2006 5th International Power Electronics and Motion Control Conference

Shanghai, China
13-16 August 2006

Volume 2 of 4

IEEE Catalog Number: 06EX1405
ISBN: 1-4244-0448-7

Copyright © 2006 by The Institute of Electrical and Electronics Engineers, Inc.
All Rights Reserved

Copyright and Reprint Permissions: Abstracting is permitted with credit to the source. Libraries are permitted to photocopy beyond the limit of U.S. copyright law for private use of patrons those articles in this volume that carry a code at the bottom of the first page, provided the per-copy fee indicated in the code is paid through Copyright Clearance Center, 222 Rosewood Drive, Danvers, MA 01923.

For other copying, reprint or republications permission, write to IEEE Copyrights Manager, IEEE Operations Center, 445 Hoes Lane, Piscataway, New Jersey USA 08854. All rights reserved.

IEEE Catalog Number:	06EX1405
ISBN:	1-4244-0448-7
Library of Congress:	2006925601

Additional Copies of This Publication Are Available from:

IEEE Service Center
445 Hoes Lane
Piscataway, NJ 08854
IEEE Service Center
445 Hoes Lane
Piscataway, NJ 08854
Phone: (800) 678-IEEE
             (732) 981-1393
Fax: (732) 981-9667
E-mail: customer-service@ieee.org

Table of Contents

Design Challenges For Distributed Power Systems 1
Fred C. Lee, Ming Xu, Shuo Wang, Bing Lu

A Smarter Grid for Improving System Reliability and Asset Utilization 16
D. Divan, H. Johal

Medium-Voltage Power Conversion Systems in the Next Generation 23
Hirofumi Akagi, Shigenori Inoue

Modern Electrical Drives: Design and Future Trends 31
R. W. De Doncker

Power Semiconductors development trends 39
L. Lorenz

Power Electronics in Wind Turbine Systems 46
F. Blaabjerg, Z. Chen, R. Teodorescu, F. Iov

Sustainable Energy and Mobility, and Challenges to Power Electronics 57
C.C.Chan

Wind farms with increased transient stability margin provided by a STATCOM 63
Marta Molinas, Jon Are Suul, Tore Undeland

A New Super Junction LDMOS with N+-Floating Layer 70
Baoxing Duan, Bo Zhang, Zhaoji Li

Unified Power Flow Controller: Comparison of Two Advanced Control Schemes and Performance Analysis for Power Flow Control 74
Liu Liming, Zhu Pengcheng, Kang Yong, Chen Jian

A New Analytical Model for the Surface Electrical Field Distribution of Double RESURF LDMOS 79
Qi Li, Zhaoji Li

A Novel Centralized HID Ballast System with Power-Bus 83
Xiaodong Lu, Bo Yang, Jiande Wu, Xiangning He

Research on a Novel Structure of SiGeC/Si Heterojunction Power Diodes 88
Liu Jing, Gao Yong, Ma Li

Gate driving of high power IGBT by wireless transmission 92
Stéphane Bréhaut, François Costa

The Characteristics of Thyristor Controlled Reactance Series Compensation by Adjustable Coupling 97
Guo-rong Zhu, Min-zu Li, Yong Kang

Dual-Side Cooled Novel IPM and Improved Capability of Inverter for Elevated-Temperature Operations 102
Jie (Jay) Chang, Changming Liao

An Improved Current-Doubler with Coupled Inductors 108
T.-F. Wu, C.-T. Tsai, W.-C. Lin, Y.-M. Chen

Monolithic Integration of Trench Power JFET with Schottky Diode 113
Yang Gao, Jie Chen, Alex. Q. Huang

Sequential Color LED Back-Light Driving System for LCD Panels 117
C.-C. Chen, C.-Y. Wu, P.-C. Lu, Y.-M. Chen, T.-F. Wu

Development of Large Capacity Programmable Harmonic Current Generator Based on Three-phase-four-wire Configuration 122
LIU Tao, ZHUO Fang, CHEN Bo, ZHAI Xi, WANG Zhao-an

A Universal Digital Platform and Software Library for Power Electronic Systems Integration 127
Haibing HU, Tianjun Jin, Wenxi YAO, Zhengyu LU, Zhaoming Qian

Table of Contents

Unipolar SiC Devices - Latest Achievements on the Way to a New Generation of High Voltage Power Semiconductors 132
Peter Friedrichs

Implementation of GA-trained GRNN for Intelligent Fast Charger for Ni-Cd Batteries 137
Panom Petchjatuporn, Noppadol Khaehintung, Khamron Sunat, Phaophak Sirisuk, Wiwat Kiranon

Modelling and Analysis of a Novel Transformer with Ability to Suppress Conducted interference 142
Zongxiang Chen, Pengsheng Ye, Junmin Pan

An Observer-Based Three-Phase Current Reconstruction using DC Link Measurement in PMAC Motors 147
Li Ying, Nesimi Ertugrul

Experiment Research of Chaotic PWM Suppressing EMI in Converter 152
R. Yang, B. Zhang , F.Li, J.J. Jiang

Emitter Size Effect in 4H-SiC BJT 157
Yan Gao, Alex Q. Huang, Sumi Krishnaswami, Anant K. Agarwal, Charles Scozzie

PSIM and SIMULINK Co-simulation for Threelevel Adjustable Speed Drive Systems 161
Zhang Yongchang, Zhao Zhengming, Baihua, Yuan Liqiang, Zhang Haitao

Three-Phase Z-Source AC-AC Converter for Motor Drives 166
Xu-Peng Fang

Construction and Application of Macro Model for ZVS Resonant Mode Controller MC34067 171
Wei Chen, Yilei Gu, Zhengyu Lu, Zhaoming Qian

Optimum Design of Hollow Conductor in Stator Winding for Large Evaporative Hydro-generator 175
Z. Wen , L. Ruan, G. Gu

Rotor Suspension Principle and Decoupling Control for Self-bearing Induction Motors 179
Tengchao Zhang, Huangqiu Zhu, Yuxin Sun

Field Oriented Control of Linear Induction Motor Considering Attraction Force & End-Effects 184
Jianqiang Liu, Fei Lin, Zhongping Yang, Trillion Q. Zheng

Series Resonant High Frequency Link Sine-wave Inverter System Modeling using Sampled Data 189
Jin Xiaoyi, Dong Wei, Sun Xiaofeng, Wu Weiyang

Maximal Power Point Tracking under Speed-Mode Control for Wind Energy Generation System with Doubly Fed Introduction Generator 194
Y. Zhao, X. D. Zou, Y. N. Xu, Y. Kang, J. Chen

Effective Mobility in Nano-Scaled n-MOSFETs 199
Yue-Hua Dai, Jun-Ning Chen, Dao-Ming Ke, Jia-E Sun

Investigation on the Factors Affecting Inrush Current of Transformers Based on Finite Element Modeling 204
M. Reza Feyzi, M. B. B. Sharifian

An Improved Support Vector Machine Method for Harmonic and Inter-harmonic Detecting 209
Ma Li, Liu Kaipei, Lei Xiao

A Common Mode and Differential Mode Integrated EMI Filter 214
Liu Nan, Yang Yugang

Electromagnetism Model and Characteristic Simulation of Novel Claw Pole Generator with Permanent Magnet Outer Rotor 219
Fengge Zhang, Haijun Bai, Shifu Zhang, Hans Pert Gruenberger, Eugen Nolle

An Improved Adaptive Filter for Voltage and Current Reference Extraction 224
A. Abedini, A. Nasiri

Simulation Analysis on Current SVM Algorithm of Matrix Rectifier 229
Xi-jun Yang, Peng-sheng Ye, Xiang Liu, Xing-hua Yang, Jian-quan Wang, Luan-guo Zhang

Table of Contents

Study of Measurement Approach of Loop Gain of Converter 236
Weiping Zhang, Yunpeng Chen, Yuanchao Liu, Dongyan Zhang, Zheng Meng

A Stand-Alone Hybrid Generation System Combining Solar Photovoltaic and Wind Turbine with Simple Maximum Power Point Tracking Control 242
Nabil A. Ahmed, Masafumi Miyatake

Design Optimization of Industrial Motor Drive Power Stage Using Genetic Algorithms 249
F. Wang, W. Shen, D. Boroyevich, S. Ragon, V. Stefanovic, M. Arpilliere

FEM Based Simulation of a Permanent Magnet Synchronous Motor Performance Characteristics 254
L. Petkovska, G. Cvetkovski

Analytical Modeling of Semiconductor Losses in Matrix Converters 259
Bingsen Wang, Giri Venkataramanan

Nonlinear Robust Sliding Mode Control for PM Linear Synchronous Motors 267
Xi Zhang, Junmin Pan

Dynamic Analysis of PWM Switching DC-DC Converters 272
Liu Jian, Wang Yuanbin

A Novel LLC Resonant Converter Topology: Voltage Stresses of All Components in Secondary Side Being Half of Output Voltage 276
Yilei Gu, Zhengyu Lu, Zhaoming Qian

On the hybrid automaton models and control synthesis of a single inductor, double output boost converter 281
Sreekumar C, Vivek Agarwal

Complex Intermittency in Voltage-Mode Controlled Buck Converter 286
Zheng-Ping Li, Yu-Fei Zhou, Jun-Ning Chen

Dual Mode Control Multiphase DC/DC Converter for CPU Power 291
Li-Wei Lin, Chung-Hsing Chang, Huang-Jen Chiu, Shann-Chyi Mou

An Analog Implementation of Pulse-Width-Modulation Based Sliding Mode Controller for DC-DC Boost Converters 296
Siew-Chong Tan, Y. M. Lai, Chi K. Tse

Low Cost Electronic Ballast with Buck Converter as PFC Stage 301
Li Xiangrong, Xu Dianguo, Zhang Xiangjun

A New Converter Architecture for Future Generations of Microprocessors 306
Dodi Garinto

A Combined ZVS Converter with Naturally Sharing Input-Current and High Voltage Gain 311
Linbing Wang, Bo Yang

Matrix Coefficient Polynomial Description Model of DC-DC Converters Based on Switched Linear Systems 316
Yongping Zhang, Bo Zhang, Zongbo Hu, Dongyuan Qiu, Guiping Du

Development of DC-DC Multiple Converter based on Push-pull Forward Topology accomplished 321
Weihao Hu, Yunqing Pei, Zhaoan Wang

Voltage Fed and Current Fed Full Bridge Converter for the Use in Three Phase Grid Connected Fuel Cell Systems 325
M. Mohr, F.-W. Fuchs

Small-Signal Modeling of Asymmetrical Half Bridge Flyback Converter 332
Tso-Min Chen, Chern-Lin Chen

A DSP Based Controller for High Power Dual-Phase DC-DC Converters 337
Xin Guo, Xuhui Wen, Ermin Qiao

v

Table of Contents

Effective Load Resistance; A New Method to Evaluate DC/DC converters Efficiency 342
Alan Elbanhawy

Calculation of Power Loss in Output Diode of a Flyback Switching DC-DC Converter 346
Jiaxin Chen, Jianguo Zhu, Youguang Guo

A Multiple Output Forward Converter Adopting Weighted Time-Sharing Control and Switch-Linear Hybrid Scheme 351
Xiaodong Liu, Songqin Hu, Sizhou Sun

A Novel Soft-Switching PWM Full-Bridge DC/DC Converter with DC Busline Series Switch-Parallel Capacitor Edge Resonant Snubber Assisted by High-Frequency Transformer Leakage Inductor 356
Khairy Fathy, Toshimitsu Doi, Keiki Morimoto, Hyun Woo Lee, Mutsuo Nakaoka

High-Efficiency Cascode Forward Converter of Low Power PEMFC System 361
Jiann-Fuh Chen, Wei-Shih Liu, Ray-Lee Lin, Tsorng-Juu Liang, Ching-Hsiung Liu

Control of Bifurcation by Fuzzy Logic Controller for Current-mode Boost Converters 368
Noppadol Khaehintung, Phaophak Sirisuk, Anantawat Kunakorn

An Improved Three-Level Soft-Switching DC/DC Converter 373
Z. L. Lou, Z. S. Wang

A Novel Soft Switching Bidirectional DC/DC Converter and Design Consideration 378
Ma Gang, Qu Wenlong, Liu Yuanyuan

State-Variable Description and Analysis of a DC-Rail ZVT Inverter Feeding a Permanent Magnet Synchronous Motor 382
Ming Zhengfeng, Zhong Yanru

Analysis, Simulations and Experiments Of A Novel ZVS -ZCS Inverter With Pulse Current Feedback Transformer Auxiliary Commutation 386
Yaogang, Mahamnad Mansoor Khan, Chenchen

A Novel Eddy-Current Based Far-Infrared Rays Radiant Planner Heater using High-Frequency ZVT-PWM Inverter 392
Hisayuki Sugimura, Bishwajit Saha, Hideki Omori, Hyun Woo Lee, Mutsuo Nakaoka

3 Phases-3 Devices AC Voltage Regulator With Quasi-Zero Switching 397
Qianzhi Zhou, Wenhua Hu, Bin Wu

Study on Power Decoupling Control of Three Phase Voltage Source PWM Rectifiers 401
Wang Jiuhe, Yin Hongren, Zhang Jinlong, Li Huade

A Fully Digital Controlled 3KW, Single-Stage Power Factor Correction Converter Based on Full-Bridge Topology 406
HANG Li-jun, YANG Yue-feng, SU Bin, LU Zheng-yu, QIAN Zhao-ming

A New ZVT Power Factor Corrected Three-Phase AC-AC Converter with Single-Phase HF Link 411
T. H. Abdelhamid, A. Sabzali

Simple Bridge-Type AC/DC Converters with Natural Input-Current-Shaper 417
Hsing-Fu Liu, Chih-Yu Wu, Chin Sun, Lon-Kou Chang

Rough Controlling TSC for Reactive Current Compensation in Traction Substations 423
Hongsheng Su, Qunzhan Li

A Digitally Controlled 4-kW Single-Phase Bridgeless PFC Circuit for Air Conditioner Motor Drive Applications 428
Yong Li, Toshio Takahashi

Optimized Electrical Design for Single Phase PFC Active IPEM 433
Qiaoliang Chen, Xu Yang, Zhao-an Wang

Table of Contents

A Novel Topology of APFC with On-Line Half-Bridge UPS Controlled by DSP 438
Xuejun Ma, Xuezhi Hu, Hongxia Wu, XuWu Chen

Nonlinear Current Control of Single-Phase PFC Suitable for Mixed-Signal IC Implementation 442
Min Chen, Anu Mathew, Jian Sun

A Novel Detection Method for Three-Phase Reactive Current 449
Zong Ming, Wang Fengxiang, Hua Funian, Sun Yidan

**Selective Harmonic Controlling for Three-Level High Power Active Front End Converter with Low
Switching Frequency** 453
Hui Zhang, Kaipei Liu,

A Unity Power Factor Three-Phase Buck Type SVPWM Rectifier Based on Direct Phase Control Scheme 458
LI Yabin, Li Heming, Peng Yonglong

3-Phase Current-Source SMES-UPS Based on TFSC and its Control Strategies Control Strategies 463
WANG Fu-sheng, LI Hong-mei

**A novel control scheme of 230kA DC power source using thyristor, Phase-shifting rectifier transformer
and On-load tap changer** 468
Qiao Shutong, Jiang Jianguo, Zuo Dongsheng, Wu Xiaojie

Research on Control Method of Double-Mode Inverter with Grid-Connection and Stand-Alone 473
Herong Gu, Zilong Yang, Deyu Wang, Weiyang Wu

Power and Energy Management of a Dual- Energy Source Electric Vehicle - Policy Implementation Issues 478
P.C.K. Luk, L.C. Rosario

**Study on Non Contact Automatic on-Load Voltage Regulating Distributing Transformer Based on Solid
State Relay** 484
Zhao-Yulin, Dong-Shoutian, Li-Jiahui, Yao-Xin, Zheng-Na, Liu-Xueli

The Principle of a Novel Arc-suppression Coil and its Implementation 489
Cheng Lu, Chen Qiaofu, Zhang Yu, Zhang Changzheng

Grid Connection to Stand Alone Transitions of Slip Ring Induction Generator During Grid Faults 494
G. Iwanski, W. Koczara

System Control of Power Electronics Interfaced Distribution Generation Units 499
D. Feng, Z. Chen

**Test Loadability of Power Systems using A Networked Power Electronic Devices Control and
Measurement System** 505
Sheng Yang, Venkataramana Ajjarapu, Bo Zhang

**Test-Bed of Doubly Fed Induction Generator for Variable-Speed Constant-Frequency Wind Power
Generation** 510
S. Y. Yang, X. Zhang, C. W. Zhang, R. X. Cao

Control strategy of Hybrid sources for Transport applications using supercapacitors and batteries 515
M.B. Camara, H. Gualous, F. Gustin, A. Berthon

Wind Generator Stabilization With Doubly-Fed Asynchronous Machine 520
Li Wu, Zhixin Wang

**Design Consideration of a Novel Digital Bidirectional Constant Current Source Used in Hybrid Electric
Vehicle** 526
Qingbo Hu, Zhengyu Lü

A Single-Phase Grid-Connected Inverter System With Zero Steady-State Error 532
Guo Xiaoqiang, Zhao Qinglin, Wu Weiyang

DC Transformer with Line Frequency Ripple Cancellation 537
Sen Dou, Wilson Wu, Annabelle Pratt, Pavan Kumar

Table of Contents

A Novel PWM Method for Stacked Flying Capacitor Inverter 542
Gangui Yan, Gang Mu, Yafeng Huang, Wenhua Liu

Study on a New Method of Voltage-Source Induction Heating Load-Matched 549
Li Jin-gang, Zhong Yan-ru, Zhao Miao

An Alternating-master-salve Parallel Control Research for Single Phase Paralleled Inverters Based on CAN Bus 554
Zhang Chunjiang, Chen Guitao, Guo Zhongnan, Wu Weiyang

Analysis and Design of a Novel Dual Secondary Winding and Dual Power Bridge High Frequency Link Inverter 559
Zhang Zhe, Zhang Chunjiang, Wu Weiyang, Gu Herong, Shen Hong

Reduction of Common Mode EMI in a Full-Bridge Converter through Automatic Tuning of Gating Signals 564
Kai Zhang, Yunbin Zhou, Yonggao Zhang, Yong Kang

Phase Multilevel Inverter Fault Diagnosis and Tolerant Control Technique 569
Wang Baocheng, Wang Jie, Sun Xiaofeng Wu Junjuan, Wu Weiyang

Microcontroller-Based Single Phase Inverter Using a New Switching Strategy 574
K. Meghriche, O. Mansouri, A. Cherifi

Study of Stability Regions in Parallel Connected Boost Converters 580
Yuehui Huang, Chi. K. Tse

A Novel Analysis and Design Method for Integrated Magnetics 585
Zheng Feng, Weihao Hu, Pei Yun-qing

Investigation on the Space Vector PWM for Large Power Three-Level DC-Link Voltage Source Inverter Equipped with IGCTs 589
Wang Chengsheng, Li Chongjian, Li Yaohua, Zhao Xiaotan

Status and Opportunities of Photovoltaic Inverters in Grid-Tied and Micro-Grid Systems 593
Xiaoming Yuan, Yingqi Zhang

Adaptive Neuro-Fuzzy Control with Fuzzy Supervisory Learning Algorithm for Speed Regulation of 4-Switch Inverter Brushless DC Machines 597
A. Halvaei Niasar, H. Moghbelli, A. Vahedi

Combined Modulation and Harmonic Suppression 602
Cheng Weibin, Zhong Yanru, Jin Shun

Application Research of Maximum Wind-energy Tracing Controller Based Adaptive Control Strategy in WECS 607
Changhong Shao, Xiangjun Chen, Zhonghua Liang

Research on Synchrodrive Control Technology for Wind Turbine Adjustable-Pitch System Based on Adaptive decoupling Control 612
Hongche Guo, Qingding Guo

Limit-Trajectory Single- and Two-Mode Overmodulation Technology 617
Shun Jin, Yan-ru Zhong

Multiphase Permanent Magnet Motor Drive System Based on A Novel Multiphase SVPWM 622
Shan Xue, Xuhui Wen, Zhao Feng

Novel Random-Harmonic Elimination PWM Technique for Single-Switch Three-Phase AC-DC Buck Converter 627
Guang-Hui Tan, Wenchuan Ma, Yanchao Ji, Hongxiang Yu, Wancai Xu

Table of Contents

FPGA Based Multichannel PWM Pulse Generator for Multi-modular Converters or Multilevel Converters 632
Liqiao Wang, Weiyang Wu

Cascaded Multilevel Converters with Non-Integer or Dynamically Changing DC Voltage Ratios 637
Shuai Lu, Keith A. Corzine

Practical Thermal Design Considerations for IPEM-based Converter 642
Qiaoliang Chen, Xu Yang, Zhao-an Wang

Realization of an FPGA-Based Space-Vector PWM Controller 647
Zhou Yuan, Xu Fei-peng, Zhou Zhao-yong

Chaotifying Control of Permanent Magnet Synchronous Motor 652
Hai Peng Ren, Chong Zhao Han

Analysis of PMLSM Direct Thrust Control System Based on Sliding Mode Variable Structure 657
Junyou Yang, Guofeng He, Jiefan Cui

Carrier-based Pulse Width Modulation for Three-Level Inverters: Neutral Point Potential and Output Voltage Distortion 662
Jang-Hwan Kim, Seung-Ki Sul

AC Current Sensorless Control of Three-Phase Three-Wire PWM rectifiers under the Unbalanced Source Voltage 669
Jia-peng Xu, Yu-peng Tang

Waveform Library Control of Converter 674
Xiaofeng Sun, Bin Wang, Meng Lingjie, Weiyang Wu

d-model Adaptive Algorithm Based on Plant-Parameterization 679
Zhao Feng, Liu Weiguo

Dynamics and Control of Electronic Cascaded Systems 684
Wen Wei, Xu Haiping, Wen Xuhui, Shi Wenqing

The Controlling Strategy for Electronic Ballast of HID Lamps 688
Weiping Zhang, Xiaohan Guan, Xusen Zhao, Hongtao Li, Zhengang Liu

Voltage Spectra of Three-Level Inverters with Three-Phase Modulation 693
S. Halász, I. Varjasi

Design of Motion Control System Used for Filter Rod Production Machine 699
Yang Qingyu, Ge Sibo, Ye Kesong, Shi Ren

Magnetic Pole Identification for PMSM at Zero Speed Based on Space Vector PWM 703
Jiangang Hu, Longya Xu, Jingbo Liu

Study on Stagewise Control of Connecting DFIG to the Grid 708
Xueguang Zhang, Dianguo Xu, Yongqiang Lang, Hongfei Ma

Generalized Control Approach for Active Power Filters 713
Xiaoyu Wang, Jinjun Liu, Chang Yuan, Zhaoan Wang

Novel Circuit Configuration for Hybrid Reactive Power Compensator 718
H.L Jou, J.C Wu, J.J. Yang, W.P. Hsu

Shunt Active Power Filter with Sample Time Staggered Space Vector Modulation Based Cascade Multilevel Converters 724
Liqiao Wang, Weiyang Wu

Shunt Active Power Filter Synthesizing Resistive Loads by Means of Adaptive Inverse Control 729
Wu Yanfeng, Wu Zhengguo, Li Hua, Li Hui

Table of Contents

Single Neutral Element Self-Adaptive PID Controller Used In SVC..734
Zeng Guang, Ke Min-qian, Su Yan-min, Fu Qi-gang

A Novel Shunt Single-Phase Active Power Filter for High Voltage Application 739
Zhang Changzheng, Chen Qiaofu, Zhao Youbin, Chen Yuda, Cheng Lu

Three-phase Active Power Filter Based on Space Vector and One-cycle Control............................744
Wang Yong, Shen Songhua, Guan Miao

Implementation of a Shunt-Series Compensator for Nonlinear and Voltage Sensitive Load..............748
Bor-Ren Lin, Chien-Lan Huang

Three-Phase Active Filter using a Single-Phase STATCOM Structure with Asymmetrical Dead-band Control..753
Seyyed Hossein Hosseini, Mehran Sabahi

Mitigation of Voltage Sag Using Adaptive Neural Network with Dynamic Voltage Restorer....................759
M. R. Banaei, S. H. Hosseini, M. Darkalee Khajee

Mitigation of Current Harmonic Using Adaptive Neural Network with Active Power Line Conditioner....................764
M. R. Banaei, S. H. Hosseini

A direct control strategy for UPQC in three-phase four-wire system....................................769
Tan Zhili, Li Xun, Chen Jian, Kang Yong, Duan Shanxu

Three-Phase Harmonic Selective Active Filter Using Multiple Adaptive Feed Forward Cancellation Method..774
Lewei Qian, David Cartes, Qiang Zhang

Reactive Power Compensation in Distribution Networks with STATCOM by Fuzzy Logic Theory Application 779
Seyyed Hossein Hosseini, Reza Rahnavard, Yousef Ebrahimi

A Distributed Fuel Cell Based Generation and Compensation System to Improve Power Quality....................784
Haimin Tao, Jorge L. Duarte, Marcel A. M. Hendrix

Parallel Control of Three-Phase Three-Wire Shunt Active Power Filters 789
Xueliang Wei, Ke Dai, Xin Fang, Pan Geng, Fang Luo, Yong Kang

Study and Design of Noninductive Bus bar for high power switching converter 794
Zhiling Qiu, Hongyan Zhang, Guozhu Chen

A New Minimum Torque-ripple and Sensorless Control Scheme of BLDC Motors Based on RBF Networks..798
Juan Wang, Hongwei Liu, Yuran Zhu, Bo Cui, Huijuan Duan

Improved Modelling and Calculation on Electromagnetic Transient of Power Transformer....................802
Chen Zhe, Wen Yuanfang, Lu Guojun

The Simulation and the Experimental Research of the Stator Bars' Evaporative Cooling System in the Three Gorges' Hydrogenerator 808
Ruan Lin, Gu Guobiao, Tian Xindong, Yuan JiaYi

An Investigation of Multi-phase Transverse Flux Permanent Magnet Machine 813
G.Q. Bao, J.K.Wang, D.Zhang, J.Z. Jiang

Suspension Principle and Digital Control for Bearingless Permanent Magnet Slice Motors 817
Huangqiu Zhu, Liang Fang

The effect of parameter variations on the performance of indirect vector controlled induction motor drive 821
A. Shiri, A. Vahedi, A. Shoulaie

Magnetic Field Analysis and Performance Calculation for New Type of Claw Pole Motor with Permanent Magnet Outer Rotor 826
Fengge Zhang, Shifu Zhang, Haijun Bai, Eugen Nolle, Hans Pert Gruenberger

x

Table of Contents

Performance Analysis of a PM Claw Pole SMC Motor with Brushless DC Control Scheme............831
Youguang Guo, Jianguo Zhu, Jiaxin Chen, Jianxun Jin

Solving Induction Motor Equivalent Circuit using Numerical Methods for an In-Service and Nonintrusive Motor Efficiency Estimation Method............836
Bin Lu, Wei Qiao, Thomas G. Habetler, Ronald G. Harley

Fault Investigation of X-by-wire Permanent Magnet Synchronous Machine............842
L. Feng, A. Binder, A. Rentschler, A. Paweletz, D. Guenther

PLC-Based Speed Control of DC Motor............847
Ashraf Salah El Din Zein El Din

H8 Control of Adjustable-Pitch Wind Turbine Adjustable-Pitch System............853
Hongche Guo, Qingding Guo

The Motion Control Algorithm based on Quaternion Rotation for a Permanent Magnet Spherical Stepper Motor............857
Qun-jing Wang, Kun Xia

Research on Restraining Thrust Force Ripple for Permanent Magnet Linear Synchronous Motor............862
Cui Jiefan, Wu Hui, Sun Qing, Zhang Yi, Zhao Lijun

Using Recurrent Fuzzy Wavelet Neural Network to Control AC Servo System............866
Yan Tang, Wei Sun, Yaonan Wang, Xiaohua Zhai

new topology of multi - level - converter for harmonic reduction............870
Frank Grundmann, Jian Xie

PWM Based Sensing and Control of Magnetic Bearings............875
Zhuliang Yeic, Flalph Vansencc

Position Sensorless Direct Torque Control of Synchronous Reluctance with Permanent Magnet Motor............880
Jiang Dong, Zhao Zhengming, Duan Yao, Guo Wei

Counter-Rotating Permanent Magnet Brushless DC Motor for Underwater Propulsion............885
Jianqi Qiu, Cenwei Shi, Mengjia Jin, Ruiguang Lin

A Special Flux-weakening Control Scheme of PMSM - Incorporating and Adaptive to Wide-Range Speed Regulation............890
Song Chi, Longya Xu

Model-based Disturbance Attenuation for Linear Motor Servo System............896
Guiqiu Liu, Qingding Guo

A Fuzzy-Wavelet-Network-Based Position Control for PMSM............899
Wang Jun, Peng Hong, Xia Ling

Stability Analysis of Magnetic Bearing with Resonance Circuit............903
Zong Ming, Wang Fengxiang, Sun Yidan, Wang Jiqiang

Flux-Weakening Characteristics of Trapezoidal Back-EMF Machines in Brushless DC and AC Modes............908
Z.Q. Zhu, J.X. Shen, D. Howe

A Cost Effective Sensorless Control Method for Permanent Magnet Synchronous Motors Based on Average Terminal Voltage............913
Cheng-Hu Chen, Wei-Chih Tai, Ming-Yang Cheng

DSP-based Discrete-Time Reaching Law Control of Switched Reluctance Motor............918
Ge Baoming, Zhao Nan

Digital Control System on Bearingless Permanent Magnet-type Synchronous Motors............923
Jianming Deng, Huangqiu Zhu, Yang Zhou

xi

Table of Contents

Practical Issues in Sensorless Control of PM Brushless Machines Using Third-Harmonic Back-EMF.......................928
J.X. Shen, Z.Q. Zhu, D. Howe

Switched Reluctance Motors Drive for the Electrical Traction in Shearer.......................933
H. Chen

Research on Three-level Inverter of Six-phase Synchronous Motor.......................937
Yao Wenxi, Hu Haibing, Lu Zhengyu, Xu Haijie

Doubly-Salient Permanent-Magnet Machine with Skewed Rotor and Six-State Commutating Mode.......................942
Yongbin Li, Chris Mi

Sensorless Control and PMSM Drive System for Compressor Applications.......................947
Dongsheng Li, Takahiro Suzuki, Kiyoshi Sakamoto, Yasuo Notohara, Tsunehiro Endo, Chikara Tanaka, Tatsuo Ando

Analysis and Experimental Study of Slot Effect in Synchronous Reluctance Permanent Magnet Motors.......................952
Wei Guo, Zhengming Zhao, Yingchao Zhang

A New BLDC Motor Drives Method Based on BUCK Converter for Torque Ripple Reduction.......................958
Zhang Xiaofeng, Lu Zhengyu

Performance Investigation of a Fault-Tolerant Brushless Permanent Magnet AC Motor Drive.......................962
Jingwei Zhu, Nesimi Ertugrul, Wen Liang Soong

Current sensorless integral variable structure controller of synchronous reluctance motor.......................967
Huann-Keng Chiang, Chien-An Chen, Bor-Ren Lin, Kai-Sheng Hsu

An Improved Sliding Mode Observer for Speed Sensorless Vector Control Drive of PMSM.......................972
K. Paponpen, M. Konghirun

Analysis of an AC fed direct converter for a switched reluctance machine in aerospace applications.......................977
S. J. Forrest, J. Wang, G. W. Jewell, C. M. Johnson, S.D. Calverley

Direct Torque Control of an Interior Permanent Magnet Synchronous Machine fed by a Direct AC-AC Converter.......................983
D. Xiao, M. F. Rahman

A Novel Modular Permanent Magnet Drive System Design.......................989
Wen Ouyang, Nicholas Lemberg, Ruoping Yao, T.A.Lipo

Research on Digital Control Systems for Large Power AC-DC-AC Converters with Synchronous Motor Load.......................995
Xiaotan Zhao, Chongjian Li, Weihui Sheng, Yaohua Li

About the Prediction of Undesired Higher Current and Torque Harmonics of Inverter Driven Motors with Numerical Methods.......................999
C. Grabner

A Method of Stator Voltage Error Compensation in MRAS Sensorless Vector Control of Induction Motor.......................1006
Wen Xuhui, Chen Guilan, Han Li

Systematic Design of Fuzzy Logic Based Hybrid On-Line Minimum Input Power Search Control Strategy for Efficiency Optimization of IM.......................1012
Zhang Liwei, Liu Jun, Wen Xuhui, Trillion Q. Zheng

Research on an AC Variable-frequency Power Dynamometer Based on PWM Rectifier and Fuzzy Direct Torque Control.......................1017
Jia-qiang Yang, Jin Huang

Characteristic Research of Bearing Currents in Inverter-Motor Drive Systems.......................1023
Xing Shancheng, Wu Zhengguo

Research on a New Motor Drive Control System for Electric Transit Bus.......................1027
SHAO Gui-xin, ZHANG Cheng-ning

Table of Contents

New Micro-Drive Series For Induction Motors & Survey of Market Trends 1032
Henrik Rosendal Andersen, Ruimin Tan, Zhang Hui

Robust Backstepping Control of Induction Motor Drives Using Artificial Neural Networks 1038
J. Soltani, R. Yazdanpanah

Robust Nonlinear Control of Linear Induction Motor taking into account the Primary End Effects 1043
J. Soltani, M.A. Abbasian

A Novel Adaptive Scheme for Stator Resistance Estimation in Sensorless Induction Motor Drives 1049
Han Li, Wen Xuhui, Chen Guilan

Ripple-Free Sampling of Current Signals in Drives with Carrier-based PWM Patterns 1054
Haihui Lu, Qiang Yin, Russel J. Kerkman, Thomas A. Nondahl

**Study of Speed Sensorless Control Methodology for Single Inverter Parallel Connected Dual Induction
Motors Based on the Dynamic Model** 1061
Shi Wei, Wang Ruxi, Wang Yue, He Yanhui, Wang Zhaoan, Liu Jinjun

ADC architecture with direct binary output for digital controllers of high-frequency SMPS 1066
Tao Zhou, Jianping Xu

Analysis and Evaluation of a High-Voltage AC Amplifier for Electrostatic Suspension 1071
F. T. Han, Q. P. Wu, K. Liu, Z. Y. Gao

Design and Development of a 50kW Z-Source Inverter for Fuel Cell Vehicles 1076
Miaosen Shen, Alan Joseph, Yi Huang, Fang Z. Peng, Zhaoming Qian

Identification and improvement of stray coupling effect in an L-C-L common mode EMI filter 1081
Junping He, Wei Chen, Jianguo Jiang

**High Step-up Converter Associated with Soft-Switching Circuit with Partial Energy Processing for
Livestock Stunning Applications** 1086
S. -Y. Tseng, S.-H. Tseng, J. -Z. Shiang

**A Computationally Intelligent Methodologies and Sliding Mode Control Based Traction control System
for in-wheel driven EV** 1091
Ming Zhengfeng, NI Guangzheng

A Low-Cost Gate Driver Design Using Bootstrap Capacitors for Multilevel MOSFET Inverters 1096
J. J. Graczkowski, K. L. Neff, X. Kou

An Effective Method to Suppress Resonance in Input LC Filter of a PWM Current-Source Rectifier 1101
Y.W. Li, B. Wu, N. Zargari, J. Wiseman, D. Xu

Topological and Modulation Design of Three-Level Z-Source Inverters 1107
P. C. Loh, F. Gao F. Blaabjerg

Investigation of Power Supplies for a Piezoelectric Brake Actuator in Aircrafts 1112
Rongyuan Li, Norbert Fröhleke, Hermann Wetzel, Joachim Böcker

A Line Power-Supply for LED Lighting using Piezoelectric Transformers in Class-E Topology 1117
F.E. Bisogno, S. Nittayarumphong, M. Radecker, A. V. Carazo, R. N. do Prado

Integrating Large Wind Farms into Weak Power Grids with Long Transmission Lines 1122
Richard Piwko, Nicholas Miller, Juan Sanchez-Gasca, Xiaoming Yuan, Renchang Dai, James Lyons

Turn-on Condition and Characteristics of Highpower Semiconductor Switch RSD 1129
Y. M. Zhou, Y. H. Yu, H. G. Chen, L. Liang

The analysis and simulation of power circuits for high voltage converter 1133
S. I. Volskiy, Y. Y. Skorokhod, V. V. Shergin

A novel IGCT-based Half-controlled Bridge Type Fault Current Limiter 1138
Wanmin Fei, Yanli Zhang

xiii

Table of Contents

Influence of Proton Irradiation dose on the Performance of Local Lifetime Controlled Power Diode with Proximity Gettering of Platinum 1143
B.D. Han, D.Q. Hu, S.S. Xie, Y.P. Jia, B.W. Kang

IMPLEMENTATION OF A HIGHER QUALITY DC POWER CONVERTER 1148
Barsoum, N.N., YII, M.L.

Design of a Digital Programmable Control IC for Single-Phase Controlled Rectifiers 1154
Ming-Fa Tsai, Fu-Jing Ke, Ying-De Lin, Jui-Kum Wang

Feasibility Study of AlGaN/GaN HEMT for Multimegahertz DC/DC Converter Applications 1159
Yang Gao, Alex Q. Huang

The Mechanism Analysis of IGBT Module Invalidation 1162
Xu Aide, Fan Yinhai, Wang Xinxin, Liu Yuanyuan

A New Injection Efficiency Controlled GTO 1167
Wang Cailin, Gao Yong, Zhang Ruliang

Implementation and Analysis of 3-phase Voltage Sourced Regenerative Rectifier 1171
Rui Chen, Qiongxuan Ge, Shijie Li

Design and Implementation of Electronic Ballast for Fluorescent Lamps with Low Lighting Flicker 1178
Yang-Sheng Lin, Chun-An Cheng, Jiann-Fuh Chen, Tsorng-Juu Liang, Wei-Shih Liu

A Floating-point Coprocessor Configured by a FPGA in a Digital Platform Based on Fixed-point DSP for Power Electronics 1183
Haibing HU, Tianjun Jin, Xianmiao Zhang, Zhengyu LU, Zhaoming Qian

An Analytical Model for 4H-SiC Super-Junction Devices 1188
L.C. Yu, K. Sheng

Architecture Implementation of Class-D Amplifiers Using Digital-Controlled Multiphase-Interleaved PWM Technique 1192
Yu-Tzung Lin, Chi-Yang Lee, Ying-Yu Tzou,

Integrated IC-like Thyristor–based Switching Structure for Pulse Current Generation to Electronic Ignition 1198
C. L. Zhang, K. S. Jeon, C. H. Ahn, J. D. Park, E. D. Kim, Na Zhi, Yong Gao

A Wide Bandwidth Current Probe Based on Rogowski Coil and Hall Sensor 1202
Dong Li, Guiyou Chen

Voltage Dip Detection Based on an Efficient Least Squares Algorithm for D-STATCOM Application 1207
Thip Manmek, Chathura P. Mudannayake, Colin Grantham

Optimal Design and Analysis on Bearingless Permanent Magnet-type Synchronous Motors Using Finite Element Method 1213
Chang Jiang , Huangqiu Zhu, Zhenyue Huang

The Restrain of Harmonic Circulating Currents between Parallel Inverters 1218
Yu Zhang, Shanxu Duan, Yong Kang, Jian Chen

Simulation of Permanent Magnet Synchronous Motor with Dual Closed Loop by Time-Stepping Finite Element Model 1223
Xinhua Liu, Jianzhong Jiang, Yu Gong, Ye Ding

Online Dynamic Parameter Estimation of Transformer Equivalent Circuit 1228
M. Reza Feyzi, Mehran Sabahi

Worst-Case Tolerance Analysis for a Power Electronic System by Modified Genetic Algorithms 1233
Toshiji Kato, Kaoru Inoue, Kazuya Nishimae

The Reduction of Force Ripples of PMLSM Using Field Oriented Control Method 1238
Yu-wu Zhu, Kun-seok Jung, Yun-hyun Cho

xiv

Table of Contents

Analysis and Design of Signal Stage AC/DC Converter with Resonant Model PFC 1243
Weiping Zhang, Liangrui Lin, Dongyan Zhang, Xusen Zhao

Low Frequency Model for the Metal Halide Lamp 1248
Weiping Zhang, Yuanchao Liu, Xiaoqiang Zhang, Hongtao Li, Wenji Liu

H8 Robust Controller Based on Local Feedback Recurrent Neural Network for Permanent Magnet Linear Synchronous Motor 1253
Junyou Yang, Naiguang Fa, Ruijuan Chen

Parameter Estimate Modeling of Electronic Transformer 1258
Jiaju Wu, Hidehiko Sugimoto, Changkun Wang

Analysis and Design of Boost DC-DC Converters for Intrinsic Safety 1267
Shu-Lin Liu, Jian Liu, Hong Mao

Modeling and Fuzzy Logic with Integrator Control for the ZVZCS PWM DC/DC Converter 1273
Shen Hong, Wan Jianru, Yang Xiaobo, Wu Weiyang, Wang Xiaohuan

ZVS DC-DC Converter with Parallel-Connected Current Doubler Rectifier 1278
Bor-Ren Lin, Shuh-Chuan Tsay, Chun-Sheng Yang, Chien-Lan Huang

Study on the Dynamical Model and Analytical Method for DC-DC Switching Converter 1283
Li-Li Wang, Yu-Fei Zhou, Jun-Ning Chen

A Novel Topology Family of Single-stage Parallel Mode Uninterruptible AC/DC Converter with PFC 1288
Xuejun Ma, Hongxia Wu, Congsheng Huang, Xuwen Huang

Analysis and Design of an Automatic-Current-Sharing Control Based on Average-Current Mode for Parallel Boost Converters 1293
Wenxun Xiao, Bo Zhang, Dongyuan Qiu

A Novel Digital Charge Control for DC-DC Converters 1298
Shi Wenqing, Xu Haiping, Wen Xuhui, Wen Wei

An Asymmetrical Switched Capacitor and Lossless Inductor Quasi-Resonant Snubber-Assisted ZCS-PWM DC-DC Converter with High frequency Link 1302
Khairy Fathy, Keiki Morimoto, Toshimitsu Doi, Hyun Woo Lee, Mutsuo Nakaoka

A Divided Voltage Half-Bridge High Frequency Soft-Switching PWM DC-DC Converter with High and Low Side DC Rail Active Edge Resonant Snubbers 1307
Khairy Fathy, Keiki Morimoto, Toshimitsu Doi, Hiroyuki Ogiwara, Hyun Woo Lee, Mutsuo Nakaoka

Dynamic Analysis of a Current Source Inductively Coupled Power Transfer System 1312
Wenqi Zhou, Hao Ma

A New Topology of Capacitor-Clamp Cascade Multilevel Converters 1318
Anees Abu Sneineh, Ming-Yan Wang, Kai Tian

Evaluation of Semiconductor Losses in Cryogenic DC-DC Converters 1323
C. Jia, A. J. Forsyth

Design and Performance Evaluation of a 10-kW Interleaved Boost Converter for a Fuel Cell Electric Vehicle 1328
G. Calderon-Lopez, A. J. Forsyth, D. R. Nuttall

Analysis of Abnormal Phenomenon in Common-Source-type Forward Converter with Self-driven Synchronous Rectifier 1333
Kentaro Fukushima, Takayoshi Hashimoto, Tamotsu Ninomiya, Takeshi Segawa

Power Quality Conditioning in Distributed Generation Systems 1338
R.K. Járdán, I. Nagy

xv

Table of Contents

Active Clamp Forward Converter Combined with Dither Voltage Generator for Poultry Stunning Applications 1343
S. -Y. Tseng, H.-T. Wen, H.-H. Chang, J. -S. Kuo

A Novel Zero-Voltage Switching Resonant Pole Inverter 1348
Sanbo Pan, Junmin Pan

Analysis of Three-Level ZVS PWM Inverter for Induction Heating Applications 1353
A. Jangwanitlert, J. Songboonkaew, W. Thammasiriroj, J.C. Balda

Dual Duty Cycle Controlled Voltage Source Soft-Switching High Frequency Inverter with AC Load Side Reverse Blocking Switched Resonant Capacitor 1358
Khairy Fathy, Ju-Sung Kang, Hiroyuki Ogiwara, Bin Eiuo, Hideki Omori, Hyun Woo Lee, Mutsuo Nakaoka

A Switched-Capacitor Lossless Inductor ZCS Snubber-Assisted Series Load Resonant High Frequency Inverter with Dual Mode Pulse Modulation Scheme 1363
Khairy Fathy, Takaaki Okude, Hideki Omori, Hyun Woo Lee, Mutsuo Nakaoka

Topologies of Switch-Linear Hybrid Power Conversion & Special Operation States 1368
Lu-sheng Ge, Qian-zhi Zhou, Wu bin

Single Reverse Blocking Switch Type Pulse Density Modulation Controlled ZVS Inverter with Boost Transformer for Dielectric Barrier Discharge Lamp Dimmer 1372
Hisayuki Sugimura, Bishwajit Saha, Hideki Omori, Hyun-Woo Lee, Mutsuo Nakaoka

PDM Controlled Series Load Resonant Soft Switching High Frequency Inverter for Induction Heated Toner Fixing Outer Roller with Inner Cylindrical Working Coil Stator 1377
Hisayuki Sugimura, Hideki Omori, Hyun Woo Lee, Mutsuo Nakaoka

Zero-Voltage and Zero-Current Switching Two-Transformer Full-Bridge Converter Using the Output-Voltage-Doubler 1382
H.K. Yoon, E.S. Choi, S.K. Han, G.W. Moon, M.J. Youn

A Single-stage Boost-Flyback PFC Converter 1387
Zhao Qinglin, Wen Yi, Wu Weiyang, Chen Zhe

Control Bifurcation in PFC Boost Converter under Peak Current-Mode Control 1392
Yi-Jing Ke, Yu-Fei Zhou, Jun-Ning Chen

Analysis and Design of One-Cycle-Controlled Dual-Boost Power Factor Corrector 1397
Yue-feng Yao, Yuan-rui Chen

A Novel Single-phase Buck PFC Converter Based on One-cycle Control 1401
Chen Bing, Xie Yun-Xiang, Huang Feng, Chen Jiang-Hui

Modeling and Simulation of Three Phase High Power Factor PWM Rectifier factor correction. 1406
Yu Fang, Yong Xie, Yan Xing

Effect of the Ripple Current on Power Factor of CRM Boost APFC 1412
A. Abramovitz

Simulated Study of Three-Phase Single-Switch PFC Converter with Harmonic Injected PWM by MATLAB 1416
Zhanlong Li, Yupeng Tang

A Simple Digital Controller for Constant Instantaneous Input Power type Three-Phase Boost Rectifier under Unbalanced System 1421
Jin Ai-Juan, Li Hang-Tian, Li Shao-Long

An Improved and Digital Current Control Strategy for One Cycle Control Based Three-Phase Boost Rectifier under Unbalanced System 1426
Li Shao-Long, Jin Ai-Juan, Li Hang-Tian

xvi

Table of Contents

Control Method for Power Quality Compensation Based on Levenberg-Marquardt Optimized BP Neural Networks..1431
Zhou Ming, Wan Jian-Ru, Wei Zhi-Qiang, Cui Jian

A Nonlinear Method for Hybrid Electromagnetic Suspension 1436
Junwei Cui, Jianhui Wang

New topology of multi - level - converter for harmonic reduction 1442
Frank Grundmann, Jian Xie

Model Reference Adaptive Control based on Neural Network for Electrode System in Electric Arc Furnace..1447
Zhang Shi-feng, Zhang Shao-De, Li Kun, Zheng Xiao

STATCOM ETO Failure Analysis..1450
Zhong Du, Bin Chen, Chong Han, Zhaoning Yang, Wenchao Song, Subhashish Bhattacharya, Alex Q. Huang

Modeling and Control of Three-phase Voltage Source PWM Rectifier 1454
Yao Chen, Xin Min Jin

Mitigation of Electric Arc Furnace Voltage Flicker Using Static Synchronous Compensator..................1458
Y.F. Wang, J.G. Jiang, L.S. Ge, X.J. Yang

Design of Distributed FACTS Controller and Considerations for Transient Characteristics..................1463
Gaidi Ning, Shijie He, Yue Wang, Lei Yao, Zhaoan Wang

A Wind-Power Generation System Having a Function of Suppressing Line Voltage Deviation..................1468
Y. Nakayama, S. Fukuda, M. Futami, M. Ichinose, S. Ohara, H. Kita

A Novel Active Islanding Detection Method of Grid-connected Photovoltaic Inverters Based on Current-Disturbing..1473
Zhang Chunjiang, Liu Wei, San Guocheng, Wu Weiyang

Grid Connection of Doubly-Fed Induction Generators in Wind Energy Conversion System..................1477
Ahmed G. Abo-Khalil, Dong-Choon Lee, Se-Hyun Lee

Active and Reactive Power Control of DFIG for Wind Energy Conversion under Unbalanced Grid Voltage 1482
Jeong-Ik Jang, Young-Sin Kim, Dong-Choon Lee

A BASIC STUDY OF FUZZY-LOGIC-BASED POWER SYSTEM STABILIZATION WITH DOUBLY-FED ASYNCHRONOUS MACHINE..1487
Li Wu, Zhixin Wang

Quantitative Analysis on Different Modes of Energy Optimal Control for Series Power Quality Controllers..1492
Huang Xinming, Liu Jinjun, Zhang Hui

Resonance inverter power system for improving plasma sterilization effect 1497
Y.M Kim, J.Y Kim, M. C Jo, S.H Lee, S.P Mun, H.W Lee, S.K Kwon, K.Y Suh

Generic optimization for SMPS design with Smart Scan and Genetic Algorithm..................1502
Heidi H.T. Yeung, N. K. Poon, Stephen L. Lai

Novel Single-Stage Isolated Buck-Boost Inverter Based on Improved SPWM Control Method 1507
Guang-Hui Tan, Fanpeng Zeng, Yanchao Ji, Xi Chen, Hua Wang

On the Effects of Voltage Loop in Paralleled Converters Under Master-Slave Current Sharing..................1512
Yuehui Huang, Chi K. Tse

Improved Control for Parallel Inverter with Current-Sharing Control Scheme 1517
Zhao Qinglin, Chen Zhongying, Wu Weiyang

A Novel Digital Controlled battery charger for High power UPS application..........................1522
Fang Luo, Yong Kang, Shan Xu Duan, Xueliang Wei

xvii

Table of Contents

A Novel High Input Power Factor Single-Stage Single-Phase AC/AC Converter 1527
Chien-Ming Wang, Chien-Yeh Ho, Maoh-Chin Jiag

Research on the Power Sharing of the Parallel Inverters without Control Interconnection Basing on Droop Characteristic 1532
Kan Jiarong, Xie Shaojun

Analysis and Design of Repetitive controlled Inverter System with High Dynamic Performance 1537
Mingzhu Li, Zhongyi He, Yan Xing

Study on a large-volume high-performance programmable voltage disturbance source 1542
Zhan Qizhi, Zhuo Fang, Dong Wenjuan, Wang Zhao'an

1 KW Dual Interleaved Boost Converter for Low Voltage Applications 1546
Heinz van der Broeck, Ibrahim Tezcan

Control of Multilevel Flying Capacitor Inverters for High Performance 1551
L. Zhang, S. J. Watkins, Duan Qi Chang

Analysis of Harmonics in Input Line Current for Matrix Converter based on Double Input Line-toline Voltages 1557
Guo Yougui, Deng Wenlang, Zhu Jianlin

Research on Neutral-point Balancing Control for Three-level NPC Inverter Based on Correlation between Carrier-based PWM and SVPWM 1560
Wenxiang Song, Guocheng Chen, Xiaoyu Ding, Mantang Shu

Instantaneous Voltage Regulated Seamless Transfer Control Strategy for Utility-interconnected Fuel cell Inverters with an LCL-filter 1566
Guoqiao Shen, Dehong Xu, Xiaoming Yuan

An Anti-windup Design Method for Internal Model Control Based on H8 Optimization 1571
Hou Yansong, Li Hua

Study on Pwm Control Strategy of Photovoltaic Grid-connected Generation System 1576
Shi-cheng Zheng, Pei-zhen Wang, Lu-sheng Ge

Robust Sliding Model Control for Regenerative Braking of Electric Vehicle 1581
Min Ye, Zhifeng Bai, Binggang. Cao

A Self-adaptive Fuzzy Control Scheme of High Frequency Link SPWM Inverters 1585
Herong Gu, Deyu Wan, Weiyang Wu

Using Automatic Frequency Shifting Techniques for LLC-SRC Output Voltage Regulation 1590
Kuo-Kai Shyu, Ching-Ming Lai, Ko-Wen Jwo, Ming-Ho Pan, Chung-Ping Ku

Design and Test of Novel Programmable Digital Three Phases SPWM Chip 1595
Yang Yuan, Gao Yong, Chen Lijie

An Improved Performance of Five-Leg Inverter in Two Induction Motor Drives 1598
Ryuji Omata, Kazuo Oka, Atsushi Furuya, Shuji Matsumoto, Yusuke Nozawa, Kouki Matsuse

Adaptive Three Dimensional Space Vector Modulation in abc Coordinates for Three Phase Four Wire Split Capacitor Converter 1603
Xiao-bo Yang, Wei-yang Wu, Hong Shen

Inverters Parallel Operation Based on CAN 1608
Yong Wu, Xianglong Jiang, Jinbang Xu, Qingyi Wang, Shuyun Wan

EMI Reduction Method for a Single-Phase PWM Inverter by Suppressing Common-Mode Currents with Complementary Switching 1613
Toshiji Kato, Kaoru Inoue, Koji Akimasa

Table of Contents

Analysis and Design of a Novel Dual Secondary Winding and Dual Power Bridge High Frequency Link Inverter 1618
Zhang Zhe, Zhang Chunjiang, Wu Weiyang, Gu Herong, Shen Hong

Research of Complex Fuzzy Control on-off Magnetism Team Motor Speed-Adjusting System 1623
Zhao Ming-fu, Chen Yan, Zhang Zhi-yuan, Dong Chun, DongYu

A New BLDC Motor Drives Method Based on BUCK Converter for Torque Ripple Reduction 1626
Zhang Xiaofeng, Lu Zhengyu

Design of Wind Turbine Generator Control System 1630
Chen Guiyou, Zhou Li, Sun Tongjing, Wang Zhongmin

Non-touching Intelligent Control System of Water Intenerating Equipment Based on Sodion Exchange 1634
Chen Guiyou, Zhang Qingfan, Zhou Li, Luo Donghua

Investigation of Hybrid Modeling and Control for DC-DC Converters 1637
Hao Ma, Feng Qi, Wenqi Zhou

Effect of Peak Current Mode Control on Transient Response for VRM Application 1641
Seiya Abe, Tamotsu Ninomiya

Modulations for Voltage Source Rectification and Voltage Source Inversion Using Direct Space Vector Approach 1646
Keping You, M. F. Rahman

Synchronization of Voltage Waveforms in Basic Topologies of Dual Inverter-Fed Motor Drives 1651
V. Oleschuk, F. Profumo, A. Tenconi, R. Bojoi, A.M. Stankovic

Research on Fast Magnetic Valve Controllable Reactor 1657
Zhang Jian-wen, Cai Xu

Study and comparison of fault tolerant shunt threephase active filter topologies 1663
H. El Brouji, P. Poure, S. Saadate

Application of GA-BP in Fault Diagnosis of Power Circuit of SVC 1669
Zeng Guang, Xi Yu-fan, Su Yan-min, Zhang Jing-Gang

The Optimization-Sliding Mode Control For Three-Phase Three-Wire DSP-based Active Power Filter 1674
Zhou Wei-ping, Liu Da-ming, Wu Zheng-guo, Xia Li, and Yang Xuan-fang

Three-Phase DVR using a Single-Phase Structure with Combined Hysteresis/ Dead-band Control 1679
Seyyed Hossein Hosseini, Mehran Sabahi

Harmonic Detection Based on the TLS Estimation Algorithm 1684
Liu Kaipei, Zhang Junmin

Control Strategy Study of Hybrid Active Power Filter 1689
Jia Zhang, Guohong Zeng

Novel Harmonic Free Single Phase Variable Inductor Based on Active Power Filter Strategy 1693
Mu Xianmin, Wang Jianze, Ji Yanchao, Wei Xiaoxia, Fu Xiangyun

A Multi-Output Series Resonant Inverter with Asymmetrical Voltage-Cancellation Control for Induction-Heating Cooking Appliances 1697
S.H. Hosseini, A. Yazdanpanah Goharrizi, E. Karimi

Capacitor Voltage Control in a Cascaded Multilevel Inverter as a Static Var Generator 1703
M. Li, J. N. Chiasson, L. M. Tolbert

DC-link Pumping-up Voltage Suppression of a Series Active Voltage Regulator With Phase Shift Control 1708
G. C. Xiao, Z. L. Hu, C. H. Nan, Z. A. Wang

The Fuzzy Soft-startup Controller of Active Power Filter 1713
He Na, Wu Jian, Xu Dianguo

xix

Table of Contents

A Novel Control Method for DSTATCOM Using Artificial Neural Network..1718
Yang Xiao-ping, Zhong Yan-ru, Wang Yan

A Detailed Analysis of Unexpected DC-side Voltage Boost in Series Power Quality Controllers 1722
Yuan Chang, Liu Jinjun, Wang Xiaoyu, Wang Zhaoan

Comparative Analysis of Popular Control Schemes for Parallel Active Power Filter and Experimental Verification...1726
Xiaoyu Wang, Jinjun Liu, Chang Yuan, Zhaoan Wang

Accurate Modeling of the Three Phase Induction Motor Including Saturation Effects................................1731
E. V. N. Souza, S. R. Naidu

A study on the reliability evaluation of driving parts for note handling units 1736
Joo Han Kim, Jung Kee Chung, Ha Kyeong Sung, Se Hyun Rhyu

Analysis on Toothless Permanent Magnet Machine with Halbach Array...1741
Xu Yanliang, Feng Kaijie

Improvement in Reliability of Doubly Salient Permanent Magnet Motor Drive...1746
Wenxiang Zhao, Ming Cheng, Xiaoyong Zhu, Wei Hua, Jianzhong Zhang

A New Approach of Modeling the Saturated Induction and Synchronous Salient Pole Machines 1751
A. Câmpeanu, M. Badica

Inductance characteristics of 3-phase fluxswitching permanent magnet machine with doubly-salient structure 1758
Wei Hua, Cheng Ming

Performance Index Evaluations of a Micro Axialflux Switched-reluctance Motor......................................1763
Cheng-Tsung Liu, Yen-Ming Chen, Da-Chen Pang

Study of Variable Frequency Operation of Induction Generator for Wind Power......................................1768
Noriyuki Kimura, Mitsuhiro Hirao, Toshimitsu Morizane, Katsunori Taniguchi

Optimal Power Control Strategy of Maximizing Wind Energy Tracking and Conversion for VSCF Doubly Fed Induction Generator System 1773
H. Li, Z. Chen, John K. Pedersen

Design and Evaluation of a Dual Mechanical Port Machine and System 1779
Longya Xu, Yuan Zhang

Characteristic Analysis on Overhang Effect in Axial Flux PM Synchronous Motors with Slotted Winding 1784
WonYoung Jo, YunHyun Cho, YonDo Chun, DaeHyun Koo

Design and Analysis of a Double-Stator Cup-Rotor Directly Driven Permanent Magnet Wind Power Generator 1788
Dong Zhang, Shuangxia Niu, K. T. Chau, J. Z. Jiang, Yu Gong

Feasibility Analysis of Accelerometer Configuration of Non-gyro Micro Inertial Measurement Unit.....................1793
Ding Mingli, Zhou Qingdong, Wang Qi, Wang Changhong

Design of Fractional-Order a PI Controller with two modes...1797
Wen Li, Yoichi Hori

Sliding Mode Robust Tracking Control Based on Learning Feedforward Compensation for High Precision Linear Servo System 1802
Zhu Guoxin, Guo Qingding, Zhao Ximei

Application of Fuzzy Self-learning Sliding Mode Variable Structure Control in Linear AC Servo System..............1806
Qing Hu, Shuo Jie, Dongmei Yu

Dynamics Research of Robot Manipulator 1811
Zhibing Shu, Caizhong Yan, Hairong Zhang

Table of Contents

Advanced Angle Control Schemes for Stator Hybrid Excited Doubly Salient Motor Drive 1815
Xiaoyong Zhu, Ming Cheng, Wenxiang Zhao, Wenguang Li

A Design Method of Reconfigurable Controller for AC Position Servo Systems 1820
Wu Qinmu, Qin Yi, Li Yesong

Position Sensorless Control of PMSM Based on a Novel Sliding Mode Observer over Wide Speed Range 1825
Song Chi, Student Member, Longya Xu,

Design of Motion Control System Used for Filter Rod Production Machine 1832
Yang Qingyu, Ge Sibo, Ye Kesong, Shi Ren

Analysis and Implementation of Sensorless Position Detection in a Permanent Magnet Generator 1836
Sebastian Rosado, Xiangfei Ma, Fred Wang, Jerry Francis, Dushan Boroyevich

Torque-Speed Characteristics of Interior-Magnet Machines in Brushless AC and DC Modes, with Particular Reference to Their Flux-Weakening Performance 1841
Y. F. Shi, Z. Q. Zhu, D. Howe

H8 Robust Control for Dual Linear Motors Servo System 1846
Zhao Ximei, Guo Qingding

Research on Linear Motor Driving System Based on Wavelet Transform 1849
Cui Jiefan, Zhao Lijun, Wang Hemin, Wan Junzhu, Jiang Lili

Study on Rotor Position Detection Error in Sensorless BLDC Motor Drives 1853
Li Qiang, Wang Ruixia

A New Scheme to Direct Torque Control of Interior Permanent Magnet Synchronous Machine Drives for Constant Inverter Switching Frequency and Low Torque Ripple 1858
Jun Zhang, M. Faz Rahman, Colin Grantham

A Modified Direct Toque Control for Interior Permanent Magnet Synchronous Motor Drive Without a Speed Sensor 1863
Yanping Xu, Yanru Zhong, Hui Yang

Direct Torque Control for Interior Permanent Magnet Synchronous Motors Using Matrix Converters 1867
D. Xiao, M. F. Rahman

A Neural Network Based Initial Position Detection Method To Permanent Magnet Synchronous Machines 1872
Mengjia Jin, P.C.K Luk, Jianqi Qiu, Cenwei Shi, Ruiguang Lin

A New Recurrent Fuzzy Neural Network Sliding Mode Position Controller Based on Vector Control of PMLSM Using SVM 1877
Junyou Yang, Ruijuan Chen, Naiguang Fa

DSP Implementation of Rotor Position Detection Method for Hybrid Stepper Motors 1882
M. Bendjedia, Y. Ait-Amirat, B. Walther, A. Berthon

An In-Wheel Switched Reluctance Motor for Electric Vehicles 1887
P.C.K. Luk, P. Jinupun

Speed Sensorless Vector Control of Induction Motor Based on Full-Order Flux Observer 1892
Shanshan Wu, Yongdong Li, Zedong Zheng

A Parameter Identification Method for General Inverter-fed Induction Motor Drive 1896
Xiaochun Jiang, Geng Yang, Yunfei Wang

Indirect Rotor Field Orientation Vector Control for Induction Motor Drives in the Absence of Current Sensors 1901
Z. S. WANG, S. L. HO

A Robust Adaptive Sliding-Mode Controller for Slip Power Recovery Induction Machine Drives 1906
J.Soltani, A. Farrokh Payam

xxi

Table of Contents

Identification of the Rotor Time Constant in Induction Machines without Speed Sensor 1912
M. Li, J.N. Chiasson, M. Bodson, L.M. Tolbert

Adaptive Control of Doubly Fed Field-Oriented Induction Machine Based On Recursive Least Squares Method Taking the Iron Loss Into account 1917
N. R. Abjadi, J. Askari, J. Soltani

Analysis and Design of PDM Converter with High Frequency Link for HEV Drive System 1922
Ma Xianmin

A Multi-Directional Power Converter for a Hybrid Renewable Energy Distributed Generation System with Battery Storage 1926
Mei Qiang, Wu Wei-Yang, Xu Zhen-lin

Four-bridge Multilevel Converters Based on Hybrid-clamped Techniques 1931
Xiaofeng Wang, Yan Deng, Xiangning He

Standardization of Input/Output Impedance Specifications of Buck Converters Based on the System Integration Concept 1936
Tao Wu, Xinbo Ruan

Research on The Magnetic Integration in Three-Level ZCS Quasi-Resonant Buck Converter 1942
Jiang Ying, Xiang Hui-jie, Yang Yu-gang, Liu Nan

Decoupling Control of Magnetically Levitated Induction Motor with Inverse System Theory 1947
Yang Zhou, Huangqiu Zhu, Tianbo Li

Fault Detection and Accommodation for Nonlinear Systems Using Fuzzy Neural Networks 1952
H. Xue, J.G. Jiang

A Novel Constant Power Control of High Frequency Electronic Ballast Applying the PLL Technique for a Metal Halide Lamp 1957
Chang-Hua Lin, Chung-Lun Ou, Tien-Shuo Liu, Ken-Chuan Hsu

The Voltage Stability Research of Ship Electric Power System 1962
Fanyinhai Zhaomin

Parasitic Gate Resistance and Switching Performance 1967
Alan Elbanhawy

PWM Rectifier with DC Reverse-Blocking Diode for High-Reliability Generating Apparatus and Its Application to Gas Heat Pump System 1971
Akio Toba, Toshihiro Maeda, Kouetsu Fujita, Tomohiko Kato

A Novel Stator Section Crossing Method of Long Stator Linear Synchronous Motor for Maglev Vehicles 1976
Qian Zhang, Fei Lin, Xiaojie You, Trillion Q. Zheng

Common Mode Current Suppression in Full-Bridge Converter Based on Simulated Annealing Algorithm 1981
Yonggao Zhang, Kai Zhang, Yunbin Zhou, Yong Kang

Summary of Distance Measurement Based on Vision in Localization Technology 1986
Handong Zhang, Gang Wang, Yuwan Cen

The studies of Single-phase Inverter Fault Diagnosis Based on D-S Evidential Theory and Fuzzy Logical Theory 1991
Wang Baocheng, Li Danhe, Sun Xiaofeng, Wu Weiyang

A Novel Single-Stage High-Power-Factor Electronic Ballast with Symmetrical Half-Bridge Topology 1995
Chien-Ming Wang, Chien-Yeh Ho

Smoothed-Power Output Supply System for Battery of Stand-alone Renewable Power System Using EDLC 2000
Y. Jia, R. Shibata, N. Yamamura, M. Ashida

Table of Contents

Supercapacitors characterization for hybrid vehicle applications 2005
F. Rafik, H. Gualous, R. Gallay, A. Crausaz, A. Berthon

Power Transfer Maximization and Di/Dt Based Extremum Tracking for a Swing Engine Based Portable Power System 2010
Satish Rajagopalan, Deepak M. Divan, Ronald G. Harley, J. Rhett Mayor

3D FEA of the Stator of the Linear Magnetic Flux Compression Generator 2015
Yanjie Cao, Chengxue Wang

The Effect of Current Control Strategies on Power Consumption of a Magnetically Levitated Turbomolecular Pump 2018
A.E. Hartavi, R.N. Tuncay, M.N. Sahinkaya

Direct Torque Control of an Interior Permanent Magnet Synchronous Machine fed by a Direct AC-AC Converter 2023
D. Xiao, M. F. Rahman

Control of Distributed Power Systems 2029
Z. Chen, Y. Hu, F. Blaaberg

xxiii

Test Loadability of Power Systems using A Networked Power Electronic Devices Control and Measurement System

Sheng Yang[*], Venkataramana Ajjarapu[*] and Bo Zhang[**]

[*] Department of Electrical and Computer Engineering, Iowa State University, Ames IA 50011, USA
[**] Electric Power College, South China University of Technology, Guangzhou, Guangdong 510641, CHINA
E-mail: [*] {shengy, vajjarap}@iastate.edu, [**] epbzhang@scut.edu.cn

Abstract—The study of maximum loadability of power systems reveals critical information of steady-state system behavior and allows engineers to develop control strategies against voltage collapse. To achieve this goal, we develop a networked power electronic devices control and measurement system which can be accessed by researchers and students over Internet. This work is the initial step of a Multi-University Research Initiative (MURI) project in terms of device development for remote and nondestructive testing and measurement.

Keywords-networked control and measurement; power electronic devices; loadability; remote and nondestructive experiments

I. INTRODUCTION

In August 2003, over 40 million American and Canadian residents were plunged into darkness [1]. This recent disaster urges researchers from academic and industrial communities to find effective methods to prevent similar blackouts from happening in the future. It is well known that the security of power systems is threatened when it is loaded near to or beyond its maximum capacity [2]. Therefore, the research work on maximum loadability of power systems may significantly help engineers design control strategies against blackouts.

Traditionally, most studies of large-scale power systems under stressed loading conditions are based on software simulations. Unfortunately software models can not accurately describe the behavior of actual power system hardware in many ways. On the other hand, it is impossible to equip every research affiliations enough hardware components to build an actual scaled system since enormous time, money, space and manpower is needed to achieve this goal. This dilemma has limited the possibility that researchers are able to develop state-of-the-art technologies against blackouts.

To improve this situation, five universities in US have agreed that power systems and power electronics laboratories from them will be interconnected via

This work is supported by the Office of Naval Research under ONR N0014-04-1-0404.

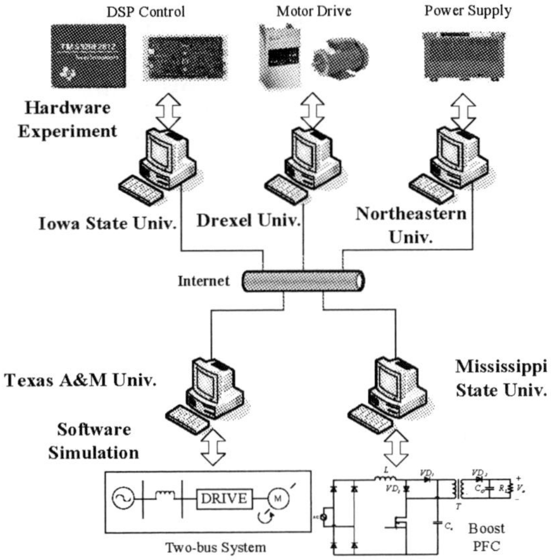

Figure 1. Interconnection of power system/power electronic laboratories at five universities

networks [3]. Therefore, all available hardware and software tools related to power systems and power electronics are combined together to create a large-scale power system laboratory. This will allow researchers at various locations to study processes that lead to system breakdowns, including system behaviors under stressed loading conditions. The Internet connection is preferred in this project because of its widespread usage, well-developed structure and affordability. The conceptual connection diagram is shown in Figure 1.

We will discuss the initial efforts at Iowa State University (ISU) related to development of the networked loadability testing system using a controllable three-phase power supply. This paper is organized as follows. In section II we derive formulas to calculate the critical active power and voltage of the load in a two-bus system. Section III and IV present the design and setup of this system, including hardware and software configurations and development. The results and analysis are provided in section V. The final section concludes this paper.

II. PROBLEM FORMULATION

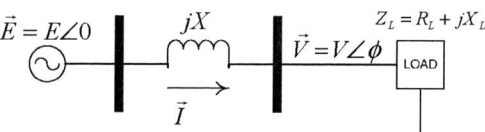

Figure 2. A two-bus system

In this section the main equations related to the maximum loadability of a two-bus system shown in Figure 2 will be introduced. The resistance of the transmission line is neglected. We assume the voltage at the generation bus is fixed and take it as the reference.

Using basic knowledge and analysis, we have

$$\overline{V} = \overline{E} - \overline{I} \cdot jX \tag{1}$$
$$\overline{S} = \overline{V}I^* \tag{2}$$
$$\overline{S} = P + jQ \tag{3}$$

where $\overline{S}$, P, Q are apparent power, active power and reactive power of the load, respectively.

Note that $\overline{E} = E\angle0$, $\overline{V} = V\angle\phi$ and (3). Solving $\overline{I}$ by (1) and substituting it into (2) lead to

$$P = -VE\sin\phi / X \tag{4}$$
$$Q = (VE\cos\phi - V^2)/X \tag{5}$$

Let $v = V/E$, $p = PX/E^2$, $q = QX/E^2$. Note that $\sin^2\phi + \cos^2\phi = 1$ and $q = p\tan\theta$, where θ is the power factor of the load. By the simplification of (4) and (5), we have

$$v^4 + (2p\tan\theta - 1)v^2 + p^2\sec^2\theta = 0 \tag{6}$$

To obtain the critical active power and voltage of the load, we differentiate (6) with respect to v and equate it to zero ($\partial p / \partial v = 0$). After the simplification of the equations, we have

$$p_{critical} = \cos\theta / (2 + 2\sin\theta) \tag{7}$$
$$v_{critical} = 1/\sqrt{(2 + 2\sin\theta)} \tag{8}$$

Consider the special case that the load is purely resistive, i.e. $\theta = 0$. We get $p_{critical} = 0.5$ and $v_{critical} = 0.707$, which imply that

$$P_{critical} = E^2 / 2X \tag{9}$$

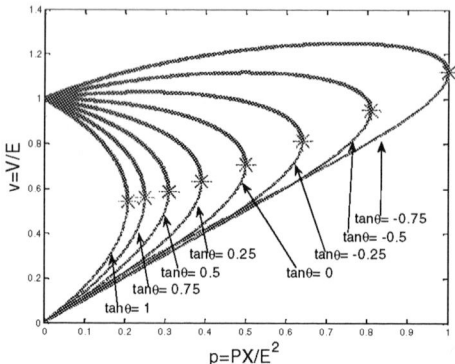

Figure 3. P-V curves with load power factor as the parameter

$$V_{critical} = E/\sqrt{2} \tag{10}$$

From (9) and (10), we may find that if the resistance of the load is equal to the inductance of the transmission line, i.e. $R = X$, the load will get the maximum active power, which is $E^2/2R$.

Figure 3 illustrates a group of P-V curves with different load power factors. The solid and dash line of each curve denote the stable and unstable operating region, respectively. The critical points are marked by stars.

III. NETWORKED TESTING AND MEASUREMENT SYSTEM

We built a scale-down hardware system to simulate the two-bus system as shown in Figure 2. The descriptions of the setup and components are detailed in the following contents.

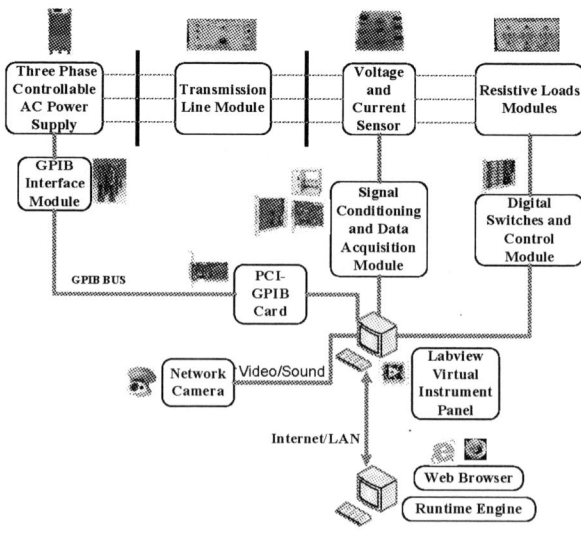

Figure 4. Networked loadability testing system setup: a sample two-bus system

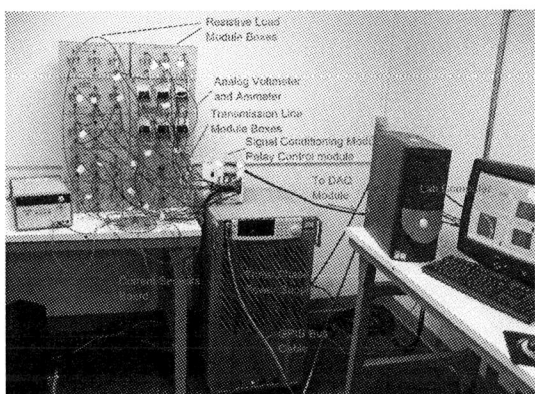

Figure 5. Actual experiment setup in local lab

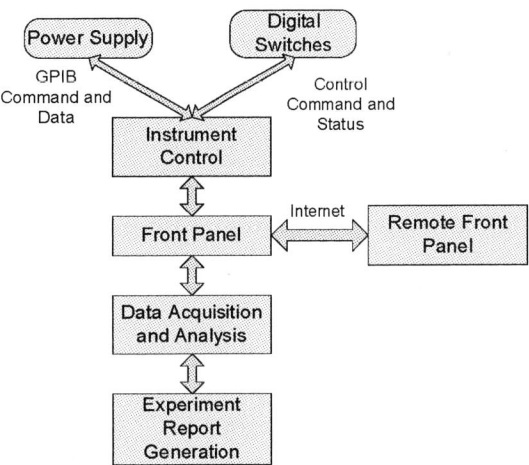

Figure 6. Software modules

A Kikusui PCR6000W2 commercial three-phase remotely controllable power supply is used as the generator shown in Figure 2 since the output voltage is assumed to be fixed. The combination of a high frequency PWM inverter and a high power factor converter inside it offers a high efficiency and a low input current, a compact and lightweight body, and high quality output.

The conceptual diagram of experiment setup is shown in Figure 4. An actual loadability testing system in the local laboratory depicted as the block diagram in Figure 4 is shown in Figure 5. Here the two bus power system consists of the three-phase controllable power supply, two three-phase Lab-Volt transmission line modules and six resistor modules as the three phase load. The power supply is equipped with a General Purpose Interface Bus (GPIB) interface module. The amplitude and frequency of output voltage is fully controlled by the lab computer over GPIB with the help of instrument drives. Three current sensors are installed between the transmission line modules and resistive loads. These current sensors can sense AC current while the linear output voltage tracks the sensed current waveform. Through the signal conditioning module and data acquisition module, load voltage and current signals are sent back to the Virtual Instrument (VI) running on a lab computer for calculation and analysis. A switch control module with sixteen digital switches is used to change the connection topology of various resistors to get different load. A network camera with pan, tilt and zoom control enables advanced remote monitoring over IP networks by delivering high quality video and audio streams.

The software system was programmed using Labview developed by National Instruments to perform device control and data acquisition, provide Human Machine Interface (HMI) to local and remote operators and communicate signals over Internet. With built-in web-based tools, Labview can publish the front panel of visual instruments on the web and provide exactly the same user interface to remote users.

IV. SOFTWARE DEVELOPMENT

The main software modules are shown in Figure 6 and explained in the following subsections.

A. Instrument Control

The manufacturer of the controllable power supply provides communication protocols based on the GPIB interface. Major functionalities of communication between the power supply and host computer are encapsulated into SubVIs. It is convenient to write control commands and read values of desired variables such as voltage amplitude, frequency, power factor, harmonic and equipment status.

A group of digital switches are operated under the control module. These switches are actually Single-Pole Double-Throw (SPDT) relays. Before connecting the common terminal and one pole, the program should check if they can be connected. After the connecting action, the program must pause until the created path has settled.

B. Data Acquisition and Analysis

There are six data input channels where waveforms of three-phase voltage and current at the load side are sampled and transferred. The program will read these data after some changes of the load, or some operations from users, depending on the type of experiment control methods, i.e. automated or user-defined, respectively. The phase difference between voltage and current waveforms is required to calculate the active or reactive power. A SubVI was developed for this purpose.

C. Experiment Report Generation

The parameters and data, such as load resistance, voltage, current and power, are stored in the software module during the process of experiments and output to some special file formats like text or excel files after

507

experiments. This provides data records for researchers and students for the future references.

D. Local and Remote Front Panel

The local front panel has a set of buttons, knobs and input fields for control inputs. Also several waveform charts, graphs and output indicators are provided for the display of equipment status and results.

The remote front panel looks exactly the same as the local one because it is generated by the web-based tools in Labview. The Labview runtime engine and a web browser are required for remote users. The runtime engine can be downloaded from the National Instruments website for free. Thus researchers and students can perform the experiment as if they were in the laboratory.

E. Experiment Procedures

The remote operating procedures for a set of specified load resistance are provided below. The front panel is shown in Figure 7.

Step 1: Open your web browser, for example, Internet Explorer. Type the address in the address bar and press "Enter".

Step 2: After the front panel is fully loaded, right click on any area of the panel except controllers, indicators and charts. A yellow rectangular box which shows "control granted" will display at the center of the front panel, which means you have gained the access to control the experiment.

Step 3: Select Operate>>Reinitialize All to Default on the menu bar.

Step 4: Select Operate>>Run or click the right arrow icon on the toolbar. The virtual instrument is now enabled.

Step 5: Select the amplitude of phase voltage and frequency of the power supply by tuning knobs or directly typing data on the input field. Select the reactance of the transmission line box. The default values of these parameters are recommended because they are selected for this set of specified load resistance. Here we set the voltage and frequency to be 50VAC and 60Hz. Click the "Save" button.

Step 6: Click the "Power Supply Output" button to enable the power supply output. You will notice the green light is on after you enable the output. Now you can begin to read and record the experiment results. In this experiment, we provide eleven different load resistances. Hence you will get eleven values of load voltage and power. "Operating Points: 1 of 11" means that resistance #1 is now implemented as the load.

Step 7: Click the "Get Voltage and Current Waveforms" button. The three-phase load voltage and current are shown on two graphs. Also, the values are displayed right to the graphs.

Step 8: Click the big "Save..." button. You will notice that the text of "Operating Points: 2 of 11" is displayed. This means resistance #2 is now implemented as the load.

Step 9: Repeat Step 7 and Step 8 until all values of 11 operating points are saved. Finally, three P-V curves for each phase will be plotted on the P-V curve graph automatically. The virtual instrument will immediately exit the running status and end this session.

Step 10: If you would like to repeat the experiment or change some parameters, go back and proceed from Step 3.

V. EXPERIMENT RESULTS

Since several hardware configurations are available for the testing system, we only provide the measurements for two sets of specified load resistances in Table I and II. One set of p-v curves are shown in the right-bottom corner of Figure 7.

TABLE I.
THREE-PHASE LOAD POWER (LINE REACTANCE=180 Ω)

No.	R1 1200Ω	R2 600Ω	R3 300Ω	R4 300Ω	Actual Load(Ω)	P(3φ) (W)
1	O	X	X	X	1200	5.99
2	X	O	X	X	600	10.63
3	O	O	X	X	400	14.17
4	X	X	X	O	300	16.02
5	O	X	O	X	240	17.48
6	X	O	X	O	200	18.22
7	O	O	O	X	171.43	18.41
8	X	X	O	O	150	18.23
9	O	X	O	O	133.33	17.84
10	X	O	O	O	120	17.33
11	O	O	O	O	109.09	16.77

TABLE II.
THREE-PHASE LOAD POWER(LINE REACTANCE=120Ω)

No.	R1 1200Ω	R2 600Ω	R3 300Ω	R4 300Ω	Actual Load(Ω)	P(3φ) (W)
1	O	X	X	X	1200	5.99
2	X	O	X	X	600	10.63
3	O	O	X	X	400	14.17
4	X	X	X	O	300	16.02
5	O	X	O	X	240	17.48
6	X	O	X	O	200	18.22
7	O	O	O	X	171.43	18.41
8	X	X	O	O	150	18.23
9	O	X	O	O	133.33	17.84
10	X	O	O	O	120	17.33
11	O	O	O	O	109.09	16.77

O: The resistor is set to ACTIVE (connected as a part of total load)
X: The resistor is set to INACTIVE (disconnected from load)

The reactance of transmission lines are selected to be different, that is, $X = 120\Omega$ and $X = 180\Omega$. According to the conclusion in section II, the load will get the maximum active power if $R = X$. In Table I, when the load resistance is 171.43Ω, which is the closest value to 180 Ω based on available resistances, the load will get the maximum three-phase power (18.41W). In Table II, when the load resistance is 120Ω, the load will get the maximum three-phase power (26.81W).

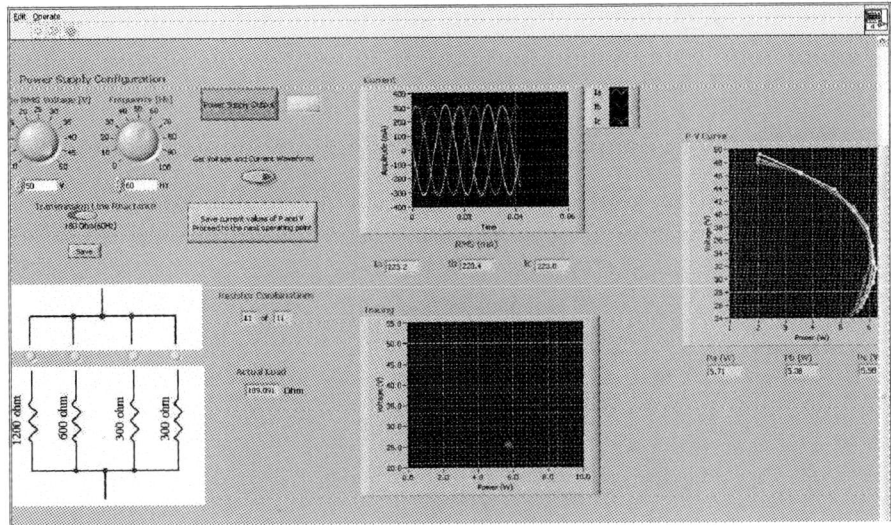

Figure 7. Remote front panel (X=180 Ω)

VI. CONCLUSION

It is believed that interconnecting power hardware systems and software simulation and control tools over network will provide a significantly enhanced platform for the study of large-scale power and power electronic systems. In this paper, we develop a networked power supply control and measurement system to test the loadability of a two-bus system. The maximum loadability of the two-bus system is derived. The design and setup of the system are presented and explained in details with respect to the hardware configuration and software development. The system provides two types of experiment operating procedures: automated and user-defined. The results of local test and demonstration between ISU and Drexel show the system can act as a learning and research platform for students and researchers who can remotely perform the loadability testing experiment.

This work is the first step of the MURI project at ISU. The next work may concentrate on the analysis on close-loop control systems when network delays and data losses are taken into account.

ACKNOWLEDGMENT

The authors would like to thank Jason Boyd for the layout design of the current sensor board.

REFERENCES

[1] U.S.-Canada Power System Outage Task Force, "Final report on the August 14, 2003 blackout in the United States and Canada: causes and recommendations," April 2004.

[2] Ajjarapu, V.; Lee, B.; "Bibliography on voltage stability," *IEEE Transactions on Power Systems,* Volume 13, Issue 1, Feb. 1998 Page(s):115 – 125.

[3] Miu. K; et al; "Device development for remote, nondestructive testing and measurement of power systems," MURI project description, Nov. 2003.

[4] Miu. K; et al, "Testing of shipboard power systems: a case for remote testing and measurement," Electric Ship Technologies Symposium, 2005 IEEE, 25-27 July 2005 Page(s):195 – 201

[5] Curran, K.; "A Web-based collaboration teaching environment," IEEE Multimedia, Volume 9, Issue 3, July-Sept. 2002 Page(s):72 – 76

[6] Sanchez, J.; Dormido, S.; Pastor, R.; Morilla, F.; "A Java/Matlab-based environment for remote control system laboratories: illustrated with an inverted pendulum", *IEEE Transactions on Education,* Volume 47, Issue 3, Aug. 2004 Page(s):321 – 329

[7] Tipsuwan, Y.; Mo-Yuen Chow; "On the gain scheduling for networked PI controller over IP network," *IEEE/ASME Transactions on Mechatronics,* Volume 9, Issue 3, Sept. 2004 Page(s):491 – 498

[8] Golder, A.; Miu, K.; Chika Nwankpa; Carullo, S.; "Remote hardware power system loading studies over the World Wide Web," Power Symposium, 2005. Proceedings of the 37th Annual North American, 23-25 Oct. 2005 Page(s):9 – 15

[9] "Kikusui PCR-W series GPIB interface operation manual," http://www.kikusui.co.jp/en

[10] "National Instruments Labview user manual, April 2003 edition," http://sine.ni.com/manuals

2006 5th International Power Electronics and Motion Control Conference

Test-Bed of Doubly Fed Induction Generator for Variable-Speed Constant-Frequency Wind Power Generation

S. Y. Yang [*], X. Zhang [*], C. W. Zhang [*] and R. X. Cao [**]

[*] School of Electric Engineering and Automation, Hefei University of Technology, Hefei, P.R.China
[**] Hefei Sunlight Power Supply Co.Ltd. Hefei, Anhui, P.R.China
e-mail eehut@mail.hf.ah.cn

Abstract—**In this paper, a test-bed built for the experiments of variable-speed constant-frequency (VSCF) wind power generation is described. The test-bed is composed of a doubly fed induction generator (DFIG), a VVVF inverter driven squirrel motor for wind turbine simulation, dual-DSP-based controllers, as well as a PC-based data-acquisition and control system. In this system, a pair of bi-directional PWM voltage-source converters in a back-to-back configuration is inserted between the rotor windings and the grid lines to control the rotor currents in order to vary the slip power thereby implementing the VSCF operation. Vector-control is used to independently control the flow of active and reactive power between the system and grid line, and to make it easy for the system to implement maximal power point tracking (MPPT) and optimal reactive power splitting in wind power plants.**

Keywords- test-bed; wind power generation; variable-speed constant-frequency; doubly fed induction generator

I. INTRODUCTION

Doubly fed induction generator (DFIG) is able to supply power at constant voltage and constant frequency while the rotor speed varies, that makes it suitable for variable-speed wind power generation [1], [2]. Another major advantage of the DFIG-based system is that the power electronic equipments only need to handle a fraction of the total system power, resulting the reduction of the power losses and the cost of the power electronic equipments [3].

A key design of DFIG-based systems is in the rotor circuits. It may be implemented with either a current-fed DC-link converter or a cyclo-converter [3], [4]. But those approaches exhibit disadvantages such as expensive DC-link chokes, extra commutation circuits, current harmonics, low power factor etc. However these disadvantages may be overcome by using back-to-back PWM converter pair in the rotor circuits [4].

Facing the system complexity and installation hardship, the overall system functionality, performance and safety should be checked and tested on ground level before installation as comprehensively as possible to

avoid a single malfunction leading to system damage, tedious reparation, schedule delay and cost increases [5]. So a test-bed is quite necessary for the development of VSCF wind power generation system.

In the literatures [3], [4], [6], [7], a DC motor is used for wind turbine simulation. The merit of this kind of test-bed system is the easiness of control; however the drawbacks are in bulk and maintenance, so that unsuitable for high power applications. To simulate high power wind power generation, an 110kW DFIG-based VSCF wind power generation test-bed has been set up with a squirrel induction motor simulating the wind turbine.

The paper describes the composition of the test-bed, the mathematical models and the control structures of both the grid-side and rotor-side converters. Then the simulation of windmills is presented. At last, the relevant experimental results to validate the design are reported

II. EXPERIMENTAL SYSTEM

DFIG-based VSCF wind power generation test-bed mainly consists of the following components: a squirrel induction machine driven by a vector-controlled inverter to simulate wind turbine, a PC with developed software simulates a virtual wind farm, a DFIG with a back-to-back PWM converter pair, and dual-DSP-based controllers. The schematic diagram of the overall system is shown in Fig.1. As described in [8], the back-to-back arrangement of two PWM converters with a common dc-link bus makes the system extremely flexible in terms of control of active and reactive power flow. Since the converters make the rotor circuit have the capability of sourcing and sinking the slip power, the generator can be operated at sub-synchronous and super-synchronous speeds.

As [9] has pointed out, the so-called crowbars is used to short circuit the rotor windings of the generator to bypass the circuit current, which makes the generator be capable of operation during grid faults.

The PC is used to monitor the system. Software, developed with Labview and run on PC, delivers a virtual wind speed and calculates the rotor speed by considering

1-4244-0448-7/06/$25.00 ©2006 IEEE

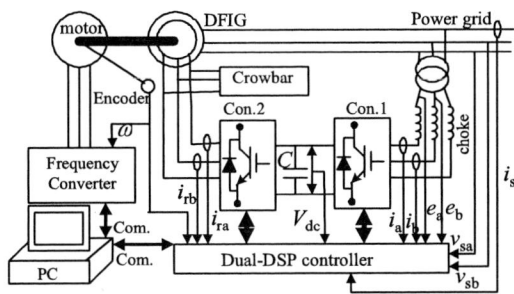

Figure 1. Schematic of experimental system

the tip-speed ratio of wind turbine, and then downloads the speed via RS-485 to the inverter to drive the squirrel motor running at the given speed. Meanwhile the maximum power point is fetched in a look-up table representing the peak power points curve and set beforehand in the software, and this value is also downloaded via RS-485 to the dual-DSP-based controller to force the DFIG output power to follow it.

III. CONTROL OF GRID-SIDE CONVERTER

The grid-side converter serves to meet two purposes: keeping the DC-link voltage constant regardless of the magnitude and direction of the rotor power [4]; joining the reactive power composition [10], [11]. The scheme of the grid-side converter is shown in Fig.2.

In the stationary three-phase coordinates, the state equation of the converter is [12]:

$$\begin{bmatrix} e_a \\ e_b \\ e_c \end{bmatrix} = R \begin{bmatrix} i_a \\ i_b \\ i_c \end{bmatrix} + Lp \begin{bmatrix} i_a \\ i_b \\ i_c \end{bmatrix} + \begin{bmatrix} v_{a0} \\ v_{b0} \\ v_{c0} \end{bmatrix} \qquad (1)$$

where L and R are the inductance and resistance of the line reactors, respectively; $p = d/dt$ is derivative operator.

To facilitate independent control of the active and reactive power flowing between the grid and the grid-side converter, (1) is transformed into a dq coordinates with its q-axis fixed on the grid voltage vector position, expressed as (2) [4].

$$\begin{bmatrix} v_q \\ v_d \end{bmatrix} = -R \begin{bmatrix} i_q \\ i_d \end{bmatrix} - Lp \begin{bmatrix} i_q \\ i_d \end{bmatrix} - \omega_e L \begin{bmatrix} -i_d \\ i_q \end{bmatrix} + \begin{bmatrix} e_q \\ e_d \end{bmatrix} \qquad (2)$$

where ω_e is the synchronous rotating speed of the grid-voltage vector; subscripts d, q indicate the vector components in the rotating (d-q) reference coordinates.

Assuming $v_q' = Ri_q + Lpi_q$, $v_d' = Ri_d + Lpi_d$ and $e_{\omega q} = -\omega_e Li_d$, $e_{\omega d} = \omega_e Li_q$, (2) can be rewritten as:

$$\begin{bmatrix} v_q \\ v_d \end{bmatrix} = -\begin{bmatrix} v_q' \\ v_d' \end{bmatrix} - \begin{bmatrix} e_{\omega q} \\ e_{\omega d} \end{bmatrix} + \begin{bmatrix} e_q \\ e_d \end{bmatrix} \qquad (3)$$

With feed forward control scheme for the grid voltage $e_{q,d}$ and compensation control scheme for the cross-coupling terms $e_{\omega q, \omega d}$, i.e. only considering the impedance voltage v_q' and v_d', the plant for the current control loops is given by (4) [4]:

$$G(s) = \frac{I_q(s)}{V_q'(s)} = \frac{I_d(s)}{V_d'(s)} = \frac{1}{Ls + R} \qquad (4)$$

Neglecting the loss and the voltage drop of inductors and considering the power balance between the AC-side and DC-side of the converter, the DC-link voltage can be approximated by:

$$\frac{dV_{dc}}{dt} = \frac{1}{C}(\frac{\sqrt{3}}{2}mi_q - i_{dc}) \qquad (5)$$

where m is the SVPWM modulation index.

From (5), it can be found that the DC-link voltage is controlled by the current i_q. Therefore the output of the voltage control loop corrector is selected as the reference of the q-axis component of the AC currents. The reference of the d-axis component is dependent on the optimal reactive power splitting [10], [11]. The space vector control scheme for the grid-side converter is shown in Fig.3.

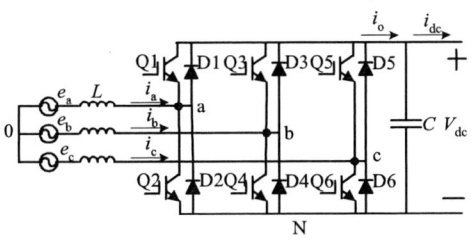

Figure2. Structure of the supply-side PWM converter

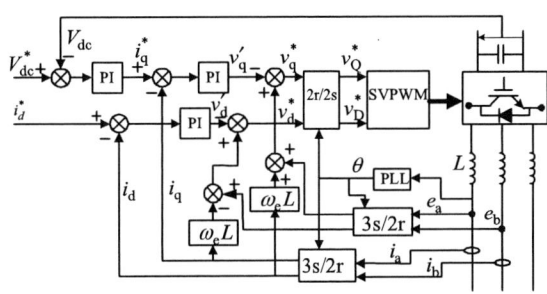

Figure3. The control structure for grid-side converter

IV. CONTROL OF DFIG

In order to improve the robustness with respect to generator parameters variation [13] and make the stability of the system immune to the value of the d-axis component of the rotor currents [14], the stator voltage vector, rather than flux vector, oriented reference coordinates is adopted for the control of the generator, i.e. the q-axis of the reference coordinates is oriented along the stator-voltage vector position, d-axis is 90° lag behind the q-axis in the direction of rotation and rotates at the power-grid voltage vector speed ω_s. In the dq reference coordinates, the decoupled control between the electrical torque component and the excitation component of the rotor currents is obtained. Since the stator resistance is quite small compared to the stator reactance ($R_s \ll \omega_s L$) and can be neglected [15], the relationship among the rotor fluxes ψ_{rdq}, the stator fluxes ψ_{sdq}, the rotor voltages v_{rdq} and the electrical torque T_e can be expressed by motor convention:

$$\begin{cases} \psi_{sd} = \psi_s = L_s i_{sd} + L_m i_{rd} = L_m i_{ms} \\ \psi_{sq} = L_s i_{sq} + L_m i_{rq} \end{cases} \quad (6)$$

$$\begin{cases} \psi_{rq} = (L_r - L_m^2/L_s) i_{rq} \\ \psi_{rd} = (L_m^2/L_s) i_{ms} + \sigma L_r i_{rd} \end{cases} \quad (7)$$

$$\begin{cases} v_{rq} = R_r i_{rq} + \sigma L_r p i_{rq} \\ \quad + \omega_{sl}((L_m^2/L_s) i_{ms} + \sigma L_r i_{rd}) \\ v_{rd} = R_r i_{rd} + \sigma L_r p i_{rd} t - \omega_{sl} \sigma L_r i_{rq} \end{cases} \quad (8)$$

$$T_e = -(1.5 n_p L_m^2/L_s) i_{ms} i_{rq} \quad (9)$$

where, i_{ms}: the equivalent stator magnetizing current, $\sigma = 1 - L_m^2/(L_s L_r)$: the leakage coefficient; ω_s, ω_{sl}: the synchronous and slip angular speed; n_p: the number of the pole pair; R_r: the rotor resistance; L_r, L_s: the stator and rotor self-inductances; L_m: the mutual inductances.

With a similar process as that for grid-side converter, (8) is rewritten as:

$$\begin{cases} v_{rq} = v'_{rq} + \omega_{sl} \sigma L_r i_{rd} + \omega_{sl}(L_m^2/L_s) i_{ms} \\ v_{vd} = v'_{rd} - \omega_{sl} \sigma L_r i_{rq} \end{cases} \quad (10)$$

where $v'_{rq} = R_r i_{rq} + \sigma L_r p i_{rq}$, $v'_{rq} = R_r i_{rq} + \sigma L_r p i_{rq}$.

In (10), the second terms are cross-coupling terms, and the third term is equivalent to a speed-dependent induced EMF term associated with the stator flux. As described in section III, with the compensation for the cross-coupling

terms and feed forward control for the induced EMF, the independent control in d,q-axis is obtained.

The control for the DFIG mainly includes two stages: connecting DFIG to power grid and grid-connected generation. For the two stages, two control structures are designed in the paper, shown in Fig.4.

The objective of no-load control for DFIG is to make the stator voltages keep up with the supply voltages, and then make the stator be softly connected to the grid. To control the stator voltages, naturally the outer loop is voltage control loop. Before being connected to the grid the stator currents equal zero. Considering (6), it can be found that the magnitude of the voltage is mainly controlled by the d-axis component of the rotor currents, so the output of the voltage corrector is selected as the reference of the d-axis component, i_{rd}^*. At the same time make the command of the q-axis component equal zero. The control structure is shown in Fig.4 (a).

In the process of grid-connected generation, the outer loop is power control loop, and the command of the active

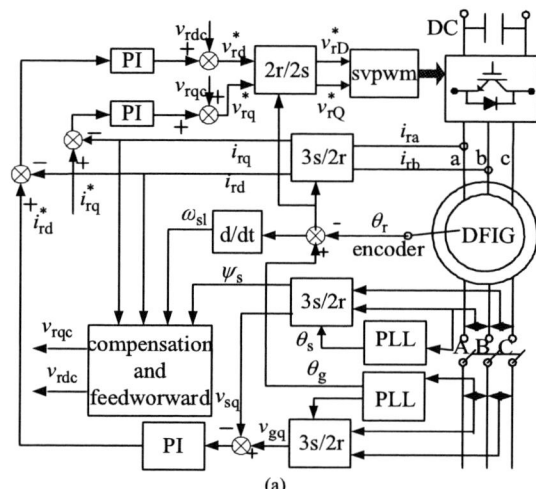

(a)

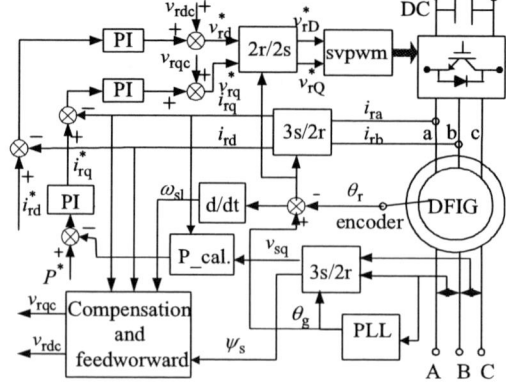

(b)

Figure4. The control structure for DFIG:(a) Structure of no-load control, (b) Structure of grid-connected control

power comes from the PC. Considering (9), at the given speed and magnetizing current the output active power is dependent on the q-axis component of the rotor currents, i_{rq}. So the output of the power corrector is selected as the reference of the q-axis component, i_{rq}^{*}. And the d-axis component of the rotor currents is up to the reactive power composition and the rotor currents up limitation. The control structure for grid-connected DFIG is shown in Fig.4 (b).

V. WIND-SPEED SIMULATION

In this experimental system, a vector-controlled VVVF inverter is used to drive a squirrel motor at a given speed set by software that is developed with Labview and runs on PC, thus the system simulates a variable-speed wind turbine under the environment of a wind farm.

A simple mathematic model for wind speed is adopted. In the model, natural wind is considered to be composed of four parts: base wind, gust wind, ramp wind and noisy wind, described as (11) [6], [16]:

$$V_{w} = V_{wb} + V_{wg} + V_{wr} + V_{wn} \qquad (11)$$

The base wind speed, V_{wb}, is usually regarded as a constant value and is set according to a given site. And the gust wind speed, V_{wg}, is expressed as:

$$V_{wg} = \begin{cases} 0, & \text{if } t \notin (T_{1g}, T_{1g} + T_{g}) \\ (G_{max}/2)\{1 - \cos[2\pi(t - T_{1g})/T_{g}]\}, \text{else} \end{cases} \qquad (12)$$

where, G_{max}: the amplitude of gust wind speed, T_{g}: the period of the gust wind, T_{lg}: the start time of the gust wind, t: the time.

The ramp wind is used to describe the gradual change characteristic of wind speed, and its mathematic model is expressed as:

$$V_{wr} = \begin{cases} 0, & \text{else} \\ R_{max}(t - T_{1r})/(T_{2r} - T_{1r}), t \in (T_{1r}, T_{2r}) \\ R_{max}, & t \in (T_{2r}, T_{2r} + T) \end{cases} \qquad (13)$$

where, R_{max}: the maximum value of the ramp-wind speed, T_{1r}, T_{2r}, and T: the start time, end time, and span time of the ramp wind.

The last term in (11) represents the nosy-wind speed, V_{wn}, expressed as:

$$V_{wn} = 2 \sum_{i=1}^{n} \left[S_{V}(\omega_{i}) \Delta \omega \right]^{1/2} \cos(2\omega_{i} t + \phi_{i}) \qquad (14)$$

where, $S_{V}(\omega_{i})$: the spectrum-density function of wind speed, $\omega_{i} = (i - 0.5)\Delta \omega$: the angular frequency of the i-th component, $\Delta \omega$: the intervals between two random

components, ϕ_{i}: randomly generated, with an uniform distribution in the domain $[0, 2\pi]$.

VI. EXPERIMENTAL RESULTS

To validate the design and verify the performance of the test-bed, an 110kW experimental prototype is set up and several relevant experimental results are reported. The main parameters are: squirrel motor 165kW, DFIG 110kW, back-to-back PWM converters 40kW, line-reactor inductance 0.7mH, and the DC-link capacitance 13.2mF. The main controllers are implemented with two DSP TMS320LF2407A.

In the procedure of connecting the stator of DFIG to grid, the technique of soft connection is adopted to reduce the impact on the grid. The results of soft connection are shown in Fig.5.

Correspondingly, the technique of soft disconnection is also adopted in the procedure of disconnecting the stator of DFIG from the grid. The stage following soft connection is grid-connected generation. The stator currents, the currents flowing from AC-side of grid-side converter, and the currents flowing from the points of common coupling (PCC) at super- and sub- synchronous speed are shown in Fig.6.

The waveform of the simulated wind speed is shown in Fig.7.

VII. CONCLUSION

Based on this test-bed, a DFIG-based VSCF wind power generation is investigated by setup of an 110kW experimental prototype, and the system design is validated by the experimental results. This test-bed provides experimental condition for high-level windmill installation and for further research of wind power generation.

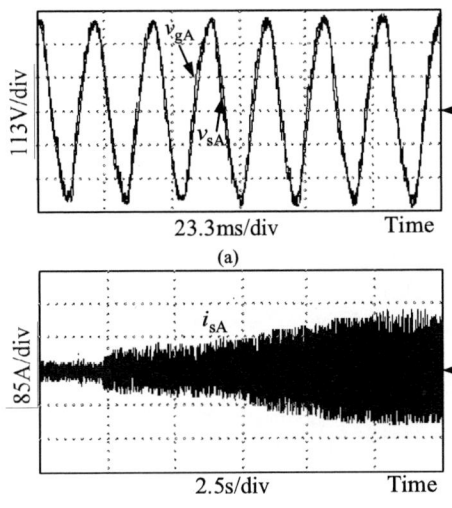

Figure5. The process of softly connecting to the grid: (a) Stator voltage and supply voltage, (b) Stator a-phase current

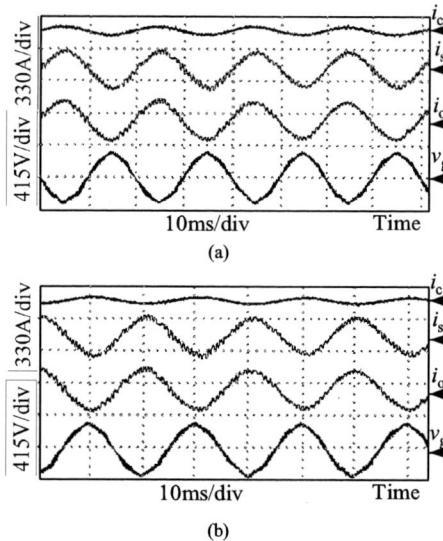

(a)

(b)

Figure6.   Experimental waveform of grid-connected generation: (a) At super-synchronous speed, (b) At sub-synchronous speed (Note: i_{c1} : current from AC-side of supply-side converter, i_s : current from stator side, i_o : current from PCC, v_g: grid voltage)

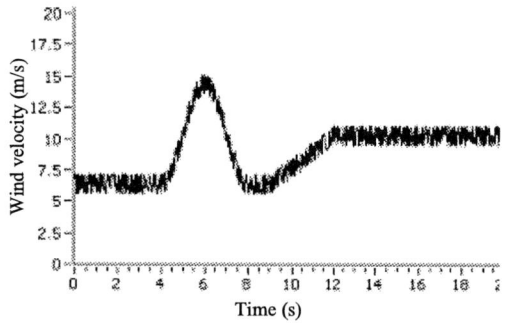

Figure7.   Waveform of the simulated wind speed

REFERENCES

[1] R. Pena, J. C. Clare, and G. M. Asher, " A doubly fed induction generation using back-to-back PWM converters supplying an isolated load from a variable speed wind turbine," *IEE Proc.*, vol.143, pp. 380-387, September 1996.

[2] M. Yamamoto and O. Motoyoshi, "Active and reactive power control for doubly-fed wound rotor induction generator," *IEEE Trans. Power Electron*, vol.6, pp.624-629, October 1991.

[3] S.Gallardo, M.J. Carrasco, E. Galvan and L.G. Franquelo, "DSP-based doubly fed induction generator test bench using a back-to-

back PWM converter," *IEEE IECON, 30th Annual Conf.* 2004, vol.2, pp.1411-1416.

[4] R. Pena, J.C. Clare and G.M. Asher, "Doubly fed induction generator using back-to-back PWM converters and its application to variable-speed wind power generation," *IEE Proc. Electric Power Applicat.*, vol.143, pp.231-241, May 1996.

[5] C. Dufour and J. Belanger, "A real-time simulator for doubly fed induction generator based wind turbine applications," *IEEE PESC, 35th*, vol.5, pp.3597-3603, June 2004.

[6] K. Zheng and Z. P. Pan, "The imitation of the fan characteristic in variable-speed constant-frequency wind-power generation system," *Mechanical & Electrical Engineering Magazine*, vol.20, pp.40-43, 2003.

[7] R. Pena, J.C. Clare and G.M. Asher, "A doubly fed induction generator using back-to-back PWM converters supplying an isolated load from a variable speed wind turbine," *IEE Proc. Electric Power Applicat.*, vol.143, pp.380-387, September 1996.

[8] R. Datta and V. T. Ranganathan, "Decoupled control of active and reactive power for a grid-connected doubly-fed wound rotor induction machine without position sensors," *IEEE IAS*, 34th, vol.4, pp.2623-2630, October 1999.

[9] J. Morren and S. W. H. de Hann, "Ridethrough of wind turbines with doubly-fed induction generator during a voltage dip," *IEEE Trans. Energy Conversion*, vol.20, pp.435-441, June 2005.

[10] B. Rabelo and W. Hofmann, "Control of an optimized power flow in wind power plants with doubly-fed induction generators," *IEEE PESC, 34th*, vol.4, pp.1563-1568, June 2003.

[11] C. Abbey and G. Joos, "Optimal reactive power allocation in a wind powered doubly-fed induction generator," *IEEE Power Engineering Society General Meeting*, vol.2, pp.1491-1495, June 2004.

[12] J. B. Xu, J. Zhao, L. luo, X. Z. Ma, and S.Y. Wan, "A new control strategy of unity power factor for three-phase PWM rectifier system," *IEEE IECON, 30th*, vol.1, pp.709-714, November 2004.

[13] S. Peresada, A. Tilli, and A. Tonielli, "Indirect stator flux-oriented output feedback control of a doubly fed induction machine," *IEEE Trans. Control System Technology*, vol.11, pp.875-888, November 2003.

[14] A. Petersson, L. Hamefors, and T. Thiringer, "Comparison between stator-flux and grid-flux-oriented rotor current control of doubly-fed induction generators," *IEEE PESC*, 35th, vol.1, pp.482-486, June 2004.

[15] B. Hopfensperger, D. Atkinson, and R. A. Lakin, "Stator-flux-oriented control of a doubly-fed induction machine with and without position encoder," *IEE Electric Power Applications*, vol.147, pp.241-250, July 2000.

[16] X. Wu, X. Zhang, Y. H. Yin, and H. Z. Dai, "Application of models of the wind turbine induction generators (WTIGs) to wind power system dynamic stability analysis," *Power System Technology*, vol.22, pp.68-72, June 1998

2006 5th International Power Electronics and Motion Control Conference

Control strategy of Hybrid sources for Transport applications using supercapacitors and batteries

M.B. Camara, H. Gualous, F. Gustin, A. Berthon

L2ES Laboratory, University of Franche-Comte
Rue Thierry Mieg, F90010 BELFORT, FRANCE
mamadou.camara@utbm.fr

Abstract—**In this article, the authors propose an approach to the problem of the power management in transport applications. The mobile experimental platform ECCE is a series hybrid vehicle which currently has three sources of energy: two thermal machines each coupled with an alternator and a lead-acid battery pack of nominal voltage 540V. The alternators are inter-connected with the DC-link by means of the rectifiers. Our contribution is focused on studying the energy coupling between this battery pack and that of supercapacitors in order to find the best compromise between dimensions of the electric power devices, the efficiency mobile energy storing devices, the energy exchanges, and the capacity of exchange of electric power. The supercapacitors module consists of a pack of 108 cells and can supply a maximum of 270V. The main objective is to be able to provide a power of 216kW by supercapacitor module to the DC-link for 20 seconds.**

Keywords- Hybrid vehicle, Supercapacitors, Buck-Boost, Control strategy, Energy storage, multi source system

I. INTRODUCTION

The energy requirements in transport applications are increasing day by day. The developments in the field of power electronics and command have made it possible to increase the performances of traction devices. However the limitations in the case of single source vehicles are due to batteries which till present are the principle source of energy. The various type batteries of traction used until now cannot satisfy the future energy requirements of the transport applications. The batteries are designed to supply power requirement in the transition period. The high valves of energy in the transition state severely affect the battery life. The battery manufacturers are not very optimistic on the fast evolution of de capacities and performances of the batteries of traction; it is therefore interesting to make the hybridization of the source, to rely less on battery during the transition phases. Among the various possible technological solutions (Supercapacitors, flywheels) for the hybrid sources, we are interested in that using of supercapacitors. But, for the
Supercapacitors to follow the energy needs of the future applications of transport and find their place with or in replacement of the batteries, it is necessary that the

productions of these devices of strong power are at reasonable cost. Software (SABER) is used for simulation and a data acquisition system with a microcontroller (PIC18F4431) is setup. The general structure of the hybrid system is represented by Figure1.

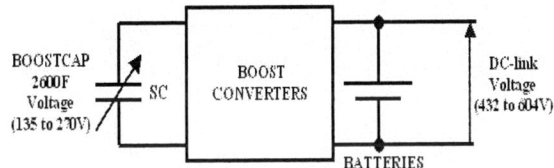

Figure1. General structure of hybrid system

II. MODELING AND COMMAND BY INVERSE BOOST MODEL

A. Converter topology

In this configuration, a boost converter will control the energy exchange between the packs of battery and supercapacitors as indicated in the Figure 2. The system consists of a power source (supercapacitors), a source of energy (battery), a load representing traction vehicle and a boost type converter [1, 4, 6].

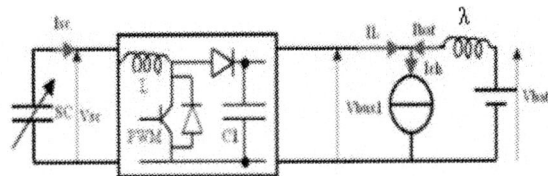

Figure2. Simple structure of the boost converter

B. Modeling of boost converter

Suppose that the semi-conductors are simple switches, the dynamic model of supercapacitors can be represented by equations (1).

$$
\begin{cases}
\dfrac{d}{dt}(i_{sc}) = \dfrac{V_{sc} - (1-\alpha) \cdot V_{bus1}}{L} \\
\dfrac{d}{dt}(i_{bat}) = \dfrac{V_{bat} - V_{bus1}}{\lambda} \\
i_c = C \cdot \dfrac{d}{dt}(V_{bus1}) \\
I_{ch} = I_L + I_{bat}
\end{cases}
\tag{1}
$$

1-4244-0448-7/06/$25.00 ©2006 IEEE   515

C. Control by inverse boost model

The control law can be represented by equation (2).

$$\alpha = 1 - \frac{V_{sc} - V_L}{V_{bat} - V_\lambda} \approx 1 - \frac{I_L}{I_{sc}} \qquad (2)$$

This voltage is compared with a triangular signal having 1V amplitude (Umax) with frequency of IGBT of 10 kHz to generate PWM signal.

The references of current for supercapacitors are obtained from power assessments between the pack and DC-link.

$$I_{scref} = \frac{V_{bus1}}{\eta \cdot V_{sc}} \cdot (I_{ch} - I_{batref}) = \frac{V_{bus1}}{\eta \cdot V_{sc}} \cdot I_{Lref} \qquad (3)$$

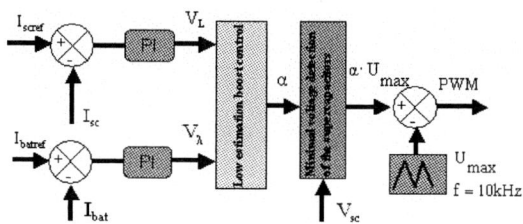

Figure3. Control strategy of the boost converter

where η is the efficiency of the boost converter

III. MODELING AND CONTROL OF PARALLEL BOOSTS

A. Converter topology

To avoid overloads with only one boost converter a parallel structure has been proposed, to satisfy the energy request of the hybrid electric vehicle during transient states [2, 3, 5]. This topology is shown on Figure4.

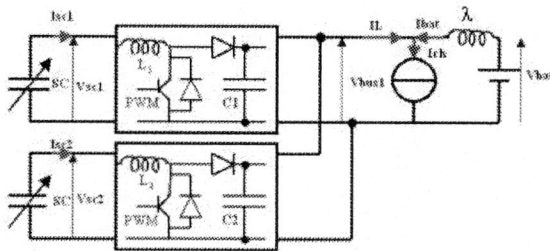

Figure4. Parallel topology of boost converters

The system consists of two sources of power (supercapacitors), a source of energy (batteries), an active load which represents vehicle traction system and two boost converters which ensures the energy exchange between the sources and the DC-bus.

B. Modeling boost converters

The model for parallel boost structure for current regulation can be represented by differential equations (4), where I_{ch} and I_L are active load and DC-link

current respectively.

$$\begin{cases} I_{ch} = I_L + I_{bat} \\ \dfrac{d}{dt}(i_{sc1}) = \dfrac{V_{sc1} - (1-\alpha) \cdot V_{bus1}}{L_1} \\ \dfrac{d}{dt}(i_{sc2}) = \dfrac{V_{sc2} - (1-\alpha) \cdot V_{bus1}}{L_2} \\ \dfrac{d}{dt}(i_{bat}) = \dfrac{V_{bat} - V_{bus1}}{\lambda} \\ i_c = C_1 \cdot \dfrac{d}{dt}(V_{bus1}) = C_2 \cdot \dfrac{d}{dt}(V_{bus1}) \end{cases} \qquad (4)$$

C. Application of control by inverse boost model

Two regulation methods have been investigated. The first method consists in controlling independently the voltage drops across inductors. The control law established for this purpose is given by the expression (5)

$$\alpha = 1 + \frac{1}{2} \cdot \frac{(V_{L1} - V_{sc1}) + (V_{L2} - V_{sc2})}{V_{bat} - V_\lambda} \qquad (5)$$

The drawback of this method is the significant number of parameters which should be optimized to control the battery current.

The second method consists in controlling the sum the voltage drops across the inductances. The control law which generates modulating signal of the PWM is given by (6).

$$\alpha = 1 + \frac{1}{2} \cdot \frac{(V_{L1} + V_{L2}) - (V_{sc1} + V_{sc2})}{V_{bat} - V_\lambda} \qquad (6)$$

The advantage of this method compared to the first is in the reduction of the number of parameters to be optimized to control the battery current.

IV. MODELING AND CONTROL OF PARALLEL BOOST STRUCTURES WITH PARALLEL SOURCES

To reduce the weight of the vehicle a parallel structure has been proposed which enables to decrease the number of smoothing inductors for supercapacitors currents. The topology of this structure of boosts is indicated on the Figure5.

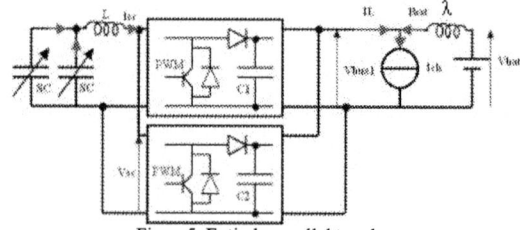

Figure5. Entirely parallel topology

$$
\begin{cases}
V_{sc1} = V_{sc2} = L \cdot \dfrac{d}{dt}(i_{sc}) + (1-\alpha) \cdot V_{bus1} \\[2mm]
V_{bus1} = V_{bat} - 1 \cdot \dfrac{d}{dt}(i_{bat}) \\[2mm]
i_c = C_1 \cdot \dfrac{d}{dt}(V_{bus1}) = C_2 \cdot \frac{d}{dt}(V_{bus1}) \\[2mm]
I_{ch} = I_L + I_{bat}
\end{cases}
\tag{7}
$$

The control strategy of this structure is same as applied to the first one.

V. SIMULATION RESULTS

Some results of simulations of various structures will be presented. The common objective is to provide to the DC-link a power of 216kW during 20 seconds or 400A under a voltage ranging between 432V and 604V. We fixed the reference of battery current at 100A.

A. Simple structure of the boost converter

From 0 to 0.5 second and 10.5 to 11 seconds the reference of battery current is equal to the current requested by the active load. In these intervals of time the boost converter is switched off. It can be noted that the control makes it possible to keep of battery current (Ibat) around 100A except at the moment of change of the active load current. These peaks of current which appear are caused by the delay of the control of the current loops. Because of the fast discharge of supercapacitors, an overlapping loop of supercapacitor current has been used, so that the reduction of the supercapacitor voltage (Vsc) is compensated by the supercapacitors current (Isc) increase.

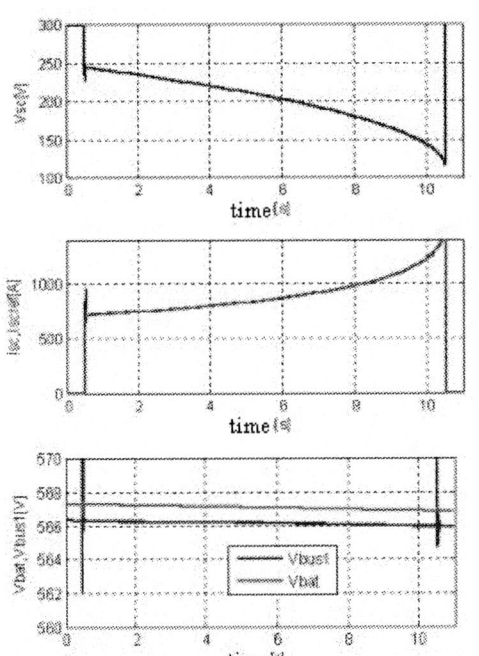

Figure6. Supercapacitors and batteries current and voltages.

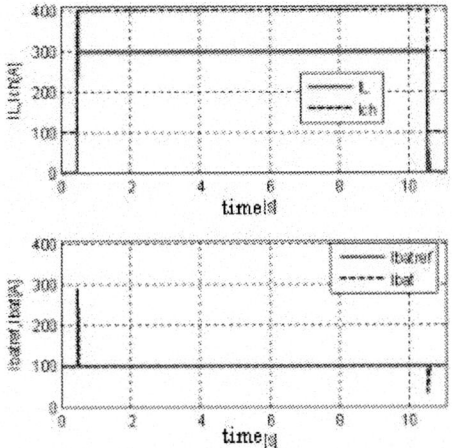

Figure7. Percentage contribution of supercapacitors and batteries

B. Parallel structure of boost converters

To check the control of the boost converters, the reference of battery current (Ibatref) is fixed at 100A and active load current is set to (Ich) 400, 200 and 400A respectively. The control of the boost converter is satisfactory except around 0.5 second and 1 second when the current loops do not have necessary time to act. The currents (Isc1) and (Isc2) output respectively by the pack1 and the pack2 are same. The voltages of two modules of supercapacitors (Vsc1, Vsc2) are same.

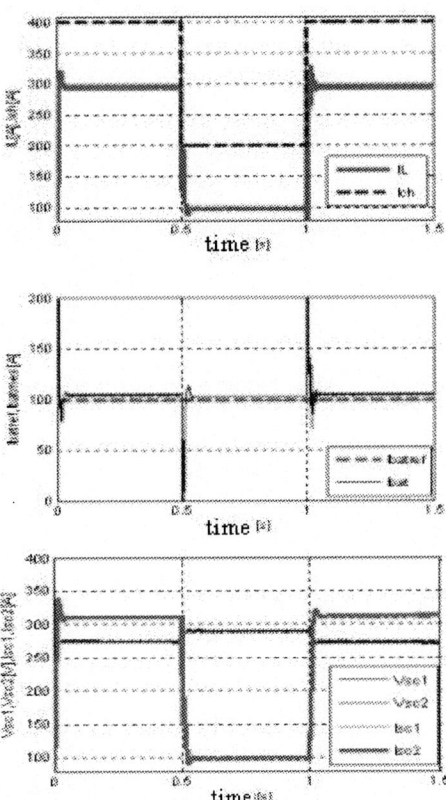

Figure8. Results for (Ich=400A-200A-400A) and Ibatref=100A

517

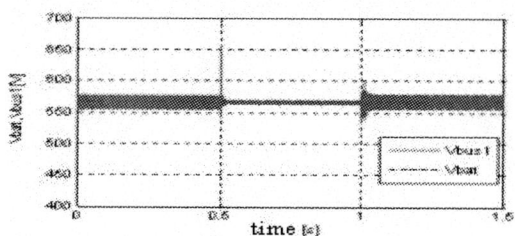

Figure9. DC-link and batteries voltages.

The reference battery current (Ibatref) always remains within 100A. Between 0-0.5 second and 20.5-22 seconds the converters are with blocked off and in the interval (0.5-20.5 seconds) 75% of current requested by the load is output by both supercapacitor packs (IL). After 20.5 seconds of discharge the control of battery current (Ibat) becomes almost impossible. The distribution of the powers is equal between the supercapacitor packs; this is why the voltage and current output by both packs are identical (Vsc1=Vsc2 and Isc1=Isc2).

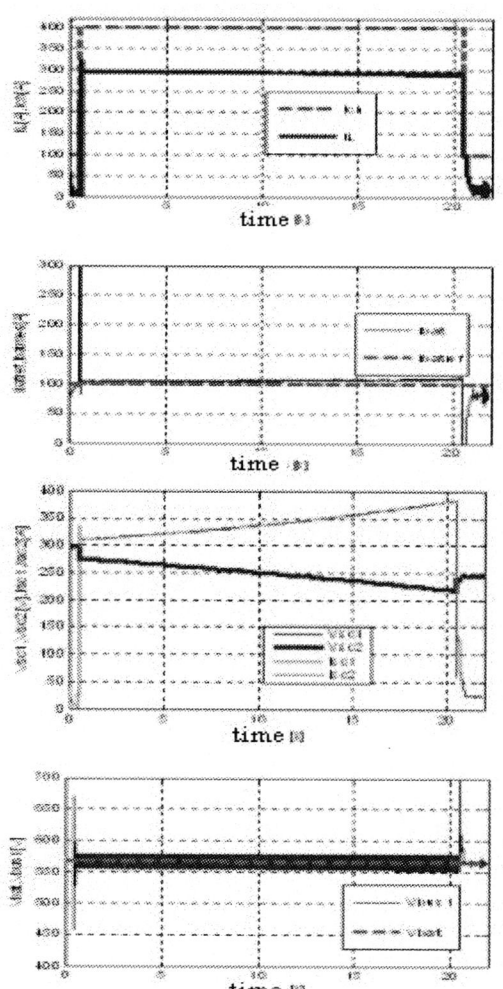

Figure10. Results for (Ich=100A-400A-100A) and Ibatref=100A

C. Entirely parallel structure

The simulation conditions are identical to the parallel structure. The peaks of battery current around 0.5 second and 20.5 seconds are due to the commutation of load current (Ich).

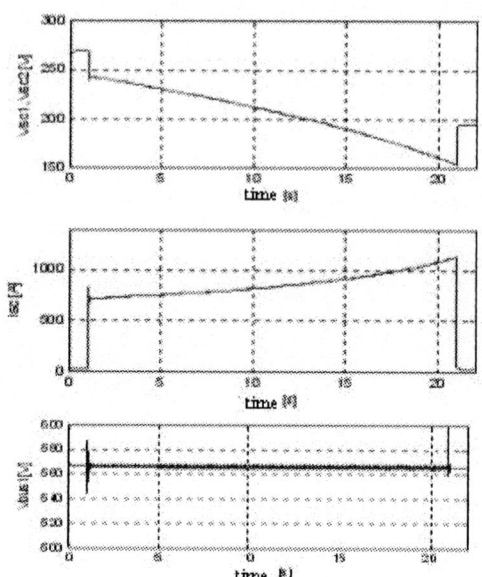

Figure11. Supercapacitors and DC-link current and voltages.

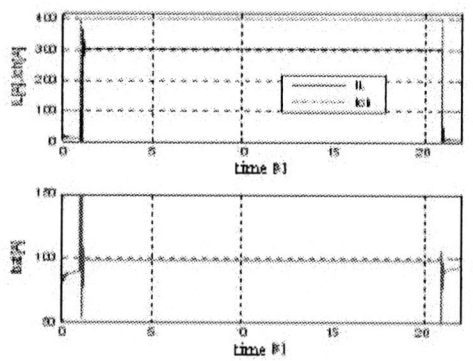

Figure12. Percentage contribution of supercapacitors and batteries

VI. DESIGN AND EXPERIMENTAL RESULTS

A reduced size experimental testbench has been designed. It includes a pack of supercapacitors made up of 11 cells of 2600F each with a maximum voltage of 2.5V, a pack of batteries of 4 elements of 12V with a capacity of 92Ah, a boost converter, a filtering capacitor of 6.8mF, inductors for smoothing supercapacitor and battery currents of values 50µH and 25µH respectively. The control of the system is ensured by a microcontroller of the type PIC18F4431 which presents 9 analog inputs and 8 PWM outputs.

A. Experimental setup

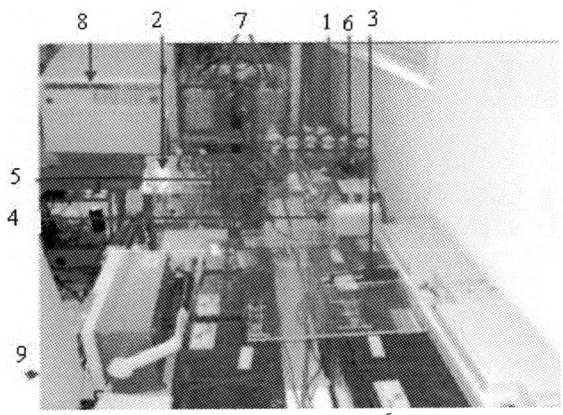

1: Supercapacitors
2: Drive and control (µcontroller)
3: Batteries
4: Inductor (25µH)
5: Boost converter
6: Capacitor (6.8mF)
7: Inductor (50µH)
8: Active load
9: PC

Figure13. Realisation of simple structure of the boost converter

B. Experimental results

The current supplied by supercapacitor pack to the continuous bus (Ilref) is fixed at 20A. The current of the load is fixed at 25A. The powers (Pbus) and (Pch) are obtained respectively by the product of the currents measured (IL and Ich) by the measured DC-link voltage and that of the battery by its current and measured voltage. The efficiency of the converter is defined by the ratio of the power of the continuous bus by that of supercapacitors. We notice that the efficiency according to the current output by the supercapacitors decreases. This is explained by the fact that, the progressive discharge of the supercapacitor pack is compensated by a progressive increase in its current; which results in higher losses by Joule effect.

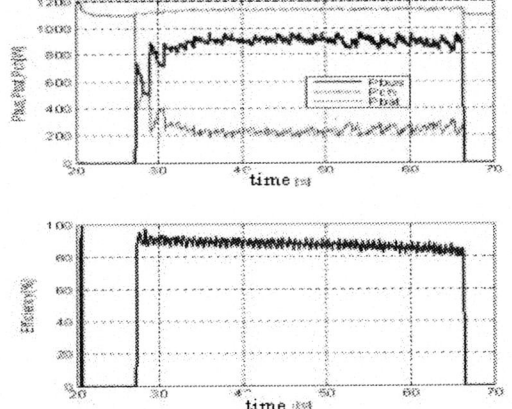

Figure14. Powers and efficiency of boost converter

VII. CONCLUSION

In this article, we have presented the modeling of various possible structures of multi source system of energy and the associated strategies of control. To control the total system, we have established laws for boost converters with a reference of battery current of 100A. The references for supercapacitors are calculated by the strategy of energy management for the vehicle. The current supplied by battery follows the reference value except in the transition region where load current varies from 100 to 400A and 400 to 100A. For reasons of simplicity and cost, the simple structure of the boost converter is most interesting but unfortunately is far from the DC-link power requested by the vehicle (216kW) during 20 seconds. To avoid the DC-link ageing of supercapacitors, we proposed the parallel structure of boost converters. What seems satisfactory for the energy requested of the hybrid electric vehicle in 20 seconds. This structure presents some disadvantages in particular the cost is higher and the weight is more significant contrary to the simple structure of the boost converter. To reduce the weight of the vehicle we proposed the parallel structure which enables us to decrease the number of smoothing inductances of supercapacitors current. The control allows to maintain battery current around the desired value (100A) in spite of the significant request for power on the DC-link.

ACKNOWLEDGMENT

This work is under the continuity of the work started within laboratory L2ES within the program framework ECCE in collaboration with Electronic and Electrical engineering Research Center of BELFORT (CREEBEL).

REFERENCES

[1] L.Solero, L. Lidozzi, J-A. Pomilio, Design of Multiple-Input Power converter for Hybrid Vehicles, IEEE Trans. On power Electronics, vol:20, N°5, September 2005

[2] P. Mestre, S.Astier, Utilization of Ultracapacitors as auxiliary power source in Electric Vehicle, EPE ,vol:4, pp4670-4673, Sept.1997

[3] A. Di Napoli, F.Crescimbini, L. Solero, F. Caricchi, F.G. Capponi, Multiple-Input DC-DC power for power-flow management in hybrid vehicles, IEEE,vol.3,pp1578-1585,oct 2002

[4] B. Michael, B.Burnett, L.J. Borle, A power system combining batteries and supercapacitors in a solar/hydrogen hybrid electric vehicle, IEEE,vol:3, Sept. 2005

[5] M. Marchesoni, C. Vacca, A New DC-DC Converter Structure for Power Flow Management in Fuel-Cell Electric Vehicles with Energy Storage Systems, 2004 35[th]Annual IEEE Conf. Power Electronics Specialists , Germany

[6] P.Thounthong,S.Raël,B.Davat, Control strategy of fuel cell/supercapacitors hybrid power sources for electric vehicle, Journal of Power Sources September 2005

Wind Generator Stabilization With Doubly-Fed Asynchronous Machine

Li Wu, Zhixin Wang

*Shanghai Jiao Tong University, 800 Dongchuan Road, 200240, P.R.China
wuli@sjtu.edu.cn wangzxin@sjtu.edu.cn

Abstract—This paper investigates the function of DASM (Doubly-fed ASynchronous Machine) with emphasis placed on its ability to the stabilization of the power system including wind generators. P(active power) and Q(reactive power) compensation from DASM can be regulated independently through secondary-excitation controlling. Simulation results by PSCAD (Power System Computer Aided Design) show that DASM can restore the wind-generator system to a normal operating condition rapidly even following severe transmission-line failures.

Keywords-DASM(Doubly–fed ASynchronous Machine)

I. INTRODUCTION

Generation of electricity using wind power has received considerable attention worldwide in recent years. It has been predicted that the annual growth of wind power between 1998 and 2040 would be between 20-30%[1]. However, the continuous trend of the increasing of the number of wind generators would influence the operation of existing utilities networks. Up to now many concerns[2]-[4] have been focused on three principal problems of the wind generator system: (problem 1): Reactive power compensation is required for the operation of the general wind generator (induction machine) in steady state, otherwise the power quality of the system will be decreased; (problem 2): When a large electrical disturbance occurs in the network , the terminal voltage of the wind farm will decrease while the rotor speed of the machine will increase tremendously, which will lead to the collapse down of the system; (problem 3):When wind power fluctuates tremendously due to wind speed variations such as wind gust, the power delivered to the network system will suffer power surge if there is no adequate amount of active power compensation.

Although various ways, such as capacitor banks have been supposed to compensate the reactive power for the wind generator in steady state, because of its fixed capacity, the reactive power supplied to the wind generator will be superfluous sometimes while insufficient sometimes, depending on the output of the generator. Controlled static VAR compensation [5] can offer the variable reactive power compensation both for steady state and transient state. When large electrical disturbances, such as 3LG (three-line-to-ground) fault, occurs in the network, terminal voltage will decrease, thus rotor speed

of the wind generator will be accelerated, and consequently much more reactive power will be consumed from the network system, which leads to the drop of the wind farm voltage. As SVC can respond to this with adequate amount of reactive power compensation, wind farm voltage, and also wind generator rotor speed can recover to the initial state. In addition, since wind farm suffer severe wind gust frequently, when wind speed fluctuates, active power supplying to the network will also fluctuate accordingly. However, at this time active power compensation from SVC is impossible. The combination of SVC and battery storage system [6] seem to be a solution to all the three problems, in which active and reactive power compensation can be supplied at a time. However, in this case, full capacity inverter is required, which makes the system costly, and harmonic is inevitable. Moreover, in the battery storage system, the output of the active power can not change rapidly due to its chemical limitation. So after all, no ways have been found to solve the above three problems at a time.

DASM(Doubly-fed Asynchronous machine) seems to be a solution to these problems. It has been emerged as a system stabilizer by injecting/absorbing adequate amount of active/reactive power quickly and instantaneously to the network system. The rotor of the DASM is fed by a three-phase, sinusoidal current of slip frequency, which enables to regulate the active and reactive power compensation for the system independently through secondary excitation controlling[7]. Moreover, as compared to the combination of SVC and battery storage system, the capacity of the inverter required for the system is about 0.3 times the DASM power rating, which makes it cheaper. In addition, less harmonic is found, and it is much quicker in its electrical action. So DASM might be a good solution to the 3 problems facing the wind generator system.

It is well known that hydro generator by using DASM has been put into practical use successfully [8]. On the other hand, there are also some reports concerning the application of flywheel generator by using DASM for the stabilization of the power system [9]. However, up to now, few discussions have been done on the stabilization of wind generator by flywheel generator using DASM.

Therefore, it is supposed that DASM, when properly controlled, can be used to provide locally the leading reactive power compensation for the wind generator in

1-4244-0448-7/06/$25.00 ©2006 IEEE

steady state; be used to stabilize the wind power system under large electrical disturbances in the network, with adequate amount of reactive power compensation; and be used to stabilize the wind power system under a severe wind gust, with adequate amount of active power compensation. In this paper, DASM, with proper control scheme, is proposed to enhance the stability of wind generator system, and various simulations are done to verify the propositions above.

Simulation studies by PSCAD (Power System Computer Aided Design) /EMTDC were performed to verify the effectiveness of the DASM on the stabilization of wind generator system when a 3LG fault occurs and when wind generators suffer severe wind gusts. Simulation studies on the effectiveness of the DASM on transient stability problems have been done on both single wind generator system and multiple wind generator system. One severe wind speed pattern is used to verify the effectiveness of DASM on smoothing power surge and terminal voltage of the wind farm during wind gust.

II. MODEL SYSTEM CONFIGURATION

The system used as a basis for this investigation is shown in Figure. 1, in which SG denotes a synchronous generator and a DASM is connected to the high-voltage side of a wind generator (induction generator, IG) through a Δ/Y transformer. Table I shows various parameters of each generator. There is a local transmission line with one circuit between the main transmission line and a transformer at the wind power station.

Though a wind power station is composed of many generators practically, it is considered to be composed of a single generator with total power capacity in the first few parts of the paper. A condenser C is connected to the terminal of the wind generator to compensate the reactive power demand for the induction generator in steady state. The value of C has been chosen so that the power factor of the wind power station becomes unity when the terminal voltage and output power are 1.0[pu].

AVR and governor control systems are considered in the synchronous generator model, and they are shown in detail in Figure.2.

The DASM used is based on a wound rotor structure, whose rotor is supplied with an alternating current of slip frequency. Therefore the resulting magnetic flux due to the impressed rotor currents rotates with slip frequency referred to the rotor and with the network frequency referred to the stator; whereby it generates the corresponding counter magnetic flux within the stator. So the amplitude and phase position of the stator current and thereby the active and reactive power consumption of the machine is determined by control of the rotor current. Although various ways[10],[11]has been proposed to control the rotor current, the simplest way of PI controlling is used in the paper, to verify the effectiveness of the DASM to the stabilization of the wind power system. Moreover, in the paper the rotating reference frame (dq

frame) to analyze the DASM is fixed on the space axis of stator voltage.

In the control scheme illustrated in Figure.3[7], when P_{WG} (active power from wind farm into the network system) and V_{WG}(wind farm terminal voltage) are detected, the P/Q compensation required from DASM is thus determined. To regulate the error between the desired and detected values of P/Q, a two-step PI controller is used, the first step of which is APR/AQR, and the second is ACR. These controllers are shown in Figure.4.Therefore the required field voltage is specified and applied to the rotor side of the DASM. In Figure.4, P_{WG} and P_{WG}^{ref} are the active power output and its reference value of wind farm; P_{DASM} and P_{DASM}^{ref} are the active power detection and its reference value of DASM ; Q_{DASM} and Q_{DASM}^{ref} are the reactive power detection and its reference value of DASM respectively. V_{WG} and V_{WG}^{ref} are detection and its reference value of the wind farm terminal voltage respectively.

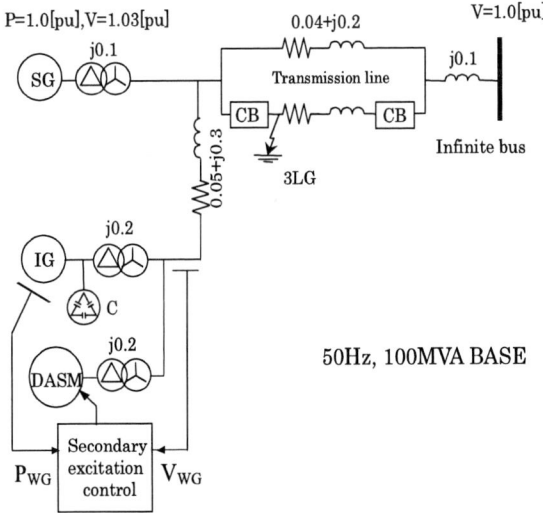

Figure.1 Power system model

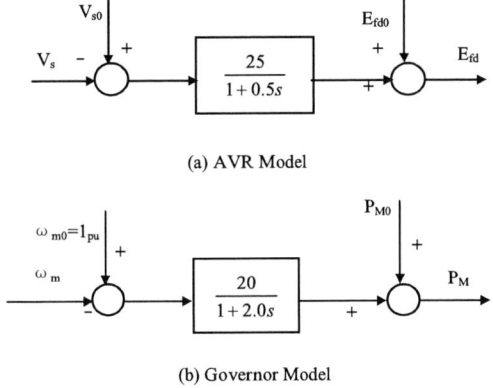

(a) AVR Model

(b) Governor Model

Figure.2 Control system models of synchronous generator

TABLE I.
NOMINAL VALUES AND PARAMETERS OF EACH GENERATOR

SG		Wind generator(IG)		DASM
Rated output (MVA)	100	Rated output (MVA)	50	50
Rated voltage(kV)	11	Rated voltage(kV)	0.69	11
r_a(pu)	0.003	r_1(pu)	0.01	0.0045
x_a(pu)	0.13	x_1(pu)	0.1	0.142
X_d(pu)	1.2	X_{mu}(pu)	3.5	2.75
X_q(pu)	0.7	r_2(pu)	0.01	0.0045
X_d'(pu)	0.3	x_2(pu)	0.08	0.142
X_d''(pu)	0.22	H(sec)	1.5	19.5
X_q''(pu)	0.25			
T_{do}'(sec)	5.0			
T_d''(sec)	0.04			
T_q''(sec)	0.05			
H(sec)	2.5			

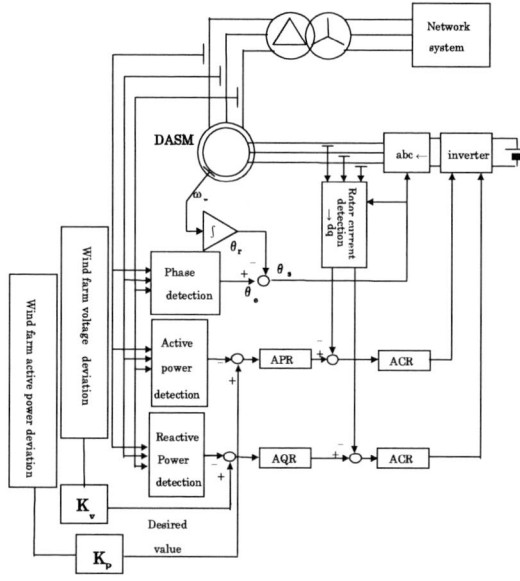

Figure.3 DASM circuit configuration

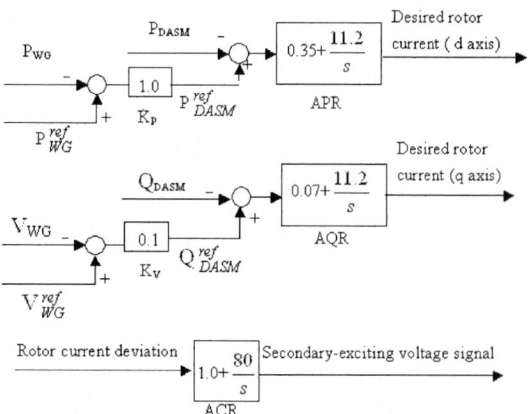

Figure.4 Control signal blocks for DASM

This paper simulates the secondary-exciting source with an ideal DC source for the sake of simplicity, and only the fundamental component of inverter output is considered.

III. SIMULATION STUDIES AND DISCUSSIONS ON THE SYSTEM TRANSIENT STABILITY PROBLEM

The general way to deal with wind generators when a fault such as a 3LG (three-line-to-ground) fault occurs in the network system is to disconnect them from the grid and then apply brake to stop them. Although there are also reports investigating the effect of the pitch control system on the transient stability recently[12], this paper proposes a new way of compensating the amount of required reactive power from DASM to recover the voltage when a 3LG fault occurs.

A. Simulation Studies on The Single Wind Generator System

To investigate the validity of the proposed method, a 3LG fault is applied in the following two kinds of circumstances, and simulation results are studied and compared:

(Case 1): Capacitor banks are connected at the terminal of the wind generator to offer the required reactive power in steady state.

(Case 2): Without the capacitor banks at the terminal of the wind generator, DASM (50MVA) is placed nearby to provide the required reactive power compensation in steady state. The mechanical input to the DASM is 0 [pu] and the output from DASM has been limited to 0. 5[pu](rated output in 100MVA base).

In the above two cases, the grid configuration is totally the same as shown in Figure.1. Figure.5(a)-(f) show various simulation results when a 3LG fault occurs at 1.1 s, the CBs(Circuit Breaker) are opened at 1.2s and re-closed at 2.0s. The initial values of the wind generator and synchronous generator are shown in Table II, in which the condenser value, 0.264 [pu] (100MVA base), is for case1. Because of the step out of the machine, the simulation result of case 1 is cut at 8 seconds. It is also assumed that the wind speed Vw is constant here.

Figure.5 (a) and Figure.5 (b) show the terminal voltage and rotor speed of the wind generator. It is clear that when fault occurs, voltage collapses and rotor speed keeps on increasing tremendously for case 1, while for case 2, it recovers to its pre-fault voltage and rotor speed in less than 3 second. Figure.5(c)-(d) show various reaction of the DASM to the accident in the system. It is clear that after the fault, DASM contributes to the stabilization of the system by offering adequate amount of reactive power compensation, and that even in the steady state DASM offer 0.26[pu](100MVA Base) reactive power, the amount of which is generally offered by the capacitor banks to maintain the system voltage. This means that DASM works effectively in providing reactive power compensation to the wind generator in steady state. While the active power output of DASM is controlled to 0[pu] in steady state, it oscillated after the

522

fault because of the cross coupling with the reactive power part. Figure.5 (e) shows synchronous generator load angle response, and it remains stable in each of the two cases. It may be concluded that the dynamic behavior of wind generators do not have so significant effect on the stability of synchronous generator.

So as far as the transient stability is concerned, it can be said that DASM can be used effectively to stabilize the power system, decreasing the transient of the system. The results also show that instead of conventional capacitor bank, the DASM works effectively in providing locally the leading reactive power compensation to the wind generator in steady state.

TABLE II.

INITIAL VALUES OF WIND GENERATOR AND SYNCHRONOUS

IG				SG	
P(pu)	V(pu)	C(pu)	V_w(m/s)	P(pu)	V(pu)
0.500	1.000	0.264	12.49	1.000	1.034

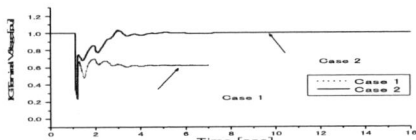

Figure.5(a) IG terminal voltage

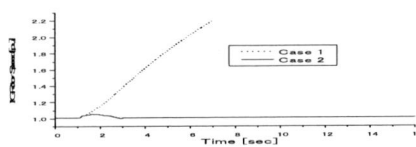

Figure.5(b) IG rotor speed

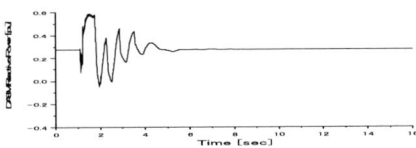

Figure.5(c) DASM reactive power output(case 2)

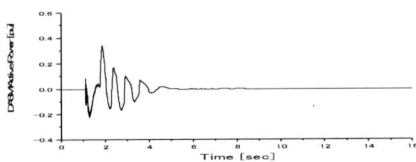

Figure.5(d) DASM active power output(case 2)

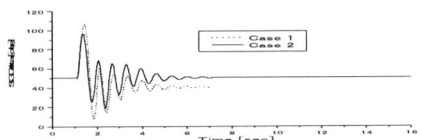

Figure.5(e) Synchronous generator load angle

B. Simulation Studies On Multiple Wind Generator System

As explained previously, simplified lumped wind generator model was considered by now, but in most cases, wind power station is practically composed of many separated generators.

In order to examine the effect of DASM on the practical multiple wind generator system, 5 wind generators(each 10 MVA power rating) with different output are used for the transient stability simulation study, and fault conditions are the same as those described above. Two cases below are considered in the simulation study:

(Case 3): with capacitor banks connected at the terminals of the wind generators.

(Case 4): with DASM(50MVA) placed nearby while without the capacitor banks at the terminals of the wind generators.

The schematic model of the wind generators is shown in Figure.6 and the rest of the model are the same as those shown in Figure.1. The initial values of each wind generator and synchronous generator are shown in Table III.

Figure.7 (a)-(e) show various simulation results of the wind generator system, and it is clear that after the fault wind generators with high output go out of step because of the high initial rotor speed. It is also clear that after the fault, for case 4 the system regains its pre-fault voltage and rotor speed due to the reactive power compensation from DASM, while the system experience voltage collapse and tremendous speed increasing for case 3. The reactive output of DASM reaches 0.23 [pu] in steady state to meet reactive power compensation requirement of the system.

It can be concluded from all the results above that DASM also work effectively on decreasing the transient of the multiple wind generator system.

Although it seems that DASM may be replaced by SVC in offering the variable reactive power when 3LG fault occurs, it is less effective because of the disability of SVC in offering active power compensation when wind power fluctuates, which will be discussed in Section IV.

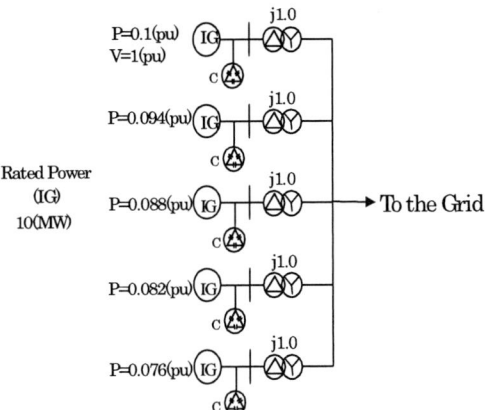

Figure.6 Wind generator part of the model system

523

TABLE III.

INITIAL VALUES OF THE WIND GENERATORS AND SYNCHRONOUS GENERATOR

IG	P(pu)	0.076	0.082	0.088	0.094	0.100
	V(pu)	1.011	1.009	1.006	1.003	1.000
	Vw(m/s)	10.95	11.35	11.74	12.12	12.49
	C(pu)	0.053 (100 MVA Base, for Case 3)				
SG	P(pu)	1.000				
	Q(pu)	0.222				
	V(pu)	1.013				
	T_G(pu)	1.003				
	E_f(pu)	1.703				
	δ (deg)	52.84				

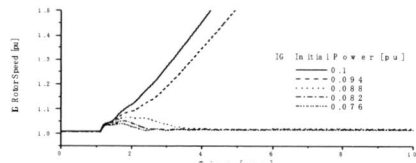

Figure.7(a) IG rotor speed (Case 3)

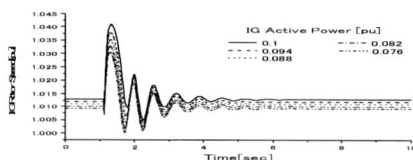

Figure.7(b) IG rotor speed (case 4)

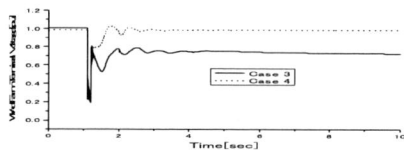

Figure.7(c) Wind farm terminal voltage

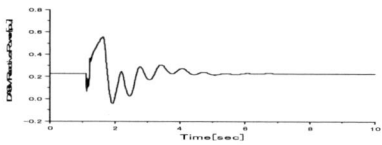

Figure.7(d) DASM reactive power(case 4)

IV. SIMULATION STUDIES AND DISCUSSIONS ON THE WIND GENERATOR POWER FLUCTUATION PROBLEMS

It is well known that the wind speed is fluctuating because of its turbulent nature and it varies on a random basis. Thereby the output of the wind generator can't be connected to the network directly since it is continuously varying. Although the output may sometimes be maintained constant by adjusting its blade pitch angle, the generator can sometimes be unstable during highly variable wind conditions. Although various ways have been tried[13],[14], a new way using DASM is proposed in this paper to stabilize the power fluctuations from the wind farm to the network system.

To investigate the effectiveness of the proposed method, simulation is done on the multiple wind generator system shown in Figure.6. The two cases (case3 and case4), system model and the initial values of each wind generator and synchronous generator are totally the same as those described in section III.B

As the wind speed is randomly changing all the time, it is important to investigate its influence to the system when the multiple generators are under typical wind gusts. Among various wind patterns, it is considered that when 4 of the 5 wind generators are under the same type of wind gust. As all the generators will be accelerated at the same time in this case, there may be a high possibility of leading to the total collapse of the system if without the adequate compensation from DASM.

Based on the expectations above, representative wind speed pattern is chosen for the simulation and superiority of DASM over capacitor banks when wind speed changes are tested. Wind speed variations shown in Figure.8 have been applied to 4 of the 5 wind generators for both cases, case3 and case4.

Simulation results are shown in Figure.9 (a) to (d). It is clear that, for case 3, wind generators go out of step, and active power from wind farm(shown in Figure9(a)) and wind farm terminal voltage(shown in Figure.9(b)) collapse totally.

On the other hand, for case 4, since DASM supply adequate amount of active and reactive power compensations(shown in Figure. 9 (c) and (d)) to the system, both the power output and terminal voltage of the wind farm (shown in Figure. 9(a) and (b)) can be maintained almost constant.

Therefore it can be said that, even in the severe wind pattern shown in Figure.8, DASM works much effectively in stabilizing the system though the wind generators go out of step if there is not DASM.

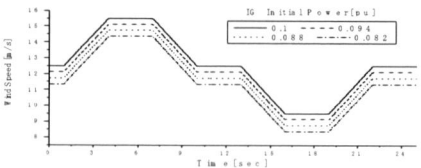

Figure.8 Wind speed variations

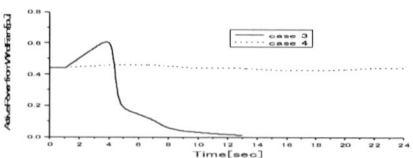

Figure.9(a) Active power from wind farm

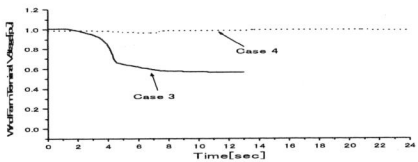

Figure.9(b) Wind farm terminal voltage

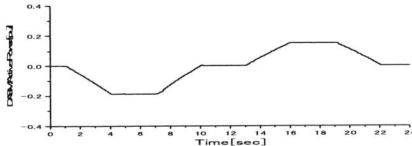

Figure.9(c) DASM active power(case 4)

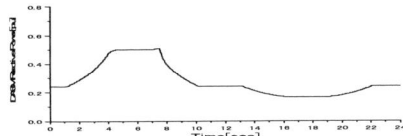

Figure.9(d) DASM reactive power (case 4)

V. CONCLUSION

The paper proposed a new way of stabilization of the wind generator system with DASM. Various simulation results by PSCAD show the effectiveness of preventing the voltage collapse of the wind generators after a 3LG fault with the reactive power compensation from DASM. It has also been found that DASM can be used to provide locally the leading reactive power compensation to the wind generators in the steady state.

Simulation has also been done on multiple wind generator system when wind power fluctuates randomly. a typical types of wind speed pattern is used in the paper. Results show the effectiveness of DASM on smoothing variations of output power and terminal voltage of the wind farm when wind power fluctuates.

All in all, simulation results show that DASM can be used to minimize the influence to the system under large electrical disturbances in the network and under a severe wind gust effectively. Simulation results also show the effectiveness of DASM in providing locally the leading reactive power compensation to the wind generator in steady state.

ACKNOWLEDGMENT

This Project was granted financial support from China Postdoctoral Science Foundation, No: 2005038435.

This Project was granted financial support from Shanghai Postdoctoral Science Foundation, No: 05R214133.

REFERENCES

[1] The European wind energy association (EWEA):"EWEA Publication,1998,"[online],Available:http://www.ewea.org/Force1 0.htm

[2] E. N. Hinrichsen et al.: Dynamics and Stability of Wind Turbine Generators, IEEE Transactions on PAS, Vol.PAS-101, No. 8, pp. 2640-2648,August 1982.

[3] C. S. Demoulias et al.: Electrical Transients of Windturbines in a Small Power Grid, IEEE Transactions on EC, Vol. 11, No. 3, pp. 636-642,September 1996.

[4] P. M. Anderson et al.: Stability Simulation of Wind Turbine Systems, IEEE Transactions on PAS, Vol. PAS-102, No. 12, pp.3791-3796, December 1983.

[5] E.S.Abdin et al: Control Design and Dynamic Performance Analysis of a Wind Turbine-Induction Generator Unit, IEEE Trans. on EC, Vol. 15, No. 1, pp. 91-96,March 2000.

[6] B.S.Borowy et al: Dynamic Response of a Stand-Alone Wind Energy Conversation System with Battery Energy Storage to a Wind Gust, IEEE Trans. on EC, Vol. 12, No. 1, pp. 73-78,March 1997.

[7] Takahashi,Tamura,Tada,Kurita, Derivation of Model of an Adjustable Speed Hydro Generator and Its Control System , Trans. IEE of Japan, Vol.124-B, No.2, pp.181-189,2004(in Japanese)

[8] C.Takanashi et al: Principal and Effect of Variable Speed Hydro Generator System, Trans. IEE of Japan, Vol.115-B, No.5, pp.447-450,1995(in Japanese)

[9] T.Chida et al: Stabilization of a Large-capacity and Long-distance Transmission System by combination of a System Damping Resistor and an Adjustable Speed Flywheel Generator, Trans. IEE of Japan, Vol.120-B, No.8/9, pp.1030-1038,2000(in Japanese)

[10] Takaya Shigetoh et al: Relationship between Robust Stability and Power System Damping Enhancement by Excitation Control System of Adjustable-Speed Generator, Trans. IEE of Japan, Vol.118-B, No.1, pp.6-13,1998.(in Japanese)

[11] Kaoru Koyanagi et al: New Analytical Method for Studies of Enhancement of Power System Dynamic Stability by Application of Adjustable-Speed Generating System, Trans. IEE of Japan, Vol.119-B, No.1, pp.74-81,1999. (in Japanese)

[12] J.Tamura et al: Transient Stability Simulation of Power System Including Wind Generator by PSCAD/EMTDC, 2001 IEEE Porto Power Tech.Conference, No.EMT-108,2001.

[13] O. Wasynczuk et al.: Dynamic behavior of a Class of Wind Turbine Generators during Random Wind Fluctuations, IEEE Transactions on PAS, Vol. PAS-100, No. 6, pp. 2837-2845, 1981.

[14] T. Matsuzaka et al: Study on Stabilization of a Wind Generator Power Fluctuation, Trans. IEE of Japan, Vol.117-B, No.5, pp.625-631,1997, (in Japanese)

2006 5th International Power Electronics and Motion Control Conference

Design Consideration of a Novel Digital Bidirectional Constant Current Source Used in Hybrid Electric Vehicle

Qingbo Hu, Zhengyu Lü

National Key Laboratory of Power Electronics, Zhejiang University, Hangzhou 310027, P R China

hqbcszz@hotmail.com

Abstract – **Two novel circuit topologies dedicating to system within storage battery as standby power supply and operating as bidirectional current source used in hybrid electric vehicle (HEV) are proposed in this paper, and both advantages and disadvantages between the two proposed topologies are investigated. Functions of both discharging process with large current from battery to the relative lower voltage load end and charging process with constant current to the reverse direction are realized by the presented circuit. Meanwhile, a strategy of digital control, which implements the control of constant current for the bidirectional circuit and furthermore realization of power management for battery is achieved, is proposed as well. In addition, an opening loop control scheme base on F2407A DSP chip is adopted when inductor current is in an intermittent condition. Characteristic of excellent current source for the very digital control circuit is verified in the experiment, and these circuit topologies and control scheme can be used in HEV system.**

Keywords – *Bidirectional constant current source; full-digital control; storage battery; intermittent current*

I. INTRODUCTION

Nowadays, storage battery with bulky capacity is adopted as equipment of energy storage in standby power supplies, which are widely used in power system, distributed power supply system, communication power supply and hybrid electric vehicle (HEV) [1-6]. As well known, a bidirectional DC/DC converter is required in application within storage battery for realizing conversion of power form. Structure frame of standby power supply used in HEV is shown in Fig.1. Storage battery is paralleled with other energy storages such as super-capacitor and fuel battery, and this energy storage discharges with large current to drive motor of the electric vehicle. Whereas, when gas engine in the HEV system starts to work, the bidirectional DC/DC circuit is served as a charger. In general, the motor of HEV system works in a constant torque mode, so it is required that an extremely large and stable current must be supplied by

Project Supported by National Natural Science Foundation of China (50237030ZD).

battery. Besides, when charging process is performed, requirement of constant current control mode must be satisfied. Therefore, usage life of electro-equipment and storage battery is directly affected by characteristic of constant current control mode adopted in bidirectional DC/DC converter.

Application of digital control technique is widely broadened in power circuit due to the improvement of digital signal processor (DSP) [7-8]. Compared with analogy control, digital control possesses the advantages of high cost effective- ness, stable performance and so on. In addition, function of circuit can be easily realized by programming. Based on the comparison of existing non-isolated bidirectional DC/DC topologies, two new topologies of bidirectional circuit, which are dedicated to solve problem of bidirectional power conversion involved in such storage equipments as battery and so on, are presented in this paper. These new circuit topologies not only achieve power flow from storage battery to motor loading, also can serve as charger used in Buck mode or Boost mode corresponding to high speed or low speed of the hybrid electric vehicle. Constant current control mode is realized by the proposed topology and furthermore problem of current sampling is successfully dealt with by the very topology as well. Excellent performance of the circuit is verified by experimental result. Therefore, the proposed circuit is a candidate adopted in such storage equipments as battery and super-capacitor to realize bidirectional power conversion, and it can be used in HEV system.

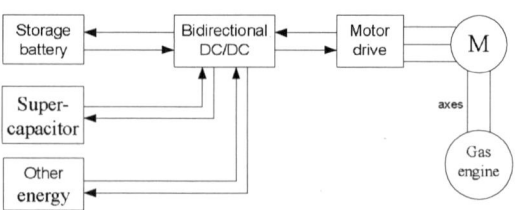

Figure.1 Drive system frame for the hybrid electric vehicle system

II. COMPARISON OF BIDIRECTIONAL TOPOLOGY

Various types of bidirectional DC/DC topologies are

researched in references [9-10] during recent years, where in two types of bidirectional Buck-Boost topologies are compared with each other. An ordinary Buck-Boost is shown in Fig.2 (a), whereas, Fig. 2 (b) shows Buck-Boost circuit connected in cascade mode. Distinction that electric stress difference between the above two topologies is not obvious provided that their operating state is Buck mode, whereas, once they operates under the state of Boost mode, performance of the later topology is superior to that of the former one, is specified in the reference. In cascade Buck-Boost, stress of inductor and capacitor is reduced by addition of power devices, therefore cost consideration of the circuit must be involved once it is adopted in industrial fields.

Both operation modes of Buck and Boost can be realized by either circuit that shown in Fig.2. Provided that the circuit is desired to operate under one mode of Boost or Buck separately, synchronous Boost or synchronous Buck is respectively preferable, as shown in Fig.3. As shown in Fig.3 (a), it is assumed that the circuit operates in the mode of Buck, and T_1 is controlled by PWM signal. Whereas, T_2 acts as synchronous switching device, energy is transferred from V_{in} to V_{out}. On the contrary, once energy is transferred from V_{out} to V_{in}, T_2 is controlled by PWM signal and that T_1 acts as synchronous switching device, and the whole circuit operates in the mode of Boost. Similarly, operation mode of synchronous Boost, as shown in Fig.3 (b), can be concluded.

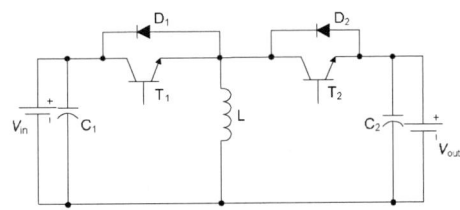

(a)Bidirectional Buck-Boost converter

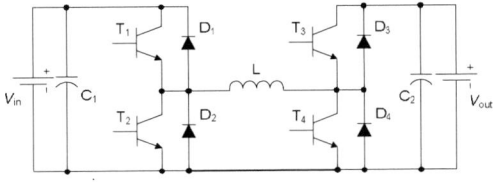

(b)Bidirectional Buck-Boost cascade converter

Figure.2 Two kinds of bidirectional Buck-Boost converter

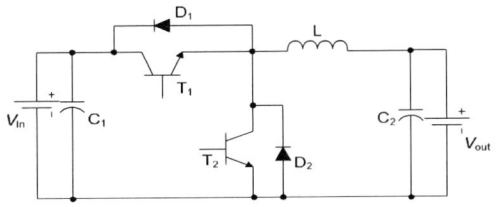

(a)Synchronous Buck converter

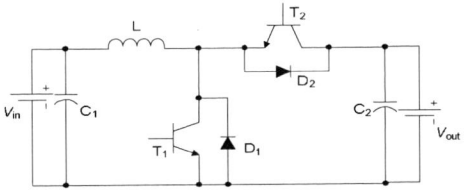

(b)Synchronous Boost converter

Figure.3 Two types of bidirectional synchronous converter

In the experimental application, within consideration of two factors that V_{in} is larger than V_{out} and Buck operation mode contributes to increase of output current on the prediction that capacity of battery is settled, synchronous Buck is adopted, as shown in Fig.3 (a). However, current difference between discharging mode that current reaches 120A and charging mode that the maximum current is 30A is extremely distinct, so the current sampling circuit and inductor design are difficult problems. In the charging mode, the relatively weak sampling signal is achieved and disturbance is easily introduced if the same current sampling circuit and inductor are adopted with the discharging circuit. Therefore, it is required to modify the circuit shown in Fig.3 (a).

III. MODIFICATION OF SYNCHRONOUS BUCK

A. Paralleled topology

Due to the large difference between discharging current with charging current, the charging current has to be in an intermittent condition adopted the same inductor as discharging mode. It is disadvantage to digital current sampling and control, so the discharging circuit and charging current may as well independence.

Topology modified from typical synchronous Buck, as shown in Fig. 3(a), is shown in Fig.4. The modified topology separates discharging loop from charging loop. As shown in Fig.4, discharging loop of Buck circuit is enclosed in the above broken frame and correspondingly charging loop is enclosed in the below frame. It can use different inductor and sampling circuit, and the inductor current can be designed on a continuous condition which is propitious to digital sampling. In addition, control requirement of constant output current must be achieved during discharging process, whereas control requirement of constant charging current must be achieved during charging process. Analyzing the current loop during discharging mode of Buck circuit, it can be found a ground loop, which is marked by a black and thick line in Fig.5, and it must be avoided.

Structure improvement of the paralleled topology is required for making the two paralleled circuits can separately operate normally, but ground loop must be avoided. Therefore modified parallel type circuit, as shown in Fig.6, is the preferable candidate. Compared

527

with topology shown in Fig.4, ground line between the input end of Buck of the charging mode and D_3 is removed from the circuit, as shown in Fig.6. Current sampling can use hall component which serially put into the branch of inductor, so it must use two hall components to sample discharging current and charging current, respectively.

In the discharge mode, T_1 is controlled by PWM signal, and T_2 and T_3 are turned off, just as Fig.7 (a) shown. In the charge mode, it can work as Buck and Boost modes. While the vehicle speed is high, the motor back electromotive force is higher than battery voltage, and the circuit works as Buck mode. Shown as Fig.7 (b), T_3 is controlled by PWM signal, and T_2 and T_1 are turned off. The circuit of Buck mode in charging is as same as discharging mode. While the vehicle speed is low, the motor back electromotive force is lower than battery voltage, and the circuit works as Boost mode as Fig.7 (c) shown. T_2 is controlled by PWM signal, and T_3 and T_1 are turned off. The real line with arrow as Fig.7 shows the switching device work loop, while broken line with arrow shows the diode work loop. In a word, this paralleled topology only works a half at any time, so the switching loss is low.

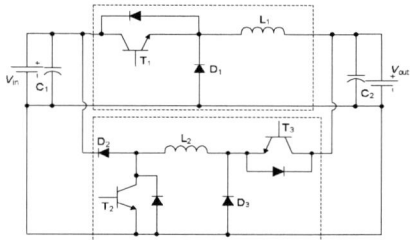

Figure.4 Modified bidirectional DC/DC converter

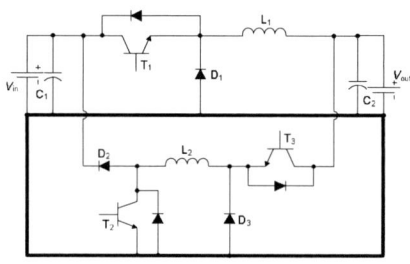

Figure.5 Ground loop of the circuit

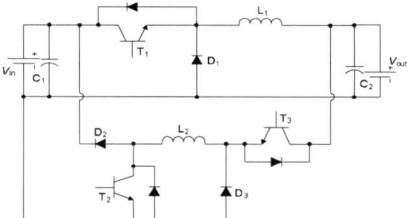

Figure.6 Modified topology of bidirectional converter

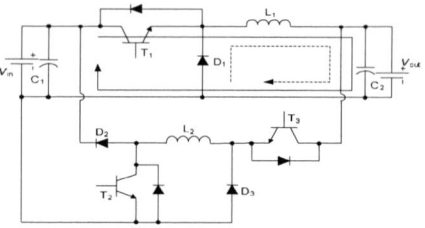

(a)Buck operation loop of discharging mode

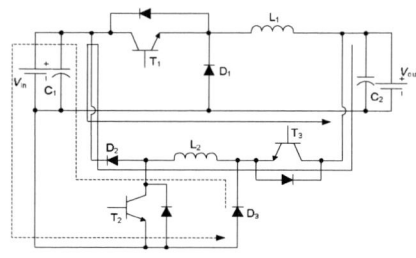

(b)Buck operation loop of charging mode

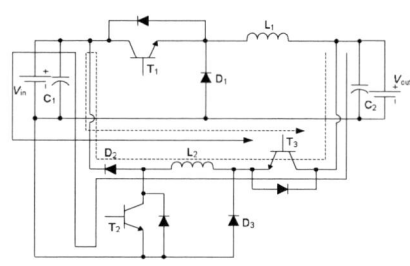

(c)Boost operation loop of charging mode

Figure.7 Operation loops of bidirectional converter

B. Combined topology

The paralleled topology of above analysis can achieve discharging mode and charging mode in HEV system, but the active devices and inductors are so many, and it makes system more complexity.

The problem that current sampling and control can not be achieved commendably used digital chip during the inductor current in an intermittent mode is resolved by an open looping control scheme just as following sector analysis. Only one inductor component can be adopted in the circuit, and one hall component which is used to sample inductor current is added. The discharging current and charging current sampling are used in different proportional coefficient of the amplifier. Fig.8 is the proposed topology.

In the discharge mode, T_1 is controlled by PWM signal, and D_3 is unforced on, while T_2 and T_3 are turned off, just as Fig.9 (a) shown. The circuit is served as Buck

528

mode. In the charge mode, it can work as Buck and Boost modes. While the vehicle speed is high, the motor back electromotive force is higher than battery voltage, and the circuit works as Buck mode. Shown as Fig.9 (b), T_3 is controlled by PWM signal, and D_1 is unforced on, while T_2 and T_1 are turned off. The circuit of Buck mode in charging is as same as discharging mode. While the vehicle speed is low, the motor back electromotive force is lower than battery voltage, and the circuit works as Boost mode as Fig.9 (c) shown. T_2 is controlled by PWM signal, and T_3 is forced on, while T_1 are turned off. The real line with arrow as Fig.9 shows the switching device work loop, while broken line with arrow shows the diode work loop.

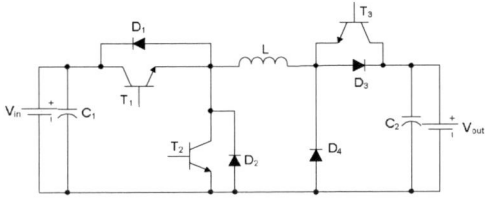

Figure.8 Proposed bidirectional converter

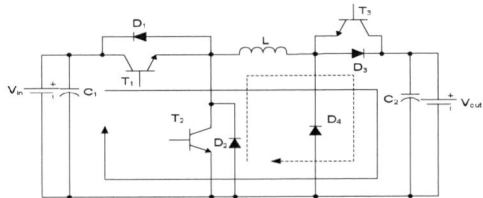

(a)Operation loop of Buck during discharging mode

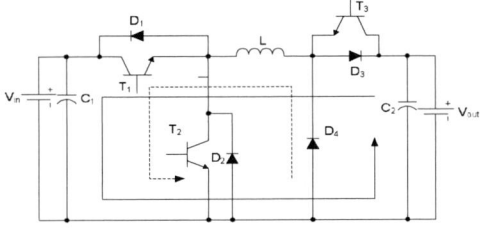

(b)Operation loop of Buck during charging mode

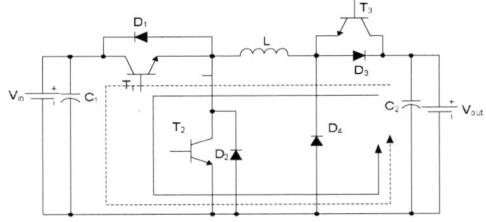

(c)Operation loop of Boost during charging mode

Figure.9 Operation loops of the proposed bidirectional converter

C. Topology analysis

Both two bidirectional topologies can achieve discharging and charging functions in HEV system, but paralleled topology is more complexity than the combined topology. Tab.1 shows the active device and passive component number between two topologies, respectively.

	Switching device	diode	inductor	capacitor
Paralleled topology	3	3	2	2
Combined topology	3	1	1	2

Table.1 Component number of the two topologies

Compared electric stress of the switching device, the voltage stress is same for all switching device, while current stress is different. Only one switching device in the paralleled topology need to work at large current, whereas, all switching device of the later one have to work at large current, so the combined circuit has more switching loss than the paralleled circuit. But this slight loss in HEV system does not affect system efficiency because the rate power of whole system is 50kW. Consideration simpleness of circuit topology, the combined topology is adopted in this paper.

IV. REALIZTION OF INTERMITTENT CURRENT CONTROL

It must sample inductor current in DC/DC converter real-time, but limiting by sampling speed of DSP, only one point can be sampled in each switching period. Fig.10 shows three sampling points in one switching period. When digital system samples inductor current in B point or C point, the value can be regarded as average current because the inductor current just locals midpoint of raise or fall. The sampling value can be regarded as average current and used in digital PI control when inductor current is continuous. While inductor current is intermittent, the sampling value is a half of ripple current ΔI. Now, this paper analyzes the intermittent current control in a Buck converter.

1) $\frac{1}{2}\Delta I > I_o$, inductor current is intermittent.

2) $\frac{1}{2}\Delta I < I_o$, inductor current is continuous.

3) $\frac{1}{2}\Delta I = I_o$, inductor current is in a critical condition.

In a Buck converter, it has following relationship as

$$\frac{1}{2}\Delta I = \frac{1}{2}\frac{(U_{in}-U_o)DT_s}{L} . \tag{1}$$

529

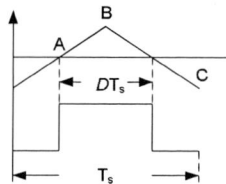

Figure.10 Sampling points in one switching period

From above formula, the ripple inductor current can be calculated by sampling input voltage U_{in} and output voltage U_o with duty cycle D, switching period T_s and inductor L. Then circuit state can be judged by comparing ripple current with average output current.

This paper only discusses the calculation method of average inductor current when current is intermittent. Fig.11 shows the inductor current in an intermittent state. The average current can be calculated as

$$I_o = \frac{1}{2}\Delta I(D + D_p) . \tag{2}$$

Wherein, $D_p = \dfrac{(U_{in} - U_o)D}{U_o}$, so the average current can be written as

$$I_L = \frac{1}{2}\Delta I \frac{U_{in}D}{U_o} . \tag{3}$$

Similarly, when the converter works in charging mode it just serves as a Boost circuit, and the average inductor current can be written as

$$I_L = \frac{1}{2}\Delta I \frac{DU_o}{U_o - U_{in}} . \tag{4}$$

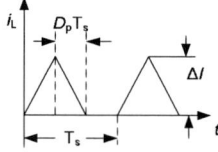

Figure.11 Inductor current is intermittent

In an intermittent condition, using the calculative average inductor current to replace the sampling value is just an opening loop control scheme. Next sector will compare transfer functions between continuous mode with intermittent mode.

In a continuous mode, the ac small signal model is shown as Fig.12, where, $e(s) = V_o / D^2$, $j(s) = V_o / R$. It has some equations as following. From Equ.(5) to Equ.(7), the inductor current transfer function can be written as Equ. (8).

$$\frac{v_o(s)}{d(s)}\Big|_{v_1(s)\,=\,0} = \frac{V_1}{LCS^2 + SL/R + 1} \tag{5}$$

$$v_o(s) = i_L(s) \times Z_o \tag{6}$$

$$Z_o = \frac{R}{RCS + 1} \tag{7}$$

$$\frac{i_L(s)}{d(s)}\Big|_{v_1(s)\,=\,0} = \frac{V_1(RCS + 1)}{LRCS^2 + SL + R} \tag{8}$$

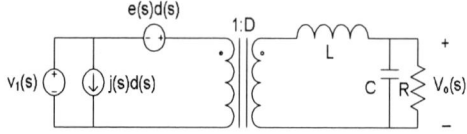

Fig.12 Ac small signal model in a continuous mode

In an intermittent mode, the ac small signal model is shown as Fig.13, where, $g_1 = \dfrac{1}{R_e}$, $j_1 = \dfrac{2(1-M)V_1}{DR_e}$, $r_1 = R_e$, $g_2 = \dfrac{2-M}{MR_e}$, $j_2 = \dfrac{2(1-M)V_1}{DMR_e}$, $r_2 = M^2 R_e$, $M = \dfrac{V_o}{V_1}$, $R_e = \dfrac{2L}{D^2 T_s}$, T_s is the switching period. The inductor current transfer function can be written as Equ.(9).

$$\frac{i_L(s)}{d(s)}\Big|_{v_1(s)\,=\,0} = \frac{r_2 j_2(SRC + 1)}{LRCS^2 + (RCr_2 + L)S + (R + r_2)} \tag{9}$$

Figure.13 Ac signal model in an intermittent mode

Contrasted Equ.(8) with Equ.(9), it has small difference. The parameters of current PI regulator can be designed as the same in the two current modes.

V. EXPERIMENTAL RESULT

Storage Battery with specification of 336V, 100Ah is used in the input port of Buck. During experimental process, constant current control of output is adopted. Switching frequency of experiment is 20kHz and output current range is set to be 0-120A. During charging process, charging mode of constant current is implemented and furthermore function of floating charge is realized. Main parameters of the circuit is listed as follows: $L = 75\mu H$, $C_1 = C_2 = 100\mu F$.

During discharging process, current waveform of inductor at 100A output is shown as Fig.14. Fig.15 shows the diode D_2 voltage waveform in a continuous mode. Furthermore, Fig.16 shows a current waveform of inductor at 10A when current is in an intermittent mode.

530

Fig.17 shows the diode D_2 voltage waveform in an intermittent mode.

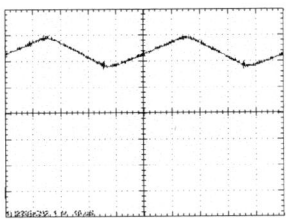

20A/div , 10μs/div

Figure.14 Current waveform in a continuous mode

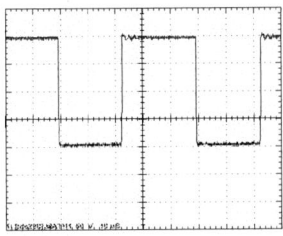

50V/div , 10μs/div

Figure.15 Voltage waveform of diode D_2 in a continuous mode

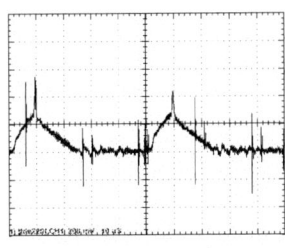

20A/div , 10μs/div

Figure.16 Current waveform in an intermittent mode

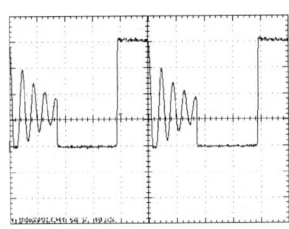

50V/div , 10μs/div

Figure.17 Voltage waveform of diode D_2 in an intermittent mode

VI.   CONCLUSTION

Two novel bidirectional topologies dedicated to power conversion of battery are presented in this paper. Requirement of power conversion is satisfied by these topologies and that constant current control during both discharging and charging process is realized. For satisfying the output requirement of low voltage and large current during the motor start, Buck circuit is adopted during discharging process. In addition, total digital control scheme of bidirectional constant current source, whose excellent control effect is verified by experiment, is presented in this paper. The proposed bidirectional DC/DC converter can be adopted on hybrid electric vehicle and so on.

REFERENCES

[1] Qun Zhao, Fred C. Lee. High-Efficiency, High Step-Up DC-DC Converter[J]. IEEE Trans on Power Electronics, 2003,vol.18, no.1,pp: 65-73.

[2] Fang Z. Peng, Fan Zhang, Zhaoming Qian. A Novel Compact DC-DC Converter for 42V Systems[C]. PESC'03, 2003, vol.1, 33-38.

[3] Xu Haiping, Sun Changfu, Ma Gang, et al. The Research on Bi-directional DC-DC Converter in FCEV[J]. Electrical Automation, 2004, 26(3):33-35.

[4] Barbosa, P, Canales, F, Lee, F.C. A Front-end Distributed Power System for High-power Applications[C]. IEEE, Industry Application Conference Rec, 2000, vol.4, pp: 2546-2551.

[5] Zhang Fang-hua, Yan Yang-guang. A Family of Forward-Flyback Hybrid Bi-directional DC-DC Converters[J]. Proceedings of the CSEE, 2005, vol.24, no.5, 157-162.

[6] Song Peng, Zhang Yi-cheng, Yao Yong-tao, et al. Application of Average Current Control Method for DC/DC Converter Used in Electric Vehicle[J]. Low Voltage Apparatus, 2004，No.5: 8-11.

[7] Xu Haiping, Wen Xuhui, Kong Li. Analysis and Design of Digitally Controlled Bi-directional DC/DC Converter[J]. Power Electronics, 2003,12, vol.37, No.6, pp: 13-17.

[8] Jinghai Zhou, Zhengyu Lu, Zhengyu Lin et al. A Novel DSP 2kW PFC Converter with a Simple Sampling Algorithm [C]. APEC, 2000, vol.1, 434-437.

[9] Caricchi. F, Crescimbini. F, Capponi. F.G, Solero. L. Study of Bi-directional Buck-Boost Converter Topologies for Application in Electrical Vehicle Motor Drives[C]. APEC'98, 1998, Conference Proceedings. Vol.1, pp:287-293.

[10] Schupbach. R.M, Balda. J. C. Comparing DC-DC Converters for Power Management in Hybrid Electric Vehicles[C]. IEMDC'03, 2003, vol.3, pp: 1369-1374.

2006 5th International Power Electronics and Motion Control Conference

A Single-Phase Grid-Connected Inverter System With Zero Steady-State Error

Guo Xiaoqiang, Zhao Qinglin and Wu Weiyang
Department of Electrical Engineering, Yanshan University
Qinhuangdao, Hebei P. R. China
E-mail: yeduming@163.com

Abstract—**A single-phase grid-connected inverter system with a novel current control strategy is presented in this paper. The controller consists of a proportional regulator and a new type of resonant regulator. Compared with traditional PI control methods, the P+Resonant(PR) control can introduce an infinite gain at the fundamental frequency and hence can achieve zero steady-state error. A more practical quasi-PR controller is used based on theoretical analysis. In this paper, a step by step procedure for designing the quasi-PR controller is proposed and verified by experiments. Theoretical analysis and experimental results of an experimental prototype verified the high performance of the proposed system in both the sinusoidal reference tracking and the disturbance rejection.**

Keywords-grid-connected inverter; quasi-PR controller; harmonic impedance; bilinear transformation

I. INTRODUCTION

In recent years, there has been an increase in the use of renewable energy due to growing concern for the gradually serious shortage of energy and environment pollution. With the rapid progress of the power electronics techniques, solar energy as an alternative energy source has been put to use such as Photovoltaic (PV) arrays. Nowadays, more and more concern of PV array has been focused on interconnecting the PV power systems with the grid. Generally, the grid-connected PV power systems mainly consist of a boost DC/DC converter unit and a DC/AC inverter unit. As the core of the whole system, the inverter unit plays an important role in grid-connected operation. Many control strategies of the grid-connected inverter have been proposed in recent works such as hysteresis control and predictive control [1]. While the hysteresis control is simple and robust, it has major drawbacks in variable switching rate, current error of twice the hysteresis band, and high frequency limited cycle operation [2]. As for predictive control, it depends on the accuracy of both the system model and the reference current prediction [3], [4]. Therefore, there is an unsolved problem of steady-state error in those current control strategies mentioned above.

In this paper, a novel current control is introduced for the grid-connected inverter to achieve zero steady-state error. Compared with the conventional PI control, the introduced PR control can overcome two well-known drawbacks of PI control: inability to track a sinusoidal reference with zero steady-state error and poor disturbance rejection capability. Due to achieving an infinite gain at the fundamental frequency and keeping the gain much reduced at other frequencies, the PR controller can achieve high performance in both the sinusoidal reference tracking and the disturbance rejection. Furthermore, another significant advantage of this controller is that it can be applied to single-phase current regulated systems as well as three-phase systems.

II. DESCRIPTIONS OF THE PROPOSED SYSTEM

Fig.1 shows the schematic diagram of the grid-connected inverter system. The proposed system mainly consists of a DC/AC SPWM inverter, an output filter inductor, a DSP control board, an IGBT drive board and two isolated transformers which enable electrical isolation as well as voltage level adjustment.

The main parameters of the proposed system shown in Fig.1 are listed in TABLE I.

TABLE I. SYSTEM PARAMETERS

Utility voltage	110V rms, 50Hz
DC bus voltage	200V
Switching frequency	20KHz
Sampling frequency	20KHz
Filter inductor	3mH
Deadbeat time	1us

A novel modulation strategy is introduced into the grid-connected inverter system [5]. It modulates the triangle carrier wave and two inverse sinusoidal modulation waves. Compared with the conventional modulation method, the novel modulation strategy has its advantages. First, it is no need of checking the pole of the output voltage to switch the modulated signals. Second, the modulation strategy is able to weaken the distortion of the output waveform on zero passage as well as to improve the stability of the closed-loop system.

1-4244-0448-7/06/$25.00 ©2006 IEEE   532

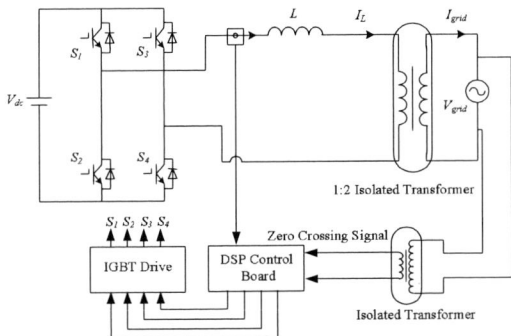

Figure 1. Schematic Diagram of the grid-connected inverter system

III. SYSTEM CONTROL SCHEME

A. System Model

Assuming the switching frequency is high enough to neglect the inverter dynamics (the effective value is 40kHz due to the novel modulation), the equivalent representation of the single-phase current regulated grid-connected inverter system is obtained as shown in Fig.3. Due to relatively high switching frequency, the PWM inverter can be represented by a gain for simplicity of analysis.

The transfer function of the output current of the inverter can be derived from Fig.3 as follows:

$$I_L = \frac{KG(s)}{sL + R + KG(s)} I_{ref} - \frac{1}{sL + R + KG(s)} V_{grid} \quad (1)$$

Where $G(s)$ is the controller of the grid-connected inverter system. The transfer functions of the PI and PR controller are given by (2) and (3) respectively [6], [7].

$$G_{PI}(s) = k_p + \frac{k_i}{s} \quad (2)$$

$$G_{PR}(s) = k_p + \frac{2k_r s}{s^2 + \omega_0^2} \quad (3)$$

Equation (1) shows that the inductor current depends on both the reference current and the grid voltage. In steady state, the PI controller has a finite gain at the fundamental frequency. So the second term of (1) can't be neglected. However, PR controller introduces an infinite gain, which means that the first term of (1) approaches I_{ref} and the second term approaches zero. That is, the zero steady-state error of the controlled inductor current can be achieved by using PR controller.

Unfortunately, it is hard to implement the PR controller in reality. Firstly, the infinite gain introduced by PR controller leads to an infinite quality factor which can't be achieved in either analog or digital system. Secondly, the gain of PR controller is much reduced at other frequencies and it is no adequate to eliminate harmonic influence caused by grid voltage.

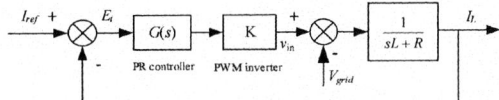

Figure 3. The control diagram of the grid-connected inverter system

To solve the problems mentioned above, a quasi-PR controller is used as follows[7]:

$$G(s) = k_p + \frac{2k_r \omega_c s}{s^2 + 2\omega_c s + \omega_0^2} \quad (4)$$

B. Quasi-PR Controller Design

In this paper, a detailed guideline for designing the quasi-PR controller is proposed in order to optimize the performance of the proposed system in the sinusoidal reference tracking as well as the disturbance rejection.

From (4), it can be seen that there are three parameters in the quasi-PR controller including k_p, k_r and ω_c. For simplicity of analysis, any two of them are assumed to be constant, and then the effect of changes in the third parameter can be easily observed.

Firstly, supposing $k_p = 0$, $\omega_c = 1$, the Bode diagram of (4) can be obtained with the variations in k_r as shown in Fig.4.

From Fig.4, it is concluded that k_r has no effect on the bandwidth but the gain of the controller. The gain increases as k_r is added.

Secondly, as shown in Fig.5, the Bode diagram of (4) can be obtained with the variations in ω_c on the assumption that $k_p = 0$, $k_r = 1$.

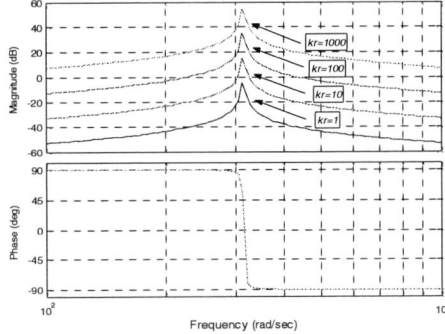

Figure 4. Bode diagram of the quasi-PR controller for variation in k_r

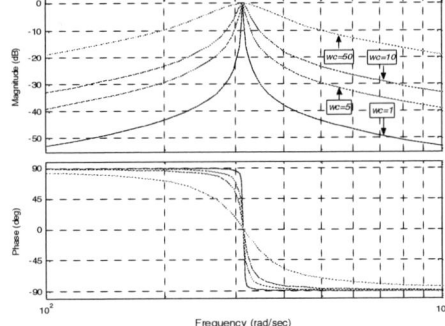

Figure 5. Bode diagram of the quasi-PR controller for variation in ω_c

Fig.5 shows that ω_c has effects on both the bandwidth and the gain of the controller. The bandwidth and gain increase as ω_c is added. However, it is worth noting that the same gain can be achieved at the resonant frequency of the quasi-PR controller with variation in ω_c .

In this paper, harmonic impedance is defined as the relationship between grid harmonic voltage disturbance and the resulting harmonic inductor current [8]. High harmonic voltage impedance will result in a relatively small harmonic current in response to a harmonic grid voltage disturbance. Hence, it is a useful measure of the system's disturbance rejection capability. The harmonic impedance of the proposed grid-connected system is given by (5). Supposing ω_c =5, k_r =100, the Bode diagram of (5) can be obtained with the variations in k_p as shown in Fig.6.

From Fig.6, it can be concluded that the harmonic impedance of the system is subjected to k_p and the harmonic impedance increases as k_p is added. That is, higher k_p can lead to a relatively low harmonic current caused by grid harmonic voltage.

In conclusion, there are three degrees of freedom in the design of the quasi-PR controller. Therefore, the quasi-PR controller can be designed following a step by step procedure according to the theoretical analysis mentioned above.

First, design ω_c to obtain an appropriate bandwidth for the controller.

Second, choose k_r to meet the requirement of the controller gain.

Finally, design k_p to make sure that the system can achieve high performance in the sinusoidal reference tracking as well as the disturbance rejection.

C. Stability Analysis

The open-loop and close-loop transfer functions of the system with quasi-PR controller are given by (6) and (7) respectively.

The characteristic polynomial is given by the denominator of (7). Hence, the characteristic equation can be obtained given by (8).

Supposing k_p varies from 1 to 10, where ω_c =5, k_r =100, the root locus of the closed-loop system is obtained as shown in Fig.7 where the non-dominant poles are ignored. It can be observed that the dominant poles move towards the imaginary axis when k_p becomes too small or too large. Therefore, in order to achieve the stable operation as well as high harmonic impedance shown in Fig.6, a tradeoff should be made to determine the value of k_p.

In this paper, the parameters of the quasi-PR controller are given as follows:

$$k_p = 4,\ \omega_c = 5,\ k_r = 100$$

Under the conditions mentioned above, the system can achieve the desired performance in the sinusoidal reference tracking as well as the disturbance rejection.

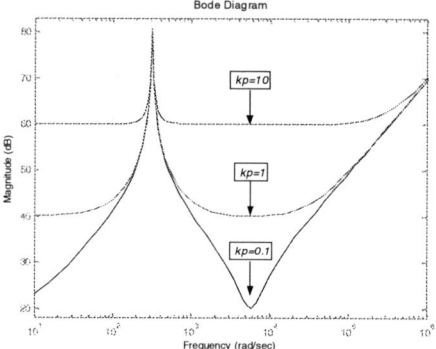

Figure 6. Harmonic impedance for the quasi-PR controller

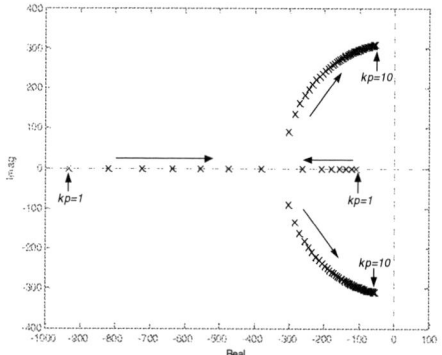

Figure 7. Root locus for the variation of k_p

$$Z_{PR} = \frac{V_{grid}}{I_L} = \frac{Ls^3 + (Kk_p + 2L\omega_c + R)s^2 + (L\omega_0^2 + 2Kk_p\omega_c + 2R\omega_c + 2Kk_r\omega_c)s + Kk_p\omega_0^2 + R\omega_0^2}{s^2 + 2\omega_c s + \omega_0^2} \tag{5}$$

$$G_{ol}(s) = \frac{I_L(s)}{E_i(s)} = K\frac{k_p s^2 + 2\omega_c(k_p + k_r)s + k_p\omega_0^2}{Ls^3 + (R + 2L\omega_c)s^2 + (2R\omega_c + L\omega_0^2)s + R\omega_0^2} \tag{6}$$

$$G_{cl}(s) = \frac{I_L(s)}{I_{ref}(s)} = \frac{G_{cl}(s)}{G_{cl}(s)+1} = K\frac{k_p s^2 + 2\omega_c(k_p + k_r)s + k_p\omega_0^2}{Ls^3 + (Kk_p + R + 2L\omega_c)s^2 + (2Kk_p\omega_c + 2Kk_r\omega_c + 2R\omega_c + L\omega_0^2)s + Kk_p\omega_0^2 + R\omega_0^2} \tag{7}$$

$$Ls^3 + (Kk_p + R + 2L\omega_c)s^2 + (2Kk_p\omega_c + 2Kk_r\omega_c + 2R\omega_c + L\omega_0^2)s + Kk_p\omega_0^2 + R\omega_0^2 = 0 \tag{8}$$

D. Digital Implementation

The discrete transfer function for the quasi-PR controller can be obtained by means of the bilinear transformation as follows [9], [10]:

$$G(z) \Leftrightarrow G(s) \qquad s = \frac{2}{T} \times \frac{1 - z^{-1}}{1 + z^{-1}} \qquad (9)$$

Where T — sampling cycle

Substitute (9) into (4), and then the discrete transfer function is given by (10).

$$G(z) = \frac{b_0 + b_1 z^{-1} + b_2 z^{-2}}{1 + a_1 z^{-1} + a_2 z^{-2}} \qquad (10)$$

Where,

$$b_0 = \frac{k_p \omega_0^2 T^2 + 4k_r \omega_c T + 4k_p \omega_c T + 4k_p}{\omega_0^2 T^2 + 4\omega_c T + 4}$$

$$b_1 = \frac{2k_p \omega_0^2 T^2 - 8k_p}{\omega_0^2 T^2 + 4\omega_c T + 4}$$

$$b_2 = \frac{k_p \omega_0^2 T^2 + 4k_p - 4k_r \omega_c T - 4k_p \omega_c T}{\omega_0^2 T^2 + 4\omega_c T + 4}$$

$$a_1 = \frac{2\omega_0^2 T^2 - 8}{\omega_0^2 T^2 + 4\omega_c T + 4}$$

$$a_2 = \frac{\omega_0^2 T^2 - 4\omega_c T + 4}{\omega_0^2 T^2 + 4\omega_c T + 4}$$

In this paper, $G(z)$ is implemented with a fixed point arithmetic TI TMS320LF2407 DSP. The discrete difference equation is given by (11).

$$u(k) = b_0 e_i(k) + b_1 e_i(k-1) + b_2 e_i(k-2) - a_1 u(k-1) - a_2 u(k-2) \quad (11)$$

Where $u(k)$ is the output of the transfer function and $e_i(k)$ is the inductor current error.

IV. EXPERIMENTAL VERIFICATION

In this paper, a 300W experimental prototype based on the quasi-PR controller was constructed. In order to verify the efficiency of the proposed system, two factors will be used as follows:

Total Harmonic Distortion of the current:

$$THD = \frac{\sqrt{\sum_{h=2}^{20} I^2(h)}}{I(1)}$$

Power Factor $\qquad PF = \frac{I(1) \cos \varphi}{I}$

Where, $I(h)$ is the RMS value of the h harmonic and φ is the angle between the fundamental current and fundamental voltage.

Fig.8 shows the steady-state waveforms of the grid voltage and the grid current based on quasi-PR control. It can be seen that the grid current waveforms are nearly

perfect sinusoids. The PF of the grid current is 0.989, and the THD is 3.737% which is well below the IEEE standard of 5% for grid-connected inverters [11].

On the other hand, experimental results based on PI control are also obtained shown in Fig.9 with THD of 5.066% and PF of 0.987.

Compared with all the experimental results mentioned above, it is concluded that the quasi-PR control scheme can achieve better steady-state performance than the conventional PI control scheme.

Fig.10 show the start-up current based on quasi-PR control. It is noted that the grid current has a slight overshoot and the transient dies out rapidly. Therefore, the quasi-PR control scheme can also achieve a good transient performance.

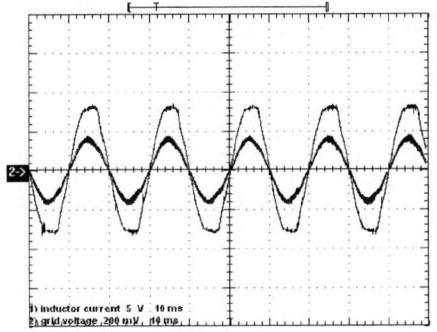

Figure 8. Grid voltage and grid current based on quasi-PR control
(Y-axis:100V/div, 5A/div, X-axis:10ms/div)

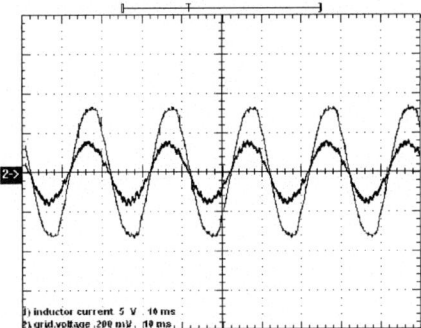

Figure 9. Grid voltage and grid current based on PI control
(Y-axis:100V/div, 5A/div, X-axis:10ms/div)

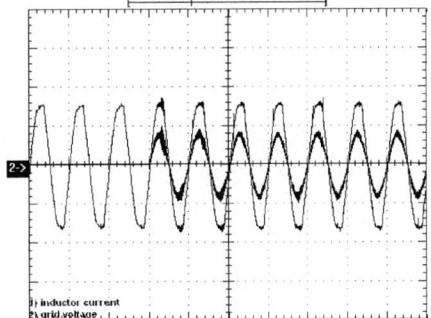

Figure 10. Grid voltage and start-up current
(Y-axis:100V/div, 5A/div, X-axis:10ms/div)

V. Conclusion

This paper presents a single-phase grid-connected inverter system with the quasi-PR control scheme. A detailed theoretical analysis has been performed to enable the proposed grid-connected inverter system to achieve high performance in both the sinusoidal reference tracking and the disturbance rejection. The proposed analysis also offers a useful guide to choose the value of the three parameters of the quasi-PR controller. The modulation and control strategy is implemented by using the TMS320LF2407A DSP. Theoretical analysis and experimental results of a 300W experimental prototype verified the high performance of the proposed grid-connected inverter system.

Acknowledgment

This work was supported by the National Natural Science Foundation of China, NO. 50237020.

References

[1] A. Kotsopoulos, J. L. Duarte, M.A.M Hendrix, "A predictive control scheme for DC voltage and AC current in grid-connected photovoltaic inverters with minimum DC link capacitance," *The 27th Annual Conference of the IEEE Industrial Electronics Society*, vol.3, pp. 1994-1999, November/December 2001.

[2] R. D. Lorenz, T. A. Lipo, and D. W. Novotny, "Motion control with induction motors," *Proc. IEEE*, vol. 82, pp. 1115–1135, August 1994.

[3] F. Kamran and T. G. Habetler, "An improved deadbeat rectifier regulator using a neural net predictor," *IEEE Trans. Power Electronics*, vol. 10, pp.504–510, July 1995.

[4] D. G. Holmes and D. A. Martin, "Implementation of a direct digital predictive current controller for single and three phase voltage source inverters," *in Proc. IEEE IAS Annu. Meeting*, 1996, pp. 906–913.

[5] Zhao Qinglin, Xu Yunhua, Jin Xiaoyi, Wu Weiyang, Cao Lingling, "DSP-based closed-loop control of bi-directional voltage mode high frequency link inverter with active clamp", *IEEE Industry Applications Conference*, 2005, pp. 928 – 933.

[6] S. Fukuda, and T. Yoda, "A Novel Current-Tracking Method for Active Filters Based on a Sinusoidal Internal Model," *IEEE Trans. on Industry Applications*, vol. 37, pp.888-895, May/June 2001.

[7] Poh Chiang Loh, D. M. Vilathgamuwa, Seng Khai Tang and H.L. Long, "Multilevel dynamic voltage restorer," *IEEE Trans. Power Electronics Letters*, vol. 2, pp.125–130, December 2004.

[8] E. Twinning and D. G. Holmes, "Grid current regulation of a three-phase voltage source inverter with an LCL input filter," *IEEE Trans. Power Electronics*, vol. 18, pp. 888–895, May 2003.

[9] F. Groutage, L. Volfson, A. Schneider, "S-plane to Z-plane mapping using a simultaneous equation algorithm based on the bilinear transformation," *IEEE Trans. on Automatic Control*, vol. 32, pp. 635 – 637, July 1987.

[10] H. Krishna, "Computational aspects of the bilinear transformation based algorithm for S-plane to Z-plane mapping," *IEEE Trans. on Automatic Control*, vol. 33, pp. 1086 – 1088, November 1988.

[11] IEEE Std. 929-2000, "IEEE Recommended Practice for Utility Interface of Photovoltaic (PV) Systems," pp. 6, April 2000.

2006 5th International Power Electronics and Motion Control Conference

DC Transformer with Line Frequency Ripple Cancellation

Sen Dou, Wilson Wu,* Annabelle Pratt and Pavan Kumar**

Intel Asia-Pacific Research & Development Ltd.
* No. 880, Zi Xing Rd, Shanghai Zizhu Science Park
Shanghai, China 200241
**Intel Corporation, Hillsboro, Oregon, 97124, USA
Sen.dou@intel.com

Abstract—In this paper, a DC transformer with line frequency ripple cancellation is proposed. The line frequency ripple cancellation scheme is discussed and applied to the Asymmetrical Half Bridge converter. An Asymmetrical Half Bridge DC transformer prototype is built to verify the proposed approach.

Keywords-DC transformer; DCX; efficiency; line frequency ripple cancellation; duty cycle; AHB converter

I. INTRODUCTION

A typical power delivery architecture within a server platform is shown in Figure 1. It consists of a Power Supply Unit (PSU) containing an AC/DC converter which provides power factor correction (PFC), electrolytic capacitors which store energy for ride-through of AC line outage events, and an isolated DC/DC converter which provides a regulated 12V output. For the majority of loads, a Voltage Regulator (VR) further converts the 12V downstream, resulting in duplicate regulation stages.

Requiring the isolated DC/DC converter in the PSU to provide regulation even during AC line outages results in design compromises which lead to lowering its efficiency [1,2,3]. Power delivery efficiency can be increased by splitting the regulation and isolation functions, replacing the regulated DC/DC converter in the PSU with a more efficient, unregulated DC/DC converter or DC transformer (DCX) [4,5,6,7] as shown in Figure 2.

The DCX output voltage varies as a function of load and input voltage. Voltage Regulators (VRs) shield the majority of the loads from these variations by providing a regulated voltage. However, loads that are supplied directly from the output of PSU, e.g. hard disk drives and

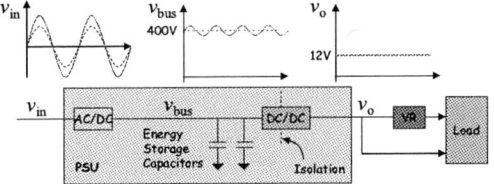

Figure 1. Power delivery within a server platform

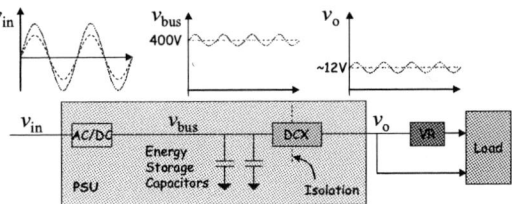

Figure 2. Power delivery and distribution of power with DCX

I/O cards, are subjected to an unregulated voltage. The variation with load is small because of the high efficiency of the DCX. The DCX input voltage varies due to AC/DC regulation inaccuracy, storage capacitor discharge during AC line outages, and line frequency ripple. The first two factors may be addressed by tightening the AC/DC regulation, and adding a hold-up time extension circuit [1,2] respectively. With these solutions, the application of a DCX only has the line frequency ripple problem.

A DC Transformer (DCX) with line frequency ripple cancellation which provides an unregulated 12V output with minimal line frequency ripple is proposed. The DCX is designed to operate over a narrow duty cycle range and the line frequency ripple cancellation is achieved by modulating the duty cycle within this range. The required duty cycle range is proportional to the magnitude of the line frequency ripple at the input of the DCX and since this is small (typically <5% peak to peak), only a small variation in duty cycle is required, so high efficiency is maintained, along with elimination of the line frequency ripple at the output of the converter.

In this paper, prototypes of a half bridge converter operating at 50% constant duty cycle and an Asymmetrical Half Bridge (AHB) with line frequency cancellation is built to verify the efficiency and ripple cancellation benefits. The proposed AHB DCX provides line frequency ripple attenuation of more than 40dB, and achieves 95.7% efficiency at full load.

II. DC/DC EFFICIENCY TRADE-OFF

It is well known that an unregulated DC transformer (DCX) can achieve higher efficiency than a regulated

1-4244-0448-7/06/$25.00 ©2006 IEEE

DC/DC converter [4,5,6,7], because a DCX always operates at an optimized condition with constant 50% duty cycle.

For a regulated DC/DC converter, the hold-up time requirement results in a wide input voltage range requirement. The input voltage of the DC/DC converter will drop to some minimum voltage V_{min} when the AC line drops out. The value of V_{min} is generally selected based on a trade-off between efficiency and energy utilization. At V_{min}, the converter achieves regulation by operating at the maximum duty cycle and for a lower V_{min}, a lower primary to secondary turns ratio n is required. Therefore, at nominal input voltage, the regulated DC/DC converter operates at a lower duty cycle condition with a lower turns ratio, resulting in higher primary current (peak and rms), and higher secondary side rectifier voltage rating. That means there will be higher conduction and switching loss in the primary side switch and the transformer. And the higher secondary side rectifier voltage rating leads to a higher forward voltage drop of a diode rectifier or conduction resistance of a synchronous rectification (SR) MOSFET, which increases the secondary side rectifier conduction loss. In summary, the regulated DC/DC converter has an obvious reduction in efficiency because of the wide input voltage range.

However, A DCX is operated close to the maximum duty cycle of 50% with a higher turns ratio. From the above analysis, the power loss of the DCX at this operating condition is lower than for the wide input range regulated DC/DC converter. Therefore, the DCX can achieve the higher efficiency.

The duty cycle of the proposed DCX with line frequency ripple cancellation only need vary over a narrow range. If the duty cycle varies over a narrow range near 50%, the efficiency of the DCX is expected to only decrease a little compared to a 50% duty cycle DCX.

This paper proposes an approach, applied to an AHB converter as shown in Figure 3, to reduce the amount of line frequency ripple allowed to propagate through the DCX stage. With the AHB converter, the DCX with line frequency ripple cancellation operates at a high nominal duty cycle, and only varies over a small range.

III. DCX WITH RIPPLE CANCELLATION

The principle of line frequency ripple cancellation is to attenuate line frequency ripple by modulating the duty cycle. There are two methods to realize this function as

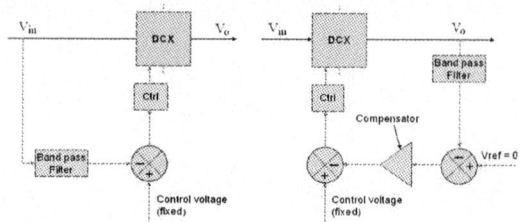

Figure 4. Line frequency ripple cancellation architecture scheme

shown in Figure 4. The first one is sensing the line ripple signal at the DCX input bus with a band pass filter (feed forward). The second one is sensing the line ripple signal at the DCX output terminal with a band pass filter (feedback). In this paper, the feed forward approach is used to realize ripple cancellation.

The design specification for the prototype AHB converter with line ripple cancellation is: input voltage range 394V to 406V (400V DC voltage plus ±6V line ripple voltage), output voltage 12V at no load, output ripple less than 120mV peak to peak, i.e. 40dB line frequency attenuation. Because of variable duty cycle close to 50% to realize ripple cancel function, it is necessary to design an appropriate turns ratio for the AHB converter. The DC gain M of AHB converter is

$$M = \frac{V_o}{V_{in}} = \frac{2}{n} \cdot D \cdot (1-D) \qquad (1)$$

According to the DC gain equation (1) of the AHB converter, the available turns ratios which can achieve close to 50% duty cycle were selected as $n=16$ and $n=15.5$. The duty cycle curve with input voltage corresponding to these two turns ratio designs are shown in Figure 5, D1 is for $n=15.5$ and D2 is for $n=16$.

It can be seen that the duty cycle curve $D1$ at $n=15.5$ is almost linear with ripple voltage but $D2$ at $n=16$ is more non-linear. The feed forward approach is used to implement the ripple cancellation function by converting the input ripple voltage signal to a duty cycle signal. When the turns ratio is $n=16$, a very complex circuit is required to realize ripple cancellation function, but for $n=15.5$, it is easier. Therefore, in this paper the feed forward ripple cancellation approach is applied to the AHB converter with $n=15.5$ turns ratio.

From the above analysis, it can be said that the required duty cycle should be very close to the duty cycle relation

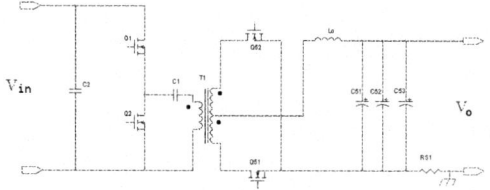

Figure 3. The AHB converter topology

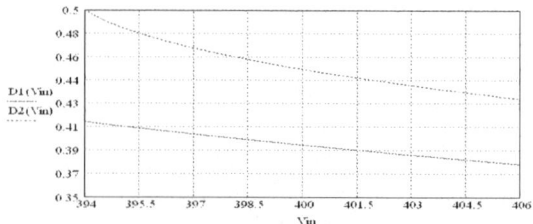

Figure 5. Duty cycle curve with input voltage at different turns ratio conditions

538

curve $D1(V_{in})$ shown in Figure 6. From the sensed line ripple signal, a linearized duty cycle curve can be created as follows

$$D3(V_{in}) = D_0 - A \cdot (V_{in} - V_{DC}) \qquad (2)$$

where D_0 is the constant value of duty cycle without line frequency ripple, V_{DC} is the input voltage DC value, i.e., V_{DC}=400V. A is the amplifier gain for adapting the slope rate of this linear line. Then the requirement of parameter D_0 and A is to achieve the smallest gap between $D3(V_{in})$ and $D1(V_{in})$ over the whole input voltage range from 394V to 406V. It is easy to get this duty cycle $D3(V_{in})$ by mathematical method as shown in Figure 6. With this duty cycle curve, the line ripple cancellation result in theory is shown in Figure 7. The output line frequency ripple voltage is about 20mV peak to peak, therefore the attenuation is 55dB, which is far better than the design goal of 120mV, or 40dB attenuation. Certainly, this is the ideal result without considering component tolerances.

A line frequency ripple cancellation circuit is proposed as shown in Figure 8. The duty cycle generation circuit is made up of two timer ICs TC555. One generates a constant frequency pulse corresponding to the switching frequency whereas another generates a duty cycle which is modulated between the maximum value and the minimum value corresponding to the magnitude of the line frequency ripple.

There are two critical points in designing the ripple sensing circuit: 1) the line frequency signal should be detected exclusively to distinguish line ripple from other transient signals; and 2) no phase shift is allowed to ensure phase cancellation can be achieved as expected.

The band pass filter permits only 100/120Hz line frequency signal to pass and attenuates the lower and the

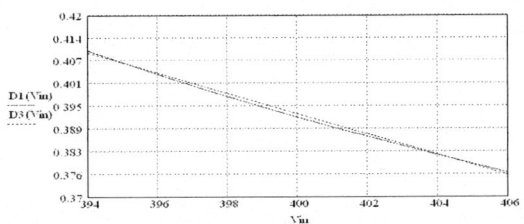

Figure 6. Required duty cycle and available duty cycle curve with input voltage

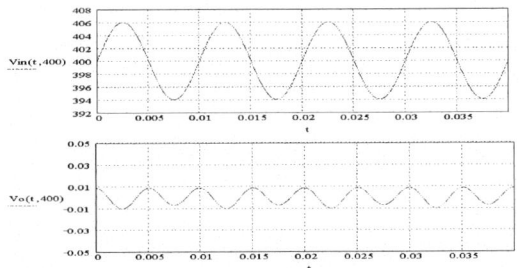

Figure 7. Theoretical line ripple cancellation result at input voltage 400VDC±6V line ripple voltage

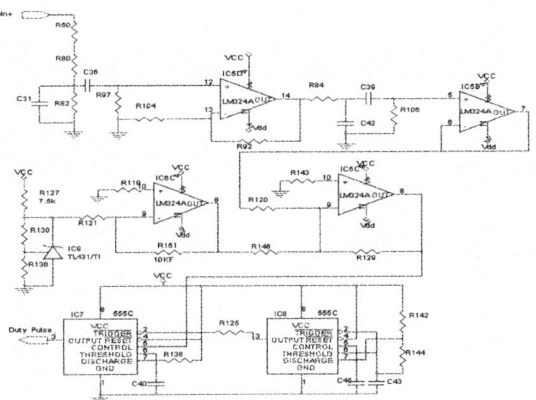

Figure 8. Line frequency ripple cancellation control circuit

higher frequency signals. So the corner frequency of band pass filter is set at 110Hz to minimize the phase shift of 100Hz and 120Hz signals. In order to eliminate the effect of low frequency bus voltage transient conditions and switching frequency ripple, the slope of the gain curve should be +40dB/Dec below 110Hz and -40dB/Dec above 110Hz as shown in Figure 9. The band pass filter is comprised of two filters with the cross over frequency of each centered around 110Hz. The phase shift at 100Hz and 120Hz is +10° and -10° respectively. This phase shift at 100/120Hz has some impact on ripple cancellation performance, but since the phase shift is very small, the output ripple requirement can still be met. Figure 10 shows the calculated output ripple voltage at +10° and -10° phase shift corresponding to 100/120Hz conditions. It can be seen that output ripple increases a lot compared to the zero phase shift condition shown in Figure 5, but the output ripple still is lower than 120mV and meets design requirement even after considering the switching frequency ripple.

The above analysis and calculation results in theory indicate that the band pass filter and duty cycle generation circuit can attenuate line frequency ripple and meet the design requirement. This control circuit was applied to the AHB converter and its performance was verified on a prototype.

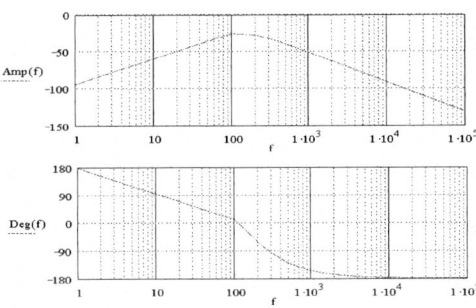

Figure 9. Bode plot of band pass filter {Sure, it is 180 degrees, just display problem, have been corrected.}

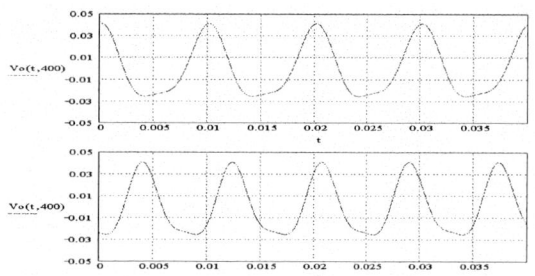

Figure 10. Output (a) 100Hz (b)120Hz line frequency ripple voltage waveforms at input voltage 400VDC±6V line ripple voltage

IV. EXPERIMENTAL RESULTS

An Asymmetrical Half Bridge DCX prototype with line frequency ripple cancellation was built to verify the described benefits. Key specifications and design parameters are provided in Table I. In order to provide a performance comparison, a 50% duty cycle, 12V output DCX prototype was also built by modifying the proposed AHB DCX design. This 50% duty cycle DCX uses the same power stage except for the changes indicated in Table I, and its duty cycle is fixed at 50%.

The 50% duty cycle DCX can use 60V synchronous rectifiers (SRs), as opposed to the 75V SRs used in the proposed DCX design. However, the on-resistance of 60V devices are not significantly lower than that of 75V devices, so it would make very little difference to the efficiency.

TABLE I
PROTOTYPE DESIGN PARAMETERS

DC input voltage	400 V
Input voltage line frequency ripple	12V peak-to-peak
Nominal output voltage	12 V
Maximum output power	240 W
Switching frequency	68 kHz
Primary switches	Infineon SPA20N60C3
Primary switch driver	IR IR21844S
Output inductor	6μH
Output inductor core	PQ26 core
Transformer core	Ferroxcube 3C96 PTS4026
Transformer turns ratios	Proposed DCX: 15.5:1
	50% Duty DCX: 16:1
Transformer primary magnetizing inductance	Proposed DCX: 500μH
	50% Duty DCX: 600μH
Synchronous rectification switches (self-driven with coupled windings)	Proposed DCX: Fairchild FDP047AN08A0
	50% Duty DCX Fairchild FDP038AN06A0

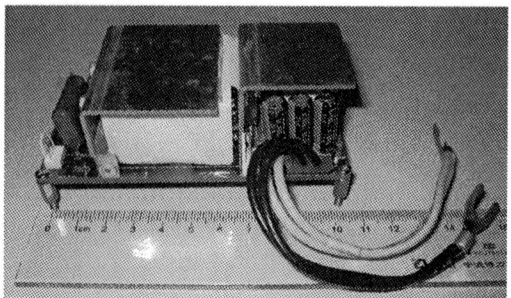

Figure 11. Prototype of AHB with line ripple cancellation

Figure 11 shows a picture of the prototype. The AHB converter DCX prototype can operate with primary switch ZVS condition and secondary side self-driven synchronous rectifier over the whole load range. The duty cycle varies from 0.38 to 0.42 when 12V peak to peak line frequency ripple is applied to the input voltage. It was experimentally verified that the applied line frequency ripple has no impact on the power conversion efficiency, therefore efficiency is measured under constant input voltage condition.

Figure 12 shows the measured full load efficiency of the proposed AHB DCX prototype and 50% duty DCX prototype as a function of the input voltage. The full load efficiency of the proposed AHB DCX is 95.7% at 400V input. It can be seen that the proposed DCX full load efficiency is only 0.4% lower than the 50% duty cycle DCX. This is a 10% increase in losses from 9.7W to 10.8W at full load.

Figure 13 shows the measured efficiency curve as a function of the output current at a constant input voltage of 400V. The highest efficiency of the proposed AHB DCX is 96.5% at an output current of 10A, i.e. at 50% load. The efficiency of the proposed AHB DCX is lower than the 50% duty DCX at heavy load conditions. At the middle and light load conditions, the efficiency of the proposed AHB DCX is higher than the 50% duty DCX. The reason for this behavior is two-fold. The transformer core loss of the 50% duty DCX is higher because the flux density variation is higher. And the SR driver loss is higher for the 50% duty DCX since the SR self-driven

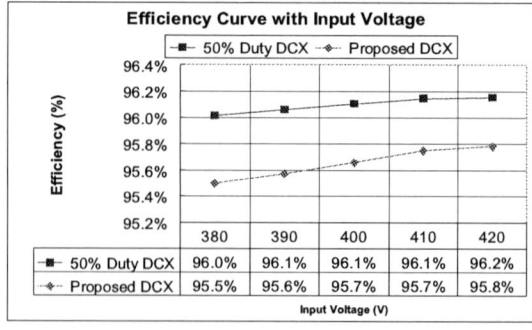

Figure 12. Full load efficiency comparison between proposed AHB DCX and 50% duty DCX

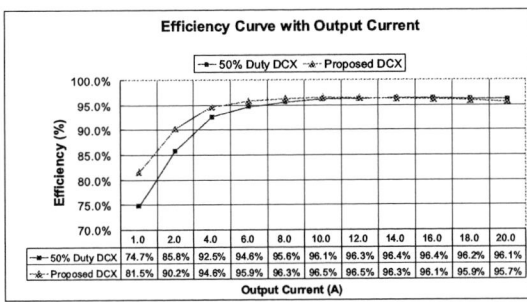

Figure 13. Efficiency comparison between current AHB DCX and 50% duty DCX at whole load range

Output Current (A)	1.0	2.0	4.0	6.0	8.0	10.0	12.0	14.0	16.0	18.0	20.0
50% Duty DCX	74.7%	85.8%	92.5%	94.6%	95.6%	96.1%	96.3%	96.4%	96.4%	96.2%	96.1%
Proposed DCX	81.5%	90.2%	94.6%	95.9%	96.3%	96.5%	96.5%	96.3%	96.1%	95.9%	95.7%

voltage is higher. These two losses are not variable with output current. Although the loss is only a small portion of total power loss at full load, it is a significant portion of total loss at light load conditions.

The measured output ripple of AHB converter without ripple cancellation is shown in Figure 14, when 12Vpp 100/120Hz line ripple is applied to the AHB converter input. Without the ripple cancellation function, the peak to peak output ripple is 383mV, which exceeds the output ripple requirement of 120mV. It can be seen that the ripple is transferred to the output and there is no phase shift.

When 12V peak-to-peak 100/120Hz line ripple is applied to the proposed AHB DCX input with ripple cancellation incorporated, the measured output ripple is shown in Figure 15. The maximum total output ripple, including line frequency ripple and switching frequency ripple, is 106mV at 100Hz and 75mV at 120Hz. The output line frequency ripple is attenuated by at least 41dB, which meets the design goal of 120mV and 40dB attenuation. It can also be clearly seen in Figure 15 that there is a small phase shift in the output signal with respect to the input voltage variation. It can be concluded that the proposed AHB DCX with line frequency ripple cancellation meets the design requirements.

CONCLUSION

In this paper, an unregulated DC/DC converter or DC transformer (DCX) with line frequency ripple cancellation is proposed to apply to a server power delivery architecture. The duty cycle of this DCX is set as high as possible and varies over a narrow range. The higher duty cycle will result in higher efficiency. The Asymmetrical Half Bridge converter is selected as the topology of DCX with line frequency ripple cancellation. The ripple cancellation circuit uses a feed forward scheme, in which the duty cycle signal is inversely proportional to the input voltage line ripple signal in order to attenuate the output voltage line ripple. A prototype of the AHB DCX with line frequency ripple cancellation is built to verify this scheme. The measured results show that it can not only

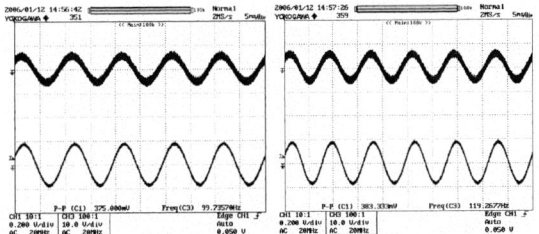

Figure 14. Measured input (top, 10V/div) and output (bottom, 0.2V/div) ripple at 100Hz (left) and 120 Hz (right) of proposed DCX without line frequency ripple cancellation

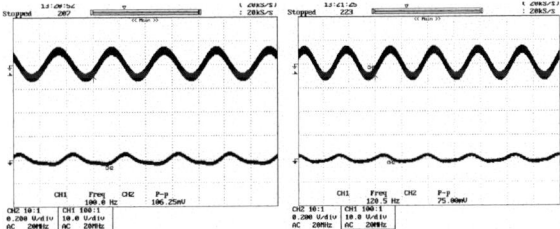

Figure 15. Measured input (top, 10V/div) and output (bottom, 0.2V/div) ripple at 100Hz (left) and 120 Hz (right) of proposed DCX with line frequency ripple

attenuate line frequency ripple to 120mV, but also achieve 95.7% efficiency at full load, which is only 0.4% lower than the 50% duty DCX without the line ripple cancellation function. At lighter loads, the proposed DCX has higher efficiency than the 50% duty cycle DCX.

REFERENCES

[1] B. Yang, P Xu, and F.C. Lee, "Range Winding for Wide Input Range Front End DC/DC converter," *IEEE Applied Power Electronics Conf. Rec.*, pp.476-479, 2001.

[2] Y. Jang, M. Jovanovic, and D.L. Dillman, "Hold-up time extension circuit with integrated magnetics", *IEEE Applied Power Electronics Conf. Rec.*, pp. 219-25, 2005.

[3] Y. Xing, et al., "A combined front end DC/DC converter", *IEEE Applied Power Electronics Conf. Rec.*, pp. 1095-99, 2003.

[4] Y. Ren, M Xu, C.S. Leu, and F.C. Lee, "A family of high power density bus converters," *PESC 2004*, vol. 1, pp. 527-32.

[5] M. Barry, "Design issues in regulated and unregulated intermediate bus converters," *IEEE Applied Power Electronics Conf. Rec.*, pp. 1389-94, 2004

[6] R. Ridley, "The Incredible Shrinking Power Supply", *Switching power magazine*, vol. 4, issue 3, pp. 25 – 30, Oct 2003.

[7] R. Ridley, "IBM's next generation : servers look to bus converters", *Switching power magazine*, vol. 5, issue 3, pp. 14 – 20, Oct 2004

[8] B. Yang, F.C. Lee, A. J. Zhang, and G. Huang, "LLC Resonant Converter for Front End DC/DC conversion," *IEEE Applied Power Electronics Conf. Rec.*, pp. 1108-1112, 2002.

A Novel PWM Method for Stacked Flying Capacitor Inverter

Gangui Yan [*], Gang Mu [*], Yafeng Huang [*], Wenhua Liu [**]

[*] Northeast China Dianli University, Chang chun Road 169, Jilin City, China, 132012
[**] Department of Electrical Power Engineering, Tsinghua University, Beijing City, China, 100084

Abstract — Allowing a significant reduction of capacitor volumes as well as scalable in terms of voltage levels and control flexibility, stacked flying capacitor inverter has gained great industrial attention recently. However, it also needs to ensure balancing flying capacitor voltages under all operation conditions. A novel PWM method is proposed for stacked flying capacitor multilevel inverter to control the balance of flying capacitor voltages, which can be readily implemented on a digital signal processor and also guarantee the switch frequency of each switching device almost same. Extensive simulation results demonstrated its validity in a 5-level stacked flying capacitor inverter.

Keywords-Power Electronics; Stacked Flying Capacitor Inverter; PWM Control; Flying Capacitor Voltage

I. INTRODUCTION

Flying Capacitor Multi-Level (FCML) inverter has been attracting wide industrial interests recently for scalable in terms of voltage levels and control flexibility. However, the increase of the voltage indirectly implies a significant increase in the size of the capacitors as well as the energy stored in the converter, and the price and volume of flying capacitors tend to become too important beyond 6kV. To solve this problem, stacked topology was proposed in [2]-[3] and allows to work on higher input voltage with the same power switches. However, it also needs to ensure balancing flying capacitor voltages under all operation conditions. Carrier-shifted PWM was proposed in [2]-[3], however, the method is a kind of open loop control with poor stability. In this paper a novel PWM method is proposed for stacked flying capacitor multilevel inverter to control the balance of flying capacitor voltages. Validity of the method is demonstrated with extensive simulation results in a 5-level stacked flying capacitor inverter.

II. A NOVEL PWM METHOD FOR STACKED FLYING CAPACITOR INVERTER

A. Stacked Flying Capacitor Inverter

In Figure 1, a $(2N+1)$-level stacked flying capacitor inverter is shown. It is composed of two same $(N+1)$ level flying capacitor inverters, which consists of N elementary commutation cells composed of two switches sa_{uk} and sb_{uk} (or sa_{dk} and sb_{dk}) with complementary states, $k=1,2,\ldots,N$. and the DC-link voltage of each inverter is equal to NE. Figure 1 also show the inverter switching statuses to produce the positive and negative half part output voltage respectively. The operation principle of the inverter can be described as below:

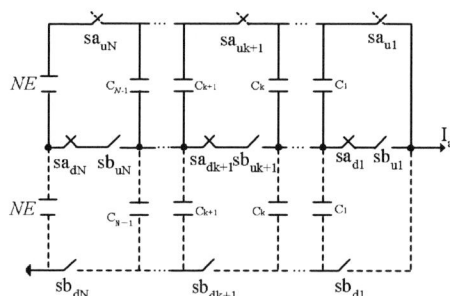

(a) When producing positive output voltage

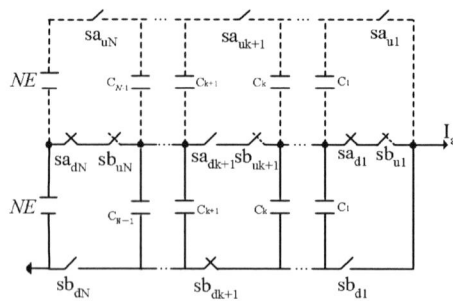

(b) When producing negative output voltage

Fig.1. Topology of (2N+1)-level stacked Flying capacitor inverter and its operation principle schemes (phase A)

1) When the commanded output voltage is positive, switches sa_{dk} are controlled in conduction mode and the top $(N+1)$ level inverter is operated in inversion mode as shown in Fig.1 (a), $(N+k)$-level output voltages can then be generated, $k=0,1,\ldots,N$.

2) when the commanded output voltage is negative, sb_{uk} are controlled in conduction mode and the underside inverter is operated in inversion mode as shown in Fig.1 (b), therefore, k-level output voltages can be obtained.

In order to obtain the full AC output voltage, two inverters should be operated alternatively, and only $(N-1)$ capacitor voltages need to balance to their reference values $V_{c,k}=kE$, $k=1,\ldots,N-1$. Respect to $2N+1$ level flying capacitor inverter (capacitor voltages are equal to $V_{c,k}=kE$, $k=1,\ldots,2N-1)$[1], capacitor voltages of this topology reduce significantly, and only half capacitor voltages need to balance. Therefore, much attention has been paid to this topology in high-voltage conversion.

B. Principle of Output Voltage Synthesizing Control

Denoting $V_{ref.x}(T_j)$ as reference voltage sampled at $t=j*T_c$ of phase x, x=a,b,c, and T_c is DSP control cycle. Then the commanded inverter output voltage is:

$$V_{inv.x}(T_j) = V_{ref.x}(T_j) + NE \tag{1}$$

Where $-NE \leq V_{ref.x}(T_j) \leq NE$, $0 \leq V_{inv.x}(T_j) \leq 2NE$. According to "volt-time equivalent principle", the commanded output voltage can be synthesized effectively by the following two adjacent level voltages:

$$\begin{cases} n_{0.x} = \text{int}\left[V_{inv.x}(T_j)/E\right] \\ n_{1.x} = n_{0.x}+1 \end{cases} \tag{2}$$

Where the int($\cdot$) function returns the nearest integer that is less than or equal to its argument. The switching times in these level voltages can be determined from:

$$\begin{cases} T_{0.x} = \left[1 - \dfrac{V_{ref.x}}{E} + \text{int}\left(\dfrac{V_{ref.x}}{E}\right)\right]*T_c \\ T_{1.x} = T_c - T_{0.x} \end{cases} \tag{3}$$

Where $T_s = T_c*2N$ is the switching cycle[5].

In order to optimize the output line voltage, the switching strategy is accomplished in an alternative manner as follows:

If $T_j = j*T_c \leq t \leq (j+1)*T_c = T_{j+1}$ $j=0,1,2,\ldots$

$$S_x = \begin{cases} n_{0.x} & jT_c \leq t \leq jT_c + T_{0.x} \\ n_{1.x} & jT_c + T_{0.x} \leq t \leq (j+1)T_c \end{cases} \tag{4a}$$

If $T_{j+1} = (j+1)*T_c \leq t \leq (j+2)*T_c = T_{j+2}$

$$S_x = \begin{cases} n_{1.x} & (j+1)T_c \leq t \leq (j+2)T_c - T_{0.x} \\ n_{0.x} & (j+2)T_c - T_{0.x} \leq t \leq (j+2)T_c \end{cases} \tag{4b}$$

Where S_x is switching function of phase x, x=a,b,c.

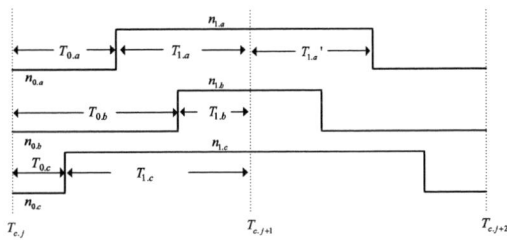

Fig.2. Diagram of output voltage synthesizing pattern

Figure 2 illustrates the example of the switching pattern. It can be seen from Fig. 2 that the switching state production of the proposed PWM algorithm is equivalent to that of Space Vector Modulation. However, the proposed PWM algorithm does not require the definition of voltage vectors and is mathematically straightforward but discrete in nature for DSP implementation, and the method can also be employed in other topology inverter for output voltage synthesizing.

In flying capacitor multilevel inverter, there are enormous redundant switch combinations. Although the output voltages are the same regardless of the redundant switch combinations used, the resulting currents in flying capacitors are widely different. Therefore the appropriate redundant combination that results in improving the balance of flying capacitor voltages should be selected.

C. Generic Rules for Switching Selection Oriented from Capacitor Voltage Balancing Control

In order to have inverse voltage constraints equally distributed among all the series devices, the flying capacitor voltages need to be balanced. Refer to Fig. 1 the blocking voltage of switch k can be described:

$$V_{cel.k}(t) = V_{c.k}(t) - V_{c.k-1}(t) \tag{5a}$$

$$\Delta V_{cel.k} = \frac{1}{C}\int_0^t (s_{k+1} + s_{k-1} - 2s_k)i_a dt \tag{5b}$$

543

Where $V_{c.k}(t) = \frac{1}{C}\int_0^t i_{ck}\,dt + V_{c.k}(0)$, $V_{c.0}(t)=0$.

$i_{c.k}=(s_{k+1}-s_k)i_a$ is charging current of C_k, s_k is switching functions of sa_{uk} and i_a output current of phase a.

Under the consideration that the control period is very small, it is assumed that the current sign is constant on that control period. Then it can be seen from (5) that

When $i_a>0$

Let $s_k=1$ then $\Delta V_{cel.k} \leq 0$

Let $s_k=0$ then $\Delta V_{cel.k} \leq 0$

While $i_a<0$

Let $s_k=1$ then $\Delta V_{cel.k} \leq 0$

Let $s_k=0$ then $\Delta V_{cel.k} \leq 0$

The principles for switching selection can then be concluded in Table I、□.

TABLE I Principles for Switching Selection oriented from Capacitor Voltage Balancing

Δn	$T_j \leq t \leq T_{j+1}$		$T_{j+1} \leq t \leq T_{j+2}$	
	$[s]_{j.0}$	$[s]_{j.1}$	$[s]_{j+1.1}$	$[s]_{j+1.0}$
0	$[s]_{j-1.1}$	$[s]_{j.0}+s_{max}$	$[s]_{j.1}$	$[s]_{j+1.1}-s_{min}$
1	$[s]_{j-1.1}+s_{max}$	$[s]_{j.0}+s_{arb.off}$	$[s]_{j.1}$	$[s]_{j.1}$
-1	$[s]_{j-1.1}$	$[s]_{j-1.1}$	$[s]_{j.1}-s_{min}$	$[s]_{j+1.1}-s_{arb.on}$

TABLE □ Characteristics of s_{max}、s_{min}、$s_{arb.off}$、$s_{arb.on}$

	i_a	Characteristics
s_{max}	>0	Among cells at state "off " and the one having the highest cell voltage
	<0	Among cells at state "off " and the one having the lowest cell voltage
s_{min}	>0	Among cells at state "on" and the one having the lowest cell voltage
	<0	Among cell at state "on" and the one having the highest cell voltage
	n_k	Characteristics
$s_{arb.on}$	>1	The cell at state "on" both during $[s]_{k-1.0}$ and $[s]_{k.1}$
	=1	The cell at state "on" during $[s]_{k.1}$
$s_{arb.off}$	<N-1	The cell at state "off" both during $[s]_{k-1.1}$ and $[s]_{k.0}$
	$n_k=N-1$	The cell at state "off" during $[s]_{k.0}$

Where $\Delta n=n_{0.x}(T_j)-n_{0.x}(T_{j-1})$, and $n_{0.x}$ is determined from (2)；When $V_{ref.x}(T_j)\geq 0$, $[s]_{j.0}$ and $[s]_{j.1}$ are switching states of upper inverter to generate $n_{0.x}$-N and $n_{1.x}$-N level

voltage during $[T_j, T_{j+1}]$ period, and their initial values are listed as below:

$$[s]_{1.0} = [\underbrace{1,\cdots 1,}_{n_{0.x}-N}\ \underbrace{0,\cdots,0}_{2N-n_{0.x}}]\text{，}\quad [s]_{1.1} = [\underbrace{1,\cdots,1,}_{n_{1.x}-N}\ \underbrace{0,\cdots,0}_{2N-n_{1.x}}]\text{；}$$

when $V_{ref.x}(T_j)<0$, $[s]_{j.0}$ and $[s]_{j.1}$ are switching states of downside inverter for generating $n_{0.x}$ and $n_{1.x}$ level voltage during $[T_j, T_{j+1}]$ period, And their initial values are listed as below:

$$[s]_{1.0} = [\underbrace{1,\cdots 1,}_{n_{0.x}}\ \underbrace{0,\cdots,0}_{N-n_{0.x}}]\text{，}\quad [s]_{1.1} = [\underbrace{1,\cdots,1,}_{n_{1.x}}\ \underbrace{0,\cdots,0}_{N-n_{1.x}}]\text{；}$$

III. Simulation RESULTS

In order to verify the proposed PWM method, extensive system simulations are carried out in stacked flying capacitor 5-level inverter using PSCAD/EMTDC software. The main parameters of the system are summarized in Table III、IV, and the simulation results are shown in Figs. 4-8 accordingly.

TABLE III   Simulation System Parameters

Inverter Parameters	4E=8.6kV, Ts=2ms, Tc=Ts/4, Ia=250A, Vcel.r=2.15kV, ΔVmax=300V, C1=C2 =500μF, third-harmonic injection to the reference wave；
Example 1	RL loads: R=19Ω, L=0.038H, cosθ=0.85
Example 2	Induction motor loads：Pn=1250kW, Un=6kV, Ia=159A, η=93.5%, pole pair=4, load torque is square to rotate speed.
Disturbances	Delay Trigger pulse 5μs to sa1 and no time delay to any other switches, and sudden change in input voltage

TABLE IV Initial Value Setting for Switching States at The Point of Operation Shifting

		The former half of basic cycle	The latter half of basic cycle
Positive zero crossing	$n_{k-1}=n_k=0$	$[s]_{k-1.0}=[0,0]$	$[s]_{k-1.0}=[0,1]$
Negative zero crossing	$n_{k-1}=n_k=2$	$[s]_{k-1.0}=[0,1]$	$[s]_{k-1.0}=[1,1]$

A. Startup steps of Stacked FCML Inverter

The task of the Starting up the inverter is to charge up all the capacitors and connect them to system in a short time. The procedure can be described as following two steps[6]:

1) Close switch group sa_{u2}、sb_{d2}, open sa_{u1} and sb_{d1}. By doing this, the DC-link capacitor and the flying capacitors in all three phases are connected in parallel and the converter becomes a three phase rectifier and these capacitors in parallel as its only load. The system will then charge these capacitors.

2) Once the voltage of the parallel capacitors reach E, open all switches and only DC-link capacitors are charged. Till the voltage reaches 2E, bypassing the inserted resistor, and the converter is put into the inversion mode.

The above steps are summarized in Table □, and the other two phase starting up steps are the same as phase a.

TABLE □ Starting up Steps for Stacked Flying Capacitor 5-Level Inverter

Vdc	Switching state of phase a
<E	$sa_{u1}=sb_{d1}=0$, $sa_{u2}=sb_{d2}=1$
(E, 2E)	$sa_{u1}=sb_{d1}=sa_{u2}=sb_{d2}=0$
>2E	Bypass the inserted charging resistor

B. Inversion Mode Simulation

B1. Example 1: with RL loads

As can be seen from Fig.3, the converter operates as rectifier before 0.1s, and the flying capacitor voltages are charged to the reference values smoothly. During the whole inversion mode, flying capacitor voltage balancing is maintained excellently even input voltage swells suddenly at 0.5s and different trigger delay exists as shown in Table III.

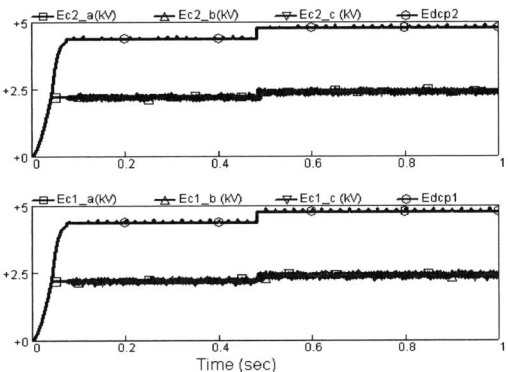

Fig.3. Voltage waveforms of Flying Capacitors

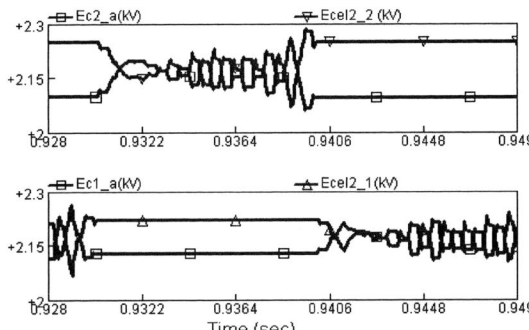

Fig.4. Cell voltage waveforms of phase a

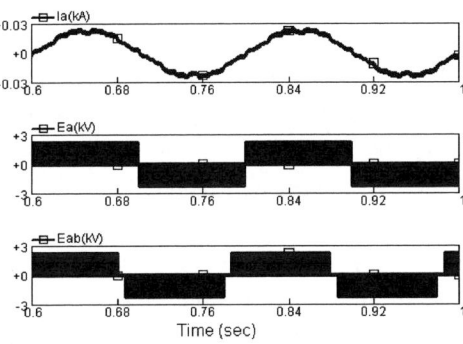

(a) Waveforms of i_a, u_a and u_{ab} when f_1=5Hz

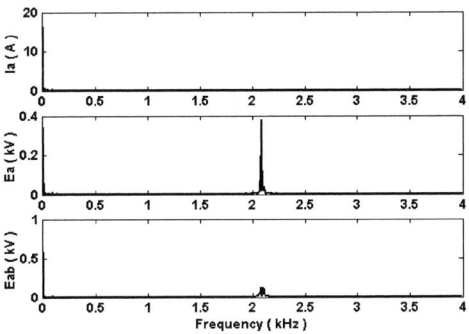

(b) Frequency Spectrums of i_a, u_a and u_{ab} when f_1==5Hz

(c) Waveforms of i_a, u_a and u_{ab} when f_1=50Hz

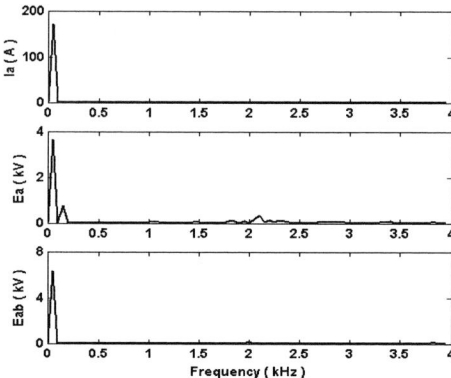

(d) Frequency Spectrums of i_a, u_a and u_{ab} when f_1=50Hz

Fig.5. Waveforms of inverter output currents、phase and line-line voltages and their frequency spectrums

The waveforms of phase A cell voltage are shown in Fig. 4 with rated value of 2.2kV. It can be seen that its variation can be controlled firmly within ±150V.

Both output current、phase and line-line voltages and their frequency spectrums of phase a are shown in Fig. 5 (a)-(d). As can be seen that the output voltages exhibits the typical 5-level wave-shape, and the phase current has exceptionally low harmonics due to the load inductance.

B2. Example 2：with induction motor load

Figure 6 shows the acceleration transients from 5Hz to 50Hz with speed-square load characteristics and the transient flying capacitor voltages. The popular volt-per-hurtz method is employed to control the motor speed, and the inverter is controlled by the proposed method.

As can be seen from Fig. 6(a), speed oscillation occurs around 25Hz and a sudden input voltage change takes place at t=8s, while voltage balancing is maintained excellently during the whole transients.

As shown in Fig.7, although the load current is distorted heavily during speed oscillation transients, the output line-line voltage still exhibits the typical 5-level wave-shape which indicates that flying capacitor voltages maintain balance perfectly.

Figure 8 shows four cell control waveforms of one phase inverter in one output fundamental period. It can be counted that the switching numbers of four power switching devices are 18、17、19、18 respectively when f_1=28Hz ，and 10、11、10.5、10 respectively when

f_1=50Hz, They are consistent with the theoretical value $1/(T_s*f_1)$ in both cases, which means that each switching device operates with uniform conditions.

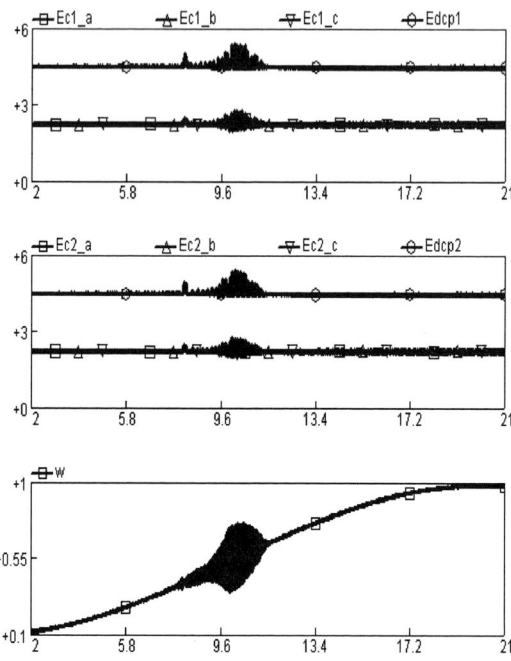

(a) Flying capacitor voltages (kV) and motor speed (p.u.) during acceleration transients

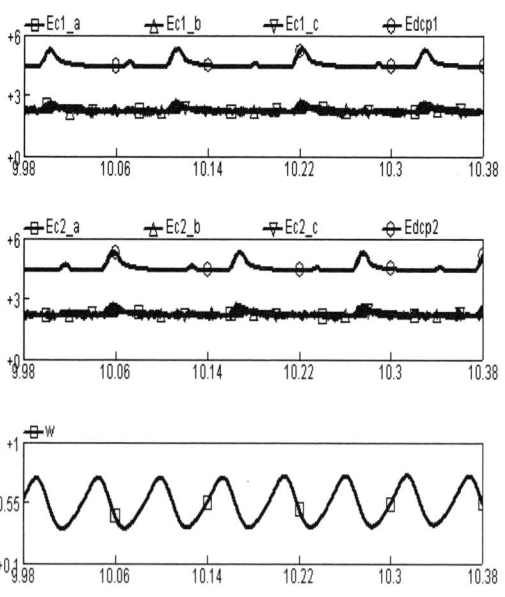

(b) Capacitor Voltages (kV) and motor speed (p.u.) during speed oscillation

Fig.6. capacitor voltages and motor speed during acceleration transients

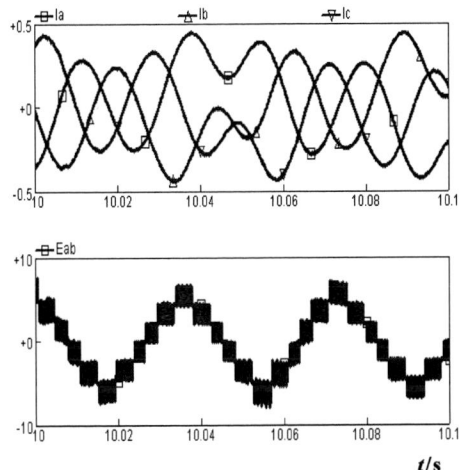

Fig.7. Output current (kA) and line-line voltage (kV) during speed oscillations

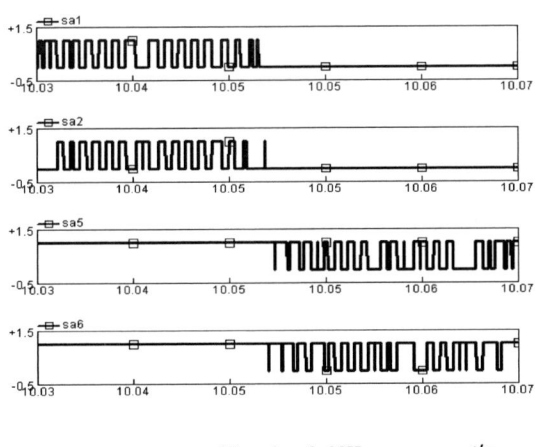

(a)　　when f_1=28Hz　　　　t/s

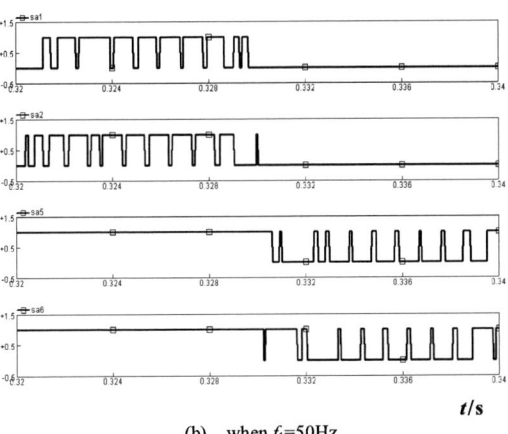

(b)　　when f_1=50Hz

Fig.8. Commutation times of each cell in one output fundamental period

IV.　Conclusions

The extensive simulations show that the novel PWM method proposed has the advantages as below:

1) The stacked inverter can be controlled to generate the volt-time equivalent output voltages with the proposed PWM method, and flying capacitor voltages of the inverter can also be controlled balance perfectively even disturbances occur both in front-end or in dynamic loads.

2) The proposed method can be easily implemented in DSP processor with less calculation time, and it is scalable in terms of voltage level.

3) Uniform operating conditions can be guaranteed even during transients.

REFERENCES

[1]. Meynard T A, Foch H. Multilevel conversion: high voltage choppers and voltage-source inverters. *Proceedings of IEEE Power Electronics Specialists Conference –PESC. 1992: 397~403.*

[2]. G. Gateau, T. A. Meynard, H. Foch. Stacked multilevel converter(SMC): properties and design. *Proceedings of IEEE Annual Power Electronics Specialists Conference,* Vol. 3 (PESC'2001). Vancouver: Jun 17-21 2001: 1583~1588

[3]. G. Gateau, T. A. Meynard, H. Foch. Stacked multicell converter (SMC): Control and natural balancing. *Proceedings of IEEE Annual Power Electronics Specialists Conference,* Vol.2 (PESC'2002). Cairns: Jun 23-27 2002: 689~694

[4]. Liu Wen Hua, Song Qiang, Yan Gangui, etal. Medium voltage drive with NPC three-level inverter using IGCTs. *Automation of Power Systems,* 2002, 26 (20): 61~65

[5]. Yan Gangui, Liu Wenhua, Chen Yuanhua, Han Ying duo. A generic PWM control method for flying capacitor inverter [J]. *Proceedings of the CSEE,* 2003, 23(6): 35-40.

[6]. Yiqiao Liang, C. O. Nwankpa. A Power-Line Conditioner Based on Flying-Capacitor Multilevel Voltage-Source Converter with Phase- Shift SPWM. *IEEE Transaction on Industry Applications* 36(4), 2000:965-971

BIOGRAPHIES

Gangui Yan was born in Jiangxi Province China in 1971. He obtained his M. E. and Ph.D from Northeast China Institute of

Electrical Power Engineering and Tsinghua University in 1997 and 2003 respectively. His research interests include FACTS technology, Wind-Energy, Power system simulation technology. He is an IEEE PELS Member.

e-mail: yangg@mail.nedu.edu.cn

Gang Mu was born in Dalian City China in 1957. He received his M. E. and Ph.D from Northeast China Institute of Electrical Power Engineering and Tsinghua University in 1982 and 1991 respectively. His research interests include Wind-Energy, Power System Transient Stability Analysis. He is an IEEE PES Member.

Yafeng Huang was born in Hubei Province China in 1979. Special interests in inverter-fed motor speed control.

Wenhua Liu was born in Hunan Province China in 1968. Special interests in FACTS control and inverter-fed motor speed control.

2006 5th International Power Electronics and Motion Control Conference

Study on a New Method of Voltage-Source Induction Heating Load-Matched

Li Jin-gang, Zhong Yan-ru，Zhao Miao

School of Automation and Information Engineering, Xi'an University of Technology, Xi'an, China

shgr@xaut.edu.cn

Abstract--**In this paper, the static characteristics of three types of circuit topology used for voltage-source induction heating static electricity induction load-matched are analyzed. The systematic analyzed results are given, with regard to the transforming range of load resistances, the circuit responding sensitivity to the variation of the inductor coils' parameters, and the voltage and current of each compensatory element enduring. According to theoretical analysis, the applied conditions of every compensatory circuit are pointed, as well as the merit and demerit of different topology. The rationality of the theoretical analysis is proved by simulation. The matching experimental results of typical induction coils are given at last. With the practicality of analysis illuminated, the consistency of simulation results with actuality is testified.**

Key words-voltage-source; induction heating; static electricity induction; load-matched

I. INTRODUCTION

The load of induction heating supply is induction coils with high quality factor (*Q*). In order to improve the power factor and efficiency of source, the reactive VAR of heating load is always compensated by series or parallel branch of capacitors, consequently forms two kinds of inverter, voltage-source and current-source. With the development of power semiconductor devices and the need of industrial production, science research and modern life, the induction heating equipments are developing towards higher frequency and larger capacity. Because of the particularity of the load of induction heating power supply, it is important to research induction heating supply load-matched in improving the efficiency and safety of source. Normal electromagnetic coupling (transformer) matching method makes high requirements, expensive cost and hard design towards transformers at high frequency and high-power. Static electricity induction matching method using many energy-stored components to form different resonant loops instead of high frequency and high-power matching transformers to achieve load-matched, which realize the matching of high efficiency and low cost. Static electricity induction matching method is especially suitable to the field as follows where the parameters of load vary a little during the heating process: quenching, welding, metal surface process, semiconductor crystal heating and heat preservation technology, and modern family heating

equipments such as electric water-heating machine and electric kitchen and so on. The voltage-source induction heating inverter is relatively simpler in structure, more easily to start and to be controlled at constant power compared with the current source inverter. The modern power semiconductor devices applied to high-frequency inverter can't subject to inverse voltage, hence it is more suitable for voltage-source inverter. The static voltage-source induction heating inverter characteristics are mainly studied in this paper, in which the three-order resonant circuit consisting of three energy-stored components is used in load-matched. It is proved that analysis results are correct and practical by simulation and experiment.

II. THE CIRCUIT TOPOLOGY

The topology of voltage-source induction heating inverter is shown as Fig.1.

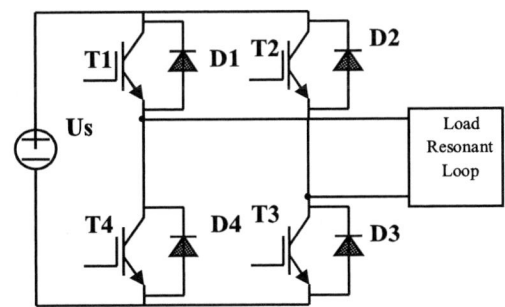

Fig.1 Voltage-source induction heating circuit

Due to the integration of induction coil with the equivalent load, three kinds of three-order circuit of voltage-source inverter are discovered shown as Fig.2.

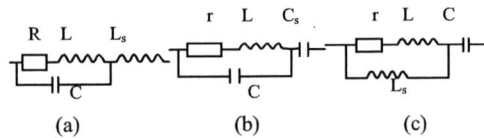

Fig.2 The topology of voltage-source three-order circuit

L is inductance of the induction coil and r is the sum of the inner resistance of induction coil and the resistance which is converted from the heated load resistance to the coil, they are indivisible as a device. Other elements in the circuits and heating coil constitute a three-order circuit, to achieve compensation and matching. Suppose

1-4244-0448-7/06/$25.00 ©2006 IEEE 549

the voltage amplitude of fundamental wave which inverter output to resonant loop is U_m.

III. STATIC STATE ANALYSIS

A. The Range of Load-Matched Resistance

Analyze the circuit as Fig.3, which is equivalent to the ones of Fig. 2(a) and Fig. 2(b).

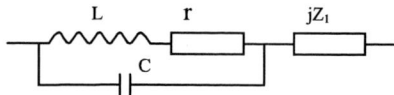

Fig.3 Equivalent configuration of the two circuits

Suppose the resonant frequency is ω, load resistance is R at rated power. Capacitor C compensates for inductive var and makes load matched. "jZ_1", reactance of another compensation component, may be capacitive or inductive depending on the different value of C.

The equivalent circuit reactance shown as Fig.3 is expressed as:

$$Z = \frac{r + j\omega(L - \omega^2 L^2 C - r^2 C)}{(1 - \omega^2 LC)^2 + \omega^2 r^2 C^2} + jZ_1 \quad (1)$$

Suppose :

$$d = \frac{r}{R} = (1 - \omega^2 LC)^2 + \omega^2 r^2 C^2 \quad (2)$$

Then:

$$C = \frac{\omega L \pm \sqrt{\omega^2 L^2 d - r^2(1-d)}}{\omega^3 L^2 + \omega r^2}$$

The quality factor of induction coil is $Q = \omega L / r$, so:

$$C = \frac{Q \pm \sqrt{(Q^2 + 1)d - 1}}{\omega r (Q^2 + 1)} \quad (3)$$

The condition which ensures the formula have real number solution is:

$$d \geq \frac{1}{Q^2 + 1}$$

Discussing with d in two ranges according to the previous expression, when d<1, formula (3) has two real solutions, respectively corresponding with the topology shown as Fig.2 (a) and (b).When d<1, it has a single solution, corresponding with the topology of Fig. 2(a). Hence the range of the matched resistance corresponding with Fig. 2(a) is: $R < (Q^2 + 1)r$ .The matched range of topology Fig. 2(b) meet: $r < R < (Q^2 + 1)r$.

The reactance expression of topology 3(c) is:

$$Z = \frac{\omega^2 L_s^2 r + j\omega(L_s r^2 + \omega^2 L^2 L_s + \omega^2 LL_s^2)}{r^2 + \omega^2(L + L_s)^2} + \frac{1}{j\omega C} \quad (4)$$

Hence when the circuit is in the resonant state:

$$R = \frac{\omega^2 L_s^2 r}{r^2 + \omega^2(L + L_s)^2}$$

So R<r.

B. The Circuit Responding Sensitivity Analysis

As the temperature-raise of the heated workpiece, it's induction coil permeability μ and resistivity ρ will change. Then the resonant frequency and load resistance vary in the process of induction heating source's working. The change of permeability manifests itself in the change of induction coil's equivalent inductance.

When the simple series resonant of electromagnetic coupling load-matched is used, resonant frequency ω_1 of circuit is expressed as:

$$G(L) = \omega_1^2 = \frac{1}{LC_1}$$

Where, C_1 is the compensating capacitor.

Thus when induction L varies, the formula below come true:

$$\frac{dG}{dL} = -\frac{1}{L^2 C_1} \quad (5)$$

In the Fig2 (a), considering Q>>1, the resonant frequency ω_2 is expressed as:

$$f(L) = \omega_2^2 = \frac{L + L_s}{LL_s C}$$

So:

$$\frac{df}{dL} = -\frac{1}{L^2 C} \quad (6)$$

When at the same rating resonant frequency the condition of $C > C_1$ is contented, the frequency change $\Delta\omega$ in topology (a) is smaller than that in simple series resonance topology when the same induction change ΔL takes place.

The load-matched resistance can be expressed approximately as:

$$R = r(\frac{Ls}{L})^2 \quad (7)$$

The transformation to the load resistance of the topology as in Fig2(a) is similar to the transformer at the ratio $n = L_s/L$.

In diathermy and smelting process, as the temperature of the heated metal rises, the load equivalent resistance r gets bigger, while equivalent inductance L gets smaller. We can conclude that R will increase greatly and cause the source output power to reduce. This is the shortcoming of topology 2(a) when it is used in the field in which the load resistance parameters vary greatly.

As for topology 2(b), also considering Q>>1, so the resonant frequency is:

$$\omega = \sqrt{\frac{1}{L(C + C_s)}} \quad (8)$$

It is obviously that the same of frequency change would be brought in the topology 2(b) as it in two-order series circuit when the same varying of L happens.

Matched load can be indicated as:

$$R = r(1 + \frac{C}{C_s})^2 \qquad (9)$$

The circuit of Fig. 2(b) is equivalent to the transformer at the ratio $n=1+C/C_s$ when is used in the function of load-matched. The variation of load inductance affect slightly on matched load R.

As for topology 2(c), resonant frequency and matched load are respectively expressed as:

$$\omega = \sqrt{\frac{L + L_s}{LL_s C}} \qquad (10)$$

$$R = r(\frac{L_s}{L + L_s})^2 \qquad (11)$$

The change amount of resonant frequency caused by coil induction variation is the same as simple series resonance, and load resistance change is similar as topology Fig.2(a).

C. Electric Parameters Analysis

When matching the load using the static induction way, matching inductance or capacitance, besides the required compensating capacitance in simple two-order series resonant circuit, the others of energy-stored components have to be appended. Therefore it is necessary to research the voltage and current by which the energy-stored components subjected. Through analysis, we can get all the advantage or disadvantage of the topology described above.

To simplify the analysis, the three kinds of circuit topology in Fig.2 can respectively be equivalent to the circuit analytic model in Fig.4 according to the circuit resonant principle.

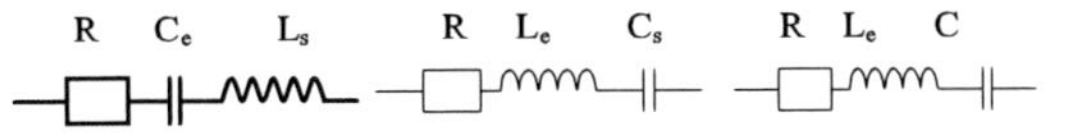

Fig.4 Analytic model of load-matched circuit

According to the Fig. 4(a), when the circuit is resonant, the amplitude of the electric current flowing through compensating inductor L_s is:

$$I_{Ls} = \frac{U_m}{R} \qquad (12)$$

It can be proved that the maximum of voltage that matching inductor subjects appears at the resonant state, and can expressed as:

$$U_{Ls} = Q\sqrt{\frac{r}{R}} U_m \qquad (13)$$

The voltage that compensating capacitor C subjects should be the sum of the ones which inductance L_e and resistance R subjected in equivalent circuit Fig.4(a). Expressed by:

$$U_C = U_m \sqrt{1 + (\frac{L}{L_s} Q)^2} = U_m \sqrt{1 + \frac{r}{R} Q^2} \qquad (14)$$

Therefore, for the compensating circuit Fig.2 (a), the maximum of the voltage applied to the compensating inductance and capacitance is related to the required R: comparing to r, the R is larger and the smaller voltage they subjected. It is benefit for design and choosing of compensating component.

After analyzing the Fig.4(b), the main electric parameters of the components in matching circuit Fig.2(b) can be got.

The voltage amplitude that compensating capacitor C_s subjected is:

$$U_{Cs} = \frac{C_s}{C + C_s} Q U_m = Q\sqrt{\frac{r}{R}} U_m \qquad (15)$$

The voltage amplitude of compensating capacitor C subjected is:

$$U_C = U_m \sqrt{1 + \frac{r}{R} Q^2} \qquad (16)$$

Obviously, when R>r, the voltage imposed on the two capacitor is smaller than it in the simple two-order resonant circuit. It is helpful for choosing the capacitor.

The voltage amplitude on the compensating capacitor C in topology Fig.2 (c) can be got from Fig.4 (c):

$$U_C = (1 + \frac{L}{L_s}) Q U_m = \sqrt{\frac{r}{R}} Q U_m \qquad (17)$$

The voltage amplitude that the compensating inductor subjects is equal to that on the load induction coil, it can be expressed as:

$$U_{Ls} = U_m \sqrt{1 + \frac{r}{R} Q^2} \qquad (18)$$

We can see, making use of the matching mode as Fig2(c), the voltage amplitude applied to the compensating capacitor, the inductor and load induction coil will all enlarge.

Suppose the current in load coil is I_r, current in resonant loop is I_R, and current in compensating inductor is I_L. Considering $Q\gg1$, then:

$$I_{Ls} = I_R \sqrt{\frac{R}{r}} = I_r \frac{L}{L_s} \qquad (19)$$

So, when current in load induction coil L is constant at rated load, the larger the discrepancy between the matched load R and load r is, the greater that I_L flowing in the compensating inductor will become, even greater than that of load induction coil.

IV. SIMULATION AND EXPERIMENTAL RESULT

In order to prove the rationality of theoretical analysis, for typical load induction coil: $r=6\Omega$, $L=0.1mH$. The simulation and experiment results are given.

It is requested that rated resonant frequency is 50kHz, i.e. Q=5. Matched load resistance is: R=30Ω in Fig. 2(a) and (b), R=3Ω in Fig. 2(c). The compensating component parameters are calculated following formula (7)-(11). They are given by the Table I.

Table I
The Parameters Designed Result

Circuit Topology	Matched Load R	Parameters Designed Result
Fig.2(a)	R=30Ω	C=0.15μF; L_s=0.22mH
Fig.2(b)	R=30Ω	C=0.056μF; C_s=0.045μF
Fig.2(c)	R=3Ω	C=0.014μF; L_s=0.24mH

The simulation results of all the load-matched topology can be got as Fig. 5, it is completely identical with the theoretical analysis.

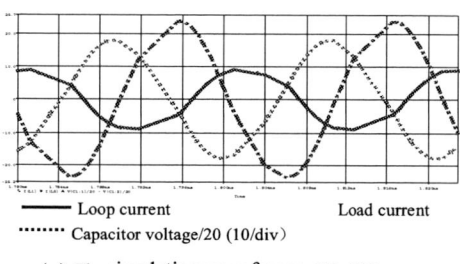

──── Loop current   ──── Load current
········ Capacitor voltage/20 (10/div)

（a）The simulation waveforms of Fig.2(a)

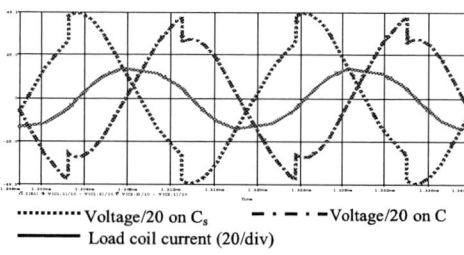

········Voltage/20 on C_s ─·─·─ Voltage/20 on C
──── Load coil current (20/div)

（b）The simulation waveforms of Fig.2（b）

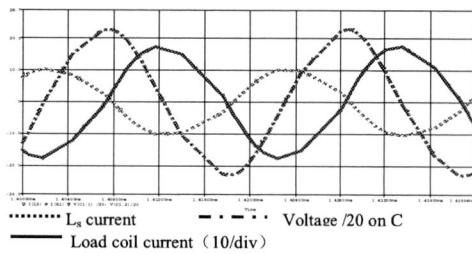

········ L_s current   ─·─·─ Voltage /20 on C
──── Load coil current（10/div）

（c）The simulation waveform of Fig.2（c）

Fig.5 Simulation waveforms

It can be proved that theoretical analysis is proper and simulation results are correct by experiment for the given induction heating coil. The experiment results of all the

topology which Fig.2 indicates are shown as Fig.6, which the square wave is inverter output voltage wave in every curve, and the sine-wave is the current waveform of the resonant loop.

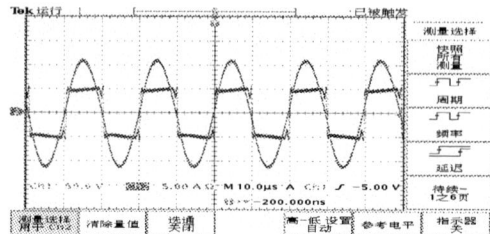

(a) The waveforms of simple series resonant

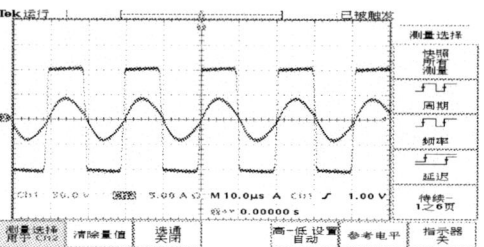

(b) The waveforms of topology Fig.2(a)

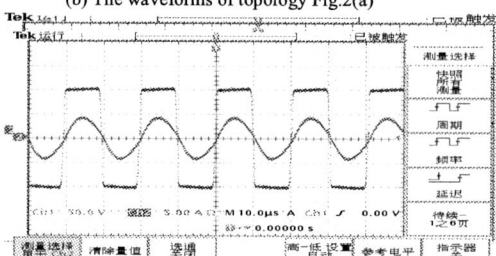

(c) The waveforms of topology Fig.2(b)

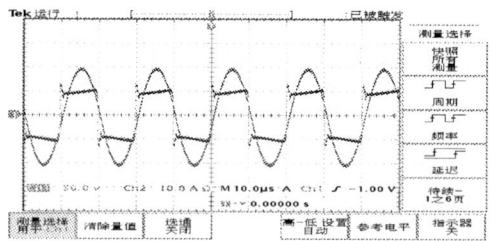

(d) The waveforms of topology Fig.2(c)

Fig.6 Experimental waveforms

V. SUMARRY

The analysis in this paper is based on the supposition Q>>1, and on the basic of considering fundamental wave only. The engineer design can be simplified with the precondition of having no affection on accuracy.

The characteristics of three kinds of topology circuit are as follows.

The topology as Fig. 2(a): the range of matched load is the widest, when R>r, the voltage that the compensating capacitor C subjected is small. The current flowing in the matched inductor L_s is also smaller than that flows in load coil L. The matched load resistance R varies great as the load coil parameter changes along with the temperature. Therefore it is applied in heating field where equivalent inductance and resistance change a little.

The topology as Fig. 2(b) : the range of matched load is $r<R<(Q^2+1)r$. It can't achieve the ending of R<r. When the power supply works near resonant frequency, the matched resistance is affected slightly by the change of the inductance of load induction coil. Furthermore, in general, the maximum of voltage that each capacitor subjected is lower than that of simple two-order resonant compensating capacitor.

The last topology, means Fig2(c): the range of matched resistance: $R< r$. Very large current may flow in the matching inductor, and compensating capacitor will subject relatively high voltage.

The three kinds of circuit have themselves advantages and disadvantages under different use, thus they should be considered synthetically according to actual requirement when they are chosen.

The reasonability of the analysis is proved by simulation and experiment.

REFERENCES

[1] Wu Zhaoling，Yuan Junguo，Yu Fei. "Load-matching of high frequency induction heating device. Power Electronics Technology"，1999，33（4）：29-32

[2] Chen Huiming, Qian Zhaoming. "The Newset development of Induction heating Power Source". Transactions of China Electrotechnical Society，1999，14（supplement）：50~53

[3] S.S. Tanavade, Mrs M.A. Chaudhari, H.M. Suryawanshi and K.L. Thakre. "Design of Resonant Converter for High Power Factor Operation Optimum Losses in Magnetic Components". PESC'2004 35th Annual conference 1818 - 1822

[4] Issa Batarseh. " Resonant Converter Topologies with Three and Four Energy Storage Elements". IEEE Trans. Power Electro，1994，9(1)：64-73

[5] S.V. Mollov, M. Theodoridis and A.J. Forsyth. High frequency voltage-fed inverter with phase-shiftcontrol for induction heating. IEE Proc.-Electr. Power Appl., Vol. 151, No. 1, January 2004: 12-18

[6] Li Jingang, Zhong Yanru and Zhang Yang, " Study on static electricity induction load-matched of voltage-source converter for induction heating". IECON 2005, 32nd Annual Conference of IEEE 1284- 1287

[7] http://www.richieburnett.co.uk/indheat.html

2006 5th International Power Electronics and Motion Control Conference

An Alternating-master-salve Parallel Control Research for Single Phase Paralleled Inverters Based on CAN Bus

Zhang Chunjiang   Chen Guitao   Guo Zhongnan, Wu Weiyang
College of Electrical Engineering, Yanshan University
Qinhuangdao, Hebei P. R. China
E-mail: zhangcj@ysu.edu.cn

Abstract—A novel alternating-master-salve control strategy for paralleling inverter is proposed, which merely make use of CAN bus to achieve the high precision of synchronization and fast current-sharing control. Only single interconnection bus is requires and simple arithmetic is used by the proposed scheme. Moreover, the paralleled inverters act as master in turn during each cycle of the output voltage and the paralleled modules are identical. There are not difference between master and slave and the parallel system will still operate normally under either paralleled module fault, so N+1 redundancy can be achieved. The experimental results verify good static and dynamic current-sharing performance by adopting the parallel strategy. And the parallel system also has a good parallel performance under the linear and nonlinear load.

Keywords-parallel; inverter; synchronization; current-sharing; DSP; CAN bus

I. INTRODUCTION

The power systems composed of multi-module parallel operation can cater for different power demands of load, high quality power supply and high reliability demands. The parallel systems also make for standardization, integration of power modules, the expedition as well as maintenance[1]. Originally inverters parallel merely aims at expanding the capacity and enhancing the reliability. With the developing of the technique and increasing of demand for power supply, the function of parallel systems are enhanced, even refer to saving electricity. That is, under the light load, one module or several modules of parallel system are removed, so the efficiency is advanced; while the "removed" modules will start-up to supply the power automatically when the load attains to a degree.

At present, there are many control schemes for inverter parallel operation, such as concentrate control [2], master-slave control [3], distributed control[4][5] and droop control[6][7]. Not all parallel strategies can achieve redundancy. The expression of redundancy is N+X, where, N is the numbers of the parallel modules; X is the max numbers of broken-down modules that parallel system still can work normally. Some existed parallel control

schemes have good paralleling performance at the cost of poor redundancy and complicated interconnection. Others parallel schemes are redundant and have intricate arithmetic or have poor current-sharing performance. On the basis of analyzing and referring the existing parallel schemes, an alternating-master-salve interactive following control strategy for paralleling inverter based CAN(Controller Area Network) bus is proposed, which synchronize the reference voltage and fast adjust the amplitude of reference voltage to share the active and reactive power by CAN bus. The proposed method adopts DSP to implement the full digital control, which needn't calculate the active power and reactive power. It has real-time characteristic, simple algorithm and only single interconnection bus is required. The experimental results demonstrate that this proposed method can greatly realize the power sharing, and has good dynamic response.

II. POWER ANALYSIS of PARALLEL SYSTEM

To facilitate the analysis, two paralleled inverters are taken as example, as shown in Fig. 1. The output voltage of inverter1 and inverter2 are u_1 and u_2, respectively, and u_0 is the parallel grid voltage. They correspond to $U_1 \angle \varphi_1$, $U_2 \angle \varphi_2$ and $U_0 \angle 0$. The phase angles between u_1, u_2 and u_0 are φ_1 and φ_2 respectively. X_1, X_2 are reactance of output filter inductance, assumed $X_1 = X_2 = X$. C_1, C_2 are output capacitance and Z is load. i_{o1}, i_{o2} are load current of inverter1 and inverter2. i_1, i_2 are inductor current of inverter1 and inverter2. The complex power of the inveter1 at load is given by:

$$S_1 = P_1 + jQ_1 = \dot{U}\dot{I}_1^*$$ (1)

The inductor current is:

$$\dot{I}_1 = \frac{U_1(\cos \varphi_1 + j \sin \varphi_1) - U_o}{jX}$$ (2)

Thus

$$S_1 = U_o[\frac{U_1(\cos \varphi_1 + j \sin \varphi_1) - U_o}{jX}]^*$$
$$= \frac{U_o U_1 \sin \varphi_1}{X} + j \frac{U_o(U_1 \cos \varphi_1 - U_o)}{X}$$ (3)

1-4244-0448-7/06/$25.00 ©2006 IEEE   554

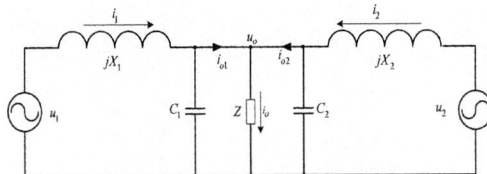

Figure 1. Equivalent circuit of two inverters parallel connection

Then the active power and reactive power of inverter1 are yielded.

$$\begin{cases} P_1 = \dfrac{U_o U_1 \sin \varphi_{12}}{X} \\ Q_1 = \dfrac{U_o (U_1 \cos \varphi_1 - U_o)}{X} \end{cases} \tag{4}$$

Similarly for the second inverter:

$$\begin{cases} P_2 = \dfrac{U_o U_2 \sin \varphi_2}{X} \\ Q_2 = \dfrac{U_o (U_2 \cos \varphi_2 - U_o)}{X} \end{cases} \tag{5}$$

Due to the angle difference between output voltage and grid voltage is very small, it can be concluded from (4) and (5) that active power balance of the inverters mainly depends on the coherence of the phase angle difference, φ_1 and φ_2, and the reactive power of balance mainly depends on the coherence of voltage amplitudes, U_2.and U_2. Therefore, changing output voltage amplitude of the inverters can control the reactive power, and changing the phase angle differences can control the active power. As far as the inverters controlled by the instantaneous voltage are concerned, the amplitudes and phases of output voltage of each inverter can reach coherence by adjusting the amplitude and phase of voltage reference, that is, it can archive the sharing of the active power and reactive power. The proposed parallel control strategy is suitable for multi-modules parallel operation.

III. THE ALTERNATING-MASTER-SLAVE PARALLEL CONTROL SCHEME

The field bus is an exoteric, digital, multipoint communication control system area network. CAN belongs to this category. CAN bus has many advantages, such as flexible configuration, and extendable, adaptive hot-swap. Using unique bit arbitration technique, CAN has strong real-time property and its transmission rate can be up to 1Mbps. The reliability is very high. So the CAN is very adapted to the inverters of parallel control. The bit rate is set to 1Mbps, each frame has 10 bytes, so lest time of a frame from transmitting to receiving is $T = 10 byte * 8bit / 1M = 80us$. Therefore, the parallel system can achieve fast control.

The DSP(TMS320LF2407A) is adopted in the proposed parallel system to implement the instantaneous voltage regulator, current regulator, generating SPWM. At the same time, it also answers for regulating the phase and amplitudes of voltage reference to achieve the control of output synchronization and current sharing. Fig. 2 is the

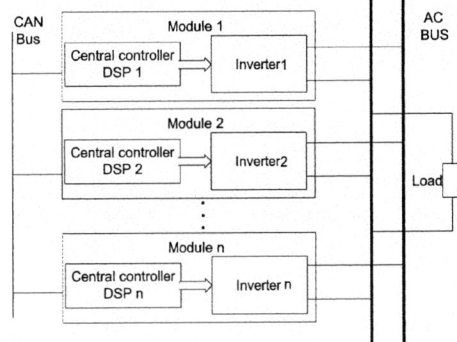

Figure 2. Parallel system block diagram

diagram of the parallel system. The parallel units of system consist of DSP controllers and inverters which adopt voltage and current dual closed-loops control as well as parallel control. And it can work either alone or in parallel mode by the control. The information of each inverter is transmitted via CAN bus which is the digital communication bus.

In this paper, twisted-pair is chosen as the CAN transport medium. SN65HVD232D is chosen as the CAN transceiver which is the interface between CAN controller in DSP and physical bus. Each CAN transceiver is designed to provide differential transmit capability to the bus and differential receive capability to a CAN controller at speeds up to 1 Mbps. Its high input impedance allows for 120 Nodes on a Bus Each end of the bus is terminated with 120Ω resistors in compliance with the standard to minimize signal reflections on the bus. The interface circuit of the CAN bus is shown in Fig.3.

A. The Synchronization Control of Paralleled Inverters

The CAN bus is multi-master serial communication bus, which supports distributed control or real-time control. The message does not include the aim address, which bases on broadcasting in whole network. One of outstanding properties is that the mode of coding to messages substitutes for the traditional mode of coding to station address. The messages are filtrated according the coding. Then each node can be connected or disconnected to network on-line. And hot-swap and multi-station receiving can be reached with CAN bus. Because of the merits mentioned above, CAN bus is very suitable for synchronization of paralleled inverters.

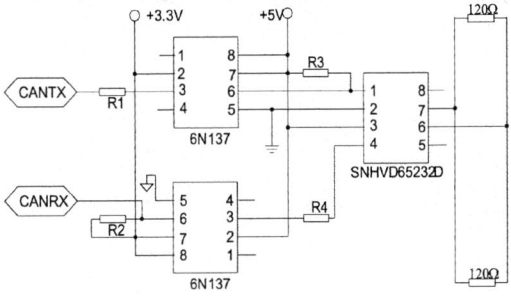

Figure 3. Interface of CAN bus

In order to solve the problem of output sine waves synchronization and keep the frequency and phase of each inverter consistent with the others to avoid the errors of active power among inverter modules, an alternating-master-salve synchronization method is proposed. The paralleled modules in the system broadcast their own synchronization information to the other parallel units by CAN bus in turn according to a certain rule during each cycle of the output voltage. The broadcasted messages included the identification (ID) and phase of each module. If one is the temporary master module, the other modules of receiving messages are the slave. The slave modules will judge the ID and the phase of the master module, and adjust their own phase in accordance with the master as soon as the messages are received.

Therefore, all paralleled modules act as master in turn during each cycle of sinusoidal output voltage, and the phase of the paralleled modules will follow each other. The parallel system will still work normally whichever modules failed. This synchronization method not only has high accuracy, but also achieves redundancy. The synchronization performance of inverters is dependent on the following accuracy.

The concrete method is that: during the each cycle of output voltage, each paralleled modules transmit a synchronization message to parallel network. And each inverter corresponds with a certain phase position; when the message is received by the temporary slave modules, the interruption will occur and the ID and phase message will be accessed, and the voltage reference of the receiving modules will be adjusted in accordance with the temporary master. The outputs of inverters follow each other and act as master in turn. The sketch map of the method is shown in Fig. 4, in which four paralleled inverters are taken as example. There is some time delay in communication, so the phase compensation is required. The high precision phase-lock in paralleled inverters can be achieved and redundant parallel can be gained by this method, which enhances the reliability of parallel system.

B. The Control of Circulating-Current

The reactive circulating-current is mainly caused by output voltage amplitude deviation among modules. As analysis above, the output voltage follows the voltage reference tightly, so the regulation of reactive power can be implemented by controlling the amplitude the voltage

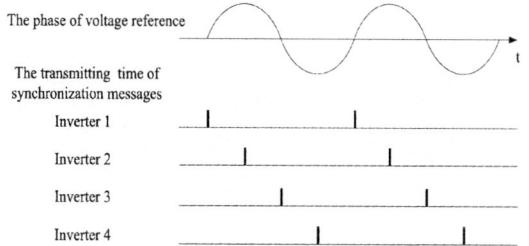

Figure 4. Principle schematic of the phase synchronization

reference. In this paper, the communication among the paralleled inverters is achieved by the use of CAN bus. A proportional loop is constructed for the current sharing control. During the cycle of the output voltage, the paralleled modules transmit the sample values of inductor current to others by CAN bus. The times of transmitting depends on the speed of the communication. The modules calculate the circulating-current according to all the inductor current of the paralleled inverters. And the voltage references are adjusted according to the circulating-current to share the reactive power.

The current sharing of specific arithmetic is as follow: The ID of inverters, the index of sine table-COUNT and corresponding inductor currents are broadcasted to other nodes by CAN bus. And the received inductor current values are saved in the memory space according to the ID and COUNT as address index when the transmitted information is received. Similarly, during each sample period, the module saves its own inductor current according to its ID and COUNT. Then each module has a table which includes the information of each paralleled inverter in the memory space. As soon as the adjusting time comes, each module calculates the current difference, Δi, between the average current and one's own current. Four modules are taken as example. As for inverter1, Δi_1 is:

$$\Delta i_1 = i_1 - \frac{i_1 + i_2 + i_3 + i_4}{4} \tag{6}$$

Define:
$$\Delta U_g = k_{vf}\Delta U = k_{vf}\Delta i_1 X \tag{7}$$

And define:
$$k_u = \frac{U_g - \Delta U_g}{U_g} \tag{8}$$

After substituting (7) into (8), k_u is yielded.

$$k_u = \frac{U_g - k_{vf}\Delta i_1 X}{U_g} \tag{9}$$

The equation of the reference voltage about time is:
$$u_i = k_u U_g \sin \omega t \tag{10}$$

Where k_{vf} is the feedback coefficient of the output voltage; U_g is the amplitude of sine wave reference signal. The voltage reference is adjusted by multiplying with the amplitude regulation coefficient k_u. And current-sharing can be achieved. The flow charts are shown in Fig. 5.

Therefore, the regulation of current-sharing is carried out frequently during each cycle of output voltage. The scheme not only avoids the insufficiency of current-sharing control by simulation circuits, such as poor anti-jamming, complicated interconnection and low reliability, but also overcomes poor instantaneity of the digital parallel to some extent. Although the advantages are achieved at the price of occupying large memory space, the memory resource is quite abundant enough in DSP, considering memory extension in TMS320LF2407A.

556

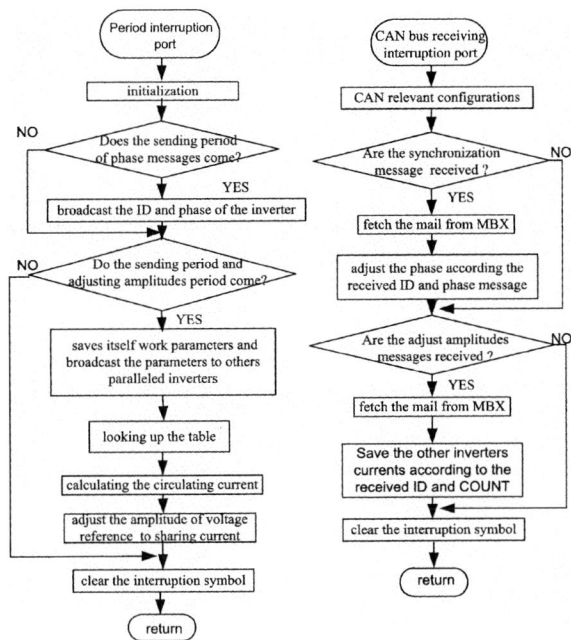

Figure 5. Flow chart of current-sharing program

IV. EXPERIMENT RESULT

Based on the analysis above, a parallel system with two inverter modules has been built to verify the proposed parallel strategy. A block diagram of the single module is shown in Fig. 6. The RMS of output voltage is 220V; the power is 1.2 kVA.

The load voltage and output current waveforms of the parallel modules under resistive load are shown in Fig. 7. Corresponding output voltage harmonic analysis is shown in Fig. 8, in which the THD is 0.867%. The results demonstrate that the parallel system has good static parallel performance and outstanding waveform control. And Fig. 9 shows output currents of two modules, i_{o1} and i_{o2} under step load change, respectively. The rapid adjustment and good current-sharing can be achieved during the dynamic. The behavior of the parallel scheme

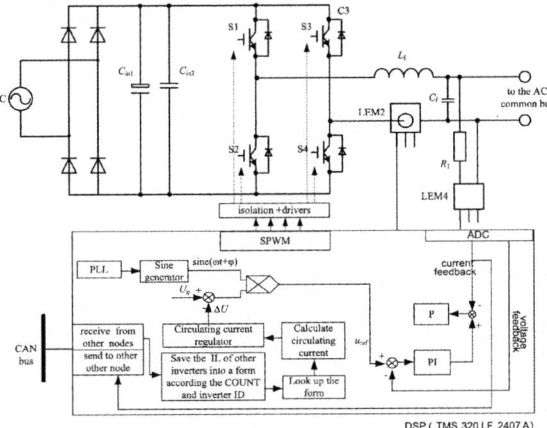

Figure 6. Block diagram of an inverter supply system

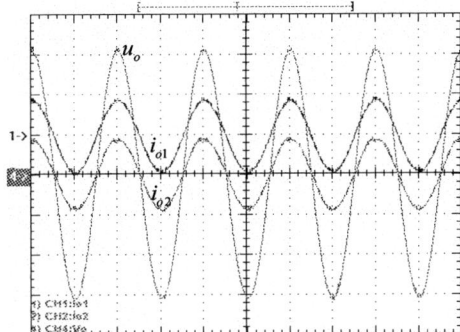

Figure 7. Voltage and current waves of the parallel system
(Y-axis: 100V/div 8A/div, X-axis:10ms/div)

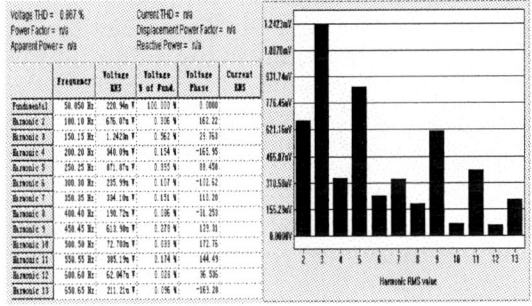

Figure 8. THD analysis of the output voltage

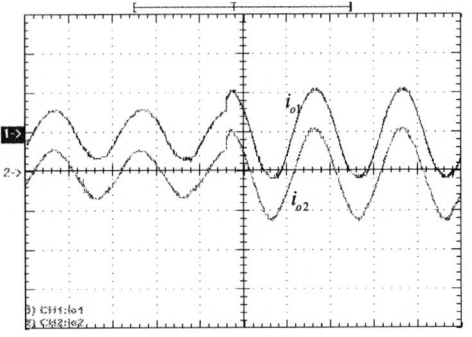

(a) Load increasing

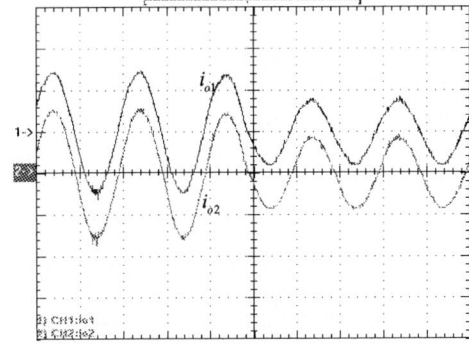

(b) Load decreasing

Figure 9. Dynamic experiment of the parallel system
(Y-axis: 4A/div, X-axis:10ms/div)

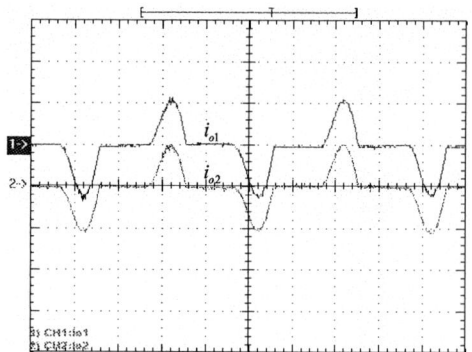

Figure 10. Output current of paralleled inverters under nonlinear load (Y-axis: 8A/div, X-axis: 5ms/div)

under rectifier load is also studied. The waveform of output current under rectifier load is shown in Fig. 10. The experiment results demonstrate that inverter parallel system designed in proposed strategy can operate reliably and realize power sharing evidently.

V. CONCLUSION

This paper presents a full digital paralleled inverter system designed by the proposed alternating-master-salve parallel scheme. DSP is adopted as the central controller to realize the operation. The high precision of synchronization and the fast regulation of current-sharing make for parallel control via CAN bus among the paralleled inverters. The proposed scheme presented in the example of two modules is also suitable for multiple inverters. The proposed parallel strategy is smart and simple and only single interconnection bus is required. It has the advantages of anti-jamming, tight configuration, high reliability and low-cost. The strategy not only keeps the advantages of digital parallel control, but also modifies real-time property compared with other digital parallel scheme. All the inverter units can operate individually or run paralleled. The parallel system operates normally even as faults occurring in whichever paralleled module. So the proposed parallel method can achieve N+1 redundancy. The experiment results demonstrate that inverter parallel system designed in proposed strategy can operate reliably and realize prominently power sharing.

ACKNOWLEDGMENT

This work was supported by the National Natural Science Foundation of China, NO. 50237020.

REFERENCES

[1] Chen daolian, "the DC-AC inverter technique and application," *Electronic Industry Press,* 1998.

[2] Xiao Lan, Hu Wenbin, Yan Yangguang, "Summary of the Control Techniques of Paralleled Inverters," *APSC'2000,* pp.164~167. October 2000

[3] Jiann-Fuh Chen, Ching-lung Chu, "Combination Voltage-Controlled and Current Controlled PWM Inverters for UPS Parallel Operation," *IEEE Trans. Power Electronics,* pp.547-558, 1995

[4] Pei Yunqing, Jiang Guibin, Yang Xu, Wang Zhaoan, "Auto-Master-slave control technique of parallel inverters in distributed AC power systems and UPS," *IEEE 35th Annual Power Electronics Specialists Conference,* Vol3, 2004, pp.2050-2053.

[5] Huang Lei, Xiao Lan, "Study of Paralleling Inverter Based on Phase synchronizing and amplitude Modulating," *Power Electronics,* Vol.3, pp.1432-1437, 2003

[6] S. J. Chiang, J. M. Chang, "Parallel Control of the UPS Inverters With Frequency-dependent Droop Scheme," *IEEE Annual Power Electronics Specialists Conference,* Vol.2, 2001, pp.957-961.

[7] Lin Xinchun, Duan Shanxu, "Modeling and Stability Analysis for Parallel Operation of UPS with No Control Interconnection Basing on Droop Characteristic," *Zhongguo Dianji Gongcheng Xuebao,* Vol.24, pp.33-38, February 2004.

2006 5th International Power Electronics and Motion Control Conference

Analysis and Design of a Novel Dual Secondary Winding and Dual Power Bridge High Frequency Link Inverter

Zhang Zhe, Zhang Chunjiang, Wu Weiyang, Gu Herong and Shen Hong
Yanshan University, Qinhuangdao, China
zhangzhe@ysu.edu.cn

Abstract—a novel dual secondary winding and dual power bridge high frequency link (HLF) inverter is discussed in this paper Single-stage power conversion, standard half-bridge connection of devices, soft-switching for all the power devices (ZVS or ZCS), low conduction loss, simple bipolar combined phase-shifted control, and high efficiency are among the salient features of the HLF inverter. The principles of circuit operation, PWM control and synthesis, topological extension are discussed. The theoretical analysis is presented and the realization of power stage and close loop control is described. The experimental results are given to validate the effectiveness of analytical result and operation principle of the circuit.

Keywords-inverter; high frequency link; soft-switching

I. INTRODUCTION

The bi-directional high frequency link inverters are used widely in the UPS system, battery-backup stand-alone inverter systems, and alternative energy systems such as photovoltaic applications, specially the power supply systems of recreational vehicle and marine boats that need the inverters having the features of higher reliability, smaller volume, higher efficiency and lower cost [1].

In the past decade, the bi-directional HFL inverters have undergone a great development wordwide. The existing HFL inverters can be classified into three major categories from the view of power processing. One is single stage bi-directional circuit, which is mainly based on cycloconverter topology [2-3]. It has the feature of just single power conversion stage with the bidirectional switches, and without the large energy storage elements. But it has a inherent problem of high voltage surge on the bidirectional switches [4]. The second one is the two stage power converter configuration, which usually has two or three power converters in cascade and needs the DC-link [5]. The more power stages relatively leads to more power loss and lower reliability. The third one is quasi-single-stage bi-directional inverter/charger proposed by Virginia Power Electronics Center in 1998, and with the voltage clamp schemes all switch devices can achieve soft-

---

[a] This paper is supported by the Key Program National Nature Science Foundation of China (No. 50237020).

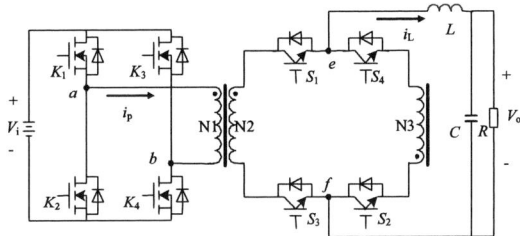

Figure 1.   Dual Secondary Winding and Dual Power Bridge HFL Inverter

switching [6].

In this paper a novel soft-switched single-stage bi-directional HFL inverter topology is discussed with the bipolar combined phase-shifted SPWM control schemes. Analysis of the operation principles in four quadrants is performed. Based on the analytical result, the design consideration is given and also illustrated in a design. The close loop control for the whole system is is also discussed and described. Finally, the experimental results from prototype (24 V dc input and 110 V ac output, 500 W output power) are presented.

II. ANALYSIS OF THE TOPOLOGY

A. Dual secondary winding and dual power bridge HFL inverter topology

The HFL inverter topology is shown in Fig.1. It consists of a full bridge high frequency inverter, two standard half-bridges to achieve the high frequency voltage polarity inversion to get the sinusoidal output voltage on the ac load and a high frequency transformer with two independent secondary windings. Because of the two secondary winding, the secondary side of the transformer increases a new degree of freedom, and the circuit structure of the secondary side is more flexible. So the proposed topology can be extended to a topology family according to the different requirements.

The proposed dual secondary winding and dual power bridge high frequency link inverter has the following salient features compared to other existing circuit topologies furnishing the same functionality:

•Simplified SPWM pattern and control;
•Simplified the circuit structure and the number of

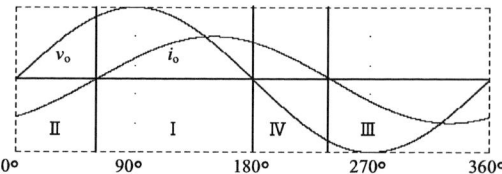

Figure 2. Four-quadrant operation

switches is small;

•Standard half-bridge circuit connection and easily achieved modularization;

•Bi-directional switches are not necessary due to two independent secondary windings;

•Easier circuit layout.

B. Operational Principles

The high-frequency operation principles of the dual secondary winding and dual power bridge HFL inverter in each operation mode will be discussed in this section. Because the load can assume any power factor, either leading or lagging, the circuit needs to operate in all the four quadrants in the V_o-I_o plane during an output line cycle as indicated in Fig. 2. In this paper the bipolar combined phase-shifted control method is used to make the inverter realize four-quadrant operation. It is clear that the output current and voltage are both positive in quadrant I, and that v_o is positive and i_o is negative. Circuit operations in quadrants III and IV are exact replicas of those in quadrants I and II. So only the operations in quadrants I and II need to be analyzed.

In the following analyses, it is assumed that all the power devices are ideal, and that the inductor of output filter, L_f, is much higher than leakage inductor of the transformer, L_k, so the inductor current, I_L, can be considered constant during circuit commutations.

1) Operating Principles in Quadrant I: The PWM pattern and the main operating waveforms are shown in Fig. 3 for operation in quadrant I. The equivalent circuits in each interval during a half high frequency cycle are drawn in Fig. 4.

[t_0~t_1]: Before t_0, switches K_1 and K_4 are on, the primary current, i_p, flows through the K_1, the primary winding of the transformer N_1 and K_4. I_L flows through the secondary winding N_2, S_1 and the antiparallel diode of S_3, D_3. The voltage on the output filter, u_{ef}, is positive, the dc source delivers the energy to the load.

At t_0, S_2 is closed under ZCS and K_1 and K_4 are turned-off. under ZVS. The u_{ab} is:

$$u_{ab} = V_i - \frac{nI_L}{C}t. \qquad (1)$$

Where $C=C_1+C_2$, and n is turn ratio of the transformer.

[t_1~t_2]: At t_1, u_{ab} reduces to zero and increases continually in negative direction. Because the I_L can reflect to the primary side, $i_p=nI_L$, is approximately constant.

[t_2~t_3]: C_1 and C_4 are charged to V_i, so C_2 and C_3 are discharged to zero at t_2. i_p starts to freewheel through D_{K2}, D_{K3} and N1. I_L flows through N_2, S_1 and D_3 so that

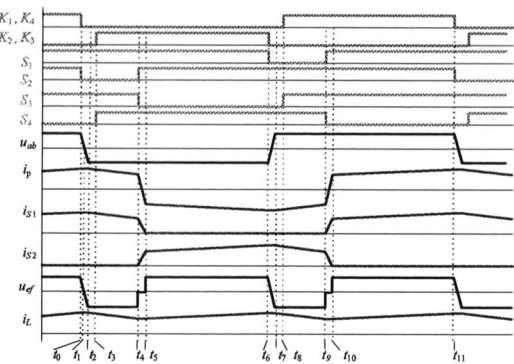

Figure 3. PWM pattern and key waveforms in quadrant I.

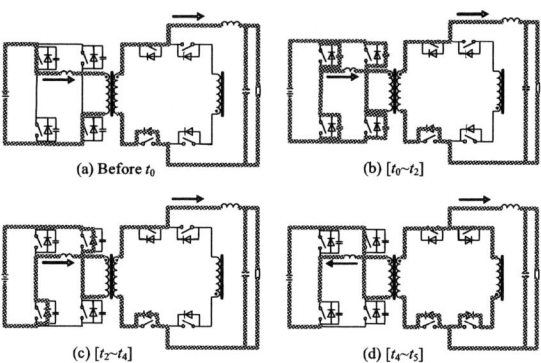

(a) Before t_0 (b) [t_0~t_2]

(c) [t_2~t_4]   (d) [t_4~t_5]

Figure 4. Equivalent circuits in a half-frequency cycle in quadrant I

$u_{ef} = nu_{ab} = -nV_i$ the load feedbacks the energy to dc source.

[t_3~t_4]: Both K_2 and K_3 are turned on under ZVS at t_3 without interruption to the freewheeling path.

[t_4~t_5]: At $t4$, S_3 is turned off under ZVZCS. At the same time S_2 is turned on under ZCS, and according to the voltage polarity of N_3 the current in the secondary side begins to commutate from S_1 to S_2. So $u_{ef}=0$, $u_{ab}=0$. The dc voltage is exerted on the transformer leakage inductance L_k oppositely and i_p starts to ramp down, described as:

$$\frac{di_p}{dt} = -V_i / L_k. \qquad (2)$$

when i_p reverses its direction, it starts to flow through K_2 and K_3.

After the current in S_1 is reduced to zero and the current in S_2 is reached to I_L the next half high frequency cycle begins.

2) Operating Principles in Quadrant II: The PWM pattern and high frequency waveforms are shown in Fig. 5, while the equivalent circuits in each interval are shown in Fig. 6.

[t_0~t_1]: Before t_0, are on, $u_{ab}=-V_i$, and i_p is negative and flows through K_2, N_1 and K_3. While I_L flows through N_2, D_1 and S_3. At t_0, K_2 and K_3 are turned off under ZVS, and i_p starts to charge C_1, C_4, and discharge C_2, C_3.

560

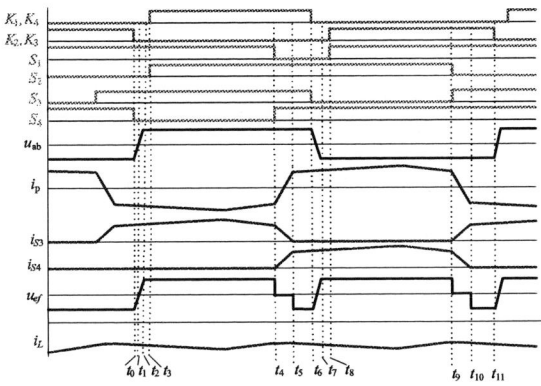

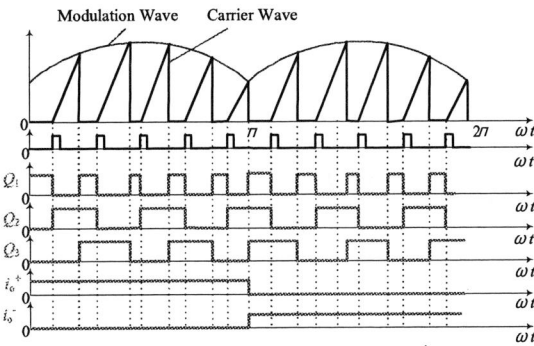

Figure 7. Bipolar combined phase-shifted control scheme

Figure 5. PWM pattern and key waveforms in quadrant II.

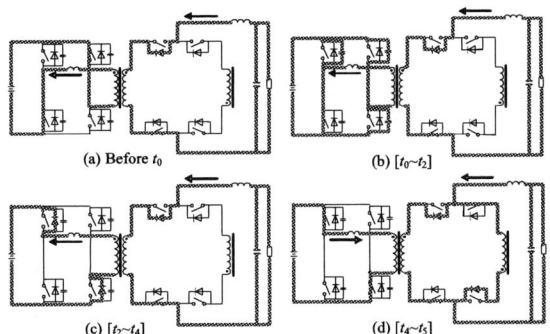

(a) Before t_0

(b) $[t_0\sim t_2]$

(c) $[t_2\sim t_4]$

(d) $[t_4\sim t_5]$

Figure 6. Equivalent circuits in a half-frequency cycle in quadrant II

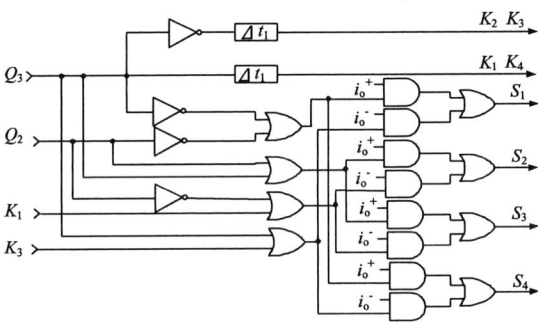

Figure 8. Synthesis of logic signals

$[t_1\sim t_2]$: At t_1, u_{ab} increased to zero and resumed to increase in positive direction. The voltage on S_2 begins to increase from zero and the current paths are not changed.

$[t_2\sim t_3]$: C_1 and C_4 are charged to V_i, so C_2 and C_3 are discharged to zero at t_2. i_p starts to freewheel through D_{K2}, D_{K3} and N1. I_L flows through N_2, S_1 and D_3 so that $u_{ef} = nu_{ab} = -nV_i$ the load feedbacks the energy to dc source.

$[t_3\sim t_4]$: Both K_2 and K_3 are turned on under ZVS at t_3 without interruption to the freewheeling path.

$[t_4\sim t_5]$: Turns off S_1 under ZVZCS. S4 is turned on under ZCS, and the current in S_4 begins to increase and the current in S_3 starts to reduce. So i_p begins to reset by the dc voltage. Finally, i_p reverses its direction, and i_p starts to flows through K_1, K_4 and N1. During this interval the secondary side of the HF transformer is equivalent to be short-circuit condition, so $u_{ef}= 0$.

After the current in S_4 is reached to I_L the next half high frequency cycle begins.

C. *SPWM Control Method*

The bipolar combined phase-shifted SPWM with the signal of output current polarity ensures transition between different quadrants as shown in Fig. 7. The modulation wave is the output of the current regulator rectified and biased, and the carrier signal is output of the

constant-frequency integration. Q_2 and Q_3 are modulated by the SPWM phase-shifted scheme. The driving signals for all the main switches can be synthesized with simple circuits as shown in Fig. 8. The signal of Q_3 is obtained by dividing trailing edge of the carrier signal and is 50% fixed duty cycle. Moreover, 50% fixed duty-cycle switching signals are applied to all the switches on the high frequency inverter bridge with proper delay, Δt_1, inserted between the complementary switch-pairs. With some simple logic operations of Q_2, Q_3, K_1 and K_3 we can generate the driving signals for switches on secondary side.

III. SYSTEM DESIGNS

The block diagram of the whole HFL inverter system is shown in Fig. 9.

A. *Designs of the Power Stage*

1) Transformer turn ratio n: For the high frequency transformer turn ratio should guarantee the voltage range. To ensure appropriate inverter output voltage scaling, the turns ratio n of the transformer needs to be selected such that $n(V_{i\min} - V_{FET})D_M \geq V_o^p = \sqrt{2}V_o$, where D_M is the maximum duty ratio, $V_{i\min}$, V_{FET}, and V_o^p are the minimum input battery voltage, on-drop voltage on the primary MOSFET switches, and peak output ac voltage, respectively. If D_M=0.9, V_{FET}=0.5 V, and V_o=110 V, then $n \geq 8$. In reality, $n = 10$ is selected.

561

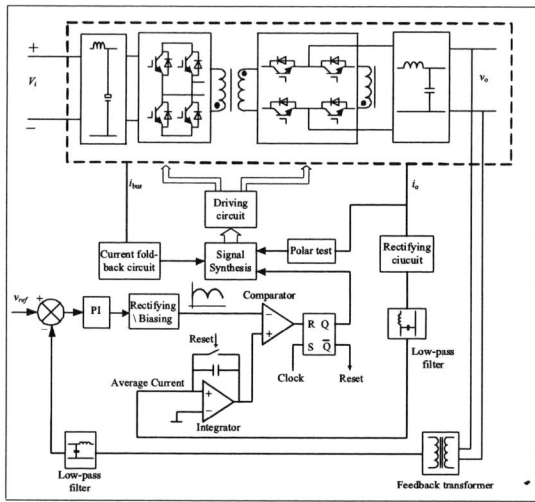

Figure 9. Block diagram for the whole system

2) Election of the switch devices: The high voltage justify the use of IGBT devices for the secondary side switches. Because of the structure of two independent secondary windings of the transformer, standard low cost half-bridge IGBT modules can be used. On the primary side, the switches are switched under ZVS and low voltage MOSFETs can be adequately used.

3) Output filter: The output filter of the power inverter is used to smooth out the waveforms generated from the DC/AC stage. The signal contains many unwanted harmonic frequencies including multiples modulation switching frequency. The cut-off frequency f_c is described as:

$$f_c = \frac{1}{2\pi\sqrt{LC}} = \frac{R}{2\pi L} \qquad (3)$$

Where $R = \sqrt{L/C}$.

So from (3) we can get:

$$L = \frac{R}{2\pi f_c}. \qquad (4)$$

$$C = \frac{L}{R^2} = \frac{1}{2\pi f_c R}. \qquad (5)$$

When the fundamental frequency of output voltage is 50 Hz, f_c is chosen to be 400~1000 Hz generally. And R is $(0.5$~$0.8)R_L$, where R_L is the load resistance. If $R = 0.5R_L = 0.5 \times 16.1 \approx 8\,\Omega$ and f_c=1 kHz, we can get: L_f=1.4 mH and C=10 μ F:

B. Close loop control scheme

This paper proposed a novel control strategy with output voltage and inductance current double loop, shown in Fig. 9. Because the one cycle control can restrain the dynamic and steady-state errors in one on-off cycle, it be used to the inner loop of the inductance current and to realize the PWM scheme. A instantaneous outer voltage feedback loop with PI controller is used to stabilize the sinusoidal voltage.

IV. EXPERIMENTAL RESULTS

A laboratorial prototype based on dual secondary winding and dual power bridge HFL Inverter is designed according to the following specifications:

Input battery voltage V_i=20~28 V (nominal value V_i=24 V);

Output ac voltage V_o = 110 V, 50 Hz;

Output capacity P_o = 500 W;

Output voltage THD<5%.

The main power stage parameters are as follows:

K_1~K_4: FQA65N20, 200V/65A, TO-3P, MOSFET;

S_1~S_4: Toshiba MG50J2YS50 600V/50A IGBT Module;

The transformer T: PC40 materials, ETD-59 core, N_1 : N_2 : N_3 = 1 : 10 : 10;

Switching frequency: f_s =30 KHz;

Primary leakage inductance L_k=1 uH.

Fig. 10 shows output of constant-frequency integrator is low level while the Q_1 is high. Fig. 11 shows the logic relationship between output of outer loop regulator and output of constant frequency integrator in a large time range. From Fig. 12 it is clear that voltage surge on collector-emitter of switch in secondary side is restrained effectively. The bipolar SPWM voltage on the output LC filter, u_{ef}, is shown in Fig. 13. The output voltage is shown in Fig. 14

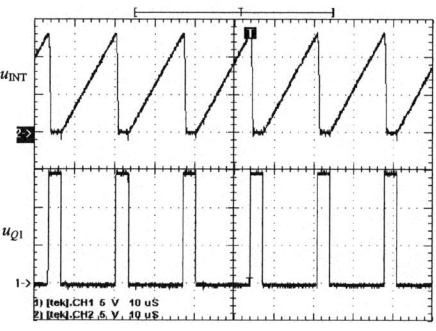

Figure 10. Waveforms of u_{INT} and u_{Q1} (5 V/div)

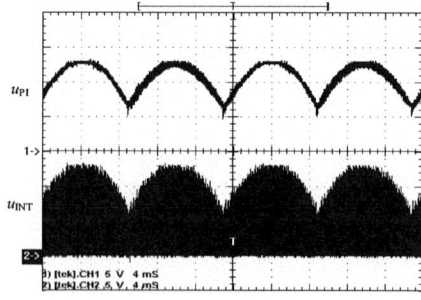

Figure 11. Waveforms of u_{PI} and u_{INT} (5 V/div)

562

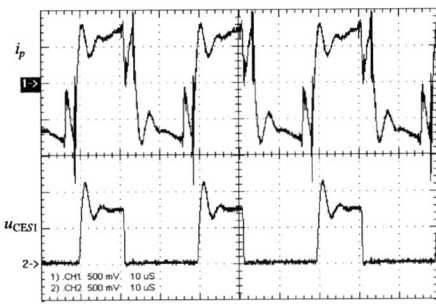

Figure 12. Waveforms of i_p and u_{CES1} (CH1: 20 A/div; CH2: 250 V/div)

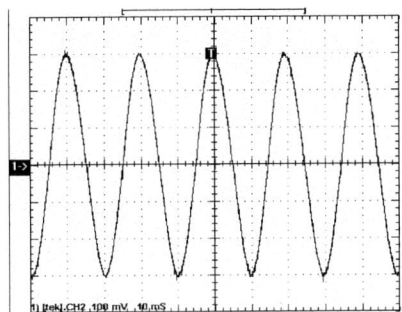

Figure 14. Waveforms of output voltage in resistant load (50 V/div)

by the experimental results on a on a laboratorial prototype.

REFERENCES

[1] W. Wu, C. Zhang, X. Sun and X. Yang, "Research on the topologies of medium and small power high frequency link inverter," Proceedings of Ninth Chinese Power Electronics Seminar, Sept., 2004, pp. 24-27

[2] I.Yamato, N.Tokunage, Y.Matsuda, H.Amano and Y.Suzuki, "High frequency link DC-AC converter for a new voltage clamper," *IEEE PESC*, 1990, pp. 749-756.

[3] H. Pinherio, P. Jain and G. Joos, "Zero voltage switching series resonant based UPS", *IEEE PESC*, 1998, pp. 1879-1885.

[4] Tazume K., Aoki T., Yamashita T., "Novel method for controlling a high-frequency link inverter using cycloconverter techniques," *IEEE PESC*, 1998, pp. 497-502.

[5] Salam Z., Ramli, Z., "A DC-DC type bidirectional high frequency link inverter using center-tapped active rectifier," *IEEE IECON, 30th Annual Conference of IEEE*, 2004, pp. 47-50.

[6] Kunrong Wang, Fred C. Lee, and Wei Dong, "A new soft-switched quasi-single(QSS) bi-directional Inverter/Charger", *Proceedings of the Sixteenth Annual VPEC Seminar*, Sept., 1998, pp. 55-62.

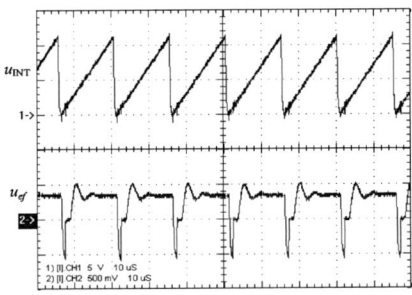

Figure 13. Waveforms of u_{INT} and u_{ef} (CH1: 5V/div; CH2: 250V/div)

V. CONCLUSION

A novel single-phase soft-switched single-stage bi-directional high frequency link inverter topology is discussed in this paper. With the bipolar combined phase-shifted control method it can realize four-quadrant operation. The main circuit structure, detailed operation modes, PWM control scheme and synthesis, the closed-loop control are investigated. The feasibility for the novel HFL inverter with the proposed control scheme has been verified

2006 5th International Power Electronics and Motion Control Conference

Reduction of Common Mode EMI in a Full-Bridge Converter through Automatic Tuning of Gating Signals

Kai Zhang[*], Yunbin Zhou, Yonggao Zhang, and Yong Kang
Huazhong University of Science and Technology, Wuhan, China
[*]Email: zhkhust@263.net

Abstract—Theoretically, a full-bridge converter under bipolar PWM should produce very little common mode noise because the two phase-legs can compensate for each other, but this feature is often disrupted by different transmission delays of the gating signals. In this paper, a method to address this problem through automatic tuning of the pulse edges of gating signals is proposed. Specifically, the CM noise current is sensed by a DSP and the noise energy is calculated, then a search algorithm is employed to tuning the pulse edges of the gating signals in order to minimize the CM noise. Experimental results show that up to 10dB reduction of the CM current can be obtained from 500kHz to 5MHz. Implementation of the this method requires little hardware expense, and the software can be synthesized into the main control DSP of the converter.

Keywords—full-bridge converter; CM noise rejection; energy; gating signal

I. INTRODUCTION

To reduce volume and weight, and improve dynamic response, there has been a continuing demand for increasing the switching frequency of power electronic converters. As a result, high-speed power devices find more and more applications. However, high *dv/dt*s associated with high speed switching can cause severe common mode electromagnetic interference (CM EMI) [1]. The latter usually constitutes a major part of total conducted EMI, especially beyond MHz level. CM EMI also acts as a major cause of radiated EMI because of the "antenna effect" of the CM current loop.

Among many countermeasures against the CM EMI problem, those based on the idea of active cancellation have drawn a lot of research interests in recent years. Some of these methods attempt to eliminate common mode voltage by means of balanced switching actions, such as the four-leg [2,3] and dual-bridge [4,5] approaches. Others focus on CM current [6-9] or voltage [10-14]

This work was sponsored by the National Natural Science Foundation of China (NSFC, under project 50407011) and Delta Electronics Ltd.

compensation with some extra circuitry.

Most of the above-mentioned methods deal with AC drive systems where the oscillation frequency of the CM noise current are relatively low due to a fairly large distributed capacitance between the motor windings and the ground. Therefore, transmission delays of gating signals are a minor factor there. However, for most power supply applications where the oscillation frequency of the CM noise current is much higher, transmission delays of gating signals can no longer be overlooked. For example, according to the generation mechanism of CM noise current, a full-bridge converter under bipolar PWM operation should produce very little CM noise because each of the two phase legs acts as a perfect active compensator for the other one. However in a real product different transmission delays of gating signals can easily disrupt this feature, rendering the CM noise higher than expected. To address this problem, a method to minimize the CM noise current through automatic tuning of the pulse edges of the gating signals is proposed in this paper. The method is validated by experiment, which shows that the CM current can be significantly reduced in a quite wide frequency range.

II. CM NOISE CURRENT IN A FULL-BRIDGE CONVERTER

Fig.1 shows the main circuit of a full-bridge converter. For reducing the thermal resistance within a power device, there is only a thin insulation layer between the device junction and the base plate, with the latter being fixed on a heat sink. The heat sink is in turn fixed to the chassis that is usually grounded for safety reasons. This gives rise to a significant parasitic capacitance (see C_{p1} and C_{p2} in Fig.1) between the midpoints of the phase legs and the ground. As the devices switch on and off with high speed, high *dv/dt*s at the midpoints generate currents through the parasitic capacitors. These currents converge into the chassis/ground, forming a major part of the CM current.

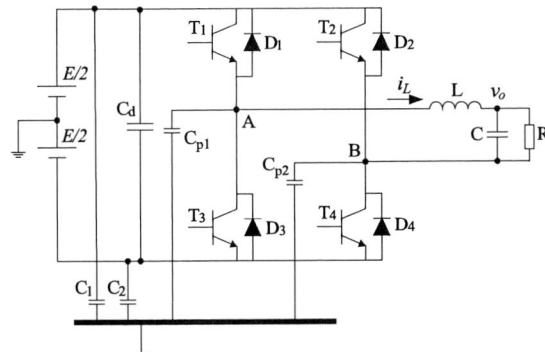

Fig.1 A full-bridge converter.

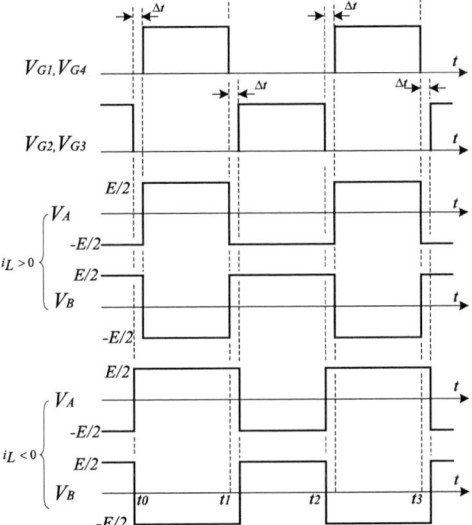

Fig.2 Gating signals and midpoint voltages under bipolar PWM operation.

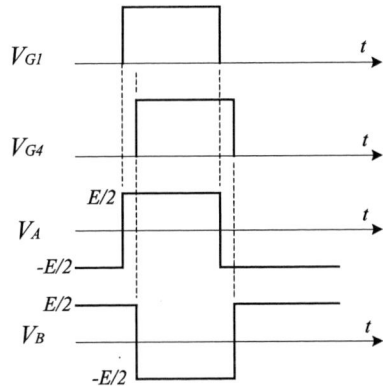

Fig.3 Time skew between midpoint voltages due to different transmission delays of gating signals (supposing $i_L > 0$).

This CM current then will find its way into the DC buses through common mode capacitors at the front end of the inverter. (These CM capacitors can be part of an EMI measurement system or part of the CM filter of the converter itself.)

With unipolar PWM, one of the midpoint voltages alternates in switching frequency and the other in fundamental frequency. Therefore, they cannot cancel out each other. With bipolar PWM, however, diagonal switch pair T1-T4 (and T2-T3) turns on and off simultaneously, while two switch pairs operate in a complementary manner. Fig.2 shows gating signals $V_{G1} \sim V_{G4}$ and midpoint voltages V_A and V_B of the phase legs for bipolar PWM taking into account the blank time Δt and different directions of the load current i_L.

As Fig.2 shows, no matter which direction the load current is in, the two midpoint voltages always change in an exact opposite way. Usually, power devices used in one converter are of the same type, so the parasitic capacitance between each phase leg's midpoint and the ground can be considered as the same. As a result, the two phase legs provide perfect CM compensation for each other and the CM current can be cancelled out automatically.

Unfortunately, in a real product the gating signals always have to go through some processing circuits before arriving at the power devices. Such circuits usually include logic gates, voltage comparators, optical couplers, and/or some application-specific driving modules. Due to inevitable deviation in signal transmission characteristics of these devices, transmission delay of each gating signal can be different. Therefore, the exact synchronism of switching actions shown in Fig.2 can be easily disrupted.

An example of this is shown in Fig.3, where it is supposed that $i_L > 0$ and transmission of T1's gating pulses (both the edges) is faster than T4's. It is seen that in this case ($i_L > 0$), V_{G1} determines V_A and V_{G4} determines V_B. Similarly, it can be found that when $i_L < 0$, voltages V_A and V_B will be determined by V_{G2} and V_{G3}, respectively. In either case, the two midpoint voltages no longer change in an exactly opposite way. Although the differences in transmission delays of the gating signals are fairly small —usually around a hundred ns—compared with the pulse widths, it can easily disrupt the CM current compensation mechanism mentioned above, because the oscillation frequency of the CM current can be as high as several mega Hz, depending on the exact parameters in the CM current path.

III. AUTOMATIC TUNING OF GATING SIGNALS

To compensate the differences between the transmission delays of the gating signals for the diagonal switch pairs T1-T4 and T2-T3, a simple solution is to fine-

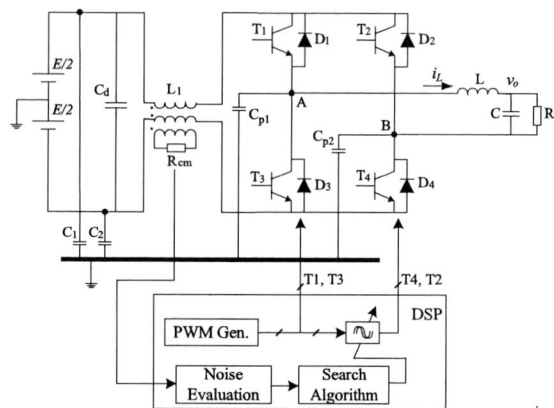

Fig.4 Automatic tuning of gating signals for CM noise reduction.

tune the gating signals before they are transmitted. Also, it would be preferable if the tuning process can be done automatically with a closed loop.

Although the problem in its nature is just to align two pulse edges, a phase-lock-loop (PLL) solution does not fit here because of the difficulty in detecting gating signals at the power device side. The phase delay introduced by the phase detecting circuit is another denying factor. Since the ultimate purpose of the tuning process is to minimize the CM noise, it would be a better choice to detect the CM current, and then use a search algorithm to find out the optimal amount of phase advance/delay of the gating signals in order to minimize it. The basic idea of this method is shown in Fig.4, where gating signals of T1 and T3 are transmitted as they are, but those of T4 and T2 are fine-tuned in their phase with respect to T1 and T3, respectively, in order to minimize the CM noise current.

It should be noted that in order to achieve phase advancement, which may be needed, it is assumed that the gating signals are pre-determined by the PWM generation algorithm one switching period ahead. This is not a difficult task for digital controllers.

A. Evaluation of CM Noise Level

From standard EMC point of view, the best description of the CM noise current is its spectrum. Since the latter needs FFT calculation, it's not feasible here. Actually, a rough idea of whether the CM noise is worsening or improving would be ample for the decision-making of the search algorithm. Therefore, the time-domain energy of the CM noise current within one switching period is chosen as the evaluation index of the CM noise level, the definition of which is:

$$ e_{CM} = \sum_{n=0}^{N-1} \left| i_{CM}[n] \right|^2, \qquad (1) $$

where N denotes the number of samplings of the CM current (i_{CM}) within one switching period.

According to Parseval's theorem, time-domain energy of a signal is equal to its frequency-domain energy, i.e.,

$$ \sum_{n=0}^{N-1} \left| i_{CM}[n] \right|^2 = \frac{1}{N} \sum_{k=0}^{N-1} \left| I_{CM}[k] \right|^2, \qquad (2) $$

where I_{CM} is the discrete Fourier transformation (DFT) of i_{CM}.

From (1) and (2) it is seen that the calculation of time-domain energy of the CM noise current is fairly simple, yet it is proportional to the squared sum of the spectral components of the CM noise current.

Another advantage of this index is that its value is not affected by the variation of duty ratio, as long as the positive- and negative-going pulses of the CM current do not merge into each other. This prevents the automatic tuning process from being interfered by the normal duty ratio regulation of the converter.

B. Automatic Search Algorithm

Typical CM current waveform generated by each phase leg can be characterized by a series of damped oscillations, as can be expected from an RLC circuit under square wave excitation. Based on this knowledge, the relationship between the total CM noise energy and the time skew between the CM currents of the two phase legs can be found through simulation. Although the exact shape of the relationship curve may vary with simulation parameters, the salient features are easy to identify. Shown in Fig.5 is a typical curve. Not surprisingly, the curve is symmetrical about the central vertical line where the time skew is zero, and it has multiple local minimums and maximums. The time skew corresponding to the first pair of local maximums is found to be a half of the oscillation cycle of the CM noise current. Within this boundary, there is only one minimum, which is also the global minimum of the curve. In many cases the initial time skew resulted from different transmission times of gating signals fall within

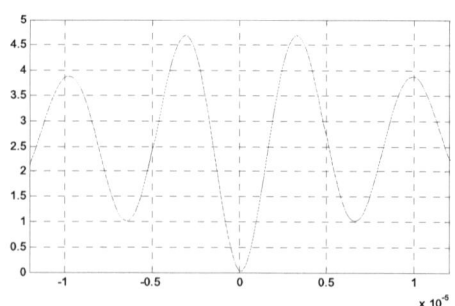

Fig.5 Relationship between total CM noise energy and time skew between CM noise currents of the two phase legs

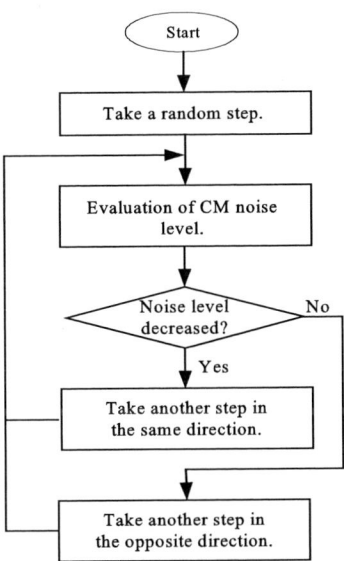

Fig.6 Flow graph of the search algorithm.

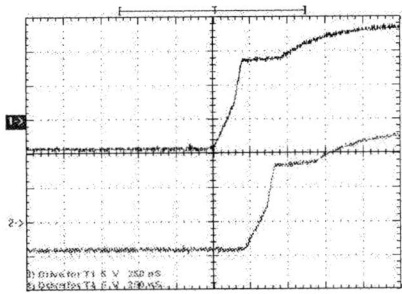

Fig.7 Gating signals of T1 and T4.

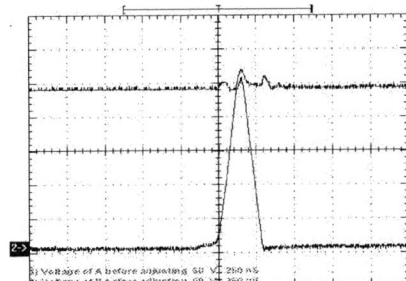

Fig.8 Midpoint voltages of the two phase legs.

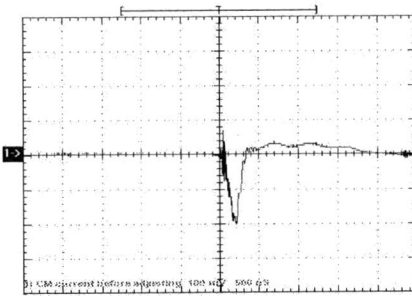

Fig.9 CM current before the tuning process.

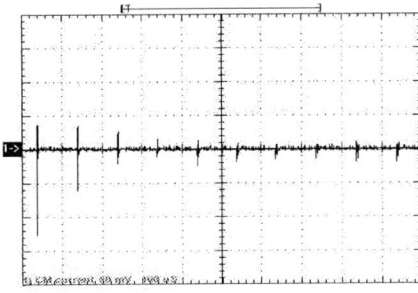

Fig.10 Reduction of CM current during tuning process.

this boundary, therefore a local search algorithm based on the evaluation of CM noise level can reach the global minimum.

The flow graph of the proposed search algorithm is shown in Fig.6. The first step is taken at random, then the direction of every following step depends on the effect (i.e. whether the CM noise level has increased or decreased) of the preceding one. Apparently this is a typical "greedy algorithm", which is unable to overcome local minimums. If such capability is required, more complex algorithms such as simulated annealing should be employed.

IV.   EXPERIMENTAL RESULTS

The main circuit of the experimental setup has been shown in Fig.4. The full-bridge converter is controlled as a single-phase SPWM inverter. The power devices are 1200V/75A dual IGBT modules from SEMIKRON. The driving modules are EXB841s from Fuji Electric. The DSP is a TMS320F2812 from Texas Instruments. The CM current is sensed by a CM transformer. Sampling of the CM current is done with the embedded A/D converter of the DSP.

Gating signals of T1 and T4 before the automatic tuning algorithm is activated are shown in Fig.7, where a 200ns time skew can be observed. The resulting midpoint voltages (with respective to the negative dc bus) of the two phase legs are shown in Fig.8. Apparently they are not exactly opposite to each other. The CM current waveform under this situation is shown in Fig.9.

After the activation of the proposed algorithm, the CM current starts to decrease, as shown in Fig.10. Fig.11 is the midpoint voltages of the phase legs in steady state. It can be seen that the time skew between the two voltages has disappeared. The CM currents before and after the tuning process are put together in Fig.12. The reduction in the amplitude is evident. Fig.13 is the spectrum of the CM noise current from 150kHz to 30MHz before and after the tuning process. The reduction of noise level is evident

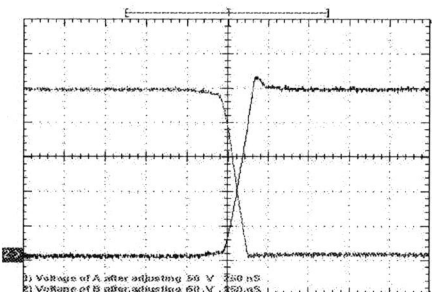

Fig.11 Midpoint voltages after the tuning process.

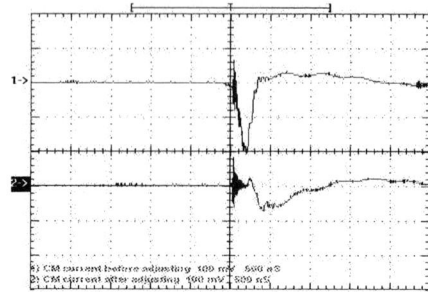

Fig.12 CM current before and after tuning process.

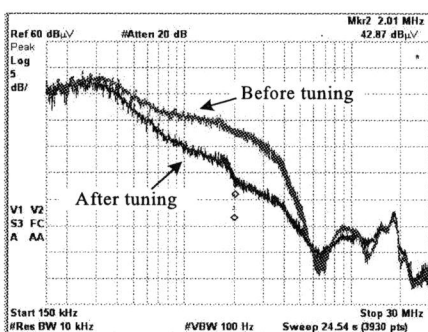

Fig.13 CM current spectrum before and after tuning process.

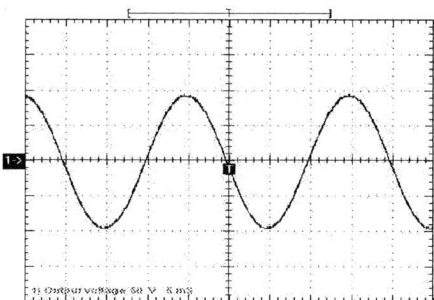

Fig.14 Output voltage of the inverter.

from 500kHz to 5MHz. The maximal reduction is nearly 10dB.

During the experiments, the same DSP also functions as the main controller of the SPWM inverter. Fig.14 is the waveform of the output voltage of the inverter.

V. CONCLUSION

A method is introduced to compensate the differences in transmission delays of gating signals, which can disrupt the CM noise cancellation mechanism of a full-bridge converter. The method features evaluation of CM noise level by its time-domain energy and a simple search algorithm. The hardware expense is minor and the control software can be synthesized into the main control DSP of the converter. Experiments show that this method can effectively reduce the CM noise level in a wide frequency range. The basic idea of this method also applies to other active CM noise rejection applications where time skew due to different signal transmission delays needs to be taken care of.

REFERENCES

[1] Laszlo Tihanyi, Electromagnetic Compatibility in Power Electronics, IEEE Press, 1995.

[2] A. L. Julian, G. Oriti and T. A. Lipo, "Elimination of Common Mode Voltage in Three Phase Sinusoidal Power Converters", *IEEE Trans. on Power Electronics*, vol.14, no.5, pp.982-989, 1999.

[3] G. Oriti, A. L. Julian and T. A. Lipo, "A New Space Vector Modulation Strategy for Common Mode Voltage Reduction", *IEEE-PESC'97*, vol.2, pp.141-1546.

[4] A. V. Jouanne and H. Zhang, "A dual-bridge Inverter Approach to Eliminating Common mode Voltages and Bearing and Leakage Currents", *IEEE-PESC'97*, vol.2, pp.1276-1280.

[5] H. Zhang and A. V. Jouanne, "Suppressing Common-Mode Conducted EMI Generated by PWM Drive Systems Using a Dual-Bridge Inverter", *IEEE-APEC'98*, pp.1017-1020

[6] I. Takahashi, A. Ogata et al, Active EMI Filter for Switching Noise of High Frequency Inverters", *IEEE Power Conversion Conference-Nagaoka 1997*, vol.1, pp.331-334, Aug. 1997.

[7] Y. C. Son and S. K. Sul, "A Novel Active Common-Mode EMI Filter for PWM Inverter" *IEEE-APEC2002*. vol.1, pp.545-549, Mar. 2002.

[8] Xin Wu, M.H. Pong, Z.Y. Lu, Z.M. Qian, "Novel Boost PFC with Low Common Mode EMI: Modeling and Design", *IEEE-APEC'00*, vol.1, pp.178-181.

[9] D. Cochrane, D. Chen, and D. Boroyevich, "Passive Cancellation of Common-Mode Noise in Power Electronic Circuits", *IEEE Trans. on Power Electronics*, vol.18, no.3, pp.756-763, May 2003.

[10] Y. Murai, T. Kubota, and Y. Kawase, "Leakage Current Reduction for a High-Frequency Carrier Inverter Feeding an Induction Motor," *IEEE Trans. on Industry Application*, vol. 28, no.4, pp. 858-863, Jul. 1992.

[11] M. M. Swamy, K. Yamada, and T. J. Kume, "Common Mode Current Attenuation Techniques for Use with PWM Drives", *IEEE Trans. on Power Electronics*, vol.16, no.2, pp.248-255, 2001.

[12] S. Ogasawara, H. Ayano, and H. Akagi, "An Active Circuit for Cancellation of Common-Mode Voltage Generated by a PWM Inverter", *IEEE Trans. on Power Electronics*, vol.13, no.5, pp.835-841, Sep. 1998.

[13] S. Ogasawara and H. Akagi, "Circuit Configurations and Performance of the Active Common-Noise Canceler for Reduction of Common-Mode Voltage Generated by Voltage-Source PWM Inverters", *IEEE-IAS'2000*, vol.3, pp.1482-1488.

[14] Y. Q. Xiang, "A Novel Active Common-Mode-Voltage Compensator (ACCom) for Bearing Current Reduction of PWM VSI-Fed Induction Motors", *IEEE-APEC'98*, pp.1003-1009.

2006 5th International Power Electronics and Motion Control Conference

Phase Multilevel Inverter Fault Diagnosis and Tolerant Control Technique

Wang Baocheng, Wang Jie, Sun Xiaofeng Wu Junjuan and Wu Weiyang
Yanshan University, Qinhuangdao City, Hebei province, China
E-mail: ldh820@tom.com

Abstract-This paper gives a solution to implement a self-diagnosis and tolerant control for a 3-level neutral point clamped inverter. Redundancies in semiconductor configurations are used to enable a continuous operation with a faulty power switch. The fault tolerance is obtained by using a SCR to substitute faulty switch. In case of failure events, the on-line diagnosis technique can fix faulty switch and take relevant tolerance measures. Through simulations and experiments, this diagnosis technique is validated.

Keywords-component; multilevel inverter; on-line fault diagnosis; digital signal processing; tolerant technique

I. INTRODUCTION

Conventional inverter has limitations in high-voltage and high-power occasions. Nowadays, multilevel inverters are widely applied in such systems. This topology improves power quality by inserting some voltage steps in the line-to-ground voltages. Then it makes trapeziform wave composed by multiple levels approach to sine wave. So lower losses and improved EMC are another advantages of multilevel converters. Generally, more numbers of levels, more output voltage waveform approaching to sine wave. At the same time, dv/dt of output voltage decreases largely [1-4]. Multilevel inverter will replace conventional 2-level inverter in more application. These characters attract a lot of experts and researchers to study its topology, control and modulation method.

A main disadvantage of multilevel converters is that they need more power semiconductors [5]. Following the increase of voltage levels, failure probability increases, inverter reliability decreases [6-7]. So fault diagnosis in multi-level inverter is a worthy study field to perfect this topology. Previous researches concerning fault diagnosis of 2-level inverter are mostly based on theory analysis, such as neural network, expert system and so on. This paper aims at single switching fault of Neutral Point Clamped (NPC) multilevel inverter. Fault types include open-circuit fault and trigger signal loss fault come from driver circuit failure. Fault makes some switching states missing. The lost switching states will reduce the utilization of the dc source, lead to inverter defeat. Through analyzing fault fore-and-aft waveforms, we propose a new on-line fault diagnosis based on digital signal analysis and processing. According to inherent features of

more switches and more levels in multi-level inverter, a fault-tolerant measure is introduced. In cases of failure events, inverter can fix on faulty switch and take tolerant measures to ensure inverter operating continuously.

II. OPERATION IN FAULT FREE MODE

NPC topology is a common multilevel format. Fig.1 shows single phase NPC 3-level voltage source inverter (VSI) topology. In conventional drive, each bridge has three switching states. By turning on two switches of upper (S1, S2), lower (S3, S4) or middle (S2, S3), the middle voltage of bridge are clamped to be positive, negative or zero respectively [8]. To single phase three-level inverter, it has 2 bridges, 3^2 switching states.

Table I shows conventional switching states and V_{AO}. Switch drivers meet:

$$S1 = \overline{S3}, \ S2 = \overline{S4}. \ S5 = \overline{S7}, \ S6 = \overline{S8}$$

fmn are switching states of three-level inverter (m-bridge number, n-switch state). Voltage between bridges (V_{AB}) has five levels (Vdc, Vdc/2, 0, -Vdc/2, -Vdc). In half period, it has three levels (Vdc, Vdc/2, 0), so called three-level inverter. In case of resistance load, drive method can be optimized as Fig.2 shows. Double carriers SPWM is used in producing switch drivers of S1, S2, S3, S4. When reference voltage (Vgiven) is positive, S7 and S8 are on, S5 and S6 are off. When Vgiven is negative, S5 and S6 are on, S7 and S8 are off. This modulation makes driver switch simple, switch actions less. It sometimes distributes dc input voltage over 3 switches in first bridge compared to conventional 3-level inverter. Experiment waveforms of Vo and V_{AB} are shown as Fig.4. Switch frequency is 10kHz.

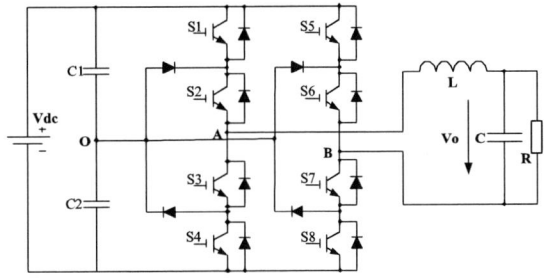

Figure 1. Single phase NPC 3-level inverter topology

1-4244-0448-7/06/$25.00 ©2006 IEEE

TABLE I. SWITCHING STATES AND V$_{AB}$

Switching state								Bridge switching states						Voltage
S1	S2	S3	S4	S5	S6	S7	S8	f_{11}	f_{12}	f_{13}	f_{21}	f_{22}	f_{23}	V$_{AB}$
1	1	0	0	1	1	0	0	1	0	0	1	0	0	0
1	1	0	0	0	1	1	0	1	0	0	0	1	0	Vdc/2
1	1	0	0	0	0	1	1	1	0	0	0	0	1	Vdc
0	1	1	0	1	1	0	0	0	1	0	1	0	0	-Vdc/2
0	1	1	0	0	1	1	0	0	1	0	0	1	0	0
0	1	1	0	0	0	1	1	0	1	0	0	0	1	Vdc/2
0	0	1	1	1	1	0	0	0	0	1	1	0	0	-Vdc
0	0	1	1	0	1	1	0	0	0	1	0	1	0	-Vdc/2
0	0	1	1	0	0	1	1	0	0	1	0	0	1	0

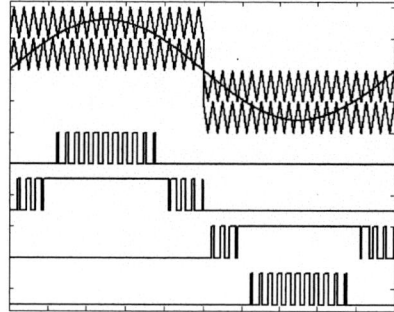

Figure 2. Simplified modulation with resistance load

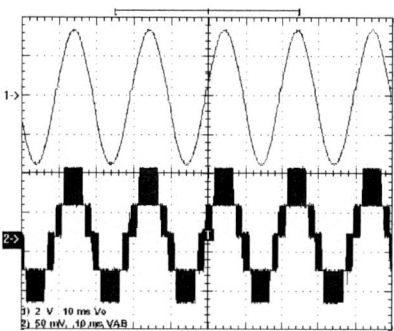

Figure 3. Experiment waveforms in fault free mode

III. FAULT DIAGNOSIS

The heart of multilevel inverter is to use a higher number of power semiconductor switches for adding multiple voltage steps on the AC side. The multiple steps voltage has several advantages, such as lower harmonic content in Vo waveform, lower dv/dt on output load, lower electromagnetic compatibility (EMC) and higher dc/ac power quality. But it brings higher possibility of power semiconductor fault, beside higher cost and more complicated control.

In power electronics field, short-circuit (SC) fault, open-circuit fault (OC) and losing drive pulse (LDP) fault are three common semiconductor switch faults. SC fault is generally caused by counterblow destroy of semiconductor switch. By adding protection actions, SC fault can be avoidable. In SC fault prevention aspect, some experts and scholars have studied deeply. OC fault comes from switch self. As we know, each switch in the inverter has self-drive circuit. If power supply or element in drive circuit is invalid, it can't cause normal trigger pulse to corresponding switch. Then the switch can't work normally. This fault is defined as losing drive pulse fault. LDP fault has the similar fault features with OC fault in output waveforms. In conventional fault diagnosis, more detected point and sensors are demanded to set up. It makes circuit complicated and cost increased; it is also easy to cause diagnosis defeat because of outside interrupt. To these fault formats, experts propose some diagnosis method. For example, detect switch voltage or current, signal transformation and so on. This paper presents a new and effective method aimed at detecting single- switching OC fault and LDP fault in 3-level NPC inverter. The main step for fault diagnosis is finding fault characters. Diagnosis should be based on analyzing fault circuit and waveforms.

A. Fault Analysis

Fault changes switching states. In 3-level inverter, each switching fault makes V$_{AB}$ levels decrease. More decreased levels, more serious output voltage Vo waveform distortion. In order to detect the fault switch immediately and accurately, we should simulate fault, then analyze fault features. Through the course, find out detected waveform and in-depth point out fault switch. Features of each switch fault are shown as Table II.

Fault makes V$_{AB}$ and Vo distort, a key difference between switch fault in 2 bridges is the effect to V$_B$. This feature occurs at the moment when fault destroy inverter. In Table II, this distortion phenomenon is analyzed deeply. When inverter operates normally, V$_B$ is square wave with 50 Hz frequency. It varies between 0 and Vdc. The given voltage can be represented as:

$$Vgiven = A\sin \omega t \qquad (1)$$

The relationship between time(t) and Vo, V$_B$ in fault free mode is shown as:

$$kT < t < kT + T/2 \qquad Vo > 0, V_B = 0 \qquad (2)$$
$$kT + T/2 < t < (k+1)T \qquad V_o < 0, V_B = Vdc/2 \qquad (3)$$

Fig.4 shows experiment waveforms in S1 fault mode to validate the fault analysis above. From waveforms, we can observe V$_{AB}$, Vo and V$_B$ varieties fore-aft S1 fault from top to down. V$_{AB}$ and Vo waveforms in positive axis appear distortion. The average value becomes negative so that it leads to inverter failure.

TABLE II. FAULT FEATURES

Faulty Switch	V_{AB} levels lost	Effect to V_B	Effect to Vo
S1	Vdc	No	Positive waveform, disappear partly
S2	Vdc, Vdc/2	No	Positive waveform, disappear completely
S3	-Vdc, -Vdc/2	No	Negative waveform, disappear completely
S4	-Vdc	No	Negative waveform, disappear partly
S5	-Vdc	Negative waveform, Vdc fall to Vdc/2	Negative waveform, disappear partly
S6	-Vdc, -Vdc/2	Negative waveform, Vdc fall to 0	Negative waveform, disappear completely
S7	Vdc, Vdc/2	Positive waveform, 0 rise to Vdc	Positive waveform, disappear completely
S8	Vdc	Positive waveform, 0 rise to Vdc/2	Positive waveform, disappear partly

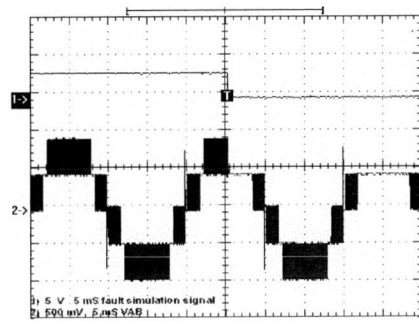

(a) V_{AB} in S1 fault mode

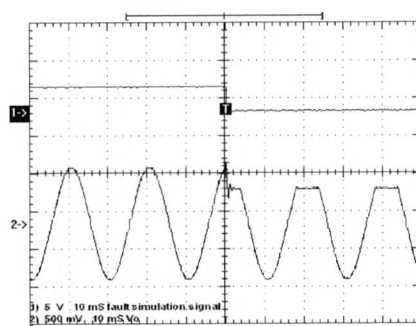

(b) Vo in S1 fault mode

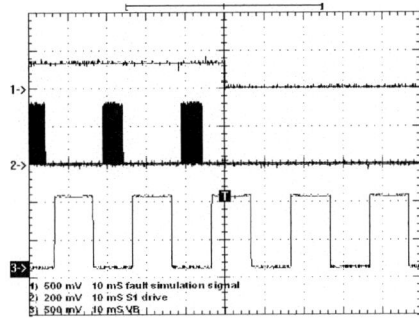

(c) S1 drive and V_B in S1 fault mode

Figure 4. Waveforms in S1 fault mode

In NPC topology, outside 4 switch faults affect V_{AB} and Vo waveforms partly. Inside 4 switch faults will make V_{AB} and Vo disappear completely. Fig.5 shows V_B waveforms in S5 and S6 fault modes. In experiment application, S5 fault makes V_B become a voltage near to Vdc/2 when Vgiven<0. When S6 fault occurs, V_B falls

from Vdc to a voltage near to zero. Because of circuit symmetry, S8 fault features are similar to S5's. S7's are similar to S6's. The only difference lies in the area where fault features show. But switch fault in first bridge has no effect to V_B as Fig.4(c) shows. On the other hand, we can judge area where fault characters occur to point out S1, S2, S7, S8 faults or S3, S4, S5, S6 faults.

Through each switch fault simulation and analysis, V_{AB}, Vo and V_B should be looked as main detected parameters. We will set about fault diagnosis on the basic of detected parameter analysis and transformation.

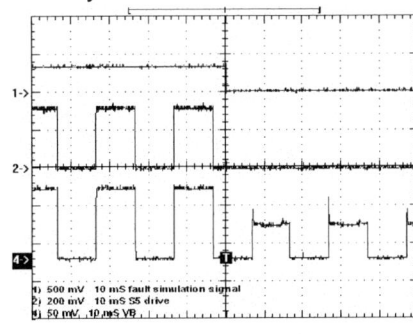

(a) V_B in S5 fault mode

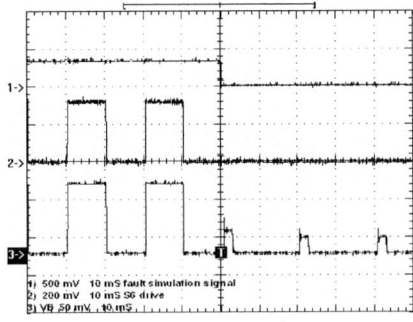

(b) V_B in S6 fault mode

Figure 5. V_B waveforms in S5 and S6 fault modes

B. Fault Diagnosis

Fault diagnosis includes these contents:

- V_B transformation;
- Vo detection;
- Region judgement;
- Fault report.

According to V_B fault feature when switch in second bridge, V_B is transformed in the circuit as Fig.6 shows. It compares V_B with a direct-current voltage. Voltage value is relative with Vdc and Vdc/2. We utilize outer detection interrupt of TMS320LF2407A to enter XINT1CR (XINT2CR) interrupt. If faulty switch lies in second bridge, program will enter into interrupt and ensure the faulty switch by judge fault area.

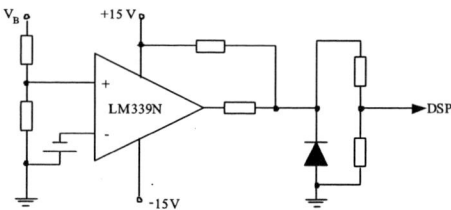

Figure 6. V_B transformation circuit

If there is no faulty switch in 2nd bridge, program continues to detect Vo. As soon as detecting Vo=0 and analyzing the fault area, it can conclude S2 or S3 fault and send relevant fault report signal. During S1 and S4 faults diagnosis, we should consider duty ratio of inverter. As Fig.7 shows, we suppose one triangle carrier peak value is 1. Double carriers' peak value is 2. Inverter duty ratio is D. So modulation wave can be expressed as:

$$f(t) = 2D \sin \omega t = 2D \sin 314 t \qquad (4)$$

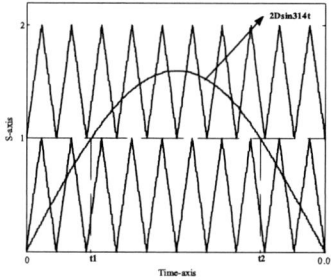

Figure 7. Modulation wave and double carrier wave

t1, t2 are two moments when modulation wave is equal to triangle carrier peak value 1. T is inverter AC output period.

$$2D \sin 314 t = 1 \qquad (5)$$

$$t1 = \frac{\arcsin \dfrac{1}{2D}}{314} \qquad (6)$$

$$t2 = \frac{T}{2} - t1 = 0.01 - \frac{\arcsin \dfrac{1}{2D}}{314} \qquad (7)$$

The region where each fault characters show is presented as Table III. It is a key fault character to point out faulty switch. We can detect modulation wave and the region with digital signal processor TMS320LF2407A to judge S1 and S4 faults. In switch fault diagnosis, the first step is to judge faulty bridge; then it detects the faulty region; at last it will ensure the faulty switch and send fault report signal. The first bridge fault diagnosis time is longer a little than the second's.

TABLE III. FAULTY SWITCH AND REGION

Faulty Switch	Faulty region
S1	kT+ t1~kT+ t2
S2	kT~kT+T/2
S3	kT+T/2 ~ (k+1)T
S4	kT+T/2+ t1~kT+T/2+ t2
S5	kT+T/2~ (k+1)T
S6	kT+T/2~ (k+1)T
S7	kT~kT+ T/2
S8	kT~kT+ T/2

C. Experiment Results

This diagnosis utilizes TMS320LF2407A to realize on-line fault detection. Some experiment waveforms are showed as Fig.8. In Fig.8(a), there are fault simulation signal, S2 fault report signal, V_B and Vo respectively. Vo damped speed affects fault diagnosis time. Vo damped speed is decided by the instantaneous parameter at fault moment and element value of inverter. We can find S2 fault has no effect to V_B. In Fig.8(b), waveforms are fault simulation signal, S5 fault report signal, XINT1 signal and V_B from top to down. If S5 fault occurs, V_B will vary from Vdc to Vdc/2. We compare V_B with a dc voltage between Vdc and Vdc/2, and then connect the comparison output to XINT1 pin of TMS320LF2407A. As soom as DSP detects the jump, it will enter interrupt to detect the region. If the detected information lies in Vgiven<0, it will send S5 fault report signal just like Fig.8(b) shows.

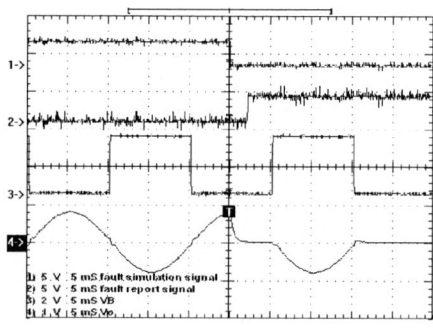

(a) S2 fault diagnosis

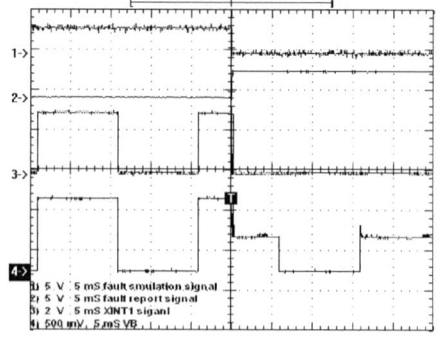

(b) S5 fault diagnosis

Figure 8. Fault diagnosis waveforms

C. Tolerant control

When faulty switch is fixed, it sends fault report signal to drive a SCR to continue inverter operation [7]. The SCR substitutes the faulty switch. Fig.9 shows fault tolerant control circuit and waveforms. Fault diagnosis time directly affects tolerant control result. We can decrease fault diagnosis time through making diagnosis program simpler and adopting higher-speed digital signal processor.

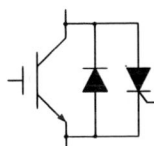

(a) Tolerant circuit

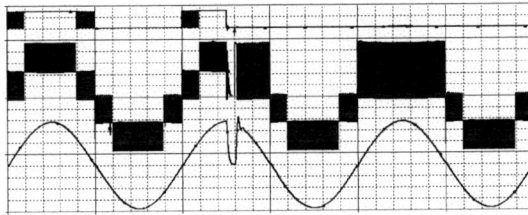

(b) Tolerant control waveform of S2 fault

Figure 9. Fault tolerant control

IV. CONCLUSION

With the development of multilevel inverter, system performance improves largely. But system reliability becomes an important question. Paper aims at diagnosing single phase NPC multilevel inverter single-switching fault of losing drive pulse fault and open-circuit fault. At first, paper presents an optimized drive mode. Secondly, paper proposes a new and effective fault diagnosis method through fault analysis and simulation. Then paper shows experiment waveforms to validate diagnosis theory. At last, a tolerant control circuit is proposed to complete inverter operation with fault.

Acknowledgment

This work was supported by the National Natural Science Foundation of China, NO. 50237020,50407012.

References

[1] A.Nabac, I.Takahashi, and H.Akagi, "A new neutral-point-clamped PWM inverter", IEEE *Trans*. On Industry Applications, Vol. 17, No. 5, p. 518-523, 1981.

[2] H.Stemmler, "Power electonics in electric traction application", Industrial Electronics, Control, and Instrumentation, 1993. Proceedings of the IECON '93, International Conference on 15-19 Vol.2, p. 707-713,1993

[3] T.A.Meynard and H.Foch, "Multi-level conversion: High voltage choppers and voltage-source inverters", Proceedings of IEEE PESC'92, Toledo.Spain, June 1992, p. 397-403

[4] S.Ogasawara and H.Akagi, "Analysis of variation of neutral point potential in neutral-point-clamped voltage source PWM inverters", 1993 IEEE-IAS Annu. meeting, p. 965-970

[5] G.sinha, C.hochgraf, R.H.Lasseter, D.M.Divan, T.A.Lipo, "Fault protection in a multilevel inverter implementation of a static condenser", Proceedings of the IEEE Industry Applications Society Conference, vol. 3, p. 2557-2564, Orlando Florida, USA, 1995

[6] B.francois, J.P., Hautier, "Design of a fault tolerant control system for a N.P.C multilevel inverter", Proceedings of the IEEE International Symposium on Industrial Electronics, vol.4, p. 1075-1080, Aquila, Italy, 2002

[7] Xiaomin. Kou, "Fault Tolerant Design For Multilevel Inverters", Doctor Degree Dissertation of Philosophy in Engineering, University of Wisconsin-Milwaukee, December, 2003

[8] Ho-In Son, Tae-Jin Kin, Dae-Wook Kang, Dong-seok Hyun, "Fault diagnosis and neutral point voltage control when the 3-level inverter faults occur", 2004 35th Annual IEEE Power Electronics Specialists Conference, p. 4558-4563, Aachen, Germany,2004

2006 5th International Power Electronics and Motion Control Conference

Microcontroller-Based Single Phase Inverter Using a New Switching Strategy

K. Meghriche*, O. Mansouri** and A. Cherifi***
University of Versailles Saint-Quentin en Yvelines (UVSQ)
Versailles Laboratory of Systems Engineering (LISV),
Integrated Systems Research Group (GRIS), IUT Mantes-en-Yvelines
7, rue Jean Hoët 78200, Mantes-La-Jolie, FRANCE
*meghrich@lisv.uvsq.fr, **mansouri@lisv.uvsq.fr, ***abderrezzak.cherifi@.uvsq.fr

Abstract—In this paper, we present a single-phase inverter using a new switching strategy based on the use of pre-calculated switching angles. A microcontroller system is used to control the inverter switching process. A passive filter using polarized filtering capacitors is used to obtain substantial reduction of the harmonic rate in the main output voltage source. Both simulation and experimental results are given verifying substantial reduction of total harmonic distortion.

Keywords-inverter; harmonics; microcontroller; PFC

I. INTRODUCTION

The performance of power systems is best enhanced using single or three phase power supplied actuators rather than dc machines [1]–[2]. In this case, the inverter switches are operated in the on-off state thus leading to harmonic distortion. In mechatronic systems, several techniques were developed in order to reduce harmonic currents and voltages [3]–[10]. Some benefits of harmonic reduction are decrease of eddy currents and hysterisis losses, and increase of the life time of the machine winding insulators.

Pre-calculated switching angles developed in [2], are used so as to minimize the harmonic distortion and improve power factor correction (PFC).

In this paper, we present a microcontroller based single phase inverter using a new switching strategy. Compared to other systems as in [11]–[15], the proposed approach allows a simpler inverter control scheme, and a reduced total harmonic distortion (THD) rate. Moreover, this system allows the use of polarized filtering capacitors and can also be used to compensate four-quadrant active and reactive power.

This paper is organized as follows: in the next section, we present the system background theory and analysis. Section III describes the microcontroller-based inverter. The obtained results are given in section IV and we draw conclusions in section V.

II. BACKGROUND AND ANALYSIS

In [2], we have proposed a novel configuration of a single-phase static PFC inverter. An LC circuit is used to filter the inverter output. It has been shown that this system, using pre-calculated switching angles strategy, exhibits substantial reduction of harmonic distortion, while keeping small components values.

Unfortunately, the system of [2] requires the use of non polarized capacitors, thus impairing design and implementation issues. To achieve single-phase reactive power compensation while using polarized small valued capacitors, a novel scheme of the single phase structure is proposed.

Figure 1 shows the new structure of the single-phase static full-bridge PFC inverter, with E being the dc input voltage. R represents the internal inductors resistance. The semiconductor switches Q_i and Q'_i (i=1,2) are operated in complementary states. V_{C1} and V_{C2} are the inverter filtered output voltages taken across capacitors C1 and C2 respectively and $Vout = V_{C1} - V_{C2}$ is the ac output voltage obtained via a dual LC filter.

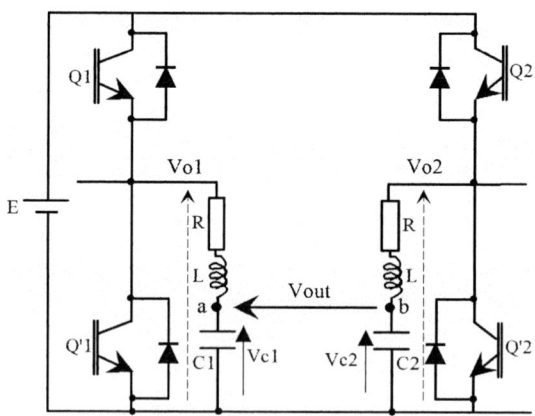

Figure 1. Single phase static PFC inverter

In terms of the dc input voltage E, the unfiltered inverter output voltages V_{O1} and V_{O2} are desired to be expressed as

$$V_{O1} = \frac{E}{2}\left(1 + \cos \omega t\right) \qquad (1)$$

$$V_{O2} = \frac{E}{2}\left(1 - \cos \omega t\right) \qquad (2)$$

for which, the harmonic coefficients are given by (3).

1-4244-0448-7/06/$25.00 ©2006 IEEE

$$a_k\left(V_{O1}\right)=\frac{2E}{k\pi}\sum_{i=1}^{N_\alpha}\sin k\alpha_i\left(-1\right)^{i+1} \qquad (3)$$

where

a_k is the kth harmonic amplitude

N_α is the number of switching angles α_i per half period

A. Precalculated Switching Angles

The objective is to directly determine the switching angles α_i ($Vo1$) so as to best possible match the inverter output V_{O1} and the reference input ac voltage. A perfect matching is achieved only when an infinite number of their harmonics is considered. In practice, the number of harmonics N that can be identical is finite. This number, to be maximized, depends on the number of switching times per period.

Dividing (3) by the dc input voltage E, we obtain the ratio r_k given by (4).

$$r_k=\frac{a_k\left(V_{O1}\right)}{E}=\frac{2}{k\pi}\sum_{i=1}^{N_\alpha}\sin k\alpha_i\left(-1\right)^{i+1} \qquad (4)$$

Using the simplex method [16], we determine the switching angles α_i, from (4) that satisfy the set of equations given by (5).

$$\begin{cases} r_1=a \\ r_3=0 \\ r_5=0 \\ r_7=0 \\ r_9=0 \\ r_{11}=0 \\ r_{13}=0 \end{cases} \qquad (5)$$

In theory, (1) and (2) can be represented as an infinite sum of harmonic components as given by (6) and (7) respectively.

$$V_{O1}=\frac{a_0}{2}+\sum_{k=1}^{\infty}a_k\cos k\alpha \qquad (6)$$

$$V_{O1}=\frac{a_0}{2}+\sum_{k=1}^{\infty}a_k\cos k\left(\alpha+\pi\right) \qquad (7)$$

where $a_0=E=2V_{O1(mean)}$, a_1 is the amplitude of the fundamental and a_k represents the amplitude of the kth harmonic component $\left(k=2,\infty\right)$.

The solution of (5) gives the switching angles α_i that allow us to determine V_{O1}. The other inverter output V_{O2} is obtained by phase shifting V_{O1} with π as illustrated in figure 2.

Considering the inverter direct output fundamental, the LC filter transfer function is given by (8).

$$\overline{T}=\frac{V_{C1}}{V_{O1}}=\frac{1}{1-LC\omega^2+jRC\omega} \qquad (8)$$

From (8), one can notice that for $\omega=0$, $\overline{T}=1$, meaning that the mean value (dc part) of the input voltage is not altered by the filter.

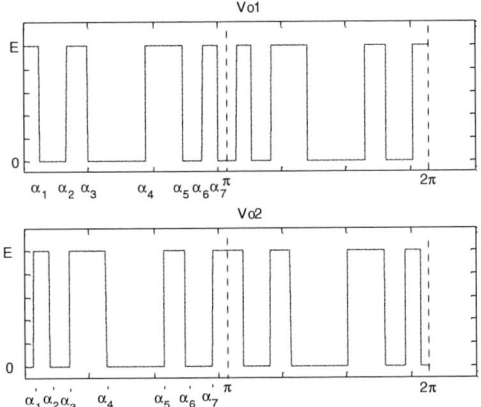

Figure 2. Inverter unfiltered output representation for $N_\alpha=7$

Letting

$$\begin{aligned} x&=\omega\sqrt{LC} \\ y&=R\sqrt{\frac{C}{L}} \end{aligned} \qquad (9)$$

we get,

$$\overline{T}=\frac{V_{C1}}{V_{O1}}=\frac{1}{1-x^2+jxy} \qquad (10)$$

For a given harmonic component k, the LC filter transfer function is obtained by replacing ω with $k\omega$ as given by (11).

$$\overline{T}_k=\left(\frac{V_{C1}}{V_{O1}}\right)_k=\frac{1}{1-LC\left(k\omega\right)^2+jRCk\omega} \qquad (11)$$

that can be rewritten as

$$\overline{T}_k=\left(\frac{V_{C1}}{V_{O1}}\right)_k=\frac{1}{1-x^2k^2+jkxy} \qquad (12)$$

Assuming that the filter L and C components are not saturated and using the superposition principle, we obtain the inverter filtered output voltages V_{C1} and V_{C2}, taken across capacitors $C1$ and $C2$ as given by (13) and (14) respectively.

$$V_{C1}=\frac{a_0}{2}+\sum_{k=1}^{N}a_kT_k\cos\left(k\alpha+\varphi_k\right) \qquad (13)$$

$$V_{C2}=\frac{a_0}{2}+\sum_{k=1}^{N}a_kT_k\cos\left[k\left(\alpha+\pi\right)+\varphi_k\right] \qquad (14)$$

with $\alpha=\omega t$

T_K and φ_K are the kth components of the modulus and phase shift of the LC filter transfer function respectively.

Each harmonic term of rank k (inverter output voltage) has a frequency of $k\omega$ and its amplitude equal to $a_k T_k$, where a_k is the amplitude of the kth harmonic of V_{O1}.

The transfer function magnitude and argument (phase) will be given by (16) and (18) respectively.

$$T_k = \frac{1}{\sqrt{\left(1 - LC\omega^2 k^2\right)^2 + (RC)^2 \omega^2 k^2}} \quad (15)$$

$$\left|T_k\right| = T_k = \frac{1}{\sqrt{\left(1 - x^2 k^2\right)^2 + y^2 x^2 k^2}} \quad (16)$$

$$\varphi_k = -\arctan\frac{RC\omega k}{1 - LC\omega^2 k^2} \quad (17)$$

that can be rewritten as

$$\varphi_k = -\arctan\frac{R\sqrt{\dfrac{C}{L}}kx}{1 - x^2 k^2} \quad (18)$$

with x and y being given by (9).

As illustrated in Fig. 3, the analysis of (16) shows that for $x = \omega\sqrt{LC} = 2\pi f \sqrt{LC} = 0$, T_1 $(T_{k=1})$ is equal to 1, meaning that the fundamental is not altered by the filter transfer function.

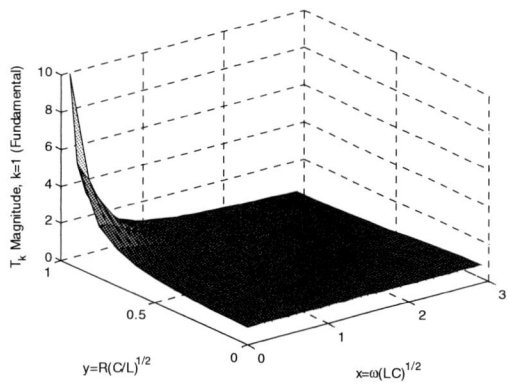

Figure 3. Behavior of the filter transfer function magnitude for the fundamental

Moreover, the maximum value of T_k is obtained when $\dfrac{dT}{dx} = 0$, in which case, $\omega_{max} = \dfrac{\sqrt{2}}{RC}\sqrt{1 - \dfrac{y^2}{2}}$, and T_{max} is given by (19).

$$T_{max} = \frac{\dfrac{1}{y^2}}{\sqrt{\dfrac{1}{y^2} - \dfrac{1}{4}}} \quad (19)$$

ω_{max} exists if and only if $y < \sqrt{2}$

If $y > \sqrt{2}$, the filter transfer function will exhibit a damped behavior.

If $y < \sqrt{2}$, the filter transfer function will show a peak value then decreases towards zero. This means that besides the fundamental, the harmonics may also be amplified leading to undesirable situation.

The analysis of (16) taking into consideration (9) and (19), shows that the optimum transfer function will be obtained when $x = 1$ and $y = \sqrt{2}$, for which figure 4 illustrates the corresponding variations for the L and C filter components in terms of the frequency.

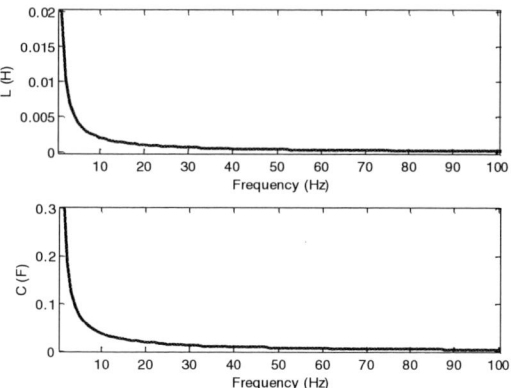

Figure 4. LC variations with respect to frequency

B. Total Harmonic Distortion Rate

Combining (1), (2) and (4), we get

$$V_{O1} = \frac{E}{2}\left(1 + \cos\omega t\right) + a_5 \cos 5\omega t + a_6 \cos 6\omega t + \ldots \quad (20)$$

$$V_{O2} = \frac{E}{2}\left(1 - \cos\omega t\right) - a_5 \cos 5\omega t + a_6 \cos 6\omega t - \ldots \quad (21)$$

Taking into consideration the filter transfer function, we get the expressions (22) and (23) for V_{C1} and V_{C2} respectively.

$$V_{C1} = \frac{E}{2}\Big[1 + a_1.T_1.\cos\left(\alpha + \varphi_1\right) + a_{15}.T_{15}.\cos\left(15\alpha + \varphi_{15}\right) + a_{17}.T_{17}.\cos\left(17\alpha + \varphi_{17}\right) + \ldots\Big] \quad (22)$$

$$V_{C2} = \frac{E}{2}\Big[1 - a_1.T_1.\cos\left(\alpha + \varphi_1\right) - a_{15}.T_{15}.\cos\left(15\beta + \varphi_{15}\right) - a_{17}.T_{17}.\cos\left(17\alpha + \varphi_{17}\right) - \ldots\Big] \quad (23)$$

with $\alpha = \omega t$.

Using (22) and (23), we get the inverter filtered output voltage expression as given by (24).

$$V_{out} = E.\left[a_1 T_1 \cos\left(\alpha + \varphi_1\right) + \sum_{\substack{k=15 \\ k=2n+1}}^{\infty} a_k T_k \cos\left(k\alpha + \varphi_k\right)\right] \quad (24)$$

The total harmonic rate is given by (25).

$$H_T = 100 \frac{1}{a_1 . T_1} \sqrt{\sum_{\substack{k=3 \\ k=2n+1}}^{\infty} \left(\frac{a_k T_k}{k} \right)^2} \, \% \qquad (25)$$

In practice, the sum in (24) and (25), is limited to the number N of harmonics considered. In this case, H_T becomes H_R and is given by (26).

$$H_R = 100 \frac{1}{a_1 . T_1} \sqrt{\sum_{\substack{k=3 \\ k=2n+1}}^{N} \left(\frac{a_k T_k}{k} \right)^2} \, \% \qquad (26)$$

III. RESULTS AND DISCUSSION

A 16-bit microcontroller is used to control the switching process. A circuit block diagram is shown in Fig. 5.

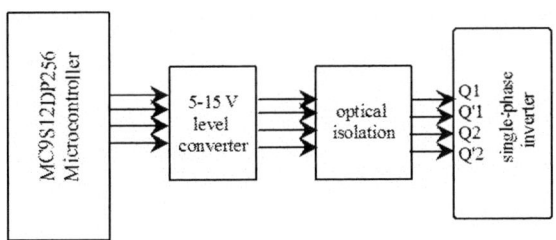

Figure 5. Microcontroller based PFC control system

Figure 6 shows the microcontroller experimental set up. It consists of an MC9S12DP256 16-bit microcontroller card, the inverter and an induction motor.

Figure 6. Microcontroller based experimental setup

For each different value a of r_1 in (5), corresponds a different solution of the nonlinear system of equation for the switching angles α_i. We have carried out simulations

for 35 different values of r_1 leading to 35 distinct families of switching angles as illustrated in Fig. 7.

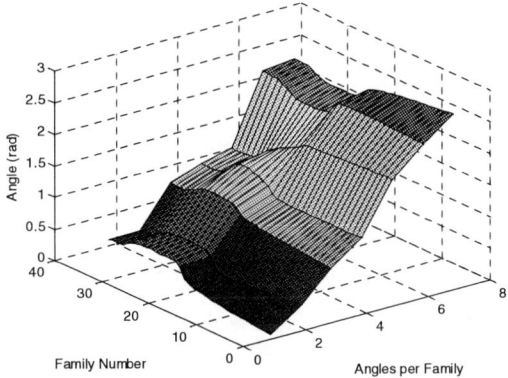

Figure 7. Families of switching angles α_i

Figure 8 represents the THD rate per switching angles family corresponding to different values of r_1.

The analysis of Fig. 8 shows that family 24 (r_1=0.54), given in Table I, and family 1 (r_1=0.08) exhibit the minimum and maximum THD rates respectively.

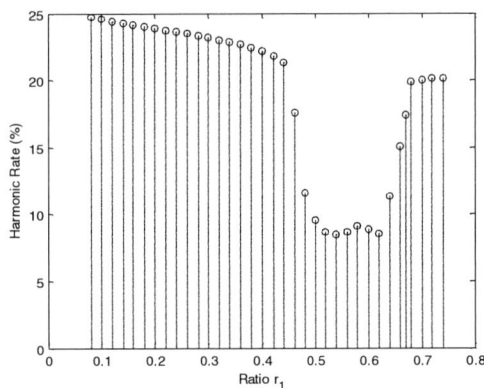

Figure 8. Total harmonic rate per switching angles family ratio r_1

TABLE I.
MINIMUM THD RATE SWITCHING ANGLES

Optimal angles family #24 (r_1=0.54)		
	radians	degrees
α_1	.5242298452415577	30.0362
α_2	.5715928478038887	32.7499
α_3	1.149189723838989	65.8437
α_4	1.41548576602197	81.1014
α_5	1.660415377441947	95.1348
α_6	2.165774550803808	124.0897
α_7	2.298212023090896	131.6778

We have used family 24 to control the inverter semiconductor switches. Taking into consideration the LC filter transfer function given in (16), we have used the parameters given in Table II, to determine the THD rate within the inverter filtered ac output voltage.

TABLE II.
SIMULATION PARAMETERS

Frequency (f)	R	$x = \omega\sqrt{LC}$	$y = R\sqrt{\dfrac{C}{L}}$
50 Hz	0.2 Ω	0 to 1	0.05 to 2.9
		incremental step=0.1	

In Table II, R represents the internal inductor resistance.

The THD rate for the inverter filtered output voltage with respect to x and y variations, is shown in Fig. 9-a and 9-b corresponding to families 24 and 1 respectively.

For x near zero, family 24 THD rate is 4 times less than family 1 THD rate.

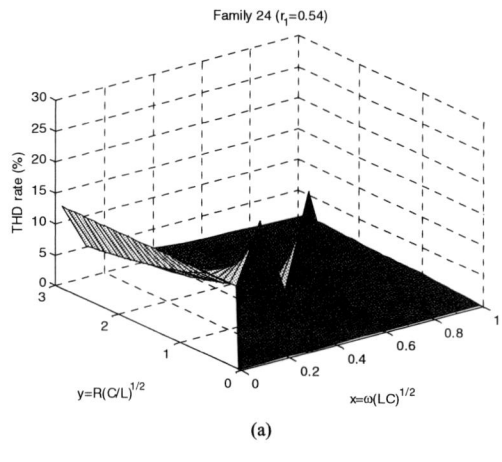

(a)

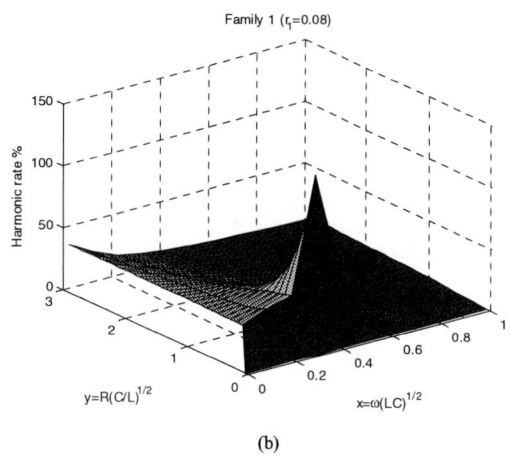

(b)

Figure 9. Total harmonic distortion rate for filtered inverter output

Comparing figures 8 and 9, one can see that for the dc component (x near 0), the obtained THD rates for both the unfiltered and filtered inverter outputs are similar.

For the ac component, apart y being near zero, the THD rate is inversely proportional to x and y.

The filter transfer function contribution in eliminating the inverter output voltage harmonics, is optimized by choosing appropriate values for x and y.

Figure 10 shows the inverter filtered output for x=0.6 and y=1.41.

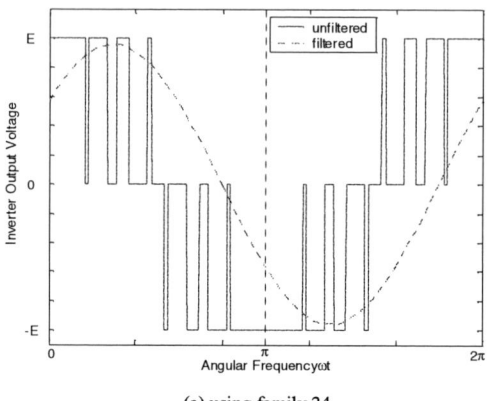

(a) using family 24

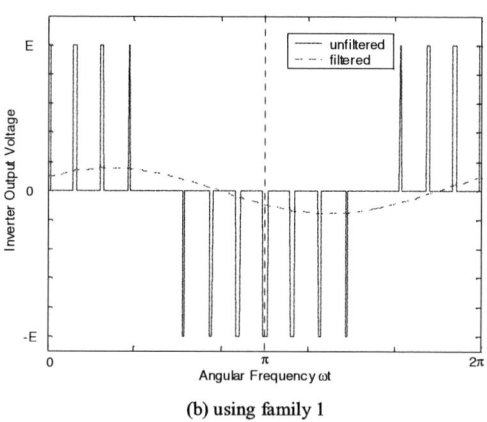

(b) using family 1

Figure 10. Inverter output ac voltage

It is worth to mention that for the selected values of x and y, the THD rate for both families is substantially reduced leading to a sinusoidal shaped output voltage.

IV. CONCLUSION

We have presented a microcontroller-based static PFC inverter using a new switching strategy.
Simulation results show that the THD rate decreases with the total number of switching angles per period.

We have tested the method on an inverter driving an induction motor. Experimental results exhibit good matching with the theoretical values.

Due to substantial reduction of harmonic distortion, the proposed scheme succeeds to balance two-quadrant reactive power using small component values, in particular polarized capacitors, thus providing more reliability and increase of system components life time.

REFERENCES

[1] K. Meghriche, F. Chikhi, A. Cherifi, "A New Switching Angle Determination Method for Three Leg Inverter", *Proc. of IEEE Mechatronics & Robotics 2004*, MechRob-2004, Aachen, Germany, pp. 378–382, 13-15 September 2004.

[2] K. Meghriche, O. Mansouri, and A. Cherifi, "On the use of pre-calculated switching angles to design a new single phase static PFC inverter," *Proc. of the 31st IEEE IECON'05*, pp. 906–911, Raleigh North-Carolina, 6-10 November 2005.

[3] T. Key and Jih-Sheng Lai, "Costs and benefits of harmonic current reduction for switch-mode power supplies in a commercial office building," *IEEE Trans. on Industry Applications*, vol. 32, no. 5, pp. 1017–1025, September/October 1996.

[4] M. Izhar, C.M. Hadzer, S. Masri, and S. Idris, "A study of the fundamental principles to power system harmonic," *Proceedings of the National Power Engineering Conference, PECon 2003*, pp. 225–232, 15-16 December 2003.

[5] F.Z. Peng, H. Akagi and A. Nabae, "A study of active power filters using quad-series voltage-source pwm converters for harmonic compensation," *IEEE Trans. Power Electronics*, vol. 5, no. 1, January 1990, pp. 9-15.

[6] O. García, M. D. Martínez-Avial, José A. Cobos, J. Uceda, J. González, and José A. Navas, "Harmonic reducer converter," *IEEE Trans. on Industrial Electronics*, vol. 50, no. 2, pp. 322–327, April 2003.

[7] S. Kim, and Prasad N. Enjeti, "A modular single-phase power-factor-correction scheme with a harmonic filtering function," *IEEE Trans. on Industrial Electronics*, vol. 50, no. 2, pp. 328–335, April 2003.

[8] Y.J. Song, and P.N. Enjeti, "A high frequency link direct dc-ac converter for residential fuel cell power systems," *in Proceedings of the 35th Annual IEEE Power Electronics Specialists*, 20-25 June 2004 Aachen, Germany, vol. 6, pp. 4755-4761.

[9] M. Ghone, M. Schubert, and John R. Wagner, "Development of a Mechatronics laboratory–eliminating barriers to manufacturing instrumentation and control," *IEEE Trans. on Industrial Electronics*, vol. 50, no. 2, pp. 394–397, April 2003.

[10] S. Paul, S.K. Basu, and R. Mandal, "A microcomputer controlled static VAr compensator for power systems laboratory experiments," *IEEE Trans. on Power Systems*, vol. 7, no. 1, pp. 371–376, February 1992.

[11] R. Mandal, S.K. Basu, A. Kar and S. P. Chowdhury, "A microcomputer-based power factor controller," *IEEE Tran. On Industrial Electronics*, vol. 41, no.3, pp. 361–371, June 1994.

[12] K. De Gussemé, D.M. Van de Sype, and J.A. Melkebeek, "Design issues for digital control of boost power factor correction converters," *Proceedings of the 2002 IEEE International Symposium on Industrial Electronics, ISIE 2002*, vol. 3, pp. 731–736, 26-29 May 2002.

[13] Z. Pan, F. Z. Peng, and S. Wang, "Power factor correction using a series active filter," *IEEE Trans. on Power Electronics*, vol. 20, no. 1, pp. 148–153, January 2005.

[14] K. De Gussemé, D.M. Van de Sype, A.P.M. Van den Bossche, and J.A. Melkebeek, "Digitally controlled boost power-factor-correction converters operating in both continuous and discontinuous conduction mode," *IEEE Trans. on Industrial Electronics*, vol. 52, no. 1, pp. 88–97, February 2005.

[15] Saul I. Gass, *Linear Programming Methods and Applications*, 5th edition, ISBN: 048643284X, 2003.

2006 5th International Power Electronics and Motion Control Conference

Study of Stability Regions in Parallel Connected Boost Converters

Yuehui Huang and Chi K. Tse
Department of Electronic and Information Engineering
The Hong Kong Polytechnic University, Kowloon, Hong Kong
Email: yuehui.huang@polyu.edu.hk, encktse@polyu.edu.hk

Abstract—This paper describes the coexisting attractors of parallel connected boost switching converters under a master-slave current sharing scheme. We present the basins of attraction of desired and undesired attractors, which provide design information on the conditions for hot-swap operations. The system employs a typical proportional-integral (PI) controller for regulation. It is shown that the system will converge to different attractors for different initial conditions with the same control parameters. Simulation results are given to illustrate the phenomenon. This study is relevant to practical design. Specifically, we show that the stability regions obtained from linear methods (i.e., considering only local stability) can be over-optimistic as the global stability regions are found to be more restrictive in the parameter space.

I. INTRODUCTION

Power supplies based on paralleling switching converters offer a number of advantages over a single, high-power, centralized power supply. They enjoy low component stresses, increased reliability, ease of maintenance and repair, improved thermal management, etc. [1], [2]. Paralleling of standardized converters is an approach used widely in distributed power systems for both front-end and load converters. Since current sharing has to be maintained among the paralleled converters, some form of control has to be used to equalize the individual currents in the converters. One widely used method for balancing currents is the *master-slave current sharing* method [3], [4].

The system under study in this paper is a parallel connected system of two boost converters. Under the master-slave scheme, one of the converters is the master and the other is the slave. Both of the converters are under peak-current-mode (PCM) control. The master consists of a typical proportional-integral (PI) control, to regulate the output voltage, and a comparator, to compare the feedback current with the reference current. The slave basically sets its current to equal that of the master via an active loop involving comparison of the currents of the two converters, as shown in Fig. 1. Previous studies of such systems have focused on pure proportional control, which is not normally used in practice [5]. The use of PI control introduces a low-pass characteristic to the feedback loop, thereby suppressing high-frequency components in the feedback signal. The resulting bifurcation and stability behavior is therefore different. In this paper we will consider practical PI control in our simulation study.

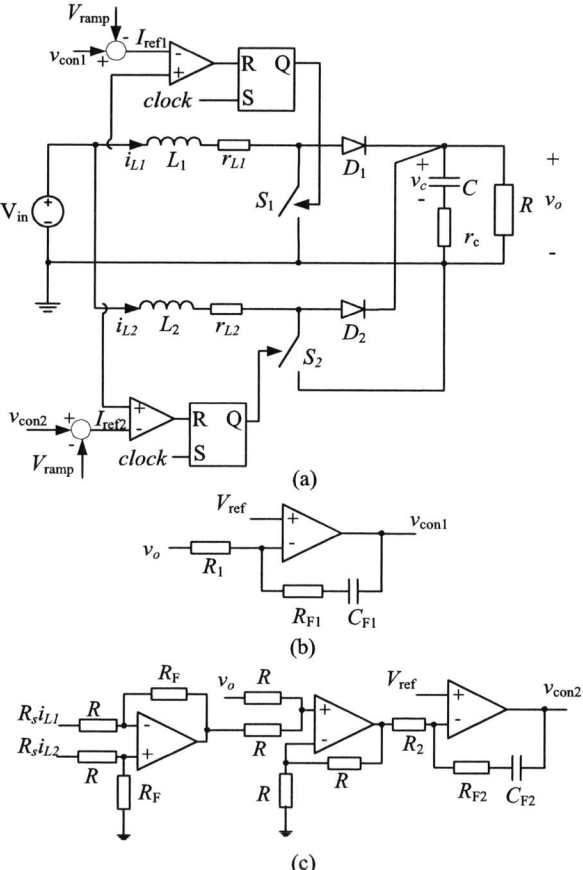

Fig. 1. Paralleled boost converters under master-slave current sharing.

Basically we find that for parallel connected boost converters, the desired operating orbit is not always reached from all initial conditions, even though the orbit has been found locally stable (e.g., from a linearized model). Depending on the initial state, the system may converge to different attractors, which can be a stable period-1 orbit, quasi-periodic orbit or chaotic orbit. In the paper, we examine two parallel connected boost converters with PCM control under master-slave current sharing. And it is easy to extend to N-paralleled converters. We show that different initial conditions may lead to different

1-4244-0448-7/06/$25.00 ©2006 IEEE   580

steady states. Thus, linear stability analysis methods, which basically evaluate the convergence of the system trajectory to the desired steady state starting from a nearby point, can be misleading.

In this paper, we report the phenomenon, present specific basins of attraction for the different attractors, and derive the critical values of control parameters for which the system loses stability of its expected operation. We generally observe that stability boundaries obtained from equivalent linear methods are over-optimistic, in that the system is actually more prone to instability. Thus, reliable stability information can only be obtained with the basin of attractions duly taken into consideration.

II. SYSTEM DESCRIPTION AND OPERATION

Figure 1 (a) shows two boost converters connected in parallel. In this circuit, S_1 and S_2 are switches, which are under peak-current-mode (PCM) control. In the PCM, The switch is set to be on by the latch at the beginning of each cycle. Then if the feedback current reaches the reference current I_{ref}, the switch will be turned off. The reference current is decided by the output of voltage regulator and the ramp compensation. The compensatory ramp signal is given by

$$V_{\text{ramp}} = V_L + (V_U - V_L) \left(\frac{t}{T_s} \bmod 1 \right) \quad (1)$$

where V_L and V_U are the lower and upper thresholds of the ramp, respectively, and T_s is the switching period. The role of ramp compensation is to stabilize the system when duty cycle exceeds 0.5 in peak current-mode-control.

The control signals v_{con1} and v_{con2} are derived from the voltage compensator, as shown in Figs. 1 (b) and (c). Here the compensator is a PI controller, e.g.,

$$\frac{V_{\text{con1}}(s)}{E(s)} = -K_p \left(1 + \frac{1}{\tau_{F1} s} \right) \quad (2)$$

where $V_{\text{con1}}(s)$ and $E(s)$ are the Laplace transforms of $v_{\text{con1}}(t)$ and $e(t)$; $e(t)$ is the error between reference and output; K_p and τ_{F1} are the control parameters. With respect to the slave, extra current sharing signal is included. We can likewise write the equation.

We assume that the converter operates in continuous conduction mode (CCM) and diodes D_1 and D_2 are always in complementary state to S_1 and S_2. Consequently, the state equations of the converter stage of Fig. 1 are

$$\begin{cases} \dot{i}_{L1} = \frac{1}{L1}[V_{in} - r_{L1}i_{L1} - (1 - q_1(t))v_o] \\ \dot{i}_{L2} = \frac{1}{L2}[V_{in} - r_{L2}i_{L2} - (1 - q_2(t))v_o] \\ \dot{v}_c = \frac{1}{C}[(1 - q_1(t))i_{L1} + (1 - q_2(t))i_{L2} - \frac{v_o}{R}] \end{cases} \quad (3)$$

where v_o can be written as

$$\begin{aligned} v_o &= v_c + r_c i_c \\ &= v_c + r_c[(1 - q_1(t))i_{L1} + (1 - q_2(t))i_{L2} - \frac{v_o}{R}] \quad (4) \end{aligned}$$

and $q_1(t)$ and $q_2(t)$ are the switching function decided by the output of controllers. They are time varying functions given by

$$q_i(t) = \begin{cases} 1, & \text{if } S_i \text{ is on,} \\ 0, & \text{if } S_i \text{ is off.} \end{cases} \quad (5)$$

Depending upon the feedback circuit in Figs. 1(b) and (c), we have

$$\frac{dv_{\text{con1}}}{dt} = -K_1 \frac{dv_o}{dt} - \frac{K_1}{\tau_{F1}} v_o + \frac{K_1}{\tau_{F1}} V_{\text{ref}} \quad (6)$$

$$\frac{dv_{\text{con2}}}{dt} = -K_2 \frac{dv_o}{dt} - \frac{K_2}{\tau_{F2}} v_o + K_2 K_i \left(\frac{di_{L1}}{dt} - \frac{di_{L2}}{dt} \right)$$

$$+ \frac{K_2 K_i}{\tau_{F2}} (i_{L1} - i_{L2}) + \frac{K_2}{\tau_{F2}} V_{\text{ref}} \quad (7)$$

where K_1 and K_2 are the proportional coefficients, τ_{F1} and τ_{F2} are the integral coefficients, K_i is the current sharing coefficient, and V_{ref} is the reference voltage (expected output voltage). In circuit terms, $K_1 = R_{F1}/R_1$, $\tau_{F1} = R_{F1}C_{F1}$, $K_2 = R_{F2}/R_2$, $\tau_{F2} = R_{F2}C_{F2}$, $K_i = R_F R_s/R$, where R_s is the current sensing resistance. Equations (6) and (7), together with (3), form the complete set of state equations of the system. It is a fifth order system.

III. BASINS OF ATTRACTION

In this section, we begin our investigation of the basins of attraction of the operation orbits. Our simulations are based on the state equations derived in the foregoing section and hence are exact cycle-by-cycle simulations. We are primarily concerned with the system stability in relation to the initial condition X_0 ($X = [i_{L1}, i_{L2}, v_c]$ refers to the converter state variables), feedback parameters of the PI controller K_1, K_2, τ_{F1}, τ_{F2} and current sharing coefficient K_i. The circuit parameters and component values are listed in Table I.

TABLE I
COMPONENT VALUES USED IN SIMULATIONS

Circuit Components	Values
Switching Period T_s	10 μs
Input Voltage V_{in}	5 V
Reference Voltage V_{ref}	10 V
Ramp Voltage V_L, V_U	0 V, 0.8 V
Inductance L_1, ESR r_{L1}	50 μH, 0.01 Ω
Inductance L_2, ESR r_{L2}	60 μH, 0.1 Ω
Capacitance C, ESR r_c	126 μF, 0 Ω
Load Resistance R	2 Ω
Current sensing Resistance R_s	0.01 Ω

Under the same controller but with different initial conditions, we find that the system will converge to stable period-1 orbit or unstable orbits as well as what we found in paralleled buck converters [6]. Again, there are more than one attractor in paralleled boost converters. The steady-state behavior of the system depends on where it starts [7]. The basins of attraction are therefore important.

In the following, we find the basin boundaries numerically in relation to initial point X_0, and determine how they are affected by the controller parameters K_1, K_2, τ_{F1} and τ_{F2}, as shown in Figs. 2, 3, 4 and 5. Figures 2 and 3 show the basins of attraction for different K_1 and K_2. We first get the boundary of stable and unstable operations in the i_{L1}–i_{L2} plane, and then extend it to a 3-D space by gathering boundaries for different v_{c0}. Figures 2 (a), (b), (c) and (d) are basins of attraction presented on the i_{L1}–i_{L2} plane for different initial v_{c0} with $K_1 = K_2 = 5$. The yellow region is the basin corresponding to the desired operating orbit (stable region), whereas the blue region is the basin corresponding to attractors other than the desired operating orbit (unstable region). Thus, if the system starts from the blue region, it will not converge to the expected operating orbit. Figure 2 (e) shows the interfaces in 3-D space for various X_0 in a cubic box. The space below the interface is the unstable region. Actually, it is clearly displayed in the slices as shown in figs. 2 (a), (b), (c) and (d). Similarly, figs. 3, 4 and 5 show the basins of attraction for different feedback parameters.

Furthermore, we observe that the yellow region diminishes as proportional coefficients K_1, K_2 increase; and vice versa.

For large K_1 and K_2, the yellow region subsides and the desired operating point is almost never stable. For small K_1 and K_2, the blue region subsides and the desired operating point is almost always stable. In practice, K_1 and K_2 determine the response speed of the system. Comparing fig. 2 and fig. 3, we clearly see the limitation on selecting K_1 and K_2 so as to maintain stability for a wider basin of attraction. In addition, there are some effects for different v_{c0}. The farther it is away from the equilibrium orbit (centered around $v_{c0} = 10$V), the smaller the basin is.

Figures 4 and 5 show the basins of attraction for different integral coefficients τ_{F1} and τ_{F2}. Obviously, $1/\tau_{F1}$ and $1/\tau_{F2}$ are the zero point in the PI controller. The general trend of the variation of the basin boundaries is similar to that of Figs. 2 and 3. As $1/\tau_{F1}$ and $1/\tau_{F2}$ increase, the system goes from being globally stable to partially stable, and eventually unstable.

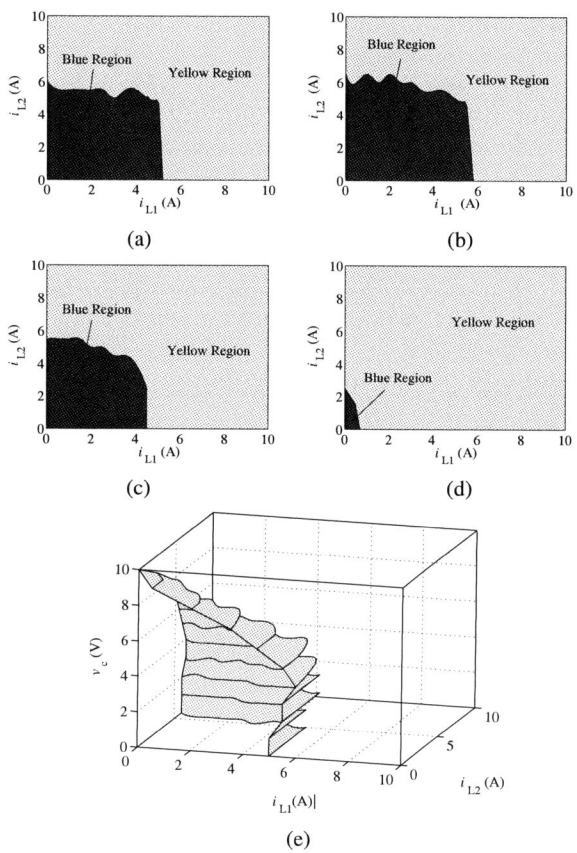

Fig. 2. Basins of attraction for $K_1 = K_2 = 5$, $1/\tau_{F1} = 1/\tau_{F2} = 12000$, $K_i = 1$. Yellow region is the basin of attraction of the desired operating orbit. Blue region is the basin of attraction of attractors other than the desired operating orbit. (a) $v_{c0} = 0$; (b) $v_{c0} = 3$; (c) $v_{c0} = 6$; (d) $v_{c0} = 9$; (e) interface in 3-D space.

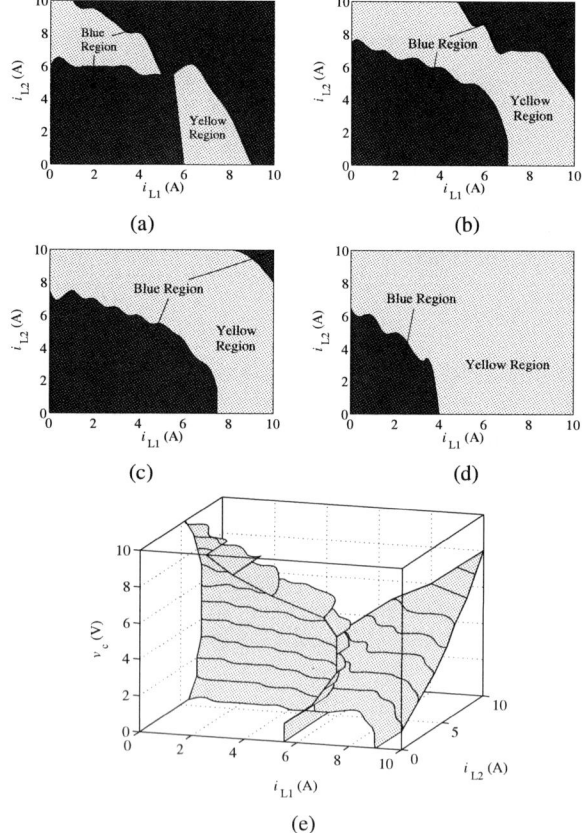

Fig. 3. Basins of attraction for $K_1 = K_2 = 6$, $1/\tau_{F1} = 1/\tau_{F2} = 12000$, $K_i = 1$. Yellow region is the basin of attraction of the desired operating orbit. Blue region is the basin of attraction of attractors other than the desired operating orbit. (a) $v_{c0} = 0$; (b) $v_{c0} = 3$; (c) $v_{c0} = 6$; (d) $v_{c0} = 9$; (e) interface in 3-D space.

IV. CAUTIONS ON STABILITY INFORMATION AND STABILITY BOUNDARIES

From the above results, an important conclusion can be made. The stability of the operating orbit cannot be determined purely from the linear model or any method that tests stability by perturbing near the operating orbit. Stability information can be unreliable since global stability is not generally guaranteed from local stability tests. In general, we can get different stability boundaries for different initial conditions.

The stability boundaries for the parallel connected boost converter system are shown in Figs. 6, 7 and 8, corresponding to two initial points. One is the origin point $X_0 = [0, 0, 0]$, and the other is a point near the equilibrium orbit, e.g., $X_0 = [5.0, 5.1, 10]$. The curve divides the parameter space into stable region (lower) and unstable region (upper). The system works in the normal stable period-1 operation when the feedback parameters are located in the stable region. Otherwise, if the parameters crosses the boundary and enters into the unstable region, the system loses stability. In

Fig. 6 (a), K_1 and K_2 decrease with $1/\tau_{F1}$, $1/\tau_{F2}$ increase. Also, the gap between the two boundaries widens as $1/\tau_{F1}$ and $1/\tau_{F2}$ increase. Within the gap, coexisting attractors exist and stability information may be unreliable. Actually, the coexisting attractors exist in single boost converters when $1/\tau_F$ is large enough, as shown in Fig. 6 (b).

Figure 7 shows the effect of the current sharing parameter K_i. Again, these two boundaries are not overlapped. Coexisting attractors exist when parameters are in the gap. In the figure, when K_i is very large, the system is easy to be unstable. Thus, the two boundaries are very close.The coexisting attractors are not obvious.

Finally, Fig. 8 shows the effects of changing the size of inductors L_1 and L_2. We fix the ratio of L_1 and L_2, and maintain the system in CCM in steady state. From the figure, we clearly observe that the coexisting attractors exist in the whole inductance range.

V. CONCLUSIONS

This paper studies the coexisting attractors in two parallel connected boost converters under master-slave current sharing

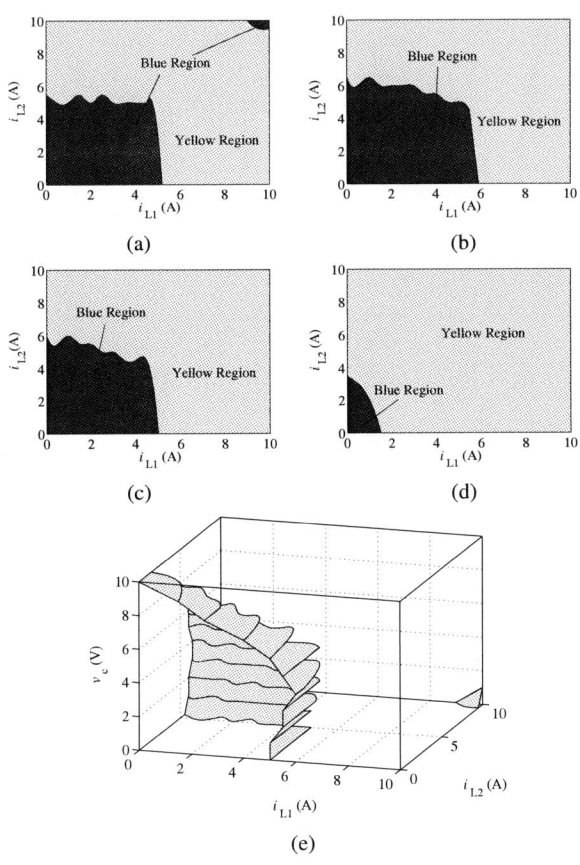

Fig. 4. Basins of attraction for $K_1 = K_2 = 5.5$, $1/\tau_{F1} = 1/\tau_{F2} = 11000$, $K_i = 1$. Yellow region is the basin of attraction of the desired operating orbit. Blue region is the basin of attraction of attractors other than the desired operating orbit. (a) $v_{c0} = 0$; (b) $v_{c0} = 3$; (c) $v_{c0} = 6$; (d) $v_{c0} = 9$; (e) interface in 3-D space.

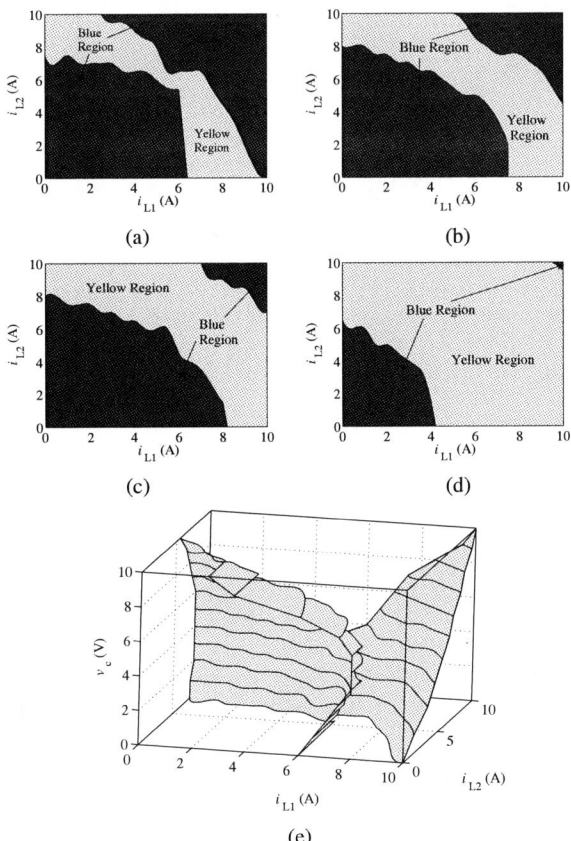

Fig. 5. Basins of attraction for $K_1 = K_2 = 5.5$, $1/\tau_{F1} = 1/\tau_{F2} = 14000$, $K_i = 1$. Yellow region is the basin of attraction of the desired operating orbit. Blue region is the basin of attraction of attractors other than the desired operating orbit. (a) $v_{c0} = 0$; (b) $v_{c0} = 3$; (c) $v_{c0} = 6$; (d) $v_{c0} = 9$; (e) interface in 3-D space.

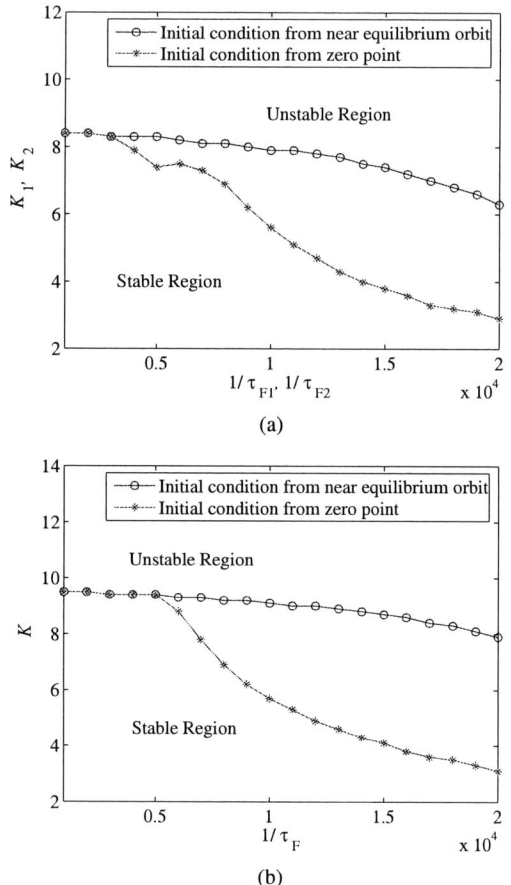

(a)

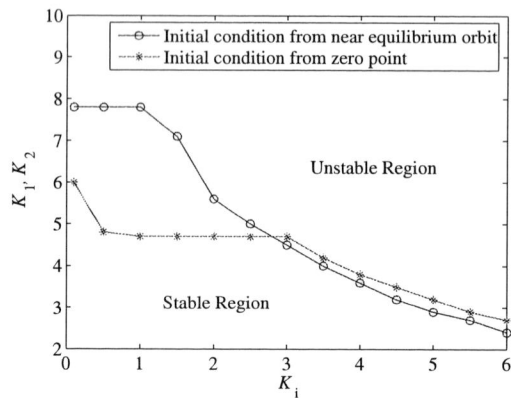

Fig. 7. Stability boundary of feedback parameters K_i versus K_1, K_2 for $1/\tau_{F1} = 1/\tau_{F2} = 12000$.

(b)

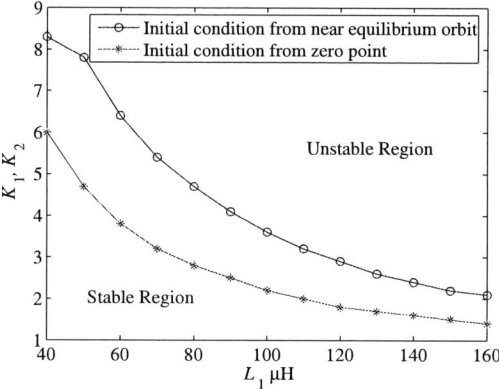

Fig. 6. Stability boundaries of feedback parameters in (a) two paralleled boost converters in $1/\tau_{F1}$, $1/\tau_{F2}$–K_1, K_2 plane for $K_i = 1$; (b) single boost converter in $1/\tau_F$–K plane.

Fig. 8. Stability boundary of feedback parameters K_1, K_2 in relation to L_1 for $1/\tau_{F1} = 1/\tau_{F2} = 12000$, $K_i = 1$.

and peak-current-mode control. The system is either stable or oscillatory depending on the initial condition and the control parameters. The implication of this finding is relevant to practical operation since stability information obtained from linear models or any method that involves perturbation around the operating orbit can be unreliable. Specifically, stability information obtained from linear methods has been shown over-optimistic. Practically, enough margins have to be considered in linear methods. In fact, the basins of attraction of an operating orbit is an important piece of design information, and stability boundaries in parameter space have to be interpreted in conjunction with the initial conditions. Different initial conditions may give rise to different stability boundaries. In this paper, we have reported the phenomenon and illustrated the effects of different parameters by presenting the numerical basins of attraction and specific stability boundaries.

ACKNOWLEDGMENT

This work was supported by Hong Kong Research Grants Council under a CERG project (Ref. PolyU 5237/04E).

REFERENCES

[1] V. J. Thottuvelil and G. C. Verghese, "Analysis and control of paralleled dc/dc converters with current sharing," *IEEE Trans. Power Electron.,* vol. 13, no. 4, pp. 635–644, July 1998.

[2] J. Rajagopalan, K. Xing, Y. Guo, F. C. Lee, and B. Manners, "Modeling and dynamic analysis of paralleled DC/DC converters with master-slave current sharing control," *Proc. IEEE APEC'96*, pp. 678–684, 1996.

[3] Y. Panov, J. Rajagopalan, and F. C. Lee, "Analysis and design of N paralleled DC-DC converters with master-slave current-sharing control," *Proc. IEEE APEC'97*, pp. 436–442, 1997.

[4] K. Siri, C. Q. Lee, and T. F. Wu, "Current distribution control for parallel connected converters: Part I and Part II," *IEEE Trans. Aerospace Electron. Syst.,* vol. 28, no. 3, pp. 829–851, July 1992.

[5] C.K. Tse, *Complex Behavior of Switching Power Converters*, Boca Raton: CRC Press, 2003.

[6] Y. Huang, C. K. Tse, "On the Basins of Attraction of Parallel Connected Buck Switching Converters," *Proc. IEEE ISCAS'06*, to appear.

[7] S. Banerjee, G. C. Verghese, *Nonlinear Phenomena in Power Electronics: Attractors, Bifurcations, Chaos, and Nonlinear Control*, New York: IEEE Press, 2001.

2006 5th International Power Electronics and Motion Control Conference

A Novel Analysis and Design Method for Integrated Magnetics

Zheng Feng, *Student Member, IEEE,* Weihao Hu, Pei Yun-qing, *Member, IEEE,*
Yang Xu, *Member, IEEE* and Wang Zhao-an, *Senior Member, IEEE*
Industrial Automation Department
Xi'an Jiaotong University, Xi'an, China

Abstract—**Integrated magnetics (IM) has been widely used in power electronics systems. A method to design IM has been proposed systematically. The rule to decide whether three and more discrete magnetic components can be integrated together is also proposed in this paper. A half bridge with current doubler converter and a single IM component are taken as an example to demonstrate the practical application of this method.**

Keywords- Integrated magnetics; magnetic cores; matrix decomposition; topology

I. INTRODUCTION

Magnetic components are one of the important factors that affect the size/cost of power electronic circuits. The integration of magnetic components of a converter leads to low component count, low cost, high power density, and high reliability [1]. Since so much benefit can be gained, more and more attention has focused on this area.

In order to serve the purpose of integrated DM components with different AC flux together, the core of IM must have multi magnetic branch. Then the methods of magnetic integration can be classified into two groups according to how to get the multiple magnetic branches. The character of first group is to get multiple branches by change the magnetic topology of core. The other is that achieving integration by using the general core such as E-E or E-I shape without any change. The method proposed in this paper is the second group.

The reluctance model of a three limbs core is illustrated in Figure 1(a). It is assumed that several DM components will be integrated and the sum of the DM ones winding number is m. Then these windings will be wounded on the three limbs and Figure 1(b) demonstrates their distributions by reluctance and MMF source model. Where N_{i1}, N_{i2} and N_{i3} (i=1,2...m) are distribution turn numbers of the ith winding on three limbs respectively. and each part of the winding are represented by a MMF source which value equals to $N_{ij}i_i$ (j=1, 2 or 3). Now the problem of how to fulfill the IM is transferred to get the value of these mX3 variables.

The principle of operation of IM is not difficult to understand, but the full design and analysis of a complex

magnetic component can be difficult.

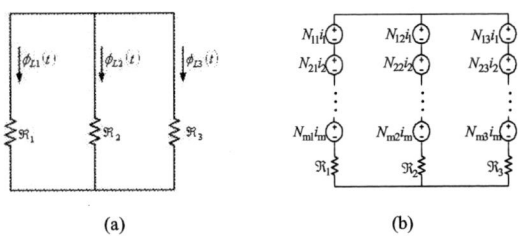

(a)   (b)

Figure 1.   Reluctance model of a magnetic component with three limbs core

Ed. Bloom has done the outstanding work [2]-[4]. He took a design of single ended forward converter as an example to introduce the method by analysis the relationship of flux in DM component. The method is very effective in small number discrete ones to be integrated. If the number is a bit larger, such as three and more, the work may need more patience and carefulness since the procedure is relied on experience and hard to be programmed. On the same time, this method cannot give the answer whether these discrete ones can be integrated or not. Dr. Chen Qianhong proposed another IM design method [5], which is based on flux decoupling and change the integration procedure easier. But a more systematical method is needed to answer whether these DM ones can or not be integrated and how many topology may be existed. In this paper, a new method to find the IM topologies are proposed, by which whether the discrete ones can be integrated is determined in the first step, then all of the possible topologies can be find.

II. INTEGRATED TWO DM COMPONENTS

If only 2 magnetic components are to be integrated, we assume that the $\phi_{L1}(t)$、 $\phi_{L2}(t)$ and $\phi_{L3}(t)$ are the flux through three limbs of core, on which all DM have been integrated. We suppose the winding of DM 1 has been divided into three parts: m_1, m_2 and m_3, each part is winded on the corresponding limb. So the flux linkage of DM 1 can be described

$$\Psi_1(t) = m_1\phi_{L1}(t) + m_2\phi_{L2}(t) + m_3\phi_{L3}(t) \qquad (1)$$

For the reason that there are only 2 independent fluxes, not losing generality, we take for granted that the third flux is composed of the others. Then the (1) can be rewritten as

---

This work is supported by National Natural Science Foundation of China Key Project (50237030) and Delta Science & Technology Educational Development Program (2005009).

$$\Psi_1(t) = z_{11}\phi_{L1}(t) + z_{12}\phi_{L2}(t) \qquad (2)$$

Just by the same way, the flux linkage of DM 2 can be expressed

$$\Psi_2(t) = z_{21}\phi_{L1}(t) + z_{22}\phi_{L2}(t) \qquad (3)$$

Combining (2) and (3), we can derive

$$\begin{bmatrix} \Psi_1(t) \\ \Psi_2(t) \end{bmatrix} = \begin{bmatrix} z_{11} & z_{12} \\ z_{21} & z_{22} \end{bmatrix} \begin{bmatrix} \phi_{L1}(t) \\ \phi_{L2}(t) \end{bmatrix} \qquad (4)$$

Let

$$A = \begin{bmatrix} z_{11} & z_{12} \\ z_{21} & z_{22} \end{bmatrix} \qquad (5)$$

Thus A, which is named coefficient matrix, has two possibilities. They are whether the matrix is nonsingular or not. Firstly we can suppose the matrix is nonsingular, and then we get

$$\begin{bmatrix} z_{11} & z_{12} \\ z_{21} & z_{22} \end{bmatrix}^{-1} \begin{bmatrix} \Psi_1(t) \\ \Psi_2(t) \end{bmatrix} = \begin{bmatrix} \phi_{L1}(t) \\ \phi_{L2}(t) \end{bmatrix} \qquad (6)$$

We can see clearly from (6) that flux through the two limbs is linear combination of flux linkage of the 2 DM components and the assumed flux waveform can be realized. Under this condition, the two DM components can be integrated.

If the coefficient matrix is singular, an elementary row operation can be taken and (4) is rewritten as

$$\begin{bmatrix} 1 & k \\ 0 & 1 \end{bmatrix} \begin{bmatrix} \Psi_1(t) \\ \Psi_2(t) \end{bmatrix} = \begin{bmatrix} 1 & k \\ 0 & 1 \end{bmatrix} \begin{bmatrix} z_{11} & z_{12} \\ z_{21} & z_{22} \end{bmatrix} \begin{bmatrix} \phi_{L1}(t) \\ \phi_{L2}(t) \end{bmatrix} \qquad (7)$$

Furthermore, (7) can be transferred as

$$\begin{bmatrix} 1 & k \\ 0 & 1 \end{bmatrix} \begin{bmatrix} \Psi_1(t) \\ \Psi_2(t) \end{bmatrix} = \begin{bmatrix} 0 & 0 \\ z'_{21} & z'_{22} \end{bmatrix} \begin{bmatrix} \phi_{L1}(t) \\ \phi_{L2}(t) \end{bmatrix} \qquad (8)$$

From the first row of (8), the following relationship can be get

$$\Psi_1(t) = -k\Psi_2(t)$$

It is clear that $\Psi_1(t)$ and $\Psi_2(t)$ are proportional. And if one flux of the three limbs is chosen as the base vector, then the two DM components may be winded on the same limb. In other words, the two DM components are also able to integrate. From above two conditions, it can be drawn that two DM components can always be integrated.

III. Integration of 3 and more DM components

When three or more DM components are to be integrated, we can choose two of them (the flux linkage of the chosen DM components should not be proportional). Just following the steps proposed, the two flux $\phi_{L1}(t)$、$\phi_{L2}(t)$ of the limbs can be expressed

$$\begin{bmatrix} \phi_{L1} \\ \phi_{L2} \end{bmatrix} = \begin{bmatrix} z_{11} & z_{12} \\ z_{21} & z_{22} \end{bmatrix}^{-1} \begin{bmatrix} \Psi_1 \\ \Psi_2 \end{bmatrix} \qquad (9)$$

If the third DM component can be integrated, its flux linkage equation must satisfy

$$\Psi_3 = \begin{bmatrix} z_{31} & z_{32} \end{bmatrix} \begin{bmatrix} z_{11} & z_{12} \\ z_{21} & z_{22} \end{bmatrix}^{-1} \begin{bmatrix} \Psi_1 \\ \Psi_2 \end{bmatrix} \qquad (10)$$

We obtain

$$\Psi_3(t) = z_{31}\phi_{L1}(t) + z_{32}\phi_{L2}(t) \qquad (11)$$

So do the remainder

$$\Psi_m(t) = z_{m1}\phi_{L1}(t) + z_{m2}\phi_{L2}(t) \qquad (12)$$

Where $m \geq 3$. The rule that decide whether one of the remainder can be integrated with the former is:

On the condition that one of the remainder can be integrated, then z_{m1} and z_{m2} must have solution. If no solution, then it can't be integrated.

This rule can be expressed in another way:

Let

$$F = \begin{bmatrix} d\Psi_{11}/dt & d\Psi_{12}/dt & \cdots & d\Psi_{1n}/dt \\ d\Psi_{21}/dt & d\Psi_{22}/dt & \cdots & d\Psi_{2n}/dt \\ \vdots & \vdots & \vdots & \vdots \\ d\Psi_{m1}/dt & d\Psi_{m2}/dt & \cdots & d\Psi_{mn}/dt \end{bmatrix} \qquad (13)$$

where m is the number of DM components that will be integrated, and n is the piece of the flux linkage waveforms. Then the conclusion can be drawn that the rank of matrix F is greater than 2, these DM cannot be integrated in three limbs core, otherwise, they can.

IV. Simulation and Experimntal Verification

From the upper procedure, we can find the key point of design an IM component is to integrate two DM components with flux waveform out of relation. Drawing from (4), finding of IM topology can be carried out in two ways: by the first one, we can figure the optimized flux waveform and solve (4) to get coefficient matrix. By the other, firstly we list the entire possible coefficient matrix, and then select the most favorable one that satisfies constraint. Finally we can get the magnetic topology by coefficient matrix. The flows of these two methods are illustrated in Figure 2.

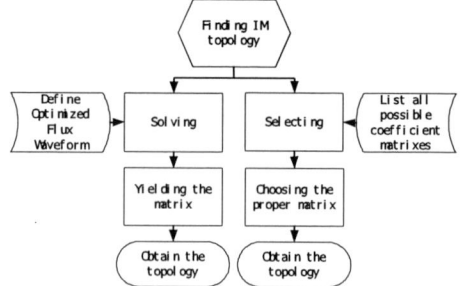

Figure 2. Procedure or proposed method to find IM topology

The half bridge with current doubler rectifier topology (it is illustrated in Figure 3), which has the merits such as its simple topology, the transformer core is excited alternately in both direction and so on, are widely used in medium or low power rating and high output current DC/DC converters [7]-[9]. But there are three DM components being used in the circuit, which baffle the converter to achieve a higher power density. To overcome this difficult, the IM is the proper choice.

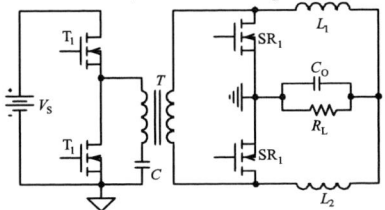

Figure 3.   Schematic of half bridge with current doubler rectifier

The switching frequency of this converter is 200kHz and its input and output voltages are 48V and 5V respectively. Following the magnetics core selection rule, an E32/6/20 planar core, which material is 3F3, is selected. For a generating of 50℃ temperature rise, the allowable material loss density is about 181mW/cm³. But considering of safe operating, the loss density is restricted under 150mW/cm³. Under this condition, the amplitude and peak of flux density are selected at 60 and 200mT respectively. The ratio of transformer is 4:3 and the duty ratio is 2/3. Then the waveforms of flux linkage of the three DM components are illustrated in Figure 4.

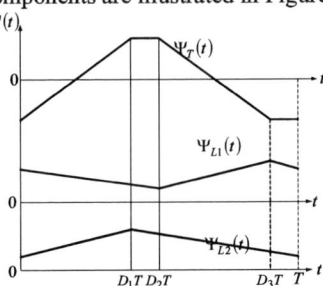

Figure 4.   Flux linkage waveform of transformer, inductor 1 and 2 respectively

From the rule just presented for the multiple DM, the following matrix can be get

$$F = \begin{bmatrix} -18 & 0 & 18 & 0 \\ 12 & -6 & -6 & -6 \\ -6 & -6 & 12 & -6 \end{bmatrix}$$

The first row in the matrix represents the derivative of the voltage on the transformer; the second row does the inductor 1 and the third one, the inductor 2. It is clearly the rank of F is equal to 2. So these three components can be integrated in a three limbs core. Now we can program a program to find the possible magnetic topology. From the result of the program, it is found three topologies for the three DM components integration, which are presented in table I, II and III respectively. It can be clearly see that the topology in table I is the familiar CDR integration method proposed by Wei Chen. But the others are novel. In order to verify the feasibility of them, a serious simulation and experiment have been carried out, and the simulation and experiment results are illustrated in Figure 5-8 respectively. The gyrator-

capacitor approach is selected to model the magnetic components. Simultaneous simulation of the electrical and magnetic circuits is performed with Pspice.

TABLE I.
POSSIBLE TOPOLOGY 1

Turns	Limb 1	Limb 2	Limb3
Transformer(s)	0	3	0
Inductor 1	-3	0	0
Inductor 2	0	0	3

TABLE II.
POSSIBLE TOPOLOGY 2

Turns	Limb 1	Limb 2	Limb3
Transformer(s)	0	3	0
Inductor 1	-3	-1	0
Inductor 2	0	-1	3

TABLE III.
POSSIBLE TOPOLOGY 3

Turns	Limb 1	Limb 2	Limb3
Transformer(s)	0	3	0
Inductor 1	-2	0	0
Inductor 2	0	-1	2

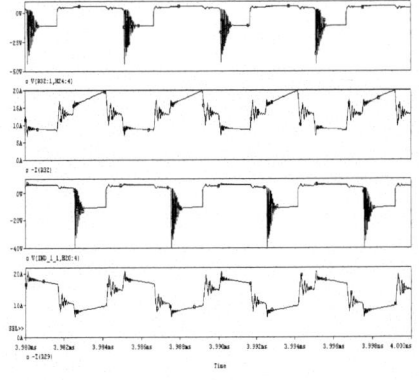

Figure 5.   Simulation waveforms of possible topology 2. Top to bottom: inductor 1 voltage and current, inductor 2 voltage and current.

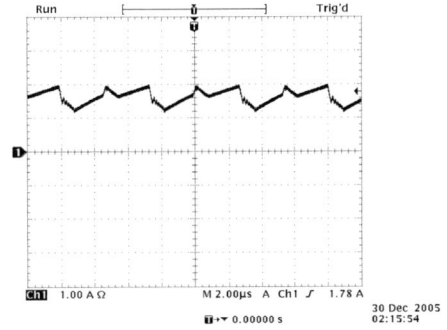

Figure 6.   Current waveform of inductor 1 in possible topology 2

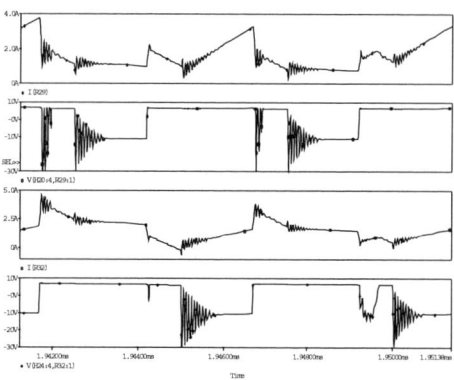

Figure 7. Simulation waveforms of possible topology 3. Top to bottom: inductor 1 voltage and current, inductor 2 voltage and current.

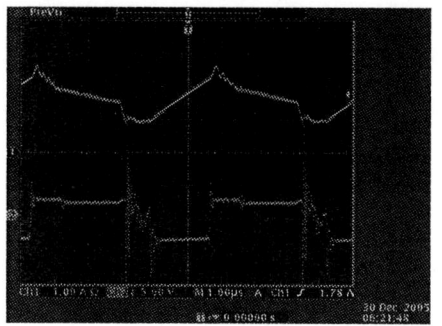

Figure 8. Current (top) and voltage waveform of inductor 2 in possible

V. CONCLUSION

This paper proposed a method to find the IM topology. The method is straightforward and simple. Moreover, the design of IM components is not relied on experience and we can almost find all of the possible magnetic topology. A rule by which we could decide whether 3 and more DM components can be integrated has been derived. Practical implementation of a half bridge with current doubler, employing only a single IM component, has been carried out to demonstrate the practical application of this method.

REFERENCES

[1] Y.S. Lee, L.P. Wong and D.K. Cheng, "Simulation and design of integrated magnetics for power converters", *IEEE Trans. Magn.,* Vol. 39, No. 2, pp. 1008-1018, Mar, 2003.

[2] R. P. Severns and G.Bloom, *Modern DC-to-DC Switchmode Power Converter Circuits,* New York: Van Nostrand Reinhold, 1985, pp. 262-324.

[3] E. Bloom, "Core selection for and design aspects of an integrated magnetic forward converter," *in Applied Power Electronics Conf. And Exposition, New Orleans, LA,* April 28-May 1, 1986, pp.141-150.

[4] E. Bloom, "New integrated–magnetic DC-DC power converter circuits and systems," *in Applied Power Electronics Conf. And Exposition, San Diego, CA,* March 6-7, 1987, pp.57-66.

[5] Q.H. Chen, "Research to the application of the magnetics-integration techniques in switching power supply," Ph.D. thesis, Nanjing University of Aeronautics and Astronautics, Nanjing, 2001(in Chinese).

[6] Jai P. Agrawal, *Power Electronics Systems –Theory and Design,* Upper Saddle River, N.J: Prentice Hall, 2001.

[7] Wei Chen, Guichao Hua, Dan Sable, Fred Lee, "Design of High Efficiency, Low Profile, Low Voltage Converter with Integrated Magnetics", *in Applied Power Electronics Conf. And Exposition, Vol.2, Atlanta GA,* Feb 23-27,1997, pp. 911-917.

[8] Peng Xu, Qiaoqiao Wu, Pit-Leong Wong and Fred C. Lee, "A Novel Integrated Current Doubler Rectifier", *in Applied Power Electronics Conf. And Exposition, Vol. 2, New Orleans, LA,* Feb. 6-10, 2000, pp. 735-740.

[9] Sergey Kornokov, Valery Meleshin, Alexey Nemchinov and Simon Fraidlin , "Small-Signal Modeling of Soft-Switched Asymmetrical Half-Bridge DC/DC Converter", *in Applied Power Electronics Conf. And Exposition, Dallas, TX,* March 5-9, 1995, pp.707 –711

2006 5th International Power Electronics and Motion Control Conference

Investigation on the Space Vector PWM for Large Power Three-Level DC-Link Voltage Source Inverter Equipped with IGCTs

Wang Chengsheng[*,**], Li Chongjian[***], Li Yaohua[*], Zhao Xiaotan[*,**]
[*] Institute of Electrical Engineering, Chinese Academy of Sciences, Beijing, China
[**] Graduate University of the Chinese Academy of Sciences, Beijing, China
[***] Automation Research and Design Institute of Metallurgical Industry, Beijing, China

Abstract—In high performance ac drive systems, the large power three-level voltage source inverters equipped with IGCTs are widely used now and have promising future. In this paper, the principle of space vector PWM (SVPWM) for three-level inverter is introduced. Some problems of SVPWM for three-level inverter are discussed. Based on the system simulation, a high power three-level inverter equipped with IGCTs is designed. The working performance of the inverter is verified with system experiment. The experiment results show that the inverter has excellent performance.

Keywords-SVPWM; three-level; inverter; IGCT

I. INTRODUCTION

In recent years, there are rapid improvements in high voltage technologies. The high performance ac drive systems are at increased power level. And high quality inverters with low harmonic loss and torque pulsation output are largely demanded. Compared with the conventional two-level inverter, the three-level neutral point clamped voltage source inverters (NPC-VSI) have many merits, such as they can block double of the voltage for a given power electronic device, reduce the voltage and current harmonics and increase the equivalent switching frequency [1][2][3][4]. They have been widely used in medium or high power drive systems, and will be promising in the near future.

The integrated gate commutated thyristors (IGCTs) are a new kind of power electronic device, which integrate the GCTs and the integrated gates together. The IGCTs have the features of high current, high voltage, high reliability, compact structure and low loss. The design feature of the IGCTs makes them have excellent performance in high-voltage and high-current field. And the high integration of the drive and the device makes it very convenient to use them [5][6]. They have been widely used in large power voltage source inverter systems now, and they have extensive application prospect.

In this paper, the principle of SVPWM for three-level inverter is introduced. Then the duration calculation of space vectors is given. Some problems of SVPWM are

also discussed, such as switching pattern selection to reduce minimum pulse width and synchronous control to limit the over-current. Then the simulation system is set up. Based on the system simulation, a high power three-level voltage source inverter equipped with IGCTs is designed, and the working performance of the inverter is verified with experiment.

II. SPACE VECTOR PWM FOR THREE-LEVEL INVERTER

A. Introduction of SVPWM for Three-level Inverter

Fig.1 shows the circuit diagram of a three-level inverter equipped with IGCTs. In the three-level inverter, each phase consists of four main power devices, four freewheeling diodes and two clamping diodes. The clamping diodes are connected to the neutral point of the two DC-link capacitors. If the voltage of each DC capacitor is E, then by selecting different switching states, the output voltage of each phase has three values +E, 0 and –E respectively. So the three phase three-level inverter has 27(3^3) switching states in total.

Fig.2 shows all the switching states generated by the three-level inverter and the corresponding space vectors. In each state, P means that the upper two main power devices are on and the lower two are off, O means that the middle two are on and the others are off, and N means the lower two are on and the others are off. The corresponding output voltages are +E, 0 and -E

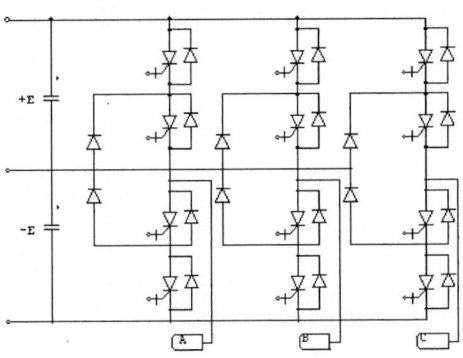

Figure1. Circuit diagram of a three-level inverter

1-4244-0448-7/06/$25.00 ©2006 IEEE

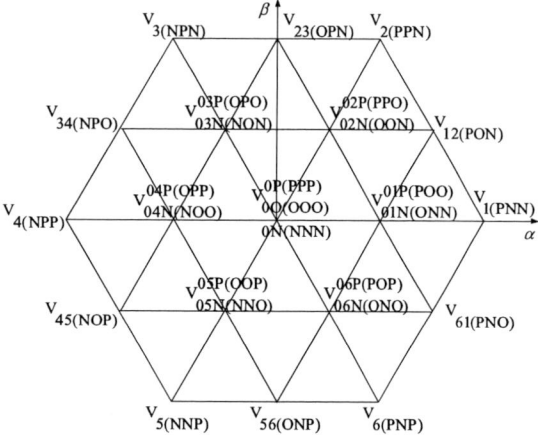

Figure 2. Voltage space vectors for three-level inverter

respectively. For example, the vector PON means that the output of phase A is the positive voltage +E, the output of phase B is the zero voltage 0 and the output of phase C is the negative voltage –E. Other vectors can also be described as above.

B. Duration Calculation of Space Vectors

In order to simplify the study, the hexagon in Fig.2 can be divided into 6 sectors when the angle changes from 0 to 360 degree and each sector occupies 60 degree. Fig.3 is one of the sectors and it is referred as sector 0. The sector is divided into 7 regions.

Generally, in standard space vector methods, only three voltage vectors corresponding to the apexes of the triangle in which the reference vector is inside are used to minimize the harmonic components of the output voltage [7][8]. In this way, the duration of the nearest three vectors in every region can be easily obtained.

Now, it is supposed that the reference voltage vector falls in the regions of 4 or 5, as shown in Fig.3, the duration time of the voltage space vectors V_{01}, V_{02}, V_{12} is t_{a0}, t_{b0}, t_{c0} respectively. Then t_{a0}, t_{b0}, t_{c0} can be calculated with the following equations:

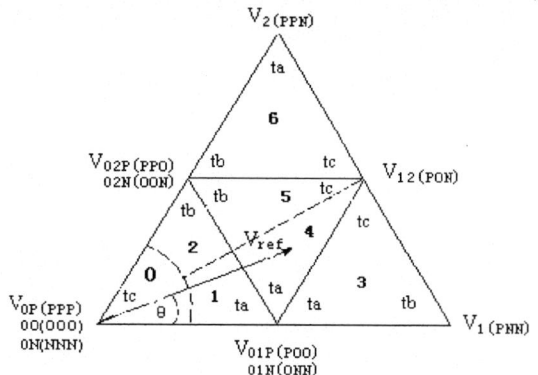

Figure 3. Space vectors and region division of sector 0

$$V_{01}t_{a0} + V_{02}t_{b0} + V_{12}t_{c0} = V_{ref}T_S \qquad (1)$$

$$t_{a0} + t_{b0} + t_{c0} = T_S \qquad (2)$$

Where T_S is the sampling period and

$$\begin{cases} V_{01} = \dfrac{2}{3}E \\[2mm] V_{02} = \dfrac{2}{3}Ee^{j\frac{\pi}{3}} \\[2mm] V_{12} = \dfrac{2\sqrt{3}}{3}Ee^{j\frac{\pi}{6}} \\[2mm] V_{ref} = Ve^{j\theta} \end{cases} \qquad (3)$$

From (1), (2) and (3), the duration of each vector can be expressed as below:

$$\begin{cases} t_{a0} = T_s(1 - 2m\sin\theta) \\[2mm] t_{b0} = T_s\left[1 - 2m\sin(\dfrac{\pi}{3} - \theta)\right] \\[2mm] t_{c0} = T_s\left[2m\sin(\dfrac{\pi}{3} + \theta) - 1\right] \\[2mm] m = \sqrt{3}V_{ref}/2E \end{cases} \qquad (4)$$

If the reference space vector falls in other regions of Fig.3, the duration can be obtained in the same way, and they are shown in TABLE I.

The duration of voltage space vector in other 5 sectors can be calculated as above similarly. In this way, the complexity of the arithmetic is greatly decreased.

C. Switching Pattern Selection to Avoid the Minium Pulse width (MPW)

If the modulation coefficient m is little enough, such as being less than 0.1, the reference voltage vector is classified to region 0. Otherwise, the reference vector voltage belongs to other regions.

In each region, there is a main vector, which means that the duration of the vector is always long enough to be much more than the minimum on/off time of the main switching device in the region. In the IGCT inverter system, the duration of the main vector should be more than 4 times of the minimum on/off time of IGCT.

TABLE I
THE DURATION OF SPACE VECTORS IN SECTOR 0

Regions of sector 0	Duration time of space vectors		
	t_a	t_b	t_c
0,1,2	$T_S - t_{b0}$	$T_S - t_{a0}$	$-t_{c0}$
3	$T_S - t_{c0}$	$-t_{b0}$	$T_S - t_{a0}$
4,5	t_{a0}	t_{b0}	t_{c0}
6	$-t_{a0}$	$T_S - t_{c0}$	$T_S - t_{b0}$

In region 0, it can be easily got that the duration of V_0 is large enough to be much more than the minimum on/off time of IGCT, so V_0 can be selected as the main vector.

In region 1, it can be got that

$$\begin{cases} m \geq 0.1 \\ V_{ref} < 2E/3 \\ 0 \leq \theta < \pi/6 \\ t_{V_{01}} = T_s \cdot 2m\sin(\frac{\pi}{3} - \theta) \end{cases} \qquad (5)$$

Then we can get

$$0.1T_s \leq t_{V_{01}} < T_s \qquad (6)$$

For $0.1T_S$ is much larger than the minimum on time of IGCT, vector V_{01} can be chosen as main vector in region 1. In the same way, V_{02} can be chosen as the main vector in region 2. If m is not too large, V_{01} can be the main vector in region 3, 4, and V_{02} can be the main vector in region 5, 6.

Then we can choose the main vector as the first vector in the sampling time T_S, and the minimum pulse width can be avoided. In region 0, we can choose the sequence of the vectors as

$$OOO \rightarrow POO \rightarrow PPO \rightarrow PPP \rightarrow PPO \rightarrow POO \rightarrow OOO$$

The output switching sequences are shown in Fig.4. In the sampling period T_S, the minimum pulse width will be $t_{V0}/4$. For V_0 is the main vector, $t_{V0}/4$ will be much larger than the MPW. So the MPW is avoided in region 0 in this way.

In other regions, similar methods can be used to solve the MPW problem.

When m is large enough to be nearly 1.0, other vector should be chosen as the main vector or the region can be divided into more sub-regions to get the main vector. On the other hand, when m is large enough, we can simply remove the MPW by setting the narrow pulse width to a certain value. In the same time, the output error is also acceptable. And the problem of MPW is not serious. So the MPW problem with a much large m is not included in this paper.

D. Synchronous Control in SVPWM

Compared with asynchronous control method, the efficiency of synchronous control is higher, and the output harmonics is much better. So in large power motor drive systems, synchronous control method is widely used.

In the large power three-level DC-link voltage source inverter, the main switching devices are IGCTs. The switching frequency of IGCTs is not very high. Normally, it is less than 1 kHz. So the subsection synchronous PWM control method is used in the inverter, which can ensure that the modulation ratio is big enough to reduce the output harmonics in low frequency, and the switching

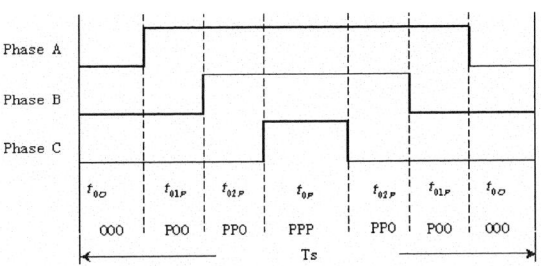

Figure 4. Switching sequence of region 0

frequency of the IGCT is not higher than its rated one in high frequency.

Normally, a threshold value of the frequency is set. When the output frequency is larger than the threshold, the modulation ratio changes from one to another. But most of the time, this will lead to the over-current and high harmonics of the output. In this paper, an identification condition is added to the control system to ensure the smooth transition of the output. When the output frequency is up to the threshold, only if the identification condition is also met, synchronous transition occurs. Otherwise, the control system will keep the original modulation ratio.

The simulation result of the synchronous transition is shown in Fig.5. From which we can see that there is almost no over-current in the transition process. It means that the synchronous transition problem is well solved.

III. SYSTEM SIMULATION AND EXPERIMENT

Based on the analysis of SVPWM, a simulation system is set up to study the working process of the IGCT inverter. Based on the simulation results, an IGCT inverter is designed, and a series of experiments are done to verify the performance of the inverter. The simulation system is shown in Fig.6 (a), the simulation and experiment output line current and voltage are shown in Fig.6 (b) and (c) respectively.

When the output voltage is high enough and the system parameters are properly selected, the output phase voltage is three-level and the output line voltage is five-level state, and the output line current is almost sinusoidal. On the same time, the voltage over IGCT is in the safe range.

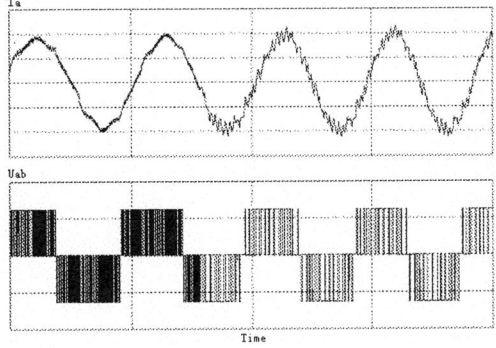

Figure 5. Simulation result of the synchronous transition process

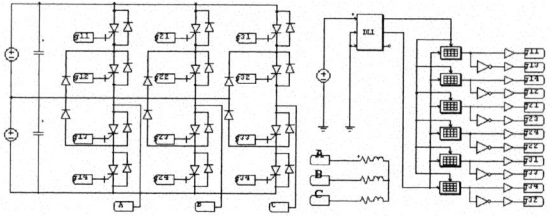

(a) The diagram of the simulation system

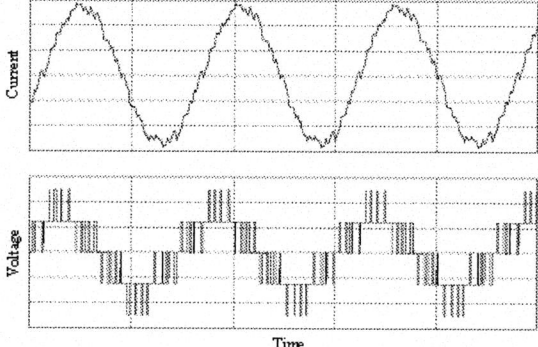

(b) Simulation output current and voltage

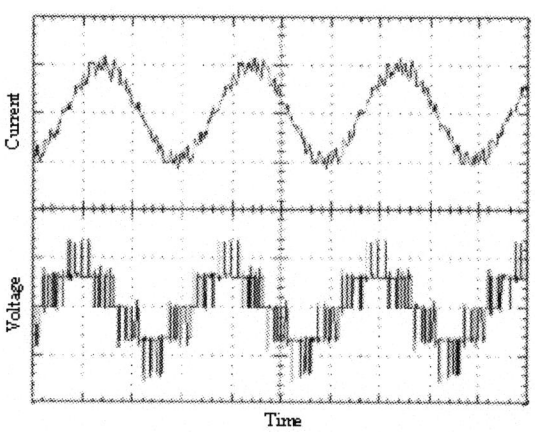

(c) Experiment output current and voltage

Figure 6. System experiment

With the system simulation results, the parameters of the inverter are determined. In the inverter, the main switch devices are a new kind of IGCTs. As to this kind IGCT, the repetitive peak off-state voltage is 4500V and the maximum controllable turn-off current is 4000A. For the operating voltage of the inverter is very high, all the control signals are transmitted through fiber in the inverter. The IGCTs are directly controlled with optical signal and they have status feedback signal, which can be used to protect the system. The closed loop water-cooling system is used in the system, and the sealing problem is properly solved. And the laminated bus technology is used in the system in order to decrease the stray inductance. Perfect protection system is also designed in the inverter. If something wrong happens in any part of the inverter, the control system will cut off the high voltage input and enable the protection system to assure the safe operation of the inverter.

After the installation of the inverter, a series of experiments were done to check its performance. In the experiment, the DC bus voltage is 5000V, and the peak value of the output current is up to 2500A. So the output power is up to about 10 MVA. And the maximum voltage over the IGCT device is still in the safe region.

From Fig.6, we can see that the experiment result fits the simulation result very well. The test results show that the design of the three-level DC-link voltage source inverter is successful, and the inverter has excellent performance.

VI. CONCLUSION

Based on the system simulation, combining with the feature of the three-level circuit, a large power three-level inverter equipped with IGCTs is designed. Many problems in the practical application of the large power inverter are properly solved. The unique stack press technology and laminated bus technology is used to decrease the stray inductance of the circuit and make the system structure compact. And there also has perfect protection system to assure the safe operation of the inverter. An excellent control system with SVPWM method is designed, and some problems such as the MPW and synchronous transition are well solved.

In the system experiment, the DC bus voltage is 5000V, and the peak value of the output current is up to 2500A. So the output power is up to about 10 MVA. The experiment results show that the high voltage large power three-level inverter system has excellent performance.

REFERENCES

[1] Steffen Bernet, "Recent development of high power inverters for industry and traction applications," IEEE Trans. Vol. 15, No. 6, pp1102-1117, Nov. 2000

[2] Thomas Bruckner, Steffen Bernet, "Investigation of a high-power three-level quasi-resonant DC-link voltage-source inverter," IEEE Trans. Vol. 37, No. 2, pp619-627, March/April, 2001

[3] A. Zuckerberger, E. Suter, Ch. Schaub, A. Klett, P. Steimer, "Design, simulation and realization of high power NPC inverters," IEEE-IAS, St Louis, Oct. 1998

[4] A. Nabae, I. Takahashi, and H. Akagi, "A new neutral-point-clamped PWM inverter," IEEE Trans. Ind. Applicat., Vol. 17, pp.518-523, Sept./Oct. 1981

[5] Eric Carroll, Sven Klaka, Stefan Linder, "Integrated gate-commutated thyristors: a new approach to high power electronics," IEMDC, Milwaukee, May 1997

[6] Peter Steimer, Oscar Apeldoorn, Eric Carroll, Andreas Nagel, "IGCT technology baseline and future opportunities," IEEE-PES. Atlanta, Oct. 2001

[7] Masato Koyama, Toshiyuki Fujii, Ryohei Uchida, Takao Kawabata, "Space voltage vector-based new pwm method for large capacity three-level gto inverter,"IEEE-IECON, 1:271-276, November 1992

[8] Yo-Han Lee, Bum-Seok Suh, and Dong-Seok Hyun, "A novel pwm scheme for a three-level voltage source inverter with gto thyristors," In IEEE Transitions on Industry Applications, volume 32, pages 260-268, March/April 1996

2006 5th International Power Electronics and Motion Control Conference

Status and Opportunities of Photovoltaic Inverters in Grid-Tied and Micro-Grid Systems

Xiaoming Yuan and Yingqi Zhang
GE Global Research – Shanghai
1800 Cailun Rd., Zhangjiang High Tech Park, 201203, Shanghai, China
xiaoming.yuan@geahk.ge.com

Abstract: **This paper reviews the status in industry and academia regarding configurations, topologies, controls, and grid connections in grid-tied and micro-grid PV inverter applications. The paper will discuss the major technical needs to address problems in bringing cost down, increasing efficiency and improving reliability/availability. The paper foresees that new grid interconnection features will have to be integrated more into the inverters, along with the widespreading use of distributed generations.**

Keywords: grid-tie PV inverter, modular inverter, Inverter efficiency, micro-grid

I. Introduction

The tremendous growth in the PV market in the recent years all over the world has been stimulating wide interests from industry researches and universities participating in the technology development. Significant efforts on new materials, device concepts and processes, and manufacturing technologies are being made in order to bring down the costs of PV cells. PV inverter which usually represents 20% of the system cost and the major reliability bottleneck in a PV system should receive due attention. Typical guarantee for PV inverter is 5 years, except for a few manufacturers who are jumping to 10 years, in comparison to 20 years for PV panels.

The key technical aspects that will drive improvements in cost, reliability and efficiency of PV inverters, which are key to success, will be addressed in this paper. Inverter configuration and topology represents the way that how DC power from PV array, small or large scale, will be extracted, and how this extracted power will be converted to AC connected to grid or fed to island load. Inverters designed with modularity and scalability will drive the cost down in large volume production due to simplification in designing, manufacturing, operation, and maintenance processes. PV inverter system with multiple inverter modules operating in parallel are also expected to improve system availability and efficiency. Inverter controls deal with optimizing the interactions during steady state or transient operation between inverter and PV array at the DC side as well inverter and grid at the grid side. Especially at the grid side, depending on how the grid authority will respond to the wide spread of PV and other dispersed generations at the low voltage distribution system, new functionalities such as grid fault transient ride through, voltage support, power control, as well as observability and controllability to grid will need to be integrated into the design, in addition to existing requirements in power quality and safety.

Grid connected PV represents about 75% of the global PV market today. Given the need for remote area electrification particularly in developing countries, PV micro-grid is starting taking off recently. Technical success with PV micro-grid will depend largely on a plug-and-play configuration of the system implying minimal required engineering in installation, commissioning, operation and maintenance, in addition to an efficient operation management model. The paper will point out a few major technical challenges in power quality assurance, fault location and protection etc toward that direction.

II. Grid-tied PV inverters

A Inverter configurations and topologies

PV inverter configurations: PV inverters can be classified as central inverters, string inverters, multi-string inverters and module-integrated PV inverters, as shown in Fig. 1.

Central inverter is generally used in large-scale 3-phase PV installations, where strings of PV modules are connected in parallel necessitating complex DC cabling. Losses due to module mismatch and shading are usually high in this configuration. To overcome the above disadvantages, string inverter and multi-string inverter were developed. Multi-string inverter accommodates different PV sizes and orientations, where each string is equipped with its own MPPT. Module-integrated PV inverters achieve highest MPPT efficiency and eliminate the DC cabling. When long lifetime (20 years) is achieved, these plug & play inverters will have a very promising future in Building Integrated PV applications.

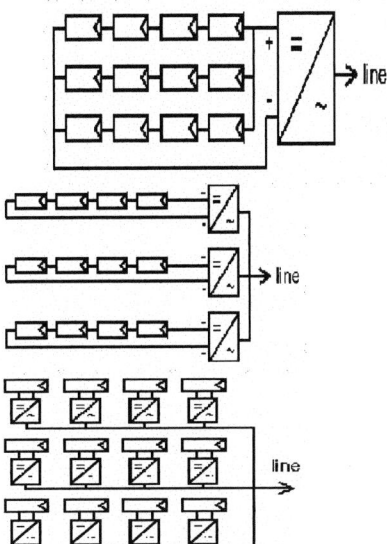

Fig. 1. Configurations of central, string, multi-string and module integrated PV inverters.

PV inverter topologies: Traditional grid-tie PV inverters are equipped with a LF-transformer to boost the input voltage, as

1-4244-0448-7/06/$25.00 ©2006 IEEE

shown in Fig. 2(a). LF-transformer provides galvanic isolation between the grid and PV array. There is no problem of DC injection to the grid. The DC/AC inverter performs MPPT and delivers low distortion output current to the grid. The inverter is heavy weight, large size, and low efficiency however. By replacing LF-transformer with a HF-transformer, as shown in Fig. 2(b), inverter efficiency can be increased by 2%. Phase-shifted DC/DC converter performs MPPT function and at the same time provides galvanic isolation. The DC injection to the grid should be controlled to avoid saturation of distribution transformers. Transformer-less inverters, as shown in Fig. 2(c) can further increase the efficiency by 2%. Boost converter is commonly used in transformer-less inverters to regulate PV array voltage, as shown in Fig. 2(d), simple H-bridge inverter or Neutral-Point-Clamped (NPC) inverter can be used as the DC/AC section. The NPC inverter so connected has no dc injection and achieves a large MPPT range.

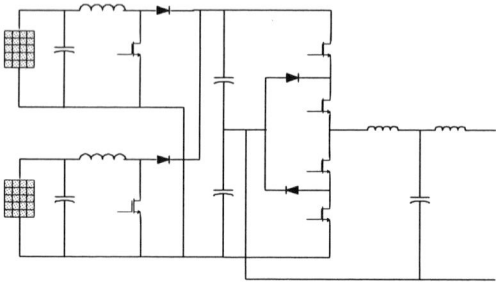

(d) Neutral-Point-Clamped (NPC) inverter

Fig. 2. Potential inverter topologies for PV applications

For large-scale PV applications, three-phase central inverters are commonly used. Several central inverters can be put in parallel with master-slave operation. This modular configuration has advantages of high availability. It's flexible to cover a large power range by parallel operation of modular inverters. Besides, under partial load conditions, extra inverters are turned off to increase the DC/AC conversion efficiency.

PV Inverter efficiency: Overall inverter efficiency is the product of MPPT efficiency and DC/AC conversion efficiency. Usually MPPT efficiency under low irradiation conditions is lower since the power curve is much flatter and the true maximum power point is more difficult to identify. Voltage ripple at the array will impact the tracking and will reduce MPPT efficiency. More attention is being paid on MPPT efficiency under partial load and dynamic irradiance conditions. Dynamic irradiance patterns are being developed for MPPT efficiency testing.

DC/AC inverter efficiency largely depends on the selected topologies and PWM schemes. DC input voltage of a DC/AC conversion can be real time optimized for efficiency improvement. A recent innovation utilizes two extra unidirectional switches to provide zero voltage at the output offering equivalent output current ripple with reduced switching of the inverter switches and therefore increasing DC/AC conversion efficiency.

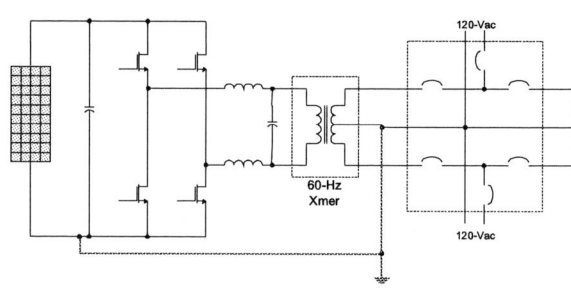

(a)   PV inverter with LF-transformer

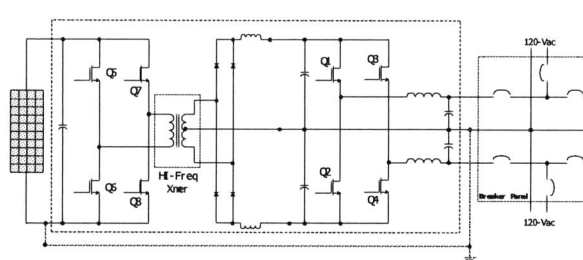

(b) PV inverter with HF-transformer

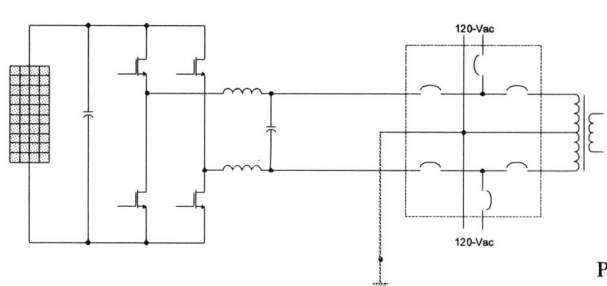

(c) Transformer-less PV inverter

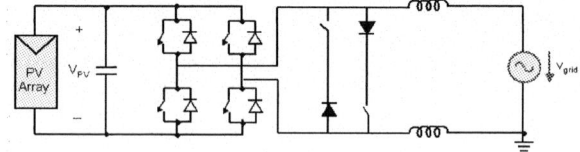

Fig. 3. Highly efficient and reliable inverter concept (HERIC).

Several string inverters can be reconfigured paralleling at the PV array sides and selectively operating a portion of the paralleled string inverters to improve the system efficiency, at a trade-off of potentially mismatch losses.

Europe is adopting a measurement of PV inverter efficiency as following taking into account of both full load and partial efficiency operation conditions and is therefore a better indication

594

of the efficiency performance across different irradiance over a day.

$$\eta_E = 0.03\,\eta_{5\%} + 0.06\,\eta_{10\%} + 0.13\,\eta_{20\%} + 0.1\,\eta_{30\%} + 0.48\,\eta_{50\%} + \eta_{100\%}$$

B PV inverter controls and grid interactions

MPPT control and array–inverter interaction: The most used MPPT algorithm is based on the concept of Perturbation and Observation (P&O), in which the PV array voltage is perturbed and the corresponding array output power is observed to determine the direction of next voltage change. Appropriate selection of the MPPT stage input capacitor will impact not only the array voltage/current ripple, but also the MPPT control loop stability. Due to the very non-linear characteristics of the PV array, certain damping in the loop will be necessary. Future research will need to be focused on tracking accuracy during low irradiance and dynamic irradiance conditions. Monitoring and diagnosis of large area PV array performance deserves also due attention.

PV inverter control and steady state power quality: Primary objective for PV inverter control is to guarantee quality of produced power. No matter whether the grid voltage is distorted or not, current flowing to the grid has to be sinusoidal without DC or harmonic components be present. Topologies with low frequency transformer at the grid side, or half bridge inverter with output neutral line returned to capacitor neutral point are free from DC current injection problem. Dedicated harmonic current control should be integrated in the PV inverter controller.

Multiple PV inverter interaction: Individual PV inverter exhibits current source behavior at certain harmonic frequencies. Electrical grid, on the other hand, will behave like a voltage source at certain harmonic frequencies. Especially for sites with high density PV inverter installation, those harmonic current or voltage sources will potentially excite resonances among the individual inverter output capacitors, cable stray effects and grid internal impedances, resulting in over-voltage across certain connection points or over-current through certain components. These resonance effects happen even when individual inverter meets the harmonic specification and it is not being considered in today's current control loop. They usually become worse in a weak grid when the grid exhibits significant impedances at certain frequencies.

Islanding detection and grid dynamic performance impact: Island occurs when a portion of the distribution system becomes electrically isolated from the reminder of the power system, which is normally a result of feeder circuit breaker, recloser, sectionalizer or fuse operation in response to a fault. Intentional island operation can be desirable to increase customer reliability, which however will require considerable engineering in communication and control functionality infrastructure especially when the island includes other distributed generations and a portion of the primary power system. Unintentional island operation presents a number of problems in maintenance personnel and public safety, power quality, islanded system protection and out-of-phase re-closing protection etc.

Non-fault induced island detection is usually done by so called Sandia Frequency Shift and Sandia Voltage Shift, or grid impedance jump measurement. Unintentional island disconnection is required to happen within 2 seconds. Fault induced island normally involves voltage or frequency transients and the detection is realized via under/over voltage/frequency protection. Required disconnect duration in this case varies from 2 cycles to 10 cycles depending on the severity of the

disturbances, in order to avoid interference with the grid auto re-closers, or damage to the PV generation itself by the re-closing. Unfortunately, with exception of the grid impedance measurement concept, other detection concepts respond to grid disturbances by destabilizing the grid through modulating the active or reactive power output in the case of active islanding detection, or stopping the output in the case of voltage/frequency detection. With high penetration of PV and other distributed generations compared to local capacity, solutions to detect unintentional islanding can aggregate local disturbances. When distributed generations become widespread in the grid, such anti-islanding detections will also impact bulk system stability.

Grid impedance measurement, due to the nature of the detection, however will not function appropriately when PV generations are densely installed at the site.

Grid transient ride-through: Within the voltage/frequency limit regulated in islanding detection, PV inverter should be designed to survive grid transients remaining connected to the grid during disturbances. In this way, unnecessary trips will be reduced and customer power reliability will be increased. Today's PV inverters on the market are widely different in this performance.

Grid voltage support and power control: As shown in Fig. 4, in areas with high-density PV installations, connection point voltage will fluctuate with varying power production from PV inverters. With the low voltage distribution system cable being highly resistive, reverse active power flow happening when excess power is produced from PV inverters will potentially raise the PV connection point voltage to the trip point in the inverter protection. Certain control actions from the PV inverters will have to be integrated to ensure power flow into the grid.

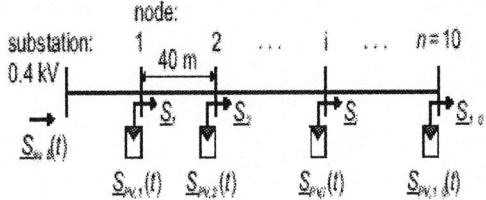

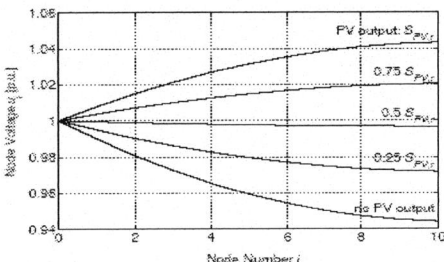

Fig. 4. Node voltage profile along the distribution line under varying PV productions.

Grid interconnection long-term strategy: Increasing penetration of distributed generations like PV in the low voltage grid imposes unprecedented challenges for grid operation and management. New functionalities that will allow for distributed generations to provide ancillary services like regulation, load following, unit commitment in different time scale to the grid will need to be embedded, close to what are being required for large generations today. Communication infrastructure will be necessary to provide required observability and controllability regarding to the grid operator, which will represent a major business opportunity in the distributed generation area. Low voltage distribution system infrastructure design will have to take up new philosophy especially in protection coordination in order for bidirectional power flow in the distribution system.

A future scenario in power system will be based on the "internet" model known today. Individual generation will take up an agreed global protocol for information exchange in the system regarding consumption and supply. Individual generation at each node listens to the rest of the network, adjusting its supply or consumption in relation to the global state of the electricity network.

III. PV Micro-grid system

Micro-grid system comprises of various energy resources from PV, wind, diesel-generation, energy storage etc mainly for electrification of areas where access to grid electricity is not viable or economically not worthwhile. A few major challenges are conceived today.

Local control and central control: The intermittent nature of PV, wind etc presents a major challenge to power quality in micro-grid, which in general has to be managed from supply control/prediction, storage management and consumption control/prediction. Fast regulation of power quality will have to be relied on individual generation reactive or active power output behavior responding to voltage and frequency variation the individual generation sees. The challenge lies in the way for PV or wind generation to be controlled with such behavior. In a longer time scale, a central control will help to maintain longer period power quality, stability and economics just like dispatching strategy for bulk power system today. For a small-scale micro-grid, economics may be realized via control distributed in the individual generation.

Fault location and protection: When most generations output power via power electronics with low overload capability, fault location in the micro-grid becomes a challenge. One expensive way to locate fault will be differential protection which principle does not rely on amplitude of fault current.

Micro-grid is usually seen in remote areas with poor access to technical skills. It has to be designed in a way to enable easy engineering upon installation, commissioning, operation and maintenance. It also has to be designed into a plug-and-play architecture that will allow for failure tolerance and capacity scalability. On the other hand, due to multiple stakeholders and users involved, an effective management model will be key to success.

In the longer term, the need for interconnections among multiple micro-grids and to conventional electrical grid will arise and strategic research has to be started. PV inverters with universal power, control and communication functionalities applicable to different applications will become reality.

IV. Conclusions

The paper concludes that modularity and scaling capability of PV inverters will be the major driving factors for cost down. Efficiency optimization has to be targeted for system including inverter efficiency at varying illuminations (partial load efficiency), in addition to maximum power tracking of PV modules. Control design to achieve robustness to electrical grid and illumination transients will be critical to system reliability growth, while ensuring fast islanding detection and equipment protection. In the long run, it is expected to see universal PV inverter coming to reality.

References:

[1] European commission, A Vision for Photovoltaic Technology, 2005.

[2] Juergen Schmid, Review of Advances in PV Systems Technology, Sixteenth European Photovoltaic Solar Energy Conference, 2001.

[3] Sigifredo Gonzalez et al, Removing Barriers to Utility Interconnected Photovoltaic Inverters, IEEE Photovoltaic Specialists Conference, 2000.

[4] Journal of Photo International, Market Survey on Inverters for Grid Tied PV Systems, May 2005.

[5] Journal of Renewable Energy World, Choosing the Right Inverter for Grid-Connected PV Systems, March-April 2004.

2006 5th International Power Electronics and Motion Control Conference

Adaptive Neuro-Fuzzy Control with Fuzzy Supervisory Learning Algorithm for Speed Regulation of 4-Switch Inverter Brushless DC Machines

A. Halvaei Niasar[*], H. Moghbelli[**] and A. Vahedi[*]

[*] Department of Electrical Engineering, Iran University of Science & Technology, Tehran, Iran
[**] Department of Electrical & Computer Engineering, Isfahan University of Technology, Isfahan, Iran

halvai@iust.ac.ir   hamoghbelli@yahoo.com   avahedi@iust.ac.ir

Abstract— **Principle of a new Adaptive Neuro-Fuzzy Inference System (ANFIS) with supervisory learning algorithm is introduced and is used to regulate the speed of a Four-Switch, Three-Phase Inverter (FSTPI) Brushless DC (BLDC) drive. The proposed algorithm has advantages of neural and fuzzy networks. To enhance of drive's performance, instead of well-known back propagation learning method, a fuzzy based supervisory learning algorithm is used. This newly developed design leads to a controller with minimum structure and improved dynamic performance. System implementation is relatively easy since it has minimum fuzzy rules and membership functions as compared with the conventional fuzzy and/or neural networks, used for electrical drive applications. In order to demonstrate the proposed ANFIS controller abilities to follow the reference speed and to reject disturbances, its performance is simulated and compared with that of a conventional PI controller.**

I. INTRODUCTION

The permanent magnet brushless DC (BLDC) motor is increasingly being used in automotive, computer, aerospace, military, industrial and household products because of its high efficiency, high power factor, high torque, simple control, and lower maintenance [1]. Moreover, to reduce the cost of BLDC motor drive, one approach is the using of minimum number of switches and to design the control algorithms in conjunction with a reduced component inverter to produce the desired speed–torque characteristics.

In other hand, in variable speed operations of BLDC motor, the PI control is still the most used control. This is because of its simplicity and ease of design. However, it has disadvantages that the performance depends to proportional and integral gains. Therefore, when operating condition changes such as disturbances, load changes and motor's parameters variations, the re-tuning process of control gains is necessary.

Recent advances in simulation methodologies have enabled the development and implementation of more complex algorithms like robust control, sliding mode and fuzzy control for drive systems. Fuzzy logic controllers have been implemented successfully in many control applications where fuzzy regulators lead to improved performance in the face of load disturbances. Fuzzy Logic Control (FLC) introduces a good tool to deal with complicated, non-linear and ill-defined systems. Artificial Neural Networks (ANN) has the powerful capability for learning, adaptation, robustness and rapidity. Here the advantages of both the FLC and ANN have been combined together to design a new controller.

This paper develops an ANFIS (or neuro-fuzzy) controller for improving the transient responses to torque disturbance and speed reference following of the BLDC motor drive. To train the proposed ANFIS controller, a fuzzy critic supervises the learning of ANFIS controller. Simulation results are used to show the abilities and shortcomings of the proposed algorithm as compared with the conventional PI controller.

II. ANALYSIS OF 4-SWITCH BLDC MOTOR DRIVE

The typical mathematical model of a three-phase BLDC motor is described by the following equations:

$$
\begin{bmatrix} v_a \\ v_b \\ v_c \end{bmatrix} = \begin{bmatrix} R & 0 & 0 \\ 0 & R & 0 \\ 0 & 0 & R \end{bmatrix} \times \begin{bmatrix} i_a \\ i_b \\ i_c \end{bmatrix} + \begin{bmatrix} L-M & 0 & 0 \\ 0 & L-M & 0 \\ 0 & 0 & L-M \end{bmatrix} \frac{d}{dt} \begin{bmatrix} i_a \\ i_b \\ i_c \end{bmatrix} + \begin{bmatrix} e_a \\ e_b \\ e_c \end{bmatrix} \quad (1)
$$

where R, L, M are the resistance, inductance and mutual inductance of stator windings and v_x, i_x and e_x are phase voltage, back-EMF voltage and phase current of each phase of stator respectively. The electromagnetic torque is expressed as

$$
T_e = \frac{1}{\omega_r}(e_a i_a + e_b i_b + e_c i_c) \quad (2)
$$

Fig. 1 shows the three waveform phase currents with the trapezoidal back-EMF of this type of BLDC motor. The shown waveforms are the ideal. To have a constant electromagnetic torque, phase currents should be regulated to be as quasi-square waveforms [1,2]. For four-switch BLDC motor drive, current regulation needs to more attention rather than 6-switch drive [3].

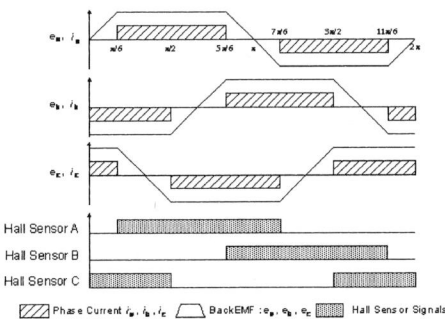

Figure 1. Signal waveforms of the BLDC motor

1-4244-0448-7/06/$25.00 ©2006 IEEE   597

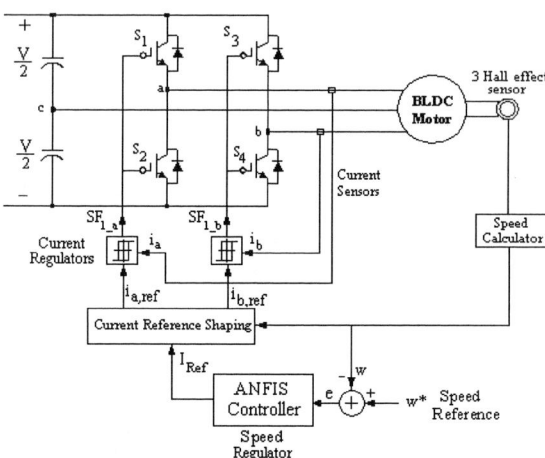

Figure 2. Configuration of BLDC motor drive system

According to Fig. 1, a BLDC motor needs quasi square current waveforms, which are synchronized with the back-EMF to generate constant output torque. Also, at every instant only two phases are conducting and the other phase is inactive. However, in the four-switch inverter, the generation of 120 conducting current profiles is inherently due to its limited voltage vectors. Therefore, in order to use the four-switch inverter topology for the three-phase BLDC motor drive, we use the direct phase current control method which regulates the i_a, i_b independently. Fig. 2 shows the overall system configuration of the 3-phase 4-switch BLDC motor drive. Phase a, b currents are controlled via hysteresis current controller. The advantages of the hysteresis current controller are fast transient response and simple implementation. The current ripple is a disadvantage of the controller.

For modeling the four-switch inverter BLDC motor drive, we use the switching function concept which has been proposed in [4,5] for a six-switch inverter BLDC motor drive and then modify it for our four-switch inverter. According to Fig. 2, there are only 4 switches and so, we have two switching functions SF_{1_a} and SF_{1_b}, to regulate i_a, i_b currents.

III. ANFIS CONTROLLER

Many researchers have recently studied intelligent controller in order to solve problems with classical methods of control, and what remarkable of all is fuzzy control using expert's knowledge or experience or linguistic variables and neural network with learning capability. The characteristics of fuzzy control are: First, approximate knowledge of plant is required. Second, knowledge representation and inference is simple with plural forms of IF-THEN. Third, its implementation is fairly easy [6]. The characteristics of neural network are: First, with learning ability, neural network can adapt itself to changing control environment and can learn just by the type of input and output. Second, it can work real time performing random data mapping by parallel distributed processing, and so any kind of nonlinear mapping is possible. Third, it does not required difficult theories of control, knowledge of system, or other environment [7].

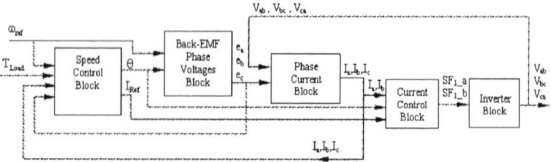

Figure 3. Switching functions concept based model of the BLDC motor drive

Fuzzy control and neural network control have many advantages as above, but fuzzy control also has a drawback that you have to set new control law and membership function every time types of system change. These types of controllers also lack the parallelism of neural controllers. And neural network has a drawback that while learning, it can easily fall onto local minimum instead of global minimum, and it take much time to make as many neurons learn as how complicated the system. In order to make up for defects, research on integration of neural network and fuzzy control is under way recently [8].

A. Structure of Neuro-Fuzzy controller

To overcome the shortcomings of fuzzy and neural networks it is wised to use the combination of both, which leads to neuro-fuzzy controllers. The basic concept of neuro-fuzzy control models is first to use structure-learning algorithms to find appreciate fuzzy logic rules and then use parameter learning algorithms to fine-tune the membership functions and other parameters. In this hybrid structure, the input and output nodes represents the input states and output control or decision signal respectively and in hidden layer there are nodes functioning as membership functions and fuzzy logic rules [9].

The basic structure of the utilized neuro-fuzzy controller in this paper takes the form of fuzzy controller, and separate elements are composed of a neural network. Therefore as shown in Fig. 4 below, the structure of this controller contains the parts of fuzzification, inference engine and the part of defuzzification [10].

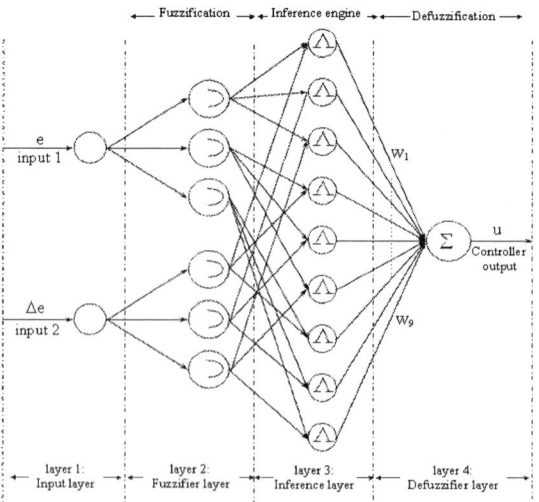

Figure 4. Neuro-Fuzzy controller structure

In this paper, neuro-fuzzy controller use an error E and change rate of error ΔE as an input signal, both of which are calculated by comparator. In order to calculate the two input signals through qualitative fuzzy, fuzzy process is carried out by means of membership function and nonlinear quantization. The membership functions are adopted as shown in Fig. 5. Because of adaptation and supervisory learning, input variables have only three membership functions as *NEG*, *ZE* and *POS*. The rule base of connectionist structure contains fuzzy IF-THEN rules of Mamdani type. The general formula is as follows:

R_i: IF E is A_i and ΔE is B_i THEN u is C_i ; for i=1,…,9.

Here, E, ΔE and u are inputs and output variable and A_i, B_i and C_i are quantitative linguistic value to each variable respectively. The final output is the weighted average of each rule's output.

To implement the neuro-fuzzy controller, firstly we define a general neuron as Fig. 6 that its input-output relation is as:

$$O_i^k = a\left(f(u_1^k, u_2^k, u_3^k, ..., u_n^k, W_1^k, W_2^k, ..., W_n^k)\right) \quad (3)$$

Which a is the node activity function, f is the node input function, u_i^k is i'th input in layer k, W_i^k is i'th weight of node in layer k and O_i^k is the output of i'th node in layer k. As shown in Fig. 4, architecture of the ANFIS has a total of 4 layers as follow:

Layer 1: or input layer; every input is scaled in limited range of input membership functions. All weights in this layer are constant and equal to unit ($W_i^1 = 1$) and we have:

$$O_i^1 = k \cdot u_i^1 \qquad \text{for} \quad i = 1, 2 \quad (4)$$

Layer 2: or fuzzification layer; every node is a constant node with triangle membership functions. Output function of each node is as

$$O_i^2 = \mu_{A_j}(u_i^2) \qquad \text{for} \quad i, j = 1, 2, 3 \quad (5)$$

Layer 3: or inference and decision layer; every node is a fixed node whose output is the product of all incoming signals. i.e., these nodes perform the fuzzy AND operation as:

$$O_i^3 = \mu_{A_j}(x) \cdot \mu_{B_k}(y) \quad \text{for} \quad i = 1, 2, ..., 9 \quad (6)$$

where $j, k = 1, 2, 3$, x, y are inputs of 3'th layer and O_i^3 represents the firing strength of i'th rule.

Layer 4: or defuzzifier layer; a single node computes the overall output as summation of all incoming signals. To defuzzify of output, mass of gravity method is used as:

$$O_1^4 = \frac{\sum_{i=1}^{9} W_i^4 u_i^4}{\sum_{i=1}^{9} W_i^4} \quad (7)$$

where W_i^4 is weight of input variables to 4'th layer. They are adaptive and training of neuro-fuzzy controller equals to adjusting these adaptive weights.

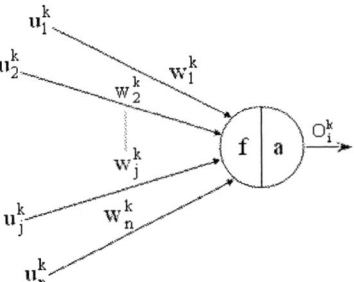

Figure 6. i'th general node in k'th layer

B. Supervisory learning in ANFIS controller

The Error Back Propagation Through Plant (EBP-TP) technique is one of the general approaches for training neural networks, that output error of the controller is passed through the plant, and updating law of the weights is achieved [11]. However, this technique has some defects, such as sensitivity to noise, disturbance and learning rate coefficient. To develop the learning, supervisory learning can be added to EBP-TP algorithm. As shown in Fig. 7, we can add a critic to ANFIS. Signal S is the output of critic and shows amount of the system stress.

For training the neuro-fuzzy system with a fuzzy critic, the criterion is selected as:

$$E(W_i^4) = \frac{1}{2} S^2 \quad (8)$$

W_i^4 should be adjusted in the direction of negative gradient of the $E(W_i^4)$. Thus, for 4'th layer, we have

$$\Delta W_i^4 \propto -\frac{\partial E}{\partial W_i^4} \quad (9)$$

By using the chain differential law

$$\frac{\partial E}{\partial W_i^4} = \frac{\partial E}{\partial S} \cdot \frac{\partial S}{\partial y} \cdot \frac{\partial y}{\partial u} \cdot \frac{\partial u}{\partial W_i^4} \quad (10)$$

or

$$\frac{\partial E}{\partial W_i^4} = S \cdot (-k) \cdot J \cdot \frac{\partial u}{\partial W_i^4} \quad (11)$$

where J is the plant gradient and always is positive. Thus,

$$\Delta W_i^4 = \eta \cdot S \cdot \frac{\partial u}{\partial W_i^4} \quad (12)$$

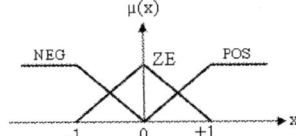

Figure 5. input/output membership function

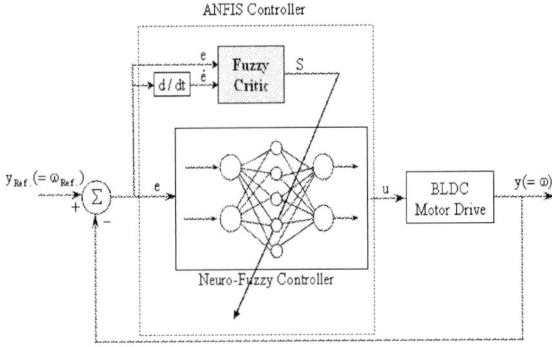

Figure 7. ANFIS controller with fuzzy supervisory learning

So the online updating law of weights is written as:

$$W_{i,new}^4 = W_{i,old}^4 + \eta \cdot S \cdot \frac{u_i^4}{\sum_{i=1}^9 u_i^4} \qquad (13)$$

where η is the learning rate coefficient of network. It is possible to generalize training to previous layers. But in the sense of practical remarks, it has some defects and therefore, we content learning only for the last layer [10]. Also, it is possible to use another critic in parallel to critic of error. This can limit the control effort. Operation of critics makes it possible to lower following error and control effort simultaneously.

C. Design of the Critic

Critic evaluates the performance of neuro-fuzzy controller and with respect to error and variation of error, it generates stress signal in range of [-1 +1]. -1 and +1 are corresponding to the worst cases in control of the plant. Critic can be described like a simple PD control system or for enhancing the training of ANFIS is considered as a simple fuzzy system. In this paper we design a fuzzy critic which its surface view of its fuzzy rules is shown in Fig. 8. It is the simplest fuzzy structure that includes of only 3 membership functions and 9 rules.

IV. SIMULATION

In this section some simulation results are used to explore the proposed ANFIS controller and compare its performance with the conventional PI controller. The entire system is modeled and simulated using the Matlab/Simulink toolbox. Fig. 9 shows the implementation of ANFIS controller in simulink. Simulations are performed using a 3-phase BLDC motor which its nominal parameters are given in the table 1. The critic in the ANFIS controller has been designed as a fuzzy system as shown in Fig. 8, and the learning rate coefficient is set to $\eta = 0.6$. PI controller is assigned by $k_P = 1$, $k_I = 10$. Usually PI controller parameters are adjusted by trial and error and are adjusted for rated speed. Also i_a, i_b currents are controlled via hysteresis current controllers.

Fig. 10 compares the performance of the BLDC motor drive with ANFIS & PI controllers in various speeds. It is obvious that speed tracking by using ANFIS controller in all speeds is better than PI controller. Also in Fig 11 ANFIS controller has rejected load disturbance better than PI controller. All simulation results indicate to fast and smoothness of ANFIS controller response.

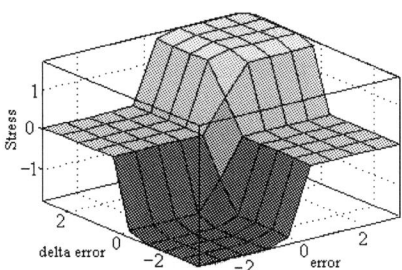

Figure 8. Surface view of fuzzy rules for cirtic

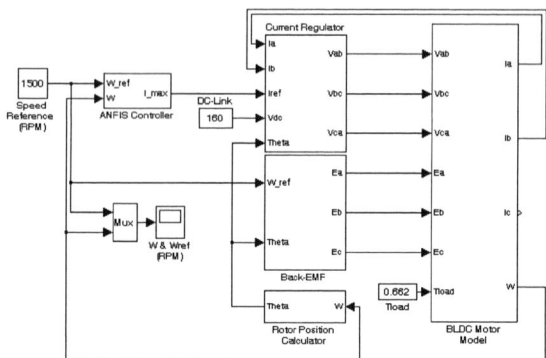

Figure 9. Block diagram of simulated BLDC motor drive in Simulink

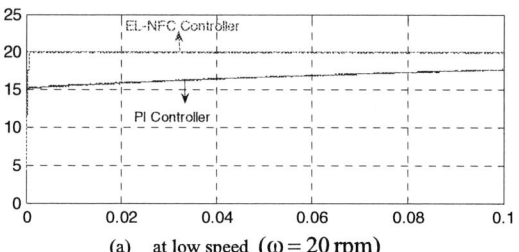

(a) at low speed ($\omega = 20\,\text{rpm}$)

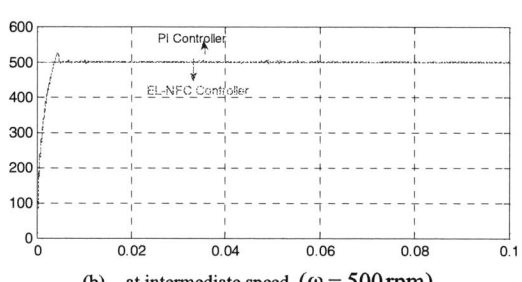

(b) at intermediate speed ($\omega = 500\,\text{rpm}$)

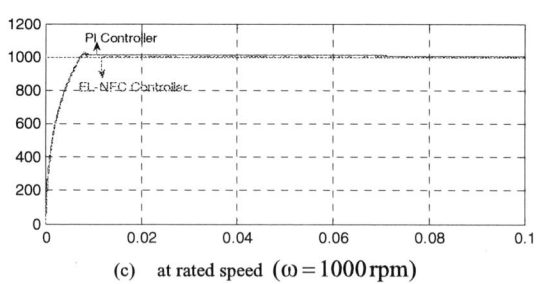

(c) at rated speed ($\omega = 1000\,\text{rpm}$)

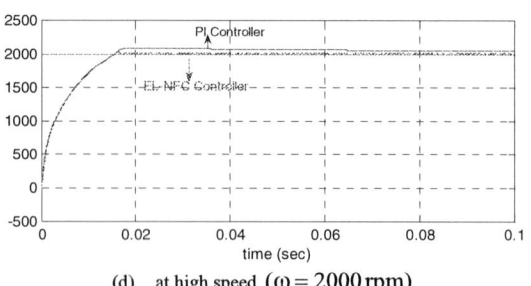

(d) at high speed ($\omega = 2000\,\text{rpm}$)

Figure 10. Comparison of speed tracking of ANFIS & PI controllers at various speeds

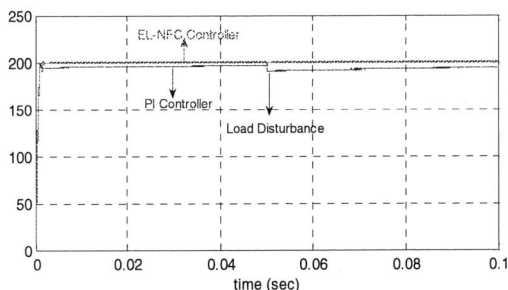

Figure 11. Disturbance rejection by ANFIS & PI controllers

TABLE 1:
RATED PARAMETER OF BLDC MOTOR

K_t	0.21 [Nm]	R	0.75 [Ω]
K_{e_LL}	0.21 [V.(rad/sec)$^{-1}$]	L-M	3.05 [mH]
J	8.2e-5 [kgm^2]	P_{rated}	1[HP]
T_L	0.66 [Nm]	ω_{rated}	3500 [rpm]

V. CONCLUSION

The purpose of this paper is to design an adaptive neuro-fuzzy inference system (ANFIS) based controller in order to solve the general problems of fuzzy systems and neural networks. Training of the proposed ANFIS controller is based on emotional learning. To improve controller performance a fuzzy critic has been defined and used to supervise the learning of neural network instead of back propagation learning method.

This controller was compared with the conventional PI controller for speed tracking and disturbance rejection of a four-switch inverter BLDC motor drive. The following is the main conclusion based on simulation results:

- ANFIS controller has the better transient and steady state responses rather than PI controller in entire of speed range.
- It is simpler than other reported adaptive neuro-fuzzy controllers in literature [12-16]. In other words it has only 18 neurons, 4 layers and 9 rules.
- It does not require an accurate model of the plant, its knowledge representation and interface description is relatively simple and therefore its construction and implementation is fairly easy.
- It doesn't require knowledge of expert man to obtain and set its rule bases since less number of adjustable parameters is involved (as compared with fuzzy and/or neural systems).

So, the proposed ANFIS controller, due to its non-model base, can be used to control a wide range of complex and nonlinear systems. In order to a practical use of this controller, we demand the research for more stability fuzzy rule.

VI. REFERENCES

[1] R. Krishnan; "Electric Motor Drives: Modeling, Analysis and Control ", *Printice-Hall of India,* New Delhi, 2003.

[2] B.K. Lee and M. Ehsani; "Advanced BLDC Motor Drive for Low Cost and High Performance Propulsion System in Electric and Hybrid Vehicles", *IEEE 2001 International Electric Machines and Drives Conference,* 2001, Cambridge, MA, June 2001, pp. 246-251.

[3] B.K. Lee, T.H. Kim, M. Ehsani; "On the feasibility of four-switch three-phase BLDC motor drives for low cost commercial applications: topology and control", *IEEE Transactions on Power Electronics,* Vol. 18, No. 1, pp. 164-172, January 2003.

[4] E. P. Wiechmann, P. D. Ziogas, V. R. Stefanovic; "Generalized Functional Model for Three Phase PWM Inverter/Rectifier Converters", *in Proc. IEEE IAS'85,* 1985, pp. 984-993.

[5] B. K. Lee, B. Fahimi, M. Ehsani; "Dynamic Modeling of Brushless DC Motor Drives", *European Conference on Power Electronics and Applications (EPE'2001), Austria.*

[6] C. Chein, " Fuzzy Logic in Control Systems: Fuzzy Logic Controller, Part 1-2", *IEEE, Trans. on Systems, Man and Cybernetics,* Vol. 20, No. 2, pp. 404-428, March/April 1990.

[7] P.K. Dash, S.K Panda, T.H. Lee, J.X. XU, A. Routray, "Fuzzy and Neural Controllers for Dynamic Systems: An Overview", *IEEE, Proc. on Power Electronics and Drive Systems,* Vol. 2, pp. 810-816, 1997.

[8] J. Jang, C. Sun, E. Mizutani, Neuro-Fuzzy and Soft Computing, *Prentice-Hall, Inc.* 1997.

[9] C. Lin, Y. Lu, "A Neural Fuzzy System with Fuzzy Supervised Learning", *IEEE, Trans. on Systems, Man and Cybernetics,* Vol. 26, No. 5, Oct. 1996.

[10] S.A. Jazbi, Development of Emotional Learning Methods for Intelligent Control and its Industrial Applications, *M.S. Thesis, Dept. of EE engineering, University of Tehran,* Tehran, Iran, spring 1998.

[11] J. Jang; "Self Learning Fuzzy Controller Based on Temporal Back-Propagation", *IEEE Trans. On Neural Network,* Vol. 3, pp. 714-723, Sep. 1992.

[12] R. Sankaran, P.S Chandramohan; "Adaptive neuro-fuzzy controller for improved performance of a permanent magnet brushless DC motor", *The 10th IEEE International Conference on Fuzzy Systems,* 2001, pp. 493-496.

[13] Khalil Shujaee, S. Sarathy, R. Nicholson; "Neuro-fuzzy Controller and Convention Controller: A Comparison", *Proceedings of the 5th Biannual World Automation Congress,* 2002, pp. 207-214.

[14] G. Bologna; "FDIMLP: a new neuro-fuzzy model", *IJCNN '01, International Joint Conference on Neural Networks,* 2001, pp. 1328-1333.

[15] S. H. Kim, L. K. Kim; "Design of a neuro-fuzzy controller for speed control applied to ac servo motor*", ISIE 2001, IEEE International Symposium on Industrial Electronics,* 2001, pp. 435-440.

[16] C. T. Su, G. R. Lii, H.R. Hwung; "A Neuro-fuzzy Method For Tracking Control", (ICIT '96), *Proceedings of The IEEE International Conference on industrial Technology,* 1996, pp. 482-486.

2006 5th International Power Electronics and Motion Control Conference

Combined Modulation and Harmonic Suppression

Cheng Weibin[1,2], Zhong Yanru[2], Jin Shun[2]

1. Xi'an Shiyou University , Xi'an , 710065 , China
2. Xi'an University of Technology , Xi'an , 710048 , China
e-mail: wbcheng@xsyu.edu.cn
zhongyr@xaut.edu.cn

Abstract—The output of pulse width modulation includes harmonics with invariable frequencies and invariable magnitudes, which are sources of conducted interference and load resonance . Based on the spectrum analysis of switching modulation and sinusoidal pulse width modulation (SPWM) , frequency modulation and low frequency amplitude modulation are combined to make a nonlinear swing for the modulated frequencies, to spread the harmonics frequencies broader , to reduce the average amplitudes of various harmonics , and the experimental results are also given out in this paper .

Keywords - Harmonic ; Spectrum analysis ; Frequency modulation ; Pulse width modulation ; Amplitude modulation

I. INTRODUCTION

Pulse width modulation is widely used in switching powers and converters, whose output includes large numbers of harmonics with constant frequencies and constant amplitudes , which often give birth to conducted interferences . In some power circuits , the harmonics maybe resonate with the eigen-frequencies of load , then destroy the load and driving circuit [1,2].In order to reduce the output harmonics , many kinds of PWM techniques are researched , such as SPWM , three level PWM[3] and multi-level PWM[4],et al. .

Based on spread-spectrum technique [5,6] , a novel combination with nonlinear frequency modulation and low frequency amplitude modulation is presented in this paper , to make the output frequency take a nonlinear swing about a central frequency , to decrease the quasi-peak amplitude and average amplitude of the output harmonics .

II. SPECTRUM ANALYSIS OF SWITCHING MODULATION MODE

Switching modulation mode in switching-mode power supply (SMPS) and converters includes pulse width modulation , pulse phase modulation , carrier frequency modulation with fixed duty (CFMFD) and carrier frequency modulation with varied duty (CFMVD)[7,8] .

A. PWM with constant frequency and constant duty

The output of the PWM wave lies on its duty , a PWM wave with a fixed period T_2 depicted in Fig.1 can be considered as odd integer multiples of PWM waves with the common period T_2 shift laterally interval T_1 in turn and superpose , and $T_2 = N \times T_1$. Suppose the duty of each single PWM wave is τ , and the amplitude is unity , then this PWM wave with period T_2 can be expressed as :

$$f_T(t) = \sum_{k=-\frac{N-1}{2}}^{\frac{N-1}{2}} f(t - kT_1) \tag{1}$$

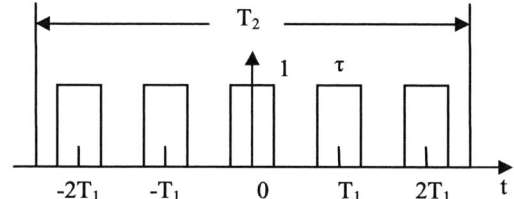

Figure 1. PWM waves superposition (suppose N =5)

Use the time-shift property , then obtain the complex coefficients of Fourier transforms as followed :

$$
\begin{aligned}
\dot{F}_n &= \frac{\tau}{T_2} \left(\sum_{k=-\frac{N-1}{2}}^{\frac{N-1}{2}} e^{jw_2 kT_1} \right) \bullet \frac{Sin\left(\frac{n\tau}{T_2}\pi\right)}{\frac{n\tau}{T_2}\pi} \\
&= \frac{\tau}{T_2} \frac{\sin\left(\frac{N\omega_2 T_1}{2}\right)}{\sin\left(\frac{\omega_2 T_1}{2}\right)} \bullet \frac{Sin\left(\frac{n\tau}{T_2}\pi\right)}{\frac{n\tau}{T_2}\pi} \\
&= \frac{\tau}{T_2} \frac{\sin\left(\frac{N\omega_2 T_1}{2}\right)}{\sin\left(\frac{\omega_2 T_1}{2}\right)} \bullet Sinc\left(\frac{n\tau}{T_2}\right) \tag{2}
\end{aligned}
$$

Where $n = 0, \pm 1, \pm 2 \cdots$, and $\omega_2 = \frac{2\pi}{T_2}$.

The spectrum of PWM wave includes discrete fundamental frequency and its harmonics , and the amplitude of the envelope changes as the sinc function $Sinc(\frac{n\tau}{T_2})$. The maximum of sinc function is unity at

1-4244-0448-7/06/$25.00 ©2006 IEEE   602

$\frac{n\tau}{T_2} = 0$. The zero crossings occur at $\frac{n\tau}{T_2} = m$ (m is integer).

For $\omega_2 = \frac{2m\pi}{T_1}$ ($m = 0, \pm1, \pm2 \cdots$), can obtain :

$$\lim_{\omega_2 \to \frac{2m\pi}{T_1}} \frac{\sin(\frac{N\omega_2 T_1}{2})}{\sin(\frac{\omega_2 T_1}{2})} = N \qquad (3)$$

The coefficient sums $\sum\limits_{k=-\frac{N-1}{2}}^{\frac{N-1}{2}} e^{jw_2 kT_1}$ come to the maximum , amounts to N multiples of the amplitude of each frequency .

For $\omega_2 = \frac{2m\pi}{N \cdot T_1}$ (m is integer , but not a integer multiple of N), can also obtain :

$$\sin(\frac{N\omega_2 T_1}{2}) = 0 \qquad (4)$$

The amplitude sums are also equal to zero .

To sum up , the output spectrums of PWM wave concentrates on fundamental frequency and its harmonics with higher amplitudes , which are the activation resources of load resonance and conducted interference .

B. Pulse phase modulation (PPM)

The output of the PPM lies on its starting phase, a PPM wave with fixed period T_2 and fixed duty τ depicted in Fig.2 can be considered as odd integer multiples of pulse modulation waves with the common period T_2 shift laterally interval $kT_1 + \alpha_k$ from the origin and superpose , then the PPM wave can be expressed as :

$$f_T(t) = \sum_{k=-\frac{N-1}{2}}^{\frac{N-1}{2}} f(t - kT_1 - \alpha_k) \qquad (5)$$

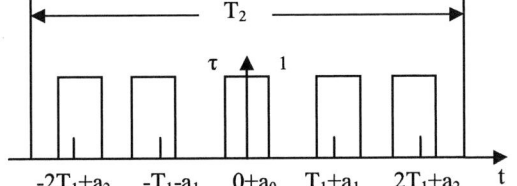

Figure 2. PPM waves superposition (suppose $N = 5$)

Then the complex coefficients of Fourier transforms can be written as followed :

$$\dot{F}_n = \frac{\tau}{T_2} \left(\sum_{k=-\frac{N-1}{2}}^{\frac{N-1}{2}} e^{jw_2(kT_1 + \alpha_k)} \right) \bullet \frac{Sin(\frac{n\tau}{T_2}\pi)}{\frac{n\tau}{T_2}\pi} \qquad (6)$$

The harmonics frequencies of PPM are consistent with those of PWM ,but those amplitudes are decreased .

C. Carrier frequency modulation with fixed duty and carrier frequency modulation with varied duty

Carrier frequency modulation with fixed duty can be considered as pulse modulation waves with varied period $T_2(t)$ and fixed duty τ ,the harmonics amplitudes are equal to those of PWM, only the frequencies are different .

Carrier frequency modulation with varied duty can be considered as pulse modulation wave with varied period $T_2(t)$ and varied duty $\tau(k)$,and the complex coefficients of Fourier transforms can be written as followed :

$$\dot{F}_n = \sum_{k=-\frac{N-1}{2}}^{\frac{N-1}{2}} \left(\frac{\tau(k)}{T_2(t)} e^{jw_2(t)kT_1} \bullet \frac{Sin(\frac{n\tau}{T_2(t)}\pi)}{\frac{n\tau}{T_2(t)}\pi} \right) \qquad (7)$$

Although the amplitudes are not decreased , and also varied as period ,but the spectrum of CFMVD wave are spread broader than those of PWM .

III. COMBINED FREQUENCY MODULATION

A. sinusoidal pulse width modulation

In order to improve the output efficiency , decrease the harmonics , the mechanism of equivalent area is used to make the area of PWM wave equal to that of sinusoidal wave , this means that the pulse width changes as sinusoidal wave , which is called sinusoidal pulse width modulation (SPWM) , and is depicted in Fig.3 , then the width of the nth pulse is expressed as :

$$\tau(n) = T_1 \sin(nT_1) \qquad (8)$$

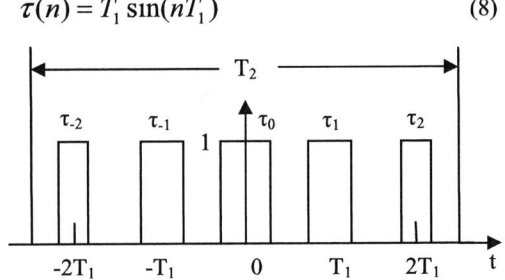

Figure 3 sinusoidal PWM waves ($N = 5$)

Then , the Eq.(2) can be written as followed :

$$\dot{F}_n = \sum_{k=-\frac{N-1}{2}}^{\frac{N-1}{2}} \left(e^{jw_2 kT_1} \frac{\tau(n)}{T_2} Sinc(\frac{n\tau(n)}{T_2}) \right)$$

(9)

Where $n = 0, \pm 1, \pm 2 \cdots$.

$$\tau(n) = T_1 \sin(nT_1) \tag{10}$$

Only at $\frac{n\tau(n)}{T_2} = m$ (m is integer) , sinc function is equal to zero , the amplitudes of envelope are also equal to zero . For $\omega_2 = \frac{2m\pi}{T_1}$ ($m = 0, \pm 1, \pm 2 \cdots$) and

$\omega_2 = \frac{2m\pi}{N \cdot T_1}$ (m is integer, but not an integer multiple of N) , the amplitudes of those frequencies spectrums superpose together , can not approach to the maximum , also can not equal to zero , but the sum amplitudes are decreased greatly .

B. Nonlinear frequency modulation

The Eq.(2) denotes that the amplitudes of each harmonics not only change with the pulse width , but also are the function of pulse period , if the frequencies are modulated , the amplitudes of each harmonics are also changed . The modulated angular velocity ω_i can be written as followed :

$$\omega_i = \omega_2 + \Delta\omega \cos(\omega_m t) \tag{11}$$

where $\omega_2 = \frac{2\pi}{T_2}$ is the central angular frequency ,

$\omega_m = \frac{2\pi}{T_m}$ is the modulation angular frequency , $\Delta\omega$ is the maximum angular displacement , and $\omega_2 = M\omega_m$.

According to the equivalent area , the modulated frequency PWM wave with period T_m , which is depicted in Fig.4 , can be considered as M pulses with common period T_m and different duty τ_i , displace interval T_1 in turn and superpose together , the duty τ_i is a function of T_2 and T_m , and decided by Eq.(12) :

$$\tau_i = f_M(i, T_2, T_m) \tag{12}$$

As a result of time-shift property , the complex coefficient of Fourier transforms is written as :

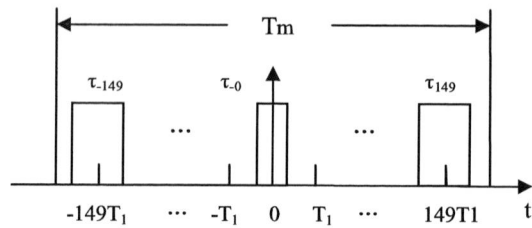

Figure 4 . Nonliear frequency modulated waves (suppose M = 299)

$$\dot{F}_n = \sum_{i=-\frac{M-1}{2}}^{\frac{M-1}{2}} \left(e^{jwnT_1} \frac{\tau_i}{T_m} Sinc(\frac{n\tau_i}{T_m}) \right)$$

(13)

Where $n = 0, \pm 1, \pm 2, \cdots$.

At $\frac{n\tau_i}{T_m} = k$ (k is integer) , there will be a single frequency with its amplitude equal to zero , but the sum amplitude is not equal to zero . For $\omega_2 = \frac{2m\pi}{T_1}$ ($m = 0, \pm 1, \pm 2 \cdots$) and $\omega_2 = \frac{2m\pi}{M \cdot T_1}$ (m is integer , but not an integer multiple of M) , the amplitudes of those frequencies spectrums superpose together , can not approach to the maximum , also are impossible equal to zero , but the sum amplitudes are decreased greatly .

As mentioned previously , the frequencies of the switching modulation are invariable , but the frequencies of the nonlinear modulation can be changed , while the modulation ratio M is not integer , so the frequencies are taking nonlinear swing in a certain extent , spread the frequency spectrums broader , then decrease averagely the activation power so as to lighten the wave interferences and avoid the load resonance .

C. Low frequency amplitude modulation

The amplitude is generally invariable in the spectrum analysis of PWM waves , now introduces a low frequency voltage amplitude modulation . The voltage amplitude is modulated as followed :

$$V(t) = V_A + \Delta V \sin(\omega_3 t) \tag{14}$$

Where V_A is the unity voltage , ΔV is the maximum modulation voltage displacement , $\omega_3 = \frac{2\pi}{T_3}$ is the angular frequency of modulation voltage .

Use the linear and time shift properties to express the complex coefficients of the Fourier transform :

$$\dot{F}_n = \sum_{i=-\frac{M-1}{2}}^{\frac{M-1}{2}} \left(V(KT_1)e^{jwnT_1}\frac{\tau_i}{T_m}Sinc(\frac{n\tau_i}{T_m}) \right)$$

(15)

The ripple voltage with period 100Hz of the power factor correction circuit can be used to modulate the output voltage amplitude .The modulation ratio can be regulated by the output filter capacitor , the less capacitor makes the higher modulation amplitude , the stronger energy the high frequency harmonics have , then load resonance maybe occurs , at same time makes the driving current wave crest factor so high as to wear down the load quickly , to shorten the lifetime of driving circuit , to increase the total harmonic distortion (THD) of the input current.The experimental result in electronic ballast is shown in Fig.5 , the modulation ratio is generally less then 2% .

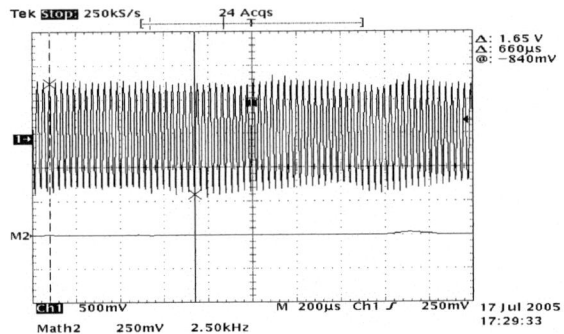

Figure 5 . Low frequency ripple voltage amplitude modulation

IV.   APPLICATION

The previous analyses show that the output frequency driven with voltage amplitude and frequency modulation , can be expressed as :

$$f = n_1 f_1 \pm n_2 f_2(t) \pm n_3 f_3 \qquad (16)$$

Where $n_1 = 0, \pm 1, \pm 2, \cdots$, $n_2 = 0, \pm 1, \pm 2, \cdots$,and $n_3 = 0, \pm 1, \pm 2, \cdots$ .This means that the output spectrums include three kinds of fundamental frequencies , the first is high frequency central frequency f_1 , the second is nonlinear swinging modulation frequency $f_2(t)$, the third is voltage amplitude modulation frequency f_3 .

The experimental waves and spectrums are shown in Fig.6 . The modulated frequency keeps taking a nonlinear swinging , so makes the output spectrums spread over a broad frequency band , and reduces evidently the average value of the amplitude of each frequency.

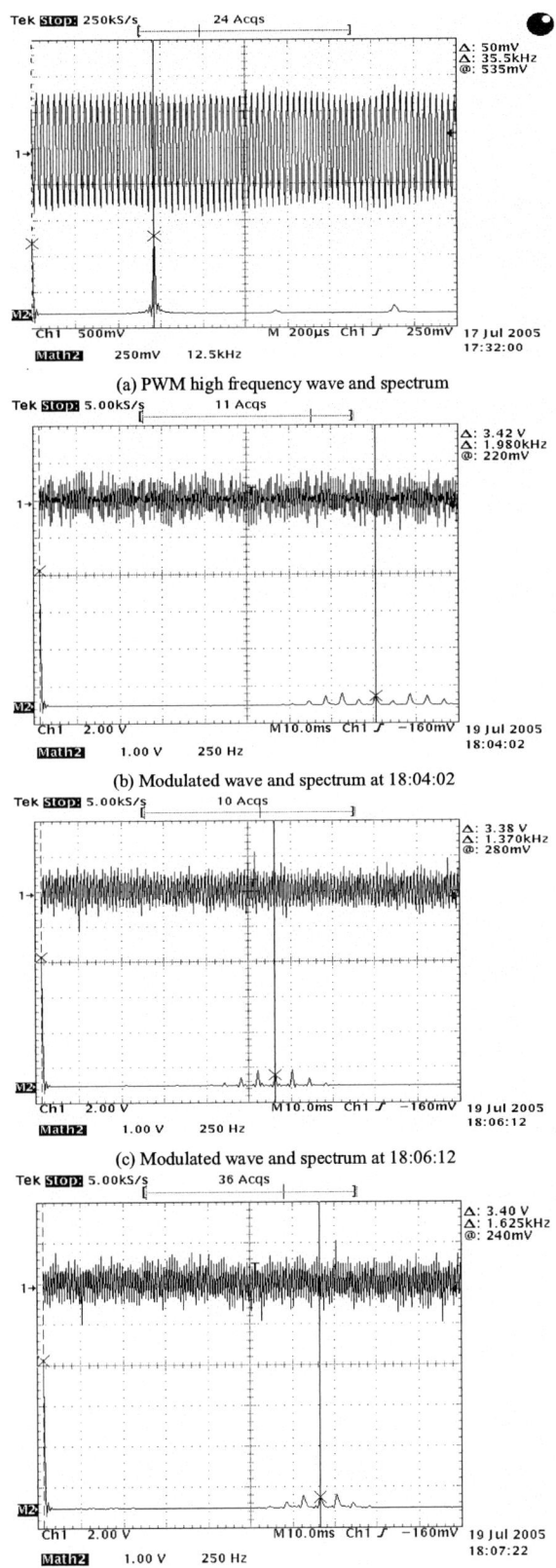

(a) PWM high frequency wave and spectrum

(b) Modulated wave and spectrum at 18:04:02

(c) Modulated wave and spectrum at 18:06:12

(d) Modulated wave and spectrum at 18:07:22

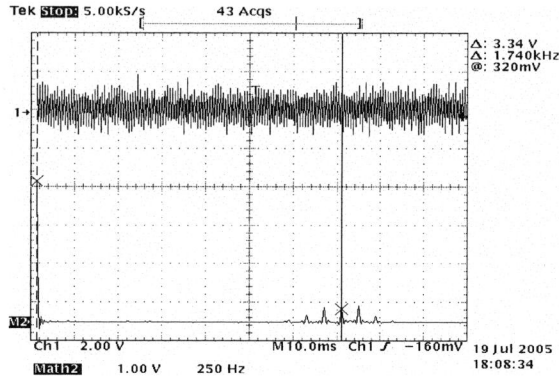

(e) Modulated wave and spectrum at 18:08:34

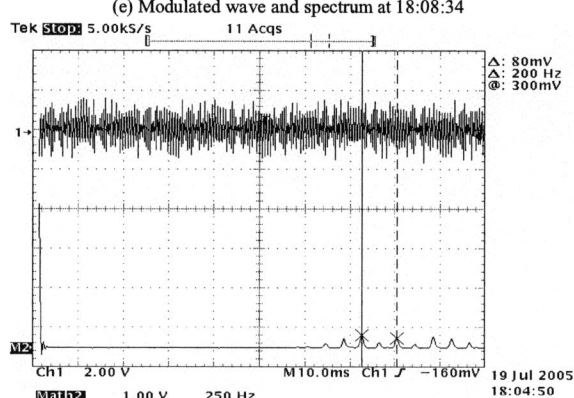

(f) 100Hz amplitude modulated wave and spectrum
Figure 6. Combined frequency modulation spectrums and waves

V. CONCLUSION

This paper analyses the frequency spectrums of switching modulation waves , in particular deduces the spectrum formula of PWM wave , and introduces a novel combination of nonlinear frequency modulation with continuous swinging center frequency and low frequency voltage amplitude modulation , to spread the frequency spectrums over a broader frequency band , to reduce the activation energies of harmonics . It is shown by experimental results that the combined modulation with the spread-spectrum technique can avoid efficiently the acoustic resonances in HID lamps with low cost .

REFERENCES

[1] C. F. Gallo , J. E Courtney, "Acoustic resonances in Modulated Xenon and Krypton Compact Arc Lamps", *Applied optics*,1967,vol.16,pp.939.

[2] S. Bhosle , M. Aubes , and M Cristea , "Acoustic resonance in HID lamps", *Light Sources 2004 - Proceedings of the Tenth International Symposium on the Science and Technology of Light Sources*,2004,pp.35-40.

[3] Zhang Yanli , Fei Yuanmin , and Lu Zhengyu , "Three-level voltage inverter selected harmonic elimination PWM", *Transactions of China Electrotechnical Society*,2004, vol.19,pp.16-20,54.

[4] Lu Qiwei ,Wang Long , "Research on space vector PWM for five-level inverter",Transactions of China Electrotechnical Society, 2005,vol.20,pp.83-88.

[5] M. Rahkala ,T. Suntio , and Kalliomaki , "Effects of switching frequency modulation on EMI of a converter using spread spectrum approach",*Conference Proceeding – IEEE Applied Power Electronics Conference and Exposition – APEC*,2002,vol.1,pp.93-99.

[6] D. A. Stone , B. Chambers , "Effects of spread-spectrum modulation of switched mode power converter PWM carrier frequencies on conducted EMI", *Electronics Letters* , 1995,vol.31,pp.769-770.

[7] Gao Jinfeng ,Wu Zhenjun , Zhao Kun, "Research on Suppressing Electromagnetic Interference Level of Buck dc/dc converter by Using Chaotic Modulation Technolgy",*Transactions of China Electrotechnical Society*,2003,vol.18,pp.23-27.

[8] K. K. Tse , Henry Shu-hung Chung , S. Y. R. Hui ,et al. , "A comparative investigation on the use of random modulation schemes for dc-dc converters",*IEEE Trans. On Industrial Electronics* , 2000, vol. 47 , pp. 253-263.

Application Research of Maximum Wind-energy Tracing Controller Based Adaptive Control Strategy in WECS

Changhong Shao *, Xiangjun Chen ** and Zhonghua Liang ***

* Products Business Department, Beijing AriTime Intelligent Control Co. Ltd, Beijing China
** Products Business Department, Beijing New Image Electronic Co. Ltd, Beijing China
*** Department of Electrical Engineering Shenyang University of Technology, Shenyang, China

Abstract—Based on the prototype of Wind Electricity Conversion System (WECS), this paper presents a self-adaptive control strategy for doubly fed induction generators (DFIG) applied to variable speed constant frequency wind generation system. The strategy mainly depends on wind speed estimation, the wind turbine and generator parameters. Wind speed is estimated by the output power and the efficiency of the generator. In the adaptive Maximum Power Point Tracking (MPPT) strategy, an estimated maximum efficiency is adopted by the maximum tip-speed ratio tracker, without the measurement of mechanical quantities. The steady and dynamic performance of the control strategy had been verified for efficacy by MATLAB/Simulink software and on experiment platform of a 2kW laboratory variable speed constant frequency wind generation system. WECS can operate at the expected points in the wide range of wind velocities. The controller may be applicable to all doubly fed generators, including brushless double fed machine.

Keywords- adaptive optimal controller, wind power, double fed generator, wind speed estimated, maximum power point tracking

I. INTRODUCTION

Development of wind electricity conversion system not merely can save the routine energy resources, but also is favor of the environmental protection and one of the effective ways of improving the makeup of the energy resources and decreasing pollution of the environment. Thus it can bring immediate economic benefits, community benefits and environment benefits [1]. Currently, the utilization ratio of the wind energy is not high in wind electricity conversion system, so how to control the electric machine and capture maximum wind energy becomes very hot topic.

In order to capture maximum wind energy as soon as possible, the control of the rotational speed of the generator connected to wind turbine is one of the most significant features characterizing the overall design of a wind power plant. The speed of the wind turbine determines the efficiency of wind electricity conversion system, for given wind speed, blades geometry and turbine orientation [2]. This mechanical energy is then transformed into electric energy by the generator driven by the wind turbine. The wind turbine is connected with generator by gear case. The speed control of the generator thus provides the means to regulate the amount of energy captured from the wind [3].

Many different topologies and control strategy [4] have been proposed to achieve the desired speed control on wind generators. [5] proposed a chopping and doubly-fed adjusting speed control strategy. In order to capture maximum wind energy, [6] presented a decoupled active and reactive power control strategy for doubly excited induction machine. The overall system power flow problem was not studied, as only the output power from the stator side is controlled. [7] studied a tracking maximum power point. The dynamic performance is not well in this method. [8] studied a fuzzy logic approach to control the electromagnetic torque for maximum power capture and dynamic performance improvement of the turbine generator system. The paper, thus, presents a self-adaptive control strategy based RBF neural network for doubly fed induction system. The aim of the control strategy is to maximize the output power, maximizing the efficiency of the energy conversion. The 2kW doubly fed asynchronous generators is used in the experiment in order to verify the control strategy presented by the paper.

II. WIND POWER AND TURBINE CHARACTERISTICS

Wind is an intermittent and variable source of energy. Wind speed is random in magnitude and direction at any site except that average wind speed and direction distribution on an annual basis is repeatable within moderate limits. Wind variations, including time variation and spatial variation, are caused by many factors such as wind turbulence, wind shear, and tower shadow and turbine rotation. However, the effect on the turbine torque can be modeled using wind profile at the hub height to produce the equivalent torque variations. So that a simple wind turbine model can be used [7].

The power of the wind is proportional to the cube of the wind speed and may be expressed as

$$P_{\text{wind}} = 0.5\rho\pi R^2 v^3 \tag{1}$$

Where ρ (kg/m^3) is the air density and R is the radius (m) of the wind turbine blade. A wind turbine, however, can only withdraw a certain percentage of the wind power, up to the maximum Betz limit of 59%. This fraction is described as the power coefficient. The value of the power coefficient is a function of the wind speed, rotation speed and pitch of the specific wind turbine. A simple equation (2) is quite often used to describe the mechanical power transmitted to the hub shaft.

$$P_{\text{turbine}} = 0.5 C_p \rho \pi R^2 v^3 \qquad (2)$$

This coefficient (C_p) has only one maximum value ($C_{P\max}$) at a fixed operating point under a certain constant wind speed. When wind speed changes, it is therefore necessary to adapt the rotational speed of the turbine to correspond to the wind speed, in order to gain the maximum value ($C_{P\max}$). The characteristics of the power coefficient are normally expressed in terms of the tip-speed ratio λ, which is defined as:

$$\lambda = \frac{v_p}{v} = \frac{\Omega R}{v} \qquad (3)$$

Where: v_p -tip-speed of turbine blade;

R-turbine rotor radius;

Ω-rotational turbine angular velocity;

v -wind speed.

Fig. 1 shows a typical C_P-λ curve under a certain constant wind speed where an optimum value of tip-speed ratio (λ_{opt}) corresponds to the maximum power coefficient ($C_{P\max}$).

Clearly the turbine speed has to be changed with wind speed so that optimum tip-speed ratio is maintained for maximum power capture. The maximum turbine power can be expressed as

$$P_{\max} = \frac{1}{2} \rho \pi R^2 \left[\frac{R\Omega_{opt}}{\lambda_{opt}} \right]^3 C_{p\max} \qquad (4)$$

or

$$P_{\max} = K\Omega_{opt}^3 \qquad (5)$$

where

$$K = \frac{1}{2} \rho \pi R^5 \frac{C_{p\max}}{\lambda_{opt}^3} \qquad (6)$$

Equation (4) indicates that the maximum extractable power by a turbine has a cubic relation with the turbine optimum speed. The optimal torque can be obtained by dividing equation (5) with optimal speed:

$$T_{opt} = K\Omega_{opt}^2 \qquad (7)$$

Equation (7) shows that the optimal torque, corresponding to the maximum power, is a quadratic function of the optimal speed. Consequently, the maximum power can be achieved by either imposing an optimal speed or an optimal torque. The paper chooses the optimal speed control based estimated wind speed.

III. CONTROLLER OF ADAPTIVE CONTROL ALGORITHM

A. An Adaptive Control Algorithm

From Fig.1 it is clear that, in order to obtain maximum power from the wind turbine, it is necessary to keep the optimal tip-speed ratio over a wide range of wind speed. As the wind speed varies, this involves appropriate speed control of generator shaft to ensure that maximum wind power is traced.

The main control blocks of this novel algorithm are as shown in Fig.2. The total output power P_T is calculated by the stator voltage and current for the MPPT control loop. The typical I_c versus P_T and the I_c versus η relation for the block of the Fig.2 is mapped by RBF neural network, which parameters are trained according to the data given by the factory of the generator (or gained by the experiments). The curve of C_p versus λ is drawn according to the data given by the factory of the wind turbine. The wind speed can be estimated by the measured output power and the estimated maximum efficiency $\eta_{\max}$. Then, according to the estimated wind speed and equation (3), the angle velocity can be calculated. At the established shaft speed, stable operation is possible for a variety of control winding current levels. For a given mechanical input power, this can be used to regulate the power flow in the stator windings and power converter. The maximize efficiency and output power can be gained. Thus, the overall optimization problem for the turbine and generator systems involves finding the maximum output power as a

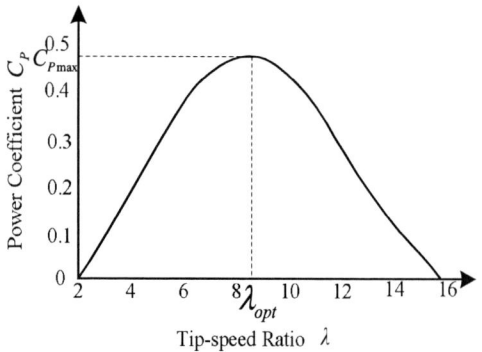

Fig. 1. Typical C_P-λ curve

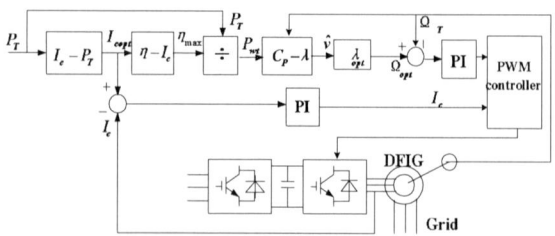

Fig.2. Adaptive control algorithm

function of both speed and control current.

B. Wind Speed Estimation Based MPPT Algorithm

Following the iterative algorithm, as expressed in Fig.2, the wind speed is estimated, which determines the optimal rotational speed Ω_{opt} to track the desired tip-speed ratio for the MMPT. Following the measured power, the optimum control winding current I_{opt} is commanded based upon information of I_c-P_T which is mapped by the RBF neural network. The maximum DFIG efficiency η_{max} is estimated at a particular control current optimized operating point utilizing a mapped η versus I_c^{opt} characteristic of the generator. DFIG input power P_{wt} is calculated from the estimated maximum efficiency η_{max} and measured output power P_T as

$$P_{wt} = \frac{P_T}{\eta_{max}}$$

Based upon information of P_{wt} and the measured angular velocity, the wind speed is estimated employing the following procedure.

The characteristics of the power coefficient of a wind turbine are normally expressed in terms of the tip-speed ratio λ, as shown in Fig.1. C_P depends on the parameters of the turbine blade and can be represented as a function in λ, such as an nth-order polynomial.

$$C_p(\lambda) = C_{p0} + C_{p1}\lambda + C_{p2}\lambda^2 + \cdots + C_{pn}\lambda^n \quad (8)$$

Power output of the wind turbine is related to the cube of the upstream wind velocity and can be expressed as (2). Substituting for the wind speed v from (3) in (2), the power delivered by the turbine can be expressed in terms of the angular velocity Ω_T and the tip-speed ratio λ as

$$P_{wt} = \frac{1}{2}\pi\rho C_p(\lambda)R^5\frac{\Omega_T^3}{\lambda^3} \quad (9)$$

Employing an iterative method for determination of the root of a polynomial, such as Newton-Raphson or bisection method, the roots of (9) can be determined. Further change the form of (6).

$$F(\lambda) = P_{wt} - \frac{1}{2}\pi\rho R^5\Omega_T^3[C_{p0}\lambda^{-3} + C_{p1}\lambda^{-2} + C_{p2}\lambda^{-1} + \cdots + C_{pn}\lambda^{n-3}] = 0$$

$$\frac{\partial F(\lambda)}{\partial \lambda} = -\frac{1}{2}\pi\rho R^5\Omega_T^3[-3C_{P0}\lambda^{-4} - 2C_{P1}\lambda^{-3}$$

$$- C_{P2}\lambda^{-2} - \cdots - (n-3)C_{Pn}\lambda^{n-4}]$$

An iterative method is then used, such that

$$\lambda^{(i)} = \lambda^{(i-1)} - \Delta\lambda^{(i-1)} \quad (10)$$

where

$$\Delta\lambda^{(i)} = [\frac{\partial F^{(i)}(\lambda)}{\partial\lambda}]^{-1}F^{(i)}(\lambda) \quad (11)$$

and the superscripts (i) and $(i-1)$ represents the ith and $(i-1)$th iterations.

Only the root that satisfies the range of λ as defined by the C_P-λ curve is valid and retained for wind speed estimation. Substituting for λ in (3), the wind speed v is estimated. Again, utilizing (3) with the estimated wind speed v and the optimal tip-speed ratio λ_{opt}, the desired angular velocity of the turbine is determined as

$$\Omega_T^{opt} = \frac{v\lambda_{opt}}{R} \quad (12)$$

Using the (9) and substituting C_P^{max} and λ_{opt} as determined by above equation (10) ~ (12), the optimal turbine output power P_{wt}^{max} is estimated. The estimated maximum output power of the electrical generator P_T^{max} is calculated using the following equation:

$$P_T^{max} = \eta_{max}^{DFIG} P_{wt}^{max} \quad (13)$$

The system is commanded to the desired optimum shaft speed Ω_T^{opt} determined from above (12). The shaft speed is regulated within a small speed range and the total output power P_T is measured repeatedly to ensure that the maximum power point P_T^{max} is tracked in different wind speed. The local regulation is utilized to determine the change of power coefficient of the turbine due to variation in parameters of the turbine over time caused.

IV. CONTROLLER IMPLEMENTATION

A. Control System Description

The maximum power tracking controller, as described in section III, was developed and implemented in a laboratory variable speed wind generation model to verity its efficacy and dynamic tracking performance. The laboratory model consists of a dc machine and doubly fed generator, as illustrated in Fig.3 and Fig.5.

The optimization controller of doubly fed generator was implemented in two TMS320LF2407A digital signal processors (DSP). One DSP mainly realizes the self-adaptive control strategy based on the estimated wind speed and completes the control of rotor power converters. Anther the same type DSP is used to complete the control of the grid power converter. Details of the system implementation are presented in the following section.

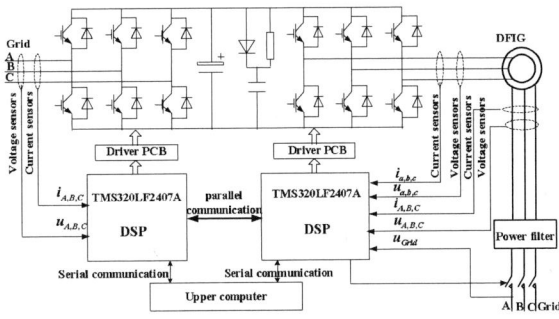

Fig. 3. Control system structure

The dc machine was set up as a motor to emulate the wind. The dc machine controller utilized the speed and current measurement signals to operate in closed-loop. The torque of the dc machine, which emulates the wind turbine torque, is controlled via the current of the dc machine. Wind turbine emulator development is presented in the following section.

B. System Controller Implementation

The system adopted dual PWM converters. The grid converter realizes to feedback the energy from the rotor.

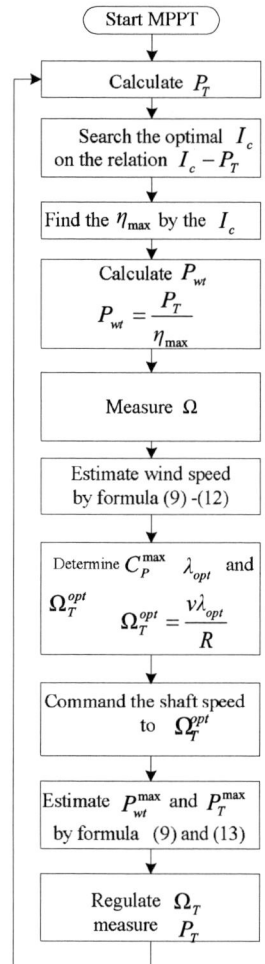

Fig. 4. System controller implementation

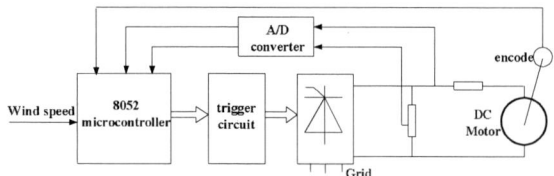

Fig. 5. Wind turbine emulator system structure

The rotor converter mainly controls the rotor speed for capturing maximal wind energy.

Fig.4 illustrates the flow of the optimization controller. The optimization controller firstly searches the maximum efficiency η_{max} , and calculates the maximum input mechanical power. Then, the controller estimates wind speed by the digital signal processor. According to the principle of optimal tip-speed ratio, the wind speed signal is used to estimate the angle velocity of wind turbine, which is adopted to adjust rotor speed after proportion.

C. Wind Turbine Emulator

As illustrated in Fig.5, the microprocessor 8051 realizes the control of the speed and torque in order to emulator the wind turbine. A dc machine torque controller was developed which could be adapted to effectively model a wind turbine operating under varying wind speeds. The wind turbine model controller utilizes the current and speed signal available form the transducer, which is transferred to the microprocessor by A/D converter. The dc machine drive employed for the purpose consists of a rectified bridge and a trigger. The rectified bridge

uses controllable semiconductor device (Thyristor). By controlling the phase angle of the thyristor gating signals, the armature voltage for the dc machine can be controlled.

The output power of the dc machine can be expressed in terms of the angular velocity and the armature current as

$$P_m = T_e \omega = C_m \Phi i_a \omega$$

While the output power of the wind turbine is expressed by equation (2). It is only related to the angular velocity of the wind turbine, as shown Fig.1. It can be obtained through the angular velocity of the wind turbine. According to the desired output power, the armature current can be calculated. Thus, controlling the armature voltage can adjust the curve of ω-i_a. In anther word, it emulates the power curve of the wind turbine.

D. Experiment Result

The adaptive control strategy for tracking maximum power point is applied to wind electricity conversion system in this paper, and this control strategy has been verified for efficacy and reliability by the experiment of 2kw wind electricity conversion system. The measured waveforms through the experiment are shown in the figure 6 and figure 7.

From the figure 6, the response time reduces to 4.5 seconds, while the response time of the conventional control strategy is 15 seconds. When the wind speed suddenly increases, wind turbine can fleetly track the maximum power point.

From the figure 7, it shows this control strategy can more efficiently trace the maximum power points than traditional controller when the wind speed varies. The given wind speed is 0.005Hz sine wave , wind turbine

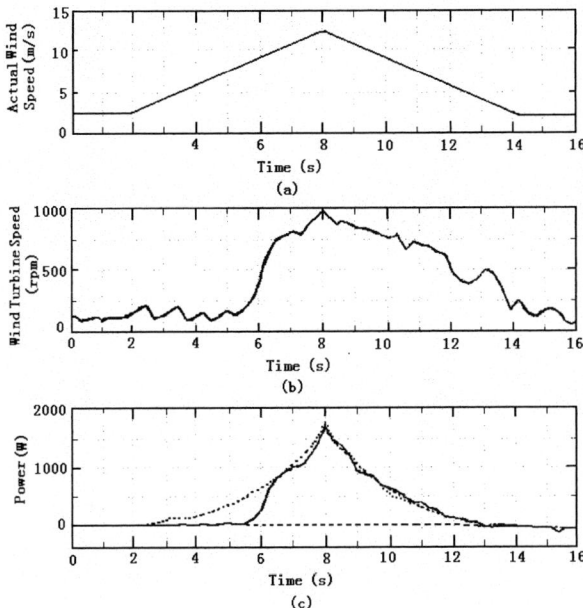

Fig. 6. Simulation results of a flurry of wind
(a) wind speed of a flurry, (b) wind turbine speed,
(c) the output power of doubly fed induction generators

speed and the output power can track the maximum power point in this experiment.

The experiment result shows that the maximum wind energy is converted into electric power by generator corresponding to the different wind speed. This strategy can ensure the optimal efficiency and output electric power is maximal.

From figures above, the system has better stability and faster response performance than traditional controller. Furthermore, actual output can well trace the maximum wind energy, and the steady error is very little.

V. CONCLUSIONS

A maximum wind energy tracking controller, used in the 2kW laboratory generator system, has been present. In the laboratory implementation, it is observed that the system controller can responded to different wind speed without any information on the wind speed. It realizes the optimal tip-speed ratio and captures maximum wind power. The controller improves the response speed form 15 seconds without controllers to 4.5 seconds in a flurry of wind. The controller may be applicable to all doubly fed configurations, including brushless double fed machine, but, at the present time, the controller was only applied in the 2kW laboratory generator system. If it can be applied in high power system, it will solve the problem of wind speed which is difficult to accurately measure it.

REFERENCES

[1] Santiago Dominguez Rubira and Malcolm D. McCulloch "Control method comparison of doubly fed wind generators connected to

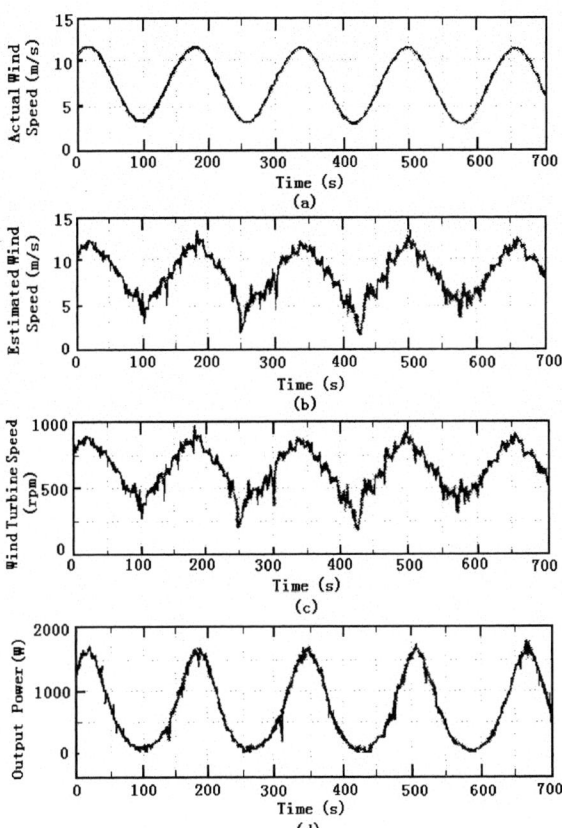

Fig. 7. Simulation results of 0.005Hz fluctuation wind speed (a) 0.005Hz fluctuation wind speed (DC motor giving speed), (b) estimated wind speed, (c) wind turbine speed (DC motor practical speed), (d) the output power of doubly fed induction generators

the grid by asymmetric transmission lines" *IEEE Trans. Ind. Applicat.*, vol. 36, pp. 986-991, July/Aug. 2000.

[2] Ye Hangye, *The Control Technique of Wind Generators.*Beijing, China: China Machine Press, 1995, pp. 150–175.

[3] T. Thiringer and J. Lindres, "Control by variable rotor speed of a fixed-pitch wind turbine operating in a wide speed range," *IEEE Trans. Energy Conversion,* vol. 8, pp. 520–526, Sept. 1993.

[4] I. Eskandarzadeh and H. Ince, "Modeling and output power optimization of a wind turbine driven doubly output induction generator," *Proc. IEEE – Electr. Power Applicat*, vol.141, no.2, pp. 33–38, March.1994.

[5] Fengxiang Wang, Chengwu Lin, and Longya Xu "A chopping and doubly-fed adjustable speed system without bi-dirctional converter," *IEEE Trans. Ind. Applicat.*, vol. 20, pp. 2393–2397, October 2002.

[6] Yifan Tang, and Longya Xu "A flexible active and reactive power control strategy for a variable speed constant frequency generating system," *IEEE Trans. Power Electronic*, vol. 10, pp. 472–478, July 1995.

[7] René Spée, Shibashis Bhowmik, and Johan H. R. Enslin, "Adaptive control strategy for variable-speed doubly-fed wind power generation system," *IEEE Trans. Ind. Applicat*, vel. 12, pp. 545–552, Jun.1994.

[8] Z. Chen, S Arnalte Gómez, and M McCormick, "A fuzzy logic controlled power electronic system for variable speed wind energy conversion systems," presented at the IEE power electronic and power drives Meeting, Sept. 18–19, 2000, conference publication No. 475.

2006 5th International Power Electronics and Motion Control Conference

Research on Synchrodrive Control Technology for Wind Turbine Adjustable-Pitch System Based on Adaptive decoupling Control

Hongche Guo and Qingding Guo
Shenyang University of Technology   Shenyang   P.R.China 110023
ghc_work@163.com

Abstract— **This paper presents a robust control design for a variable-speed adjustable-pitch wind turbine system. An adaptive decoupling controller is designed to drive the turbine speed to extract invariable power from the wind and to allow pitch adjustment to power regulation. An adaptive decoupling controller is designed to drive the turbine speed to extract maximum power from the wind and to allow pith adjustment to constant power regulation. The simulations show the robustness of the controlled system.**

Keywords-wind turbine; adjustable-pitch system; decoupling control; adaptive control

I. INTRODUCTION

Globally enthusiastic research has been focused on the exploitation of wind power as a source of renewable energy. Benefited from the ceaseless development of information technology and control technique, research on grid-connected wind power generation system has been in fast progress, with commercial application from several kilowatts up to more than one megawatts per unit, and category of controls from stall type to pitch regulation type up to the state-of-the-art varying speed type.

Because variable-speed wind turbines have the potential for increased energy capture, controller design has become an area of increasing interest. Blade-pitch regulation provides means for initiating rotation, varying rotational speed to extract power at low wind speeds, and maintaining power production at a maximum level. Controllers must be designed to meet each of these objectives, but this study pertains only to constant power production.

The power regulation regime is entered when the turbine reaches the design rotor speed for maximum power production. Under these conditions, rotational speed is constrained to a specified maximum value through blade-pitch regulation. Fluctuations in wind speed are accommodated to prevent large excursions from the desired rotational speed. Thus the power production is also constrained to a relatively constant level. In addition

to maintaining a constant rotational speed, actuator movement must be restrained to prevent fatigue and overheating. The combination of maintaining a constant rotational speed and minimizing actuator motion are the control objectives specified for the power regulation regime.

The controlled system must be robust to parameters uncertainties in the plant model. It is considered a turbine that allows pitch angle adjustment to power regulation above maximum speed and to help start. The controller is only active above rated wind speed. Since the wind frequently crosses rated wind speed, the controller must operate without a switching transient. The main objective of the control system designed in this paper is regulate pitch demand such that power is kept at its rated value. A change in the operating point of the plant was simulated for a disturbed one showing the robustness of the controlled system.

II. BACKGROUND

Considering the system of Fig.1, the turbine torque is given by [1]-[3]:

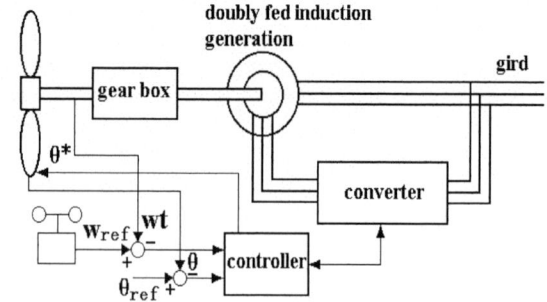

Figure 1.   Complete system

$$T_t = \frac{1}{2}C_q\rho\upsilon^2\pi R^3 \qquad (1)$$

where ρ is the air density, υ is the wind velocity, R is the turbine radius and C_q is a turbine characteristic:

$$C_q = f(\omega_r, \upsilon, \theta) \qquad (2)$$

been ω_r the turbine angular velocity and θ the pitch angle.

The torque coefficient C_q has the characteristics shown in Fig.2. In high speed, $C_{q\max}$ occur with $\theta = 0$ rad. A change in the pitch angle decrease C_q making the behavior of the system worst (1). In a low wind speed situation, the pitch angle can be used to improve C_q. The pitch can also be used in high wind speeds to limit the power by reducing the turbine aerodynamics.

There are a direct relation between C_q and the power coefficient C_p:

$$C_p = \lambda C_q \qquad (3)$$

where λ is the pitch velocity and is given by

$$\lambda = \frac{\omega_r R}{\upsilon} \qquad (4)$$

The power is given $P_t = \dfrac{1}{2} C_p \rho A \upsilon^3$ by, thus:

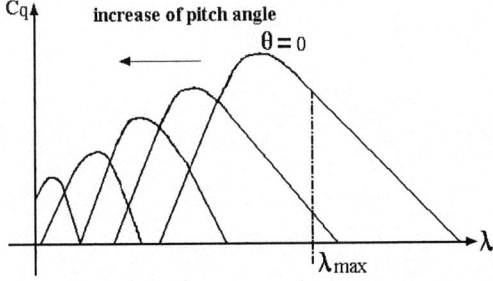

Figure 2. Torque coefficient characteristics

$$P_t = \frac{1}{2} C_p \rho A \frac{R^3}{\lambda^3} \omega_t^3 = K_T \omega_t^3 \qquad (5)$$

where, A is the rotor disk area. This equation gives the points of maximum power transfer with K_T calculated for $\lambda_{\max}$. Therefore, the torque to maximum power transfer is:

$$T_t = K_T \omega_t^2 \qquad (6)$$

for $\lambda_{\max}$.

So, the maximum power transfer could be reached acting on the resistant torque of the generator so that the turbine speed is as near as possible from that corresponding to $\lambda = \lambda_{\max}$.

The turbine speed dynamics behave as a low pass filter given by:

$$\tau \Delta \dot{\omega} + \Delta \omega_t = \mu \Delta \upsilon \qquad (7)$$

with: $\tau = \dfrac{J}{T_r'(\omega_{t0}) - K R \upsilon_0 C_q'(\lambda_0)}$

$$\mu = \frac{2 K \upsilon_0 C_q(\lambda_0) - K \omega_{r0} R C_q'(\lambda_0)}{T_r'(\omega_{t0}) - K R \upsilon_0 C_q'(\lambda_0)}$$

$$K = \frac{1}{2} \rho \pi R^3$$

where J is the inertia of the system and $C_r'(\lambda_0)$ and $T_r'(\omega_{t0})$ denote derivatives with respect to λ and ω_t respectively.

Therefore, the system can not recover all energy from the wind because ω_t can not respond immediately to turbulent winds without high torque. Thus, there is a trade-off between maximum energy extraction on turbulent winds and torque limits. Because the torque coefficient is related to the power coefficient C_p, through the (3) relation manipulation of the torque coefficient using λ and θ will result in manipulation of the power produced by the turbine. The block diagram in Fig.3 illustrates the simulation logic.

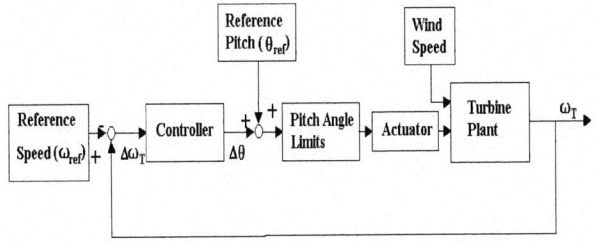

Figure 3. Simulation block diagram

III. MODELING OF ADJUSTABLE-PITCH SYSTEM

Sketch of hydraulic actuators is showed in Fig.4, where, 1 is synchro-disk, 2 is blade, 3 is drive rod and 4 is hydro cylinder.

Transfer Function of Hydraulic actuators that adjust to the blade-angle may be approximately represented by third-order polynomial [4]:

613

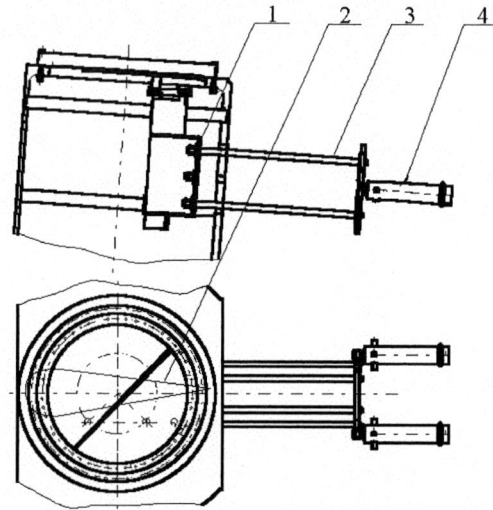

Figure 4.   Sketch of hydraulic actuators

$$G(s) = \frac{K_a}{s\left(\dfrac{s^2}{\omega_h^2} + \dfrac{2\zeta_h}{\omega_h^2}s + 1\right)} \qquad (8)$$

with: $K_a = (1+K)K_uU - \dfrac{K_{ce}}{A^2(1+K^2)}(1+\dfrac{CS}{K_{ce}})F_l$

$$\omega_h = \sqrt{\frac{A^2(1+K^2)}{mC}}$$

$$\zeta_h = \frac{1}{2}\frac{K_{ce}m + bC}{2\sqrt{mC}\sqrt{A^2(1+K^2)}}$$

where, K_a is open loop gain of system, ω_h is natural frequency of system and ζ_h is damping coefficient.

Through analysis, open loop transfer Function of hydraulic position actuators may be simplified by:

$$G(s) = \frac{K}{s(Ts+1)} \qquad (9)$$

IV.   DESIGN OF THE DECOUPLING CONTROLLER

Dual Hydraulic actuators of adjust the blade-angle have the same drive signal, but the synchronous error between two actuators still generate because of non-balanced forces adding to the actuators and other moving parts, position variation of blade carrier and various uncertainty disturbances during the working process.

For the servo system return to the synchronous state, to remove coupling effect of controlled object, must adopt decoupling method. After decoupling that dual variable control system, the control system becomes single variable system. In this paper, adopt ideal decoupling theory [5]. Block diagram of decoupling control is showed

in Fig.5, where, **C** is controller, **N** is decoupling controller and **G** is controlled object.

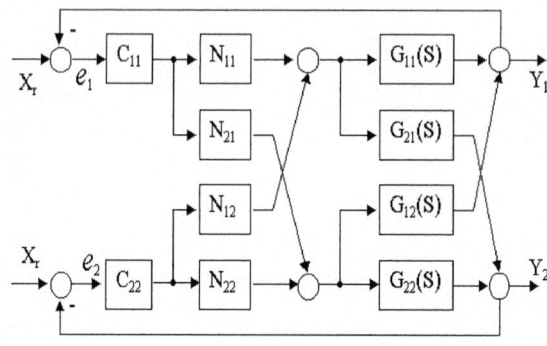

Figure 5.   Block diagram of decoupling control

The characteristic of system is:

$$\mathbf{Y} = (\mathbf{I} + \mathbf{GNC})^{-1}\mathbf{GNCX} \qquad (10)$$

After decoupling it is:

$$\frac{Y_1}{X_1} = \frac{G_{11}C_{11}}{1 + G_{11}C_{11}} \quad , \quad \frac{Y_2}{X_2} = \frac{G_{22}C_{22}}{1 + G_{22}C_{22}} \qquad (11)$$

then,

$$(\mathbf{I} + \mathbf{GNC})^{-1}\mathbf{GNC} = \begin{bmatrix} \dfrac{G_{11}C_{11}}{1 + G_{11}C_{11}} & 0 \\ 0 & \dfrac{G_{22}C_{22}}{1 + G_{22}C_{22}} \end{bmatrix} \qquad (12)$$

$$= \begin{bmatrix} \dfrac{1}{1 + G_{11}C_{11}} & 0 \\ 0 & \dfrac{1}{1 + G_{22}C_{22}} \end{bmatrix} \begin{bmatrix} G_{11} & 0 \\ 0 & G_{22} \end{bmatrix} \begin{bmatrix} C_{11} & 0 \\ 0 & C_{22} \end{bmatrix}$$

for decoupling:

$$\mathbf{GN} = \begin{bmatrix} G_{11} & 0 \\ 0 & G_{22} \end{bmatrix} \qquad (13)$$

so:

$$\mathbf{N} = \mathbf{G}^{-1}\begin{bmatrix} G_{11} & 0 \\ 0 & G_{22} \end{bmatrix} \qquad (14)$$

namely:

$$\begin{bmatrix} N_{11} & N_{12} \\ N_{21} & N_{22} \end{bmatrix} = \begin{bmatrix} \dfrac{G_{11}G_{22}}{G_{11}G_{22} - G_{12}G_{21}} & -\dfrac{G_{22}G_{12}}{G_{11}G_{22} - G_{12}G_{21}} \\ -\dfrac{G_{11}G_{21}}{G_{11}G_{22} - G_{12}G_{21}} & \dfrac{G_{11}G_{22}}{G_{11}G_{22} - G_{12}G_{21}} \end{bmatrix}$$

$$= \frac{G_{11}G_{22}}{G_{11}G_{22} - G_{12}G_{21}} \begin{bmatrix} 1 & -\dfrac{G_{12}}{G_{11}} \\ -\dfrac{G_{21}}{G_{22}} & 1 \end{bmatrix} \qquad (15)$$

So the decoupling system recovered the original control system from the dynamic state. After decoupling system becomes double single variable system, but can carry out independent control.

V. DESIGN OF THE ADAPTIVE CONTROLLER

Although the above-mentioned decoupling controller carries out the decoupling design to the system, but this design is an establishment on foundation of accurate mathematics model. Therefore have the necessity very much to go together with the SISO adaptive controller [6][7], to carry out each control loop independently accurate follow the track of input change. We can thinks controlled object that is change with time, various disturb is change of model. So in this paper, same dual structure is used in which no merely proceed position control to single hydraulic actuator. When the one actuator suffers from the disturbance so as to produce the position synchronous error, the speed given of the other one actuator will follow the variation. The system structure is showed in Fig.6.

State equation of controlled object is:

$$\dot{\mathbf{x}}_p = \mathbf{A}_p \mathbf{x}_p + \mathbf{B}_p \mathbf{u} \qquad (16)$$

namely,

$$\begin{bmatrix} \dot{x}_1 \\ \dot{x}_2 \end{bmatrix} = \begin{bmatrix} 0 & 1 \\ -\dfrac{R_s D}{L_q M} & -\dfrac{MR_s + DL_q}{L_q M} \end{bmatrix} \begin{bmatrix} x_1 \\ x_2 \end{bmatrix} + \begin{bmatrix} 0 \\ \dfrac{K_T}{l_q M} \end{bmatrix} u$$

$$y = \begin{bmatrix} 1 & 0 \end{bmatrix} \begin{bmatrix} x_1 \\ x_2 \end{bmatrix} \qquad (17)$$

State equation of reference model is:

$$\dot{\mathbf{x}}_m = \mathbf{A}_m \mathbf{x}_m + \mathbf{B}_m \mathbf{u} \qquad (18)$$

In this paper, system is used forward feed controller $\mathbf{G}$ and state back feed controller $\mathbf{F}$ that parameter is adjusted. The control action of controlled object u is given by [6]:

$$\mathbf{u} = \mathbf{G}\mathbf{x}_r + \mathbf{F}\mathbf{x}_p \qquad (19)$$

where, $\mathbf{G}$ is plus matrix of forward feed controller; $\mathbf{F}$ is plus matrix of state back feed controller

When the model is suited completely, thus:

$$\begin{aligned} \mathbf{A}_m &= \mathbf{A}_p + \mathbf{B}_p \mathbf{F}^* \\ \mathbf{B}_m &= \mathbf{B}_p \mathbf{G}^* \end{aligned} \qquad (20)$$

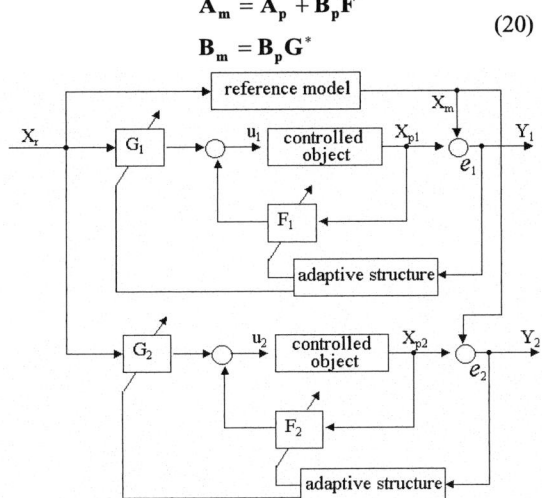

Figure 6.   Block diagram of adaptive control

with, sign* means the model is suited completely.

Thus rule of controller may be represented by:

$$\begin{aligned} \lim_{t \to \infty} \mathbf{G} &= \mathbf{G}^* \\ \lim_{t \to \infty} \mathbf{F} &= \mathbf{F}^* \end{aligned} \qquad (21)$$

If generalized error is:

$$\mathbf{e} = \mathbf{x}_m - \mathbf{x}_p \qquad (22)$$

$$\dot{\mathbf{e}} = \mathbf{A}_m \mathbf{e} + (\mathbf{A}_m - \mathbf{A}_p - \mathbf{B}_p \mathbf{F})\mathbf{x}_p + (\mathbf{B}_m - \mathbf{B}_p \mathbf{G})\mathbf{x}_r \quad (23)$$

$$\dot{\mathbf{e}} = \mathbf{A}_m \mathbf{e} + \mathbf{B}_m (\mathbf{G}^*)^{-1}[\mathbf{F}^* - \mathbf{F}]\mathbf{x}_p + \mathbf{B}_m (\mathbf{G}^*)^{-1}[\mathbf{G}^* - \mathbf{G}]\mathbf{x}_r$$

$$= \mathbf{A}_m \mathbf{e} + \mathbf{B}_m (\mathbf{G}^*)^{-1}\widetilde{\mathbf{F}}\mathbf{x}_p + \mathbf{B}_m (\mathbf{G}^*)^{-1}\widetilde{\mathbf{G}}\mathbf{x}_r \quad (24)$$

where, $\widetilde{\mathbf{F}} = [\mathbf{F}^* - \mathbf{F}]$, $\widetilde{\mathbf{G}} = [\mathbf{G}^* - \mathbf{G}] \qquad (25)$

It is represented error that matrix $\mathbf{F}$ and $\mathbf{G}$ with model is suited completely.

thus:

$$\dot{\mathbf{e}} = \mathbf{A}_m \mathbf{e} + \mathbf{B}_m (\mathbf{G}^*)^{-1}(\widetilde{\mathbf{F}}, \widetilde{\mathbf{G}}) \begin{pmatrix} \mathbf{x}_p \\ \mathbf{x}_r \end{pmatrix} \qquad (26)$$

Last equation is the equation of equivalent error system.

The asymptotic stability condition of equivalent error system is:

$$\widetilde{\mathbf{F}} = -\mathbf{R}_1 [\mathbf{B}_m (\mathbf{G}^*)^{-1}]^T \mathbf{P}\mathbf{e}\mathbf{x}_p^T \qquad (27)$$

$$\widetilde{\mathbf{G}} = -\mathbf{R}_2 [\mathbf{B}_m (\mathbf{G}^*)^{-1}]^T \mathbf{P}\mathbf{e}\mathbf{x}_r^T \qquad (28)$$

and $\mathbf{X}_r$ frequency spectrum is enough abundance.

Where, $\mathbf{R}_1$, $\mathbf{R}_2$ and $\mathbf{P}$ is symmetric positively definite matrix and $\mathbf{P}$ is explain of Lyapunov equation (23):

$$\mathbf{A}_m^T \mathbf{P} + \mathbf{P}\mathbf{A}_m = -\mathbf{Q} \qquad (29)$$

where, $\mathbf{Q} = \mathbf{Q}^T > 0$.

The control rule of adaptive is solved from the (25), (27) and (28):

$$\mathbf{F} = \int_0^t \mathbf{R}_1 [\mathbf{B}_m (\mathbf{G}^*)^{-1}]^T \mathbf{P}\mathbf{e}\mathbf{x}_p^T + \mathbf{F}(0) \qquad (30)$$

$$\mathbf{G} = \int_0^t \mathbf{R}_2 [\mathbf{B}_m (\mathbf{G}^*)^{-1}]^T \mathbf{P}\mathbf{e}\mathbf{x}_r^T + \mathbf{G}(0) \qquad (31)$$

So we change the adaptive rule into:

$$\mathbf{F} = \int_0^t \mathbf{R}_1' \mathbf{P}\mathbf{e}\mathbf{x}_p^T + \mathbf{F}(0) \qquad (32)$$

$$\mathbf{G} = \int_0^t \mathbf{R}_2' \mathbf{P}\mathbf{e}\mathbf{x}_r^T + \mathbf{G}(0) \qquad (33)$$

with, $\mathbf{R}_1' = \mathbf{R}_1 [\mathbf{B}_m (\mathbf{G}^*)^{-1}]^T \qquad (34)$

$\mathbf{R}_2' = \mathbf{R}_2 [\mathbf{B}_m (\mathbf{G}^*)^{-1}]^T \qquad (35)$

Last matrix parameter is given by simulation.

VI. SIMULATION RESULTS

The simulation study was performed to verify the proposed control strategy. The air density ρ is 1.2kg/m^3. The specification of the wind turbine is given in the Table1.

TABLE I.

Adjustable-Pitch Wind Turbine Parameters

Allowable rotor speed	13~25rpm
Generator output power	1000kW
Cut-in wind speed	3m/s
Rate wind speed	12.5m/s
Furling wind speed	25m/s
Rotor diameter	62m
Adjust pitch rate	7.5° /s
Generator voltage	690V

With the proposed control scheme, we get the tracking performance as illustrated by the following Figures:

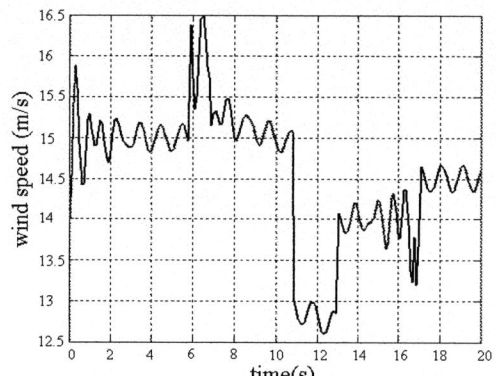

Figure 7.   The wind speed curve

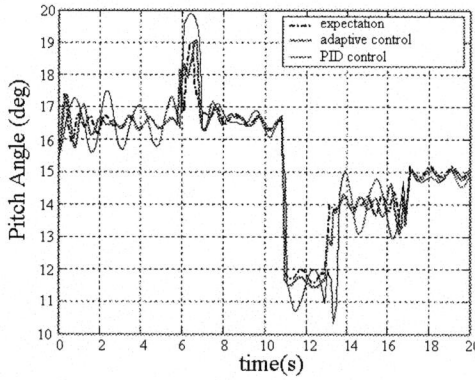

Figure 8.   The pitch angle curves

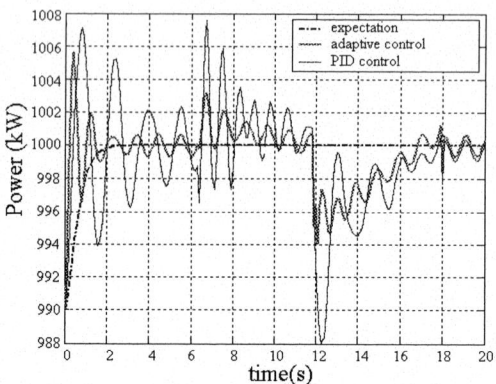

Figure 9.   The output power curves

VII.   CONCLUSION

The adaptive decoupling algorithm is base on mechanical dynamics. The diagrams of analysis and simulation show that the proposed method is able to achieve smooth and satisfactory speed tracking and the motion of the actuator is desired.

REFERENCES

[1] Freris, L. L., Wind Energy Conversion Systems, Prentice Hall, 1990.

[2] W. E. Leithead and B. Connor, "Control of Variable Speed Wind Turbine with Induction Generator", Control'94, Conference Publication, No.389, March 1994.

[3] Y. D. Song, "Control of Wind Turbines using Memory-Based Method", American Control Conference, June 1998.

[4] Wang Zhanlin, Electric Hydraulic Pressure Servo Control of the Time, 2005, Beijing

[5] Wang yongchu, Decoupling Control System, 1985, Chengdu

[6] Landau ID. Adaptive Control-Model Reference Approach. New York: Dekker, 1979

[7] Dambrosio L. Fortunato B. One step ahead adaptive control of a wind –driven, synchronous generator system, Energy, 1999,vol. 24, pp.9-20.

2006 5th International Power Electronics and Motion Control Conference

Limit-Trajectory Single- and Two-Mode Overmodulation Technology

Shun JIN and Yan-ru ZHONG
Xi'an University of Technology, Xi'an, 710048 China
shunjin@126.com

Abstract—**This paper extends the limit-trajectory two-mode overmodulation ideal [1] into single mode. The harmonic characteristics of them are analyzed and compared. Based on one TI TMS320LF 2407A DSP, the proposed method is verified with an open-loop V/F controlled diode-clamped three-level model inverter. With the test result, it is confirmed that single-mode limit-trajectory overmodulation technology is correct and effective.**

Key Words-PWM; overmodulation; three-level inverter

I. INTRODUCTION

By using Limit-trajectory two-mode overmodulation technique brought forward by N. V. Nho[1], linear control (output fundamental component has unit gain over reference voltage) in overmodulation region is successively achieved. Unlike other methods, one can gain the desired reference voltage vector, which can achieve the goal of unit gain, by analytical calculation. So, the additional look-up table operation which is needed by other methods can be avoided. In addition, the achieved analytical relationship between reference vector and output fundamental component has also great theoretical value. In this paper, on the basis of introduction of the limit- trajectory ideal, the method proposed by [1] was analyzed with method developed by Lee D. C. [2] firstly. Then the ideal of limit trajectory is extended into single mode overmodulation scheme. Finally, the output waveforms produced by single- and two-mode limit-trajectory overmodulation techniques are compared in regard to harmonic characteristics.

II. IDEAL OF LIMIT- TRAJECTORY CONTROL[1]

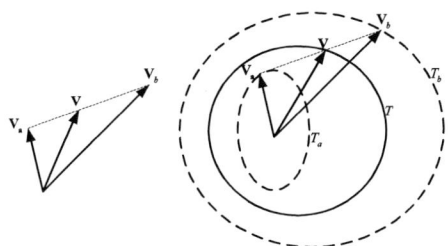

Fig.1 Ideal of limit-trajectory control [1]

Let $\mathbf{V}$, $\mathbf{V_a}$, $\mathbf{V_b}$ be three space voltage vectors, and $\mathbf{V}$ lie on the line connecting $\mathbf{V_a}$ and $\mathbf{V_b}$ (as shown in Fig.1). Vector $\mathbf{V}$ can be expressed as linear combination of two vectors $\mathbf{V_a}$ and $\mathbf{V_b}$:

$$\mathbf{V} = (1-\eta)\mathbf{V_a} + \eta\mathbf{V_b} \qquad (1)$$

Where η varies in (0, 1), when η varies from 0 to 1, $\mathbf{V}$ varies from $\mathbf{V_a}$ to $\mathbf{V_b}$.

Assuming that $\mathbf{V_a}$ and $\mathbf{V_b}$ rotate along trajectory T_a and T_b, and $\mathbf{V}$ along T, then the fundamental component of $\mathbf{V}$ can be calculated as follows:

$$
\begin{aligned}
V_1 &= \frac{1}{2\pi}\int_0^{2\pi}[(1-\eta)\cdot\mathbf{V_a} + \eta\cdot\mathbf{V_b}]\cdot e^{-j\theta}d\theta \\
&= \frac{(1-\eta)}{2\pi}\int_0^{2\pi}\mathbf{V_a}\cdot e^{-j\theta}d\theta + \frac{\eta}{2\pi}\int_0^{2\pi}\mathbf{V_b}\cdot e^{-j\theta}d\theta
\end{aligned}
\qquad (2)
$$

And thus we can get:

$$V_1 = (1-\eta)V_{a1} + \eta V_{b1} \qquad (3)$$

In (3), V_{a1} and V_{b1} represent the corresponding fundamental component amplitude of T_a and T_b. When $\eta = 0$, $\mathbf{V}=\mathbf{V_a}$, and thus $V_1=V_{a1}$; when $\eta = 1$, $\mathbf{V}=\mathbf{V_b}$, $V_1=V_{a2}$. Through changing the η in (0, 1), we can control the fundamental component within the scope $V_{a1}<V_1<V_{b1}$.

The linear control in overmodulation region means that the fundamental component of $\mathbf{V}$ satisfies the following equation:

$$V_1 = MI\frac{2V_{dc}}{\pi} \qquad (4)$$

with (3), one can get:

$$\eta = \frac{MI - M_a}{M_b - M_a} \qquad (5)$$

Where M_a, M_b represents the modulation index of V_{a1}, V_{b1} respectively:

$$M_a = \frac{V_{a1}}{2V_{dc}/\pi}, \qquad M_b = \frac{V_{b1}}{2V_{dc}/\pi} \qquad (6)$$

Now, an important conclusion can be achieved: if vector $\mathbf{V}$ is controlled with linear function (1) and (5), then, in the scope $M_a<MI<M_b$, the fundamental component (V_1) of $\mathbf{V}$ satisfies (4). In other words, the fundamental component of output voltage has unit gain

1-4244-0448-7/06/$25.00 ©2006 IEEE

over reference voltage.

III. TWO-MODE LIMIT-TRAJECTORY OVERMODULATION SCHEME AND ITS ANALYSIS

The literature [1], like [3], divides the modulation index into 3 regions: linear region (0<MI<0.907), overmodulation region I (0.907<MI<0.9514) and overmodulation region II (0.9514<MI<1). In linear region, $\mathbf{V_a}$ is set to zero vector, thus $M_a=0$, $V_{a1}=0$; the trajectory of $\mathbf{V_b}$ is set to be $\mathbf{V_b} = \dfrac{V_{dc}}{\sqrt{3}} \cdot e^{j\theta}$, which is the inscribed circle of vector diagram and the upper limit of linear modulation. So, $M_b = \pi/2\sqrt{3}$, $V_{b1} = V_{dc}/\sqrt{3}$. With (5), we can get $\eta = \dfrac{2\sqrt{3}}{\pi} MI$, and then the desired reference voltage vector $\mathbf{V}$ can be achieved by replacing this formula into (1).

In overmodulation region I, the trajectory of $\mathbf{V_a}$ is set to be $\mathbf{V_a} = \dfrac{V_{dc}}{\sqrt{3}} \cdot e^{j\theta}$, and thus $M_a = \pi/2\sqrt{3}$, $V_{a1} = V_{dc}/\sqrt{3}$; The trajectory of $\mathbf{V_b}$ is set to be the boundary of vector diagram, and thus $M_b = 0.9514$, $V_{b1} = \dfrac{\sqrt{3}V_{dc}}{\pi} \ln 3$. By replacing it into (5), η can be calculated as: $\eta = \dfrac{2\sqrt{3}MI - \pi}{3\ln 3 - \pi}$.

In overmodulation region II, the trajectory of $\mathbf{V_a}$ is set to be the boundary of vector diagram, and thus $M_a = 0.9514$, $V_{a1} = \dfrac{\sqrt{3}V_{dc}}{\pi} \ln 3$; The trajectory of $\mathbf{V_b}$ is set to be the corresponding 6 discrete vectors of the 6 vertexes of vector diagram, and thus $M_b = 1.0$, $V_{b1} = \dfrac{2V_{dc}}{\pi}$. By replacing it into (5), η can be calculated as: $\eta = \dfrac{2\sqrt{3}MI - 3\ln 3}{2\ln 3 - 3\ln 3}$.

For analysis of the output voltage with the method introduced by Lee D. C. [2], the vector $\mathbf{V}$ need to be represented by its amplitude and phase angle. Considering the 1/6 period symmetry, only sector 1 is considered in the following. In overmodulation region I, it is easy to calculate the amplitude of reference vector as follows:

$$|\mathbf{V}| = \frac{(1-\eta_{12}) \cdot V_{dc}}{\sqrt{3}} + \frac{\eta_{12} \cdot V_{dc}}{\sqrt{3}\cos(\theta - \frac{\pi}{6})} \qquad (7)$$

Where, $\eta_{12} = (m - \dfrac{\pi}{2\sqrt{3}}) \Big/ (\dfrac{\sqrt{3}}{2}\ln 3 - \dfrac{\pi}{2\sqrt{3}})$, $m \in [0.907, 0.952]$, and m is MI indeed(Fig.2 b). θ is the vector angle of inner or outer vector.

Bold real line in Fig.2a shows the trajectory of reference voltage vector gained with (7). As can be seen in Fig.2b, the phase voltage waveform can be separate into 2 parts, and their analytical expression can be as follows:

$$\begin{cases} f_1 = R_1 \cdot V_{dc} \cdot \sin\phi & ,0 \le \phi < \pi/6 \\ f_2 = R_2 \cdot V_{dc} \cdot \sin\phi & ,\pi/6 \le \phi < \pi/2 \end{cases} \qquad (8)$$

Where, $R_1 = (1-\eta_{12})/\sqrt{3} + \eta_{12}/(\sqrt{3}\cos\phi)$, $R_2 = (1-\eta_{12})/\sqrt{3} + \eta_{12}/[\sqrt{3}\cos(\phi - \pi/3]$

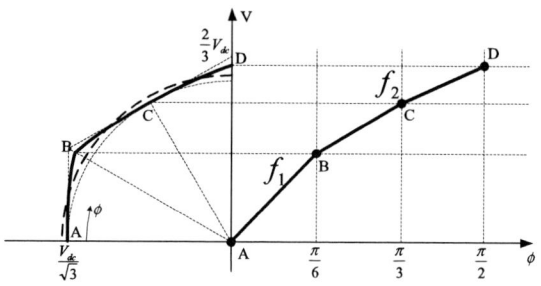

(a) Trajectory of voltage vector and modulated phase voltage waveform

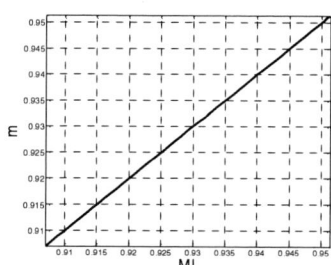

(b) Relationship of MI and m
Fig. 2 Analysis of TMLT mode I

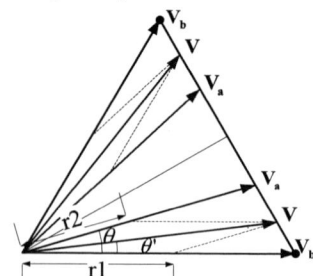

Fig. 3 Geometrical relationship between $\mathbf{V_a}$, $\mathbf{V_b}$ and $\mathbf{V}$ in overmodulation region II

In overmodulation region II, $\mathbf{V_a}$, $\mathbf{V_b}$ and $\mathbf{V}$ satisfy the relation shown in Fig.3. The following expression can be easily gain by the geometrical relation in Fig.3:

$$|\mathbf{V}| = \frac{V_{dc}}{\sqrt{3}\cos(\pi/6 - \theta')} \qquad (9)$$

618

$$\theta' = \begin{cases} \arctan(\dfrac{r_2 \cdot \sin\theta}{r_2 \cdot \cos\theta + r_1}) & ,0 \le \theta < \pi/6 \\ \dfrac{\pi}{3} - \arctan[\dfrac{r_2 \cdot \sin(\pi/3 - \theta)}{r_2 \cdot \cos(\pi/3 - \theta) + r_1}] & ,\pi/6 \le \theta < \pi/3 \end{cases} \quad (10)$$

Where $\qquad r_1 = 2\eta_{23} \cdot V_{dc}/3 \qquad$,

$r_2 = (1 - \eta_{23}) \cdot V_{dc}/[\sqrt{3}\cos(\theta - \pi/6)] \qquad$,

$\eta_{23} = \dfrac{m - \sqrt{3}\ln 3/2}{1 - \sqrt{3}\ln 3/2}$. θ is the phase angle of inner

vector. Where control parameter $m \in [0.952, 1]$, and m is MI indeed(Fig.4b).

As can be seen from Fig.4a, phase voltage waveform can be divided into 3 parts, and can be as follows:

$$\begin{cases} f_1 = R_2 \cdot \sin\phi + R_1/2 & ,0 \le \phi < \pi/6 \\ f_2 = R_3 \cdot \sin\phi + R_1/2 & ,\pi/6 \le \phi < \pi/3 \\ f_3 = R_3 \cdot \sin\phi + R_1 & ,\pi/3 \le \phi < \pi/2 \end{cases} \quad (11)$$

Where $\qquad R_1 = 2\eta_{23} \cdot V_{dc}/3 \qquad$,

$R_2 = (1 - \eta_{23}) \cdot V_{dc}/\sqrt{3}\cos\phi \qquad$,

$R_3 = (1 - \eta_{23}) \cdot V_{dc}/\sqrt{3}\cos(\pi/3 - \phi)$.

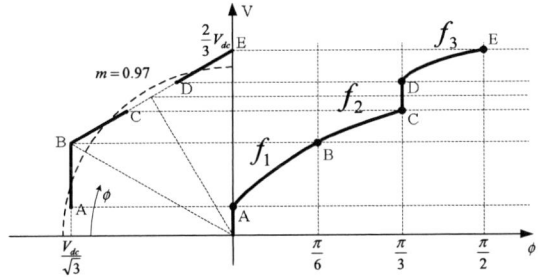

(a) Trajectory of voltage vector and modulated phase voltage waveform

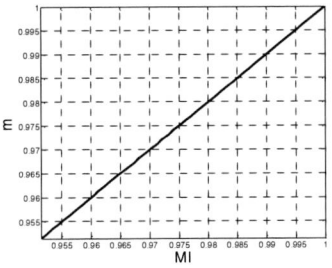

(b) Relationship of MI and m
Fig. 4 Analysis of TMLT mode II

IV. SINGLE-MODE LIMIT-TRAJECTORY OVERMODULATION SCHEME

Indeed, the ideal of limit trajectory can also be used in single mode overmodulation scheme. In modulation region $MI \in (0.907, 1)$, the inner trajectory is set to be the inscribed circle:

$$\mathbf{V_a} = \frac{V_{dc}}{\sqrt{3}} \cdot e^{j\theta} \quad (12)$$

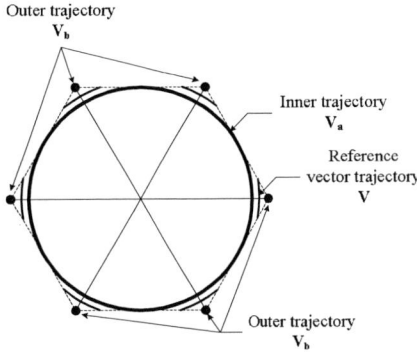

Fig.5 Single mode limit trajectory overmodulation

And the outer trajectory is set to be the corresponding 6 discrete base vectors of the 6 vertexes of vector diagram (Fig.5). Thus, we have: $M_a = \pi/2\sqrt{3}$, $V_{a1} = V_{dc}/\sqrt{3}$; and $M_b = 1.0$, $V_{b1} = \dfrac{2V_{dc}}{\pi}$. By replacing it into (5), we can get:

$$\eta = \frac{MI - \pi/2\sqrt{3}}{1 - \pi/2\sqrt{3}} \quad (13)$$

By considering (1), we can get the desired vector (in sector 1):

$$\mathbf{V} = \begin{cases} (1-\eta)\dfrac{V_{dc}}{\sqrt{3}} \cdot e^{j\theta} + \eta \cdot \dfrac{2V_{dc}}{3} & , \quad 0 \le \theta < \pi/6 \\ (1-\eta)\dfrac{V_{dc}}{\sqrt{3}} \cdot e^{j\theta} + \eta \cdot \dfrac{2V_{dc}}{3} \cdot e^{j\pi/3}, & \pi/6 \le \theta \le \pi/3 \end{cases} \quad (14)$$

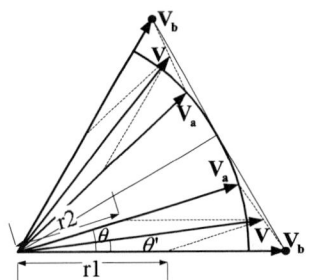

Fig.6 Geometrical relationship between $\mathbf{V_a}$,$\mathbf{V_b}$ and $\mathbf{V}$

According to the geometrical relation of $\mathbf{V_a}$, $\mathbf{V_b}$ and $\mathbf{V}$ shown in Fig.6, the amplitude and phase angle of reference voltage vector can be expressed as follows (for the 1/6 period symmetry, only sector 1 is considered):

$$r' = \begin{cases} r_1 \cdot \cos\theta' + r_2 \cdot \cos(\theta - \theta') & ,0 \le \theta < \pi/6 \\ r_1 \cdot \cos(\pi/3 - \theta') + r_2 \cdot \cos(\theta - \theta') & ,\pi/6 \le \theta < \pi/3 \end{cases}$$

$$\theta' = \begin{cases} \arctan(\dfrac{r_2 \cdot \sin\theta}{r_2 \cdot \cos\theta + r_1}) & ,0 \le \theta < \pi/6 \\ \dfrac{\pi}{3} - \arctan[\dfrac{r_2 \cdot \sin(\pi/3 - \theta)}{r_2 \cdot \cos(\pi/3 - \theta) + r_1}] & ,\pi/6 \le \theta < \pi/3 \end{cases} \quad (15)$$

Where $r_1 = 2\eta \cdot V_{dc}/3$, $r_2 = (1-\eta) \cdot V_{dc}/\sqrt{3}$, $\eta = (MI - \pi/2\sqrt{3})/(1 - \pi/2\sqrt{3})$.

As can be seen from Fig.7a, phase voltage waveform can be divided into 2 parts, and can be expressed as:

$$\begin{cases} f_1 = r_2 \cdot \sin\phi + r_1/2 & ,0 \leq \phi < \pi/3 \\ f_2 = r_2 \cdot \sin\phi + r_1 & ,\pi/3 \leq \phi < \pi/2 \end{cases} \quad (16)$$

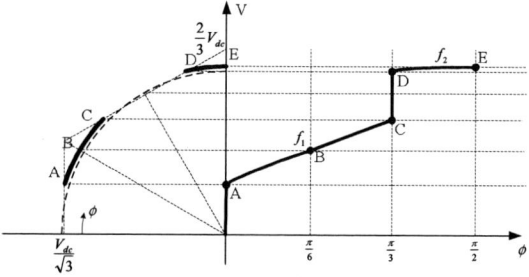

(a) Trajectory of voltage vector and modulated phase voltage waveform

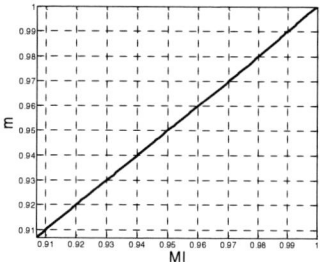

(b) Relationship of MI and m
Fig. 7 Analysis of SMLT

V. PERFORMANCE COMPARISON BETWEEN SINGLE- AND TWO-MODE LIMIT-TRAJECTORY OVERMODULATION SCHEME

THD is defined as follows:

$$THD = \frac{\sqrt{\left(V_r^2 - V_1^2\right)}}{V_1} \quad (17)$$

Where V_r, V_1 are rms values of output voltage and its fundamental component respectively. The THD of the two methods mentioned above is drawn in Fig.8 (method of [1] is referred as TMLT and the one introduced by this paper is referred as SMLT). For comparison purpose, THD of the method of [3] (referred as TM1) and [4] (referred as SM) are also drawn in the same diagram. It is easy to conclude that THD of two mode method TMLT, TM1 is smaller than that of single mode ones. Additionally, SMLT is better than SM in regard to THD. Index THD reflects the distortion degree of voltage. But for inductive machine, the distortion of current is the fact that we indeed care about. Accordingly, WTHD is used bellow to evaluate the output waveform in the situation of inductive load.

The full name of WTHD is weighted total harmonic distortion. It is induced from THD by weighted disposal of each harmonic component. It is defined as:

$$WTHD = \frac{1}{V_1}\sqrt{\sum_{i=2}^{\infty}\left(\frac{V_i}{i}\right)^2} \quad (18)$$

WTHD reflect the current distortion on inductive load. Fig.9 shows the WTHD of all the overmodulation mentioned above. As can be seen from it, two-mode scheme is still better than single- mode. The important truth is SM if better than SMLT in regard to WTHD, which is reverse in regard to THD. This exemplifies the necessary of WTHD.

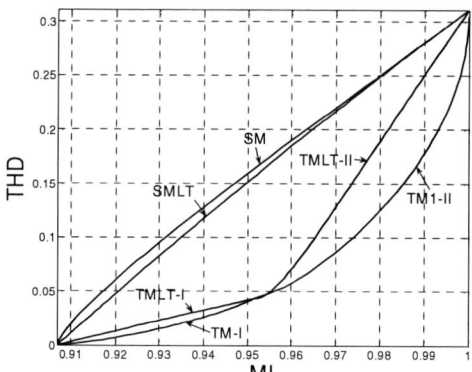

Fig.8 Comparison of THD

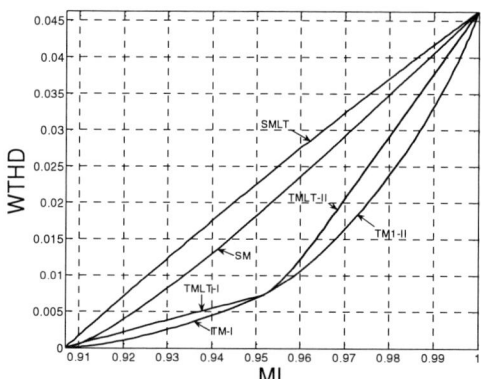

Fig.9 Comparison of WTHD

VI. EXPERIMENT

The 4 overmodulation schemes were testified on an open-loop diode-clamped V/F controlled three-level model inverter with one TI TMS320LF 2407A DSP as controller[5][6] (Fig.10, TABLE I.). The U phase voltage is calculated online and output by D/A. As can be seen from Fig.11, all the tested waveform is identical to analytical results. When MI=1, all schemes can produce the 6 step output. Fig.12 shows line voltage and current of the single-mode limit-trajectory overmodulation scheme.

TABLE I.
EXPERIMENT CONDITION

DC-Link Voltage	DC100V
Switching Frequency	1kHz
DC-Link Capacitance	1000uF
Switching Device	MOSFET 10A/400V
AC Motor	Squirrel Cage 380V/1.1kW
Controller	TMS320LF 2407A DSP
Inverter Capacitance	1kVA
Dead Time	4us
D/A Precision	12Bit

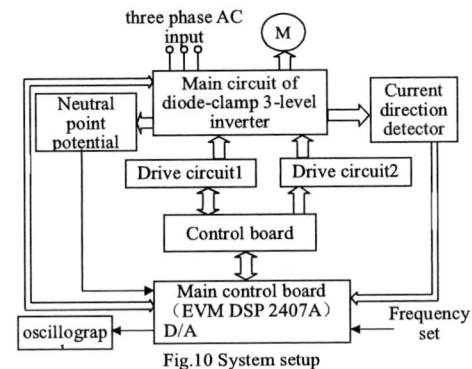

Fig.10 System setup

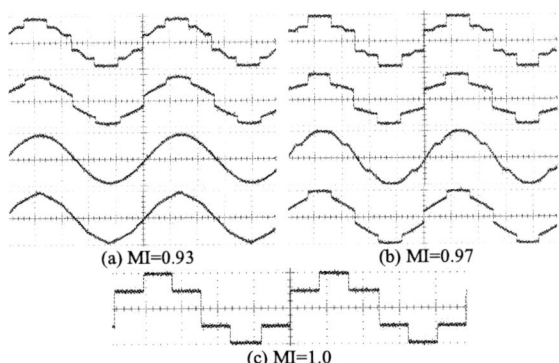

(a) MI=0.93 (b) MI=0.97

(c) MI=1.0

Fig.11 Out-put phase voltage, from top to bottom: SM, SMLT, TM1, TMLT

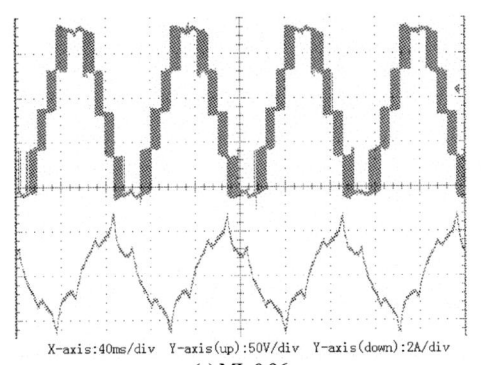

X-axis:40ms/div Y-axis(up):50V/div Y-axis(down):2A/div
(a) MI=0.96

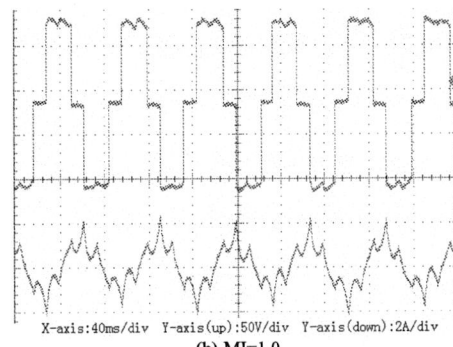

X-axis:40ms/div Y-axis(up):50V/div Y-axis(down):2A/div
(b) MI=1.0

Fig.12 Line voltage and current of single-mode limit-trajectory overmodulation scheme.

VII. CONCLUSION

It is feasible to extend the ideal of limit trajectory control into single mode overmodulation. By doing this, single analytical formula can be achieved in the whole overmodulation region, by which the desired voltage vector can be calculated according to modulation index directly and thus the fundamental component of output voltage can be controlled accurately. Despite the degraded harmonic characteristics the single-mode limit-trajectory overmodulation scheme is valuable for calculation –resource-scarce applications such as three-level inverter.

REFERENCE

[1] N ho N V. Two-Mode Overmodulation in Two-Level Voltage Source Inverter Using Principle Control Between Limit Trajectories [C]. Proceedings of PEDS2003. 2003, 1274 - 1279.

[2] Lee D C. A Novel Overmodulation Technique for Space Vector PWM Inverters [J]. IEEE trans. on PE. 1998, 13 (6): 1144 - 1151.

[3] Holtz J. On Continuous Control of PWM Inverters in the Over-Modulation Range Including the Six- Step Mode [J]. IEEE Trans. on PE. , 1993, 8(4): 546 - 553.

[4] Bolognani S. Novel Digital Continuous Control of SVM Inverters in the Overmodulation Range [J]. IEEE Trans. on Industry Application. 1997, 33(2): 525 - 530.

[5] Shun JIN, Zhong Yan-ru, et al. A three-level PWM method of neutral-point balancing and narrow-pulse elimination [J]. Proceedings of the CSEE, 2003, 23(10):114-118. (in Chinese)

[6] Shun JIN, Zhong Yan-ru. A novel three-level SVPWM algorithm considering neutral-point control and narrow-pulse elimination and dead-time compensation [J]. Proceedings of the CSEE, 2005, 25(6):60-66.

2006 5th International Power Electronics and Motion Control Conference

Multiphase Permanent Magnet Motor Drive System Based on A Novel Multiphase SVPWM

Shan Xue, Xuhui Wen, Zhao Feng

Institute of Electrical Engineering Chinese Academy of Sciences, Beijing, China

Abstract—**Multiphase motor drives are multi-dimensional systems, so multiphase vector control and SVPWM must be implemented in a multi-dimensional vector space. In this paper, multi-dimensional multiphase vector control is developed based on a novel multiphase SVPWM strategy, which could synthesize the required voltage vectors not only in d-q subspace but also in other subspaces. Therefore, all voltage and current harmonics, which contribute to the torque positively, are controllable. Throughout this paper, a five-phase permanent magnet motor drive system is developed as an example, and simulation and experimental results verify these proposed strategies. Moreover, these strategies can be easily extended to other multiphase drive systems with any number of phases.**

Keywords-multiphase permanent magnet motor; vector control; multiphase SVPWM; orthogonal vector space; harmonics; equivalent circuit

I. INTRODUCTION

The use of multiphase motors to implement high power is an alternative to reduce the current rating of the inverter power switches. Also, multiphase motors drives show other advantages over three-phase motors drives such as reducing the amplitude and increasing the frequency of torque pulsations, higher reliability, and lowering the dc link current harmonics [1-8]. Therefore, multiphase motor drives are very suitable for the applications of high power, high reliability, and low dc bus voltage, such as electrical vehicles, aerospace applications, ship propulsion etc.

Multiphase motor drives are multi-dimensional systems, which require multiphase vector control and PWM must be multi-dimensional. In addition, most multiphase motors are designed to have nearly rectangular phase back-emf in order to increase the torque density [7][10]. Therefore, conventional vector control and PWM, which are implemented only in two-dimensional d-q subspace and aims to realize a sinusoidal phase voltage, is improper for multiphase motors control.

Especially, multiphase PWM techniques are of key importance and much research has been done on them. In [1] and [2], two space vectors are selected to realize SVPWM in d-q subspace and minimize the switching losses. They simply extend the conventional three-phase SVPWM and do not consider the influence of these vectors in other subspaces, so the stator harmonic currents could be surprisingly large [3][4]. In [5] and [6], multiphase SVPWM based on vector space decomposition is developed. It is effective only for motors with sinusoidal windings, and the switching losses increase unfortunately. Kelly et al. developed *n*-phase SVPWM by extending a three-phase Unified PWM [9]. However, this method

focused on how to realize a sinusoidal phase voltage. Toliyat et al. developed the five-phase motors drives based on the hysteresis-band current PWM [7][8].

In this paper, novel multi-dimensional multiphase vector control and multiphase SVPWM will be presented based on the concept of orthogonal multi-dimensional vector space. Therefore, voltage vectors both in d-q subspace and in other subspaces can be synthesized to satisfy control requirements. At the same time, appropriate vectors conducting sequence are chose to minimize switching losses. A five-phase permanent magnet motor drive system is developed as an example, and simulation and experiment are presented.

II. VECTOR SPACE, MODELING AND VECTOR CONTROL

A. Orthogonal Vector Space And Modeling

Since a five-phase motor with star connection is basically a four-dimensional system except for zero sequence subspace, two two-dimensional subspaces should be used to define it. By applying the well known transformation (1), the original five-dimensional space can be mapped to a new vector space spanned by the new frame—d-q-z1-z2-n reference frame. The transformation possesses the following properties:

1. The transformation has the pseudo orthogonal property.

$$\mathbf{T}^{-1} = 5/2 \cdot \mathbf{T}^T$$

2. The fundamental component of the motor variables and the kth harmonics with $k = 10m \pm 1, (m = 1, 2...)$ are all mapped into d-q subspace.

3. The kth harmonics with $k = 5m \pm 2, (m = 1, 3, 5...)$ are all mapped into z1-z2 subspace.

That is, the new vector space consists of three orthogonal subspaces: d-q subspace, z1-z2 subspace and zero sequence subspace. This concept lays the foundation for the control of current or voltage harmonics in each subspace respectively.

$$\mathbf{T} = \frac{2}{5} \begin{bmatrix} 1 & \cos\alpha & \cos 2\alpha & \cos 3\alpha & \cos 4\alpha \\ 0 & \sin\alpha & \sin 2\alpha & \sin 3\alpha & \sin 4\alpha \\ 1 & \cos 3\alpha & \cos 6\alpha & \cos 9\alpha & \cos 12\alpha \\ 0 & \sin 3\alpha & \sin 6\alpha & \sin 9\alpha & \sin 12\alpha \\ 1/2 & 1/2 & 1/2 & 1/2 & 1/2 \end{bmatrix} \quad (1)$$

Similarly, thirty non-zero space voltage vectors of a five-phase inverter could be projected to d-q subspace and z1-z2 subspace as shown in Fig.1. In d-q subspace, the thirty vectors are composed of three sets of different

1-4244-0448-7/06/$25.00 ©2006 IEEE

amplitude vectors, and divide d-q subspace into 10 sectors. The amplitudes of these voltage vectors are $V_{min}=0.2472V_{dc}$, $V_{mid}=0.4V_{dc}$, $V_{max}=0.6472V_{dc}$, and the ratio of the amplitudes is $1:1.618:1.618^2$. It is the similar situation in z1-z2 subspace.

Especially, the fundamental and third harmonics of motor variables is transformed to be dc components after using synchronously rotating transformation (2). This idea is very important for multi-dimensional vector control of multiphase motors.

$$
\mathbf{C}_{s/r} = \begin{bmatrix}
\cos\theta & \sin\theta & 0 & 0 & 0 \\
-\sin\theta & \cos\theta & 0 & 0 & 0 \\
0 & 0 & \cos3\theta & \sin3\theta & 0 \\
0 & 0 & -\sin3\theta & \cos3\theta & 0 \\
0 & 0 & 0 & 0 & 1
\end{bmatrix} \quad (2)
$$

As shown in Fig.2, a two-pole, five-phase, star connection, non-salient pole permanent magnet motor with the concentrated winding will be discussed here. In order to simplify the modeling, only the fundamental and the third harmonics of the winding function are taken into account. By applying the above transformation (1) and (2), the equations of the five-phase permanent magnet motor in the synchronously rotating reference frame can be obtained:

Stator voltage equations

$$
\begin{aligned}
u_{ds} &= r_s i_{ds} + p\lambda_{ds} - \omega\lambda_{qs} \\
u_{qs} &= r_s i_{qs} + p\lambda_{qs} + \omega\lambda_{ds} \\
u_{z1s} &= r_s i_{z1s} + p\lambda_{z1s} - 3\omega\lambda_{z2s} \\
u_{z2s} &= r_s i_{z2s} + p\lambda_{z2s} + 3\omega\lambda_{z1s}
\end{aligned} \quad (3)
$$

Stator flux linkage equations

$$
\begin{aligned}
\lambda_{ds} &= L_{ds} i_{ds} + \lambda_{m1} \\
\lambda_{qs} &= L_{qs} i_{qs} \\
\lambda_{z1s} &= L_{z1s} i_{z1s} + \lambda_{m3} \\
\lambda_{z2s} &= L_{z2s} i_{z2s}
\end{aligned} \quad (4)
$$

The electromagnetic torque can be obtained using the magnetic co-energy method.

$$
T_e = \frac{\partial W_{co}}{\partial \theta_r} = \frac{5}{2} P\left(\lambda_{m1} i_{qs} + 3\lambda_{m3} i_{z2s}\right) \quad (5)
$$

where u_{ds}, u_{qs}, u_{z1s}, u_{z2s} are stator voltages in d, q, z1, z2 axes respectively. i_{ds}, i_{qs}, i_{z1s}, i_{z2s} are stator currents in these rotating axes. λ_{ds}, λ_{qs}, λ_{z1s}, λ_{z2s} are the transformed stator fluxes. λ_{m1}, λ_{m3} are the fundamental and third harmonic components of the flux linking stator coils due to the permanent magnets. ω is synchronous speed. L is stator inductance. P is number of pole pairs.

B. Multiphase Vector Control

From (3)~(5), the improvement in the torque due to the third harmonic current can be noticed, and thus control of

five-phase motors requires we can synthesize the required voltage vectors in z1-z2 subspace to realize the injection of the third harmonic current. Furthermore, Torque is only dependent on i_{qs} and i_{z2s}, so i_{ds} and i_{z1s} should be equal to zero. Fig.3 shows the control block diagram of the five-phase motor drive system, where $\mathbf{T}(\theta) = \mathbf{C}_{s/r}\mathbf{T}$ and $\mathbf{C}_{r/s} = \mathbf{C}_{s/r}^{-1}$. Clearly, the key for the multi-dimensional vector control is to implement multiphase SVPWM.

III. NOVEL MULTI-DIMENSIONAL SVPWM

The main idea of the novel multi-dimensional SVPWM is to synthesize the required voltage vectors both in d-q subspace and in other subspaces. Therefore, it can control all voltage and current harmonics which contribute to the

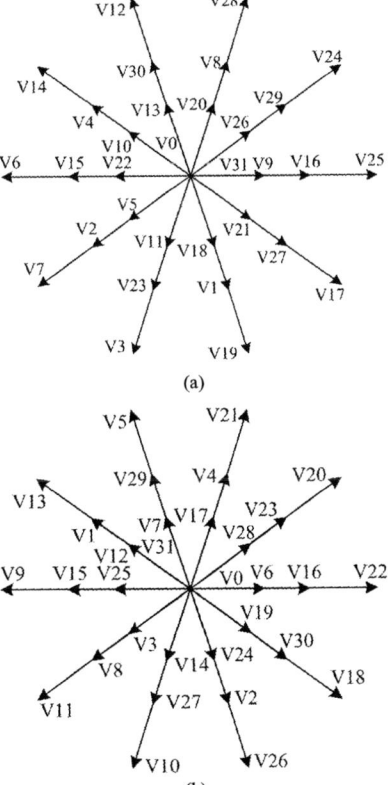

(a)

(b)

Fig. 1. Five-phase voltage vectors
(a) in d-q subspace (b) in z1-z2 subspace

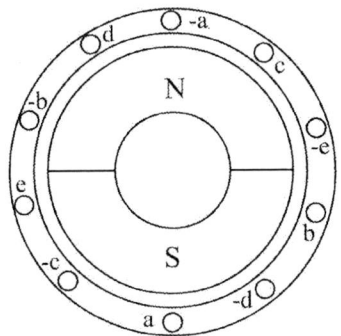

Fig. 2. Two-pole, five-phase, non-salient pole PMSM

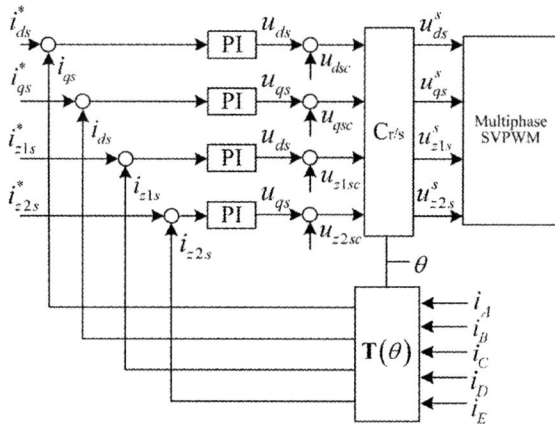

Fig. 3 Control block diagram of the five-phase motor drive system

torque positively. For a five-phase motor drive system, it can accomplished by (6), where a set of five voltage vectors are chosen to guarantee unique solution. Obviously, there are many ways to choose voltage vectors to realize (6). A novel method for choosing the voltage vectors and deciding these vectors conducting sequence will be developed.

$$
\begin{bmatrix} u_{ds}^{s*}T_s \\ u_{qs}^{s*}T_s \\ u_{z1s}^{s*}T_s \\ u_{z2s}^{s*}T_s \\ T_s \end{bmatrix} = \begin{bmatrix} V_{ds}^1 & V_{ds}^2 & V_{ds}^3 & V_{ds}^4 & V_{ds}^5 \\ V_{qs}^1 & V_{qs}^2 & V_{qs}^3 & V_{qs}^4 & V_{qs}^5 \\ V_{z1s}^1 & V_{z1s}^2 & V_{z1s}^3 & V_{z1s}^4 & V_{z1s}^5 \\ V_{z2s}^1 & V_{z2s}^2 & V_{z2s}^3 & V_{z2s}^4 & V_{z2s}^5 \\ 1 & 1 & 1 & 1 & 1 \end{bmatrix} \cdot \begin{bmatrix} T_1 \\ T_2 \\ T_3 \\ T_4 \\ T_0 \end{bmatrix} \quad (6)
$$

Since switching states determine different load equivalent circuit configurations of a VSI, switching states can be divided into different sets according to the load equivalent circuit configurations. For example, in a seven-phase inverter, when the switching state is (1111000), the equivalent circuit configuration is four parallel impedances in series with other three parallel impedances and this type of configuration is defined as C34. Accordingly, all switching states which determine C34 belong to set {C34}. Furthermore, {C34} can be divided into several subsets according to the adjacent relation of "1" in switching states, for example, (1111000) and (0111000) have the same adjacent relation of "1", while (1111000) and (1101100) do not. Among these subsets, the maximum magnitude subset {C34$_{Max}$} is typical, since the polygon formed by it encloses all polygons formed by other subsets in d-q subspace. The voltage vectors in {C34$_{Max}$} are selected in the novel SVPWM. In the same way, the other two possibilities are {C16$_{Max}$}, {C25$_{Max}$}.

In a five-phase inverter, there are only two types of basic equivalent circuit configurations: C23 and C14, as shown in Fig.4. Accordingly, the switching states can be divided into two sets: {C23} and {C14}. In Fig.1(a), the

10 outmost voltage vectors and the 10 inmost voltage vectors belong to {C23}, and 10 mid-magnitude voltage vectors belong to {C14}. Furthermore, in {C23}, the 10 outmost voltage vectors which form {C23$_{Max}$} subset have the same adjacent relation. In {C14}, there is only one adjacent relation, thus {C14$_{Max}$} subset is equal to {C14}. Voltage vectors that belong to {C23$_{Max}$} and {C14$_{Max}$} are selected to realize the novel four-dimensional five-phase SVPWM. For u_{dqs}^{s*} shown in Fig.5, as an example, $\vec{V}_{16}$, $\vec{V}_{24}$, $\vec{V}_{25}$, and $\vec{V}_{29}$ are selected, and the fifth vector is chosen from zero vectors. From (6), the time intervals T_1, T_2, T_3, T_4 and T_0 can be obtained by solving (7). Similarly, we can calculate time intervals when reference voltage u_{dqs}^{s*} locates in other sectors.

In order to minimize the total switching losses, the vector conducting sequence is shown in Fig.6. For example, when u_{dqs}^{s*} locates in the first sector, the vector sequence will be $\vec{V}_0 \rightarrow \vec{V}_{16} \rightarrow \vec{V}_{24} \rightarrow \vec{V}_{25} \rightarrow \vec{V}_{29} \rightarrow \vec{V}_{31}$ for half of the PWM period, and the witching pattern is illustrated in Fig.7. A device switches only once during each period.

IV. SIMULATION AND EXPERIMENTAL RESULTS

To verify the validity of these proposed strategies, experiments have been carried out using a 5kW five-phase Surface Permanent Magnet Motor (SPM), which was used as a wheel motor for an electrical vehicle. The electrical specifications of the motor are given in Table I. The control is implemented by using TMS320 LF2407A DSP. The motor drive system is also simulated using the technical data in Table I.

Especially, in order to prove the effectiveness of the proposed SVPWM strategy, the amplitude ratio between the fundamental and third harmonic currents is commanded to be different.

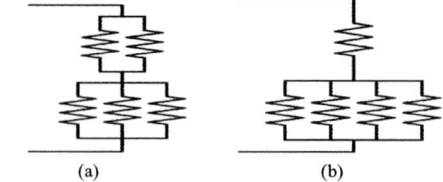

Fig. 4. Basic equivalent circuit configuration (a) C23 (b) C14

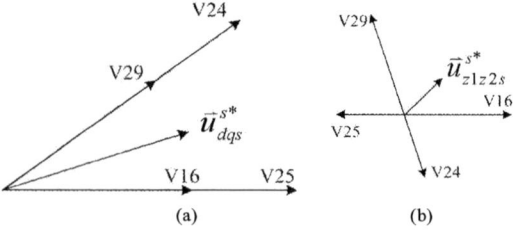

Fig. 5. Switching states selection when the reference voltage vector locates in the first sector (a) in d-q subspace (b) in z1-z2 subspace

$$\begin{bmatrix} u_{ds}^{s*}T_s \\ u_{qs}^{s*}T_s \\ u_{z1s}^{s*}T_s \\ u_{z2s}^{s*}T_s \end{bmatrix} = \begin{bmatrix} V_{mid} & V_{max}\cos\pi/5 & V_{max} & V_{mid}\cos\pi/5 \\ 0 & V_{max}\sin\pi/5 & 0 & V_{mid}\sin\pi/5 \\ V_{mid} & V_{min}\cos(\pi+3\pi/5) & V_{min}\cos\pi & V_{mid}\cos3\pi/5 \\ 0 & V_{min}\sin(\pi+3\pi/5) & 0 & V_{mid}\sin3\pi/5 \end{bmatrix} \cdot \begin{bmatrix} T_1 \\ T_2 \\ T_3 \\ T_4 \end{bmatrix} \qquad (7)$$

$$T_0 = T_s - T_1 - T_2 - T_3 - T_4$$

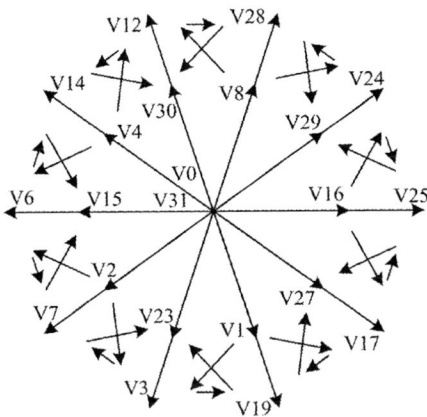

Fig. 6. Vectors conducting sequence

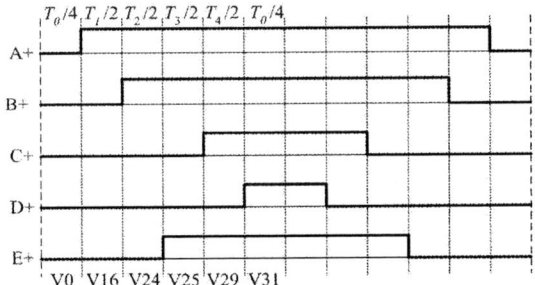

Fig. 7. Switching pattern when the reference voltage vector locates in the first sector

TABLE I
MOTOR DATA

Rated output power	5kW
Rated speed	550r/min
Rated torque	86N.m
Rated phase voltage	70V
Rated phase current	21A
Number of pole pairs	8
r_s	0.11Ω
L_{ds}	3.17mH
L_{qs}	3.17mH
L_{z1s}	1.4mH
L_{z2s}	1.4mH
λ_{m1}	0.142Wb
λ_{m3}	0.008Wb

Fig.8 shows the simulation results and experimental results of A-phase current using the proposed SVPWM,

when the amount of the injected third harmonic current is commanded to be 25% of the fundamental current. i_{qs} is commanded to be 15A, and i_{z1s} is commanded to be $0.25i_{qs}$. Fig.9 shows A-phase current without injection of the third harmonic current. i_{qs} is commanded to be 10A, and i_{z1s} is commanded to be zero. Fig.10 shows torque vs. phase current for $i_{ds} = i_{z1s} = i_{z2s} = 0$. From these results above, it is seen clearly that by using the proposed multi-dimensional vector control and SVPWM, we can control the fundamental and third harmonics currents and regulate them without steady error.

For the purpose of comparison, vector control based on conventional SVPWM just like in [1] and [2] is also tested. Fig.11 shows the A phase current using the conventional SVPWM. Obviously, the third harmonic current is uncontrollable using the conventional method.

V. CONCLUSIONS

Conventional vector control implemented only in two-dimensional d-q subspace is improper for multiphase motors drives, since they are multi-dimensional. This paper proposed multi-dimensional vector control of multiphase motor drives based on a novel multi-dimensional multiphase SVPWM, which is based on the concept of orthogonal multi-dimensional vector spaces. It can synthesize the required voltage vectors in every subspace, and thus controls stator current harmonics that contribute to torque positively according to control's requirements. Moreover, the proposed SVPWM strategy can be easily extended to other multiphase drives with any number of phases.

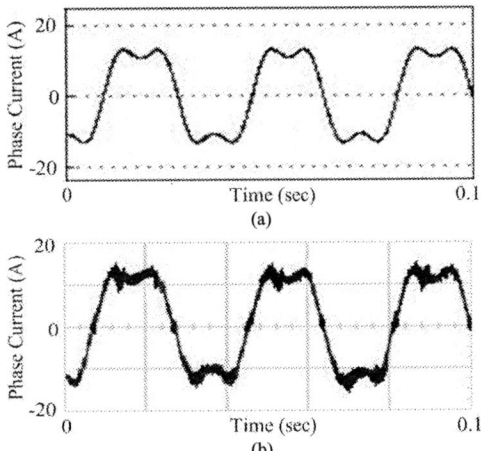

Fig. 8 A-phase current with the combined fundamental and third harmonic current (a) simulation result (b) experimental result

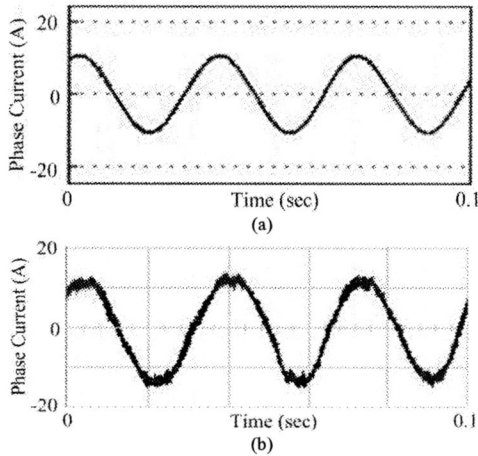

Fig. 9 A-phase current without the injection of the third harmonic current (a) simulation result (b) experimental result

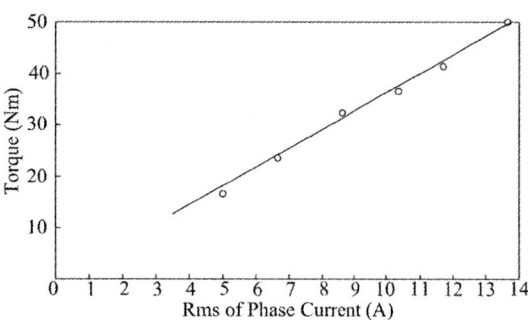

Fig. 10 Torque vs. phase current for $i_{ds} = i_{z1s} = i_{z2s} = 0$

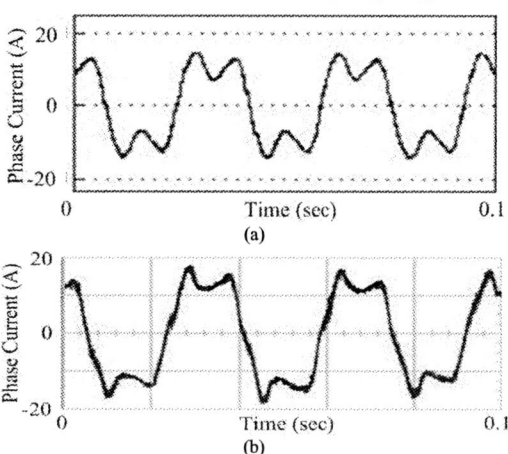

Fig. 11 A-phase current using the conventional SVPWM (a) simulation result (b) experimental result

REFERENCES

[1] Fei Yu, Xiaofeng Zhang, Huaishu Li, Zhihao Ye, "The space vector PWM control research of a multi-phase permanent magnet synchronous motor for electrical propulsion", Electrical Machines and Systems (ICEMS), Vol.2, pp.604-607, Nov. 2003.

[2] Ruhe Shi, H.A.Toliyat, "Vector Control of Five-phase Synchronous Reluctance Motor with Space Pulse Width Modulation for Minimum Switching Losses", Industry Applications Conference, 36th IAS Annual Meeting. Vol.3, pp.2097-2103, 30 Sept.-4 Oct. 2001.

[3] M.A.Abbas, R.Christen, T.M.Jahns, "Six-phase Voltage Source Inverter Driven Induction Motor", IEEE Trans. on IA, Vol.IA-20, No.5, pp.1251-1259, 1984.

[4] E.E.Ward, H.Harer, "Preliminary Investigation of an Inverter fed 5-phase Induction Motor", IEE Proc, June 1969, Vol.116(B), No.6, pp.980-984, 1969.

[5] Y.Zhao, T.A.Lipo, "Space Vector PWM Control of Dual Three-phase Induction Machine Using Vector Space Decompositon", IEEE Trans. on IA, Vol.31, pp.1177-1184, 1995.

[6] Xue S, Wen X.H, "Simulation Analysis of A Novel Multiphase SVPWM Strategy", 2005 IEEE International Conference on Power Electronics and Drive Systems(PEDS), Proc., pp:756~760, 2005

[7] Parsa L, H.A.Toliyat, "Multiphase permanent magnet motor drives", Industry Applications Conference, 38th IAS Annual Meeting. Vol.1, pp.401-408, 12.-16 Oct. 2003.

[8] H.Xu, H.A.Toliyat, L.J.Pertersen, "Five-Phase Induction Motor Drives with DSP-based Control System", IEEE Trans. on IA, Vol.17, No.4, pp.524-533, 2002.

[9] John W. Kelly, Elias G. Strangas, John M. Miller, "Multiphase space vector pulse width modulation", IEEE Trans. on Energy Conversion, Vol.18, No.2, pp.259-264, 2003.

[10] Hyung Min Ryu, Ji Woong Kim, Seung Ki Sut, "Analysis of multiphase space vector pulse width modulation based on multiple d-q spaces concept", International Power Electronics and Motion Control Conference (IPEMC), Vol.3, pp.1618-1624, Aug. 2004.

[11] D.C.White, H.Woodson, *Electromechanical Energy Conversion*. New York: Wiley, 1959.

2006 5th International Power Electronics and Motion Control Conference

Novel Random-Harmonic Elimination PWM Technique for Single-Switch Three-Phase AC-DC Buck Converter

Guang-Hui Tan, Wenchuan Ma, Yanchao Ji, Hongxiang Yu and Wancai Xu
Department of Electrical Engineering, Harbin Institute of Technology, Harbin, China
hearpc@163.com

Abstract—**A novel PWM technique for single-switch three-phase AC-DC buck converter is proposed to reject input voltage fluctuation, obtain high quality output voltage and realize open-loop operation mode. By controlling the duty cycle of the switch which tracks the change of input voltage, the energy supplied by ac mains and the energy consumed in load can keep invariable in every switching cycle. So the output dc voltage maintains at a constant value and random-harmonic voltage contained in input stage is eliminated. Based on the principle of energy conservation, the system operation is analyzed and the control strategy is introduced in detail both in discontinuous conduction mode (DCM) and continuous conduction mode (CCM). Experimental results are provided to show its validity and feasibility.**

Keywords—buck converter; harmonic elimination; energy conservation; discontinuous conduction mode; continuous conduction mode

I. INTRODUCTION

Buck converter as a basic DC-DC converter is widely used in modern power supplies. In the past dozens of years, control methods of buck converter are well researched [1]-[5]. However, when analyzing its harmonic characteristic, most researchers consider the input voltage of the buck converter as a constant. To achieve this, large smooth capacitors are employed in practice. For example, diode rectifiers and phase-controlled rectifiers are employed to convert ac voltage. The rectifier can be followed by a front-end LC filter, which provides low-ripple dc voltage, and a PWM DC-DC converter (generally buck converter), which is used to achieve the output voltage regulation [6], [7]. The front-end filter in this conventional AC-DC converter results in high cost, big size and mass harmonic current in supply. Recently some PWM techniques and control methods have been developed to cancel the bulky front-end filter and improve the harmonic performance, yet involve difficulties of complicated control strategy and degraded precision under condition of random-harmonic voltage contained in input stage [8], [9].

In order to solve the above-mentioned problems, we propose a novel random–harmonics elimination technique based on the principle of energy conservation for single-switch three-phase AC-DC buck converter. By controlling the duty cycle of the switch which tracks the change of input dc side voltage, the input energy can keep invariable that the load also consumes the same energy in every switching cycle. So the output voltage can maintain invariable and harmonics are suppressed. The proposed harmonics elimination technique is robust for the change of input voltage and load, and fulfilled by simple open-loop mode both in DCM and CCM. The validity and feasibility of the proposed control technique are supported by experiments.

II. BUCK CONVERTER SYSTEM ANALYSIS

The buck converter is shown in Fig. 1. It consists of a power controllable switch S, an inductor L, a diode D, a filter capacitor C, and a load resistor R. The switch is turned on and off at a switching frequency of $f_s = 1/T_s$. The duty cycle of the switch is $d = t_{ON}/T_s$, where t_{ON} is the time interval when the switch is on.

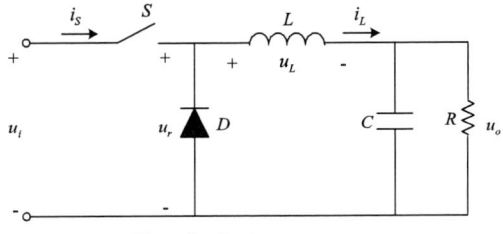

Figure 1. Buck converter

The analysis of the buck converter in Fig. 1 is based upon the following assumptions: (1) The power switch and diode are ideal switches, and switching losses are assumed zero; (2) Passive components are linear, time-invariant, and frequency-independent; (3) The output impedance of the input voltage source is zero for both dc and ac components; (4) The switching frequency is much higher than the line frequency f_o (50Hz). In Fig. 1, assuming the input voltage u_i contains random harmonic component, which can be given as

1-4244-0448-7/06/$25.00 ©2006 IEEE   627

$$u_i = U_{DC} + \sum_{k=1}^{\infty} U_K \sin(kwt + \varphi_k) \qquad (1)$$

where U_{DC} and $\sum_{k=1}^{\infty} U_K \sin(kwt + \varphi_k)$ are average and total harmonic component, respectively.

According to the assumption (4), input voltage is assumed to be almost unchanged in one switching cycle, therefore, the operation principle and characteristic of buck converter can also be analyzed in one switching cycle. Operation of buck converter is characterized by the conduction mode of the inductor current, either continuous or discontinuous. The sketch map of distinguishing work mode is shown in Fig. 2, where d is the duty cycle of the switch, I_{LC} is the average value of the current through the inductor, I_{LCM} is the peak inductor current, and I_o is the load current. When output voltage u_o is constant, the load current $I_o = u_o / R$ can also be considered as constant.

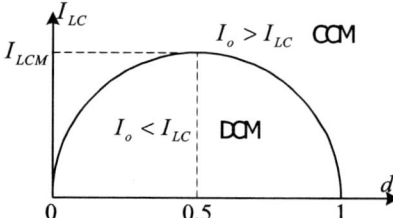

Figure 2.   Sketch map of distinguishing work mode

On condition that the load current I_o is less than the average value of inductor current I_{LC}, the buck converter operates in DCM. The waveforms of inductor voltage and current in DCM are shown in Fig. 3.

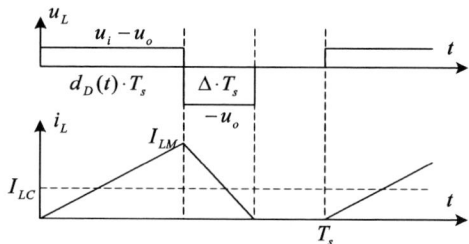

Figure 3.   Voltage and current of inductor in DCM

When the switch is on, the current through the switch i_{sD} is equal to the inductor current i_{LD}, which is expressed by

$$i_{sD} = i_{LD} = \frac{u_i - u_o}{L} t \qquad (2)$$

On condition that the load current I_o is bigger than the average value of inductor current I_{LC}, the buck converter operates in CCM. The waveforms of inductor voltage and current in CCM are shown in Fig. 4, where we see an initial inductor current I_o which is nonzero in every switching cycle. So besides the energy stored in the output capacitor in the entire switching cycle, there is always some energy stored in the inductor in CCM buck

converter, which is a remarkable difference between CCM and DCM.

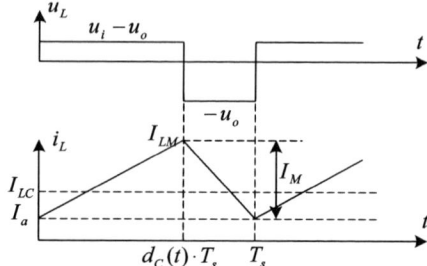

Figure 4.   Voltage and current of inductor L in CCM

According to the waveform in Fig. 4 and the characteristic of buck converter in CCM, the following expression is obtained.

$$I_a = I_o - \frac{1}{2} I_M = I_o - \frac{u_o T_s}{2L}(1 - \frac{u_o}{u_i}) \qquad (3)$$

where I_M is the peak-to-peak ripple current through the inductor in the switching cycle.

Similarly, when the switch is on, the current through the switch i_{sC} is equal to the inductor current i_{LC}, which is expressed by

$$i_{sC} = i_{LC} = i_a + \frac{u_i - u_o}{L} t \qquad (4)$$

Combination of (3) and (4) yields

$$i_{sC} = I_o - \frac{u_o T_s}{2L}(1 - \frac{u_o}{u_i}) + \frac{u_i - u_o}{L} t \qquad (5)$$

III.   CONTROL STRATEGY

In buck converter, almost all the ac component of the inductor current flows through the output capacitor, therefore, the output voltage value u_o will inevitably change. From (6) which shows the relation between the capacitor voltage ripple Δu_o and the capacitor current i_C in a switching cycle, the capacitance of the capacitor C is big enough and the switching cycle T_s is small enough, that Δu_o can be considered as zero. That is, we think output voltage is an invariable value U_o in a switching cycle.

$$\Delta u_o = \frac{1}{C} \int_0^{T_s} i_C dt \qquad (6)$$

While assuming that the output voltage keeps invariable, the energy stored in the output capacitor also keeps invariable, therefore, we neglect the transformation of energy stored in output capacitor. The current through the inductor rises when the switch is on, and the energy stored in it increases. During this state, the inductor acquires energy. When the switch is off, the energy stored in the inductor falls and releases to load. So in every switching cycle, the voltage source u_i provided energy just when the switch is on. The inductor and capacitor only play a role in storing and shifting energy, not consuming energy. Only load consumes the energy that the power offers.

Based on the principle of energy conservation, the energy supplied by the power when the switch is on is equal to the energy consumed in load in a switching cycle, that is,

$$E_{in} = \int_0^{d(t)\cdot T_s} u_i i_s dt = E_{Load} \qquad (7)$$

The expression (7) is the foundation of theory analysis, simulation and experiment in this paper, and it reflects the application of energy conservation principle to buck converter. By controlling the duty cycle which tracks the change of input voltage, the input energy can keep invariable that the load also consumes the same energy in every switching cycle.

The energy consumed in load in a switching cycle can be given by

$$E_{Load} = \frac{U_o^2}{R} T_s \qquad (8)$$

Constant frequency control is employed in practice and the load is constant on steady state condition, and from (8), the output voltage U_o only relates to the energy consumed in load E_{Load}. Therefore, by controlling the duty cycle of the switch, the energy consumed in load can keep invariable in every switching cycle that the desired output voltage can maintain and harmonics are eliminated.

Combination of (2) and (7) yields the duty cycle of the switch in DCM as

$$d_D(t) = \sqrt{\frac{2LE_{Load}}{u_i(u_i - u_o)T_s^2}} \qquad (9)$$

Combination of (5) and (7) yields the duty cycle of the switch in CCM as

$$d_C(t) = \frac{-B + \sqrt{B^2 - 4AC}}{2A} \qquad (10)$$

where

$$A = \frac{u_i(u_i - u_o)}{2L} T_s^2$$

$$B = u_i \left[\frac{E_{Load}}{u_o T_s} - \frac{u_o T_s}{2L}(1 - \frac{u_o}{u_i}) \right] T_s$$

$$C = -E_{Load}$$

The expression (9) and (10), respectively, gives the control equations based on the principle of energy conservation in DCM and CCM to eliminate input harmonics and keep output voltage constant, where the duty cycle $d_D(t)$ and $d_C(t)$ are both related to the inductor L, the energy consumed in load E_{Load}, the output voltage u_o, input voltage u_i and switching cycle T_s. After structure of the buck converter and switching frequency are confirmed, L and T_s is also confirmed. And after the load and the expectable output voltage are confirmed, u_o and E_{Load} are also confirmed accordingly. So the duty cycle are only related to input voltage u_i, which is a one-one relation between the two. That is, the control of buck converter can be implemented by open-loop mode.

According to the assumption (4), input voltage is assumed to be almost unchanged in one switching cycle. So long as the input voltage is measured in every switching cycle, substitution of which into (9) or (10) yields the duty cycle to control the power switch, thus the input energy keeps invariable that the energy consumed in load keeps invariable in a switching cycle, and that the output voltage can maintain at a constant value.

The block diagram of overall control scheme is shown in Fig. 5, which consists of a PI controller, main controller, triangle-wave generator, and comparator, etc. The objective of the voltage feedback loop is to maintain the desired output voltage level under normal operating conditions, as well as under sudden change of input voltage or load. After the circuit enters steady state, the PI output must be the energy consumed in the load. The input voltage is measured that the duty cycle can be obtained by (9) or (10) to control the power switch. Thus the change of input voltage is tracked and the harmonics contained in input stage can be suppressed. In addition, through the given voltage, the output dc voltage can be regulated conveniently and quickly.

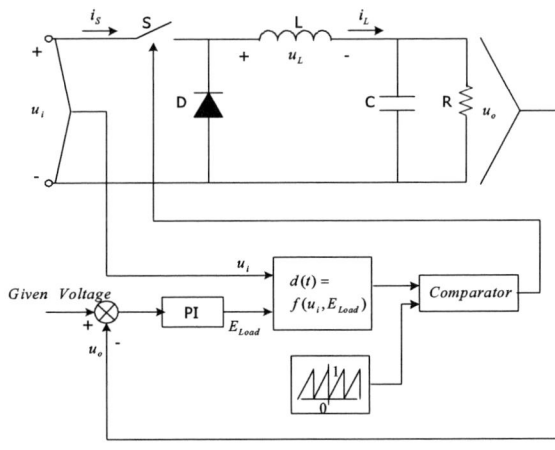

Figure 5. The block diagram of overall control scheme

IV. RANDOM-HARMONIC ELIMINATION PRINCIPL

A. Analysis in DCM

In buck converter, almost all the ac component of the inductor current flows through the output capacitor and the dc component flows through the load. To maintain the output voltage u_o invariable, it is necessary to keep the load current $I_o = u_o / R$ invariable, that is, the average value of inductor current I_{LC}, which is equal to the load current I_o, must be kept invariable in every switching cycle.

Fig. 3 shows the waveforms of inductor voltage and current in DCM, where assuming the triangular area enclosed by the inductor current and time axle in a switching cycle is S, therefore, I_{LC} is equal to S/T_s. To keep the output voltage u_o constant, S must be constant in every switching cycle, which can be called the equal area principle to eliminate harmonics in buck converter.

When the switch is off, the slope of the above-mentioned triangle is always $-U_o/L$, that is, all the down edges of the triangles are parallel. Based on this characteristic of the triangles, when input voltage changes, the slope of the up edge varies with it too. Therefore, in order to maintain the area S invariable, the peak inductor current I_{LM} and the duty cycle $d_D(t)$ must change accordingly in operation.

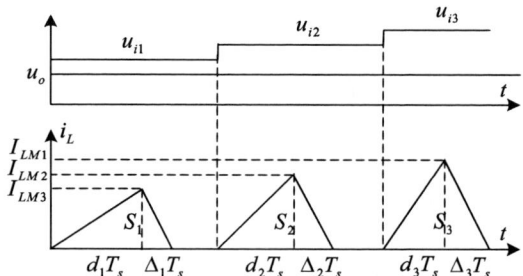

Figure 6. Waveform of inductor current under different input voltages condition in DCM

Under three different input voltages condition, the waveform of inductor current in DCM is shown in Fig. 6. The bigger u_i is, then the faster i_L rises, the bigger I_{LM} is, the smaller $d_D(t)$ is and the longer the discontinuous conduction time of the inductor maintains, but which always complies with the above-mentioned equal area principle and the principle of energy conservation.

B. Analysis in CCM

When the input voltage fluctuates, the initial values of inductor current in those switching cycles are not all equal. But the duty cycle $d_C(t)$ can be obtained from (10) to control the power switch, and the average value of the inductor current in every switching cycle can keep invariable.

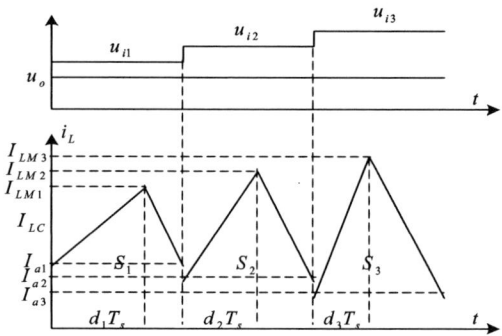

Figure 7. Waveform of inductor current under different input voltages condition in CCM

Under three different input voltages condition, the waveform of inductor current in CCM is shown in Fig. 7. Similarly, the equal area principle is satisfied, which reflects the principle of energy conservation in buck converter.

Considering the pentagons enclosed by the inductor current and the time axle, the slopes of down edges in which are all $-u_o/L$, that is, they are parallel, and the slopes of up edges are $(u_i-u_o)/L$, which varies with the change of input voltage. The bigger u_i is, then the faster i_L rises, the bigger I_{LM} is, the smaller I_a and $d_D(t)$ are, but which always complies with the above-mentioned equal area principle and energy conservation principle.

V. EXPERIMENTAL RESULTS

In order to verify the system performance, an 800W laboratory prototype is built and investigated. Fig. 8 gives the modules description of the experimental AC-DC buck system without front-end filter. Reference of the output voltage is set as 200V. The voltage signals are sensed using Hall-effect sensors (turn ratio: 0.05). The rectifier output voltage u_C is shown in Fig. 9, the dominant harmonic of $6f_0$, is inherent in diode bridge rectification, and the harmonic of $2f_0$ resulted from an unbalanced input supply is rather large.

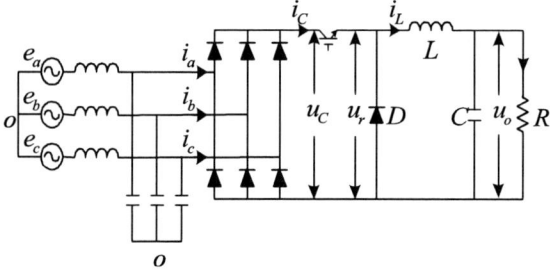

Figure 8. The experimental AC-DC buck system without front-end filter

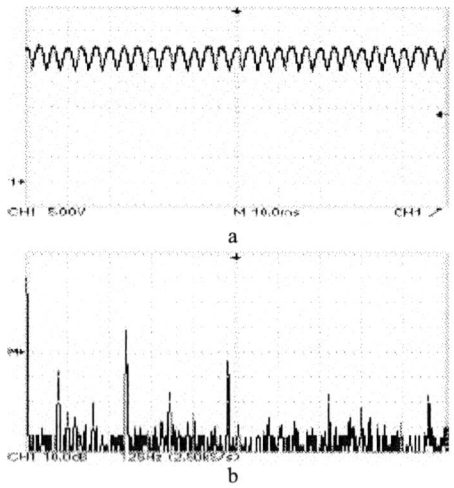

Figure 9. Waveform and spectrum of the rectifier output voltage (a:100V/div;b:10dB/div)

The circuit parameters are following:
In DCM

$$f_s = 8\text{kHz} \quad f_0 = 50\text{Hz}$$
$$L = 1\text{mH} \quad C = 220\,\mu\text{F}$$
$$R = 50\,\Omega \quad e_{a,b,c} = 220\text{V(rms)}$$

In CCM

$$f_s = 8\text{kHz} \quad f_0 = 50\text{Hz}$$
$$L = 10\text{mH} \quad C = 220\,\mu\text{F}$$
$$R = 50\,\Omega \quad e_{a,b,c} = 220\text{V(rms)}$$

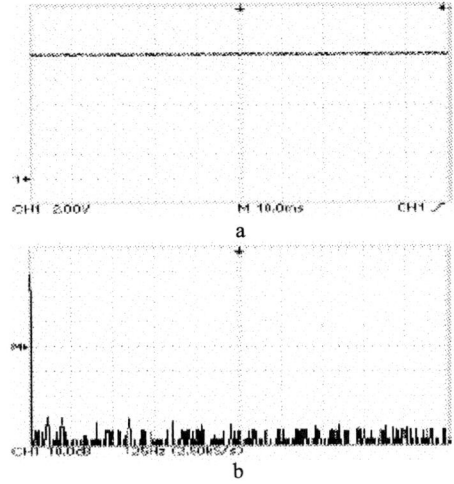

Figure 10. Waveform and spectrum of the output voltage in DCM
(a:100V/div;b:10dB/div)

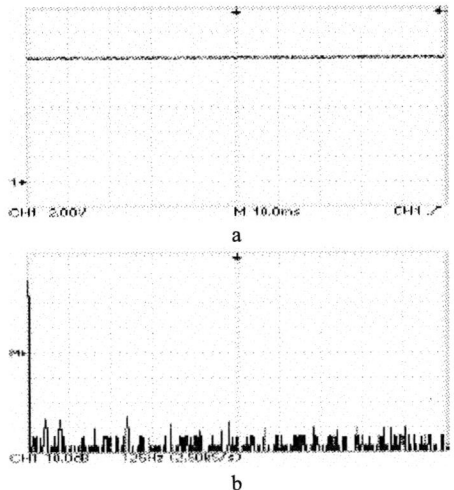

Figure 11. Waveform and spectrum of the output voltage in CCM
(a:100V/div;b:10dB/div)

Fig. 10 and Fig.11, respectively, show the experimental results of output voltage u_o in DCM and CCM. Under the unbalanced system, the ripple of three-phase rectifier output voltage u_C is relatively large as shown in Fig. 9, but the one of output voltage u_o in DCM or CCM is rather small as shown in Fig. 10 and Fig. 11. This verifies the validity of the proposed harmonics elimination technique.

VI. CONCLUSION

The proposed harmonics elimination technique based on the principle of energy conservation has the excellent performance of suppressing random harmonics. The method is simple and feasible, and it can be applied to buck converter both in DCM and CCM. In addition, the output dc voltage can be regulated conveniently and quickly, and it is robust for the change of input voltage and load. The experimental results of an 800W laboratory prototype prove the validity of the analysis and the feasibility of the proposed harmonics elimination technique.

REFERENCES

[1] K. M. Smedley and S. Cuk, "One-cycle control of switching converters," *IEEE Trans. Power Electron.*, vol. 10, pp. 625-633, Nov. 1995.

[2] F. Carfalo, P. Marino, S. Scala, and F. Vasca, "Control of DC-DC converters with linear optimal feedback and nonlinear feedforward," *IEEE Trans. Power Electron.*, vol. 9, pp. 607-615, Nov. 1994.

[3] H. Sira-ramirez and M. Garcia-estebaba, "Dynamical, adaptive pulsed-width-modulation control of DC-to-DC power converters: a backstepping approach," *INT. J. Control*, vol. 65, pp. 205-222, 1996.

[4] F. Chen and X. S. Cai, "Design of feedback control laws for switching regulators based on the bilinear large signal model," *IEEE Trans. Power Electron.*, vol. 5, pp. 236-240, Apr. 1990.

[5] A. G. Perry, G. Feng, Y.–F. Liu, and P. C. Sen, "A new design method for PI-like fuzzy logic controllers for DC-to-DC converter," *Proc. IEEE-PESC'04*, Jun. 20-25, 2004, Aachen, Germany, pp. 3751-3757.

[6] B. Dewan, "Optimum input and output filters for a single-phase rectifier power supply," *IEEE Trans. Ind. Applicat.*, vol. 17, pp. 282-288, 1981.

[7] B. K. Bose, "Modern power electronics—A technology review," *Proc. IEEE*, vol. 80, pp. 1303-1334, 1992.

[8] J. W. Kolar, H. Ertl, and F. C. Zach, "A novel three-phase single-switch discontinuous-mode AC-DC buck-boost converter with high-quality input current waveforms and isolated output," *IEEE Trans. Power Electron.*, vol. 9, pp. 160-173, Mar. 1994.

[9] Y. C. Ji and M. W. Shan, "A novel three-phase AC/DC converter without front-end filter based on adjustable triangular-wave PWM technique," *IEEE Trans. Power Electron.*, vol. 14, pp. 233-245, Mar. 1999.

[10] B. Singh, B. N. Singh, A. Chandra, K. Al-Haddad, A. Pandey, and D. P. Kothari, "A review of single-phase improved power quality AC-DC converters," *IEEE Tran. Ind. Electron.*, vol. 50, pp. 962-981, Oct. 2003.

[11] R. E. Strawser, B. T. Nguyen, and M. K. Kazimierczuk, "Analysis of a Buck PWM DC-DC converter in discontinuous conduction mode," *Proc. IEEE-NAECON'94*, May 23-27, 1994, pp. 35-42.

2006 5th International Power Electronics and Motion Control Conference

FPGA Based Multichannel PWM Pulse Generator for Multi-modular Converters or Multilevel Converters

Liqiao Wang, Weiyang Wu

Institute of Electric Engineering, Yanshan University, Qin Huangdao, P. R. China

Abstract- **Familiar special microprocessor with PWM pulse generator cannot provide enough PWM pulses for multi-modular converters or multilevel converters. Field programmable gate array (FPGA) is applied to implement the multichannel PWM pulse generator with no synchronization and communication trouble due to the single chip structure of FPGA compared with the multi-microprocessor scheme. A three-phase voltage source 5-level cascade inverter prototype with sample time staggered space vector modulation is accomplished to verify the correctness of the proposed multichannel PWM pulse generator.**

Keywords- multichannel PWM pulse generator; multi-modular converters or multilevel converters; sample time staggered space vector modulation

I. INTRODUCTION

Pulsewidth modulation (PWM) is widely used in power converter. Many new PWM strategies are brought out for large capacity equipments like multi-modular converters or multilevel converters [1][2][3]. The digital implementation of these PWM strategies encounters obstacles.

Familiar special microprocessors with PWM pulse generator are generally designed for two-level converters, which can always provides a six-channel PWM pulse generator only enough for a three-phase voltage source six-switch converter. Some latest products can provide two six-channel PWM pulse generators. However, it is not yet enough for multi-modular converters or multi-

level converters. For example, a three-phase five-level voltage source converter composed of 24 power devices needs 24 channels PWM pulses. Familiar special microprocessors cannot provide enough channels PWM pulse generator for it.

Some schemes with paralleled multi-microprocessor are presented to solve this problem. However, the synchronization and communication of every microprocessor are much difficult to realize due to separate clock and disperse parameters. Field programmable gate array (FPGA) is a new technique for very large scale integration. Multichannel PWM pulse generator can be conveniently implemented by FPGA with no synchronization and communication trouble due to its single chip structure. Therefore, the FPGA scheme is very promising to multi-modular converters and multilevel converters.

In this paper, a multichannel PWM pulse generator by FPGA is designed for multi-modular converters or multilevel converters. Sample time staggered space vector modulation (STS-SVM) is applied as switch modulation strategy. A three-phase voltage source 5-level cascade inverters prototype is accomplished through this PWM pulse generator. The experimental results verify the correctness of the design.

II. SAMPLE TIME STAGGERED SPACE VECTOR MODULATION

Space vector modulation (SVM) is a regular modulation strategy in much common use [4]. STS-SVM is the extended application of SVM to multi-modular convert-

1-4244-0448-7/06/$25.00 ©2006 IEEE

ers or multilevel converters [5]. To be brief, the principle of STS-SVM is to stagger the sampling time for each converter unit. Instanced by voltage sourced multi-modular converters (shown as Fig.1), each three-phase six-switch voltage source converter unit is modulated by traditional two-level SVM under the same modulated index and the sample time of the contiguous converter units is staggered by: $2\pi/(N \cdot \omega_c)$, where N is the amount of converter units and ω_c is the sample frequency. With such modulation, the equivalent sample frequency is improved by N times.

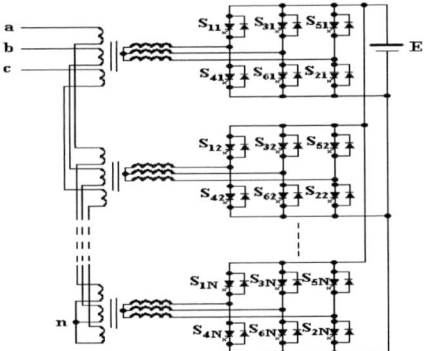

Fig.1 Voltage source STS-SVM multi-modular converters

STS-SVM can be also applied to cascade multilevel converters. A (N+1)-level cascade converter is equivalent to an N-unit multi-modular converter.

STS-SVM [6] is of much good performance such as excellent harmonic feature as well as wide transmitting bandwidth at very low device switching frequency, balanced switching load of every power device, high voltage utilization ratio and much easy digital algorithm. STS-SVM is much promising to large capacity power electronic equipments.

III. MULTICHANNEL PWM PULSE GENERATOR BASED ON FPGA

A. Realization of two-level SVM

For a three-phase N-unit voltage source multi-modular converter with STS-SVM, each converter unit is modulated by traditional two-level SVM. The modulation al-

gorithm of traditional two-level SVM is much regular. By the traditional two-level SVM, there are three pairs of complementary pulses for each converter unit. In this paper, the symmetrical regular sample modulation is applied. The waveform of one pair of complementary pulses in one sample cycle is shown as Fig.2, where T_w is the pulsewidth, T_s is the sample cycle and T_D is the dead time.

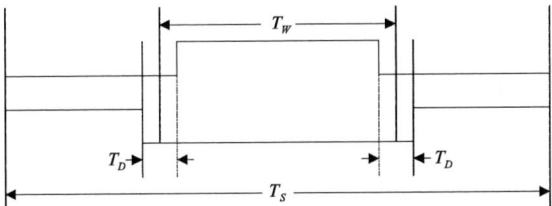

Fig.2 One pair of complementary pulses in one sample cycle

To generate the waveform shown as Fig.2, a programmable counter is needed. The counter value can be overwritten at anytime. The initial value of the counter can be discretional. The counter is corresponding with a cycle register, two comparative registers and two output pins. The cycle register is used for registering $T_s/2$. The value of $(T_s-T_w+T_D)/2$ is loaded in the comparative register of the upper pulse, and the value of $(T_s-T_w-T_D)/2$ is loaded in the comparative register of the lower pulse. The counter will be reset when its value is equal to zero. The counter value is set by the cycle register value when it is reset. The counter is controlled by a counting-enable bit. When the counting-enable bit is set to high level, the counter starts to work as a plus counter. From this moment, the counter will not stop until the counting-enable bit is set to low level. When the counter starts to work, the output pin for the upper pulse is set to low level, and the output pin for the lower pulse is set to high level. The counter value is compared with the value of each comparative register continuously. When the counter value is greater than $(T_s-T_w-T_D)/2$, the output pin for the lower pulse is set to low level. When the counter value is greater than $(T_s-T_w+T_D)/2$, the output pin for the upper pulse is set to high level. When the counter value is equal to the value of $T_s/2$, the counter goes on to work as a minus counter. When the counter value is less than $(T_s-T_w+T_D)/2$, the output pin for the upper pulse is set to low level. When the counter value is less than $(T_s-T_w-T_D)/2$, the output pin for the lower pulse is set to high level. When the counter value is reduced to zero, the counter is reset. The values of the cycle register and comparative registers will also be updated. The next sample cycle starts. The pulsewidth T_w is variable subjected to algorithm of SVM, and then the PWM pulse of one pair of power devices is realized. The other two pairs of the PWM pulses can be also generated by the same way like the former pair.

The process above-mentioned is only a simple operating logic of the PWM pulse generator for one converter unit. Actually, there many logic gates needed to realize the process. This PWM pulse generator for one converter unit can be packaged as a PWM module.

B. Realization of STS-SVM

For the three-phase N-unit voltage source multi-modular converter, N PWM modules should be constructed. Each PWM module has a counter. The cycle register value of each counter is the same as $T_s/2$. The initial value of each counter should be staggered equably by T_s/N in turn. The initial value of the first counter can be assigned with 0. The initial value of the second counter will be assigned with T_s/N. The initial value of the second counter will be assigned with $2T_s/N$. The initial value of other counters can be attained by analogy. In this way, the multichannel PWM pulse generator with STS-SVM is realized.

The FPGA of the proposed multichannel PWM pulse generator is only used to generate PWM pulses. The system control strategy should be realized by another microprocessor such as a digital signal processor (DSP).

IV. EXPERIMENTAL RESULTS

A three-phase voltage source 5-level cascade inverter prototype with STS-SVM is accomplished to verify the correctness of the proposed multichannel PWM pulse generator. There are 24 channels of PWM pulses needed for this converter. The multichannel PWM pulse generator is accomplished by a chip of ACEX1K30. The system control is accomplished by a chip of TMS320F240. The link diagram of the FPGA and DSP is shown as Fig.3.

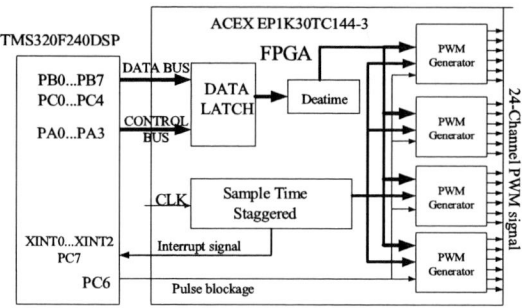

Fig.3 Link diagram of the FPGA and DSP

The main circuit structure is shown in Fig.4. MOSFET is chosen for the power device and the load is resistance. The modulated waveform frequency is 50Hz. The switching frequency is 1050Hz. The amplitude modulation index is 0.9.

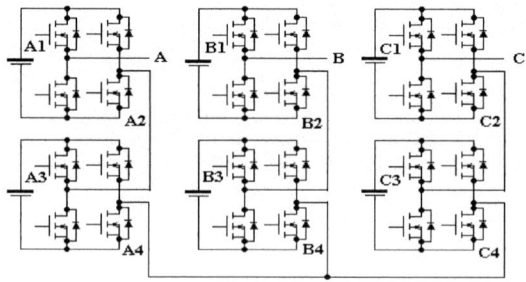

Fig.4 Three-phase voltage source 5-level cascade inverters

The four trigger waveforms A1, A2, A3, A4 of phase A is shown in Fig.5. The line voltage waveform is shown in Fig.6 (a), together with its spectrum shown in Fig.6 (b). From the spectrum, the harmonic feature of the output voltage is in agreement with theoretical analysis.

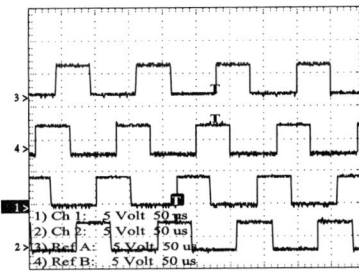

Fig.5 Four trigger waveform of phase A

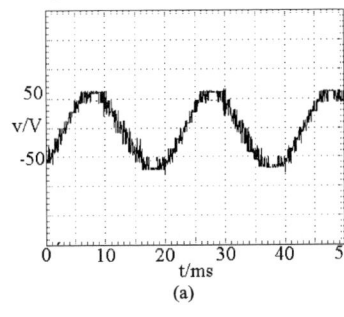

(a)

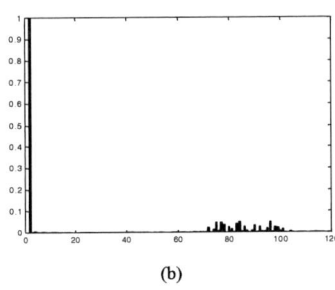

(b)

Fig.6 Line voltage waveform and its spectrum (a) Voltage waveform of phase (b) Spectrum of (a)

V. DISCUSS

Almost 200,000 logic gates applied for the 24-channel PWM pulse generator in the experimental FPGA. This 24-channel PWM pulse generator can provide enough PWM pulses for a three-phase voltage source 5-level cascade inverter. The total control can be implemented by two microprocessors, one FPGA and one DSP. This control can also be implemented with the paralleled multi-microprocessor scheme. Select DSP with two six-channel PWM pulse generators as control IC, three microprocessors are needed with multi-microprocessor scheme. If the synchronization and communication problem is considered, FPGA scheme is of obvious superiority.

According to the proportion 200,000:24, an FPGA with one million logic gates can provide 120-channel PWM pulse generator that is suitable for a three-phase voltage source 21-level cascade inverter. The total control can be implemented by this FPGA plus a DSP, again two microprocessors. If the multi-microprocessor scheme is selected, at least ten microprocessors are needed without consideration of the synchronization and communication problem. It can be seen that the more the voltage levels are, the more superior the FPGA scheme is.

VI. CONCLUSIONS

Familiar microprocessors cannot supply enough PWM pulse generators for multi-modular converters and multi-level converters. A multichannel PWM pulse generators based on FPGA is a good choice. According to the principle of STS-SVM, a multichannel PWM pulse generator by FPGA is designed. Through a three-phase voltage source 5-level cascade inverters prototype, the designed multichannel PWM pulse generator is verified. Experimental results prove that the waveform generated by this PWM pulse generator is correct. Other modulation strategies, such as carrier phase-shifted PWM [7] and carriers-based PWM [8], can also be accomplished by this way. This is of much significance to the further application of multi-modular converters and multilevel converters.

ACKNOWLEDGEMENT

The authors wish to thank the National Natural Science Foundation of China (50237020) and the Doctor Found of Yanshan University (B155) for their important help.

REFERENCES

[1] X. Wang and B. T. Ooi. Unity PF current-source rectifier based on dynamic trilogic PWM IEEE Transaction on PE, Vol.8, No.3, 1993, pp288-294.

[2] Zhang Zhongchao, Boon-Teck Ooi. Forced commutated HVDC and SVC based on phase-shifted multi-converters. IEEE Trans. on Power Delivery, Vol.8, NO.2, pp.712 − 718, 1993

[3] B.P. McGrath, D.G. Holmes. A comparison of multicarrier PWM strategies for cascaded and neutral point clamped multilevel inverters. Power Electronics Specialists Conference, 2000. PESC 00. 2000 IEEE 31st Annual, vol.2, pp. 674 −679, 2000

[4] H. W. Van. Der. Broeck, et al. Analysis and realization of a pulse width modulator based on voltage space vector. Proc, IEEE IAS Annu. Meeting, Denver, CO, pp.244~251,1986

[5] Wang Liqiao, Zhang Zhongchao, et al. Cascade Multi-Level Converters with Sample-Time-Staggered Space Vector Modulation. IEEE Applied Power Electronics Conference and Exposition - APEC, v 1, pp.268-273, 2003

[6] Wang Liqiao, et al. Harmonic characteristic of sample time staggered space vector modulation for multi-modular or multilevel Converters. Proceedings of IEEE PESC04, pp:4554~4557, 2004

[7] Wang Changyong, Zhang Zhongchao. Phase-shifted SPWM technique based cascade converter and its application to active power filter. Automation of Electric Power Systems, 25(1): 28~30, 2001

[8] B.P. McGrath, D.G. Holmes. A comparison of multi-carrier PWM strategies for cascaded and neutral point clamped multilevel inverters. Proceedings of PESC'00, 674-679, 2000

2006 5th International Power Electronics and Motion Control Conference

Cascaded Multilevel Converters with Non-Integer or Dynamically Changing DC Voltage Ratios

Shuai Lu and Keith A. Corzine

Electrical and Computer Engineering Department, University of Missouri – Rolla, Missouri, USA

SL4xf@umr.edu and Keith@corzine.net

Abstract- **Multilevel converters made of cascaded cells (MCCC) offer a high number of voltage levels with a given switch count. Many variants of the MCCC topology have been introduced over years. Mostly, only integer dc voltage ratios between the cascaded converter cells have been applied to its PWM modulation. This is mainly due to the non-uniform distribution of voltage vectors with the non-integer dc ratios. It poses difficulty to the conventional SVM modulators relying on locating the equilateral "triangle" made of three nearest vectors enclosing the reference vector. In this paper, an original modulation method is introduced to operate with any dc voltage ratio between cascaded converter cells, integer or non-integer; and even with a dynamically varying ratio; while the normal PWM output fast-average is not disrupted. The space vector analysis and the detailed simulations verify the modulation method.**

The new concept of MCCC PWM with non-integer dc voltage ratio offers great flexibility in its practical operation, particularly in MCCC with single dc source where the auxiliary inverter cell dc-link capacitor voltage can be regulated at arbitrary value online. Its wide range of practical applications is briefly discussed in this paper.

I. INTRODUCTION

In summary, there are three major types of topologies for multilevel converter with cascaded converter cells (MCCC). Figs. 1 (a), (b) and (c) show the cascaded H-bridges, cascaded multilevel converters through split neutral load and the multilevel converter in series with H-bridge cells, respectively. Given the same dc voltage ratio

between the cascaded cells, all three topologies can provide the equivalent voltage levels with same IGBT switch count.

Recent developments in MCCC control methods witnessed two major trends. First is the hybrid operation between converter cells [1, 4, 5 and 8], where the main and auxiliary inverter cells operate at fundamental and PWM frequencies respectively. The other progress is in single dc source operation [4, 5 and 8], with only capacitor sources in auxiliary inverter. This greatly simplifies the converter dc front end complexity. Therefore, the topologies (b) and (c) are particularly preferable since only one isolated dc source is needed in the main converter cell (suppose diode clamped topology is used), while the topology (a) requires three of them.

The MCCC output performance (directly related to the number of voltage levels or layers of the voltage vectors in the hexagon patterns) depends on the dc-link voltage ratio between the main and auxiliary converter cells. The Fig. 2 shows the examples when two three-level converters are cascaded. As the voltage ratio increases, fewer voltage vectors overlap and there're more voltage vector positions available for the modulator, until the 4:1 ratio results in discontinuous vector pattern on the edge of the plot. So the voltage ratio 3:1 is called "maximal distension" [6, 7], and its total voltage levels (layers of hexagon rings in plot) are the product of the level numbers of all cascaded converter cells. In this example, the maximal distension offers 9-level performance.

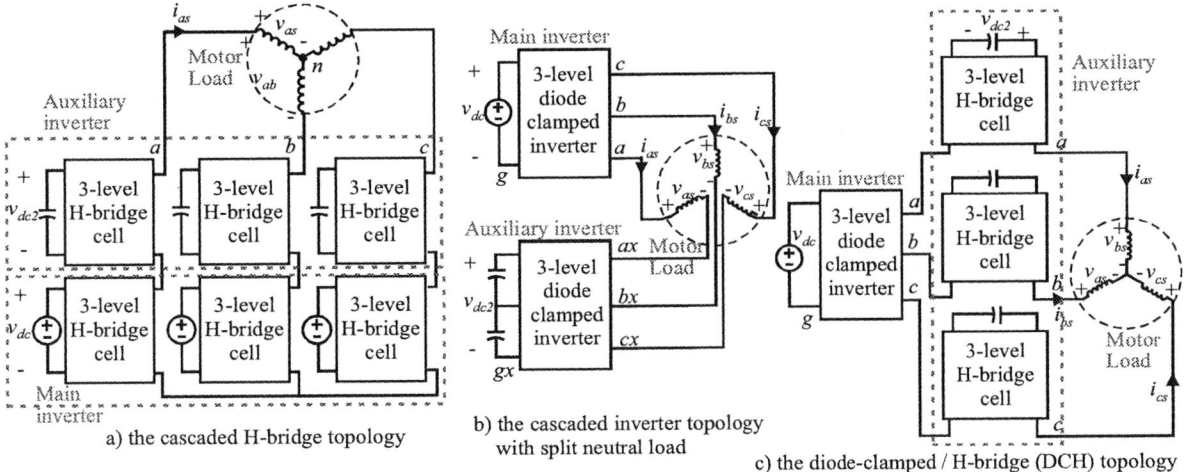

a) the cascaded H-bridge topology

b) the cascaded inverter topology with split neutral load

c) the diode-clamped / H-bridge (DCH) topology

Figure 1. The three major MCCC topologies.

1-4244-0448-7/06/$25.00 ©2006 IEEE

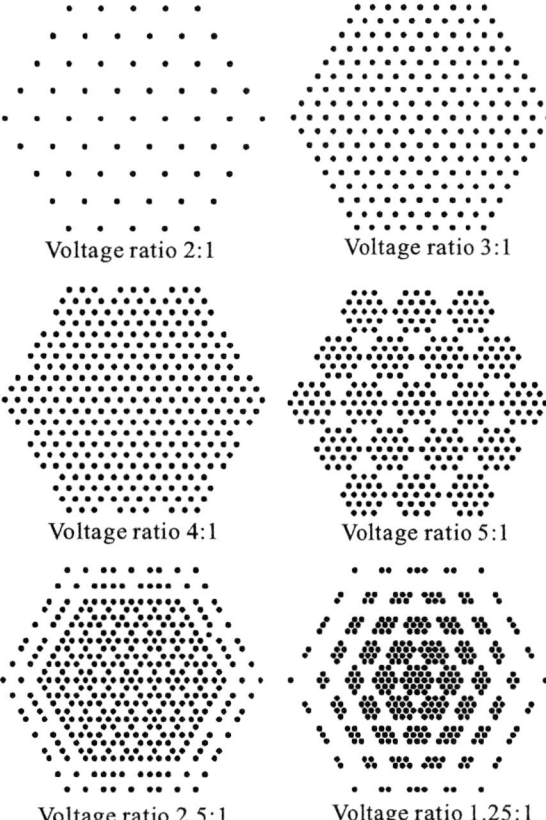

Voltage ratio 2:1   Voltage ratio 3:1

Voltage ratio 4:1   Voltage ratio 5:1

Voltage ratio 2.5:1   Voltage ratio 1.25:1

Figure 2. Vector plot with different dc voltage ratio

All integer dc ratios at or below maximal distension result in evenly distributed voltage vectors constituting the meshes of equilateral triangles. The 4:1 ratio is called "over-distension" [7], and for the m-index lower than the discontinuous region, it offers higher output resolution than "maximal distension". At the voltage ratio beyond "over-distension", like 5:1, the vector pattern becomes isolated "islands", hence an invalid option for modulator.

The non-integer voltage ratios between the MCCC cells can be used in the fundamental frequency switching (block switching) with Selective Harmonics Elimination (SHE) to achieve extra orders of harmonics elimination [9 and 10]. However, the non-integer ratio creates non-uniformly distributed vectors as in Fig. 2. The PWM modulator locating the enclosing equilateral triangle for the reference vector seems to be impossible. That is most likely the reason why there's no previous report on the PWM with non-integer voltage ratio in MCCC.

The "hierarchical modulation" introduced in the next section solves the problem. It can always locate the equilateral triangles out of the unordered vector patterns. Moreover, this method works even with the online changing of the voltage ratio, i.e. dynamic changing vector patterns. The normal MCCC output will not be disrupted during the transient.

II. PROPOSED MODULATION METHOD

Various multilevel converter modulation schemes have been introduced over years. The book [13] provides a comprehensive review for recent developments in modulation techniques. Particularly in [11, 12], the new coordinate system for SVM is introduced and then the equivalence between the SVM and the carrier based modulation (natural sampling) is proven. The conventional carrier based multilevel modulation uses multi-layer carrier comparison with the reference signal in each phase. The resulting switching states are then decomposed into the switching states for each converter cell using lookup table. In SVM perspective, this method automatically locates the enclosing triangle and the vectors on its vertex; it can also generate the optimal switching sequence and duty ratios when the proper common mode reference signal is applied [12]. As stated previously, this conventional practice will not handle the non-uniform vector patterns when non-integer dc voltage ratio is applied to MCCC.

The "hierarchical modulation" method is proposed to address this problem. The voltage vector plot of MCCC can be readily represented by the hierarchical organization of the switching states of the main and auxiliary inverter cells. Fig. 3(a) shows such an example with two cascaded 3-level converters at 3:1 dc voltage ratio. The heavy vector dots form a 3-level vector pattern representing all switching states of the main converter. Each vector is then cascaded with complete enumeration of the auxiliary converter switching states, which forms a sub-hexagon designated by dotted line. The sub-hexagon dimension is 1/3 of the main hexagon. As voltage ratio changes (integer or non-integer value), the size of the sub-hexagon and the vectors distribution changes accordingly. As illustrated in Fig. 3(b) and (c), beneath the seemingly convoluted vector patterns at non-integer ratio, there're still cascaded sub-hexagons made of meshes of equilateral triangles.

As in Fig. 3(c), any reference voltage vector v_{ref} can be decomposed into the combination of the nearby main cell vector v_{main} and the "relative reference vector" $v_{relative}$ with the origin at v_{main}. The problem is then hierarchically reduced to synthesizing $v_{relative}$ within the 3-level sub-hexagon. Still in Fig. 3(c), suppose the v_{ref} and V_{dc} remains fixed; as the voltage ratio (or V_{dc2}) changes, it could fall into a different triangle with its size changed as well. To trace its enclosing triangle with changing size, all the vector computations must use their actual values (instead of m-index) and the resulting $v_{relative}$ is then transformed into the a-b-c coordinate to obtain the reference signal $v_{relative\text{-}phase\_x}$ per-phase. The instantaneous auxiliary inverter dc voltage V_{dc2} measurement is updated every PWM cycle to compute the duty ratios for each phase leg according to the illustration in Fig. 3(d). Herein, the point 'p' is the a-phase reference value in certain PWM cycle. It is synthesized by 15% of state 1 and 85% of the state 0 in the auxiliary inverter a-phase (three level inverter with switching state 0, 1 and 2 in each phase).

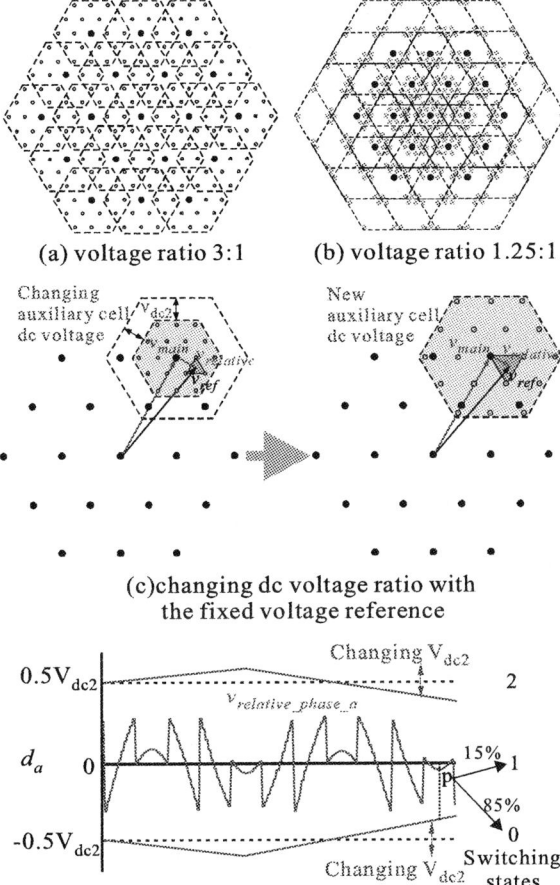

(a) voltage ratio 3:1   (b) voltage ratio 1.25:1

(c)changing dc voltage ratio with
the fixed voltage reference

(d) auxiliary converter duty ratio computation

Figure 3. Hierarchical modulation illustration

In the space vector view, the above modulation process automatically locates the enclosing equilateral triangle with its size changing at the same ratio as the V_{dc2}. Then the duty ratio of each vector on its vertex is computed.

The above description of generating the reference signal per-phase is a simplistic version of the real process. The 3-phase reference signal will be further adjusted with common mode reference to balance the voltages between the capacitors in the auxiliary cell as analyzed in [8]. In topologies in Fig. 1(a), (c), the balance is between the three phase capacitors, and in Fig. 1(b), the balance is between the upper/lower capacitors.

The complete process of the hierarchical modulation also needs to determine which main inverter cell vector is to be applied at given reference vector position along its locus. This process determines the main inverter vector traversal pattern, i.e. all the main inverter vectors to be used in a fundamental cycle and their time duration. This traversal pattern directly affects the power flow distribution between the main and auxiliary inverter cells as in [5]. For example, when MCCC operates at single dc source mode, the net power into the auxiliary cell is to be maintained at zero by properly adjusting the main

inverter vector traversal pattern, so as to maintain its overall capacitor charges and V_{dc2}.

III. THE NON-INTEGER VOLATGE RATIO OPERATION SPACE VECTOR ANALYSIS

Fig. 4 provides an insightful look at the non-integer voltage ratio PWM output using the proposed modulation method. The ratio 1.25:1 is used in the example. Its complete vector plot is in Fig. 3(b). The Fig. 4(a), (b) and (c) are the zoom-in illustrations of the modulation when the main inverter uses switching states 200, 210 and 220, respectively. The red circle represents the locus of the reference vector. The larger grey dots represent all switching states of the main inverter. Only the sub-hexagons affiliated with 200, 210 and 220 are shown here for clarity.

The line to line output voltage v_{ab} is usually shown as the indicator of the multilevel output performance. The resulting v_{ab} with the non-integer ratio has interesting pattern as shown in Fig. 4(d). In space vector view, v_{ab} is the ab-axis projections of all the voltage vectors used along the modulation path. The labels ('a' through 'g') along the ab-axis as shown in Fig. 4 (a), (b) and (c) represent such projections and also correspond to the labels in Fig. 4(d). The highlighted triangles are the ones used to synthesize the reference vector. From Fig. 4(a) to (c), there're totally 11 such triangles used. When triangle 1 (belonging to the sub-hexagon 200) is used, the v_{ab} switches between level 'a' and 'b'; then it switches between 'b' and 'c', when triangle 2, 3, 4 are used. At certain point when the reference is going across the triangle 4, the main inverter switches to 210 and the triangle 5 in the new sub-hexagon will be used to synthesize the reference vector. The projections of triangle 5 and subsequent triangle 6 onto the axis-ab are between 'd' and 'e', then the triangle 7 is between 'e' and 'f'. These are corresponding to the ['d', 'e'], ['e', f] segments in v_{ab} time domain waveform.

At certain point when the reference is in triangle 7, the main vector jumps to 220, and the triangle 8 is then used. Hence the v_{ab} will switch from segment ['e', 'f'] to ['g', 'h'] during this transition. The triangle 9, 10 and 11 are then used subsequently. The same space vector analysis is applicable for the whole fundamental circle of the reference vector.

The multilevel PWM output of MCCC with non-integer ratio does not have clear cut number of levels in its line to line voltage as in the case when the integer voltage ratio is used. v_{ab} waveform of non-integer ratio appears to have more levels. However, the THD is not lowered by more observable levels. The dominant PWM frequency harmonics magnitudes are approximately proportional to the size of the equilateral triangle.

IV. OPERATIONS WITH DIFFERENT M-INDEX OR DYNAMICALLY CHANGING VOLTAGE RATIO

The shape of the v_{ab} evolves with the main inverter vectors traversal pattern. In MCCC single DC source operation, the m-index directly affects the traversal pattern [5] and hence the v_{ab} output shape. A simulation

639

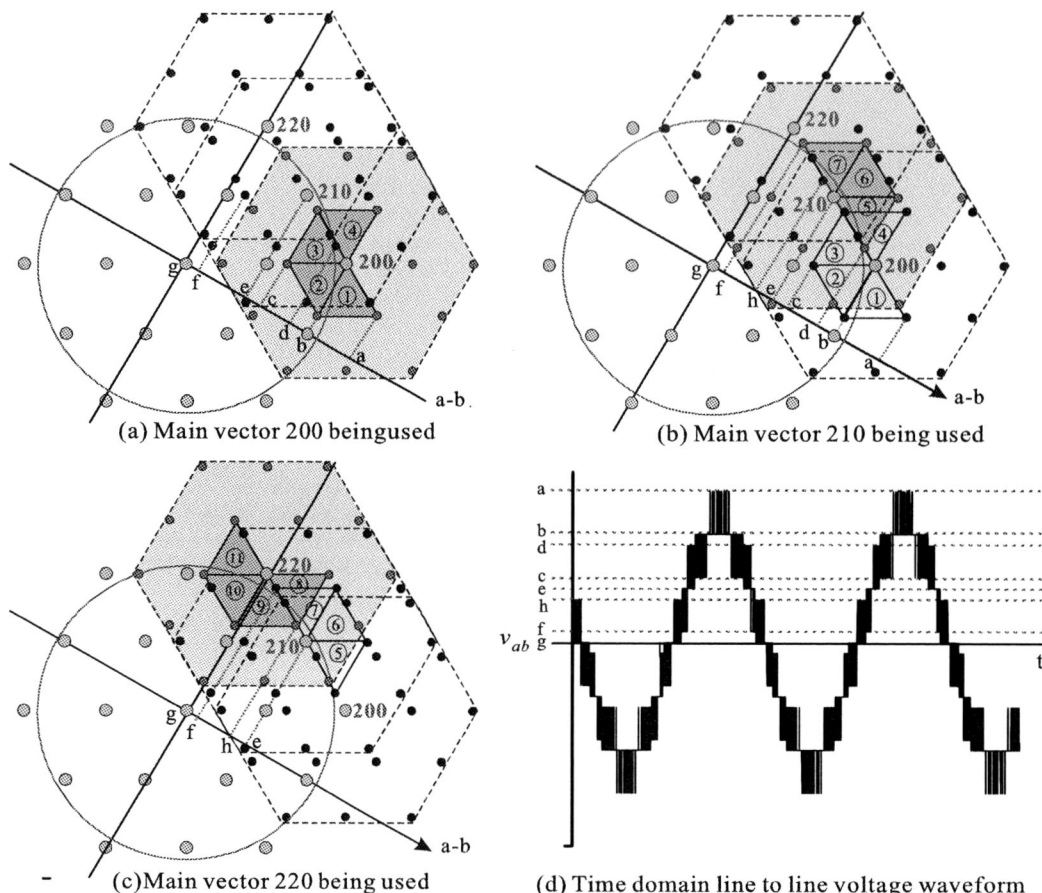

(a) Main vector 200 beingused

(b) Main vector 210 being used

(c)Main vector 220 being used

(d) Time domain line to line voltage waveform

Figure 4. Space vector analysis of non-integer ratio

is given in Fig. 5 where the m-index is lower than the previous case and the main inverter vectors follow the star-shaped pattern as in Fig. 5(a). The resulting voltage waveforms as shown in Fig. 5(b) are different than the previous case. As the main inverter vector jumps from 201 to 110, the triangles labeled 1(in red) and 2(in blue), which belongs to the sub-hexagons originating from the vector 201 and 110, are used. Their projections onto the ab-axis are highlighted in v_{ab}. Despite the evolving shape

of the MCCC output voltage envelope, its fast average is always correctly synthesized with the proposed modulation method, as verified by the sinusoidal phase voltage v_{ffas} after the PWM filter.

The Fig. 6 shows the effectiveness of the proposed modulation method during the voltage ratio dynamic change. Herein, the voltage ratio is dynamically changed from 2:1 to 3:1 by commanding net power out of the auxiliary inverter which has only capacitor source. The

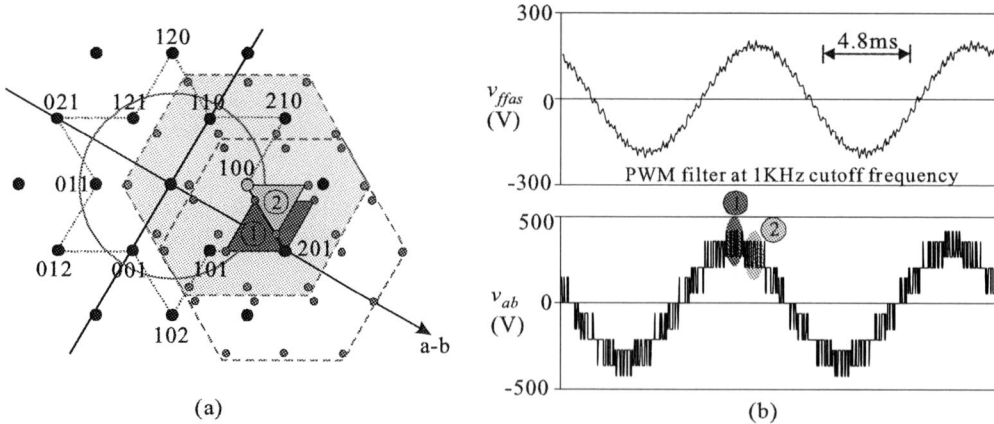

(a)

(b)

Figure 5. Space vector analysis with dc ratio at 1.25:1 (m-index is smaller)

640

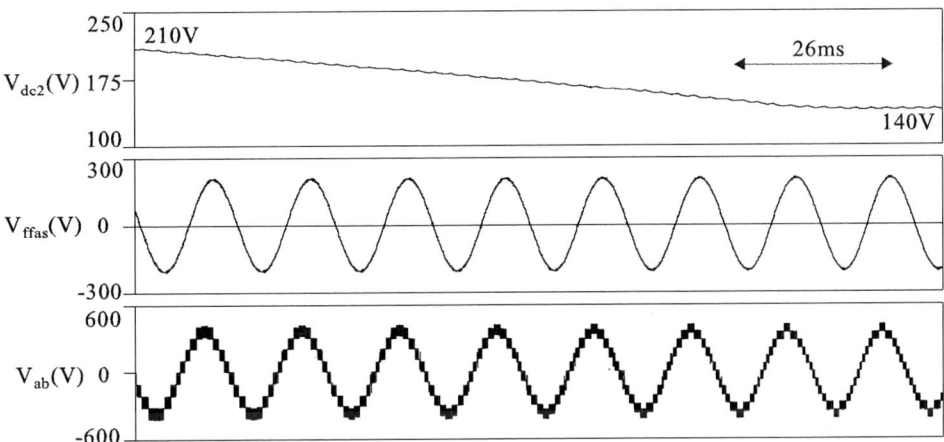

Figure 6. The dynamic voltage ratio change from 2:1 to 3:1 (V_{dc2} from 210V to 140V)

m-index (sinusoidal reference magnitude) is kept constant during the process. Initially, v_{ab} has 5-levels with the ratio 2:1. As V_{dc2} decreases to 1/3 of the main inverter V_{dc}, the levels in v_{ab} increases to seven. It is instructive to observe the gradual transition of the v_{ab} waveform. The first trace in Fig. 6 shows the V_{dc2}, which is decreased from 210V to 140V (V_{dc} is 420V). The phase voltage fast average (trace 2 in Fig. 6) after PWM filter remains sinusoidal and not disrupted during dynamic voltage ratio transition.

V. CONCLUSION

In this paper, a new concept of the non-integer dc voltage ratio operation in MCCC PWM operation was introduced. The modulation method proposed offers the unique feature which enables the online change of the auxiliary inverter capacitor (dc-link) voltage without disrupting the MCCC output voltage. Obviously, this offers the wide range of practical application possibilities.

First, the new concept can be applied to the large hybrid electric vehicle or electric vehicle propulsion [14], where MCCC is used in the electric power train; the ultra-capacitor banks are directly installed across the auxiliary converter cell dc-link, and its state of charge can be regulated online at any value. This offers a unique way to manage the vehicle regenerative braking energy without using a dc-dc converter to interface the ultra-capacitor. Many technical details to implement and integrate this general concept with the vehicle motor drive are addressed in the literature [14].

Similarly, the concept can be easily adapted to other large motor drive systems with frequent braking and accelerations. Hereby, the regenerative braking energy can be stored for subsequent motor acceleration, which avoids a complicated inverter regenerative front-end design or braking resistor energy loss.

Furthermore, for any type of application using one of the topologies of the multilevel converter with cascaded cells, the proposed new concept will provide an efficient and simplified energy storage methodology to directly place the energy storage devices across the dc-link since the wide dc-link voltage variation will be transparent to the voltage output.

REFERENCE

[1] M.D. Manjrekar, P.K. Steimer, and T.A. Lipo, "Hybrid Multilevel Power Conversion System: A Competitive Solution for High-Power Applications," *IEEE Transactions on Industry Applications*, volume 36, number 3, pages 834-841, May/June 2000.

[2] P. Steimer, Operating a Power Electronic Circuit Arrangement Having Multiple Power Converters, U.S. Patent Number 6,009,002, Assigned to ABB, December 1999.

[3] K.A. Corzine, *Cascaded Multi-Level H-Bridge Drive*, U.S. Patent Number 6,697,271, Assigned to Northrop Grumman Corporation, February 2004.

[4] K.A. Corzine, M.W. Wielebski, F.Z. Peng, and J. Wang, "Control of Cascaded Multi-Level Inverters," *IEEE Transactions on Power Electronics*, volume 19, number 3, pages 732-738, May 2004.

[5] S. Lu, K.A. Corzine, and T.H. Fikse, "Advanced Control of Cascaded Multilevel Drives Based on P-Q Theory," *Proceedings of the IEEE International Electric Machines and Drives Conference*, May 2005.

[6] K. A. Corzine, S. D. Sudhoff, and C. A. Whitcomb, "Performance Characteristics of a Cascaded Two-Level Converter," *IEEE Transactions on Energy Conversion*, volume 14, pages 433-439, September 1999.

[7] X. Kou, K.A. Corzine, M.W. Wielebski, "Over-distention operation of cascaded multilevel inverters," *IEEE International Electric Machines and Drives Conference*, volume 3, pages 1535-1542, June 2003.

[8] M. Veenstra and A. Rufer, "Control of a Hybrid Asymmetric Multilevel Inverter for Competitive Medium-Voltage Industrial Drives," *IEEE Transactions on Industry Applications*, volume 41, number 2, pages 655- 664, March/April 2005.

[9] Q. Jiang and T.A. Lipo, "Switching angles and DC link voltages optimization for multilevel cascade inverters," *International Conference on Power Electronic Drives and Energy Systems for Industrial Growth*, volume 1, pages 56-61, December 1998.

[10] L.M. Tolbert, J.N. Chiasson, Zhong Du, K.J. McKenzie, "Elimination of harmonics in a multilevel converter with nonequal DC sources," *IEEE Transactions on Industry Applications*, volume 41, number 1, pages 75-82, January/February 2005.

[11] N. Celanovic and D. Boroyevich, "A Fast Space-Vector Modulation Algorithm for Multilevel Three-Phase Converters," *IEEE Transactions on Industry Applications*, volume 37, pages 637-641, March/April 2001.

[12] B.P. McGrath, D.G. Holmes and T. Lipo, "Optimized space vector switching sequences for multilevel inverters," *IEEE Transactions on Power Electronics*, volume 18, pages 1293-1301, Nov. 2003.

[13] D.G. Holmes, T.A. Lipo, *Pulse Width Modulation for Power Converter - Principles and Practice*, IEEE press, 2003.

[14] S. Lu, K.A. Corzine and M. Ferdowsi, "An Unique Ultracapacitor Direct Integration Scheme in Multilevel Motor Drives for Large Vehicle Propulsion," *Proceeding of the IEEE Industrial Application Society Conference*, October 2006.

2006 5th International Power Electronics and Motion Control Conference

Practical Thermal Design Considerations for IPEM-based Converter

Qiaoliang CHEN, Xu YANG and Zhao-an WANG
School of Electrical Engineering
Xi'an Jiaotong University
Xi'an, Shaanxi, CHINA, 710049
Email: QLCHEN@ieee.org

Abstract— **In this paper, a 2kW PFC active IPEM (Integrated Power Electronics Modules) employing CoolMOS and SiC diode is developed. The thermal model of IPEM is developed and the power dissipation of every power devices in the active IPEM is calculated through the measurement-based power loss graph in datasheet. With the requirements of uniform temperature distribution and slight thermal interaction between power dice, the variation of temperature gradient distribution with the heat transfer coefficient of heatsink and die position is analyzed. The thermal interaction distance of power die is investigated and the tradeoff between thermal and electrical design is considered. The analysis is verified by the simulation results.**

Keywords- integrated power electronics modules; thermal;

I. INTRODUCTION

Today, there is a growing trend toward power electronics integration. Generally, the use of active IPEM and passive IPEM will allow systems assembly for customized applications with relative ease instead of having to design and build the systems from the component level [1]. It will reduce the cost and size of power electronic converters as well as to improve electrical performance.

The single PFC converter can be divided into active IPEM and passive IPEM. The active IPEM with half brick dimension consists of the common rectifier diodes dice, current sensor, MOSFET dice and SiC Schottky output diode dice, which are soldered on one side of DBC (Direct Bonded Copper).

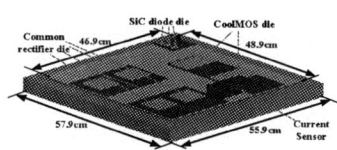

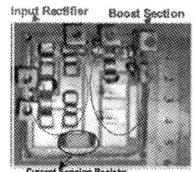

Fig.1. Components demonstration in PFC IPEM

Fig.2. The final fabricated module

The operating temperatures at the device junction have a great impact on long-term reliability and device operation. Over-temperature and power cycling are the two main reasons for the IPEM (Integrated Power Electronics Module) failures. Fig.1 shows the established power cycling characteristics relating Tm to both over-temperature and power cycling limits. It shows that the junction temperature dominates the reliability of power semiconductors [2].

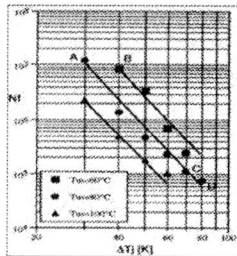

Fig.3 Power cycling results

So many switching devices which are the main heat source for the module is integrated on a DBC. Therefore the thermal design is a primary concern for a successful module. The temperature of devices is dependent on device spacing due to thermal interaction between the devices [3]. The temperature distribution of DBC should be as uniform as possible. Otherwise, that will weaken the ability of thermal dissipation of power module. At the same time, if the temperature is much localized in some area and this may lead to worse thermo-mechanical stress concentration [4]. On the other hand, the optimized electrical design is strongly related to the thermal consideration.

In this paper, a 2kW PFC active IPEM employing CoolMOS and SiC diode is developed. For specific application, the power loss of switching devices can be calculated through power loss curve. For the given thermal transfer conditions, the temperature profile of IPEM which provides graphical insight to the IPEM can be derived. A case study illustrating improvement of DBC surface temperature uniformity and improved power dissipation ability of IPEM is carried out concerning the optimized electrical layout design.

II. THERMAL TRANSFER MECHANISM AND THERMAL MODEL FOR IPEM

A. Theoretic Fundamentals of Heat Transfer

The understanding of thermal transfer mechanism in the module is the basis for the practical thermal design considerations. It is well known that thermal energy can

1-4244-0448-7/06/$25.00 ©2006 IEEE

be transferred in three fundamental fashions: through conduction, convection, and radiation [5]. In conduction transfer, thermal energy is transferred through a stationary medium through the vibratory motion of atoms and molecules. In convection transfer, thermal energy is transferred through mass movement (such as a gas or liquid) which is flowing around the heat generating object. In radiation transfer, the thermal energy is converted into electro-magnetic radiation, which is then absorbed by the surrounding environment. In electronic packaging, radiation effects are seldom sufficient to cause a noticeable change [5]. Therefore, conduction and convection will be more important for power module.

The heat conduction equation can be written in three dimensions for the steady state situation as follows.

$$\lambda \nabla^2 T + \dot{q} = 0 \qquad (1)$$

Where, $\dot{q}$ is density of thermal source, W/m³; λ is thermal conductivity, W/ (m·K); T is the temperature, K.

The thermal convection can be governed in steady state by the following equality equations.

$$q = Ah(T_w - T_f) \qquad (2)$$

Where, q is the heat flow rate, W; A is the surface area where thermal convection occurs, m²; T_w is the temperature of surface, K; T_f is the temperature of liquid or gas, K; h is the average heat transfer coefficient, W/(m²·K).

If the temperature dependency of thermal conductivity and average heat transfer coefficient is not considered, equation (1) and (2) are both linear equations. Therefore, superposition theorem can be used to solve the temperature distribution in the module. If the thermal source $q_1(x, y, z)$ leads to the resulted temperature distribution $T_1(x, y, z)$, and $q_2(x, y, z)$ results in $T_2(x, y, z)$, and then thermal $q_1(x, y, z) + q_2(x, y, z)$ will lead to the resulted temperature distribution $T_1(x, y, z) + T_2(x, y, z)$. This principle will be important in the following analysis.

B. Thermal Model Development of Active IPEM

Fig.4 shows the cross section view of active IPEM. The bare dice of power devices, such as CoolMOSFET and SiC Schottky diode, are soldered to direct bond copper (DBC) substrates. In turn, this assembly is soldered to a relatively thick copper base plate. In DBC substrate fabrication, a copper foil is bonded to the ceramic substrate, such as Al2O3 and AlN, by forming a thin oxide on the copper and then bringing the copper into intimate contact with the ceramic at an elevated temperature [6]. Typical DBC substrates are comprised of a sandwich structure as shown in Fig.4, with the upper copper layer etched to form pads and traces and the bottom copper layer providing not only a solderable surface but also an aid in forming a mechanically balanced flat piece. For mounting the module on a metal heat sink, thermal grease is used to eliminate thin air layers which would cause thermal isolation.

Table □ lists the corresponding thermal properties of individual components of IPEM. The full thermal model is shown in Fig.5. The model includes a full packaging structure with heat spreader and heat sink. The area of Al heat sink is 86×86 mm², with gap between adjacent plates of 10 mm, height of plate from base of 30 mm and plate thickness of 2 mm. Forced air cooling is available in this situation and a mean heat transfer convection coefficient will assigned to the surface of the heatsink in the thermal simulation.

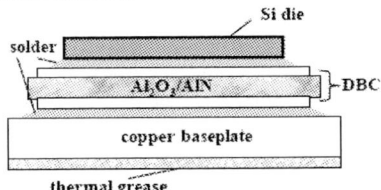

Fig.4 Cross section of DBC based IPEM

TABLE I THERMAL PROPERTIES OF IPEM COMPONENTS

	Thermal Conductivity (W/m*K)	Thickness (mm)
Silicon	120	0.35
Solder	65	0.05
Copper (Top)	395	0.25
Al₂O₃	24	0.38
Copper (Bottom)	395	0.25
Solder	65	0.05
Copper plate	395	4.0
Thermal Grease (adhesive)	1	0.05
Al (Heat sink)	140	

Fig.5 Thermal model of PFC IPEM

The dimension of power dice employed in the IPEM is illustrated in Fig.6. The input rectifier diode die is IXYS DWN50, sized in 7.1×7.1 mm². The CoolMOS die is Infineon SIPC69N60C3 sized in 10.5×6.5 mm² and SiC Schottky diode die Infineon SIDC19D60SIC3 with 1.38×1.38 mm². The maximum operation junction of power semiconductor die is roughly limited below 120□.

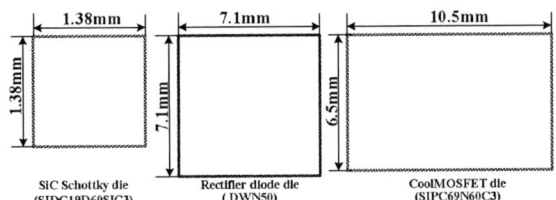

Fig.6 Dimensions illustration of power dice

III. POWER LOSS EVALUATION OF PFC ACTIVE IPEM

The power loss in switching device die is the main heat source. The power dissipation in these components is the most important parameter in the thermal design concerning heat source distribution and selection of heat sink. Therefore, for certain electrical specifications and PFC active IPEM, the power loss of power dice, including rectifier diode, CoolMOS and SiC diode, must be evaluated.

For facilitating the calculation of power dissipation, the current-voltage characteristics of diode is often represented by a threshold voltage V_{Th} and a differential resistance r_d at the initially assumed operating junction temperature, 120□ in this case. As for the on-state power dissipation of CoolMOS, the RMS current through it can be derived according to the operation waveform. Due to the strong temperature dependency of on-state resistance, the value used in the calculation is on-state resistance at 120□, which is roughly 2 times that of 25□.

Among those, the switching loss evaluation of power loss is most complex. The switching loss can be calculated through the measurement-based power loss graph in datasheet [7]. Fig.7 demonstrates the dependence of turn-on and turn-off loss versed drain current and external gate resistor. Particularly, the E_{on} involves commutated loss by freewheeling diode. The inductive load condition and SiC diode SDP06S60 are used in the test bed of datasheet for CoolMOS SPW47N60C3 [8], that is similar to practical operating condition of IPEM-based PFC. Therefore, the value in loss graph can be applied in the power loss evaluation.

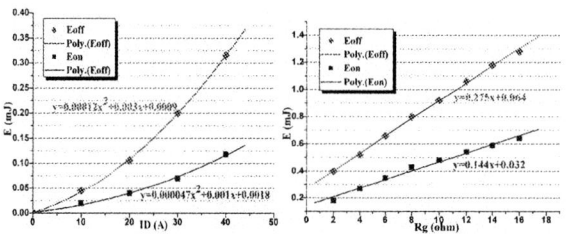

(a) With drain current   (b) With gate drive resistor
Parameters (a) Vds=380V, Vgs=0/+13V, Rs=1.8Ω,Inductive Load , Tj=125℃
Parameters (b) Vds=380V, Vgs=0/+13V, Id=47A, Inductive Load, Tj=125℃

Fig.7. The measurement-based fitted power loss graph versed drain current and gate drive resistor

From the loss curve in Fig.7, the power losses value can be numerated at the corresponding drain current value and driver resistor in specific case. However, for PFC converter, the input current is sinusoidal. There is not a stable operation point for PFC converter, the curve fitting method of the power loss graph is necessary for determining the switching loss expediently. The curve can be interpolated with the second order polynomials in order to simplify the calculations. For the IPEM PFC converter, the intermediate DC bus voltage is 385V and the gate driver resistor is 7.5 ohm. In order to calculate the power loss correctly, a corrective factor into the equations of switching losses should be implemented as following.

In this active PFC IPEM, two CoolMOS dice are used

in parallel and three SiC diode dice in parallel, and the input power is 2.1kW. Fig.8 shows the variations of total power loss with the input voltage and Fig.9 illustrates the power loss of single power die. From the loss chart in Fig.8, the maximum power dissipation occurred at 180V input voltage will be considered in following analysis.

$$E_{on}\left(i_d, v_{ds(on)}, R_g\right) = E_{on}\left(i_d\right) \cdot \frac{E_{on}\left(V_{ds(on)}\right)}{E_{on}\left(V_{ds(test)}\right)} \cdot \frac{E_{on}\left(R_{gate}\right)}{E_{on}\left(R_{gate(test)}\right)} \quad (1)$$

$$E_{off}\left(i_d, v_{ds(off)}, R_g\right) = E_{off}\left(i_d\right) \cdot \frac{E_{off}\left(V_{ds(off)}\right)}{E_{off}\left(V_{ds(test)}\right)} \cdot \frac{E_{off}\left(R_{gate}\right)}{E_{off}\left(R_{gate(test)}\right)} \quad (2)$$

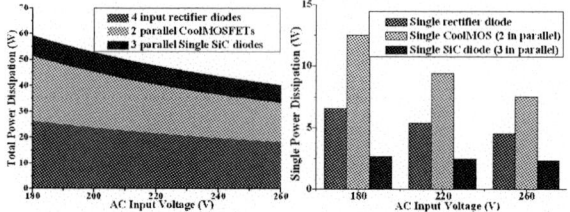

Fig.8. Total power loss of IPEM   Fig.9. Power loss of single component of IPEM

IV. THERMAL DESIGN CONSIDERATIONS

For a good thermal design process, the following requirements must be considered.

1) total power dissipation as low as possible;
2) temperature distribution as uniform as possible;
3) thermal interaction between power dice as slight as possible;
4) more tradeoff considerations in connection with optimized electrical design;

According to the superposition theorem mentioned above in section □, the temperature distribution of full IPEM can be derived through the summation of temperature distribution which individual power die results in. Therefore, the resulted temperature distribution of individual die, with certain die dimension, location and power dissipation, will be investigated in this section.

A. Temperature distribution with heat transfer coefficient

In this case, the rectifier die with power dissipation of 10W will be investigated. The die, sized in 7.1mm×7.1mm, is located in the center of the DBC, illustrated in Fig.10. The ambient temperature is 27□ and the area of upper DBC copper foil equals that of ceramic layer of DBC.

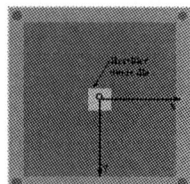

Fig.10. thermal model as single rectifier diode die considered

Fig.11 shows the X-axis direction temperature distribution on the top surface of DBC Al$_2$O$_3$ layer for different heat transfer coefficient of heatsink. The temperature gradient can be calculated when the edge point in x direction of Al₂O₃ layer is defined as referenced temperature. The results indicate that heat transfer coefficient does affect the average temperature, but not influence the temperature gradient.

Fig. 12 shows the temperature gradient distribution when single SiC diode die and single CoolMOS die are considered, respectively. The 10W power dissipation is loaded. For the SiC diode, due to the smaller die dimension, the temperature is localized more seriously and the gradient is much more. For the 10W power loss, the difference between maximum temperature point on SiC die and referenced temperature point on Al₂O₃ layer is as much as 28.3□. The temperature gradient value becomes very small when the x coordinate (distance) is slightly larger than the half of size of power die. That means the temperature interaction between power dice will be slight in DBC-based IPEM. In other words, the reduced distance between devices does not affect the junction temperature of devices so much. Therefore the distance between power dice can be reduced for reducing the parasitic inductance of critical high di/dt branch.

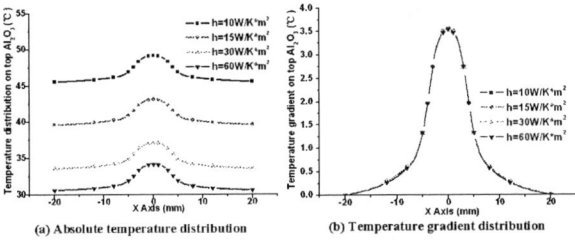

(a) Absolute temperature distribution   (b) Temperature gradient distribution

Fig.11. absolute temperature and temperature gradient distribution for different heat transfer coefficient

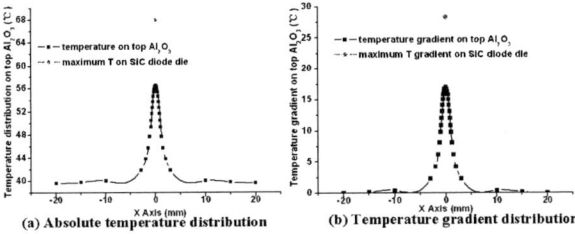

(a) Absolute temperature distribution   (b) Temperature gradient distribution

Fig.12. absolute temperature and temperature gradient distribution for SiC diode die

B. Temperature distribution with power die location

In order to construct the desired circuit topology, the power device should be placed at the desired location. Therefore, it is necessary to investigate the variation of device location with temperature distribution. In this case, the rectifier die with power dissipation of 10W is investigated. The ambient temperature is 27□ and heat transfer coefficient of heatsink is 15 W/(m²·K). Fig.13 illustrates the relation between device location and temperature distribution. The results indicate that the die position variation within certain range affects slightly the maximum temperature gradient. That will be true for SiC

die and CoolMOS die too.

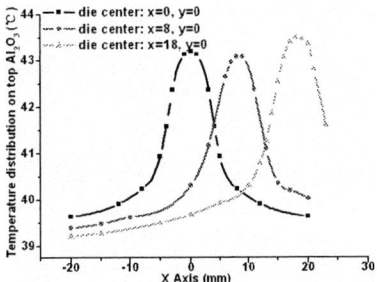

Fig.13. the variation of device position with temperature distribution

C. Temperature distribution with copper track area

From the optimized electrical design point of view, the area of DBC copper track, with high dv/dt and thin ceramic layer, is required as small as possible for reducing the common EMI current. From the thermal design consideration, the larger area of DBC copper track, on which power die is soldered, is beneficial to thermal spreading, and then leads to lower junction temperature and more uniform temperature distribution. Therefore, there will be a tradeoff between these two aspects.

Fig.14 shows the relationship between temperature distribution and copper track area as SiC diode die and rectifier diode die are considered, respectively. The center of power die and copper track is localized in the axis origin O, illustrated in Fig.10. The power dissipation are both 10W, heat transfer coefficient of heatsink of 15 W/(m²·K) and the ambient temperature 27℃.

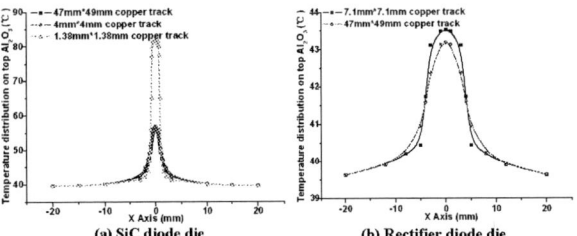

(a) SiC diode die   (b) Rectifier diode die

Fig.14. the relationship between temperature distribution and copper track area

Due to the smaller area of SiC diode die, when the copper track area is the very close to the die area, the temperature of on SiC die is localized seriously. When the copper track area equals to 4mm×4mm, the situation is improved dramatically. For the die area as much as that of rectifier diode, the copper track area, which is always larger than die area at least, does not affect the temperature gradient so much. Therefore, the area of copper track, on which CoolMOS drain with high dv/dt is solder, can be reduced as small as possible.

D. Compoents selction for more uniform temperature distribution and reducing power loss

The reduction in power loss of IPEM is the most effective method to improve thermal management of IPEM. The use of MOSFETs in parallel can reduce the total on-state power dissipation on MOSFET. That is true for diodes in parallel. At the same time, the thermal source can also be decentralized. Especially for SiC diode of small die area, the high power dissipation will lead to large temperature gradient. In order to ensure safe junction temperature of SiC diode, the lower temperature of copper case must be guaranteed. That will weaken the ability of thermal dissipation of IPEM. Fig.15 shows the power dissipation comparison charts. The total power loss is reduced as devices are used in parallel. Furthermore, the less total power dissipation can be distributed on more power dice effectively.

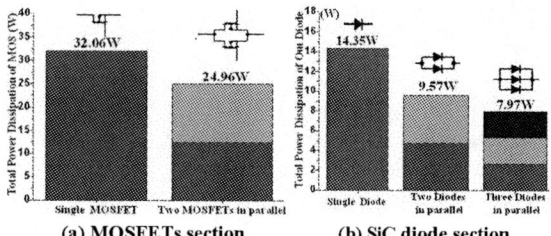

(a) MOSFETs section   (b) SiC diode section

Fig.15. the power loss comparison for MOSFETs and SiC diodes

Fig.16 shows the thermal simulation results of IPEM for two cases. One case is where only single CoolMOS and SiC diode are used. The other is where two CoolMOSFETs in parallel and three SiC diodes in parallel are employed. The total loading power dissipation is that of 180 input voltage derived in section □, average heat transfer coefficient of heatsink of 15 W/(m²·K) and the ambient temperature 27□. From the results, the SiC junction temperature in former case exceeds the 120□ so much, and the junction temperature in latter case is below the defined maximum operating temperature.

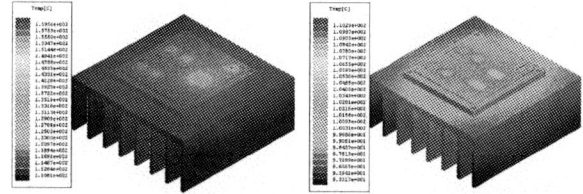

(a) Case 1: single MOS and single SiC diode   (b) Case 2: two MOSFETs in parallel and three SiC diodes in parallel

Fig.16. the thermal simulation of IPEM in two cases

TABLE II JUNCTION TEMPERATURE OF POWER DEVICES

	Case 1 (□)	Case 2 (□)
Rectifier diode	121.2	104.0
CoolMOSFET	129.4	105.7
SiC Schottky diode	157.6	110.0

V. CONCLUSIONS

In this paper, a 2kW PFC active IPEM employing CoolMOS and SiC diode is developed. The thermal model of IPEM is developed. And the power dissipation of every power devices in the active IPEM is calculated through the measurement-based power loss graph in datasheet.

The uniform temperature distribution and slight thermal

interaction between power dice is required for good thermal design. Considering the superposition theorem, the resulted temperature distribution of individual die with certain dimension and power dissipation in some cases is investigated. The heat transfer coefficient of heatsink does affect the average temperature, but does not influence the temperature gradient. The distance between devices does not affect the junction temperature of devices so much. But the smaller area of power die, more seriously the temperature is localized.

The die location variation within certain range affects the maximum temperature gradient slightly. For the die area as much as that of rectifier diode, the copper track area does not affect the temperature gradient so much. Therefore, the area of copper track with high dv/dt can be reduced as small as possible. The use of MOSFETs and SiC diode in parallel can reduce the total power dissipation on them. Furthermore, the less total power dissipation can be distributed on more power dice effectively. The thermal simulation results of IPEM are demonstrated to verify the analysis.

ACKNOWLEDGEMENT

This work is supported by Key Project of National Natural Science Foundation of China under award number 50237030. The author would like to acknowledge Infineon Technologies providing the power CoolMOS and SiC diode dice samples.

REFERENCES

[1] J. D. van Wyk, F. C. Lee, Z. X. Liang, R. G. Chen, S. Wang, B. Lu, "Integrating Active, Passive and EMI-Filter Functions in Power Electronics Systems: A Case Study of Some Technologies", *IEEE Trans. Power electronics*, vol. 20, no. 3, pp. 523-536, May. 2005.

[2] M. held, P. Jacob, G. Nicoletti, P. Scacco, M.-H. Poech, "Fast power cycling test for IGBT modules in traction application," *Proc. IEEE Power Electronics and Drive Systems Conf.*, 1997, vol. 1, pp. 425-430.

[3] Shatil Haque, et al. "Thermal management of power electronics modules packaged by a stacked-plate technique", *Microelectronics Reliability*, vol. 39, pp. 1343-1349, 1999.

[4] N. ZHU, J.D. VAN WYK, Z.X. LIANG, "Thermal-Mechanical Stress Analysis in Embedded Power Modules", *35th Annual IEEE Power Electronics Specialists Conference*, pp. 4503-4508, 2004.

[5] A.B. Lostetter, F. Barlow, A. Elshabini, "An overview to integrated power module design for high power electronics packaging", *Microelectronics Reliability*, vol. 40, pp. 365-379, 2000.

[6] Jun He, Vivek Mehrotra and Michael C. Shaw, "Thermal Design and Measurements of IGBT Power Modules: Transient and Steady State", *IEEE IAS*, pp. 1440-1444, 1999.

[7] Infineon Technologies Application Note, "How to Select the Right CoolMOS and its Power Handling Capability", Jan. 2002. AN-CoolMOS-03.

[8] Infineon Technologies, Datasheet for SPW47N60C3, 2002.

2006 5th International Power Electronics and Motion Control Conference

Realization of an FPGA-Based Space-Vector PWM Controller

Zhou Yuan[*], Xu Fei-peng[**] and Zhou Zhao-yong[**]

[*] Department of Automation, Tianjin University of Technology and Education, Tianjin, China
[**] Department of Electrical Engineering, Harbin Institute of Technology, Harbin, China
E-mail: Candy_zhy@hotmail.com

Abstract—This paper presents a design of space-vector PWM (SVPWM) integrated circuit. The proposed strategy is implemented and verified on a single Altera's Cyclone™ series of FPGA (EP1CF400C7). Experimental results show that this controller can present an excellent drive performance and its switching frequency, as well as its dead time, is adjustable. The design can be used in vector control of ac servo drive system or three-phase ac-voltage regulation systems with a little modification.

Keywords- SVPWM; FPGA; Cyclone™

I. Introduction

Variable-frequency ac drives are increasingly used for various applications in industry and traction. Due to the improvement of fast-switching semiconductor power devices, voltage source inverters with pulse-width-modulated (PWM) control find particularly growing interest. Although most ac drives in use today adopt microprocessor-based digital control strategy, implementation of current control loop and PWM control are still tied to analog control circuitry. This kind of control scheme possesses the advantage of fast dynamic response, but suffers the disadvantages of complex circuitry, limited functions, and difficulty in circuit modification.

PWM dc–ac converters may serve a wide range of applications in ac motor drives and ac power conditioning systems. The PWM strategy plays an important role in the minimization of harmonics and switching losses in these converters, especially in three-phase applications. In the past two decades, various PWM strategies, control schemes, and realization techniques have been developed [1]–[5]. These PWM strategies were realized either by analog circuit or microprocessor-based software control techniques. However, with the advance of high-frequency switching power devices, complex modulation schemes can no longer be realized, even employing the most advanced digital signal processors, because of the high-speed switching requirement. In recent years, motor control and power conversion IC's employing ASIC/FPGA technology are receiving increased attention [6]–[10]. Employing FPGA to realize PWM strategies provides advantages such as rapid prototyping, simple

hardware and software design, higher switching frequency, and relieving the computation load of microprocessors [11].

In this paper an FPGA-based SVPWM (Space-vector PWM) controller is designed and realization in a single FPGA. The designed SVPWM IC may serve either for ac motor drives or three-phase ac-voltage regulation systems. The rest of this paper is organized as follows. Section II briefly introduces the principle of the space-vector PWM method. Section III discusses developing a strategy for FPGA-based SVPWM IC and gives a detailed description of the digital circuit realization scheme for SVPWM. Section IV describes the simulation and experimental results. Section V is the conclusion.

II. Principle of Space-Vector PWM

SVPWM is one of the most popular technologies in modern high performance ac servo drive systems. Compared with sinusoidal PWM (SPWM), which is another useful modulation strategy, the linear range of the SVPWM is 15% higher.

The principle of SVPWM is based on the space vector representation of the voltages in the a, β plane. The a, β components are found by Park transform, where the total power, as well as the impedances, remains unchanged. The basic ideas are summarized as follows.

The motor stator voltage vector can be expressed as a combination of the inverter output-phase voltage u_U, u_V and u_W, which can be expressed in vector form as

$$\vec{U}_s = u_{UN} + \alpha u_{VN} + \alpha^2 u_{WN}, \alpha = e^{j\frac{2}{3}\pi}. \quad (1)$$

where $\begin{cases} u_{UN} = V_m \sin \omega t \\ u_{VN} = V_m \sin(\omega t - 120°) \\ u_{WN} = V_m \sin(\omega t + 120°) \end{cases}$

and V_m is the amplitude of the fundamental component. There are eight basic switching configurations of the three-phase PWM inverter. (See also [2] and [5].) Their corresponding voltage vectors are depicted in Fig. 1, expressed as

1-4244-0448-7/06/$25.00 ©2006 IEEE

$$\begin{cases} \vec{V}_n = \dfrac{2}{3}U_{dc}e^{j\frac{(n-1)\pi}{3}}, n=1,2....6 \\ \vec{V}_7 = \vec{V}_8 = 0 \end{cases} \quad (2)$$

where U_{dc} is the dc-link voltage.

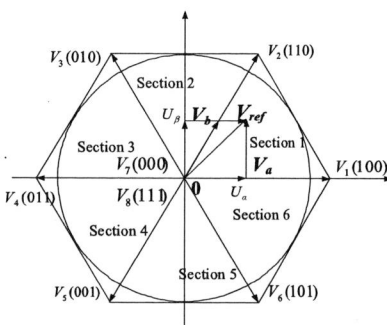

Figure 1. Voltage Space Vectors for a three-phase voltage source inverter

The stator voltage vector can be decomposed into two orthogonal components in a two-axis coordinate or as a combination of two basic vectors, as Fig. 1 indicates. The SVPWM strategy aims to minimize harmonic distortion in the current by selecting the appropriate switching vectors and determining of their corresponding dwelling widths.

If the reference vector is located in sector I, then it is composed of voltage vectors $\vec{V}_1$, $\vec{V}_2$, and zero voltage vectors $\vec{V}_7$, $\vec{V}_8$.

We assume that t_1 and t_2 are the dwelling times for $\vec{V}_1$ and $\vec{V}_2$, respectively and T_s is the switching period. In one switching period, besides t_1 and t_2, the time left is the dwelling time for zero vectors $\vec{V}_7$ and $\vec{V}_8$, which are usually taken as the same, represented by t_0. Then, it can be express as follows

$$t_0 = (T_s/2 - t_1 - t_2)/2 \quad (3)$$

On considering of minimal in flux ripple and minimum number of switching, we get the optimal synthesis of the flux vector, show in Fig. 2, which indicates the PWM switching patterns.

According to the principle that the composed vector be the same, we get equation as follows, that

$$\frac{2t_1}{T_s}\frac{2}{3}U_{dc}\begin{bmatrix}1\\0\end{bmatrix} + \frac{2t_2}{T_s}\frac{2}{3}U_{dc}\begin{bmatrix}\dfrac{1}{2}\\\dfrac{\sqrt{3}}{2}\end{bmatrix} = U_\alpha\begin{bmatrix}1\\0\end{bmatrix} + U_\beta\begin{bmatrix}0\\1\end{bmatrix} \quad (4)$$

Working it out, then we get

$$t_1 = \frac{3T}{4}\left(\frac{U_\alpha}{U_{dc}} - \frac{U_\beta}{\sqrt{3}U_{dc}}\right) \quad \text{and} \quad t_2 = \frac{3T}{4}\frac{2U_\beta}{\sqrt{3}U_{dc}} \quad (5)$$

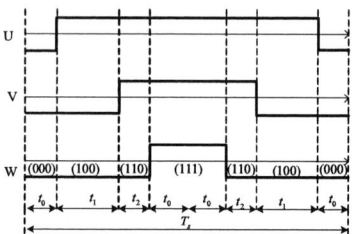

Figure 2. PWM waveforms in sector I

Dwelling times of the nonzero vectors in the other five sections may be worked out using the same method. A general equation can be written as

$$\begin{cases} t_k = \dfrac{\sqrt{3}T_s}{U_{dc}}\left[U_\alpha \sin\left(\dfrac{k\pi}{3}\right) - U_\beta \cos\left(\dfrac{k\pi}{3}\right)\right] \\ t_{k+1} = \dfrac{\sqrt{3}T_s}{U_{dc}}\left[U_\beta \cos\left(\dfrac{(k-1)\pi}{3}\right) - U_\alpha \sin\left(\dfrac{(k-1)\pi}{3}\right)\right] \end{cases}$$
$$k = 1, 2, ..., 5 \quad (6)$$

where t_k and t_{k+1} are the dwelling times for two adjacent state of the inverter $\vec{V}_k$ and $\vec{V}_{k+1}$, respectively. And k stands for the number of the section, which is determined by the turn angle.

III. DESIGN OF THE SVPWM CONTROLLER USING FPGA TOOLS

The main structure of the Space-Vector PWM Controller can be divided into six parts. They are Synchronizing Control Signal unit, Sine Table Memorizer unit, Section Calculation unit, Switching Time Calculation unit, Vector Generating unit, and Dead Time Regulation unit. Fig..3 is the structural diagram of the SVPWM module based on FPGA. Each one of the modules is described by Verilog HDL and synthesized with QuartusII software.

A. Sine Table Memorizer

In real-time operation, the input signals of a SVPWM controller are two orthogonal voltage vectors U_α and U_β. While simulation, the sine and cosine input of SVPWM controller can be supplied by looking-up the ROM table. Both the sine and cosine value shares one table, only staggered by $\pi/2$ electrical degree.

B. Synchronizing Control Signal unit

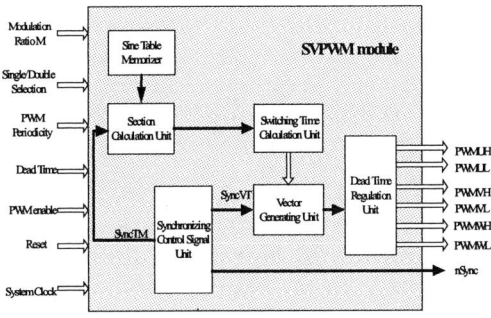

Figure 3.   The structural diagram of SVPWM module

Each sampling of Synchronizing Control Signal unit will generate a SyncTM signal, a SyncVT signal and an nSYNC signal. According to different mode of modulation, one or two times of sample occur during a PWM period. Fig. 4 shows time sequence among the three signals. The SyncTM acts as a control signal of Section Calculation unit. To ensure the accomplish of calculation before refresh of vector, SyncTM should be sent at least 30 clock periods ahead of SyncVT, which represent by T_m. The width of SyncTM signal is one clock period. At each sampling, nSYNC turns into high level after keeping low level for a time represent by T_n. And nSYNC signal can be provided to a processor as a control signal.

C. Section Calculation unit

On reception of SyncTM signal, the Section Calculation unit looks up the sine table according to current degree.

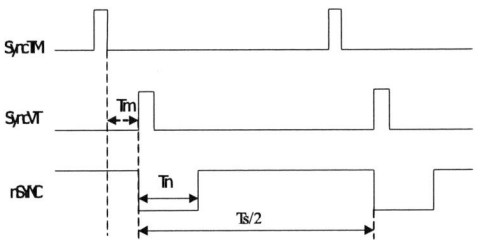

Figure 4.   Time Sequence Among the Synchronous Signals

First, two values are calculated as follows

$$\begin{cases} \dfrac{U_\alpha}{U_{dc}} = \dfrac{k * \cos(\theta)}{U_{dc}} \\ \dfrac{U_\beta}{\sqrt{3}U_{dc}} = \dfrac{k * \sin(\theta)}{\sqrt{3}U_{dc}} \end{cases} \qquad (7)$$

where $k = M * \dfrac{1}{\sqrt{3}U_{dc}}$, $M \in [0,1]$.

And then, the section of the reference voltage vector is determined according to thereinafter regulation, that

If $0 < U_\beta \le \sqrt{3}U_\alpha$, V_{ref} is in Section I;

If $U_\beta \ge \sqrt{3}U_\alpha$ and $U_\alpha \ge 0$, or if $U_\beta > -\sqrt{3}U_\alpha$ and $U_\alpha < 0$, V_{ref} is in Section II;

If $0 < U_\beta \le -\sqrt{3}U_\alpha$, V_{ref} is in Section III;

If $\sqrt{3}U_\alpha < U_\beta \le 0$, V_{ref} is in Section IV;

If $U_\beta \le \sqrt{3}U_\alpha$ and $U_\alpha \le 0$, or if $U_\beta < -\sqrt{3}U_\alpha$ and $U_\alpha > 0$, V_{ref} is in Section V;

If $-\sqrt{3}U_\alpha < U_\beta \le 0$, V_{ref} is in Section VI.

D. Switching Time Calculation unit

The value of $\dfrac{U_\alpha}{U_{dc}}, \dfrac{U_\beta}{U_{dc}}$ and number of the section are sent to Switching Time Calculation unit so as to calculate t_1 and t_2 according to (6). Meanwhile, the dwelling time for zero vectors can be calculated from (3).

E. Vector Generating unit

Before the generation of SyncVT signal, the calculation of t_1, t_2 and t_0 have already finished. SyncVT signal acts as a control signal of Vector Generating unit, which aims at manipulation of switchover among the vectors according to the value of dwelling time t_1, t_2 and t_0. Well, the sequence is

$$V_7 \to V_k \to V_{k+1} \to V_8 \to V_{k+1} \to V_k \to V_7 .$$

F. Dead Time Regulation unit

This module divided each of the three phase modulation signals into two parts, acting as driving signals of upper and lower arm path, respectively and stagger with each other by the setting dead time. The regulation of dead time is realized by changing the clock frequency of Dead Time Regulation unit.

G. Treatment of Overmodulaiton

When we get the result $t_0 < 0$ during calculation, it indicates that overmodulation occurs. In practical application, it is possible to take place especially when the motor system gets a sudden change in rotational speed so that the torque takes a great alteration. To ensure the operation of modulation under such condition, t_1 and t_2 are calculated as follows.

If $t_1 > t_2$,

$$t_1{}' = \frac{t_1}{t_1 + t_2} \cdot \frac{T_s}{2}, \text{ and } t_2{}' = \frac{T_s}{2} - t_1{}' \qquad (8)$$

else

$$t_2{}' = \frac{t_2}{t_1 + t_2} \cdot \frac{T_s}{2}, \text{ and } t_1{}' = \frac{T_s}{2} - t_2{}' \qquad (9)$$

IV. SIMULATION AND EXPERIMENTAL RESULTS

The object device is Cyclone FPGA EP1CF400C7, one of Altera's Cyclone series of FPGA, which claims to be an FPGA with the highest cost performance ever. After compilation, the Resource Usage Summary is shown as Fig. 5. We may learn that the usage of LE is 1560, which take only 7% of the total resource. Thus plenty of resources are left for other modules such as angular measurement, and coordinate conversion etc. so as to form a single-chip control system with further functions.

	Resource Usage Summary	
	Resource	Usage
1	Logic cells	1,560 / 20,060 (7 %)
2	Registers	567 / 21,850 (2 %)
3	User inserted logic cells	0
4	I/O pins	9 / 301 (2 %)
5	-- Clock pins	1 / 2 (50 %)
6	Global signals	3
7	M4Ks	8 / 64 (12 %)
8	Total memory bits	32,768 / 294,912 (11 %)
9	Total RAM block bits	36,864 / 294,912 (12 %)
10	Global clocks	3 / 8 (37 %)
11	Maximum fan-out node	nRESET
12	Maximum fan-out	567
13	Total fan-out	6561
14	Average fan-out	4.16

Figure 5.   Resource usage summary

Fig. 6 shows the result of timing simulation waveform under QuartusII circumstance. The frenquency of simulation clock is 20MHz and swiching time of power device is 20kHz. The output signals are six PWM signals with mutative duty ratio. The result of timing simulation validates the design.

Figure 6.   Timing simulation waveforms diagram

Downloading the design to FPGA device and filtering the harmonic wave, we get the resulting waveforms. Fig. 7 shows two of the three-phase PWM waves using double-sideband modulation. Fig. 8 shows the waveform of line voltage. The result indicates that it is regular sine wave.

Cutting out the Sine Table Memorizer and with a little modulation, the designed IC could receive signals from DSP and output real-time SVPWM signals. The output acts as an exterior interrupt source to the DSP. At each nSYNC signal, the DSP chip would sample the current rotor position and work out the value of U_α and U_β.

V. CONCLUSION

This paper presents the design and realization of a programmable SVPWM IC for high performance ac servo drives and three-phase ac-voltage regulation systems. The proposed control strategy was implemented and tested using an FPGA technology. The designed IC may interface with other IP cores or DSP to form a control system. The maximal output current of Cyclone FPGA is 24mA, large enough to drive power devices directly such as EXB841. The SVPWM IC has a wide range from linear to overmodulation. Besides, the source codes can be easily replanted to different FPGA without any change. Simulation and experimental results show that this IC can acquire excellent operating performance.

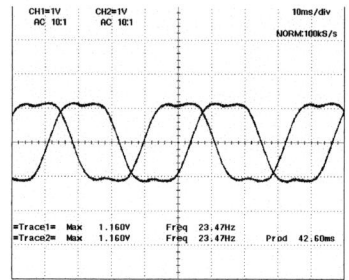

Figure 7.   The experimental result of phase voltage waveforms

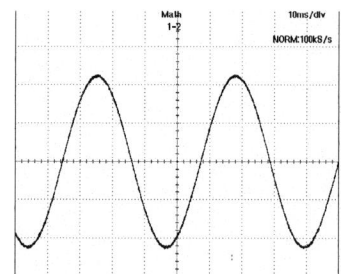

Figure 8.   The experimental result of line voltage waveform

REFERENCES

[1] J. Holtz, "Pulse width modulation-A survey," *IEEE Trans. Ind. Electron*, vol 39, no. 5, pp. 410-420, 1992.

[2] H. W. Van Der Broeck, H. Skudelny, and G. V. Stanke, "Analysis and realization of a pulsewidth modulator based on voltage space vector," *IEEE Trans. Ind. Applicat.*, vol. 24, no. 1, pp. 142–150, 1988.

[3] T. G. Habetler, "A space vector-based rectifier regulator for ac/dc/ac converters," *IEEE Trans. Power Electron.*, vol. 8, no. 1, pp. 30–36, 1993.

[4] M. Morimoto, S. Sato, K. Sumito, and K. Oshitani, "Single-chip microcomputer control of the inverter by the magnetic flux control PWM method," *IEEE Trans. Ind. Electron.*, vol. 36, no. 1, pp. 42–47, 1989.

[5] Yang Gui-jie, Sun Li, Cui Nai-zheng, Lu Yong-ping, "Study on method of the space vector PWM", *Proceedings of the CSEE, China,* vol.21, no.6, 2001, pp.79-83.

[6] M. G. Egan, J. M. Murphy, E. J. Heffernan, S. U. Lidholm, and M.L. McGrath, "An ASIC-based PWM waveform generator for AC motor control applications," in *IEEE Int. Symp. on Circuits and Systems. Proc.,* vol. 2, 1988, pp. 1369–1372.

[7] T. C. Green, M. Mirkazemi-Moud, J. K. Goodfellow, and B. W. Williams, "Field-programmable gate-arrays and semi-custom designs for sinusoidal and current-regulated PWM," in *IEE Colloquium on ASIC Technology for Power Electronics Equipment,* 1992, pp. 4-1/4-4.

[8] M. Mirkazemi-Moud, T. C. Green, and B. W. Williams, "Use of ASIC technology in the design of two novel PWM generators," in *IEE 4th Int. Conf. on Power Electronics and Variable-Speed Drives,* 1990, pp. 347–532.

[9] J. M. Retif, B. Allard, X. Jorda, and A. Perez, "Use of ASIC's in PWM techniques for power converters," in *Proc. IEEE IECON Conf. Rec.,* vol. 2, 1993, pp. 683–688.

[10] Mauricio Tonelli, Pedro Battaiotto and Maria I.Valla, "FPGA implementation of an universal space vector modulator", IECON'01: The 27th Annual Conference of the IEEE Industrial Electronics Society, 2001, pp. 1172-1177.

[11] Ying-Yu Tzou, Hau-Jean Hsu, "FPGA Realization of Space-Vector PWM Control IC for Three-Phase PWM Inverters," *IEEE Trans. Power Electron.,* Vol. 12, No. 6, pp. 953-963. 1997

2006 5th International Power Electronics and Motion Control Conference

Chaotifying Control of Permanent Magnet Synchronous Motor

Hai Peng Ren[1,2], Chong Zhao Han[1]

[1]School of Electronic and Information Engineering, Xi'an Jiaotong University, Xi'an, China

[2] School of Automation and Information Engineering, Xi'an University of Technology, Xi'an, China

Email: renhaipeng@xaut.edu.cn

Abstract—**Purposely generating chaos—chaotifying control, when it is useful or beneficial, becomes one of the focuses in chaos engineering. In this paper, the direct time delay feedback is proposed for chaotifying control of Permanent Magnet Synchronous Motor (PMSM). The direct-axis or quadrature-axis stator voltage is used as manipulated variable, and the direct-axis or quadrature-axis current is used as time delay state feedback in the control law. This method can be physically realized and is simple in comparison to indirect time delay feedback method. The proposed controller has the same form of the controller proposed by Pyragas for eliminating chaos, this investigation, together with the previous research, will show that the direct time delay feedback control can generate or enhance chaos when it is useful, and eliminate chaos when it is harmful, which will give more flexibility for the control engineer. Simulation results show its effectiveness.**

Keywords-PMSM, Chaotifying control, Direct Time delay feedback

I. INTRODUCTION

In the field of power electronics, Kuroe and Hayashi[1] introduced chaos phenomena in motor drive systems on the 20th IEEE power electronics specialist conference in 1989. It was pointed out that motor drives would behave in a chaotic manner, when parameters of motor lie in certain area. The further research work were done in some kinds of motor drive systems[2-5]. Li Zhong et al. described the bifurcation and chaos in Permanent Magnet Synchronous Motor (PMSM)[6]. When a PMSM fall into chaos, the torque of the motor changes in a "random" manner. The speed would oscillate in a wide range. This situation is not desirable in most application, in such case, chaos has to be controlled or eliminated. Therefore, some chaos control method is suggested for chaos control in PMSM[7-8]. However, recently research work has shown that chaos is beneficial in some time- and /or energy-critical applications where chaos provides a system designer with a variety of special properties, richness of flexibilities, and a cornucopia of opportunities [9]. Examples include liquid mixing[10], secure communication[11], abrasive machine[12], and road roller[13]

It is expected that chaos research in engineering will eventually reach the point at which it will lead to improved and refined procedures, enabling an engineer to design a system to be either chaotic or nonchaotic as desired.

To generate chaos, Chen has suggested some methods for the discrete systems[14][15]. Chaos synthesis from the continuous systems has also been investigated using time delay feedback with sinusoidal form and feedback linearization techniques[9,16,17]. A switching piecewise-linear controller has been employed to generate a chaotic attractor in the linear systems[18]. Generating chaos via jerk function has been studied in [19]. Maintenance of chaos by weak harmonic perturbation is introduced in [20].

Motivated by the work in [9], a direct delay feedback control method is proposed for generating chaos in PMSM. Compared with the method in [9], this method is sample and easy to be implemented for a practical plant PMSM by removing the sinusoidal function in the controller and linearization operation. Moreover, it is well known that the direct time delay feedback can be used for eliminating chaos[21-23]. In this paper, we will show that it can also be used for chaotifying control. By this way, in a same framework of control system, the designer can design a system to be either chaotic or nonchaotic as desired.

This paper is organized as follows: In Section 2, chaotifying control of PMSM via the direct time delay feedback is described. In Section 3, the simulation results are given to show the effectiveness of the proposed method. Some conclusion remarks are given in Section 4.

II. CHAOTIFYING CONTROL OF PMSM VIA DIRECT TIME DELAY FEEDBACK

The transformed model of PMSM with the smooth air gap can be described as follows [6]:

$$
\begin{cases}
\dfrac{d\tilde{i}_d}{dt} = -\tilde{i}_d + \tilde{\omega}\tilde{i}_q + \tilde{u}_d \\[2mm]
\dfrac{d\tilde{i}_q}{dt} = -\tilde{i}_q - \tilde{\omega}\tilde{i}_d + \gamma\tilde{\omega} + \tilde{u}_q \\[2mm]
\dfrac{d\tilde{\omega}}{dt} = \sigma\left(\tilde{i}_q - \tilde{\omega}\right) - \tilde{T}_L
\end{cases}
\tag{1}
$$

Where $\widetilde{i}_d$, $\widetilde{i}_q$ are the transformed direct- and quadrature-axis stator current respectively, $\widetilde{\omega}$ is the transformed angle speed of the motor, $\widetilde{u}_d$, $\widetilde{u}_q$ are the transformed direct- and the quadrature-axis stator voltage components respectively, $\widetilde{T}_L$ is the transformed external load torque, σ and γ are system parameters.

The bifurcation and chaotic behaviors of Eq.(1) with different parameter set are investigated in [6], for example, when $\widetilde{u}_d = 0$, $\widetilde{u}_q = 0$, $\widetilde{T}_L = 0$, initial condition $\left(\widetilde{i}_d(0), \widetilde{i}(0)_q, \widetilde{\omega}(0) \right) = \left(20, 0.01, -5 \right)$, parameter $\sigma = 5.46$, $\gamma = 20$, system (1) will demonstrated chaotic motion, the chaotic attractor is shown in Fig.1.

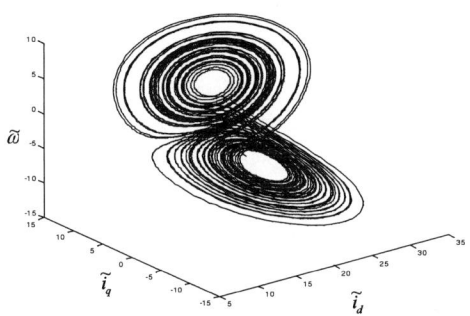

Figure 1.   Strange attractor in PMSM

For more bifurcation and chaotic behavior of system (1) with the change of parameter, one can refer to [6]. In this paper, our aim is to make system (1) chaotic by an external control force, when the system parameters are not belonging to the chaotic set. To implement this aim, the direct time delay feedback is employed.

The proposed controller scheme is given in Eq.(2)

$$\widetilde{u}_d = K_d \left(\widetilde{i}_d(t) - \widetilde{i}_d(t - \tau_d) \right) \tag{2}$$
$$\widetilde{u}_q = K_q \left(\widetilde{i}_q(t) - \widetilde{i}_q(t - \tau_q) \right)$$

The controlled system is given as

$$\begin{cases} \dfrac{d\widetilde{i}_d}{dt} = -\widetilde{i}_d + \widetilde{\omega}\widetilde{i}_q + K_d \left(\widetilde{i}_d(t) - \widetilde{i}_d(t - \tau_d) \right) \\[2mm] \dfrac{d\widetilde{i}_q}{dt} = -\widetilde{i}_q - \widetilde{\omega}\widetilde{i}_d + \gamma\widetilde{\omega} + K_q \left(\widetilde{i}_q(t) - \widetilde{i}_q(t - \tau_q) \right) \\[2mm] \dfrac{d\widetilde{\omega}}{dt} = \sigma \left(\widetilde{i}_q - \widetilde{\omega} \right) - \widetilde{T}_L \end{cases} \tag{3}$$

It can be seen from Eq.(3) that the proposed controller has the same form as the delay feedback control method proposed by Pyragas for eliminating chaos[22]. In the following section, we will show that this controller can be used for chaotifying control.

III.   SIMULATION RESULTS

When the parameter $\sigma = 5.46$, $\gamma = 3$, $\widetilde{T}_L = 0$, $\widetilde{u}_d = 0$, $\widetilde{u}_q = 0$, the system has three equilibriums, (0,0,0) is saddle point, the other equilibriums are focuses, therefore the unforced system (3) is asymptotically stable. The three dimensional phase plot of unforced system (3) starting from the initial condition $\left(\widetilde{i}_d(0), \widetilde{i}(0)_q, \widetilde{\omega}(0) \right) = (20, 0.01, -5)$ is shown in Fig.2. For the unforced system, the trajectory of the state will stabilize at one of the two stable equilibriums according to the basin where the initial condition locates.

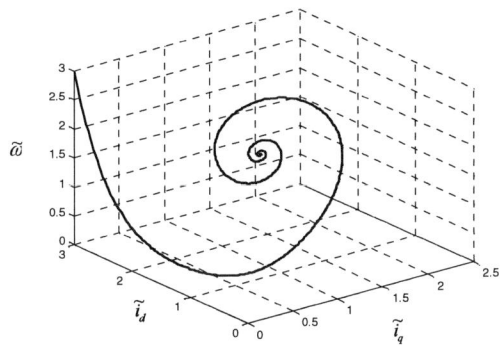

Figure 2. three dimensional phase plot of the unforced system starting from initial condition (3,0.01,3)

The aim of the chaotifying control is to generate chaos from the originally stable system via the proper control force. Set controller parameter $K_d = 1$, $K_q = -0.1$, $\tau_d = 0.8$, $\tau_q = 0.9$, the system parameter $\sigma = 5.46$, $\gamma = 3$, $\widetilde{T}_L = 0$, the three dimensional phase plot is shown in Fig.3.

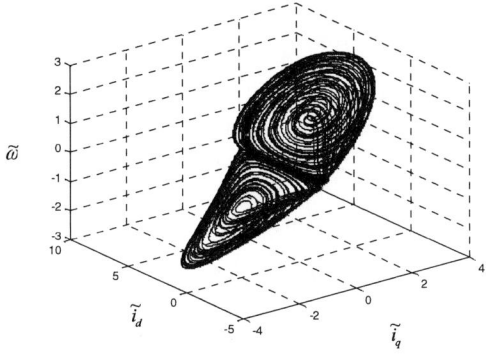

Figure 3.   the 3D phase plot of the chaotic attractor generated by direct delay feedback

The time domain waveform of the transformed speed output and the power spectrum of the speed output is shown in Fig.4 (a) and Fig.4(b) respectively. The largest Lyapunov exponent of the strange attractor is 0.014. The phase plot, time domain waveform and the power spectrum are congruously indicating the chaotic state of system (3).

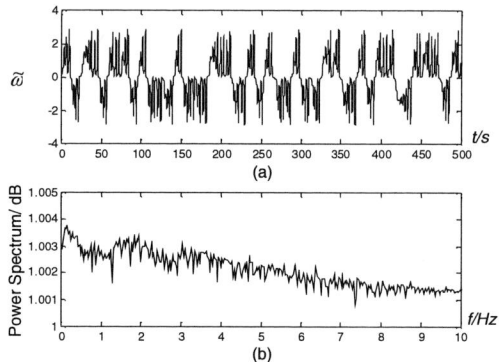

Figure 4. The time domain waveform of the transformed speed output and the power spectrum of the speed output

From the Fig.3 and Fig.4, we can see the chaos is generated by the proposed method. To probe the relationship of the controller parameters and the system state, we can derive the bifurcation diagram of the controller parameter using the Poincaré map and numerical methods given in [24].

If the parameters are set as $K_d = 1$, $\tau_d = 0.9$, $\tau_q = 0.8$, K_q is used as bifurcation parameter, the bifurcation diagram is shown in Fig.5. The parameters are set as $K_q = -0.1$, $\tau_d = 0.9$, $\tau_q = 0.8$, K_d is used as bifurcation parameter, the bifurcation diagram is shown in Fig.6. The Parameters are set as $K_d = 1$, $\tau_d = 0.9$, $K_q = 1$, τ_q is used as bifurcation parameter, the bifurcation diagram is shown in Fig.7. $K_d = 1$, $\tau_q = 0.8$, $K_q = 1$, τ_d is used as bifurcation parameter, the bifurcation diagram is shown in Fig.8.

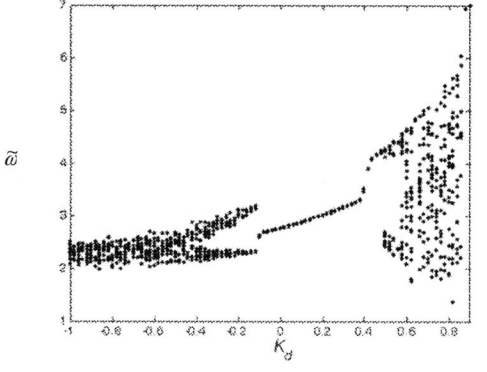

Figure 5. Bifurcation diagram with respect to K_d $K_q = -0.1$, $\tau_d = 0.9$, $\tau_q = 0.8$

From the bifurcation diagram Fig.5 - Fig.8, we can see clearly that the system behavior will undergo periodic

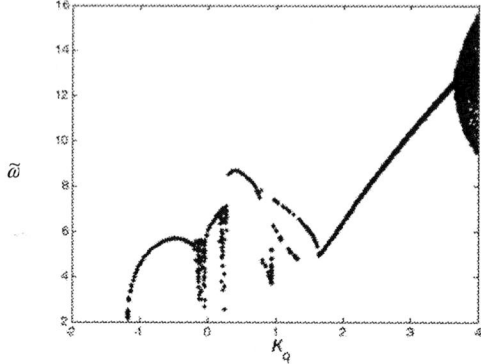

Figure 6. Bifurcation diagram with respect to K_q with $K_d = 1$, $\tau_d = 0.9$, $\tau_q = 0.8$

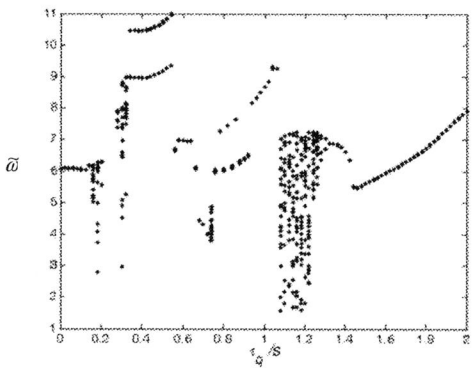

Figure 7. Bifurcation diagram with respect to τ_q with $K_d = 1$, $\tau_d = 0.9$, $K_q = 1$

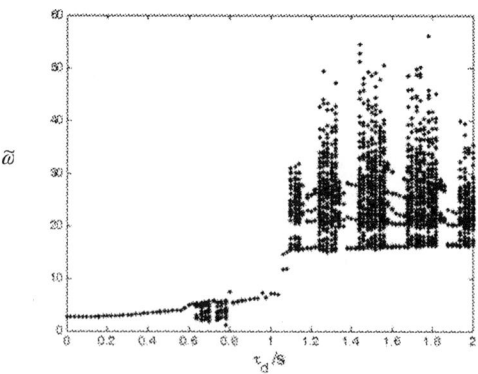

Figure 8. Bifurcation diagram with respect to τ_d with $K_d = 1$, $\tau_q = 0.8$, $K_q = 1$

operation and chaos as the parameters change smoothly. There might be more than one chaotic regime as parameter changing. The amplitude of the different chaotic regime is different. The designer can choose the parameter according to the required amplitude of chaotic motion. In Figs.9-10, the chaotic attractor, the time sequence and the power spectrum are given, when $K_d = 1$, $K_q = 4.3$, $\tau_d = 0.8$, $\tau_q = 0.9$. It can be calculated that the Lyapunov exponent of the chaotic attractor is 0.02377.

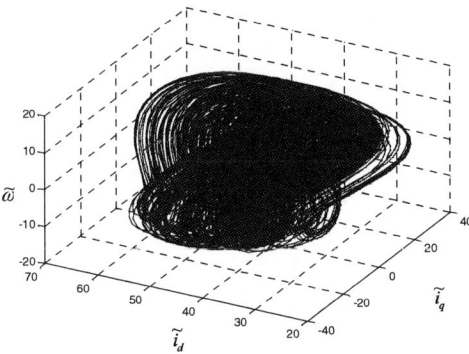

Figure 9. Chaotic attractor derived by time delay feedback with $K_d = 1, K_q = 4.3, \tau_d = 0.8, \tau_q = 0.9$

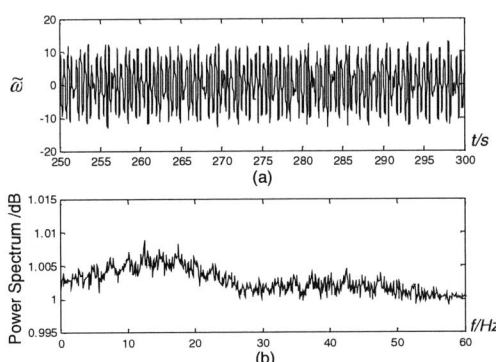

Figure 10. Time sequence and power spectrum of the chaotic motion corresponding to Fig.9

In the other parameter regime of chaotic motion, the similar result can be observed to shown the generation of chaos by the direct time delay feedback control. When the external torque is not zero, the similar results can be derived, which also shows the bifurcation and chaos will occur with the change of the parameter.

From the bifurcation diagram, we can also have there exists periodic operation under the control force of the direct time delay feedback.

The work in this paper, together with the previous work [8], shows that one can eliminate chaos or generate chaos by using the direct time delay feedback control according

to the practical demand, which will give a great flexibility to the designer.

IV. CONCLUSION

The direct time delay feedback control method is investigated for chaotifying control of PMSM. This method is proved to be feasible and effective by simulation results. Compared with the indirect time delay feedback control method, the proposed method is simple, because the sinusoidal function is removed and no linearization is needed. This investigation also indicates that the direct time delay feedback control can be used for generating chaos or eliminating chaos according to the practical demand.

This paper just investigates the effectiveness of the proposed method. The deepen research is still open for further research to probe the underlying the theoretical aspect of the direct time delay feedback for both generating chaos and eliminating chaos.

ACKNOWLEDGMENT

This work supported in part by Specialized Research Fund for the Doctoral Program of Higher Education (20020698026), and Natural Science Foundation of Shaanxi Province (2003F028). H P Ren thanks Prof. G Chen of City university of Hong Kang for his helpful suggestions.

REFERENCES

[1] Y.Kuroe and S.Hayashi, Analysis of Bifurcation in Power Electronic Induction Motor Drive System [C], IEEE Power Electronics Specialists Conference Rec., 1989:923-930

[2] Hemail N., strange attractors in brushless DC motor [J], IEEE Transactions on Circuits and System- I : Fundamental Theory and Application, 1994, 41(1): 40-45.

[3] Z. Suto, I. Nagy and E. Masada, Avoiding Chaotic Processes in Current Control of AC Drive [C], IEEE Power Electronics Specialists Conference Rec., 1998, 255-261.

[4] J H Chen, K. T. Chau and C. C. Chan, Analysis of chaos in current-mode-controlled DC drive systems, IEEE Trans. on Circuit and Systems I , 2000, 47(2): 67-76

[5] K. T. Chua and J. H. Chen, Modeling, analysis and experimentation of chaos in switched reluctance drive system, IEEE Trans. on Circuit and Systems I , 2003, 50(3): 712-716

[6] Zhong Li, Jin Bae Park, etal. Bifurcation and Chaos in a Permanent –Magnet Synchronous Motor [J]. IEEE Trans. on Circuits and Systems- I , 2002, 49(3): 383-387

[7] Zhang Bo, Li Zhong, Mao Zongyuan, Entrainment and Migration Control of Permanent-Magnet Synchronous Motor System, Control Theory and Applications, 2002,19(1):53-56

[8] REN Haipeng, Liu Ding, Li Jie, Delay Feedback Control of Chaos in Permanent Magnet Synchronous Motor, Proceeding of the CSEE, 2003, 23(6), pp.175-178

[9] X F Wang, G Chen and X Yu, anticontrol of chaos in continuous time system via time delay feedback, Chaos, Vol.10(4), 2000, pp.771-779

[10] J. M. Ottino et al., "Chaos, symmetry, and self-similarity: Exploiting order and disorder in mixing processes," Science 1992 257, 754–760.

[11] Andreas Abel, Wolfgand Schwarz , Chaos communications— principles, schemes, and system analysis, Proceeding of IEEE, Vol.90(5), 2002, 691-710

[12] Ito Shunji, Narikiyo Tatsuo, Abrasive machine under wet condition and constant pressure using chaotic rotation, Journal of the Japan Society for Precision Engineering, 1998, 64(5): 748-752

[13] Yunjia Long, Yong Yang, Congling Wang, Road roller engineering based on chaotic vibration mechanics, Chinese Engineering Science, 2000, Vol.2(9), 76-79

[14] G Chen, D Lai. Feedback anticontrol of discrete chaos [J]. Int. J. of Bifurcation and Chaos, 1998, 8: 1585-1590.

[15] X F Wang, Chen G. Chaotification via arbitrarily small feedback controls: theory, method and applications. Int. J. Bifurcation and Chaos, 2000, 10(3): 549-579.

[16] Xiao Fan Wang et al, Generating chaos in Chua's circuit via time delay feedback, IEEE Trans. On Circuit and Systems Ⅰ , 2001, 48(9): 1151-1156

[17] Tianshou Zhou, Guanrong Chen, Qigui Yang, A simple time-delay feedback anticontrol method made rigorous. Chaos, 2004, 14(3): 662-668

[18] J Lu, T Zhou, etal. Generating chaos with a switching piecewise-linear controller [J]. Chaos, 2002, 12(2): 344-349

[19] X F Wang, G Chen. Generating topologically conjugate chaotic systems via feedback control [J]. IEEE Trans. on Circuits and Systems 1, 2003, 50(6): 812-817

[20] Ricardo Chacón, Maintenance and suppression of chaos by weak harmonic perturbations: a unified view, Physics Review Letters, 2001, 86(9), 1737-1740

[21] K Pyragas, Continuous Control of Chaos by Self-controlling Feedback [J], Phys. Lett. A, 1992, 170(6):421-428

[22] K Pyragas, Control of chaos via extended delay feedback, Phys. Lett. A, 1995, 206(10):323-330

[23] Kazuyuki Yagasaki and Yoshiyuki Tochio, Experimental control of chaos by modifications of delayed feedback, Int. J. of Bifurcation and Chaos, 2001,11(12): 3125-3132.

[24] Thomas S Parker, Leon O Chua, Practical numerical algorithms for chaotic systems, Springer-Verlag, New York, 1993

2006 5th International Power Electronics and Motion Control Conference

Analysis of PMLSM Direct Thrust Control System Based on Sliding Mode Variable Structure

Junyou Yang, Guofeng He, Jiefan Cui

School of Electrical Engineering, Shenyang University of Technology, Shenyang, 110023, China
junyouyang@sut.edu.cn

Abstract—The permanent magnet linear synchronous motor (PMLSM) direct thrust control (DTC) systems using space vector modulation (SVM) and sliding mode variable-structure are presented and analyzed. Firstly, simulation results of the classical DTC system and the system based on SVM and predictive PI controller are compared. Both systems based on SVM can keep the fast response characteristics of the conventional DTC and apparently reduce the mover flux and thrust force ripples, and furthermore, it can improve the inverter's switching frequency at low speeds. Under the same conditions, simulation results show that the systems using SVM produce almost the same mover flux and thrust ripples. But the system using sliding mode variable structure controller is more robust to the mover flux estimated angle error caused by the change of the mover resistance. The conventional DTC and the proposed system have been realized on the experiment platform based on TMS320LF2407. Experiment results show that the system has good performances.

Keywords- permanent magnet linear synchronous motor; direct thrust control; space vector modulation; sliding mode variable structure

I. INTRODUCTION

Permanent magnet linear synchronous motors (PMSMs) have been widely applied to the industrial and servo drives. With appropriate control strategies, they can result in significant energy-savings and high performances. Direct torque control was introduced in the mid-1980s [1,2], which has been widely used for induction motor drives. Besides its simplicity, direct torque control is able to produce fast torque and flux control and, if the torque and flux can be estimated correctly, is robust with respect to motor parameters and perturbations. However, it is well known that direct torque control presents some disadvantages, such as the variable switching frequency behavior of the inverter, and notable torque, flux and current ripples. Space vector modulation (SVM) technique is widely used nowadays in inverter controls. Comparing with the sinusoidal pulse width modulation

(SPWM), the SVM is more suitable for digital implementation and is able to increase the maximum obtainable output voltage. Using SVM-based direct torque control can apparently reduce torque and stator flux ripples. Lai YS, Chen JH proposed proportional integral (PI) controllers for the stator flux and the torque regulations [3], but the PI controllers limit the dynamic response. The PI predictive controllers adopted by these SVM-based direct torque control schemes limit the transient performances and reduce the robustness by contrasting classical direct torque control scheme. Variable-structure control strategy is an effective, high frequency switching control for nonlinear systems with uncertainties. It features simple implementation, disturbance rejection, strong robustness and fast response. Wang Huangang [4] and Zhuang Xu [5] proposed that the stator flux and the electrical torque were directly controlled by variable-structure controllers, and stator voltage vectors calculated by the variable-structure controllers were applied to the induction motor and interior PMSM respectively. In this study, this method is applied to the direct thrust control (DTC) system. The amplitude of mover flux linkage and the thrust are directly controlled by sliding mode variable-structure (SMVS) controllers. Three simulation systems including the classical DTC, the DTC based on predictive PI controller and the proposed system by authors are set up, and the simulation results are compared and analyzed. The conventional DTC and the new system have been realized on the experiment platform basing on TMS320LF2407. Experiment results show that the proposed system has good performances.

II. MODELING OF PMLSM

When neglecting the longitudinal end effect, the mathematical model of PMLSM under the d-q reference frame can be described as,

$$\begin{bmatrix} v_d \\ v_q \end{bmatrix} = \begin{bmatrix} R + pL_m & -\omega_e L_m \\ \omega_e L_m & R + pL_m \end{bmatrix} \begin{bmatrix} i_d \\ i_q \end{bmatrix} + \begin{bmatrix} 0 \\ \omega_e K_E \end{bmatrix}, \quad (1)$$

where v_d, v_q, i_d, i_q are mover voltages and currents, p is differential operator, K_E is EMF constant, R is

---

Supported by National Natural Science Foundation of China (50375102) and Program for Liaoning Excellent Talents in university (RC-04-14)

the mover resistance, L_m is the mover induction.

$$\omega_e = \pi p_n v / \tau \; , \qquad (2)$$

where ω_e is the mover electrical angle speed converted, p_n is the number of pole pairs and τ is the pitch, and v is the mover line speed.

Transform (1) into the static $\alpha - \beta$ coordinate, then

$$\begin{bmatrix} v_\alpha \\ v_\beta \end{bmatrix} = \begin{bmatrix} R+pL_m & 0 \\ 0 & R+pL_m \end{bmatrix} \begin{bmatrix} i_\alpha \\ i_\beta \end{bmatrix} + \omega_e K_E \begin{bmatrix} \sin\theta_e \\ \cos\theta_e \end{bmatrix}, \qquad (3)$$

where

$$\theta_e = \int \omega_e dt + \theta(0), \qquad (4)$$

θ_e is the converted mover angle, $\theta(0)$ is the initial value of θ_e. Let

$$e = \begin{bmatrix} e_\alpha \\ e_\beta \end{bmatrix} = \begin{bmatrix} -\omega_e K_E \sin\theta_e \\ \omega_e K_E \cos\theta_e \end{bmatrix}. \qquad (5)$$

From the new model of (2), the PMLSM can be described by state equations as (6)-(8):

$$\begin{bmatrix} \dot{i}_\alpha \\ \dot{i}_\beta \end{bmatrix} = \begin{bmatrix} -R/L_m & 0 \\ 0 & -R/L_m \end{bmatrix} \begin{bmatrix} i_\alpha \\ i_\beta \end{bmatrix} + \begin{bmatrix} 1/L_m & 0 \\ 0 & 1/L_m \end{bmatrix} \begin{bmatrix} e_\alpha \\ e_\beta \end{bmatrix}$$

$$+1/L_m \begin{bmatrix} v_\alpha \\ v_\beta \end{bmatrix}, \qquad (6)$$

$$\begin{cases} \dot{\lambda}_\alpha = v_\alpha - R i_\alpha \\ \dot{\lambda}_\beta = v_\beta - R i_\beta \end{cases}, \qquad (7)$$

λ_α and λ_β are mover flux linkages.

$$F = \frac{3}{2}\frac{\pi}{\tau} p_n (\lambda_\alpha i_\beta - \lambda_\beta i_\alpha) = Mv + Bv + F_L, \qquad (8)$$

where F is estimated thrust, M is the inertia, B is the friction factor, F_L is the load force.

$$\lambda = \lambda_\alpha^2 + \lambda_\beta^2, \qquad (9)$$

λ is the square of mover flux linkage norm.

III. TWO DTC SYSTEMS BASED ON SVM

A. The Diagram of the Two DTC Systems

Block diagram of the DTC is shown in Fig.1. The outer

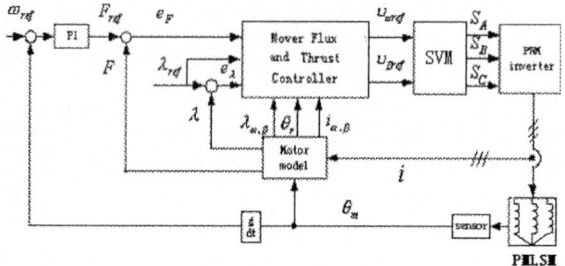

Fig. 1. Block diagram of the two direct thrust control systems.

loop contains a speed PI controller, which produces the reference thrust command. The inner loop includes the mover flux and thrust controllers, which generates the most appropriate mover voltage vectors for thrust and mover flux to track the references. The control mover voltage signals have been limited before proceeding to SVM module. The space vector modulation provides a solution for high-resolution voltage control and constant inverter switching frequency. The main difference between the two DTC systems is in the inner loop. One uses predictive PI thrust and flux controller, the other adopts sliding mode variable structure controller.

B. Design of Predictive PI Controller

The predictive PI thrust and mover flux controller diagram is shown in Fig. 2, in which input signals are the mover flux reference and the thrust error and outputs are reference voltage vectors.

The voltage vectors calculating equations are as (10) and (11) [4].

$$v_{\alpha ref} = \frac{\lambda_{ref}\cos(\theta_e + \Delta\delta) - \lambda\cos\theta_e}{T_p} + R i_\alpha, \qquad (10)$$

$$v_{\beta ref} = \frac{\lambda_{ref}\sin(\theta_e + \Delta\delta) - \lambda\sin\theta_e}{T_p} + R i_\beta, \qquad (11)$$

where $\Delta\delta$ is the mover flux predictive angle variance.

C. Design of Sliding Mode Variable Structure Controller

The control objectives are to track desired thrust and mover flux trajectories. The sliding surface is set as

$$S = \begin{bmatrix} S_1 & S_2 \end{bmatrix}^T, \qquad (12)$$

where

$$S_1 = e_F(t) + K_1 \int_0^t e_F(\tau)d\tau - e_F(0), \qquad (13)$$

$$S_2 = e_\lambda(t) + K_2 \int e_\lambda(\tau)d\tau - e_\lambda(0), \qquad (14)$$

where $e_F = F_{ref} - F$, $e_\lambda = \lambda_{ref} - \lambda$, e_F and e_λ are the

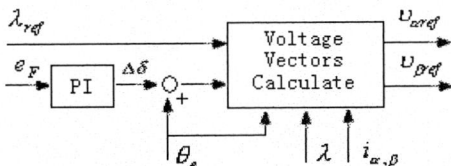

Fig.2. The diagram of the predictive PI controller

errors between the references and estimated values of thrust and the mover flux, K_1 and K_2 are positive control gains.

Select switching control law,

$$\underline{V}_{ref} = \begin{bmatrix} v_{\alpha ref} \\ v_{\beta ref} \end{bmatrix} = -\underline{D}^{-1} \begin{bmatrix} \mu_1 & 0 \\ 0 & \mu_2 \end{bmatrix} \begin{bmatrix} sign(S_1) \\ sign(S_2) \end{bmatrix}, \quad (15)$$

where

$$\underline{D} = \begin{bmatrix} -\dfrac{3}{2}\dfrac{\pi}{\tau} p_n (i_\beta - \dfrac{\lambda_\beta}{L_m}) & -\dfrac{3}{2}\dfrac{\pi}{\tau} p_n (\dfrac{\lambda_\alpha}{L_m} - i_\alpha) \\ -2\lambda_\alpha & -2\lambda_\beta \end{bmatrix}. \quad (16)$$

In variable-structure control systems, the control signals can be switched ideally at an infinite frequency. This is not possible due to the finite sample times of microprocessor implementation. The switching non-idealities can cause a chattering phenomenon. The standard solution is to introduce a boundary layer (BL). Replace the switching function by a continuous function in the sliding surface neighborhood, that is,

$$sign(S_i) = \begin{cases} 1, & S_i > \lambda_i \\ -1, & S_i < -\lambda_i \\ \dfrac{S_i}{\lambda_i}, & |S_i| < \lambda_i \end{cases}, \quad (17)$$

where $\lambda_i > 0$ is a smoothing factor, $i = 1, 2$.

Let

$$F_1 = -\frac{3}{2}\frac{\pi}{\tau}\left[\lambda_\alpha(-\frac{R}{L_m}i_\beta - \frac{1}{L_m}e_\beta) - \lambda_\beta(-\frac{R}{L_m}i_\alpha - \frac{1}{L_m}e_\alpha)\right] + K_1 e_F \quad (18)$$

$$F_2 = K_2 e_\lambda + 2\lambda_\alpha R i_\alpha + 2\lambda_\beta R i_\beta, \quad (19)$$

when $\mu_1 > |F_1|$ and $\mu_2 > |F_2|$, the proposed DTC system becomes asymptotically stable [4][6].

IV. VOLTAGE VECTOR GENERATOR AND SVM

According to (α, β) reference vector, the SVM modulator produces the inverter control signals. The SVM principle is based on the switching between two adjacent active vectors and zero vectors during one modulation period T_p. If, for instance, the reference voltage vector $\underline{V}_{ref}$ is as in Fig. 3, the vector can be approximated by two adjacent active vectors $V_1(100)$, $V_2(110)$ and zero vectors $V_0(000)$, $V_7(111)$.

$$\begin{aligned} \underline{V}_{ref} T_p &= V_1 T_1 + V_2 T_2 \\ T_1 &= T_p \frac{2a}{\sqrt{3}} \sin(\frac{\pi}{3} - \theta_{us}) \\ T_2 &= T_p \frac{2a}{\sqrt{3}} \sin \theta_{us} \\ T_0 &= T_7 = \frac{T_p - T_1 - T_2}{2} \end{aligned} \quad , \quad (20)$$

where $a = \underline{V}_{ref} / V_{dc}$, and V_{dc} is the DC link voltage.

V. SIMULATION RESULTS

Three Matlab/Simulink simulation systems are designed to examine the different controller schemes. The mover flux and thrust force responses are respectively presented in Fig.4, Fig.5 and Fig.6. Parameters of PMLSM in the simulation are $R = 2.1\Omega$, $L_m = 0.0163H$, $\lambda_f = 0.211$Wb, $p_n = 6$, $M = 25kg \cdot m^2$, $B = 8.00Ns/m$, $\tau = 0.016m$. The speed PI controller parameters are $K_p = 6$, $K_i = 5.33$, $F_{max} = 300$, $F_{min} = 0.01$. The predictive PI controller parameters are $K_p = 10$, $K_i = 8$, $T_{max} = 0.01$, $T_{min} = -0.01$, where K_p is the proportional coefficient, K_i is the integral coefficient, F_{max} is the upper-limit of the thrust reference, F_{min} is the lower-limit, T_{max} is the upper-limit of the mover flux rotation angle, T_{min} is the lower-limit. The sliding mode variable-structure controller parameters are $K_1 = 20$, $K_2 = 10$, $\mu_1 = 400$, $\mu_2 = 3$, $\lambda_1 = 3$, $\lambda_2 = 0.03$.

These three control systems are compared under the same operating conditions. Speed reference is 2m/s, and the load force is 200N. These three figures show that the

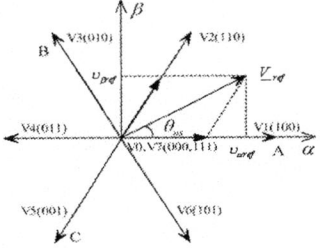

Fig. 3. The principle behind reference voltage vector SVM

last two systems' thrust dynamic responses are as fast as classical DTC and have much smaller thrust and mover flux ripples than the classical DTC system. Compared Fig.5 with Fig.6, the two systems' mover flux ripples are not evidently different.

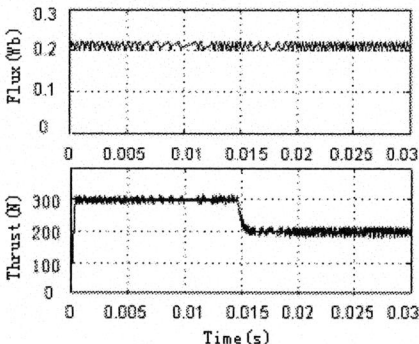

Fig. 4. The response curve of classical DTC system

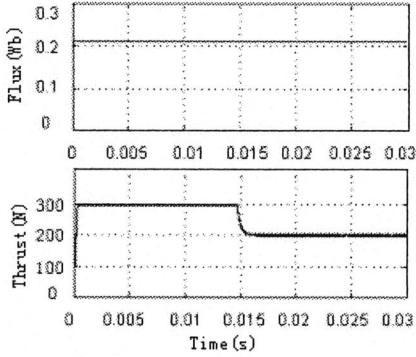

Fig. 5. The response curves of the system using predictive PI controller

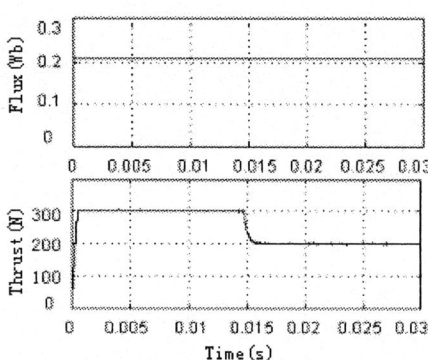

Fig. 6. The response curves of the system

using sliding mode controller

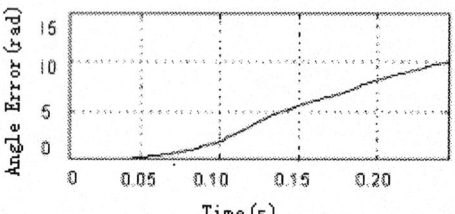

Fig.7. The estimate error curve of flux linkage of the system using predictive PI controller

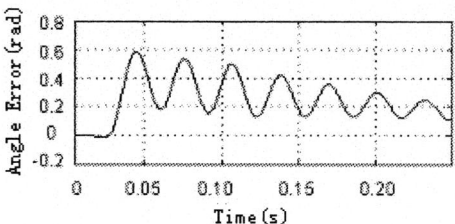

Fig. 8. The estimate error curve of flux linkage of the system using sliding mode controller

Fig.7 and Fig.8 show the robustness of the two systems using SVM. When the mover resistance has varied from normal value to 3 Ω, simulation result of the system using sliding mode variable structure controller shows that angle error between real and estimated mover fluxes that are caused by the variation of the mover resistance gradually converges to zero. The mover flux regulation is robust with respect to the variation of mover resistances, however, the system using predictive PI mover flux controller's angle error gradually becomes large.

VI. EXPERIMENTAL RESULTS

The classical DTC and the proposed system are implemented on a TMS320LF2407 digital signal processor with a clock speed of 30MHz. The mover position and speed are obtained from linear raster ruler, and 6MBP50RA060 is used for the voltage inverter, which is supplied at a DC link voltage of 110V. Real time control software is coded using assemble language. One PWM period is 132 μs. This figure is directly drawn by DSP, which 1000 data are acquired from the DARAM. The motor and controller parameters are same as the simulation, speed reference is 2m/s. Fig.9 and Fig.11 show the conventional DTC mover flux and thrust responses. Fig.10 and Fig.12 show the system mover flux and thrust responses based on sliding mode variable structure controller. These figures illustrate the mover flux and thrust ripples of the proposed system are much smaller than the classical DTC system.

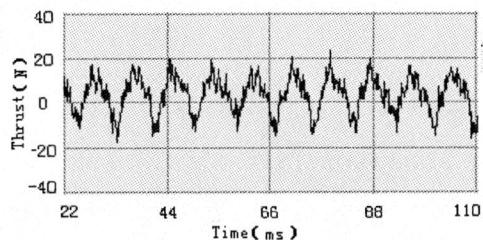

Fig. 9. The thrust response curve of the conventional DTC system

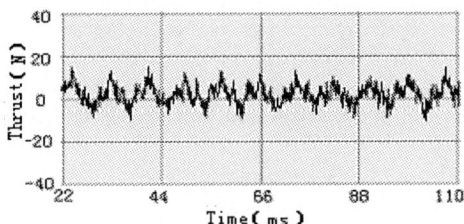

Fig. 10. The thrust response curve of the system using sliding mode controller

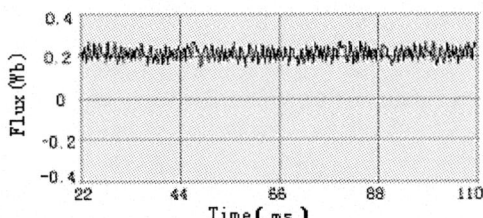

Fig. 11. The mover flux linkage response curve of the conventional DTC system

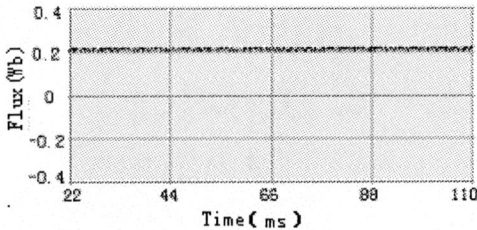

Fig. 12. The mover flux linkage response curve of the system using sliding mode controller

VII. CONCLUSIONS

The conclusions can be got from simulation and experiment results.

(1)   The DTC systems using SVM can apparently reduce the mover flux and thrust ripples.

(2)   The system based on sliding mode variable structure controller is more robust than the system using PI controller.

REFERENCES

[1]   M. Depenbrock, "Direct self-control of inverter-fed machine," *IEEE Trans. Power Electron*, vol. 3, pp. 420–429, Oct. 1988.

[2]   I. Takahashi and T. Naguchi, "A new quick-response and high-efficiency control strategy of an induction motor," *IEEE Trans. Ind. Applicat*, vol. IA-22, pp. 820–827, Sept./Oct. 1986.

[3]   Lai Y S. Chan J H, "A new approach to direct torque control of induction motor drives for constant inverter switching frequency and torque ripple reduction," *IEEE Trans. Energy*, vol. 16, no. 3, pp. 220-227, 2001.

[4]   Zhuang Xu, M.F. Rahman, "A variable structure torque and flux controller for a DTC IPM synchronous motor," *Proceedings of the 35th Annu. IEEE Power Electronics Specialists Conference*, Aachen, Germany, pp.445-450, 2004

[5]   Wang Huangang, Xu Wenli, Yang Geng and Li Jian, "A new approach to direct torque control of induction machines," *Proceedings of the CSEE*, vol.24, no.1, pp.107-111, Jan. 2004. (in Chinese)

[6]   Yang Junyou, He Guofeng, Cui Jiefan, "Sliding Mode Variable-structure Direct Thrust Control of PMLSM using SVM," *Proceedings of the 6th. ICEMS Conference*, Nanjing, China, pp.1655-1658, 2005

2006 5th International Power Electronics and Motion Control Conference

Carrier-based Pulse Width Modulation for Three-Level Inverters: Neutral Point Potential and Output Voltage Distortion

Jang-Hwan Kim and Seung-Ki Sul

Electrical Engineering and Computer Science
Seoul National University, Seoul 151-744, Republic of Korea

Abstract- **Inherent problem in three-level neutral-point-clamped voltage source inverters (NPC VSI) is the variation of the neutral point potential. This paper proposes a carrier-based PWM technique, which can balance the neutral point potential and minimize the output voltage distortion. Additionally, the modulated voltage error caused by the unbalance of the neutral point potential is analyzed and the method to eliminate the error is described. The validity of the proposed PWM method is verified by computer simulation and experimental results.**

Keywords-neutral point-clamped voltage source inverter, neutral point potential, carrier-based PWM, output voltage distortion

I. INTRODUCTION

Multi-level neutral-point-clamped(NPC) voltage source inverters(VSI) have been widely used in medium/high power applications shown in Fig.1 (a). Practical problem in the NPC VSI is that the neutral point potential varies due to the neutral point current, which depends on the output currents and switching of the power devices. Many PWM strategies have been suggested to solve this problem [1]-[4],[10]. Technically these strategies manipulate the redundant voltage vectors using their property that the neutral point currents are different even though the effective output voltage is same. In space-vector PWM schemes, average neutral point current in each switching period can be restrictedly kept zero adjusting the duties of the respective redundant voltage vectors according to the modulation index and power

factor [1]-[3]. This function is similarly implemented in carrier-based PWM schemes [4] by adding the zero sequence offset voltage to the output voltage references. It is understandable that the magnitude of the zero sequence offset voltage may be associated with the selection of redundant voltage vectors and the dwell time of those vectors as well. This paper analyzes the relationship between the space-vector PWM and the carrier-based PWM method, and also proposes a novel carrier-based PWM strategy to balance the neutral point potential. The modulation voltage error caused by the unbalance of the neutral point potential is analytically described and a method to eliminate the voltage error is proposed.

II. PWM METHOD FOR THREE-LEVEL VSI [5]-[8]

A. Space vector PWM scheme [5]

The standard space-vector PWM method requires three-nearest voltage vectors which can create a desired reference vector. As the level of the inverter increases, the combination of the three-nearest voltage vectors, that is, the sectors to be identified become numerous. Once the sector is identified, that is, three-nearest voltage vectors are found, duties of the respective voltage vectors can be obtained by the linear combination of those vectors as follows.

$$\mathbf{V_{ref}} = d_x \cdot \mathbf{V_x} + d_y \cdot \mathbf{V_y} + d_z \cdot \mathbf{V_z}, \text{ where } d_x + d_y + d_z = 1 .$$

The dwell times of respective switching states within the PWM cycle should be calculated by arranging the three-nearest voltage vectors with the calculated duties. This procedure gives the freedom to select one of

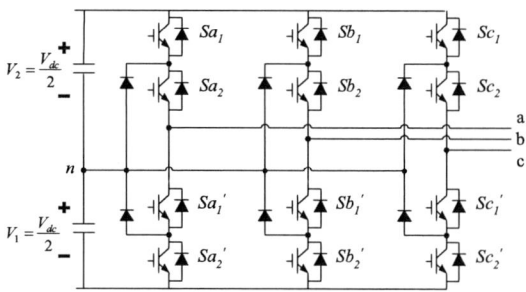

a) Circuit configuration

b) Output space in dq frame

Fig.1 Three-phase three-level VSI

1-4244-0448-7/06/$25.00 ©2006 IEEE

redundant vectors if necessary, and how to place the redundant voltage vector within the PWM cycle allows the additional degree of freedom in optimizing the switching sequence.

B. Optimized PWM shemes [6]-[8]

There is only one redundant voltage vector in two-level PWM inverter, that is, a zero vector. It is a well accepted method to minimize the output distortion that the zero vector is split into two so as to occupy the first and last state of the switching sequence with the same duty [9], [10]. Contrary to the two-level inverter, there are 7 redundant voltage vectors in the three-level inverter. It is natural that the redundant voltage vectors can be used for minimizing the output current distortion [6]-[8]. The redundant voltage vectors except zero vector influence the neutral point potential according to the respective switching states, thus they have been exploited for balancing the neutral point voltage instead of reducing the output distortion [1]-[4]. The vertices of redundant voltage vectors are marked with 7 dots in Fig 1. (b).

Space-vector PWM method is quite comprehensive, but it requires a lot of computational burden such as sector identification, the sequence table according to the sectors, even though the balancing of the neutral point potential is not considered in the modulation strategy.

The carrier-based PWM method in [4] has improved harmonic distortion in a low modulation index, but it still didn't completely use the three-nearest voltage vector, and also requires the RSS (Redundant Switching Selection) table. The next section will describe the carrier-based PWM method to balance the neutral point potential, which is exactly equivalent to the SVPWM method, since it uses the three-nearest voltage vectors and their duties are exactly same with the results from the space vector PWM method.

III. STRATEGY TO BALANCE THE NEUTRAL POINT POTENTIAL AND VOLTAGE SYNTHESIS

A. Carrier-based PWM method with optimizing output voltage harmonics

The output voltage space in a three-phase inverter can be depicted on the dq plane, since it has two degree of freedom in voltage synthesis. Therefore the output reference vector is expressed by $\mathbf{V_{dq}}^*$ but it is not useful if the carrier-based PWM method is adopted to implement the PWM strategy. It is transferred to a stationary reference frame as the phase voltage references (V_{as}^* V_{bs}^*, V_{cs}^*). Once the offset voltage (V_{sn}^*) is decided, pole voltage references (V_{an}^*, V_{bn}^*, V_{cn}^*) can be expressed using the phase voltage references and the offset voltage as follows.

$$V_{an}^* = V_{as}^* + V_{sn}^*$$
$$V_{bn}^* = V_{bs}^* + V_{sn}^* \qquad (1)$$
$$V_{cn}^* = V_{cs}^* + V_{sn}^*$$

The comparison between the co-phasal multi-carrier and the pole voltage references generates the switching states of the respective legs. The pole voltages are physically bounded by $-V_{dc}/2$ and $V_{dc}/2$, if the neutral point potential is balanced by half of the dc bus voltage (V_{dc} is the total dc bus voltage). Therefore the references of the pole voltage should exist between $-V_{dc}/2$ and $V_{dc}/2$, so that V_{sn}^* has a physical limitation as follows.

$$-V_{dc}/2 - V_{min} < V_{sn}^* < V_{dc}/2 + V_{max} \qquad (2)$$

V_{max} and V_{min} are defined as a maximum and a minimum value among V_{as}^*, V_{bs}^*, and V_{cs}^* respectively. If the V_{sn}^* is set to $-0.5 \cdot (V_{max} + V_{min})$, the modulation index can be increased by 15% compared to Sine PWM method. Besides, the output distortion can be minimized in case of two- level VSI system [9]. Due to the property of the multi-carriers in multi-level inverter, the switching instants are not the same with the case of single carrier in two-level inverter. Single virtual carrier was used for generating PWM pattern [6], [7], [8] as shown in Fig.2. The switching instant in a virtual carrier can be expressed by the modulus function, MOD of the pole voltage references [7], [8].

modulus function $\equiv \mathrm{MOD}(x, y) = x - y \cdot \mathrm{int}(x / y)$

To minimize the output voltage harmonics in three-level VSI, the redundant voltage vectors can be placed at the

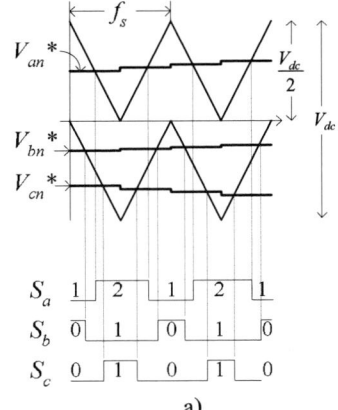

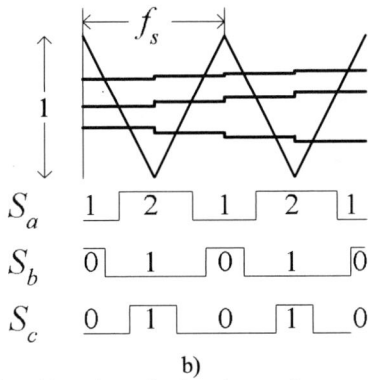

a) Switching state generation from the comparison between the co-phasal multi-carrier and pole voltage references,
b) Conceptual view in a virtual single carrier
Fig.2 Conceptual view in a virtual single carrier (S_a, S_b, S_c in (5))

first and last in the switching sequence using the extra offset voltage, which can be expressed as follows.

$$V_{extra} = \left[-\frac{D_{\max} + D_{\min}}{2} + 0.5 \right] \cdot \frac{V_{dc}}{2} \tag{3}$$

where $D_{\max}$ is the maximum and $D_{\min}$ is minimum value among $\text{MOD}(V_{an}^*, V_{dc}/2)$, $\text{MOD}(V_{bn}^*, V_{dc}/2)$, and $\text{MOD}(V_{cn}^*, V_{dc}/2)$.

Additionally the range where the extra offset voltage can vary is physically given by (4). Note that V_{extra} in (3) is the mid-point of this interval.

$$-D_{\min} \cdot V_{dc}/2 \le V_o \le V_{dc}/2 - D_{\max} \cdot V_{dc}/2. \tag{4}$$

B. Strategy to balance the neutral point potential

The pole voltage in three-level inverter according to three switching states can be expressed by (5).

$$V_{an} = \frac{V_{dc}}{2}(S_a - 1), \quad S_a = 0,1,2 \quad \text{,(fixed number)} \tag{5}$$

By averaging the pole voltage in (5) during half of carrier period ($T_s = 0.5/f_s$), S_a^* can be introduced as (6) that is related to the pole voltage reference.

$$\frac{1}{T_s} \int^{T_s} S_a dt \equiv S_a^* = 2 \cdot \left(\frac{V_{an}^*}{V_{dc}} + \frac{1}{2} \right) = \frac{V_{an}^* + V_{dc}/2}{V_{dc}/2} \tag{6}$$

, where $S_a^* = 0 \sim (n-1)$, (real number)

S_a^* can be obtained by this procedure that the pole voltage reference is added to $V_{dc}/2$ and then it is normalized by $V_{dc}/2$. S_a^* has an important characteristic that decimal part of S_a^* is directly associated with the dwell time of the upper switching state as (7). Therefore it is exactly equivalent to the carrier-based method.

$$T_a\_ON = T_s \cdot \text{MOD}(S_a^*, 1) \tag{7}$$

Using this modulus function, average neutral point current can be obtained by (8). Because there are three pole voltage references in three-phase three-level VSI, total average neutral point current can be written by (9).

$$\bar{I}_{np\_a} = \begin{cases} \text{MOD}(S_a^*, 1) \cdot I_a & 0 < S_a^* < 1 \\ [1 - \text{MOD}(S_a^*, 1)] \cdot I_a & 1 \le S_a^* < 2 \end{cases}, \tag{8}$$

$$\bar{I}_{np} = \bar{I}_{np\_a} + \bar{I}_{np\_b} + \bar{I}_{np\_c} = C_2 \frac{dV_2}{dt} - C_1 \frac{dV_1}{dt} = \bar{i}_2 - \bar{i}_1 \tag{9}$$

On the other hand, the total dc bus voltage is the sum of the V_1 and V_2 shown in Fig.1 (a).

$$V_{dc} = V_1 + V_2 \tag{10}$$

Assuming that their capacitances are same, the total energy stored in the capacitors can be expressed by (11).

$$E = \frac{1}{2}C \left[\sum_{k=1}^{2} V_k^2 \right] = \frac{1}{2}CV_1^2 + \frac{1}{2}CV_2^2 \tag{11}$$

Under the given condition (10), the necessary and sufficient condition .to minimize the total energy stored in the capacitor is given by (12).

$$V_1 = V_2 \tag{12}$$

It means that the minimization of the total energy stored in the capacitors permits one to perform the balancing control of the neutral point potential. Thus, the derivative of this energy related to the neutral current should be negative or zero as follows.

$$\begin{aligned} dE/dt &= V_1 i_1 + V_2 i_2 \\ &= \frac{1}{2}(V_2 - V_1)(i_2 - i_1). \\ &= \frac{1}{2}I_{np}(V_2 - V_1) \le 0 \end{aligned} \tag{13}$$

The average neutral point current in next half of PWM cycle can be predicted using (9) from the pole voltage references at this moment. Note that the pole voltage references can be modified to satisfy the condition (13) by the addition of the offset voltage, V_o in (4). In order to meet the condition (13), V_o should be selected differently from V_{extra} in (3). It is equivalent to the fact that the duty of the redundant voltage vector is manipulated in order to balance the neutral point potential in SVPWM method. V_o can be simply found by one of $-D_{\min} \cdot V_{dc}/2$ and $V_{dc}/2 - D_{\max} \cdot V_{dc}/2$ in order to control neutral point potential, and also can be set to V_{extra} in order to minimize the voltage distortion.

C. Implementation of the proposed PWM method

Under the unbalance condition of the neutral point potential, the voltage vectors represented on dq-space are relocated according to the unbalance factor defined by (14).

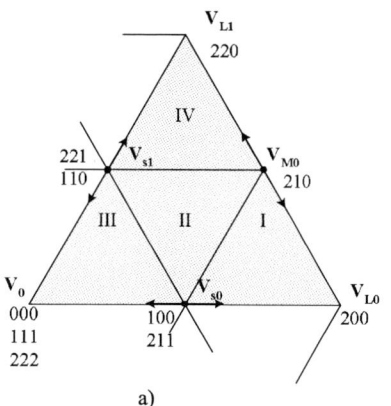

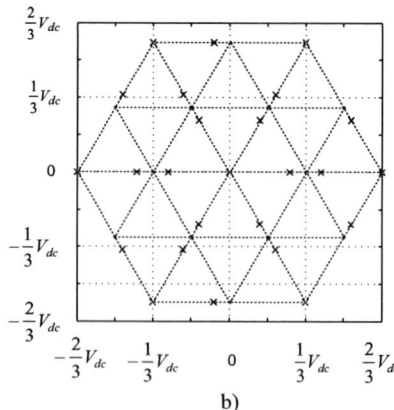

a) Switching vectors movement b) Switching vector movement under the condition that $K_b = 0.2$, x: relocated voltage vectors, •: voltage vectors on the balanced neutral point potential

Fig.3 Movement of the voltage vectors, which is caused by the unbalance of the neutral point potential

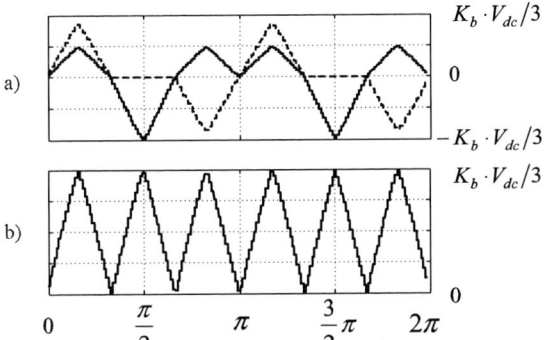

a) D-q axis voltage errors (solid: d-axis voltage error, dotted: q-axis voltage error), b) Magnitude of the voltage error (distance from the reference and averaged output voltage)
Fig.4 Output voltage error due to the movement of the voltage vectors without the compensation scheme, MI =1

$$K_b \equiv \frac{V_2 - V_1}{V_{dc}} \qquad (14)$$

Assuming that K_b = 0.2, $V_1 = V_{dc}/2 - K_b \cdot V_{dc}/2$, $V_2 = V_{dc}/2 + K_b \cdot V_{dc}/2$ and the variation of the voltage vectors are shown in Fig. 3. In space vector PWM method, the unbalance in neutral point potential causes the modulation error, if there is no proper compensation scheme. The magnitude of the modulated voltage errors is shown in Fig.4, under the condition that the modulation index is 1.

The information of the unbalance factor should be taken into account when the ON time of the upper switching states in (7) and average neutral point current in (8) are calculated, thus they can be modified as follows.

$$T_{a\_ON}$$

$$= \begin{cases} T_s \cdot MOD\left(\dfrac{S_a{}^*}{1-K_b},1\right) & S_a{}^* < 1-K_b \\[3mm] T_s \cdot MOD\left(\dfrac{S_a{}^* - (1-K_b)}{1+K_b},1\right) & S_a{}^* > 1-K_b \end{cases} \qquad (15)$$

$$\bar{I}_{np\_a}$$

$$= \begin{cases} MOD\left(\dfrac{S_a{}^*}{1-K_b},1\right) \cdot I_a & 0 < S_a{}^* < 1-K_b \\[3mm] \left[1 - MOD\left(\dfrac{S_a{}^* - (1-K_b)}{1+K_b},1\right)\right] \cdot I_a & 1-K_b \le S_a{}^* < 2 \end{cases} \qquad (16)$$

The modulation voltage error can be eliminated by this modification, which is equivalent to the carrier adjustment in the carrier-based PWM method as shown in Fig. 5.b). Fig.5.a) shows the voltage vector movement due to the neutral point potential variation under the K_b = 0.2 condition. If V_{ref} is given as a reference, three-nearest switching vectors are changed. It can be seen in Fig.5.b) that (110) switching vectors appears in right figure due to the relocation of the switching vectors in (a) under the same references condition.

The block diagram for the practical implementation of the proposed PWM method is shown in Fig.6. The method to place the redundant voltage vector within the PWM cycle can be selected to optimize the voltage

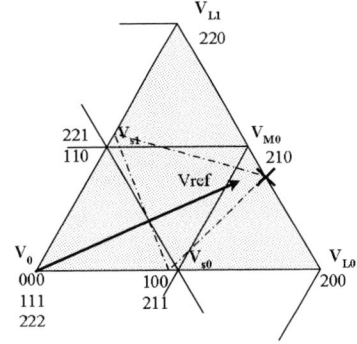

a) Voltage vectors variation due to the neutral point potential variation

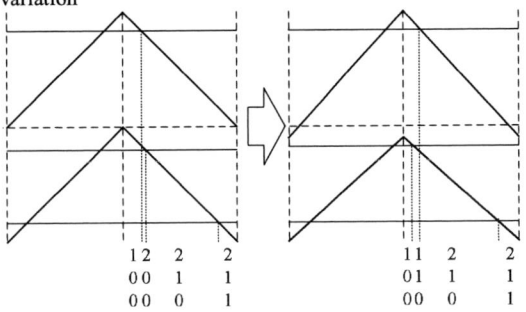

b) Carrier adjustment according to V_1, V_2 variation and switching states transition, $K_b = 0.2$
Fig.5 Conceptual view of the method to eliminate the voltage error

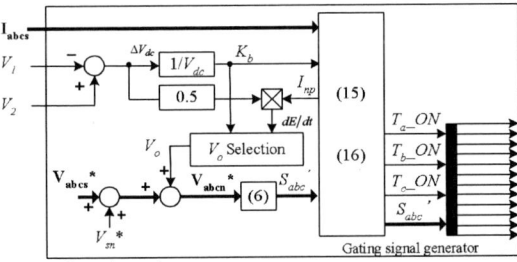

Fig.6 Implementation of the proposed PWM method

harmonics and also to balance the neutral point potential. The proposed PWM strategy is focused on the minimization of the output distortion within the range where the system is capable of accepting the variation of the neutral point potential, and if K_b is out of that range, the balancing control is performed instead of minimizing the output voltage harmonics.

IV.   SIMULATION RESULTS

The simulations have been carried out under the various operating condition using Simulink and PLECS. Total dc bus voltage is given by 1800[V], and total dc capacitance is 6[mF]. The carrier frequency (f_s) is set to 1[kHz]. Fig. 7 is the simulation result which shows modulated line-to-line voltage (V_{ab}) and its averaged voltage. It can be seen that the proposed PWM method synthesizes the output voltage accurately even in the unbalanced condition. Fig.8

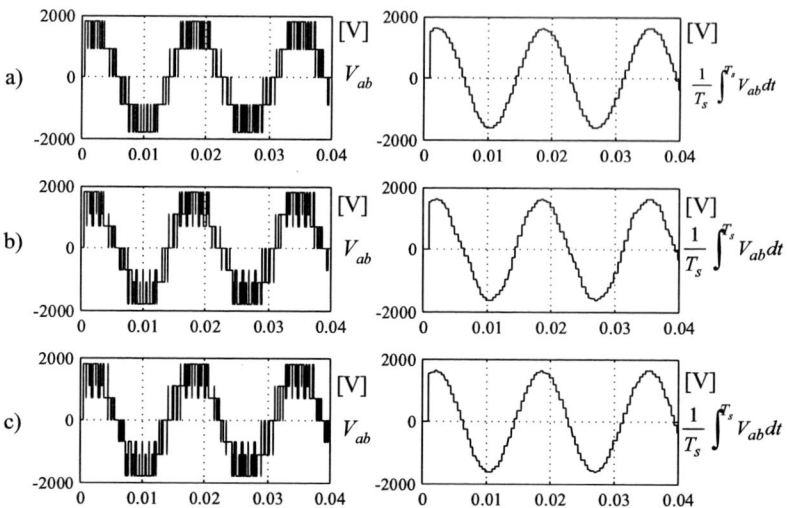

a) balanced condition, b) unbalanced condition without any compensation scheme, c) unbalanced condition with the proposed method

Fig. 7 Modulation performance, MI =0.9

shows the simulation results, which depict the line-to-line output voltage (V_{ab}), the difference between the V_2 and V_1, neutral point current (I_{np}), line currents (I_{abcs}), the pole voltage reference ($V_{an}*$) and additional offset voltage (V_o). The simulation condition is that the initial unbalance factor is given by 0.2, the modulation index is 0.9, the

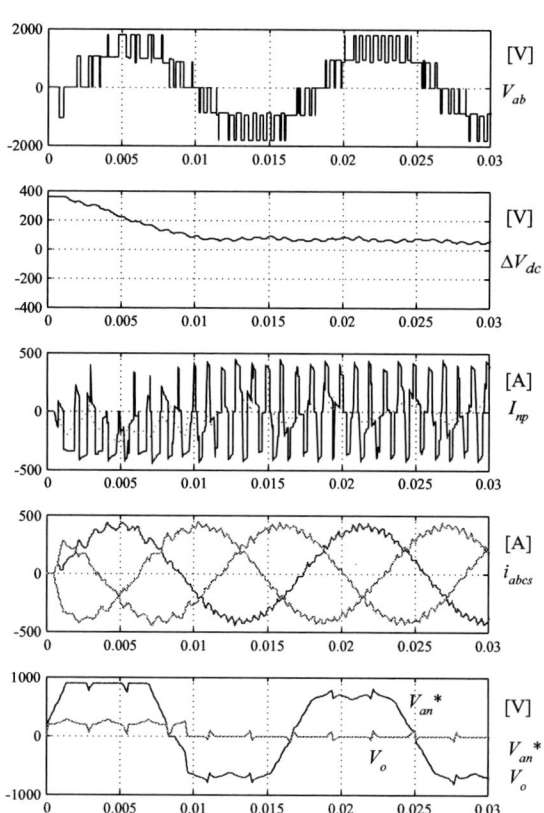

Fig. 8 Voltage balance and output voltage synthesis by the proposed PWM method

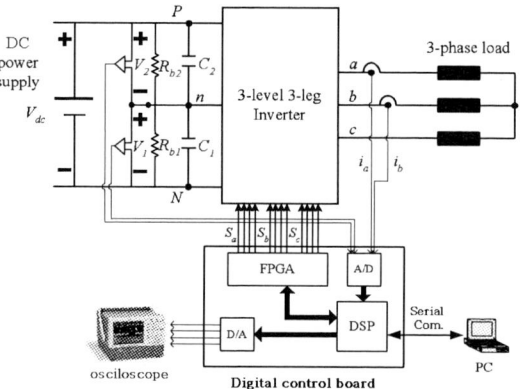

Fig.9 Experimental setups

power factor is poor about 0.26 due to the RL load (1Ω, 10mH) and the range where the system is capable of accepting the variation of neutral point potential is set to 50[V]. It can be seen in Fig.8 that the neutral point potential is maintained in the acceptable range, and the average neutral point current is negative during the balancing control. After the neutral point potential is maintained in the acceptable range, the offset voltage is modified to minimize the output distortion.

V.   EXPERIMENTAL RESULTS

Experiments have been carried out to verify the proposed PWM method. The experimental setup is shown in Fig.9, composed of prototype the 3-level 3-leg PWM inverter and digital controller. Fig.10 depicts the experimental results. Experimental conditions for the test are summarized at table I. The neutral point potential is initially unbalanced, since balancing resistors (R_{b1}, R_{b2}) are intentionally selected differently. It can be seen that the neutral point potential is stabilized by the proposed method. The pattern of the output voltages (V_{ab}) is uneven

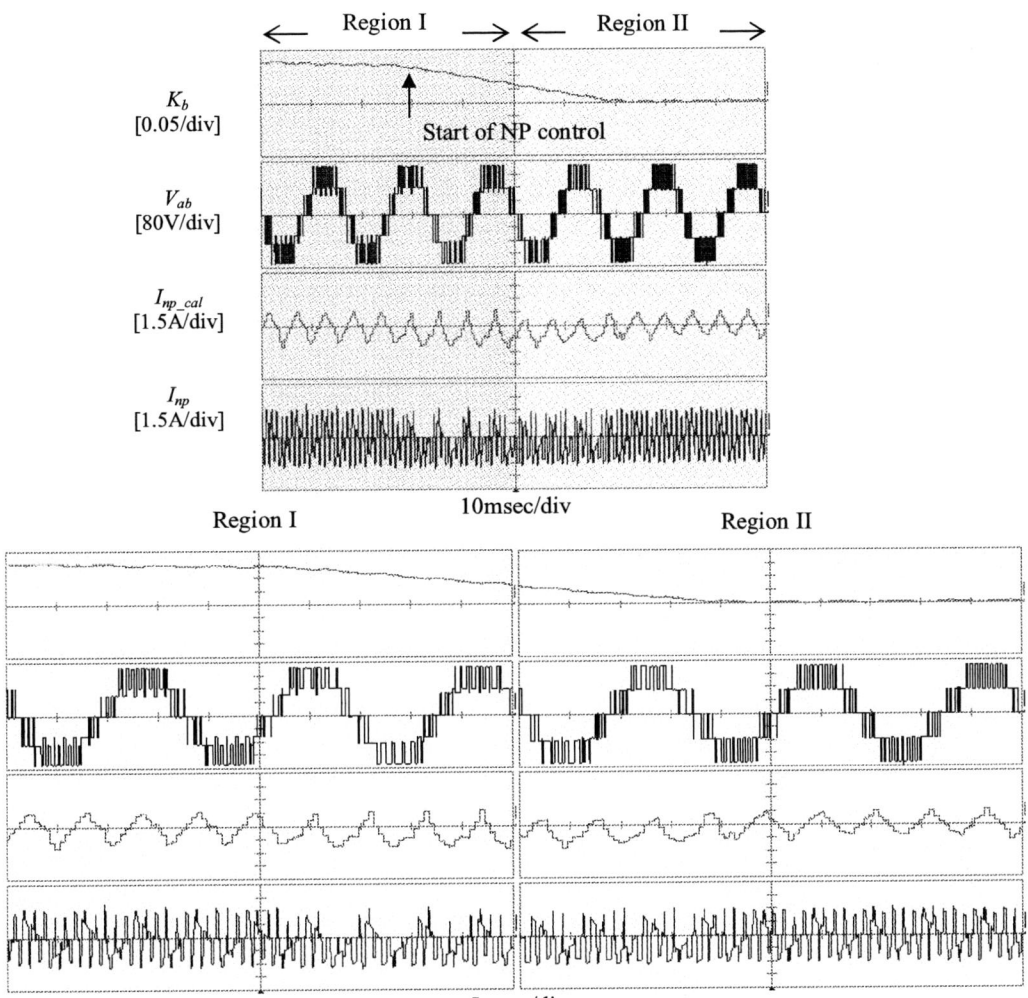

Fig. 10 Experimental results

within the region I due to the unbalance of the neutral point potential compared to that in the region II. Once the NP potential control starts, note that the average neutral point current, which is calculated by (9) and (16), becomes negative to balance the neutral point potential, reducing the magnitude of K_b. I_{np} is the measured neutral point current, which is just for monitoring.

TABLE I TEST CONDITIONS

V_{dc}	300 [V]
C_1, C_2	500 [uF]
Carrier frequency	1 [kHz]
3phase R-L load	R = 40 [Ω]
	L = 50 [mH]
R_{b1}	33[kΩ]
R_{b2}	20[kΩ]

VI. CONCULSION

This paper analyzes the relationship between the space-vector PWM and the carrier-based PWM method, and also proposes a novel carrier-based PWM strategy to balance the neutral point potential. The voltage error of the modulation caused by the unbalance of the neutral point potential is analytically described and an implementation method to reduce the unbalance is proposed. The validity of the proposed method has been verified by the simulation and experimental results.

REFERENCES

[1] N. Celanovic, and D. Boroyevich, "A comprehensive study of neutral-point voltage balancing problem in three-level neutral-point-clamped voltage source PMW Inverter" *IEEE Trans. Power Electron.*, Vol. 15, No. 2, March 2000.

[2] S. Busquets-Monge, J. Bordonau, D. Boroyevich, and S. Somavilla, "The nearest three virtual space vector PWM – A modulation for the comprehensive neutral-point balancing in the

three-level NPC inverter," *IEEE Power Electron Lett.*, Vol. 2, No. 1, March 2004.

[3] K. Yamanaka, A.M. Hava, H. Kirino, Y. Tanaka, N. Koga, and T. Kume, "A novel neutral point potential stabilization technique using the information of output current polarities and voltage vector," *IEEE Trans. Ind. Appl.*, Vol. 38, No. 6, Nov./Dec. 2002.

[4] R. M. Tallam, R. Naik, and T. A. Nondahl, "A carrier-based PWM scheme for neutral-point voltage balancing in three-level inverters," *IEEE Trans. Ind. Appl.*, Vol. 41, No. 6, Nov./Dec. 2005.

[5] N. Celanovic, and D. Boroyevich, "A fast space-vector modulation algorithm for multilevel three-phase converters," *IEEE Trans. Ind. Appl.*, Vol. 37, No. 2, March/April 2001.

[6] J. K. Steinke, "Switching frequency optimal PWM control of a three-level inverter," *IEEE Trans. Power Electron.*, Vol. 7, No. 3, July 1992.

[7] B.P. McGrath, D.G. Holmes and T. Lipo, "Optimized Space Vector Switching Sequence for Multi-level Inverters," *IEEE Trans. Power Electron.*, Vol. 18, No.6, Nov. 2003.

[8] J.-H. Kim, S.-K. Sul and P. N. Enjeti, "A Carrier-based PWM method with optimal switching sequence for a multilevel four-leg voltage source inverter," in Proc. IAS Annual meeting Vol.1, Oct. 2005.

[9] J.-S. Kim and S.-K. Sul, "A Novel Voltage Modulation Technique of the Space Vector PWM," in Conf. Rec., IPEC'95 Yokohama, pp742-747, 1995, also appeared in *IEEJ Trans.* Vol.116-D, pp820-825 1996.

[10] H.-J. Kim, H-D. Lee, and S.-K. Sul, "A New PWM Strategy for Common-Mode Voltage Reduction in Neutral-Point-Clamped Inverter-Fed AC Motor Drives," *IEEE Trans. Ind. Appl,* Vol.37, No.6, Nov./Dec., 2001.

2006 5th International Power Electronics and Motion Control Conference

AC Current Sensorless Control of Three–Phase Three-Wire PWM rectifiers under the Unbalanced Source Voltage

Jia-peng Xu , Yu-peng Tang

Department of o Electrical Engineering , Beijing Jiaotong University

Abstract—**This paper proposes an AC current sensorless control strategy for a PWM rectifier under the unbalanced source voltage. In the unbalanced system, if we use the traditional control scheme, it will yield large input current harmonics and dc-link voltage ripples due to the effect of the negative components. Meanwhile, in the ordinary control methods, at least two current sensors are needed to test the phase current, which increases the system's cost and reliability. This paper's control objective is to balance the input current and minimize the dc-link voltage ripples of the three-phase three-wire boost rectifier without the AC current sensors in the case of the unbalanced situation. The control is implemented in the positive SRF, and notch filters will be applied. The two AC current sensors will be replaced by only one current sensor, which is expected to test the dc-link current. The feasibility of the method has been confirmed through computer simulations.**
Key words: PWM rectifier   unbalanced   sensorless

NORMENCLATURE

Superscript p and n	Positive and negative sequence components
Subscript I and E	Current and voltage variables
Superscript*	Reference command
SRF	Synchronous reference frame
P	Active power
S	Reactive power
Q	Apparent power
w	Angular frequency

I. INTRODUCTION

Recently, the PWM rectifier(Fig.1) has been more and more widely used, for its distinct advantages, such as sinusoidal input currents, unity input power factor and bi-directional power flow, and lots of controlling methods have been proposed.[1]-[3]. But most of the schemes are based on the assumption of the balanced source voltage. If the source voltage is unbalanced, the control methods will be invalid. However, it's known to us that the three–phase voltage of power network may be unbalanced in practice.[4]In the unbalanced condition, the traditional way will have large input current harmonics and dc-link voltage ripples, for the existence of the negative components in the voltage. In addition,normally, the three-phase rectifiers require at least 2 current sensors to test the phase current[5],which largely increase the cost of the system , furthermore, in some cases, it's rather difficult to test the phase current.

This paper presents the AC current sensorless control for the 3-phase PWM rectifier, which can solve the problem listed above. First, the method to balance the current is studied, which is achieved by eliminating the negative components of the voltage in the positive SRF in the control algorithm. The relationship among the network voltage, the input current and the switching function is discussed. In this scheme, the input current commands are given by a set of the equations, which appear as dc in the frame. Hence, just one PI controller is needed to adjust the current. Besides, the way that how to control the rectifier without AC current sensors in the unbalanced condition will be given, as well as the flowchart of the program

Computer simulation results are provided to verify the validity of the proposed control scheme.

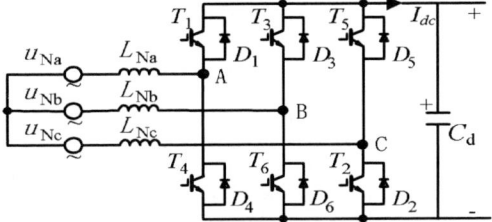

Figure1 configuration of three-phase PWM rectifier

II. ADUSTMENT OF INPUT CURRENT

We suppose in this experiment that the input voltages though may be unbalanced are sinusoidal. and no zero components exist in the network voltage,. therefore, in this paper, the voltage only includes positive sequence components and the negative sequence components,. which can be devoted in the stationary frame that

$$E_{dqs} = E_{dq}{}^{p}e^{jwt} + E_{dq}{}^{n}e^{-jwt} \qquad (2\text{-}1)$$

1-4244-0448-7/06/$25.00 ©2006 IEEE   669

where

$$E_{dq}{}^s = 2/3[E_a + E_b e^{j(2\pi/3)} + E_c e^{-j(2\pi/3)}]$$

$$E_{dq}{}^p = I_d{}^p + jI_q{}^p, \quad E_{dq}{}^n = I_d{}^n + jI_q{}^n,$$

multiplying I_{dqs} by e^{-jwt}, the equation can be devoted that

$$e^{-jwt}E_{dqs} = E_{dq}{}^p + E_{dq}{}^n e^{-2jwt} \tag{2-2}$$

Here w=50Hz.Then a 100Hz notch filter can eliminate the negative components of the voltage. The method can be shown in figure 2

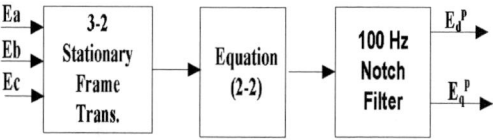

Figure2.elimination of the negative components of the voltage

After that, the frame has been converted to the positive SRF. Hence the input current will only be determined by the positive components of the voltage. However, only in the control algorithm, can the negative components of the input voltage be eliminated , while in the main circuit, the negative components of the voltage can't be removed. And the direct control object of the algorithm is the switches of the main circuit. Then the associated question is whether the input current of the main circuit can be adjusted by controlling the switches. Although the answer may be obvious, we illustrate it by taking phase a as an example. The equation can be devoted that :

$$L\frac{dI_a}{dt} + I_a R = E_a(t) - V_a(t) = E_a{}^p(t) + E_a{}^n(t) - V_{dc}(t)S_a(t) \tag{2-3}$$

where $S_a(t)$ is the switching function..
Letting $I_{dqs}(0)=0$ for convenience, we assume that for $t \ge 0$

$$I_a(t) = \frac{E_a{}^p(t) + E_a{}^n(t) - S_a(t)V_{dc}(t)}{jwL} \tag{2-4}$$

From (2-3) and (2-4),it can be known that the input current is related to the switching function, which can be controlled by the control algorithm. Hence, the input current of the main circuit can be made balanced as long as the switches are controlled properly. In the other word, the negative components of the input current can be eliminated by controlling switches.

III.DETERMINATION OF THE COMMANGD CURRENT

In the unbalanced system, the apparent power can be devoted that

$$S = P + jQ = \left(e^{jwt}E_{dq}{}^p + e^{-j\theta}E_{dq}{}^n\right)\left(\overline{e^{jwt}E_{dq}{}^p + e^{-jwt}E_{dq}{}^n}\right) \tag{3-1}$$

By solving it , the equation can be expressed that

$$\begin{cases} P(t) = P_0(t) + P_{c2}\cos(2wt) + P_{s2}\sin(2wt) \\ Q(t) = Q_0(t) + Q_{c2}\cos(2wt) + Q_{s2}\sin(2wt) \end{cases} \tag{3-2}$$

Where

P_0 the average value of the active power
P_{c2} the cosine component of the 2nd harmonic of the active power
P_{s2} the sine component of the 2nd harmonic of the active power separately
Q_0 the average value of the reactive power,
Q_{c2} the cosine component of the 2nd harmonic of the reactive power .
Q_{s2} the sine component of the 2nd harmonic of the reactive power

From equation (3-1) to (3-2), the components , $P_0, Q_0, P_{c2}, Q_{c2}, P_{s2}, Q_{s2}$, are determined by the eight components, $I_d{}^p, I_d{}^n, I_q{}^p, I_q{}^n, E_d{}^p, E_d{}^n, E_q{}^p, E_q{}^n$. Since $E_d{}^p, E_d{}^n, E_q{}^p, E_q{}^n$ are determined by the network, and in the last part, the negative components have been eliminated to balance the input current, which means both $I_d{}^n$ and $I_q{}^n$ are equal to zero , only two variables , $I_d{}^p$ and $I_q{}^p$, can be controlled. So only two components concerned with the power can be determined. Usually ,it is asked to control the P_0 to a specific value, and Q_0 to the zero. So , the equation can be devoted that

$$\begin{cases} P_0 = 1.5(E_d{}^p I_d{}^p + E_q{}^p I_q{}^p) \\ Q_0 = 1.5(E_d{}^p I_d{}^p - E_q{}^p I_q{}^p) \end{cases} \tag{3-3}$$

and that

$$\begin{cases} I_q{}^{p*} = \dfrac{2P_0{}^* E_q{}^p}{3D} \\ I_q{}^{p*} = \dfrac{2P_0{}^* E_q{}^p}{3D} \end{cases} \tag{3-4}$$

where

$$D = \sqrt{(E_d{}^p)^2 + (E_q{}^p)^2} \neq 0$$

From (3-3) and (3-4), to get the value of the command current, it is important to get the value of the P*,which is determined by the following method:.

The difference error between the dc capacitor voltage command V_d* and its actual value V_d is delivered to a PI regulator. The output of PI regulator generates the magnitude I*.Then multiplying I* by V*, the value of the P* can be determined. After that , using the equation (3-4), the value of the command current can be calculated. The method is shown in the figure 3.

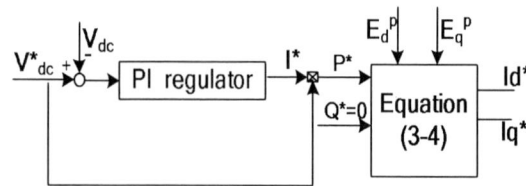

Figure3. Detailed control block diagram of the determination of the command current.

670

IV. AC CURRENT SENSORLESS CONTROL SCHEME

As talked above, the zero component of the voltage is not considered in three-phase three-wire system. So, the equations,

$$I_a + I_b + I_c = 0 \quad (4\text{-}1) \quad E_a + E_b + E_c = 0 \quad (4\text{-}2)$$

are still valid. Then, the estimator and observer used in the balanced system[6]-[7] can still be applied in this unbalanced situation, the analysis of which had been detailed.[7].

In each switching state, the value of the dc-link current is equal to one phase current accordingly. The value of phase current, hence, can be reconstructed by the value of the dc-link current. The relationship between the value of the dc-link current and that of the phase current can be shown in table 1

TABLE 1
RELATIONSHIP BETWEEN SWITCHING STATES AND THE DC-LINK CURRENT

S_a	S_b	S_c	i_{dc}
1	0	0	i_a
1	1	0	$-i_c$
0	1	0	i_b
0	1	1	$-i_a$
0	0	1	i_c
1	0	1	$-i_b$
1	1	1	0
0	0	0	0

The observer and the estimator can be expressed by the equations as follows [5]:

$$I_a(k) = I_{dc}(\tau_1) + \frac{T_2 + \dfrac{T_0}{2}}{L}[E_a - V_{an} - RI_{dc}(\tau_1)] \quad (4\text{-}3)$$

$$I_c(k) = I_{dc}(\tau_2) + \frac{T_0}{2L}[E_c - V_{cn} - RI_{dc}(\tau_2)] \quad (4\text{-}4)$$

$$I_b(k) = -(I_a + I_c) \quad (4\text{-}5)$$

where V_{an} is the averaged a-phase voltage, which is applied from τ_1 to the end of the sampling period, and V_{cn} is the averaged c-phase voltage, which is applied from τ_2 to the end of the sampling period.

The predictive state observer equation can be given as follows[5]

$$X(k+1) = F\hat{X}(k) + GU(k) + K(Y(k) - \hat{Y}(k)) \quad (4\text{-}6):$$

Where

K   the observer gain

$\wedge$ the estimated quantity.

$$F = \begin{bmatrix} e^{-\frac{R}{L}T} & e^{-wT} \\ e^{wT} & e^{-\frac{R}{L}T} \end{bmatrix}$$

$$G = \begin{bmatrix} \dfrac{1}{R}(1 - e^{-\frac{R}{L}T}) & \dfrac{1}{\omega L}(1 - e^{-wT}) \\ \dfrac{1}{\omega L}(e^{wT} - 1) & \dfrac{1}{R}(1 - e^{-\frac{R}{L}T}) \end{bmatrix}$$

To determine the switching state, the out pins of the PWM pulse of he DSP are tested by designing software. The flowchart of the program can be shown in the figure 4.

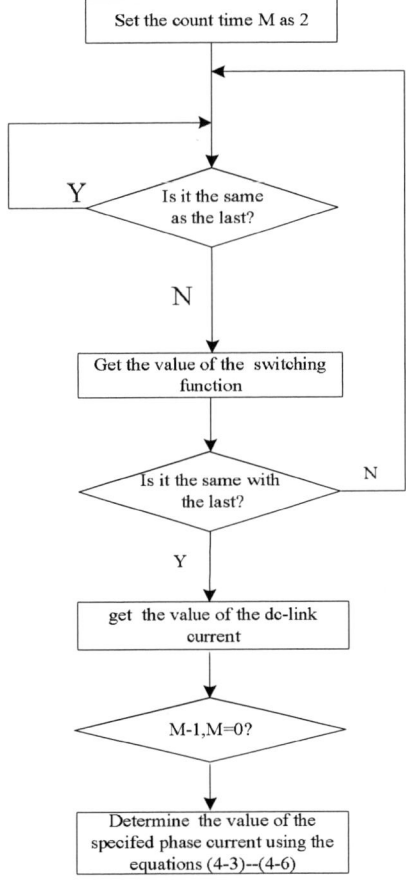

Figure4 The block of the caculation of the phase current

V. SIMULATION RESULTS

To prove the feasibility of the proposed control scheme, computer simulations were conducted by the software MATLAB.(Figure5).

In this simulation ,we used notch filters with 100-Hz notch frequency,. The output dc-link voltage is 750V, the frequency of the input fundamental wave is 50HZ and the output power is 15kw.Figure 5,shows the simulation results of the dc-link voltage, input current response, contrasted to the result of the traditional control way in the unbalance system.

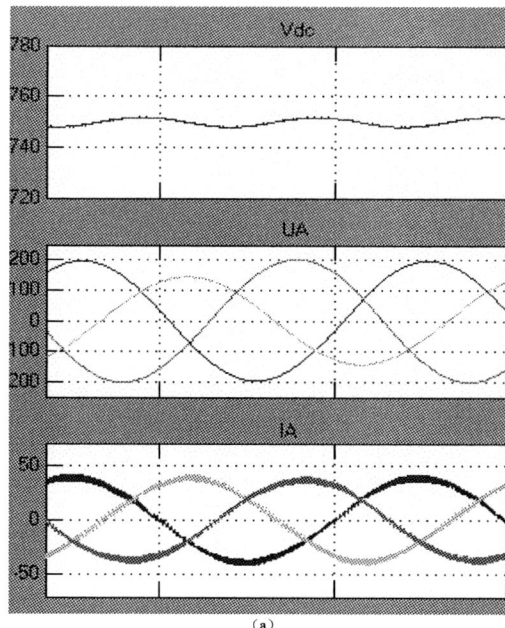

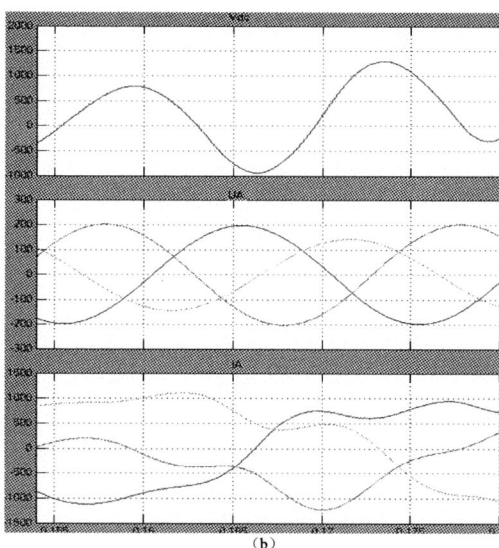

Figure 5 simulation result of the dc-link voltage, input current under the unbalanced system.(a)the proposed scheme.(b)the traditional scheme.

VI..CONCLUSION

In this paper, we proposed a control scheme for the rectifier without AC current sensors in the unbalanced system. First, the input currents were balanced by using a 100Hz notch filter to eliminate the negative component in the positive SRF, and got the current commands by a set of equations concerned with the input current and power. In addition , one current sensor is taken to test the dc-link current instead of the two AC current sensors. To solve the time delay, an estimator was

remolded, which was synthesized by a control software routine .The computer simulations proved the feasibility of the method

VII.REFERENCE

[1] Rusong Wu, Shashi B and Dewan, "A PWM AC-to-DC Converter with Fixed Switching Frequency", IEEE Transactions on Industry Applications, VOL, 26 .NO.5,Sep./Oct. 1990

[2] D.M.Brod and D.W.Novotny, "Current control of VSI-PWM inverters ,"in Conf. Rec.IEEE-IAS,1984,pp.418-425

[3] H.R.Mayer and G.Pfaff, "Direct control of induction motor currents-Design and experimental results," in Conf .Rec,IEEE-IAS,1982,PP.692-697

[4] Ana Vladan Stankovic and Thomas A.Lipo. "A Novel Control Method for Input Output Harmonic Elimination of the PWM Boost Type Rectifier Under Unbalanced Operating Conditions."IEEE Transactions On Power Electronics,VOL.16. NO.5,SEP.2001.

[5] H.Song,K.Nam, "Dual Current Control Scheme for PWM Converter Under Unbalanced Input Voltage Conditions", IEEE Transactions on Industrial Electronics,Vol.46,No.5,Oct.199.953-959. .

[6] Hansen, S., Malinowski, M., Blaabjerg, F. ,and Kazmierkowski, M.P "Sensorless control strategies for PWM rectifier."; in IEEE APEC 2000 pp832-838

[7] Dong-Choon Lee and Dae-Sik Lim, "AC Voltage and Current Sensorless Control of Three-Phase PWM Rectifiers" IEEE Transactions On Power Electronics VOL17, Issue6 , Nov..2002 PP:883-890

 **Jiapeng Xu** was born in Qingdao. He received the B.A.Sc degree in automation engineering from East China Jiaotong University, Nanchang, Jiangxi, China. He is currently studying as a postgraduate candidate in power electronics in the department of electrical engineering in Beijing Jiaotong University ,Beijing ,China. His main interest is power converter/inverter systems

2006 5th International Power Electronics and Motion Control Conference

Waveform Library Control of Converter

Xiaofeng Sun*, Bin Wang, Meng Lingjie, and Weiyang Wu

Yanshan University, Qinhuangdao City, Hebei province, China

*E-mail: sxf123@263.net

Abstract—**A novel control method which named Waveform Library Control is presented in this paper. This control method is based on the rated open loop waveforms instead of mathematical model as usual. The proposed control method is easy to be realized because it only needs to collect the typical system response waveform without the mathematical model. As an exploration for power electronic circuit waveform control, the research is important to improve the control performance and decrease the design cost. In the paper, we present two control schemes of the method, and the feasibility of the method is verified by the simulation and experiment results.**

Keywords- wave-form library, converter

I. INTRODUCTION

Most of the exited control methods of the power electronic circuit are based on mathematical model[1~3]. In general, Average signal model and small signal model is often used in controller design. But some system information, such as switch delay, measure delay, died time and effect of nonlinear magnetic cell all will be omitted in mathematical model. So there is some parameter uncertain in mathematical model. Some control theory[4~8], such as Hinf theory is tried to be used in power electronic circuit to overcome the parameter uncertain problem. But it is complex to design the controller. Repetitive control method is a kind of inner model control. Mathematical model is not need in its basic method. Repetitive control can present an excellent steady-state performance under severe nonlinear cyclic loads, but they do not present a good dynamic performance under non-periodic disturbances, such as load changed.

Not only the system time constant, gain, frequency, phase, amplitude, but also those omitted ingredients in mathematical model such as switch died time, snub circuit, switch delay, measure delay, can be embodied in system waveforms. This paper presents a control method which is based on typical waveform measuring. The control method is based on the rated open loop waveforms instead of mathematical model as usual. It is easy to be realized because it only needs to collect the typical system response waveform without the mathematical model. The control method can reduce some existing control method

This work was supported by the National Natural Science Foundation of China, NO. 50407012.

drawbacks, such as mathematical model accuracy problem for the control method based on mathematical model, complexity for some intelligent and nonlinear control such as neural networks control and fuzzy control.

II. THE WAVEFORM CONTROL

Fig 1(a) shows some step response waveforms for three buck converters. Fig 1(b) shows step response waveforms and a step load change response waveform from 3 to 6 at t=0.008. Only switch conduction resistance is different in these three converters. The parameters are:L=2.1mH, C=41.7uF, R=3ohm,Vin=28V, $d = 0.5$, conduction resistance of converter1 is 0.1ohm, conduction resistance of converter2 is 0.3 ohm, conduction resistance of converter 3 is 0.5ohm.

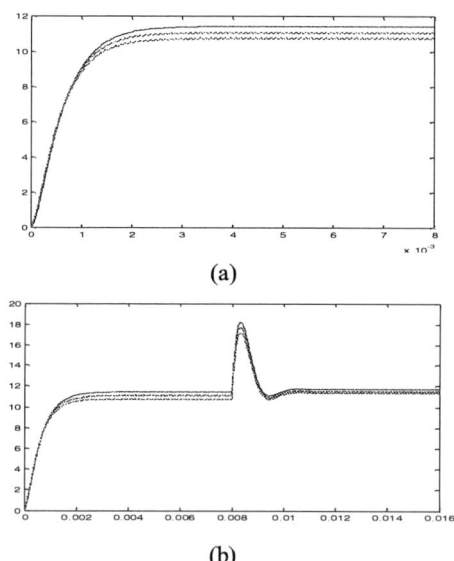

(a)

(b)

Blue: converter1; green : converter 2; red: converter 3

Fig 1.(a) Step response of the one order inertial system

(b) System response when the input is changed

From Figures parameter difference can be shown clearly in their waveforms. With the small signal model we can design control modulator. But for a boost converter the small signal model will changing with different rated ratio. The bode diagram of boost converter are shown in Fig 2.

Now try to use the sampling data to form ratio to control the converter. For a converter, a open loop step response voltage output of system at rated duty can be written as:

$$v = v_1 z^{-1} + v_2 z^{-2} + v_3 z^{-3} + \cdots \qquad (1)$$

Where: v is the output vector of system and $v_n = v(nT)$, and it is output at sample moment t=kT. A library, waveform library, can be build up by cell of the vector. Waveform library is an aggregate of waveform vectors.

Sigle signal waveform-sampling-library: A waveform library which is composed by a typical system output waveform

Muti- signal waveform-sampling- library: A waveform library which is composed by a typical system output waveform.

System characteristic can be composed in these waveforms aggregates. This paper will only propose the control scheme based on single-waveform waveform library. With the waveform library, switch ration and output voltage can be written as:

$$d(kT) = \Delta d_0 \partial(t-0) + \Delta d_1 \partial(t-1) + \cdots \qquad (2a)$$

$$v(kT) = \frac{v_k}{D}\Delta d_0 + \frac{v_{k-1}}{D}\Delta d_1 + \cdots = \sum_{i=0}^{k} \frac{v_{k-i}}{D}\Delta d_i \quad (2b)$$

Where $\Delta d(k)$ is input increment of the system at sample moment t=kT.

From (2), not only the system output can be predicted, but also control variable can be gotten if there is a reference of next sample moment.

This control scheme can be divided into two parts, too. One part is to tracking reference without considering the disturbance; the other is to decrease effect of disturbance. The controller is able to make the output of the system track the reference at the next sample moment or the time given, and to eliminate the disturbance having been detected at the next sample time (or the time given).

Duty ratio increment can be calculated by reference, converter output evaluation at t=(k+1)T and waveform library,

$$\Delta d_{k+1} = \frac{D(V_{ref}(kT+1) - V_{evl}(kT+1))}{v_0}$$

$$= \frac{D(V_{ref}(kT+1) - \sum_{i=0}^{k} v_{k+1-i}\Delta d_i)}{v_0} \qquad (3)$$

Where t=kT is currently sample moment. t=(k+1)T is next sample moment in future. $v_{ref}(k+1)$ is the reference at next sample moment. $v_{evl}(k+1)$ is the evaluation at t=(k+1)T which not considering the influence of this duty ratio increment, gotten by waveform library and duty ratio increment.

A compensating duty ratio will be given to decrease the evaluation error as (5).

$$\Delta \hat{d}_{k+1} = \frac{D(V_{evl}(kT) - V_{out})}{v_0}$$

$$= \frac{D(\sum_{i=0}^{k} v_{k-i}\Delta d_i - V_{out})}{v_0} \qquad (4)$$

The sum of the two increments is the real increment used to control the main switch.

$$\Delta d = \Delta d_{k+1} + \Delta \hat{d}_{k+1} \qquad (5)$$

The control ratio can be written as:

$$\Delta d = \frac{D(V_{ref}(kT+1) - V_{out} - \sum_{i=0}^{k}(v_{k+1-i} - v_{k-i})\Delta d_i)}{v_0}$$

$$(6)$$

As transient finished, converter output becomes stable.

$$v_{i+1} - v_i = 0 \quad \text{when } i \geq n \qquad (7)$$

So calculation in (6) can be reduced a lot. The middle part of the numerator in (6) only reflects the dynamic progress of the converter which equals one order difference or impulse response which is gotten by waveform library through some data procession, so it also is an inner model control to some extent.

III. WAVEFORM LIBRARY CONTROL IN A BUCK CONVERTER

The scheme is realized in a buck converter which is shown in Fgi.3.The buck converter parameters of the simulation is:L=1mH,C=100uF, R=3 Ohm, Vin=28v, 0<D<1. Figure 4 shows the step response of the open loop that is used to build a waveform library. Fig. 5 shows the close loop response and there is step change of the load at 0.005 second with waveform lib control. A single signal waveform library is used here.

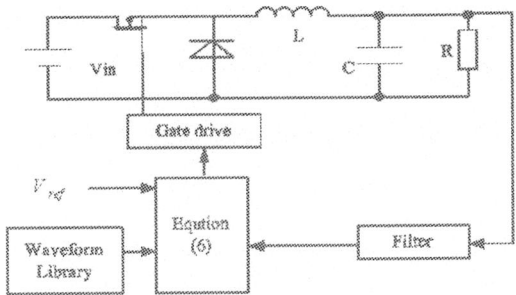

Fig. 3 Diagram of waveform library control system for a buck converter

675

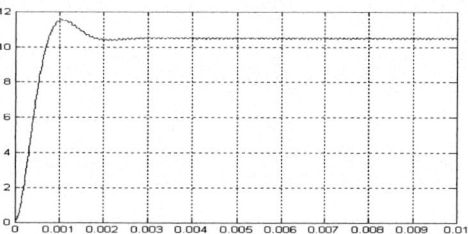

Fig. 4 Open response of the system

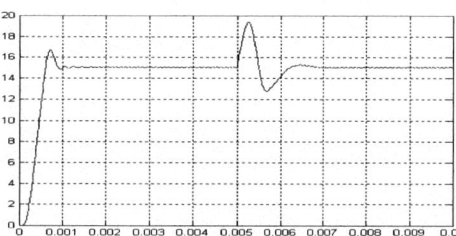

Fig. 5 Close loop response

The complete control arithmetic was implemented in a DSP with a clock speed of 40MHz and a sampling rate of 10KHz. The figures below are all experiment results. Fig.6 is a BUCK typical step response which is used to built a waveform library.

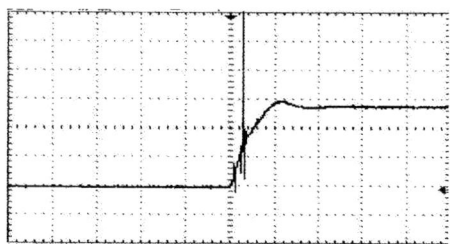

Fig.6 Typical step response

Fig.6 shows the step response waveform and Fig 7 shows the sampling data. Figure 8 show the capability of resisting disturbance. From the experimental results, waveform library control can be used in voltage control with out mathematical analysis.

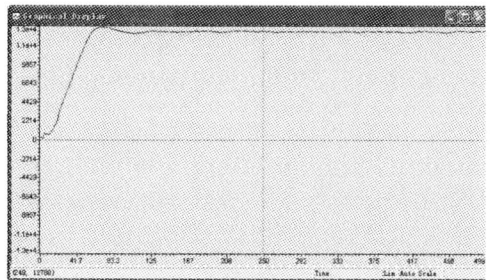

Fig. 7 The data sampled by DSP

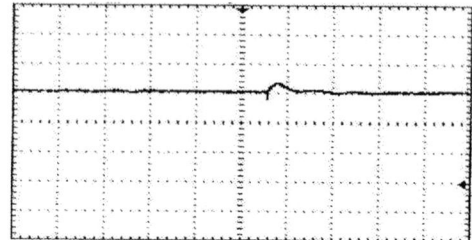

Fig8 Wave-form with a step change load

IV. WAVE-FORM LIBRARY CONTROL IN A BOOST CONVERTER

Because of the non-linear characteristic of the controlled object, the object is hard to be controlled perfectly to track the reference as well as the inner model when converter power changed. A multi library is used to enhance the control ability which is verified by a DC/DC boost converter and AC/DC three phase converter.

The boost converter waveform library control system is shown in Figure 9. The boost converter parameters of the simulation is: Vin=50v, L=3mH,R_L=1.5,Rc=0.3,R=70. and the rated output is 80v.Figure 10 shows the close loop response and there is step change of the load at 0.005 second with waveform lib control. A single signal waveform library is used here. And sample data number is 400. the laod is changed form 70 Ohm to 140 Ohm.

Figure 11 shows that reference output step change form 80 to 88v. The next curve shows system output response when the input voltage source step change form 50 to 55. From these experimental results, waveform library control can be used in boost converter voltage control with out mathematical analysis.

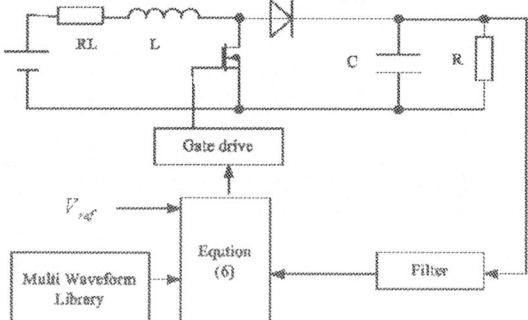

Fig. 9 Diagram of waveform library control system for a boost converter

IV. WAVE-FORM LIBRARY CONTROL IN A AC/DC CONVERTER

A three-phase ac/dc converter is presented with multi waveform library control to prove the DC voltage quality without mathematical model analyses as out loop controller which is shown in Fig10 and Fig11.

The current reference can be written as:

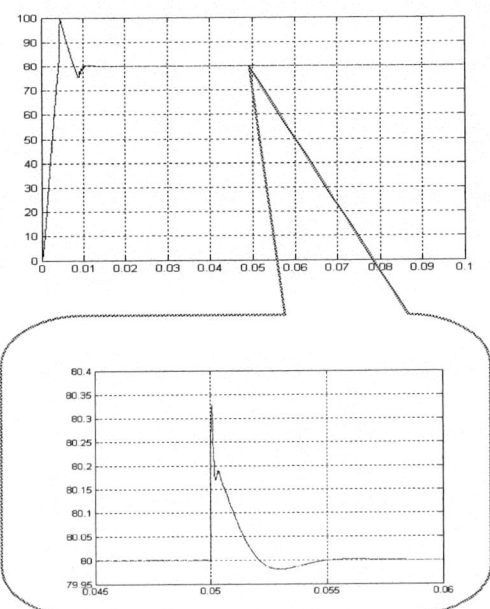

Fig. 10 Simulation waveform with a step change load

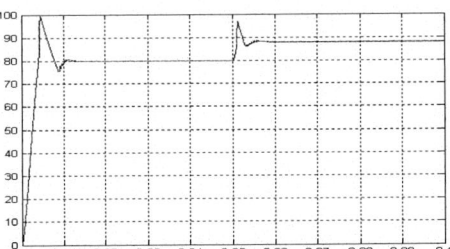

Fig. 11 Simulation waveform with a step change load

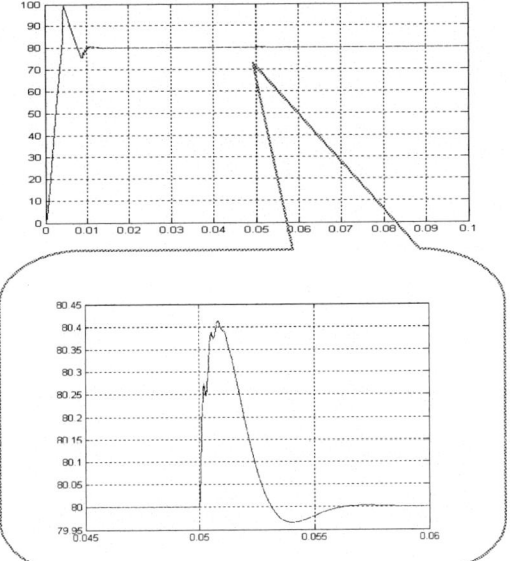

Fig. 12 Simulation waveform with a step change load

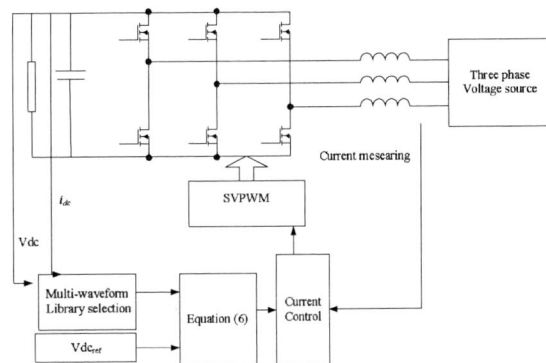

Fig.13 Diagram of control system

$$\Delta I_{rd}^* = \frac{I_u\left(U_{ref}(kT+1) - U_{out} - \sum_{i=0}^{k}(v_{k+1-i} - v_{k-i})\Delta I_i\right)}{v_0} \quad (12)$$

Fig.14-Fig.17shows the typical step response with different loads. Current step in Fig 14 is 1.27A to3.8A with a 60 Ohm load. Current step in Fig 15 is 1.27A to3.8A with a 138 Ohm load. Current step in Fig 16 is 0.63A to1.9A with a 200 Ohm load. Current step in Fig 17 is 0.63A to1.27A.

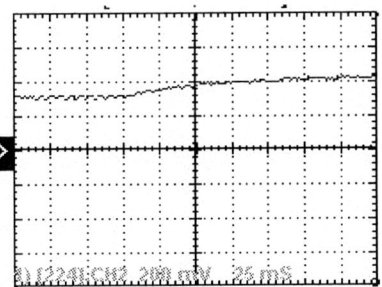

Voltage:100V/div, time:25ms/div

Fig.14 Response of DC voltage under load of 60Ω

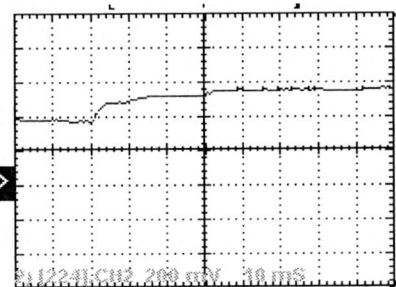

Voltage:100V/div, time:25ms/div

Fig.15 Response of DC voltage under load of 138Ω

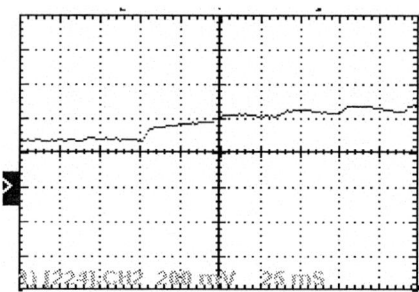

Voltage:100V/div, time:25ms/div

Fig.16 Response of DC voltage under load of 200Ω

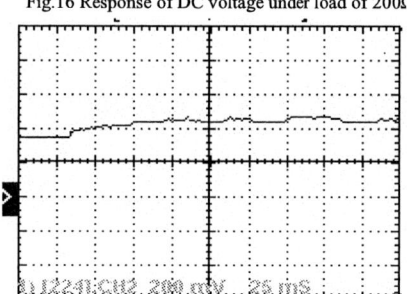

Voltage:100V/div, time:25ms/div

Fig.17 Response of DC voltage under load of 276Ω

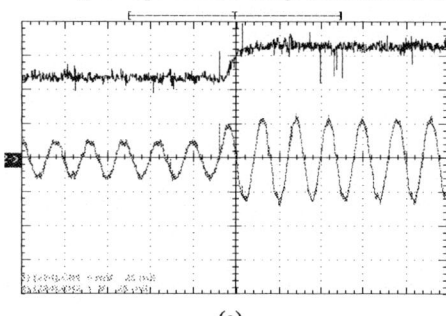

(a)

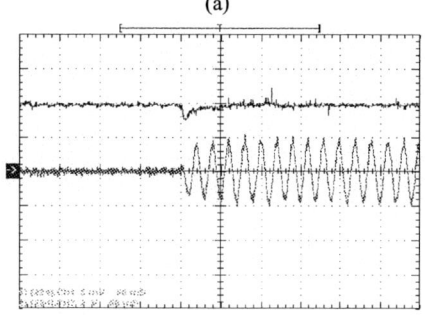

(b) time:25ms/div

From top to bottom: DC stage voltage,100V/div; Rectifier current 7A/div

Fig 18 system response waveform

Fig 18 shows the experimental results of the AC/DC converter. From top to bottom are: DC stage voltage, 100V/div; Rectifier current 7A/div. Figure 18(a) shows the DC voltage reference step response and Fig18 (b) shows the load step from no load to rated load. All the experimental results shows that the waveform lib control can control the Dc link voltage.

IV. CONCLUSION

System information and characteristic are consisted in its response waveform. The dynamic response data build the Wave-Form library. The Wave-Form Library control method is one of these controls based on dealing with the data sampled from the step response or other typical response waveforms of the system. Waveform library control can control system without mathematical model. The calculation in this scheme is less than inner model control. But this arithmetic is also limited by measure precision. Wave-Form library control has the effect of compensating poles and zeros of the controlled objects. Because the Wave-Form library is built by actual response waveform, the built inner model can reflect the poles and zeros better then mathematic model. The essential of this control scheme is a non-linear control method, the effect of the control can achieve the blur control, and the close loop control can be stable only if the open loop system is stable. And the system is under increment control, so the output can track the reference without error. Because of the difference of the control method, the effects of tracking reference and resisting disturbance are different. The effect of resisting disturbance is not as good as that of tracking, but is much better then that of normal PI control

REFERENCES

[1] Middlebrook, R. D.; Cuk, Slobodan. "Source Modeling and analysis methods for DC/DC switching converters". *Digest of Papers - Semiconductor Test Symposium,* 1977, p 90-111

[2] Li, Hui; Peng, Fang Z . "Modeling of a new ZVS bi-directional DC-DC converter", *IEEE Transactions on Aerospace and Electronic Systems,* v 40, n 1, January, 2004, p 272-283

[3] Niculescu; Elena Iancu; Eugen-Petrisor "Functional block diagram of the fourth-order PWM converters with DCM" ,*Computers and Computational Engineering in Control, Computer and Computational Engineering in Control,* 1999, p 102-106

[4] Basak, A. Demirci, R. "Adjustable position controller for PM DC linear motor". *Proceedings of the Universities Power Engineering Conference,* v 2, 1997, p 918-921

[5] Liu, Gang; Li, Sheng-Yi; Fan, Da-Peng "Studies of predictive functional control arithmetic in electromechanical servo system". *Guofang Keji Daxue Xuebao/Journal of National University of Defense Technology,* v26, n2, April, 2004, p89-93

δ-model Adaptive Algorithm Based on Plant-Parameterization

Zhao Feng[*], Liu Weiguo[*]

[*] Northwestern Polytechnical University Automation Department, Xian, China
Northwestern Polytechnical University PO box352, Xian710072, China
e-mail:1978.zhaofeng@163.com

Abstract— **Focusing on some adaptive control problems of nonlinear industrial processes under given conditions. This paper proposes a modified iterative scheme of the closed-loop identification and designs δ-model adaptive controller based on plant-parameterization. The modification enables to identify the whole plant using only one coefficient and without the necessity of reducing the order of a new plant model. Moreover, instead of using a least squares algorithm, only a simple formula for identification is used. In addition, introduction of δ-models helps to cope with numerical instabilities of discrete models occurring when a sampling interval is being shortened. Digital simulation demonstrates that the proposed algorithm brings about good control results.**

Keywords- adaptive algorithm, Youla-Kucera parameter, parameter identification, and least squares algorithm, iterative methods

I. INTRODUCTION

The nearly several for ten years, swiftly and violently develop along with the computer technology the auto-adapted control, the Robust-control and so on a series of modern control theories are widely applied to do not determine the system the non-linearity controls (Ref. [3]). However, in the actual hardware operating system (for example PC machine, labor control machine as well as programmable logical controller) carries out before the corresponding task, all algorithms must express is the separate form, therefore the system first should choose the reasonable sampling gap. In general, the parameter-sampling gap regards is controlled the process the dynamic characteristic decides. On the one hand, the system does not determine and nonlinear response request sampling gap as far as possible short, but simultaneously reduces the sampling gap possibly can cause the discrete model the numeral not to be stable. Introduced the δ model from this article, it could effectively avoid this kind of question the production. Its biggest merit is when the sampling gap reduces, delta the operator restrains also may lead, and namely sampling time infinite short time, delta the model is similar to continuously time model processing. This article unifies the actual application and the research achievement which the recent years auto-adapted control and the Robust- controlled, This article

unifies the actual application and the research achievement which the recent years auto-adapted control and the Robust-controlled, the use iteration algorithm debates knows the closed-loop system and the realization control system designs. This method based on receives controls the image parameter, therefore is called " matches the Youla-Kucera parameter "This algorithm biggest advantage is, when estimates when one new model needs not to reduce this model step, moreover, entire is controlled the object only needs the few parameters may realize debates the knowledge.

II. CONTROL ALGORITHM

A. Basic principle

Typical reaction control structure diagram like fig. 1 shows. y, u, e, r separately is controlled the output signal, the control input, the control error and the reference signal, n expresses the attachment input noise, d is the unwanted signal. Here defines P is receives controls the object the transfer function, C is the controller transfer function, namely:

$$P = \frac{N}{D} \qquad C = \frac{X}{Y} \qquad (1)$$

Previous type N, D and X, Y separately is co prime the transfer function, P, C all is stable moreover the strict real rational fraction. Through co prime, May then according to the M. Vidyasaga theory to ask to take the stable controller, it is the Diophantine equation characteristic solution:

$$D \cdot Y + N \cdot X = 1 \qquad (2)$$

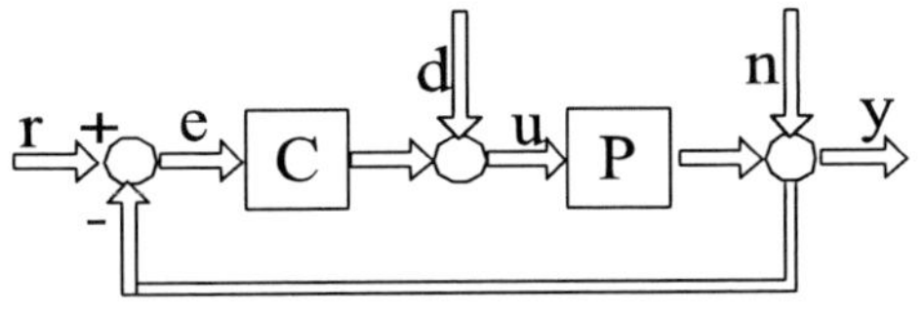

Figure 1. Model reaction control structure diagram

The maintenance is controlled the object P stables all controllers the function expression general formula is:

$$C_s = \frac{X + D \cdot S}{Y - N \cdot S} \quad (3)$$

S is the free parameter, C_s is stable moreover the strict real rational fraction, this method is called "the Youla-Kucera parameter ", it simply causes the controller design to be convenient. Solves one stable controller after this method, may find all maintenances using the antithesis question is controlled the object stable controller. Reconsiders (1) the formula describes is controlled the object and the controller points expression, conceives the Bezout, (2) restrains effectively, then possesses is controlled as follows the object the model expression description:

$$P_M = \frac{N + Y \cdot S}{D - X \cdot S} \quad (4)$$

S is the free parameter, P_M is stable moreover the strict real rational fraction, when is controlled the object by the parameter, this method is called "matches the Youla-Kucera parameter", it is similar the closed loop auto-adapted recognizes in the noise system.

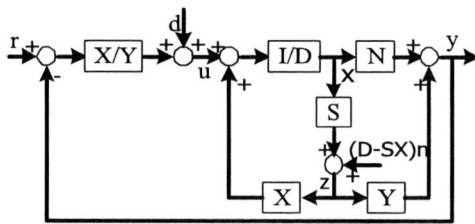

Figure 2.   Closed loop feedback structure diagram

According to the (4) choices the object model, shows with Fig. 2 the closed loop feedback structure diagram replaces Fig.2. Substitute's u, n, is easy to prove is controlled outputs y may express as follows:

$$y = P_M \cdot u + n \quad (5)$$

This article main discussion in closed-loop system, from by in noise pollution observed value u, y is recognized the real non-linearity controlled object, the essential target is the realization recognizes to parameter S but is not to PM coefficient recognize. Thin observation Fig. 2 may know, this is one standard split-ring recognizes the question, the signal x, z satisfies the following relations:

$$x = Y \cdot d + X \cdot r = Y \cdot u + X \cdot y \quad (6)$$
$$z = D \cdot y - N \cdot u \quad (7)$$
$$z = S \cdot x + (D - S \cdot X) \cdot n \quad (8)$$

The (8) formula in recognizes in the process is extremely important According to the (6) and (7), the signal x and z in the closed loop is observable. If n independently to r and d, then x and n is two mutually

independent processes, also S is stable. Therefore, parameter S recognizes is one standard split ring recognizes the question.

B.   Auto-adapted control algorithm

First time separate, the k time is controlled the object the control input by the controller C^k actuation, from this produces is controlled object P the nominal model P^k. Make P^k=P, then S=0, Therefore auto-adapted control algorithm iteration step as follows:

1. Regarding receives controls object model P^k, According to (2) computation stable state controller:

$$D^k \cdot Y^k + N^k \cdot X^k = 1 \quad (9)$$

2. The cueing signal x, the z reduction is:

$$\begin{aligned} x &= Y^k \cdot u + X^k \cdot y \\ z &= D^k \cdot y - N^k \cdot u \end{aligned} \quad (10)$$

3 . Above the signal uses in parameter S is least squares method recognizes, output error the expression is:

$$\varepsilon = z - \hat{S} \cdot x \quad (11)$$

Here, x is the input signal; z is the output signal.

4. Using (4), recognizes the model the new transfer function computation may result in:

$$P_M^{k+1} = \frac{N^k + Y^k \cdot \hat{S}}{D^k - X^k \cdot \hat{S}} \quad (12)$$

5. Make:

$$P^k = P_M^{k+1} \quad (13)$$

6. After obtains is controlled the object model new transfer function P^k, returns 1 continues the new turn of iteration computation to the step.

In fact, whether is estimated the model the parameter approaches to the real process, this very easily distinguishes from in the parameter S recording.

If this parameter coefficient restrains to 0, then recognizes the model and the real system tallies. Therefore, parameter S actual represented is estimated the model and is really controlled between the object the difference.

C.   The algorithm revises

The revision algorithm directly should use in to control recognize of the object model P_M co prime factor.

The cueing signal x may survey in the closed-loop system also is not related with the noise signal.

680

According to Fig.2, supposition n=0.Then has:

$$y = (N + Y \cdot S) = N_M \cdot x$$

$$u = (D - X \cdot S) = D_M \cdot x \qquad (14)$$

Here NM and DM is receives controls the object model (4) co prime the factor.

$$P_M = \frac{(N + Y \cdot S)}{(D - X \cdot S)} = \frac{N_M}{D_M} \qquad (15)$$

(7), (8) the formula unifies with (14) the formula, then obtains the following equality:

$$S = D \cdot N_M - N \cdot D_M \qquad (16)$$

The previous type in to the original algorithm revision process in is the very essential one step, besides the 4th step, all computations step all and the original algorithm is same.

Using the (16) recognizes the model new transfer function computation is:

$$\hat{S} = D^k \cdot N_M^{k+1} - N^k \cdot D_M^{k+1},$$
$$P_M^{k+1} = \frac{N_M^{k+1}}{D_M^{k+1}} \qquad (17)$$

D. δMethod of portrayal

When the sampling gap reduces, in order to avoid the model the digital instability, Middleton and Goodwin proposed one kind of special discrete model, its definition was:

$$\delta = \frac{z-1}{T_0} \qquad (18)$$

Here z the expression plural number z- transformation, T_0 is the sampling gap.

This model main merit is when the sampling gap reduces, the operator δ may differentiable also restrain, namely, and is easy to prove all models form may define as the following expression:

$$\gamma = \frac{z-1}{\lambda \cdot T_0 \cdot z + (1-\lambda) \cdot T_0}, \qquad (19)$$
$$0 \le \lambda \le 1$$

From this obtains delta the operation about differential coefficient arithmetic operators restrains the characteristic,

most commonly used several kinds of expressions as follows:

$$\delta = \frac{z-1}{T_0}(\lambda = 0); \sigma = \frac{1-z^{-1}}{T_0}(\lambda = 1);$$
$$\gamma = \frac{2}{T_0}\frac{z-1}{z+1}(\lambda = 0.5) \qquad (20)$$

All these models all are called "the δ model". In this article, the initial definition type (18) uses in to realize and the improvement based on receives controls the image parameter auto-adapted control algorithm.

III. CONTROL SYSTEM DESIGN AND SIMULATION

A. The controlled system

The control structure diagram of controlled the object model like Fig.3 shows, the control system is one reservoir, the input variable q_0 expression liquid current capacity, liquid horizontal plane highly h is controlled the state variable, q_1 is the liquid flow output. Therefore, the control system mathematical model is:

$$F \cdot \frac{dh}{dt} + q_1 = q_0; \quad q_1 = c_1 \cdot h + c_2 \cdot \sqrt{h} \qquad (21)$$

Among them, F is the reservoir plan area, c_1 and c_2 is the constant, regards the actual control system decides. The definition deviation variable is as follows:

$$y = h - h^s; \quad u = q_0 - q_0^s \qquad (22)$$

Here h^s is the reservoir stable state horizontal plane altitude, q_0^s is stabile state flow inputs, is actual is controlled the object the related constant as follows: $c_1 = 1.53322 \times 10^{-3} \mathrm{dm^2/s}$, $c_2 = 3.31142 \times 10^{-3}$ $\mathrm{dm^{5/2}/s}$, $F = 1.44$ $\mathrm{dm^2}$, $h^s = 1.5$ dm, $q_0^s = 0.006359$ $\mathrm{dm^3/s}$, the change scope of the liquid pours into speed is [-0.006359dm³/s, 0.004161dm3/s] 。

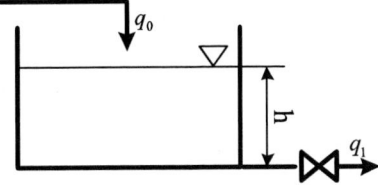

Figure 3. The controlled object model control structure

B. The control algorithm designs

The controlled object the continual system model may express as follows:

$$P_M(s) = \frac{n_0}{s + d_0} \qquad (23)$$

Among them, constant n_0, d_0 initial estimate cost gains from the control system liberalized model, in this article establishes n_0 (0) =0.6944dm^{-2}, d_0 (0) =0.0020 s^{-1}.Using the equality (2), the stable controller transfer function is:

$$C(s) = \frac{x_1 \cdot s + x_0}{s} \qquad (24)$$

x_1 and x_0 formula as follows:

$$x_1 = \frac{(2 \cdot \alpha - d_0)}{n_0}, \quad x_0 = \frac{\alpha^2}{n_0} \qquad (25)$$

Here α is the constant, uses in to guarantee the controller strong stability. Youla-Kucera parameter S uses in to the controlled system recognizes its expression as follows:

$$S(s) = \frac{s_0}{s + \alpha} \qquad (26)$$

The cueing signal of recognizes system parameter S: x, z expression is:

$$x = Y \cdot u + X \cdot y = \frac{s}{s + \alpha} \cdot u + \frac{x_1 \cdot s + x_0}{s + \alpha} \cdot y$$

$$z = D \cdot y - N \cdot u = \frac{s + d_0}{s + \alpha} \cdot y - \frac{n_0}{s + \alpha} \cdot u \qquad (27)$$

In addition, according to the (16), computation of the controlled object model coefficient equations as follows:

$$\hat{S} = D^k \cdot N_M^{k+1} - N^k \cdot D_M^{k+1} \Rightarrow$$
$$\frac{\hat{s}_0}{s + \alpha} = \frac{s + d_0^k}{s + \alpha} \cdot \frac{n_0^{k+1}}{s + \alpha} - \frac{n_0^k}{s + \alpha} \cdot \frac{s + d_0^{k+1}}{s + \alpha} \qquad (28)$$

C. Algorithm application

In order to propose the algorithm applies in the actual control system, using the (18) transforms the transfer function continual form for the separate form.

Thus, the auto-adapted control method may transform as a simpler easy programmable pattern, its concrete step as follows:

1. Regarding the (22) describes controlled the system model, substitute the formula (23) calculates to the (24) the stable controller, the concrete execution realizes by the below equation (29).

Also may obtain from the chart 2, the constant α choice may affect smoothness of the control input curve.

$$u(t) = u(k-1) + x_1 \cdot e(k) + (x_0 \cdot T_0 - x_1) \cdot e(k-1) \,(29)$$

2. Cueing signal x, z computation expression as follows:

$$x(k) = (1 - \alpha \cdot T_0) \cdot x(k-1) + u(k) - u(k-1) +$$

$$+ x_1 \cdot y(k) + (x_0 \cdot T_0 - x_1) \cdot y(k-1)$$

$$+ (d_0 \cdot T_0 - 1) \cdot y(k-1) - n_0 \cdot T_0 \cdot u(k-1). \qquad (30)$$

3. Above the signal will use in parameter S simply recognize.

$$\hat{s}_0 = \frac{z(k) + (\alpha \cdot T_0 - 1) \cdot z(k-1)}{T_0 \cdot x(k-1)} \qquad (31)$$

4. The substitution recognizes parameter s_0 and the (20), new model coefficient computation as follows:

$$n_0^{k+1} = \hat{s}_0 + n_0^k; \quad d_0^{k+1} = \frac{d_0^k \cdot n_0^{k+1} - \hat{s}_0 \cdot \alpha}{n_0^k} \qquad (32)$$

5. Substitutes these computations obtained coefficients in the new turn of iteration computation and the parameter estimate. In here constant T_0 expression sampling gap, its choice should as far as possible short, guarantee with the continual system nonlinear response tallies. In the (30), first should guarantee x (k-1) $\neq 0$.

D. Simulation result

In the simulation experiment, the establishment sampling gap is 6s, the auto-adapted control iteration cycle as T_A=10T_0=60s. The constant α originates from model coefficient d_0, decides the control system and recognizes the process the dynamic characteristic.

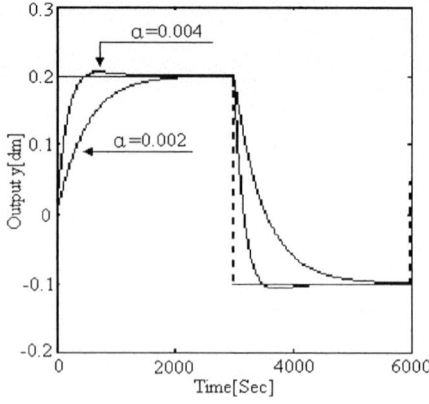

Figure 4. Controlled output signal y the response curve

Parameter S as necessary change curve Fig. 4 is the expression the controlled output signal response curve.

From the chart 4 may see influence the constant alpha to the controlled output signal y. Fig. 5 demonstrated this article designed stable controller control input response.

When constant α=0.002, 6 described Youla-Kucera which recognized the parameter as necessary change curve.

Pays attention parameter s_0 very quickly draws close to 0, this means in the very short time, the model coefficient which recognizes is approached to is really controlled the object.

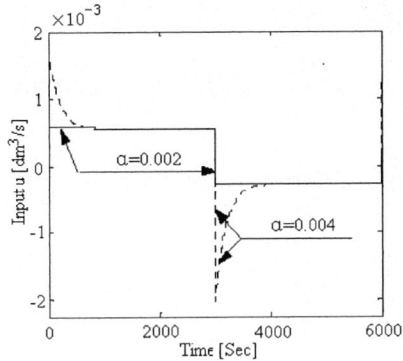

Figure 5.   The stable controller control inputs u the response

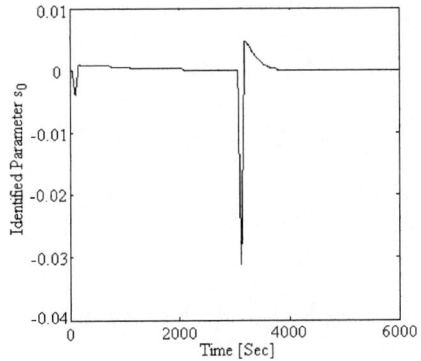

Figure 6.   Youla-Kucera which recognizes

IV. CONCLUSIONS

This article proposed one use iteration method delta model adaptive algorithm, this was one kind of effective system recognizes the method. The digital simulation result demonstrates, this method obtained the good control effect under the certain condition.

In the controlled process, only needs one coefficient and completes some simple computations, may realize entirely to be controlled the object recognizes.

Because the constant α choice may affect to the entire control and recognizes the process, therefore, α may as far as possible reduce the sampling gap through the adjustment constant but cannot initiate the entire algorithm the digital instability.

Therefore, the next step of research direction and the main duty, is in under the model coefficient initial estimate cost not definite situation, how maintains the entire the stability and the astringency of iteration algorithm.

REFERENCES

[1] Qian Kun, Dong Xinmin, Qu Zhihong. Research of Multi-level Optimization Algorithm in Large-Scale System [J]. In CD-ROM Proceedings of 11th Mediterranean Conference on Control and Automation, Rhodes, Greece. 2003. 6.

[2] R. H. Middleton and G. C. Goodwin, Digital Control and Estimation [J]. A Unified approach. Englewood Cliffs, New Jersey: Prentice Hall, 1989. S.

[3] Xiao jian . The modern control system synthesizes and designs M].Bejin: Chinese railroad publishing house. 2000.2

[4] B. D. O. Anderson, From Youla-Kucera to identification, adaptive and nonlinear control [J], Automatica, no. 34, pp. 1485–1506, 1998.

[5] S. Kozka and J. Mikles, An identification based on the Youla-Kucera parameterisation without model reduction[J], in CD-ROM Proceedings of 13rd Int. Conference on Process Control, Strbske Pleso, Slovakia, 2001.

[6] F. Gazdos and P. Dostal, Adaptive control of technological processes based on dual Youla-Kucera patameterization[J], in Proc. IFAC Workshop on Adaptation and Learning in Control and Signal Processing, Cernobbio-Como, Italy, 2001, pp. 467–471. M.

[7] M. Vidyasagar, Control Systems Synthesis: A factorization Approach [M]. Cambridge, MA: MIT Press, 1985.

Dynamics and Control of Electronic Cascaded Systems

Wen Wei[*,**], Xu Haiping[*], Wen Xuhui[*], Shi Wenqing[*,**]
[*] Institute of Electrical Engineering, Chinese Academy of Sciences, Beijing, China
[**] Graduate School of Chinese Academy of Science

Abstract—Electronic Cascaded Systems are complex and extensively systems integrated with different converters. In general, there are two kinds of loads in these systems, constant power loads and constant voltage loads. The potential instability effected by constant power loads is analyzed with the power system of Electric Vehicle(EV) by Impedance Ratio Criterion. The stability qualifications are given in this paper. Furthermore, a nonlinear control method is described to stable the systems with complex loads. The control effects are proved by Simulations.

Keywords-Electronic cascaded system, constant power load, impedance ratio criterion, one-circle control

I. INTRODUCTION

Electronic Cascaded systems are complex systems integrated by different power electronic converters. This topology has the advantage of flexibility, augment ability and redundancy. In recent years, such design are wide used in automotive power systems, Electric and Hybrid Electric Vehicles, advanced industrial electrical systems, telecommunications, terrestrial computer systems, International Space Station, spacecraft and More Electric Ship power systems and so on. Usually each module is designed based on its own stand-alone operation. As a result, after system integration, the interaction between subsystems may cause performance degradation, and even system instability.

Power electronic converters, when tightly regulated, behave as constant power loads. [1][2] Constant power Loads have negative impedance characteristic. This destabilizing effect for the system is known as negative impedance instability. Classical linear control methods, which are often used to design controllers for DC/DC converters, have stability limitations around the operating points. Therefore, new stabilizing control methods are required to ensure system stability.

There are two kinds of typical loads in these systems. One is constant power load and the other is constant Voltage load. In this paper, the power system of Electric Vehicles is analyzed by Impedance Ratio Criterion. The stability qualification and some improve methods are given, then a nonlinear controller for Negative impedance is designed. The stability improvement is obtained by Nyquist criterion. Simulations prove the conclusions

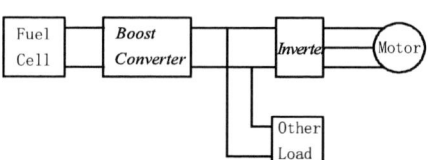

Figure1. Power System of EV

II. POWER SYSTEM OF EV AND CONSTANT POWER LOAD CHARACTERISTIC

Fig.1 shows the power system of Electric Vehicles. The Interface of the cascaded system is very important for the performance of the integral system, thus will be the key point of the research.

The motor drive controller is to keep constant speed and the motor has one-to-one torque-speed characteristic, namely, the motor output $P = T\omega$ of each operating point is a constant power. So the inverter system is also a constant power load. Besides, DC/DC converter which feeds an electric load and tightly regulates the voltage when the electric load has one-to-one voltage current characteristic is also a constant power load. According to the DC/DC converter in this system, there are mainly two types of loads: constant voltage loads and constant power loads.

To get the constant input power when there is small disturbance in system.

$$(V + \Delta u)(I + \Delta i) = VI \qquad (1)$$

Ignoring the higher infinitesimal:

$$\Delta v / \Delta i = -V / I \qquad (2)$$

As shown in Fig 2, the instantaneous resistance of the load is $R = V / I > 0$, while the small signal resistance is $\Delta r = \Delta v / \Delta i = -R < 0$. The negative impedance char-

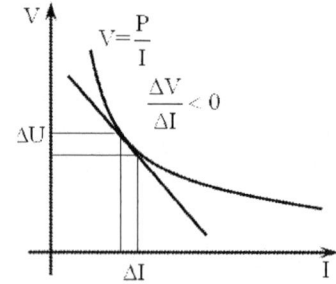

Figure 2. Constant Power Load Characteristic

acteristic might have impact on the power quality and the system stability. [1][2]

III. IMPEDANCE RATIO CRITERION

Figure 3 shows a source and a load subsystem connected together. Each subsystem is assumed to be stable and well designed for stand-alone operation. While they are integrated, the total system transfer function is given by (3). [3][6]

$$F = V_{o2} / V_{i1} = GH / (1 + T_m) \quad (3)$$

Where $T_m = Z_o / Z_{in}$

The term $(1 + T_m)$ represents the loading effect caused by integrating the two subsystems. So poles of (3) must have negative real part to get system stability. The term T_m can be viewed as the system equivalent loop gain and can be used to determine system stability. The integrated system stability can be determined by checking if the Nyquist plot encircles (-1,0) point.

IV. NEGATIVE IMPEDANCE INSTABILITY ANALYSIS

There might be instability in such system because not meet the impedance ratio criteria. The boost converter operating in continuous conduction mode and the equivalent small signal circuit of the power system is depicted in Fig.4. [3][4]

A. Cascaded system

Equivalent constant power and constant voltage loads of the Boost converter represented by P and R, respectively. Using the impedance ratio criteria, it is necessary to calculate output impedance of the source module and input impedance of the load system

$$Z_o = sL_e / (s^2 L_e C + 1) \quad (4)$$

$$Z_{AC} = V_o^2 / (V_o^2 / R - P)$$
$$Z_{DC} = V_o^2 / (P + V_o^2 / R) \quad (5)$$

Z_{AC} is input impedance and Z_{DC} is instantaneous impedance. So the term:

$$1 + T_m = \frac{s^2 L_e C + sL_e (1/R - P/V_o^2) + 1}{s^2 L_e C + 1} \quad (6)$$

According to the impedance ratio criterion, if the zero of the equation (6) has a negative real part, the system is stable. Then can get (7) which is the same as the conclusion of the conventional analytical method. This shows that the applicability of the impedance analytical method.

$$1/RC - P/CV_o^2 > 0 \Rightarrow P < V_o^2 / R \quad (7)$$

Meeting (7) can be regarded as that the equivalent input impedance of the load subsystem is positive, which is

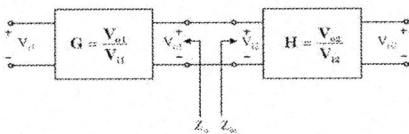

Figure 3. Two section cascaded subsystems

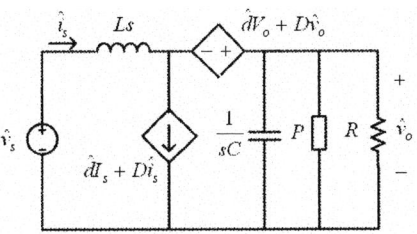

Figure 4. Small signal model of Boost Converter

similar to the situation with resistance loads. With this condition, using conventional PI control method can keep the system steady and provide constant output power and voltage. Otherwise, the system would present instability

B. Distribution system

In a distribution system with the damp of the Boost converter, the stability margin is improved. The new $(1 + T_m)$ can be calculated as (8).

$$1 + T_m = \frac{s^2 L_e C + sL_e k_1 + k_2}{s^2 L_e C + sr_e C + 1} \quad (8)$$

Where $L_e = L / D_0^2$, $r_e = r / D_0^2$,

$k_1 = r e C + L_e / R - L_e P / V_o^2$, $\quad k_2 = 1 + r_e / R - r_e P / V_o^2$

Since r_e is very small, generally, the constraint $k_2 > 0$ is satisfied. Therefore, $k_1 > 0$ is the most important constraint. The effect of the distribution system and its resistance is shown in the first term at the left hand side of (9). This is an additional term compared to (7). It is related to the distribution system, filters, and the DC/DC converter parameters as well as the output nominal voltage.

$$k_1 > 0 \Rightarrow rCV_o^2 / L + V_o^2 / R > P \quad (9)$$

In order to stabilize the system, based on the relation (9), there are four different possible solutions. First one is Increasing r_e which is not practical due to the power loss. The second one is decreasing L_e which is not feasible also. However, in the design stage of the system, it should be considered to have minimum inductance as possible. The third method is increasing C which is easily possible by adding a filter. In fact, adding passive filters is the solution, which has been proposed to maintain the stability of the system. Last method is increasing V_o In most cases, there is no control over the nominal voltages of the system. However, it shows systems with higher base voltages have better negative impedance stability than systems with low base voltage

Some solutions have been tempted to solve the instability problem such as designing media filter and changing input impedance of load block. This paper advised a nonlinear control method to stabilize the system.

V. NEGATIVE RESISTEANCE STABILITY CONTROLL METHOD

One-cycle control [5], which has been reported in recent years, is a nonlinear control method. Using this

685

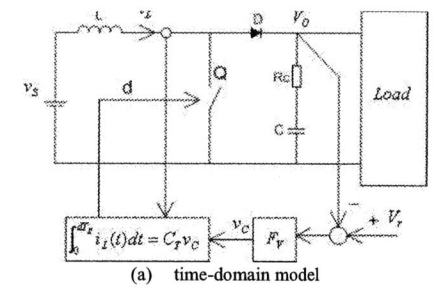

(a)   time-domain model

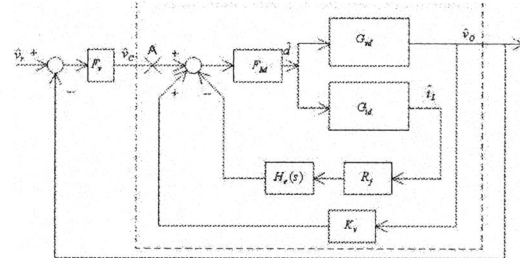

(b) small signal analysis

Figure 5. Model of One-cycle Control

method, the perturbations of the load and the line changes can be eliminated rapidly, theoretically in one switch cycle.

The active switch in the power stage turns on at the beginning of each cycle, and the switch current is integrated to obtain its total charge. When the voltage across C_T reaches the control voltage v_C, the power stage turns off, and the switch across capacitor turns on to discharge C_T. It should be totally discharged before the next switching cycle starts. Hence the total charge of the switch current in one cycle, which is proportional to the average value of the switch current, is controlled.

The dissipation resistance of output capacitor R_C is considered to analyze the system's transfer function G_{vd} and G_{id}.

$$G_{vd} = \frac{\hat{v}_o}{\hat{d}} = \frac{V_s(1+sCR_c)(1-sL_e/Z_{DC})}{D_0^2\left(1+s/Q\omega_0+s^2/\omega_0^2\right)} \quad (10)$$

$$G_{id} = \frac{\hat{i}_L}{\hat{d}} = \frac{V_oCs+I_o\left(1-Z_{DC}/Z_{AC}\right)}{D_0^2\left(1+s/Q\omega_0+s^2/\omega_0^2\right)} \quad (11)$$

Where $Q = Z_{AC}/\left[\omega_0\left(Z_{AC}CR_c-L_e\right)\right]$ $\omega_0 = 1/\sqrt{L_eC}$

Analyzing the small signal model of the system as Fig.5(b), F_m is the PWM modulator gain, and R_j, $H_e(s)$ represent the inductor current feedback. R_j is the equivalent current gain and $H_e(s)$ represents the sampled-and-hold effect. K_r represent the effect of the perturbation of output voltage on the duty-cycle.

$$H_e(s) = 1+s/\omega_n Q_n+s^2/\omega_n^2 \quad (12)$$

$$R_j = I_{LP}L/C_TV_s \quad (13)$$

$$F_m = C_T/I_{LP}T_s \quad (14)$$

$$K_r = R_jD_0^2T_s/2L \quad (15)$$

Where $\omega_n = \pi/T_s$, $Q_n = -2/\pi$, $C_T = I_LDT_s/v_C$

From Fig.5, it is easy to get the input to output function for the current loop.

$$G_{oc} = \frac{\hat{v}_O}{\hat{v}_C} = \frac{F_MG_{vd}}{1+R_jH_e(s)F_MG_{id}-K_rF_MG_{vd}} \quad (16)$$

Applying (12)-(15) to (16), and using some approximations and defining M as:

$$M = T_s/2L_e+(Z_{AC}-Z_{DC})/\left(D_0Z_{DC}Z_{AC}\right) \quad (17)$$

$$G_{oc} \approx K\left(\frac{1+sCR_c}{1+s/\omega_{pL}}\right)\frac{(1-sL_e/Z_{DC})}{1+s/\omega_HQ_H+s^2/\omega_H^2} \quad (18)$$

Where $K = 1/R_jM$, $\omega_{pL} = MD_0/C$, $\omega_H = \pi/T_s$, $Q_H = -1/\left[\pi(D_0-0.5)\right]$

Design voltage compensator as:

$$F_V = k_m(1+s/\omega_z)/\left[s(1+s/\omega_p)\right] \quad (19)$$

where $1/\omega_p = L_e/R_{DC}$, $\omega_z < 1/\sqrt{L_eC}$

According to Fig.6, the close-loop output impedance of the converter can be expressed by:

$$Z_o = Z_{oo}/\left(1+F_vG_{oc}H\right) \quad (20)$$

To express the operation point, it is necessary to see the load as constant current, namely, let $Z_{AC} = \infty$. This condition brings great convenience to the practical applications and the output impedance can be calculated only with the output power. So can rapidly determine the stability of cascade system with various loads using the impedance ratio criteria, as well as design the controller parameters. Since this method can solve the problem that conventional methods depend on the load characteristic, which is now wide used in stability analysis of distribute power system.

Example: $V_S = 250V$, $V_O = 400V$, $P_{cpl} = 20kW$, $v_C = 5V$, $T_s = 50\mu s$, $P_{cvl} = 10kW$, $L = 300\mu H$, $C = 1500\mu F$, $R_C = 0.02\Omega$.

Design $1/\omega_z = 1.1\times10^{-3}$, $1/\omega_p = 1.44\times10^{-4}$. When $485 < k_m < 5250$, the Nyquist plot of T_m don't encircle (-1,0), the system's poles have negative real part. If $k_m = 2300$, the controller can impact the output impedance as Fig.7. It reduces the amplitude of output impedance at low and middle frequency with a phase change about $90°$. The Nyquist plot of T_m shows the designed system is stable from Fig.8.

Injecting a identify step perturbation $\hat{i}_o$ at $t = 0.1s$ on the dc bus, the dynamics of bus voltage depict in Fig.9. The result shows the system have enough stability margin.

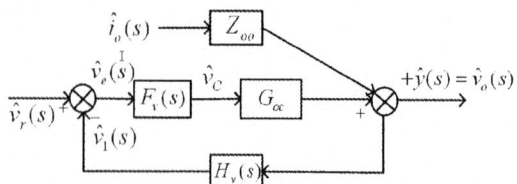

Figure 6. Close-loop output impedance of converter

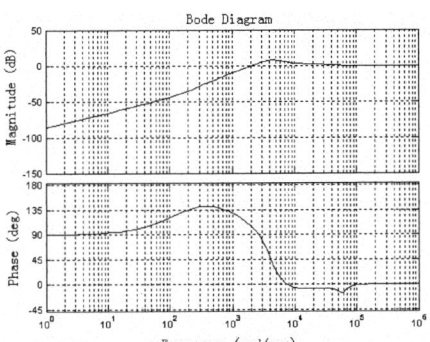

Figure 7. Effect of control to output impedance

VI. CONCLUSION

In this paper, electronic cascaded system's instability effected by constant power loads is analyzed by

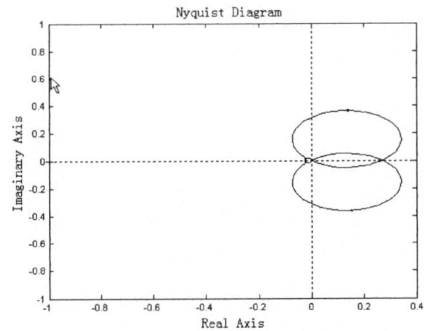

Figure 8. Nyquist plot of T_m when $k = 2300$

impedance ratio criterion with power system of Electric Vehicle. It is illustrated that the stability is relative to the input impedance of equivalent load. Furthermore, the necessary and sufficient conditions of stability for such systems are explained. Through small signal analysis, guidelines of stabilizing such systems are presented. In addition, one-cycle control method is used to stabilize the system with the mixed load. Nyquist plot of T_m shows the stability and simulations prove the conclusions

REFERENCES

[1] A. Emadi, "Modeling, analysis, and stability assessment of multi-converter power electronic system," Ph. D. Dissertation, Texas A&M University, College Station, TX,Aug.2000.

[2] V. Grigore, J. Hatonen, J. Kyyra, and T. Suntio, "dynamics of a Buck Converter with a Constant Power Loads" in IEEE Power Electronics Specialists Conference Record, vol.1, pp. 72-78, 1998.

[3] L. R. Lewis, B. H. Cho, F. C. Lee and B. A. Carpenter, "Modeling and Analysis of Distribute Power Systems," PESC 1989.

[4] R. D. Milddlebrook, and Slobodan Cuk, "A general unified approach to modeling switching-convert power stages," in IEEE PESC, june 8-10 1976, pp. 18-34

[5] Wei Tang, F. C. Lee, R. B. Ridley, and Isaac Cohen, "Charge control: modeling, analysis, and design," in IEEE Transactions on Power Electronics, vol.8, no.4, October 1993.

[6] R. D. Middlebrook, "Input Filter Consideration in Design and Application of Switching Regulators", Proc. IEEE Industrial Application Society Annual Meeting, 1976.

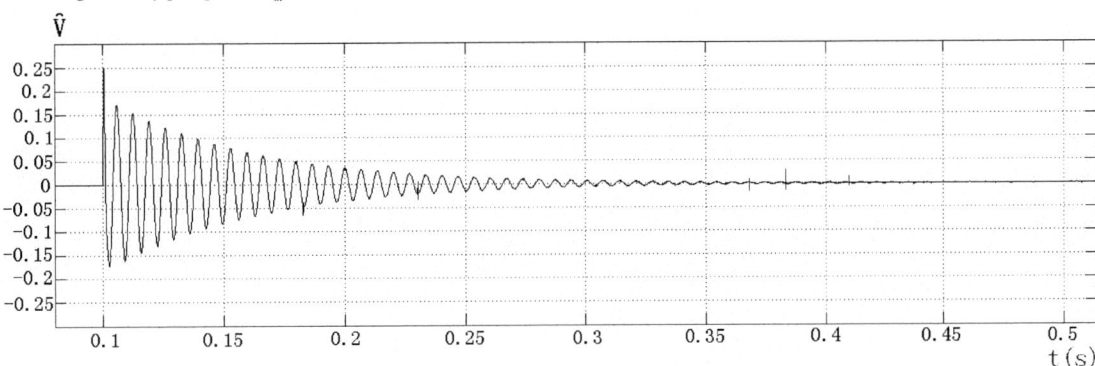

Figure 9. Dynamics of bus voltage after perturbation

2006 5th International Power Electronics and Motion Control Conference

The Controlling Strategy for Electronic Ballast of HID Lamps

Weiping Zhang, Xiaohan Guan, Xusen Zhao, Hongtao Li and Zhengang Liu
North China University of Technology/Lab of Green Power & Energy System, Beijing, China
Email: zwp@ncut.edu.cn

Abstract—Based on the electric characteristics of HID (High-Intensity-Discharge) lamps, novel control strategy and implement method for electronic ballasts of HID lamps have been investigated in this paper. The main contributions are as the followings: (1) To ensure to ignite the HID lamps safely, the auto-tracking technique, which can offer proper igniting voltage, has been put forward. (2) To make sure to warm up the HID lamps safely, the start control regulation has been proposed, and the principle of this strategy is that the maximum current must be limited in the start moment, once the lamp starts conducting, the current through lamp should be exponential or approximatively exponential to decrease, and dissipation power should be increased as time is increasing . (3) In steady stage, HID lamps should operate in constant power state. Maintaining the product of the voltage and current of the lamp tracking to equal the rating power, a constant-power-control technique, which takes the current as the main regulation variable, has been proposed in this stage.

Keywords- HID lamps; electric characteristics; control strategy; closed-loop feedback

I. INTRODUCTION

The HID lamps have negative incremental impedance characteristics. As a result, gas discharge lamps cannot be directly connected to a voltage source. Certain impedance must be placed between the discharge lamp and the voltage source as a means to limit lamp current. Therefore, a current-limiting device called ballast is necessary to ensure lamp's stable operation. The first generation ballast, called electromagnetic ballast, is just an inductor virtually. Thanking to mature technology, this conventional type of ballast perform dependable stability, and the output maximum power can get to 24KW, but this ballast which operates at a low frequency (50 or 60 Hz) is not the most practical solution due to its large size, heavy weight, low power factor, high no-load loss, consumed cost of iron and copper, and low efficiency. The second generation ballast, belonged to electric ballast, is based on switching PWM converter, it adopts three-level hiberarchy combined with PFC、DC/DC converter and DC/AC inverter, operates in low frequency(less than 300Hz) and the input power is constant. Its merits include

Project supported by Beijing Natural Science foundation (No. 4052011)

no acoustic resonance, high power level (about 24KW) and no frequency-flickering. However, this ballast have some innate disadvantages: overmany sects of circuit, complicated topology structure, high product costs, low efficiency (about 85%) and high EMI, so it is only suitable for some special lighting applications(such as cinema lighting and military airfield lighting), in addition, it is hard to bring about industrialization.

The third generation ballast is high-frequency electronic one. It is based on the principles of resonant converter, consists of PFC and DC/AC high-frequency resonant converter—two-level structure, and utilize high-frequency sine wave(50kHz~150kHz) to supply power. Its block diagram is as showed in Fig.1, where,

(1) Controller① adopts PWM control mode, the PFC circuit can work in the mode of both CCM and DCM. When the input voltage of PFC circuit range from 90V to 270V, it is realized that the output DC voltage is fixed at 400V;

(2)DC/AC resonant inverter provides high-frequency sine signal to drive the lamp, the controlling parameter of controller② is a current sampling, and the constant-power control in steady state can be achieved by high-frequency regulation.

(3)When the high voltage trigger is started up, it can provide very high voltage to ignite the gas inside of lamp to ionize and induce arc discharge. There is commonly no special high-voltage trigger in the medium and low level power electronic ballasts, the igniting-voltage will be provided by DC/AC resonant inverter in the process of start-up.

The research about third generation ballast has regarded as one of the study hotspot at home and abroad. The main research outcome reported are as followings: (a)the maximum output power is 400W;(b)the maximum efficiency is 93%;(c)the luminous efficiency is improved by 10-20%;(d)the power density is doubled compared with the low-frequency electronic ballast. In conclusion, the third generation ballast can almost overcome all the

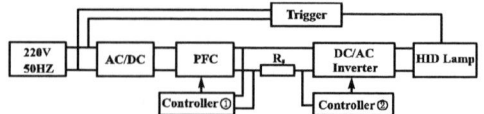

Figure 1. Block diagram of high-frequency resonant inverter system for HID lamp

shortages of the first and second generation ballasts, but it brings new flaws: hard to maintain the stability of system working, can not eradicate acoustic resonance, attaching many difficulties in industrialization.

The strategy of constant-power control is adopted in the third generation ballast, as showed in Fig.1. The control signal of controller② is sampling from the input current of DC/AC inverter. Because the output DC voltage of PFC circuit is constant, the output power of DC/AC inverter will be also invariable if the input quantity of DC/AC resonant inverter is changeless. Therefore, the constant-power control can be realized. The control strategy can be suitable only when the HID lamp is in the steady state. The core of the control circuit is an integrator. The characteristics of it are as followings: (a) high static-control precision;(b)slow response speed;(c)difficult to be up to particular request of HID lamp in various operation condition;(d)low-frequency oscillation occurring probably. Therefore, it is needed that special control strategy accorded to different operation condition of HID lamp must be study to make the device work reliably, no matter what the HID lamp is in any one condition.

The low-frequency electrical characteristics of MH (Metal Halide) lamp is presented in Fig.2, where, the lamp produced by OSRAM with rating power 2500W. Observing Fig.2 reveals the following information: (1) The electrical characteristics of MH lamps are felicitously divided to warm-up stage and steady stage. (2) MH lamps behave time-varying nonlinear electrical characteristics. (3)MH lamps can be regarded as a quasi-steady-state in warm-up stage, because the period of warm-up is correspondingly longer and no saltation.

Based on the low-frequency characteristics of MH lamp shown in Fig.2, both working principle of HM lamp and electrical characteristics of ballast would be investigated in this paper. In order to better study the behaviors of lamp, according to Fig.2, the electrical characteristics of MH lamp are characterized by three distinct periods: the pre-ignition period (t<0s), the warm-up period (0s<t<4s) and the steady-state period (t>4s).

The electrical characteristics of every period and the corresponding control strategies will be studied as following.

II. Control Strategy of Operation Periods

In pre-ignition period, lamp's equivalent resistance is as big as opening, therefore the voltage endured by lamp is just the output voltage of ballast, and both current and power of lamp are zero. For a given lamp, its igniting-voltage is a relatively fixed value, but in fact, the igniting voltage cannot keep a constant in different ambient, so, the output voltage of ballast must be rectified with the change of lamp's operation condition, namely, the ballast must be able to offer auto-tracking igniting voltage [1].

The strategy presented in this paper is auto scanning and gradually increasing the output voltage until that the igniting-voltage supplied by ballast is equal to or a little higher than the lamp' s igniting-voltage, this method is short for auto-tracking to offer igniting-voltage technique in this paper.

As shown in Fig 3, at the moment of HID lamp is ignited, the electrical characteristics present two obvious traits. First, the voltage across the lamp fall from the ignition voltage (very high voltage) to 20%～30% of the rating operation voltage instantaneously; second, the current carried by the lamp rise from 0 to 140%～150% of the rating current rapidly. The co-operation of the two traits mentioned above can be used to judge whether the HID lamp is ignited.

During warm up, with conducting-channel's diameter dwindling and the carrier concentration of channel falling, the voltage cross the lamp will escalate, the current through the lamp will step down, and the lamp's equivalent resistance will increase correspondingly. In addition, the dissipation power of lamp will be continuously raised. In order to maintain the arc discharge as well as to evenly raise the temperature of lamp's tube, the proper control strategy of start-up period must be studied [1].

The control strategy of warm-up period presented in this paper is as following: to limit lamp's maximum current in the moment of start-up, to decrease the current of lamp gradually with exponential or approximatively exponential law, and increase the lamp's dissipation power. The practical means are that the lamp's current is supposed to be restricted at 140%~150% of rating value at the moment of lamp' s ignition, simultaneously,

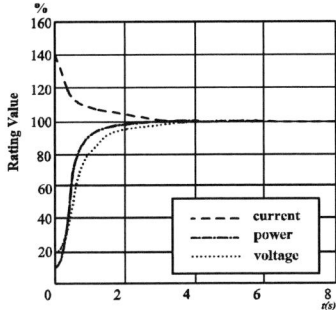

Figure 2. Typical electrical characteristics of HID lamp
(2500W, OSRAM)

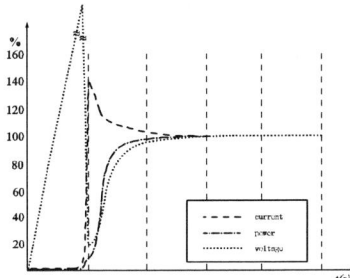

Figure 3. Electrical characteristics of HID lamp's igniting process

confine the dissipation power to 20%~30% of rating value, and after some time, the steady state will be reached by degrees.

When the lamp's performance enters steady state, the gas ionization velocity is still slightly higher than recombination one, so there has enough carriers inside of lamp, but the quantity of carrier transferred electric energy is decided by the ballast. However, it is not able to realize constant-power control only by to stabilify the current of lamp, because that the voltage of lamp is not a constant value even if the current through the lamp is stable. Therefore, a novel constant-power control strategy must be adopted in steady state [1].

The technique of constant-power control presented in this paper is that to take the current as the main regulation parameter and keep the product of the lamp's voltage and the current carried by lamp tracking to equal the rating power. That is to say, the regulation of lamp's current is primary to implement steady power output when the lamp's power fluctuated.

III. ACTUALIZATION OF CONTROL STRATEGIES

The block diagram of control circuit is shown in Fig. 4. The main circuit is a DC/AC resonate inverter, the variable-frequency control need be adapted to drive it. In Fig.4, VFO is just a voltage-controlled oscillator, which output frequency is in direct ratio to the input voltage. Inside of T trigger, the VFO's output signals frequency divided by two and produce two-channel reciprocal signals-- Q and $\overline{Q}$, which are amplified by driver 1 and driver 2 to control the two switches of main circuit. The processor of the lamp's voltage is comprised of voltage sampling unit, rectifier and filter, its input is the lamp's voltage U_{lamp} --high-frequency AC signal, and the output voltage U_{01}, being in direct ratio to root-mean-square value of U_{lamp}, is approximative to DC voltage. The processor of the lamp's current consists of current sampling unit, rectifier and filter. Its input is the lamp's current I_{lamp}-- high-frequency AC signal, and the output voltage U_{o2} is in direct ratio to root-mean-square value of I_{lamp}.

The A controller has two-channel input signals--U_{01} and U_{o2}, and the corresponding output are P_1 and P_2. At the moment of boot-strap, the switch S is set to location①. When a negative saltation of the lamp's voltage is detected by controller A and a negative current saltation occurs, the controlling signal from P_2 is sent out

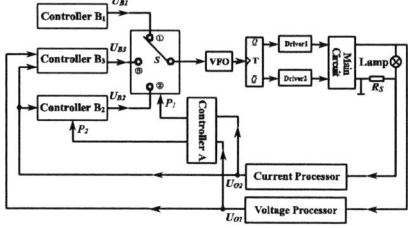

Figure 4. Block diagram of control circuit

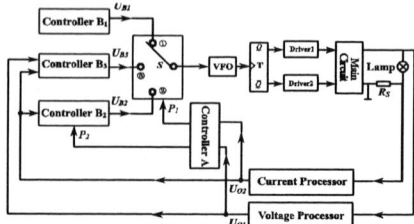

Figure 5. Block diagram of control circuit

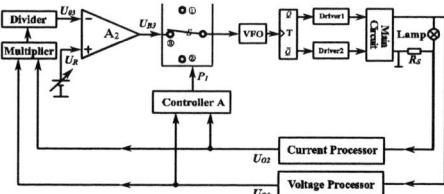

Figure 6. Controller B3 （period of steady state）

to direct the exponent-voltage generator to run, at the same time, the signal from P_1 turns switch S to location②. When controller A find that the lamp's voltage and current get to their rating value, drive signal via P_1 is send out and the switch will be located to position③.

Controller B_1 is a voltage-sweep generator. At the moment of boot-strap, it start to operate, its output voltage descend linearly, the maximum and the minimum are U_H and U_L respectively. If U_{B1} is equal to U_L, VFO's output frequency is the highest; and if U_{B1} and U_L are equal, VFO offers the lowest frequency.

Controller B_2 is shown in Fig.5. The A_1 is an error amplifier. The control signal P_2 is used to decide the exponent-voltage generator to operate or not, if P_2 is at high-level voltage, the generator will go into operation and decrease the output voltage according to exponent rule, in addition, it also provide an reference level to error amplifier.

Controller B_3 is shown as in Fig.6. The function of divider is to divide the product of U_{01} and U_{o2} with a constant voltage and obtain an output signal U_{o3}. Obviously, the U_{o3} is in direct ratio to the product of U_{01} and U_{o2}, namely, the relationship of both U_{o3} and the output power is in direct proportion.

IV. OPERATION TIME SERIES OF CONTROL CIRCUIT

The time series of operation about control circuit is shown as in Fig.7. Time Period① indicate auto-swept gradually to increase the output voltage until that it is equal to the igniting-voltage of HID lamp. At the moment of igniting, Time Period① ends and Time Period② will begin.

The relationship between operation frequency and the ballast's output voltage U_{lamp} of pre-ignition period is depicted in Fig 8. Observing Fig.4 reveals the following information, at the moment of start-up, namely t=0,

690

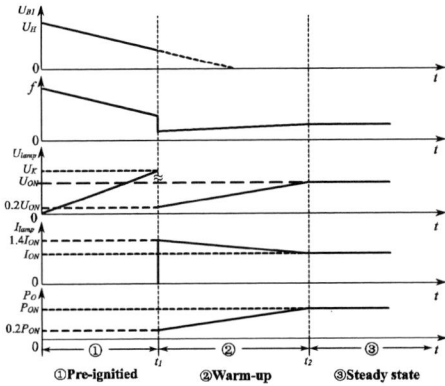

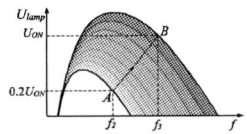

Figure 9. Relationship of lamp's voltage and operation frequency in Pre-ignited

power is under control of the inner parameter U_R of controller B_3 completely.

V. CONCLUSION

Because the increment impendence of MH lamp is negative, a ballast device must be placed between the discharge lamp and the voltage source to ensure lamp's stable operation. Observing the low-frequency electrical characteristics of HID lamps, the following information can be revealed: (1)The electrical characteristics of MH lamps are felicitously divided to start-up stage and steady stage. (2)MH lamps behave time-varying nonlinear electrical characteristics. (3)MH lamps can be regarded as quasi-steady-state in start-up stage, because the period of start up is correspondingly longer and not a saltation.

Ballast and the load--HID lamp compose a time-varying nonlinear system. Through studying the system, some conclusions can be obtained as followings: (1) The igniting-voltage is not a constant in different ambient even if the power level of HID lamps are the same, so, the output voltage of ballast must be rectified with the change of lamp's operation condition, therefore, the technique of auto-tracking to offer igniting voltage is adopted to ensure the lamp to be ignited safely; (2) During warm up, with conducting-channel diameter dwindling and the carrier concentration of channel falling, the voltage cross the lamp will escalate, the current through the lamp will step down, and the lamp's equivalent resistance will increase correspondingly. In addition, the dissipation power of lamp will be continuously raised. Based on these characteristics, the start control regulation has been proposed, and the principle of this strategy is to restrict the maximum in the start moment, decrease the current flowed in lamp by exponential or approximatively exponential law, and increase the lamp's dissipation power, this strategy can ensure to start the HID lamps safely; (3)When the lamp's performance enters steady state, the gas ionization velocity is still slightly higher than recombination one, so there has enough carriers inside of lamp, but the quantity of carrier transferred electric energy is decided by the ballast. Because the voltage of lamp is not a constant quantity, it is not able to realize constant-power control only by to stabilify the current of lamp. The technique of constant-power control presented in this paper is that to keep the product of the lamp's voltage and current tracking to equal the rating power and take the current as the main regulation variable to confirm the lamp to operate in steady state. In this paper, some practical control strategies are offered for electronic ballast of HID lamps and the conclusion drew

Figure 7. Time Series of controlling

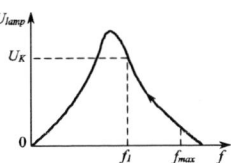

Figure 8. Relationship of lamp's voltage and operation frequency in Pre-ignited

switch S is connected to controller B_1, then, the output voltage U_{B1} of controller B_1 begins to fall linearly from the maximum U_H, this results in the input frequency of VFO to decrease linearly. So, the operation frequency of circuit will fall step by step, correspondingly, the input voltage of main circuit increases automatically. If the voltage endured by lamp is raised to be equal to or a little higher than the igniting voltage, the gas inside of lamp will occur electron avalanche, and the arc discharge will be established.

In Fig.9, the relationship between operation frequency and the ballast's output voltage in warm-up period is illustrated. At the instant of igniting, the equivalent resistance is very little, so it corresponds to a lower output voltage. When lamp enters the steady state, its equivalence resistance becomes greater and the corresponding output voltage will get higher. After lamp is ignited, controller A send out two signals P_1 and P_2, the signal P_1 informs the switch S to turn to position②, and signal P_2 will start the exponent generator inside of controller B_2 start to operate, synchronously, the current-feedback loop begin to work as shown in Fig.5, the operation frequency of control circuit is marked f_2. With lamp's current falling, the voltage across lamp will rise by degrees, and operation frequency will fulfill a transition from f_2 to f_3. If the loop-gain of the closed-loop current control circuit is infinite, the output current will change with exponential law. In Figs. 7 and 9, I_{ON} and U_{ON} are the rating current and rating voltage respectively.

As shown in Fig.6, when lamp arrives at the steady state, under the direction of signal P_1 from controller A, switch S will be turn to position③, thence, both output voltage and output current take part in feedback, a closed-loop feedback loop will be created. Then, the output

in this paper can be regard as an integrated theory reference for designing high-power electronic ballast.

REFERENCES

[1] Jinji Yang. Gas discharge (the first edition). *Science press*, 1983. P.183

[2] Dahua Chen (translated). *Lamps and lighting (the forth edition). Fudan press*, 2000/UK. (J.R.Coaton), (A.M.Marsden).

[3] E. Rasch and E. Statnic, ' Behavior of Metal Halide Lamps with Conventional and Electronic Ballasts' , *J. of the IES*, pp.88-96, Summer, 1991

[4] Harald L. Witting ' Acoustic Resonance in Cylindrical High-Pressure Arc Discharges' , *J. Appl. Phys.* Vol.49, No.5, pp.2680-2683, 1978

[5] Chi-Hwan Lee, ' Electronic Ballasts for 400W Metal Halide Lamps' , *Ph.D Dissertation*, Uiduk University, 2002

[6] G. A. Trestman, D. L. Bay, " Apparatus and Method for Operating a High Intensity Gas Discharge Lamps Ballast" , *U.S. Patent*, NO.US 6, 181, 076 B1, 2003

[7] X.Cao, W. Yan, S.Y.Hui and H. Chung, "Lamps Arc Resistance Modelling of High—Intensity—Discharge (HID) Lamps" , *IEE Proc.—Sci. Meas. Technol.* Vol.149, No.I, January, 2002

[8] G. A. Trestman, D. L. Bay, "Apparatus and Method for Operating a High Intensity Gas Discharge Lamps Ballast", *U.S. Patent*, NO.US 6,181,076 B1, 2003

2006 5th International Power Electronics and Motion Control Conference

Voltage Spectra of Three-Level Inverters with Three-Phase Modulation

S. Halász*, I. Varjasi**

*Electric Power Engineering Department,
**Automation and Applied Informatics Department,
Budapest University of Technology and Economics, Budapest, HUNGARY
shalasz@eik.bme.hu, varjasi@aut.bme.hu

Abstract—**In the area of high voltages and powers, the high quality inverter-fed ac drive is more easily achieved by the use of three-level inverters. For three-level inverters three different PWM control regions can be introduced where different three-phase PWM strategies may be applied. This paper deals with the analysis of that three-phase PWM method of three-level inverters which ensures the lowest value of motor harmonic losses. However, the realization of this PWM method is not so simple as the space vector PWM. At the same time in the low and middle voltage area the space vector PWM method gives bigger harmonic losses than the optimal PWM method. The motor voltage spectrum derivation for both the optimal version and the space vector PWMs are given and the results are checked by straight harmonic computations and by simulations.**

Keywords-component; three-level inverter, three-phase PWM techniques, voltage spectra, one and two carrier waves.

I. INTRODUCTION

In the region of high voltages and powers, the high quality inverter-fed ac drive is more easily reached by the use of three-level inverters [1-4] instead of two-level ones. Several PWM strategies for three-level inverters have been elaborated in the last years where the different two-phase (discontinuous) PWM strategies and the different three-phase ones are applied. This paper deals with the derivation and the analysis of the voltage spectra of three-phase PWM methods which ensures the lowest value of motor harmonic losses. The comparison of the different realization methods of these modulations is given.

II. STATE OF THE ART

In Fig. 1 the configuration and in Fig. 2 the possible voltage vectors of the three-level inverter are presented for 1/6 of the fundamental cycle. Owing to the symmetry it is usually enough to investigate the first half of this time duration where $0 \leq W_1 t \leq \pi/6$. In case of the space vector PWM technique those three voltage vectors are applied which are nearest to the wanted fundamental voltage vector. From Fig. 2 it is clear that in the normal linear modulation area there are three (I-II-III) different regions, while region IV belongs to the overmodulation region.

One of the important quality indexes of the PWM technique is the inverter-fed motor harmonic loss. It can be characterized by the generalized loss-factor [4-5] which for the pure inductive load is given as follows:

$$G = f_t^2 \Delta \Psi^2, \qquad (1)$$

where: $\Delta \Psi^2 = \sum_{\nu > 1}^{\infty} \Psi_\nu^2$, $\Psi_\nu = L' I_\nu$ and I_ν, Ψ_ν are the harmonic current and the harmonic flux of the order ν, $\Delta \Psi$ is the rms. value of the harmonic flux, L' is the stator transient inductance, f_t is the switching frequency of one transistor (GTO), and all the values are in p.u. system.

In equation (1) it is assumed that sampling frequency

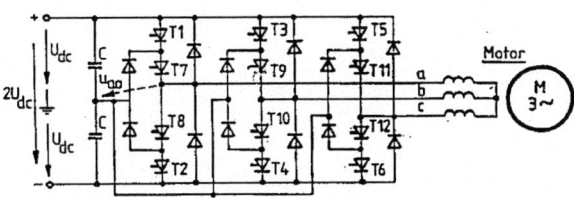

Fig. 1. Three-level inverter configuration

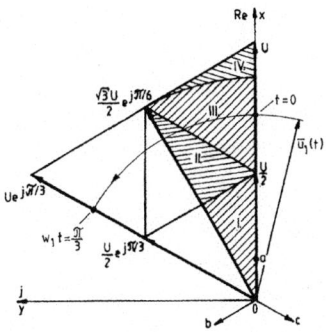

Fig. 2. Voltage vectors and control regions of three-level inverters

1-4244-0448-7/06/$25.00 ©2006 IEEE

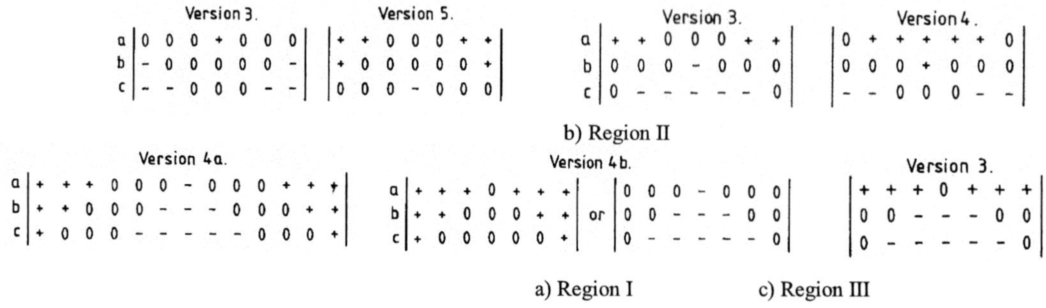

```
              Version 3.              Version 5.                              Version 3.              Version 4.
          a | 0 0 0 + 0 0 0 |     | + + 0 0 0 + + |            a | + + 0 0 0 + + |     | 0 + + + + + 0 |
          b | - 0 0 0 0 0 - |     | + 0 0 0 0 0 + |            b | 0 0 0 - 0 0 0 |     | 0 0 0 + 0 0 0 |
          c | - - 0 0 0 - - |     | 0 0 0 - 0 0 0 |            c | 0 - - - - - 0 |     | - - 0 0 0 - - |

                                                              b) Region II

                  Version 4a.                              Version 4b.                              Version 3.
      a | + + + 0 0 0 - 0 0 0 + + + |      a | + + + 0 + + + |   | 0 0 0 - 0 0 0 |      | + + + 0 + + + |
      b | + + 0 0 0 - - - 0 0 0 + + |      b | + + 0 0 0 + + | or| 0 0 - - - 0 0 |      | 0 0 - - - 0 0 |
      c | + 0 0 0 - - - - - 0 0 0 + |      c | + 0 0 0 0 0 + |   | 0 - - - - - 0 |      | 0 - - - - - 0 |

                                  a) Region I              c) Region III
```

Fig.3. Motor phase connections to dc bars, 3 phase PWMs

$f_s \to \infty$, so the T_s sampling (carrier) period must be infinitely small. In the above equations it is assumed that the rated voltage is equal to the maximal inverter voltage $U_{1max}=4U_{dc}/\pi$, where $2U_{dc}$ is the inverter input voltage and the rated frequency f_r corresponds to this voltage.

The motor fundamental voltage vector path is in region I only for $0 \le U_1/U_{dc} \le 1/\sqrt{3}$ and it is in regions I and II when $1/\sqrt{3} \le U_1/U_{dc} \le 2/3$. The PWM strategies of region I use $U/2, Ue^{j\pi/3}/2$ and zero voltage vectors and are very similar to the PWM strategies of the two-level inverters. The motor fundamental voltage in region II is $1/\sqrt{3} \le U_1/U_{dc} \le 2/\sqrt{3}$ and voltage vectors $U/2$, $Ue^{j\pi/3}/2$ and $U\sqrt{3}e^{j\pi/6}/2$ are applied. In region III the motor fundamental voltage is $2/3/ \le U_1/U_{dc} \le 2/\sqrt{3}$ and voltage vectors $U/2, U$ and $U\sqrt{3}\,e^{j\pi/6}/2$ are used.

The two-phase PWMs of versions 0, 1 and 2 are investigated in [9-11] respectively. The possible three-phase PWM versions 3-5 are presented in Fig. 3 for all the three voltage control regions. It was shown [7] that version 3 insures the lowest motor harmonic losses (except the region 1 where the version 4 and 4a gives the slightly lower value harmonic losses in $0 \le U_1/U_{dc} \le 0.37$), therefore further on the version 3 is analyzed.

III CARRIER-BASED REALIZATION OF VERSION 3

The realization of version 3 in region I is given in [9]. There it was shown that in case of the equal time sharing of the voltage vector $\bar{u}=U/2$ between the cycle sides and the middle part of the cycle the reference wave becomes discontinuous one as it is seen in Fig.4.

The PWM strategies of region III during a sampling period T_s uses $U/2$ vector for the time

$$t_1/T_s = 2 - \sqrt{3}\,A\sin(W_1t + \pi/3),\qquad(2)$$

$Ue^{j\pi/3}/2$ vector for the time

$$t_2/T_s = \sqrt{3}\,A\sin(W_1t),\qquad(3)$$

and U voltage vector is used for $t_3/T_s =1-t_1-t_2$ time:

$$t_3/T_s = -1 + \sqrt{3}\,A\sin(\pi/3 - W_1t).\qquad(4)$$

The dc component of reference waves - related to U_{dc} - in case of the equal time sharing of the voltage vector $\bar{u}=U/2$ - after substitution (2-4) - will be:

$$U_0 = T_s(-t_1 - t_3 + t_1/2)/3 = -0.5A\sin(\pi/6 - W_1t).$$

Therefore the phase reference waves are:

$$U_{ar} = A\cos(W_1t) - 0.5A\sin(\pi/6 - W_1t)$$
$$= 0.5\sqrt{3}\sin(\pi/3 + W_1t);$$
$$U_{br} = A\cos(W_1t - 2\pi/3) - 0.5A\sin(\pi/6 - W_1t)$$
$$= -1.5A\sin(\pi/6 - W_1t);$$
$$U_{cr} = A\cos(W_1t - \frac{4\pi}{3}) - 20.5A\sin(\pi/6 - W_1t)$$
$$= -0.5\sqrt{3}A\sin(W_1t + \pi/3).$$

The reference wave of region II is determined similarly. These reference waves are presented in Fig.5. It is seen that the reference wave of region II is discontinuous. At the same time the reference wave of region III is continuous and coincides with the reference wave of the traditional space vector PWM. It should be noted that these reference waves provide the equal time sharing of the voltage vector that is used on the both sides and in the middle of a sampling cycle. Usually, the minimum of harmonic losses is related to the unequal time sharing

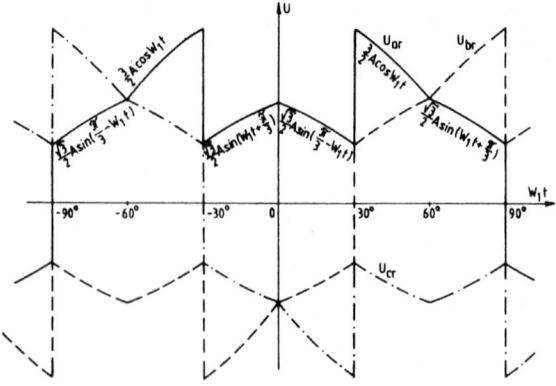

Fig. 4. Reference waves of region I

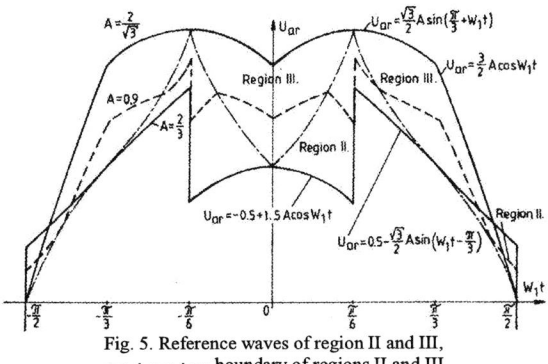

Fig. 5. Reference waves of region II and III,
— · — · — boundary of regions II and III

however the losses in this case differ from the minimum point in the worst case only by 1.2% [12].

From the practical point of view this carrier-based realization is not too simple, especially in region II-III where for $2/3 \leq A \leq 2/\sqrt{3}$ inside each $W_1 t = 30°$ both regions are used as it is shown in Fig. 5 in case $A = 0.9$. The modulation process is presented in Fig. 6 where the positive and negative carrier waves are shifted by $T_s/2$ since the correct voltage vector sequence of version 3 can be realized only in this case.

For practical application the simplification of PWM is expected. One possible simplification is the use of two-phase modulation [10-11]. In this case the reference wave in all control regions has the same shape but the harmonic losses in a considerable part of control area are bigger than for version 3 as it is shown in Fig. 7. The second possibility to simplify the PWM control is the use of the region III reference wave for whole voltage control area. The generalized loss-factor of this PWM is also given in Fig. 7 (space vector PWM). The space vector PWM uses the same voltage vector sequences as version 3 however except the region III this space vector modulation cannot provide the equal sharing times of voltage vector that is used on the both sides and in the middle of the sampling cycle. This leads to the considerable increase of motor harmonic losses in regions I and II. As it is shown later this result is explained by very low amplitudes of voltage harmonics of m side orders in the version 3.

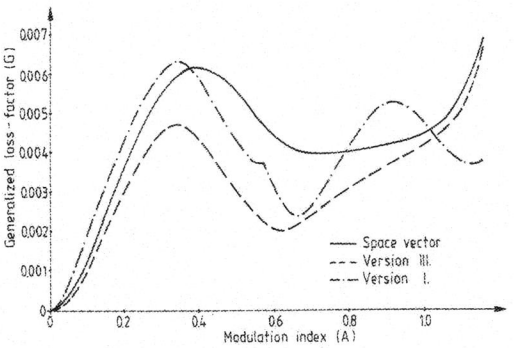

Fig. 7. Generalized loss-factor,
— · — two-phase PWM, - - - and —— three-phase PWMs

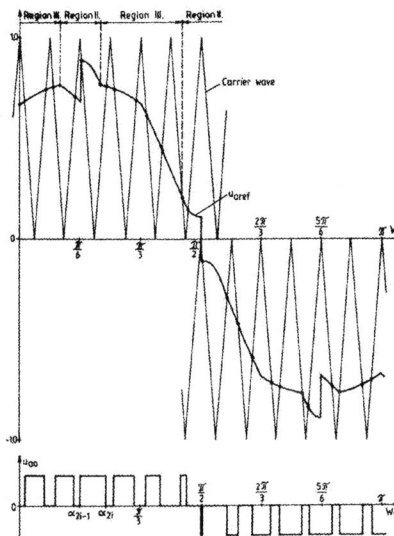

Fig. 6. Modulation process in region II-III and phase-end voltage to the middle point of dc ($A = 0.9$)

IV. VOLTAGE SPECTRA

The voltage spectra equation derivation of three-phase PWMs of three-level inverters will be given in Appendix (in final paper). The voltage spectrum equations of two-phase PWMs are derived in [10-11].

The order of the harmonics can be given as follows:

$$\nu = km \pm n,$$

where $m = f_c / f_1$, and f_1 is the fundamental frequency, f_c is the carrier one, k and n are positive integers.

The motor voltage to the middle of the dc circuit u_{a0} (Fig. 1) can be decomposed in Fourier series (Fig. 6).

$$u_{a0} = a_0 + \sum_{\nu=1}^{\infty} (a_\nu \cos \nu\alpha), \qquad (5a)$$

where

$$U_\nu = \frac{a_\nu}{U_{dc}} = \frac{2}{\pi} \operatorname{Re} \int_0^\pi u_{a0}\, e^{j\nu\alpha}\, d\alpha =$$
$$-\frac{2}{\pi\nu} \operatorname{Im} \sum_i (e^{j\nu\alpha_{2i}} - e^{j\nu\alpha_{2i-1}}) sign(U_{dc}) \qquad (5b)$$

and $a_0 = 0$ usually, α_{2i} and α_{2i-1} are the modulation angles (Fig. 6). In [5] it was shown that the harmonic amplitudes do not depend on the phase angle between the reference and carrier waves therefore it is reasonable to derive the voltage harmonics for the simplest case when $m/12$ is integer and the phase angle is as in Fig. 6. The derivation method is based on the summation of harmonic amplitudes for all parts of the reference wave with duration 30° or 60°.

A.   Space vector PWMI

In this case the reference wave of region III is used. The result is presented in Appendix but for the case $m \to \infty$ the voltage harmonics are:

695

$$\frac{U_v}{U_1} = \frac{4}{\pi^2 kA} \sum_p \left\{ J_p\left(\frac{\sqrt{3}}{2}\pi Ak\right) \cdot \cos\left(\frac{5\pi n}{6}\right) \sin\left((p+n)\frac{\pi}{6}\right) \right.$$

$$\left. + J_p\left(\frac{3}{2}\pi Ak\right) \cos\left((p+n)\frac{5\pi}{12}\right) \sin\left[(p+n)\frac{\pi}{12}\right] \right\} \quad (6)$$

$$\left((-1)^k - (-1)^n\right) \sin\left(p\frac{\pi}{2}\right)/(p+n),$$

where usually it is enough to take $p = -21, -19, ...19, 21$.

B.   Version 3 , region I-II

The border between region I and region II at $0 \le W_1 t \le \pi/6$ is $\alpha^* = \pi/6 - a\cos(1/A/\sqrt{3})$.

The voltage harmonic equation for $m \to \infty$ is:

$$\frac{U_v}{U_1} = \frac{4}{\pi^2 kA} \sum_p \left\{ B \cdot \sin\left(p\frac{\pi}{2}\right) \cdot \sin\left[\frac{p+n}{2}\alpha^*\right] \right. \quad (7)$$

$$\left. + C \sin\left(p\frac{\pi}{2} + k\frac{\pi}{2}\right) \sin\left[\frac{p+n}{2}\left(\frac{\pi}{6} - \alpha^*\right)\right] \right\} \frac{(-1)^k - (-1)^n}{p+n},$$

where $\alpha^* = \pi/6$ if $A \le 1/\sqrt{3}$ and:

$$B = \sum_p \left\{ J_p\left(\frac{\sqrt{3}}{2}\pi Ak\right) \cdot \left[\cos\left(p\frac{\pi}{6} + (p+n)\frac{\alpha^*}{2}\right) - \right.\right.$$

$$\left. -\cos\left(p\frac{5\pi}{6} + (p+n)(\frac{\pi}{3} + \frac{\alpha^*}{2})\right) \right]$$

$$\left. + J_p\left(\frac{3}{2}\pi Ak\right) \cos\left[(p+n)(\frac{\pi}{3} + \frac{\alpha^*}{2})\right] \right\}, \quad (8)$$

$$C = \sum_p \left\{ 2J_p\left(\frac{\sqrt{3}}{2}\pi Ak\right) \cdot \cos\left(p\frac{\pi}{2} + n\frac{\pi}{3}\right) + \right.$$

$$\left. J_p\left(\frac{3}{2}\pi Ak\right)(-1)^k \right\} \cos\left((p+n)(\frac{\pi}{12} + \frac{\alpha^*}{2})\right). \quad (9)$$

In this equation p can be also even: $p = -21, -20, ... 21$.

C.   Version 3 , region II-III

The border between region II and region III at $0 \le W_1 t \le \pi/6$ is $\alpha^{**} = a\cos(1/A/\sqrt{3}) - \pi/6 = -\alpha^*$. The voltage harmonic equation for $m \to \infty$ is:

$$\frac{U_v}{U_1} = \frac{4}{\pi^2 kA} \sum_p \left\{ D \cdot \sin\left(p\frac{\pi}{2}\right) \cdot \sin\left[\frac{p+n}{2}\alpha^{**}\right] + \right. \quad (10)$$

$$\left. C \sin\left(p\frac{\pi}{2} + k\frac{\pi}{2}\right) \sin\left[\frac{p+n}{2}(\frac{\pi}{6} - \alpha^{**})\right] \right\} \frac{(-1)^k - (-1)^n}{p+n},$$

and α^{**} must be substitute also in (9) for C while

$$D = \sum_p \left\{ 2J_p\left(\frac{\sqrt{3}}{2}\pi Ak\right) \cos\left((p+n)(\frac{\pi}{6} - \frac{\alpha^{**}}{2})\right) \cos\frac{n}{6} + \right.$$

$$\left. J_p\left(\frac{3}{2}\pi Ak\right) \cos\left((p+n)(\frac{\pi}{3} + \frac{\alpha^{**}}{2})\right) \right\}. \quad (11)$$

The voltage harmonics as function of modulation index are presented in Figs. 8-11. From that it is clear that the version 3 produces lower value of motor harmonic losses owing to lower amplitudes of m-side voltage harmonics ($m \pm 2$, $m \pm 4...$ orders) while the $2m$-side harmonics ($2m \pm 1, 2m \pm 5, 2m \pm 7...$ orders) virtually have approximately the same amplitudes.

V.   SIMULATION RESULTS

The motor current and torque as function of time are presented in Fig. 12 for the switching frequency of one transistor $f_t \cong 715Hz$, and no-load conditions. The no-load current is $I = 0.5 p.u.$ and the transient reactance is $L' = 0.12 p.u.$ Two different operation points are investigated for both the version 3 and the space vector PWM: a) and b): $A = 0.4, m = 72$, c) and d): $A = 0.8, m = 36$. It is seen that in agreement with derived voltage harmonic equations and Fig. 7 the motor current pulsations is bigger in case of space vector PWM and this difference for $A = 0.4, m = 72$ is more sensible than for the case $A = 0.8, m = 36$. At the same time the motor torque ripples have approximately the same amplitudes

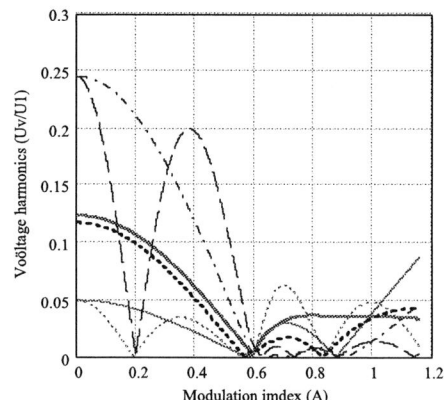

Fig.8. Version 3, voltage harmonics of order
---- $m \pm 2$, - - - $3m \pm 2$, * * $m \pm 4$, ∼∼∼ $3m \pm 4$,
——— $m \pm 8$, - - - $m \pm 10$

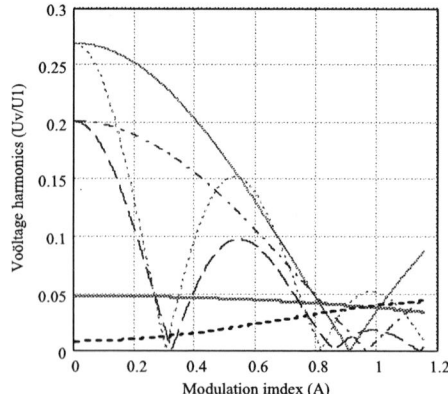

Fig. 9. Space vector PWM, voltage harmonics of order
---- $m \pm 2$, - - - $3m \pm 2$, * * $m \pm 4$, ∼∼∼ $3m \pm 4$,
——— $m \pm 8$, - - - $m \pm 10$

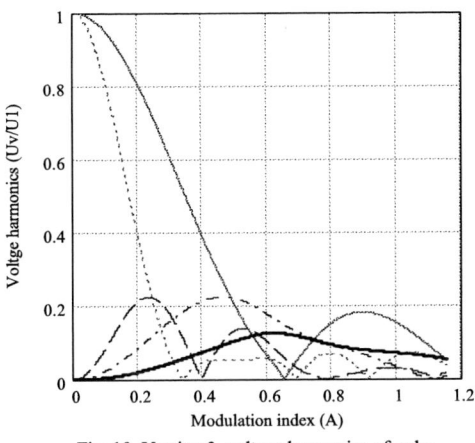

Fig. 10. Version 3, voltage harmonics of order
──── $2m\pm1$, - - - $4m\pm1$, ─··─ $2m\pm5$, ─── $4m\pm5$,
••• $2m\pm7$

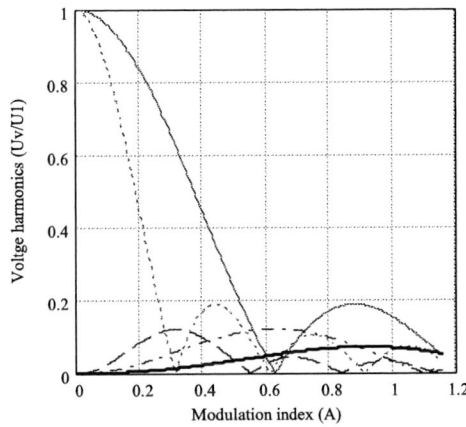

Fig.11. Space vector PWM, voltage harmonics of order
──── $2m\pm1$, - - - $4m\pm1$, ─··─ $2m\pm5$, ─── $4m\pm5$,
•• $2m\pm7$

since these amplitudes are mainly determined by the real component of current harmonic vector which in both cases has approximately the same values (Fig. 13). It should be noted that for $m < 36$ the voltage harmonic of order $\nu = 2$ reaches the sensitive value. The effect of this current harmonic leads to asymmetry of the current shape and for the appearance of the torque harmonic of order 3 with amplitude of a few percent (Fig.12).

The derived voltage harmonic equations were checked by straight voltage harmonic computations for selected operational points. The direct analytical computer calculation in case e.g. $A = 0.8$ gives:

Version 3:
$U_{2m-1} = -0.1557, U_{2m+1} = -0.1554 \ (m = 36)$;
$U_{2m-1} = -0.1553, U_{2m+1} = -0.1554 \ (m = 192)$.

Space vector PWM:
$U_{2m-1} = -0.1735, U_{2m+1} = -0.1734 \ (m = 36)$;
$U_{2m-1} = -0.1735, U_{2m+1} = -0.1735 \ (m = 192)$;

The derived equations give:

Version 3: $U_{2m\pm1} = -0.1554 \ (m = \infty)$;

Space vector PWM: $U_{2m\pm1} = -0.1735 \ (m = \infty)$;

It is seen that the derived equations for $m \rightarrow \infty$ give a very accurate result.

VI. CONCLUSIONS

It is shown that the modulation of three-level inverters according to version 3 and space vector PWM can be realized by two carrier PWMs. The analytical expressions of voltage spectra of these PWM techniques are given and it is proved that these expressions ensure a good accuracy of computations. It is shown that the modulation according to version 3 and the space vector produces approximately the identical values of motor harmonic losses for high voltage area while for low voltage area these losses are considerably lower in case of version 3. However the realization of version 3 is not so simple as the space vector PWM. The computation and the simulation result is in agreement with the theoretical results.

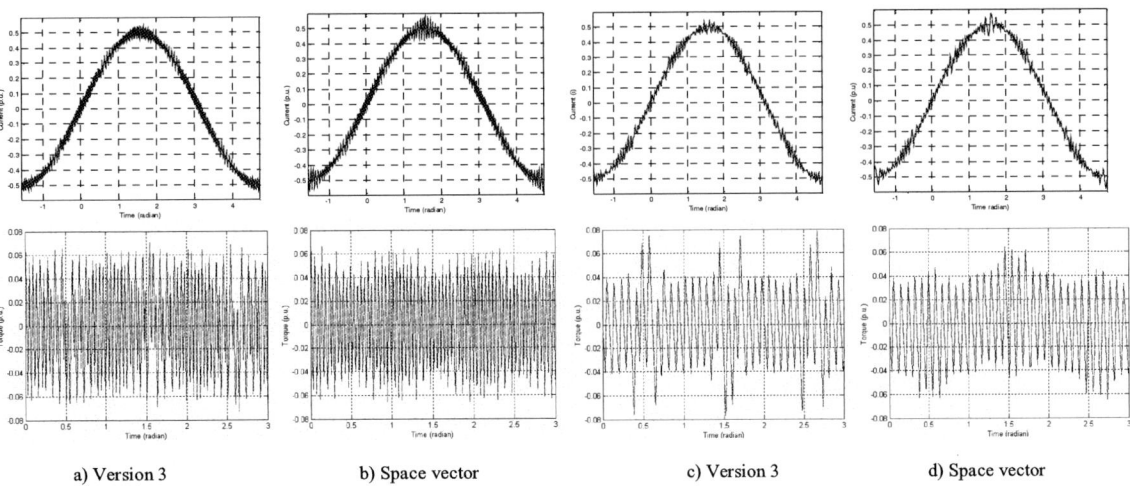

a) Version 3   b) Space vector   c) Version 3   d) Space vector

Fig. 12. Motor current and torque vs. time, no-load, a) and b): $A = 0.4, m = 72$; c) and d): $A = 0.8, m = 36$

ACKNOWLEDGMENT

The research was supported by the Hungarian National Scientific Fund (OTKA #T 042866) for which authors express their sincere gratitude.

REFERENCES

[1] A. Bellini, S. Bifaretti, S. Constantini: A space vector modulation technique for NPC inverter", *in Proceedings of 3rd European Conference . on Power Electronics and Applications,* 2003, Toulouse, France, pp1-10.

[2] C. Cecati, A . Dell'Aquila, A. Lecci, M. Liserre, V. G. Monopoli: A discontinuous carrier-based multilevel modulation for multilevel inverters, *The 30th Annual Conference of the IEEE Industrial Electronics Society,* Nov. 2-6, 2004, Busan, Korea, CD ROM.

[3] A.D Pizzo, M. Pasquariello, A. Perfetto: An optimized PWM control technique of three-level inverters for induction motor drives, *IPEC-Tokyo, Tokyo, Japan,* 2000, pp.124-129.

[4] S. Halász, A. Zacharov: PWM strategies of three-level inverter-fed ac drives, *37th Annual Meeting of IEEE Ind. Application Society,* 13-17 October 2002, Pittsburgh, USA, CD ROM:

[5] H. van der Broeck: Analysis of the harmonics in voltage fed inverter drives caused by PWM schemes with discontinuous switching operation, *in Proceedings of 3rd European Conference on Power Electronics and Applications,* Florence, Italy, 1991, pp.3-261-266.

[6] S. Halász, B.T.Huu: Generalized harmonic loss-factor as a novel important quality index of PWM techniques, *PCC-Nagaoka, Japán,* Aug. 3-6. 1997. pp. 787-792.

[7] S. Halász, I. Varjasi: Small vector PWM strategies of three-level inverters, *in Proceedings of the International Symposium on Industry Electronics,* May, 4-7, 2004,Ajaccio, France, CD ROM,pp. 1273-1278.

[8] Halász S., Huu B. T, Zacharov: Two-phase modulation technique for three-level-fed ac drives, *IEEE Trabsactions on Ind. Electronics,* Vol. 47, No.6, pp.1200-1211, December 2000.

[9] S. Halász, A. Zacharov: Analysis of two-phase PWM techniques in inverter fed ac drives, *in Proceedings of 8th European Conference on Power Electronics and Application,* Lausane, 1999, pp. p1-p9.

[10] S. Halász: Analysis of discontinuous PWM strategies of three-level inverters (II), *The Sixth International Conference on Power Electronics and Drive Systems (PEDS2005),* 28 Nov - 1 Dec 2005 Kuala Lumpur, Malaysia, CD ROM.

[11] S. Halász:: Analysis of discontinuous PWM strategies of three level inverters (I), *IEEE International Conference on Industry Technology (ICIT2005),* 14-17 Dec. 2005, ,Hong Kong, CD ROM.

[12] S. Halász, I. Varjasi: Analysis of tree phase PWM strategies of three-level inverters, *Intrnational Symposium on Power Electronics, Electrical Drives, Automation and Motion,* Taorrmina, Italy, 2006,, in press.

APPENDIX

The derivation time of the voltage spectrum expression is $0 \le W_1 t \le \pi$ and in case version 3 contains according to each 30° (Fig. 6) six terms. Since the reference waves have 90° symmetry only cosine components of Fourier series components are possible with non zero harmonic of

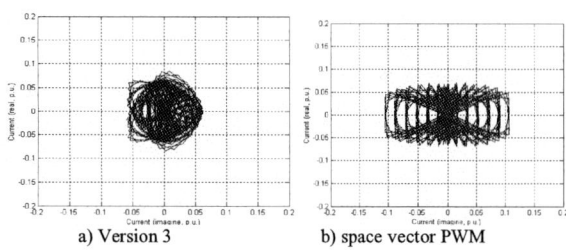

a) Version 3   b) space vector PWM

Fig. 13. Current harmonic vector in synchronous coordinate system no-load, $A = 0.4$, $m = 72$.

order $v^* = 1 \pm 3k^*$ where $k^* = 0, 1, 2 \cdots$.

Let us derive the first term of the voltage spectra expression for $0 \le \alpha \le 30°$ and $2/3 \le A \le 2/\sqrt{3}$. The intersection points in Fig. 6 in region III are as follows:

$$\alpha_{2i-1} = \gamma i - \gamma(A\sqrt{3}/2\sin(\alpha_{2i-1} + \pi/3) \qquad (A1)$$
$$\alpha_{2i} = \gamma i + \gamma(A\sqrt{3}/2\sin(\alpha_{2i} + \pi/3),$$

where $\gamma = \pi/m$, $i=1,3,5...i^*$ and i^* is the total number of the intersection points α_{2i} in $0 \le \alpha \le \alpha^{**}$. The number of the intersection points α_{2i-1} equals in this case $i^{**} = i^*$ or $i^{**} = i^* + 1$. After rearrangement of (A1):

$$\gamma i + 4\pi/3 = \alpha_{2i-1} + 4\pi/3 - \gamma A\sqrt{3}/2\sin(\alpha_{2i-1} + 4\pi/3)$$
$$\gamma i + \pi/3 = \alpha_{2i} + \pi/3 - \gamma A\sqrt{3}\sin(\alpha_{2i} + \pi/3) \qquad (A2)$$

and taking into account (5) and expressing the exponential function [10] by the first-kind Bessel function $J_\rho(x)$ of the ρ order the result is expressed as:

$$U_v^* = \frac{2}{\pi} \text{Im} \sum_{\rho=1}^{\infty} \{\frac{1}{\rho} J_{\rho-v}(\gamma\sqrt{3}/2A\rho)$$
$$[e^{j\rho(\pi/3)} e^{-jv\pi/3} \sum_{i}^{i^*} e^{j\rho\gamma i} - e^{j\rho(4\pi/3)} e^{j4v\pi/3} \sum_{i}^{i^{**}} e^{j\rho\gamma i}]\}.$$

In the above expression the Bessel functions of order $\rho + v$ have been neglected (for small m these orders can affect only the fundamental voltage). After rearranging and changing the order of the summation the result with simplification $i^{**} = i^*$ can be written as follows:

$$U_v^* = \frac{4}{\pi} \sum_{\rho=1}^{\infty} \{\frac{1}{\rho} J_{\rho-v}(\gamma\sqrt{3}/2A\rho) \frac{\sin(\rho i^* \pi/m)}{\sin(\rho\pi/m)}$$
$$\cos[\rho i^* \pi/m + (\rho - v)5\pi/6]\sin[(\rho - v)\pi/2]\}. \qquad (A3)$$

In region II the intersection points are as follows:

$$\alpha_{2i-1} = \pi/6 - \gamma i - \gamma(-0.5 + A\ 1.5\cos(\alpha_{2i-1}) \qquad (A4)$$
$$\alpha_{2i} = \pi/6 - \gamma i + \gamma(-0.5 + A\ 1.5\cos(\alpha_{2i})$$

and, after similar computation like that for region III, the result is:

$$U_v^{**} = \frac{4}{\pi} \sum_{\rho=1}^{\infty} \{\frac{1}{\rho} J_{\rho-v}(\gamma 1.5A\rho) \frac{\sin(\rho i^* \pi/m)}{\sin(\rho\pi/m)}$$
$$\cos[\rho i^* \pi/m + (\rho - v)5\pi/6]\sin[(\rho - v)\pi/2]\}.$$

The final result for the first term is:

$$U_v^1 = U_v^* + U_v^{**}$$

and the similar computation is necessary for the rest five terms. In case of $m \to \infty$ after substitution $p = \rho - v$ the simpler equations (7-11) are obtained.

The derivation of voltage spectrum expressions for space vector modulation is considerably simpler and the result for $m \to \infty$ is given in (6).

2006 5th International Power Electronics and Motion Control Conference

Design of Motion Control System Used for Filter Rod Production Machine

Yang Qingyu, Ge Sibo, Ye Kesong and Shi Ren
Department of Automation, Xi'an Jiaotong University, Xi'an, P.R.China
yangqingyu@mail.xjtu.edu.cn

Abstract—The control system of filter rod production machine has characters of high control precision and high reliability, its performance affects quality of filter rod directly. How to improve its integrative control level is the key problem of control system design. Therefore, a suit of motion control system based on Profibus is presented in this paper. In this control system, some new technologies such as fieldbus and AC servo drive control are employed. System structure and software framework are issued. The key technologies are researched in depth. This system has already been implemented on the product line in a large-scale tobacco group successfully. The performance results show that the motion control system improves outputs and quality of products and integrative control level, saves energy, and obtains favorable economic and social benefits.

Keywords-filter rod production machine; Fieldbus; motion control; PLC

I.   INTRODUCTION

Filter rod production machine used to make cigarette filter tip is the important device in tobacco industry. Carbonic acid fibre is incised into cigarette filter tip through a series of technics including slack control, insufflating glycerin, packaging papers, circle control, and so on. In this process, slack control performance of Carbonic acid fibre affects "draw resistance" directly, which is the key parameter used to judge the quality of filter rod and content of tar in cigarette. Today, the traditional method is mechanism drive when controlling filter rod production machine in tobacco industry in china. But this method has disadvantages of large noise, high energy wasting, and unstable quality of products. Recent years, along with the development of cigarette technologies and high demand of automation level and management information integration in cigarette industry, more higher and stricter control demand about control of filter rod production machine are brought forward[1].

Therefore using some new technologies, a suit of motion control system based on Profibus that is used for filter rod production machine is presented in this paper, and the key technologies are researched. This paper is structured as follows. Section II introduces shortly the technics and control demand of filter rod production machine. Section III describes the network architecture

and software framework of the motion control system. Section IV is the detailed design, such as AC servo drive control, data communication and advanced program design, etc. Section V is implementation results. Section VI summarizes this paper.

II.   TECHNICS AND CONTROL DEMAND OF FILTER ROD PRODUCTION MACHINE

The technics process of filter rod production machine is shown as Fig.1. Normal length of the filter rod is 120mm. Firstly carbonic acid fibre is pulled into fibre rod with same thickness. In order to control slack performance, the fibre rod passes through slack roller 1-2 and output roller 3. Then glycerin is insufflated into the fibre rod, and the content of glycerin is 49mg/120mm. Moreover the fibre rod is packaged with twist papers after insufflating glycerin. Finally, the fibre rod passes through cooling, circle control and length incising to form the filter rod with the length of 120mm.

Control demands of filter rod production machine include three points. Firstly, the speed of main motor can be controlled according to multi-modes, which decides the production speed of the whole machine. The running mode of the main motor includes starting speed, normal production speed and meeting paper speed. Secondly, the speed of slack motors and glycerin motor must correspond with the ones of the main motor. In order to change glycerin content of filter rod, the speed of glycerin motor should be adjusted independently. There is a ruler that speed of the other motors cannot be changed as long as the speed of main motor is invariable, which is to keep the reliability of glycerin content. The most important control portion is slack roller 1-2 and output roller 3, which affects the slacking effects of fibre and the draw resistance of filter rod directly. Finally, the whole control system must have high reliability.

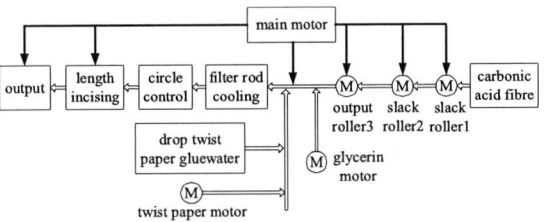

Fig. 1 Technics process of filter rod production.

1-4244-0448-7/06/$25.00 ©2006 IEEE   699

III. NETWORK ARCHITECTURE AND SOFTWARE FRAMEWORK

Network architecture of control system not only affects the exertion of system performance, but also decides the integration automatic level of the system in some sense. How to design reasonable system architecture is the chief problem during developing control system.

A. Network Architecture of Motion Control System

In order to satisfy the demand of integration automation, the motion control system designed for filter rod production machine adopts the structure of distributed control system, and is composed of operator stations (OS), local control stations (CS) and sensors/actuators. The three levels communicate with digital buses including fieldbus and RS485. The network architecture of the whole motion control system is depicted in Fig.2.

As shown in Fig.2, the OS is the center of the control system, its mission is to monitor the whole production line. Furthermore the OS provides an interface for operators to set parameters. According to the demand of high reliability, two suits of operator station are redundant and achieve redundant control of slack control segment. In order to save invest, multiple function panel MP370 is used as main station, and ordinary touch panel TP170A is used as redundant station. Communication between two redundant stations and PLC is realized through Profibus-DP. Process monitoring and parameters setting are implemented on MP370. TP170As is used as adjusting of slack motor, which can control two slack rollers, one output roller and glycerin motor independently.

CS responds for complicated control functions, and realizes the communication between OS and transducers at the same time. Siemens S7-300 PLC is chosen as the CS, which has the advantages of high reliability and high respond speed. The signals from sensors/actuators enter into I/O modules of PLC directly.

The control of main motor adopts technology of AC frequency conversion. CT/UNI transducer is adopted. The control mode of slack rollers change from mechanism drive to AC servo drive, and Germany LENZE drivers are selected. Panasonic NAIS driver is used to control glycerin motor. All the drivers and transducers communicate with PLC through RS485 protocol.

B. Software Framework

We use the thinking of modularization software design when realizing monitoring program. We design PLC program using Step7 language, and design OS HMI software using ProTool-SIMATIC HMI configuration language[2]. The functions on OS have device and process monitor, parameters setting, report and data statistic, fault alarming, providing correlative information for users, and so on. The software framework is represented in Fig.3. If there were alarm and fault in control system, the current menu changes to alarm and fault display menus automatically. After alarm and fault are eliminated, the current menu will redisplay automatically.

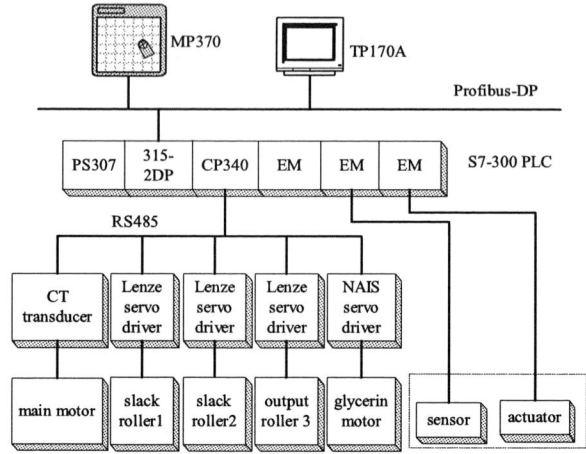

Fig. 2 Network architecture of motion control system.

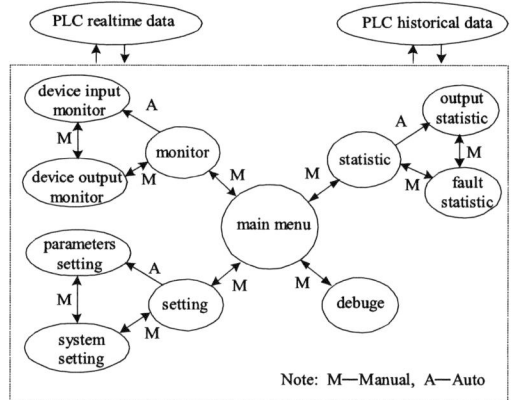

Fig. 3 Software framework of motion control system.

IV. KEY TECHNOLOGIES OF SYSTEM REALIZATION

Besides PLC control program and general HMI menus, the key technologies of the control system used for filter rod production machine are AC drive control, communication among all parts and advanced program design, which are also difficult and important. The realization of these key technologies is described as follows.

A. AC Servo Drive Control

As shown in Fig.1, the most important portion in the control of filter rod production machine is speed control of motors. The speed relation among main motor, slack motors and glycerin motor must be proportional strictly. Therefore we choose transducers and AC servo drive control subsystem to realize speed control.

Among these motors, slack and output roller motors have the highest control precision, which affect draw resistance directly. The speed of roller motors is very high, and the unit of speed changing is 1 rpm. The speed of the three motors has also proportional relation strictly, which is 892:1258:1146 (r/m) during normal production. Therefore three Lenze drivers are chosen to control slack roller 1-2 and output roller 3. Lenze drivers can realize complicated control functions and precise position control through embedded phase controller. The three drivers are connected into the form of digital frequency degree[3][4].

The control precision of the glycerin motor is lower relatively. The speed of this motor is not high and about 80 rpm. In stead of expensive Lenze driver, economic Panasonic driver is chosen to control glycerin motor. The configuration and setting of Panasonic driver are easier than Lenze driver.

The main motor should provide two suits of normal production speed and meeting paper speed at the same time according to the technics demand. So CT transducer is selected to control the main motor. CT transducer has abundant inside parameters and functions, and can realize complicated control strategy and high precise digital setting[5]. In addition, CT transducer has four work modes, which are open loop, closed vector loop, rebirth and user interface. According to control demand, we configure the transducer into open loop mode in this motion control system.

B. Data Communication

Data communication includes between not only PLC and drivers/transducers, but also PLC and OS. Reliability of communication affects the system performance directly. When designing the control system, we try our best to make data communication easy and reliability.

1) Data Communication Between Drivers/Transducers and PLC: RS485 protocol is used to realize communication between drivers/transducers and PLC. On PLC we use CP340 RS422/485 communication processor to extend RS485 interface. Because Lenze drivers and CT transducers haven't RS485 interface, UD71 and RS232/485 module 2102 are used to realize RS485 communication respectively. Reliable communication between PLC and drivers/transducers is achieved via steps above. Please note that the configuration data is downloaded to drivers using RS232 protocol, using RS485 protocol when system running.

2) Data Communication Between OS and PLC: The OS MP370 and TP170A communicate with PLC through Profibus-DP. For PLC and OS, communication configuration must be done in Step7 and Protool software respectively. The key configuration portion is the setting of baud rate and device bus address. All the devices on one Profibus must have the same baud rate and different address.

C. Advanced Program Design -Eembedded VB Script

When developing HMI using ProTool on MP370[6], some complicated functions such as the computation of production efficiency can not be developed using general function block of Protool. So we use embedded VB script to realize these functions.

We design the computation of production efficiency using embedded VB script according to real-time production data. The data includes outputs, runtime and stop-time because of fault. In addition, the VB script program must be trigged by some concrete event. Because real-time production is variable, the change of production is selected as trigger. The production efficiency will be computed one time as long as production varieties, then the efficiency display will be refreshed. Adopting this method, we design the computation subprogram of production efficiency and other complicated function modes. Please note the relationship between project variables and temporary ones in scripts when designing VB scripts to realize complicated control algorithm.

V. IMPLEMENTATION RESULTS

The motion control system that is presented in this paper has already been implemented on the product line in a large-scale tobacco group successfully since 2004. The control system has abundant functions. The speed of slack motors can be decreased or increased in 1 rpm unit. In addition, control mode of the system has changed from mechanism drive to servo drive, and decreased noise greatly. Table I and Table II is draw resistance of filter rod with the length of 120 mm before and after adopting the motion control system respectively. Here normal range of draw resistance with is 380 ± 30 mmH$_2$O.

Table I and II show that quality of draw resistance is improved greatly, at the same time stability and control precision increase. According to the results and documents from consumer, change range of draw resistance is very large, and even reaches 60 mmH$_2$O before the motion control system is implemented. Furthermore operators must verify and adjust system parameters every 10 minutes manually. After the motion control system is implemented, change range of draw resistance decreases to 20 mmH$_2$O, and the quality and output of production increase greatly. In addition, total automation is achieved without operator's participation, and good economic benefits are obtained.

TABLE I
DRAW RESISTANCE FROM OLD CONTROL SYSTEM (BEFORE)

No. Unit	1	2	3	4	5	6	7	8	9	10
mmH$_2$O	395	401	375	362	368	352	412	385	392	378

TABLE II
DRAW RESISTANCE FROM MOTION CONTROL SYSTEM (NOW)

No. Unit	1	2	3	4	5	6	7	8	9	10
mmH$_2$O	375	392	372	388	380	383	387	385	392	375

VI. Conclusions

Using some new technologies such as fieldbus and AC servo drive control, a suit of motion control system based on Profibus that is used for filter rod production machine is presented in this paper, and the key technologies are researched in depth. The performance results show that the motion control system improves output and quality of products and integrative control level, and saves energy and invest. Design thinking and application experience of the motion control system are useful for us to establish fieldbus control system in other industrial process, such as slack control system in spin industry.

References

[1] X.C. Gu, "Application of AC servo drive system in tobacco production machine," *China west science and technology*, no.10, pp.12-14, 2004.

[2] ProTool User Configuration Manual Based On Windows. Germany Siemens Ltd., 1999.

[3] *Lenze Manual.* Germany Lenze GmbH & Co KG, 1998.

[4] Global Driver Control Getting Started. Germany Lenze GmbH & Co KG, 2000.

[5] *Unidrive Modle No.1-5 User Manual.* England Control Techniques Drives Ltd., 1999.

[6] Multi Panel MP370 Device Manual. Germany Siemens Ltd., 2001.

2006 5th International Power Electronics and Motion Control Conference

Magnetic Pole Identification for PMSM at Zero Speed Based on Space Vector PWM

Jiangang Hu, Longya Xu, *Fellow IEEE* and Jingbo Liu
Department of Electrical and Computer Engineering
Columbus, OH 43210, USA
Email: hu.158@osu.edu, xu.12@osu.edu

Abstract— **This paper contributes to an improved magnetic pole identification based on space vector PWM for an arbitrary initial rotor position of PM Synchronous Machines. The principle of magnetic pole identification is discussed and the N-S pole identification method based on space vector PWM presented. The improved identification method is verified by experiment results.**

Keywords-Magnetic pole identification; space vector PWM; initial rotor position; sensorless control; PM synchronous machine

I. INTRODUCTION

Significant attention has been paid to sensorless control of permanent magnet (PM) synchronous machines recently. For sensorless control of PM synchronous machines at zero speed, both the initial d-axis position and the polarity of the magnetic pole are to be identified. Although a conventional high frequency injection (HFI) method for initial rotor position estimation can identify the axis of the magnets but not the magnetic polarity. The conventional method for the North-South (N-S) pole is determined based on the magnetic saturation [1, 2]. The commonly used scheme is to inject pilot voltages and then detect the corresponding currents to determine the pole polarity.

In this paper, we propose a simple yet effective method to identify the N-S pole based on space vector PWM suitable for arbitrary initial rotor position at zero speed.

The principle of the magnetic pole identification is briefly discussed and the proposed method introduced. The validity of the proposed identification method is verified by experiment results.

II. IDENTIFING PERMANENT MAGNET POLARITY

The principle of identifying the magnet pole based on magnetic saturation effects has been discussed in literature and is illustrated in Figs.1 (a), (b) and (c). Initially, the flux increases in direct proportion to the increase of currents. Further increases in currents result in progressively smaller increases in flux because of magnetic saturation. As shown in Fig.1 (c), with the permanent magnet excitation alone, the original operating point is "A". However, with the stator excitation voltage applied in the same polarity with respect to that of the permanent magnet, the stator currents will increase from i_0 to i_1, and operating point from "A" to "B" due to the deep saturation. On the other hand, when the applied stator excitation voltage and magnetic pole are in the opposite directions, the magnetic path will not be saturated. The stator current will change from i_0 to i_2. At the same time, the operating point moves accordingly from point A to C. In these two cases, the applied volt-seconds are the same but different in polarity. Due to core saturation nonlinearity, the magnitudes of current variations are different. Therefore, we can detect the polarity of permanent magnets mounted on the rotor.

When the magnetic pole is in an arbitrary initial position as shown in Fig.1 (a), we have to detect the axis

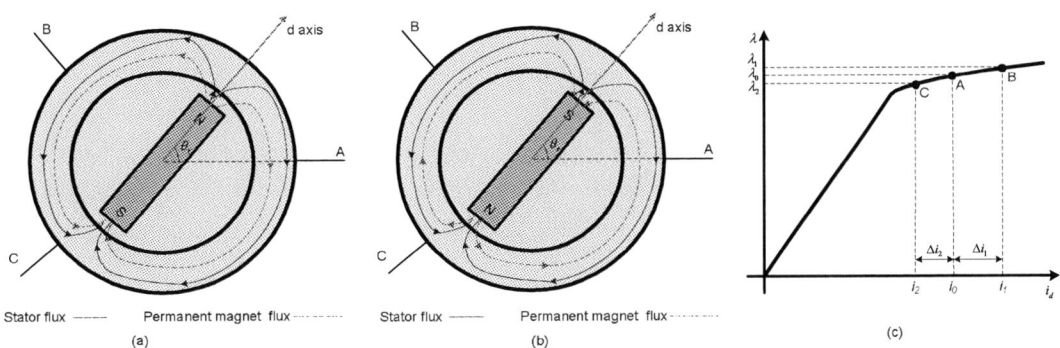

Figure 1.   Magnetic pole identification based on saturation effects

1-4244-0448-7/06/$25.00 ©2006 IEEE

of the permanent magnet first, normally achieved by the high frequency voltage injection method but the polarity of the magnet is left unknown. In the second step, according to the axis direction of the permanent magnet, positive and negative pilot voltages are applied sequentially in a controlled mode such that the volt-seconds are the same in both cases. When the positive voltage is applied, the current incremental is $+|i_1|$ and when negative the current incremental is $-|i_2|$. In the third step, we compare the magnitudes of the current incrementals. If $|i_1|$ is larger than $|i_2|$, the magnetic pole that corresponds to $|i_1|$ is the North and the rotor position in the range from 0 to π. Similarly, when the magnetic

pole is in the orientation as shown in Fig.1 (b), the corresponding current ($|i_2|$) with the positive stator excitation applied will be smaller than current ($|i_1|$) with the negative stator excitation applied. Therefore, the magnetic pole that corresponds with $|i_1|$ is the North Pole. In this case, the rotor position angle is actually an angle between π and 2π and the angle will be $\theta_r + \pi$.

III. IDENITFYING POLE POLASRITY BASED ON SPACE VECTOR PWM

In the proposed magnetic poles identification scheme, the principles discussed above are applied but the voltages are based on the space vector PWM that is readily

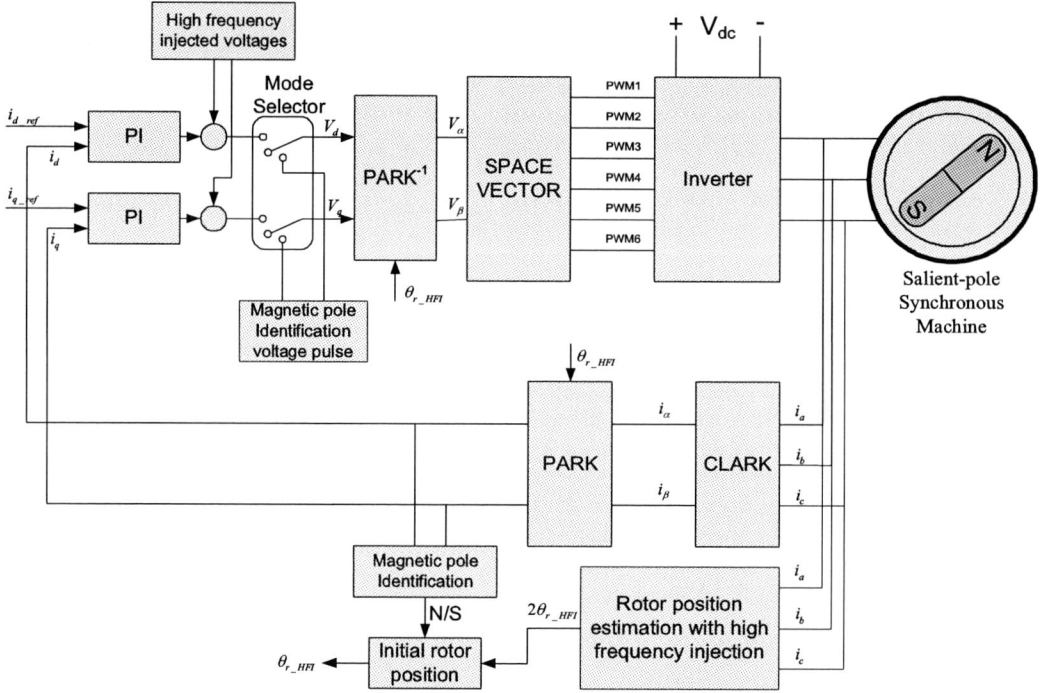

Figure 2. System diagram for rotor position estimation

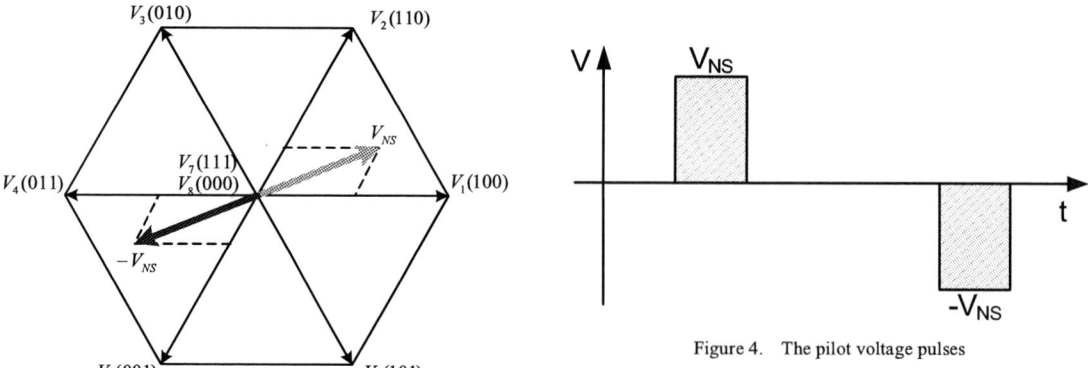

Figure 3. Space vectors

Figure 4. The pilot voltage pulses

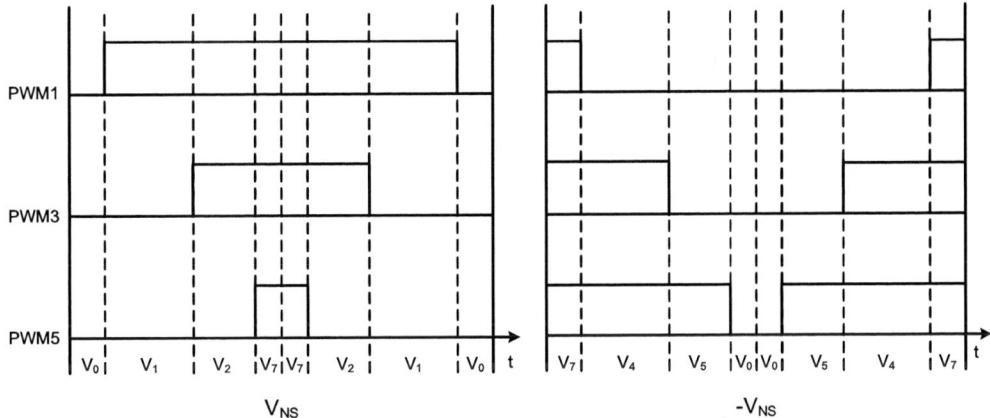

Figure 5. Pilot voltage vectors in one PWM cycle for magnetic pole identification: V_{NS} and $-V_{NS}$

available in the controller hardware and software. The system diagram for rotor position estimation with magnetic pole identification is shown in Fig. 2. The space vector is defined as shown in Fig. 3. When we know the d-axis direction of the rotor, we can conveniently figure out the phase windings closest to the rotor poles. Then, a group of pilot voltage vectors will be applied to the phase windings closest to the rotor axis to detect the magnetic poles. The pilot voltages are composed of one positive pulse and one negative pulse alternatively. Normally, the rotor's initial position could be arbitrary and the magnetic pole is not perfectly aligned with any phase windings. As an example, the V_{NS} and $-V_{NS}$ shown in Fig. 4 are one pair of pilot voltage vectors corresponding to an arbitrary

initial rotor position. The applied SVPWM switching signals are shown in Fig. 5. The sensed three phase currents through A/D circuits will be transformed into the synchronous reference frame first to obtain the relative information on the polarity of magnetic pole. If the current in the synchronous reference frame is larger when V_{NS} applied than the current when $-V_{NS}$ is applied, the magnetic pole must be the North pole. Note that the applied voltage vectors should be sufficiently large to cause the magnetic saturation.

One special case is that the rotor magnetic pole is aligned with one of phase windings. For example, only vectors V_1 and V_4 (shown in Fig. 3) would be sufficient to detect magnetic pole provided that rotor magnetic pole is

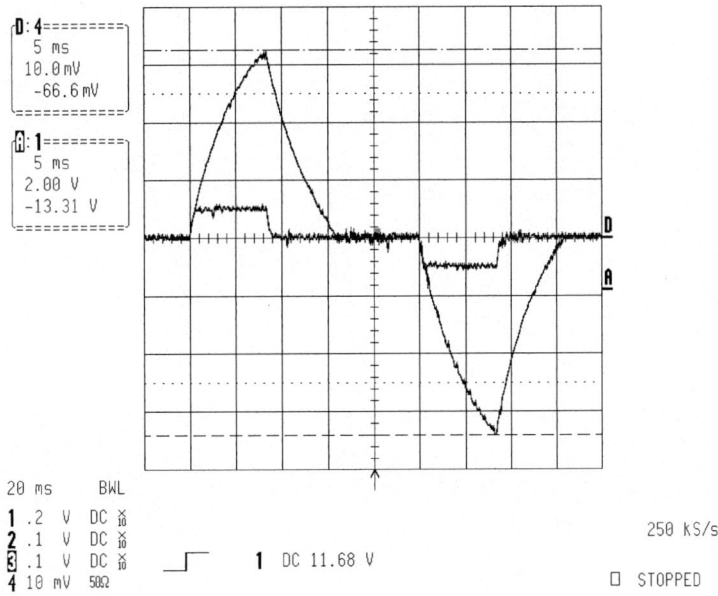

Figure 6. Magnetic pole identification (Channel A: applied voltage pulses; Channel D: The current in synchronous reference frame)

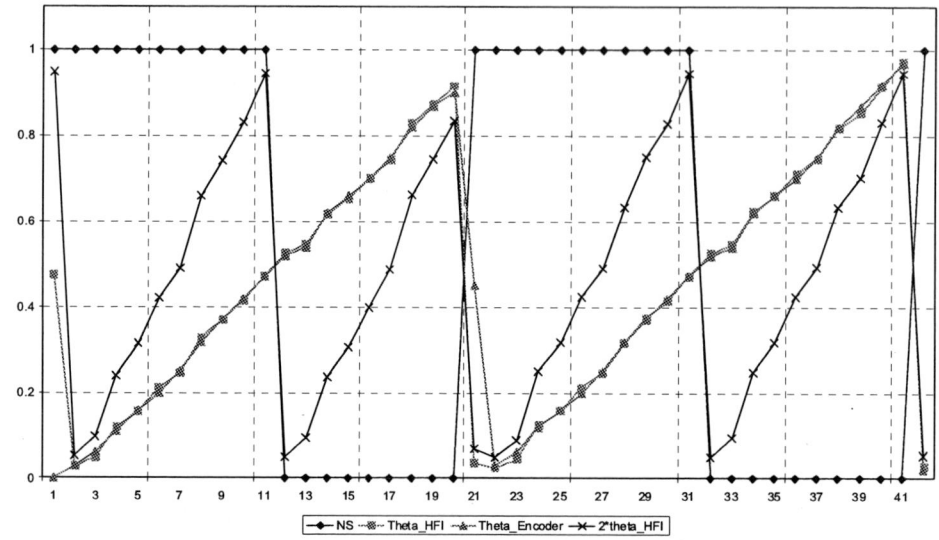

Figure 7. The estimated rotor position θ_{r_HFI} with magnetic pole detection (Calibrated by a 60-pulse encoder ($\theta_{r\_Encoder}$))

aligned with Phase "A" axis. In this case, positive and negative flux linkage is produced only with Phase "A" winding and therefore, we can obtain the N-S pole information by checking Phase "A" current peak value only.

IV. EXPERIMENTAL RESULTS

Experiments have been done based on a 5 HP salient pole permanent magnet synchronous machine. One of the experimental results is shown in Fig. 6. The waveform shown in Channel A is the applied pilot voltage pulses and Channel D the stator current that transformed into the synchronous reference frame at zero speed (10A/div). The magnitude of positive current is 32.5A and the magnitude of negative current is 34A with the aforementioned pilot voltage vectors based on SVPWM (V_{NS} and $-V_{NS}$). Obviously, the magnetic pole corresponding to applied negative current is North pole and accordingly the rotor position angle is between π and 2π.

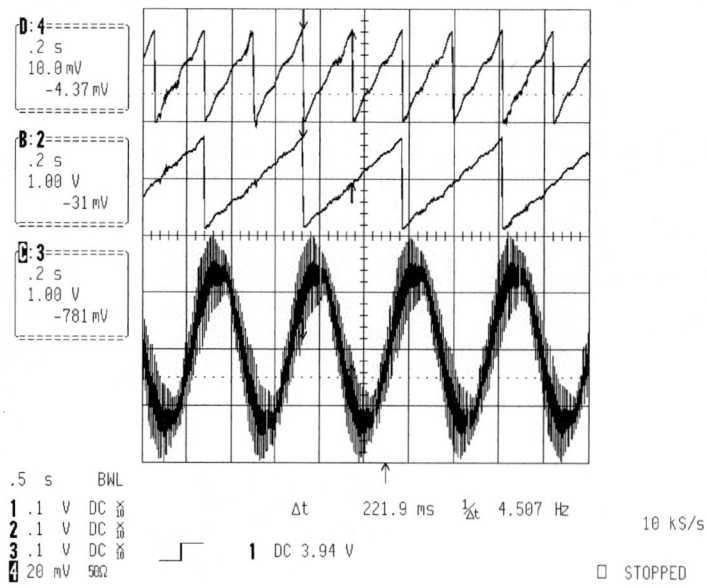

Figure 8. Low speed (2.25Hz) with fundamental current waveforms (top trace is $2\theta_{r\_HFI}$, middel is θ_{r_HFI} and bottom trace is the phase A current)

An encoder mounted on the rotor shaft is used to calibrate the N-S pole detection method. The detected magnetic pole information is applied with $2\theta_{r\_HFI}$ to generate the estimated rotor position θ_{r_HFI} as shown in Fig. 7. The θ_{r_HFI} matches the real rotor position $\theta_{r\_Encoder}$ very well. Fig. 8 shows the estimated rotor position waveforms of $2\theta_{r\_HFI}$, θ_{r_HFI} and phase current in a constant low speed range with regulated currents. The results show that the proposed magnetic pole detection scheme works very well for the estimation of the arbitrary initial rotor position.

V. SUMMARY

In this paper, an improved magnetic pole identification method based on space vector PWM for the arbitrary initial rotor position estimation and sensorless control of PM synchronous machines is presented. Through applying vector controlled pilot voltages by SVPWM, the N-S pole can be identified at any initial rotor positions at any including zero rotor speed without rotor alignment actions. The proposed method can be combined with the conventional initial rotor position and sensorless control scheme to ensure the effective estimation of initial magnetic pole position. The validity of the proposed identification method has been proven through experimental results based on a 5-HP salient-pole synchronous machine drive system.

REFERENCES

[1] Chuanyang Wang and Longya Xu, "A novel approach for sensorless control of PM machines down to zero speed without signal injection or special PWM technique", IEEE Trans. on Power Electronics, Vol. 19, No. 6, pp. 1601-1607, Nov 2004.

[2] Takashi Aihara, Akio Toba, Takao Yanase, Akihide Mashimo and Kenji Endo, "Sensorless torque control of salient-pole synchronous motor at zero-speed operation", IEEE Trans. on Power Electronics, Vol. 14, No. 1, pp. 202-208, Jan 1999.

2006 5th International Power Electronics and Motion Control Conference

Study on Stagewise Control of Connecting DFIG to the Grid

Xueguang Zhang, Dianguo Xu, *IEEE member*, Yongqiang Lang, Hongfei Ma
Department of Electrical Engineering
Harbin Institute of Technology, Harbin 150001,China
zxghit@126.com

Abstract—A novel control scheme in grid-voltage-orient frame for connecting doubly fed induction generator (DFIG) to the grid is proposed. The model of DFIG in grid-voltage-orient frame is analyzed. A PI controller to compensate the rotor position initial error is adopted in this design Stagewise control strategy based on the DFIG model is proposed. The method to design the controllers is described. The grid synchronization procedure is analyzed in detail. The proposed techniques have been implemented on a 4.2kW experimental DFIG system. Experimental results are presented to illustrate the effectiveness of the proposed techniques.

Keywords- DFIG, rotor position initial error compensation, stagewise control, grid connection

I. INTRODUCTION

Many countries have now recognized the wind power as a sustainable source of energy. Institutional and governmental support on wind energy conversion technology has led to a fast development of wind power generation in recent years. Several benefits can be achieved by letting the turbine speed varying with the wind fluctuation. A wind turbine with variable speed can be realized with a DFIG system [1-3].

In the scheme of DFIG the stator is directly connected to the grid and the rotor is supplied by a conventional AC-DC-AC converter, which consists of two voltage source PWM converter back-to-back connected. Direct access to the rotor circuit of DFIG by the converter provides a means to control the phase and amplitude of the currents flowing through the stator circuit [4].

Grid voltage oriented vector control have been popularly used for DFIG[5], in this case, the rotor position relative to the stator A phase winding is needed for converting the rotor currents into the synchronous frame. Due to the mechanical installation of the incremental encoder, the rotor initial position can not be known beforehand. One method is that experimentally determining the rotor position at the point of the positioning pulse (rising or falling edge) from the encoder. Driven by shaft coupled wind turbine, DFIG rotates before control strategy being implemented. Control system of the converter will not start until the positioning pulse is detected. However, experimentally determining

the initial rotor position is not convenient each time the encoder has to been reinstalled.

A compensation strategy is proposed in this paper which can compensate the initial error of the rotor position automatically. With being applied to the process of connecting DFIG wind power generation system to the grid, a stagewise control strategy is proposed.

II. MODEL OF DFIG

The model of DFIG in synchronous frame includes the following equations:

$$\begin{cases} u_{sd} = p\psi_{sd} - \omega_1\psi_{sq} - R_s i_{sd} \\ u_{sq} = p\psi_{sq} + \omega_1\psi_{sd} - R_s i_{sq} \end{cases} \quad (1)$$

$$\begin{cases} u_{rd} = p\psi_{rd} - \omega_2\psi_{rq} + R_r i_{rd} \\ u_{rq} = p\psi_{rq} + \omega_2\psi_{rd} + R_r i_{rq} \end{cases} \quad (2)$$

$$\begin{cases} \psi_{sd} = -L_s i_{sd} + L_m i_{rd} \\ \psi_{sq} = -L_s i_{sq} + L_m i_{rq} \end{cases} \quad (3)$$

$$\begin{cases} \psi_{rd} = L_r i_{rd} - L_m i_{sd} \\ \psi_{rq} = L_r i_{rq} - L_m i_{sq} \end{cases} \quad (4)$$

$$T_{em} = 1.5pL_m(i_{sq}i_{rd} - i_{sd}i_{rq}) \quad (5)$$

III. INITIAL ERROR OF ROTOR POSITION COMPENSATION

3.1 Definition of rotor position initial error

Applying grid voltage oriented vector control, d component of stator voltage will be control to Us, which is rated value for stator voltage and equals to the grid voltage magnitude; q component of stator voltage will be control to zero.

There two important angles for grid voltage oriented vector control of DFIG: grid voltage phase angle (θ_s) and rotor position angle (θ_r). Grid voltage phase angle can be determined easily. Rotor position angle is important to calculate the slip angle which is used to achieving vector transformation for rotor current. Rotor position angle is usually got from an incremental encoder. Due to the mechanical installation, the rotor initial position can not be known beforehand and is usually set to zero or a certain angle in software which defiantly causes error which we define as the initial error of rotor

1-4244-0448-7/06/$25.00 ©2006 IEEE

position (θ_{error}). Without any compensation method, control performance of DFIG will deteriorate.

With initial error of rotor position, the orientation on the rotor side will be deviated from the synchronous frames (d-q), here we define deviated frames as d'-q' frame. The angle difference between two frames is θ_{error} (Fig.1).

3.2 Compensation controller of initial error of rotor position

Compensation PI controller use q axis component of stator voltage only (u_{sq}) as input:

$$\theta_c = \left(K_p + \frac{K_i}{s}\right)\left(0 - u_{sq}\right) \qquad (6)$$

The output of the controller here we name it compensation angle (θ_c), is added to the rotor position angle measured by the incremental encoder. After the compensation, the actual error of rotor position becomes:

$$\theta'_{error} = \theta_{error} + \theta_c \qquad (7)$$

So the objective is to compensate the dynamic angle error to be zero ($\theta'_{error} \to 0$).

The condition for the compensation controller is that only rotor current closed loop is applied with stator open circuit. Using PI closed loop controller, rotor currents can be controlled to the reference in the deviated synchronous frame ($d' - q'$). When the current references are:

$$\begin{cases} i_{rqref} = -\dfrac{u_{sdref}}{\omega_1 L_m} \\ i_{rdref} = 0 \end{cases} \qquad (8)$$

According to the mathematical model of DFIG, induced stator voltage in deviated frame will be $u'_{sd} = u_{sdref}$, $u'_{sq} = 0$. From Fig.1, stator voltage in synchronous frame can be got:

$$u_{sq} = u'_{sd} \sin \theta_{error} \qquad (9)$$

$$u_{sd} = u'_{sd} \cos \theta_{error} \qquad (10)$$

Look-up table based on (9) and (10) is the most straightforward method to get θ_{error}. However, 2-D table has to be used which occupied much hardware resource and its implementation is complex and time consuming, so this method is rarely adopted.

3.3 Principle of compensation

Substituting (9) into (6), θ_c can be expressed as following:

$$\theta_c = K \sin \theta_{error} \left(K_p + \frac{K_i}{s}\right) \qquad (11)$$

Where: $K = -u'_{sd}$, $K < 0$.

In order to analyze the compensation process, it is necessary to divide the whole plane into three regions: the

first region is the right half plane, including vertical axis; the second region is the sum of the second and third quadrant exclusive of both axes; the third region is the minus part of the horizontal axis.

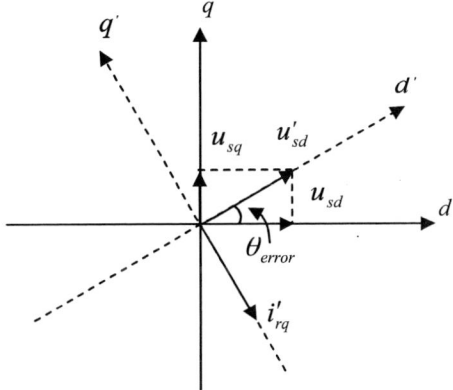

Fig.1 Grid voltage vector oriented synchronous frames

(1) $\theta_{error} \in [-\dfrac{\pi}{2}, \dfrac{\pi}{2}]$

When the initial error of rotor position falls into this region, because the compensation angle θ_c is proportional to the sinusoidal value of θ_{error}, a series of curves of θ_c are got with the variety of K, as shown in Fig.2. The slope coefficient of the straight line is -1, and the point of intersecting with cures of θ_c is the desired state of compensation.

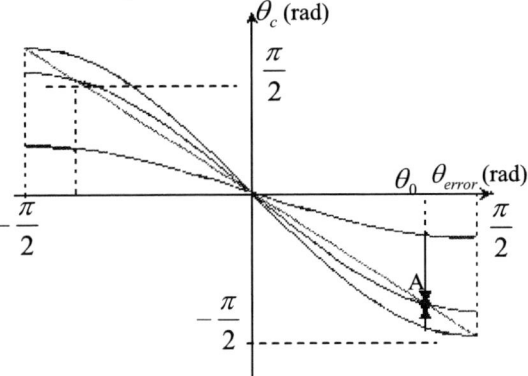

Fig.2 Curves of compensation angle when initial error falls into the first region

When $\theta'_{error} > 0$, from (9) it is got that through rotor currents close-loop control, u_{sq} will be positive ($u_{sq} > 0$) and then the compensation controller (6) outputs negative θ_c, from (7), the dynamic error θ'_{error} decreases. When $\theta'_{error} < 0$, the compensation controller outputs a positive θ_c, so θ'_{error} increases. Finally, θ'_{error} gradually approaches the zero point, at the same time the desired state, intersecting point A (Fig.2) is reached.

(2) $\theta_{error} \in (\frac{\pi}{2}, \pi)$ or $\theta_{error} \in (-\pi, -\frac{\pi}{2})$

In this region, the compensation process is similar to the previous one. The absolute value of θ'_{error} decreases gradually and then definitely will be pushed into the first region. However, from (9), it is obvious that when θ'_{error} is in this region, the magnitude of u_{sq} increases gradually which is different from that in the first region.

(3) $\theta_{error} = \pi$ (or $\theta_{error} = -\pi$)

It is unfortunate that the compensation controller can not deal with the point ($\theta_{error} = \pi$). However, by detecting such a state, the compensation angle θ_c can be set to $\theta_c = -\pi$, or by adding θ_{noise} ($\theta_{noise} \neq \pi$) to θ_{error}, θ_{error} is pushed into the previous two region.

IV. STAGEWISE CONTROL OF CONNECTING DFIG TO THE GRID

The stagewise control strategy is proposed to connect DFIG to the grid. The process is composed of three stages in sequence: the stage of compensating initial error of rotor position, the stage of establishing stator voltage and the stage of connecting DFIG to the grid.

4.1 The stage of compensating initial error of rotor position

In this stage the initial error of rotor position is compensated by the compensation controller, and only rotor currents close-loop control is adopted. The current reference is determined by (8) for two purposes: Firstly, with the stator side open circuited the induced stator voltage is proportional to rotor current, so the compensation precision will be improved using as large stator voltage as possible. Secondly, when the compensation is finished, the stator voltage will approach its rated value which can make the second stage as short as possible.

The rotor current controller is as follows:

$$\begin{cases} u_{rdref} = \left(K_{ip} + \dfrac{K_{il}}{s} \right)\left(i_{rdref} - i_{rd} \right) - \omega_2 L_r i_{rq} + \omega_2 L_m i_{sq} \\ u_{rqref} = \left(K_{ip} + \dfrac{K_{il}}{s} \right)\left(i_{rqref} - i_{rq} \right) + \omega_2 L_r i_{rd} - \omega_2 L_m i_{sd} \end{cases}$$
(12)

When the stage ends, the compensation angle θ_c will maintain its last value and the second stage follows.

4.2 The stage of establishing stator voltage

In the second stage, same as in the first stage, the stator is open and the magnetizing current of the generator is supplied by the power converter at the rotor terminals. The stator voltage is indirectly controlled by adjusting the rotor current to meet the requirement of the grid connection: amplitude, frequency and phase angle of the stator voltage have to be as same as possible with those of

the grid voltage which can make the grid connection process smooth, i.e. without large inrush current. In this stage stator voltage close-loop control is adopted [5].

The stator voltage controller is as follows:

$$\begin{cases} i_{rdref} = \left(K_{up} + \dfrac{K_{ul}}{s} \right)\left(u_{sqref} - u_{sq} \right) \\ i_{rqref} = -\left(K_{up} + \dfrac{K_{ul}}{s} \right)\left(u_{sdref} - u_{sd} \right) \end{cases}$$
(13)

Because grid voltage-oriented control is adopted here, $u_{sdref} = U_s$, $u_{sqref} = 0$.

At the end of the second stage, the requirement of the grid connection is met, then ac contactor between the stator and the grid will be triggered, DFIG will be connected to the grid smoothly.

4.3 The stage of connecting DFIG to the grid

At the end of the previous stage the stator of DFIG is connected to the grid directly. Assuming that the grid is strong enough, the stator voltage is determined by the grid voltage. In this stage, the power control strategy is applied to achieve the decoupled control of active and reactive power. The reference of active power tracks the wind velocity to maximize the wind energy conversion; the reference of the reactive power is determined by the grid demand. The relationships between rotor currents and the stator side power can be got according to the mathematical model of DFIG [6,7]:

$$\begin{cases} i_{rd} = \dfrac{L_s}{L_m} \dfrac{2P_a}{3U_s} \\ i_{rq} = -\dfrac{L_s}{L_m} \dfrac{2P_r}{3U_s} - \dfrac{U_s}{\omega_1 L_m} \end{cases}$$
(14)

Obviously the active power and reactive power at stator side are decoupled from each other.

The power controllers are:

$$\begin{cases} i_{rdref} = \left(K_{pp} + \dfrac{K_{pl}}{s} \right)\left(P_{aref} - P_a \right) \\ i_{rqref} = -\left(K_{pp} + \dfrac{K_{pl}}{s} \right)\left(P_{rref} - P_r \right) - \dfrac{U_s}{\omega_1 L_m} \end{cases}$$
(15)

4.4 Problems with control strategy switch

The most important task of stagewise control is to tackle the strategies switch which may cause inrush in currents: rotor and stator. Obviously, within three stages the inrush is caused mainly by large sudden change in rotor current references.

In the stage of compensating initial error of rotor position, the rotor current references are selected according to (9) which equals to the steady rotor currents in the following stage. At the end of the stage, induced stator voltage will approach its rated value, the proportional parts of the voltage controller are almost zeros, and so the inrush can be avoided by initializing the integral components of the controller as (9). When the

710

stage of establishing stator voltage switches to the power control stage, because of the existence of the term ($-\dfrac{U_s}{\omega_1 L_m}$) in reactive power controller (15), the continuity of rotor current reference is achieved and the inrush can be avoided too.

V. EXPERIMENTAL RESULT

The experimental platform of variable speed constant frequency DFIG wind power generation system has been established in the lab. The Scheme of VSCF doubly fed induction machine generating system experimental platform is shown in Fig.3.

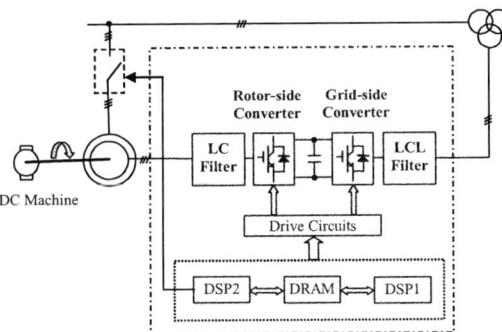

Fig.3 Scheme of VSCF doubly fed induction machine generating system experimental platform

In this design, the AC-DC-AC converter composed of two voltage PWM inverters connected back-to-back in rotor circuit supply the excited current. The stator side is connected to the grid directly. The control hardware comprises two fix point DSP controllers. Grid side inverter is controlled by DSP1 and rotor side inverter is controlled by DSP2. Some data is shared through dual-port RAM between two DSP controllers. DSP1 performs the vector control of the DFIG and it is also in charge of the grid connection control. To switch the control strategy, the statement of the grid connection is detected by DSP1 in real-time. The generator is a 4.2kW wound induction generator. the windmill is simulated by a DC motor.

The compensation strategy and stagewise control proposed have been verified experimentally. The experiment results are shown in Fig.4 to Fig.8.

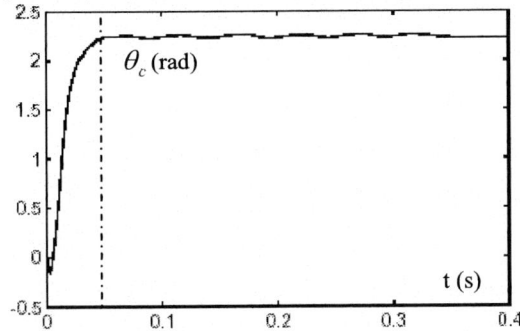

Fig.4 Response of compensation angle during the first stage

Fig.4 shows the curve of θ_c recorded by DSP in the stage of compensating initial error of rotor position. It can be seen that θ_c reaches the expected value at nearly 50ms. The compensation process takes about 0.35s. Afterwards, θ_c keeps its expected value.

Fig.5 shows the waveforms of both grid voltage and stator voltage during the first stage. With initial error of rotor position, there is obvious phase difference between the two voltages. When θ_c reaches the expected value first time (50ms), the phases of the two voltages are approximately the same. The phase difference disappears gradually and the compensation is achieved at about 0.35s.

Fig.6 shows waveforms (steady state) of both grid and stator voltage during the second stage. The requirement of grid connection is met at the end of the stage.

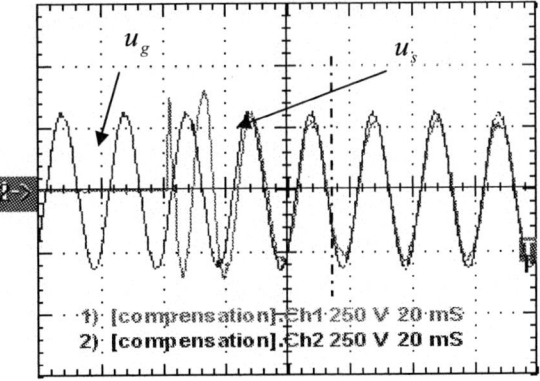

Fig.5 Waveforms of both grid voltage (u_g) and stator voltage

(u_s) during the first stage

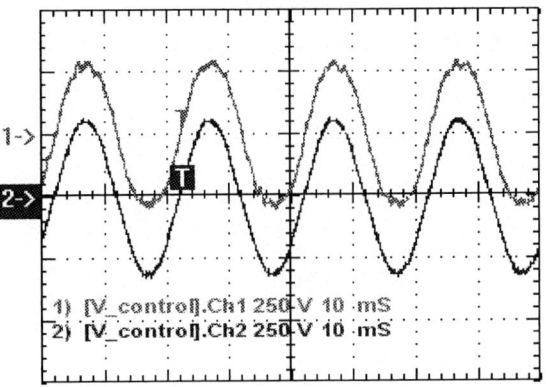

Fig.6 Waveforms of both grid voltage (u_g) and stator voltage

(u_s) during the second stage

Fig.7 shows the waveforms of both grid voltage and stator voltage at the grid connection instant. When the requirement of the grid connection is met, DFIG is connected to the grid through an ac contactor controlled by the controller. Afterwards, the stator voltage is the same as the grid voltage.

Fig.8 shows the waveforms of both grid voltage and stator current at connecting instant. It is obvious that the inrush current is very small. The system fast got into steady state controlled by the power controller.

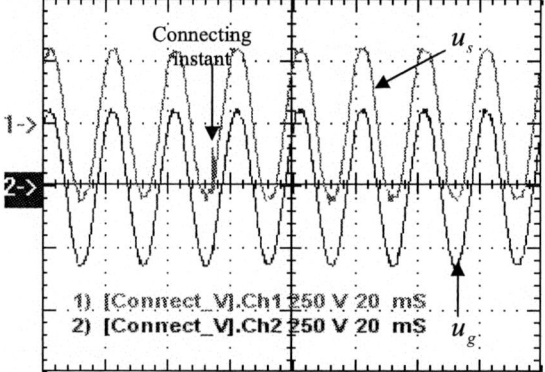

Fig.7 Waveforms of both grid voltage (u_g) and stator voltage

(u_s) at the grid connecting instant

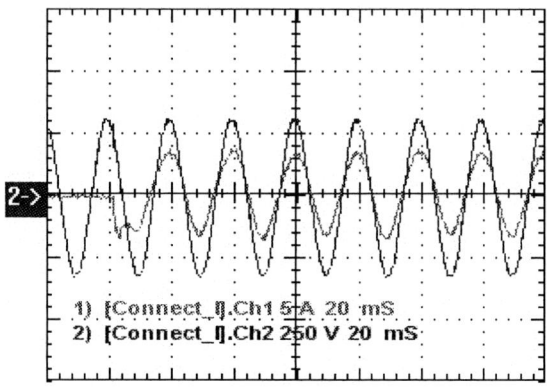

Fig.8 Waveforms of both grid voltage (u_g) and stator current

(i_g) at the grid connecting instant

VI. CONCLUSION

A compensation method for initial error of rotor position is proposed in this paper. Incorporating it into the control application of DFIG, a stagewise control strategy of connecting the generator to the grid is proposed, and design method of the controllers is given. The strategy proposed has been validated on the laboratory platform of VSCF DFIG wind power generation system. The experimental results show that the rotor position initial error can be compensated and soft grid connection of the DFIG was achieved under the stagewise control strategy.

REFERENCES

[1] Muller S, Deicke M, De Doncker R W. Doubly fed induction generator systems for wind turbines [J]. IEEE Industry Applications Magazine, 2002, 8(3): 26-33.

[2] Muller S, Deicke M, De Doncker R W. Adjustable speed generations for wind turbines based on doubly-fed induction machines and 4-quadrant IGBT converters linked to the rotor [C]. IEEE Proceeding of IAS, 2000（4）: 2249-2254

[3] R Hoffmann, P Mutschler. The influence of control strategies on the Energy capture of wind turbines [C]. IEEE Proceeding of IAS, 2000（2）: 886-893.

[4] Richard Piwko, Xiaoming Yuan, Nicholas Miller, Renchang Dai, James Lyons [C]. Proceedings of 2005 IEEE/PES Transmiss -ion and Distribution Conference & Exhibition: Asia and Pacific, Dalian(China), 2005: 1193-1198.

[5] Pena R, Clare J C, Asher G M. Doubly fed induction generator using back-to-back PWM converter and its application to variable-speed wind-energy generation [J]. Electric Power Applications, IEE Proceedings, 1996, 143(3): 231-241.

[6] Sergei Peresada, Andrea Tilli, Alerto Tonielli. Power control of a doubly fed induction machine via output feedback. Control Engineering Practice, 2004(12): 41-57.

[7] A.Tapia, G.Tapia, J.X.Ostolaza. Reactive power control of wind farms for voltage control applications. Renewable Energy, 2004, 29: 377-392.

2006 5th International Power Electronics and Motion Control Conference

Generalized Control Approach for Active Power Filters

Xiaoyu Wang, Jinjun Liu, Chang Yuan, and Zhaoan Wang

School of Electrical Engineering
Xi'an Jiaotong University
28 West Xianning Road, Xi'an, Shaanxi, China
e-mail: xywang@ieee.org

Abstract—**Active Power Filter (APF) is one of the most effective solutions to compensate the current harmonics and to enhance the power quality in industrial application. Control is one of the key elements to their performance. This paper proposes a generalized control approach for APF, which provides a systematical way not only to analysis the regular control schemes but also develop new control schemes. Based on the simplified model of power stage, the development of generalized control is illustrated. With properly designed gains of detected currents, different control aims and compensation objectives can be achieved. Regular control scheme is reviewed and several new practical implementations of the generalized control approach are illustrated in detail. Simulation and experimental results are provided to verify the approach and analysis.**

Keywords-active power filter, control strategy

I. INTRODUCTION

Power quality problems in industrial application have been drawing more and more attentions these years, especially with the development of modern electronics industry. Active Power Filter (APF) is an effective solution to compensate the current harmonics and to enhance the power quality in industrial application [1, 2]. Parallel Active Power Filter (PAPF) is one of the most popular APFs and has been put into many field installations especially in Japan.

Control is one of the key elements to the performance of APF. The literature of APF contains reports of many control approaches, plus several reviews. To simplify the discussion, the control method for PAPF is focused only, and can normally be applied to a series compensation case using the ideas of duality.

Generally, PAPF is controlled as current source, and these control strategies consist of three parts: calculation of the compensating current reference; the current control of the voltage-source inverter (VSI); and the control of the DC-side capacitor voltage. Fig. 1 a) shows the equivalent circuit, in which the PAFP is substituted with a controlled current source for simplicity.

Since different control strategies deal with the same power stage, this paper combined together the reference calculation and current control of VSI of the previous control strategies. Different with regular current-mode control, the VSI is controlled as voltage source here, shown as Fig. 1b).

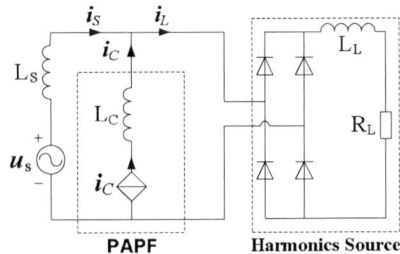

a) Regular current-mode control.

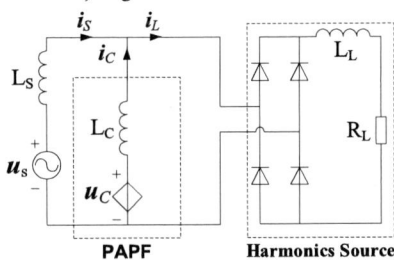

b) Generalized voltage-mode control.
Figure 1. Control algorithm of Parallel Active Power Filter (PAPF).

Generally, the fundamental function of any control approach is to calculate the duty cycle d of the VSI from the detected variable. This paper proposed a generalized control approach after modeling of the power stage. When proper gains in the proposed approach are designed, different control aims and compensation objectives can be achieved. In section IV, several practical implementations of the generalized control approach are discussed in detail.

The proposed generalized control approach provides a systematical way not only to analysis the regular control schemes but also develop new control schemes. Simulation and experimental results are provided to verify the approach and analysis.

1-4244-0448-7/06/$25.00 ©2006 IEEE

II. Generalized Control Approach

Fig. 2 shows the general system diagram of APF, where u_s represents the source voltage, i_S, i_C and i_L indicate the source current, compensate current and load current separately, and u^*_c is the reference to control the PAPF. In the following discussion, the subscript letter f indicates the fundamental, the subscript letter h is the harmonics, and the subscript letter s, c and L indicate the variables of the source side, the compensation and the load side separately.

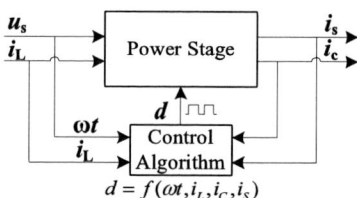

Figure 2 General control diagram of PAPF.

Since different schemes detect the similar system variables and control the same power stage, which is voltage-source inverter (VSI), the whole system can be divided into two parts as Fig. 2 in system level: power stage and control algorithm. The power stage is governed by the basic circuit laws, and the control strategies are represented by the abstract control function f. Accordingly, to analyze the system with APF and design its control strategies generally follows these two steps: first, to calculate the relationship between the system variables and sources; then to design the control algorithm according to the compensation objectives.

A. Model of Power Stage

Fig. 3 shows the linear model of the power stage of PAPF, where u_S represents the source, and u_C represents the power electronics inverter in PAPF. Typically, the harmonics load, that PAPF are used to compensate, has const input current and can be view as harmonics current source. In Fig. 3, i_{Lh} represents this kind of current-source-type harmonics-generating load with a const harmonics current source.

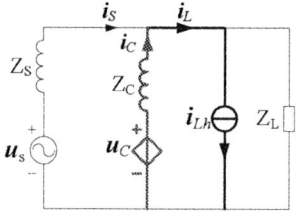

Figure 3 Equivalent circuit of power stage.

There are two independent sources u_S, i_{Lh}, which are not controllable, and one controllable voltage source u_C. These three sources stimulate the power stage, which is governed by Kirchoff's Voltage Law (KVL) and Kirchoff's Current Law (KCL) as the following equations in Laplace form:

$$\begin{cases} U_S = I_S \cdot Z_S + U_L \\ U_C = I_C \cdot Z_C + U_L \\ I_S + I_C = I_{Lh} + \dfrac{U_L}{Z_L} \end{cases} \quad (1)$$

By analysis these equations that govern the power stage, the relationship between the output currents (source current and compensated current) and the input (three independent voltage sources) can be got as (2).

$$\begin{bmatrix} I_S \\ I_C \end{bmatrix} = \begin{bmatrix} \dfrac{1}{Z_S + Z_C} & \dfrac{Z_C}{Z_S + Z_C} & -\dfrac{1}{Z_S + Z_C} \\ -\dfrac{1}{Z_S + Z_C} & \dfrac{Z_S}{Z_S + Z_C} & \dfrac{1}{Z_S + Z_C} \end{bmatrix} \cdot \begin{bmatrix} U_S \\ I_{Lh} \\ U_C \end{bmatrix} \quad (2)$$

Fig. 4 illustrates this simple model with control diagram. Two independent sources u_S, i_{Lh}, and one controlled voltage source are the input of power stage model. Source current i_S and compensated current i_C are the output.

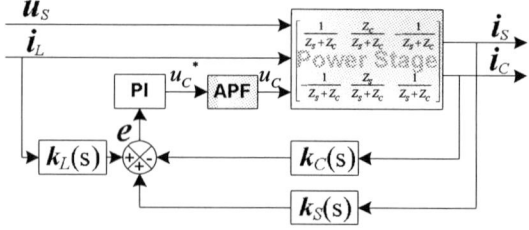

Figure 4. Simplified model of power stage.

B. Generalized Control Approach

The fundamental function of any control approach is to calculate the duty cycle d from the detected system variable. Theoretically, the control reference u^*_c to the VSI can be obtained from lots of system variables, yet for practical purpose in most application, the reference u^*_c comes from the currents, which may be source current, load current and compensated current. So any control function in Lapalace Transformation form can be represented as (3):

$$U^*_C(s) = K_S(s) \cdot I_S(s) - K_C(s) \cdot I_C(s) + K_L(s) \cdot I_L(s) \quad (3)$$

Here K_S, K_C and K_L are the feedback gain of corresponding detected current, which may be not only simple as proportional index, but also complicated as a low-pass filter (LPF) or high-pass filter (HPF). Fig. 5 shows the whole system diagram combining the generalized control function to the power stage of APF.

Figure 5 Whole system diagram of PAPF with generalized approach.

Equivalent (3) provides a generalized approach to analysis and control PAPF, by which not only regular

714

control schemes can be reviewed in a systemic way, but also new control schemes can be obtained.

If proper gains of K_L, K_S and K_C are designed carefully, different control aims and compensation objectives can be achieved. By substituting the control function (3) to the power stage (2), the final system response can be obtained as (4).

$$
\begin{cases}
I_S = \dfrac{1+\frac{Z_C+K_C}{Z_L}}{(Z_S+K_S)+(Z_C+K_C)+\frac{Z_S\cdot(Z_C+K_C)}{Z_L}} \cdot U_S + \dfrac{Z_C+K_C-K_L}{(Z_S+K_S)+(Z_C+K_C)+\frac{Z_S\cdot(Z_C+K_C)}{Z_L}} \cdot I_{Lh} \\[2em]
I_C = -\dfrac{1-\frac{K_S}{Z_L}}{(Z_S+K_S)+(Z_C+K_C)+\frac{Z_S\cdot(Z_C+K_C)}{Z_L}} \cdot U_S + \dfrac{Z_S+K_S+K_L+\frac{Z_S}{Z_L}K_L}{(Z_S+K_S)+(Z_C+K_C)+\frac{Z_S\cdot(Z_C+K_C)}{Z_L}} \cdot I_{Lh}
\end{cases}
\tag{4}
$$

In the application of APF, the compensation object is to obtain pure sinusoidal I_S and pure harmonics I_C. So if $K_S = Z_L$, $K_C = -Z_C$ and $K_L = 0$, the final system response can be obtained as (5), which means in this control approach, the source current I_S is only determined by the linear interaction between the source voltage and the load, and the compensated current is just equal to the load current harmonics as expected.

$$
\begin{cases}
I_S = \dfrac{1}{Z_S+Z_L} \cdot U_S \\[1em]
I_C = I_{Lh}
\end{cases}
\tag{5}
$$

But in practice, it is difficult to detect the exact impedance of Z_L and Z_C, several practical implementations of the generalized control approach are proposed in section IV.

III. REVIEW OF REGULAR CONTROL

With proper feedback gain of K_L, K_S and K_C, regular control schemes of PAPF can be derived. When $K_c = 1$, $K_L = HPF$ and and $K_S = 0$, the generalized control algorithm can be simplified to

$$
U^*_C(s) = HPF(s) \cdot I_L(s) - I_C(s) \tag{6}
$$

in s domain (with Lapalace Transformation), or

$$
u^*_C(t) = i_{Lh}(t) - i_C(t) \tag{7}
$$

in time domain.

Fig. 6a) shows the whole system diagram with this control scheme, and Fig. 6b) shows the corresponding control diagram.

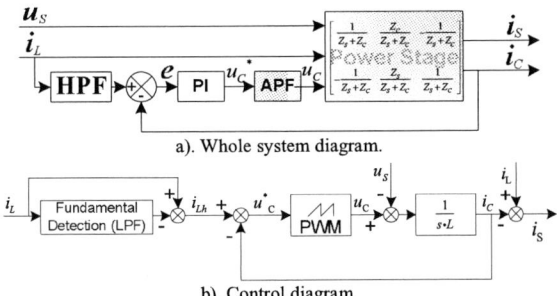

a). Whole system diagram.

b). Control diagram.

Figure 6. Control diagram of regular control scheme.

When the control system is fast enough, the compensated current i_c could follow i_{Lh}, thus the VSI is controlled as current source with i_{Lh} as current reference. Actually, the control diagram also represents the regular current-mode control. So in this case the equivalent circuit of whole system can be shown as Fig. 1a).

IV. PRACTICAL IMPLEMENTATIONS

With different feedback gain of K_L, K_S and K_C, new control schemes of PAPF can be derived. In this section, several new practical implementations of the generalized control approach are discussed in detail.

A. Detecting Both Source Current and Load Current

When $K_S = k_S$, $K_L = -k_L \cdot LPF(s)$, and $K_C = 0$, the control algorithm can be simplified to

$$
U^*_C(s) = k_S \cdot I_S(s) - k_L \cdot LPF(s) \cdot I_L(s) \tag{10}
$$

in s domain (with Lapalace Transformation), or

$$
u^*_C(t) = k_S \cdot i_S(t) - k_L \cdot i_{Lf}(t) \tag{11}
$$

in time domain.

Fig. 7 shows the corresponding control diagram. In this scheme, the source current i_S is controlled to follow only the fundamental part of load current, thus the other part of load, which is the load harmonics current, can't flow through the source.

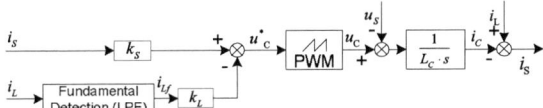

Figure 7. Control diagram when detecting the source current and compensated current

B. Detecting Both Source Current and Compensated Current

When $K_S = k_S \cdot HPF(s)$, $K_C = k_C \cdot LPF(s)$ and $K_L = 0$, the control algorithm can be simplified to

$$
U^*_C(s) = k_S \cdot HPF(s) \cdot I_S(s) - k_C \cdot LPF(s) \cdot I_C(s) \tag{8}
$$

in s domain (with Lapalace Transformation), or

$$
u^*_C(t) = k_L \cdot i_{Sh}(t) - k_C \cdot i_{Cf}(t) \tag{9}
$$

in time domain.

Fig. 8 shows the corresponding control diagram.

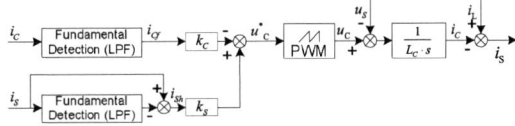

Figure 8. Control diagram when detecting source current and compensated current.

Fig. 9 illustrates this with equivalent circuit diagrams. In this case, the APF functions as a capacitor connected with a resistor at low frequency shown as Fig. 9a). While

715

in high frequency, the APF is a transconductor as the one shown in Fig. 9b).

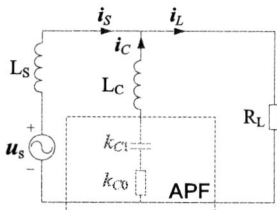

a) Fundamental equivalent circuit.

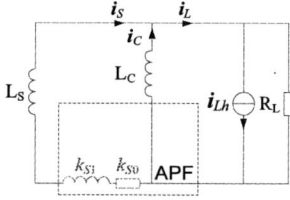

b) Harmonics equivalent circuit.

Figure 9. Equivalent circuits of PAPF when detecting source current and compensated current.

C. Only Detecting Compensated Current

The proposed generalized control approach is effective even when only compensated current is detected. In this case, less current sensors are required; hence the cost of APF is reduced evidently which is a key issue to widely application of APF.

When $K_S=0$, $K_L=0$ and $K_C = -k_C \cdot LPF(s)$ the generalized control algorithm can be simplified to

$$U^*_C(s) = -k_C \cdot LPF(s) \cdot I_C(s) \qquad (12)$$

in s domain (with Lapalace Transformation), or

$$u^*_C(t) = -k_C \cdot i_{Cf}(t) \qquad (13)$$

in time domain.

In this way, the APF works just like frequency-selective impedance. It presents zero impedance for the harmonics frequency and acts as a resistor with high resistance of k_C [Ω] for fundamental frequencies. If the ratio $k_C[\Omega]$ is large enough, the compensated current will be close to the harmonics current of the load.

To get a better compensation result, the power stage of APF can be modified to enlarge the source impedance. Fig. 10 shows the equivalent circuit of the modified power stage, in which a inductor of L_{APF} is inserted in the source.

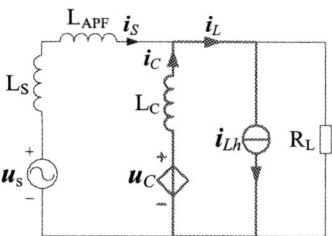

Figure 10. Modified equivalent circuit when only detecting compensated current.

D. Only Detecting Source Current

The proposed control approach is also effective even when only source current is detected.

When $K_S=0$, $K_L=0$, and $K_S = k_S \cdot HPF(s)$, the control algorithm can be simplified to

$$U^*_C(s) = k_S \cdot HPF(s) \cdot I_S(s) \qquad (14)$$

in s domain (with Lapalace Transformation), or

$$u^*_C(t) = k_S \cdot i_{Sh}(t) \qquad (15)$$

in time domain.

In this way, the APF is a transconductor similar to Fig. 8 (b), which presents high impedance for the harmonics. To get a better performance, a capacitor C_{APF} is inserted into the shunt branch of the APF in order to block fundamental current, shown as Fig. 11.

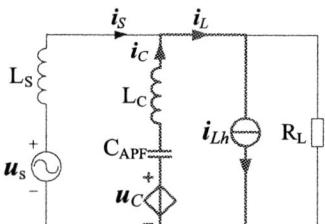

Figure 11. Modified equivalent circuit when only detecting source current.

V. SIMULATION VERIFICATION

Plenty of simulations were held to verify the proposed control approach and different practical implementations. The simulation results with each control scheme of PAPF in three-phase circuit are shown as Fig. 12.

VI. EXPERIMENTAL VERIFICATION

A 3kV three-phase experimental prototype was built and comparative experimental was conducted to verify the previous study. Fig. 13 is one snapshot of the experimental results of different implementations of the proposed control approach. For comparative study, the same compensators are used in all of the control schemes, with the same power stage.

VII. CONCLUSION

Active Power Filter is one of the most effective solutions to compensate the current harmonics and to enhance the power quality in industrial application. This paper proposes a generalized control approach for APF, which provides a systematical way not only to analysis the regular control schemes but also develop new control schemes.

Since different control strategies deal with the same power stage, this paper combined together the reference

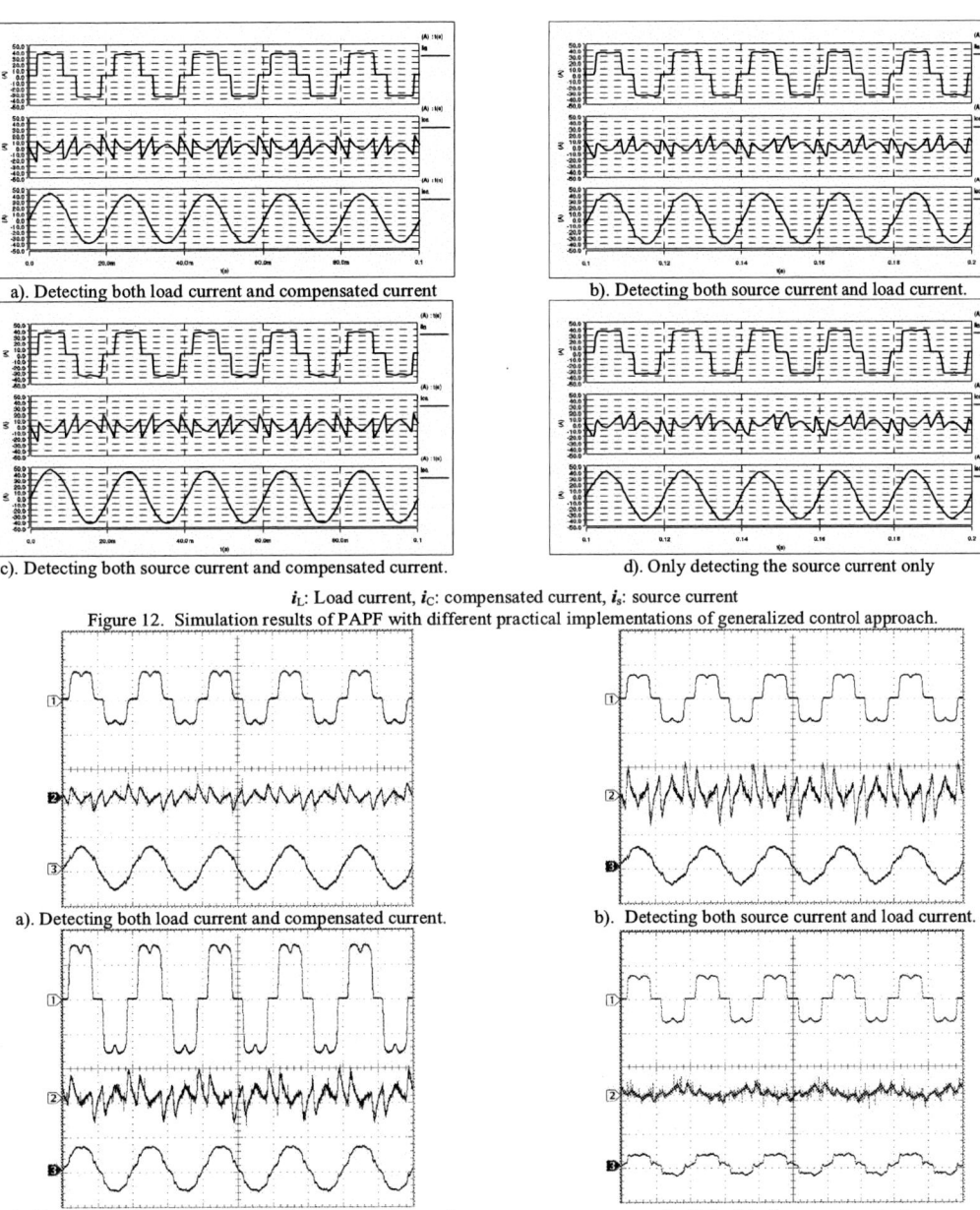

a). Detecting both load current and compensated current

b). Detecting both source current and load current.

c). Detecting both source current and compensated current.

d). Only detecting the source current only

i_L: Load current, i_C: compensated current, i_s: source current

Figure 12. Simulation results of PAPF with different practical implementations of generalized control approach.

a). Detecting both load current and compensated current.

b). Detecting both source current and load current.

c). Detecting both source current and compensated current.

d). Only detecting source current.

(i_L: Load current, i_C: compensated current, i_s: source current)

Figure 13. Experimental of PAPF with different practical implementations of generalized control approach.

calculation and current control of VSI in regular current-mode control strategies. Since the fundamental function of any control approach is to calculate the duty cycle d of the VSI from the detected variable, this paper proposed a generalized control approach, in which the reference to VSI comes from the currents for practical purpose. If proper gains of detected current are well designed, different control aims and compensation objectives can be achieved. Later on, several practical implementations of the generalized control approach are discussed in detail. Simulation and experimental results are provided to verify the approach and analysis.

ACKNOWLEDGMENT

The authors would like to pray their thanks to the graduate student Mr. Guopeng Zhao and Ms. Jing Li in Xi'an Jiaotong University for their great help in the experiments on hardware prototypes.

REFERENCES

[1] Akagi, H., "New trends in active filters for power conditioning," *Industry Applications, IEEE Transactions on* , vol.32, no.6pp.1312-1322, Nov/Dec 1996.

[2] Akagi, H., "Active harmonic filters," *Proceedings of the IEEE* , vol.93, no.12pp. 2128-2141, Dec. 2005.

Novel Circuit Configuration for Hybrid Reactive Power Compensator

H.L Jou*, J.C Wu**, J.J. Yang* and W.P. Hsu***

* Department of Electrical Engineering, National Kaohsiung University of Applied Sciences
** Department of Electrical Engineering, Kun Shan University of Technology
*** UIS Abler Electronics Co., Ltd.

Abstract—A novel circuit configuration to compensate for the reactive power is proposed in this paper. This hybrid reactive power compensator includes an AC capacitor set serially connected to a small-capacity power converter. The AC capacitor set can avoid the damage caused by the power resonance and harmonic current injection after using this hybrid reactive power compensator. Because the AC capacitor set can effectively block the DC voltage generated from the power converter to the utility, only a two-arm structure without using the split DC capacitors is required for the power converter in the three-phase three-wire distribution power system. Consequently, the required number of power electronic devices for the power converter is reduced. A prototype is developed and tested to verify the performance of the proposed hybrid reactive power compensator. The tested results show that the proposed hybrid reactive power compensator has the expected performance.

Keywords: hybrid; reactive power compensator; harmonic

I. INTRODUCTION

AC power capacitor set was connected in parallel to the distribution power system, so as to compensate for a lagging reactive power to increase the power factor [1]. However, it has some disadvantageous. The AC capacitor set merely provides with a fixed reactive power that cannot be adjusted in response to the load variation. This may cause over-voltage in the load side due to over-compensating the reactive power in the condition of light load. In consequence, the over-voltage caused by the over- compensating reactive power may damage the other power facilities. Besides, the AC capacitor is frequently damaged by harmonics. Meanwhile, this may result in the power resonance between the AC capacitor and the reactance of the distribution power system. Thus, the AC capacitors may be damaged due to over-voltage or over-current [2, 3]. The power resonance also may damage the neighboring electric power facilities and even result in public accidents.

The power converter based reactive power compensator has been developed [4-6] to solve the above

This work was supported by the UIS Abler Electronics Co., Ltd. and the National Science Council under NSC-94-2622-E-151-023-CC3.

problems. The power converter can provide with either of a leading reactive power or a lagging reactive power by means of controlling the operation of the power converter. This means that the reactive power can be adjusted in linear. Advantageously, no power resonance problem occurs between the power converter and the distribution power system, and no problem of harmonic current injection due to neighboring nonlinear load. However, the power converter is employed to provide with the entire reactive power compensation. Hence, the capacity of the power converter must be larger to provide with the entire reactive power in response to the full range load variation. Consequently, it may increase technical difficulty and manufacturing cost that limits the wide application.

A low capacity power converter serially connected to the AC capacitor set to protect AC capacitor set was proposed by authors [7, 8]. This power converter can avoid the harmonic resonance generated between power capacitor and the impedance of power system. However, the supplied reactive power is fixed.

This paper proposed a novel hybrid reactive power compensator employing an AC capacitor set serially connected to a power converter to eliminate the problems of power resonance and the injection of neighboring harmonic current. The proposed hybrid reactive power compensator has the advantages that the capacity of power converter can be reduced and the required number of power electronic devices in the power converter can be reduced. To verify the performance of the proposed hybrid reactive power compensator, a three-phase prototype is developed and tested.

II. CIRCUIT CONFIGURATION

Figure 1 shows the circuit configuration of the proposed hybrid reactive power compensator applied to the three-phase three-wire distribution power system. Generally, the proposed hybrid reactive power compensator consists of an AC capacitor set serially connected to a power converter. The AC capacitor set provides with a fundamental reactive power and also uses to withstand the major fundamental component of the utility voltage that may reduce the capacity of the *power converter. Additionally, the AC capacitor set can also block the DC voltage generated from the power converter to the utility.

The power converter is used to solve the power resonance and the harmonic current injection problems of AC capacitor, and it permits the proposed hybrid reactive power compensator to provide with a compensation reactive power that can be adjusted within a predetermined range in response to the load variation. The power converter consists of a DC capacitor, a power electronic device set and a high-frequency ripple filter. The DC capacitor acts as an energy buffer, and provides with a DC voltage for normally operating the power converter. The power electronic switch set is connected to the DC capacitor switching the DC voltage to generate a desired compensation current. Because the AC capacitor set can effectively block the DC voltage generated from the power converter to the utility, only two-arm bridge structure is required for the power converter in the three-phase three-wire distribution power system. In consequence, it permits one of the phase power lines of the utility to directly connect to the negative DC terminal of the power electronic switch set through a capacitor of the AC capacitor set without passing any power electronic devices in the power electronic switch set. The high-frequency ripple filter is an inductor set adapted to filter the high-frequency ripple current due to the switching operation of the power electronic switch set.

Due to the existence of the AC capacitor set, the operating voltage of the DC capacitor and the capacity of the power converter can be lower and two power electronic devices can be saved. Thereby, the manufacture cost of the hybrid reactive power compensator applied in the distribution power system is reduced.

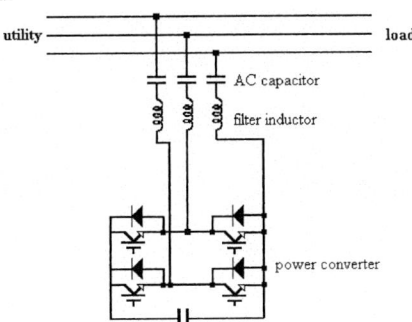

Figure 1.   The proposed circuit configuration for hybrid reactive power compensator.

III.   SYSTEM ANALYSIS

Because a DC capacitor is connected to the DC bus of power converter, this power converter is regarded as a voltage-source power converter. The voltage-source power converter is controlled by the pulse width modulation (PWM) strategy in which a modulation signal is compared with a high frequency carrier. In ideal

PWM operation, the output voltages of power converter can be represented as:

$$v_{cona}(t) = V_{dc}/2 + k_{con}v_{ma}(t) + v_{rpa}(t) \tag{1}$$

$$v_{conb}(t) = V_{dc}/2 + k_{con}v_{mb}(t) + v_{rpb}(t) \tag{2}$$

where $V_{cona}(t)$ and $V_{conb}(t)$ are the output voltages of power converter, V_{dc} is the DC bus voltage of power converter, $v_{rpa}(t)$ and $v_{rpb}(t)$ are the switching ripple voltages of power converter, $v_{ma}(t)$ and $v_{mb}(t)$ are modulation signals of power converter, and k_{con}: the gain of power converter.

The gain of power converter can be represented as:

$$k_{con} = \frac{V_{dc}}{2\hat{V}_{car}} \tag{3}$$

where $\hat{V}_{car}$ is the amplitude of high frequency carrier.

The frequency of switching ripple voltages is around the integer multiple times of carrier frequency. Since the switching frequency of power converter is very high compared with the harmonic frequency of power system, it can be filtered out effectively by the filter inductor. Hence, the switching frequency of power converter can be neglected in the following discussion. Consequently, only two components of power converter output voltages, DC component and low frequency AC components, are considered.

Figure 2 is the DC equivalent circuit of the proposed hybrid reactive power compensator. Since the utility voltages contain no DC component, the utility voltages are regarded as a short circuit. The power converter contains two DC voltage sources in phase a and phase b. As seen in Figure 2, the DC voltage component appeared in the AC capacitor set can be derived as:

$$V_{pa0} = V_{pb0} = \frac{1}{6}V_{dc} \tag{4}$$

$$V_{pc0} = -\frac{1}{3}V_{dc} \tag{5}$$

where V_{pa0}, V_{pb0} and V_{pc0} are the DC voltage components appeared in the AC capacitor set. This means that the DC component of converter output voltage can be blocked effectively by the AC capacitor, and no DC current will be generated by the power converter and injected into the distribution power system.

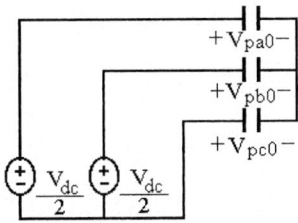

Figure 2.   DC equivalent circuit.

Figure 3 is the fundamental frequency equivalent circuit of the proposed hybrid reactive power compensator. The power converter output voltages are

$v_{cona1}(t)$ and $v_{conb1}(t)$. The amplitudes of $v_{cona1}(t)$ and $v_{conb1}(t)$ are proportional to the fundamental frequency components of the modulation signals. This figure shows that the power converter only generates two AC output voltages. In order to obtain the balanced three-phase compensation reactive current, the voltages across the AC capacitor set must be derived under losing one phase AC voltage of power converter. Using the principle of superposition, the effects of the utility voltages and the power converter output voltages to the AC capacitor set can be analyzed separately. The method of symmetrical components is used in the following analysis. Since the utility voltages contain only positive-sequence component for the ideal three-phase power system, the AC capacitor voltages caused by the utility voltages contain only positive-sequence component. The AC capacitor voltages ($V_{pa1,c}$, $V_{pb1,c}$ and $V_{pc1,c}$) caused by the power converter output voltages can be derived as:

$$V_{pa1,c} = (2V_{cona1} - V_{conb1})/3 \qquad (6)$$

$$V_{pb1,c} = (-V_{cona1} + 2V_{conb1})/3 \qquad (7)$$

$$V_{pc1,c} = (-V_{cona1} - V_{conb1})/3 \qquad (8)$$

The symmetrical components of AC capacitor voltages caused by the power converter output voltages can be derived as:

$$\begin{bmatrix} V_{pa1,c}^{(0)} \\ V_{pa1,c}^{(1)} \\ V_{pa1,c}^{(2)} \end{bmatrix} = \frac{1}{3} \begin{bmatrix} 1 & 1 & 1 \\ 1 & a & a^2 \\ 1 & a^2 & a \end{bmatrix} \begin{bmatrix} V_{pa1,c} \\ V_{pb1,c} \\ V_{pc1,c} \end{bmatrix} \qquad (9)$$

where $V_{pa1,c}^{(0)}$, $V_{pa1,c}^{(1)}$ and $V_{pa1,c}^{(2)}$ are the zero-sequence, positive-sequence and negative-sequence components of AC capacitor voltages caused by the power converter output voltages. The operator, a, is represented as:

$$a = 1\angle 120^o \qquad (10)$$

Because no zero-sequence current flows in the three-phase three-wire distribution power system, the zero-sequence components of AC capacitor set voltages caused by the power converter output voltages can be neglected. In order to obtain the balanced compensation current of hybrid reactive power compensator, the negative-sequence component $V_{pa1,c}^{(2)}$ of AC capacitor voltages caused by the power converter output voltages must be zero. Hence, the following equation can obtain:

$$\frac{1}{3}(V_{pa1,c} + a^2 V_{pb1,c} + a V_{pc1,c}) = 0 \qquad (11)$$

Substituting Eqs. (6)-(8) into Eq. (11), the relationship of the power converter output voltages can be derived as:

$$V_{conb1} = -aV_{cona1} \qquad (12)$$

Eq. (12) indicates that V_{conb1} lags V_{cona1} by 60°. As seen in Eqs.(6)-(9), the positive-sequence component of AC capacitor voltage caused by the power converter output voltages can be derived as

$$V_{pa1,c}^{(1)} = (V_{cona1} + aV_{conb1})/3 \qquad (13)$$

Substituting Eq. (12) into Eq. (13), then

$$V_{pa1,c}^{(1)} = (V_{cona1} - a^2 V_{cona1})/3$$
$$= V_{cona1}\angle 30^o / \sqrt{3} \qquad (14)$$

Considering the effects of both the utility voltages and the power converter output voltages, the positive-sequence component of AC capacitor set voltages can be derived as:

$$V_{pa1}^{(1)} = V_{sa1} - V_{cona1}\angle 30^o / \sqrt{3} \qquad (15)$$

To generate a fundamental reactive current, the positive-sequence component of power converter output voltage must be in phase with that of utility voltage. Equation (15) shows that the power converter output voltage of phase a must lag the utility voltage of phase a by 30°. Consequently, the power converter output voltage of phase b must lag the utility voltage of phase a by 90o. The compensation reactive power can be adjusted by controlling the amplitude of fundamental component of power converter. Because the power converter output voltage can be controlled to be positive and negative, the three-phase reactive power (Q_h) supplied from the hybrid reactive power compensator can be derived as:

$$Q_h = 3\omega C V_{SA1}\left(V_{sa1} - V_{cona1}/\sqrt{3}\right) \qquad (16)$$

where V_{sa1} and V_{cona1} are the RMS values of the fundamental component for the utility voltage and the power converter output voltage. The maximum RMS value ($V_{cona1,max}$) of the fundamental voltage generated by the power converter without over-modulation is dependent on the DC bus voltage of power converter, and it can be represented as [3]:

$$V_{cona1,max} = \frac{1}{2\sqrt{2}}V_{dc} \qquad (17)$$

Then, the minimum and maximum compensation reactive power can be derived by substituting Eq. (17) into Eq. (16). Hence, the proposed hybrid reactive power compensator can adjust the supplied reactive power between the minimum and maximum compensation reactive power linearly. If the variation range of the reactive power demanded by the load is pre-known, the voltage of dc bus and the capacitance of AC capacitor can be determined. If the load is very light or there is no load, the hybrid reactive power compensator can be switched away from the distribution power system.

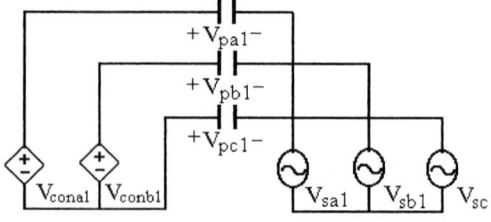

Figure 3.   fundamental frequency equivalent circuit.

IV. HARMONIC SUPPRESSION OF COMPENSATION CURRENT

The AC capacitor set used for power factor correction provides with a low impedance path for harmonic current under the polluted distribution power system. The AC capacitor is frequently damaged by harmonics. If a damping resistor is connected in series with the AC capacitor set, the problem of harmonic injection can be alleviated. However, the conventional passive resistor will result in power dissipation not only at harmonic frequencies but also at the fundamental frequency. Since the fundamental current component in the AC capacitor set for power factor correction is significant, this will result in a significant power loss and the problem of heat dissipation. Moreover, the power efficiency will be degraded. Since a power converter is connected with the AC capacitor set in the proposed hybrid reactive power compensator, the power converter can act as a damping resistor. The damping resistor can be controlled to respond to the harmonic frequency only. Because the damping resistor is not a practical resistor, it can be regarded as a virtual harmonic resistor. Meanwhile, the energy consumed by the virtual harmonic resistor is stored in the dc capacitor of power converter, and the stored energy will be regenerated to the utility at the fundamental frequency. Hence, the use of power converter can solve the problem of power loss caused by the practical resistor. Due to this purpose, the power converter must generate a voltage that is proportional to the harmonic components of the compensation current generated by the hybrid reactive power compensator.

V. CONTROL BLOCK DIAGRAM

Figure 4 shows the control block diagram of phase a for the proposed hybrid reactive power compensator. The voltage-mode control is used in the proposed hybrid reactive power compensator. The control block contains three parts, a fundamental reactive component S1, a fundamental real component S2 and a harmonic component S3. The fundamental component S1 is used to adjust the compensation reactive power, and the harmonic component S3 is used to block the injecting harmonic current. To balance the power loss of power converter and the operation of virtual harmonic resistor, the fundamental real component S2 is required. Therefore, the modulation signal of the power converter contains these three signals.

VI. EXPERIMENTAL RESULTS

In order to demonstrate the performance of the proposed hybrid reactive power compensator, a three-phase prototype is developed. The major parameters of the prototype are shown in Table I. The utility power is supplied by a three-phase three-wire utility system with 380V and 60Hz.

Figure 5 shows the experimental results of the proposed hybrid reactive power compensator under the normal condition. As seen in Figure 5, the load is inductive, and the compensating current of the hybrid reactive power compensator is leading with the utility voltage to supply a compensating reactive power. The utility current is nearly in phase with the utility voltage after compensated by the proposed hybrid reactive power compensator. Figure 6 shows the utility voltage and three-phase voltages across the AC capacitor set. As seen in Figure 6(c), there is a DC voltage dropped on AC capacitor of phase R. This proves that the AC capacitor can block the DC voltage generated by the proposed hybrid reactive power compensator. Figure 7 shows the proposed hybrid reactive power compensator under the transient response of step-on load. As seen in Figure 7, the compensating current is increased when the load is increased. Since the distribution power system is often distorted due to the use of nonlinear load, the AC capacitor is more sensitive to the harmonic distortion. Figures 8 and 9 show the experimental results under the distorted utility before and after applying the power converter respectively. A three-phase inductor is inserted between the utility and hybrid reactive power compensator to increase the system impedance, and a three-phase diode rectifier is applied behind the three-phase inductor to result in the distorted utility voltage in the experimental process. Before applying the power converter, the THD % of the utility voltage is 16.47% and the THD% of the capacitor current is 93.97%. As seen in Figure 8, the currents of AC capacitor set are distorted seriously due to the distorted utility before applying the power converter. As seen in Figure 9, the THD% of the utility voltage and the capacitor currents after applying the power converter are 4.65% and 7.4%. This proves the proposed power converter can improve the distortion of the utility voltage and the capacitor currents significantly. This shows that the performance of proposed hybrid reactive power compensator is excellent even under the distorted utility condition. Moreover, the distortion of utility voltage is improved because the harmonic amplification of AC capacitor is suppressed.

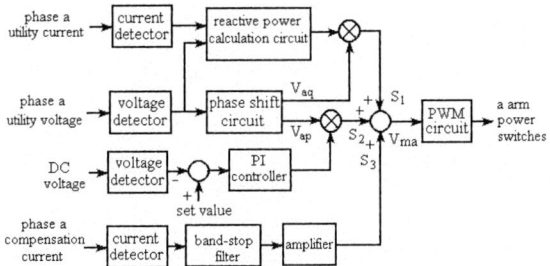

Figure 4. the control block diagram of phase a for the proposed hybrid reactive power compensator.

TABLE I.
MAIN PARAMETERS OF PROTOTYPE

utility	380V, 60Hz
DC capacitor	4700uF
AC capacitor	180uF
DC bus voltage	370V
Switching frequency	20KHz
high-frequency ripple filter	0.19mH

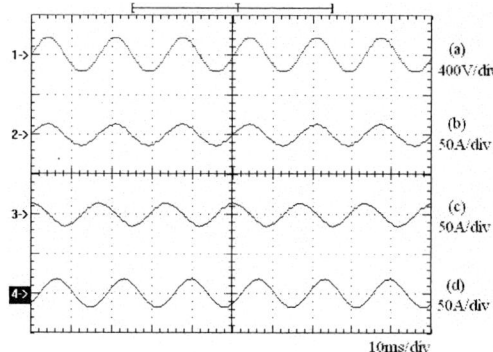

Figure 5   Experimental result of phase R under normal condition, (a) utility voltage, (b) utility current, (c) compensating current, (d) load current.

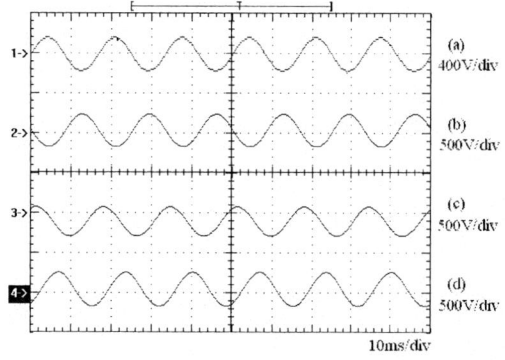

Figure 6.   Experimental result of phase R under normal condition, (a) utility voltage, (b) phase R capacitor voltage, (c) phase S capacitor voltage, (d) phase T capacitor voltage.

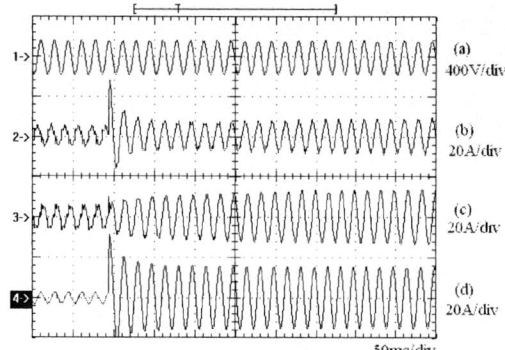

Figure 7. experimental result under the transient of step-on load, (a) utility voltage, (b) utility current, (c) hybrid reactive power compensator, (d) load current.

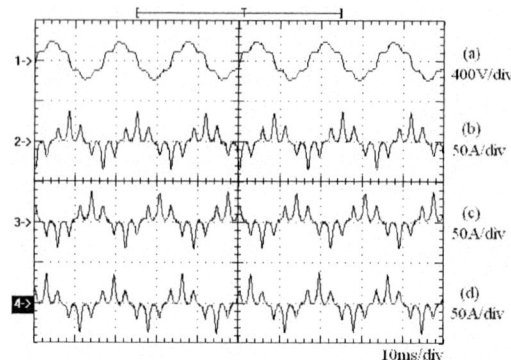

Figure 8. experimental result under the distorted utility condition, (a) utility voltage, (b) phase R capacitor current, (c) phase S capacitor current, (d) phase T capacitor current.

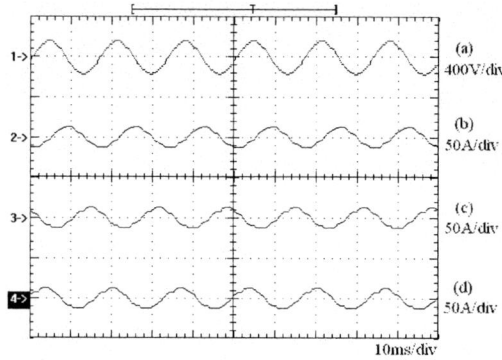

Figure 9.   Experimental result under the distorted utility, (a) utility voltage, (b) phase R capacitor current , (c) phase S capacitor current, (d) phase T capacitor current.

VII.   CONCLUSION

A novel circuit configuration for the three-phase three-wire hybrid reactive power compensator is proposed in this paper. The salient advantages of the proposed hybrid reactive power compensator are
1.   the power converter uses only a two-arm bridge structure without split DC capacitors,
2.   the harmonic damage of the AC capacitor set can be avoided,
3.   the compensating reactive power can be adjusted linearly in a predetermined range, and
4.   the power rating of power converter can be reduced.
The experimental results of the developed prototype verify that the proposed hybrid reactive power compensator has the expected performance.

ACKNOWLEDGEMENT

The authors would like to express their acknowledgement to financial support of UIS Abler Electronics Co., Ltd. and National Science Council under NSC-94-2622-E-151-023-CC3.

REFERENCES

[1] C. Wang, T. C. Cheng, G. Zheng, Y. D, L. Mu, B. Palk and M. Moon, "Failure Analysis of Composite Dielectric of Power Capacitors in Distribution Systems," IEEE Trans. Dielectrics and Electrical Insulation, Vol. 5, No. 4, pp.583-588, Aug. 1998,.

[2] D. A. Gonzalez and J. C. Mccall, "Design of Filters to Reduce Harmonic Distortion in Industrial Power Systems," IEEE Trans. on Industry Applications, 1987, Vol. 23, No. 3, pp.504-511, May/June, 1987.

[3] H. Moore, "Application of Power Capacitors to Electrochemical Rectifier Systems," IEEE Trans. on Industry Applications, Vol. 13, No. 5, pp.399-406, Sep./Oct. 1977.

[4] A. Tahri, A. Draou and M. Benghanem, "A Fast Current Control Strategy of a PWM Inverter Used for Static VAR Compensation," IEEE IECON, Vol. 1, pp. 450-455,1998.

[5] L. Xu, V. G. Agelidis and E. Acha, "Development Considerations of DSP-Controlled PWM VSC-based STATCOM," IEE Proc.-Electr. Power Appl., Vol. 148, No. 3, pp.449-455, May 2001.

[6] J. E. Hill and W. T. Norris, "Exact Analysis of a Multipulse Shunt Converter Compensator or Statcon. I. Performance," IEE Proc. Gener. Transm. Distrib., Vol.144 , No.2 , pp.213-218, March 1997.

[7] J.C. Wu, H.L. Jou, K.D. Wu and N.T. Shen, "Power Converter Based Method for Protecting Three-Phase Power Capacitor from Harmonic Destruction, " IEEE Transactions on Power Delivery, Vol.19, No.3, pp.1434-1441, 2004.

[8] J.C. Wu, H.L. Jou, K.D. Wu, Y.T. Kuo, Y.J. Chang, "Power converter based method for suppressing power capacitor harmonic current, " IEE Proceedings of Generation, Transmission and Distribution, Vol.151, No.3, pp.341-346, 2004.

2006 5th International Power Electronics and Motion Control Conference

Shunt Active Power Filter with Sample Time Staggered Space Vector Modulation Based Cascade Multilevel Converters

Liqiao Wang, Weiyang Wu

Institute of Electric Engineering, Yanshan University, Qin Huangdao, P. R. China

Abstract- **Sample time staggered space vector modulation (STS-SVM) is a kind of new modulation strategy for large-capacity power electronic equipments, which is of excellent harmonic characteristic, wide transmitting bandwidth, average switching load and easy digital algorithm. With STS-SVM, cascade multilevel converters can achieve a high equivalent switching frequency effect at very low real device switching frequency, which is much suitable for large capacity APF. A shunt active power filter experimental prototype with STS-SVM based cascade multilevel converters is established. Experimental results verify that this APF can compensate the harmonic and reactive current correctly and validly.**

Keywords- sample time staggered space vector modulation; shunt active power filter; cascade multilevel converter

I. INTRODUCTION

Active power filter (APF) is a kind of power electronics equipment applied to restrain the harmonics and compensate reactive power dynamically. APF can perform real-time compensation for nonlinearly variable harmonic and reactive power, which is of much importance to stabilize the electric power system and improve the power quality [1].

APF is a kind of large-capacity power electronic equipment. At present, large-capacity power electronic equipments always select multilevel converters as main circuit structure [2, 3]. Multilevel converters can output high voltage directly with no use of transformers or reactors to realize the link of AC side. Multilevel convert-

ers can apply pulsewidth modulation technique (PWM) and possess perfect input/output feature as well as excellent harmonic characteristic. Compared with other kinds of multilevel converters, cascade multilevel converters are of many advantages: 1) the amount of the devices is the least, 2) the DC side voltage is relatively easy to balance due to the separate DC supply structure, 3) each converter unit has the same structures and is easy to design and assembly modularized. The drawback of cascade multilevel converters is that there must be separate DC supply in the occasion in need of active power. Clearly, this drawback does not exist in APF.

There are several familiar modulation strategies for cascade multilevel converters such as fundamental frequency modulation [4], carriers-based PWM [5, 6], multilevel space vector modulation [7, 8, 9], carrier phase shifted PWM [10, 11] and sample time staggered space vector modulation (STS-SVM) [12, 13]. STS-SVM can achieve a high equivalent switching frequency effect at very low real device switching frequency. STS-SVM is with perfect harmonics characteristic, wide transmitting bandwidth, balanced switching load, high voltage utilization ratio and easy digital modulation algorithm.

In this paper, STS-SVM is introduced into cascade multilevel converters, which combines their advantages together. A shunt APF prototype with STS-SVM based cascade multilevel converters is presented. The experimental results verify that the shunt APF is of good performance to compensate harmonics and reactive power.

1-4244-0448-7/06/$25.00 ©2006 IEEE

II. SAMPLE TIME STAGGERED SPACE VECTOR MODULATION BASED CASCADE MULTILEVEL CONVERTERS

A. Sample time staggered space vector modulation

For the three-phase voltage source six-switch converter, there are eight switch modes called fundamental voltage space vectors, which are six valid vectors and two zero vectors, shown as Fig.1. Any reference vector V^* in Fig.1 can be approximated by two nearest valid vectors and zero vectors. The corresponding PWM pulses can be attained by appropriately placing the two nearest valid vectors and zero vectors in a sample cycle, which is the principle of space vector modulation (SVM) [14].

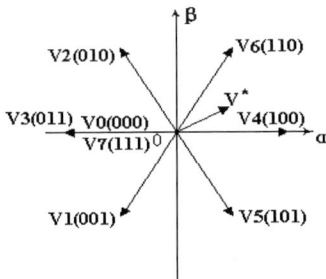

Fig.1 Diagram of fundamental voltage space vectors

STS-SVM is the extended application of SVM to multi-modular converters. Voltage source multi-modular converters in series are composed of three-phase voltage source six-switch converters, shown as Fig.2.

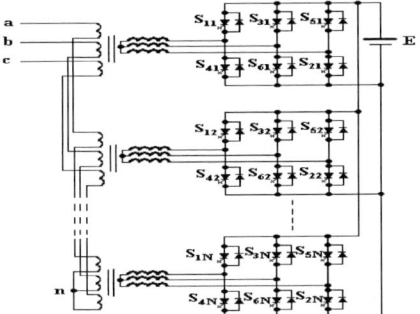

Fig.2 Voltage sourced multi-modular converters in series

Each three-phase voltage source six-switch converter is considered as one converter unit. Each converter unit is modulated by the same SVM under the same modu-

lated index and the sample time of the contiguous converter unit is staggered by: $2\pi/(N \cdot \omega_c)$, where N is the amount of converter units and ω_c is sample frequency. This is so-called STS-SVM.

STS-SVM is of excellent performance [14] that is much promising to APF such as:

1) The amplitude of the total fundamental component line voltage is N times as that of one converter unit, i.e. there is no fundamental component loss.

2) The lowest harmonic cluster is near $N\omega_c$, i.e. the equivalent switching frequency is improved by N times. In another word, STS-SVM can achieve perfect harmonic characteristic and wide transmitting bandwidth with low device switching frequency.

3) Every converter unit is modulated under the same traditional two-level SVM. The switching load of each device is average. The modulated algorithm is much easier than that of other multilevel SVM strategies.

B. Application of STS-SVM to cascade multilevel converters

The Y-link three-phase cascade multilevel converters are shown as Fig.3, which are composed of three-phase converter modules shown as Fig.4. The three-phase converter module itself is composed of three single-phase full bridge converters (SPFBs). The three-phase converter module shown as Fig.4 can be considered as two three-phase voltage source six-switch converters in series, where the three left half-bridges compose one and the three right half-bridges compose the other. Then, cascade multilevel converters composed of N three-phase converter modules, shown as Fig.3, are equivalent to the voltage source multi-modular converters in series composed of 2N three-phase voltage source six-switch converters. For each three-phase converter module shown as Fig.4, the three left half-bridges and the three right half-bridges are modulated by SVM under the same modulated index, the sample time of the left half-bridges between the right ones is staggered by: π / ω_c. The same side half-bridges of contiguous modules staggers their sample time by $\pi/(2N \cdot \omega_c)$. Then, STS-SVM is accomplished in cascade multilevel converters.

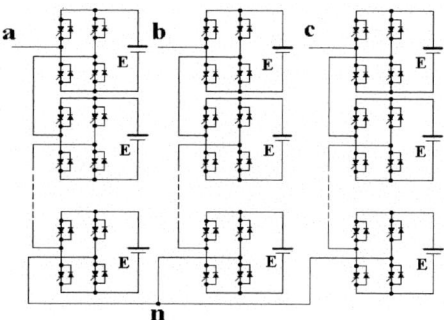

Fig.3 Y-link three-phase cascade multi-level converters

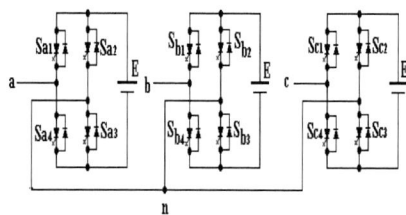

Fig.4 Three-phase converter module composed of three SPFBs

An STS-SVM based three-phase five-level cascade converter prototype is implemented. In the experiment, the amplitude modulation index is 1, the fundamental frequency is 50Hz and the sample frequency is 1050Hz.The output line voltage is a nine-level waveform, shown as Fig. 5a. Its harmonics spectrum is shown as Fig. 5b. The lowest harmonic cluster is near 4200Hz, i.e. the equivalent switching frequency is improved by 4 times.

III. SHUNT APF WITH STS-SVM BASED CASCADE MULTILEVEL CONVERTERS

The circuit topology of a shunt APF based on five-level cascade converters is shown as Fig.6. The nonlinear load is a current source diode rectifier. The cascade converters are directly paralleled into the AC source. In Fig.6, i_S is the AC source current, i_L is the load

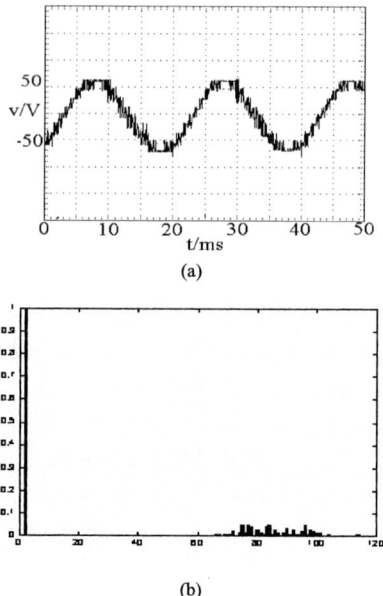

(b)

Fig.5 Output line voltage experimental waveform and its spectrums of five-level STS-SVM converter (a) line voltage waveform, (b) harmonics spectrum of (a)

current, i_C is the compensation current from APF and U_{xy} is the DC voltage of the y-th converter unit of x phase ($x=a$, b, c; $y=1,2$). It is clear that $i_S=i_L+i_C$.

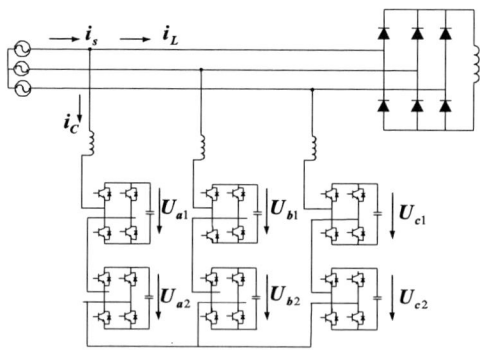

Fig.6 Shunt APF based on five-level cascade converters

The operating principle of APF is to derive the harmonic and reactive component of the load current i_L through harmonic and reactive detection circuit and reverse it as the reference compensated current i^*_C, and then construct the AC side feedback control to make the AC side current i_C of APF correctly trace i^*_C. After this operation, the compensation for the harmonic and reactive component of i_L is accomplished to let the source current i_S be close to sinusoidal and correct the source power factor to unity.

In addition, the DC voltage of each converter unit should be balanced to ensure the system to work at normal state. At meantime, APF should adopt a little active power for the source to compensate the loss of system itself. Therefore, the DC side feedback control is necessary. The overall system control diagram is shown in Fig.7.

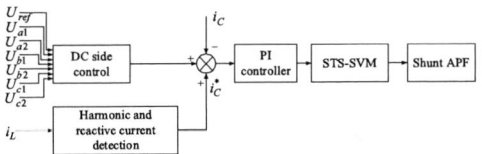

Fig.7 Overall system control diagram

In this paper, a PI controller is applied to the AC side current control. To improve the dynamic response rate, some better control strategy can be introduced such as dead-beat control [15], pole-placement control [16] and variable structure control [17].

The DC voltage balance is easy to implement due to the separate DC link of each converter module. Instanced by five-level cascade converters, the DC side control diagram is shown in Fig.8.

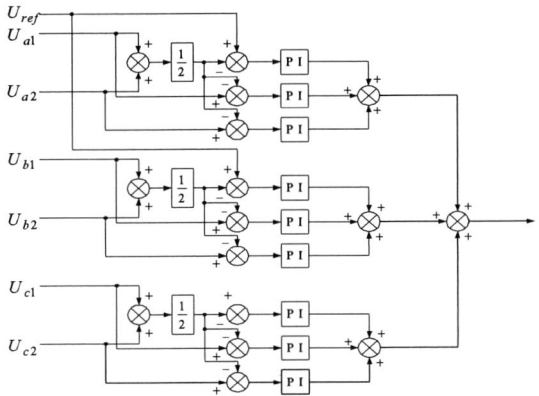

Fig.8 DC side control

As for Fig.7, it can be seen that the overall control system is a typical two closed-loops system. To avoid the interference of AC side and DC side, the time con-

stant of DC side should be much bigger than that of AC side. In another word, the response rate of DC side should be much slower than that of AC side.

IV. EXPERIMENTAL RESULTS

A 2.5kVA experimental prototype of shunt APF with STS-SVM based five-level cascade converters is established. The experimental setup is implemented by a DSP-CPLD controller. The circuit parameters are: AC source voltage 220V (rms.), AC source frequency 50Hz, AC side inductor 3.5mH and DC link capacitor 3000µF. IGBT is selected as power device with its switching frequency at 1050Hz. For five-level cascade converters, the equivalent switching frequency of the whole system reaches 4200Hz.The load current, compensation current and AC source current after compensation waveforms are shown as Fig.9.

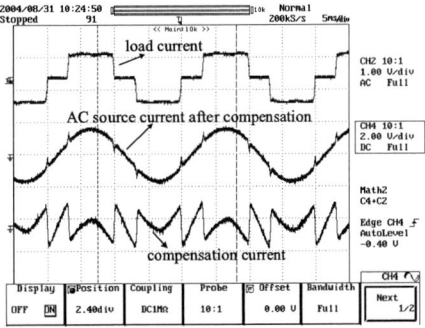

Fig.9 Experimental waveforms of the prototype

Analyze the load current and AC source current after compensation by FFT respectively, and the harmonic spectrums are attained shown as Fig.10.

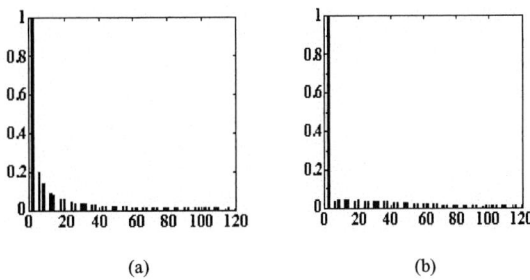

(a)   (b)

Fig.10 Harmonics spectra of load current and AC source current after compensation (a) harmonics spectrum of load current, (b) harmonics spectrum of AC source current after compensation

It can be seen that the proposed APF system can achieve good compensation effect at very low device switching frequency. The APF prototype is of remarkable function to restrain low order harmonics.

V. CONCLUSIONS

Cascade multilevel converters can directly output high voltage with many advantages such as the least number of devices needed, separate DC link easy to balance DC voltage and uniform structure of each converter unit to design and assemble modularized. Especially for the occasion with no need of active power like APF, cascade multilevel converters are of much superiority. Under STS-SVM, cascade multilevel converters can achieve a high equivalent switching frequency effect at very low real device switching frequency, decrease harmonic component and exalt transmitting bandwidth efficiently, which is much suitable for large capacity APF. In this paper, it is introduced how to apply STS-SVM in cascade multilevel converters. A shunt APF prototype with STS-SVM based cascade multilevel converters is implemented. The APF system has good compensating performance due to experimental results, which is with perfect prospect in engineering application.

ACKNOWLEDGEMENT

The authors wish to thank the National Natural Science Foundation of China (50237020) and the Doctor Found of Yanshan University (B155) for their important help.

REFERENCES

[1] H.Fujita, H.Akagi. A Practical Approach to Harmonic Compensation in Power Systems --- Series Connection of Passive and Active Filters, IECON'90, pp1107~1112, 1990

[2] Jih-sheng Lai, Fang Zheng Peng. Multilevel converter--A new breed of power converters. IEEE Trans. IA, Vol.32, No.3, pp.509–517, 1996

[3] Fang Zheng Peng, Jih-sheng Lai, John W Mckeever. A multilevel voltage source inverter with separate DC source for static var

generation. IEEE Trans. IA, Vol.32, No.5, pp.1130–1137, 1996.

[4] R. Teodorescu, F. Blaabjerg, J. K. Pedersen, et. al. Multilevel converters-a survey. EPE'99 CDROM

[5] B.P. McGrath, D.G. Holmes. A comparison of multi-carrier PWM strategies for cascaded and neutral point clamped multilevel inverters. Proceedings of PESC'00, 674-679, 2000

[6] R. Tallam, et al. A Carrier-Based PWM Scheme for Neutral-Point Voltage Balancing in Three-Level Inverters. Proceedings of APEC04, 1675-1681, 2004

[7] N. Celanovic, D. Boroyevich. A fast space vector modulation algorithm for multilevel three-phase converters. Proceeding of APEC'99, 1173-1177, 1999

[8] S. Wei; B. Wu; Wang Qianghua. An improved space vector PWM control algorithm for multilevel inverters Proceedings of IPEMC 2004, 1124 – 1129, 2004

[9] N. P. Filho, J. P. Pinto, B. K. Bose. A neural-network-based space vector PWM of a five-level voltage-fed inverter. Proceedings of IAS2004, 2181 – 2187, 2004

[10] Zhang Zhongchao, Boon-Teck Ooi. Forced commutated HVDC and SVC based on phase-shifted multi-converters. IEEE Trans. on Power Delivery, Vol.8, NO.2, pp.712 – 718, 1993

[11] Wang Changyong, Zhang Zhongchao. Phase-shifted SPWM technique based cascade converter and its application to active power filter. Automation of Electric Power Systems, 25(1): 28~30, 2001.

[12] Wang Liqiao, Zhang Zhongchao, et al. Cascade Multi-Level Converters with Sample-Time-Staggered Space Vector Modulation. IEEE Applied Power Electronics Conference and Exposition - APEC, v 1, pp.268-273, 2003

[13] Wang Liqiao Research on sample time staggered space vector modulation technique. Dissertation submitted to Zhejiang University for Ph.D. Degree of Engineering. 2002.12

[14] Wang Liqiao, et al. Harmonic characteristic of sample time staggered space vector modulation for multi-modular or multilevel Converters. Proceedings of IEEE PESC04, pp:4554~4557, 2004

[15] Simone Buso, et. al. Design and fully digital control of parallel active power filter for thyristor rectifiers to comply with IEC-1000-3-2 standards. IEEE Trans on Industry Applications, Vol.34, No.3, pp. 508~517, 1998

[16] Yan Guo, Xiao Wang, B. T. Ooi. Pole-placement control of voltage-regulated PWM rectifiers through real-time multiprocessing. IEEE Trans. on Industrial Electronics, Vol.41, No.2, pp.224~230,1994

[17] D. M. Vilathgamuwa, et. al. Variable structure control of voltage sourced reversible rectifier. IEE Proc Electric Power Application, Vol.143, No.1, pp. 18~24, 1996

2006 5th International Power Electronics and Motion Control Conference

Shunt Active Power Filter Synthesizing Resistive Loads by Means of Adaptive Inverse Control

Wu Yanfeng[1], Wu Zhengguo[1], Li Hua[2], Li Hui[1]

1. Electrical and information Engineer academy, Naval University of Engineering, Wuhan, China

2: Administrative office of training, Naval submarine academy, Qingdao China

ABSTRACT- **The power active filter(APF) is used to suppress the harmonics and compensate the reactive power which are produced by the nonlinear loads. The traditional method of the sinusoidal current synthesis(SCS) would deteriorate the distortion level in the common coupling, thus result in the harmonic propagation through the power distribution system. The method of the resistive load synthesis(RLS) can compensate the mains current in phase with the mains voltage in spit of the mains voltage distort or not. The character of the system is a resistive load. The adaptive inverse control is a novel intelligent control method. The adaptive inverse control is used to control the APF to synthesis the resistive load. The whole power distribution system is regarded as a generalized active power filter. The inverse of the plant model is obtained by using an off-line identification technique. The inverse of the model is used to be the controller to control the action of the active power filter so that the output of the voltage of the dc capacitor to follow the reference input. The action of the active power filter makes the system to synthesize a resistive load. The harmonic propagation can be damped out and the harmonics and reactive power can be compensated. The power factor can be corrected to unity. The simulation and experiment results validate the proposed algorithm.**

Keywords- active power filters; sinusoidal current synthesis; resistive load synthesis; adaptive inverse control

I. INTRODUCTION

In recent years, harmonic problems have been serious in industrial and utility power distribution systems. One of the most serious problems in utility power distribution systems is the so-called "harmonic propagation," which contributes to a significant amplification of voltage harmonics in a distribution line[1].

Hirofumi Akagi etc have proposed a shunt active filter based on harmonic-voltage detection which is intended to be installed by electric utilities. The active filter is characterized by behaving like a resistor for harmonic frequencies, making it possible to damp out harmonic propagation throughout a whole distribution line[2]-[7].

Teresa Esther Núñez-Zúñiga present that if the APF works synthesizing a resistive load, which means that the final line current will present the same waveform of the voltage, the damping effect remains unchanged.

Additionally, the power factor is corrected to unity. This implies minimum RMS current for constant active power being delivered to the load[8]. The APF action maximizes the power factor, thus minimizing the transmission losses, because this situation corresponds to the minimum RMS current that is required to deliver the necessary active power to the load.

Takaharu Takeshita etc have proposed the distorted current waveform whose harmonic components are in phase with the terminal voltage harmonics, by which the harmonics both in the terminal voltage and the source current can be suppressed. The apparent power of the converter system using the proposed current waveform hardly increased [9].

The work of adaptive inverse control was initiated in the 60 by Widrow, and is slowly gaining popularity with the power of neural networks. A recent textbook explains the principles of adaptive inverse control[10]. In its simplest form, adaptive inverse control seeks to model the inverse of the plant. The controller appears in series with the plant. The command input is fed to the controller and provides also the desired response. Hence, when the error is small the controller transfer function is the inverse of the plant.

This paper presents a resistive load synthesis method by means of adaptive inverse control. The supply system with the load and the APF are regarded as the generalized active filter and the inverse of its model is used as the controller to control dc capacitor voltage[11]. The actual model express of the APF is needn't known exactly. The proposed active filter can compensate the harmonic and suppress the passive power. In addition, the compensated mains current have the same wave with the mains voltage to suppress the terminal voltage harmonics. After compensation the power factor will be correct to unity. The distort level at the point of the common coupling is reduced significantly.

II. WAVEFORM CONTROL AND RESISTIVE LOAD SYNTHESES

The action of the active power filter produces compensating currents that once injected into the grid, result sinusoidal line currents in phase with the respective voltages. However, this theory does not provide good results if the line voltage is not sinusoidal. If the voltages

1-4244-0448-7/06/$25.00 ©2006 IEEE

are sinusoidal and balanced, sinusoidal and balanced final currents are obtained. Otherwise, the final line currents will be distorted with a harmonic content other than that of the voltage and the power factor will not be corrected to unity.

The presence of resonant circuits in voltage-distorted grids can cause resonance phenomena. In such situations the damping effect is provided by the resistive part of the circuit. As the physical resistance is usually low, the load provides the effective damping. If the local load voltage is distorted and the APF action provides a sinusoidal line current, only the fundamental frequency is available at the load. The load is an open circuit for the harmonics, as no current will flow in these frequencies. Considering the role played by the load damping, the system will lose its ability to damp any eventual resonance that may produce a severe voltage distortion at the point of the common coupling. On the other hand, if the APF works synthesizing a resistive load, which means that the final line current will present the same waveform of the voltage, the damping effect remains unchanged. Additionally, the power factor is corrected to unity.

Fig.1 shows a simple model of a power distribution system. The voltage source v_s is a distorted three-phase voltage. R_s, L_s are the sum of the source and line resistance, inductance, respectively. Generally, the PCC voltage v_t of the PWM converter system is distorted.

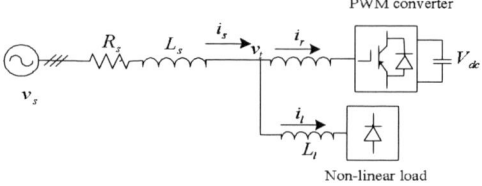

Figure 1 model of power distribution system

For a given value of the active power, a purely resistive load minimizes the apparent power and the RMS current, thus minimizing the losses in the transmission system. Defining $u(t)$ and $i(t)$ as periodical quantities associated with the voltage and current in an electrical circuit, their respective RMS values are

$$U = \sqrt{\frac{1}{T} \int_0^T u^2(t)dt} \qquad (1)$$

$$I = \sqrt{\frac{1}{T} \int_0^T i^2(t)dt} \qquad (2)$$

The active power and apparent power are, respectively

$$P = \frac{1}{T} \int_0^T u \cdot i \, dt \qquad (3)$$

$$S = U \cdot I \qquad (4)$$

By using the Schwartz inequality

$$\left[\int_a^b f(x) \cdot g(x) \cdot dx \right]^2 \le \int_a^b f^2(x) \cdot dx \cdot \int_a^b g^2(x) \cdot dx \qquad (5)$$

The condition of the equation in (5) is that

the $f(x)$ and $g(x)$ is linear correlation, which means that:

$$g(x) = k \cdot f(x) \qquad (6)$$

So that

$$S \ge P = \lambda \cdot U \cdot I \qquad (7)$$

The equality in (7) applies only if the ratio $f(x)/g(x)$ is constant, which characterizes the voltage per current ratio is constant. So that $u(t)$ and $i(t)$ have the same waveform, which characterizing a resistive load. The parameter λ is a constant that, in this case, represent the power factor.

III. ACTIVE POWER FILTER SYNTHESIZE RESISTIVE LOAD BY MEANS OF ADAPTIVE INVERSE CONTROL

A plant can track an input command signal if it is driven by a controller whose transfer function approximates the inverse of its transfer function[12]. A dc capacitor located in the dc bus of the voltage-source converter has two functions, one is to maintain a dc voltage with small ripple in the steady state, the other is to serve as an energy storage element for supplying the real power difference between the mains and the loads in the transient state for a shunt active power filter[13]-[14]. For a lossless active power filter in the steady state, the active power supplied from the mains should be equal to the real load demanded, where no active power passes through the power converter into the dc capacitor. The average dc capacitor voltage can be thus maintained at the reference voltage level. Once the mains voltage has varied or the load condition has changed, the real power difference between the mains and load is thus required compensated by the aid of the dc capacitor. This scenario drives the average voltage of dc capacitor away from the reference voltage. The average voltage of the dc capacitor can supply the real power flow information and the amplitude control of the mains current can be obtained by using a voltage regulation circuit of the dc capacitor. By controlling the average voltage of the dc capacitor as the reference voltage, it can effectively complete the amplitude control of mains current.

Because that the average voltage of the DC capacitor can supply the real power flow information and the amplitude control of the mains current can be obtained by using a voltage regulation circuit of the dc-side capacitor. The inverse control technique can be applied to the active power filter. The parts of the main circuits including the supply, the loads and the APF are defined as a generalized active power filter. An equation could be defined as follows:

$$V = f(i) \qquad (8)$$

Where V is the voltage of the dc-side capacitor, and i is the amplitude of the mains current. The f is the mapping function between V and i. It is needn't to know the exactly express of f. The adaptive plant modeling or identification can be done with an adaptive FIR filter according to equation (1), the plant input signal is the input of the adaptive filter, here, which is V, the plant output signal is the desired response, here, which is i. The adaptive algorithm, N-LMS, causing the model to be a best least squares match to the plant for the given input signal. The inverse model $i = f^{-1}(V)$ of the generalized active power filter is obtained by using an off-line identification technique. This inverse model is implemented by means of an FIR transversal adaptive filter, which coefficients are adapted by the N-LMS algorithm for computational simplicity. This FIR will be referred as "inverse model". Then another FIR transversal adaptive filter, referred as "control model", which coefficients are obtained by copying the coefficients of the "inverse model filter". This "control model" filter is connected to the APF input, as shown in Fig. 2. Provided that the inverse model is exact, the output voltage of the DC-side capacitor of the active power filter will be equal to the reference signal, due to the computation process. Therefore the output of the generalized APF, that is , the dc-side capacitor voltage, will follow the input of the "control" block, the reference voltage.

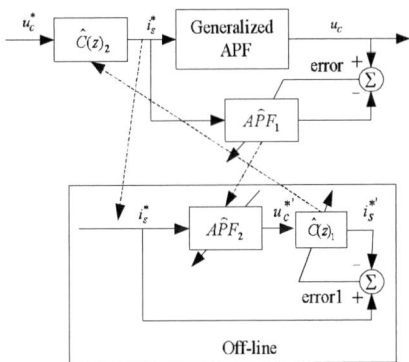

Figure 2   Block diagram of shunt active power filter based on adaptive inverse control

In this figure, the $A\hat{P}F_1$ has the same dynamical response with the generalized APF. And the inverse model $\hat{C}(z)_1$ of the off-line process is achieved by the model $A\hat{P}F_2$ whose weights are copied from the adaptive filter $A\hat{P}F_1$. $\hat{C}(z)_2$, whose weights are copied from the inverse model $\hat{C}(z)_1$ works as the controller to control the generalized active power filter. Thus, when adaptation has converged, the cascade of plant and adaptive inverse behave like a unit gain. Also, the cascade of $A\hat{P}F_2$ and $\hat{C}(z)_1$ behave like a unit gain too. So the output of the generalized APF is obtained by applying a reference voltage to the input of the control block $\hat{C}(z)_2$.

During the modeling process, both the process of the plant model and the inverse model are adaptive. The exact express of the generalized active power filter needn't to be known. The only important point is that the plant model has the same response with the generalized active power filter. If the length of the filter is not so long the normalized least mean square adaptive algorithm would converge fast.

Fig.3 presents the principle of the proposed method. The output of the dc side capacitor voltage is the desired response, the target signal for the filter output. The controller is the copy of the plant inverse which is generated by the off-line process that delivers new values of tap almost instantaneously with new values of the plant model. The reference dc voltage is the controller input signal.

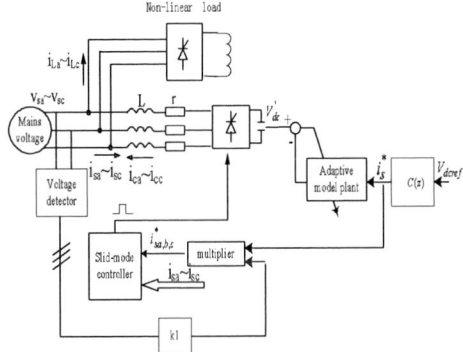

Figure 3   Block diagram of the proposed active filter

The adaptive plant modeling or identification of the generalized active power filter is an important function. The adaptive algorithm, LMS, which converge slowly, is not properly here. The N-LMS algorithm is used to cause the model to be a best match to the plant. The inverse plant identification is another important function. The inverse plant is generalized in the process of the offline. The plant inverse is to be used as a controller to provide a driving function for the plant. Three voltage detector and two current detectors are used in the proposed algorithm. The three voltage detectors are used to detect the mains voltage and dc capacitor voltage. The two current detectors are used to detect the mains current. The output of the inverse controller is seen as the amplitude of the desired mains current. The detected mains voltage is normalized to be unity with the same waveform of the mains voltage. The multiplication of the output of the inverse controller and the unity voltage signal is them performed to obtain the desired mains current, which has the same waveform as the mains voltage. The desired mains current is compared with the detected mains current. The compares error is then sent to a waveform controller. Finally, the output of the waveform controller is delivered to a PWM modulator in order to produce appropriate switching signal for the control of power devices in the power converter, where the waveform controller is implemented by slide mode controller.

If the proposed algorithm is work well and both the two adaptive process converge rapidly, then the compensated mains current can follow the desired mains current well, which has the same waveform as the mains voltage in spite of the mains voltage distort or not. The mains voltage per mains current ratio is constant, characterizing a resistive load. The power factor will be correct to unity.

IV.   SIMULATION AND EXPERIMENTAL TESTS

A.   *Simulation Tests*

Table I lists the simulation parameter of the proposed algorithm. Both of the adaptive filters are transversal FIR filter. The LMS algorithm is too slowly to be used, so the N-LMS algorithm can meet the need. The filter length is two.

TABLE I THE PARAMETERS OF THE SIMULATION MODEL CIRCUIT

Mains voltage resistance	0.3Mh，1 Ω
Filter inductance	5mh
Mains voltage	220V
Step-size parameter	μ=0.001
Dc capacitor	2200 μ F
Switching frequency	20KHz
Mains frequency	50Hz
Load parameter	50mh, 30 Ω

The following three-phase mains voltage are used in the simulation:

$$U_{sa} = U_1 \sin(\omega t) + U_2 \sin(5\omega t)$$
$$U_{sb} = U_1 \sin(\omega t - 120°) + U_2 \sin(5\omega t - 120°) \quad (9)$$
$$U_{sc} = U_1 \sin(\omega t + 120°) + U_2 \sin(5\omega t + 120°)$$

Where U_1 is the amplitude of the fundamental wave, U_2 is the amplitude of the fifth order harmonic.

Fig. 4 depicts the mains voltage and the desired mains current. In this figure, it shows that the desired current and the mains voltage have the same waveform. If the compensated current follows the desired current well, then the proposed active filter can synthesize a resistive load.

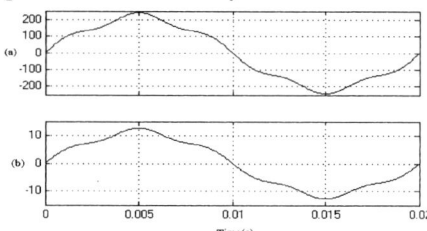

Figure 4 The mains voltage and the expected mains currents

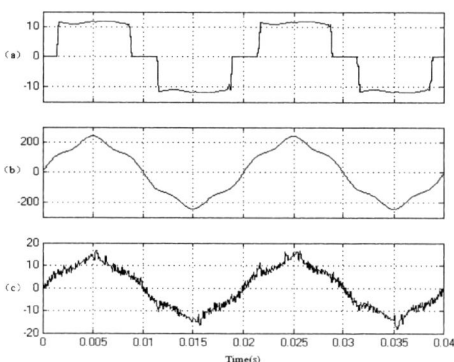

Figure 5 Simulation results

Fig. 5 depicts the simulation results of the proposed algorithm under the distort mains voltage(a phase). In this figure, the load current, the mains voltage, and the compensated current are demonstrated, where the compensated current has nearly the same waveform with the mains voltage. The power unity can be corrected to unity.

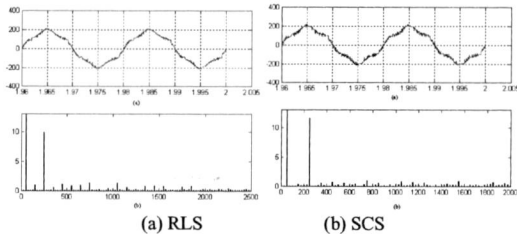

(a) RLS   (b) SCS

Figure 6　The contrast of the PCC voltage based on both method

Fig. 6 depicts a comparative situation of the voltage and their spectrum at the point of the PCC under the RLS method and the SCS method. In Fig. 6(a), the system is connected to an ideal APF which operates synthesizing a resistive load. The case of sinusoidal current synthesis is shown in Fig. 6(b). Notice that the voltage distortion at the PCC becomes more severe under SCS method. Under RLS the voltage distortion at the PCC becomes alleviate. Therefore, the action of the active power filter by means of adaptive inverse control can make the system synthesizing a resistive load. This system can damp any eventual resonance at the PCC under distort mains voltage.

B.　Experimental Test

To verify the effectiveness of the proposed algorithm, a laboratory prototype of three-phase three-wired active power filter was developed and tested in the laboratory. Table II lists the mains parameters used for the buildup of this laboratory prototype. The control algorithm is implemented through a TMS320F240 digital signal processor. Mains voltage and mains current are sampled for one cycle on the occurrence of zero crossing interrupt. For the ease of implementation, reference current is estimated for phase A and phase B assuming a balanced system. A negative adding circuit is used to obtain phase C current. Estimated reference current and actual source currents are then processed in slide mode controller, to obtain the switching signals and applied to the PWM converter after proper isolation and amplification.

TABLE II THE PARAMETERS OF THE SIMULATION MODEL CIRCUIT

Mains voltage	110V
Filter inductance	7mh
Step-size parameter	N=2,μ=0.001
Dc capacitor	1650μF
Switching frequency	10KHz
Mains frequency	50Hz
Load parameter	196mh, 26 Ω

Fig. 7 depicts the experimental results of the proposed algorithm. (a) is the distort mains voltage whose THD is less than 3%. (b) is the mains current before compensated, whose THD is 28%. (c) is the compensated mains current. The figure indicates that the compensated mains current has the same waveform with the mains voltage. The system can be operated as a resistive load. The power factor can be corrected to unity.

Figure 7   experimental results:(a) mains voltage and (b) load current and (c) mains current

V.   CONCLUSIONS

Due to the distort or unbalanced mains voltage, the traditional action of the active filter to synthesize the sinusoidal current would make the voltage distort level at the point of the common coupling more severe. The system couldn't damp out the harmonic propagation effectively. The adaptive inverse control is a novel intelligent control method. The action of the active power filter by means of the adaptive inverse control to synthesize a resistive load can damp out the harmonic propagation in the power distribution system. The PCC voltage distort becomes alleviate. Simulation and experiment validated the proposed algorithm.

REFERENCES

[1] Pichai J, Hideaki F, Hirofumi A etal. Implementation and performance of automatic gain adjustment in a shunt active filter for harmonic damping throughout a power distribution system[J]. IEEE Transactions on Power Electronics. 2002, 17(1):438-447

[2]Keiji W, Hideaki F, Hirofumi A. Considerations of a shunt active filter based on voltage detection for installation on a long distribution feeder[J]IEEE Transactions on industrial applications. 2002, 38:1123-1130

[3]Pichai J, Hideaki F, Hirofumi A etal. Implementation and performance of cooperative control of shunt active filters for harmonic damping throughout a power distribution system[J]. IEEE Transactions on industry applications. 2003,39:556-564

[4]Pichai J, Hideaki F, Hirofumi A. Control and performance of a fully-digital-controlled shunt active filter for installation on a power distribution system[J]. IEEE Transactions on power electronics. 2002, 17:132-140

[5]Hirofumi A, Hideaki F, Keiji W. A shunt active filter based on voltage detection for harmonic termination of a radial power distribution line[J]. IEEE Transaction on Industry Applicaltions. 1999, 35:638-645,

[6] Pichai J, Hideaki F, Hirofumi A etal. Implementation and performance of automatic gain adjustment in a shunt active filter for harmonic damping throughout a power distribution system[J]. IEEE Transactions on Power Electronics. 2002, 17:438-447,

[7]Hirofumi A. Control strategy and site selection of a shunt active filter for damping of harmonic propagation in power distribution systems[J]. IEEE Transactions on Power Delivery. 1997, 12: 354-363

[8]Teresa E N, José A P. Shunt active power filter synthesizing resistive loads[J]. IEEE Transactions on power electronics. 2002,17(2):273-278

[9]Takaharu T, Nobuyuki M. Current waveform control of PWM converter system for harmonic suppression on distribution system[J]. IEEE Transactions on Industrial Electronics. 2003, 50(6):1134-1139

[10] Widrow B, Wallach E. Adaptive Inverse Control. Prentice Hall,1996

[11] Wu Yanfeng, Wu Zhengguo, Li Hua etc. The Novel Detection Approach of Shunt Active Filter Based on Adaptive Inverse Control[c]. The Six International Conference on Power Electronics and Drive Systems. Kuala Lumpur, Malaysia, December 2005:124-127

[12] Bernard Widrow, Michel Bilello. Adaptive inverse control[C]. Proceedings of the 1993 International Symposlum on Intelligent Control Chicago, USA, August 1993:1-6

[13] J C Wu, H L Jou, "Simplified control method for the single-phase active power filter", in Proc. IEE Electr. Power Appl., vol. 143, pp.219-224, 1996.

[14] S J Huang, J C Wu, "A control algorithm for three-phase three-wired active power filters under nonideal mains voltages", in IEEE Trans. Power Electron., vol.14, pp753-759, 1999

Single Neutral Element Self-Adaptive PID Controller Used In SVC

Zeng Guang*, **, Ke Min-qian**, Su Yan-min* and Fu Qi-gang***

* Xi'an Jiaotong University, Xi'an, China;
** Xi'an University of Technology, Xi'an, China
***Shanghai Surpass Sun Electric Co.Ltd. ,Shanghai ,China
Email: kmq.xaut@163.com

Abstract—In order to overcome the poor adaptability of conventional PID control method which over depend on precise mathematical model of the control object, a kind of control algorithm called single neutral element self-adaptive PID control is designed. The controller can modulate PID parameters on-line and dynamically optimize them what conventional PID controller cannot live up to. It can improve the control robust and self-adaptive of the system. The controlling effect is perfect by using the single neutral element self-adaptive PID controller to SVC nonlinear system. Experimental results indicate the superiority of the control algorithm.

Keywords-mathematical model; single neural element; robustness; self-adaptive PID control

I. INTRODUCTION

Power system must be carefully controlled in order to obtain an acceptable power supply quality, a reliable and economical power network. Adjusting the reactive power timely and effectively is very important which is in favor of the voltage stabilization, the reasonable distribution of tidal current and the restriction of the over voltage. The SVC which can compensate the reactive power and decrease the voltage imbalance degree is widely applied in transmitting and distributing electricity at present. It may be a perfect choice in the compensator market for high performance and low price.

Conventional proportional-integral-derivative (PID) controllers have been well developed and are extensively used for industrial automation and process control today. However, with the controlling desire advancing constantly and nonlinear characters and time-varying of controlled object, conventional PID controller generally does not work very well any more. Tuning of PID parameters becomes difficult and adaptive ability falls. Control strategies combining conventional PID with other intelligent controlling algorithm for improving the performance with better self tuning ability become one hotspot recently. Single neutral element self-adaptive PID control algorithm independent of the precise mathematical model of the control object, with self learning feature of neuro, it can solve the problem that

conventional PID controller can not easy to modulate the PID parameters on-line. The robust and self adaptive performance can be improved that it can more easily to adapt the complicated working condition and satisfy the high precision demand.

Because of the nonlinear character of SVC system and the fierce undulation of load especially arc furnace, both the self adaptive and controlling precision of conventional PID controller becomes worse, the settling time becomes long. In this paper, single neutral element self-adaptive PID control algorithm is used into SVC system, experimental results satisfactorily verified that the proposed control strategy has better adaptability and doughty robustness. It reveals good performance when applied to the experimental prototype.

II. CONTROLLER DESIGN

A. Conventional PID Controller

The output of conventional PID controller is given by:

$$u(t) = K_P \left[e(t) + \frac{1}{T_I} \int_0^t e(t)dt + T_D \frac{de(t)}{dt} \right] \tag{1}$$

where K_P is the proportional gain, T_I the integral time constant, T_D the derivative time constant, and e(t) is the tracking error signal. This equation can be transformed into the following well-known discrete version:

$$u(k) = u(k-1) + K_P \left[e(k) - e(k-1) \right] + K_I e(k) + K_D \left[e(k) - 2e(k-1) + e(k-2) \right] \tag{2}$$

It is defined: $K_I = K_P \dfrac{T}{T_I}$ and $K_D = K_P \dfrac{T_D}{T}$. There K_I and K_D are the integral gain and the derivative gain, and T>0, is the sampling period.

B. Single Neutral Element Self-Adaptive PID Controlle

Single neutral element self-adaptive PID control diagram is shown as fig.1.

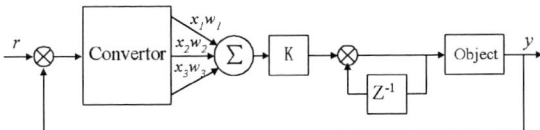

Figure 1. Single neutral element self-adaptive PID control

The input of convertor is $r(t)$ and $y(t)$, and the output of it is $x_1(k)$, $x_2(k)$ and $x_3(k)$ by which the neuro used to self learning. K is the gain of the neuro. The main learning algorithm of weight $w_i(k)$ are as follows:

1) Supervised Delta learning algorithm

$$w_i(k+1) = w_i(k) + \eta_i u(k) x_i(k) \qquad (3)$$

2) Nonsupervised Hebb learning algorithm

$$w_i(k+1) = w_i(k) + \eta_i e(k) x_i(k) \qquad (4)$$

Integrating supervised Delta learning algorithm with nonsupervised Hebb learning algorithm, we can get:

$$w_i(k+1) = w_i(k) + \eta_i e(k) u(k) x_i(k)$$
$$u(k) = u(k-1) + K \sum_{i=1}^{3} w_i(k) x_i(k) \qquad (5)$$

where $w_i(k)$ is weight of $x_i(k)$. Function of state variable $x_i(k)$ is shown as follows:

$$\begin{cases} x_1(k) = r(k) - y(k) = e(k) \\ x_2(k) = e(k) - e(k-1) \\ x_3(k) = e(k) - 2e(k-1) + e(k-2) \end{cases} \qquad (6)$$

From Eq.(5) we know, the tuning of weight $w_i(k)$ is related to state variable $x_i(k)$, learning speed $\eta_i(k)$ and output error $e(k)$. Single neutral element self-adaptive PID controller is capable of self learning and self adaptive by tuning the weight $w_i(k)$ on-line. To assure the astringency and robustness of single neutral element self-adaptive PID control diagram, the arithmetic can be standardization as the follow equations:

$$u(k) = u(k-1) + K \frac{\sum_{i=1}^{3} w_i(k) x_i(k)}{\sum_{i=1}^{3} |w_i(k)|} \qquad (7)$$

where:

$$\begin{cases} w_1(k+1) = w_1(k) + \eta_I e(k) u(k) x_1(k) \\ w_2(k+1) = w_2(k) + \eta_P e(k) u(k) x_2(k) \\ w_3(k+1) = w_3(k) + \eta_D e(k) u(k) x_3(k) \end{cases} \qquad (8)$$

η_P、 η_I、 η_D express learning speed of proportional, integral and derivative respectively. We can know from above-mentioned, single neutral element self-adaptive PID control algorithm can be regard as self tuning PID control. We can look on weights of w_1, w_2, w_3 as proportional coefficient, integral coefficient and differential coefficient. This learning formula used in single neutral element self-adaptive PID controller is propitious to improve the learning capability, flexibility and controlling ability which makes the controlling system be easy to control on real time.

The performance of the above single neutral element self-adaptive PID learning algorithm contact with the selected η_P、 η_I、 η_D、 K. By experimental debugging of SVC system, we can summarize the following parameter tuning principle:

1) The select of initiative weights $w_1(0)$, $w_2(0)$, $w_3(0)$ is important to the system. It will make the system emanative if the select of weights is ill-suited. The choice of initiative weight value is based upon an expert's experience and good understanding of how a controller should operate. We should select different initiative weight value according to different control object. Because the weights w_1, w_2, w_3 are equal to proportional factor, integratiag factor and derivative factor respectively, so we can regard the parameter of PID as initiative weights.

2）K is the neruo gain, it will cause the system transient process to become long if it is too small. But too great selection of K may cause the system overshoot becoming bigger and enlarge the system steady-state error.

3）According to the different values of proportional P, integral I and derivative D, we can choice different learning speed η_P, η_I, η_D in order to tuning the parameters for different weights. When K is decided, if the overshoot is too big we can decrease η_P, η_D, or if settling time is too long we can properly decrease η_I.

III. SVC CONTROLLING ALGORITHM

A Static Var Compensator (SVC) consists of capacitors and reactors connected in shunt, which can be quickly controlled by thyristor switching. The block model for SVC system is given in fig.2. In this case the antiparallel thyristor is connected in series with the reactor. By phase angle control of thristor, the flow of current through the reactor is varied. In three phase arrangement, TCR is normally Delta connected. The TCR control is stepless control and used with fixed capacitor (FC) which acts as harmonic filter constituting the dynamic compensation.

735

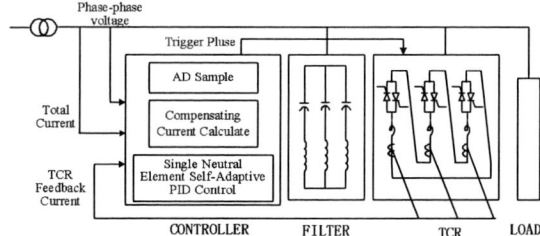

Figure 2. Block model for SVC system

The controller is the center part of the SVC system. The closed loop controlling strategy is adopted in the controller in this paper. The controller includes the following parts:

1) AD sample;
2) Compensating current calculate;
3) Single neutral element self-adaptive PID control.

Fig.3 is the detail drawing of controller:

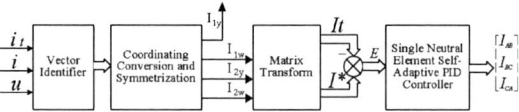

Figure 3. Detail drawing of controller

From fig.3 we know, instantaneous values of the phase-phase voltage u, total current i and thyristor feedback current i_t are sampled, calculating by vector identifier(VI), coordinating conversion and symmetrization, then we can get positive-sequence active power current I_{1y}, positive-sequence reactive power current I_{1w}, negative-sequence active power current I_{2y}, negative-sequence reactive power current I_{2w}. I_{1w} brings reactive power, I_{2y} and I_{2w} make voltage fluctuate. All these current should be eliminated. The relationship of I_{1w}, I_{2y}, I_{2w} with the total current is shown as follows(Re represents real part and Im meals imaginary part.):

$$\begin{bmatrix} I_{1W} \\ I_{2Y} \\ I_{2W} \end{bmatrix} = \begin{bmatrix} 0 & \dfrac{1}{3} & \dfrac{1}{2\sqrt{3}} & -\dfrac{1}{6} & -\dfrac{1}{2\sqrt{3}} & -\dfrac{1}{6} \\ \dfrac{1}{3} & 0 & -\dfrac{1}{6} & \dfrac{1}{2\sqrt{3}} & \dfrac{1}{6} & -\dfrac{1}{2\sqrt{3}} \\ 0 & \dfrac{1}{3} & -\dfrac{1}{2\sqrt{3}} & -\dfrac{1}{6} & \dfrac{1}{2\sqrt{3}} & -\dfrac{1}{6} \end{bmatrix} \begin{bmatrix} \mathrm{Re}\,I_A \\ \mathrm{Im}\,I_A \\ \mathrm{Re}\,I_B \\ \mathrm{Im}\,I_B \\ \mathrm{Re}\,I_C \\ \mathrm{Im}\,I_C \end{bmatrix}$$

(9)

and

$$\begin{bmatrix} I_{AB}^{*} \\ I_{BC}^{*} \\ I_{CA}^{*} \end{bmatrix} = \frac{1}{\sqrt{3}} \begin{bmatrix} -1 & \sqrt{3} & -1 \\ -1 & 0 & 2 \\ -1 & -\sqrt{3} & -1 \end{bmatrix} \begin{bmatrix} I_{1w} \\ I_{2y} \\ I_{2w} \end{bmatrix}$$

(10)

By matrix transform, if U_A is chosen as reference phasor, the calculation reference current is shown as following equation:

$$\begin{bmatrix} I_{AB}^{*} \\ I_{BC}^{*} \\ I_{CA}^{*} \end{bmatrix} = \begin{bmatrix} \dfrac{1}{3} & -\dfrac{2}{3\sqrt{3}} & \dfrac{1}{6} & \dfrac{5}{6\sqrt{3}} & -\dfrac{1}{6} & -\dfrac{1}{6\sqrt{3}} \\ 0 & \dfrac{1}{3\sqrt{3}} & -\dfrac{1}{2} & -\dfrac{1}{6\sqrt{3}} & \dfrac{1}{2} & -\dfrac{1}{6\sqrt{3}} \\ -\dfrac{1}{3} & -\dfrac{2}{3\sqrt{3}} & \dfrac{1}{6} & \dfrac{1}{6\sqrt{3}} & \dfrac{1}{6} & \dfrac{5}{6\sqrt{3}} \end{bmatrix} \begin{bmatrix} \mathrm{Re}\,I_A \\ \mathrm{Im}\,I_A \\ \mathrm{Re}\,I_B \\ \mathrm{Im}\,I_B \\ \mathrm{Re}\,I_C \\ \mathrm{Im}\,I_C \end{bmatrix}$$

(11)

Then the calculation reference current I^* is compared with the thyristor feedback current I_t and the error signal E is given to the single neutral element self-adaptive PID controller. By self tuning and adaptive control of the controller, we can get the final thyristor fundamental compensating current: $[I_{AB}\ I_{BC}\ I_{CA}]$.

Linearization is provided to line the relationship between the thyristor fundamental compensating current and the trigger pulses. In the case of unbalanced load, each phase of TCR trigger angle can be independently controlled and regulated.

IV.   EXPERIMENTAL RESULT

In order to test the effectiveness of the proposed single neutral element self-adaptive PID controller for SVC system, a prototype was designed and applied into the power system where a lot of experiments had been conducted. Fig.4 gives the main circuit diagram. In the main circuit, the primary side of the transformer is connected to Delta and the secondary windings to Star which transforms the 380V AC voltage to 40V phase voltage by transformer T1. The three-phase power filter and the thyristor controlled reactor (TCR) join into the 40V phase voltage one after the other. The power filter tuned to the 4th, 5th and 7th filters which can be switched into the circuit separately. The three-phase resistor can be modulated independently simulating the unbalanced load.

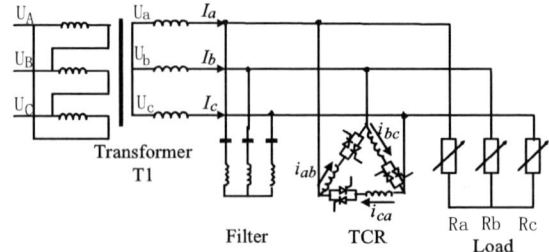

Figure 4. Main circuit diagram

The waveforms of the voltage and current snatched with single neutral element self-adaptive PID under different working condition are shown as follows. Contrastively, the parameters were chose finally as follows: η_P=0.3、 η_I=0.5、 η_D=0.5、K=0.8.

Fig.5 is the waveform of fore-and-aft compensate of voltage and current of Phase A under balanced load condition (the A, B, C three-phase load is all 250Ω) with 4th, 5th, 7th harmonic filters. We can see from fig.5(a), the phase angle of voltage and current is staggered where the current led the voltage. With the compensating of the TCR, the current and the voltage are with one phase in fig.5(b). Fig.5 reveals that SVC system with single neutral element self-adaptive PID controller plays a good compensating performance with high precision.

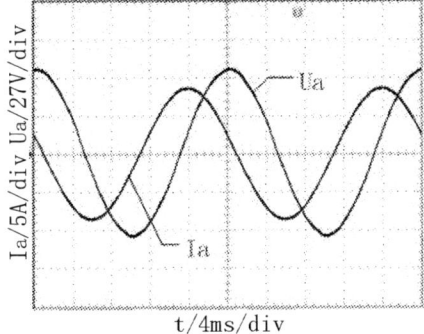

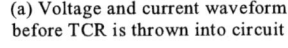

t/4ms/div

(a) Voltage and current waveform
before TCR is thrown into circuit

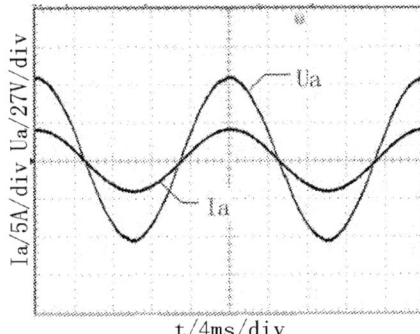

t/4ms/div

(b) Voltage and current waveform
before TCR is thrown into circuit

Figure 5. Waveform of fore-and-aft compensate of voltage and current of Phase A under balanced load condition

Under balanced load condition (the A, B, C three-phase load is all 250Ω), first throws the 4th harmonic filter, latter throws the 5th harmonic filter, we can use the single sequence trigger of oscilloscope to gain the voltage and current waveform of phase A. Fig.6 is obtained under fuzzy-PID and PID. We may see from the fig.6, before the t1 time, only throws the 4th harmonic filter, the voltage and current are in one phase because of the compensating of SVC. In the t1 time, the 5th harmonic filter is switched into the circuit, the capacitive current

increases, the voltage and current phase are displaced, the current leading voltage. After the adjustment of TCR, in the t2 time the amplitude of current finally achieves stably, the current and the voltage are in one phase. Entire adjustment process is finished about in 4.5 cycles. Fig.7 is the waveform of dynamic compensate of voltage and current from one unbalanced load condition to anther unbalanced load condition (Firstly phase A is 50Ω and B,C are both 250Ω; then phase A and B hold the line, phase C plays down to 50Ω). We can also use the single sequence trigger of oscilloscope to gain the waveform. The experimental condition is that the 4th and 7th harmonic filters are thrown into the circuit. Before the t1 time, we may see from the fig.6, the voltage and current is with one phase because of the compensating of SVC. In the t1 time, part load of Phase C is shorted. In the t1 time, we can find the phase of the current lagged to the voltage. After the adjustment of TCR, in the t2 time the amplitude of current finally achieves stably, the current and the voltage are with one phase. Entire adjustment process is finished about in 4 cycles.

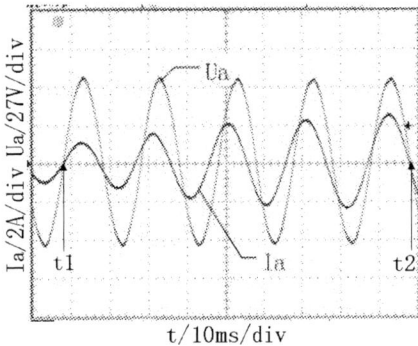

t/10ms/div

Figure 6. Waveform of dynamic compensate of voltage and current of Phase A under balanced load condition

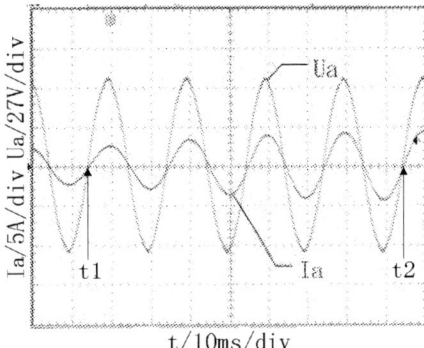

t/10ms/div

Figure 7. Waveform of dynamic compensate of voltage and current of Phase A under unbalanced load condition

V. EXPERIMENTAL RESULT

This paper presents a novel approach using single neutral element self-adaptive PID control algorithm in allusion to nonlinear character of SVC system. The control algorithm integrating supervised Delta learning algorithm with nonsupervised Hebb learning algorithm needn't set up the precise referenced model to tuning the parameter as conventional self-adaptive control. Self tuning and adaptive control make the system more robust. Experimental results show that the control algorithm has preferable controlling precision and robust character.

REFERENCES

[1] M.Zaheer-uddin, N.Tudoroiu, "Neuro-PID tracking control of a discharge air temperature system", *Energy Conversion and Management*. 2004, pp. 2405-2415.

[2] HanPu, SunHairong, ZhouLihui, "Application of self-adaptive single-neuron PID controller in superheated steam temperature control", *Journal of North China Electric Power University*. Sep.2005, Vol.32, No.5, pp. 62-65.

[3] Yuyao, DengFeiqi, "Simulation of PID Controller Based on Artificial Neural Cell and Its Improved Algorithms", *Control Engineering of China*. May.2005, Vol.12, S_0, pp. 46-48.

[4] PengJianchun, HuangChun,WangYaonan, "Intelligent Adaptive PID Controller Design for Static Var Compensator.", *Journal of Hunan University(Natural Sciences Edition)*. Oct.1999, Vol.26, No.5, pp:0-55.

[5] LiQi, LiShihua, "Analysis and Improvement of a Kind of Neural Networks Intelligent PID Control Algorithm", *CONTROL AND DECISION*. July.1998, Vol.13, No.4, pp. 311-316.

[6] WanJianru,ZhanHaibo,CaoCaikai, "Drive System based on Single Neuron PID Controller", *Power Electronics*. Feb.2005, Vol.39, No.1, pp. 75-77

[7] Dusan Ranonic, Ljubisa Stankvic, Dusan Petranovic, "ADAPTIVE CONTROL SYSTEM FOR HYBRID SVC", *IEEE*, 1995, pp:378-382.

[8] P.K.Muttik, P.Wang, M.W.H.Minchin, "Detailed Simulation of SVC Transient Performance Using PSCAD", *IEEE*, 2000, pp. 685-690.

[9] Fumitoshi Ichikewa, Kenichi, Tatsuhito Nakajima, Shoichi Irokawa, Tadayuki Kitahara, "Development of Self-Commutated SVC for Power System", IEEE, 1993, pp. 609-614.

[10] M.Parniani, H.Mokhtari, M.Hejri, "Effects of Dynamic Reactive Compensation in Arc Furnace Operation Characteristics and its Economic Benefits", *IEEE*, 2002, pp. 1044-1049.

[11] Athula Rajapakse, Anawat Puangpairj, Surapong Chirarattananon, "Harmonic Minimizing Neural Network SVC Controller for Compensating Unbalanced Fluctuating Loads", *IEEE*, 2002, pp. 403-408.

2006 5th International Power Electronics and Motion Control Conference

A Novel Shunt Single-Phase Active Power Filter for High Voltage Application

Zhang Changzheng, Chen Qiaofu, Zhao Youbin, Chen Yuda, and Cheng Lu

Dept. of Electric Machinery
College of Electrical and Electronic Engineering
Huazhong University of Science and Technology
Wuhan, Hubei, 430074, P. R. China
E-mail: longmarch_zhang@sohu.com

Abstract—In this paper, a novel shunt single-phase active power filter is proposed for high voltage application, which is based on the harmonic impedance control of a special linear transformer with multiple secondary windings. The transformer's primary winding is shunted with harmonic-producing loads, while its multiple secondary windings are connected with inverters. The transformer's secondary currents that produced by the inverters are proportional to its primary harmonic current. When the harmonic current compensation condition is satisfied, the transformer can really exhibit nearly zero impedance to harmonic current and primary self-impedance to fundamental current. Thus, the harmonic currents in high voltage power distribution systems can be led to flow into the transformer branch. The validity of the novel principle and excellent filtering characteristics are verified by the experimental results.

Keywords- Active power filter; high voltage; large capacity; linear; multiple windings; transformer

I. INTRODUCTION

The proliferation of nonlinear loads such as static power converters and arc furnaces has led to serious harmonic contamination in power distribution systems, which becomes a severe problem for both utilities and customers. In order to keep the harmonic contamination within acceptable limits as well as to increase the power factor, passive LC filters have been used conventionally. However, in practical applications passive LC filters present many disadvantages. As a result, attention has been focused on active power filters. Various active power filter configurations with their respective control strategies have been proposed during the last decade [1]. Most of the researches are restrained in low voltage utilities due to the constraint of power semiconductor switches. So far, there is still no generally accepted scheme for high voltage grid. Nowadays, studies on high voltage active power filters are mainly focused on two ways. One is the hybrid cascade multilevel inverter [2], [3]. The other is the hybrid active power filter, which

This work is supported by the National Natural Science Foundation of China (50477047).

combines active and passive compensation techniques [4], [5]. Problems exist in these schemes: (a) the security and reliability is low considering the possible direct short of the lines or loads; (b) their configurations and control schemes are rather complicated.

In order to realize harmonic suppression, controllable impedance is needed in power systems. A novel practical principle of series hybrid active power filter based on the fundamental magnetic flux compensation of a series transformer is proposed in [6]. In this scheme, the series transformer can exhibit primary leakage impedance to fundamental and magnetizing impedance to harmonics under the control of power electronic inverters. In order to increase the capacity of the series hybrid active power filter, a novel series transformer structure with multiple secondary compensation windings is proposed in [7] and a 100kVA prototype is developed successfully. Another novel principle of adjustable reactor based on magnetic flux controllable is proposed in [8]. Combining the new principle with the idea of multiple compensation windings, a 180kVar automatic resonant arc-suppressing coil is developed and tested successfully [9].

On the basis of the references [6]-[9], a novel shunt single-phase active power filter is proposed for high voltage application, which is based on the transformer's harmonic impedance control. A specially designed linear transformer with air gap and multiple secondary windings is used. The transformer's primary winding is shunted with harmonic-producing loads, while its secondary windings are respectively connected with corresponding PWM voltage source inverters. The transformer's primary harmonic current is detected, and then tracked by the inverters with fixed harmonic compensation coefficient to produce harmonic compensation currents. When the primary harmonic current and the secondary harmonic compensation currents satisfy the harmonic current compensation condition, the transformer exhibits nearly zero impedance to harmonic current and primary self-impedance to fundamental current. Thus, the harmonic currents in power systems can be led to flow into the transformer branch. Active power filters implemented in terms of the new principle can be used in 6kV~35kV alternate power distribution systems.

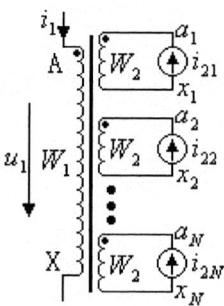

Figure 1. Configuration of the linear transformer.

II. THE COMPENSATION PRINCIPLE

Fig.1 shows the configuration of a linear transformer with air gap and multiple secondary windings. The turns of the primary winding and each secondary winding of the transformer are W_1 and W_2, respectively. The turns ratio is represented by $k = W_1 / W_2$. Assume that the primary winding AX is shunted with harmonic-producing loads. Then the transformer's primary current consists of no-load current $i_1^{(1)}$ and harmonic current $\sum i_1^{(n)}$, i.e., $i_1 = i_1^{(1)} + \sum i_1^{(n)}$. The harmonic component $\sum i_1^{(n)}$ is detected, and then multiplied by a harmonic current compensation coefficient to get a reference current. The reference current is tracked by PWM voltage source inverters to produce multiple harmonic compensation currents, which are respectively injected into the corresponding multiple secondary windings.

The nth-order harmonic voltage equation of the transformer's primary winding can be obtained in phasor form

$$U_1^{(n)} = (r_1 + j\omega_n L_{11})I_1^{(n)} + j\omega_n M_{11}I_{21}^{(n)} + ... + j\omega_n M_{1N}I_{2N}^{(n)} \quad (1)$$

Where, ω — angular frequency;

r_1 — resistance of primary winding;

L_{11} — self-inductance of primary winding;

M_{1P} — mutual inductance between the primary and Pth secondary winding. Here, P =1, 2…N.

Assume that the transformer's secondary windings are designed with identical structures. Then, the mutual inductances are almost the same

$$M_{11} = M_{12} = ... = M_{1N} = M \quad (2)$$

The secondary harmonic compensation currents are also the same, which can be expressed as

$$i_{21} = i_{22} = ... = i_{2N} = -\alpha \cdot \sum i_1^{(n)} \quad (3)$$

Where, α is the harmonic current compensation coefficient.

Then equation (1) can be reformulated to

$$U_1^{(n)} = r_1 \cdot I_1^{(n)} + (j\omega_n L_{11} - j\omega_n \alpha NM)I_1^{(n)} \quad (4)$$

If the harmonic current compensation coefficient α satisfies

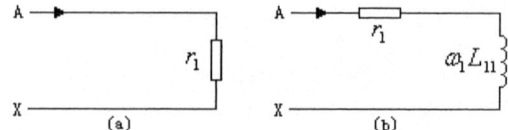

Figure 2. The equivalent circuits of the linear transformer. (a) The harmonic equivalent. (b) The fundamental equivalent circuit.

$$\alpha = L_{11} / N \cdot M \quad (5)$$

Then, from terminals AX, the equivalent impedance of the transformer at the nth-order harmonic frequency is derived

$$Z_{AX}^{(n)} = U_1^{(n)} / I_1^{(n)} = r_1 \quad (6)$$

The fundamental voltage equation of the transformer's primary winding can be obtained in phasor form

$$U_1^{(1)} = (r_1 + j\omega_1 L_{11})I_1^{(1)} \quad (7)$$

Then, from terminals AX, the equivalent impedance of the transformer at the fundamental frequency is derived

$$Z_{AX}^{(1)} = U_1^{(1)} / I_1^{(1)} = r_1 + j\omega_1 L_{11} \quad (8)$$

Note that equation (5) is the harmonic current compensation condition.

The equivalent circuits of the transformer to the fundamental and the harmonics are shown in Fig.2.

The resistance of the transformer's primary winding is nearly zero. Therefore, when the harmonic current compensation condition is satisfied, the transformer exhibits nearly zero impedance to harmonic current and primary self-impedance to fundamental current. The PWM voltage source inverters in nature behave as harmonic current controlled harmonic current sources. Before compensation, the primary harmonic current is very weak. When the compensation system starts to work, positive feedback occurs, which results in sharp increasing of the primary harmonic current. The harmonic currents in power systems are led to flow into the transformer branch.

A specially designed linear transformer with multiple secondary windings is adopted to increase the capacity of the active power filter and ensure that the power semiconductor switches work in relatively low voltage. Saturation must be avoided in the transformer in order to obtain excellent filtering performance. To make sure that the transformer works in linear range, i.e. to reduce the magnetic flux density, air gap is necessary in the core.

In order to reduce the reactive power produced by the transformer, i.e. to reduce the no-load current $i_1^{(1)}$, the only approach is to increase the primary self-inductance. However, too large primary self-inductance brings two disadvantages: (a) higher precise control system is needed; (b) the bandwidth of the positive feedback loop is increased, which may cause the invalidation of the positive feedback in view of the error of harmonic current detection and compensation. Therefore, the primary self-inductance should be moderate.

The transformer's secondary windings are identical in

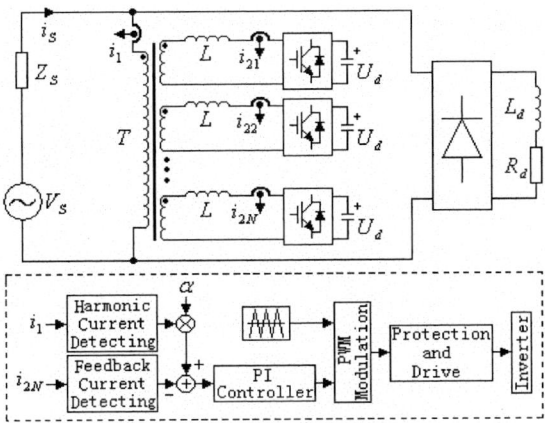

Figure 3. Configuration of the proposed active power filter.

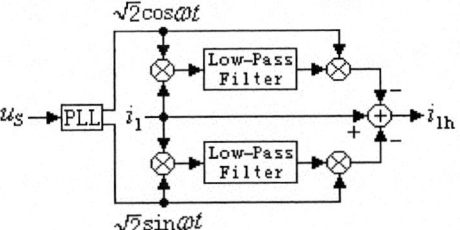

Figure 4. Block diagram of harmonic current detection method.

turns, lead cross section, parallel wound lead number, and structural dimension including inner diameter, outer diameter and height. Therefore, the mutual inductances between the primary winding and each of the secondary windings are almost the same. The self-inductances of the multiple secondary windings have small discrepancy due to the difference of their distributing position in the core, i.e. the leakage inductances of the multiple secondary windings are different. The discrepancy has negligible influence on filtering performance since each secondary winding is connected with a harmonic current source, which is formed by a PWM voltage source inverter.

III. SYSTEM CONFIGUTION AND CONTROL STRATEGY

Fig.3 shows the system configuration of the proposed active power filter. V_S and Z_S represent the source voltage and the system impedance, respectively. The transformer's primary winding is shunted with a harmonic-producing load, while the secondary windings are connected with PWM voltage source inverters. L represents the output inductance of each inverter, which is negligible if the corresponding secondary leakage impedance is relatively large. U_d represents the voltage of the DC side of each inverter. A typical diode full-bridge rectifier with resistive and inductive (RL) load is used as the current type harmonic-producing load.

The control scheme of the proposed active power filter includes the harmonic reference current generator and the harmonic compensation current controller. The harmonic reference current generator first detects the primary harmonic current of the transformer. Then, the primary harmonic current is multiplied by the harmonic current compensation coefficient, the result of which is the reference current. The harmonic compensation current controller compares the reference current with the feedback current of the transformer's secondary winding, the error of which is regulated by a proportional and integral (PI) regulator. Compared with the triangular waveform, the output of the PI regulator is modulated to

get PWM driving pulses, which drive the power semiconductor switches. The multiple secondary windings use one controller to get PWM driving pulses since the required harmonic compensation currents are the same.

Fig.4 shows the transformer's primary harmonic current detection method, which is based on the single-phase d-q transformation. In the proposed block diagram, $\cos \omega t$ is synchronized with the phase-to-neutral source voltage by implementing the phase loop lock (PLL). Low pass filters (LPFs) are used to achieve the rms value of the fundamental active current and reactive current.

Assume that the transformer's primary current is

$$i_1(t) = \sqrt{2} \sum_{n=1}^{\infty} I_n \cos(n\omega t + \varphi_n) \qquad (9)$$

Multiplying $i_1(t)$ by $\sqrt{2} \cos \omega t$, the following expressions can be obtained

$$\sqrt{2} \cos \omega t \cdot i_1(t) = \sum_{n=1}^{\infty} I_n \{\cos[(n+1)\omega t + \varphi_n] + \cos[(n-1)\omega t + \varphi_n]\} \qquad (10)$$

The rms value of the fundamental active current can be achieved by a low pass filter

$$I_{1p} = I_1 \cos \varphi_1 \qquad (11)$$

Multiplying $i_1(t)$ by $\sqrt{2} \sin \omega t$, the following expressions can be obtained

$$\sqrt{2} \sin \omega t \cdot i_1(t) = \sum_{n=1}^{\infty} I_n \{\sin[(n+1)\omega t + \varphi_n] - \sin[(n-1)\omega t + \varphi_n]\} \qquad (12)$$

The rms value of the fundamental reactive current can be achieved by a low pass filter

$$I_{1q} = -I_1 \sin \varphi_1 \qquad (13)$$

Then, the fundamental current can be obtained

$$I_{1p} \cdot \sqrt{2} \cos \omega t + I_{1q} \cdot \sqrt{2} \sin \omega t = \sqrt{2} I_1 \cos(\omega t + \varphi_1) \qquad (14)$$

The harmonic current can be obtained by subtracting the fundamental current from $i_1(t)$.

741

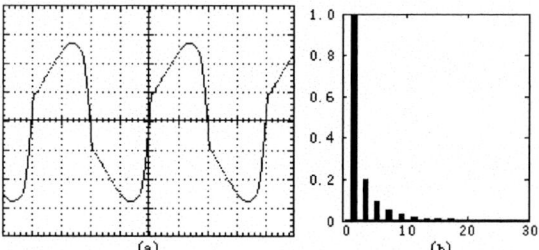

Figure 5. Experimental source current waveform and its frequency spectrum without filter. (a) Source current waveform (10A/div and 5ms/div). (b) Frequency spectrum of source current.

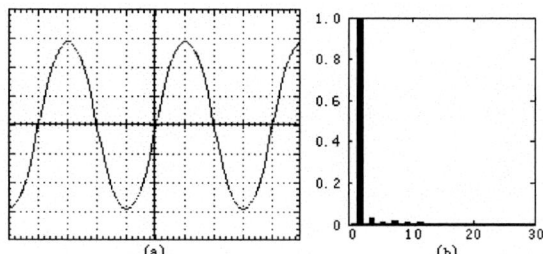

Figure 6. Experimental source current waveform and its frequency spectrum with active power filter. (a) Source current waveform (10A/div and 5ms/div). (b) Frequency spectrum of source current.

TABLE I
MAIN HARMONIC CONTENTS OF EXPERIMENTAL SOURCE CURRENT

	Without filter	With filter
Fundamental	1	1
3rd	20.15%	2.99%
5th	9.30%	1.35%
7th	5.36%	1.49%
9th	3.04%	0.92%
11th	1.63%	0.56%
13th	1.08%	0.41%
15th	0.88%	0.30%
17th	0.72%	0.25%
19th	0.65%	0.18%
21st	0.50%	0.15%
THD	23.17%	3.82%

IV. EXPERIMENTAL RESULTS

To demonstrate the validity of the novel principle of the proposed active power filter, a 5kVA prototype on the basis of the system configuration shown in Fig.3 is built. The rms value of the source voltage V_S is 220V; the source frequency f_S is 50Hz. A linear transformer with air gap and two secondary windings is adopted. The primary self-inductance L_{11} is 150mH; the mutual inductance between the primary winding and each secondary winding M is 74.88mH; the turns ratio k is 2. The harmonic current compensation coefficient α is 1. SEMIKRON's NPT type IGBT SKM300GB123D is used to be the switching device; the switching frequency is 20kHz. The output inductance of each inverter L is 0.6mH. The DC voltage of each inverter U_d is 180V.

Experimental waveforms are recorded by Tek2002

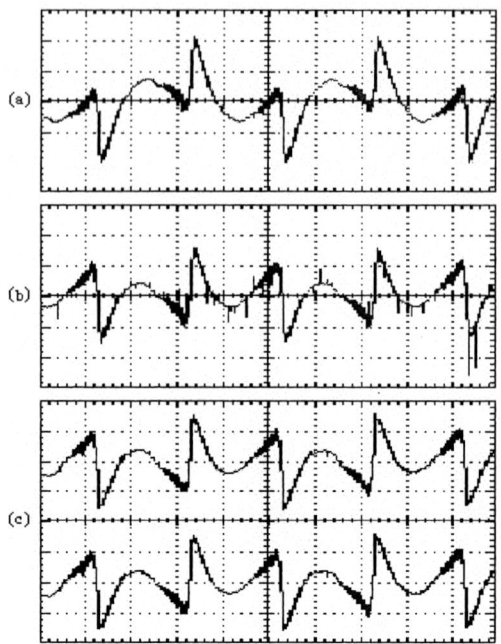

Figure 7. Experimental current waveforms of linear transformer. (a) Primary current waveform (10A/div and 5ms/div). (b) Reference current waveform (10A/div and 5ms/div). (c) Compensation current waveforms of two secondary windings (10A/div and 5ms/div).

digital oscilloscope. Fig.5 shows the experimental source current waveform and its frequency spectrum without filter. Fig.6 shows the experimental source current waveform and its frequency spectrum after the active power filter is under operation. The experimental source currents in two cases are analyzed into Fourier series. The contrast of the main harmonic contents is shown in table I. The total harmonic distortion (THD) is 23.17% before compensation and 3.82% after the active power filter is used. Fig.7 shows the experimental current waveforms of the linear transformer.

A three-phase active power filter can also be implemented in terms of the same principle. Assume that the three-phase disturbing loads are balanced, and a three-phase transformer is added. The magnetic flux in each phase is symmetrical, and does not influence each other, which is equivalent to three independent single-phase transformers. In this case, the configuration of a three-phase transformer with three-phase inverters can be adopted, and the configuration of three single-phase transformers with single-phase inverters can also be adopted. If the three-phase disturbing loads are unbalanced, especially seriously unbalanced, the magnetic flux in each phase is asymmetrical, and influences the magnetic flux of others. In this case, only the configuration of three single-phase transformers with single-phase inverters can be adopted. Each single-phase transformer can respectively exhibit nearly zero impedance to harmonic current under the control of corresponding single-phase inverters. The imbalance of the three-phase loads will not affect the filtering performance.

V. CONCLUSION

This paper presents a novel principle of active power filter for high voltage power distribution systems application. The active power filter in nature provides a low impedance path for harmonic currents in power systems. The linear transformer with air gap and multiple secondary windings is used to realize the low impedance path under the control of PWM voltage source inverters. In terms of the characteristics of the linear transformer and the law of superposition, when the harmonic current compensation condition is satisfied, the linear transformer exhibits nearly zero impedance at each harmonic frequency and primary self-impedance at the fundamental frequency. Thus, the harmonic currents in power systems can be led to flow into the linear transformer branch. The validity of the novel principle and excellent filtering characteristics are verified by the experimental results.

In the active power filter, the primary harmonic current of the linear transformer needs to be detected and tracked. Since the no-load current of the linear transformer is almost constant, the primary harmonic current can easily be detected. A harmonic current detection method based on the single-phase d-q transformation is adopted. Excellent filtering performance can easily be obtained by a simple current control scheme.

REFERENCES

[1] H. Akagi, "New trends in active filters for power conditioning," *IEEE Trans. Ind. Applicat.*, vol. 32, pp. 1312-1322, Nov./Dec. 1996.

[2] M. G. Lopez, L. T. Moran, J. C. Espinoza, J. R. Dixon, "Performance analysis of a hybrid asymmetric multilevel inverter for high voltage active power filter applications," in *Proc. 2003 IEEE Annual IECON Conf.*, pp. 1050-1055.

[3] G. Wang, Y. Li, X. You, "A novel control algorithm for cascade shunt active power filter," in *Proc. 2004 IEEE Annual PESC Conf.*, pp. 771-775.

[4] J. Hafner, M. Aredes and K. Heumann, "A shunt active power filter applied to high voltage distribution lines," *IEEE Trans. Power Delivery*, vol. 12, pp. 266-272, Jan. 1997.

[5] A. Luo, Q. Fu, L. Wang, P. Mellors, "High-capacity hybrid power filter for harmonic suppression and reactive power compensation in the power substation," *Proceedings of the CSEE*, vol. 24, pp.115-123, Sep. 2004.

[6] D. Li, Q. Chen, Z. Jia, J. Ke, "A novel active power filter with fundamental magnetic flux compensation," *IEEE Trans. on Power Delivery*, vol. 19, pp. 799-805 , Apr. 2004.

[7] Q. Chen, D. Li, Y. Xiong, C. Zhang, "A series active power filter with large capacity," *Automation of electric power systems*, vol. 29, pp. 73-76, Jan. 2005.

[8] D. Li, Q. Chen, Z. Jia, "A novel principle of adjustable reactor based on magnetic flux controllable," *Proceedings of the CSEE*, vol. 23, pp. 116-120, Feb. 2003.

[9] J. Sheng, Q. Chen, Y. Xiong, Y. Zhang, Z. Jia, "A new type automatic resonant arc-suppressing coil based on controllable magnetic flux," *Transactions of China Electrotechnical Society*, vol. 20, pp. 88-93, Feb. 2005.

2006 5th International Power Electronics and Motion Control Conference

Three-phase Active Power Filter Based on Space Vector and One-cycle Control

Wang Yong, Shen Songhua, Guan Miao

School of Automation Science and Electrical Engineering, BeiHang University, Beijing, China
Email: wylc@sina.com

Abstract—In this paper, a three-phase active power filter based on space vector and one-cycle control was proposed. According to the ideal of space vector, the three-phase voltage is divided to six regions. In each region, the model of one-cycle control is established. The proposed control method has some advantages, such as it only needs to detect the three-phase electrical power's voltage, current, neutral current, and DC side capacitance voltage of APF. It eliminates the multipliers and the calculating of harmonics and reactive current is not need. The proposed controller needs only one integrator with reset, three comparators, two flip-flops and some linear components. In each region one leg works under low frequency and other two legs work under high frequency. The proposed controller is simple, robust, reliable, and high efficiency. The one-cycle belong to nonlinear and unified constant-frequency control method, so it has features of high precision, effective compensation, and desirable for industrial applications. The analysis, modeling, and simulations are carried out with the three-phase active power filter. The simulation results verify that the proposed three-phase three-legs active power filter can dynamically and effectually compensate the harmonics, reactive and zero sequence current.

Keywords-active power filter; space-vector; one-cycle control; harmonic suppression; reactive power compensation

I. INTRODUCTION

In recent years, the usage of modern electronic equipment is increasing rapidly. The utility ac mains suffer increased harmonic pollution due to the increased number of nonlinear loads and distributed power sources connected to the grid, such as diode rectifiers, thyristor converters, power inverters, and electronic appliances. These nonlinear loads/sources generate harmonic and reactive current, which leads to low power factor, low energy efficiency, low power capacity, and harmful disturbance to other appliances. Power-factor-correction (PFC) techniques and active power filters (APFs) are viable solutions to eliminate the harmonics and improve the power factor. With the PFC approach, it processes all the power and corrects the current to unity power factor. As a contrast, an APF provides only the harmonic and reactive power. Most previously reported control approaches need to sense the load currents and then

calculate their harmonic and reactive components in order to generate the reference for controlling the power converters. Those methods require fast and real-time calculation; therefore, a high-speed digital microprocessor and high performance A/D converters are necessary, which yields high cost, low reliability, and high complexity. The one-cycle control is one of the nonlinear control strategies, it was proposed by Keyue M. Smedley and Slobodan Cuk. The control idea is that in one switch period, the average of switch variable is equal to or proportional to the reference signal, though controlling the duty ratio of switches. So it can eliminate the steady state error and the transient error. Based on one-cycle control, a three-phase PWM active power filter was introduced in [4], which represents by far the simplest APF controller with high performance. In that circuit, all switches are triggered with switching frequency, therefore, the switching losses are relatively higher that of the space vector active power filters.

II. THE THREE-PHASE APF WITH SPACE VECROT AND ONE-CYCLE CONTROL

A typical power stage of a three-phase APF is composed of a voltage-source converter that is connected in parallel with a nonlinear load as show in Fig.1. The three-phase voltage waveforms v_a, v_b, and v_c of the grid is shown in Fig.2. The three-phase voltage is divided to six regions by across zero-voltage of each voltage waveforms.

A. The Selection of Space Vector

The voltage-source converter is operated in CCM mode and the driver signals to the switches in each arm are set to be complementary, i.e. the duty ratios of switches S_{an}, S_{ap} in phase A are d_{an} and $d_{ap}=1-d_{an}$ respectively.

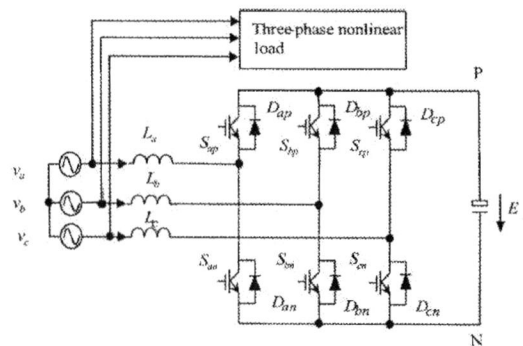

Figure 1.　Three-phase APF with six-switch bridge

1-4244-0448-7/06/$25.00 ©2006 IEEE　　744

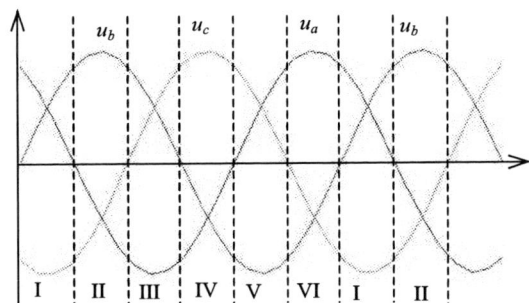

Figure 2. Three-phase voltage waveforms and the regions

In Fig.1 the average node voltages for node A, B, C referring to the bridge "N" can be write as,

$$
\begin{cases}
u_{AN} = d_{ap}E \\
u_{BN} = d_{bp}E \\
u_{CN} = d_{cp}E
\end{cases}
\tag{1}
$$

Under the state of rectifier and the symmetrical state of three-phase system,

$$
u_{NO} = -\frac{1}{3}E(d_{ap} + d_{bp} + d_{cp})
\tag{2}
$$

Since the switching frequency is much higher than the line frequency,

$$
\begin{cases}
u_a \approx d_{ap}E - \frac{1}{3}E(d_{ap} + d_{bp} + d_{cp}) \\
u_b \approx d_{bp}E - \frac{1}{3}E(d_{ap} + d_{bp} + d_{cp}) \\
u_c \approx d_{cp}E - \frac{1}{3}E(d_{ap} + d_{bp} + d_{cp})
\end{cases}
\tag{3}
$$

According to the space vector and the equation (3), the space vector and the rule of control in each region is in the table 1.

B. The Realize of one-cycle Control

When the power factor is unity, the phase of the phase-voltage and the phase-current is same. The impedance of APF and the loads are equivalent with the resistance. For the three-phase APF, the control goal is to achieve unity power factor, ie.

$$
u_i = R_e \cdot i_i \quad (i = a, b, c)
\tag{4}
$$

Where R_e is the emulated resistance that reflects the real power of the load.

According to the equation (4), the control model of one-cycle control in each region can be derived. The table 2 shows the control model of one-cycle control.

In the table 2, R_s is current sensing resistance and V_m is given by

$$
V_m = R_s \cdot \frac{E}{R_e}
\tag{5}
$$

From table 2 the one-cycle key equation is derived as

$$
\begin{cases}
V_m \begin{bmatrix} 1 - d_p \\ 1 - d_n \end{bmatrix} = R_s \begin{bmatrix} 2 & 1 \\ 1 & 2 \end{bmatrix} \begin{bmatrix} i_p \\ i_n \end{bmatrix} \\
d_t = 1
\end{cases}
\tag{6}
$$

Where d_p, d_n is the duty ratio of each switch, i_p, i_n is the phase current. Table 3 shows the relation d_p, d_n with the duty ratio of switch and i_p, i_n with the phase current.

TABLE I.
VOLTAGE SPACE VECTOR SELECTED AND FORMULAS OF THE DUTY RATIO IN EACH REGION

Region	Vector	Duty ratio
I	V_0 V_4 V_6	$1 - d_{an} = \dfrac{2u_a + u_b}{E}$ $1 - d_{bn} = \dfrac{u_a + 2u_b}{E}$ $1 - d_{cn} = 0$
II	V_7 V_6 V_2	$1 - d_{ap} = -\dfrac{2u_a + u_c}{E}$ $1 - d_{bp} = 0$ $1 - d_{cp} = -\dfrac{2u_c + u_a}{E}$
III	V_0 V_2 V_3	$1 - d_{an} = 0$ $1 - d_{bn} = \dfrac{2u_b + u_c}{E}$ $1 - d_{cn} = \dfrac{2u_c + u_b}{E}$
IV	V_7 V_3 V_1	$1 - d_{ap} = -\dfrac{2u_a + u_b}{E}$ $1 - d_{bp} = -\dfrac{u_a + 2u_b}{E}$ $1 - d_{cp} = 0$
V	V_0 V_1 V_5	$1 - d_{an} = \dfrac{2u_a + u_c}{E}$ $1 - d_{bn} = 0$ $1 - d_{cn} = \dfrac{2u_c + u_a}{E}$
VI	V_7 V_5 V_4	$1 - d_{ap} = 0$ $1 - d_{bp} = -\dfrac{2u_b + u_c}{E}$ $1 - d_{cp} = -\dfrac{2u_c + u_b}{E}$

TABLE II.
THE CONTROL MODEL OF ONE-CYCLE CONTROL

Region	Duty ratio
I	$V_m(1-d_a')=(2i_a+i_b)R_s$ $V_m(1-d_b')=(2i_b+i_a)R_s$ $V_m(1-d_c')=0$
II	$V_m(1-d_a)=-(2i_a+i_c)R_s$ $V_m(1-d_b)=0$ $V_m(1-d_c)=-(2i_c+i_a)R_s$
III	$V_m(1-d_a')=0$ $V_m(1-d_b')=(2i_b+i_c)R_s$ $V_m(1-d_c')=(2i_c+i_b)R_s$
IV	$V_m(1-d_a)=-(2i_a+i_b)R_s$ $V_m(1-d_b)=-(2i_b+i_a)R_s$ $V_m(1-d_c)=0$
V	$V_m(1-d_a')=(2i_a+i_c)R_s$ $V_m(1-d_b')=0$ $V_m(1-d_c')=(2i_c+i_a)R_s$
VI	$V_m(1-d_a)=0$ $V_m(1-d_b)=-(2i_b+i_c)R_s$ $V_m(1-d_c)=-(2i_c+i_b)R_s$

Fig.3 shows the diagram of one-cycle controller for the three-phase APF and Fig.4 shows the operation waveform of one-cycle controller.

In the beginning of each switching cycle, the clock sets the two flip-flops. The currents i_p and i_n from the input multiple is linearly combined to form an input to each of the two comparators. At other input of the two comparators is the value of V_m minus the integrated value of V_m. Signal V_m- $V_m t/T_s$ is compared with $R_s(2i_p+i_n)$ in the upper comparator and is compared with $R_s(i_p+2i_n)$ in the lower comparator as show in Fig.3. When the two inputs of a comparator meet as show in Fig.4, the comparator changes its state, which resets the correspondent flip-flop. As a result, the correspondent switch is turned off. Therefore, the duty ratios d_p and d_n are determined for the correspondent switch in each switching cycle.

III. THE SIMULATION RESULTS

Simulation using the control construction in Fig.3 was conducted using Saber. In the simulation, the nonlinear load is a three-phase rectifier. The others simulation conditions are:

Input AC voltage amplitude:

$v_a = v_b = v_c = 115V$ (RMS value)

Frequency: $f = 400Hz$

DC bus voltage: $E = 400V$

DC bus capacitance: $C_{dc} = 470\mu F$

Input inductance: $L = 0.2mH$

Switch frequency: $f_s = 50Kz$

Resistance of the nonlinear load: $R = 15\Omega$

Inductance of the nonlinear load: $L = 1mH$

Capacitance of the nonlinear load: $C = 220\mu F$

The waveforms in Fig.5 show the three-phase voltages and the three-phase currents, when the three-phase APF is connected with the nonlinear and the AC main line. The waveforms in Fig.6 show the input inductance current of three-phase APF. The waveforms in Fig.7 show the load currents. The waveforms in Fig.8 show the upper switch driver signals of three-phase APF.

From the Fig.5, Fig.6 and Fig.7, we can see, the AC main line currents is almost in-phase, and they are almost sine, so the power factor is almost unity. The harmonious current of load is compensated by APF. From the Fig.8, we can see, there are 120° the switch works under the state of low frequency in one cycle of AC main line, so the switch loss is more low.

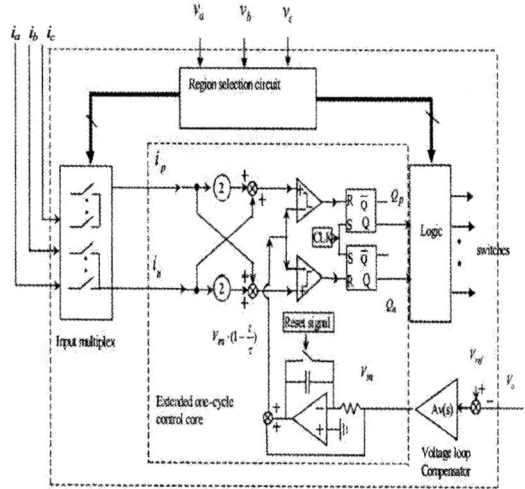

Figure 3. The diagram of one-cycle controller for the three-phase APF

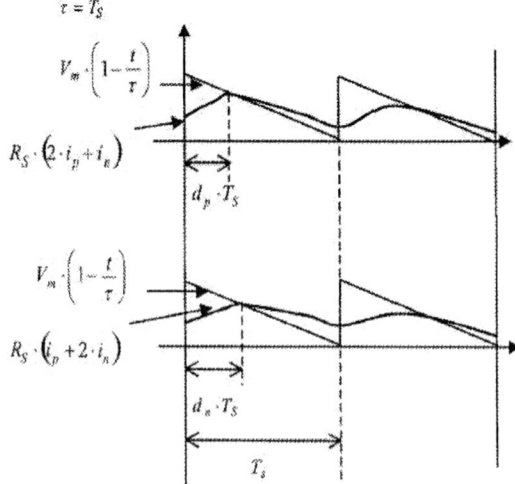

Figure 4. The operation waveform of one-cycle controller

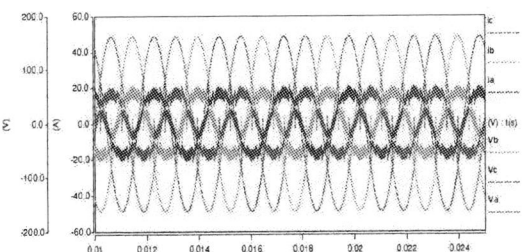

Figure 5.   The AC main line currents and voltages after APF compensation

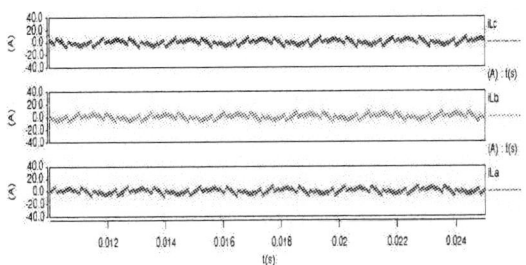

Figure 6.   The input inductance currents of the three APF

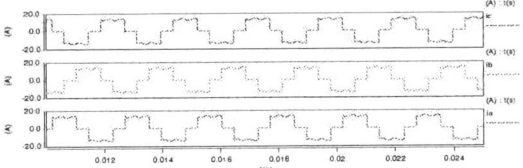

Figure 7.   The AC main line currents before APF compensation

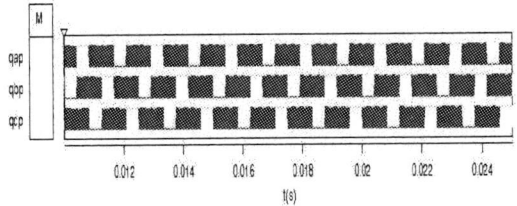

Figure 8.   The upper switch drive signal of three APF

IV.   CONCLUSION

In this paper, a three-phase APF with space vector and one-cycle control has been proposed. By this method the controller of APF is more simply. It is a cost-effective and reliable solution for power quality control and efficient energy use. This kind APF can be plugged in at the entrance of a building, a factory, a ship, an airplane, or a substation to realize overall sinusoidal current draw the mains.

REFERENCES

[1] Chongming Qiao, Keyue Ma Smedley, "Three-Phase Bipolar Mode Active Power Filter," *IEEE Trans on Industry Applications*, 2002, 38(1), pp. 149-158.

[2] Yong Wang, Songhua Shen, "Research on One-cycle Control for Switching Converters," *WCICA'2004*, Hangzhou, China, June. 6(1), pp.74-77.

[3] C.Qiao and K. M. Smedley "Three-phase Active Power Filter With Unified Constant-frequency Control," *IPEMC'2000*, Beijing, China, Aug, pp.15-18.

[4] Chongming Qiao, Keyue M. Smedley and Franco Maddaleno, "A Single-phase Active Power Filter With One-Cycle Control Under Unipolar operation," *IEEE Trans on Circuits and Systems- I: Regular papers*, 2004, 51(8), pp.1623-1630.

[5] Zhou Lin, Shen Xiao-li, Zhou Luo-wei, et al, "Application of One-Cycle Control Method to Active Power Filter," *Power Electronics*, 2004, 38(4), pp.11-13.

[6] Liu Zhiqiang, Lü Hongli, Wang Yong, "Based on space-vector pulse-width modulated reversible rectifier," *Proceedings of the fifth international conference on electrical machines and systems*, Shenyang, China, 2001, pp.518-521.

[7] Aredes M, Hafner J, Heumann K, "Three-phase four-wire shunt active filter control strategies", *Power Electronics, IEEE Transactions on*, 1997, 12(2), pp. 311-318.

[8] Zhou Xiaojun, Zhou Lin, Shen Xiaoli, "One-cycle control three-phase four-wire active power filter with four-arm", *Journal of Chongqing University*, 2004, 27(3), pp. 77-80.

[9] Zhou Lin, Jiang Jianwen, Zhou Luowei, Ye Yilin, "Three-phase four-wire active power filter with one-cycle control", *Proceedings of the Chinese Society for Electrical Engineering*, 2003, 23(3), pp. 85-88.

[10] Wang Yong, Shen Songhua, Lü Hongli, et al, "Study of three-phase inverters based on simplified voltage space-vector method", Transactions of China electrotechnical society, 2005, 20(10), pp: 25-29.

[11] Zhou Fang, Yang Jun, Hu Junfei, Wang Zhaoan, "Main circuit structure and control of the active power filter for three-phase four-wire system", *Advanced Technology of Electrical Engineering and Energy*, 2000. Vol.2.

[12] H Y Kanaan, A Marquis, K Al-Haddad, "Small-Signal Modeling and Linear Control of a Dual Boost Power Factor Correction Circuit", IEEE Power Electronics Specialists Conference (PESC), 2004, pp: 3127-3133.

[13] Wang Zhaoan, Yang Jun, Liu Jinjun, "Harmonic reduction and reactive power compensation", Beijing: China Machine Press, 1998.

2006 5th International Power Electronics and Motion Control Conference

Implementation of a Shunt-Series Compensator for Nonlinear and Voltage Sensitive Load

Bor-Ren Lin, *Senior Member, IEEE,* and Chien-Lan Huang

National Yunlin University of Science and Technology, Yunlin 640, Taiwan, China

Email: linbr@yuntech.edu.tw

Abstract—This paper presents the system analysis and circuit implementation of single-phase shunt-series compensator to improve voltage and current quality at the load side. The series compensator is adopted to compensate the voltage disturbance including voltage sag, swell, flicker and harmonics. The shunt compensator is used to supply the necessary active current to main the constant dc-link voltage of the inverter and to improve the power quality of utility source including the reactive current and harmonic current. The full-bridge inverters are used in the series and shunt compensator with the common dc-link. The full-bridge inverter has less voltage stress of power semiconductor compared with the voltage stress of switching devices in the half-bridge inverter. There is an active power flow between the series and shunt compensators to main the dc-link voltage constant when voltage disturbance is detected at the point of common coupling. The system analysis and operational principle of the adopted system is presented. Some experimental results of a scale-down prototype circuit are presented to verify the effectiveness and validity of the proposed control scheme.

Keywords – series compensator, shunt compensator, harmonic, voltage sag.

I. INTRODUCTION

Power quality problems have been become serious problem due to a large number of nonlinear load used in the modern industry products such as switching mode power supplies, electronic fluorescent lamps, industrial motor drives and uninterruptible power supplies. High quality source current and stable source voltage are generally welcome in the utility and load sides. To overcome the above problems high quality ac compensator operated with low harmonic current, high power factor, low total harmonic distortion and high reliability have been developed. The shunt compensator or shunt active filters [1-3] were used to improve the power quality of the ac source. The shunt compensator is operated as a controllable current source to supply the currents that are equal to the harmonic and reactive components of nonlinear load. The utility system only supplies active current component to the load such that the system power factor is close to unity. A clean and stable ac source is required in the voltage-sensitive equipments such as computers, telecommunication

systems and biomedical instrumentations. Voltage regulators [4-5] and uninterruptible power supplies (UPS) [6-7] have capability to provide the stable sinusoidal voltage that is independent of the mains voltage to the critical load. However the cost of the UPS system is very expensive. The series-connected compensators or series active filters [8-9] were proposed to protect the sensitive load against the voltage disturbances due to short-term abnormal voltage conditions. The technique of series compensator is an effective and cost competitive approach to improve voltage quality at the load side.

A shunt-series compensator is presented in this paper to compensate current quality at the utility side and voltage quality at the load side for the nonlinear and voltage sensitive load. Two full-bridge inverters with a common dc bus are used in the adopted shunt-series compensator. The shunt compensator can supply the harmonic and reactive components of nonlinear load current and draw or inject the active current from or to the system under the voltage sag or swell condition. Therefore the dc-link voltage can be maintained at the constant value. The series compensator can maintain the load voltage at the desired root mean square under the abnormal utility conditions. When voltage disturbance is detected at the load side, the series compensator is operated as a controllable voltage source to protect the voltage sensitive load against abnormal voltage disturbance. The single-phase shunt-series compensator based on full-bridge inverter topology for low power application is studied to improve the current and voltage qualities. The system operation and control approach are presented. The experimental results are presented to verify the validity and effectiveness of the control algorithm.

II. CONFIGURATION OF SHUNT-SERIES COMPENSATION

Fig. 1(a) gives the conventional voltage compensator. The ac/dc converter is used to achieve power factor correction and maintain the dc link voltage at the constant value. The dc/ac inverter is adopted to generate the stable load voltage to the voltage sensitive load. Therefore the load voltage is insensitive to the voltage disturbance. However the power rating of conventional ac/dc/ac converter is equal to the rated load power. The cost of this compensator is very high. Fig. 1(b) gives the system configuration of adopted shunt-series compensator for nonlinear and voltage sensitive load. Two voltage source inverters are used in the compensator. The series

1-4244-0448-7/06/$25.00 ©2006 IEEE

compensator is connected in series between the utility and load through an isolation transformer. The series compensator is operated in the voltage-controlled mode to inject or extract the compensating voltage to against the voltage disturbance such as voltage sag or swell at the load side. Therefore the load voltage is maintained at the stable root mean square value. The shunt compensator is connected in parallel with the load. The shunt compensator is operated in the current-controlled mode to supply the necessary active current for dc bus voltage regulation as well as to compensate the harmonic and reactive currents generated from the nonlinear load such that the power factor at the utility side is close to unity. The power rating of the adopted voltage compensator is only one part of rated load power. Therefore the cost of the adopted compensator is lower than the cost of conventional ac/dc/ac compensator. Fig. 2 gives the equivalent circuit of shunt-series compensator for current and voltage quality compensation.

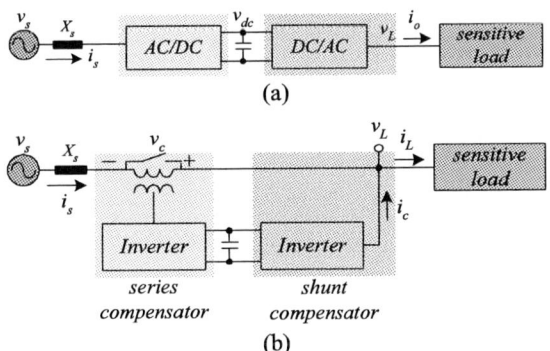

(a)

(b)

Fig. 1 (a) Conventional AC/DC/AC voltage compensator (b) adopted shunt-series voltage compensator.

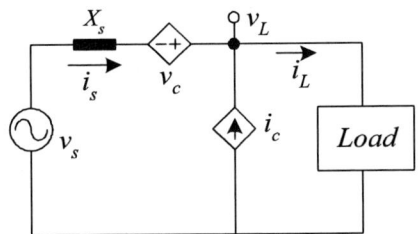

Fig. 2 Equivalent circuit of the shunt-series compensator.

The circuit topologies of shunt-series configuration can be two half-bridge inverters or two full-bridge inverters connected with a common dc link. Fig. 3 gives the shunt-series compensator based on two half-bridge inverters with the reduced number of switching devices and control functions for low power system. Two switches and two split capacitors are used in each inverter to generate the bipolar pulse-width modulation (PWM) waveforms on the ac side. Fig. 4 illustrates the circuit configuration of single-phase shunt-series compensator based on two full-bridge inverters. The shunt-connected inverter consists of four switching devices and one dc bus capacitor to achieve unipolar PWM operation. The full-

bridge based topology has better voltage utilization and less harmonic content of compensating voltage and current compared with that of half-bridge based topology.

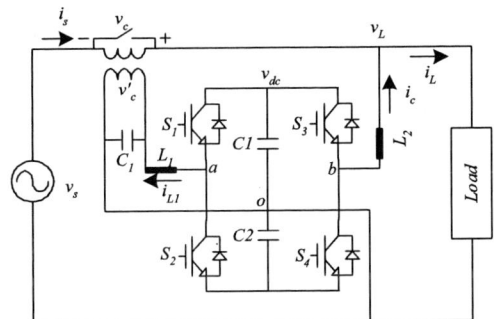

Fig. 3 Circuit configuration of single-phase shunt-series compensator based on two half-bridge inverters.

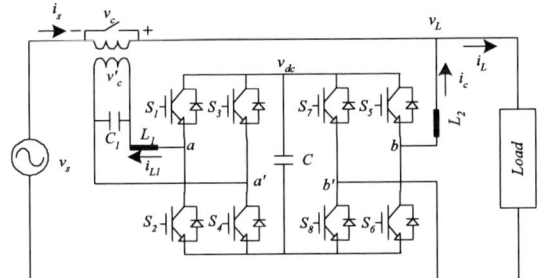

Fig. 4 Circuit configuration of single-phase shunt-series compensator based on two full-bridge inverters.

III. SYSTEM ANALYSIS AND OPERATION PRINCIPLE

The adopted shunt-series compensator is controlled to perform the following goals: 1) sinusoidal line current, 2) unity power factor, 3) constant dc link voltage, and 4) stable load voltage. Before analysis of the proposed converter, the power switches are assumed ideal and two switches in each inverter leg are complementary each other to avoid the power switches conducting at the same time. The relationships between the ac side voltages and the dc bus voltage of the proposed compensator are given as

$$v_{aa'} = (S_1 S_4 - S_2 S_3) \cdot v_{dc}, \quad v_{bb'} = (S_5 S_8 - S_6 S_7) \cdot v_{dc} \quad (1)$$

where $v_{aa'}$ and $v_{bb'}$ are the ac side voltages of the series and shunt compensators, respectively, v_{dc} is the dc bus voltage. Based on the on/off state of each switching devices, three voltage levels v_{dc}, 0 and $-v_{dc}$ are generated on the ac terminal voltages $v_{aa'}$ and $v_{bb'}$. The compensating voltage of series compensator is based on the desired load voltage command and the measured load voltage. In the series compensator, the voltage level v_{dc} and 0 are generated on $v_{aa'}$ during the positive compensating voltage. On the other hand, another two voltage levels $-v_{dc}$ and 0 are generated on voltage $v_{aa'}$ during the negative compensating voltage. For the shunt compensator, the compensating current is based on the reference line current command and the measured line current. During the positive load voltage, two voltage

749

levels v_{dc} and 0 are generated on the $v_{bb'}$ to control the compensating current to follow the current command. On the other hand another two voltage levels $-v_{dc}$ and 0 are generated on ac terminal voltage $v_{bb'}$ to track the reference compensating current.

The ac and dc side equations of the adopted compensator are expressed as follows:

$$\frac{di_c}{dt} = \frac{(S_5 S_8 - S_6 S_7) \cdot v_{dc}}{L_2} - \frac{v_L}{L_2} - \frac{r_{L2} i_c}{L_2} \qquad (2)$$

$$\frac{di_{L1}}{dt} = \frac{(S_5 S_8 - S_6 S_7) \cdot v_{dc}}{L_1} - \frac{1}{L_1} v'_C \qquad (3)$$

$$\frac{dv'_c}{dt} = \frac{i_{L1}}{C_1} + \frac{i_s / n}{C_1} \qquad (4)$$

where $n = v'_C / v_C$ is the turn ratio of isolation transformer. Fig. 5 gives the equivalent circuit of the shunt-series compensator based on the above analysis. The above equations of the adopted system can be used to to achieve high input power factor, regulate DC bus voltage and provide the stable sinusoidal output voltage using the computer simulation software.

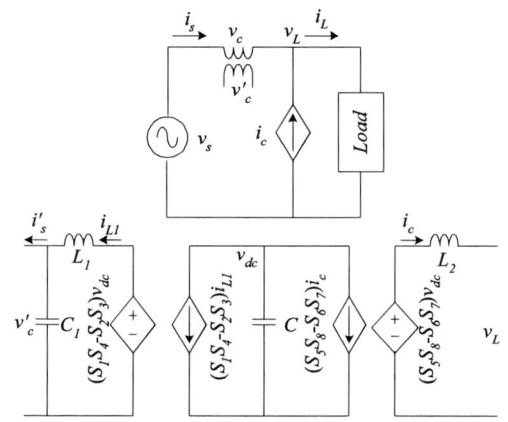

Fig. 5 Equivalent circuit of the adopted shunt-series compensator.

IV. CONTROL STRATEGY

Based on the proposed control scheme the adopted compensator can achieve power factor correction, draw a sinusoidal line current, regulate dc bus voltage, and provide a stable load voltage.

A. Shunt Compensator

Three valid operation modes are used to generate three voltage levels on the ac terminal of the shunt compensator. Based on the load voltage and the line current error, the appropriate switching state of $S_5 \sim S_8$ can be selected to generate proper voltage level on the ac terminal and to track the line current command. The actual line current will track the line current command with the current slope of $(v_L - v_{bb'})/L_2$ by controlling the voltage $v_{bb'}$. During the positive and negative load voltage, one high voltage level and one low voltage level are used to control the line current. During the positive load voltage, two voltage levels 0 ($S_6 = S_8 = 1$) and v_{dc}

$(S_5 = S_8 = 1)$ are generated on the voltage $v_{bb'}$. Power switch S_8 is always turned on in the positive load voltage. High voltage level v_{dc} generated on the ac side voltage $v_{bb'}$ will decrease line current. Low voltage $v_{bb'} = 0$ will increase line current in the positive half cycle of mains voltage. During the negative half cycle of load voltage, another two voltage levels 0 ($S_5 = S_7 = 1$) and $-v_{dc}$ ($S_6 = S_7 = 1$) are generated on the ac side and to control line current. For each half cycle of load voltage, the low voltage level is selected to increase the line current and high voltage level is used to decrease the line current. Fig. 6 shows a control block diagram of the shunt compensator. To improve the dynamic response and eliminate the steady-state error of dc bus voltage, a proportional-integral voltage controller is used in the inner control loop to determine the active source current command I^*_s. When mains voltage or load voltage variation is detected, the real power between the load and utility is not be sustained. The real source power is changed by adjusting the line current command to match the real power variation of the load. A phase-locked loop (PLL) circuit is adopted to produce a reference sinusoidal wave in phase with mains voltage. The current error between the source current command and measured current is sent to the current controller. The hysteresis current comparator is used in the current control loop to track the source current command i^*_s. Three voltage levels ($v_{dc}/2$, 0, and $-v_{dc}/2$) can be generated on the ac side of the rectifier.

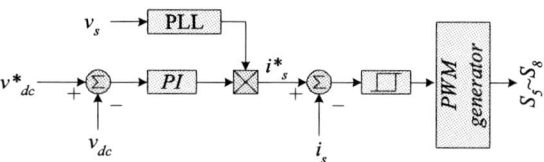

Fig. 6 Control block diagram of the shunt compensator for power factor correction.

B. Series Compensator

In the adopted series compensator, four power switches $S_5 \sim S_8$ are used. Inductor L_1 and capacitor C_1 are used to eliminate the voltage harmonics for generate the compensating voltage waveform to protect the sensitive load. The series compensator can achieve both directions of voltage regulation such as voltage sag and swell. Under the voltage sag condition, the series compensator generates the compensating voltage at the isolation transformer such that the load voltage is a sinusoidal voltage in phase with source voltage. Fig. 7(a) shows the phasor diagram of the series compensator under the source voltage sag condition. Normally the phase angle of load current is less than 90 degree. Therefore the series compensator should supply active power to the load under the voltage sag condition. This active power is supplied from the shunt compensator by regulating the dc-link voltage. An active current from the load side is flowing through the shunt compensator to dc bus. Under the voltage swell condition, the generated voltage at the output side of the series compensator should be out of phase of the source voltage. Fig. 7(b) gives the phasor diagram of the adopted series compensator operating at the voltage swell condition. To keep the dc bus voltage at the desired value, an active power absorbed by the series compensator should be delivered into the load side through the shunt compensator. Therefore an active current is supplied by the shunt compensator and flows to the load side.

Fig. 8 shows the control block diagram of the series compensator under the voltage disturbance condition. First the root mean square value of load voltage is measured and compared with the desired voltage command. The dc voltage controller based on a proportional plus integral control is used to obtain the amplitude of the compensating voltage command V^*_c. The phase locked loop circuit is used to generate the unit sinusoidal wave in phase with source voltage. The compensating voltage command and the measured compensating voltage are compared to obtain the voltage error. A proportional-integral (PI) voltage controller is used to obtain the compensating current to improve the output voltage regulation. Therefore, the calculated reference inductor current i^*_{L1} is equal to

$$i^*_L = i'_s + i_{comp} = i'_s + k_p \Delta v'_c + k_i \int \Delta v'_c \, dt \qquad (5)$$

where i'_s is the current at the output side of the series compensator, k_p and k_i are the proportional and integral gain of the voltage controller. The inner inductor current controller is adopted to improve the dynamic response of the load change. The carrier-based PWM schemes are used to generate the switching signals of power switches $S_1 \sim S_4$. Based on the switching signals of power switches, a unipolar PWM voltage waveform is generated on the ac side of the series compensator.

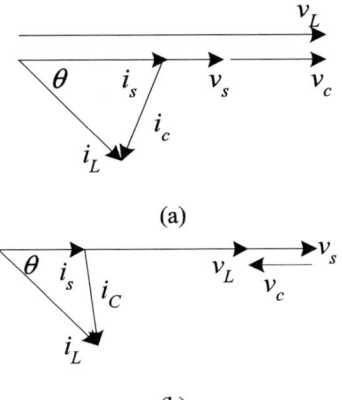

(a)

(b)

Fig. 7 Phasor diagram of the inverter (a) under voltage condition (b) over voltage condition

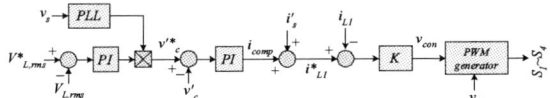

Fig. 8 Control block diagram of the series compensator for load voltage regulation.

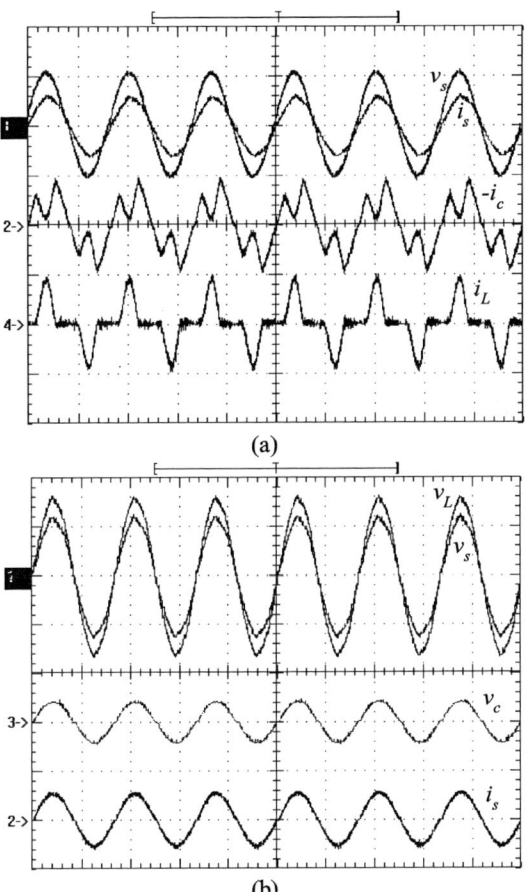

(a)

(b)

Fig. 9 Experimental results under the voltage sag condition (a) source voltage, line current, shunt compensating current and load current (b) source voltage, load voltage, series compensating voltage and line current [v_s, v_L, v_C: 100V/div; i_s, i_L, i_C: 10A/div; time:10ms/div].

V. EXPERIMENTAL RESULTS

The adopted shunt-series compensator is verified by the experimental results. The source voltage is $110V_{rms}/60Hz$. The circuit parameters are $L_1=1mH$, $L_2=0.2mH$, $C_1=36\mu F$, and $C=2200\mu F$. The experimental results of the adopted system under the source voltage disturbance and nonlinear load are shown in Figs. 9 and 10. The measured waveforms of the shunt compensator under the voltage sag are given in Fig. 91(a). The shunt compensating current will generate the harmonic and reactive components of load current into the system in order to improve the system power factor. At the same time an active current will be absorbed by the shunt compensator to compensate the dc-link voltage due to the generated in phase series compensating voltage v_C. Fig. 9(b) illustrates the measured waveforms of the series compensator under the voltage sag condition. The in phase compensating voltage v_C will be generated on the out side of the series compensator to make the load voltage at the desired value. Fig. 10 gives the measured waveforms of the shunt-series compensator at the voltage swell and nonlinear load case. The shunt compensator will eliminate the harmonic and reactive currents of the nonlinear load. The series compensator will generate the series compensating voltage that is out of phase of the load voltage to make the load voltage at the desired root mean square voltage. Based on the measured waveforms shown in Figs. 9 and 10, the control scheme of the adopted shunt-series compensator can effectively improve the voltage and current qualities for the nonlinear and voltage sensitive load.

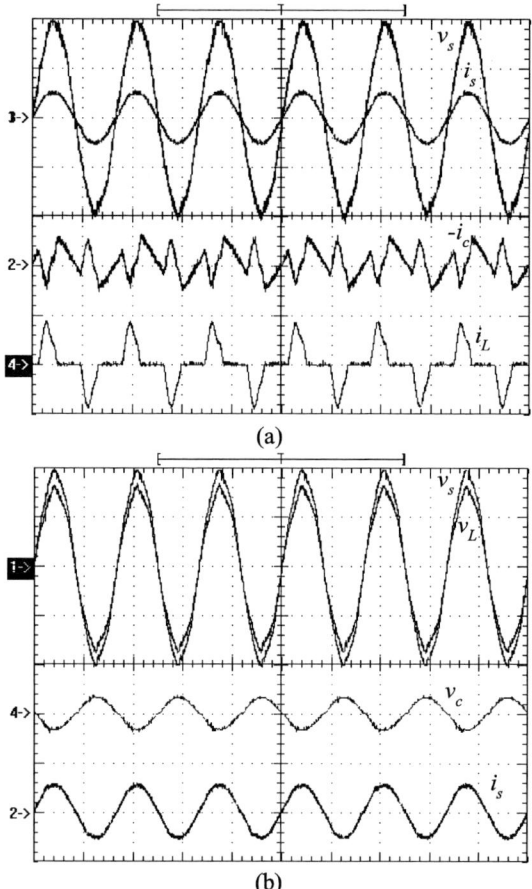

(a)

(b)

Fig. 10 Experimental results under the voltage swell condition (a) source voltage, line current, shunt compensating current and load current (b) source voltage, load voltage, series compensating voltage and line current [v_s, v_L, v_C: 100V/div; i_s, i_L, i_C: 10A/div; time:10ms/div].

VI. CONCLUSION

A shunt-series compensator is adopted in the single-phase system to improve power quality for nonlinear and voltage sensitive load. The full-bridge inverters are used in the system to generate unipolar voltage waveforms on the ac side of the inverters. The series compensator connected between the ac source and load will compensate source voltage disturbance. The in-phase or out-of-phase compensating voltage will be generated on the output side of series compensator to make the load voltage at its desired value. The shunt compensator will generate the necessary compensating currents to improve the system power factor and regulate the dc-link voltage. Based on the simulation and experimental results, the adopted control scheme can make the load voltage insensitive to source voltage disturbance and power factor of ac source is close to unity.

REFERENCES

[1] V. M. Moreno, A. P. Lopez, and R. I. D. Garcias, "Reference current estimation under distorted line voltage for control of shunt active power filters", IEEE Transactions on Power Electronics, vol. 19, no. 4, pp. 988-994, 2004.

[2] J. W. Dixon, J. J. Garcia and L. Moran, "Control system for three-phase active power filter which simultaneously compensates power factor and unbalanced loads", IEEE Transactions on Industrial Electronics, vol. 42, no. 6, pp. 636-642, 1995.

[3] P. Verdelho and G. D. Marques, "An active power filter and unbalanced current compensator", IEEE Transactions on Industrial Electronics, vol. 44, no. 3, pp. 321-328, 1997.

[4] Jang, D. H. and Choe, G. H., "Step-up/down AC voltage regulator using transformer with tap changer and PWM ac chopper", IEEE Transactions on Power Electronics, 1998, 45, (6), pp. 905-911

[5] Chen, C. C. and Divan, D.M., "Simple topologies for single phase AC line conditioning", IEEE Transactions on Industry Applications, 1994, 30, (2), pp. 406-412

[6] Su, G. J. and Ohno, T., "A new topology for single phase UPS systems", In Proceedings of the IEEE Industry Applications Annual Meeting, IAS'97, 1997, pp. 1376-1382

[7] Jensen, U. B., Enjeti, P. N. and Blaabjerg, F., "A new space vector based control method for ups systems powering nonlinear and unbalanced loads", IEEE Transactions on Industry Applications, 2001, vol. 37, no. 6, pp. 1864-1870

[8] Y. S. Kim, J. S. Kim and S. H. Ko, "Three-phase three-wire series active power filter, which compensates for harmonics and reactive power", IEE Proceedings - Electric Power Applications, vol. 151, no. 3, pp. 276-282, 2004.

[9] M. T. Tsai, "Analysis and design of a cost-effective series connected ac voltage regulator", IEE Proceedings – Electric Power Applications, vol. 151, no. 1, pp. 107-115, 2004.

2006 5th International Power Electronics and Motion Control Conference

Three-Phase Active Filter using a Single-Phase STATCOM Structure with Asymmetrical Dead-band Control

Seyyed Hossein Hosseini and Mehran Sabahi

Department of Electrical Engineering, Islamic Azad University of Tabriz, Tabriz, Iran

hosseini@tabrizu.ac.ir

Abstract. **This paper proposes a parallel active filter with a single phase STATCOM structure to compensate unbalance currents and reactive power of the power line due to unbalanced non-liner three phase loads. Therefore the number of required switches of conventional parallel active filters is reduced from six to four. A new control routine is applied to eliminate requirements for fast data samplers. In addition a new asymmetrical dead-band control is used to modify line currents and reactive power compensation for asymmetrical non-liner unbalanced loads. A circuit model has been simulated using PSCAD/EMTDC software and the results are presented which verify the analysis.**

Keywords- Active filter –Unbalanced asmymetric non-liner load – Ssingle phase STATCOM

I. INTRODUCTION

Nowadays the requirements for power quality become more and more important to keep safety of the electric devices and costumer satisfaction. The growth of the nonlinear loads like the devices with switching power supplies have increased the current harmonics, EMI problems, unnecessary reactive power and power losses. The static VAR compensators are presently used in the power systems to improve power quality but they haven't fast enough time responding to the instantaneous load variations. Fast switching devices like CMOS or IGBT transistors provide implementation of full bridge inverters to serve as a real time parallel compensators by bidirectional energy flow to control and compensate reactive power and current harmonics. Several circuits have been proposed to act as a dynamic active filter (AF) [1-5] and different control algorithms. In [6] an active filter configuration by means of digital signal processor is presented and a synchronous reference frame based controller is compared to an instantaneous reactive power based controller. Another DSP based indirect current controller STATCOM is presented in [7]. Also control of active power filter by using of DC bus voltage is an attractive control strategy [8]. Recently Discrete Fourier transform (DFT) algorithm to control of an active filter is also applied [9].

In this paper the main goal is proposing a cheap single phase circuit topology to act as an active filter for three phase non-liner unbalanced loads by a control manner like

hysteresis band control routine with some difference. Further more this method is useful to modify line currents and to compensate reactive power due to relatively *asymmetrical* unbalanced non-liner loads. Required data is achieved from capacitors DC voltage, zero crossing detector of the input power for reference synchronization and current comparators which maybe contain some fast Op-Amps. There are not any mathematical computation blocks or DSP processors. Main tuning parameters are obtained by means of the conventional PI controllers. A circuit model has been simulated with PSCAD – EMTDC software [10]. The results show the good performance of proposed circuit using asymmetrical control routine and verify the analysis.

II. CIRCUIT CONFIGURATION

Fig.1 illustrates the power circuit diagram of the proposed AF based on the single phase STATCOM structure. The load has not null connection and the middle point of capacitors is connected to phase C. Middle point of capacitors C_1 and C_2 acts as a reference point to the inverter circuit so compensation of the currents $i_{as}(t)$ and $i_{bs}(t)$ cause to compensate $i_{cs}(t)$. Main condition to proper operation of the circuit is the amount of charged voltage on the DC link capacitors, so to have bidirectional current flow of $i_{ai}(t)$ and $i_{bi}(t)$ the following condition should be satisfied:

$$v_{c1}(t) \geq \sqrt{2}V_{LL} \qquad (1)$$

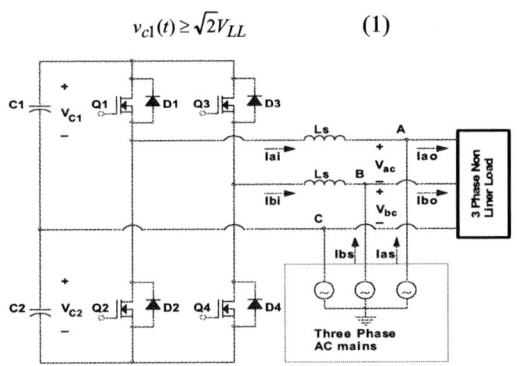

Figure 1. Proposed AF circuit diagram

1-4244-0448-7/06/$25.00 ©2006 IEEE

$$v_{c2}(t) \geq \sqrt{2} V_{LL} \qquad (2)$$

Term V_{LL} is effective line to line voltage amount. The operation modes are as follows.

a) Boost mode:

In this case switches and inductances act like a boost converter to transfer the stored energy in L_s to C_1 (or C_2). To create this mode two dead-bands between gate pulses of transistors pair Q_1 - Q_2 and Q_3 – Q_4 are required. The width of the dead band determines dead time interval of the transistors and hence determine the amount of the energy pumping to the capacitors. It is possible to consider two separate dead bands to independent control of the energy flow to C_1 or C_2. First case in the boost mode control is shown by (3).

$$i_{ai} < 0 \quad ; \quad D_1 \text{ is conducting} \quad ;$$
$$v_{c1}(t^+) = v_{c1}(t^-) - \frac{1}{C_1} \int_{t^-}^{t^+} i_{ai}(t)dt \qquad (3)$$

By assuming constant amount of the inductor current during the switching interval as (4):

$$i_{ai}(t) = Iai \qquad (4)$$

$$v_{c1}(t^+) \approx v_{c1}(t^-) + \frac{1}{C_1} I_{ai}(t^+ - t^-) \qquad (5)$$
$$\Rightarrow v_{c1}(t^+) > v_{c1}(t^-)$$

Second case in the boost mode control is similar to (3) and (4) which is shown by (6):

$$i_{bi} < 0 \quad ; \quad D_3 \text{ is conducting} \quad ;$$
$$v_{c1}(t^+) \approx v_{c1}(t^-) + \frac{1}{C_1} I_{bi}(t^+ - t^-) \qquad (6)$$
$$\Rightarrow v_{c1}(t^+) > v_{c1}(t^-)$$

Third and 4th cases of the boost mode control are applied to charge capacitor C_2 with similar equations which are given briefly by (7) and (8):

$$i_{ai} > 0 \quad ; \quad D_2 \text{ is conducting} \quad ;$$
$$v_{c2}(t^+) \approx v_{c2}(t^-) + \frac{1}{C_2} I_{ai}(t^+ - t^-) \qquad (7)$$
$$\Rightarrow v_{c2}(t^+) > v_{c2}(t^-)$$

Also

$$i_{bi} > 0 \quad ; \quad D_4 \text{ is conducting} \quad ;$$
$$v_{c2}(t^+) \approx v_{c2}(t^-) + \frac{1}{C_2} I_{bi}(t^+ - t^-) \qquad (8)$$
$$\Rightarrow v_{c2}(t^+) > v_{c2}(t^-)$$

b) Compensation mode:

In this mode of operation both diodes and transistors may be conducting relative to amount and direction of compensation current. First case in compensation mode is given by (9):

$$\text{to have } \Delta i_{ai} > 0 \Rightarrow \text{if } v_{c1}(t) > v_{ac}(t) \text{ then } Q_1 \text{ turns on}$$
$$L_s \frac{di_{ai}}{dt} = v_{c1}(t) - v_{ac}(t) \qquad (9)$$

Due to small switching interval, the amount of compensation current is approximated by (10):

$$\Delta i_{ai} \approx \frac{1}{L_s}[v_{c1}(t^-) - v_{ac}(t^-)] > 0 \qquad (10)$$

Second case of the compensation mode is given similarly by (11):

$$\text{to have } \Delta i_{ai} < 0 \Rightarrow \text{if } v_{c2}(t) < v_{ac}(t) \text{ then } Q_2 \text{ turns on}$$
$$\Delta i_{ai} \approx \frac{1}{L_s}[v_{c2}(t^-) - v_{ac}(t^-)] < 0 \qquad (11)$$

Third and 4th cases of the compensation mode control are applied to tune the current of phase B with similar equations which are given briefly by (12) and (13):

$$\text{to have } \Delta i_{bi} > 0 \Rightarrow \text{if } v_{c1}(t) > v_{bc}(t) \text{ then } Q_3 \text{ turns on}$$
$$\Delta i_{bi} \approx \frac{1}{L_s}[v_{c1}(t^-) - v_{bc}(t^-)] > 0 \qquad (12)$$

And,

$$\text{to have } \Delta i_{bi} < 0 \Rightarrow \text{if } v_{c2}(t) < v_{bc}(t) \text{ then } Q_4 \text{ turns on}$$
$$\Delta i_{bi} \approx \frac{1}{L_s}[v_{c2}(t^-) - v_{bc}(t^-)] < 0 \qquad (13)$$

After proper charging of capacitors C_1 and C_2 the minimum amounts of the compensation currents for each phase are given by (14):

$$\Delta I_{\min 1} = \frac{\Delta V_1}{L_s(t^+ - t^-)} \; ; \Delta I_{\min 2} = \frac{\Delta V_2}{L_s(t^+ - t^-)} \qquad (14)$$

Where $\Delta V_1 = V_{c1} - \sqrt{2} V_{LL} > 0$, $\Delta V_2 = V_{c2} - \sqrt{2} V_{LL} > 0$

Also maximum values of the compensation currents are given by (15):

$$\Delta I_{\max 1} = \frac{2V_{LL} + \Delta V_1}{L_s(t^+ - t^-)} \; ; \Delta I_{\min 2} = \frac{2V_{LL} + \Delta V_2}{L_s(t^+ - t^-)} \qquad (15)$$

It is clear that to correct control of compensation current proper values should be chosen for capacitor voltages, inductors and time interval.

III. ESTIMATION OF REFERENCE CURRENT VALUE

The energy equilibrium of the whole AF system may be presented by (16):

$$W_S = W_{Load} + W_{Loss} + W_{C1} + W_{C2} \qquad (16)$$

Where W_s is the input energy from the power system, W_{Loss} is wasted energy of the inverter elements due to the losses, and W_C is the stored energy in the capacitors. During specific time interval, it is possible to suppose average amounts of W_{Load} and W_{Loss} as constant values, so during long time interval an approximate relation as (17) is obtained:

$$\Delta W_S[avr] = \Delta W_{C1}[avr] + \Delta W_{C2}[avr] \qquad (17)$$

Or

$$\Delta P_S[avr] \approx \frac{1}{2\Delta T} C_1 \left[V_{c1}(T^+)^2 - V_{c1}(T^-)^2 \right]$$
$$+ \frac{1}{2\Delta T} C_1 \left[V_{c2}(T^+)^2 - V_{c2}(T^-)^2 \right] \quad (18)$$

In (18), ΔT presents long time interval duration. If I_s is assumed as the effective value of the main harmonic amplitude of the input current, average power is obtained as follows:

$$\Delta P_S[avr] = \sqrt{3} I_s V_{LL} \cos \varphi$$
$$\Rightarrow \Delta P_S[avr] = k \Delta I_s \quad (19)$$

Comparing (18) and (19) yields to drive (20):

$$\Delta I_s \propto \Delta V_{cap1} + \Delta V_{cap2} \quad where$$
$$\Delta V_{cap1} = V_{c1}(T^+)^2 - V_{c1}(T^-)^2 \quad ; \quad (20)$$
$$\Delta V_{cap2} = V_{c2}(T^+)^2 - V_{c2}(T^-)^2$$

It is possible to arrange a conventional PI controller by considering (20) to track sum of the capacitors voltage variations regarding a specified DC value, as input signal to PI controller and effective value of estimated reference current as output signal. In the stable conditions with proper amount of reference current value, the variations of the capacitor voltages tends to zero.

IV. ASYMMETRICAL DEAD BAND CONTROLLER

In this method a dead zone for switches is considered. Line currents i_{as} and i_{bs} are compared with a reference dead band and proper gate pulses are provided from a logic circuit, for example if the current of phase A takes place between dead zone then the gate pulses of Q_1 and Q_2 are stopped. The width of dead zone and its symmetry relative to reference current determine boost mode time intervals to C_1 and C_2. The following equations offer detailed explanation. By considering a positive sequence voltage in phase A with respect to null point equals with $V_{an} = V_s \cos(\omega t)$, to proper compensation of reactive power and current harmonics, the reference current may be defined by (21):

$$I_{refA} = \hat{I}_{mref} \cos(\omega t)$$
$$I_{refB} = \hat{I}_{mref} \cos(\omega t - \frac{2\pi}{3}) \quad (21)$$

In the above equation I_{mref} is the estimated reference peak current which obtains from PI controller output, which is explained on the pervious section. Upper and lower dead-band zone limits are given by (22):

$$I_{upA} = I_{refA} + \alpha = \hat{I}_s \cos(\omega t) + \alpha \quad ; \quad \alpha > 0$$
$$I_{downA} = I_{refA} - \beta = \hat{I}_s \cos(\omega t) - \beta \quad ; \quad \beta > 0$$
$$I_{upB} = I_{refB} + \alpha = \hat{I}_s \cos(\omega t - \frac{2\pi}{3}) + \alpha \quad (22)$$
$$I_{downB} = I_{refB} - \beta = \hat{I}_s \cos(\omega t - \frac{2\pi}{3}) - \beta$$

Logic of the gate pulses is obtained from (23):

$$\begin{aligned}
&if \quad i_{as} > I_{upA} \; ; Positive \; current \; injection \\
&\Rightarrow Q_1 = ON \quad ; \quad Q_2 = OFF \\
&if \quad i_{as} < I_{downA} \; ; Negative \; current \; injection \\
&\Rightarrow Q_1 = OFF \quad ; \quad Q_2 = ON \\
&if \quad I_{downA} < i_{as} < I_{upA} \; ; Boost \; mode \\
&\Rightarrow Q_1 = OFF \quad ; \quad Q_2 = OFF \\
&if \quad i_{bs} > I_{upB} \; ; Positive \; current \; injection \\
&\Rightarrow Q_3 = ON \quad ; \quad Q_4 = OFF \\
&if \quad i_{bs} < I_{downB} \; ; Negative \; current \; injection \\
&\Rightarrow Q_3 = OFF \quad ; \quad Q_4 = ON \\
&if \quad I_{downB} < i_{bs} < I_{upB} \; ; Boost \; mode \\
&\Rightarrow Q_3 = OFF \quad ; \quad Q_4 = OFF
\end{aligned} \quad (23)$$

Fig.2 shows dead-band diagram for phase A. To avoid of high frequency chattering of the transistor switches on the border lines, a mono-stable logic is used to hold high the gate pulses of the transistors within a specific minimum time interval. Furthermore values α and β determine the boosting intervals of C_1 and C_2 respectively. Boosting intervals increase by increasing α and β values and cause more charging of capacitors. To determine proper values for α and β, the difference between capacitors voltages is used with two PI controllers. For symmetric loads α and β becomes equal. Final values of them are depended on the inverter losses and boost mode required time interval. Ideal conditions are yield when width of dead-band limits to zero.

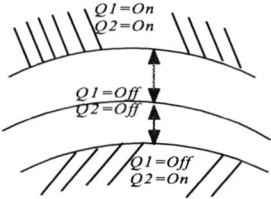

Figure 2. Asymmetric dead-band control for phase A

Figure 3. Control diagram for asymmetrical dead-band routine

Fig. 3 illustrates control diagram of the proposed circuit. It is clear the used parts in this diagram can be provided easily and with cheap structure. Also there are not fast analog to digital converters or DSP processors. All of the required functions may be obtained by conventional analog circuits, CMOS or TTL logic gates. So it is possible to obtain fast online responding of the proposed AF system. E_{SPS} in Fig. 3 is given by (24):

$$E_{SPS} = 2V_{LL} + \Delta V_1 + \Delta V_2 \qquad (24)$$

V. THE RESULTS OF SIMULATION

The following values are selected for simulation of the proposed circuit. Switching frequency is considered about 20 [KHz], V_{LL} =380V RMS, $C_1=C_2$=3300 [μF], L_S=50 [mH], a non-linear load with an active power consumption about 2800 [W] and E_{SPS}=1150 [V]. Fig. 4 shows wave forms of the line currents without compensation and Fig. 5 shows harmonic spectrum of the line currents respectively. In the first simulation, the load is assumed to have symmetrical condition which is clear from Fig. 4 so the capacitors voltages have same variations as is shown in Fig. 6. Next figure, Fig. 7 shows estimated peak reference current I_{mref} which is obtained from PI controller output. Fig. 8 shows the width values of the asymmetrical dead-band controller. It is clear in the steady state conditions for symmetrical load, α will be equal with β.

The optimum values of PI controller maybe obtained from try and error and experimental results. Fig. 9 shows the reference current I_{RefA} and compensated three phase currents I_{as}, I_{bs} and I_{cs} respectively and Fig. 10 shows harmonic spectrum of line currents with compensation. Simulation results for compensated three phase's currents illustrate the proper operation of the proposed AF circuit. Table I compares the total harmonic distortion values, THD, for line currents and compensated reactive power (respect to positive sequence value of the main harmonic of line current) to each phase of power lines without and with applying proposed active filter respectively for symmetric unbalance nonlinear load. In the next step a non-liner unbalanced load with asymmetrical currents is considered. Fig. 11 and 13 show the wave forms and harmonic spectrum of the load currents respectively.

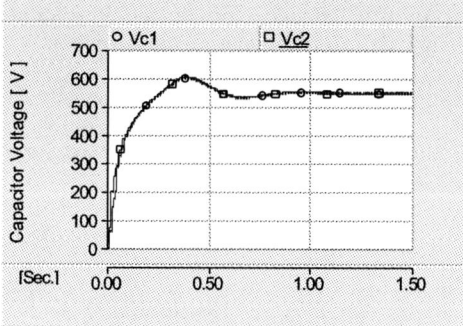

Figure 6.   Voltage variation of capacitors

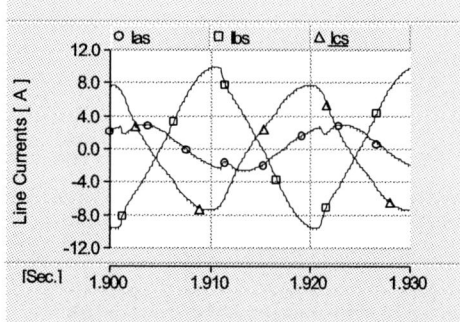

Figure 4.   Line currents of a non-liner unbalanced three phase load

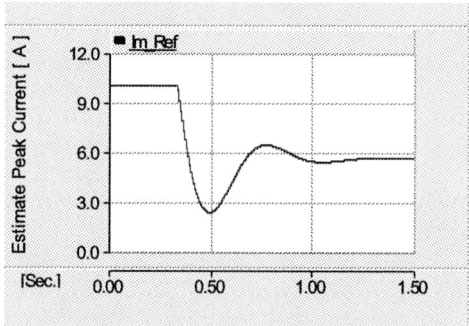

Figure 7.   Estimated reference current I_{mref}

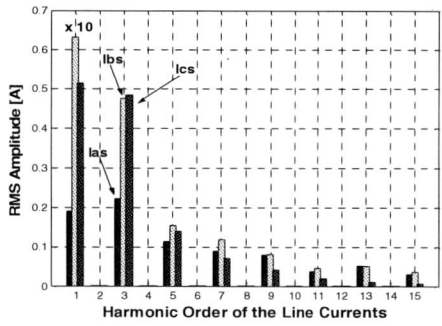

Figure 5.   Harmonic spectrum of line currents before compensation

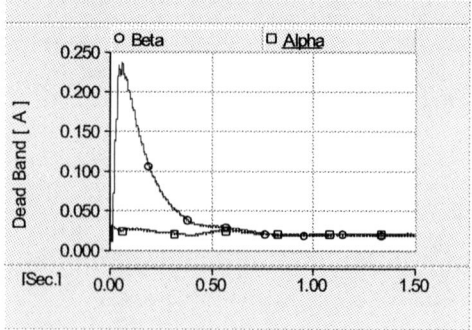

Figure 8.   Variations of the dead-band width values α and β

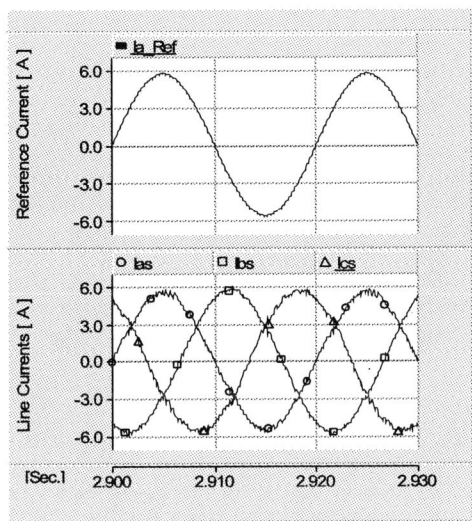

Figure 9. Compensated phase currents

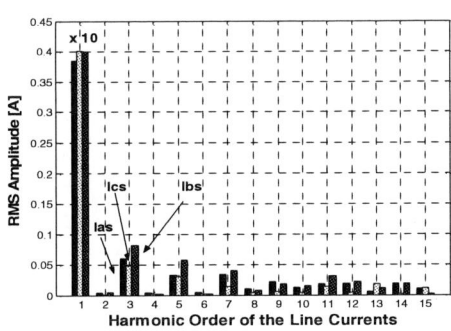

Figure 10. Harmonic spectrum of line currents after compensation

TABLE I.
Comparing results for each phase for symmetric load

Non liner unbalanced symmetric load					
Without Compensation			With Compensation		
THD$_A$	THD$_B$	THD$_C$	THD$_A$	THD$_B$	THD$_C$
0.15	0.083	0.1	0.023	0.016	0.03
Q$_A$[VAR]	Q$_B$	Q$_C$	Q$_A$	Q$_B$	Q$_C$
298	700	286	6.41	23	15.3

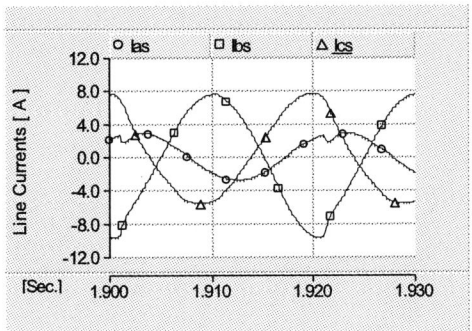

Figure 11. Currents of a asymmetrical non-liner three phase load

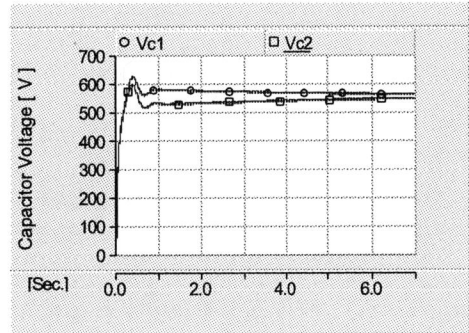

Figure 12. Voltage of capacitors for asymmetric conditions

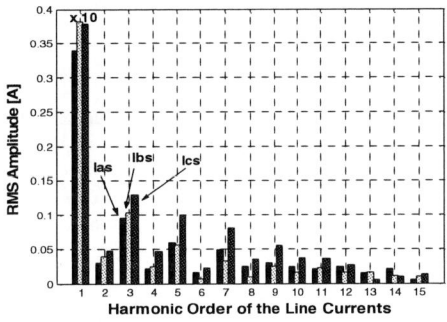

Figure 13. Harmonic spectrum of load currents for asymmetric load

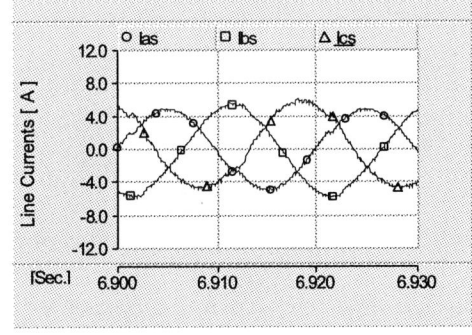

Figure 14. Compensated phase currents for asymmetric conditions

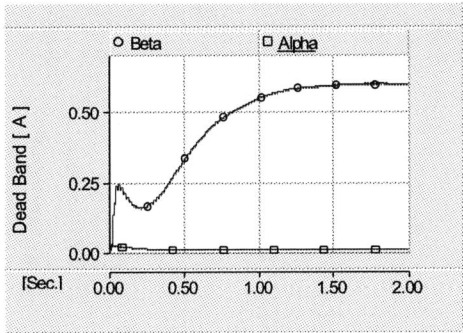

Figure 15. Variations of α and β for asymmetric load currents

757

Fig. 12 shows the difference between V_{C1} and V_{C2} because of the load asymmetry and Fig. 14 illustrates compensated phase currents which are disturbed corresponding to the previous results in Fig. 9. Fig. 15 illustrates width values of the asymmetrical dead-band controller.

Table II compares the total harmonic distortion values, THD, for line currents and compensated reactive power (with respect to main harmonic of the line current) to each phase of power lines without and with applying proposed active filter respectively for asymmetric unbalance nonlinear load. Finally Fig. 16 presents harmonic spectrum of the line currents with compensation.

VI. CONCLUSION

A single phase STATCOM topology, as a parallel active filter to compensation of three phase non-liner unbalanced symmetry or relatively asymmetric loads, has been proposed. The number of the required switches of the conventional parallel active filters is reduced from six to four.

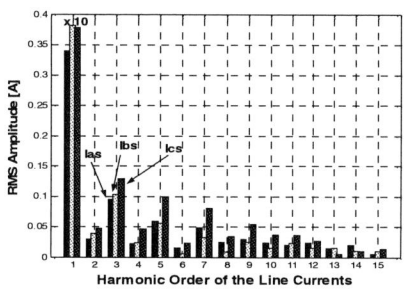

Figure 16. Harmonic spectrum of line currents after compensation

TABLE II.
Comparing results for each phase for asymmetric load

Non liner unbalanced Asymmetric load					
Without Compensation			With Compensation		
THD$_A$	THD$_B$	THD$_C$	THD$_A$	THD$_B$	THD$_C$
0.107	0.088	0.11	0.042	0.036	0.057
Q$_A$[VAR]	Q$_B$	Q$_C$	Q$_A$	Q$_B$	Q$_C$
301	628	225	25.7	69.4	29

Unsymmetrical dead-band control method is applied to modify line currents and reactive power compensation of low level asymmetric loads, sufficient charging of the DC link capacitors and required optimum reference current effective value estimation which causes minimum energy flow between power system and active filter circuit. Furthermore proposed control system maybe structured with conventional cheap devices with fast online time response. A circuit model has been simulated and the results verify the analysis. It is shown that the proposed AF circuit can be compensated symmetrical unbalanced non-linear loads and also approximately compensate low level asymmetric loads and improves line currents wave forms.

REFRENCES

[1] Don A.G. Peedder, A.D. Brown, ''A parallel-connected active filter for the reduction of supply current distortion'' ,IEEE Trans. Ind. Elect.., Vol. 47, No. 5, pp 1108-1117, Oct. 2000

[2] A. Dasfan, V.J.Gosbell, D. Platt, " Control of a new active power filter using 3-D vector control", IEEE Trans. On Pow. Elect. Vol. 15, No. 1, pp 5-12, Jan. 2000

[3] M.El-Habrouk, M.K.Darvish, P.Mehta, "Analysis and design of a novel active power filter configuration", IEE Proc.-Electr. Power Appl., Vol. 147, No. 4, pp 320-328, July 2000

[4] J.H.Sung, S.Park, K.Nam, "New hybrid parallel active filter configuration minimising active filter size"; IEE Proc.-Electr. Power Appl., Vol. 147, No.2, pp 93-97, March 2000

[5] J.M. Carrasco, E. Galvan, M.Perales, G.Escobar, A.M. Stankovic, P.Mattavelli, " Direct current control: A novel control strategy for harmonic and reactive compensation with active filters under unbalanced operation", IECON'01, 27th Annual Conference of the IEEE Ind. Electr. Society, pp 1138-1143, 2001

[6] L.R.Limongi, M.C.Cavalcanti, F.A.S. Neves, G.M.S. Azevedo, "Implemention of a digital signal processor-controlled shunt active filter", PELINCEC, International Conf. Power Electr. And Intelligent Control for Energy Conservation, Poland, Oct. 2005

[7] B.N.Singh, A.Chandra, K.Al-Haddad, " DSP- based indirect-current-controlled STATCOM , Part 1: Evaluation of current control techniques", IEE Proc.-Electr. Power Appl., Vol. 147, No.2, pp 107-112, March 2000

[8] M. Tarafdar Haque, Sh.Sarlak, N. Mahboobifar, " Control of active power filter using DC bus voltage", PELINCEC, International Conf. Power Electr. And Intelligent Control for Energy Conservation, Poland, Oct. 2005

[9] K.P. Sozanski, "Harmonic compensation using the sliding DFT algorithm for three-phase active power filter", PELINCEC, International Conf. Power Electr. And Intelligent Control for Energy Conservation, Poland, Oct. 2005

[10] Manitoba HVDC research center Inc., "PSCAD/EMTDC Version 4", 2003

2006 5th International Power Electronics and Motion Control Conference

Mitigation of Voltage Sag Using Adaptive Neural Network with Dynamic Voltage Restorer

M. R. Banaei*, S. H. Hosseini** and M. Darkalee Khajee ***

*Electrical Engineering, Faculty of Engineering, Azarbaijan University of Tarbiat Moallem, Tabriz, Iran
**Tabriz University, Faculty of Engineering, Department of Electrical &Electronics Engineering, Iran
***Young Researcher Club of Tabriz, Iran
e-mail: banaei_mohamad@yahoo.com and m.banaei@azaruniv.edu

Abstract— **A dynamic voltage restorer is a power quality (custom power) device used to correct the voltage disturbances by injecting voltage as well as power into the system. The compensation capability of a dynamic voltage restorer (DVR) depends primarily on the maximum voltage injection ability and the amount of stored energy available within the restorer. In this paper a simple structure feed forward neural network is presented for the separate of negative sequence components from fundamental sequence components in unbalance voltage sag. In addition, the strategies of correcting the supply voltage sag in a distribution feeder are presented. In this paper a new control strategy based adaptive neural network is proposed to inject minimum energy for DVR during compensation. Simulation results carried out using PSCAD/EMTDC, verify the effectiveness of the proposed control strategy.**

Keywords- neural network; minimal energy strategy; DVR

I. INTRODUCTION

The increment of voltage-sensitive load equipment has made industrial processes much more vulnerable to degradation in the quality of power supply. Voltage deviations, often in the form of voltage sags, can cause severe process disruptions and result in substantial economic loss. Therefore, cost-effective solutions, which can help such sensitive loads, ride through momentary power supply disturbances, have attracted much research attention. Among the several novel custom power devices, the dynamic voltage restorer (DVR) for application in distribution systems is a recent invention [1]. Voltage sags may cause equipment tripping, shutdown for the domestic and industrial equipment, and miss operation to the drive systems [2]. When a fault occurs in the system, the customer voltage drops below its nominal value on one or more phases. Voltage sags of down to 70% (remaining voltage is 70% of nominal voltage) are much more common than complete outages [3]. Dynamic voltage restorer (DVR) with energy storage can be used to correct the voltage sag at distribution system [4-7]. A DVR is basically a controlled voltage source that is connected in series with the network. It injects a voltage on the system in order to compensate any disturbance affecting the load voltage. The compensation capacity of a particular DVR depends on the maximum voltage injection ability and the

active power that can be supplied by the DVR. When DVR restorer's voltage disturbances, active power or energy should be injected from DVR to distribution system. If the capability of energy storage of DVR were infinite, DVR could maintain load voltage unchanged during any kind of faults [4]. Energy storage devices, such as batteries, super conducting magnetic energy storage (SMES), are required to provide active power to the load when voltage sags occur. Therefore, it is necessary to minimize energy injection from DVR. D-STATCOM similar to DVR can be used to correct the supply voltage sags. However the amounts of apparent power injection required by a D-STATCOM to correct a given voltage sag is much higher than that of a DVR.

Voltage unbalance spreads widely in the medium and low voltage distribution systems due to the increasing of nonlinear loads with different loading conditions. Unbalance voltage sages, generate negative and positive sequence voltages. The high pass filter (HPF) commonly has been used to block fundamental sequence. In [8] adaptive neural network is used for the detection of harmonic components generated by nonlinear current loads. In this paper the strategies of restoring voltage sag such as in-phase voltage and minimal energy is mentioned. In addition, a new concept of restoring techniques based adaptive neural network is proposed to inject minimum energy for DVR during compensation. The simulations carried out using PSCAD/EMTDC [9] to show capability of proposed control strategies.

II. POWER CIRCUIT CONFIGURATION OF DVR IN DISTRIBUTION SYSTEM

Power circuit of a DVR in a distribution system is shown in Fig.1. The main function of a DVR is the protection of sensitive loads from voltage sags coming from network. Therefore, the DVR is located on approach of sensitive loads. If a fault occurs on other feeders, DVR inserts series voltage, V_{dvr} and compensates load voltage to pre fault value. Distribution systems commonly use a delta-star or a star-star transformer. If delta-star transformer is used in distribution system, zero-sequence voltages will not propagate through the transformer when earth faults occur on the higher voltage level. Therefore, restoration of positive sequence and compensation of negative voltage are necessary. The main elements of the

1-4244-0448-7/06/$25.00 ©2006 IEEE

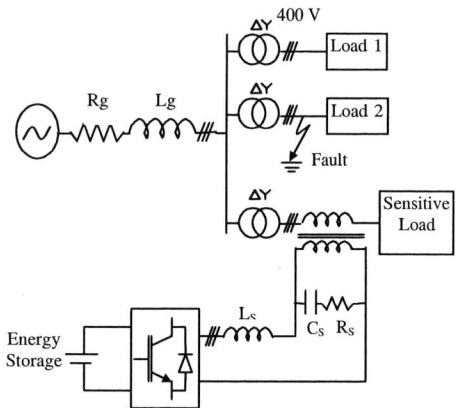

Figure 1. Power circuit of a DVR

DVR are the energy storage system, the voltage source converter, the LC filter and the coupling transformers. The sinusoidal pulse width modulation technique (SPWM) is commonly used to control forced-commutated converters. This method has been used in this paper too.

III. CONTROL STRATEGY

The characteristics of the sensitive load determine the control method and the compensation strategy for the DVR. The linear loads are not sensitive to phase angle jump and only magnitude of voltage is dominant. The control techniques should consider the limitations of the DVR such as the voltage injection capability (converter and transformer rating) and minimizing of exchanged real power from energy storage to the network. Several control techniques have been proposed for voltage sag compensation such as pre-sag method, in-phase method and minimal energy control [4], [6].

A. Minimal energy strategy

For a given load if V_{dvr} deviates to upward so that magnitude of load voltage is 1pu, injected active power will decrease. When the injected voltage, V_{dvr} is kept in quadrant with I_L, real power injection is not required to restore the voltage by the DVR.

Fig.3 shows the phasor diagram to explain the minimal energy control. α, β are the angle of V_L and the angle of

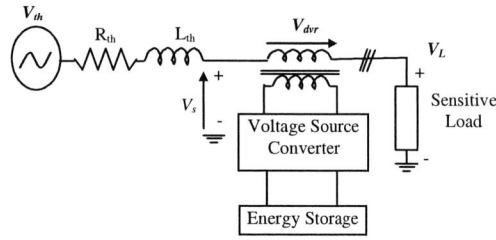

Figure 2. Equivalent circuit of power system

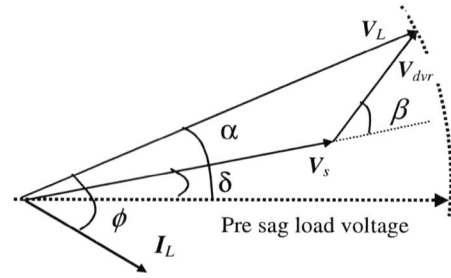

Figure 3. Phasor diagram of Minimal energy control

V_{dvr} respect to V_s, respectively. In this case, α be obtained as:

$$\beta = \frac{\pi}{2} - \phi + \alpha - \delta \qquad (1)$$

The value of δ is obtained as:

$$\alpha = \phi + \delta - \cos^{-1}\left(\frac{V_1 \cdot \cos(\phi)}{V_s}\right) \qquad (2)$$

If the supply voltage parameters satisfy the condition of (3) then the value of δ is feasible.

$$V_L . \cos(\varphi) \le V_s \qquad (3)$$

Inequality (3) means that the level of voltage sag is shallow sag. Therefore, injected real power of DVR is zero and the optimum β is obtained from (1). If inequality (3) is not satisfy then level of voltage sag will be deep sag and injected active power is not zero.

B. Proposed control method Using Adaptive Neural Network Equations

As mentioned earlier, for shallow sag injected power, P_{dvr} is zero. However, for deep sag P_{dvr} is not zero. For a given load and left side voltage of DVR, V_s with increasing β in Fig.3 magnitude and angle of V_{dvr} increases and controls P_{dvr}. Fig.4 shows the injection power according to injection voltage in several cases of voltage sags for the given load with 0.8 power factor.

It is considered that for the 0.2 pu voltages sag the minimum value of P_{dvr} is zero. While, for the shallow sag (less than 0.2 pu) minimum value of P_{dvr} is negative and for the deep sag (more than 0.2 pu) minimum value of P_{dvr} is positive. The negative P_{dvr} gives no economical advantage because DVR requires additional storage facility for the power absorption [4]. Proposed control method makes P_{dvr} zero during shallow sag and controls DVR so that P_{dvr} is minimized during deep sag according to Fig.5. It is obvious from Fig.5, which the negative value of minimum injected active power is assumed zero and the positive value of minimum injected active power is estimated as follows:

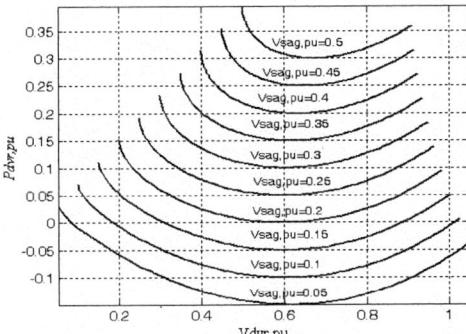

Figure 4. Injected active power according to injection voltage in several cases of voltage sags

$$P_{dvr} = V_{sag} - 0.2 \quad (4)$$

The voltage sags mentioned above were balance three-phase. For unbalance voltage sages, the three-phase instantaneous values of V_s, i.e., (V_{sr}, V_{ss} and V_{st}) generate negative, positive voltages. The synchronous d-q reference frame algorithm is applied to extract the reference voltage for the DVR i.e., ($V_{dvrr,ref}, V_{dvrs,ref}$ and $V_{dvrt,ref}$) from the measured V_s (or V_{sr}, V_{ss} and V_{st}).

For the unbalance V_s, which contain negative components, the transformation to the d-q axes results in:

$$\begin{bmatrix} V_{sd} \\ V_{sq} \end{bmatrix} = \sqrt{\frac{2}{3}} T \begin{bmatrix} V_{sr} \\ V_{ss} \\ V_{st} \end{bmatrix} = \begin{bmatrix} V_{sd,dc} \\ V_{sq,dc} \end{bmatrix} + \begin{bmatrix} V_{sd,ac} \\ V_{sq,ac} \end{bmatrix} \quad (5)$$

Where T is:

$$T = \begin{bmatrix} cos(\omega t) & cos(\omega t - \frac{2\pi}{3}) & cos(\omega t + \frac{2\pi}{3}) \\ sin(\omega t) & sin(\omega t - \frac{2\pi}{3}) & sin(\omega t + \frac{2\pi}{3}) \end{bmatrix} \quad (6)$$

The fundamental frequency components of V_s are transformed into $V_{sd,dc}$, $V_{sq,dc}$ (DC quantities) and negative sequence components are converted to AC components, $V_{sd,ac}$ and $V_{sq,ac}$. The V_s is expressed in the time domain as:

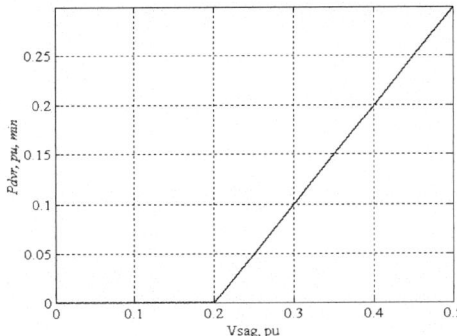

Figure 5. The minimum of injected active power according to fundamental component of voltage sag

$$V_{s,d}(t) = A_0 + A_2 \sin\left(2\omega_0 t + \phi_2\right) \quad (7)$$

Where A_0, A_2, ϕ_2 are the DC component, the magnitude and the phase of AC component in $V_{s,d}$, respectively. Equation (7) can be written as:

$$V_{s,d}(K) = A_0 + A_2 \cos\Phi_2 \sin\frac{4\pi k}{N_s} + A_2 \sin\Phi_2 \cos\frac{4\pi k}{N_s} \quad (8)$$

Where K is the iteration number, N_s is the sampling rate given by $N_s = f_s / f_0$, f_s=sampling frequency, f_0=nominal power system frequency.

For the on-line estimation of the signal features, the input to the adaptive perceptron is defined by:

$$X = \begin{bmatrix} 1 & \sin\frac{4\pi k}{N_s} & \cos\frac{4\pi k}{N_s} \end{bmatrix}^T \quad (9)$$

The structure of neural network to estimation of $V_{sd,dc}$ is shown in Fig.6. Neural network key part is a feed forward NN with three inputs (the components of X) and one output (or $V_{sd,estimated}$). The inputs are sampled uniformly and one sample is taken at a time. The output of neural network, $V_{sd,estimated}$ is calculated as follows:

$$V_{sd,estimated} = WX \quad (10)$$

NN training is aimed at minimizing J cost function:

$$J = \frac{1}{2}e^2, \quad e = V_{sd,desired} - V_{sd,estimated} \quad (11)$$

Training is accomplished by changing the NN weights according to back propagation algorithm. The weight changes are expressed as:

$$\Delta W_i = -\eta \frac{\partial J}{\partial W_i} \quad (12)$$

Where η is the learning rate. W_i is the generic weight which is connecting i^{th} input to output neuron. The weight derivate of cost function can be described as:

$$\frac{\partial J}{\partial W_i} = \frac{\partial J}{\partial e}\frac{\partial e}{\partial W_i} \quad (13)$$

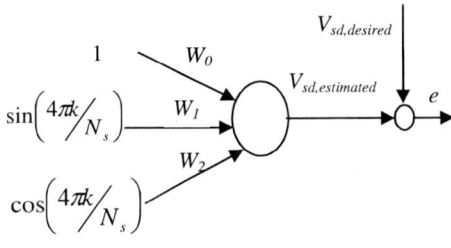

Figure 6. The structure of feed forward neural network

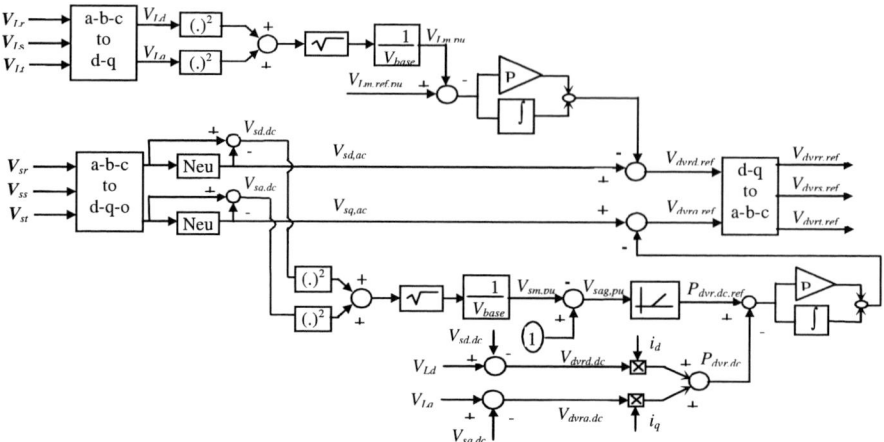

Figure 7. The proposed control strategy

After NN training W_0 is equal $V_{sd,dc}$. To estimate $V_{sq,dc}$ a neural network similar Fig.6 is necessary.

As shown in Fig.7 the adaptive neural network has been used to block DC quantities and the difference between the per unit magnitude of load voltage, $V_{Lm,pu}$ has been calculated by (14) and the per unit magnitude of reference load voltage, $V_{Lm,ref,pu}$ which is commonly 1 pu, passes through the PI controller and the output of PI controller is added to the d axis AC component of voltage.

$$V_{Lm,pu} = \sqrt{V_{Ld}^2 + V_{Lq}^2} / V_{base} \qquad (14)$$

V_{Ld} and V_{Lq} are d-q components of load voltage and V_{base} is base voltage. In the case of full compensation of load voltage for unbalance voltage sags, DVR injects DC component of active power $P_{dvr,dc}$ which restores fundamental components of voltage sags and in the same time injects AC component of active power, $P_{dvr,ac}$ which restores negative and zero sequence components of voltage sags. The DC component of active power, $P_{dvr,dc}$ can be calculated as follows:

$$P_{dvr,dc} = V_{dvrd,dc} i_d + V_{dvrq,dc} i_q \qquad (15)$$

Where $V_{dvrd,dc}$ and $V_{dvrq,dc}$ are:

$$V_{dvrd,dc} = V_{Ld} - V_{sd,dc}$$
$$V_{dvrq,dc} = V_{Lq} - V_{sq,dc} \qquad (16)$$

The per unit magnitude of fundamental components of voltage sags, $V_{sm,pu}$ can be obtained by the following equation:

$$V_{sm,pu} = \sqrt{V_{sd,dc}^2 + V_{sq,dc}^2} / V_{base} \qquad (17)$$

As shown in Fig.7, the per unit magnitude of fundamental components of voltage sags, $V_{sag,pu}$, has been applied to nonlinear function, shown in Fig.5 and the output of nonlinear function is the reference DC component of injected active power, $P_{dvr,dc,ref}$.

The difference between reference values, $P_{dvr,dc,ref}$ and the feedback value, $P_{dvr,dc}$ passes through the PI controller and the output of the PI controller is added to the q axis AC component of V_s.

IV. SIMULATION RESULTS

Simulation results carried out to verify the efficiency of mentioned control method for system shown in Fig.1 by the PSCAD/EMTDC. The system parameters are listed in Table.1.

TABLE I. PARAMETERS OF THE SYSTEM

Parameter	Value
Load voltage	400 V
Load power	125 kVA
Load power factor	0.8
Base power	125 kVA
Energy storage voltage	300 V
L_s	1 mH
R_s	0.07 Ω
C_s	300 μF

It is assumed that the voltage magnitude of the load bus is maintained at 1 pu during the voltage sag conditions.

At start-up, the NN weights are initialized to zero. During to simulation, they are changed at every sampling time. After a trial and error simulation process, satisfactory responses for the left side voltage have been achieved for $\eta = 0.1$ and the sampling rate is 0.2ms.

The results of the most important simulations are represented in Fig.8 to Fig.9.

Left side DVR voltage, injected voltage, load voltage and injected active power were shown in Fig.8 (a), (b), (c), (d) for 0.3 pu balanced three-phase voltage sag at t=0.05 sec. This fault clears at t=0.2sec.

It is observed that for deep sag the magnitude of injected voltage is 0.6 pu (according to Fig.4) and injected active is 12.5 kW equal 0.1 pu. While, injected reactive power is 73.875 kVAR equal 0.591 pu.

Fig.9 (a) shows left side voltage for two phase earth fault in phase 'r', 's' at t=0.05sec. This fault clears at t=0.2sec. Voltage sag in phase 'r', 's' is 0.24 pu. Fig.9 (b), (c) show injected voltage and load voltage, respectively. Fundamental component of voltage sag is 0.15 pu. In this case, DC component of injected active power is zero and reactive power is 0.31 pu.

These simulation results show that based on the suggested control strategy, DVR consumes zero injection power during shallow sag and minimizes injection power during deep sag.

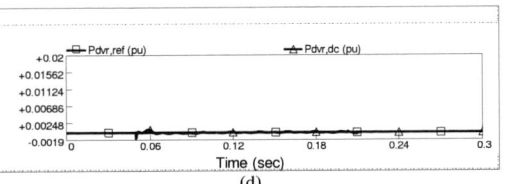

(d)

Figure 9. (a) Left side DVR voltage, V_s (b) Injection voltage, V_{dvr} (c) Load voltage, V_L (d) DC component of injected active power, P_{dvr}, for 0.24 pu sag in phase 'r', 's' for two phase earth fault

V. CONCLUSION

In order to compensate voltage sag it is possible to use dynamic voltage restorer (DVR) in distribution system for a sensitive load. Due to the limit of energy storage capacity of DC link, it is necessary to minimize energy injection from DVR. In this paper a simple structure neural network is suggested for separating the DC component of left side voltage. In addition, a new concept of restoration technique is proposed to inject minimum energy in unbalance sags.

Proposed control method makes zero injection power during shallow sag and controls DVR so that injection of power is minimized during deep sag. Simulation results carried out by the PSCAD/EMTDC shows that the proposed method using adaptive neural network can minimize the injected active power of DVR

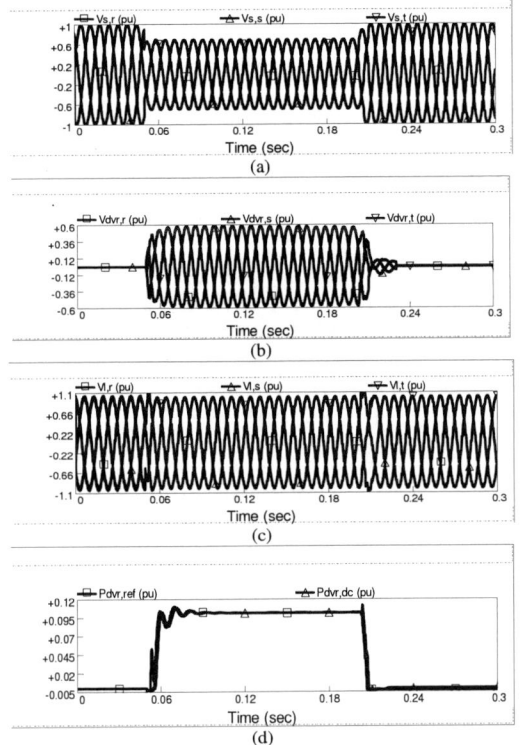

Figure 8. (a) Left side DVR voltage, V_s (b) Injection voltage, V_{dvr}
(c) Load voltage, V_L (d) DC component of injected active power P_{dvr}, for 0.3 pu three-phase balanced voltage sag

REFERENCES

[1] S. S. Choi, B. H. Li, and D. M. Vilathgamuwa, "Design and analysis of the inverter-side filter used the dynamic Voltage Restorer," *IEEE Transactions on Power Delivery*, vol. 17, no. 3, July 2002.

[2] A. Elnady and M. M. A. Salama, "New functionalities of the unified power quality conditioner," *Transmission and Distribution Conference and Exposition*, vol. 1, IEEE 2001.

[3] X. Lei, D. Retzmann and M. Weinhold, "Improvement of power quality with advanced power electronic equipment," *Electric Utility Deregulation and Restructuring and Power Technologies*, 4 -7 April, IEEE 2000.

[4] Il-Yop Chung, Dong-Jun Won, Sang-Young Park, Seung-Il Moon, Jong-Keun Park, "The DC link energy control method in dynamic voltage restorer system," *Electrical Power and Energy Systems*, vol. 25, 2003, 525-531.

[5] Hongfa Ding, Shu Shuangyan, Duan Xianzhong, Gao Jun, "A novel dynamic voltage restorer and its unbalanced control strategy based on space vector PWM," *Electrical Power and Energy Systems*, vol. 24, 2002, 693-699.

[6] S. H. Hosseini, M. R. Banaei, "Minimal energy control of the DC link energy in dynamic voltage restorer" *Proceedings of the Fourth IASTED International Conference Power and Energy Systems*, pp: 175-179, Rhodes, Greece, June 28-30, 2004.

[7] Changjiang Zhan, Atputharajah Arulampalam, and Nicholas Jenkins, "Four-Wire dynamic voltage restorer based on a three-dimensional voltage space vector PWM algorithm," *IEEE Transactions on Power Electronics*, vol. 18, no. 4, July 2003.

[8] M. Rukonuzzaman, Changjiang Zhan, Mutsuo Nakaoka, "Adaptive neural network based harmonic current compensation in active power filter," IEEE 2001.

[9] Manitoba HVDC Research Center, PSCAD/EMTDC: Electromagnetic transients program including dc systems, 1994.

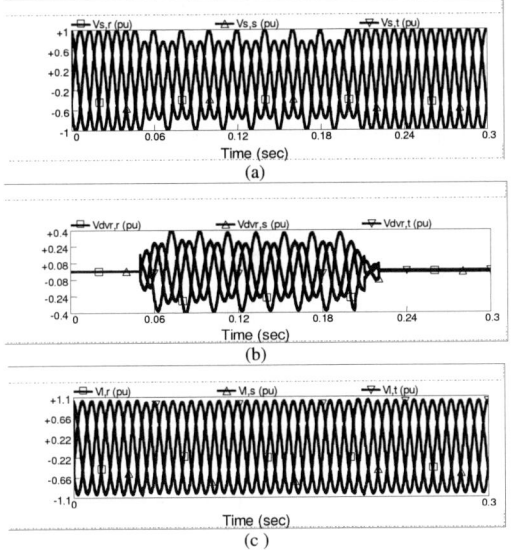

2006 5th International Power Electronics and Motion Control Conference

Mitigation of Current Harmonic Using Adaptive Neural Network with Active Power Line Conditioner

M. R. Banaei[*], S. H. Hosseini[**]

[*] Electrical Engineering, Faculty of Engineering, Azarbaijan University of Tarbiat Moallem, Tabriz, Iran
[**] Tabriz University, Faculty of Engineering, Department of Electrical &Electronics Engineering, Iran
e-mail: banaei_mohamad@yahoo.com and m.banaei@azaruniv.edu

Abstract—**This paper deals with special active power line conditioner (APLC), which is based, the same as UPQC, on integration of series and shunt power converters sharing a common DC link. In this paper the main purpose of the shunt inverter is to compensate the reactive power and current harmonics caused by nonlinear load. In addition, a simple structure feed forward neural network is presented for the separate of harmonic current components in nonlinear current load. It is known that loss reduction can be proposed in the parallel paths and also parallel transformers too. The aim of series power converter of APLC is to control power flow in order to reduce losses of distribution system. suggested APLC not only performs the functions of UPQC but also can effectively reduce power losses in parallel paths. Simulation results, carried out by PSCAD/EMTDC, show that the proposed strategy using adaptive neural network can effectively reduce power losses and eliminate harmonic currents for sensitive nonlinear loads.**

Keywords-adaptive neural network; APLC; harmonic current; loss reduction

I. Introduction

Active power line conditioner (APLC) is an advanced concept in the field power quality control [1]. APLC includes two voltage source inverters (VSIs) that connected to a DC energy storage capacitor. One of these two VSIs is connected in series with AC line while the other one is connected in shunt with load bus. Active power line conditioner can be applied in power systems for unbalance and harmonic compensation (UPQC) [2-3] and in flexible AC transmission systems (FACTS) [4], for power flow control. Other novel custom power devices can be dynamic voltage restorer (DVR) [5], for improving the quality of power supply, and distribution static compensator (DSTATCOM) for current unbalance and harmonic compensation of nonlinear load, and combined SVC, and DSTATCOM [6], for fast reactive power generator and good load balancer. In recent years, supply reliability can be proposed as power quality parameter. The parameter influencing supply reliability can be

resulted by interruption of supply and may be solved by feeding from two sides for sensitive load. Reconfiguration at the power distribution system with dispersed generation (DG), proposed at [7], for loss reduction. The high pass filter (HPF) commonly has been used to block fundamental sequence of nonlinear load current. In [8] adaptive neural network is used for the detection of harmonic components generated by nonlinear current loads.

In this paper the main purpose of series power converter that works as a voltage source is to control of power flow at the looped distribution system for loss reduction. The aim of shunt inverter, which works as a current source, is to compensate for reactive power and harmonics of current caused by nonlinear load using adaptive neural network. The simulations carried out using PSCAD/EMTDC [9] to show capability of proposed control strategies.

II. Looped Power Distribution System

Active power flow and reactive power flow of transmission line can be controlled by a series inverter in the UPFC [4]. A radial power distribution system has only one power source for a group of customers. A power failure, short-circuit, or downed power line would interrupt power in the entire line. A loop system, as the name implies, loops through the service area and returns to the original point. The loop is usually tied into an alternate power source. By placing switches in strategic locations, the utility can supply power to the customer from either direction. The loop system provides better supply reliability than the radial system. In [7] mentioned that locally loop can be obtained from reconfiguration at the power distribution system with dispersed generation (DG). In Fig.1 the sensitive nonlinear load supplied from two lines. A combined series and shunt inverter can be inserted close to the nonlinear load and in series with the line 1, such that the series compensating voltage V_{se} controls the active and reactive power of line 1 for loss reduction. The shunt inverter compensates the reactive power of load and harmonics of load current and controls voltage of DC link capacitor.

1-4244-0448-7/06/$25.00 ©2006 IEEE

764

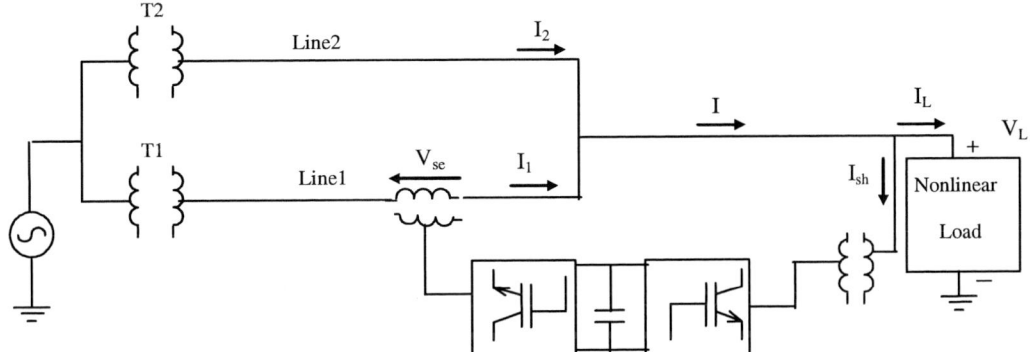

Figure 1. Looped power distribution system

III. THE SHUNT INVERTER CONTROL

The synchronous d-q reference frame algorithm is applied to extract the reference current for the shunt inverter (I_{aref}, I_{bref}, I_{cref}) from the measured load current (I_{La}, I_{Lb}, I_{Lc}) in Fig.2.

If the three-phase load currents are unbalanced and contain harmonics, the transformation to the d-q axes results in:

$$
\begin{bmatrix} I_{Ld} \\ I_{Lq} \end{bmatrix} = \sqrt{\frac{2}{3}} T \begin{bmatrix} I_{La} \\ I_{Lb} \\ I_{Lc} \end{bmatrix}
$$

$$
\equiv \begin{bmatrix} I_{Ldp} \\ I_{Lqp} \end{bmatrix} + \begin{bmatrix} I_{Ldn} \\ I_{Lqn} \end{bmatrix} + \begin{bmatrix} \sum_{k=2}^{\infty} I_{Ldk} \\ \sum_{k=2}^{\infty} I_{Lqk} \end{bmatrix} \quad (1)
$$

Where T is:

$$
T = \begin{bmatrix} \cos(\alpha) & \cos(\alpha - \frac{2\pi}{3}) & \cos(\alpha + \frac{2\pi}{3}) \\ \sin(\alpha) & \sin(\alpha - \frac{2\pi}{3}) & \sin(\alpha + \frac{2\pi}{3}) \end{bmatrix} \quad (2)
$$

Where fundamental frequency components of load currents are transformed into I_{Ldp}, I_{Lqp} which are DC quantities and negative sequence and harmonic quantities converted to AC components.

The fundamental frequency components of I_{Ld} are transformed into $I_{Ld,dc}$, $I_{Lq,dc}$ (DC quantities) and harmonic components are converted to AC components, $I_{Ld,ac}$ and $I_{Lq,ac}$. The I_{Ld} is expressed in the time domain as:

$$
I_{Ld}(t) = \sum_{n=0}^{N} A_n \sin(n\omega t + \phi_n) \quad (3)
$$

Where A_n and ϕ_n are the amplitude and phase of any term, respectively. N is the total number of terms.

Equation (3) can be written as:

$$
I_{Ld}(K) = I_{Ld,dc} + \cdots + A_N \cos \Phi_N \sin \frac{2N\pi k}{N_s}
$$

$$
+ A_N \sin \Phi_N \cos \frac{2N\pi k}{N_s} \quad (4)
$$

Where K is the iteration number, N_s is the sampling rate given by $N_s = f_s / f_0$, f_s=sampling frequency, f_0=nominal power system frequency.

For the on-line estimation of the signal features, the input to the adaptive perceptron is defined by:

$$
X = \begin{bmatrix} 1 & \cdots \cdots & \sin \frac{2N\pi k}{N_s} & \cos \frac{2N\pi k}{N_s} \end{bmatrix}^T \quad (5)
$$

The structure of neural network to estimation of $I_{Ld,dc}$ is shown in Fig.2. Neural network key part is a feed forward NN with $2N+1$ inputs (the components of X) and one output (or $I_{Ld,estimated}$). The inputs are sampled uniformly and one sample is taken at a time. The output of neural network, $I_{Ld,estimated}$ is calculated as follows:

$$
I_{Ld,estimated} = WX \quad (6)
$$

NN training is aimed at minimizing J cost function:

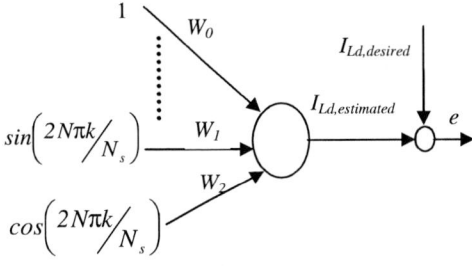

Figure 2. The structure of feed forward neural network

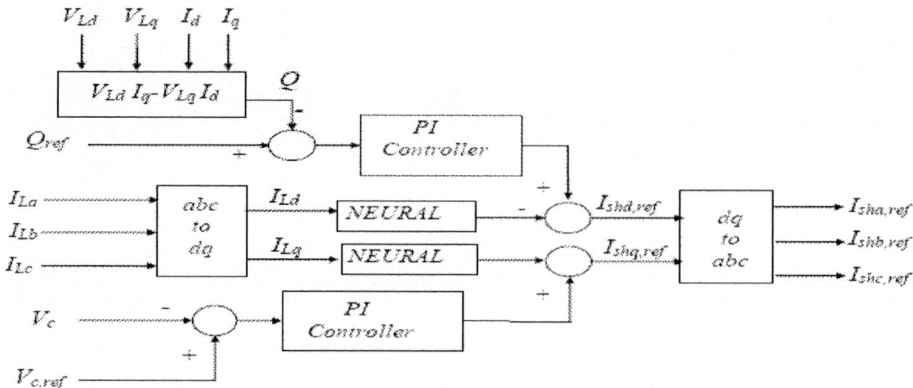

Figure 3. Reference current generator for the shunt inverter

$$J = \frac{1}{2}e^2, \quad e = I_{Ld,desired} - I_{Ld,estimated} \quad (7)$$

Training is accomplished by changing the NN weights according to back propagation algorithm. The weight changes are expressed as:

$$\Delta W_i = -\eta \frac{\partial J}{\partial W_i} \quad (8)$$

Where η is the learning rate. W_i is the generic Weight which is connecting i^{th} input to output neuron. The weight derivate of cost function can be described as:

$$\frac{\partial J}{\partial W_i} = \frac{\partial J}{\partial e} \frac{\partial e}{\partial W_i} \quad (9)$$

After NN training W_0 is equal $I_{Ld,dc}$. To estimate $I_{Lq,dc}$ a neural network similar Fig.2 is necessary.

As shown in Fig.3 the adaptive neural network has been used to block DC quantities. The reactive currents produced by loads in a three-phase distribution feeder are often dynamic in nature [6]. In order that reduction of current load and line loss minimization, it is necessary to reactive current component eliminated by the shunt active filter. In Fig.3, the difference between load instantaneous reactive power Q_L obtained from (10) and the reference value Q_{Lref} which is commonly zero, passes through PI controller and the output of PI controller is assumed with the d axis value of the AC component of current.

$$Q_L = V_{Ld}I_{Lq} - V_{Lq}I_{Ld} \quad (10)$$

Because of the energy loss due to conduction and switching power losses associated with the diodes and IGBTs of the inverter, the DC side voltage of inverter should be controlled and kept at a constant value to maintain the normal operation of inverter, which tend to reduce the value of V_c across capacitor C. A feedback voltage control circuit needs to be incorporated into the inverter for this reason.

The difference between the reference values, V_{cref} and the feedback value, V_c applied to PI controller and the output of the PI controller is assumed with the q axis value of the AC component of current. Therefore, extra fundamental components are added to load harmonic current.

IV. THE SERIES INVERTER CONTROL

Active power flow and reactive power flow of transmission line can be controlled by a series converter of a UPFC [1,4]. A radial power distribution system has only one power source for customers. A system failure, short-circuit, or line outage would interrupt the power in the part of system. By placing switches in strategic locations of the distribution system, the utility can supply power to the customer from both directions.

A parallel path is an alternate power source for sensitive loads, which requires high reliability. It is obvious that the parallel path system provides better supply reliability than the radial system. Also two parallel transformers can present the simplest parallel path in distribution system. The case study of this paper is shown in Fig. 1. As it can be seen the sensitive nonlinear load is supplied by two feeders. A combined series and shunt converter can be inserted close to the nonlinear load and in series with the line 1, such that the series compensating voltage V_{se} controls the active and reactive power of line 1 and distribution system losses.

If the APLC of Fig. 1 is not active, the line 1 current is:

$$I_1 = \frac{Z_2}{Z_1 + Z_2} I \quad (11)$$

Where Z_1, Z_2 are lines 1 and line 2 impedance, respectively. Power loss of line 1 and line 2 can be formulated as follows:

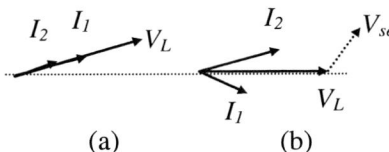

(a)   (b)

Figure 4. Phasor diagram
(a) With series compensation
(b) Without of series compensation

$$P = R_1 I_1^2 + R_2 I_2^2 \qquad (12)$$

By assumption that load current, I is constant, the line 1 current, I_1 which minimize power loss in (12) can be calculated as:

$$I_1 = \frac{R_2}{R_1 + R_2} I \qquad (13)$$

It is considered from (11) that line 1 current, I_1 is in the same phase with compensated load current, I as shown in Fig. 4. By perfect compensation of load reactive power with shunt inverter, line 1 active power flow can be obtained as:

$$P_1 = \frac{R_2}{R_1 + R_2} P \qquad (14)$$

Where P is fundamental component of load active power. It is observed from Fig.4 that line 1 reactive power flow should be regulated to zero.

Series converter inserts series voltage, V_{se} so that (14) satisfies.

Fig. 5 shows the control system of series active filter. The difference between line 1 reactive power flows, Q_1 and the reference value, Q_{1ref} that is zero, passes through the PI controller and the output of the PI controller is the q axis series inverter reference voltage, V_{qref}. The active power flow of line 1, P_1 in order to loss reduction should be controlled. The difference between P_1 and the reference value, P_{1ref} from (14), passes through the PI controller and the output is d axis series inverter reference voltage, V_{dref}.

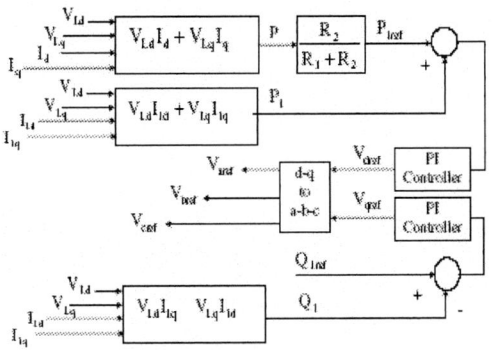

Figure 5. Control of series converter.

V. SIMULATION RESULTS

Simulation results carried out by PSCAD/EMTDC for the case study shown in Fig. 1. A three-phase thyristor rectifier with RL load is simulated as a nonlinear load and in parallel a three-phase symmetrical RL load is used as a reactive load.

The parameters of the distribution system and APLC are listed in Tables 1.

At start-up, the NN weights are initialized to zero. During to simulation, they are changed at every sampling time. After a trial and error simulation process, satisfactory responses for the left side voltage have been achieved for $\eta = 0.1$ and the sampling rate is 0.2ms.

The results of the most important simulations are presented in Figs. 6–10.

In order to study the capability of shunt inverter for reactive load compensation, the symmetrical RL load are connected to the system at t = 0 sec. The voltage of DC capacitor is controlled to be 600 V. In all cases parallel inverter is active. Fig. 6(a) shows the load reactive power Q_L and the compensated load reactive power Q and Fig. 6(b) shows the load voltage, V_L and compensated load current, I. The results indicate that shunt inverter can effectively compensate the load reactive power.

It must be noted that at t = 0.3 s the nonlinear RL load are connected to the system. Fig. 7(a) and (b) shows the waveforms of the load current, I_L and the compensated load current using adaptive neural network, I. The content of harmonics in supply current, without and with current harmonic compensation is presented in Table 2. The results show that the proposed strategy using neural network is capable of significantly reducing the current harmonics. The THD factor is in the acceptable range after compensation. The proposed loss reduction control strategy shown in Fig. 5 is used during these simulations. Fig. 8(a) reports that the line 1 reactive power is equal to zero and in the Fig. 8 (b), the phase _a_ line 1 current, I_{1a} is in the same phase with V_{La}. Fig. 9(a) shows the active power of the load, P, and the line 1 reference active power

TABLE I.   PARAMETERS OF THE DISTRIBUTION SYSTEM

Parameter	Value
Parallel RL Load	1.27+j0.628 Ω
Rectifier RL Load	3+j18.84 Ω
%Z_1	1.5+j8
%Z_2	2.25+j4
Base power	100 kVA
V_L	380 V

TABLE II.   CONTENT OF HARMONICS IN SUPPLY CURRENT

Order of harmonics	%Without compensation	%With neural network compensation
3th	11.4	0.05
5th	8.9	0.98
7th	6	1.02
9th	1.5	0.12
11th	3.2	0.92
THD	0.21	0.025

(mentioned in Eq. (14)), P_{1ref}, and the line 1 active power, P_1. It is obvious that in this figure after variation of P, P follows P_{1ref} and good power regulation can be obtained by the suggested load control strategy. In the case of full reactive load compensation without and with series inverter (for 180 kW load), line 1 active power, P_1 is almost 64 kW and 108 kW, respectively. Fig. 9(b) shows the system losses of the case study. It is obvious that for t > 0.3 sec the maximum losses occurs which is equal to 2.85 kW. In the case of the deactivation of APLC series inverter, the distribution system losses were 3.7 kW. As a result the suggested strategy can minimize the losses too.

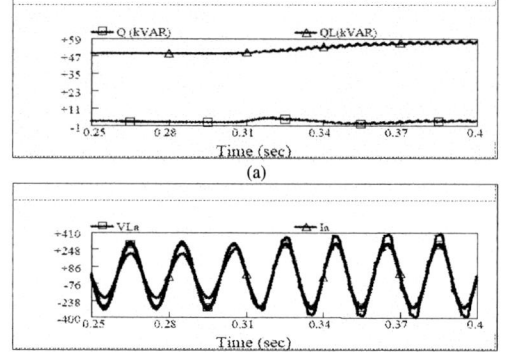

(a)

(b)

Figure 6. (a) Line 1 reactive power (b) Line 1 current and load voltage

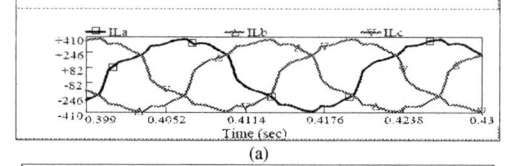

(a)

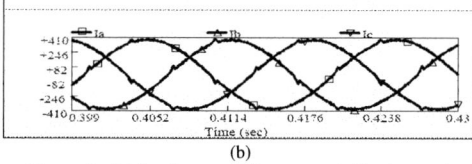

(b)

Figure 7. (a) Load current (b) compensated load current

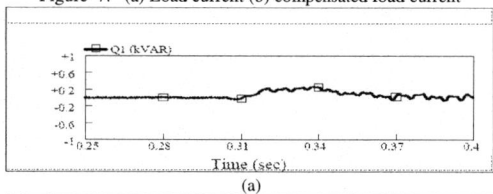

(a)

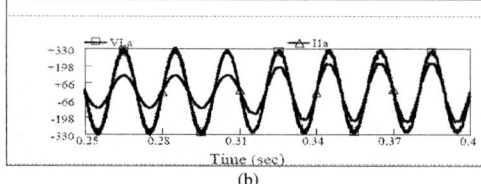

(b)

Figure 8. (a) Line 1 reactive power (b) Line 1 current and load voltage

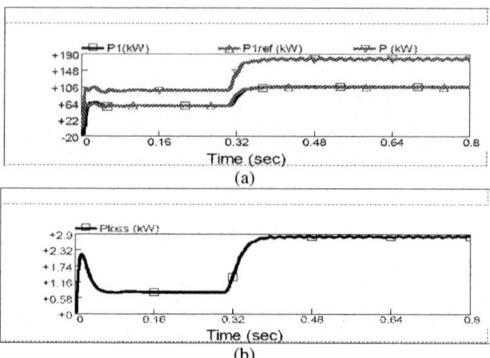

(a)

(b)

Figure 9. (a) Active power of load, P and line 1 reference active power, P_{1ref} and line 1 active power, P_1 (b) Power loss

VI.   CONCLUSION

The aim of shunt inverter in APLC is not only to compensate for reactive power but also to mitigate the harmonics produced by nonlinear load.

In this paper a simple structure neural network is suggested for separating the DC component of load current. In addition, a control strategy proposed for loss reduction based on power flow controls. In this case series inverter controls the power flow for loss reduction. The simulation results carried out by PSCAD/EMTDC show that proposed strategy can effectively reduce power losses and eliminate harmonic current using adaptive neural network.

RERENCES

[1] Mauricio Aredes, Klemens Heumann and Edson H. Watanabe, "An universal active power line conditioner" *IEEE Transactions on Power Delivery*, vol. 13, no. 2, April 1998.

[2] Arindam Ghosh, Gerard Ledwich, "A unified power quality conditioner (UPQC) for simultaneous voltage and current compensation" *Electrical Power and Energy Systems,* vol. 59, 31 August 2001, 55-63.

[3] Hideaki Fujita, Hirofumi Akagi, "The unified power quality conditioner: the Integration of series-and shunt-active filters" *IEEE Transactions on Power Electronics,* vol. 13, no. 2, March 1998.

[4] M. Vilathgamuwa, X. Zhu, S.S. Choi, "A robust control method to improve the performance of a unified power flow controller" *Electrical Power and Energy Systems*, vol. 55, pp. 103-11, 2001.

[5] Hongfa Ding, Shu Shuangyan, Duan Xianzhong, Gao Jun, "A novel dynamic voltage restorer and its unbalanced control strategy based on space vector PWM," *Electrical Power and Energy Systems,* vol. 24, pp. 693-699, 2002.

[6] San-Yi Lee, Chi-Jui Wu, "Combined compensation structure of a static var compensator and an active filter for unbalanced three-phase distribution feeders with harmonic distortion" *Electrical Power and Energy Systems,* vol. 46, pp. 243-250 September 1998.

[7] Joon-Ho Choi, Jae-Chul Kim, "Network reconfiguration at the power distribution system with dispersed generations for loss reduction" *Power Engineering Society Winter Meeting,* vol. 4, pp. 2363-2367, IEEE 2000.

[8] M. Rukonuzzaman, Changjiang Zhan, Mutsuo Nakaoka, "Adaptive neural network based harmonic current compensation in active power filter," IEEE 2001.

[9] Manitoba HVDC Research Center, "PSCAD/EMTDC: Electromagnetic transients program including dc systems", 1994.

2006 5th International Power Electronics and Motion Control Conference

A direct control strategy for UPQC in three-phase four-wire system

Tan Zhili[*,**], Li Xun[*], Chen Jian[*], Kang Yong[*], Duan Shanxu[*]

*School of Electrical & Electronics Engineering ,Huazhong University of SCI&TECH,Wuhan, China
**School of Applied Geophysics & Space Information, China University of Geosciences,Wuhan,China

Abstract— **Based on the *p-q-r* theory, this paper presents a direct control strategy of unified power quality conditioner (UPQC) used in the nonlinear and unbalance three- phase four-wire system .A algorithm of calculating the series compensation current and shunt compensation voltage in the proposed system is introduced. An analysis of the proposed control strategy and its schematic diagram is described in particular .Simulation results using MATLAB/ SIMULINK show that the harmonic current, reactive power of load as well as neutral current are compensated well when unbalance and nonlinear occur in load current or unbalance and sag in source voltage, load voltage get balanced and rated and power factor of power source is about unity.**

Keywords-p-q-r theory; three-phase four-wire system; UPQC; direct control strategy

I. INTRODUCTION

When unbalance and nonlinear load are applied to a three-phase four- wire system, quiet a large amount of current flows into the neutral-line and the transmission /distribution system, which depresses the power quality. On the other hand, in order to assure of its operational reliability, load side needs balance and sinusoidal voltage even if system voltage is in transient state, transient interrupt, sag or flicker. Employment of UPQC (unified power quality conditioner)could decrease impact on transmission and distribution harmonics and neutral-line current caused by unbalance and nonlinear load, enhance custom power quality meanwhile supply balance and

sinusoidal voltage to load and enhance power distribution reliability [1],[2],[6].

Fig.1 shows the circuit configuration of the studied three-phase four-wire UPQC.Usually there are two control scheme of UPQC,one is most used ,known as indirect control strategy, in which series compensator work by way of voltage source, and shunt compensator as current source. The other is direct control strategy，in which series compensator work as sinusoidal current source，shunt compensator as sinusoidal voltage source. Employing this strategy, series compensator isolate the voltage disturbance between power line and load，as well as shunt compensator prevent the reactive power, harmonic and neutral current on the load side into power line .Additionally, another benefit from the direct control strategy is that it is not necessary to change the work mode when power line dumping or restoring， for shunt compensator all along is controlled as sinusoidal voltage source [8]-[10].

This paper firstly introduces the *p-q-r* theory, and then discusses the calculating method of compensating current and voltage by using it. Based on that, *p-q-r* coordinates reference wave and control schematic diagram in details.Finally, simulation result was gave out, that shows this scheme is feasible and effective.

II.   *p-q-r* THEORY AND ITS POWER DEFINETION

Voltage at three-phase a-b-c coordinates can be transformed to α-β-0 coordinate as (1) [2],[3],[5]:

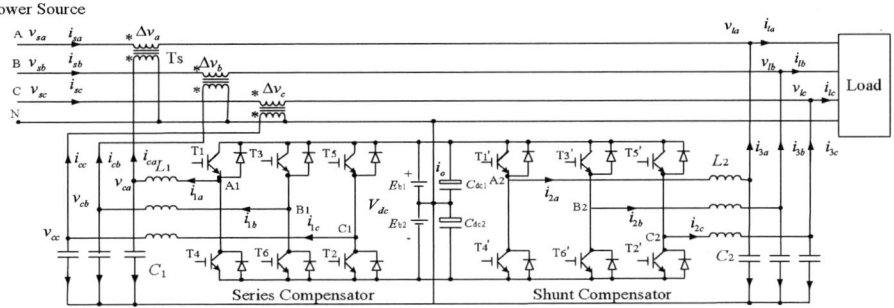

Fig.1 Circuit configuration of a three-phase four-wire UPQC

$$\begin{bmatrix} e_\alpha \\ e_\beta \\ e_0 \end{bmatrix} = \sqrt{\frac{2}{3}} \begin{bmatrix} 1 & -\frac{1}{2} & -\frac{1}{2} \\ 0 & \frac{\sqrt{3}}{2} & -\frac{\sqrt{3}}{2} \\ \frac{1}{\sqrt{2}} & \frac{1}{\sqrt{2}} & \frac{1}{\sqrt{2}} \end{bmatrix} \begin{bmatrix} e_a \\ e_b \\ e_c \end{bmatrix} = C_{\alpha\beta 0} \begin{bmatrix} e_a \\ e_b \\ e_c \end{bmatrix} \tag{1}$$

For the rotating voltage or current space vector $\vec{x}$, if choosing e_a, e_b and e_c as the coordinate reference wave, the $\vec{x}$ components in p-q-r coordinates defined by (2),

$$\begin{bmatrix} x_p \\ x_q \\ x_r \end{bmatrix} = \begin{bmatrix} \dfrac{e_\alpha}{e_{\alpha\beta 0}} & \dfrac{e_\beta}{e_{\alpha\beta 0}} & \dfrac{e_0}{e_{\alpha\beta 0}} \\ -\dfrac{e_\beta}{e_{\alpha\beta}} & \dfrac{e_\alpha}{e_{\alpha\beta}} & 0 \\ -\dfrac{e_0 e_\alpha}{e_{\alpha\beta} e_{\alpha\beta 0}} & -\dfrac{e_0 e_\beta}{e_{\alpha\beta} e_{\alpha\beta 0}} & \dfrac{e_{\alpha\beta}}{e_{\alpha\beta 0}} \end{bmatrix} \begin{bmatrix} x_\alpha \\ x_\beta \\ x_0 \end{bmatrix} \tag{2}$$

Where $e_{\alpha\beta} = \sqrt{e_\alpha^2 + e_\beta^2}$, $e_{\alpha\beta 0} = \sqrt{e_\alpha^2 + e_\beta^2 + e_0^2}$

If the system voltage $\vec{v}$ is chosen as the coordinate reference and the system current is $\vec{i}$, the instantaneous active power p, instantaneous reactive power q_q on the q axis and instantaneous reactive power q_r on the axis can be described as

$$\begin{bmatrix} p \\ q_q \\ q_r \end{bmatrix} = \begin{bmatrix} v_p i_p \\ -v_p i_r \\ v_p i_q \end{bmatrix} \tag{3}$$

Its apparent power $s = \sqrt{p^2 + q_q^2 + q_r^2}$

In this case, i_q and i_r are orthogonal with the p axis and reference voltage $\vec{v}$, having no effect on active power. The q axis and r axis relate to reactive power. The q axis lies on α-β coordinates, having something to do with phase-shifted angle with the reference voltage and harmonic wave deviated from it, other than the r axis relates to zero sequence components, such as zero sequence voltage and or the neutral current. Generally speaking, to a measured vector $\vec{x}$, such as the system voltage and current vector, x_p and x_q include the dc and ac components, while x_r include only ac components. The dc components come from the positive sequence component of $\vec{x}$, as well as the ac components from negative sequence and harmonic components of it. The value of dc components of x_q is decided by its phase-shifted angle with the positive sequence component of reference voltage.

Especially, if p-q-r coordinate reference $e_{\alpha ref}$, e_{bref}

and e_{cref} are sinusoidal and balanced, the reference voltage $\vec{e}_{ref}$ has no 0 axis components, (1)and(2)can be simplified as(4)and(5), $\vec{x}$ can be transformed to p-q-r coordinates as shown in Fig.2 [2],[5].

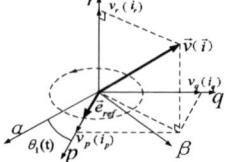

Fig.2 Physical meaning of the p-q-r transformation when e_{aref}, e_{bref} and e_{cref} are sinusoidal and balanced

$$\begin{bmatrix} e_{\alpha ref} \\ e_{\beta ref} \end{bmatrix} = \sqrt{\frac{2}{3}} \begin{bmatrix} 1 & -\frac{1}{2} & -\frac{1}{2} \\ 0 & \frac{\sqrt{3}}{2} & -\frac{\sqrt{3}}{2} \end{bmatrix} \begin{bmatrix} e_{aref} \\ e_{bref} \\ e_{cref} \end{bmatrix} \tag{4}$$

$$\begin{bmatrix} x_p \\ x_q \\ x_r \end{bmatrix} = \begin{bmatrix} \dfrac{e_{\alpha ref}}{e_{\alpha\beta ref}} & \dfrac{e_{\beta ref}}{e_{\alpha\beta ref}} & 0 \\ -\dfrac{e_{\beta ref}}{e_{\alpha\beta ref}} & \dfrac{e_{\alpha ref}}{e_{\alpha\beta ref}} & 0 \\ 0 & 0 & 1 \end{bmatrix} \begin{bmatrix} x_\alpha \\ x_\beta \\ x_0 \end{bmatrix} = C_{pqr} \begin{bmatrix} x_\alpha \\ x_\beta \\ x_0 \end{bmatrix} \tag{5}$$

III. CALCULATION OF COMPENSATION CURRENT AND VOLTAGE

A. The choose of coordinate reference voltage

This paper proposes the direct control strategy, in which the series compensator is controlled as current source, outputting three-phase balanced sine current and compensating the factor of power line unity, the shunt compensator work as voltage source, offering three-phase balanced and sinusoidal voltage to the load. In fact, because the current of power line is compensated as the sine, the reactive and harmonic current is still supplied by shunt compensator.

From the analysis above, it can be found out that the source current and loads voltage have the same phase angle with the positive sequence component of source voltage after compensation. For this compensation purpose, the balanced and sinusoidal voltage vector $\vec{e}_{ref}$ which has the same direction with positive sequence component of source voltage $\vec{v}_s^+$ must be found out. As it seen in equation (4),when v_{sa}^+, v_{sb}^+ and v_{sc}^+ transform into α-β-0 coordinate, $v_{s\alpha}^+$ and $v_{s\beta}^+$ are sinusoidal and orthogonal, the value of 0 axis components is zero. Thus if $e_{\alpha ref}$ and $e_{\beta ref}$ has the same direction as $v_{s\alpha}^+$ and $v_{s\beta}^+$ respectively, they can be chosen as the coordinate reference wave. To get the $e_{\alpha ref}$ and $e_{\beta ref}$, the reference

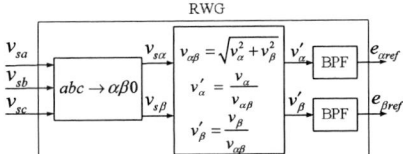

Fig.3 block for reference wave generator (RWG)

wave generator (RGW) proposed in[3][5]was used in this paper. The block diagram of the RGW is shown in Fig.3.By this algorithm, the reference wave $e_{\alpha ref}$ and $e_{\beta ref}$ can be obtained. As $\sqrt{v'_\alpha + v'_\beta} = 1$, the $e_{\alpha ref}$ and $e_{\beta ref}$ have the amplitude unity, although they have the same direction as $v_{s\alpha}^+$ and $v_{s\beta}^+$.

B. Calculation of series compensation current

The control purpose to series compensator is to supply the power line of balanced and sinusoidal current with same phase as the source voltage, so the power line only offers active power to the load. as described in (6)

$$p_s = p_{lp} = v_{lp}.i_{lp} = v_{sp}.i_{sp} \qquad (6)$$

Because v_s and i_l are unbalanced or distort, then

$$p_s = v_{sp}.i_{sp} = I_{spdc}.[V_{spdc} + \sum_{n=1}^{\infty}\sqrt{2}V_{spn}\sin(n\omega t - \varphi_{sn})] \quad (7)$$

$$p_l = v_{lp}.i_{lp} = V_{lpdc}.[I_{lpdc} + \sum_{n=1}^{\infty}\sqrt{2}I_{lpn}\sin(n\omega t - \varphi_{ln})] \ (8)$$

According to the control strategy, the power line current can only has the dc components on p axis after compensation. As known in(7) and (8),average active powers of power line are $P_s = I_{spdc}.V_{spdc}$,as well as average active powers of load are $P_l = V_{lpdc}.I_{lpdc}$,so the dc components of source current on p axis are $I_{spdc} = (V_{lpdc}.I_{lpdc})/V_{spdc}$. Because the value of C_1 is very small, the current flowing through it as well as the series transformer magnetization current can be neglected. Considering the transformer ration $N_2 : N_1 = N$,the compensation current of series compensator can be shown as(9),(10)and(11).

$$i_{cp}^* = \frac{V_{lpdc}.I_{lpdc}}{N.V_{spdc}} + \Delta I_{cp} \qquad (9)$$

$$i_{cq}^* = 0 \qquad (10)$$

$$i_{cr}^* = -(\frac{e_0}{e_{\alpha\beta ref}}).i_{cp} = 0 \qquad (11)$$

Where ΔI_{cp} is used for compensating the active powers that the converter and condenser consume, which can get by measuring the voltage of condenser. All mentioned dc components can obtain from the LPF.

C. Calculation for compensation voltages of shunt compensator

Because the shunt compensator offer three phase

balanced voltage of sine with same phase as source voltages and adopt $e_{\alpha ref}$ and $e_{\beta ref}$ as the coordinate reference wave, under the ideal situation, when the load voltages are transformed into the p-q-r coordinates, only there are the dc components on p axis, the components on the q axis and r axis are zero, so choose the compensation voltage as shown in (12),(13)and(14)

$$v_{cp}^* = v_l^* \qquad (12)$$

$$v_{cq}^* = 0 \qquad (13)$$

$$v_{cr}^* = -(\frac{e_0}{e_{\alpha\beta ref}}).v_{cp} = 0 \qquad (14)$$

Where v_l^* is rated voltages of load transformed on the p axis, its value is $\sqrt{3}$ time of effective value of rated phase voltage.

IV. THE CONTROL SYSTEM

Supposing that

$$X = [x_a \ x_b \ x_c]^{\mathrm{T}} \qquad (15)$$

$$X_{\alpha\beta 0} = [x_\alpha \ x_\beta \ x_0]^{\mathrm{T}} \qquad (16)$$

$$X_{pqr} = [x_p \ x_q \ x_r]^{\mathrm{T}} \qquad (17)$$

$$X_{\alpha\beta 0} = C_{\alpha\beta 0}.X \qquad (18)$$

$$X_{pqr} = C_{pqr}.X_{\alpha\beta 0} \qquad (19)$$

Where x may be voltage or current in the system, X is their vectors, such as $V_S = [v_{sa} \ v_{sb} \ v_{sc}]^{\mathrm{T}}$。

For the series compensator shown in Fig.1, its output voltages are

$$V_1 = V_C + L_1\frac{\mathrm{d}\,I_1}{\mathrm{d}t} + R_{L1}\,I_1 = V_C + \frac{L_1}{N}\frac{\mathrm{d}\,I_s}{\mathrm{d}t} + \frac{R_{L1}}{N}\,I_s \quad (20)$$

From (18) and (20)

$$V_{1\alpha\beta 0} = V_{C\alpha\beta 0} + L_1\frac{\mathrm{d}\,I_{1\alpha\beta 0}}{\mathrm{d}t} + R_{L1}\,I_{1\alpha\beta 0}$$

$$= V_{C\alpha\beta 0} + \frac{L_1}{N}\frac{\mathrm{d}\,I_{S\alpha\beta 0}}{\mathrm{d}t} + \frac{R_{L1}}{N}\,I_{S\alpha\beta 0} \qquad (21)$$

The control block diagram for the series compensator can be deduced from(21),as shown in Fig.4(b).By comparing the compensation current order i_{cp}^* , i_{cq}^* and i_{cr}^* to the actual current i_{1p} , i_{1q} and i_{1r} ,we can get the voltages order of L_1 after regulating their difference by current regulator CR1.Then transforming them to α-β-0 coordinate ,adding them to the voltages of secondary winding of series transformer $v_{c\alpha}$, $v_{c\beta}$ and v_{c0} ,we can get the output voltages order of the series compensator $v_{1\alpha}^*$, $v_{1\beta}^*$ and v_{10}^* .The control signals of the switch devices can be got from the PWM generator by calculating the voltages orders.

For the shunt compensator, its output voltages and current can be described as

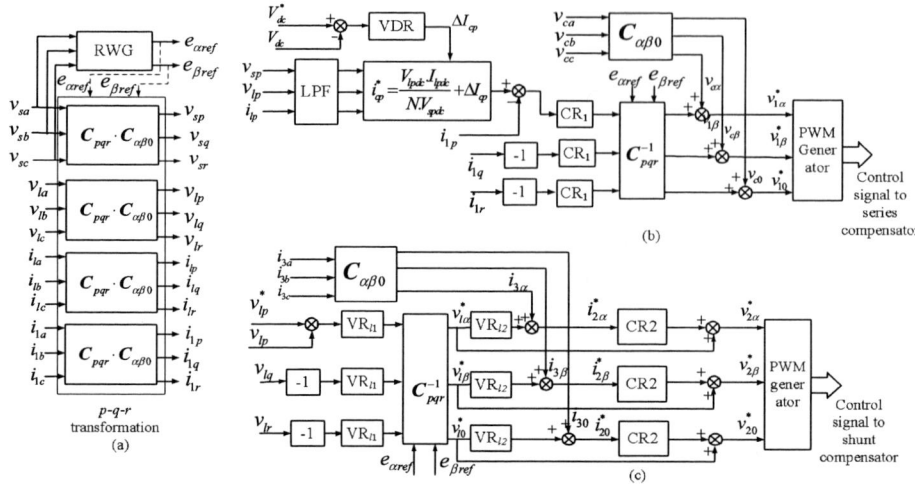

Fig.4 Block diagram for the proposed control system

$$V_2 = V_1 + L_2 \frac{d I_2}{dt} + R_{L2} I_2 \tag{22}$$

$$I_2 = C_2 \frac{d V_1}{dt} + I_3 \tag{23}$$

From(18),(22)and(23)

$$V_{2\alpha\beta0} = V_{1\alpha\beta0} + L_2 \frac{d I_{2\alpha\beta0}}{dt} + R_{L2} I_{2\alpha\beta0} \tag{24}$$

$$I_{2\alpha\beta0} = C_2 \frac{d V_{1\alpha\beta0}}{dt} + I_{3\alpha\beta0} \tag{25}$$

The control block diagram for the shunt compensator can be deduced from(24)and (25),as shown in Fig.4(c). The compensation voltage orders v_{cp}^* , v_{cq}^* and v_{cr}^* described in (12),(13)and (14)are compared to the actual load voltages v_{lp}, v_{lq} and v_{lr} .The compensation voltages $v_{l\alpha}^*, v_{l\beta}^*$ and v_{l0}^* can be got by regulating the difference of them by the voltage regulator VR$_{l1}$. Regulating them by voltage regulator VR$_{l2}$, the results plus the current $i_{3\alpha}$, $i_{3\beta}$ and i_{30} are current orders $i_{2\alpha}^*$, $i_{2\beta}^*$ and i_{20}^* calculated by (25).Output voltage orders $v_{2\alpha}^*, v_{2\beta}^*$ and v_{20}^* are the sum of $v_{l\alpha}^*, v_{l\beta}^*$ and v_{l0}^* and the regulating results of current regulator CR1.The control signals of the switch devices can be got from the PWM generator by calculating the voltages orders $v_{2\alpha}^*, v_{2\beta}^*$ and v_{20}^* .

V. SIMULATION RESULTS

To evaluate the various circuit conditions such as unbalance condition, nonlinear load ,a simulation model is set up in Matlab\Simulink according to the system structure illustrated in Fig.1. $L_1 = 5.8$mH , $N_2 : N_1 = 3.464$. $L_2 = 2$mH and $C_2 = 20\mu F$ .The dc capacitor $C_{dc1} = C_{dc2}$

$= 3300\mu F$ and dc source $E_{b1} = E_{b2} = 384V$ .The loads adopt three phase controllable rectifier bridge with L-R load and three phase unbalance resistance-inductance load .The dc active power is about 21 kW .

Fig.5 shows the coordinate reference wave $e_{\alpha ref}$ and $e_{\beta ref}$ generated by the proposed RWG. $e_{\alpha ref}$ and $e_{\beta ref}$ are sinusoidal and orthogonal ,whose peak value are always maintained unity.

Fig.6 shows the simulation waves when unbalance occurred in source voltage , as well as unbalance and distort occurred in loads.Fig.6(a) shows the waves of voltages of power line, the amplitude of three-phase voltages are 358V , 311V and 264V. Fig.6 (b) shows the waves of load current. i_{lp} include dc and ac components, i_{lq} include the two parts, too. i_{lr} is ac component. The compensation current order of series compensator is dc value, as shown in Fig.6(c).Fig.6(d) shows the power line current wave after compensating. Obviously, the three-phase current is balanced and sinusoidal, zero sequence current is about 0. Fig.6(e) shows v_{sp} , v_{sq} and v_{sr} . v_{sp} include ac components and v_{sr} is not zero, whose frequency is 2 time of fundamental wave. v_{sq} is zero because of the zero phase-shift between source voltages and coordinate reference voltages. The compensation voltage orders of the shunt compensator v_{lp}^* is permanent value (380V), so the change of the voltages of the power line does not influence the value of the load voltages, as Fig.6(f) shown.Fig.6(g) shows the a phase

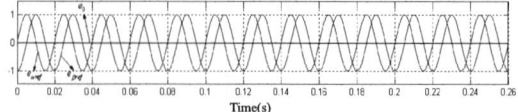

Fig.5 coordinate reference waves

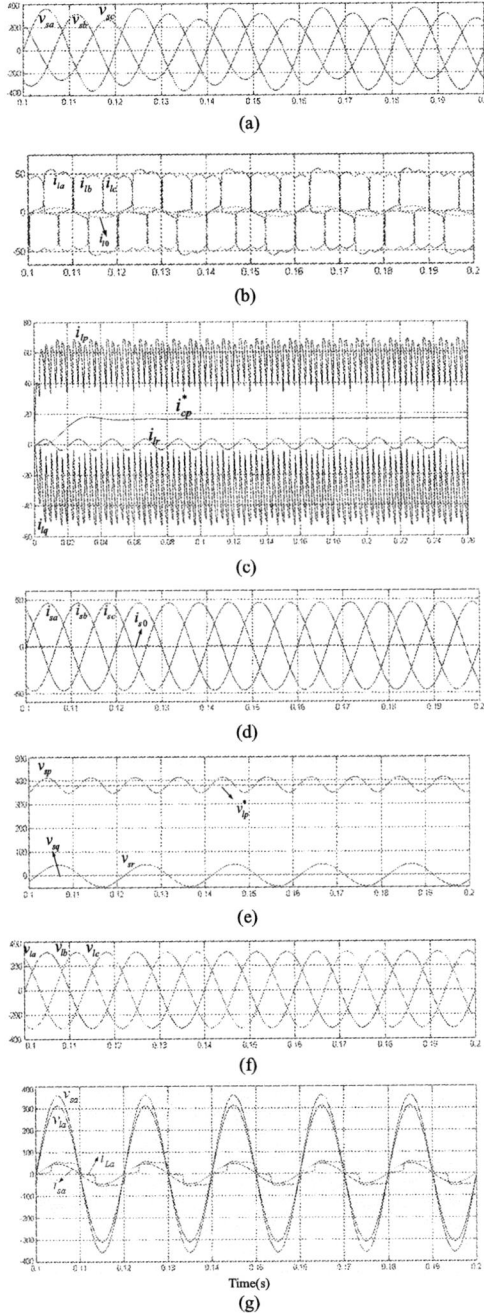

Fig.6 Simulation waves when unbalance occurred in source voltage, unbalance and distort occurred in loads

voltage v_{sa} and v_{la} as well as current i_{sa} and i_{la} as an example. It can be found out that the load current is unbalance, distort and phase-shifted with v_{sa}. i_{sa} and v_{la} are sinusoidal and in-phase with v_{sa}.

VI. CONCLUSIONS

This paper introduce a algorithm of calculating for compensation current and voltage in UPQC based on *p-q-r* theory, and apply it in the direct control strategy. The block diagram of control system is proposed. The simulation results shows that, when unbalance and nonlinear occur in load current or unbalance and sag in source voltage ,the above control algorithms eliminate the impact of distortion and unbalance of load current on the power line, making the power fact of it unity. Meanwhile, the series compensator isolate the loads voltages and source voltage, shunt compensator provide three-phase balanced and rated voltages of sine for loads. All above mentioned have realized the function of UPQC adopting direct control strategy.

REFERENCES

[1] Hyosung Kim, F.Blaabjerg, B.Bak-Jensen, and Jaeho Choi, "Instantaneous power compensation in three-phase systems by using p-q-r theory," *IEEE Trans. On Power Electronics*, vol.17, pp.701-710,Sept. 2002

[2] Hyosung Kim，Sang-Joon Lee,and Seung-Ki,Sul."A calculation for the compensation voltages in dynamic voltage restorers by use of PQR power theory,"*the 19th Annual of IEEE on Applied Power Electronics Conference and Exposition,*2004,pp.573-579

[3] Fan Ng,Man-Chung Wong,and Ying-Duo Han,"Analysis and control of UPQC and its DC-link power by use of p-q-r instantaneous power theory," *Proceedings of 1th International Conference on Power Electronics Systems and Applications,*Nov. 2004 ,pp.43-53

[4] Sang-Joon Lee; Hyosung Kim; Seung-Ki Sul;F.Blaabjerg,"A novel control algorithm for static series compensators by use of PQR instantaneous power theory,"*IEEE Trans. on Power Electronics,* vol.19,pp.814-827, May 2004

[5] Hyosung Kim,F.Blaabjerg ,and B.Bak-Jensen,"Spectral analysis of instantaneous powers in single-phase and three-phase systems with use of p-q-r theory,"*IEEE Trans. on Power Electronics,* vol.17,pp.711-720,Sept. 2002

[6] ZHU Pengcheng, LI Xun,KANG Yong,and CHEN Jian,"Study of control strategy for a unified power quality conditioner," *Proceedings of the CSEE,*vol.24,pp.67-73,Aug.2004.

[7] Chen Jian,Dai Ke,LI Xun,ZHU Pengcheng,and Liu Peiguo, "Series-Parallel Compensated UPS with Double Converters," *Journal of Power Supply* vol.1,pp.262~271, Jan. 2003.

[8] Xun Li, Pengcheng Zhu, Yinfu Yang, and Jian Chen."A New Controlled Scheme for Series-Parallel Compensated UPS System".*IEMDC'03*. Madison WI, USA. 2003. vol.2, pp. 1133 ~1136

[9] da Silva,S.A.O., Donoso-Garcia, P.F.,Cortizo, P.C., and Seixas P.F.,"A comparative analysis of control algorithms for three-phase line-interactive UPS systems with series-parallel active power-line conditioning using SRF method," *PESC00*. 18-23 June 2000.vol.2 ,pp.1023~1028

[10] Monteiro, L.F.C.; Aredes, M.,and Moor Neto, J.A.,"A control strategy for unified power quality conditioner" IEEE International Symposium on Industrial Electronics(ISIE '03) ,9-11 June 2003 , vol.1,pp.391~ 396

2006 5th International Power Electronics and Motion Control Conference

Three-Phase Harmonic Selective Active Filter Using Multiple Adaptive Feed Forward Cancellation Method

Lewei Qian*, Student Member, IEEE, David Cartes*, Member, IEEE, and Qiang Zhang**

* Center for Advanced Power Systems, Florida State University, Tallahassee, FL, USA
** Power Electronics Research Center, Hefei University of Technology, Hefei, Anhui, China
Qian@caps.fsu.edu, Dave@caps.fsu.edu, archer0402@sina.com

Abstract— This paper proposes a three phase harmonic selective active filter with multiple adaptive feed forward cancellation (MAFC) for harmonic reference generation. The MAFC method is derived from previously used adaptive feed forward cancellation (AFC) method used in acoustics. The adaptive law is derived by an SPR (Strictly Positive Real) Lyapunov function. This adaptive method is simple and effective in extracting fundamental and harmonic current information from nonlinear load current. The effectiveness of this method is verified by testing experimentally a nonlinear load current. The extracted harmonic currents are used as the reference signals for the active filter harmonic selective cancellation control. The active filter feature's harmonic selective ability reduces the active power filter rating, bandwidth requirements and enhances its flexibility. Simulation results are used to validate the control of this harmonic selective active filter system.

Key Words—Multiple Adaptive Feed Forward Cancellation, Harmonic Selective, SPR-Lyapunov, Active Filter

I.   INTRODUCTION

In today's power distribution systems, many semiconductor based switches are used to transfer controlled electric power to electrical loads. However, switch based nonlinear loads, like adjustable speed drives, medical equipment, etc., deteriorate power quality by drawing harmonic currents from the system. As a solution to these problems, shunt active filters (AF) are used in power systems. But additional active filters increase system cost and space. Therefore, the idea of using a drive's active front end to have the ability to active filter current harmonics has been proposed by [1][2]. The front end converter of the drive can be reconfigured to function as an active filter to mitigate harmonic current effects.

A reconfigurable power conversion system is shown in Fig.1. The induction motor drive is used to realize the reconfigurable active filter's function. To realize it, active filter's function must be fully implemented first. Considering that the motor drive's power rating typically exceeds the active filtering power required, it should have functionality of harmonic selective ability to make the system flexible and with low power rating limit.

The harmonic selective concept of the active filter has been proposed by [3]-[7]. The harmonic selective feature has several advantages such as controllable harmonic cancellation, low filter rating, and bandwidth requirements reductions. In [3]-[6], a three-phase active filter (dominant harmonic active filter) can be used for specific harmonic cancellation. The control method proposed in [7] uses d-q transformations in respective harmonic synchronous frames. Recently, adaptive detecting methods are gaining interest. Examples of adaptive detection methods are Kalman filter [8], adaptive filters [9], and neural network methods [10][11]. In this paper, the previously used AFC [12]is extended using an SPR Lyapunov method for active filter application. The derived control structure of the algorithm is found to be similar to that used in [10][11]. The missing procedure derivation in [10][11] is fully discussed in this paper. A three-phase active filter with MAFC is simulated and results presented to verify the application of this method.

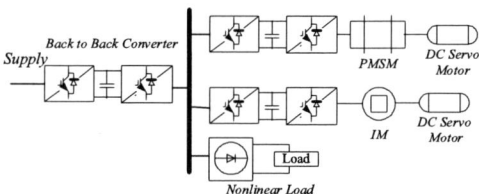

Fig.1 Proposed test system

II.   MAFC ALGORITHM

Adaptive feed forward cancellation (AFC) is a control method by which the disturbance is cancelled at the input of the plant by adding the negative of its value at all time[12] for a given frequency signal. In this section, the MAFC design process based on an SPR Lyapunov function and its stability analysis are fully discussed.

Assume the load current has the following form:

$$i_L = I_{L1}\sin(\omega_1 t + \varphi_1) + \ldots\ldots + I_{Ln}\sin(\omega_n t + \varphi_n) + I_{dc}e^{-\sigma t} \ (1)$$

In (1), $n = 2k+1, k = 0,1,2,\ldots\ldots$, however for a three-phase system, the current harmonic component with order of $n = 3k, k = 1,2,3,\ldots\ldots$ may not exist. Additionally, a decaying dc current may exist. The fundamental has frequency ω_1 as $2\pi \cdot 60 rad/s$ in USA. The phase φ_n and amplitude I_{Ln} are unknown. The current i_L is measured

1-4244-0448-7/06/$25.00 ©2006 IEEE

through a sensor with transfer function $W(s)$ to attenuate any possible higher frequency noise present. Therefore, we have:

$$z = W(s)i_L = W(s)[I_{L1}\sin(\omega_1 t + \varphi_1) + \ldots\ldots + I_{Ln}\sin(\omega_n t + \varphi_n) + I_{ed}e^{-\sigma t}] \quad (2)$$

In equation (2), $W(s)$ is an SPR transfer function.

Using first two terms of the Taylor series expansion of dc component, we can transform (2) to the linear parametric model form as:

$$i_L = I_{L11}\sin\omega_1 t + I_{L12}\cos\omega_1 t + \ldots\ldots + I_{Ln1}\sin\omega_n t + I_{Ln2}\cos\omega_n t + A_{dc} - A_{dc}\sigma t \quad (3)$$

where $\begin{array}{l} I_{L11} = I_{L1}\cos\varphi_1, I_{L12} = I_{L1}\sin\varphi_1, \ldots\ldots \\ I_{Ln1} = I_{Ln}\cos\varphi_n, I_{Ln2} = I_{Ln}\sin\varphi_n \end{array}$.

Therefore, equation (2) can be rewritten in the form of $z = W(s)\theta^{*T}\phi$. In above equation, $\theta^* = [I_{L11}, I_{L12}, \ldots\ldots, I_{Ln1}, I_{Ln2}, A_{dc}, \sigma A_{dc}]^T$ and $\phi = [\sin\omega_1 t, \cos\omega_1 t, \ldots\ldots, \sin\omega_n t, \cos\omega_n t, 1, -t]^T$.

It becomes a standard estimation problem and from the estimate $\theta = [\hat{I}_{L11}, \hat{I}_{L12}, \ldots\ldots, \hat{I}_{Ln1}, \hat{I}_{Ln2}, \hat{A}_{dc}, \hat{\sigma}\hat{A}_{dc}]^T$ of θ^*, the estimates of $[I_{L1}, I_{L3}, \ldots\ldots, I_{Ln}]$ and $[\varphi_1, \varphi_3, \ldots\ldots, \varphi_n]$ can be calculated by the following equations:

$$\hat{I}_{Ln}(t) = \sqrt{\hat{I}_{Ln1}^2(t) + \hat{I}_{Ln2}^2(t)}, \hat{\varphi}_n(t) = \cos^{-1}\left(\frac{\hat{I}_{Ln1}(t)}{\hat{I}_{Ln}(t)}\right) \quad (4)$$

The problem now is to derive the adaptive law to get the estimates. The derivation process is based on [13] and is summarized here. The estimate error ε_1 is generated as $\varepsilon_1 = z - \hat{z}$, where $\hat{z}$ is the estimate result of z and the normalized estimation error is: $\varepsilon = \varepsilon_1 - W(s)\varepsilon_1 n_s^2$. In this application, $\phi = [\sin\omega_1 t, \cos\omega_1 t, \ldots, \sin n\omega_1 t, \cos n\omega_1 t, 1, -\sigma t]^T \in \ell_\infty$, we have $\varepsilon = \varepsilon_1$.

The parameter error can be written as $\tilde{\theta} = \theta - \theta^*$ and the error becomes:

$$\varepsilon = -W\tilde{\theta}^T\phi \quad (5)$$

Since W is strictly proper, (5) can be rewritten in the following state space representation [13]:

$$\dot{e} = A_c e + B_c(-\tilde{\theta}^T\phi) \quad (6)$$
$$\varepsilon = C_c^T e$$

In (6) A_c, B_c and C_c are the matrices associated with a state space representation that has a transfer function $W(s) = C_c^T(sI - A_c)^{-1}B_c$.

Now considering the following Lyapunov like function of equation (6):

$$V(\tilde{\theta}, e) = \frac{e^T P_c e}{2} + \frac{\tilde{\theta}^T \Gamma^{-1}\tilde{\theta}}{2} \quad (7)$$

In equation (7), $\Gamma = \Gamma^T > 0$ is a constant matrix and

$P_c = P_c^T > 0$ satisfies the algebraic equations:

$P_c A_c + A_c^T P_c = -qq^T - vL_c$ for some vector q, matrix $P_c B_c = C_c$
$L_c = L_c^T > 0$ and a small constant $v > 0$.

Considering the time derivative $\dot{V}$ along the solution of equation (6), we have:

$$\dot{V}(\tilde{\theta}, e) = -\frac{1}{2}e^T qq^T e - \frac{1}{2}ve^T L_c e - e^T P_c B_c \tilde{\theta}^T\phi + \tilde{\theta}^T\Gamma^{-1}\dot{\tilde{\theta}}$$

Since $P_c B_c = C_c \Rightarrow e^T P_c B_c = e^T C_c = \varepsilon$, the above equation can be rewritten as $\dot{V}(\tilde{\theta}, e) = -\frac{1}{2}e^T qq^T e - \frac{1}{2}ve^T L_c e - \varepsilon\tilde{\theta}^T\phi + \tilde{\theta}^T\Gamma^{-1}\dot{\tilde{\theta}}$. And choose

$$\dot{\theta} = \dot{\tilde{\theta}} = \Gamma\varepsilon\phi \quad (8)$$

the time derivative $\dot{V}$ becomes:

$$\dot{V}(\tilde{\theta}, e) = -\frac{1}{2}e^T qq^T e - \frac{1}{2}ve^T L_c e \le 0 \quad (9)$$

Therefore equation (8) is the adaptive law chosen for estimation to guarantee stability.

The adaptive gain Γ can be chosen as $\Gamma = diag(\gamma)$ for some $\gamma > 0$, then equation (8) can be rewritten as:

$$\dot{I}_{L11} = \gamma_{11}\varepsilon\sin\omega_1 t, \dot{I}_{L12} = \gamma_{12}\varepsilon\cos\omega_1 t$$

$$\ldots\ldots$$

$$\dot{I}_{Ln1} = \gamma_{n1}\varepsilon\sin n\omega_1 t, \dot{I}_{Ln2} = \gamma_{n2}\varepsilon\cos n\omega_1 t \quad (10)$$

$$\dot{A}_{dc} = \gamma_{dc}\varepsilon, \qquad \dot{\hat{\sigma}\hat{A}}_{dc} = \gamma_{dc2}\varepsilon t$$

Equation (10) gives the detailed form of different harmonic estimation formulas. The convergence of θ to θ^* respectively is guaranteed as by proof in [13]. With adaptive laws the proposed method control block is easily derived and is shown in Fig.2.

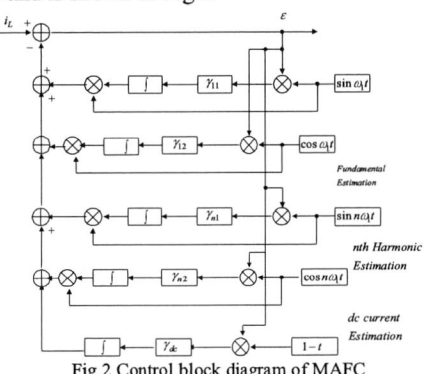

Fig.2 Control block diagram of MAFC

III. ALGORITHM SIMULATION AND EXPERIMENTAL TESTS

The adaptive law derived in previous section is used to cancel the harmonics in the input signals. A very important feature of this algorithm is that fundamental and harmonic signal waveforms can be rebuilt individually.

A. Simulation Test

To test the algorithm's effectiveness, an input signal with the fundamental frequency of 60Hz and harmonic signals up to 11th (660Hz) is used for estimation process. The parameters of the fundamental and the harmonics are shown in Table I. All the parameters are chosen randomly. The second row is the magnitude information (unitless) and the third row is the phase information in radians. DC component has a decaying factor of $-5s^{-1}$.

TABLE I
PARAMETERS OF INPUT SIGNAL

1st	3rd	5th	7th	9th	11th	dc
100	50	30	25	15	10	20
$3\pi/4$	$\pi/2$	$\pi/4$	$\pi/6$	$\pi/7$	$2\pi/3$	-5

Fig.3 shows the estimation results of the fundamental and harmonics magnitudes and phases (to 7th). All the magnitudes and phases are estimated correctly and quickly. The estimation error goes to zero in about 0.05s. The dc component is also estimated well although there is some minor estimation error during to approximation.

In real applications, the fundamental frequency may change. Therefore the sine and cosine items in Fig.2 cannot be preset items, they need the frequency information. Two methods are available: one is using a PLL for tracking the fundamental frequency and the other is using the following adaptive law for fundamental frequency tracking [11].

$$\dot{\omega} = -\alpha\varepsilon\left(\sum_{n=1}^{N}(\hat{I}_{Ln1}\cos n\omega t - \hat{I}_{Ln2}\sin n\omega t)\right) \quad (20)$$

In (20), α is the adaptive gain. In this paper, the second method is used for frequency tracking and simulation results verify this method as shown in Fig.4. In this simulation, at 0.3 second, the frequency is jumping from 60 Hz to 70 Hz and at 0.6 second, the frequency is jumping from 70 Hz to 50 Hz. Fig. 4 (a) shows the estimated frequency of the input signal. It clearly demonstrates that the frequency is tracked precisely and quickly. Fig.4 (b) shows the estimation error of the input signal and estimated signal, and the error goes to zero quickly with very litter error after convergence. Fig.4 (c) gives an estimation example of the 3rd harmonic from which we can find that estimated 3rd harmonic matches the input 3rd harmonic precisely and quickly after first frequency jump at 0.3 second.

B. Experimental Test

To verify the method's effectiveness for real world signals, the load current of a thyristor drive for a dc motor under no load situation is analyzed using this method in this paper. The current is shown in Fig.5 (a), where great dc offset and significant harmonics exist. Without any pre-filtering work, the (phase A) load current is fed into the algorithm for identification and the real current and estimated current are shown in Fig.5 (b) for comparison. It clearly shows that the estimated current matches the real current well after the first cycle. Considering the fact that no filtering existed when $t < 0$ and initial conditions were unknown, the performance is good. There is still some mismatch between the real the estimated signal, and that is because of high order frequency harmonics existing and only up to 7th harmonic identification modules are realized. It can be improved by adding no phase shift filter, proper initial conditions and more identification modules to get even better performance. The fundamental current estimation and dc offset estimation are shown in Fig. 5 (c), the identification process is fast and the algorithm can also work for time varying signals well.

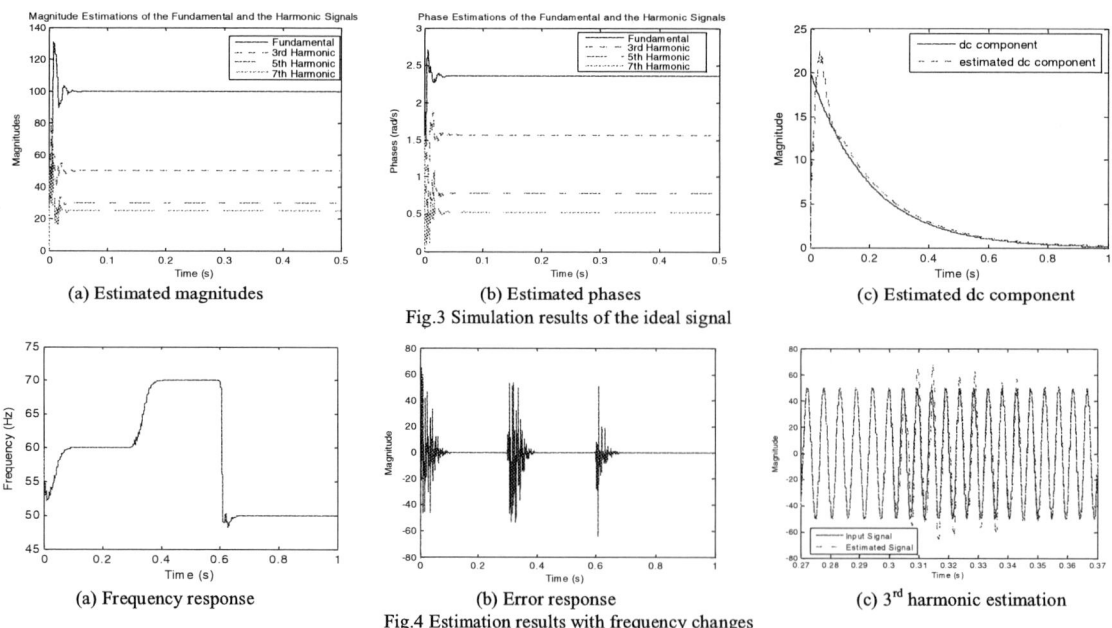

(a) Estimated magnitudes   (b) Estimated phases   (c) Estimated dc component

Fig.3 Simulation results of the ideal signal

(a) Frequency response   (b) Error response   (c) 3rd harmonic estimation

Fig.4 Estimation results with frequency changes

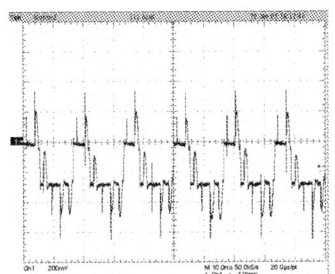

(a) Experimental current (100mV/A, 200mV/Div)

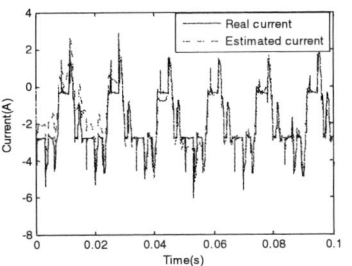

(b) Experimental and estimated load current

Fig.5 Experimental Test

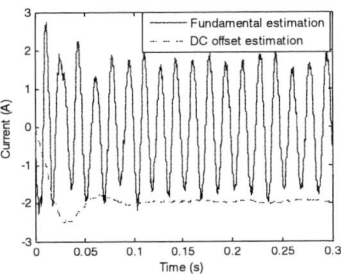

(c) Fundamental and dc offset estimations

IV. ACTIVE FILTER SIMULATIONS

The proposed current harmonic selective active filter with MAFC method is tested simultaneously in MATLAB and PSIM simulating environments. The active filter circuit part is implemented in PSIM. PSIM receives the control signals from MATLAB. The control parts are realized in MATLAB and the measured current signals are from PSIM. Fig.5 shows the control system of the proposed harmonic selective active filter. The currents i_s, i_L and i_{AF} are the source current, load current and active filter output current respectively. The three-phase currents i_{LA}, i_{LB} and i_{LC} are measured for harmonic current extraction by three MAFC blocks respectively. The current i_{AF} is measured and fed back for current tracking by hysteresis controller.

For dc bus regulation, the measured dc bus voltage V_{dc} is compared with the reference voltage V_{dc}^*. The error signal is regulated by a PI controller to get the d axis reference. After inverse dq transformation, the reference current signal for the dc voltage is obtained. The reference dc voltage is $800V$. The system parameters are:

Supply: Three-phase balanced voltage source with line-line voltage RMS magnitude of 480V at the frequency of 60 Hz. *Load:* Ideal three-phase diode bridge with the load capacitance of 7.8mF and the load resistance of 10Ω. *Active Filter:* A three-phase voltage source IGBT converter. Other parameters are shown in Table II.

TABLE II
PARAMETERS OF TEST SYSTEM

R_s	L_s	R_{LD}	L_{LD}	R_{AF}	L_{AF}	C_{dc}
0.05Ω	$0.1mH$	0.5Ω	$0.1mH$	$2m\Omega$	$2mH$	$2mF$

Fig.6 shows the source current without active filter control, where the source current is heavily distorted and its spectrum shows that significant low order harmonics exist with THD (Total Harmonic Distortion) near 85%, which is very high. To verify the effectiveness of this adaptive method in this active filter, two cases are studied. One is the single harmonic selective cancellation and the other one is the multiple harmonic selective cancellation.

A. Single Harmonic Selective Cancellation

In this case, only one harmonic is cancelled for the situation where only one dominant harmonic exists. Fig.7 shows the simulation results with the 5th harmonic cancellation. The source current waveform is shaped and its spectrum clearly demonstrates that the 5th harmonic is almost zero with other components nearly unchanged. The THD is reduced to about 48% as shown in Fig.7 (b) The active filter output current and its spectrum shown in Fig.7 (c) and (d) clearly shows that 300Hz frequency current is injected to the power system for 5th harmonic cancellation.

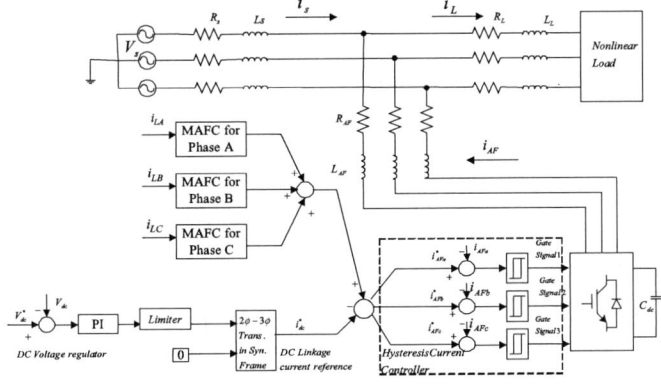

Fig.5 Harmonic selective active filter under test with control diagram

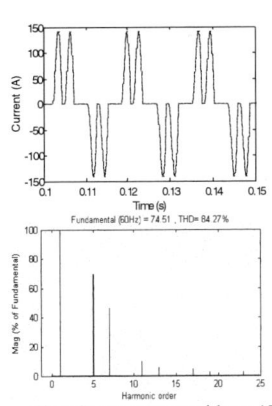

Fig.6 Source current without AF control and its spectrum

B. *Multiple Harmonic Selective Cancellation*

To get even better performance, in application, two or three main harmonic references can be combined for multiple harmonics selective cancellation. An example simulation with 5th, 7th, 11th, and 13th harmonic cancellation is given in Fig.8, where the source current is reshaped to a nearly sinusoidal signal. The THD is reduced dramatically to about 6%. The simulation results show that adding more harmonic cancellation blocks to the system can further reduce the THD to about 2%. In our simulation, the load current is unchanged and the dc voltage is also well controlled as shown in Fig.9.

V. CONCLUSIONS

This paper proposes a multiple adaptive feed forward cancellation method for harmonic reference generation in a three-phase harmonic selective active filter. This detection method is both tested under ideal and experimental situations. The adaptive laws are derived by the SPR Lyapunov method. This method can be easily applied to unbalanced systems because each phase's harmonics information is extracted individually. With the knowledge of fundamental current's phase, reactive power compensation is also applicable. The three-phase harmonic selective active filter with this MAFC method has advantages of simplicity and flexibility.

REFERENCES

[1] Brogan P. and Yacamini R., "Harmonic control using an active drive," *IEE Proceedings of Electric Power Applications*, Volume 150, Issue 1, Jan. 2003 Page(s): 14-20.

[2] Lewei Qian and David Cartes, "A Reconfigurable Induction Motor Drive with Harmonic Cancellation Feature," *Proceeding of 2005 IEEE Electric Ship Technologies Symposium*, Page(s): 93-98.

[3] Po-Tai Cheng, Bhattacharya S., and Divan D., "Operations of the dominant harmonic active filter (DHAF) under realistic utility conditions," *IEEE Transactions on Industry Applications*, Volume: 37, Issue: 4, July-Aug. 2001, Pages: 1037 – 1044

[4] Po-Tai Cheng, Subhashish Bhattacharya and Deepak M. Divan, " Control of Square-Wave Inverters in High-Power Hybrid Active Filter Systems," *IEEE Transactions on Industry Applications*, Vol.34, NO.3, May/June 1998.

[5] Po-Tai Cheng, Subhashish Bhattacharya and Deepak D. Divan, " Line Harmonics Reduction in High-Power Systems Using Square-Wave Inverters-Based Dominant Harmonic Active Filter," *IEEE Transactions on Power Electronics*, Vol. 14, NO.2, March 1999.

[6] Po-Tai Chen, Subhashish Bhattacharya and Deepak Divan, "Experimental Verification of Dominant Harmonic Active Filter for High-Power Applications," *IEEE Transactions on Industry Applications*, Vol.36, NO.2, March/April 2000.

[7] Paolo Mattavelli, " A Closed-Loop Selective Harmonic Compensation for Active Filters," *IEEE Transactions on Industry Applications*, Vol. 37, NO.1, Jan./Feb., 2001

[8] Julio Barros and Enrique Perez, " An Adaptive Method for Determining the Reference Compensating Current in Single-Phase Shunt Active Power Filters," *IEEE Transactions on Power Delivery*, Vol.18, No.4, Oct. 2003.

[9] Shiguo Luo and Zhencheng Hou, " An Adaptive Detecting Method for Harmonic and Reactive Currents," *IEEE Transactions on Industrial Electronics*, Vol.42, No.1, Feb. 1995

[10] Ramadan EI Shatshat, Mehrdad Kazerani and M.M.A. Salama, " Modular Active Power-Line Conditioner," *IEEE Transactions on Power Delivery*, Vol.16, No.4, Oct. 2001

[11] L.H.Tey, P.L.So and Y.C. Chu, " Improvement of Power Quality Using Adaptive Shunt Active Filter," *IEEE Transactions on Power Delivery*, Vol.20, NO.2, April 2005.

[12] Alexei Sacks, Marc Bodson and Pradeep Khosla, " Experimental Results of Adaptive Periodic Disturbance Cancellation in a High Performance Magnetic Disk Drive," *ASME, Journal of Dynamic Systems Measurement Control, 118, 416-424, 1996.*

[13] Petros A. Ioannou and Jing Sun, " *Robust Adaptive Control"*, Prentice Hall PTR, 1996, Chapter.4

[14] Marks J.H. and Green T.C., "Predictive transient-following control of shunt and series active power filters," *IEEE Transactions on Power Electronics*, Volume: 17, Issue: 4, July 2002, Pages: 574 – 584

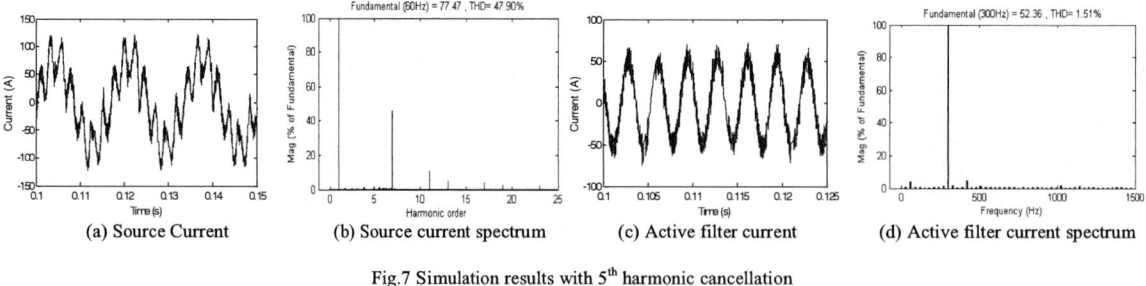

(a) Source Current   (b) Source current spectrum   (c) Active filter current   (d) Active filter current spectrum

Fig.7 Simulation results with 5th harmonic cancellation

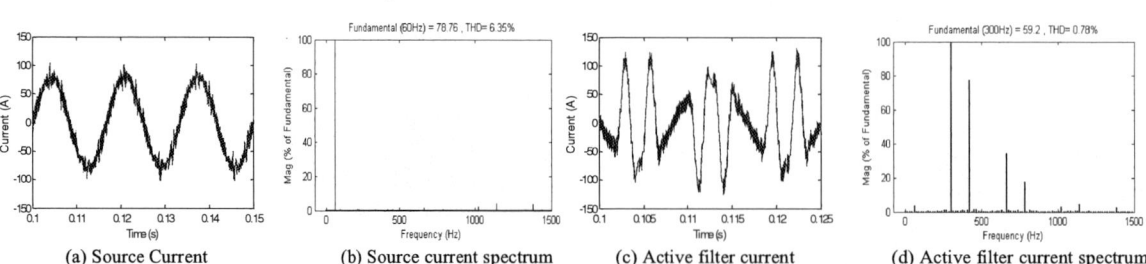

(a) Source Current   (b) Source current spectrum   (c) Active filter current   (d) Active filter current spectrum

Fig.8 Simulation results with 5th, 7th, 11th, and 13th harmonic cancellation

2006 5th International Power Electronics and Motion Control Conference

Reactive Power Compensation in Distribution Networks with STATCOM by Fuzzy Logic Theory Application

Seyyed Hossein Hosseini[*, ***], Reza Rahnavard[**, ***], Yousef Ebrahimi[***]
[*] Department of Electrical Engineering, Faculty of Engineering, Islamic Azad University of Tabriz
[**] Azarbyjan Regional Electric Power Company, Tabriz, Iran.
[***]University of Tabriz, Tabriz, Iran
email: hosseini@tabrizu.ac.ir
email: rerahnavard @gmail.com

Abstract— **In this paper a STATCOM using fuzzy logic controller to compensate reactive power in distribution networks is developed. The advantage of fuzzy control is that it is based on a linguistic description and does not require a mathematical model of the system. The performance of the fuzzy logic controller is compared with a conventional PI controller and results are presented under steady state and transient condition. The $\pm$ 3MVAR STATCOM is connected to the 20kV B_2 bus through a $20^{kV}/2^{kV}$ Wye/Delta transformer. On this 20 MVA inductive network, the $\pm$ 3MVAR STATCOM can correct voltage dip of $\pm$ 12%. The secondly voltage is synthesized by a pulse width modulation inverter (PWM) using a 2.5 kHz chopping frequency.**

Keywords- STATCOM ; fuzzy logic ; distribution network

I.   INTRODUCTION

The problems due to rapid variations of reactive power by loads are well known. Several solutions have been suggested for reactive power compensation. STATCOM is one such promising solution. The STATCOM consists mainly PWM inverter connected to the network through a transformer (Fig.1). The dc link voltage is provided by capacitor C while power taken from network the control system ensures the regulation of the voltage and the dc link voltage. The STATCOM is to regulate the voltage by absorbing or generating reactive power to the network, like a thyristor static compensator. This reactive power transfer is done through the leakage reactance of the coupling transformer by using a secondary voltage in phase with the primary voltage (network side), this voltage is provided by a voltage-source PWM inverter. In the control system of STATCOM, the conventional PI controller was used for generation of a reference current template. The PI controller requires precise linear mathematical models, which are difficult to obtain and fails to perform satisfactorily under parameter variation, nonlinearity, load disturbance. Recently, Fuzzy Logic Controller (F.L.C) have generated a good deal of interest in certain application, the advantages of F.L.Cs over conventional controller are that they do not need on accurate mathematical model. They can work with imprecise inputs, can handle none-linearity and they are more robust than conventional controllers. [1], [2]

II.   BASIC COMPENSATION PRINCIPLE

The STATCOM operation is illustrated by the phasor diagrams shown in Fig2. When the secondly voltage (VD) is lower than the bus voltage (VB) the STATCOM acts like an inductance absorbing reactive power from the bus. When the secondly voltage (VD) is higher than the bus voltage (VB), the STATCOM acts like a capacitor generating reactive power to the bus.

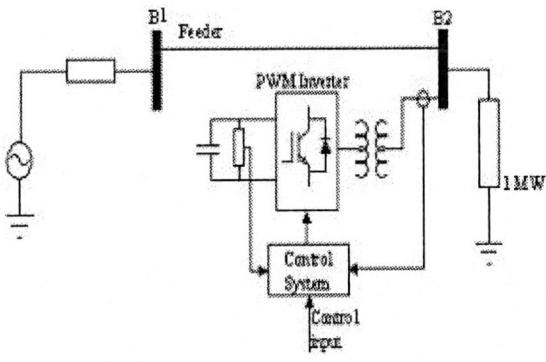

Figure.1. Schematic Diagram of a STATCOM

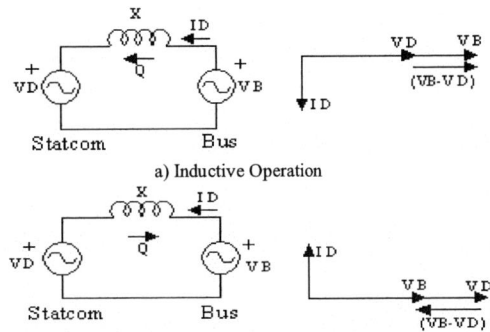

a) Inductive Operation

b) Capacitive operation

Figure2.   Phasor Diagram of STATCOM Operation.

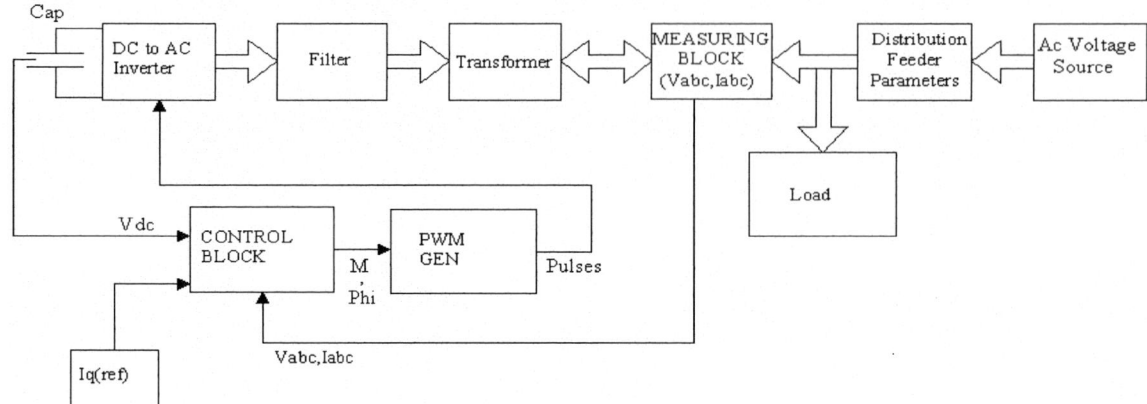

Figure 3. Schematic Diagram of a STATCOM on a Distribution Network.

III. PROPOSED FUZZY CONTROLLER

Fig.3 shows the schematic diagram of the control algorithm. Fig.4 and 7 shows the block diagram of the proposed fuzzy logic schematic diagram of the control algorithm. In order to implement the control algorithm of a STATCOM in closed loop, the DC side capacitor voltage is sensed and then compared with a reference value. The obtain error e= (vdc, ref– vdc, act) and change of error signal ce (n) = e (n) – e (n-1) at the nth sampling instant are used as inputs for the fuzzy processing. The out put of the fuzzy controller after a limit is considered as the amplitude of the reference current or voltage.

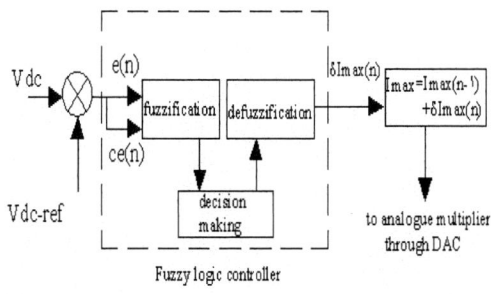

Figure 4. Block Diagram of Fuzzy Logic Controller.

This current or voltage signal takes care of the active power demand of load and the losses in the system. The short circuit at 20kv B_2 bus is 20 MVA inductive with an angle this current or voltage signal takes care of the active power demand of load and the losses in the system. The short circuit at 20kv B_2 bus is 20 MVA inductive with an angle of 56 degrees. The $\pm$ 3MVAR STATCOM is connected to the 20kV B_2 bus through a $20^{kV}/2^{kV}$ Wye/Delta transformer. On this 20MVA inductive network, the$\pm$ 3MVAR STATCOM can correct voltage

dip of $\pm$ 12%. The secondly voltage is synthesized by a pulse width modulation inverter (PWM) using a 2.5 kHz chopping frequency. In order to minimize harmonic frequencies (around multiple of 2.5 kHz) generated by the PWM, the output voltage of the inverter is filtered before being sent to the secondly of the transformer. In a fuzzy logic controller, the control action determined from evaluation of a simple linguistic rules. The development of the rules requires a through understanding of the process to be controlled, but it does not require a mathematical model of the system. Here, the error e and change of error ce are used as numerical variables into linguistic variables, the following seven fuzzy levels or sets are chosen as:

NB (Negative Big), NM (Negative Medium), NS (Negative Small), ZE (Zero), PS (Positive Small), PM (Positive Medium), and PB (Positive Big).

In Table.1 fuzzy rules are summarized.

Table.1. Control Rule Table

de \ e	NB	NM	NS	ZE	PS	PB	PM
NB	NB	NM	NS	ZE	PS	PM	PB
N M	NB	NB	NB	NB	NM	NS	ZE
NS	NB	NB	NB	NM	NS	ZE	PS
ZE	NB	NB	NM	NS	ZE	PS	PM
PS	NB	NM	NS	ZE	PS	PM	PS
PB	NB	NM	NS	ZE	PS	PM	PB
PM	NM	NS	ZE	PS	PM	PB	PB

Fig.5 determined normalized triangular membership functions.

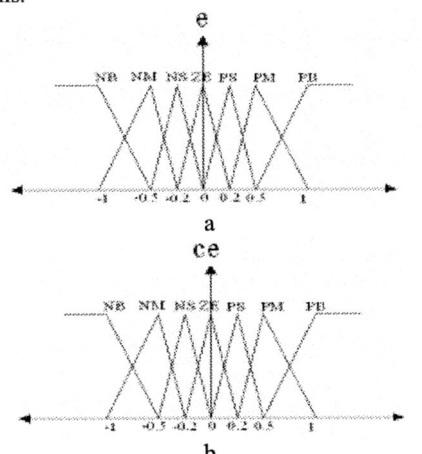

Figure 5. Normalized Triangular Membership Functions used in Fuzzification.

a. Membership Functions for e.

b. Membership Functions for change of e

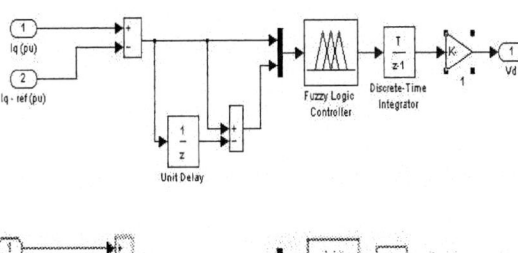

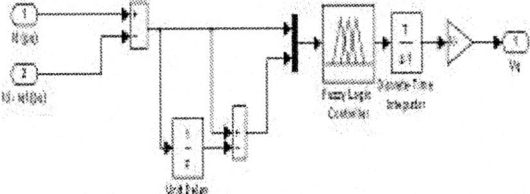

Figure 6. Block of the Fuzzy Controller

Function for e and change of e used in fuzzification block. In the control's block, Vdc and Vdc (ref) input of fuzzy controller for Vdc regulator.

- The Iq current is regulated by a fuzzy controller, (The Iq regulator follows the reference value which can be adjusted between + 1pu (capacitive) and –1 pu (inductive).
  - The Id current is also regulated by a fuzzy controller. (The Id current correspond to small active power following into the STATCOM the regulator follows a reference imposed by the DC bus voltage regulator.
  - The modulation index and phase angle of the voltage to be generated by the inverter are obtained

by $m = \sqrt{vq^2 + vd^2}$ and φ (deg) $= tag^{-1} \dfrac{vd}{vq}$.

Fig.7 shows block diagram of the STATCOM control circuit and Fig.6 shows block of the fuzzy controller.

IV. MODELING AND SIMULATION RESULTS

In this test, a step change is applied to the current reference input in order to observe the dynamic response when the linguistic changes from full inductive to capacitive operation. When the simulation starts, the DC capacitor starts charging. This requires a Id component corresponded to the active power absorbed by the capacitor.When the DC voltage reaches its reference value, the component drops to a value very close to zero and the Iq Component stays at the 1 p.u reference value (3 MVAR inductive). At t=0.1[sec] Iq reference current change to + 1p.u. when STATCOM changes from inductive to capacitive, a 180 degrees phase shift of the current Ia with respect to voltage Va is observed.

Figure 8 shows that the modulation index increases when the STATCOM changes from inductive with fuzzy controller as well as PI controller (Fig.9) but in the fuzzy controller we does not require a mathematical model of the system results of simulation .[3]

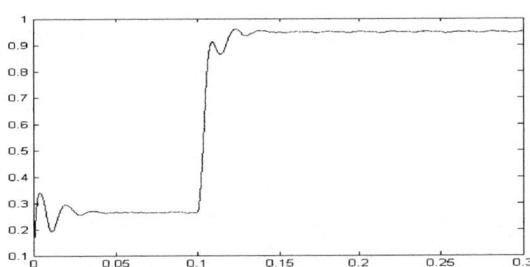

Figure 8. Modulation index for a Change at t=0.1[sec] With Fuzzy Controller.

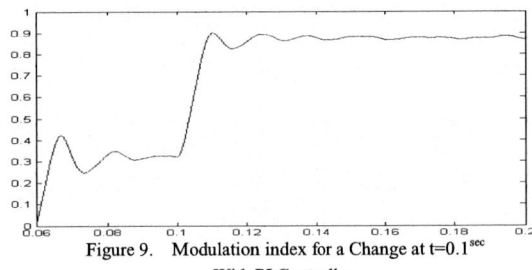

Figure 9. Modulation index for a Change at t=0.1[sec] With PI Controller.

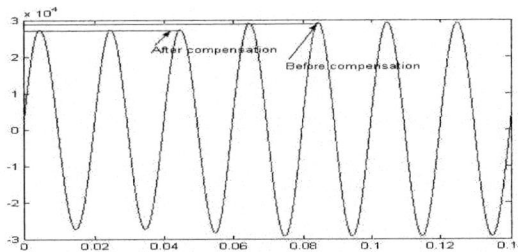

Fig.10. Voltage Increasement of the Bus for Compensation of Voltage Decreasement of the Load

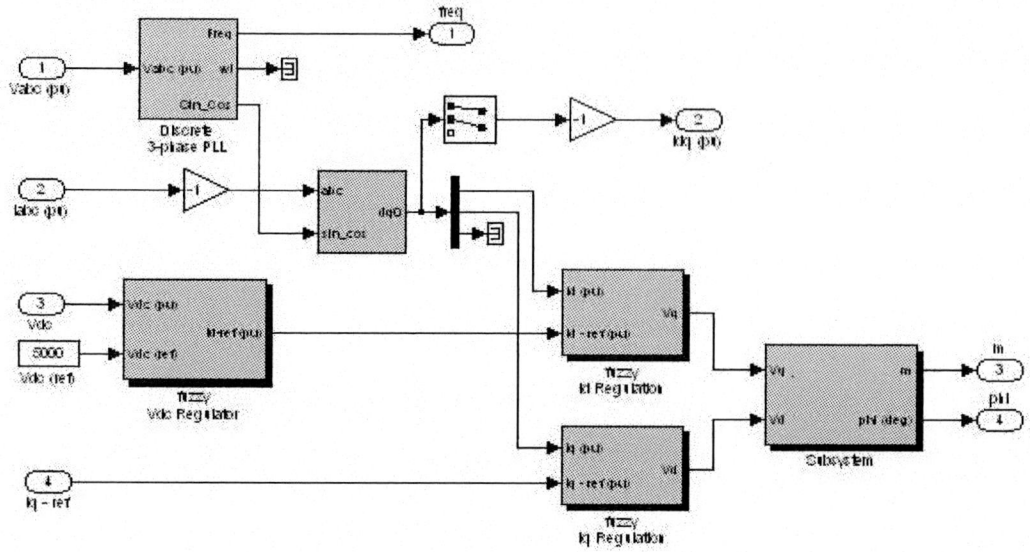

Figure 7. Block Diagram of the STATCOM Control Circuit

When inverter voltage increases for a change at t=0.1sec (Fig.11), the modulation index increases too (Fig.8). Figures 12 and 13 show, before a change at 0.1 sec the current has a lag shift phase with voltage and after t=0.1sec the current has a lead shift phase with voltage.[4]

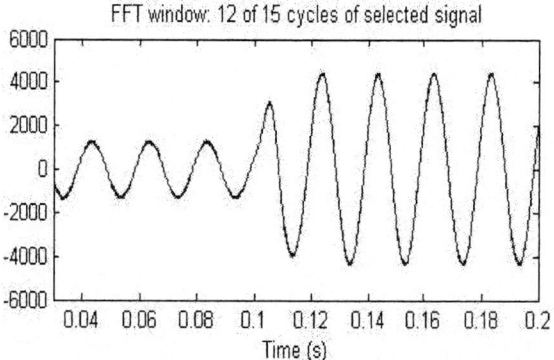

Figure 11. Inverter Output Voltage During the change at t=0.1^{sec}

Iq and Id signals during the change at t=0.1sec is are illustrated in Fig.14.

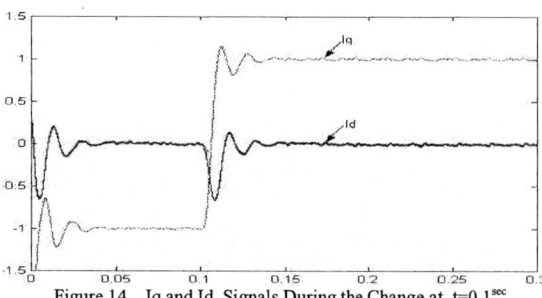

Figure 14. Iq and Id Signals During the Change at t=0.1^{sec}

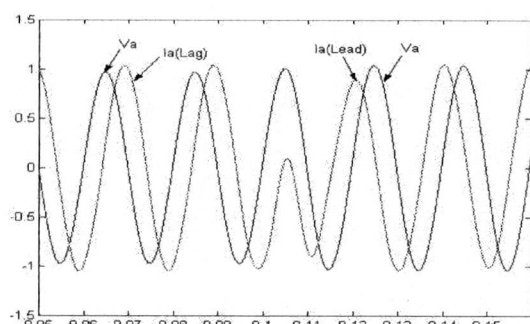

Figure 12. Voltage and Current Waveforms during the Change from Inductive Operation at t=0.1^{sec} with Fuzzy Controller

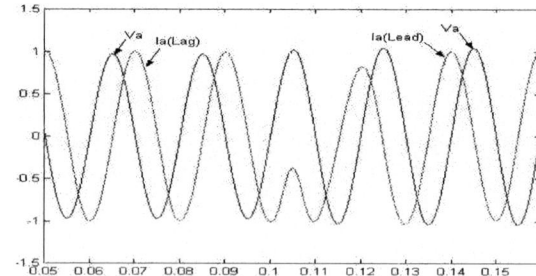

Figure 13. Voltage and Current Waveforms During the Change from Inductive Operation at t=0.1^{sec} with PI Controller

The harmonic analyzes is carried out by using FFT function and on each figure. The T.H.D value and the order of harmonics are presented. Figures 15 and 16 show that STATCOM isn't generated low order harmonics and amount of high order harmonics are low. Figures17, 18, 19 and 20 show that all harmonics have been reduced by the filters. [5]

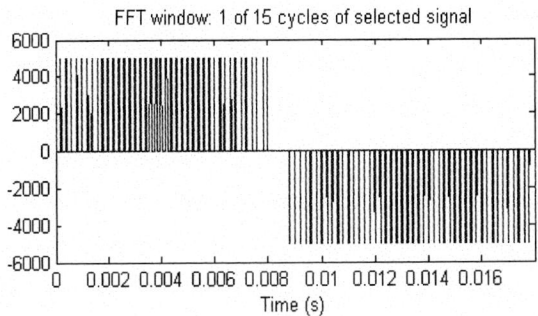

Figure15.   Output Voltage of Inverter.

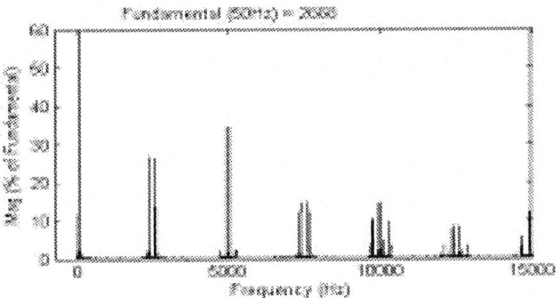

Figure 16.   Harmonic Spectrum of Output Voltage of Inverter.

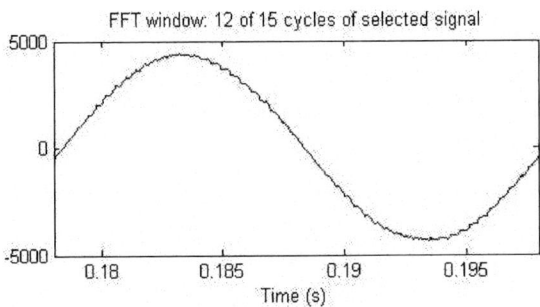

Figure 17.   Output Voltage of Filter.

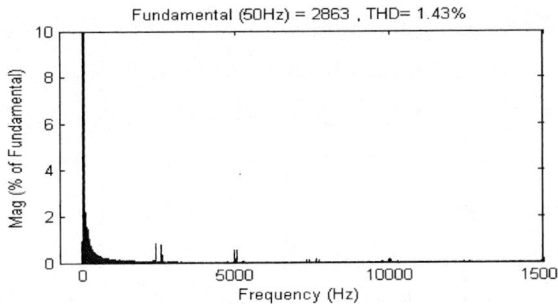

Figure 18.   Harmonic Spectrum of Output Voltage of Filter.

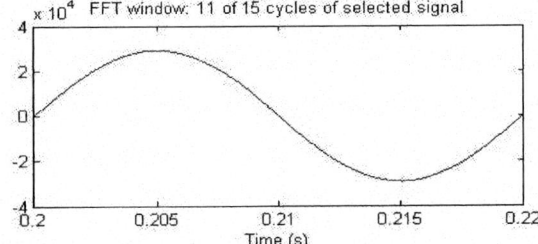

Figure 19.   Voltage at 20kV Bus.

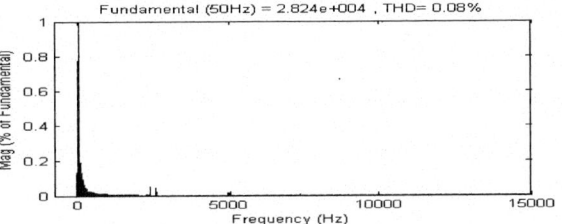

Figure 20.   Harmonic Spectrum of Voltage at 20kV Bus

V.   CONCLUTION

A fuzzy logic controlled STATCOM has been simulated to compensating reactive power in distribution networks and compared results of simulation with the PI conventional controller, we can see fuzzy logic controller compensate as well as PI controller, but design of fuzzy controller with linguistic variable is very simple and in the fuzzy controller we does not require a mathematical model of the system.

REFERENCES

[1] Hochgraf, C.; Lasseter, R.H,"STATCOM controls for operation with unbalanced voltages".Power Delivery, IEEE Transactions on:Volume: 13,   Issue 2, April 1998 Page(s):538 - 544

[2] Wei-Neng Chang; Kuan-Dih Yeh,"Design of D-STATCOM for fast load compensation of unbalanced distribution systems"Power Electronics and Drive Systems, 2001. Proceedings. 2001 4th IEEE International Conference on Volume: 2, 22-25 Oct. 2001 Page(s):801 - 806 vol.2

[3] Han-Xiong Li; Gatland, H.B,"A new methodology for designing a fuzzy logic controller"Systems, Man and Cybernetics, IEEE Transactions on Volume 25, Issue 3, March 1995 Page(s):505 – 512

[4] Chin-Ming Hong; Huan-Wen Tzeng; Shun-Yuan Wang; Chi-Wu Huang,"Design of a static reactive power compensator using fuzzy sliding mode control".Decision and Control, 1994., Proceedings of the 33rd IEEE Conference on Volume 4, 14-16 Dec. 1994 Page(s):4142 - 4145 vol.4

[5] Chia, B.H.K.; Morris, S.; Dash, P.K,"A feedback linearization based fuzzy-neuro controller for current source inverter-based STATCOM".Power Engineering Conference, 2003. PECon 2003. Proceedings.   National   15-16   Dec.   2003

2006 5th International Power Electronics and Motion Control Conference

A Distributed Fuel Cell Based Generation and Compensation System to Improve Power Quality

Haimin Tao, Jorge L. Duarte, and Marcel A. M. Hendrix
Department of Electrical Engineering
Eindhoven University of Technology
5600 MB Eindhoven, The Netherlands
Email: h.tao@tue.nl

Abstract—A small single-phase fuel cell based energy generation and compensation system is proposed in this paper. The power conditioning unit of the system comprises a grid-interfacing inverter and a three-port bidirectional converter which connects a fuel cell and supercapacitor to the inverter. The system can operate in both island and grid-connected modes. By taking advantage of the transient storage capability offered by the supercapacitor, the function of reactive power compensation is integrated into the system. Simultaneously, the inverter is operated as a shunt active power filter and, by means of a proposed control strategy, compensates for reactive and harmonic current demanded by local loads. The system is suitable for residential applications and can improve the quality of a weak power grid. Simulation and experimental results show the validity and feasibility of the proposed system.

Keywords- active power filter; fuel cell; supercapacitor; UPS

I. INTRODUCTION

Electrical generation systems based on small sustainable energy sources are gaining popularity due to their high operation efficiencies and low CO_2 emission levels. Distributed generation (DG) offers peak shaving to reduce the cost of energy by generating during peak load hours [1]. It also provides standby generation during grid outages. The use of fuel cell technology for electricity and heat generation for residential applications has generated lots of interest [2]. However, due to the slow fuel supply process, a fuel cell system requires energy storage elements to improve its system dynamics.

As a grid-interactive distributed generator, a fuel cell system is usually designed to be able to operate in both island and grid-connected modes according to grid conditions. In this paper, the system is also utilized to operate as a shunt active power filter. This idea has been applied to some line-interactive UPS systems [3]-[5]. The reactive power compensation is beneficial for the power system network. This system can minimize real/reactive power imbalances that can affect the surrounding power system and thus increases the transmittable power in the ac system. The function of active filtering is integrated into the system and realized by making use of the transient energy storage – the supercapacitor. A corresponding control strategy for the inverter is developed to achieve this

objective. The power conditioning unit of the system consists of a standard grid-interfacing inverter and a newly proposed three-port bidirectional dc-dc converter [6], [7]. This paper shows that a supercapacitor in a fuel cell system can serve as an active and reactive energy storage and buffer periodical low frequency ripple in the requested power. This minimizes the dc-link capacitance and makes it possible to use a small non-electrolytic type and thus improves the lifecycle and reliability of the system.

II. SYSTEM DESCRIPTION

Fig. 1 shows the structure of the fuel cell generation system for residential applications. A three-port converter is used to manage the power flow between the fuel cell, inverter, and supercapacitor. The second part of the system is a grid-interfacing inverter whose output is connected to the grid through a static transfer switch (STS). Local loads are coupled to the inverter output. The connecting point is usually called the point of common coupling (PCC).

A. Dc-Dc Stage

Unlike conventional power conversion equipments, the dc-dc stage for a fuel cell system features multiple inputs. A family of three-port converters [8] can be employed. In this paper, the three-port triple-active-bridge (TAB) converter [6] was chosen. The converter topology comprises a three-winding transformer and three active bridges. This topology is bidirectional due to the active bridges at all three ports. In addition to galvanic isolation, a major advantage of this converter is the ease of matching the different voltage levels of the ports. The resulting leakage inductances are an integral part of the circuit. Each bridge generates a high frequency square-wave voltage with a controlled phase angle. Moreover, this circuit can also be operated with soft-switching by duty cycle control when the port voltage varies [7]. The control scheme for the dc-dc stage aims to regulate the dc-link voltage and the fuel cell power simultaneously with the two phase shifts as control variables. The controller employs two PI controllers which are devoted to the regulation of V_{DC} and P_{FC}, respectively [7]. In addition, the supercapacitor sinks/sources the power difference between the inverter and the fuel cell. This is an autonomous system matching the variations of the power drawn by the inverter while the power of the fuel cell is kept at the same level.

1-4244-0448-7/06/$25.00 ©2006 IEEE

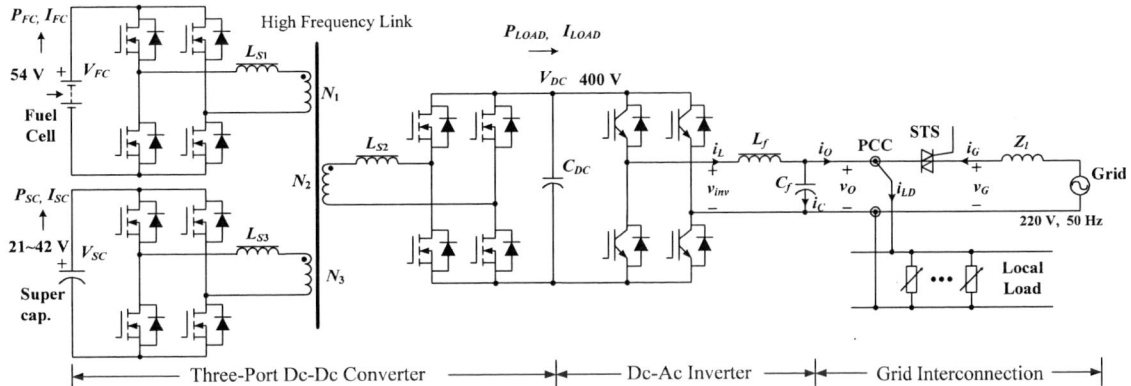

Fig. 1. A single-phase distributed fuel cell generation and compensation system.

B. Dc-Ac Stage

A standard voltage source PWM inverter is employed for dc-ac power conversion and grid-interfacing. The main function of the inverter is to maintain a regulated output voltage to power the local loads in the island mode, and to inject reactive and harmonic current required by the loads along with a constant real power into the PCC in the grid-connected mode. The proposed system combines a generation system with the function of an active power filter. Therefore it eliminates the need for an extra compensator to deal with nonlinear loads. In addition, low voltage energy storage devices are utilized because their voltages can be matched in the dc-dc stage. However, for handling both real and reactive power, the inverter must have sufficiently high power capacity with rated voltage in order to continuously condition the routine energy supplied by the fuel cell. Since the system is for residential applications, it is considered to be single-phase. The idea proposed in this paper could also be applied to a larger scale three-phase application.

III. MODES OF OPERATION AND CONTROL STRATEGIES

A. Island Mode of Operation

In this mode, the grid is disconnected from the system. Fig. 2 shows the power flow. As the transient energy storage, the supercapacitor plays two important roles. First, energy stored in the supercapacitor is utilized to handle a short-time overload or dump in load to overcome the slow dynamics of the fuel processor. The second role of the supercapacitor is to deal with instantaneous and periodical power fluctuations. The power drawn by the inverter from the dc-link capacitor is periodical with a $2*f_{grid}$ component present in it. A common way to deal with this low frequency ripple in power is to use a large electrolytic capacitor at the dc-link for power decoupling. However, the dc buffer capacitor of an inverter is often the limiting factor in terms of reliability and lifespan. In this paper, a new approach is proposed to decouple the periodical reactive power. In the dc-dc stage, since the supercapacitor serves as an energy buffer, the need for power decoupling at dc-link can be eliminated as long as the control loop has enough bandwidth. As a result, the dc-link capacitor is only used to filter out the high frequency switching ripple of the dc-dc stage. Therefore a small non-electrolytic type capacitor can be employed. Consequently, the lifecycle and reliability of the converter can be improved.

In the island mode of operation, the inverter is voltage-controlled. The ac output voltage is regulated by the inverter controller with a standard double-loop control strategy. As shown in Fig. 3(a), a PI controller is employed for the voltage loop and a proportional controller for the current loop. Other control schemes such as capacitor current feedback and resonant controller as seen in Fig. 3(b) could also be used. Since the grid voltage is absent, the controller is only dedicated to the regulation of the output voltage. Only one standalone fuel cell system is considered in this paper. If more generation units are paralleled to power the load, an external power management controller should be included in the system to share the real and

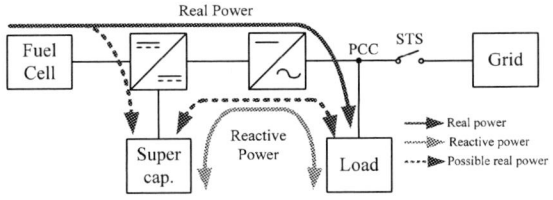

Fig. 2. Power flow in the island mode of operation.

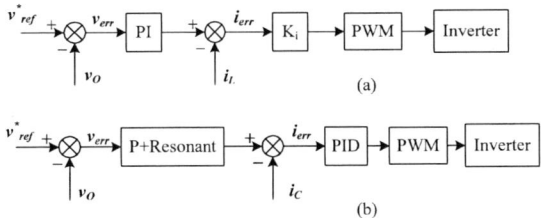

Fig. 3. Control scheme for the island mode of operation (a) inductor current feedback, (b) capacitor current feedback.

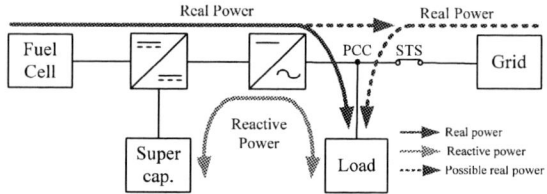

Fig. 4. Power flow in the grid-connected mode of operation.

reactive power among those generators, for instance, the commonly used voltage and frequency droop technique.

B. Grid-Connected Mode of Operation

The power flow in the grid-connected mode of operation is shown in Fig. 4. The grid is a virtually unlimited buffer for real power. In this mode of operation, we propose that the fuel cell system not only injects real power into the PCC, but also compensates for the reactive and harmonic current demanded by the load. Fig. 5 shows the proposed control scheme. It consists of an active current estimator, an amplitude current estimator, a current controller, and a phase locked loop (PLL) for generating an in-phase sinusoid. Without compensation, the line current is distorted by the power factor and harmonics of the characteristics of the load. The unwanted currents are mainly the reactive current and harmonic currents. The grid should only supply the in-phase current. To achieve this, a common way is to subtracting the active current component $i_P(t)$ from the measured load current. Then the current to be compensated, $i_F(t)$, is that

$$i_F(t) = i_{LD}(t) - I_P \cos \omega t \qquad (1)$$

where $i_{LD}(t)$ is the load current, I_P is the amplitude of the in-phase current and $\cos \omega t$ is the in-phase sinusoid. Simultaneously, the inverter should inject energy into the connecting point complying with the power sourced by the fuel cell. To do this, a in-phase reference current, $i^*_{PFC}(t)$, is superposed to $i_F(t)$ and this yields

$$i^*_O(t) = i_F(t) + i^*_{PFC}(t) \qquad (2)$$

where $i^*_O(t)$ is the reference current for the inverter. The reference $i^*_{PFC}(t)$ is determined according to the fuel cell

power P^*_{FC} and the line fundamental voltage U_G (U_G equals U_O in grid-connected mode).

$$i^*_{PFC}(t) = I^*_{PFC} \cos \omega t = \sqrt{2} \frac{P^*_{FC}}{U_G} \cos \omega t. \qquad (3)$$

where U_G is a rms value. Therefore the inverter reference current is given by

$$i^*_O(t) = i_{LD}(t) + \left(\sqrt{2} \frac{P^*_{FC}}{U_G} - I_P \right) \cos \omega t. \qquad (4)$$

The voltage source inverter can be controlled as a current source by means of a PWM signal to follow the reference current. Many control techniques such as hysteresis control, ramp comparison, predictive control, etc., have been proposed in literature.

The active current estimator shown in Fig. 5 uses a feedback loop and an integral gain block to calculate the accurate in-phase current [9]. This method is frequency-independent and does not need precise components. The principle of operation is quite simple. Because of the integrator presenting in the signal loop, the output $i_F(t)$ will not contain any in-phase component since it has been subtracted from the load current $i_{LD}(t)$. After multiplication with the in-phase sinusoid, no dc component will be present. This method requires a low-distortion sinusoid with good phase tracking with respect to the line voltage. For this purpose, a PLL block with error filtering and integration is employed to generate a line voltage in-phase sinusoid with zero phase error in steady state.

C. Transition Between the two modes

A seamless transition between the two operation modes should be guaranteed in order not to disturb the operation of critical and sensitive loads in the event of a fault on the grid. Most time the system operates in grid-connected mode. At the occurrence of a fault, the equipment may disconnect the fuel cell system as a protective measure. The procedure to shift from the grid-connected mode to the island mode can be described as follows. First, the control circuit detects a fault on the grid. Then the control circuit disconnects the separation device STS by shutting off the gate signal to the STS and monitors the magnitude and phase of the load voltage at the instant of disconnection. The inverter is then switched to voltage-mode operation with the load voltage at that moment as the reference.

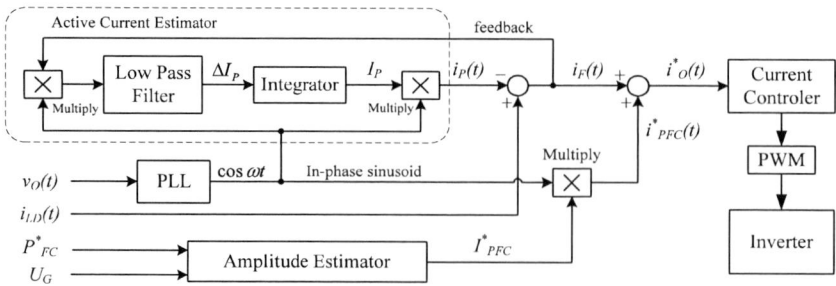

Fig. 5. Control scheme for the grid-connected mode of operation.

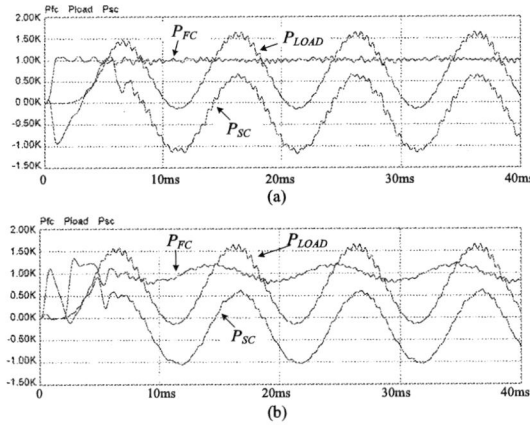

(a)

(b)

Fig. 6. The power flow in the dc-dc stage with an inverter load: (a) sufficient bandwidth for P_{FC}, (b) insufficient bandwidth for P_{FC}.

Finally, by ramping up the magnitude of the load voltage to the normal rated value, the transition procedure finishes [1]. The whole process could take a few line cycles to ensure a smooth transient.

On the other hand, when the fault on the grid is cleared, a synchronization process should shift the inverter from the island mode to the grid-connected mode. First the grid voltage is inspected and verified to be within the tolerance limits of the sensitive loads. Once a nominal grid voltage is detected, the control algorithm adjusts the load voltage to match the magnitude and phase of the grid voltage. When the voltages at the both sides of the separation switch are locked both in magnitude and phase angle, the STS can be turned on. At the instant of reconnection, the inverter is turned to current control mode.

IV. SIMULATION AND EXPERIMENTAL RESULTS

A. Power Decoupling

To verify the effectiveness of the proposed system, the dc-dc stage and control scheme was simulated with PSIM6.0. The simulation and design parameters are listed in TABLE I. The three-port converter was simulated with a standard inverter as load and the ac load is a resister in series with an inductor. Fig. 6 shows the simulation result. In (a), it can be seen that the fuel cell power is kept approximately constant at 1 kW while the ripple in the

TABLE I.   SIMULATION AND DESIGN PARAMETERS

Description	Parameter	Value	Unit
Fuel cell voltage	V_{FC}	54	V
Dc link voltage	V_{DC}	400	V
Supercapacitor voltage	V_{SC}	21 ~ 42	V
Fuel cell power	P_{FC}	1000	W
Load power	P_{LOAD}	0 ~ 2000	W
Switching frequency	f_s	20	kHz
Transformer turns ratio	$N1{:}N2{:}N3$	5:38:4	
Inductance	L_{S1}	1.0	μH
	L_{S2}	54	μH
	L_{S3}	0.61	μH

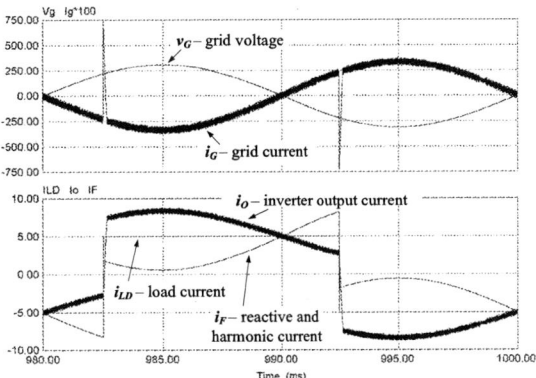

Fig. 7. Simulation waveforms of the inverter (square-wave load current).

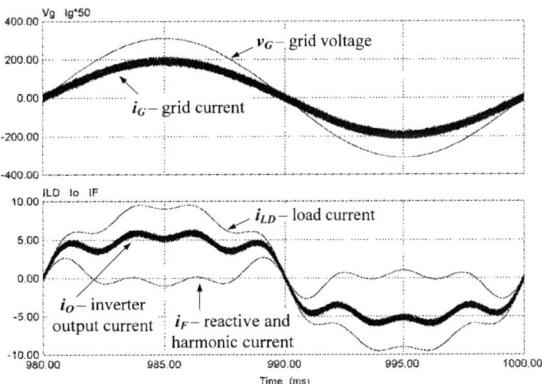

Fig. 8. Simulation waveforms of the inverter (harmonic currents).

power P_{LOAD} is buffered by the supercapacitor. However, this is on condition that the fuel cell power control loop has a sufficient bandwidth. Otherwise, there will be low frequency ripple in the power drawn from the fuel cell, as illustrated in Fig. 6(b). Although the bandwidth of the fuel cell power control loop is deliberately tuned low to avoid the interaction with the control loop that regulates the dc output voltage [7], it is normally sufficient to omit the double-line-frequency ripple. In summary, the presence of the supercapacitor as storage in the system leads to many benefits. Not only can the transients in the load be compensated for by the supercapacitor, the periodical reactive power resulting from the inverter can be dealt with in the same way. This advantage eliminates any other energy buffers in the system as long as a sufficient bandwidth is guaranteed.

B. Inverter Operation

Fig. 7 shows the key waveform in the grid-connected mode of operation. A hysteresis controller is employed to control the inverter. First, the nonlinear load is simulated by a square-wave current source which lags the grid fundamental voltage by 45 degrees. It is shown that the inverter supplies the reactive and harmonic current required by the load while injecting real power. In this simulated case, the fuel cell supplies more power than the

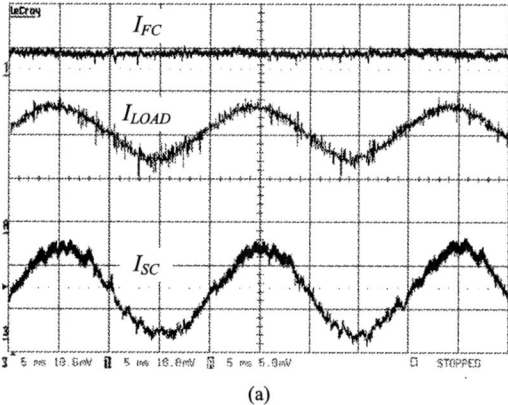

(a)

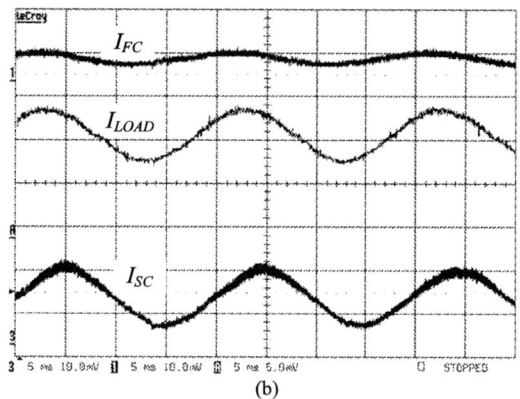

(b)

Fig. 9. Experimental result of the power flow in the system with an inverter load: (a) sufficient bandwidth for P_{FC}, (b) insufficient bandwidth for P_{FC} In both figures: I_{FC} (10 A/div), I_{LOAD} (0.2 A/div), I_{SC} (2.5 A/div), at a time base of 5 ms/div.

load demand. Therefore the surplus power is injected into the grid as seen in the figure. The grid current is an in-phase sinusoidal current and the spikes seen in the grid current are resulted from the hysteresis control and extremely high di/dt of the load current. Fig. 8 shows the case that the load current contains 3rd, 5th and 7th harmonics. As shown, the harmonic currents are compensated for by the inverter and the grid only supplies fundamental current. In this example, the load consumes more than the power generated by the fuel cell so that the grid also powers the load.

C. Experimental Results

A scaled prototype is rated at 2 kW maximum power and the control scheme is implemented with a TMS320F2812® DSP. Because of the resolution of the digitally implemented phase shift, the dc-dc stage operates at 20 kHz switching frequency. A PEM fuel cell (1 kW, 54 V) and a 145 F supercapacitor are used as the generator and storage. The measurement results of the power flow at different bandwidths for the fuel cell power control loop are shown in Fig. 9 where I_{FC}, I_{LOAD}, and I_{SC} are the averages of the fuel cell current, current drawn by the inverter, and the supercapacitor current, respectively. They

are representative for power since the voltages at all the ports remain unchanged. As shown, the periodical reactive current in I_{LOAD} is compensated for by the supercapacitor.

V. CONCLUSION

A distributed fuel cell generation and compensation system has been proposed in this paper. The system comprises a three-port converter and a grid-interfacing inverter. It can operate in both island and grid-connected modes. In the grid-connected mode the system injects a constant power to the grid, and at the same time compensates for the reactive and harmonic current required by nonlinear loads. The function of operating as a shunt active power filter is integrated into the system and realized by the proposed control scheme. Furthermore, the benefits of using a supercapacitor have been shown. It compensates for the instantaneous power fluctuations, overcomes the slow dynamics of the fuel processor, and handles periodical low frequency ripple in power.

REFERENCES

[1] R. Tirumala, N. Mohan, and C. Henze, "Seamless transfer of grid-connected PWM inverters between utility-interactive and stand-alone modes," in *Proc. 17th Annu. IEEE Applied Power Electronics Conf. Expo. (APEC'02)*, vol. 2, 2002, pp. 1081–1086.

[2] W. Choi, P. Enjeti, and J. W. Howze, "Fuel cell powered UPS system: design considerations," in *Proc. IEEE Power Electronics Specialist Conf. (PESC '03)*, vol. 1, Jun. 2003, pp. 385–390.

[3] S. A. O. D. Silva, P. F. Donoso-Garcia, P. C. Cortizo, and P. F. Seixas, "A three-phase line-interactive UPS system implementation with seriesparallel active power-line conditioning capabilities," *IEEE Trans. Ind. Appl.*, vol. 38, no. 6, pp. 1581–1590, Nov./Dec. 2002.

[4] F.-S. Pai and S.-J. Huang, "A novel design of line-interactive uninterruptible power supplies without load current sensors," *IEEE Trans. Power Electron.*, vol. 21, no. 1, pp. 202-210, Jan. 2006.

[5] P. G. Barbosa, L. G. B. Rolim, E. H. Watanabe, and R. Hanitsch, "Control strategy for grid-connected DC-AC converters with load power factor correction," *IEE Proc. Gener. Transm. Distrib.*, vol. 145, no. 5, pp. 487–491, Sep. 1998.

[6] M. Michon, J. L. Duarte, M. Hendrix, and M. G. Simoes, "A three-port bi-directional converter for hybrid fuel cell systems," in *Proc. IEEE Power Electronics Specialists Conf. (PESC'04)*, Jun. 2004, Aachen Germany, pp. 4736–4742.

[7] H. Tao, A. Kotsopoulos, J. L. Duarte, and M. A. M. Hendrix, "A soft-switched three-port bidirectional converter for fuel cell and supercapacitor applications," in *Proc. IEEE Power Electronics Specialists Conf. (PESC'05)*, June 2005, Recife, Brazil, pp. 2487–2493.

[8] H. Tao, A. Kotsopoulos, J. L. Duarte, and M. A. M. Hendrix, "Family of multi-port bidirectional dc-dc converters," *IEE Proceeding Electric Power Applications*, to be published.

[9] J. S. Tepper, J. W. Dixon, G. Venegas, and L. Moran, "A simple frequency-independent method for calculating the reactive and harmonic current in a nonlinear load," *IEEE Trans. Industrial Electronics*, vol. 43, no. 6, pp. 647–654, Dec. 1996.

2006 5th International Power Electronics and Motion Control Conference

Parallel Control of Three-Phase Three-Wire Shunt Active Power Filters

Xueliang Wei, Ke Dai, Xin Fang, Pan Geng, Fang Luo and Yong Kang

College of Electrical and Electronic Engineering, Huazhong University of Science and Technology, Wuhan, China

Email: sdslxy@163.com

Abstract—A three-phase three-wire shunt active power filter (APF) suitable for parallel operation is proposed in this paper to overcome difficulties in the field of high power applications. This APF can operate independently and compensate the load harmonic current according to its own capacity-limitation. Furthermore, the output current of each APF is optimized so that the APF with large capacity compensates more current and the one with small capacity compensates less current. In this way, the feasibility and security of modular APF can be guaranteed. Based on this, the parallel control strategy for modular APF is also given, which can ensure that each APF switches on/off according to actual load harmonic current and has little impact on the whole system. In the end, simulation and experiment results are shown to testify the validity of proportion compensation strategy based on capacity-limitation and the feasibility of coordination control strategy for parallel running of modular APF.

Keywords-shunt active power filter; capacity-limitation; parallel running; coordination control

I. INTRODUCTION

Nowadays, the widely use of non-linear power electronic devices, by injecting the harmonic and reactive current into the electrical network, has seriously threatened the secure operation and normal utilization of the utility and other related apparatus. Therefore, the filtering and compensation of harmonic and reactive current become an important research topic in many fields such as power electronics, electrical system, industrial electronics, and industry application.

APF can dynamically restrain and compensate the harmonic and reactive current, which is variable both in frequent and magnitude. Compared with passive power filter, it is an ideal device with better compensation characteristics for harmonic and reactive control. However, due to its cost, complexity and especially the problem of capability, the use of single APF in high voltage or high power field is limited. Even though great progresses have been made in the research of high power APF such as cascade multilevel APF and hybrid APF, some problems like complex circuit structures, intricate control algorithm, poor dynamic performance etc, exist in above technical solutions [1] [2].

Another way to break the capability limit of single APF is to assemble modularized small capacity blocks with

parallel control function into a large capacity APF system by switching on or off. But very few articles have illustrated the technical details on how to control APF in parallel and how to exchange information among them [3]. In this paper, a simple and secure means is given, which can make APF run in parallel. And this method is also an available way to have single APF with low capability work in combination with each other for high power applications, making the promise of secure operation and enhanced compensation effect. Simulation and experimental results have proved the practicability.

II. PARALLEL CONTROL STRATEGY

A. Load proration based on limit of capability

When the needed compensating current is beyond the available output current of APF, limitation must be taken into consideration based on the APF itself. Obviously, it's unwise to use "Magnitude Limitation" to simply cut the part over current. That is because this method will generate some undesired harmonics of new spectrum and perhaps deteriorate the harmonic pollution of network. A better way to solve this problem is using "RMS Limitation", which is aiming at limiting the output of APF to the rated range according to its compensating capability. That is to say, the output current i_c of APF should be within its rated current i_r.

The output current of APF can be expressed as follows,

$$i_c = \sum_{j=2}^{n} i_{cj} . \tag{1}$$

i_{cj} is the jth harmonic compensation value of the output of the APF.

The RMS value of the output current of APF can be expressed as follows,

$$I_c = \sqrt{I_{c2}^2 + I_{c3}^2 + \cdots + I_{cn}^2} . \tag{2}$$

The RMS value of the rated current of APF is,

$$I_r = S_c / (3 \cdot V_s) . \tag{3}$$

S_c is the rated capacity of the APF and V_s is the utility voltage.

Thus, the rated output current limit of the APF is,

$$I_c \leq I_r . \tag{4}$$

1-4244-0448-7/06/$25.00 ©2006 IEEE

If the RMS value of the compensation current reference $I_c^{'}$ is higher than that of the rated current of APF I_r, the final current reference can be got by multiplying a limited output coefficient μ. Otherwise, the final current reference will be itself. The coefficient expression is,

$$\mu = \begin{cases} I_r/I_c^{'}, I_c^{'} > I_r \\ 1, I_c^{'} \leq I_r \end{cases} . \tag{5}$$

B. Parallel running of two APFs

Parallel operation can enhance the compensating capability. Ordinarily, there are two ways of parallel connection based on their bus structure, which are referred as "Public Bus Mode" and "Non public Bus Mode" [3] [4] [5].

Here, "the non public bus mode" parallel method shown

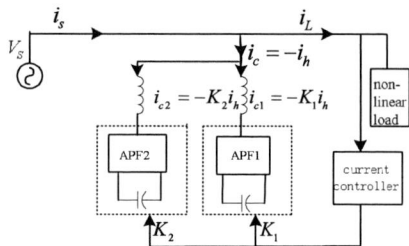

Figure 1.   Two APFs with centralized parallel control

in Fig.1 is used in the coordinated control of two APFs.

In order to realize the modularization of parallel operation, two APFs must be the same in system structure, hardware circuit, and software controller design. In Fig.1, i_{c1} and i_{c2} respectively represent the compensating current of APF1 and APF2. While the sum of these two APFs' output compensating current i_c can completely offset the harmonic current i_h of the non-linear load to make the utility current i_s only contain the fundamental active component i_1, each APF should afford its compensating current as,

$$\begin{cases} K_1 = S_1/(S_1 + S_2) \\ K_2 = S_2/(S_1 + S_2) \end{cases}, \quad I_c^{'} \leq I_r . \tag{6}$$

S_1 and S_2 is the rated capability of APF1 and APF2 respectively.

Since the utility voltage is the same, the above formula can be expressed as follows,

$$\begin{cases} K_1 = I_{r1}/(I_{r1} + I_{r2}) \\ K_2 = I_{r2}/(I_{r1} + I_{r2}) \end{cases}, \quad I_c^{'} \leq I_r . \tag{7}$$

I_{r1} and I_{r2} is the rated current of APF1 and APF2 respectively.

When the sum of compensating output current i_c of the two APFs cannot completely offset the harmonic current i_h of the non-linear load, each APF should afford its compensating proportion as,

$$\begin{cases} K_1^{*} = K_1 \mu_1 = \dfrac{I_{r1}}{I_{r1} + I_{r2}} \cdot \dfrac{I_{r1}}{I_{c1}^{'}} \\ K_2^{*} = K_2 \mu_2 = \dfrac{I_{r2}}{I_{r1} + I_{r2}} \cdot \dfrac{I_{r2}}{I_{c2}^{'}} \end{cases}, \quad I_c^{'} > I_r . \tag{8}$$

μ_1 and μ_2 is the coefficient of limited output current of APF1 and APF2 respectively

If $I_{r1} = I_{r2}$, then $K_1^{*} = K_2^{*} = K_1 \mu_1 = K_2 \mu_2 = 0.5 \mu_1 = 0.5 \mu_2$.

1) Switch off one APF from the parallel system

If APF2 has some trouble and needs repair, the system only needs APF1 in operation. APF2 can be switched off or unloaded from it. The compensating capability given by APF2 will be offered by APF1. The procedure of load transfer will be fulfilled in linear control way. The compensating coefficient of APF1 and APF2 can be expressed as,

$$\begin{cases} K_1^{'} \mu_1 = K_1^{*} + (t/T_D)K_2 \mu_1 \\ K_2^{'} \mu_2 = (1 - t/T_D)K_2^{*} \end{cases}, \quad 0 \leq t \leq T_D \tag{9}$$

$K_1^{'} \mu_1$ and $K_2^{'} \mu_2$ is the compensating proportion coefficient of APF1 and APF2 after the load transfer. K_1^{*} and K_2^{*} is the compensating proportion coefficient of APF1 and APF2 before the load transfer. T_D is the load transfer time.

If $t > T_D$, then $K_1^{*} = K_1 \mu_1 = \mu_1$ and $K_2 = 0$. That is to say all the harmonic and reactive current of the non-linear load will be offered by APF1.

Just as the same, if APF1 is switched off or unloaded, the compensating capability given by it will be offered by APF2 gradually.

$$\begin{cases} K_1^{'} \mu_1 = (1 - t/T_D)K_1^{*} \\ K_2^{'} \mu_2 = K_2^{*} + (t/T_D)K_1 \mu_2 \end{cases}, \quad 0 \leq t \leq T_D \tag{10}$$

2) Switch on one APF into the parallel system

If the capability of the non-linear load becomes larger and APF1 can't completely offer the compensating capability, APF2 can be put into the system to operate in parallel with APF1, so as to offer a part of compensation of the harmonic and reactive current. The procedure of load transfer will be fulfilled in linear control way. The compensating coefficient of APF1 and APF2 can be expressed as,

$$\begin{cases} K_1^{'} \mu_1 = \mu_1 - \dfrac{t}{T_D}\left(\dfrac{I_{r2}}{I_{r1} + I_{r2}}\right)\mu_1 \mu_2 \\ K_2^{'} \mu_2 = \dfrac{t}{T_D}\left(\dfrac{I_{r2}}{I_{r1} + I_{r2}}\right)\mu_2 \end{cases}, \quad 0 \leq t \leq T_D \tag{11}$$

Just as the same, if APF1 is switched on into the parallel system, it will offer part of compensation which is offered by APF2.

$$\begin{cases} K_1^{'}\mu_1 = \dfrac{t}{T_D}(\dfrac{I_{r1}}{I_{r1}+I_{r2}})\mu_1 \\ K_2^{'}\mu_2 = \mu_2 - \dfrac{t}{T_D}(\dfrac{I_{r1}}{I_{r1}+I_{r2}})\mu_1\mu_2 \end{cases} , \quad 0 \le t \le T_D . \quad (12)$$

C. Parallel running of more than two APFs

1) More than two APFs steadily run in parallel

As shown in fig.2, the system parallel running of more than two APFs works in the way of centralized parallel control without common DC-side bus.

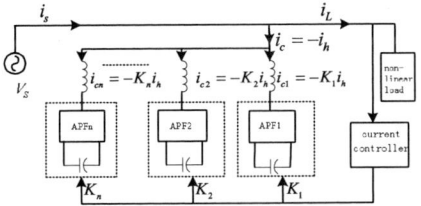

Figure 2. n APFs with centralized parallel control

The symbol n means the number of APF which should be put into the system. In order to make the system work more efficiently, n should make the following formula available.

$$\sum_{j=1}^{n-1} I_{rj} < I_H < \sum_{j=1}^{n} I_{rj} . \quad (13)$$

I_H is the total harmonic current which is injected into the utility. I_{rj} is the rated current value of the jth APF.

If there is some trouble or any APF needs repair, the sum of compensating current i_c can't completely offset the harmonic component i_h of the non-linear load's current i_L, the compensation coefficient of each APF should be,

$$K_i = \dfrac{I_{ri}}{\sum\limits_{j=1}^{n} I_{rj}} \dfrac{I_{ri}}{I_{ci}^{'}} = \dfrac{I_{ri}}{\sum\limits_{j=1}^{n} I_{rj}} \mu_i , i=1,2,\cdots,n . \quad (14)$$

If the compensating capability of all the APF is of the same, the above formula can be simplified as,

$$K_i = (1/n)\mu_i , i=1,2,\cdots,n \quad (15)$$

2) Switch off More than two APF from the parallel system

If the load capability decreases suddenly, some APFs will be switched off so as to make a full use of other APFs. We can do it by examining the variation of harmonic current of the load at any time. If the variation is more or less than the current of the APF which has the lowest capability, the number of APF and the proportion compensation coefficient of each APF need to be recalculated [4].

When the load capability decreases, the number of APF also needs to be reduced in order to restrain all harmonic

current. The number $n^{'}$ of APF which should be into the system is,

$$\sum_{j=1}^{n-1} I_{rj} < I_H^{'} < \sum_{j=1}^{n^{'}} I_{rj} . \quad (16)$$

$I_H^{'}$ is the load harmonic current after the load variation.

Therefore, the number of APF which should be switched on into the system is $m=n-n^{'}$.

If the sum of compensating current i_c offered by the system cannot completely offset the harmonic current i_h of non-linear load current i_L because of some repair or trouble, the compensating current proportion coefficient of each working APF should be,

$$K_i^{'}\mu_i = K_i\mu_i + \dfrac{t}{T_D} \dfrac{I_{ri}}{\sum\limits_{j=1}^{n^{'}} I_{rj}} \sum_{j=n^{'}+1}^{n} K_j\mu_i \sum_{j=n^{'}+1}^{n} \mu_j . \quad (17)$$

Where $i=1,2,\cdots,n$ and $0 \le t \le T_D$

The compensating current proportion coefficient of newly paralleled APF is,

$$K_i^{'}\mu_i = (1-\dfrac{t}{T_D})K_i\mu_i . \quad (18)$$

Where $i=n^{'}+1, n^{'}+2,\cdots,n$ and $0 \le t \le T_D$

If the capability of all the working APF is of the same, the above formula can be simplified as,

$$K_i^{'}\mu_i = \dfrac{1}{n}\mu_i + \dfrac{t}{T_D} \dfrac{1}{n^{'}} \dfrac{m}{n}(m-1)\mu_i^2 , i=1,2,\cdots,n, 0 \le t \le T_D \quad (19)$$

$$K_i^{'}\mu_i = (1-\dfrac{t}{T_D})\dfrac{1}{n}\mu_i , i=n^{'}+1, n^{'}+2,\cdots,n, 0 \le t \le T_D \quad (20)$$

While $t>T_D$, the parallel system gets into a new working state, we can make $n=n^{'}$ to replace the number of the paralleled APF.

3) Switch on more than APF from the parallel system

If the load capability increases suddenly when more than two APFs are working in parallel, there will be a need to switch on more APFs into the system in order to avoid overloading [4][6]. We can do it by examining the variation in harmonic current of the load at any time. If the variation is more or less than the current of the APF which has the lowest capability, the number of APF and the compensation current proportion coefficient of each APF need to be recalculated.

If the load capability increases, in order to make the system work more efficiently, the number $n^{'}$ of working APF can be shown as,

$$\sum_{j=1}^{n-1} I_{rj} < I_H^{'} < \sum_{j=1}^{n^{'}} I_{rj} \quad (21)$$

$I_H^{'}$ is the total load harmonic current after the load variation.

At this moment, the number of working APF which should be switched on into the system is $m=n^{'}-n$.

The compensation current proportion coefficient of each working APF in parallel is,

$$K_i'\mu_i = K_i\mu_i - \frac{t}{T_D}\frac{I_{ri}}{\sum\limits_{i=1}^{n}I_{rj}}\sum\limits_{j=n+1}^{n'}K_j'\mu_i\sum\limits_{j=n+1}^{n'}\mu_j \ . \tag{22}$$

Where $i = 1,2,\cdots,n$ and $0 \le t \le T_D$

Because of the load decrease, in the procedure of load transfer, the compensation proportion coefficient of the unloaded APF is,

$$K_i'\mu_i = \frac{t}{T_D}\frac{I_{ri}}{\sum\limits_{j=1}^{n'}I_{rj}}\mu_i, i = n+1,n+2,\cdots,n',0 \le t \le T_D \tag{23}$$

If the capability of all working APF is of the same, the above formula can be simplified as,

$$K_i'\mu_i = \frac{1}{n}\mu_i - \frac{t}{T_D}\frac{1}{n}\frac{m}{n}(m-1)\mu_i^2, i = 1,2,\cdots,n,0 \le t \le T_D$$

$$\tag{24}$$

$$K_i'\mu_i = \frac{t}{T_D}\frac{1}{n}\mu_i, i = n+1,n+2,\cdots,n',0 \le t \le T_D. \tag{25}$$

While $t>T_D$, the parallel system gets into a new working state, we may make $n=n'$ to replace the number of the paralleled APF [5] [6] [7].

III. SIMULATION RESULTS

A. Configuration of the paralleled APF system

The main circuit structure of the paralleled APF system is shown in Fig.3.

At the AC side, three-phase balanced utility voltage is 220/380V whose frequency is 50Hz. Two sets of three-phase three-wire APF are established by three controllable IGBT arms, whose rated output current is 100A and output inductance is 0.3mH.

At the DC side of each APF, the DC-bus is capacitor whose capacitance is 20mF/900V and rated voltage is 715V.

The non-linear load is three-phase uncontrollable diode rectifier with resistor (10.5Ω) and inductor (0.5mH) at the DC side. The harmonic current generated by the diode rectifier will be offered by two paralleled APF1 and APF2.

B. Simulation content

1) Verifying the compensation characteristic in steady state

If the capacity of each APF are of the same, the compensating proportion K_1^* and K_2^* of each APF are both $0.5\mu_1$.

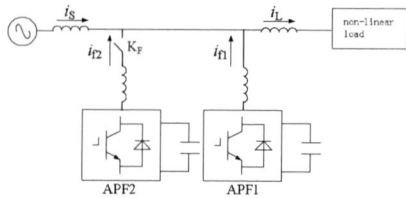

Figure 3. Configuration of the paralleled APF system

Fig.4 shows the simulation result of the paralleled APF system in steady state. From this figure, it can be learned that the output current of each APF i_{af1} and i_{af2} are almost the same, they offer 50% compensation capacity respectively.

2) Verifying the control strategies for switching on APF into the system

If the load transfer time T_D is 20ms the same as the utility period. When APF2 is switched on into the system, the compensation proportion offered by it will increase from 0 to 0.5 gradually, at the same time half of compensation capability of APF1 will be transferred to APF2, as shown in Fig.5. The whole procedure, from switching on APF2 into the system to each APF offering 50% compensation capability steadily, may need time almost a period of the utility.

Based on the above simulation results, not only the feasibility of the paralleled APF system can be verified, but also their compensation capacity can be proportioned according to the capacity limit of each APF.

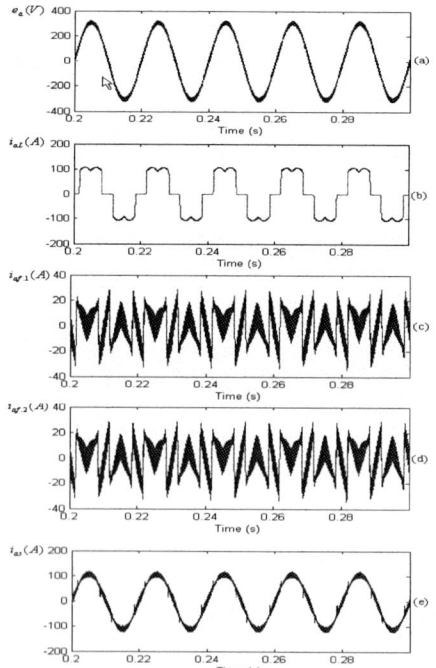

Figure 4. Simulation results of the compensation characteristic of the paralleled APF system in steady state. (a) a-phase utility voltage e_a; (b) a-phase load current i_{aL}; (c) a-phase output current of APF1 i_{af1}; (d) a-phase output current of APF2 i_{af2}; (e) a-phase utility current i_{as}.

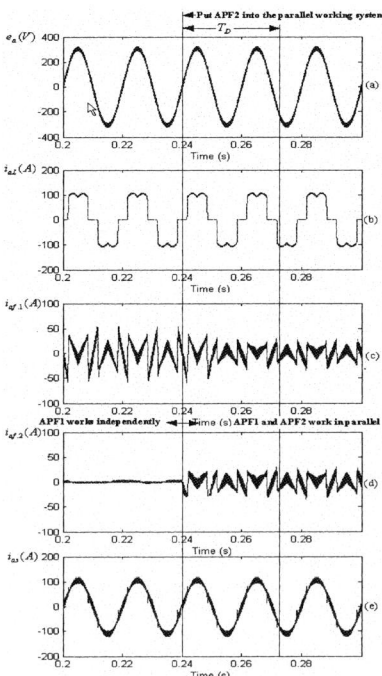

Figure 5. Simulation results of putting APF2 into the parallel APF system. (a) a-phase utility voltage e_a; (b) a-phase load current i_{aL}; (c) a-phase output current of APF1 i_{af1}; (d) a-phase output current of APF2 i_{af2}; (e) a-phase utility current i_{as}.

IV. EXPERIMENTAL RESULTS

A paralleled APF system made up of two 66 kvar modules with the same compensation capability is established in our laboratory to verify the correctness of above theoretical analyses and computer simulation results. The experiment conditions are the same as that of simulation. That is to say, the utility voltage is 220/380V, the DC side capacitor at each APF is of the value 20mF/900V and the output inductance of each APF is 0.3mH.

Fig.6 shows the experimental results of the paralleled APF system, which consist of the waveforms of non-linear load current i_{aL}, the output current of each APF $i_{af}1$ and i_{af2}, and the utility current i_{as}.

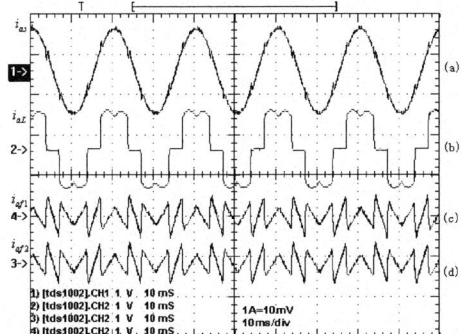

Figure 6. Experiment result of compensation characteristic of the paralleled APF system (1) a-phase utility current i_{as}; (2) a-phase load current i_{aL}; (3) a-phase output current of APF2 i_{af2}; (4) a-phase output current of APF1 i_{af1}.

From this figure, it can be learned that the compensation current of each APF are almost the same, they offer 50% compensation capacity respectively. And the utility current i_{as} close to perfect sine waveform and the utility phase voltage e_a are almost in the same phase. The experimental results verified the feasibility of the paralleled APF system. And their compensation capacity can be proportioned according to the capacity limit of each module.

V. CONCLUSION

From the above theoretical derivation, simulation and experimental results, we can come to the following conclusion.

According to the load proportion principle based on capacity limit, this paper proposes a three-phase three-wire APF which can compensate harmonic current according to its capacity and makes a study on the output current of the APF which is optimized under capacity limitation. When the load harmonic capacity is larger than the output capacity of APF, it can restrain the utility harmonic current according to its capacity. Under the coordination control of current distribution controller, the compensation current will be proportioned according to the capacity of each APF, so that the APF with large capacity compensates more current, and the one with small capacity compensates less current. In this way, we can make full use of each APF and ensure the security of the whole paralleled APF system.

Therefore, the control strategy proposed in this paper has high feasibility in the aspect of more than two modules operating in parallel and the coordination control of switching on APF into or switching off APF from the system. It also has some instructive value on the actual parallel operation of the whole APF system.

REFERENCES

[1] Wang Zhao'an, Yang Jun, Liu Jinjun. Harmonic Suppression and Var Compensation (in Chinese). *China Machine Press*, 1998.

[2] Chen Jian. Power Electronics: Converter Circuit and Control Technology (in Chinese). *Higher Education Press*, 2004.

[3] Dai Ke, Research on Control Techniques for Three-Phase Series-Parallel Compensated UPS with Double Voltage-Source PWM Converters (in Chinese). *Huazhong University of Science and Technology*, 2003.

[4] Ke Dai, Peiguo Liu, Jian Chen et al. "Practical Approaches and Novel Control Schemes for a Three-Phase Three-Wire Series-Parallel Compensated Universal Power Quality Conditioner "(APEC'04). Califorina, USA. 2004, pp. 601 – 606.

[5] S.J .Chiang and J. M. Chang, "Parallel operation of shunt active power filters with capacity limitation control," *IEEE Trans. on Aerospace and Electronic Systems*, vol. 37, no. 4, Oct. 2001, pp.1312-1320.

[6] S. J. Chiang and W. J. Ai, "Parallel operation of three-phase four-wire active power filters without control interconnection," *Proc. IEEE PESC*, 2002, pp. 1202-1207.

[7] Zhuo Fang, He yihong, Li chunyu, Wang Zhao'an. Study on the implementation of large volume active power filter (in Chinese), *POWER ELECTRONICS*, 2001, 4(2):13-15.

2006 5th International Power Electronics and Motion Control Conference

Study and Design of Noninductive Bus bar for high power switching converter

Zhiling Qiu, Hongyan Zhang and Guozhu Chen
College of Electrical Engineering, Zhejiang University, China 310027
gzchen@zju.edu.cn   Tel/FAX: 86057187953985

Abstract—To reduce the stray inductance, consequently the turn-off voltage spike is a key factor for the implementation of high power switching converter. A structure of low/noninductive bus bar with two conductive layers (positive and negative pole of the DC bus), piled together with one high-level insulating epoxy layer, several connection and support parts, and easy to machining is studied in detail. The basic theory/principle of loop inductance calculation and design method in detail of this kind of bus bar are also given. With the demonstration put into a 50 kVA shunt Active Power Filter, the designed bus bar with low stray inductance and voltage spike are achieved, therefore the principle is validated and the structure is feasible and promising for high power converters.

Index terms — *Noninductive bus bar; stray inductance; transient voltage spike; high power converter*

I .INTRODUCTION

As known, all conductors have stray inductance. In power electronic circuits, especially high power hard switching converters, e.g. AC driving inverters and Active Power Filter (APF) built in converters. When switch turns off, high di/dt on the DC bus stray inductance can cause high voltage spike. If the voltage is too high, the power switch and/or other components of the converter will be destroyed. This greatly threatens the safe operation of converters. Therefore, besides of semiconductor component and circuit topology, how to reduce stray inductance and the turn-off voltage spike has actually become a real bottle-neck especially in high power converter implementation.

In small or medium power application, snubber circuit or soft switching technique is usually adopted to constrain turn-off transient spike. Snubber circuits such as C, RC or RCD type [1] have been widely used. Essentially, snubber circuits clamp the DC bus voltage with converting magnetic energy of inductor to electric field energy of capacitor. However, in high power converters, these methods are usually found unfeasible because the energy of the DC bus stray inductor is very large, and results in large size snubber capacitor. What's more, large value capacitor with good frequency and dissipation performance is hard for manufacture.

For the reason that the energy stored in DC bus stray inductor is proportional to the inductance value, reducing the stray inductance is feasible way to restrain the excessive transient voltage in high power cases. Loop area is the main factor determining circuit stray inductance. In conventional bus using wire connection, reducing the length of power circuit is a useful, but challenge occurs in high power converters because the tradeoff between physical size and thermal dissipation requirement. Therefore, the traditional DC bus structure is not easy to achieve low stray inductance.

Another choice to reduce the circuit loop area is to reduce the distance from positive to negative bus. Following this idea, laminated copper plates piled very closely with thin insulating layer inserted in is promising and practical. A bus bar using this structure is built and put into a 50 kVA shunt APF for test. Experiment results show the stray inductance and turn-off voltage spike characteristics are very good.

II. PRINCIPLE ANALYSIS OF NOINDUCTIVE BAUSBAR

A circuit stray inductance is mainly determined by loop area. It can be explained below. For simplicity, the inductance of uniform magnetic field is illustrated firstly, which is tenable for each small element in a large nonuniform field.

Figure 1 shows a rectangular one-turn air inductor. The inductance between point A and B can be calculated as,.

$$L = \frac{\Psi}{I} = \frac{NBS}{I} \tag{1}$$

Where, L is the inductance, I is the current, Ψ is the flux linkage, N is the turn number, B is the magnetic flux density, S is the crossing area

Due to the magnetic loop Ohm equation

$$F = NI = R_{m0}\Phi = \frac{l_0}{\mu_0 \cdot S}\Phi = \frac{B}{\mu_0}l_0 \tag{2}$$

where, F is the magnetic potential, R_{m0} is the magnetic reluctance, Φ is the magnetic flux, l_0 is the length of magnetic path, and μ_0 is air magnetic permeability.

Then, equation (3) can be deduced,

$$B = \frac{NI\mu_0}{l_0} \tag{3}$$

1-4244-0448-7/06/$25.00 ©2006 IEEE

According to equation (2) and (3)

$$L = \frac{N\mu_0}{l_0} S \quad (4)$$

In Figure 1, $N = 1$, then

$$L_{AB} = \frac{\mu_0}{l_0} S \quad (5)$$

The approximate formula to calculate L_{AB} is [2]

$$L_{AB} = \frac{\mu_0}{\pi} \left(l \cdot \ln\left(\frac{2d}{w}\right) + d \cdot \ln\left(\frac{2l}{w}\right) \right) \quad (6)$$

According to equation (5), the circuit inductance L_{AB} is proportional to the loop area. The stray inductance of traditional DC bus using wire connection is large because the distance between the positive and negative bus is long,

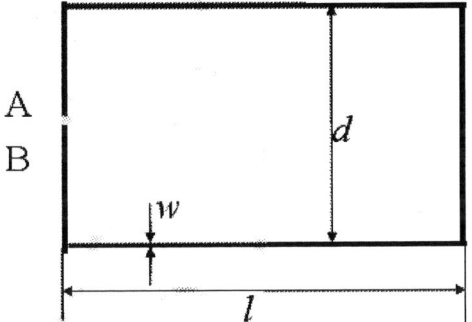

Figure.1 Rectangle loop and dimensional quantities

i.e. the power circuit loop area is difficult to be reduced.

In order to obtain low bus inductance required by high current applications, special bus structures are demanded. Therefore laminated bus bar with very low inductance is under hot study. A kind of laminated bus structure is shown in Figure2, which contains two wide copper plates and one thin epoxy plate. The copper plates act as the positive and negative bus, while the epoxy one poses an insulating layer. Because the insulating layer is very thin, low DC bus stray inductance can be achieved.

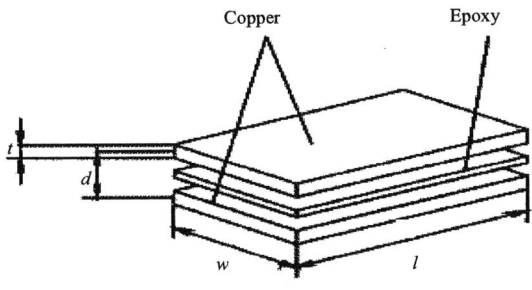

Figure.2 Dimensional quantities of laminated bus

Where, l is the length, w is the width, t is the thickness of the copper plates, and d is the distance between the two copper plates of the laminated bus bar.

The inductance of this bus bar consists of two portions: ① the internal inductance (L_i) which flux only linkages within one copper plate; ② the external inductance (L_e) which flux flows through the space area surrounding by the two copper plates.

The internal inductance depends on frequency. At high frequency, the inductance can be ignored. For low frequency, the value can be calculated as [3]

$$L_i = \frac{\mu_0 \mu_r}{8\pi} l \quad (7)$$

Where μ_r is relative permeability.

According to equation (7), the internal inductance (L_i) is proportional to the bus length (l) and not related to the distance between the two copper plates. It is indicated that although in the case of this laminated bus reducing its circuit length is important.

An exact equation to calculate the external inductance is difficult to be achieved. Under the conditions of $d \ll 2t$ and $d \ll t+w$, an approximate inductance value [4] can be calculated, i.e.

$$L_e = \frac{2\mu_0 \mu_r l}{\pi} \ln\left(1 + \frac{t}{t+w}\right) \quad (8)$$

If $t \ll w$, equation (8) can be simplified as

$$L_e = 2\mu_0 \mu_r \frac{tl}{\pi(t+w)} \quad (9)$$

Because the epoxy plate layed used in this laminated bus is very thin, formula (9) has an acceptable precision. From Equation (7) and (9) it can be concluded that a laminated bus bar, as shown in Figure 2, with short power circuit length (l), small distance (d), and/or large bus width (w) is possible to minimum both internal and external inductance, results in small loop inductance.

III. DESIGN AND IMPLEMENTATION OF THE NONINDUCTIVE BUS BAR

There are several ways to design the noninductive bus bar. Generally, they differ from two or three-conductor layers, plain or three-dimension mounted structure, a whole punched unit or consisting of connected parts. For simplicity to machining in laboratory, a two-conductor-layer (positive and negative of the DC bus), pile-on-plain mounted, connected with several blocking parts structure noninductive bus bar is chosen in this paper. Top view and connection structure are shown in Figure 3 (a) and (b) respectively.

According to the analysis in previous section, larger copper plate width (w) is better, which is chosen as 200 mm here. Considering DC bus current density and rigidness of the laminated bus, the thickness of copper plate is 1.2 mm. Power circuit length should be as short as possible so that the power modules and dc bus capacitance are piled closely to each other, as shown in Figure 3. Considering IPM width and dc bus capacitance diameter, the laminated bus length is 470 mm.

To minimize dc bus loop area, the thickness of insulation layer should be thin. In literature [5], a kind of

795

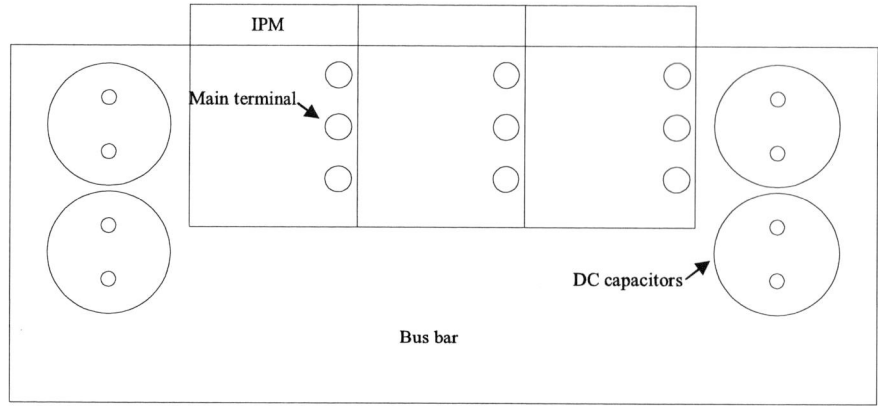

(a) Main circuit layout (top view)

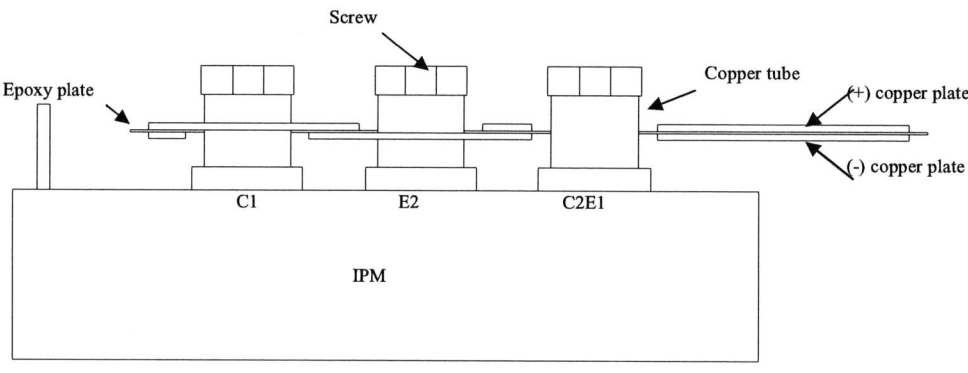

(b) Side view of bus bar connection (each module)

Figure.3 Connection structure of designed bus bar

insulation material Kapton HN with thickness of $25\mu m$ is used. This material is very thin but it requires the copper plate surface very smooth. In this paper, epoxy plate with a thickness of 0.4mm is used. The main advantages of epoxy plate include good insulation level, easy obtained, cheap and outstanding mechanical characteristics, therefore guarantee the circuit reliability.

On the other hand, insulation is apparently very important to bus bar's operation. Besides of sufficient insulation level provided by the epoxy layer, the creepage/clearance distance should be carefully considered. The edge of the epoxy plate is designed to stretch in 3 mm from the copper ones. Outside surface of the two copper plates and the connection copper tubes (used as IPM screw terminal gaskets and boots) all are covered with insulation glue.

Although the laminated bus structure seems simple, challenge still exists including the mechanical precision. In Figure 3 (b), the machining precision of the holes' location, the length of copper tubes, the flatness of the end surfaces of the tubes, and their consistency do dominate reliable electrical connection. Digital punch machine, lathe etc. are used to drill, cut, grind or mill the parts in the laboratory.

Using formula (7) and (9), internal and external inductance can be calculated separately

$$L_i = \frac{\mu_0 \mu_r}{8\pi} l = 23.5\text{nH} \tag{10}$$

$$L_e = 2\mu_0 \mu_r \frac{tl}{\pi(t+w)} = 0.45\text{nH} \tag{11}$$

Apparently, the stray inductance of the laminated bus bar is very small.

IV. EXPERIMENTAL RESULTS

The designed noninductive bus bar is built and demonstrated in a 50 kVA shunt active power filter (APF) industrial prototype. Because the output frequency range of the APF converter is generally much higher than most other ones, e.g AC driving inverter, bus bar with lower stray inductance is more necessary. Main intelligent power module (IPM) is Mitsubishi PM300DVA120. Snubber capacitor, $3.3\,\mu F$ for each leg, is still used to absorb the small stray inductance energy of the converter.

The bus bar pictures without and with other components are shown in Figure 4 and 5 respectively.

Figure 6 and 7 show the turnoff transient voltage spike of the conveter using trditional DC bus and designed noninductive laminated bus bar respectively.

796

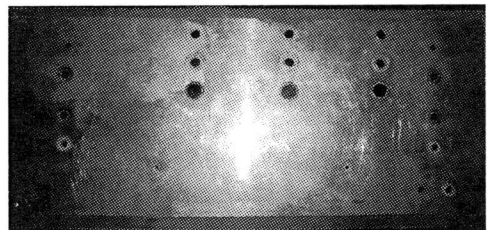

Figre.4 Laminated bus under construction

Figure.5 Complete converter

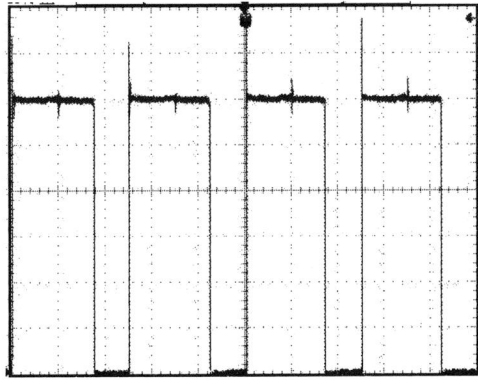

Figure.6 Turn-off voltage spike on IPM using traditional DC bus.
Voltage scale: 100V/div Time scale: 40 μs /div

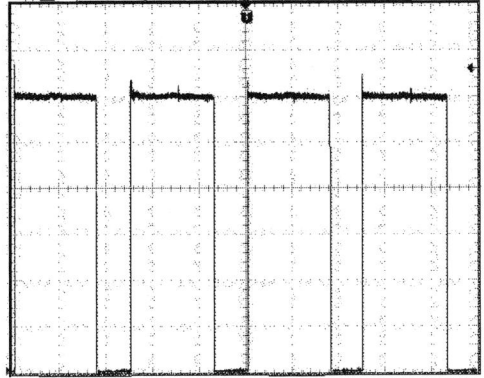

Figure.7 Turn-off voltage spike on IPM using the designed
noninductive bus bar
Voltage scale: 100V/div Time scale: 40 μs /div

It can be seen: in the case of wire connection DC bus, maximal voltage spike is 180 V additional to the DC bus volatage. However, with noninductive bus bar, maximal voltage spike is only 60V, which is 30% of the previous one. The voltage spike suppression using the designed

bus bar technique for high power converter is dramatically.

V .CONCLUDION

Low/non-inductive bus bar is important for the implementation of high power switching converter because of the reduction of turn-off voltage spike.

A structure of noninductive bus bar with two conductive copper layers, one insulating epoxy layer, and easy to machining and assemble is analyzed and designed. With the demonstration put into a 50 kVA shunt Active Power Filter (APF) industrial prototype, this kind of bus bar has the characteristics of low stray inductance ($\approx$ 25nH) and acceptable voltage spike (60 V i.e. 30% of the contrast one), which makes the APF prototype practically successful.

The basic theory/principle and the simple structure of the designed noninductive laminated bus bar could be adopted in most high power converter's application.

REFERENCES

[1] Mitsubishi Electric, *The 5th Generation IGBT Modules & IPM Modules Application Notes*, 2005, pp. 36

[2] F. B. J. Leferink, "Inductance calculations; methods and equations," in Symposium Rec. 1995 IEEE Int. Symposium on Electromagnetic Compatibility, pp. 16-22

[3] C. R. Paul, *Introduction to Electromagnetic Compatibility*. New York: Wiley, 1992, pp. 247–250.

[4] L. J. Giacoletto, *Electronics Designers' Handbook*, 2nd ed. New York: McGraw-Hill, 1977, pp. 3–44.

[5] Marco Chiadò Caponet, Francesco Profumo, Rik W. De Doncker, and Alberto Tenconi, "Low Stray Inductance Bus Bar Design and Construction for Good EMC Performance in Power Electronic Circuits", IEEE TRANSACTIONS ON POWER ELECTRONICS, VOL. 17, NO. 2, 2002, pp.225~231

2006 5th International Power Electronics and Motion Control Conference

A New Minimum Torque-ripple and Sensorless Control Scheme of BLDC Motors Based on RBF Networks

Juan Wang, Hongwei Liu, Yuran Zhu, Bo Cui, Huijuan Duan
Hebei University of Science & Technology / Science & Engineering of Information, Shijiazhuang, China
Email: wjzyq@126.com

Abstract—In this paper, a new method based on adaptive Radical Basis Function (RBF) networks is proposed to deal with the issues of rotor position requirement and high torque ripple production in a brushless DC (BLDC) motor. Two RBF networks are trained offline with a self-adjustment growing and pruning (GAP) algorithm, and all the train samples are obtained from experimental results to express the nonlinear characters of the machine more accurately. One of the trained networks is used to realize a nonlinear mapping between external voltages, phase currents and rotor position of the motor, at the same time the other is applied for estimation of phase current references with a desired torque. Actual phase currents are adjusted according to the references; therefore the torque ripples generated by non-ideal current waveforms is minimized for a BLDC motor without position sensors. Simulation results show the efficiency of this proposed method.

Keywords-BLDC motor; sensorless; torque-ripple; RBF

I. INTRODUCTION

Because of their high efficiency and power density, easy control and low maintenance, permanent magnet BLDC motors have been widely used in a variety of applications in industrial automation and consumer electric appliances. However, there are two drawbacks of this kind of motor drives. First, in order to run a BLDC motor successfully, the information about the rotor position is necessary to properly perform phase commutation. Position information is generally provided by encoders or resolvers, which increase the costs and reduce the ruggedness and simplicity of the motor. Second, the characteristic torque ripple caused by commutations of current from one phase to another, non-trapezoidal back-EMF, and cogging also limits its applications in high accurate servo systems. To be dead against these drawbacks, many researches about sensorless control and torque ripple minimization have been done [1-7].

In some papers, the researchers give their effort to eliminate rotor position sensors [1-3], and others do their best to improve the torque character of BLDC motors [4-7]. However, researches about both of them are very few.

Artificial neural networks offer an alternative to conventional controllers. They are able to adaptively model or identify a non-stationary nonlinear MIMO process/plant on-line while the process is changing, and thereby yield information that can be used by another neural network to control the process. Radical Basis Function (RBF) networks as a new kind of neural networks have gained much popularity in recent times due to their ability to approximate complex nonlinear mappings directly from the input–output data with a simple topological structure [8-10].

In this paper, a novel control strategy is presented in order to overcome the two drawbacks of BLDC motors synchronously. After analysis of the sensorless control and torque ripple generation principles, two RBF neural networks trained with a self-adjustment growing and pruning (GAP) algorithm are employed to realize a nonlinear mapping from voltages and currents to rotor positions of the motor and a mapping from torque references and rotor positions to reference currents respectively. The double RBF networks control strategy is validated by a BLDC motor with eight poles, and experimental results show efficiency of the scheme.

II. TORQUE RIPPLE ANALYSIS

The main causations, which generate torque ripples in a BLDC motor are cogging, electromagnetic factor and commutations of the phase currents. The cogging can be eliminated by skewing, and the result is very well. The other reasons are consist of the main part of torque ripples and they are both caused by the un-ideal currents, which are the focus in this research.

A. Mathematic Model of the Machine

The BLDC motor in this paper is a three-phase, eight-pole machine. The voltage equation of the brushless dc motors is given by

---

This work is supported by the Tianjin Natural Science Fund #013800811

1-4244-0448-7/06/$25.00 ©2006 IEEE

$$v_x = Ri_x + (L-M)di_x/dt + e_x \qquad x = a,b,c \qquad (1)$$

where v_x, i_x and e_x are the x phase voltage, current and EMF respectively; R, L, M are the resistance, inductanceand mutual inductance of every phase winding respectively.

The electromagnetic torque of the BLDC motor is as follows:

$$T = (e_a i_a + e_b i_b + e_c i_c)/\omega_m \qquad (2)$$

where ω_m is the mechanical angle speed of the machine.

From (2), it is seen to keep a constant torque the sum of $e_a i_a$ $e_b i_b$, and $e_c i_c$ must be a constant at a certain speed. Therefore, if the waveforms of phase back EMF and phase current are perfectly matched, torque ripples are minimized.

B Torque Ripple Caused by Electromagnetic Reasons

On the assumption that the airgap magnet field of the machine has an ideal trapezoidal waveform, the waveform of phase currents must be an ideal rectangle with same phase-angle as EMF to keep a constant electromagnetic torque. Whereas, there are inductances in the armature windings, so a sudden change of the currents is impossible, and phase currents have to vary according to a waveform like a trapezoid. Then torque ripples generate unavoidably.

C Commutation Torque Ripple

This part of torque ripples is caused by current pluses of the phase that doesn't commutate.

In the 3-phase BLDC motor, there are two phases excited at any time. Suppose that the conducting phases are a and c, then

$$i_c = -i_a, i_b = 0, e_c = -e_a$$

And the formula (2) can be expressed as:

$$T = 2e_c i_c / \omega_m \qquad (3)$$

define $\quad i_c = I, e_c = E, T_0 = T = 2EI/\omega_m$.

Suppose the commutation is between phase a and b, a is turned off and synchronously b is turned on. For i_a can't clear away immediately, the fading current has to flow through the flowing diodes. In this period, i_b increases from zero to I, and following equation is satisfied at any time:

$$i_a + i_b + i_c = 0 \qquad (4)$$

After the commutation, $i_a = 0$ and $i_c = -i_b$. The waveforms of the currents during commutations in different cases are shown in Fig.1.

From (3), T (torque) can be considered directly proportional with $e_c i_c$. During the commutation, if $t_1 > t_2$ then the amplitude of i_c is bigger than I, as in

Fig.1(a); If $t_1 = t_2$, the amplitude of i_c is keeping I, as in Fig.1(b); if $t_1 < t_2$, the amplitude of i_c is smaller than I as in Fig.1(c). Then the following conclusions can be drawn during the commutation, where u and E is the external voltage and the EMF of the windings respectively.

1) If $u > 4E$, then $t_1 > t_2$ and $T > T_0$ (see Fig.1 (a));
2) if $u = 4E$, then $t_1 = t_2$ and $T = T_0$ (see Fig.1(b));
3) if $u < 4E$, then $t_1 < t_2$ and $T < T_0$ (see Fig.1(c)).

Under the condition $u = 4E$, the change of i_c can be avoided, and therefore commutation torque ripples would not exist. However, this state isn't steady as the machine is accelerating, and EMF is getting larger and larger with an increasing speed so that the external voltage u will be smaller than $4E$. In a word, commutation torque ripples exist at a steady state and vary with the rotor speed.

III. TORQUE RIPPLE MINIMIZATION STRATEGY

A Proposed Self-adjustment GAP-RBF Algorithm

RBF network is a three-layer network. In this kind of networks the input values are each assigned to a node in the input layer and passed directly to the hidden layer without weights. The nonlinear hidden layer nodes with a Gaussian density function as an activation function are called RBF units. There are linear weights between the hidden layer and output layers. So this kind of networks has a good whole character with a very simple structure.

In a general way, for a one-output RBF network with a Gaussian function as its activation function, the output of the kth hidden unit is as follows:

$$\Phi_k(X_i) = \exp(-\|X_i - c_k\|^2 / 2\sigma_k^2) \qquad (5)$$

Where, X_i is the i th input sample and c_k, σ_k is the center and width of the kth hidden unit. The overall input-input mapping f:

$$y_i = \sum_{k=1}^{n} \omega_k \Phi_k(X_i) \qquad (6)$$

And ω_k is the weight between the output layer and the kth-hidden unit and n is the number of the hidden neurons.

In the RBF network employed here, there are no hidden neurons at first, and they increase according to a self-adjustment GAP (growing and pruning) algorithm during the offline training, at the same time the existing hidden units which have very small outputs for a certain number of sequential inputs will be deleted to keep a simple and efficient structure. The adaptive algorithm is given below.

For every new sample (X_i, t_i)

1) Calculate $\varphi_k(X_i)$ and y_i according to (5) and (6) respectively;

Calculate the error e_i, where

$$\|e_i\| = \|t_i - y_i\| \qquad (7)$$

799

and the distance between the sample and the existing hidden neurons is

$$d_j = \left\| X_i - C_j \right\| \quad j=1,2...n \qquad (8)$$

where n is the number of existing hidden units.

Define $d_{\min} = \min(d_j)$, and define ε as the expect error of the network and $\lambda(i)$ as the approximate accuracy of the network with the ith input, which decreases from $\lambda_{\max}$ to $\lambda_{\min}$ during the process of learning.

2)If $\left\| e_i \right\| > \varepsilon$, $d_{\min} > \lambda(i)$ where $\lambda(i) = \max(\lambda_{\max}\gamma^i, \lambda_{\min})$, and γ is the decline coefficient which value is between 0 and 1. Then add a new hidden unit to the existing network,and its parameters are defined as follows:

$$C_k = X_i \qquad (9)$$

$$\sigma_k = \frac{1}{p}(\sum_{j=1}^{p} \left\| X_i - C_j \right\|^2)^{\frac{1}{2}} \qquad (10)$$

where C_j are the centers of the hidden units which are the nearest with the input sample, and here $p = 2$;

3) Else, the weights of the network are adjusted with a recursive lease square (RLS) algorithm;

4) If all of the n sequential input samples are satisfied the following inequality,

$$\left\| w_k \varphi_k (X_i)/ y_i \right\| \le \delta \qquad (11)$$

where, δ is a constant predefined. Then delete the kth hidden neurons (here $n = 100$);

5) Input a new sample and repeat the first step.

B Control Scheme Based on Double RBF Networks

The control system is shown in Fig. 2. In this scheme, one RBF network is used to realize the nonlinear mapping from the external voltages and phase currents of the machine to the rotor position information, and the network is trained according to the given algorithm offline.

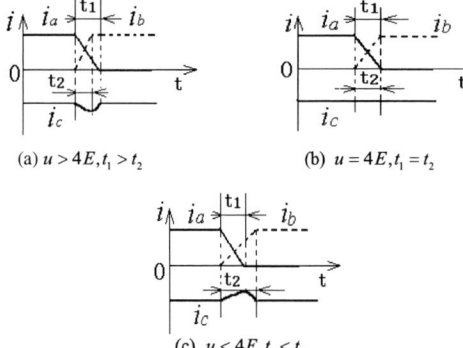

(a) $u > 4E, t_1 > t_2$ (b) $u = 4E, t_1 = t_2$

(c) $u < 4E, t_1 < t_2$

Fig.1. Waveforms of phase currents during commutation in different cases

All the training samples are obtained from the experimental results. Therefore it is believed that the trained network is the exact model of the motor and it can be used for the rotor position estimation on line. The other RBF network is trained with the same algorithm to obtain a mush simple and tight structure. It is employed here to realize the nonlinear mapping from the rotor position and the referent torque to the referent currents. As it is known, in a 3-phase, 6-state and Y connected BLDC motor, there are only two phases excited at any time. If suppose that the waveform of the EMF is ideal trapezoidal, the EMF varies with the rotor position and its rate of change with time, i.e. rotor speed. The reference currents without torque ripples can be calculated at the condition that the torque is given and the motor is running with a max torque so that the currents without torque ripple are a function of the torque and the rotor position, and this function can be realized by a RBF network. Comparing the actual phase currents with the references, and the errors are processed with a PI adjustor to drive the invertors. So the currents in the stator armatures change with the references, and torque pulses are reduced.

IV. EXPERIMENTAL RESULTS

To validate the performance of the proposed control scheme an experimental setup is employed, which is shown in Fig.3. In the experimental system, A DSP (TMS320LF2407) is used to implementation of the proposed althorighm. The parameters of the machine are given below:

U_N=36V; I_N=6A; T_N =0.4N.m; P_N=150W;
n_N = 3600 r/min; $R = 0.66\Omega$;
$L- M$=1.4mH; K_e=0.067V/(rad/s);
J=1.57×10^{-5}kg.m^2 ; p=4

The experimental result of a phase current with one RBF network to estimate the rotor position is shown in Fig.4. And the torque generated under this condition is shown in Fig.5.In this case, the motor is running properly without position sensors. However, the torque ripple is significant, and the amplitude of the pulse almost is 30 percent of the average torque.

Fig.6 shows the phase current with double RBF networks. It is seen that the waveform of the current is improved and in this case, the amplitude of the torque ripple is reduced to 4 percent of the average torque as in Fig.7.

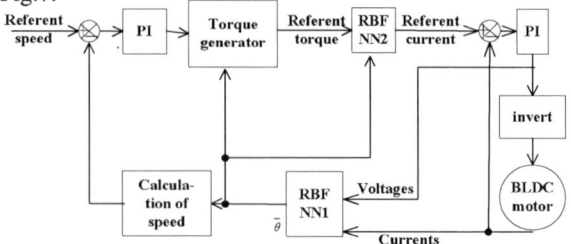

Fig.2. Control scheme with two RBF networks

800

V. CONCLUSIONS

In this work, two RBF networks are employed to realize the sensorless control and estimate the referent currents according to the reference torque at max-torque running. Then the armature currents are adjusted in accordance with the references, and the torque ripples are reduced. Therefore, the two drawbacks of the BLDC motors are overcome at a certain extent, and the performance of the motor is improved. The RBF networks are trained with an adaptive growing and pruning algorithm to obtain a much simple and tight structure.

The simulation results show the efficiency of the control scheme given in this paper. It is seen that with the given scheme, the BLDC motor can get a good torque character without position sensors, and the applications of the motor are enlarged. However, the original rotor position can't be obtained before starting the motor, so the starting performance is limited with this control scheme, and further research to ameliorate the starting performance is much needed.

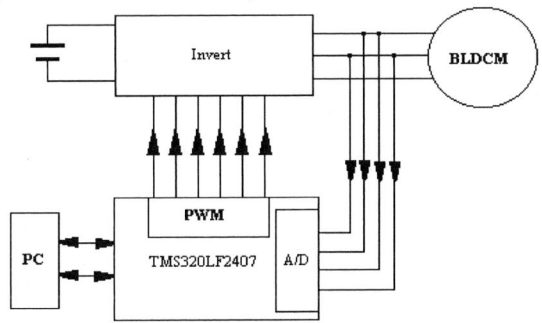

Fig.3. Experimental system

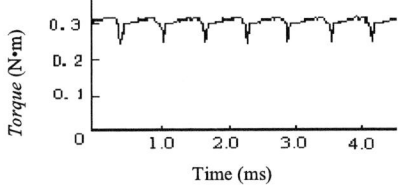

Fig.4. Current waveform of phase A without estimation of reference currents at the rated speed

Fig.5. Torque waveform without estimation of reference currents at the rated speed

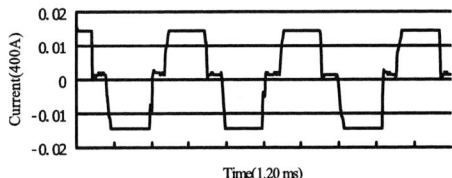

Fig.6. Current waveform of phase A with estimation of reference currents at the rated speed

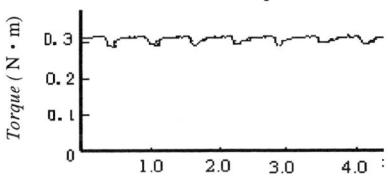

Fig.7 Torque waveform with estimation of reference currents

REFERENCES

[1] Zhiqian Chen, Mutuwo Tomita, Shinji Doki, Shigeru Okuma. "New adaptive sliding observers for position- and velocity-sensorless controls of brushless DC motors," *IEEE Trans. on Industrial Electronics*, 2000, 47(3), pp: 582-591.

[2] Changliang Xia, Juan Wang, Tingna Shi, Wei Chen, Shaohui Xu, Rong Yang. "Direct control of currents based on adaptive RBF neural network for brushless dc motors," *Proceedings of the CSEE,* vol 23, No.6, pp: 123-127,2003.

[3] Changliang Xia, De Wen, Fan Juan, Yang Xiaojun, "Based on RBF neural network position sensorless control for brushless DC motors," *Transactions of China electrotechnical society*, 2002,17(3), pp: 26-29(76).

[4] F.Bodin, S.Siala, "A new reference frame for brushless DC motor drive," *IEE Conference Publication* n 456, Sep 21-23, 1998.

[5] R.Carlson, M.Lajoie-Mazenc, J.Fagundes, "Analysis of torque ripple due to phase commutation in brushleses DC machines," *IEEE Trans. on Industry Application*, vol.28, No.3,pp. 632-638, 1992.

[6] Y.Kim, Y.Kook, Yo Ko, "A new technique of reducing torque ripples for BDCM drives," *IEEE Trans. on Industrial Electronics*, vol 44, No.5, pp.735-739, 1997.

[7] Changliang Xia, De Wen, Juan Wang, "A new approach of minimizing commutation torque ripple for brushless DC motor based on adaptive ANN," *Proceedings of the CSEE*, vol22, No.1, pp. 54-58, 2002.

[8] Shiqian Wu, Meng Joo Er, "Dynamic fuzzy networks----a novel approach to function approximation," *IEEE Trans. on System*, 2000, 30(2): 358-364.

[9] Erkan Mese, David A. Torrey, "An approach forsensorless position estimation for switches reluctance motors using artificial neural networks," *IEEE Trans. on Power Electronics*, 2002,17(1): 66-75.

[10] Guang-Bin Huang, Narasimhan Sundararajan, "A Generalized Growing and Pruning RBF (GGAP-RBF) Neural Network for Function Approximation," *IEEE Trans. on Neural Networks,* vol. 16, No. 1, pp: 57-67, January 2005.

2006 5th International Power Electronics and Motion Control Conference

Improved Modelling and Calculation on Electromagnetic Transient of Power Transformer

Chen Zhe [1*], Wen Yuanfang [1], Lu Guojun [2]

[1] College of Electrical and Electronic Engineering, Huazhong University of Science and Technology
Wuhan 430074, P. R. China
[2] Guangzhou Power Supply Company, Guangzhou 510620, P.R. China
* E-mail : chenzhe311@eyou.com

*Abstract--***Two novel methods, digital calculation based on Callahan algorithm and dynamic simulation based on MATLAB/SIMULINK software, are proposed in this paper to study the electromagnetic transient process caused by asynchronous energization of electric power transformer while taking non-linearity into consideration. The effectiveness and accuracy of the results are verified by each other. Then regularity of the inrush current generation is found and harmonics analysis by Fourier Transform is done to provide relay protection criterion. Furthermore, different conditions influencing inrush current such as the saturation characteristics, residual flux and switching on time are investigated. Finally, measures to eliminate the inrush current are suggested to avoid similar electrical breakdown accident. This paper is of great value for the studies on the dynamic process within the power system, the precise design and smooth operation of high voltage apparatus, and the proper selection for operation values of relay protection equipment for the electrical facilities.**

*Index Terms--*Callahan numerical algorithm, Dynamic simulation, Inrush current, Electromagnetic transient process

I . INTRODUCTION

Electric power transformers play an important role in the stable operation of power system, so it is quite important to guarantee that no heavy electric accidents and no giant damages happen to cause their break down. An extensive research on transient process related to them has already been done on this subject[1]-[2](as shown in Fig.1), however, most of them are based on the practical tests and on-spot experiments, so numerical calculation and digital simulation based on modern methods are also necessary to be carried out to avoid plenty of labour and vast of money spent on dynamic analogue test.

Thus it is the main aim of this paper to propose two novel methods—one is the Callahan digital calculation method, the other is the MATLAB/SIMULINK dynamic simulation method; to resolve the existing problem of prediction and prevention of disaster. Each method will be illustrated in detail to show their advantage and efficiency. Moreover, some regularities concerned with inrush current generation will also be

focused on.

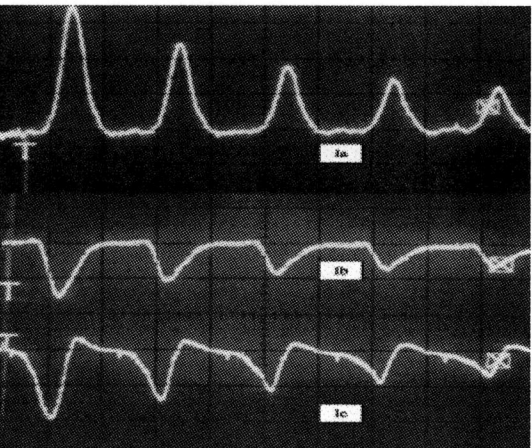

Fig.1 Test results for inrush currents[1]

II . PRINCIPLE OF INRUSH CURRENT GENERATION

When an unloaded transformer is operated to connect to the electric bus bar, the topology of the equivalent electrical circuit is therefore changed and it will lead to the state mutation. From the classical equation below (consider the winding inductor L is linear):

$$\phi(t) = \phi_m \sin(\omega t + \alpha) + (\phi_r - \phi_m \sin \alpha)e^{-\frac{R_1}{L_1}t}$$

we may get to realize, that the magnetic flux imposed by the remnant magnetism ϕ_r can go as high as $2\phi_m + \phi_r$ on condition that the initial energization angle $\alpha = 90°$, which will make the transformer inductors extremely saturated and will result in the high magnitude sharp-top inrush current. The generation principle can also be comprehended through Fig.2 We can conclude that the magnetic inrush current is composed of high order harmonics, such as the 3rd harmonic, the 4th, and so on. On the other hand, there are some countermeasures to eliminate them, such as

(1) Install over-current relay protection on the primary side of transformer to avoid the most destructive effect of the inrush current;

(2) Connect in series a certain resistance R to reduce the shock of the inrush current and accelerate the decay while switching on but remove the R when the current

1-4244-0448-7/06/$25.00 ©2006 IEEE

goes into the steady state.

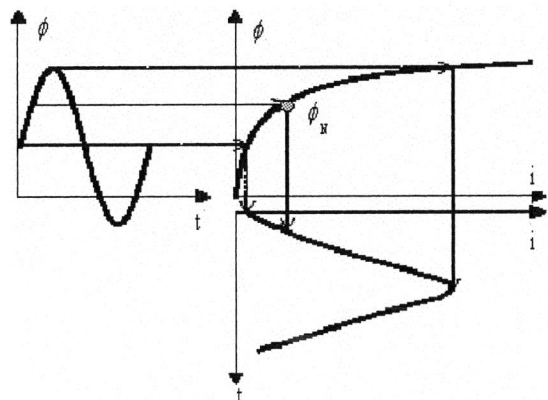

Fig.2 Schematic diagram of generation of magnetizing inrush current

III MODELS AND METHODS

In order to study the detailed process described above, concrete models by two novel methods are introduced afterwards:

A. Callahan Method

On careful parameters calculation on the electrical source system, transmission line, non-linear inductance, suction capacitance and the distributed capacitance according to the practical electrical circumstance, we can obtain the numerical calculation model shown in Fig.3

Usually the magnetic curve for the transformer inductor is simply divided into two parts or two lines to be piece-wisely represented, but it is less accurate, sometimes even unacceptable. Also, there are some precise mathematical functions to fit the saturated $\phi - $ i curve:

$$i = A \sinh B\phi$$

$$i = \sum_{k=0} a_{2k+1} \phi^{2k+1}$$

$$i = \frac{\phi}{a\phi + b}$$

$$i = A\phi + B \tan^{-1}(\frac{\phi}{k})$$

The hyperbolic sine function $i = AshB\phi$ see Fig.4 is chosen to precisely simulate the transformer inductor in this model for two reasons:

(1) only two coefficients A and B have to be set, meanwhile the hyperbolic sine function is monotonically rise, which make the fitting and programming easier.

(2) the results show the fitted curve is closer to the real magnetic curve, thus makes the computed digital results

closer to practical values.

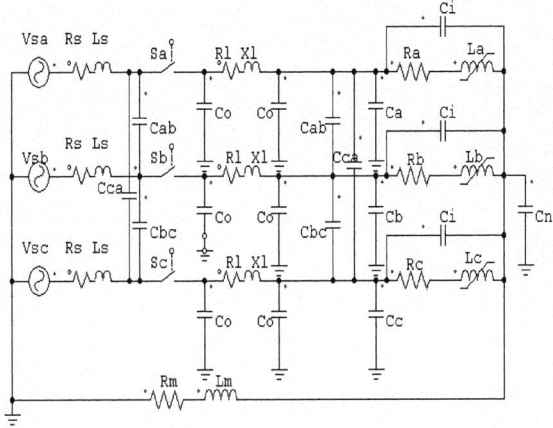

Fig.3 Equivalent circuit for three-phase circuit numerical calculation

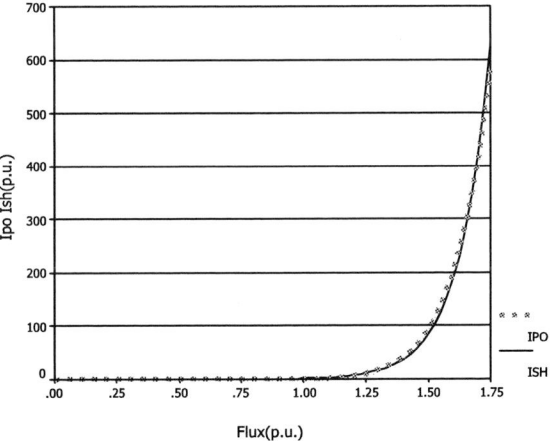

Fig.4 $I - \phi$ fitted curves by interpolation and Least-Square methods respectively

The adopted Callahan algorithm in this paper is a semi-inplicit, noniterative but explicitly recursive numerical analysis method, and it is highly stable and fit for solving high-order nonlinear circuit. The formula is as follows:

$$\begin{cases} \dfrac{dy}{dt} = f(y,t) \\ y(t_0) = y_0 \end{cases}$$

803

$$y_{n+1} = y_n + 0.75k_n + 0.25l_n \; ;$$

$$k_n = \Delta t*[I - \Delta t*a*(\frac{\partial f}{\partial y})_n]^{-1}*[f(y_n, t_n) + a*\Delta t*(\frac{\partial f}{\partial t})_n]$$

$$l_n = \Delta t*[I - \Delta t*a*(\frac{\partial f}{\partial y})_n]^{-1}*[f(y_n + b*k_n, t_n + b*\Delta t) + a*\Delta t*(\frac{\partial f}{\partial t})_n]$$

where, I, the unit matrix,
a=0.788675134595,
b=−1.15470053838

n represents values at the time t_n ,

n+1 represents values at the time t_{n+1} ,

Δt signifies the step length,

$\frac{\partial f}{\partial y}$ is partial derivative of f, named Jacobi matrix.

It can be approved that the Callahan algorithm is a three-order numerical method, which can fully meet the demands of engineering calculation. If associated with the Richardson's extrapolation method, the main error term can be deleted, then it is one order higher. What is more, the step length does not affect the result much, so it is quite suitable for solving non-linear state equations, compared with the Runge-Kutta algorithm and the others.

Regarding the randomness of the switch-on time, original source angle and time intervals of the three phases, it is supposed in this paper that phase A is firstly switched on at 0.05s, while phase B and phase C are energized 0.01s afterwards. By Callahan method, we can program to get the computed results for the three-phase inrush currents shown in Fig.5

In an easy and visualised way, a simulation model is constructed in Fig.6 and the corresponding solution is demonstrated in Fig.7, which is quite consistent to Fig.5 To point out, the three-phase saturated inductors are piece-wisely treated as a two-segment linear inductor, that is to say, when the current is low, the linear inductance equal to L_1 is very high; on the other hand, it is extraordinary low equal to L_3 when the current is high and the inductor is deeply saturated; L_2 is omitted. This effect can be understood from the rate of slope in the $\phi - i$ relationship diagram in Fig.8

Besides the simulation for the three-phase inrush currents, MATLAB/SIMULINK simulation can also simulate the harmonic components of the inrush current. Investigations of system resonance are helpful in industrial systems and transformer energizations. High over-voltages could result is the system is sharply tuned to a harmonic that is being excited by the transformer inrush current, 2nd, 3rd, 4th, or 5th harmonic, which can lead to plant shut down[2]. Fig.9 illustrates, 2nd, 3rd, 4th, and 5th harmonics are the most significant harmonics compared with the others, which are 114, 63, 42 and 34 times of the peak value of the no-load current I_0 respectively, and they descend as the whole inrush current damps oscillatory. It can be seen that the 2nd harmonic is the most predominant part and the 3rd harmonic is high than the 4th at the first period of energization but it also decays faster. Therefore, it becomes imperative to design some electrical filter (Analogue or digital), such as the Butterworth filter, Chebyshev filter and etc., accurately to eliminate the undesired harmonics to ensure the power quality of the electric network and guarantee the smooth operation of the transformer, otherwise it may crumble to result in huge loss for the industrial production and people's living.

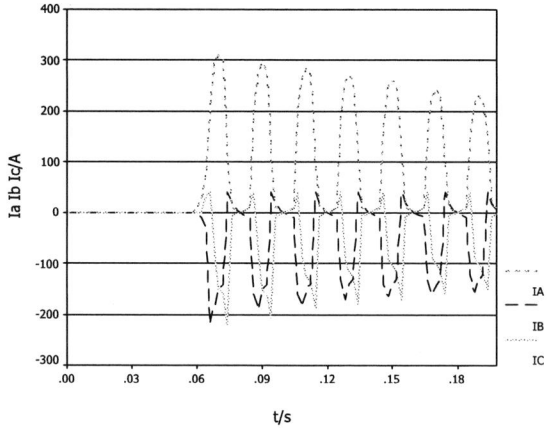

Fig.5 Computed inrush current waveforms

B. MATLAB/SIMULINK Simulation Method

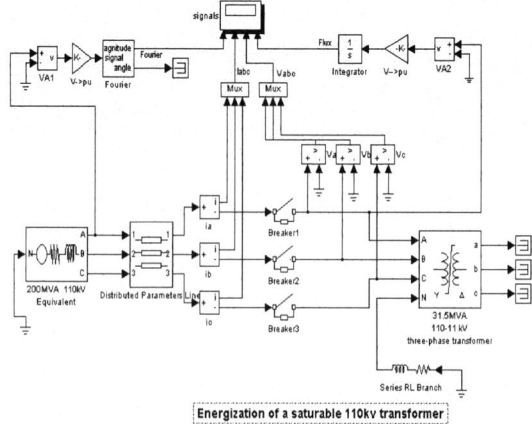

Fig.6 The equivalent simulation circuit of the unloaded transformer

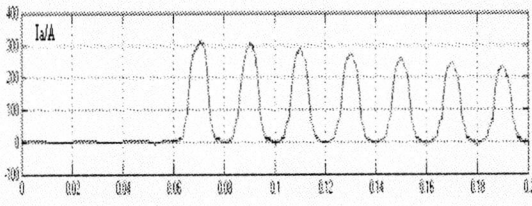

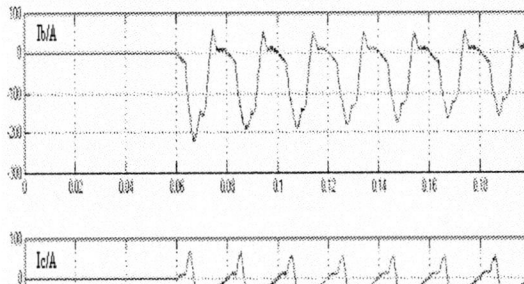

Fig.7 Simulated inrush current waveforms

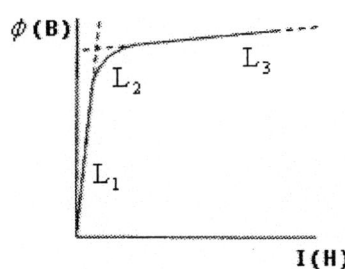

Fig.8 Typical saturation characteristic curve

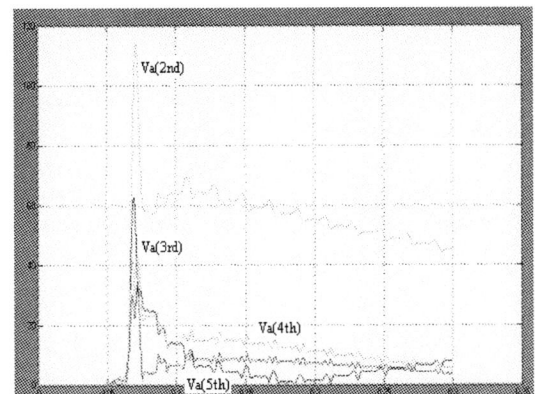

Fig.9 2nd, 3rd, 4th, 5th harmonics of phase A current

IV. RESULT ANALYSIS

Make a good comparison of the computed result and the simulated result, we can get the relationship between them, demonstrated in Tab.1. We can see that the

relative error is within the scope of 2%, and the two kinds of results are very close to each other in spite of the system error. That means, the effectiveness and rightness of one method is authenticated by the other one, and the two methods are equally right.

Furthermore, to show their respective characteristics more clearly，the merits and demerits of the above two methods are elaborated in Tab.2 in this paper.

Tab.1 Comparison of the two kinds of results

Absolute peak value	Ia/A	Ib/A	Ic/A
Callahan algorithm	320.61	-223.21	-227.19
MATLAB/SIMULINK simulation	316.43	-220.18	-223.68
Relative error(%)	1.30	1.36	1.54

Tab.2 Comparison of the two methods

Method	Callahan algorithm	MATLAB/SIMULINK simulation
Merits	Precise simulation parameters Closer to reality	Easy and visualized Powerful simulation Save time
Demerits	Modeling and programming time- costing	No precise non-linearity; certain deviation exists

V． FACTORS INFLUENCING INRUSH CURRENT

As the magnetizing inrush current has a close relationship with inductor non-linearity and hysteresis[5] of the iron core, different operating conditions such as residual flux and switching on angle, it is investigated below to study how the factors each will affect the inrush current.

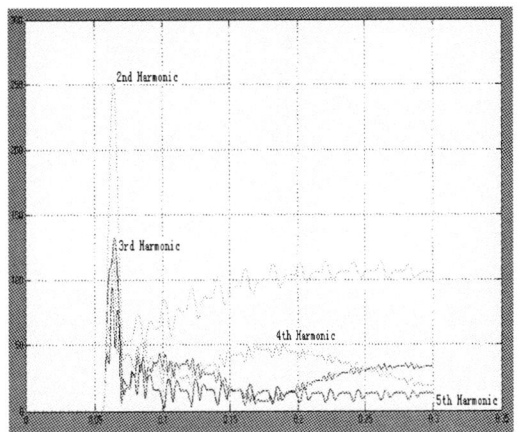

Fig.10 2nd, 3rd, 4th, 5th harmonics of phase C current

805

Tab.3. Typical magnitudes of energization inrush current harmonics

Harmonic	Magnitude
2nd Harmonic	251.25 times of $I_{No\,Load}$
3rd Harmonic	132.21 times of $I_{No\,Load}$
4th Harmonic	122.03 times of $I_{No\,Load}$
5th Harmonic	94.76 times of $I_{No\,Load}$

A. Saturation Characteristic

Circuit model of a saturable transformer is shown in Fig.11 The non-linear magnetizing inductor L_{sat} can be fitted by piece-wise linearization method, shown in Fig.12 The points 1~4 have to be accurately chosen to simulate the saturation characteristic curve of transformer. Considering the practical transformer parameters, the points (i_i, ϕ_i) are set to be [0 0; 0.00 1.20; 1.50 1.52]. Of course they may be reset flexibly to simulate various types of transformer saturation. Knowing the limiting B-H characteristic a normal distribution Preisach distribution density function has been assumed (refer to Ref. [1])

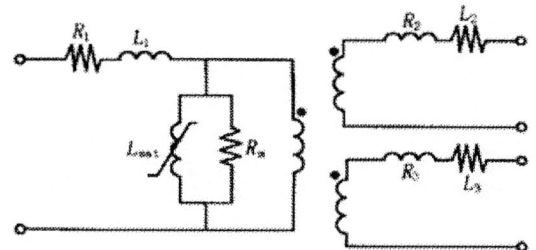

Fig.11 Circuit model of saturable transformer

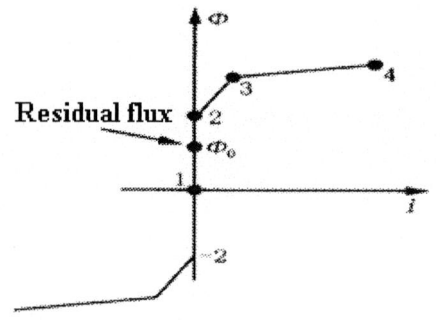

Fig.12 Saturation characteristic curve of transformer

B. Residual Flux

Residual flux has a close relationship with the magnitude of inrush current. In accordance with the practical performance of Chinese power transformers, residual flux of lots of medium and high voltage transformers is between 0.5~0.7 B_m, higher value has

to be chosen during simulation for strictness sake. In this paper, several cases of different phase C residual flux are studied to see how the residual flux matters, the relationship between phase C residual flux ϕ_{rc} and Ic is shown in Tab.4.

Tab.4. Relationship between ϕ_{rc} and Ic

Residual flux	0	$0.3\,B_m$	$0.6\,B_m$	$0.9\,B_m$	$1.2\,B_m$
Ic/A	200	300	500	600	700

Tab.5.   Relationship between switching time and Ic

Switching time	0	$\frac{1}{4}T$	$\frac{1}{2}T$	$\frac{3}{4}T$	T
Ic/A	254.16	372.18	-252.78	-371.18	254.89

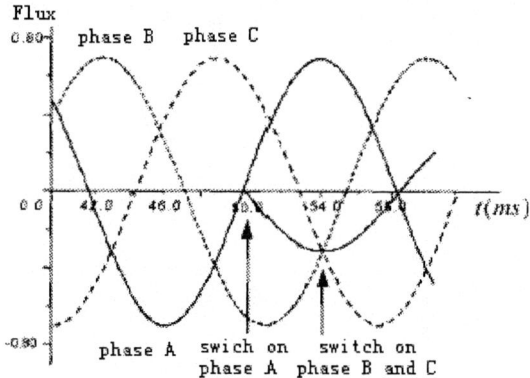

Fig.13 Three-phase flux after switching on phase A

C. Switching Time

Usually three phases' asynchronous switching time will cause distinct magnitudes of three-phase inrush current. When the saturation characteristic points (i_i, ϕ_i) are set to be [0 0; 0.00 1.20; 1.50 1.52], and three-phase residual fluxes are all zero, that is to say [phi0A phi0B phi0C]=[0 0 0], Tab.5 gives us a clear view of how much switching angle affects the inrush current. It can be easily seen that Ic is highest at the closing time of 0.25T (T=0.02s) after phase A switching on; if phase C is switched on at 0.75T, then the magnitude of Ic is almost the same but opposite. The reason of this is shown in Fig.13 When phase A is energized, phase B and C are coupled to have a same flux, which is half value of ϕ_A but opposite, because the sum of three-phase flux is zero at any time. The higher the flux ϕ_c is, the greater value Ic will be. In a word, 0.25T is a critical time for phase B and C to switch on.

VI. MEARSURES TO ELIMINATE INRUSH CURRENT

As the magnitude of inrush current has close relationship with the closing resistance, the switching phase angle, the residual flux, the transformer capacity and so on, several countermeasures are proposed to reduce it accordingly:

(1) Closing resistance plays an important role in reducing inrush current, shown in Fig.14 The effectiveness is in direct portion to the value of resistance. Tab.6 shows a $300\,\Omega$ resistance can reduce the peak value of inrush current by about 30%, while a $600\,\Omega$ resistance can reduce it by 40% at most. What's more, the elimination effect under residual flux conditions is more evident, the inrush current can be reduced to a low level. Therefore, it has to be careful to consider the inrush current effect when the closing resistors are to be replaced by MOA for over-voltage protection's sake.

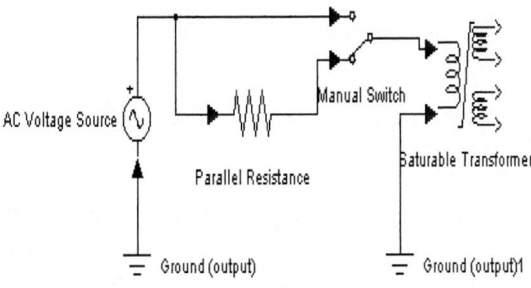

Fig.14 Diagram of switching on single phase transformer with closing resistance

Tab.6. Effect of closing resistance on inrush current

R/Ω	300	400	500	600
Elimination effect/%	28.94	32.96	35.89	39.82

Note: 110kV/11kV Transformer, S_N=35MVA, I_N=165.3A, I_0=0.29%, I_{inrush}=-227.1A(=1.37I_N) if no switching resistor is installed.

(2) Choose proper closing time. If phase A is switched on at first, then flux B and C is half but negative value of flux A, so phase B and C ought to be switched on after 1/4 cycle of phase A's closing to guarantee the flux is not so high, the principle is demonstrated in Fig.13

(3) It is suitable to install appropriate capacitance in parallel with the transformer secondary side to distort the inrush current so that the characteristics of sharp top current are eliminated. It has to take care that the capacitor should be chosen right under precise simulation result. Following this method can generate a flux that is negative to the high voltage side, therefore,

the winding flux is prevented from going into saturation.

VII. CONCLUSION

Following paper [4], it is studied in this paper based on two novel methods--Callahan algorithm and MATLAB/SIMULINK simulation, the electro magnetic transient process while energizing a transformer, which is equivalent to high-order nonlinear circuit. Regularity of the inrush current generation and harmonics analysis by Fourier transform is done. Different factors influencing inrush current such as the saturation characteristics, residual flux and switching on time are investigated. Measures to eliminate the inrush current are also suggested to guarantee safe operation of power system.

REFERENCES

[1] A. A. Adly, H. H. Hanafy and S. E. Abu-Shady, "Utilizing Preisach models of hysteresis in the computation of three-phase transformer inrush currents", *Electric Power Systems Research*, vol. 65, Issue. 3, pp. 233-238, 2003.

[2] J. F. Witte, F. P. DeCesaro, S. R. Mendis, "Damaging long-term over-voltages on industrial capacitor banks due to transformer energization inrush currents", *IAS Annual Meeting (IEEE Industry Applications Society)*, vol.2, pp. 1543-1551, 1993

[3] Chen Zhe, Wen Yuanfang, Lu Guojun. "Two novel methods for modelling and digital simulation on electric power transformer", in Proceedings of 2004 International Postgraduate Conference on Electrical and Electronic Engineering (IPCEEE 2004), Xi'an, China, pp. 131–134, 2004.

[4] Lin Xiangning, Liu Pei. "Studies for identification of the inrush current based on improved correlation algorithm", in Proceedings of Chinese Society for Electrical Engineering, vol.21, no.5, pp.56–60, 2001.

[5] I. D. Mayergoyz, Mathematical Models of Hysteresis, Springer Verlag, New York, 1991

2006 5th International Power Electronics and Motion Control Conference

The Simulation and the Experimental Research of the Stator Bars' Evaporative Cooling System in the Three Gorges' Hydrogenerator

Ruan.Lin   Gu.Guobiao   Tian.Xindong   Yuan.JiaYi

Institute of Electrical Engineering

Chinese Academy of Sciences   Beijing P.R.China 100080

Abstract—This paper presents a kind of simulation software for computing the characteristics of flowing and heat transfer in the stator bars' evaporative cooling system of large hydrogenerators. A model experiment using the real stator bar of Three Gorges' hydorgenerator was introduced also. Under experimental conditions, Temperature distributions and utmost thermal load are calculated using the simulation software. Comparison between simulation and experiment results shows that the calculating results are accord with the experimental measurement to about 5 percent relative error. On the whole, the altering of the inner water-cooling system to the evaporative cooling system can completely meet the cooling demand; the simulation software is effective and reliable and can be used to the design of the evaporative cooling system for large hydro generators in the applied engineering field.

Keywords- Three Gorges; hydro-generator; evaporative cooling system; simulation software; model experiment

I. INTRODUCTION

Evaporative cooling of the stator bar of hydro-generators is a kind of highly effective inner-cooling method. When phase change takes place, the coolant flowing inside the hollow conductors of the stator bar takes large quantity of heat generated by stator's copper losses away. The most key point is that it is much more reliable than direct water-cooling, it will benefit to the long-time safe operation of the large hydrogenerator. So far as the theory, there is no problem that the evaporative cooling technology can be applied to the Three Gorges' hydro-generators. In order to validate this technology can substitute the cooling method in engineering application and get the actual operational effect, the China Three Gorges Project Corp (CTGPC) planed to make one set inner water-cooling hydro-generator installed on the left bank of the dam altered to the evaporative cooling hydro-generator. This inner water-cooling hydro-generator designed by Siemens and use VGS stator bar. So, we must know that through altering whether the evaporative cooling system can meet the cooling demand of hydrogenerator and whether the temperature level is still comparative to that of the water-cooling. Therefore, we validate the feasibility

through two aspects of simulation and the model experiment. The simulation software was independently-developed professional software specially used for designing evaporative cooling system of generators. The model experiment using the real stator bar of Three Gorges' hydrogenerator was done and a lot of experimental results were acquired. We choose the temperature distribution and the utmost thermal load as the criteria to testify the cooling effect of the evaporative cooling system. Comparison between simulation and experimental results shows that the calculating results better accord with the experimental measurement within the average 5 percent errors. Results show the altering of the inner water-cooling system to the evaporative cooling system can completely meet the cooling demand and the level of temperature differences is comparative to that of the water-cooling. And the simulation software can be used confidently to the real engineering design for the Three Gorges' evaporative cooling.

II. THE SIMULATION OF THE EVAPORATIVE COOLING SYSTEM OF STATOR BARS

A. The operational principle

The operational principle of the CLSC evaporative cooling system is as follows: when the hydro generator works with load, the stator's bars produce heat because of the stator's copper losses; the coolant inside the hollow conductors is gradually raising its temperature; when the temperature of the coolant reaches the saturation temperature, boiling results; a large quantity of heat will be absorbed when phase change happen, then, the heat will be transferred to the second cooling water when the two-phase coolant flow through the condenser and the two-phase coolant will be cooled to one-phase liquid and flow back to the liquid-down tube, next circulation begins. This cooling system takes the advantage of the gravity difference between one-phase liquid and two-phase of liquid and gas to realize self-circulation without pump, so it is self-adaptive.

B. Theoretical basis of the simulation

The theoretical bases of the simulation are energy conversation, momentum conversation and mass conversation. Energy conversation means the heat generated by the copper losses must be take away by the coolant inside the hollow conductors. Momentum conversation means the flowing pressure head must equals to the total flowing resistances in the circulating loop. Mass conversation means the mass of gas-phase coolant plus the mass of liquid-phase coolant must equal to the mass of all coolant infused into the circulating system.

C. Flow chart of the simulation

The flow chart of the simulation is shown in Fig.1. For the detail calculating method and the selection of the parameters, please look over the reference [1][2].

D. UI of the simulation software

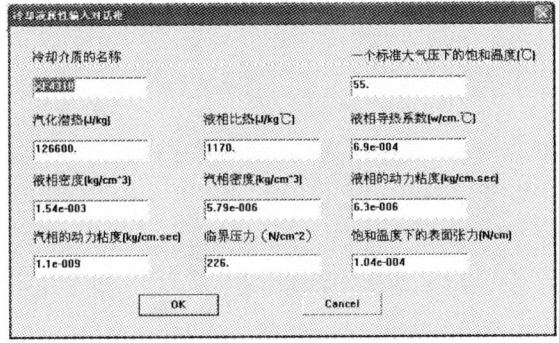

Figure 2.   Input of the characteristics of the new coolant

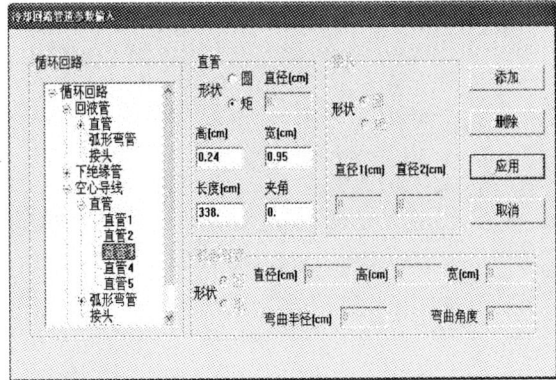

Figure 3. The objective of the VGS stator bar's evaporative cooling system

E. Result of the simulation

Under the following precondition : 1) The dimension of the stator bar is on the basis of VGS stator bar designed by Siemens; 2)Use the characteristics of the new environmental friendly coolant HFC-4310; 3) Assuming that the gauge pressure is zero; 4)Assuming that the coolant temperature at the entrance of the hollow conductor is 45℃; We got the computing results shown in Fig.4-6 and Table 1-2.

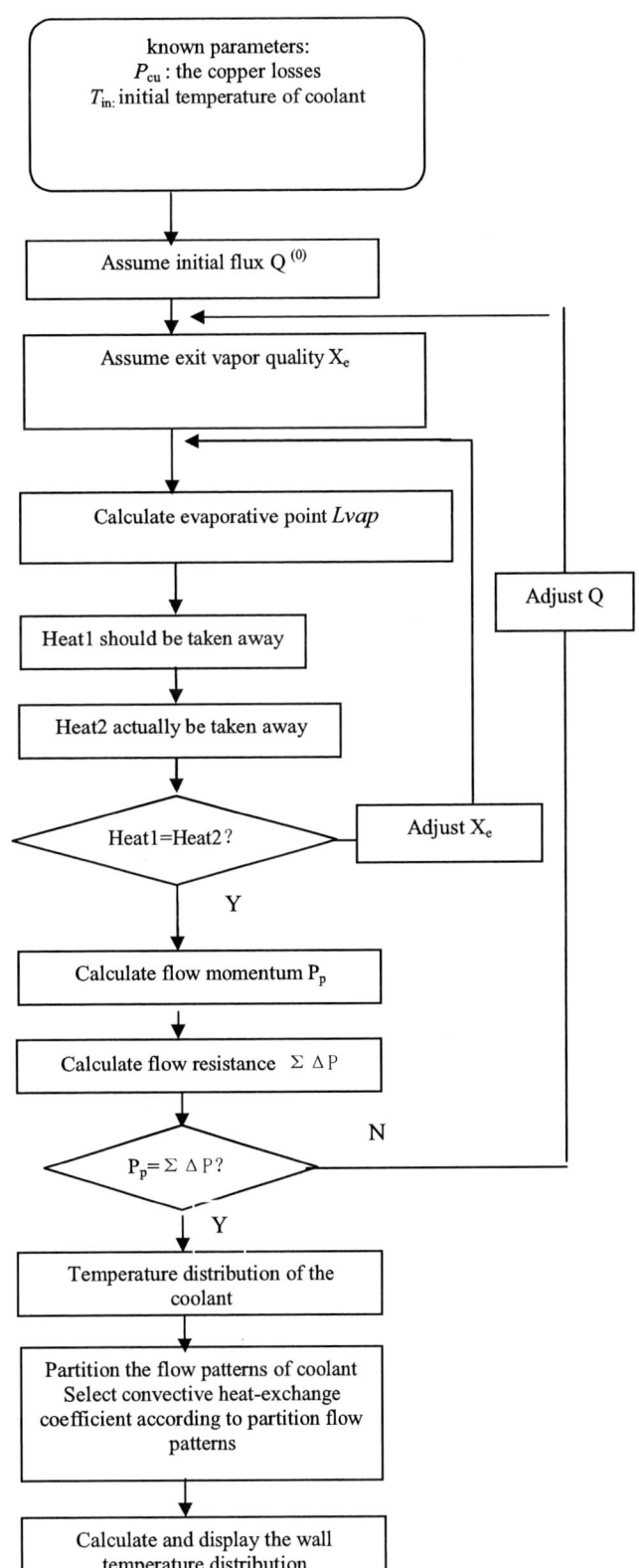

Figure 1.   Flow chart of the simulation

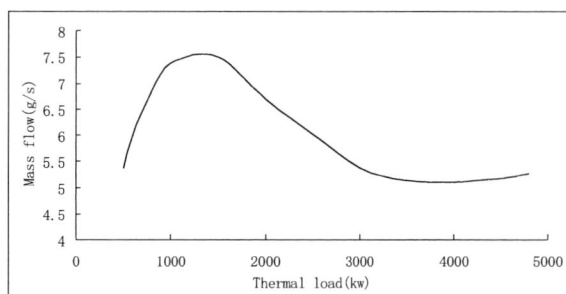

Figure 4. The relation between the mass flow in the hollow conductor with the thermal load

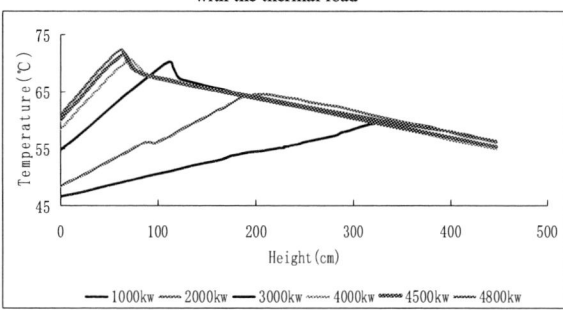

Figure 5. The temperature distribution at different thermal load

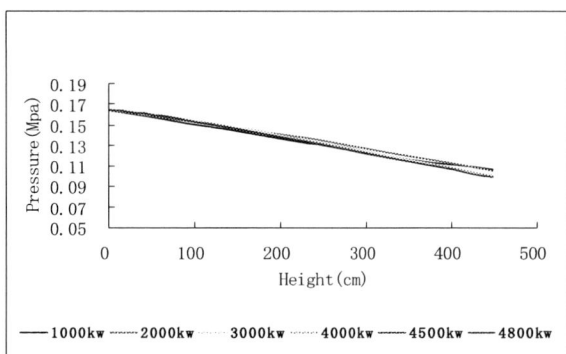

Figure 6. The pressure distribution at different thermal load

TABLE 1
THE UTMOST THERMAL LOAD OF THE VGS STATOR BAR'S EVAPORATIVE COOLING SYSTEM

Coolant	F113	HFC-4310
Rated current (A)	4490	4490
Hollow conductor number	6	6
Solid conductor number	24	24
Computing ultimate thermal load (KW)	4500	4900
Relating current Imax (A)	5433	5662
Capability of overload Imax/I	1.21	1.26

TABLE 2
THE INFLUENCE OF THE HEIGHT OF COOLANT ON THE THERMAL LOAD OF THE EVAPORATIVE COOLING SYSTEM

The height of coolant (cm)	268	380	435	477
The ultimate thermal load (KW)	2700	4200	4900	5500
Relating current (A)	4230	5276	5699	6037
Relating capacity (MVA)	680	848	916	974
Relating load (MW)	612	764	824	877

III. INTRODUCTION OF THE EXPERIMENTAL MODEL

The picture of the experimental platform is shown in Fig.7. This experimental model was completed in Dongfang electrical machinery works. It is composed of 18 stator bars that are same as ones used in the Three Gorges' hydrogenerator installed on the left bank. The coolant used in the model experiment is HFC-4310. It has been approbated by the consumers.

Figure 7. The experimental platform

IV. COMPARISON BETWEEN SIMULATION AND EXPERIMENTAL RESULT

A. Comparision of maximum temperature and average temperature

TABLE 3
COMPARISION BETWEEN EXPERIMENT AND CALCULATION

Input parameters				Experimental results		Calculating results		Relative errors (%)	
I (A)	TL(KW)	K	H(cm)	Tmax (℃)	Tave (℃)	Tmax (℃)	Tave (℃)	Tmax	Tave
4427	3232	85%	439	65.8	60.8	67.7	59.73	2.81	1.79
5209	4497	100%	505	70.7	65.5	71.55	63.05	1.19	3.89
5469	4834	105%	505	70.7	65.7	72.33	63.93	2.25	2.77
5260	3976	92.6%	514	68.5	63.48	69.72	60.48	1.75	4.96
5546	4513	100%-105%	517	69.2	64.4	71.36	62.13	3.03	3.65
3975	2801	80%	447	63.9	58.4	66.3	57.55	3.62	1.48
4823	3953	95%	510	68.9	63.93	69.87	61.14	1.39	4.56
5250	4624	100%	510	71.5	65.5	71.81	63.09	0.43	3.82

I—Current;
TL—Thermal Load;
K—Relating capacity (base on 840MVA);
H—The height of coolant;
Tmax—The max temperature;
Tave—The average temperature;

B. Comparision of temperature distributions

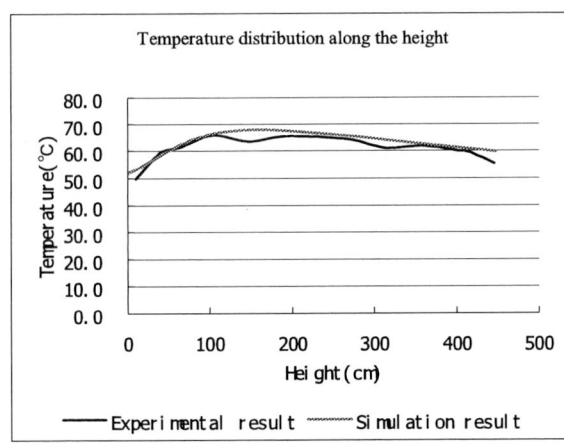

Figure 8.   Input current is 4427A

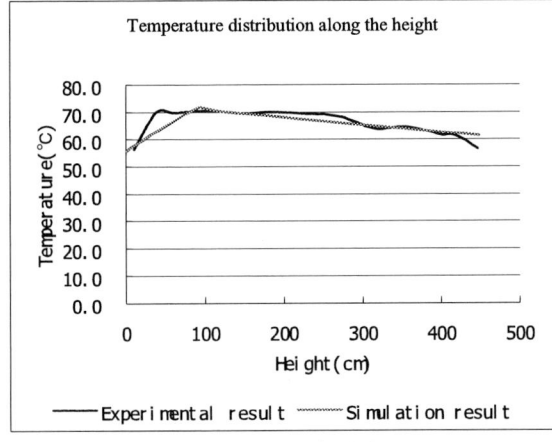

Figure 9.   Input current is 5209A

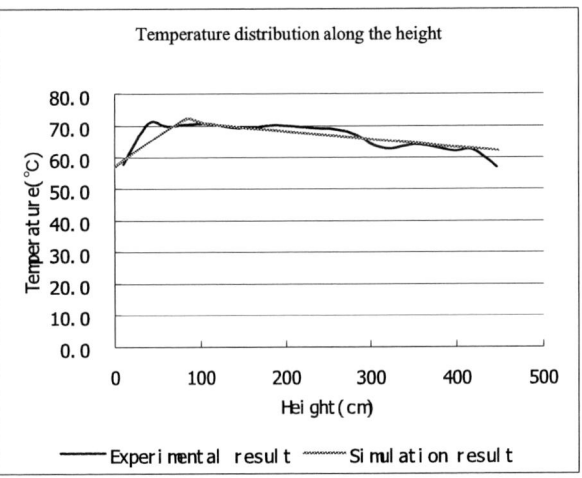

Figure 10.   Input current is 5469A

C. Comparison of Thermal load

As for the stator bar evaporative cooling system, the utmost thermal load capability is affected by the height of the coolant. According to the calculation, when the height of coolant equals to the height of stator bar, The Imax is 5662A; the model experimental result is 5546A, which shows the assumption of the utmost thermal load by the simulation software is correct. The operational height of coolant increase, the utmost thermal load capability can also be promoted, that means the system still having potential of over-generate electricity ability.

V.   CONCLUSION

Through the simulation and the model experiment, we know that the evaporative cooling system can meet the cooling demand of hydrogenerator and the temperature level is still comparative to the water-cooling.

The 1:1 model experiment carried on in the DongFang Electrical machinery works can make sure that the following points are believable:

1) The alteration can make the hydrogenerator with 700 MW (778MVA) capacity has a long-term over-loading capability by at least 115%;

2) The temperature of the stator bar is within 75℃ and there is no over-heat point;

3) The liquid-detached connectors are very reliable .

The simulation results provide the following reference foundation for the alteration of the cooling system of the Three Gorges hydrogenerator:

1) The flowing characteristic parameters;

2) The utmost thermal load capability of the stator bar's evaporative cooling system;

3) The temperature distributions of the coolant and the stator bar.

The comparison between the calculating results and the experimental results shows the simulation of the circulating system is correct and it validated the experimental results on the model experimental platform. The simulation software can be used confidently to the real engineering design for the Three Groges' evaporative cooling

REFERENCES

[1] L.Ruan, G.Gu, X.Tian "Numerical simulation for circulating systems and experimental comparison of the closed-loop,self-circulating evaporative cooling of hydro-generators" *Electrical engineering* Vol86 No.3, pp. 127–134, 2004.

[2] Ruan lin, Gu Guobiao, Tian Xindong. "Research of the temperature distribution of the hollow conductor in the evaporative cooling hydro-generators" Electrical power components and systems, Vol.33 No.2, pp.145-158, 2005.

[3] G. B. Gu, L.Ruan "The application of the evaporative cooling technology to the Three Gorges Project" *ICEMS'2004, Korean.*

[4] S. N . Ding, *Heating and Cooling of Large Electrical Machines,* Beijing: Science Press, 1992.

[5] TONG LS (1975) *Boiling heat transfer and two-phase flow.* New York, Robert E. Krieger Publishing Company. Translated by Wang Menghao, Xu Rende (1980) Beijing, China Machine Press, pp 136-141 (in Chinese).

2006 5th International Power Electronics and Motion Control Conference

An Investigation of Multi-phase Transverse Flux Permanent Magnet Machine

G.Q. Bao[1,2], J.K.Wang[2] , D.Zhang[2] and J.Z. Jiang[2]

[1] Dept.of Electrical Engineering, Lanzhou University of Technology, Lanzhou, China
[2] Dept.of Automatic Control, Shanghai University, Shanghai, China
Guangqing.bao@gmail.com

Abstract—**Based on the characteristics of transverse flux permanent magnet machine (TFPM), a novel multi-pole and multi-phase TFPM topology with assembled stator and concentrated flux rotor is presented in this paper. The associated electromagnetic fields of the prototype are investigated by means of three dimension equivalent magnetic network method (3DEMNM). It is shown the unique multi-phase structure of TFPM is valid for power generation enhancement as well as cogging torque reduction, which is vital for applications of low speed, high torque and direct drive in industrial and martial area. And the experimental measurement of four-phase prototype meets with the simulation results completely when the motor operates with a brushless DC drive.**

Keywords-Transverse flux; Permanent magnet machine; Poly-phase machine; Torque ripple

I. INTRODUCTION

Over the last decades, with the rapid developments in power electronics, permanent magnetic materials and modern control theories, the innovative designs, manufactures and applications of permanent magnet motors have made significant progress. Among them, the transverse flux permanent magnet (TFPM) machine with relatively high torque density and low speed, avoiding gearing configurations, is especially suited for direct drive such as full electric ship, electric vehicle, industrial robots etc [1]. Introduced by professor H.Weh, TFPM is different from conventional machines in many aspects [2].

•It has unique three-dimensional flux pattern which leads to the decoupling of the space requirement of the flux carrying core iron path and the space occupied by the armature winding.

•It has higher force density for the flux concentration design.

•It has a great number of degrees of freedom for design flexibility.

Every coin has its two sides. Except for advantages mentioned above, TFPM has a certain extend of disadvantages coming from its special topology.

•It is expensive to manufacture for complicated topology.

•It generates torque ripples due to the current switching and cogging torque produced by the magnetic attraction between the rotor-mounted permanent magnets and the teeth of the stator magnetic-circuit.

•It has lower power factor for the severe flux leakage.

With the recent developments of new topologies of converter fed machines, TFPM has superiority in high-power direct propulsion area. So to minimize torque ripple becomes one of the essential design considerations. The torque behavior of electric machines can be shaped by two ways. One is electrical control scheme. Namely, the harmonic content of torque output can be suppressed by optimized voltage and /or current waveform through complex control methods. The other is constructional design of machine. Constructional design can eliminate torque ripple by suitable arrangement of stator slots and/or rotor poles, individual phases and so on. To get a smooth steady torque multi-phase configuration of TFPM is adopted in this paper, by which a certain harmonic contents can be cancelled and the overall torque ripple can be reduced significantly. Except for torque ripple restraining this kind of multi-phase configuration can also improve the error tolerance ability of machine. Even when lacking several of phases, it can still work properly. The rest of the paper is organized as follows. In Section II, we present a structural design of multi-phase TFPM. In Section III, we discuss the influence of cogging torque to electro-magnetic torque in TFPM. In Section IV, we analyze the electrical phase relationship for multi-phase arrangement. In Section V, we develop the experimental and simulation measurement of four-phase prototype. In Section VI, we list our conclusions and suggest possible future work.

II. STRUCTURAL DESIGN

A large number of different TFPM topologies have been studied in many presentations. Many R&D projects are focused on efficient, cost effective and high power TFPM deigns. A new multi-phase TFPM with tooth-like and combined stator arrangement is introduced here. The multiple phases are available by means of multiple stacks and/or multiple phases per stack. Fig.1 is a 20-phase

Sponsored by Chun-hui Plan from Ministry of Education of the PRC (Z2005-1-62002)

1-4244-0448-7/06/$25.00 ©2006 IEEE

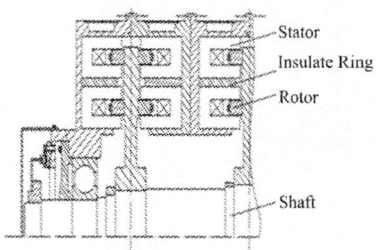

Figure 1. Cross-sectional view of TFPM

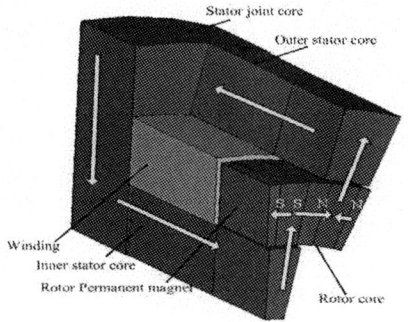

Figure 2. Structure of one pole pair TFPM with combined stator

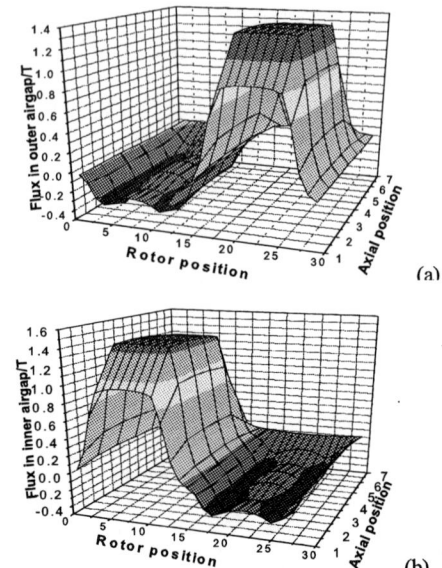

Figure 3. 3D air-gap flux distribution of TFPM at no-load

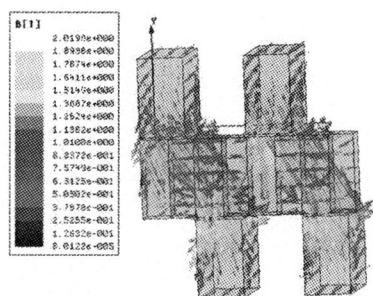

Figure 4. Magnetic flux vectors of two pole pairs by 3DFEA

TFPM, which is made up of five stacks and each stack includes four phases. For the structural similarity of each stack and each phase, one pole pair configuration with laminated pole pieces and magnets arranged is shown in Fig. 2. The rotor adopts concentrating flux design which the magnetic irons are magnetized tangentially so as to produce a high no load air gap flux density. The total stator magnetic circuit is made up of U-cores embracing the winding. Stator cores characterizing combined configurations include outer core, inner core and joint core. This arrangement deserves encouragement for the easy assembly which associated to the simple mechanical structures. To Avoid the complex curled slice and incision techniques [3], iron structures are built of thin sheets of laminations of magnetic materials. The stator circuit use sheet-steel material of silicon-iron alloy, for instance JFE Super Core, which has higher magnetic permeability and low iron loss under high frequency. These laminations, which are aligned in the direction of the field lines, are insulated from each other by an oxide layer on their surfaces or by a thin coat of insulating enamel. The rotor magnetic circuit includes permanent magnets(PMs) mounted in a flux concentration arrangement, alternated by JFE Super Core.

As no dimension is really larger than the others, the magnetic circuit geometry of a TFPM is typically three dimensional (3D) and a 3D electromagnetic analysis is required. An equivalent magnet-network analysis designed for 3D electromagnetic numeric calculation (3DEMNM) has been successfully used in TFPM design [2]. In 3DEMNM, the current winding is assumed as virtual permanent magnet, thus the scalar magnetic potential equation is satisfied, which simplifies the calculation procedure and saves time greatly. Fig.3(a) and Fig.3(b)give the outer and inner air-gap flux respectively. The flat peak induction shows the rotor is in a steady position as rotor core and stator core are concentric and air-gap reluctance gets to minimum. Fig.4 is magnetic flux vectors of two pole pairs obtained from 3D finite element algorithm (3DFEA). Noticeably, the inner and outer air gap flux get to1.5T, higher than conventional machines, so the high torque density can be achieved. An electric machine with high torque density has important commercial advantages, for it can be made in a smaller size to fit limited space and require less active material.

III. THE INVESTIGATION OF COGGING FORCE

Just as common permanent magnet machine, cogging force in TFPM is gaining increasing importance with the demand for high performance direct drive system. Cogging force is a kind of circumferential attractive force attempting to maintain the alignment between the stator cores and the rotor permanent magnets. It has been found that TFPM exhibits high cogging force especially with flux concentrating arrangement design of permanent magnets. So the impact of cogging torque can not be

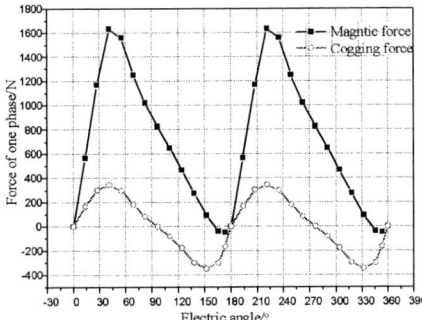

Figure 5. Electromagnetic and cogging force comparison of one-phase TFPM

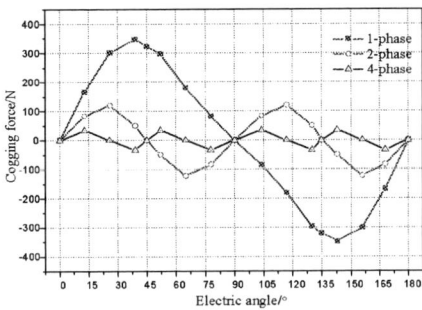

Figure 6. Magnetization as a function of applied field. Note how the caption is centered in the column

overlooked in TFPM design. Some specific features of cogging torque in multi-phase TFPM are to be emphasized [4].

• For there is no common shared air-gap, each phase produce cogging force locally.

• The multi-phase combinations which are axially and radially shifted can minimize torque ripple by the cogging force cancellation.

• For the number of rotor pole pairs is high and is equal to the number of stator teeth, the frequency of cogging force is high.

According to Fig.5, one can notice the high amplitude of the cogging force produced by a single phase and 90 poles TFPM instance, reaching almost 350N, possesses 22% of the rated force. Fig.6 is the comparison of resultant cogging force between 1-phase, 2-phase and 4-phase. Obviously, the 4-phase resultant cogging force is less than 50, representing 3% of the electro-magnetic force. Consequently increasing the number of phases would lead to lower cogging force. However, if the cost of the drive system is taken into consideration, maybe a 3-phase design is a suitable choice, where the standard converters can be used.

IV. THE ANALYSIS OF ELECTRIC RELATION

The electrical characteristics of the TFPM are similar to those of conventional synchronous machines, the same control systems and converters can be applied. Due to the

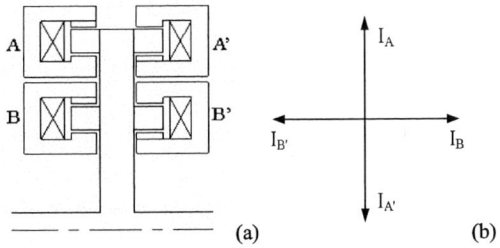

Figure 7. Structure and electrical phase relationship of four-phase TFPM column

trapezoid back-EMF waveform in our investigation, a brushless DC drive scheme is used.

Fig. 7(a) is electrical relation of ring winding. Fig. 7(b) shows the cross-sectional view of a prototype with two axially-arranged discs and 4-phases in all. There are two possible designs of 90°-electrical angle shift between the phases as discussed as follows.

• The stator teeth are aligned, the rotor permanent magnets of four adjacent phases are shifted by a 90°-electrical angle.

• The rotor permanent magnets are aligned, the stator teeth of four adjacent phases are shifted by a 90°-electrical angle.

There is no difference in electromagnetic performance between the two alternatives. The effect of cogging force cancellation is almost equivalent. Nevertheless, it is favorable to select the first design. It is much easier in assembling process.

From the analysis of electromagnetic force, the force for phase A is given by:

$$T = T_0(1 + \alpha_2 \cos 2\theta + \alpha_4 \cos 4\theta + \alpha_6 \cos 6\theta + ...) \quad (1)$$

The phase B, A' and B' have same form and magnitudes of coefficients T_0, a_2, a_4, a_6 etc, but with phase displacement of $90°$, namely being $(\theta + 90°)$, $(\theta + 180°)$ and $(\theta + 270°)$ respectively. The resultant torque for four phases becomes:

$$T = 4T_0(1 + \alpha_4 \cos 4\theta + \alpha_8 \cos 8\theta + \alpha_{12} \cos 12\theta + ...) \quad (2)$$

Altogether, all $2n$ harmonics where n is odd can be restrained, which leading to a significant reduction of torque ripple.

V. THE VERIFICATION OF SIMULATION AND EXPERIMENT

In order to investigate the drive system and to optimize the controller, the prototype and its converter have been elaborated for computer based simulations, which is implemented by MATLAB-SIMULINK. Fig.8(a) and Fig.8(b) show the current comparison of measured and simulated current of prototype. The result is seen to be good. Small differences exist due to the inaccuracies of modeling the machine, measurement and so on. The

815

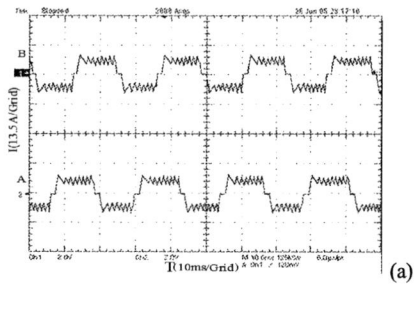

(a)

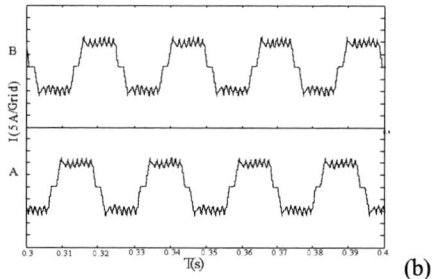

(b)

Figure 8. Phase current of measurement and simulation

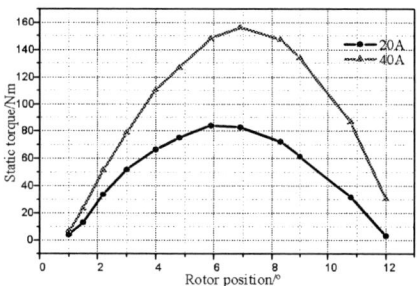

Figure 9. Static torques versus position and ampere-turns

measurements of static total torques versus angle and different currents are also provided by Fig.9.

VI. CONCLUSION

The presented transverse flux motor topology confirms the merit of high torque density on going. The skillful combined stator core configuration makes mechanical manufacture easier, advantages over conventional TFPM. Noticeably, the unique multi-phase structure of TFPM is valid for power generation enhancement as well as cogging torque reduction, which is vital for applications of low speed, high torque and direct drive area. Our future work will involve investigating the effects of other components on torque ripple, especially the factors of saturation and leakage fields since it is not taken into consideration in this paper.

REFERENCES

[1] H.Weh，H May. Achievable force densities for permanent magnet excited synchronous machine with new configurations[C]. Proceedings of ICEM, 1986

[2] G.Q.Bao, J.H.Shi, and J.Z.Jiang. A new transverse flux permanent motor for direct drive application[C]. Power Electronics and Motion Control Conference, 2004. IPEMC 2004. The 4th International Volume 1, 2004 Page(s):260 - 263 Vol.1.

[3] AJ Mitcham, Transverse Flux Motors for Electric Propulsion of Ships. IEE Colloquium on New Topologies for Permanent Magnet Machines, 18 June 1997 Page(s):3/1-3/6.

[4] A.Masmoudi and A.Elantably, A simple assessment of the cogging torque in a transverse flux permanent magnet machine. *IEEE Inter.Electric Machines and Drive Conf.,Massachusetts*, pp. 754–759, UAS,June 2001.

2006 5th International Power Electronics and Motion Control Conference

Suspension Principle and Digital Control for Bearingless Permanent Magnet Slice Motors

Huangqiu Zhu and Liang Fang
Jiangsu University/School of Electrical and Information Engineering, Zhenjiang 212013, China
zhuhuangqiu@ujs.edu.cn

Abstract—As bearingless motors have all advantages of magnetic bearings, they play an important role in solving difficult problems of special electric drives. The simplest and most versatile form of bearingless motor is the bearingless slice rotor motor. This kind of motor with centrifugal pump has a wide range of applications in life science, chemical processing industry, semiconductor manufacturing industry, foodstuff processing industry, and so on. In the paper, a 3-phase permanent magnet slice rotor motor is set up, and suspension principle of bearingless permanent magnet slice rotor motors is introduced. Theoretical formulas of driving system and radial suspension forces are given, and then the control system for the 3-phase bearingless permanent magnet slice rotor motor is designed and presented. An experimental platform of digital control system is developed to meet the need, and experiments including core loss measure, speed regulation and suspension force control are performed. The experiment results are presented under unload and load. The experiments have shown that the vibration amplitude in *x*- and *y*-direction is less than 200 μm under unload and load, while rotor speed are changed from 0 r/min to 8 000 r/min.

Keywords-bearingless motor; slice rotor; configuration; suspension principle; digital control

I. INTRODUCTION

In 1988, a disc type motor with axial force generation adjusting exciting motor current was proposed by R. Bosch, and the word "bearingless motor" was used for the first time by R. Bosch [1]-[2]. Bearingless motor does not mean the lack of bearing forces, which are necessary in any case to stabilize the rotor, but the missing of physical contact bearing components. The principle of the bearingless motor is based on the contactless magnetic bearing of the rotor. Contrast to conventional magnetic levitated drives, the bearing forces in the bearingless motor are not built up in separated magnetic bearings, which are placed on the left- and right-hand side of the motor block, but in the motor itself [1]-[6].

The simplest and most versatile form of bearingless motor is bearingless slice motor. When the length of the

rotor is small compared to the diameter, 3 degrees of freedom can be stabilized passively. The remaining 3 degrees of freedom are controlled actively. Firstly, the currents in bearing windings control radial forces. Secondly, the currents in driving windings control torque. When the rotor, which carries an impeller and is levitated within the hermetically closed pump housing, turns with a high rotational speed, liquid is pumped from the inlet to the outlet. This type of magnetically levitated centrifugal pump system has a wide range of applications in scopes of life science, chemical processing industry, semiconductor manufacturing industry, foodstuff processing industry and so on [4]-[6]. In this paper, the 3-phase permanent magnet slice rotor motor is set up, theoretical formulas of driving system and radial suspension forces are given, and then the control system for the 3-phase bearingless permanent magnet slice rotor motor is designed and presented. An experiment platform of digital control system with DSP controller is designed to meet the need, and experiments including core losses measure, speed regulation and suspension force control are performed.

II. PRINCIPLE OF BEARINGLESS MOTORS

The concept of a bearingless motor is shown in Fig. 1. Two partial motors and one axial magnetic bearing are necessary for active control of a bearingless motor 's 5 degrees of freedoms.

This topology is rather complex for economic realizations. The topology in Fig. 1 is simplified by reducing the length of the rotor compared to its diameter resulting in a slice rotor. In this case it is possible to stabilize three spatial degrees of freedom passively and only one motor/bearing part is needed to control the other 3 spatial degrees of freedom. The functional principle of the passive stabilization is shown in Fig. 2 and Fig. 3. Fig. 2 shows an axial displacement of the rotor. The

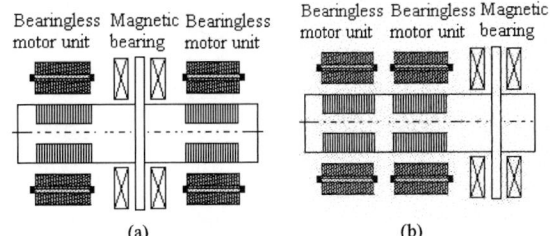

Figure 1. Basic structure of bearingless motors

The project was sponsored by the High Technology Research of Jiangsu Province (BG2005027), and SRF for ROCS, SEM.

1-4244-0448-7/06/$25.00 ©2006 IEEE

Figure 2. Passive stabilization of an axial
displacement of slice rotor

Figure 3. Passive stabilization of an angular displacement
of slice rotor

displacement results in attractive magnetic forces, which act against the displacement and therefore stabilize the axial position of the rotor. Fig. 3 shows a tilting of the rotor, which results in a torque against the displacement. The angular position is also controlled by passive magnetic forces, which turn the slice rotor back to a horizontal position.

III. CONFIGURATION OF BEARINGLESS SLICE ROTOR MOTOR WITH CENTRIFUGAL PUMP

As shown in Fig. 4, the system of bearingless permanent magnet slice motor consists of a motor, power-board, DSP-board, sensors, and RS232 interface. The motor consists of stator iron cores, stator back iron, permanent magnet slice rotor, bearing windings, and driving windings. There is one pole pair of permanent magnet slice rotor and two pole pair of bearing fields. The two sets of windings are designed to be a 3-phase symmetrical structure. The central processing unit (CPU) used in DSP-board is digital signal processor (DSP) TMS320F240. Its timing frequency is 20MHz, and it has 12 compare/pulse-width modulation (PWM) channels, 16 dual 10-bit A/D conversion modules. The instruction cycle time is 50 ns. The DSP system collects the signals of rotor positions and speed, and output PWM signal to control the bearing windings and the driving windings.

The outputs of sensors are sent to DSP-board. The RS232 interface communicates between DSP-board and a computer. The control parameters can be adjusted by a computer with special debugging software designed by a Company in Switzerland. The bearingless permanent magnet slice motor is controlled by directional control of

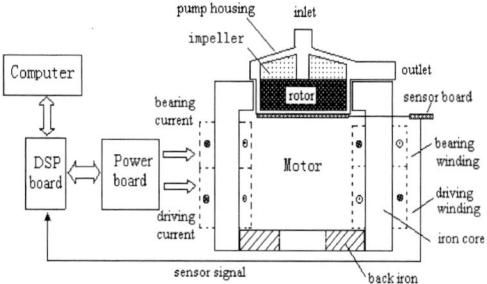

Figure 4. Configuration of bearingless slice motors

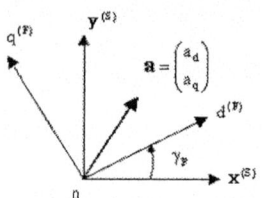

Figure 5. Connection of two kinds of coordinate systems

the rotor magnetic field.

IV. MODELS OF SLICE MOTOR

A. Coordinate Systems

A stator coordinate system is doq$^{(S)}$ and a rotating coordinate system that rotates synchronously to the rotor-flux-vector is doq$^{(F)}$, as shown in Fig. 5. Suppose vector a in the stator coordinate system, then

$$a^{(F)} = C(\gamma_F) \cdot a^{(S)} \tag{1}$$

$$C(\gamma_F) = \begin{pmatrix} \cos(\gamma_F) & \sin(\gamma_F) \\ -\sin(\gamma_F) & \cos(\gamma_F) \end{pmatrix} \tag{2}$$

$$a^{(S)} = C(\gamma_F)^{-1} \cdot a^{(F)} = C(-\gamma_F) \cdot a^{(F)} \tag{3}$$

where γ_F is field angel, $C(\gamma_F)$ is rotating matrix.

B. Basic Formulas of Driving System

As the magnetic field produced by the permanent magnetic rotor in air gap distributes according to the sine wave, define B_P as the air gap magnetic induction in the motor, ignore armature reaction, the magnetic induction B_P in the air gap of the motor can be written as

$$B_P(\varphi) = B_{Pm} \cos(p_1(\varphi - \gamma)) \tag{4}$$

where B_{Pm} is amplitude of the air gap magnetic induction, p_1 is pole pairs of the permanent magnet rotor or driving windings, φ is geometrical angle of the stator coordinate system, and r is rotating angel of the rotor.

The total magnetic flux in the air gap produced by the permanent magnetic rotor

$$\Phi_P = 2 \cdot l \cdot r \cdot B_{Pm} \tag{5}$$

where Φ_P is the total magnetic flux in the air gap, l is length of the rotor, and r is radius of the rotor.

The linked flux of each phase can be written as

$$\psi_P = \frac{N}{p_1} \cdot \xi \cdot \Phi_P \tag{6}$$

where ψ_P is linked flux of each phase, p_1 is number of pole pairs of windings, N is total number of phase turns, and ξ is winding factor.

818

Adopting the field oriented control of the permanent magnetic rotor, we can gain the voltage component formulas of motor windings in the rotating coordinate system [3]

$$\begin{cases} u^{(F)}_{1d} = 0 \\ u^{(F)}_{1q} = p_1 \cdot \psi_P \cdot \omega_m = c_u \cdot \omega_m \end{cases} \tag{7}$$

where ω_m is radian frequency of the rotor, and c_u is voltage-speed constant.

The motor torque equations can be written as [3], [5]

$$T = \frac{m\,p_1}{2} i^{(F)}_{1q}\psi_P = c_m \cdot i^{(F)}_{1q} \tag{8}$$

where m is the number of phases, and c_m is torque-current constant.

The field angel and speed of the rotor can be calculated by measuring the leakage flux density of the permanent magnetic rotor.

$$\gamma_F = \arctan(\frac{B_{sq}}{B_{sd}}) \tag{9}$$

where B_{sq} is q-coordinate magnetic induction value in the stator coordinates, and B_{sd} is d-coordinate magnetic induction value in the stator coordinates.

$$\omega_m = \frac{1}{p_1} \cdot \frac{d\gamma_F}{dt} \tag{10}$$

C. Basic Formulas of Suspension System

The radial forces of the rotor can be written as follow [3], [5]

$$\begin{cases} F_{ix} = (K_M \pm K_L) \cdot (i_{2d} \cdot \psi_{1d} + i_{2q} \cdot \psi_{1q}) \\ F_{iy} = (K_L \pm K_M) \cdot (i_{2q} \cdot \psi_{1d} - i_{2d} \cdot \psi_{1q}) \end{cases} \tag{11}$$

where $K_M = \frac{1}{2} \cdot \frac{\pi\, p_1 p_2 L_{m2}}{4 l r \mu_0 N_1 k_{N1} N_2 k_{N2}}$, $K_L = \frac{1}{2} \cdot \frac{m p_1 N_2 k_{N2}}{2 r N_1 k_{N1}}$.

F_{ix}, F_{iy} are the radial suspension forces, which are composed of Maxwell forces and Lorentz forces. K_M is Maxwell forces constant. K_L is Lorentz forces constant. i_{2d}, i_{2q} are current components of bearing windings. ψ_{1d}, ψ_{1q} are the airgap flux linkages components of motor. l is the length of rotor iron core. r is the radius of the stator inner surface. L_{m2} is the mutual inductance of bearing windings. N_1, N_2 are the number of turns of driving windings and bearing windings. k_{N1}, k_{N2} are winding factor of driving windings and bearing windings. m is the phase number of driving windings and bearing windings.

V. Control System of Slice Rotor Motor

Fig. 6 shows the system configuration of the permanent magnet slice motor. In the motor speed controller, the rotational angular speed ω_m is detected by the Hall sensors. The Proportional-Integral (PI) motor speed controller generates the q-coordinate current command $i^{(F)*}_{1q}$ in the rotor coordinates. The d-coordinate current command $i^{(F)*}_{1d}$ in the rotor coordinates equals zero. The 2-phase practical current $i^{(S)}_{1u}$ and $i^{(S)}_{1v}$ in the stator coordinates are measured by current sensors. After accounting the w-phase current $i^{(S)}_{1w}$, the 3-phase motor practical current $i^{(S)}_{1u}$, $i^{(S)}_{1v}$ and $i^{(S)}_{1w}$ are transformed from 3-phase to 2-phase, then transformed into practical current $i^{(F)}_{1d}$ and $i^{(F)}_{1q}$ in the rotor coordinates. The current commands $i^{(F)*}_{1d}$ and $i^{(F)*}_{1q}$ are compared to practical current $i^{(F)}_{1d}$ and $i^{(F)}_{1q}$, then differences are amplified by PI motor current controllers. The 2-phase motor voltage command $u^{(F)*}_{1d}$ and $u^{(F)*}_{1q}$ in the rotor coordinates are transformed into motor voltage command $u^{(S)*}_{1d}$ and $u^{(S)*}_{1d}$ in the stator coordinates, and then transformed from 2-phase to 3-phase. The 3-phase motor voltage commands $u^{(S)*}_{1u}$, $u^{(S)*}_{1v}$ and $u^{(S)*}_{1w}$ are modulated by PWM, then amplified by IPM-modules, and at last motor winding voltages $u^{(S)}_{1w}$, $u^{(S)}_{1v}$ and $u^{(S)}_{1w}$ are generated. In Fig.

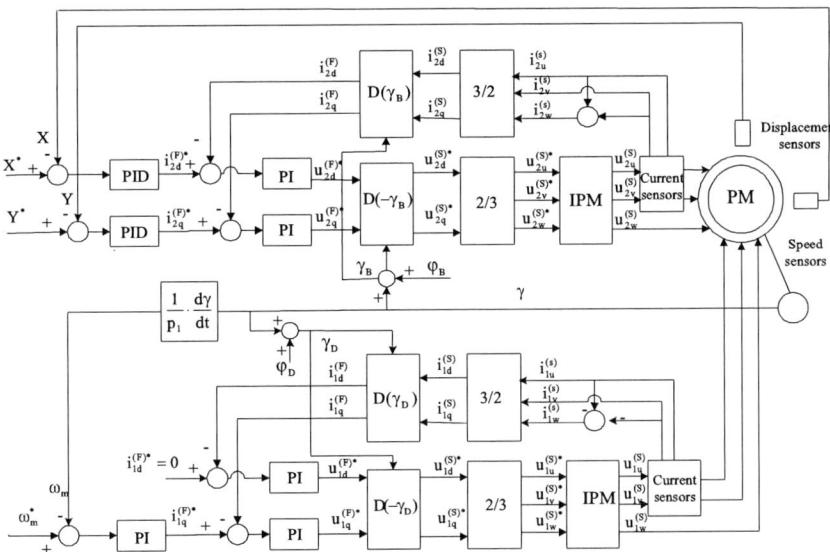

Figure 6. Sketch of control system for the bearingless permanent magnetic slice motor

6, the angel φ_D is the preliminary angel of motor winding in the stator coordinates.

Rotor shaft displacements in x and y-direction are detected by eddy current sensors. These displacements are compared to references X^* and Y^*, then the differences are adjusted by Proportional-integral-derivative (PID) controllers, and the radial force current commands $i_{2d}^{(F)*}$ and $i_{2q}^{(F)*}$ in the rotor coordinates are generated. The bearing winding voltages $u_{1u}^{(S)}$, $u_{1v}^{(S)}$ and $u_{1w}^{(S)}$ are generated like control part of the motor winding. In Fig. 6, the angel φ_B is the preliminary angel of the bearing winding in the stator coordinates.

VI. Experiments ang Results

The DSP system collects the signals of rotor positions and speed, and output PWM signals to control the bearing winding and driving winding. The power-board includes six power channels with a maximal current of 40 A. The bearing windings and driving windings are operated by 3-phase control. Each 3-phase power channel is built with one IPM-module. The converter operates with a supply voltage of 3-phase AC and is inserted in a compact case 210×200×330 mm.

The parameters of rotors are shown in TABLE I.

According to formulas of core losses [3]-[4], the total core losses will increase when the magnetic induction in air gap is enhanced, and will increase completely when the speed of the motor goes up. For two kinds of rotors, the curves of core losses are shown in Fig. 7. As shown in Fig. 7, the core losses will increase as the speed of the motor increases; and the core losses of the rotor HR633 is higher than the rotor HR510.

The amplitude of vibration of rotor in radial direction is shown in Fig. 8. Over a wide flow range the differential pressure between outlet and inlet depends only on the rotational speed of the impeller. Fig. 8 shows the rotor vibration amplitude of the pump in unload and load with

TABLE I.
PARAMETERS OF ROTORS

Rotor	Thickness (mm)	Outside Diameter (mm)	Inner Diameter (mm)	Residual magnetic induction B_r(T)
HR510	20	80	30	1.2~1.3
HR633	25	78	30	1.3~1.4

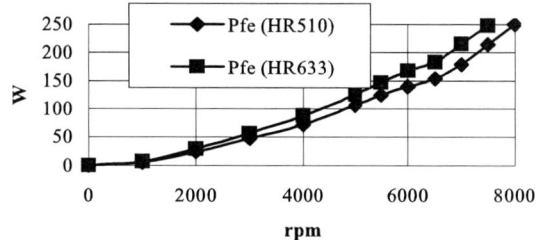

Figure 7. Core losses for two kinds of rotors

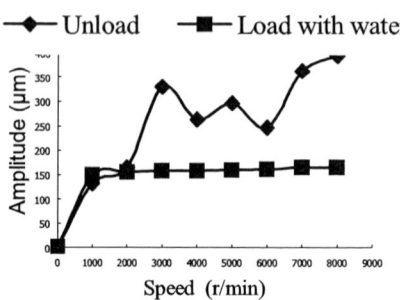

Figure 8. The rotor vibration amplitude in radial direction

water. When the bearingless motor pump system works at load, the vibration amplitudes are small because of well-proportioned liquid medium in the pump.

VII. Conclusions

The prototype of a bearingless permanent magnet slice motor with the power around 4 kW has been set up. An experimental platform of digital control system with TMS320F240 DSP controller is designed to meet the need. Experiments including speed regulation and suspension force control are performed. The experiment results have shown the good dynamic performance of the designed digital control system. The steady suspension of the rotor is realized and the speed of the motor can be continuously adjusted within the range of 0-8 000 r/min, and using the rotor HR510 Core losses are small. It is appropriate for high purity applications in the semiconductor industry, the chemical industry and so on.

Acknowledgment

The project was supported by the Laboratory of Electrical Engineering Design of Swiss Federal Institute of Technology (ETH) and GmbH in Zurich. The author was instructed by professor J. Hugel at ETH and worked together with Dr. Pascal N. Boesch in 2003.

References

[1] R. Bosch, "Development of a bearingless electrical machine," *ICEM*, pp. 373-375, 1988

[2] A. Salazar, A. Chiba, T. Fukao, "A review of developments in bearingless motors, " in *Proc. 7th Int. Symp. Magnetic Bearings*, Zurich, pp. 335-400, Aug. 2000.

[3] J. Bichsel, "Beitraege zum Lagerlosen Elektromotor", Ph.D. dissertation, ETH Zurich, Switzerland, 1990.

[4] N. Barletta, "Der lagerlose Scheibenmotor", Ph.D. dissertation, ETH Zurich, Switzerland, 1998.

[5] M. Neff, N. Barletta, R. Schoeb, "Bearingless centrifugal pump for highly pure chemicals, " in *Proc. 8th Int. Symp. Magnetic Bearings*, Mito, pp. 283-287, Aug. 2002.

[6] M. Neff, "Magnetgelagertes Pumpsystem fuer die Halbleiterfertigung", Ph.D. dissertation, ETH Zurich, Switzerland, 2003.

The effect of parameter variations on the performance of indirect vector controlled induction motor drive

A. Shiri*, A. Vahedi and A. Shoulaie
Department of Electrical Engineering
Iran University of Science and Technology, Tehran, Iran
E-mail*: abbas_shiri@yahoo.com

Abstract: In this paper, analytical functions for output torque and rotor flux of indirect vector controlled induction motor are defined and analyzed. According to the defined functions, the effect of parameter variations over outputs of vector controller can be determined. Matlab/simulink based simulations is carried out to confirm the results of the analytical functions.

Key words- induction motor; vector control; parameter variation; sensitivity analysis

I. INTRODUCTION

In recent decades, many investigations have been done by researchers to control ac motors similar to that of separately-excited dc machines that lead them to vector control theory [1]. Vector control made the ac drives equivalent to dc drives in the independent control of flux and torque. The major disadvantage of the indirect vector control scheme is that it is machine parameter dependant, since the model of the motor is used for flux estimation. The machine parameters are affected by variations in the temperature and the saturation levels of the machine. Any mismatch between the parameters in the motor and that instrumented in the vector controller will result in the deterioration of performance in terms of steady state error and transient oscillations of rotor flux and torque. These types of oscillations are not desired for some exact uses. Regarding the importance of sensitivity of vector control drive to the motor parameters, many investigations have been carried out in this field. In [2] the effects of rotor resistance and mutual inductance variations on output torque and rotor flux have been discussed qualitatively. In the other work the effect of the machine parameters variations on its outputs, referring to simulation results has been investigated and two techniques for rotor resistance estimation have been described [3]. Krishnan in [4] has derived approximate equations for parameter sensitivity of indirect vector control; and finely in some references, motor parameter estimation and compensation techniques and their effect on machine outputs have been described [5]-[9]. In this paper exact equations of parameter sensitivity have been derived. Using derived equations, the effect of parameter variations on the outputs of the machine can be determined. In the next sections, first the basic equations of indirect vector

control are presented, and then sensitivity analysis of this type of control carried out and analytical functions for output torque and rotor flux sensitivity are derived.

II. INDIRECT VECTOR CONTROL

In this section, the indirect vector controller is derived from the dynamic equations of the induction machine in the synchronously rotating reference frames. The rotor equations of the induction machine are given by:

$$R_r i_{qr}^{\ e} + p\lambda_{qr}^{\ e} + \omega_{sl}\lambda_{dr}^{\ e} = 0 \tag{1}$$

$$R_r i_{dr}^{\ e} + p\lambda_{dr}^{\ e} - \omega_{sl}\lambda_{qr}^{\ e} = 0 \tag{2}$$

where:

$$\omega_{sl} = \omega_s - \omega_r \tag{3}$$

$$\lambda_{qr}^{\ e} = L_m i_{qs}^{\ e} + L_r i_{qr}^{\ e} \tag{4}$$

$$\lambda_{dr}^{\ e} = L_m i_{ds}^{\ e} + L_r i_{dr}^{\ e} \tag{5}$$

In this equations, the various symbols denote the following: R_r, the referred rotor resistance per phase; L_m, the mutual inductance per phase; L_r, the stator referred rotor self inductance per phase; $i_{dr}^{\ e}$ and $i_{qr}^{\ e}$, the referred direct and quadrature axes currents, respectively; p, the differential operator; ω_{sl}, slip speed in rad/sec, ω_s and ω_r are synchronous speed and electrical rotor speed both in rad/sec, and $\lambda_{dr}^{\ e}$ and $\lambda_{qr}^{\ e}$ are rotor direct and quadrature axes flux linkages, respectively. The resultant rotor flux (λ_r) is assumed to be on the direct axis, to reduce the number of variables in the equations. Hence, aligning the d axis with rotor flux phasor yields:

$$\lambda_r = \lambda_{dr}^{\ e} \tag{6}$$

$$\lambda_{qr}^{\ e} = 0 \tag{7}$$

$$p\lambda_{qr}^{\ e} = 0 \tag{8}$$

Substituting equations (6) to (8) in (1) and (2) and using equations (4) and (5), the followings are obtained for i_f and ω_{sl}:

$$i_f = \frac{1}{L_m}[1 + pT_r]\lambda_r \tag{9}$$

$$\omega_{sl} = \frac{L_m}{T_r}\frac{i_T}{\lambda_r} \tag{10}$$

Where

$$i_f = i_{ds}{}^e \qquad (11)$$

$$i_T = i_{qs}{}^e \qquad (12)$$

$$T_r = \frac{L_r}{R_r} \qquad (13)$$

The q and d axis currents are labeled as torque (i_T) and flux (i_f) producing components of the stator current phasor, respectively. T_r, denotes the rotor time constant. Also using equations (6) to (8), we can summarize the induction machine torque equation as:

$$T_e = \frac{3}{2}\frac{p}{2}\frac{L_m}{L_r}(\lambda_r i_{qs}) = \frac{3}{2}\frac{p}{2}\frac{L_m}{L_r}\lambda_r i_T = K_{te}\lambda_r i_T \qquad (14)$$

Where, K_{te} is torque constant and is equal to:

$$K_{te} = \frac{3}{2}\frac{p}{2}\frac{L_m}{L_r} \qquad (15)$$

Note that the torque is proportional to the product of the rotor flux linkages and the stator q axis current. This resembles the torque expression of dc motor, which is proportional to the product of the field flux linkages and the armature current. If the rotor flux linkage is kept constant, then the torque is simply proportional to the torque producing component of the stator current (i_T), as in the case of the separately excited dc machine, where the torque is proportional to the armature current when the field current is constant. The rotor flux linkage and torque equations given in (9) and (14), respectively, complete the transformation of the induction machine into an equivalent separately excited dc machine from a control point of view.

III. SENSITIVITY ANALYSIS

A mismatch between the vector controller and induction motor occurs as a result of the motor parameters changing with operating conditions such as temperature rise and saturation or of the wrong instrumentation of the parameters in the vector controller. The mismatch produces a coupling between the flux and torque producing channels in the machine and degrades the performance of the controller.

In order to exact study of effect of machine parameters variation on drive outputs, the mathematical analyses are employed. For doing this, the machine equations which were derived in previous section are used. The motor electromagnetic torque from the equation (14) is equal to:

$$T_e = K_{te}\lambda_r i_T \qquad (16)$$

Replacing λ_r and K_{te} from equations (9) and (15) in equation (16), respectively we get:

$$T_e = \frac{1}{K_{it}}\frac{L_m}{L_r}\frac{L_m}{1+pT_r}i_f i_T \qquad (17)$$

Where $K_{it} = (\frac{2}{3})(\frac{2}{p})$. Referring to the figure 1, following equations are achievable:

$$\tan\theta_T = \frac{i_T}{i_f} \qquad (18)$$

$$i_T = i_s \sin\theta_T \qquad (19)$$

$$i_f = i_s \cos\theta_T \qquad (20)$$

And replacing equations (18) to (20) in equation (17) produces:

$$T_e = \frac{1}{K_{it}}\frac{L_m}{L_r}\frac{L_m}{1+pT_r}i_s{}^2 \sin\theta_T \cos\theta_T \qquad (21)$$

Also the slip speed derived from equation (10) as:

$$\omega_{sl} = \frac{L_m}{T_r}\frac{i_T}{\lambda_r} = \frac{L_m}{T_r}\frac{i_T}{\frac{L_m i_{ds}}{1+pT_r}} = \frac{1+pT_r}{T_r}\frac{i_T}{i_f} = \frac{1+pT_r}{T_r}\tan\theta_T \qquad (22)$$

Rearranging the above equation for $\tan\theta_T$ we can get:

$$\tan\theta_T = \frac{\omega_{sl}T_r}{1+pT_r} \qquad (23)$$

From which sine and cosine of the torque angle are defined as:

$$\sin\theta_T = \frac{(\omega_{sl}T_r)/(1+pT_r)}{\sqrt{1+(\frac{\omega_{sl}T_r}{1+pT_r})^2}} \qquad (24)$$

$$\cos\theta_T = \frac{1}{\sqrt{1+(\frac{\omega_{sl}T_r}{1+pT_r})^2}} \qquad (25)$$

Replacing $\sin\theta_T$ and $\cos\theta_T$ from (24) and (25), respectively in equation (21) we have:

$$T_e = \frac{1}{K_{it}}\frac{L_m{}^2}{[R_r(1+pT_r)^2][1+(\frac{\omega_{sl}T_r}{1+pT_r})^2]}\omega_{sl}i_s{}^2 \qquad (26)$$

Where $T_r = L_r/R_r$. Similar to the motor torque equation, commanded torque can be defined as:

$$T_e^* = \frac{1}{K_{it}}\frac{L_m{}^{*2}}{[R_r{}^*(1+pT_r{}^*)^2][1+(\frac{\omega_{sl}{}^*T_r{}^*}{1+pT_r{}^*})^2]}\omega_{sl}{}^* i_s{}^{*2} \qquad (27)$$

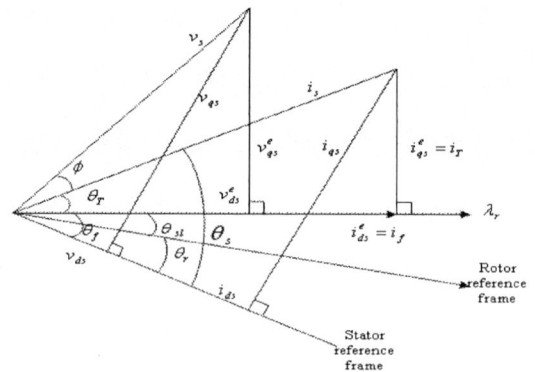

Figure 1.vector control phaser diagram

The command values are denoted with asterisks.
For rotor flux linkages following equation can be derived:

$$\lambda_r = \frac{L_m}{(1+pT_r)\sqrt{1+(\frac{\omega_{sl}T_r}{1+pT_r})^2}}i_s \qquad (28)$$

Similarly its command value is:

$$\lambda_r^* = \frac{L_m^*}{(1+pT_r^*)\sqrt{1+(\frac{\omega_{sl}^* T_r^*}{1+pT_r^*})^2}} i_s^* \quad (29)$$

Now by having the equations of torque and flux, we can define sensitivity functions in respect to the machine parameters. The ratio of torque to its command value is obtained:

$$\frac{T_e}{T_e^*} = \frac{L_m^2}{L_m^{*2}} \frac{R_r^*(1+pT_r^*)^2[1+(\frac{\omega_{sl}^* T_r^*}{1+pT_r^*})^2]}{R_r(1+pT_r)^2[1+(\frac{\omega_{sl} T_r}{1+pT_r})^2]} \frac{\omega_{sl}}{\omega_{sl}^*} \frac{i_s^2}{i_s^{*2}} \quad (30)$$

Also the ratio of actual to command rotor flux linkages is:

$$\frac{\lambda_r}{\lambda_r^*} = \frac{L_m}{L_m^*} \frac{1+pT_r^*}{1+pT_r} \frac{\sqrt{1+(\frac{\omega_{sl}^* T_r^*}{1+pT_r^*})^2}}{\sqrt{1+(\frac{\omega_{sl} T_r}{1+pT_r})^2}} \frac{i_s}{i_s^*} \quad (31)$$

Derived sensitivity functions are applicable for both transient and steady states. In steady state, p=0 and we can suppose $i_s = i_s^*$ and get:

$$\frac{T_e}{T_e^*} = \alpha\beta \frac{[1+(\omega_{sl}^* T_r^*)^2]}{[1+(\alpha\omega_{sl} T_r^*)^2]} \frac{\omega_{sl}}{\omega_{sl}^*} \quad (32)$$

$$\frac{\lambda_r}{\lambda_r^*} = \beta \frac{\sqrt{1+(\omega_{sl}^* T_r^*)^2}}{\sqrt{1+(\alpha\omega_{sl} T_r^*)^2}} \quad (33)$$

Where

$$\frac{T_r}{T_r^*} = \alpha \text{ and } \frac{L_m}{L_m^*} = \beta$$

And the following approximation was accepted (leakage inductance is supposed to be negligible compared to the mutual inductance): $\frac{L_r^*}{L_r} \approx \frac{L_m^*}{L_m}$

Ranges of α and β

Working conditions of the machine such as temperature rise, magnetic saturation and operation of the induction motor drive in the linear portion of the iron B-H characteristics can cause changes of α and β in the following manner [4]:

$$0.5 \le \alpha \le 1.5 \text{ and } 0.8 \le \beta \le 1.2$$

The torque and rotor flux sensitivity curves versus α and β are shown in figures (2) and (3). In figure (2-a) as it is seen, increasing the value of α causes a decrease in the ratio of torque to its command value, and increasing the value of β causes an increase in the ratio of rotor flux to its command value(the slip speed kept constant). In figure 2-(b), β is kept constant and torque variations are plotted as a function of α and ω_{sl}. It is evident from figure 2-(a) that the torque is not affected by the

variations of ω_{sl}. Also figure 3 shows that increasing of α and β causes a decrease and increase in rotor flux, respectively.

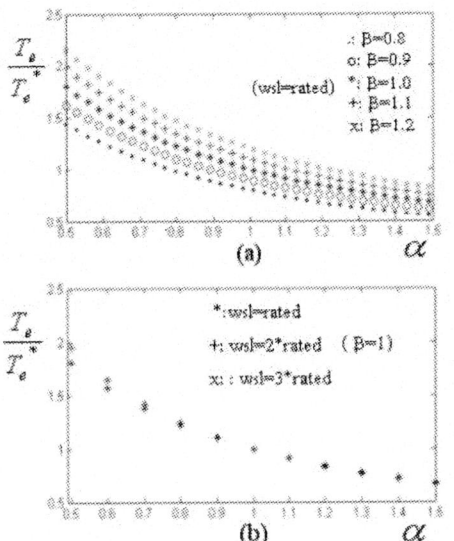

Figure 2. Torque variations as a function of α and β : (a) - α and β variations, (b) - α and ω_{sl} variations

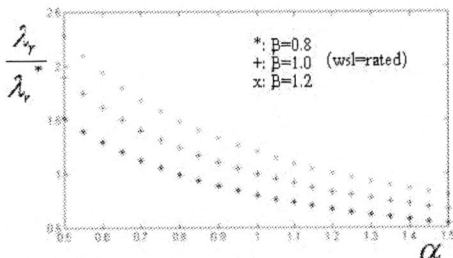

Figure 3. Rotor flux variations as a function of α and β

$$(\omega_{sl} = rated)$$

IV. SIMULATION RESULTS

To confirm the results of the analytical functions, vector controlled drive system and induction motor have been simulated in simulink/matlab environment. The motor parameters are presented in the appendix. Figure 4 illustrates the simulated drive system. Figure 5 gives the characteristics of torque, rotor flux and speed. In this figure all parameters are kept constant and the rotor resistance is doubled (α is halved). It is clear from this figure that decreasing α, increases both the torque and the rotor flux (rotor speed tracks the torque variations), and this is the same result as concluded in equations (32) and (33). Mutual inductance is doubled in figure 6 (β increases), making the torque and rotor flux increase (and confirm the results of equations (32) and (33)). Figure 7 depicts a reduction in the values of torque and rotor flux as the stator resistance increases; it is because of the voltage drop over stator resistance. The change of rotor resistance is shown as a step function; but practically it has a steep form. Figure 8 illustrates these kinds of rotor resistance variations whose results for torque and rotor flux coincide that of previous ones. Figures 9 and 10 show the curves representing the variations in torque and rotor flux as a function of rotor resistance and mutual inductance, respectively. The figures 5 – 10 indicate that the maximal effect on the torque and rotor flux is caused by variations of the rotor resistance. For a 100% increase in rotor resistance, the torque and rotor flux increase

1.2% and 66.6%, respectively, while for the same percent of increase in mutual inductance (L_m), the increase of torque and flux are only 0.43% and 8.3%, respectively. It should be mentioned that although the effect of variation of L_m on torque and flux is negligible, it causes oscillations in torque and flux that can cause resultant oscillations in rotor speed and this is not desirable for some specific applications.

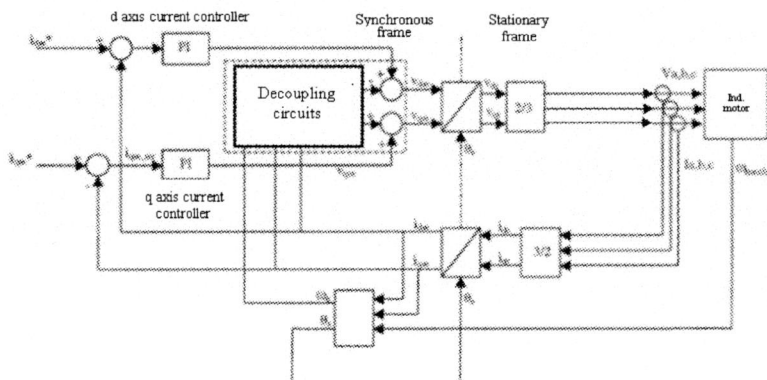

Figure 4. vector controller block diagram

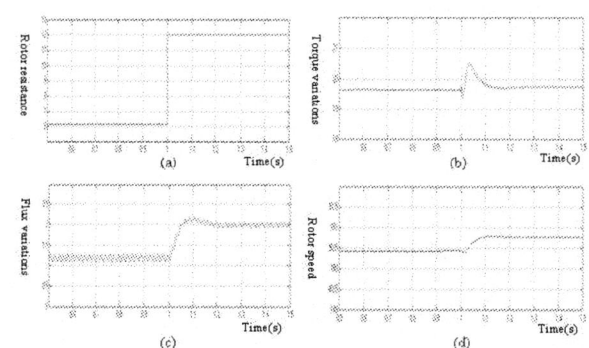

Figure 5. Torque, rotor flux and speed variations as rotor resistance change
(a) Rotor resistance (Ω), (b) torque variations (N.m), (c) rotor flux, (d) rotor speed

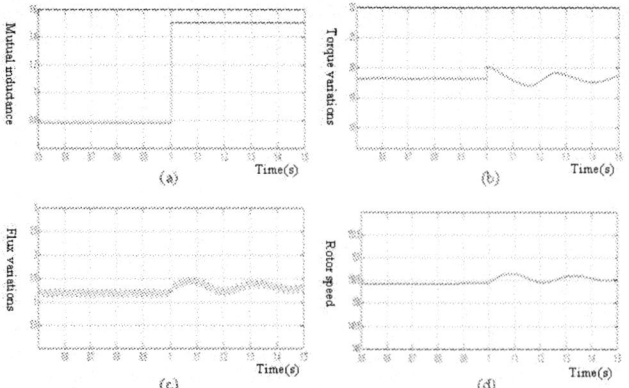

Figure 6. Torque, rotor flux and speed variations as mutual inductance change
(a)   Mutual inductance, (b) torque variations (N.m), (c) rotor flux, (d) rotor speed

824

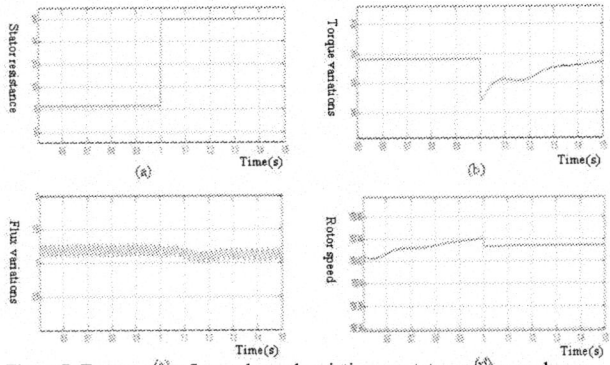

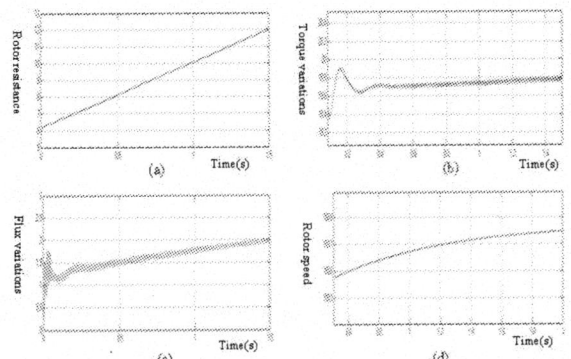

Figure 7. Torque, rotor flux and speed variations as stator resistance change
(a) Stator resistance (Ω), (b) torque variations (N.m), (c) rotor flux,
(d) rotor speed

Figure 8- Torque, rotor flux and speed variations as rotor resistance change
(a) Rotor resistance (Ω), (b) torque variations (N.m), (c) rotor flux,
(d) rotor speed

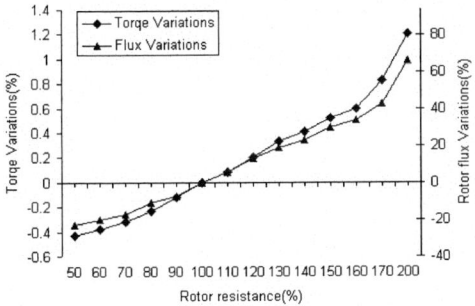

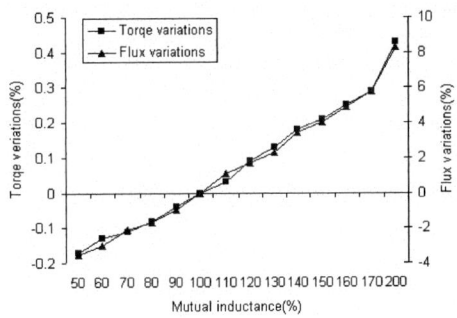

Figure 9. Variations in torque and rotor flux as a function of rotor resistance

Figure 10. Variations in torque and rotor flux as a function of mutual inductance

V. CONCLUSION

Analytical functions have been derived for the torque and rotor flux sensitivity. The results of the simulations show that variations of the motor parameters affect the outputs of the system. It has been concluded that variations of the rotor resistance in comparison with other parameters, cause the maximum deviation in torque and flux from their reference values. Changing the value of mutual inductance has a negligible effect on torque and flux values but it causes oscillations on them. These results coincide with the results of the analytical functions.

APPENDIX

The motor parameters are as:

$n_s = 1500 rpm$, $n_m = 1440 rpm$, $p_n = 4hp$, $r_r = 6.16\Omega$,

$r_s = 8.28\Omega$, $L_m = .7774H$, $L_{lr} = L_{ls} = 31.6mH$,

$J = 0.04 kg.m^2$.

REFERENCES

[1] F. Blaschke, "The principle of field orientation as applied to the new transvector closed loop control system for rotating machinery", Siemens Rev., 1972, 34, pp. 217 -220.
[2] R. Krishnan and A. S. Baharadwaj "A Review of Parameter Sensitivity and Adaptation in Indirect Vector Controlled Induction Motor Drive Systems" IEEE Tran. On Power Electronics, Vol. 6, No. 4, pp: 695-703, October 1991.

[3] B. Karanayil, M. F. Rahman and C. Grantham "PI and Fuzzy Estimators for On-line Tracking of Rotor Resistance of Indirect Vector controlled Induction motor drive" IEEE Conference On Electric Machines and Drives, pp: 820-825, 2001.
[4] R. Krishnan " electric motor drives: modeling, analysis & control" Prentice Hall pp. 411-477, 2001.
[5] A. Dittrich, "Parameter sensitivity of procedures for on-line adaptation of the rotor time constant of induction machines with field oriented control", IEE Proc., Electr. Power Appl., Vol. 141, No. 6, pp: 353-359, November 1994.
[6] R.W. De Doncker, "Parameter sensitivity of indirect universal field-oriented controllers", IEEE Trans. on Power Electronics, Vol. 9, No. 4, pp: 367-376, July 1994.
[7] H.A. Toliyat, E. Levi and M. Raina' "A review of RFO induction motor parameter estimation techniques" IEEE Tran. on Energy Conversion, Vol. 18, No. 2, pp: 271-283, June 2003.
[8] A. Ba-Razzouk, P. Sicard and V. Rajagopalan, "A simple on-line method for rotor resistance updating in indirect rotor flux orientation", IEEE Trans. on Power Electronics, Vol. 1, pp:317-322, Nov. 2002.
[9] R.S. Pena and G.M. Asher, "Parameter sensitivity studies for induction motor parameter identification using extended Kalman filters" The Fifth European Conference on Power Electronics Application, Vol. 4, pp: 306-311, Sep. 1993.

2006 5th International Power Electronics and Motion Control Conference

Magnetic Field Analysis and Performance Calculation for New Type of Claw Pole Motor with Permanent Magnet Outer Rotor

Fengge Zhang*, Shifu Zhang*, Haijun Bai*, Eugen Nolle**, Hans Pert Gruenberger**

* Shenyang University of Technology, Shenyang, 110023, China
** Esslingen University of Applied Sciences, Esslingen, 73732, Germany
e-mail: zhangfg@sut.edu.cn

Abstract—**This paper introduces the structure and operation principle of claw pole motor with permanent magnet outer rotor. The three dimensional magnetic field distribution at no-load condition has been analyzed by using the scalar magnetic potential. Moreover, the operation characteristic curves have been obtained at different speeds. The results of the paper are helpful to the computation of performance and optimal parameters.**

Keywords-claw pole; permanent magnet; outer rotor; synchronous machine

I. INTRODUCTION

Electrical machines with electrically excited claw pole DC magnetic field have been manufactured in mass production for many years. The main advantage of claw pole machines is that they can produce much higher torque density compared to the conventional machines. The number of poles can be increased while maintaining the magnetomotive force (MMF) substantially constant and this results in high torque density. These machines have been widely used for decades as automobile generators [1][2]. Because it has quite simple excitation coil and pole systems producing the excitation magnetic fields, hence, the cost of manufacturing is low [3][4].

The claw pole of claw pole motor (CPM) with electrically excited coil usually consists of steel laminations. The excessive eddy currents in the commonly used solid steel core, limit the motors to very small sizes and low speeds, and result in low efficiency [5].

In contrast to CPM, permanent magnetic claw pole motor (PMCPM) uses permanent magnet to replace excitation coil. Although the claw pole structure of the CPM and PMCPM is similar, the CPM has two sets of the brush and slip ring on the rotor and the PMCPM is brushless structure. Therefore, the PMCPM would have smaller volume and higher reliability than the CPM. Because the permanent magnet material is utilized in the design, the exciting power is not provided by winding and their current, the PMCPM would have higher efficiency and power factor than the CPM. Because of the outer rotor structure, the machine has the large moment of inertia.

The PMCPM has also the advantages of the CPM. This machine can widely obtain applications of high performance and direct drive fields.

Because of the complex structure of the claw pole, it is very difficult to fabricate the claw poles utilizing electrical steel laminations. With the improvement of soft magnetic composite (SMC), electrical machine design is no longer limited to the traditional iron lamination technology [6][7]. The SMC material can be readily pressed into complex geometries. SMC provides a new material choice where the component is manufactured as a solid piece composed of individually insulated particles and stimulates research into claw pole machines for a variety of applications.

SMC material has isotropic magnetic property, in spite of the low relative permeability value, they are suitable for applications in which the magnetic field is 3 dimensional such as claw pole machines and transverse flux machines.

Because the claw pole cores and parts of the PMCPM can be directly pressed, the further machining would be reduced. Moreover, SMC iron particles are insulated by its surface coating adhesive, the eddy current loss is much smaller than laminated steels over a wide frequency range, especially at higher frequencies. The total loss mainly is the hysteresis loss and its total loss has smaller than that of the CPM using lamination steels.

It can be seen from the analysis of the CPM and PMCPM, PMCPM has many advantages over CPM. Therefore, it is necessary to investigate the PMCPM.

II. THE STRUCTURE AND PRINCIPLE OF CLAW POLE MACHINE

A. Structure of claw pole machine

The topology of the PMCPM investigated here is illustrated in Fig. 1, where (1) is the shaft, (2) are the bearings, (3) is the thrust washer, (4) are the aluminum end plates, (5) are the screws, (6) is the mild steel yoke of the rotor, (7) are the claw poles of the stator, (8) is the stator core, (9) are the 12 permanent magnets, (10) is the stator winding or armature winding. The claw pole motor is the structure of outer rotor. The rotor is similar to that of a common PM machine, which includes a cylinder

1-4244-0448-7/06/$25.00 ©2006 IEEE

yoke, the 12 permanent magnets mounted on the inner surface of outer rotor and two aluminum end plates. The stator has three stacks and each consists of soft magnetic material claw pole pieces. The number of two claw discs is equal, which is equal to half the numbers of the pole. In assembly, left and right claw versus each other and it is uniform topology along the circumference.

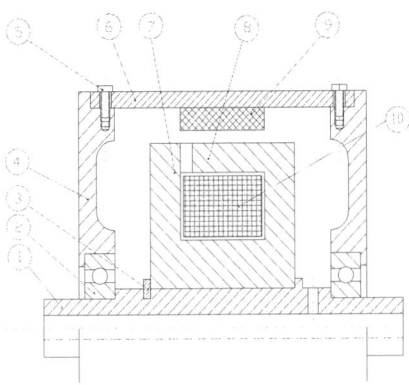

Figure 1. Schematic diagram of the PMCPM

The armature winding is located between two claw pole pieces. Each phase winding is a single concentrated coil and the manufacturing cost is relatively low because of the simple winding structure. The unique structure allows a considerably higher winding fill factor to be achieved. Moreover, the three stacks of the PMCPM are shifted each other by certain electrical degrees ($90°$ for two phases or $120°$ for three phases), the cogging and ripple torque produced by the three stacks at motor mode can compensate each other. The resultant pulsation torque would be very small, specially, the cogging torque can essentially cancel each other.

B. Principle of claw pole machine

When this machine operates, its structure must be two- or three-phases (stack), otherwise, it can not self-start. However, one-phase one-stack structure is the special example of multiple-phase multiple-stack machine, so the PMCPM of one-phase one-stack structure is considered as following analysis. The obtained theory or calculation results can extend to the PMCPM of the multiple-phase multiple-stack structure. In the PMCPM, the part with the PM is outer rotor, and the part with claw pole and single-phase winding is stator. Because the winding inductance is not related to the rotation of outer rotor, the electric-magnetic relationship of the PMCPM is similar to a conventional single-phase permanent magnet synchronous machine with the conceal pole rotor. The reluctance and reactance of d-axis is the same as that of q-axis.

In the PMCPM, the stator winding is supplied by a single phase inverter system. In this case the inverter system would convert DC voltage to AC sinusoidal voltage. The motor operates in synchronous mode by transistor bridge drive circuit.

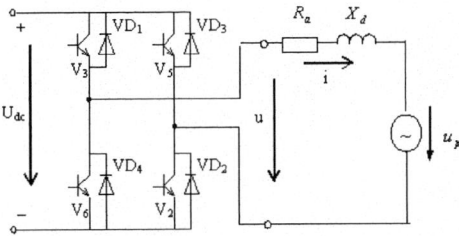

Figure 2. Schematic diagram of the PMCPM system

The schematic diagram of permanent magnet claw pole motor system is shown in Fig. 2, where R_a is the stator winding resistance, X_d is the synchronous reactance of the PMCPM. When the outer rotor of the PMCPM rotates, the flux linking the stator winding varies and the EMF (electromotive force), which is presented by u_p in Fig. 2, will be produced in single phase stator winding (the claw pole winding). Its instantaneous value can be written as following equation,

$$u = \sqrt{2}U_p \cos(\omega_1 t) \qquad (1)$$

$$U_p = 4.44 f\varphi N_1 k_{N1} \qquad (2)$$

where U_p is the effective value of the EMF. The EMF frequency depends on the rotor speed while the EMF waveform is determined by the waveform for the main magnetic flux. N_1 is the number of turns of the stator coil and f is the EMF frequency computed by the rotor speed.

$$f = \frac{pn}{60} \qquad (3)$$

Where p is the number of pole pairs of the machine and n the rotor speed in rev/min. The total magnetic flux flowing through the stator yoke and shaft can be considered as the total flux linking the stator coil.

The voltage balanceable equation of the PMCPM stator winding may be expressed as

$$U = U_p + IR_a + jI_d X_d \qquad (4)$$

Where U and U_p are the source voltage and the EMF produced by outside permanent magnets respectively, I is the armature current flowing through the stator winding.

When three phase AC current flow through stator windings, two sets of stator claws present alternately North pole or South pole. This variational magnetic field can bring the rotor with PM pieces to rotate. Thus, the electricity energy is converted into mechanical energy. However, it should be taken notice of the single phase PMCPM can not generate start-torque (the torque may be zero at some position), therefore, this kind of machine must be designed to two-phase or three-phase structure.

827

III. THE ANALYSIS OF 3-D MAGNETIC FIELD

Compared with conventional electrical machine, where analysis can be conducted in a two dimensional plane perpendicular to the machine shaft, the claw pole machine must be considered for tangential, axial and radial flux components.

Due to the complex structure of claw pole motor, which results in a truly three dimensional magnetic field in the machine, it is very difficult or almost impossible to obtain the accurate results of parameters and performance of the electrical machine by conventional equivalent magnetic circuit method [8][9]. It can not be accurately analyzed by a 2D method and a 3D method is necessary.

Finite element method is the commonest method in three dimensional magnetic field analysis. The solution of magnetic field problems is commonly obtained using potential functions. Two kinds of potential functions, the magnetic vector potential (MVP) and the magnetic scalar potential (MSP) are used depending on the problem to be solved.

The MVP formulation has more (three) degrees of freedom per node than the scalar method: AX, AY, and AZ, the magnetic vector DOFs in the X, Y, and Z directions. The MSP formulation has no direction and has the magnitude. hence, the MVP is three times than the MSP in the equation quantity. The MVP employed often occupies the vast computer resource. For current-conducting regions (including the permanent magnet region), it should be treated properly. Therefore, it can be reduced the time and improved the efficiency by using the scalar magnetic potential.

In this paper, in order to obtain accuracy of finite element analysis, 3-D magnetic field distribution has been investigated based on ANSYS for new type of claw pole motor with permanent magnet outer rotor. A model of the motor is modelled with the aid of the finite element package. The whole geometry is shown in Fig. 3.

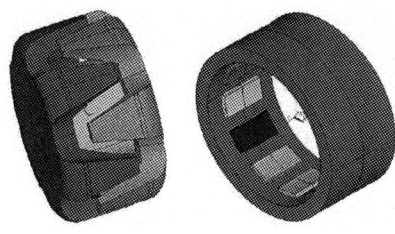

Figure 3. Geometry of the PMCPM

Because of the symmetrical structure, it is only required to analyse the magnetic field in one pole pitch. At the two radial boundary planes of one pole pitch, the magnetic scalar potential follows the half periodical boundary conditions:

$$\varphi_m(r, \Delta\theta, z) = -\varphi_m(r, -\Delta\theta, -z) \tag{5}$$

Where, $\Delta\theta = 15°$ is the half angle of one pole pitch.

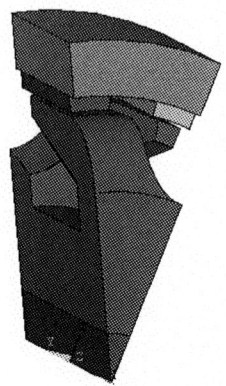

Figure 4. The model of solution

By using the period boundary condition, the model of the PMCPM is studied in one pitch as shown in Fig.4.

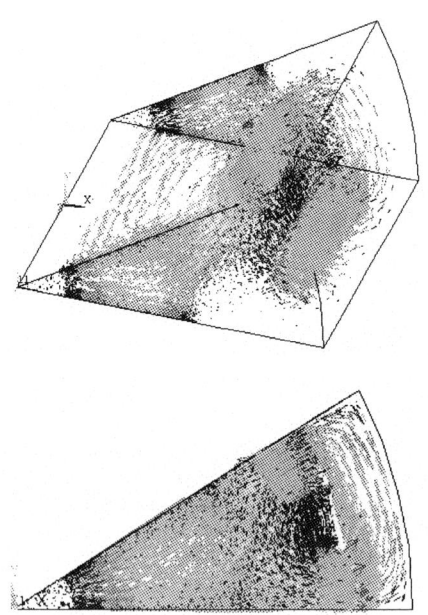

Figure 5. Magnetic flux density vector at no-load

Fig. 5 shows the magnetic field distribution at no-load condition. It is shown that the major path for the magnetic flux produced by permanent magnet is along one of the permanents, the main air gap, one of the claw pole, the stator yoke, another claw pole, main air gap, another permanent magnet, the rotor yoke to form a closed loop. The claw pole changes axial flux into radial flux.

828

IV. THE DETERMINATION OF DIMENSIONS AND PARAMETERS OF THE PMCPM

A. The choice of PM material

Compared the size and performance of PM depend on the PM material properties to very large degrees. Generally, the following factors need to be considered.

(1) In order to obtain the high power density and the high efficiency of the machine, the PM material should good magnetic property including remanence B_r, coercive force H_c and maximum energy product $(BH)_{max}$.

(2) The magnetic properties of the PM material should be stable and the demagnetization curve should be linear within the operating temperature range. An important issue is to ensure permanent magnet not to demagnetize.

Combining the above considerations, the permanent magnet of remanence (1.26T) and coercive force (925 kA/m) have been chosen. The type of the PM is VACODYM655 TP275/95.

B. The shape of the claw pole

The shape of claw pole has three types: equal wide claw pole, trapezia claw pole, sinusoidal claw pole. In this paper, the pole surface of claw pole is equilateral trapezia, the longtitudinal section plane along the pole body is the cuniform, and the cross section plane is the sector. Because of the complex geometry of claw pole, the leakage flux between the claws is rather difficult to compute. For the convenience of magnetic field analysis, the claw pole is divided into several segments along both circumferential directions as shown in Fig. 6.

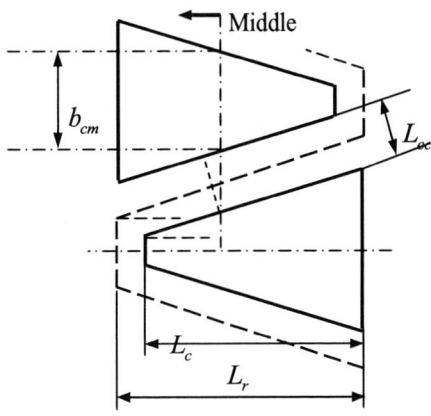

Figure 6.   Schematic diagram of the stator claw

Where L_c is the length of rotor claw pole along the axis,

L_r is the length of rotor along the axis, b_{cm} is the mean length of claw along the circumference and L_{cc} is the parallel length between adjacent claws along the circumferential directions.

C. The dimensions and parameters of the PMCPM

Table I lists the dimensions and the major parameters of machine designed

TABLE I.
THE DIMENSIONS AND MAJOR PARAMETERS OF THE PMCPM

Dimensions and parameters	Quantities
Number of phases	1
Number of pole pair	6
Number of permanent magnet	12
Remanence (T)	1.26
Coercive force (kA/m)	925
Stator core material	SMC
The length of air gap (mm)	0.7
Rotor outer radius (mm)	60
Rotor inner radius (mm)	48
The outer rotor yoke (mm)	10
The rotor iron core (mm)	37
Rated voltage (v)	12
the length of rotor claw pole L_c (mm)	26
the length of rotor L_r (mm)	31

V.   PERFORMANCE CALCULATION

When the PMCPM operates with load, the armature current would form armature magnetic field. After considering the effect of the armature reaction, the air gap compositive magnetic field would be different from no-load magnetic field. The magnetic flux and saturation degree of the magnetic circuit all are changed. And the iron core loss, the inductance parameter and armature current will also be varied. So the magnetic circuit of the PMCPM with the load should also be calculated again.

Fig. 7 and Fig. 8 show the relationship of core loss against frequency at no-load and load condition. It can be noted that the core loss of electrical machine is similar to linear rise with the frequency increases. The PMCPM has a large armature reaction field at load. Therefore, the armature reaction has already been taken into account in the design.

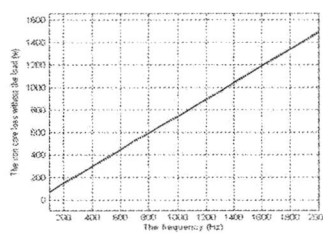

Figure 7. The curve of iron loss vs. frequency at no-load

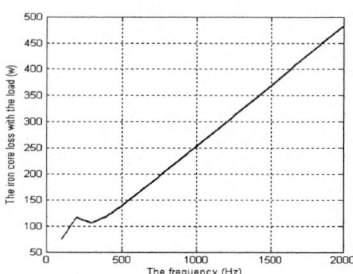

Figure 8. The curve of iron loss vs. frequency at load

The equivalent circuit of the synchronous motor can be used for the claw pole motor to predict the performance. The performance of the PMCPM has been simulated by the equivalent circuit model as shown in Fig. 2.

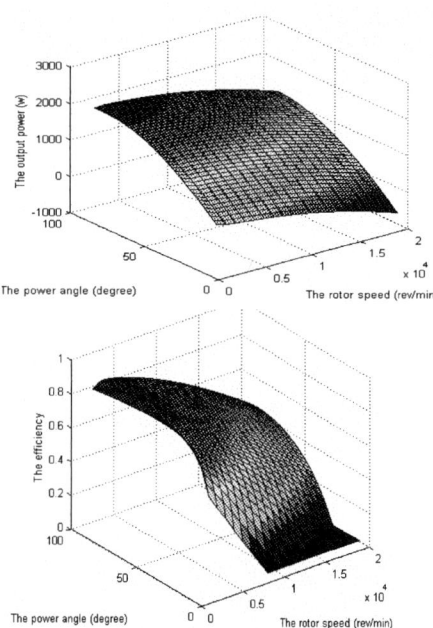

Figure 9. The performance of the PMCPM

Fig. 9 illustrates the characteristic curves of the PMCPM. It shows the relationship of the efficiency or output power against the power angel at different speeds.

VI. CONCLUSION

Because of the three dimensional structure of claw pole machine, it is very difficult to obtain accurate result. This paper analyzes the electromagnetic field distribution of the PMCPM at no-load by utilizing the scalar potential method. According to the geometry of machine, it has simplified for finite element model. It obviously shows that 3D finite element method is feasible to the field of the claw pole machine and computed time is greatly reduced. The output power and efficiency curves show the operation characteristics of the PMCPM at different power angle and speeds. The results obtained provide theory basis and important value for the optimization of parameter and performance.

REFERENCES

[1] Landmark, S.T. and Hamdi, E.S.(2004), "Design of claw-pole machines using SMC cores," WSEAS Transactions on Circuits and Systems, vol. 3, pp.1729-1934.0

[2] Landmark, S.T., et al, "Effect of pole face profile on performance of a class of claw-pole motors," Norfa Summer Seminar, Tallinn, Estonia.

[3] Y.G. Guo, J.G. Zhu, and W. Wu, "Design and analysis of electric motors with soft machine composite core," IEEE Transactions on Industry Applications, vol. 39, pp. 1658–1664.

[4] Guo Y G., Zhu J G. "Magnetic field calculation of claw pole permanent magnet machines using magnetic network method," J.Electron. ENG., vol. 22, pp. 69–75.

[5] Y.G.Guo, J.G. Zhu, P. A. Watterson, "Improved design and performance analysis of a claw pole permanent magnet smc motor with sensorless brushless DC drive," PEDS '2003, pp. 704–709.

[6] Wen Ouyang, Surong Huang, Anne Good, T.A. Lipo, "Modular permanent magnet machine based on soft magnetic composite," unpublished.

[7] A.G. Jack. "Experience with the use of soft magnetic composites in electrical machines," Proceeding of International Conference on Electrical Machines, Istanbul, Turkey, 1998, pp.1441–1448.

[8] Ramesohl I, Henneberger G, Kuppers S, et al, "Three dimensional calculation of magnetic forces and displacements of a claw pole generator," IEEE Transaction on Magnetics, vol. 32, pp. 1685–1687.

[9] Vlado Ostovic, John M. Miller, Vijay K. Garg, et al, "A magnetic-circuit-based performance computation of a lundell alternator," IEEE Transactions on Industry Applications, 1999, vol. 35, pp. 825–830.

2006 5th International Power Electronics and Motion Control Conference

Performance Analysis of a PM Claw Pole SMC Motor with Brushless DC Control Scheme

Youguang Guo[*], Jianguo Zhu[*], Jiaxin Chen[*,**], and Jianxun Jin[***]

[*]Faculty of Engineering, University of Technology, Sydney, P.O. Box 123, Broadway, NSW 2007, Australia
[**]College of Electromechanical Engineering, Donghua University, Shanghai 200051, China
[***]School of Automation Engineering, U. of Electronic Sci. & Tech. of China, Chengdu, Sichuan 610054, China
youguang@eng.uts.edu.au, joe@eng.uts.edu.au, chjiaxin@dhu.edu.cn, jxjin@uestc.edu.cn

Abstract—**Thanks to its unique properties, such as isotropic magnetic and thermal properties and low eddy current loss, the soft magnetic composite (SMC) material is suitable for application in electrical machines, especially those with complex structures and three-dimensional (3D) magnetic fluxes. This paper presents the performance analysis of a three-stack permanent magnet (PM) claw pole motor with an SMC stator core. 3D finite element magnetic analysis and improved formulations are applied to accurately compute the motor parameters, such as the back electromotive force, incremental inductance, cogging torque, and core loss. An equivalent electrical circuit is derived to predict the motor's steady-state performance under a brushless DC control scheme. Because of the large winding inductance of this type of motors, the control of the output torque and speed can be difficult. To verify the motor controllability, a Matlab/Simulink-based simulation model is compiled to simulate the motor dynamic and steady-state performances. Experiments are conducted on the motor prototype, validating the theoretical computations and analyses.**

Keywords-soft magnetic composite; claw pole motor; brushless DC control; finite element magnetic field analysis; performance simulation

I. INTRODUCTION

Compared to the laminated steels commonly used in electrical machines, the soft magnetic composite (SMC) material possesses a number of advantages, such as isotropic magnetic and thermal properties, low eddy current loss and relatively low total core loss at medium and higher frequencies, net-shape fabrication process with smooth surface and good finish (without need of any further machining), and prospect of very low cost mass production [1]. Therefore, SMC materials have a great potential for electrical machine applications, especially for those with complex topologies and three-dimensional (3D) magnetic fluxes, such as claw pole and transverse flux motors [2].

Due to its powdered nature, SMC is naturally magnetically isotropic, and this creates key design benefits [3]. The magnetic circuits can now be designed with 3D flux paths, and different radical topologies can be exploited to achieve high motor performances, for the

reason that the magnetic field does not have to be restricted in the two-dimensional (2D) plane of laminated steels.

To investigate the application potential of SMC, a three-stack permanent magnet (PM) claw pole motor with an SMC stator core has been developed by taking advantage of the unique properties of the material, as shown in Fig. 1 [4]. In this motor, the fluxes generated by the rotor PMs and the stator windings are 3D. For example, the flux produced by the PMs passes the air gap and flows into the stator claw poles via both the face and the side surfaces.

The three phases of the motor are stacked axially with an angular shift of 120° electrical from each other. Each stator phase has a single coil (not shown in the figure for clarity) around an SMC core, which is molded in two halves. The outer rotor comprises a tube of mild steel with an array of magnets for each phase mounted on the inner surface. Mild steel is used for the rotor because the flux density in the yoke is almost constant.

The major dimensions include: 94 mm for the outside diameter, 93 mm for the active axial length, 1.0 mm for the main air gap, and 81 mm for the average diameter of the airgap, etc. The motor is designed to operate under a brushless DC control scheme, delivering a torque of 2.65 Nm at 1800 rpm.

Figure 1.   Magnetically relevant parts of the claw pole SMC motor

1-4244-0448-7/06/$25.00 ©2006 IEEE

Due to the complicated structure, 3D numerical field analysis is required for accurate computation of the motor parameters. In this paper, the 3D magnetic field finite element analysis (FEA) is performed to calculate the key parameters, such as the winding flux, back electromotive force (*emf*), inductance and core losses. To predict the motor characteristics, an equivalent electrical circuit is derived under the optimal brushless DC control condition, i.e. the back *emf* is in phase with the stator current.

Generally, the claw pole machine has a large winding inductance, which may cause difficulty in the control of the output torque and speed. To verify the controllability of the motor, especially at high speed operation, a Matlab/Simulink-based model is complied to simulate the dynamic and steady-state performances. The experimental results on the motor prototype validate the theoretical analyses.

II. PARAMETER CALCULATION BY 3D FEA

A. 3D Magnetic Field FEA

By considering the detailed structure and dimensions of the motor and the non-linearity of ferromagnetic materials, the magnetic field FEA can accurately determine the magnetic field distribution and hence the parameters. Due to the almost magnetic independence and structural symmetry between stacks, only one pole region of one stack is needed for FEA, as illustrated in Fig. 2. On the two radial boundary surfaces, the magnetic scalar potentials obey the half-periodical conditions as

$$\varphi_m\left(r, \Delta\theta/2, z\right) = -\varphi_m\left(r, -\Delta\theta/2, -z\right) \tag{1}$$

where $\Delta\theta = 18°$ is the angle of one pole pitch.

Fig. 3 illustrates the no-load magnetic flux density vectors produced by the rotor PMs. It can be seen that the major path of the PM flux is along one of the PMs – the main air gap – half of the SMC claw pole stator core disk – the stator yoke – another half of SMC claw pole stator core disk – main air gap – another PM and then – the mild steel rotor yoke to form a closed loop. The magnetic field in the armature is really complex and SMC is an ideal candidate as the core material.

B. Back emf

The PM flux, defined as the flux linking a stator phase winding produced by the rotor PMs, can be obtained from the no-load magnetic field distribution (Fig. 3). The flux waveform is calculated by rotating the rotor magnets for one pole pitch in 12 steps. As plotted in Fig. 4, this flux waveform is almost perfectly sinusoidal versus the rotor position.

When the rotor rotates, the PM flux varies and an *emf* is induced in the stator winding. The *emf* frequency depends on the rotor speed, while its waveform is determined by the profile of the flux versus the rotor position. The *emf* constant is 0.2594 Vs/rad, by

$$K_E = \frac{p}{2} N_1 \frac{\phi_1}{\sqrt{2}} \tag{2}$$

where $p=20$ is the number of poles, $N_1=75$ is the number of turns of a phase winding, and ϕ_1 the magnitude of the sinusoidal flux waveform.

C. Winding Inductances

The behavior of an electrical circuit is governed by the incremental inductance rather than the secant inductance [5]. In this paper, the phase winding incremental inductance of the claw pole motor is calculated by a modified incremental energy method [6], which includes the following steps: (1) For a given rotor position, θ, conduct a non-linear field analysis considering the saturation due to the PMs to find out the operating point of the motor, and save the incremental permeability in each element; (2) Set the remanence of PMs to zero, and conduct a linear field analysis with the saved permeabilities under a perturbed stator current excitation, Δi; (3) Find out the co-energy; and (4) Calculate the self incremental inductance by

$$L_{11}(\theta) = \frac{2W_c\left(\Delta i, \theta\right)}{\left(\Delta i\right)^2} \tag{3}$$

The mutual inductance between phase windings can be considered as zero due to the independent magnetic circuit of each stack. Fig. 5 shows the computed self incremental inductance (L_{inc}) of one phase winding at different rotor positions. For comparison, the computed secant inductance (L_{sec}) and the measured inductance (L_{mea}) by the AC voltage-current method, are also plotted in the figure.

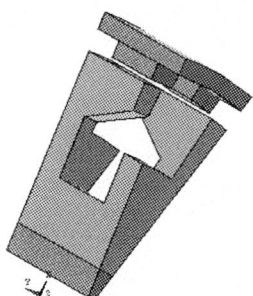

Figure 2. Region for magnetic field FEA

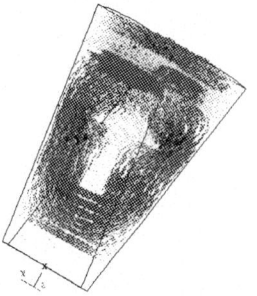

Figure 3. Plots of magnetic flux density vectors at no-load

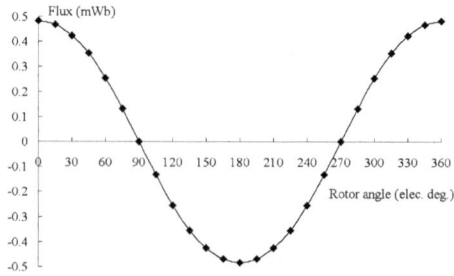

Figure 4. Per turn no-load flux of a phase winding

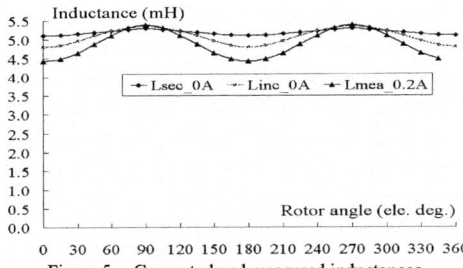

Figure 5. Computed and measured inductances

D. Core Losses

The core loss is caused not only by alternating but also by rotational magnetic fields, and should be properly considered in the motor design and performance analysis [7]. In this paper, an improved method is applied to predict the core losses in the 3D flux SMC motor [8]. Different formulations are used for core loss prediction with purely alternating, purely circular rotating, and elliptically rotating flux density vectors, respectively. A series of 3D FEAs are conducted to determine the flux density locus in each element when the rotor rotates.

It is found that the core loss increases almost linearly with respect to the rotor speed, due to the dominant hysteresis loss component in SMC. At the rated speed of 1800 rpm, the core loss is calculated as 58 W at no-load, and will increase by about 20% at the rated load due to the effect of armature current.

III. STEADY-STATE PERFORMANCE CALCULATION BY EQUIVALENT ELECTRIC CIRCUIT

When running in synchronous mode, the motor's steady-state performance can be predicted by the equivalent circuit model as shown in Fig. 6, where E_1 is the induced stator *emf*, R_1 the stator winding resistance, ω_1 the angular frequency, and L_1 the synchronous inductance of the phase winding. The motor is assumed to operate in the optimum brushless DC mode, i.e. I_1 in phase with E_1, so that the electromagnetic power and torque can be obtained by

$$P_{em} = mE_1I_1 \tag{4}$$

$$T_{em} = \frac{P_{em}}{\omega_r} = K_TI_1 \tag{5}$$

where ω_r is the rotor speed in rad/s, $K_T=mK_E$ is the torque constant, and $m=3$ is the number of phases. The rms value of the back *emf* is determined by $E_1=K_E\omega_r$.

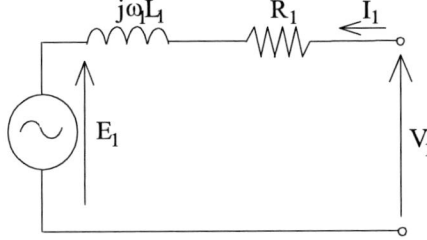

Figure 6. Per-phase equivalent electrical circuit

For a given terminal voltage, V_1, the relationship between the rotor speed and electromagnetic torque is determined by

$$(K_E\omega_r + \frac{R_1T_{em}}{K_T})^2 + (\frac{p\omega_rL_1T_{em}}{2K_T})^2 = V_1^2 \tag{6}$$

The output power, output torque, input power, and efficiency are calculated by

$$P_{out} = P_{em} - P_{Fe} - P_{mec} \tag{7}$$

$$T_{out} = P_{out}/\omega_r \tag{8}$$

$$P_{in} = P_{em} + P_{cu} \tag{9}$$

$$P_{cu} = 3I_1^2R_1 \tag{10}$$

$$\eta = P_{out}/P_{in} \tag{11}$$

where P_{Fe} is the core loss, P_{mec} the mechanical loss including windage and friction, and P_{cu} the copper loss.

IV. PERFORMANCE ANALYSIS BY A MATLAB/SIMULINK-BASED SIMULATION MODEL

The large winding inductance of claw pole motors has effect on the rise rate of the stator current, which may cause difficulties in motor control and limit the motor output toqrue, especially when the motor operates at high speeds. In this paper, the output capacity, such as the maximum steady-state speed that the motor can reach for a given load torque and inverter voltage, is investigated by a Matlab/Simulink-based simulation model. The presented model can also be employed to simulate the dynamic characteristics, such as the curves of speed, current and torque during the start-up or transients when the load or power supply varies.

A. Modeling of Brushless DC Motor with Sinusoidal Waveform Back emf

For simplification, the phase winding inductance can be considered as a constant, e.g. the average value over a variation cycle. From Fig. 5, the average self-inductance of a phase winding is L=4.9 mH. The mutual inductance between two phase windings is negligible, i.e. M=0. The voltage equations of the three phase windings can be written as

$$\begin{bmatrix} V_a \\ V_b \\ V_c \end{bmatrix} = \begin{bmatrix} R_1 & 0 & 0 \\ 0 & R_1 & 0 \\ 0 & 0 & R_1 \end{bmatrix} \begin{bmatrix} i_a \\ i_b \\ i_c \end{bmatrix} + \begin{bmatrix} L & M & M \\ M & L & M \\ M & M & L \end{bmatrix} \frac{d}{dt} \begin{bmatrix} i_a \\ i_b \\ i_c \end{bmatrix} + \begin{bmatrix} E_a \\ E_b \\ E_c \end{bmatrix} \tag{12}$$

where V_a, V_b, and V_c are the voltages of three phase windings, i_a, i_b, and i_c the three phase currents, and E_a, E_b, and E_c the three phase back *emfs*.

For the symmetrically distributed three phase windings with star connection, the three phase currents obey

$$i_a + i_b + i_c = 0 \tag{13}$$

The three phase back *emfs* are:

$$\begin{cases} E_a = E_m \sin(\omega t) \\ E_b = E_m \sin(\omega t - 120^\circ) \\ E_c = E_m \sin(\omega t - 240^\circ) \end{cases} \qquad (14)$$

where $E_m = \sqrt{2} K_E \omega_r$ is the magnitude of the sinusoidal back *emf*, ω the angular frequency, and $\omega = (p/2)\omega_r$.

The electromagnetic torque is calculated by

$$T_{em} = \frac{E_a i_a + E_b i_b + E_c i_c}{\omega_r} \qquad (15)$$

The motion equation is

$$\frac{d\omega_r}{dt} = \frac{T_{em} - T_L - \delta_0 \omega_r}{J} \qquad (16)$$

where T_L is the load torque, δ_0 the friction coefficient, and J the total inertia of the rotating parts.

B. Power Electronic Drive Circuit

Fig. 7 illustrates the schematic diagram of the typical drive circuit of brushless DC motors, from which one can work out the relationship between the phase voltages and terminal potentials (voltages) as

$$\begin{bmatrix} V_a \\ V_b \\ V_c \end{bmatrix} = \begin{bmatrix} U_a - U_N \\ U_b - U_N \\ U_c - U_N \end{bmatrix} \qquad (17)$$

where U_a, U_b, U_c, and U_N are the electrical potentials (voltages) of terminals a, b, c and N (the neutral point), respectively, and U_{dc} is the DC link voltage of the inverter.

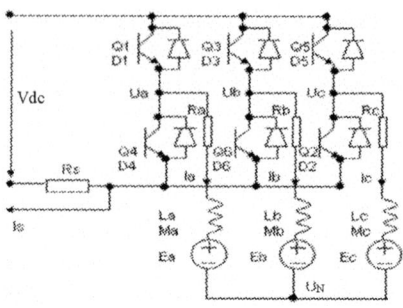

Figure 7. A typical drive circuit for brushless DC motor

Assuming that the hard switching is applied, at the moment when phase a is positively excited and phase b is negatively excited, the following equations can be obtained:

$$U_N = \begin{cases} \dfrac{\sum\limits_{k=a}^{c}(U_k - E_k)}{3} & i_c \neq 0 \\[2mm] \dfrac{\sum\limits_{k=a}^{b}(U_k - E_k)}{2} & i_c = 0 \end{cases} \qquad (18)$$

C. Performance Simulation

According to (12)-(18), a Matlab/Simulink-based simulation model is built as shown in Fig 8. The basic design requirement for the motor drive system is that for an output torque of 2.65 Nm, the steady-state speed can reach 1800 rpm when the applied voltage is $V_{dc}=165$ VDC. By using the proposed model, the motor drive system is simulated under these conditions and some results are plotted in Figs. 9-11, showing that the motor can meet the design requirements.

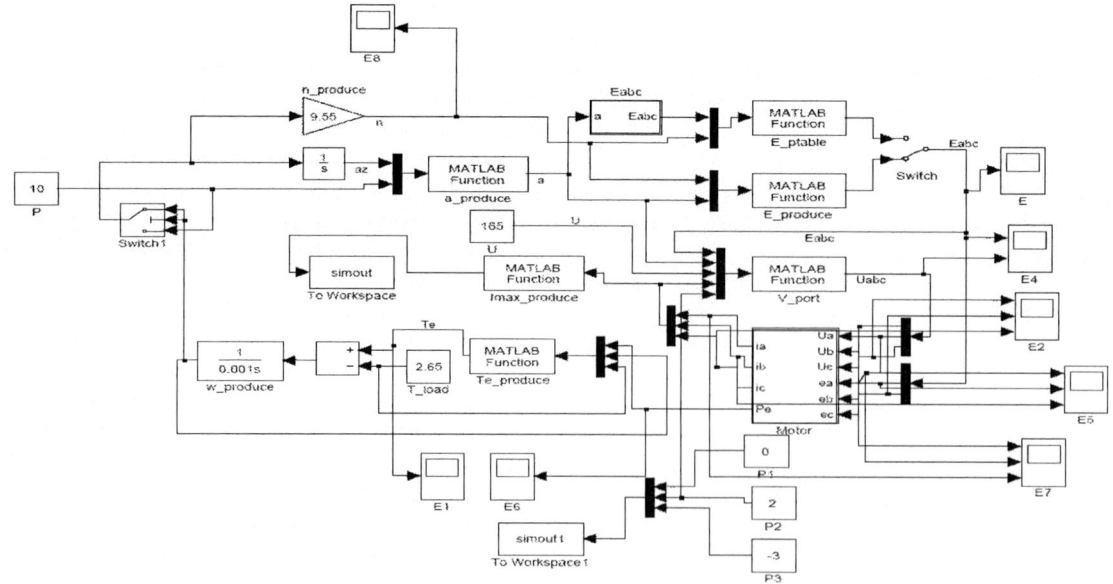

Figure 8. Matlab/Simulink-based simulation model of the brushless DC motor with sinusoidal back *emf*

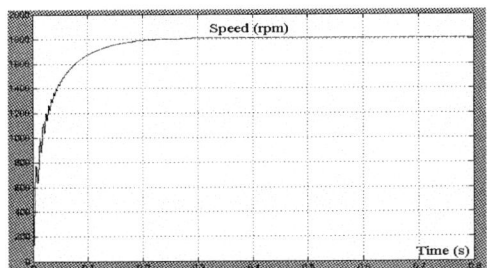

Figure 9. Speed curve during the start-up with a load of 2.65 Nm when the inverter voltage Vdc=165 V

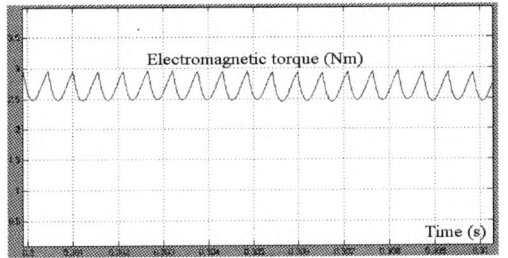

Figure 10. Steady-state electromagnetic torque when Vdc=165 V

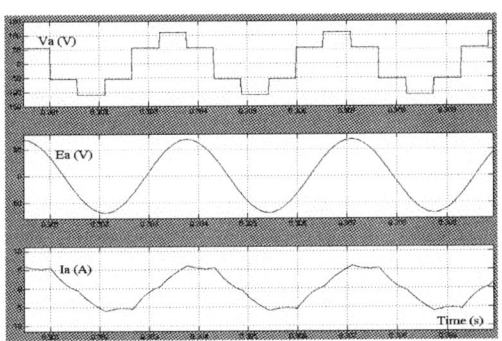

Figure 11. Voltage, back *emf* and current of a phase winding

V. EXPERIMENTAL VALIDATION

The motor prototype has successfully operated with a sensorless brushless DC scheme, delivering a toque of 2.65 Nm at 1800 rpm when the inverter DC link voltage is 165 V. Fig. 12 plots the measured torque/speed curves with different inverter voltages. It can be seen that the theoretical analysis agrees well with the experiments.

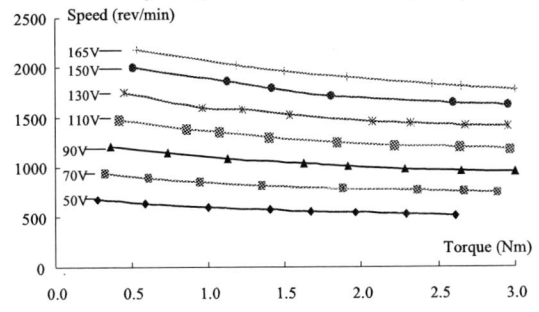

Figure 12. Measured mechanical characteristics

VI. CONCLUSION

SMC materials have a great application potential in electrical machines, particularly those with complex topologies and 3D fluxes. This paper presents the performance analysis of a three-stack PM claw pole motor with an SMC stator core by an equivalent electrical circuit and a Matlab/Simulink-based simulation model.

For accurate computation of the motor parameters, the 3D finite element magnetic field analysis is conducted. By the numerical analysis, the rotor position dependence of the back *emf* and inductance has been determined and can be considered in the simulation model for more accurate analysis.

The calculations and simulations have been validated by the experimental results on the claw pole SMC motor prototype.

REFERENCES

[1] "The latest development in soft magnetic composite technology," *SMC Update, Reports of Höganäs AB*, Sweden, 1997-2005.

[2] Y. G. Guo, J. G. Zhu, P. A. Watterson, W. Wu, "Comparative study of 3-D flux electrical machines with soft magnetic composite core," *IEEE Trans. on Industry Applications*, Vol. 39, No. 6, pp. 1696-1703, Nov. 2003.

[3] A. G. Jack, "Experience with the use of soft magnetic composites in electrical machines," in *Proc. Int. Conf. on Electrical Machines*, Istanbul, Turkey, Sept. 1998, pp. 1441-1448.

[4] Y. G. Guo, J. G. Zhu, P. A. Watterson, and W. Wu, "Development of a claw pole permanent magnet motor with soft magnetic composite stator," *Australian J. Electrical & Electronic Engineering*, Vol. 2, No. 1, pp. 21-30, 2005.

[5] M. Gyimesi and D. Ostergaard, "Inductance computation by incremental finite element analysis," *IEEE Trans. Magn.*, Vol. 35, pp. 1119-1122, 1999.

[6] Y. G. Guo, J. G. Zhu, H. W. Lu, R. Chandru, S. H. Wang, and J. X. Jin, "Determination of winding inductance in a claw pole permanent magnet motor with soft magnetic composite core," in *Proc. Australasian Univ. Power Eng. Conf.*, Hobart, Australia, Sept. 2005, pp. 491-496.

[7] Y. G. Guo, J. G. Zhu, J. J. Zhong, and W. Wu, "Core losses in claw pole permanent magnet machines with soft magnetic composite stators," *IEEE Trans. Magn.*, Vol. 39, No. 5, pp. 3199-3201, Sept. 2003.

[8] Y. G. Guo, J. G. Zhu, J. J. Zhong, P. A. Watterson, and W. Wu, "An improved method for predicting magnetic power losses in SMC electrical machines," *Int. J. Applied Electromagnetics and Mechanics*, Vol. 19, pp. 75-78, 2004.

2006 5th International Power Electronics and Motion Control Conference

Solving Induction Motor Equivalent Circuit using Numerical Methods for an In-Service and Nonintrusive Motor Efficiency Estimation Method

Bin Lu, Wei Qiao, Thomas G. Habetler, and Ronald G. Harley
School of Electrical and Computer Engineering
Georgia Institute of Technology
Atlanta, GA 30332, USA
{binlu, weiqiao, thabetler, rharley}@ece.gatech.edu

Abstract — Motor efficiency evaluation enables the energy savings in industry. However, because of the uninterrupted characteristic of industrial processes, traditional methods defined in IEEE Std-112 cannot be used for these in-service motors. A novel nonintrusive method for in-service motor efficiency estimation based on a modified induction motor equivalent circuit has been developed by the authors. A highly nonlinear and 4-dimensional system of equations needs to be solved to obtain the parameters of the motor equivalent circuit and finally the motor efficiency. This paper continues this topic and presents an in-depth discussion on solving these motor parameters using three numerical methods under various conditions. Newton's method exhibits the best suitability in this application because of its simplicity and fast convergence. In the rare cases where Newton's method does not converge, the particle swam optimization and simulated annealing methods are used. Finally, the proposed motor efficiency estimation method is verified by the experimental results from a 4-pole 7.5 hp TEFC induction motor. The performances of these three numerical methods are evaluated and compared.

Keywords — *efficiency estimation; in-service testing; induction motors; equivalent circuit; numerical methods; Newton's method; particle swam optimization; simulated annealing method*

I. INTRODUCTION

Motor efficiency evaluation enables the energy savings in industry. However, because of the uninterrupted characteristic of industrial processes, traditional methods defined in IEEE Std-112 cannot be used. Nonintrusive motor efficiency estimation methods have to be developed for in-service motor testing.

Induction motor equivalent circuit based methods are one of the least intrusive categories of motor efficiency estimation methods. Over the years, many methods have been developed based on induction motor equivalent circuit. The IEEE Std-112 F method is the standard equivalent circuit method [1]. Although this method is expected to be quite accurate, the required no-load, variable voltage, removed-rotor, and reverse rotation tests make it impossible to be used in in-service testing. Later, the standard 112-F method is modified by Ontario Hydro by eliminating the variable voltage test [2]. However, a no load test and a full load test both under rated voltage are still required. In addition, direct stator resistance measurement is also needed. To further reduce the intrusion levels, a modified equivalent circuit based method is developed by Oak Ridge National Lab in [2]. It is a low-intrusion method, however, the parameters of the equivalent circuit are solved from imaginary rated load condition and locked rotor condition, which completely rely on motor nameplate information and may have up to 20% inaccuracies according to NEMA MG-1 [3]. Another interesting method calculates the motor parameters using two different motor operating points [4]. However, it requires rather intrusive measurements of stator resistance and stator winding temperature. Besides, the solution of motor parameters requires the actual value of stator leakage reactance, which is not available for in-service testing.

To overcome the problems in these traditional methods, a novel nonintrusive method for in-service motor efficiency estimation based on a modified induction motor equivalent circuit is proposed by the same authors in [5]. A highly nonlinear and 4-dimenstional system of equations needs to be solved to obtain the parameters of the motor equivalent circuit and finally the motor efficiency.

1-4244-0448-7/06/$25.00 ©2006 IEEE

This paper continues this topic and presents an in-depth discussion on solving motor parameters using numerical methods under various conditions. Section II briefly reviews the nonintrusive motor efficiency estimation method proposed in [5]. Section III suggests the three numerical methods to solve the motor equivalent circuit parameters: Newton's method, particle swam optimization (PSO), and simulated annealing method. Finally in section IV, the performances of three methods are compared using experimental results from a 4-pole 7.5 hp TEFC induction motor.

II.   A NONINTRUSIVE EQUIVALENT CIRCUIT METHOD

In [5], the same authors propose a nonintrusive method for in-service motor efficiency estimation based on a modified induction motor equivalent circuit using only motor terminal quantities and nameplate information. Only a few cycles of line voltages and currents from two different operating points and motor nameplate information are required to develop the equivalent circuit.

A modified induction motor equivalent circuit is used in this method, as shown in Fig. 1. To simplify the problem, only the positive sequence equivalent circuit is considered here. V_1 is the stator phase voltage phasor. I_1 and I_2 are the stator and rotor phase current phasors, respectively. R_1, R_2, and R_C are the stator, rotor, and core resistances, respectively; and X_1, X_2, and X_m are the stator leakage, rotor leakage, and magnetizing reactances, respectively.

The rotor stray-load loss, W_{LLr}, is defined in [1] as

$$W_{LLr} = \left(\frac{I_2}{I_{2\_rated}}\right)^2 W_{LLr\_rated} = 3\left|\tilde{I}_2\right|^2 R_{LL}$$

where, I_2 is the magnitude of the rotor current, and the subscript "rated" denotes the rated load condition.

Therefore, an equivalent stray-load resistor, R_{LL}, can be added in series with the rotor circuit

$$R_{LL} = \frac{W_{LLr\_rated}}{3I_{2\_rated}^2} = const.$$

Since the stray-load loss is primarily determined by the rotor current, the rotor stray-load loss, W_{LLr_rated}, can be estimated using the assumed stray-load values defined in IEEE Std-112, as in Table I.

TABLE I
ASSUMED VALUES FOR STRAY LOAD LOSS IN IEEE STD-112

Machine Rating		Stray load loss percent of rated output
1-125 hp	1-90 kW	1.8%
126–500 hp	91-375 kW	1.5%
501-2499 hp	376-1850 kW	1.2%
2500 hp and up	1851 kW and up	0.9%

Considering that the stator resistance, R_1, and slip, s, can be estimated sensorlessly from motor voltages and currents as discussed in [5], and the stator and rotor leakage reactances have a specific ratio ($k = X_1/X_2 = 1.0$, 0.67, or 0.43) for a certain NEMA design [3], the input impedance can be expressed as a function F in term of only four independent unknown variables: X_1, R_C, X_m, and R_2,

$$Z_{in} = Z_1 + Z_{ag} = Z_1 + Z_m \mathbin{//} Z_2$$
$$= Z_1 + \frac{Z_m Z_2}{Z_m + Z_2} = F(X_1, R_C, X_m, R_2) \qquad (1)$$

where, Z_1, Z_2, and Z_m are the stator, rotor, and magnetizing impedances, respectively. Z_{in} is the total motor input impedance.

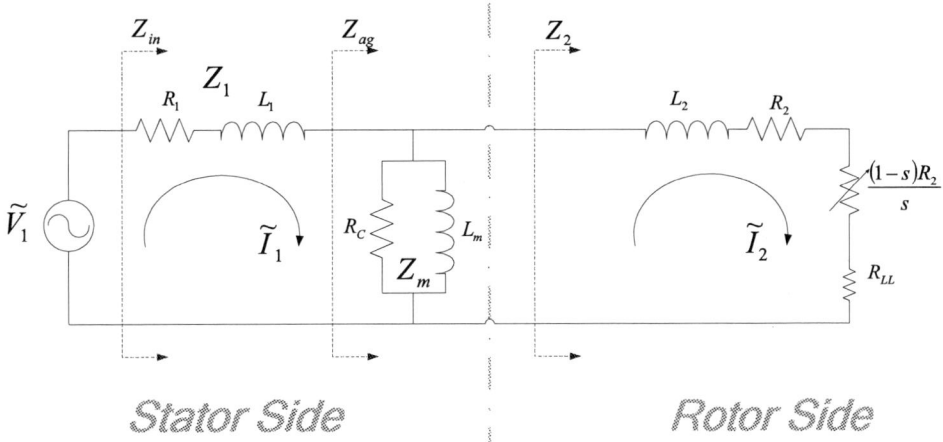

Fig. 1. A modified induction motor positive sequence equivalent circuit with an added equivalent stray-load resistor.

Splitting the real and imaginary parts in (1), two independent equations can be obtained at each motor load point.

$$\begin{cases} real\left(\dfrac{\widetilde{V}_1}{\widetilde{I}_1}\right) = real(Z_{in}) = real[F(X_1, R_C, X_m, R_2)] \\ imag\left(\dfrac{\widetilde{V}_1}{\widetilde{I}_1}\right) = imag(Z_{in}) = imag[F(X_1, R_C, X_m, R_2)] \end{cases} \quad (2)$$

Expanding (2), two independent equations can be obtained at each load points as (3).

In order to solve four independent unknowns, a set of four independent equations are developed from two carefully selected load points. The details on solving such a highly nonlinear and multi-dimensional system of equations are given in section III.

After the parameters of the equivalent circuit are solved, the motor efficiency at any load can be simply computed as

$$\eta = \frac{P_{output}}{P_{input}} = \frac{3\left|\widetilde{I}_2\right|^2 \dfrac{(1-s)}{s} R_2 - W_{fw}}{3\left|\widetilde{I}_1\right|^2 real(Z_{in})}$$

$$= \frac{3\left|\widetilde{I}_1\left(\dfrac{Z_m}{Z_m + Z_2}\right)\right|^2 \dfrac{(1-s)}{s} R_2 - W_{fw}}{3\left|\widetilde{I}_1\right|^2 real(Z_{in})}$$

where, the friction and windage loss, W_{fw}, is taken as a constant percentage of the rated motor horse power, *e.g.*, 1.2% for 4-pole motors below 200 hp, as suggested by many statistical motor efficiency estimation methods.

III. SOLVING MOTOR PARAMETERS USING NUMERICAL METHODS

In order to solve four independent unknowns
$$x = [X_1, R_c, X_m, R_2],$$
a set of four independent nonlinear equations
$$g(x) = [g_1(x); g_2(x); g_3(x); g_4(x)]$$
are developed from two carefully selected load points according to (3).

Too close load points will result in ill-conditioned equations; while too distant conditions will result in additional errors caused by the parameter variations due to temperature change, flux saturation, *etc.*

Solving such a highly nonlinear and multi-dimensional system of equations is not trivial. Three numerical methods (Newton's method [6], PSO [7], [8], and simulated annealing method [6]) have been studied and implemented. Newton's method is very simple and fast. But its convergence highly depends on the initial guess. When it converges, it finds the solution in only a few iterations. In the rare cases where Newton's method does not converge, the PSO and simulated annealing methods can be used. Both of them target on global optimization and can converge from a general zero initial condition, but their converging speeds are much slower.

A. Newton's Method

Newton's method is a very fast root finding method based on approximating $g(x)$ locally with a two-term Taylor series. It is the most widely used root finding method in engineering applications because of its simplicity and fast quadratic convergence. Since this problem is a 4-dimensional nonlinear algebraic system, an extended Newton's vector method is used. It requires the computation of a 4×4 Jacobian matrix

$$J(x) = \partial g(x) / \partial x.$$

In this problem, fortunately, the explicit expression of the Jacobian matrix $J(x)$ can be obtained offline, and the computation of partial differentiations is not needed at each iteration. The procedure of Newton's vector method is available in [6].

Like any non-bracketing method, Newton's method is not guaranteed to converge in all cases. This is the major disadvantage of this method. Its convergence highly depends on the initial point x_0. If an initial point is chosen to be close to the solution, it converges to the solution very rapidly. Fortunately, in this problem, a reasonable initial point can be always obtained from the following rough estimations.

$$\begin{cases} real\left(\dfrac{\widetilde{V}_1}{\widetilde{I}_1}\right) = R_1 + \dfrac{R_C X_m\left(\frac{R_2}{s} + R_{LL}\right)\left[R_C X_m + \frac{R_C X_1}{k} + \left(\frac{R_2}{s} + R_{LL}\right)X_m\right] - \frac{R_C X_m X_1}{k}\left[R_C\left(\frac{R_2}{s} + R_{LL}\right) - \frac{X_m X_1}{k}\right]}{\left[R_C\left(\frac{R_2}{s} + R_{LL}\right) - \frac{X_m X_1}{k}\right]^2 + \left[R_C X_m + \frac{R_C X_1}{k} + \left(\frac{R_2}{s} + R_{LL}\right)X_m\right]^2} \\[4ex] imag\left(\dfrac{\widetilde{V}_1}{\widetilde{I}_1}\right) = X_1 + \dfrac{\frac{R_C X_m X_1}{k}\left[R_C X_m + \frac{R_C X_1}{k} + \left(\frac{R_2}{s} + R_{LL}\right)X_m\right] + R_C X_m\left(\frac{R_2}{s} + R_{LL}\right)\left[R_C\left(\frac{R_2}{s} + R_{LL}\right) - \frac{X_m X_1}{k}\right]}{\left[R_C\left(\frac{R_2}{s} + R_{LL}\right) - \frac{X_m X_1}{k}\right]^2 + \left[R_C X_m + \frac{R_C X_1}{k} + \left(\frac{R_2}{s} + R_{LL}\right)X_m\right]^2} \end{cases} \quad (3)$$

It has been experimentally established that the stator leakage reactance X_1 and the magnetizing reactance X_m obey the relation

$$X_1 = \alpha \cdot X_m \qquad (4)$$

where, the ratio α is a constant ranging from 0.02 to 0.07 for a specific motor.

The magnitude of no load current I_{1_NL} can be estimated as a certain percent of full load current, e.g., 20%. Then, the magnitude of the no load input impedance $|Z_{in\_NL}|$ can be obtained by

$$\left|Z_{in\_NL}\right| = \frac{V_{1\_NL}}{I_{1\_NL}} \qquad (5)$$

Considering the core loss resistance R_C is very large and its contribution in $|Z_{in\_NL}|$ is negligible, the no load reactance can be roughly estimated by

$$X_1 + X_m \approx \sqrt{\left|Z_{in\_NL}\right|^2 - R_1^2} \qquad (6)$$

From (4) to (6), the initial guess of X_1 is

$$X_1 = \frac{\alpha}{\alpha+1} \sqrt{\left(\frac{V_{1\_NL}}{I_{1\_NL}}\right)^2 - R_1^2}$$

The initial guess of R_C can be set as one order of magnitude larger than X_m. The rotor resistance R_2 can be set in the same order of magnitude of R_1. It has been observed from experiments that for almost all cases, the above initial guess can result in final convergence.

B. Particle Swarm Optimization

The PSO method is an evolutionary computational algorithm inspired by the paradigm of birds flocking [7]-[8]. It has been successfully used for both continuous nonlinear and discrete binary optimization [8]. The PSO algorithm searches for the optimal solution from a population of moving particles. Each particle represents a potential solution. It has a position represented by a position vector X_i and a moving velocity represented by a velocity vector V_i in the problem space. Each particle keeps track of its coordinates in the problem space, which are associated with the best position it has achieved so far. This position is called individual best position $X_{i,pbest}$. Furthermore, the best position among all the particles obtained so far in the population is kept track of by all particles as X_{gbest}, which is called swarm best position. The PSO algorithm is implemented in the following iterative procedure to search for the optimal solution.

(i) Initialize a population of particles with random positions and velocities of M dimensions in the problem space.

(ii) Define a fitness measure function to evaluate the performance of each particle.

(iii) Compare particle's present position X_i with particle's $X_{i,pbest}$ based on the fitness evaluation. If current position X_i is better than $X_{i,pbest}$, then set $X_{i,pbest}$ equal to the current position X_i.

(iv) If $X_{i,pbest}$ is updated, then compare particle's $X_{i,pbest}$ with the swarm best position X_{gbest} based on the fitness evaluation. If $X_{i,pbest}$ is better than X_{gbest}, then set X_{gbest} equal to the current position $X_{i,pbest}$.

(v) At iteration k, a new velocity for particle i is updated by

$$V_i(k+1) = wV_i(k) + c_1\varphi_1(X_{i,pbest} - X_i(k)) + c_2\varphi_2(X_{gbest} - X_i(k)), \quad i = 1, 2, \cdots, N \qquad (6)$$

(vi) Based on the updated velocity, each particle then changes its position according to the following equation,

$$X_i(k+1) = X_i(k) + V_i(k+1), \quad i = 1, 2, \cdots, N \qquad (7)$$

(vii) Repeat steps (iii)-(vi) until a criterion, usually a sufficiently good fitness or a maximum number of iterations is achieved. The final value of X_{gbest} is regarded as the optimal solution of the problem.

In (6), c_1 and c_2 are positive constants representing the weighting of the acceleration terms that guide each particle toward the individual best and the swarm best positions $X_{i,pbest}$ and X_{gbest}, respectively; φ_1 and φ_2 are random numbers in the range [0, 1]; w is a positive inertia weight developed to provide better control between exploration and exploitation; N is the number of particles in the swarm.

In this problem, the fitness measure function $f(x)$ for each particle is defined as

$$f(x) = \|g(x)\|$$

where $\| \cdot \|$ represents the Euclidean norm.

In this problem, the values of c_1 and c_2 in (6) are chosen as 2; the number of particles N is chosen as 20; the inertia constant w starts with 0.9 and linearly decreases to 0.4 when the iteration number reaches a pre-specified maximum number during the simulation.

C. Simulated Annealing Method

Similar to PSO, the simulated annealing method is also a statistical optimization technique. It is based on an analogy with the annealing of metal and searches for a global minimum of the objective function.

The simulated annealing method has two stages to reach the global optimization. The first stage is a random global search based on simulated annealing. As the temperature of metal is gradually lowered from above its melting point, the atoms lose thermal mobility and decay to lower energy states. Eventually, the atoms settle into global energy state minimum. The simulated annealing methods simulate this annealing process by gradually lowering an artificial temperature T. The energy states of artificial atoms are associated with x. Their changes Δx are described using a Gaussian distribution with zero mean. The convergence speed of simulated annealing is often accelerated by considering lower and upper bonds on x and $f(x)$. A localization parameter, $0 < \gamma < 1$, is set empirically to adjust the scope of searching local minima.

The second stage is an efficient local search. Many local minimum optimization methods can be utilized, such as the conjugate-gradient method and penalty function method. The detailed procedure of the simulated annealing method is available in [6].

In this problem, the root finding process for the nonlinear 4-dimentional system of equations can be converted to an optimization process using the simulated annealing method as

Minimize: $\quad f(x) = \|g(x)\|^2 = g(x)'g(x)$

Subject to: $\quad a < x < b$

A localization parameter of $\gamma = 0.3$-0.5 provides best convergence speed. The lower and upper bonds vectors, a and b, can be roughly computed from the motor nameplate data.

IV. EXPERIMENTAL RESULTS

The proposed motor efficiency estimation method has been verified by both computer simulations and motor experiments. The parameters of the motor used in the experiments are listed in Table II.

In the experimental setup, the three-phase induction motor is line-connected to a 230-volt mains supply. A dc generator connected to resistor boxes serves as the dynamometer. The unbalances in the voltages and currents are negligible ($V_-/V_+ < 1\%$ and $I_-/I_+ < 3\%$). The line voltages and currents are sampled at 2 kHz and collected using a NI LabVIEW data acquisition system. The actual efficiency is directly calculated from the shaft torque, measured by an in-line rotary torque transducer.

The motor equivalent circuit is solved using data from two load conditions: (1) 19.09% rated load, 1775 rpm, and (2) 71.26% rated load, 1694 rpm. The motor parameters are solved using all three methods. As discussed in section III, Newton's method is regarded as the major solver for this application. The experimental results validate its fast convergence. Using the initial guess calculated from motor nameplate data, this algorithm converges in only 6 iterations. The results of Newton's method is summarized in Table III, which shows the iterations number k, the

current estimate x, and the norm of $g(x)$ at the current estimate.

Compared with Newton's method, the PSO and simulated annealing methods converge much more slowly. However, these two methods can converge from a general zero initial guess. The same equations are solved using these three methods on a computer with Pentium 4 3.4 GHz processor and 512 Mb RAM. The iterations and CPU time required by each method to reach final solution are compared in Table IV. Because both the PSO and simulated annealing methods are statistical techniques, their iterations and computation time can vary from time to time. In Table IV, the iterations and CPU time used by the PSO and simulated annealing methods are the average of 10 repeated experiments.

TABLE II. PARAMETERS OF THE INDUCTION MOTOR IN EXPERIMENT.

Brand	GE	Volts	230/460 V
HP	7.5	F.L.AMPS	18.2/9.1 A
CAT. NO.	S231	RPM	1755
Design	NEMA-A	Nom. PF	0.865
Enclosure	TEFC	Nom. Eff.	0.895

TABLE III. NEWTON'S VECTOR METHOD RESULTS.

k	x				$\|g(x)\|$
	X_1 (Ω)	R_C (Ω)	X_m (Ω)	R_2 (Ω)	
0	0.907794	70.949235	19.53566	0.500000	17.457290
1	3.41516	92.305626	55.92357	0.534489	7.0520789
2	1.77525	244.494693	86.55143	0.497332	3.0326519
3	1.93057	1325.90180	84.4203	0.520816	0.6147128
4	1.938864	824.305199	80.9109	0.520560	0.0258162
5	1.938487	846.149846	80.8698	0.520614	0.0000644
6	1.938487	846.112922	80.86943	0.520614	0.0000000

TABLE IV. COMPARISON OF THREE NUMERICAL METHODS.

Methods	Convergence	Iterations	CPU Time
Newton's Vector	Local	6	0.185 s
PSO	Global	2601	2.432 s
Simulated Annealing	Global	3377	17.123 s

It can be observed from Table IV that when the Newton's method converges, there is no doubt that it is the best method. In the rare cases where the Newton's method does not converge using the estimated initial guess, both the PSO and the simulated annealing methods can be used. The PSO method is faster, but requires relatively more configuration parameters (c_1, c_2, w, N) to be tuned. While, the simulated annealing method is about 10 times slower, but it just needs one localization parameter ($0 < \gamma < 1$) to be tuned.

Using the solved motor parameters, the motor efficiency at any load levels can be estimated. Fig. 2 compares the estimated and measured motor efficiencies versus load

840

percentage, when the load changes continuously from almost no load to full load conditions. The estimated motor efficiencies show good agreement (within 2-3% errors) with the measured efficiencies during the normal motor operations (load ranges from 30% to 90% of rated load). The errors under very low load conditions (less than 30% of rated load) are slightly larger (within 10%), but usually under such low load levels, there is no need to estimate motor efficiencies. The errors are caused by many factors, such as motor parameter variations under different load levels, stator resistance and speed estimation errors, motor nameplate information inaccuracies, *etc*. The agreement between the estimated and measured efficiencies validates the solved motor parameters and the proposed nonintrusive motor efficiency estimation method.

V. CONCLUSIONS

This paper continues the proposal of a novel nonintrusive method for in-service motor efficiency estimation based on a modified induction motor equivalent circuit using only motor terminal quantities and nameplate data. Only a few cycles of line voltages and currents from two different motor operating points and motor nameplate information are required to develop the equivalent circuit. The parameters are obtained by solving a highly nonlinear and 4-dimensional system of equations. As the focus of this paper, three numerical methods (Newton's method, PSO, and simulated annealing method) are adopted to solve the motor parameters under various conditions. Newton's method is suggested as the major solver of this application because of its simplicity and fast convergence. In the rare cases where Newton's method does not converge, the PSO and simulated annealing methods can be used using a general zero initial guess. All three suggested numerical methods have been

tested on a 4-pole 7.5 hp TEFC induction motor. Experimental results validate that the motor efficiencies estimated using the solved motor parameters agree with the measured efficiencies within 2-3% errors.

ACKNOWLEDGMENT

This work is financially supported by U.S. Department of Energy and Eaton Corporation.

REFERENCES

[1] *IEEE Standard Test Procedure for Polyphase Induction Motors and Generators*, IEEE Standard 112-2004, Nov. 2004.

[2] J. D. Kueck, M. Olszewski, D. A. Casada, J. Hsu, P. J. Otaduy, and L. M. Tolbert, "Assessment of methods for estimating motor efficiency and load under field conditions," *Oak Ridge National Laboratory report*, ORNL/ TM-13165, 1996.

[3] NEMA – MG 1 Standard, 2003.

[4] Y. El-Ibiary, "An accurate low-cost method for determining electric motors's efficiency for the purpose of plant energy management," *IEEE Trans. Industrial Applications*, vol.39, no. 4, July/Aug. 2003, pp. 1205-1210.

[5] B. Lu, T. G. Habetler, and R. G. Harley, "A nonintrusive efficiency estimation method for in-service motor testing using a modified induction motor equivalent circuit," in *Proc. of the 37th IEEE Power Electronics Specialist Conference (PESC'06)*, June 2006.

[6] R. J. Schilling and S. L. Harris, *Applied Numerical Methods for Engineers using MATLAB and C*. Pacific Grove, CA: Thomson, 2000.

[7] J. Kennedy and R. C. Eberhart, "Particle swarm optimization," in *Proc. of IEEE International Conference on Neural Networks*, vol. 4, Nov./Dec., 1995, pp.1942-1948.

[8] J. Kennedy and R. C. Eberhart, *Swarm Intelligence*. San Mateo, CA: Morgan Kaufmann, 2001.

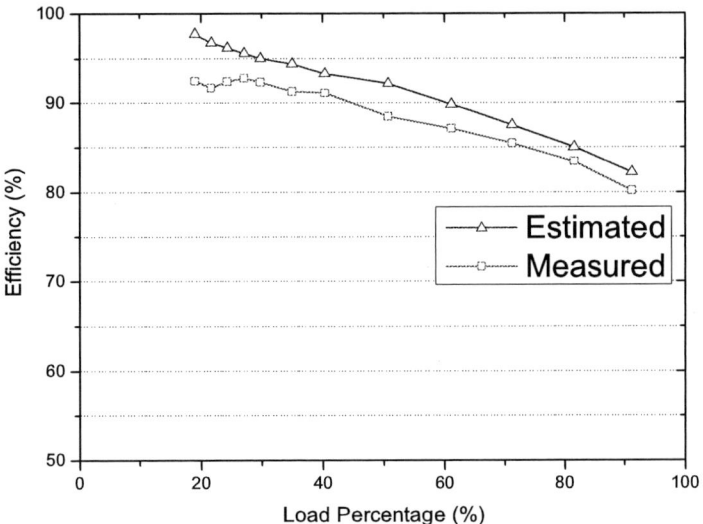

Fig. 2. Efficiency v.s. load curve of the 7.5 hp GE TEFC motor.

2006 5th International Power Electronics and Motion Control Conference

Fault Investigation of X-by-wire Permanent Magnet Synchronous Machine

L. Feng*, A. Binder*, A. Rentschler*, A. Paweletz** and D. Guenther**

* Dept. of Electrical Energy Conversion, Darmstadt University of Technology, Darmstadt, Germany
** Corporate Sector Research and Advanced Engineering, Robert Bosch GmbH, Gerlingen-Schillerhoehe, Germany

Abstract — **Permanent Magnet Synchronous Machines (PMSM) are used in x-by-wire systems, because of the high efficiency and its good dynamic properties. To investigate the performance of the machine at fault situations, a 3-phase, 16-pole prototype (rated state torque 1.7 Nm, overload torque 20 Nm for impulse torque actuation) with tooth coils was built with the possibility for measuring different kinds of winding short circuit failures. Field calculation of the PMSM was done analytically and numerically. A dynamic model of the drive system, which consists of the motor, the PWM inverter, a PI cascade speed-current controller and a simplified mechanical load, was developed to simulate the faulty operation of the x-by-wire system. The numerical calculation shows, that winding short circuit fault of the PMSM results in motor speed, torque and current ripple with twice the stator frequency. At rated load steady state, the electromagnetic torque ripple for the fault case is 111% of the rated torque, if three adjacent tooth coils of one phase are short circuited. This is about 1.6 times bigger than in the fault case, where one tooth coil is hot-wired. Short circuit fault results in big currents flowing in the stator winding, which harm the motor thermally.**

Keywords-permanent magnet synchronous machine (PMSM); x-by-wire; finite element calculation; winding short circuit fault; modeling; drive system

I. INTRODUCTION

Currently autonomous electromechanical actuators inside of "x-by-wire" systems are increasingly investigated because of their advantages. For example, "steer-by-wire" system is able to stabilize the control of the car trajectory in dangerous situations, like slippery road. The steering of cars in this case is accomplished by an electronically connected driver's wheel, acting on an electromechanical actuator, which will give the steering forces on the car's wheel set. It substitutes the mechanical coupling between the wheel set and the driving wheel. The drive system consists of a MOSFET Pulse Width Modulation (PWM) inverter, feeding the windings of the actuator. The actuator angle is positioning mostly via a gear the wheel around vertical axis. Fault-tolerance is a very important feature of such a system. As a failure of one part must not lead to malfunction, safe operational performance must be guaranteed in case of one fault condition. Faults may occur in electronic control system, in feeding inverter electronics, in the actuator or in the mechanical drive chain. This paper focuses on the actuator, which is a rotary permanent magnet excited synchronous machine, while in [1] also inverter faults are discussed. In the first stage of the project failure models due to winding faults have been investigated by simulation as preparation for future measurement and meanwhile constructed test bench (Fig. 1).

II. THE PERMANENT MAGNET SYNCHRONOUS MACHINE

This paper introduces a prototype Permanent Magnet Synchronous Machine (PMSM), with stator tooth coils, which has been selected due to high overload capability, high efficiency and very good dynamic performance ([2]). For motor design and motor parameter calculation, the analytical part of the motor design software SPEED® and numerical finite element calculation software FEMAG® were used. Comparison between both calculation methods shows, that finite element calculation achieves more accurate results especially due to the consideration of the magnetic saturation at high overload. A dynamic simulation model of the drive system is established in MATLAB/Simulink®, which includes the PMSM, the PWM inverter, a speed cascade controller and a simplified mechanical model for the load. The PMSM is mathematically described by a set of differential equations. Short circuit fault of different numbers of tooth coils in a phase can be approximately simulated with this simulation model. The torque and speed cascade controller parameters are optimized by the amplitude optimum and symmetrical optimum ([3]). In a real x-by-wire system in most cases the motor just rotates several radians at each actuator motion, while in the simulation it is simplifying assumed that it rotates with the mechanical speed of 375 rpm at steady state. This speed will be used for failure testing (equivalent 50 Hz stator frequency of the $p = 8$ pole-pair machine). The deformations of the currents, the electromagnetic torque and the mechanical speed of the machine are analyzed at fault situations. Under the restriction "One fault at one time" situation, two kinds of winding faults are presented: short circuit of a) one tooth coil, b) 3 series connected tooth coils.

1-4244-0448-7/06/$25.00 ©2006 IEEE

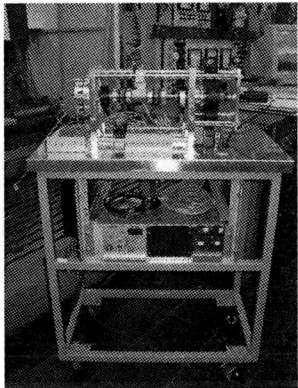

Figure 1. The test bench

TABLE I.
RATED DATA OF THE TEST MOTOR

$2p$	16	$slots\ Q_s$	18
U_N / V	8.85, Y	I_N / A	5
M_N / Nm	1.7	n_N / rpm	375
outer stator radius / mm	56.8	axial length l_{Fe} / mm	25
outer rotor radius / mm	39.2	air gap width / mm	0.7
magnets	NdFeB		

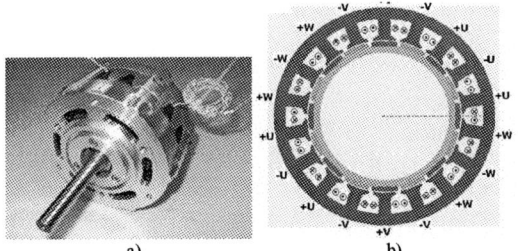

Figure 2. a) The prototype PMSM   b) Cross section with surface mounted rotor magnets and 18 stator tooth coils

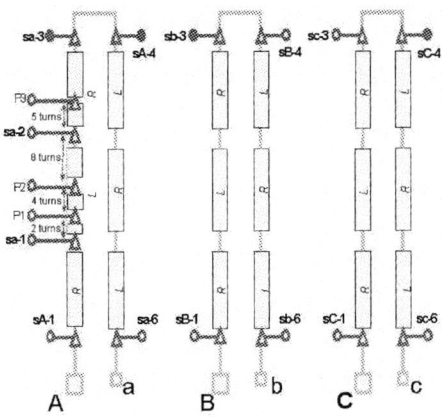

Figure 3.   The 3-phase winding scheme (phases A,B,C) of the prototype PMSM with special contacts to create short circuit

The data of the test motor are given in Table I. The geometry of the PMSM is shown in Fig. 2b) with a three phase winding and 18 tooth coils. Each phase consists of six tooth coils in series. To simulate different kinds of winding short circuit failures, the windings of the motor are specially arranged (Fig. 3). Some parts of the windings have connections with pins in a special designed motor terminal box, which are set on the motor's surface. With this the short circuit of certain numbers of turns can be done by directly connecting two corresponding pins on the outside of the machine by shortest way. This concept allows different options of the inter-turns, coils and phase to phase short circuit and can be set to measure the behaviour of the system at faulty situations.

III. FIELD CALCULATION OF THE PMSM

In the following, the three phases U=A, V=B, W=C are investigated more detailed concerning faults. For that, the healthy steady state operation is simulated first for motor design.

Fig. 4a) shows the calculated magnetic flux density distribution with finite element program FEMAG® for feeding the machine with instantaneous currents,

$$I_U = 6.333\ A, \quad I_V = 0, \quad I_W = -6.333A,$$

which results in a r.m.s q-current of

$$I_q = 5.171\ A.$$

Fig. 4b) gives the flux lines for this impressed current feeding. The corresponding calculated electromagnetic torque by FEMAG® is

$$M = 1.75\ Nm.$$

Fig.5 shows the stationary M/I_q curve calculated a) with motor calculation program package SPEED® and b) with finite element program FEMAG®, with significant overload of 6-times rated torque, which is typical for "steer-by-wire" operation.

Due to the high current loading at peak torque, for an air gap field density of 1 T, the magnetic field density in

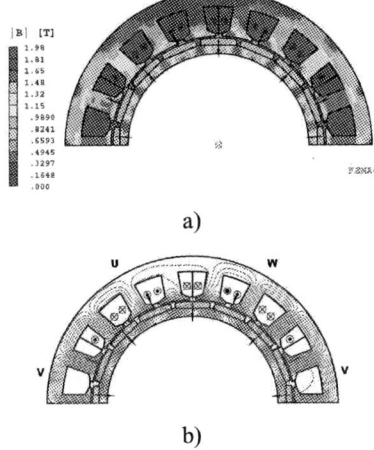

Figure 4.   a) Calculated magnetic flux density distribution with FEMAG® (absolute value)   b) Calculated flux lines with FEMAG®

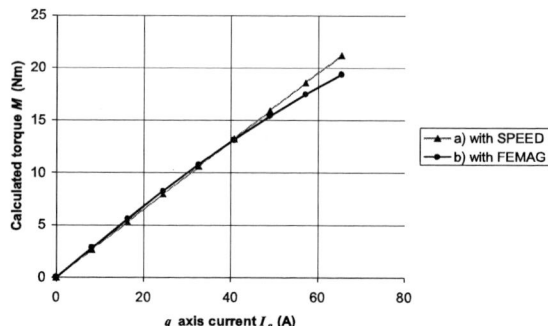

Figure 5. Stationary M/I_q curve (calculated with

a) SPEED®, b) FEMAG®)

the teeth reaches values as high as 1.8 T to 2 T. Above typically 1.5 T the iron saturates, and the magnetomotive force of the stator winding is not only used for magnetisation of the air gap, but also of the iron paths in the teeth and also partly of the stator and the rotor yoke. Therefore, the magnetic field in the air gap B_δ and hence the electromagnetic torque M increase sub-proportionally with increasing of current I_q. With the finite element software, these results are obtained due to taking into account the effect of iron saturation.

IV. DYNAMIC MODEL OF THE DRIVE SYSTEM

A. Mathematic model of the PMSM

For dynamic simulation of electrically symmetrical synchronous machine usually the d-q reference frame equations are used to build the PMSM model ([4]). In the case of winding faults, the three phases are unsymmetrical. The voltage equations for a machine with magnetically coupled tooth windings are derived to investigate different phase asymmetries. The mathematical model considers the mutual flux linkage between the 18 stator tooth coils and between stator and rotor by coil inductances. So the flux linkage in each coil comes from the flux produced by the coil current itself, by the other coil currents, and by the rotor permanent magnets. For rotor permanent magnet flux linkage only the fundamental wave is considered. For the i-th coil, $i = 1, 2, \ldots 18$, the coil flux linkage is :

$$\Psi_{c,i} = c_{s,i} k_p \hat{\Psi}_{pc} \cdot \cos\left[p\Omega_m t + p\gamma_0 - (i-1)p\alpha_Q\right] + L_{ch}i_{c,i} + \sum_{\substack{j=1,\ldots18 \\ j \neq i}} (-c_{s,i})c_{s,j} M_{ch}i_{c,j}$$
$$+ \left(L_{cQ} + L_{cb}\right)i_{c,i} + (-c_{s,i})c_{s,j-1} M_{cQ}i_{c,j-1} + (-c_{s,i})c_{s,j+1} M_{cQ}i_{c,j+1} \quad (1)$$

So the total induced voltage in the coil is:

$$u_{c,i} = -c_{s,i} k_p \hat{\Psi}_{pc} \cdot p\Omega_m \cdot \sin\left[p\Omega_m t + p\gamma_0 - (i-1)p\alpha_Q\right] + \left(L_{ch} + L_{cQ} + L_{cb}\right)\frac{di_{c,i}}{dt}$$
$$+ (-c_{s,i})M_{ch}\sum_{\substack{j=1,\ldots18 \\ j \neq i}} c_{s,j}\frac{di_{c,j}}{dt} + (-c_{s,i})M_{cQ}\left(c_{s,j-1}\frac{di_{c,j-1}}{dt} + c_{s,j+1}\frac{di_{c,j+1}}{dt}\right) + R_c i_{c,i} \quad (2)$$

The electrical torque is derived from internal power:

$$M_e = \sum_{i=1,\ldots18}\left(-c_{s,i}k_p\hat{\Psi}_{pc}\cdot p\cdot\sin\left[p\Omega_m t + p\gamma_0 - (i-1)p\alpha_Q\right]\cdot i_{c,i}\right) \quad (3)$$

The symbol $c_{si} = \pm 1$ showing the winding sense of each coil is given in Table II.

The electric and dynamic equations of the healthy

TABLE II.
WINDING SCHEME OF THE TEST MOTOR

i	1	2	3	4	5	6	7	8	9
$c_{s,i}$	+1	-1	+1	+1	-1	+1	+1	-1	+1
phase	+U	-U	+U	+V	-V	+V	+W	-W	+W

i	10	11	12	13	14	15	16	17	18
$c_{s,i}$	+1	-1	+1	+1	-1	+1	+1	-1	+1
phase	+U	-U	+U	+V	-V	+V	+W	-W	+W

i: ordinal number of coil $c_{s,i}$: coil winding sense

PMSM are:

$$v_U - v_N - u_{c,1} - u_{c,2} - u_{c,3} - u_{c,10} - u_{c,11} - u_{c,12} = 0 \quad (4)$$
$$v_V - v_N - u_{c,4} - u_{c,5} - u_{c,6} - u_{c,13} - u_{c,14} - u_{c,15} = 0 \quad (5)$$
$$v_W - v_N - u_{c,7} - u_{c,8} - u_{c,9} - u_{c,16} - u_{c,17} - u_{c,18} = 0 \quad (6)$$
$$i_U + i_V + i_W = 0 \quad (7)$$
$$J_M\ddot{\gamma}_m = M_e - c\cdot(\gamma_m - \gamma_L) \quad (8)$$
$$J_L\ddot{\gamma}_L = c\cdot(\gamma_m - \gamma_L) - M_L \quad (9)$$

Note, that coil currents are

$$i_{c,1} = i_{c,2} = i_{c,3} = i_{c,10} = i_{c,11} = i_{c,12} = i_U \quad (10)$$

according to Table II, and accordingly for i_V and i_W. This corresponds to series connection of all 6 coils per phase (number of parallel winding branches per phase $a = 1$).

If in fault case e.g. the first k tooth coils in phase U are short circuited, equations are changed:

$$\sum_{i=1,\ldots k} u_{c,i} = 0 \quad (11)$$
$$i_{c,1} = i_{c,2} = \ldots = i_{c,k} = i_{sc} \quad (12)$$

For simulation, the set of equations (1)-(12) is implemented in MATLAB/Simulink. Equations (1)-(12) contain: phase potential v_U, v_V, v_W and potential of the star point v_N. The current $i_{c,i}$ in i-th coil is assigned either to phase current i_U, i_V, i_W, or short circuit current i_{sc}. The fundamental flux linkage of rotor permanent magnets with stator winding per coil is expressed by Ψ_{pc}. In the stator, L_{ch} is coil self inductance due to air gap field, M_{ch} is mutual inductance between two stator coils due to air gap field, L_{cQ} is self inductance due to slot stray field, L_{cb} is the self inductance due to field of winding overhang, M_{cQ} is mutual coil inductance of two coils in a slot due to slot field, and R_c is the coil resistance. The number of poles is $2p$, k_p is pitch factor of the stator winding due to fundamental flux linkage of rotor field, and α_Q is the slot angle. The initial rotor angular position to the first stator coil is given by angle γ_0, and Ω_m is the rotor angular velocity. In the electromagnetic torque equation (3), M_e is the electromagnetic torque, and γ_m is the rotor angular position. The mechanical system (8), (9) is modeled with an elastic coupling ([5], [6]), defined by torsion spring coefficient c to simulate dominant vibration resonances. The angle γ_L is the load angular position, M_L is the load torque, J_M is the rotor inertia and J_L is the load inertia.

According to Fig. 2b) three coils of one phase are adjacent due to the winding arrangement of $q = 3/8$ coils

844

per pole and phase. So it is likely, that e.g. 1 coil per phase or 2 or 3 coils are bridged by fault. Here, the short circuit of one tooth coil and three tooth coils is investigated.

The electromagnetic coupling is strongly changed, if winding short circuit occurs. Here the simple case of short circuit of complete tooth-coils is simulated. Inter-phase and inter-turn short circuits need a more complicated model for voltage equation and will be presented in the future.

B.   The PMSM Drive System

The dynamic model of the drive system is established in MATLAB/Simulink, which includes the machine, the three leg PWM inverter, a PI speed-current cascade controller with q-current operation, and a torsional two-mass-oscillator simplified mechanical model. The torsion resonance at

$$f = \frac{1}{2\pi}\sqrt{\frac{c \cdot (J_L + J_M)}{J_L \cdot J_M}} = 185 Hz$$

of the elastic coupling between the motor and the driven load is considered with the torsion spring coefficient $c = 489.07$ Nm/rad. The block diagram of the drive system is shown in Fig. 6.

The three-phase inverter model consists of three modules. Each module comprises of one inverter leg. From the input DC link voltage and control signal of the leg the output potential is calculated. The inverter model is combined with the machine model by applying the three potentials of the three legs to the three phase terminals U, V, W of the PMSM as v_U, v_V, and v_W.

C.   Fault Simulation Results of the Drive Model
  a) One Tooth Coil of the PMSM Short Circuited

The MOSFET PWM inverter has a switching frequency of 20 kHz. The motor starts up at no load from standstill to a reference mechanical speed of 375 rpm, corresponding to a stator frequency of 50 Hz. After start up is finished, at $t = 0.05$ s, the machine is loaded with rated torque $M_L = 1.7$ Nm. When the system is in steady state after loading, at $t = 0.15$ s, the first tooth coil in phase U is assumed to be short circuit. The simulation results of the mechanical speed, torque, currents i_d and i_q, the three phase currents and the short-circuit coil current i_{sc} are shown in Fig. 7, 8 and 9. Winding fault causes asymmetrical motor operation, generating an electric inverse current system, which causes torque oscillation

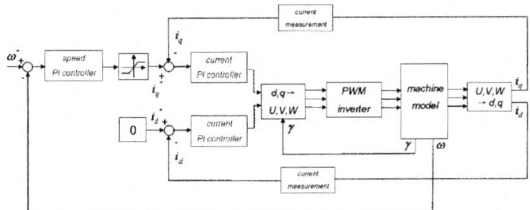

Figure 6.   Block diagram of the PMSM drive system

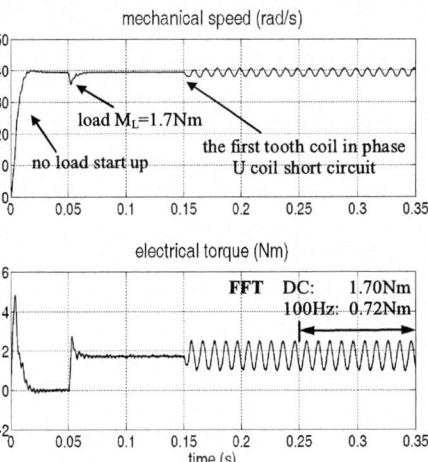

Figure 7.   Calculated speed and torque vs. time at fault a)

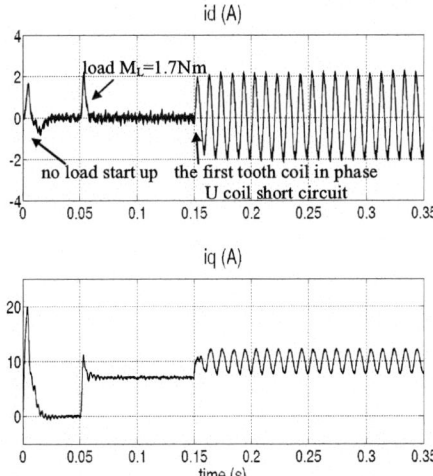

Figure 8.   Calculated i_d and i_q vs. time at winding fault a)

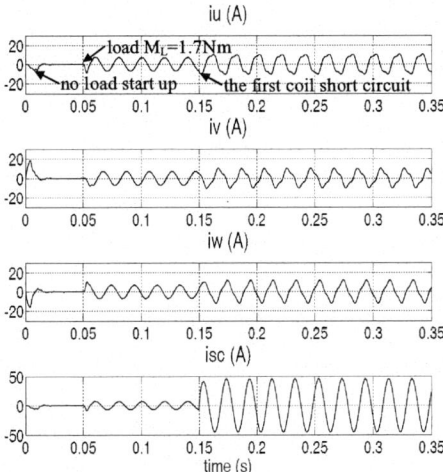

Figure 9.   Calculated phase currents and short-circuit coil current i_{sc} vs. time at winding fault a)

845

with twice stator frequency. The electromagnetic torque ripple at fault case is 42% of the rated torque. During healthy condition ($t \leq 0.15$s), the coil current of coil 1 is identically i_U, the phase current. It rises at coil short circuit up to 4-times, which means 16 times I^2R-losses and will endanger the coil thermally.

b) Three Adjacent Tooth Coils of the PMSM Short Circuited

In this simulation all basic data are the same as in case a). Now it is assumed that three adjacent tooth coils in phase U of the motor are short circuited at load steady state. The calculated speed, torque and currents characteristics are shown in Fig. 10, 11 and 12. Again, winding short circuit fault results in the speed, torque and current ripples with twice stator frequency (100 Hz), due to motor winding asymmetry, causing inverse magnetic system. The torque ripple is 111% of the rated torque, and about 1.6 times bigger than in fault case a). In this fault case not only the current in short circuit coils, but also the current in other coils, rise up to more than 2-times after fault, bringing very big copper losses in stator winding.

V.   CONCLUSIONS

A specially made x-by-wire PMSM prototype with tooth coil winding was introduced, which has the possibility to simulate different kinds of winding short circuit failure. Short circuit of one coil or of three adjacent coils of motor winding was simulated in the dynamic model of the PMSM drive system including the motor, the inverter, the speed and torque controller and the mechanical coupling parts. For "one fault at one time" failure condition, torque, speed and current ripple with twice stator frequency occurs. At rated load steady state, the electromagnetic torque ripple for the fault case, if three adjacent tooth coils are short circuited, is 111% of the rated load, and about 1.6 times bigger than in the fault case, where only one tooth coil is hot-wired. Short circuit fault results in big currents flowing in partial stator winding, which imposes big thermal threat to the motor.

REFERENCES

[1]   L. Feng, A. Rentschler, A. Binder , and A. Paweletz, "Fault Model of an Inverter-fed PM Motor for X-by-wire Systems," in press.

[2]   K. Reichert, "PM-Motors with Concentrated, non Overlapping Windings, Some Characteristics," 16th International Conference on Electrical Machines, Poland, 2004, 4 pages, CD-Rom.

[3]   B. Chen, *Control System of Drives*, Peking: China Machine Press (CMP), 1996, pp. 48 – 86.

[4]   G. Mueller, *Elektrische Maschinen*, Berlin: VEB Verlag Technik, 1990, pp. 384 – 398.

[5]   E. Wiedemann and W. Kellenberger, *Konstruktion elektrischer Maschinen*, Berlin / Heidelberg: Springer-Verlag, 1967, pp. 510 – 514.

[6]   W. Leonhard, *Control of Electrical Drives*, 3rd ed., Berlin, Heidelberg, NewYork: Springer-Verlag, 2001, pp.17–19.

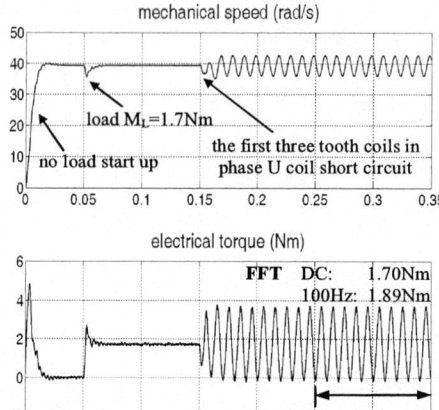

Figure 10. Calculated speed and torque vs. time at fault b)

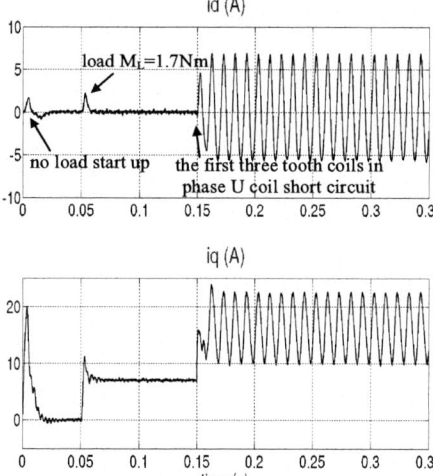

Figure 11. Calculated i_d and i_q vs. time at winding fault b)

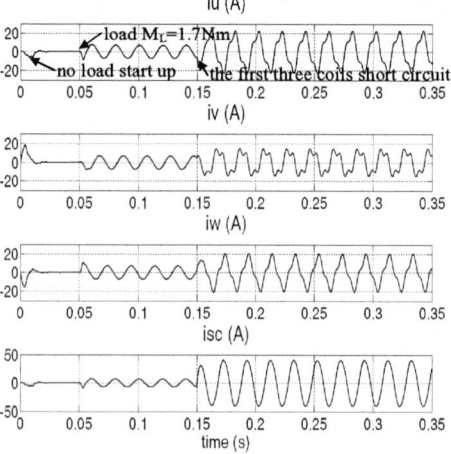

Figure 12. Calculated phase currents and short-circuit coil current i_{sc} vs. time at winding fault b)

2006 5th International Power Electronics and Motion Control Conference

PLC-Based Speed Control of DC Motor

Ashraf Salah El Din Zein El Din

Department of Electrical Engineering, Faculty of Engineering, Shebin El Kom,
Minoufiya University, Egypt, http://www.menofia.edu.eg
e-mail: ashrafzeineldin2002@yahoo.com , Tel. 20-48-2220395, Fax.20-48-2235695

Abstract- In this paper, a simplified approach for speed control of a separately excited DC motor using Programmable Logic Controller (PLC) is presented. This approach is based on providing a variable dc voltage to armature circuit of dc motor from a fixed dc supply voltage via a PLC which is used as a dc/dc chopper. Pang-Pang control method is used for switching on or off power to dc motor (armature circuit) depending on the reference (command) speed. It is easy, fast and effective by this method of control to vary motor speed from 0 to 100% of rated speed. The proposed system is suitable for different industrial applications such as subway cars, trolley buses, or battery-operated vehicles.

Keywords- PLC, DC motor, Speed Control, Pang-Pang Control

I. INTRODUCTION

The methods of speed control of DC motors are normally simpler and less expensive than that of ac drives. Due to the commutators, dc motors are not suitable for very high speed applications and require more maintenance than ac motors. Controlled rectifiers provide a variable dc output voltage from a fixed ac voltage, whereas choppers can provide a variable dc voltage from a fixed dc voltage. Due to their ability to supply a continuously variable dc voltage, controlled rectifiers and dc choppers made a revolution in modern industrial control equipment and variable-speed drives [1]. Many industrial drives and processes take power from dc voltage sources. In most cases, conversion of the dc source voltage to different levels is required. For example, subway cars, trolley buses, or battery-operated vehicles take power from a fixed dc source. However, their speed control requires conversion of a fixed voltage dc source to a variable voltage dc source for the armature of the dc motor [2,3,4]. A Programmable Logic Controller (PLC) is an industrially hardened computer-based unit that performs discrete or continuous control functions in a variety of processing plant and factory environments. Originally, it was intended as a relay replacement equipment for the automotive industry [5,6]. In the world of control, the use of PLC is ever increasing. Industrial process control is one of the very important areas where the PLC is extensively used. The flexibility offered by the system to implement various control laws is great. This paper proposed speed control of dc motor system using PLC, which is used as dc to dc chopper. The proposed system gives smooth variation over a wide range of control. This system avoids the time derivatives dv/dt or di/dt of power transistors like MOSFETs or

IGBTs if classical chopper is used. This system may be used in different industrial applications such as trolley buses, subway cars or battery-operated vehicles.

II. A CLASSICAL CHOPPER FED DC MOTORS

A dc chopper is connected between a fixed-voltage dc source and a dc motor to vary the armature voltage. Also, a dc chopper can provide regenerative braking of the motor and can return energy back to the supply. The circuit arrangement of a chopper-fed dc separately excited motor is shown in Fig.1. The motor current is either continuous or discontinuous depending on Mark space ratio (Ton/T) and the inductance of the armature circuit. Figure.2 illustrates the waveforms of the motor current in continuous and discontinuous operation. There are three possible modes of operation: Switch Q conducts, free-wheeling action takes place through the free-wheeling diode, and motor coasts. The governing equations for different modes of operation are as follows [2,3]:

Mode (1)
Switch Q is turned on, and motor is connected   to supply during the interval of
$0 < t < t_1$ and $ia = i_1$
The voltage equation is;
$$V_{dc} = R_a i_1 + L_a \, di_1/dt + K\omega \qquad (1)$$
The torque equation is
$$T_d = K \, i_1 = Jd\omega/dt + B\omega + T_L \qquad (2)$$

Mode (2)
Switch Q is turned off, and free-wheeling action is presented during the interval $t_1 < t < t_2$ and $ia = i_2$
The system equations are
$$0 = R_a i_2 + L_a \, di_2/dt + K\omega \qquad (3)$$
$$T_d = K \, i_2 = Jd\omega/dt + B\omega + T_L \qquad (4)$$

Mode(3)
Motor coasts during the interval $t_2 < t < T$
$Ia = 0$
$$0 = Jd\omega/dt + B\omega + T_L \qquad (5)$$
In order to determine the performance of the system, numerical method can be used.

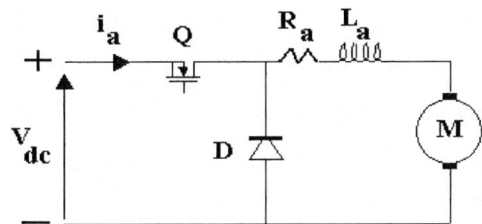

1-4244-0448-7/06/$25.00 ©2006 IEEE   847

Figure 1 A classical chopper-fed dc motor

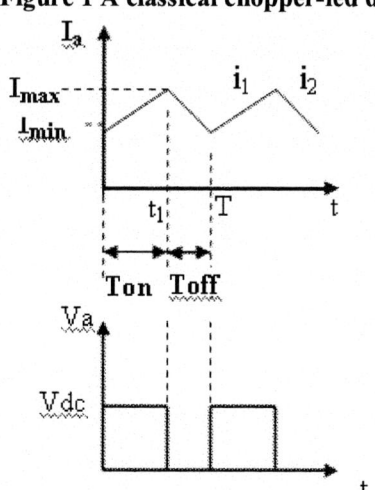 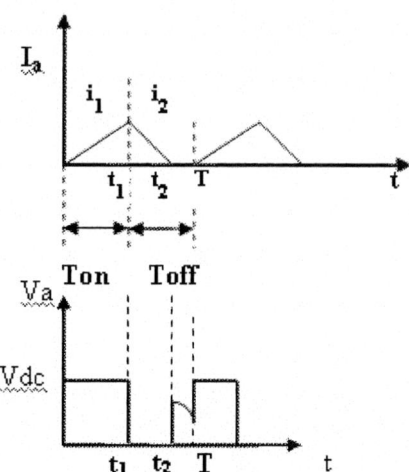

Figure 2 Continuous and discontinuous motor current at steady state operation

III. SYSTEM DESCRIPTION

The proposed system as shown in Fig.3 consists of a separately excited dc motor loaded by eddy current break. Its armature circuit is controlled via a d-c to dc chopper which is represented by a Programmable Logic Controller (PLC). There is a pair of relays interfaced to the PLC output terminals. The first relay is connected in the series with armature circuit of dc motor. On the other hand, the second relay is connected in parallel with the armature circuit of dc the motor. This combination of the two relays acts as a dc to dc chopper. A feedback signal voltage from the tacho-generator is connected as an input to an operational amplifier comparator. A reference voltage corresponding to the command speed is also an input to the comparator. The output signal of the comparator is either ± 0.9 Vcc which represents the polarity of the difference error between the reference voltage and the feed back voltage. The input interface of the PLC used in this research operates in the range of an input voltage between 8 and 30 volts. From this point, a Pang-Pang controller is suitable in this application. As the input voltage to the PLC is +0.9Vcc ≈ 13.5 volt, the PLC is operated and hence the output relay (1) is switched On, but the output relay (2) is switched off. This means that dc power is transferred to the dc motor. On the other hand, if input voltage to PLC is -0.9Vcc, output relay (1) is switched off, while output relay (2) is switched on. This means that a freewheeling period of is started. The average value of motor voltage is varied according to the variation of duty cycle, which is controlled by adjusting a reference voltage corresponding to command motor speed. An input switch is interfaced to the PLC for use in an emergency. The advantages of the proposed system may be summarized as: wide range of speed control (from 0% to 100% of rated speed) can be obtained, less complex circuit

and low size memory used program consisting of three rungs (statements), as shown in Figs.4 and 5 where both 011 and 0111 are addresses of the two output relays 1 and 2. 701 is an internal relay address which is used temporarily during the execution of the program and switch 01 is an emergency switch.

IV. SIMULATION AND EXPERIMENTAL RESULTS

The proposed system is practically designed and implemented using PLC 100 (Allen Breddly) to verify the developed model of the system. The behavior of this system under transient and steady state operations determined by solving the non-linear system equations (1-5) using Rung Kutta fourth-order method with a suitable interval time of calculation that equals 0.01 sec. It is clear that there is two region of motor current, continuous or discontinuous, it depends on the reference motor voltage and the load. Figure 6 shows the computed and experimental responses of motor speed, terminal voltage and current; due to positive change of reference speed (from 1000 to 1200 r.p.m.), at half-load. It is clear that the motor speed follows the desired speed reference and the motor current is continuous during the period of changing of the reference speed and is discontinuous otherwise. This means that the proposed controller is effective and accurate. On the other hand, as reference speed is negatively changed (from 1000 r.p.m. to 500 r.p.m.) and the motor is half-loaded, the motor speed follows the reference speed as shown in Fig.7, but the frequency of both voltage and current of the motor vary during the cycle in proportional inversely with the reference speed. Figures 8-9 show the motor voltage and current as the load torque is varied from initial steady state value, to another steady value, then returned back to its initial steady state value where the reference speed is kept constant. It is clear that the motor current and voltage are affected when moving from continuous to discontinuous current modes or vice versa, according to the load value.

V. CONCLUSION

The paper proposed a simple, effective and accurate speed controller of separately excited dc motor using PLC, which acts as a dc to dc chopper. Using PLC instead of classical controlled switches such as Thyristors, MOSFETs and IGBTs in the proposed system achieves a compact system and avoids the variation of dv/dt and di/dt of the controlled switches. The proposed system gives speed control in a wide range from 0 to 100% of the rated speed. A pang-pang controller is used. It is effective and fast, but there is a fluctuation of motor speed appears in the steady mode operation. A filter may be used for eliminating the distortion of the motor speed. The start-up and steady-state behaviors of the motor are predicted using the proposed simulation. Comparison between the simulation and experimental results for closed loop systems proved they are in line with each other.

VI. REFERENCES

[1]Mohamed Harunur-Rashid,"Power Electronics Circuits, Devices, and Applications", Prentice Hall International, 1990.

[2]David Finney, "The Power Thyristor and its Applications", McGraw-Hill Book Company (UK) Limited, 1980

[3]P.C.Sen, "Thyristor DC drives", Krieger Pub. Co. (07/01/1991)

[4]Petru Livinti and Gheorghe Livint, "The Synthesis of Automatic Controllers For Electrical Drive Systems With D.C. Motors", 9th International Research/Expert Conference TMT2005, Antalya, Turkey, 26-30 September 2005.

[5] Bolton, W. "Programmable Logic Controllers", Elsevier Science & Technology Books, San Diego, USA, 2003

[6]Tomas Sysala and Peter Dostal, "Adaptive Control Algorithm Implemented Into Programmable Logic Controller", 9th International Research/Expert Conference TMT2005, Antalya, Turkey, 26-30 September 2005.

VII. APPENDIX

The test motor is a separately excited dc motor, 50volt. 50 watt, 1 ampere, 3000 r.p.m. having the following parameters:
R_a= 10.5 ohm, L_a= 0.06 H, R_f = 550 ohm, B= 1E-05 N.m./(rad/sec.), K= 0.127 Volt/(rad./sec.), J=1.2E-04 Kg.m^2.

VIII. SYMBOLS:

L_a, R_a: Armature inductance and resistance
Rf: Field resistance
B: Viscous friction coefficient
J: Moment of inertia.
K:Back e.m.f. coefficient.
ω: Moror speed (rad./sec.)

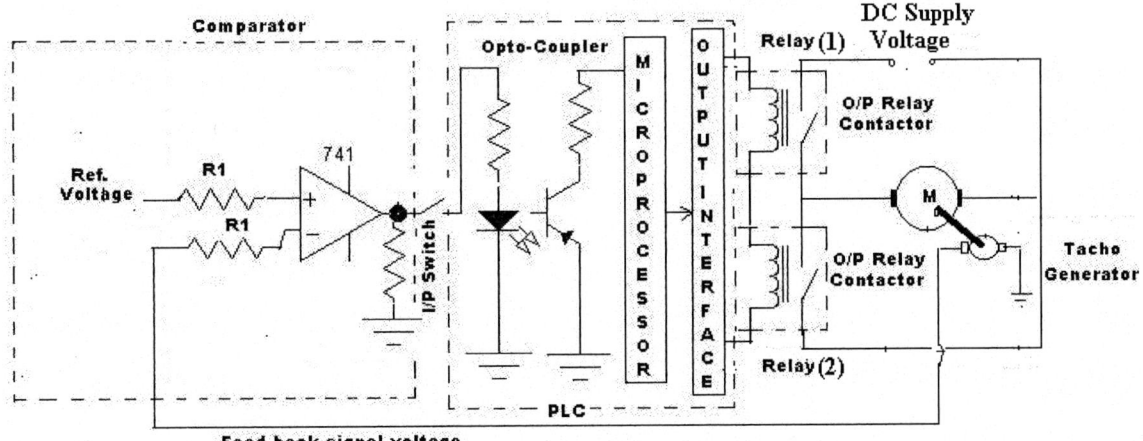

Figure 3   The proposed system

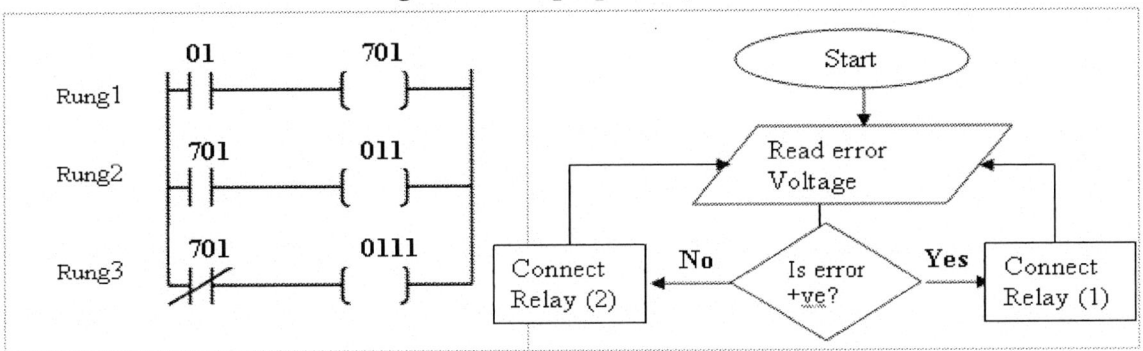

Fig. 4 A ladder diagram program   **Fig.5 A flow chrat of a program**

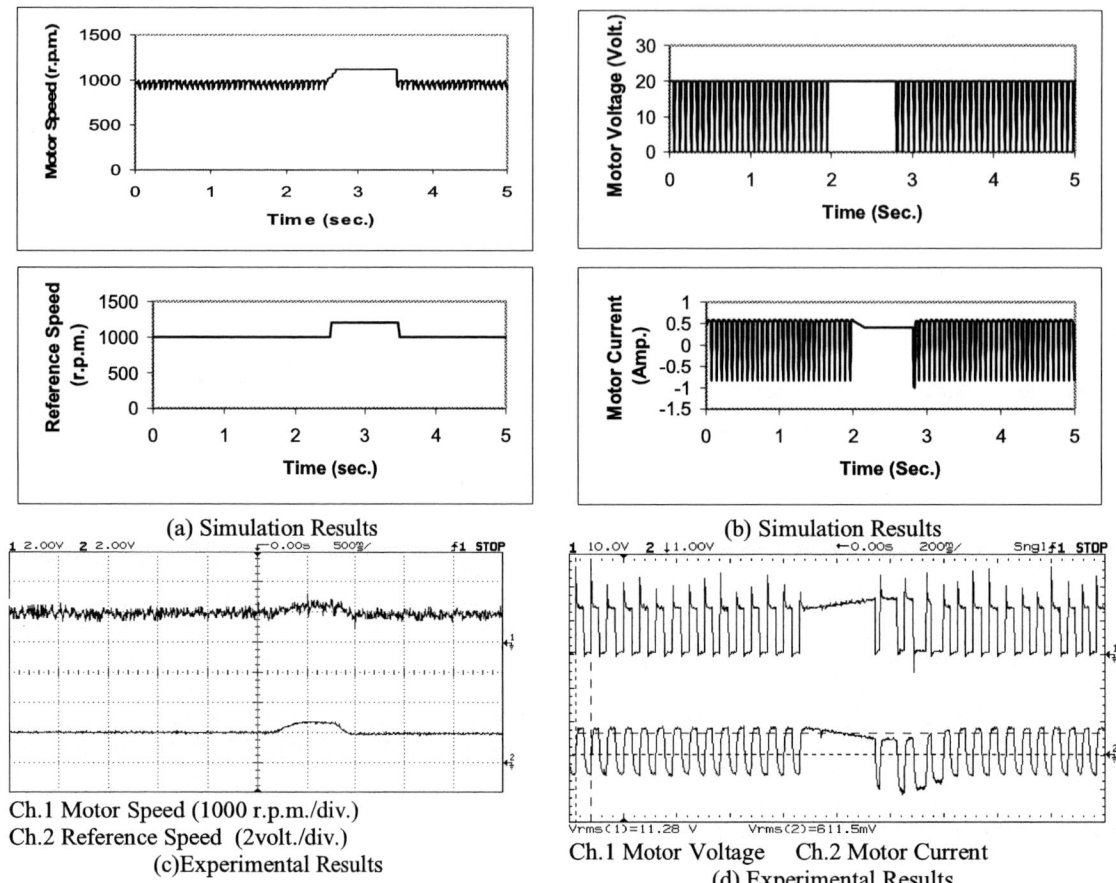

(a) Simulation Results

(b) Simulation Results

Ch.1 Motor Speed (1000 r.p.m./div.)
Ch.2 Reference Speed (2volt./div.)
(c)Experimental Results

Ch.1 Motor Voltage   Ch.2 Motor Current
(d) Experimental Results

Figure 6   Motor performance due to positive step change of reference speed at half load

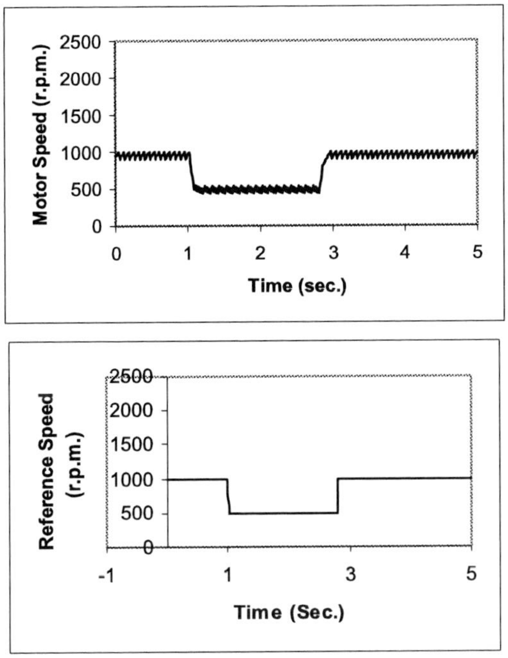

(a) Simulation Results

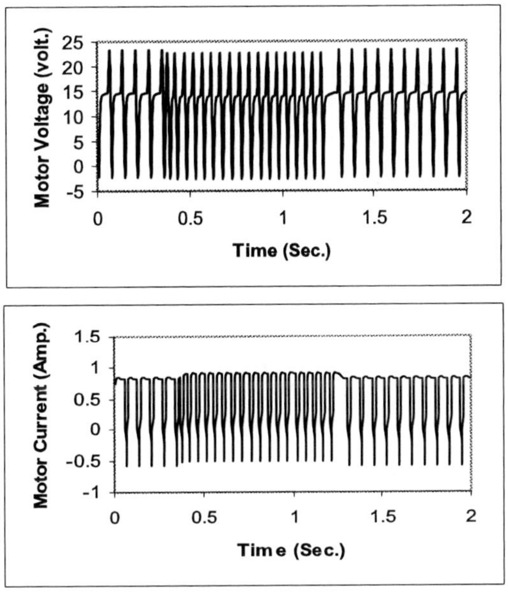

(C)Simulation Results

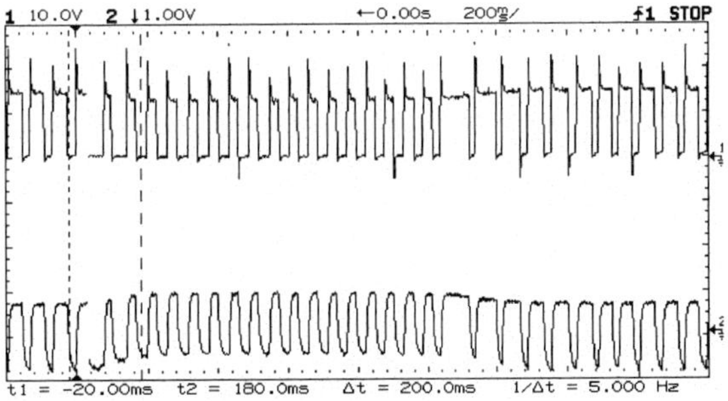

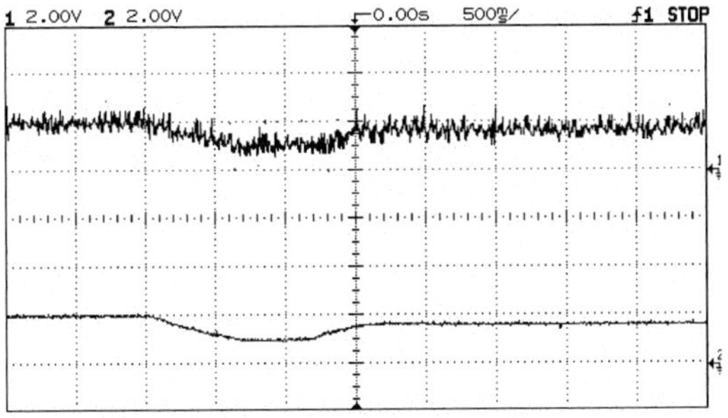

Ch.1 Motor Speed (1000 r.p.m./div)
Ch.2 Reference Speed (2volt./div.)
(b) Experimental Results

Ch.1 Motor Voltage   Ch.2 Motor Current
(D) Experimental Results

Figure 7   Motor performance due to negative step change of reference speed at half load

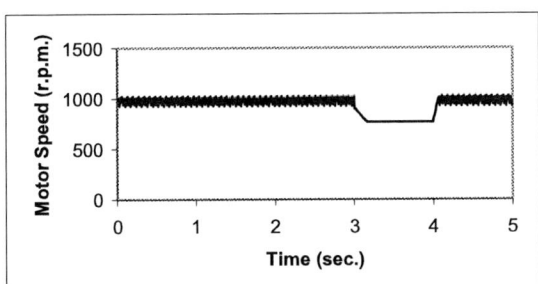

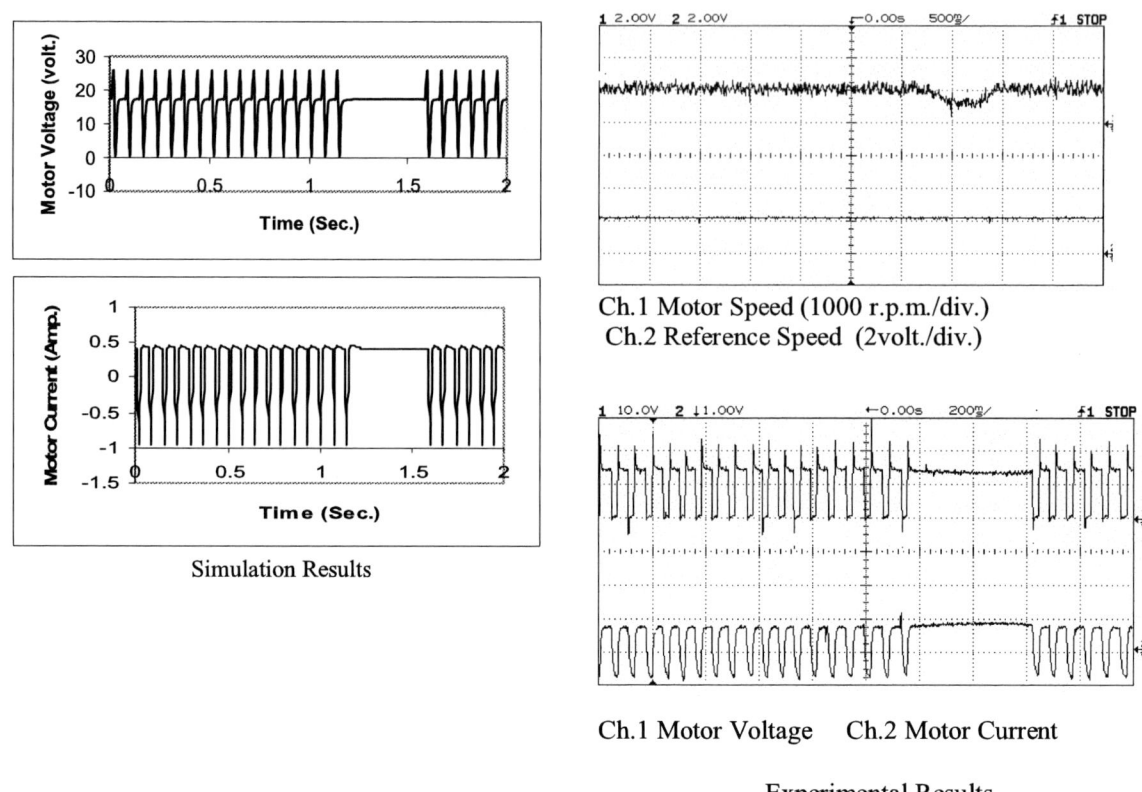

Ch.1 Motor Speed (1000 r.p.m./div.)
Ch.2 Reference Speed (2volt./div.)

Simulation Results

Ch.1 Motor Voltage   Ch.2 Motor Current

Experimental Results

Figure 8   Motor performance due to change of load from no load to full load then return back to no load at of reference speed=1000 r.p.m.

2006 5th International Power Electronics and Motion Control Conference

H_∞ Control of Adjustable-Pitch Wind Turbine Adjustable-Pitch System

Hongche Guo and Qingding Guo

Shenyang University of Technology Shenyang P.R.China 110023
ghc_work@163.com

Abstract—**Active control of a horizontal axis wind turbine via pitching of the rotor blades is considered in this paper. It presents a robust control design for a variable-speed adjustable-pitch wind energy conversion system with an uncertain model. A $H\infty$ controller is designed to drive the turbine speed to extract invariable power from the wind and to allow pitch adjustment to power regulation. It is the purpose of this paper to test and validate a $H\infty$ controller for a wind turbine by simulation.**

Keywords: wind turbine; $H\infty$ control; adjustable-pitch system

I. INTRODUCTION

The most important underlying motivation for wind energy research for large-scale energy production is the aim to reduce the price of produced electrical energy. In the area of wind and other power generation systems, where the input resource power varies considerably, variable-speed generation (VSG) is more attractive than fixed-speed systems.

Because variable-speed wind turbines have the potential for increased energy capture, controller design has become an area of increasing interest. Blade-pitch regulation provides means for initiating rotation, varying rotational speed to extract power at low wind speeds, and maintaining power production at a maximum level. Controllers must be designed to meet each of these objectives, but this study pertains only to constant power production.

The power regulation regime is entered when the turbine reaches the design rotor speed for maximum power production. Under these conditions, rotational speed is constrained to a specified maximum value through blade-pitch regulation. Fluctuations in wind speed are accommodated to prevent large excursions from the desired rotational speed. Thus the power production is also constrained to a relatively constant level. In addition to maintaining a constant rotational speed, actuator movement must be restrained to prevent fatigue and overheating. The combination of maintaining a constant rotational speed and minimizing actuator motion are the control objectives specified for the power regulation regime.

The controlled system must be robust to parameters uncertainties in the plant model. It is considered a turbine

that allows pitch angle adjustment to power regulation above maximum speed and to help start. The controller is only active above rated wind speed. Since the wind frequently crosses rated wind speed, the controller must operate without a switching transient. The main objective of the control system designed in this paper is regulate pitch demand such that power is kept at its rated value. A change in the operating point of the plant was simulated for a disturbed one showing the robustness of the controlled system.

II. BACKGROUND

Considering the system of Fig.1, the turbine torque is given by [1]-[3]:

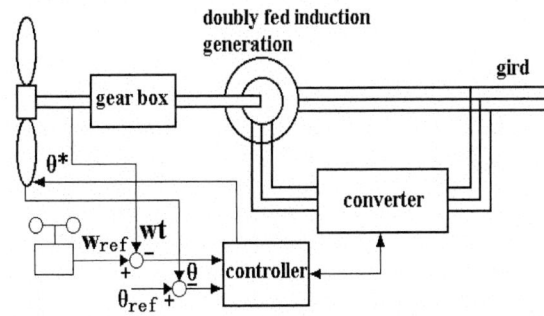

Figure 1. Complete system

$$T_t = \frac{1}{2} C_q \rho \upsilon^2 \pi R^3 \tag{1}$$

where ρ is the air density, v is the wind velocity, R is the turbine radius and C_q is a turbine characteristic:

$$C_q = f(\omega_r, \upsilon, \theta) \tag{2}$$

been ω_r the turbine angular velocity and θ the pitch angle.

The torque coefficient C_q has the characteristics shown in Fig.2. In high speed, $C_{q\max}$ occur with $\theta = 0$ rad. A change in the pitch angle decrease C_q making the behavior of the system worst (1). In a low wind speed situation, the pitch angle can be used to improve C_q. The pitch can also be used in high wind speeds to limit the power by reducing the turbine aerodynamics.

1-4244-0448-7/06/$25.00 ©2006 IEEE

There are a direct relation between C_q and the power coefficient C_p :

$$C_p = \lambda C_q \qquad (3)$$

where λ is the pitch velocity and is given by

$$\lambda = \frac{\omega_r R}{\upsilon} \qquad (4)$$

The power is given by $P_t = \frac{1}{2} C_p \rho A \upsilon^3$, thus:

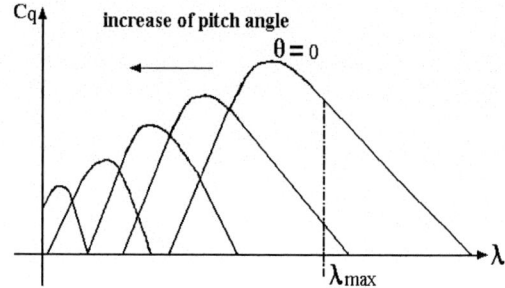

Figure 2. Torque coefficient characteristics

$$P_t = \frac{1}{2} C_p \rho A \frac{R^3}{\lambda^3} \omega_t^3 = K_T \omega_t^3 \qquad (5)$$

where, A is the rotor disk area. This equation gives the points of maximum power transfer with K_T calculated for λ_{max}. Therefore, the torque to maximum power transfer is:

$$T_t = K_T \omega_t^2 \qquad (6)$$

for λ_{max}.

So, the maximum power transfer could be reached acting on the resistant torque of the generator so that the turbine speed is as near as possible from that corresponding to $\lambda = \lambda_{max}$.

The turbine speed dynamics behave as a low pass filter given by:

$$\tau \Delta \dot{\omega} + \Delta \omega_t = \mu \Delta \upsilon \qquad (7)$$

with: $\tau = \dfrac{J}{T_r'(\omega_{t0}) - KR\upsilon_0 C_q'(\lambda_0)}$

$$\mu = \frac{2K\upsilon_0 C_q(\lambda_0) - K\omega_{r0} R C_q'(\lambda_0)}{T_r'(\omega_{t0}) - KR\upsilon_0 C_q'(\lambda_0)}$$

$$K = \frac{1}{2}\rho \pi R^3$$

where J is the inertia of the system and $C_r'(\lambda_0)$ and $T_r'(\omega_{t0})$ denote derivatives with respect to λ and ω_t respectively.

Therefore, the system can not recover all energy from the wind because ω_t can not respond immediately to turbulent winds without high torque. Thus, there is a trade-off between maximum energy extraction on turbulent winds and torque limits. Because the torque coefficient is related to the power coefficient C_p, through the (3) relation manipulation of the torque coefficient using λ and θ will result in manipulation of the power produced by the turbine. The block diagram in Fig.3 illustrates the simulation logic.

III. MODELING OF SYSTEM AND DESIGN OF THE H_∞ CONTROLLER

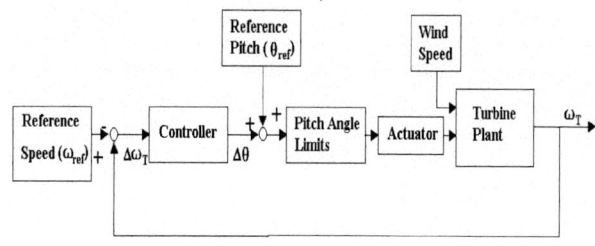

Figure 3. Simulation block diagram

Transfer Function of Hydraulic actuators that adjust to the blade-angle may be approximately represented by third-order polynomial [4]:

$$G(s) = \frac{K_a}{s\left(\dfrac{s^2}{\omega_h^2} + \dfrac{2\zeta_h}{\omega_h^2}s + 1\right)} \qquad (8)$$

with: $K_a = (1+K)K_u U - \dfrac{K_{ce}}{A^2(1+K^2)}\left(1 + \dfrac{CS}{K_{ce}}\right)F_l$

$$\omega_h = \sqrt{\frac{A^2(1+K^2)}{mC}}$$

$$\zeta_h = \frac{1}{2}\frac{K_{ce}m + bC}{2\sqrt{mC}\sqrt{A^2(1+K^2)}}$$

where, K_a is open loop gain of system, ω_h is natural frequency of system and ζ_h is damping coefficient.

Through analysis, open loop transfer Function of hydraulic position actuators may be simplified by:

$$G(s) = \frac{K}{s(Ts+1)} \qquad (9)$$

For design of the H_∞ robustness velocity controller, the velocity loop of hydraulic actuators is considered as controlled plant [5]. In this structure the controlled plant is equal to the standard plant $G_0(s)$:

854

$$G_0(s) = \frac{K}{Ts+1} \tag{10}$$

Consider a block diagram of the control system which is shown in Fig.4. If the system possesses external disturbance, we need the controller K_0 that makes the system with disturbance input which has the robust stability and servo performance (if the given input $r(t)$ is a unit step, then steady state error $e(t)$ of system output $y(t)$ trends to zero). From inner model theory, the full and

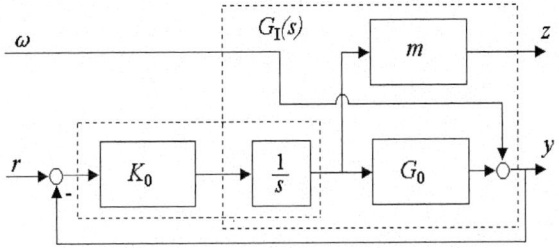

Figure 4.   System design of step input

necessary condition of the system with unit step input $r(t)$ which error $e(\infty) = 0$ is that open loop transfer matrix includes an integral factor at least. Let us now assume that system is a single variant, thus the controller may be represented by:

$$K(s) = \frac{K_0(s)}{s} \tag{11}$$

So, we may add integral factor $1/s$ to controlled object of generalized multiplicative. We have set:

$$G_I(s) = \begin{bmatrix} O & \dfrac{1}{s}m(s) \\ I & -\dfrac{1}{s}G_0(s) \end{bmatrix} \tag{12}$$

Then we intend to find the solution $K_0(s)$ to the standard design of H_∞ from $G_I(s)$. The controller which makes the system robust stability and meats the requirement of servo performance is given by (11). Putting the description reference input signal model into the controlled object of generalized multiplicative, and obtained H_∞ control system with servo performance from working out the solution of H_∞ standard design have the general sense.

Les us consider the system that was shown in Fig.5. To controlled object of generalized multiplicative $G(s)$, we hope to find out the controller $K(s)$ that makes the inner close loop system stable and $\left\| T_{z\omega}(s) \right\|_\infty < 1$. At the same time, the inner model of close loop system, under the reference input $r(t)$, makes the track error $e(\infty) = 0$

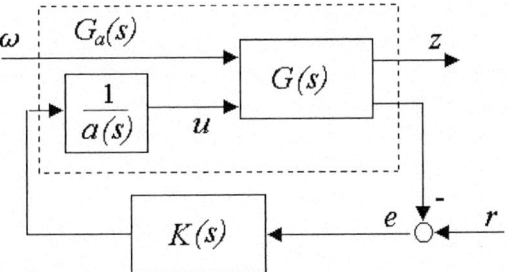

Figure 5.   Block diagram of the inner model theory and servo problem

If we assume input $r(t)$ meets the differential equation below:

$$\frac{d^m r}{dt^m} + \alpha_1 \frac{d^{m-1}r}{dt^{m-1}} + \alpha_2 \frac{d^{m-2}r}{dt^{m-2}} + \ldots + \alpha_{m-1}\frac{dr}{dt} + \alpha_m r = 0 \tag{13}$$

or

$$r(s) = \frac{1}{s^m + \alpha_1 s^{m-1} + \cdots + \alpha_{m-1}s + \alpha_m} = \frac{1}{\alpha(s)} \tag{14}$$

set

$$K(s) = \frac{1}{\alpha(s)}K_0(s) \tag{15}$$

The controlled object of generalized multiplicative is defined by:

$$G_a(s) = \begin{bmatrix} G_{11}(s) & \dfrac{1}{\alpha(s)}G_{12}(s) \\ G_{21}(s) & \dfrac{1}{\alpha(s)}G_{22}(s) \end{bmatrix} \tag{16}$$

Then we are to work out the solution $K_0(s)$ to the H_∞ standard design corresponding to $G_a(s)$. Thus, with original controlled object, the controller of meeting demanding would be given by the (15).

IV.   SIMULATION RESULTS

The real controlled object model that we choose is $G_0(s) = \dfrac{208.3}{41.7s+1}$. Given the Δp disturb-model's maximum is $m(s) = \dfrac{104.2}{41.7s+1}$. The controlled object of generalized multiplicative is solved from the (12):

$$G_I(s) = \begin{bmatrix} 0 & \dfrac{104.2}{s(41.7s+1)} \\ I & -\dfrac{208.3}{s(41.7s+1)} \end{bmatrix} \tag{17}$$

the form is written in the state equation:

855

$$\begin{cases} \dot{x} = \begin{bmatrix} 0 & 1 \\ 0 & -0.024 \end{bmatrix} x + \begin{bmatrix} 0 \\ 1 \end{bmatrix} u \\[2mm] z = \begin{bmatrix} 2.5 & 0 \end{bmatrix} x \\[2mm] y = \begin{bmatrix} -5 & 0 \end{bmatrix} x + \omega \end{cases} \qquad (18)$$

We learn from the (18), the controlled object of generalized multiplicative doesn't meet the solved condition at this time, so we change the object into:

$$\begin{cases} \dot{x} = \begin{bmatrix} 0 & 1 \\ 0 & -0.024 \end{bmatrix} x + \begin{bmatrix} 0.01 \\ 0.01 \end{bmatrix} \omega + \begin{bmatrix} 0 \\ 1 \end{bmatrix} u \\[2mm] z = \begin{bmatrix} 2.5 & 0 \\ 0 & 0 \end{bmatrix} x + \begin{bmatrix} 0 & 0.01 \end{bmatrix} u \\[2mm] y = \begin{bmatrix} -5 & 0 \end{bmatrix} x + \omega \end{cases} \qquad (19)$$

We learn from the reference [6], the change is reasonable.

The system has been simulated on a PC using the Matlab software and its toolbox packages. We gain $K_0(s) = \dfrac{120.844s + 17.436}{s^2 + 22.958s + 263.431}$, velocity controller

$$K(s) = \frac{1}{s} K_0(s) = \frac{120.844s + 17.436}{s(s^2 + 22.958s + 263.431)}$$

To evaluate the ability of the system which follows a reference, the hydraulic servo system was simulated by applying step signals. The result is shown in Fig.6 and Fig.7. Simulation results proved the robustness and fast response of the designed H_∞ controller.

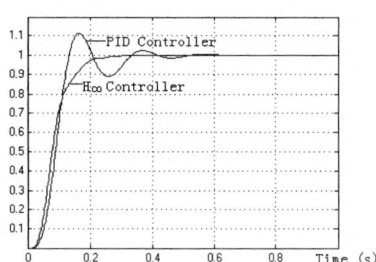

Figure 6.   Step input

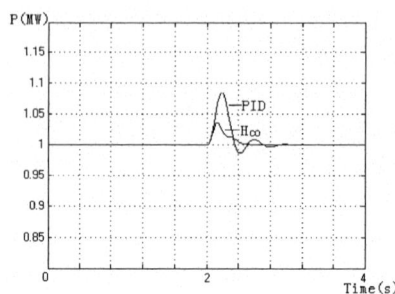

Figure 7.   Rated speed (12m/s) with 2m/s wind speed variation

V. Conclusion

The controller presented has been verified by software. Results which attain the desired control target not only demonstrate the H_∞ control strategy provided in this paper but also demonstrate that H_∞ control strategy possesses advanced ability, so we can find a new way to enhance the performance of system. This technique applied to a wind turbine maintains the stability and reaches the desired performance even in wind speed perturb.

REFERENCES

[1]   Freris, L. L., Wind Energy Conversion Systems, Prentice Hall, 1990.

[2]   W. E. Leithead and B. Connor, "Control of Variable Speed Wind Turbine with Induction Generator", Control'94, Conference Publication, No.389, March 1994.

[3]   Y. D. Song, "Control of Wind Turbines using Memory-Based Method", American Control Conference, June 1998.

[4]   Wang Zhanlin, Hydraulic Pressure Control of the Time, 1997, Beijing

[5]   Sheng Tielong, H_∞ Control Theory and application, 1996, Beijing

[6]   Marcelo L. Lima, Jose L. Silvino, Peterson de Resende, " H_∞ Control For a Variable-Speed Adjustable-Pitch Wind Energy Conversion System", ISIF'99-Bled: 556~561, Slovenia

2006 5th International Power Electronics and Motion Control Conference

The Motion Control Algorithm based on Quaternion Rotation for a Permanent Magnet Spherical Stepper Motor

Qun-jing WANG, Kun XIA

School of Electrical Engineering & Automation, Hefei University of Technology, Hefei, CHINA
wqunjing@sina.com, galaxyx@sina.com

Abstract—This paper presented the motion control algorithm based on quaternion rotation for a permanent magnet spherical stepper motor (PMSSM) designed by Gregory S. Chirikjian and David Stein et al. The movement of the rotor was achieved by controlling the current states of the stator coils with relays. While one arbitrary point on the surface of the rotor was expected to move to another appointed one, combining one-step motion into a piecewise continuous trajectory was necessary. The paper provided a control algorithm about kinetic trajectory divided into sections. Thus each segment corresponded one arc and circled around one fixed axis individually, and the continuous movement of the rotor was feasible. The simulation of the motion control algorithm and the error analysis were then presented in the following.

Keywords-permanent magnet spherical stepper motor; quaternion; motion control algorithm

I. INTRODUCTION

The spherical stepper motor combines pitch, roll and yaw motions in a single joint, which has significant potentials in applications where the demand on workspace is low but three degree-of-freedom (DOF) is required. Typical applications are robotic wrist, camera actuators for computer vision, omnidirectional wheels for mobile robots and so on. Kok-Meng Lee et al have studied the design and implementation of variable reluctance spherical motor (VRSM) for a number of years [1]~[10]. Chirikjian and Stein inherited Lee's design concept, developed a new permanent magnet spherical steeper motor (PMSSM) with the symmetry of the rotor pole arrangement. This paper addresses the motion control algorithm builds on the prototype of the PMSSM.

The prototype of the PMSSM designed by Gregory S. Chirikjian and David Stein et al is shown in Fig.1 [11] ~[13]. A 30.48-cm-diameter (12 inch) hollow plastic sphere is used as the rotor. Eighty 1.905-cm-diameter (0.75 inch) cylindrical rare-earth permanent magnets are placed along the inside surface of the rotor. If semi-

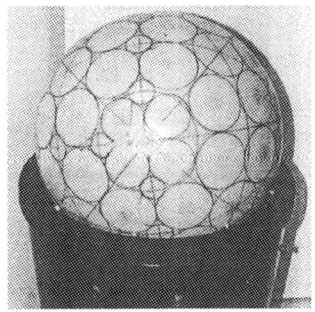

Fig.1 The permanent magnet spherical stepper motor (PMSSM)

regular packing is used as mentioned in paper [11], the principal moments of inertia of the rotor will be equal ($I_{xx}=I_{yy}=I_{zz}$) so as to eliminate the internal torque. A tapered pedestal, a magnet saddle, a ringing housing, sixteen stator electromagnets and eight miniature ball-castors make up the stator. The stator electromagnets chosen are 5.08-cm-diameter (2 inch) by 4.1275-cm-high (1.625 inch). Each electromagnet is positioned on the outside of the magnet saddle by using semi-regular packing too. Normally, the air gap between the rotor and the faces of the stator magnets is 0.0127 cm (0.005 inch). Due to the fact that the symmetry of the rotor pole arrangement is different than that of the stator poles, the fields created by energizing stator coils provide a torque that changes the orientation of the rotor. The accurate position of each magnet and electromagnet in space could be calculated according to the geometric relationship.

The PMSSM with simple structure, relative small volume and wider range of motion is a relatively new kind of spherical motor. A key feature of the PMSSM is that the stator only overlaps a very small area of the rotor, leading to design that has a much wider range of unhindered motion. However, it describes the continuous three-dimensional space with discrete-state. So the approximate error will always prevail. Furthermore, the principle of operation of the three-DOF PMSSM differs significantly from that of the conventional step motor. Two active rotor/stator pairs are always needed to hold the rotor ball in a stable configuration. The explanation for this is that one rotor/stator pair fixes the rotor and produces the fixed axis. The other rotor/stator pair

Project Supported by National Natural Science Foundation of CHINA (50377010).

1-4244-0448-7/06/$25.00 ©2006 IEEE

constrains the rotation round the fixed axis. The fixed axes can only be derived within the scope of the stator cap even more stator coils were energized.

The remainder of this paper is broken down into several sections. Section II investigates the kinetic trajectory divided into sections; Section III presents quaternion rotation representation; Section IV gives the motion control algorithm and an example of simulation; Section V analyzes the errors.

II. TRAJECTORY DIVIDED INTO SECTIONS

When the conventional step motor rotates round its axis, either one-step motion or multi-step motion must be stable. Otherwise the phenomena of step-out will occur. Similarly to the PMSSM, as mentioned before, two or more stator coils should be energized at the same time and at least one rotor/stator pair must be used as the fixed axis during each one-step drive mode to ensure the mechanical stability of the motion control. To simplify the model, this paper discusses only two active rotor/stator pairs drive mode. On this occasion as shown in Fig. 2, one rotor/stator pair produces the fixed axis, and the other one makes the rotor rotate round it.

It is obviously that only sixteen stable axes could be derived in this two active rotor/stator pairs drive mode because of the structure of the stator. It should also be noticed that sometimes one arbitrary point on the surface of the rotor could not move to another appointed surface point in any case if only one fixed axis could be derived. But, two or more individual fixed axes could resolve this problem. That is the whole kinetic trajectory could be divided into sections, each segment corresponds one arc and circles around one fixed axis individually. An example is shown in Fig. 3. While point P could not move to point P' in any case if only one fixed axis could be derived, the kinetic trajectory from point P to point P' could be divided into two sections which arc PQ circles axis $\mathbf{r1}$ and arc QP' circles axis $\mathbf{r2}$ individually. Where point Q just divides the trajectory into two sections P-Q-P' is defined as joint point. Four different trajectories can describe the rotation round two fixed axes $\mathbf{r1}$ and $\mathbf{r2}$. The arcs correspond smaller central angles less than 180 degrees are chosen for the development of computer program. While the trajectory is divided into three sections, three fixed axes are derived. It could be

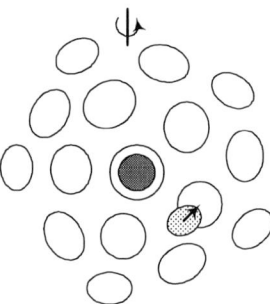

Fig.2 The two active rotor/stator pairs drive mode

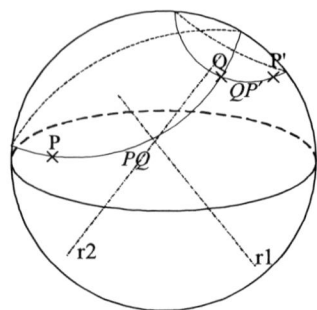

Fig. 3 The kinetic trajectory divided into sections

concluded from correlative knowledge that if there are two perpendicular axes in these three axes, it is always possible to divide the kinetic trajectory so as to let point P move to point P'.

III. QUATERNION ROTATION

It can be known from the theorem of quaternion rotation that the regular quaternion $\mathbf{A}$ contains the essential geometric factors such as rotation axis and rotation angle when quaternion representation is used. Unlike Euler angles that have some unpleasant analytical properties at some orientations, quaternion representation is free of singularities and easy to get those essential geometric factors. Moreover, to regular quaternion $\mathbf{A}$, the inverse vector equals to the conjugated one because the modulus of $\mathbf{A}$ equals 1. It can be expressed as

$$\|\mathbf{A}\| = 1, \qquad (1)$$

so

$$\mathbf{A}^{-1} = \widetilde{\mathbf{A}}. \qquad (2)$$

This can significantly simplify the quaternion rotation. Choose the center of the sphere determined by the rotor of the PMSSM as the origin of the Cartesian coordinate, S as unit sphere. So each discrete point on the spherical surface $P(a,b,c)$ corresponds a vector $\mathbf{OP}$ one to one. While sphere S rotates round some arbitrary fixed axis $\mathbf{r}$ that corresponds a random point $E(e_1,e_2,e_3)$ for a certain angle m, the rotation can be expressed by the regular quaternion representation as

$$\mathbf{A} = (\cos\frac{m}{2} + \mathbf{r}\sin\frac{m}{2}). \qquad (3)$$

It is depicted in Fig. 4, where one great arc passes through i_1, P and E, the other great arc through i_2, P' and E. The great arc i_1i_2 represents angle m. Obviously, point P on the spherical surface rotates round axis $\mathbf{r}$ for a certain angle to a new surface point P'. This new position P' could be determined by quaternion rotation accurately. We set point P representing the active rotor permanent magnet or the initial point of the trajectory, and the axis $\mathbf{r}$ representing the fixed axis round which the rotor rotates. Then each step and the whole trajectory can be described by quaternion rotation.

IV. MOTION CONTROL ALGORITHM AND SIMULATION

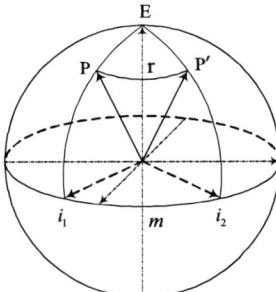

Fig. 4 Quaternion rotation

The magnetic field in the air gap of the PMSSM had been calculated in our previous work [14] with the leakage flux, the fringing flux and the eddy current losses neglected. Later we got the torque-angle characteristic by using Maxwell Stress Tensor method [15]. We set a maximum angle of magnet attraction as a threshold value. This angle is derived from the torque-angle characteristic by measuring the maximum angle where the rotor/stator pairs exhibit an adequate attraction. Angles between the active stator coils and the rotor permanent magnets smaller than this maximum angle can produce a valid torque to make the rotor rotate. All these rotor/stator pairs are defined as candidate rotor/stator pairs. However, when the rotor rotates to some position, the active stator coil may attract several rotor permanent magnets at the same time, and this complicates the motion planning. This situation should be avoided when developing the computer program.

The flow chat of the motion control algorithm is shown in Fig. 5. In this algorithm, while one arbitrary point on the surface of the rotor is expected to move to another appointed point, the kinetic trajectory is divided into two sections. The variables in the flow chat are the following: the arbitrary initial point, the expected end point, and the actual end point are defined as P, E and F individually. The spherical surface equation is defined as **S**. The first step "judgment for the position of rotor" and the second step "motor start" will be discussed in detail in other papers. This paper focuses on the motion planning in the following segment. Obviously, we just have to choose two different fixed axes from sixteen candidate axes when the trajectory is divided into two sections and the motor works in two active rotor/stator pairs drive mode. After the motor started, one rotor/stator pair comes into its working state and produces the first fixed axis **r1**. The following step is to calculate every possible joint points between point P and E. Construct plane **S1** perpendicular to the first fixed axis **r1** with point P in it. So **r1** is the normal vector of **S1**. Similarly construct plane **S2** perpendicular to one of the remainder fixed axes with point E in it. So we can easily get the candidate joint points. One joint point Q could be derived from these candidate joint points by minimizing the error that occurs at the joint point Q when the first fixed axis switches to the second one. Then point P moves continuously so as to form the piecewise continuous trajectory on the surface

of the sphere **S**. Next step is to set the two active rotor/stator pairs in each step. Obviously, the rotor/stator pair produces the fixed axis is one active rotor/stator pair. Paying attention to the stator coils that in peculiar positions as mentioned before can't be chosen, we could get the second desired rotor/stator pair from the candidate rotor/stator pairs that can produce the valid torque making the rotor rotate round the fixed axis. As the second desired active rotor/stator pair is chosen from the candidate rotor/stator pairs, two principles could be followed. One is principle of least steps, which means the line of the attraction vector is approximately perpendicular to the plane formed by the fixed axis and the imaginary torque arm. The other is principle of least angle, which can minimize the error of rotation. These two principles unite in our MATLAB program. When the rotor starts to rotate at first, it follows the principle of least steps because the initial point P is still far from the joint point Q. When the initial point P is nearing the joint point Q, the principle of least angle starts to work to ensure the control precision. This is repeated between point Q and point E after the first fixed axis switches to the second one. We could get the rotation angle and the stable state of each step by using quaternion rotation representation mentioned before to calculate the new stable position of the rotor. The whole motion control algorithm finally stops at the point F that is the nearest one to the expected end point E. The final results about the position of actual end point, the rotor position, the errors and so on are derived in the end.

It should be noticed that this relative simple two-two algorithm (The trajectory is divided into two sections and the motor works in two active rotor/stator pairs drive mode.) has dead band in motion. That is some arbitrary points on the surface of the rotor could not move to the appointed point E in any case by using this two-two algorithm. So the trajectory is in need of being divided into three sections as mentioned in section II. And the three or more active rotor/stator pairs drive mode is required to form two perpendicular axes. The work of developing this kind of "three-multi" algorithm to minimize or eliminate the effect of the dead band is underway. All of this needs the help of the calculation of the magnetic field and optimizing the parameter or the structure of the PMSSM.

An example of simulation is shown in Fig. 6. The coordinate of the initial point P is (0,0,1) and the expected end point E is (0,1,0). We could get the results by the calculation of the simulation procedure that the rotor rotates round the eighth stator coil at first, then round coil eleven. Finally, the initial point P moves to the actual end pint F (0.0084,1.0000,0.0025) on the spherical surface with thirty steps forming one piecewise continuous trajectory.

V. ERROR ANALYSIS

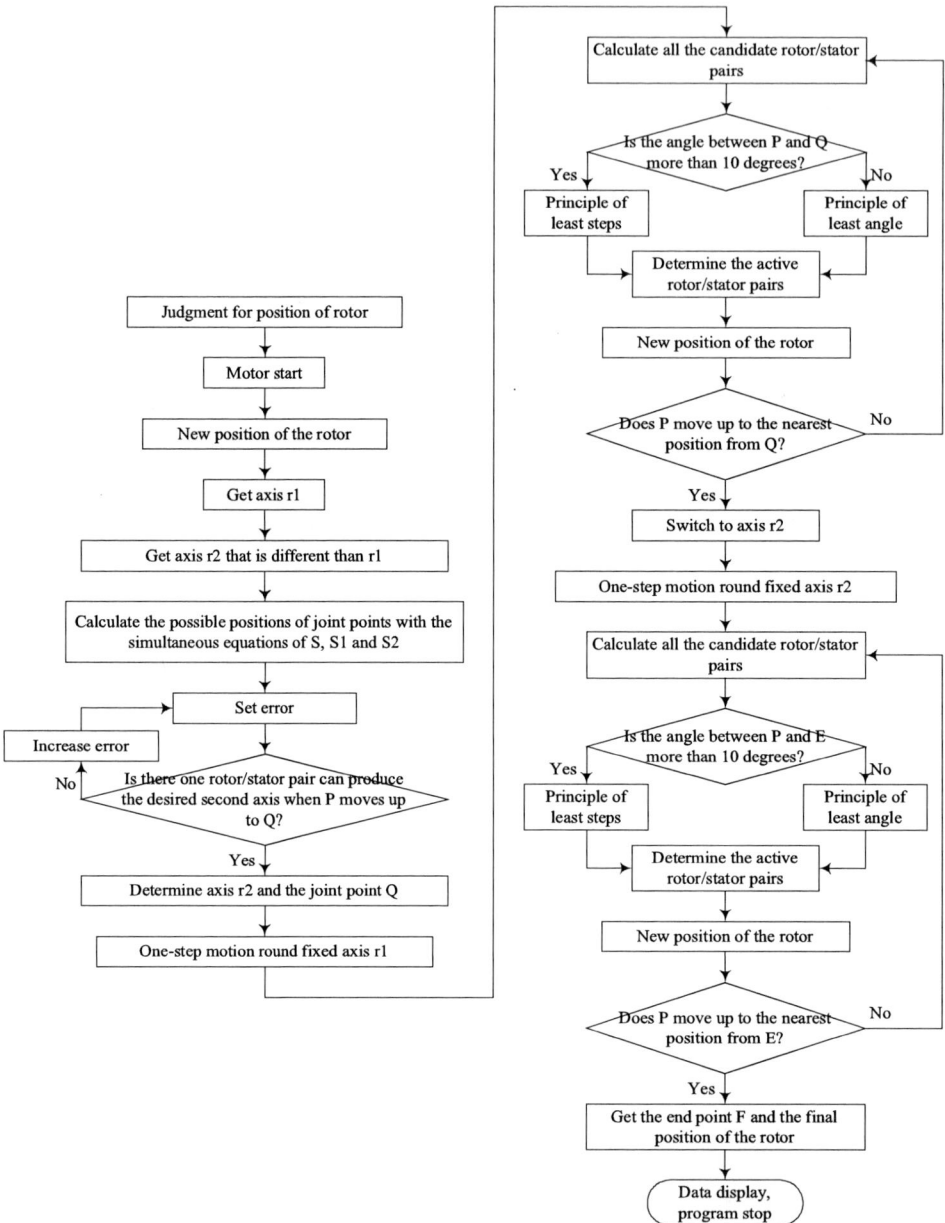

Fig. 5 The flow chat of the motion control algorithm

The errors between the expected point E and the actual end point F chiefly come from the structure of the motor and the motion control algorithm what we have used. The first one and the principal one is the commutation from the first fixed axis to the second fixed axis at the joint point Q. The second one occurs at the point where the initial point P is moving up to the joint point Q and to the expected end point E. These two errors related to the structure of the PMSSM belong to the inherent error. The last one is the calculation error. We could evaluate the error distribution of the inherent error by changing the initial position of the start point P. Without loss of generality, we choose 80 different initial points that are approximately evenly spaced on the surface of the rotor. It is shown in Fig. 7 that when the initial point varies without changing the expected end point E, the errors of the motion control algorithm are the following. There are eight points couldn't move to the expected end point E because the problem of dead band leading to destruction of the algorithm. When the algorithm is in work, the error between point E and point F always prevails. The error of first kind varies from $0.5157°$ to $7.4485°$, the average value is $2.6165°$. The error of second kind varies form 0 to $8.4673°$ with the average is $1.5639°$. These two kinds of errors sum up to the resultant error with the

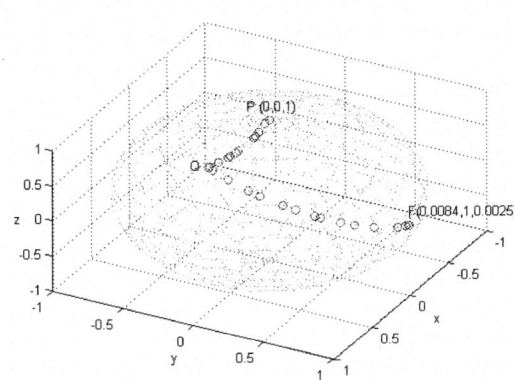

Fig. 6 The example of simulation

average value $4.1804°$ and the maximum resultant error is $11.3246°$. The resultant error below $5°$ is 70 percent.

VI. CONCLUSIONS

The two-two algorithm for the PMSSM presented in this paper has relative simplified method with feasibility and accuracy. It simplifies the calculation of the rotation and the orientation of the rotor by using quaternion representation. More significantly, it is possible to drive one arbitrary point on the surface of the rotor to another appointed surface point by using the method of kinetic trajectory divided into sections with the stator only overlaps a very small area of the rotor. Work underway is advancing three-multi algorithm to minimize or eliminate the effect of the dead band. Associated research task is exploring closed-loop control method to suppress the rotor oscillation and minimize the error of orientation. Further explorations are always needed to optimize the parameter or the structure of the PMSSM.

REFERENCES

[1] Kok-Meng Lee, George Vachtsevanos and ChiKong Kwan, "Development of a spherical stepper wrist motor," *IEEE Int. Conf. on Robotics and Automation, Philadelphia, Pennsylvania, USA*, pp. 267–272, April 1988.

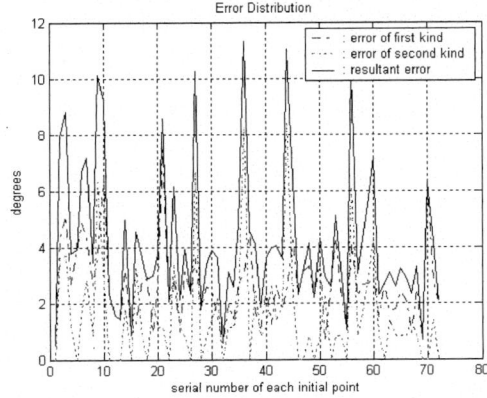

Fig.7 Error distribution

[2] Kok-Meng Lee and Chi-kong Kwan, "Design concept development of a spherical stepper for robotic applications," *IEEE Trans. on Robotics and Automation*, vol. 7, No. 1, pp. 175–181, February 1991.

[3] Kok-Meng Lee and Jianfa Pei, "Kinematic analysis of a three degree-of-freedom spherical wrist actuator," *the 5th Int. Conf. on advanced Robotics, Pisa, Italy*, pp. 72–77, June 1991.

[4] Zhi Zhou and Kok-Meng Lee, "Real-time motion control of a multi-degree-of-freedom variable reluctance spherical motor," *IEEE Int. Conf. on Robotics and Automation, Minneapolis, Minnesota, USA*, pp. 2859–2864, April 1996.

[5] Dan E. Ezenekwe and Kok-Meng Lee, "Design of air bearing system for fine motion application of multi-DOF spherical actuators," *IEEE/ASME Int. Conf. on Advanced Intelligent Mechatronics, Atlanta, USA*, pp. 812–818, September 1999.

[6] Raye A. Sosseh and Kok-Meng Lee, "Finite element torque modeling for the design of a spherical motor," *the 7th Int. Conf. on Control, Automation and Robotics and Vision, Singapore*, pp. 390–395, December 2002.

[7] Kok-Meng Lee and Raye A. Sosseh, "Effects of fixture dynamics on back-stepping control of a VR spherical motor," *the 7th Int. Conf. on Control, Automation and Robotics and Vision, Singapore*, pp. 384–389, December 2002.

[8] Kok-Meng Lee and Debao Zhou, "A real-time optical sensor for simultaneous measurement of three-DOF motions," *IEEE/ASME Trans. on Mechatronics*, vol. 9, No. 3, pp. 499–507, September 2004.

[9] Kok-Meng Lee, Jeffry Joni and Hungsun Son, "Design method for prototyping a cost-effective VR spherical motor," *IEEE Conf. on Robotics, Automation and Mechatronics, Singapore*, pp. 542–547, December 2004.

[10] Kok-Meng Lee, Zhiyong Wei and Jeffry Joni, "Parametric study on pole geometry and thermal effects of a VRSM," *IEEE Conf. on Robotics, Automation Mechatronics, Singapore*, pp. 548–553, December 2004.

[11] Gregory S. Chirikjian and David Stein, "Kinematic design and commutation of a spherical stepper motor," *IEEE/ASME Trans. on Mechatronics*, vol. 4, No. 4, pp. 342–353, December 1999.

[12] David Stein and Gregory S. Chirikjian, "Experiments in the commutation and motion planning of a spherical stepper motor," *ASME 2000 Design Engineering Technical Conf. and Computers and Information in Engineering Conf., Baltimore, Maryland, USA*, September 2000.

[13] David Stein, Edward R. Scheinerman and Gregory S. Chirikjian, "Mathematical models of binary spherical-motion encoders," *IEEE/ASME Trans. on mechatronics*, vol. 8, No. 2, pp. 234–244, June 2003.

[14] Wu Li-jian, Wang Qun-jing, Du Shi-jun and NI You-yuan, "Integral equation method for simulation of magnetic field of permanent magnetic spherical stepper," *Proceedings of the CSEE*, vol. 24, No. 9, pp. 192–197, September 2004.

[15] Qunjing Wang, Zheng Li, Youyuan Ni and Kun Xia, "3D magnetic field analysis and torque calculation of a PM spherical motor," *Proceedings of the 8th Int. Conf. on Electrical Machines and Systems*, vol. 3, pp. 2116–2120, September 2005.

2006 5th International Power Electronics and Motion Control Conference

Research on Restraining Thrust Force Ripple for Permanent Magnet Linear Synchronous Motor

Cui Jiefan[*], Wu Hui[*], Sun Qing[**], Zhang Yi[**] and Zhao Lijun[*]

[*] School of Electrical Engineering, Shenyang University of Technology, Shenyang 110023, China
[**] Higher Vocational and Technology Institute, Shenyang Pharmaceutical University, Shenyang 110026, China
e-mail: cuijf2001cn@yahoo.com.cn

Abstract—Direct thrust force control will be applied to control permanent magnet linear synchronous motor. Thrust force ripple because of special end-effect of linear motor is bigger problem in the linear motor direct driving system without any conversion block. There are different controls methods like traditional hysteresis ring control and based on space voltage vector control to analyze the linear motor direct thrust force control in the paper. Its theory is first discussed in detail. Experiment and simulation results show us the torque ripple for rotating motor and thrust force ripple for linear motor also. It is proved that the thrust force ripple is obviously reduced and restrained by suitable voltage supply in different ways.

Keywords- permanent magnet linear synchronous motor; direct thrust force control; hysteresis ring control; space voltage vector modulation

I. INTRODUCTION

As we know the direct torque control for rotating motor is widely concerned by many researchers now because of its fast response and easy to implement in fact [1-7]. The control method is now extended to control thrust force of permanent magnet linear synchronous motor (PMLSM), i.e. direct thrust force control. Because direct driving system combining with linear machine hasn't any conversion block, the system control performances are affected by various disturbances which act on it. There exits also special end effect for the linear motor on the side. Both reasons above mentioned cause bigger thrust force ripple problem comparing with rotating electrical machine. The shortcoming limits the linear motor application area, particularly for precision machine tools. Therefore it is important project that thrust force ripple for permanent magnet linear synchronous motor should be restrained for its direct driving application.

II. MATHEMATICS DESCRIPTION

A. PMLSM Equations in d-q Reference Frame

In order to analyze steady and dynamic performances of synchronous motor, two axes theory in rotating d-q

reference frame is normally chosen to obtain simple linearization equations from the complex nonlinear ones. The basic equations for PMLSM in d-q frame are following

$$
\begin{cases}
u_d = R_s i_d + \dfrac{d\psi_d}{dt} - \dfrac{\pi}{\tau} v_e \psi_q \\[2mm]
u_q = R_s i_q + \dfrac{d\psi_q}{dt} + \dfrac{\pi}{\tau} v_e \psi_d \\[2mm]
\psi_d = L_d i_d + \psi_f \\[2mm]
\psi_q = L_q i_q
\end{cases}
\tag{1}
$$

Power calculation is following

$$
p_{em} = \frac{3}{2} \frac{\pi}{\tau} v_s \left(\psi_d i_q - \psi_q i_d \right)
\tag{2}
$$

Thrust force can be derived as following

$$
F_t = \frac{p_{em}}{v_s} = \frac{3}{2} \frac{\pi}{\tau} \left[\psi_f + \left(L_d - L_q \right) \cdot i_d \right] \cdot i_q
\tag{3}
$$

There, u_d, u_q and i_d, i_q are respectively voltage and current components in d-q axes. ψ_d, ψ_q are magnetic linkage components in d-q axes. L_d, L_q are d-q axes inductance, ψ_f is permanent magnet linkage, R_s is winding resistance, v_s is linear motor velocity ($v_e = p v_s$), p is the number of pole pair, τ is pole pitch.

B. Thrust Force Equations in x-y References Frame

Relation between the d-q frame and the x-y frame is drawn in Fig.1. Let us define the x-axis direction along resultant magnetic field and the y-axis ahead of it 90 degree along the moving direction. The angle between the x-axis and the d-axis is marked in δ. According to the reflection of two frames, the conversion matrix can be written as (4)

$$
\mathbf{Z}_{dq}^{xy} =
\begin{bmatrix}
\cos\delta & \sin\delta \\
-\sin\delta & \cos\delta
\end{bmatrix}
\cdot
\tag{4}
$$

[a] The project is sponsored by the Province Education Department(2004D049) and Province Nature Science Foundation (20052040)

1-4244-0448-7/06/$25.00 ©2006 IEEE

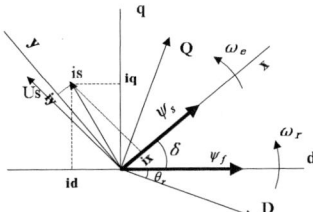

Fig. 1 x-y Reference frame

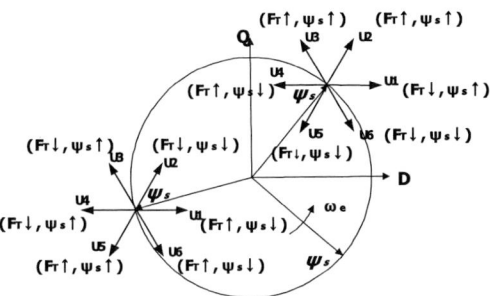

Fig. 2 Voltage vectors, magnetic linkage and thrust force

Equation (4) is same action for current, voltage and magnetic linkage. Therefore it is easy to get thrust force expression (5) combining with (3)

$$F_T = \frac{3}{2}\frac{\pi}{\tau}|\psi_s| i_y \quad .$$ (5)

Current relation is following,

$$\begin{bmatrix} i_d \\ i_q \end{bmatrix} = \left[\mathbf{Z}_{dq}^{xy}\right]^{T}\begin{bmatrix} i_x \\ i_y \end{bmatrix} \quad .$$ (6)

The general thrust force equation is further derived as (7).

$$F_T = \frac{3}{4}\frac{\pi}{\tau}\frac{1}{L_d L_q}|\psi_s|\left[2\psi_f L_q \sin\delta - |\psi_s|(L_q - L_d)\sin 2\delta\right]$$ (7)

If $L_d = L_q = L_s$, then

$$F_T = \frac{3}{2}\frac{\pi}{\tau}\frac{1}{L_s}|\psi_s||\psi_f|\sin\delta$$ (8)

C. Voltage Vectors and Magnetic Linkage

It is clear in (8) that winding magnetic linkage ψ_s decides thrust force F_t when other parameters keep the same. And voltage vectors also determine the magnetic linkage such as (9).

$$\psi_s = \int_0^{t_1}(\boldsymbol{u}_s - R_1 \boldsymbol{i}_s)dt + \psi_{s0}$$ (9)

In other words the linkage ψ_s depends on voltage $\boldsymbol{u}_s$, the force F_t depends on the linkage ψ_s. Therefore it is important to choose suitable voltage vectors in order to control the PMLSM thrust force. Illustration in Fig.2 is just the voltage vectors in different sectors effect on magnetic linkage and thrust force.

III. DIRECT THRUST FORCE CONTROL OF THE PMLSM

A. Hysteresis Ring Direct Thrust Force Control

There are 8 voltage vectors, which include 6 effective voltage vectors and 2 zero vectors in inverter supply. These limited voltages supply the linear motor with discontinuous voltages, which causes thrust force ripple for traditional hysteresis ring control. It is a serious problem for having special end effect linear motor. Based on MATLAB simulink, simulation results are partly presented in Fig. 3 to Fig. 5.

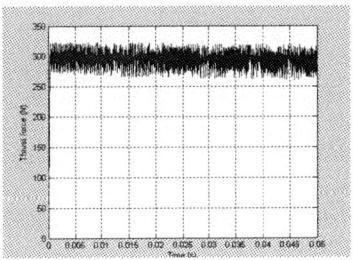

Fig. 3 Traditional direct thrust force control

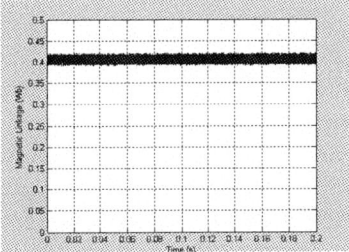

Fig. 4 Magnetic linkage magnitude

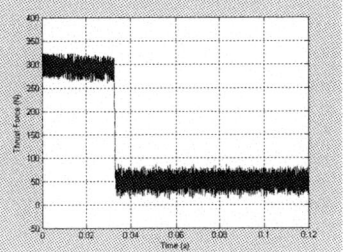

Fig. 5 a Thrust force simulation

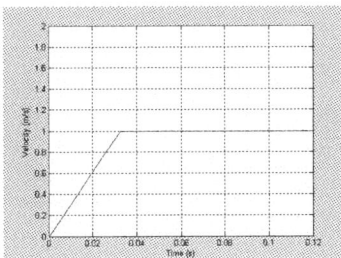

Fig. 5 b Velocity simulation with the force

863

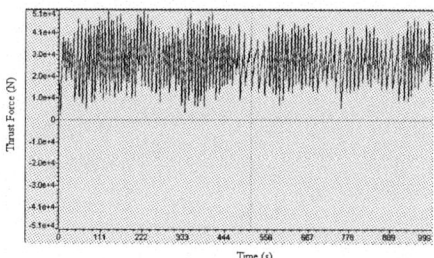

Fig.6   Traditional direct thrust force control experiment

Simulation curves of the thrust force and velocity are pictured in Fig. 5a and Fig. 5b respectively. An elementary experiment thrust force is shown in Fig. 6. Comparing Fig.3 with Fig. 6 both analysis results show us the ripple phenomena.

B.   Space Voltage Vector Modulation Supply

As we know the magnetic linkage changes with voltage. If the linkage is controlled by continuous voltages, it is possible to restrain the thrust force ripple. Considering the voltages relation in Fig. 7 it is easy to understand how to realize any continuous voltage by means of the 8 basic voltage vectors. And general relations of the space voltage vector modulation (SVM) are following,

$$
\begin{cases}
\mathbf{u}_s = \dfrac{1}{T_s}(x_i T_s \mathbf{u}_i + y_j T_s \mathbf{u}_j + z_k T_s \cdot 0) \\
\mathbf{u}_s = x_i \mathbf{u}_i + y_j \mathbf{u}_j \\
\hat{\mathbf{u}}_s e^{j\theta_s} = x u_{dc} e^{j0} + y u_{dc} e^{j\pi/3} \\
x_i + y_j + z_k = 1
\end{cases}
\tag{10}
$$

Here, x_i, y_j are the ratio of the effective voltage working time to sample period. z_k is the ratio of zero voltage working time to sample period. i, j are subscription of two neighboring voltage vectors which have 60^0 angle difference.

Simulation of magnetic linkage and thrust force is presented in Fig. 8 and Fig. 9a. The velocity tracking is drawn in Fig. 9b.

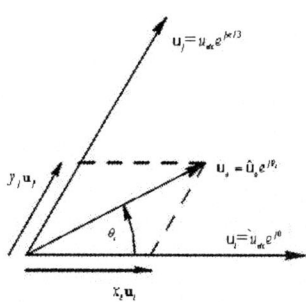

Fig. 7 Resultant of space voltage vector

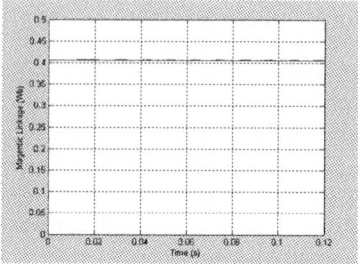

Fig. 8 Magnetic linkage magnitude wish SVM

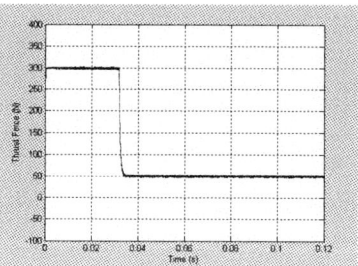

Fig. 9a Thrust force with SVM

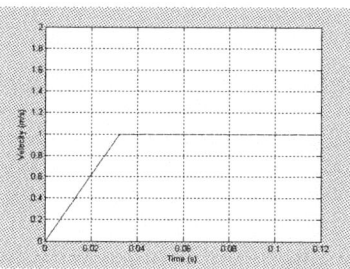

Fig. 9b   Velocity simulation with SVM

IV.   Restraining Thrust Force Ripple Analysys

It is obvious that the performance of the direct thrust force control method based on SVM is much improved than another one. In other words continuous voltage vectors aim at restraining the magnetic linkage ripple and make the thrust force ripple reduction. Thrust force ripple scope is about ± 20N, it is nearly 40 percent of given force (50N) in Fig. 10, but only ± 5 N which is 10 percent of given same value in Fig. 11. Let us observe the magnetic linkage scope in Fig. 12 and Fig. 13. The latter varying range is only 0.0005 Wb, but the former is as big as 0.02Wb. The velocity responses in Fig.5b and Fig. 9b still keep the same performance for the different methods.

V.   Conclusions

The paper has realized direct thrust force control of permanent magnet linear synchronous motor in different

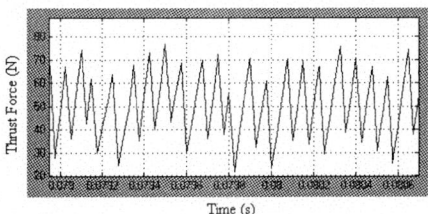

Fig 10. Enlarge display of thrust force

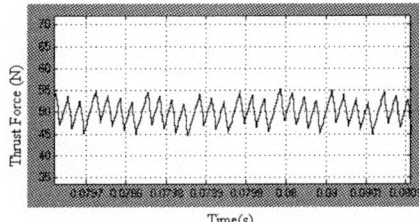

Fig 11. Enlarge display of thrust force with SVM

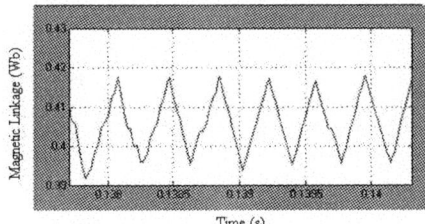

Fig 12. Enlarge display of magnetic linkage

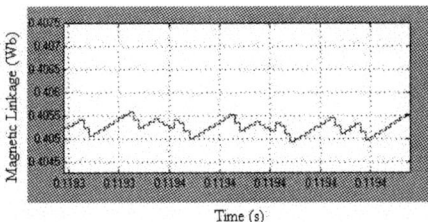

Fig 13. Enlarge display of magnetic linkage with SVM

methods. And detail descriptions about these control method have been discussed. The simple experiment result and many simulation results have been achieved also. Comparing the thrust force test with simulation as well as these simulating results, the direct thrust force control system happen serous ripple which is effective restrain by SVM method.

REFERENCES

[1] M. Depenbrock, "Direct self-control (DSC) of inverter-fed induction machine," *IEEE Trans. on Power Electronics*, 1988, 3(4) , pp. 420–429.

[2] I. Takahashi, T. Naguchi, "A New Quick-Response and High-Efficiency Control Strategy of an Induction motor," *IEEE Trans. on Industry Applications,* 1986, 22(5), pp.820-827.

[3] L. Zhong, M.F.Rahman and W. Y. Hu, "Analysis of direct torque control in permanent magnet synchronous motor drives," *IEEE Trans. on Power Electronics*,1997, 12(3), pp. 528-536.

[4] Yen-Shin Lai, Jian-Ho Chen, "A new approach to direct torque control of induction motor drives for constant inverter switching frequency and torque ripple reduction," *IEEE Trans. on Energy Conversion*,2001, 16(3), pp. 220-227.

[5] D. Telford, M. W. Dunnigan, B. W. Williams, " A novel torque-ripple reduction strategy for direct torque control," *IEEE Trans. on industrial Electronics.* 2001, 48(4) , pp. 867-870.

[6] Chun Tian, Hongping Li, Yuwen Hu, " A novel scheme of direct torque control in the permanent magnet synchronous motor drive," The 7th Workshop on Power Electronics in Transportation. 2002, USA, pp.47-52.

[7] Dariusz Swierczynski Marian P. Kazmierkowski Frede Blaabjerg, "DSP based direct torque control of permanent magnet synchronous motor (PMSM)using space vector modulation," ISIE 2002, Italy,pp723-727.

2006 5th International Power Electronics and Motion Control Conference

Using Recurrent Fuzzy Wavelet Neural Network to Control AC Servo System

Yan Tang, Wei Sun, Yaonan Wang and Xiaohua Zhai

College of Electrical and Information Engineering, Hunan University, Changsha, China

luoluo681031@sina.com

Abstract—**A kind of recurrent fuzzy wavelet neural network (RFWNN) is constructed by using recurrent wavelet neural network (RWNN) to realize fuzzy inference. In the network, temporal relations are embedded in the network by adding feedback connections on the first layer of the network, and wavelet basis function is used as fuzzy membership function. An adaptive control scheme based on RFWNN is proposed, in which, two RFWNN are used to identify and control plant respectively. The proposed adaptive control scheme is applied on AC servo control problem, and simulation results are given.**

Keywords- Recurrent fuzzy neural network; Wavelet; Adaptive control; AC servo

I. INTRODUCTION

Recently, much research has been done on using neural networks (NN) to identify and control dynamic systems [1-3]. NN can be classified as feed forward neural networks and recurrent neural networks. Recurrent neural network[4-7] can capture the dynamical response of a system with its internal feedback loop. It is a dynamic mapping and demonstrates good performance in the presence of uncertainties, such as parameter variations, external disturbance, unmodeled and nonlinear dynamics.

Recurrent fuzzy neural network (RFNN) [8,9] is a modified version of recurrent neural network, which uses recurrent network for realizing fuzzy inference. It is possible to train RFNN using the linguistic experience of human operators, and interpret the knowledge acquired from training data in linguistic form. And it is very easy to initialize the structure and parameters of RFNN from linguistic rules. Moreover, with its own internal feedback connections, RFNN can temporarily store dynamic information and cope with temporal problems efficiently.

In this paper, a recurrent fuzzy wavelet neural network (RFWNN) is proposed. In the network, the temporal relations are embedded by adding feedback connections on the first layer of fuzzy neural network, and wavelet basis function is used as fuzzy membership function. Back propagation algorithm is used to train the proposed RFWNN. For control problem, an adaptive control scheme is developed, in which, two proposed RFWNN are used to identify and control plant respectively. Finally, the

proposed adaptive control scheme is applied on AC servo control problem to achieve high servo performance.

II. CONSTRUCTION OF RFWNN

The structure of the proposed RFWNN is shown in Fig.1, which comprises n input variables, m term nodes for each input variable, l rule nodes, and p output nodes. Using u_i^k and O_i^k to denote the input and output of the ith node in the kth layer separately, the signal propagation and the operation functions of the nodes in each layer are introduced as follows.

Layer 1 (Input Layer): This layer accepts input variables. Its nodes transmit input values to the next layer. Feedback connections are added in this layer to embed temporal relations in the network.

$$u_i^1(k) = x_i^1(k) + w_i^1 O_i^1(k-1), \ O_i^1(k) = u_i^1(k), \qquad (1)$$

where $i = 1, 2, \cdots, n$; k is the number of iterations; w_i^1 is the recurrent weights.

Layer 2 (Membership Layer): Nodes in this layer represent the terms of respective linguistic variables. Each node performs a wavelet basis membership function.

$$u_{ij}^2 = \frac{O_i^1 - a_{ij}}{b_{ij}}, \ O_{ij}^2 = h(u_{ij}^2) \ , \qquad (2)$$

where $i = 1, 2, \cdots, n$, $j = 1, 2, \cdots m$. $h(\cdot)$ is a mother wavelet used in this paper, which is defined as:

$$h(x) = \cos(0.25 \cdot x) \exp(-x^2) . \qquad (3)$$

a_{ij} and b_{ij} in (2) are the dilation and translation parameters of the wavelet membership function, the subscript ij indicates the jth term of the ith input variable.

Layer 3 (Rule Layer): This layer forms the fuzzy rule base and realizes the fuzzy inference. Each node is corresponding to a fuzzy rule. Links before each node represent the preconditions of the corresponding rule, and the node output represents the "firing strength" of corresponding rule.

The qth node of layer 3 performs the AND operation in qth rule. It multiplies the input signals and output the product. Using $O_{iq_i}^2$, $q_i \in \{1, 2, \cdots, m\}$ to denote the membership of x_i to its corresponding linguistic term in qth rule, then the input and output of qth node can be described as:

1-4244-0448-7/06/$25.00 ©2006 IEEE

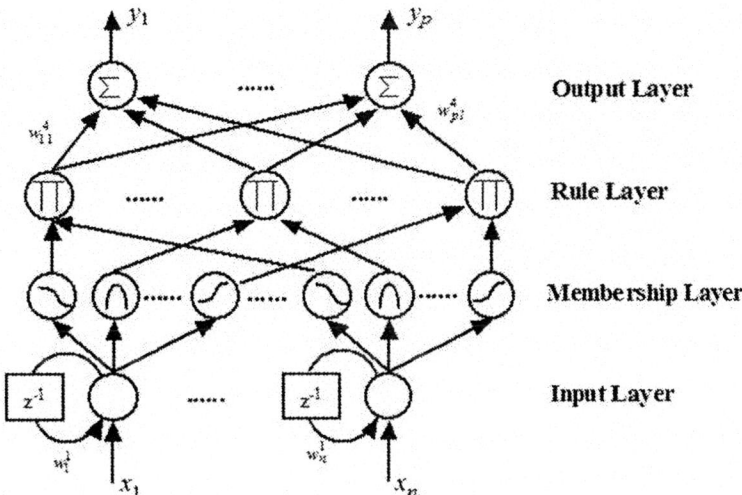

Figure 1. Structure of four-layer RFWNN

$$u_q^3 = \prod_{i=1}^n O_{iq_i}^2 \,, \; O_q^3 = u_q^3 \,, \; q = 1, 2, \cdots, l \,. \qquad (4)$$

Layer 4 (Output Layer): Nodes in this layer perform the defuzzification operation. The input and output of sth node can be calculated by:

$$u_s^4 = \sum_{q=1}^l w_{sq}^4 O_q^3 \,, \; O_s^4 = u_s^4 \Big/ \sum_{q=1}^l O_q^3 \,, \qquad (5)$$

where $s = 1, 2, \cdots, p$, w_{sq}^4 is the weight, which represents the output action strength of the sth output associated with the qth rule.

From the above description, it is clear that the proposed RFWNN is a fuzzy logic system with memory elements in first layer. Since a fuzzy system has clear physical meaning, it is very easy to choose the number of nodes in each layer of RFWNN and determine the initial value of weights.

III. ADAPTIVE CONTROL BASED ON RFWNN

The block diagram of the adaptive control system based on RFWNN is shown in Fig. 2. In this scheme, two RFWNN are used as controller (RFWNNC) and identifier (RFWNNI) separately. The plant is identified by RFWNNI, which provides the information about the plant to RFWNNC. The inputs of RFWNNC are $e(k)$ and $\dot{e}(k)$. $e(k)$ is the error between the desired output $r(k)$ and the actual system output $y(k)$. The output of RFWNNC is the control signal $u(k)$, which drives the plant such that $e(k)$ is minimized. Since the temporal relations are embedded in RFWNN, only $y(k\text{-}1)$ and $u(k)$ need to be fed into RFWNNI for identifying the model of plant [8].

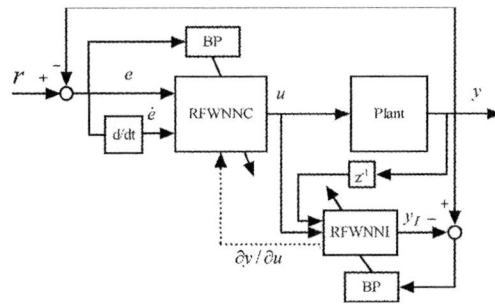

Figure 2. Adaptive control system based on RFWNN

Both RFWNNI and RFWNNC are trained by BP algorithm. For training the RFWNNI in Fig. 2, the cost function is defined as follows:

$$J_I(k) = \tfrac{1}{2} \sum_{s=1}^p (e_{Is}(k))^2 = \sum_{s=1}^p (y_s(k) - y_{Is}(k))^2 \,, \qquad (6)$$

where $y_s(k)$ is the sth output of the plant, $y_{Is}(k)$ is the sth output of RFWNNI, and $e_{Is}(k)$ is the error between $y_s(k)$ and $y_{Is}(k)$ for each discrete time k.

Then the weights of the RFWNNI can be adjusted by

$$W_I(k+1) = W_I(k) + \Delta W_I(k)$$
$$= W_I(k) + \eta_I \left(-\frac{\partial J_I(k)}{\partial W_I(k)} \right), \qquad (7)$$

where η_I represents the learning rate and W_I represents the tuning weights, in this case, which are w_{Isq}^4, a_{Iiq_i}, b_{Iiq_i}, and w_{Ii}^1. Subscript I represents RFWNNI.

For training RFWNNC in Fig. 2, the cost function is defined as

$$J_C(k) = \frac{1}{2}\sum_{s=1}^{h}(e_s(k))^2 = \sum_{s=1}^{h}(r_s(k) - y_s(k))^2 , \qquad (8)$$

where $r_s(k)$ is the sth desired output, $y_s(k)$ is the sth actual system output and $e_s(k)$ is the error between $r_s(k)$ and $y_s(k)$. Thus the parameters of the RFWNNC can be adjusted by

$$W_C(k+1) = W_C(k) + \Delta W_C(k)$$
$$= W_C(k) + \eta_C\left(-\frac{\partial J_C(k)}{\partial W_C(k)}\right). \qquad (9)$$

In above equation,

$$\frac{\partial J_C}{\partial W_C} = \sum_s \frac{\partial J_C}{\partial y_s} \cdot \frac{\partial y_s}{\partial W_C}$$

$$= \sum_s\left\{-e_s(k)\sum_t\left[\frac{\partial y_s(k)}{\partial u_t(k)} \cdot \frac{\partial u_t(k)}{\partial W_C}\right]\right\}, \qquad (10)$$

where u_t is the tth control signal, which is also the tth output of RFWNNC. In (9) and (10), η_C represents the learning rate, and W_C represents the tuning weights of RFWNNC. W_C includes $w_{C\,sq}^4$, $a_{C\,iq_i}$, $b_{C\,iqi}$, and $w_{C\,i}^1$. Subscript C represents RFWNNC.

Note that the convergence of the RFWNNC cannot be guaranteed until $\partial y_s(k)/\partial u_t(k)$ is known. Obviously, the RFWNNI can provide this information to RFWNNC.

IV. AC Servo Control

Dynamics of AC servo system are highly nonlinear and may contain uncertain elements such as friction and load. Many efforts have been made in developing control schemes to achieve the precise servo control of AC motor. In this paper, the proposed adaptive control scheme is applied on AC servo control system.

As shown in Fig. 3, Our AC servo system consists of two feedback loops: position control loop and speed control loop. The proposed RFWNNC is used as position controller. Its two inputs are position error e and error variety speed $\dot{e}$. Its output is the desired speed of motor. The speed of the motor then is controlled by the speed control loop. In our design, the whole speed control loop is regarded as the plant in Fig. 2, which is identified by RFWNNI and controlled by RFWNNC.

In Fig.3, θ_d denotes the desired AC motor position, θ is the actual position. ω_d is the desired motor speed, which is calculated by RFWNNC. ω is the actual speed. T_{ed} is the desired torque of motor, which is the output of speed controller. In this paper, we use PID controller as speed controller, and use direct torque control (DTC) method to control the torque of motor.

V. Simulation Experiments

The AC motor used for our simulation has following parameters: rated power P_n is 2.2KW, rated voltage U_n is 220V, rated current I_n is 5A, rated rotate speed n_n is 1440r/min, resistance of stator r_s is 2.91Ω, resistance of rotor r_r is 3.04 Ω, self-induction of stator l_s is 0.45694H, self-induction of rotor l_r is 0.45694H, mutual inductance between stator and rotor l_m is 0.44427H, rated electromagnetic torque T_{en} is 14N·m, the number of polar pairs n_p is 2, the inertia J is 0.002276kg·m², rated flux ψ_n is 0.96wb. And the Sampling frequency of the system is 10KHZ.

Simulation results of the proposed adaptive AC servo control system are shown in Fig. 4 and Fig. 5, and are compared with fuzzy control system. Fig. 4 shows the step response of system without load. Fig.5 shows the disturbance response under the condition that motor is added a load of 15N·m suddenly when system is stable.

From simulation results, it is obvious that the proposed

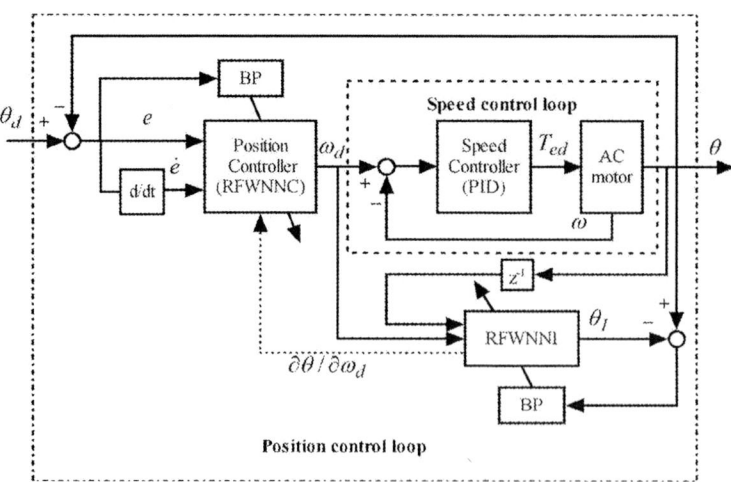

Figure 3. Adaptive control of AC servo system

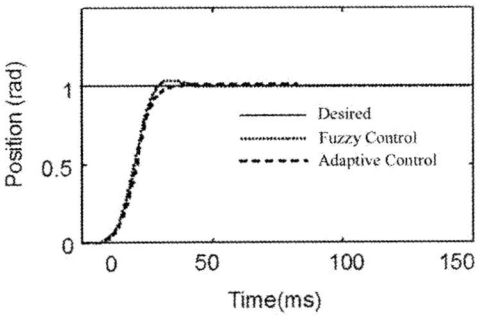

Figure 4. Step response of AC servo system

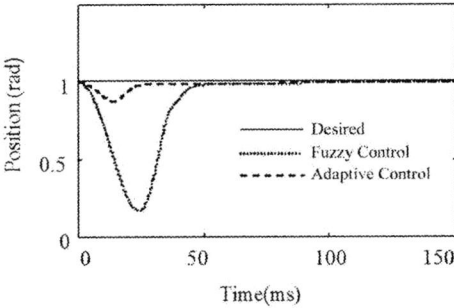

Figure 5. Disturbance response of system

adaptive control scheme can control the AC servo system very well.

VI. CONCLUSION

This paper proposed an RFWNN for realizing fuzzy inference using the dynamic fuzzy rules. The proposed RFWNN consists of four layers and the feedback connections are added in first layer. Wavelet basis function is used as fuzzy membership function. The proposed RFWNN can be used for the identification and control of dynamic system. For identification, RFWNN only needs the current inputs and most recent outputs of plant as its inputs. For control, two RFWNN are used to constitute an adaptive control system, one is used as identifier and another is used as controller. Finally, in this paper, the proposed adaptive control scheme based on RFWNN is used to control the AC servo system and simulation results verified its effectiveness.

REFERENCES

[1] Y. M. Park, M. S. Choi, and K. Y. Lee, "An optimal tracking neuro-controller for nonlinear dynamic systems," *IEEE Trans. Neural Networks*, vol. 7, no. 5, pp. 1099-1110, Sept. 1996.

[2] K. S. Narendra and K. Parthasarathy, "Identification and control of dynamical systems using neural networks," *IEEE Trans. Neural Networks*, vol. 1, no. 1, pp. 4-27, March 1990.

[3] M. A. Brdys and G. J. Kulawski, "Dynamic neural controllers for induction motor," *IEEE Trans. Neural Networks*, vol. 10, no. 2, pp. 340-355, March 1999.

[4] C. C. Ku and K. Y. Lee, "Diagonal recurrent neural networks for dynamic systems control," *IEEE Trans. Neural Networks*, vol. 6, no. 1, pp. 144-156, Jan. 1995.

[5] P. Campolucci, A. Uncini, F. Piazza, and B.D. Rao, "On-line learning algorithms for locally recurrent neural networks," *IEEE Trans. on Neural Networks*, Vol. 10, no. 2, pp. 253 – 271, March 1999.

[6] M. K. Sundareshan and T. A. Condarcure, "Recurrent neural-network training by a learning automaton approach for trajectory learning and control system design," *IEEE Trans. Neural Networks*, vol. 9, no. 3, pp.354-368, May 1998.

[7] X. B. Liang and J. Wang, "A recurrent neural network for nonlinear optimization with a continuously differentiable objective function and bound constraints," *IEEE Trans. Neural Networks*, vol. 11, no. 6, pp. 1251-1262, Nov. 2000.

[8] C. H. Lee and C. C. Teng, "Identification and control of dynamic systems using recurrent fuzzy neural networks," *IEEE Trans. on Fuzzy Systems*, vol. 8, no. 4, pp. 349-366, Aug. 2000.

[9] C. T. Lin, C. L. Chang, and W. C. Cheng, "A recurrent fuzzy cellular neural network system with automatic structure and template learning," *IEEE Trans. on Circuits and Systems I*, vol. 51, no. 5, pp. 1024-1035, May 2004.

2006 5th International Power Electronics and Motion Control Conference

new topology of multi - level - converter for harmonic reduction

Frank Grundmann and Jian Xie, University of Ulm
Frank.Grundmann@uni-ulm.de, Jian.Xie@uni-ulm.de

Abstract— This paper deals with a new topology for a multi - level - converter without an input transformer. Multi - level - converter with an increased number of voltage steps can be realised. The number of valves is equal or smaller than the one of today present topologies.

Index Terms— Multi - level - converter, topology

I. INTRODUCTION

TODAYS converter have to provide more than a suitable fundamental. The harmonics of the power consumption and production have to be taken more and more into account. Additional, the switching frequency has to be as low as possible to reduce the switching losses and to increase the live time.

The necessary quality is actually realised through high switching frequencies (by using standard two - point - converter) or very complex multi - level - converter. With the here presented topology a multi - level - converter can be build without additional switching elements. Thereby, neither a high switching frequency nor a complex multi - level - converter are necessary to achieve the intended quality.

Multi - Level - Inverter [1], [2], [3] can today be realised with for example NPC - topologies [4], flying capacitors [5] or cascaded inverters [6], [7], [8]. The results (spectrum) of the different multi - level - converter (with the same number of voltage level) are limited by the used calculation algorithm for the switching points [1], [9], [10]. The same algorithm and therefor the same switching points with the same spectrum can be used in most of the topologies. The difference lays within the complexity and thereby the costs of the circuit. All these topologies need special input transformers with additional secondary windings or complex circuit structures. The cascaded inverters work with separate DC voltages, which were separated by transformer windings with following inverter. The NPC - topologie needs either for each capacitor a separate secondary winding or a complex circiut structure.

The here presented topology only needs an transformer if the voltage level has to be changed (low voltage to high voltage transformer). In this case a standard transformer can be used (one winding per phase, not per capacitor). Otherwise only an inductor is necessary for decoupling and current control. Thereby the costs of the inverter can be slashed.

The number of valves is another cost aspect. Most of the today practicable topologies combine standard two or three - point - converter to reach a multi - level - converter. Thereby the number of valves will be multiplied. The here presented topology is an all in one design, which uses the valves in

different configuration. Thereby the number of valves can be reduced.

II. TOPOLOGY

The origin of the analysis is the known three - point - converter (figure 1)[11]. A symmetric intermediate circuit (voltage ratio 1:1) is used there, wherefor the potential $\frac{U_d}{2}$ can be actuated with two different switching states. In the presented new topology a asymmetric intermediate circuit is used. The advantage is, that with the same topology like the three - point - converter, a four - point - converter can be designed (figure 2).

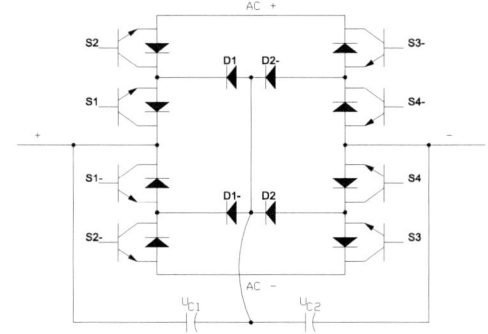

Fig. 1.   three - point - converter

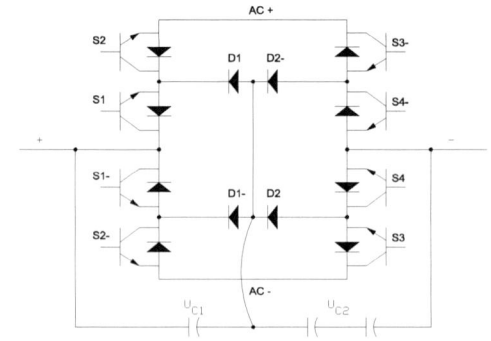

Fig. 2.   four - point - converter

With this modification an improvement in the harmonic behaviour can be achieved without additional switching elements or

1-4244-0448-7/06/$25.00 ©2006 IEEE   870

an increase in the switching frequency. A comparison between the today reachable number of voltage steps with symmetric intermediat circuit and the number of steps with asymmetric intermediate circuit is shown in table I.

TABLE I

COMPARISON OF NUMBER OF VOLTAGE STEPS WITH SYMMETRIC AND ASYMMETRIC INTERMEDIATE CIRCUIT

stage of expansion	today	new topology
0 (two - point)	2	2
1 (three - point)	3	4
2	4	7
3	5	10
⋮	⋮	⋮
7	9	23

For the reverse blocking voltage, there are several cases were switched off valves create an undefined voltage division. Through the tolerances in production, termerature or time depencence the voltage division can't be calculated exactly. Such a case is shown in figure 3, where the valves S_{1n} and S_{2n} from figure 2 are switched on and are therefor represented as a short circuit. The reverse blocking voltages of D_2, S_3 and S_4 are thereby undefined (defined through the parasitic properties of the elements). By using high ohmic resistors like in figure 4 the voltage of all valves can be dimensioned to $\frac{U_d}{3}$ or $\frac{2U_d}{3}$.

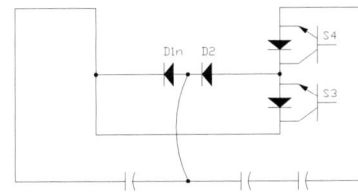

Fig. 3. undefined voltage division

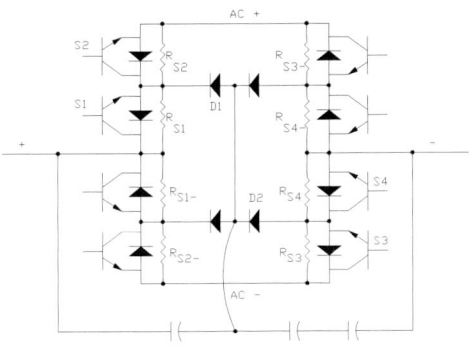

Fig. 4. four - point - converter with ohmic voltage stabilisation

The resistances have to be calculated in a relation of 1 : 2 (R_{S_4} = 2*R_{S_3}, for four - point - converter) to achieve the intended voltage distribution of $\left(\frac{1}{3}U_d\right)$ and $\left(\frac{2}{3}U_d\right)$. The blocking voltage of the diode D_2 will be forced to sero. The other resistances

can be calculated in the same way. The resistances can be high ohmic (e.g. >100 kΩ) to reduce or nearly neglect the power losses in the resistances.

The reverse blocking voltage of the valves of a four - point - converter devides in an ration of $\frac{U_d}{3}$ and $\frac{2U_d}{3}$. The ratio of an three - point - converter is always $\frac{U_d}{2}$. If the voltage reserve is large enough the three - point - converter can be easily changed in a four - point - converter by adding a capacitor (or by changing the connection of the existing). If the voltage reserve is not large enough an addional valve has to be installed. If the converter is constructed through serial connections of smaller valves, the change of the distribution can solve the problem of the reverse blocking voltage as well. The sum of the reverse blocking voltages is still the same. The reverse blocking voltages of the different elements are shown in table II.

TABLE II

VOLTAGE OF ELEMENTS, FOUR - POINT - CONVERTER

element	voltage
S_1	$\frac{1}{3}U_d$
S_2	$\frac{2}{3}U_d$
S_3	$\frac{1}{3}U_d$
S_4	$\frac{2}{3}U_d$
D_1	$\frac{1}{3}U_d$
D_2	$\frac{2}{3}U_d$
sum	$3\,U_d$

III. SIMULATION AND MEASUREMENT

The first expansion stage of the circuit is shown in figure 2. The DC - voltage (intermediate circuit) is connected to the + and - clamps. The clamps AC+ and AC- represent the connection point with the load or the net. In table III are the switching configurations for the different potentials presented. The simulation of the circuit results in figure 5.

TABLE III

SWITCHING CONFIGURATION, FOUR - POINT - CONVERTER

U_{AC}	$[S1, S2, S3, S4]$
0	$[1,1,0,0]$
$\frac{1}{3}U_d$	$[1,1,1,0]$
$\frac{2}{3}U_d$	$[0,1,1,1]$
U_d	$[1,1,1,1]$

The experimental solution is shown in figure 6. All expected voltage steps are shown. The THD (5th to 49th harmonic, without harmonics of 3rd order or order 3n) of the measured chart is about 0,13 % which is mainly reasoned by the switching behaviour of the used IGBT - modules (t_r and t_f ≈ 1 μs). An optimised pulse sequence [10] with 18 switching points per quarter periode was used. The theoretical THD is smaller than 0.01 %.

The simulation and measurement of the blocking voltages reaches values like the one shown in table II by using the mentioned parallel resistors. The measured blocking voltages of the elements S_1 and S_2 are shown in figures 8 to 11 as

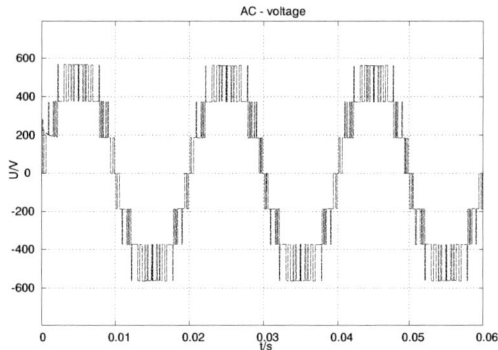

Fig. 5. AC - voltage, simulation, four - point - converter

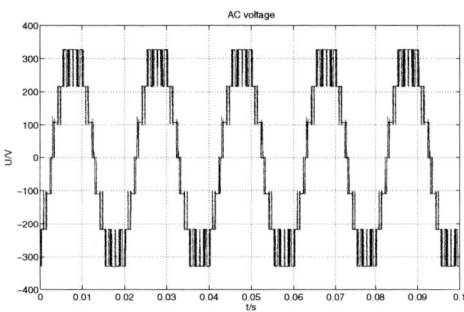

Fig. 6. AC - voltage, measured, four - point - converter

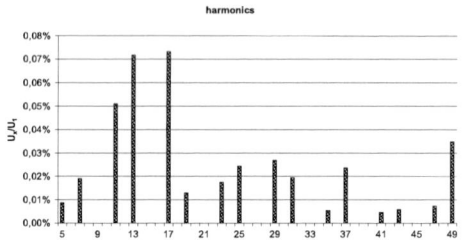

Fig. 7. amplitude of harmonics (measured)

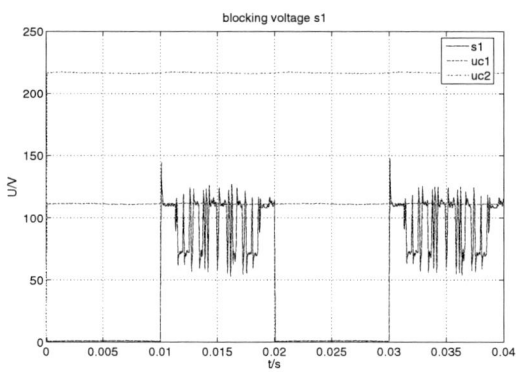

Fig. 8. blocking voltage S_1 with parallel resistors

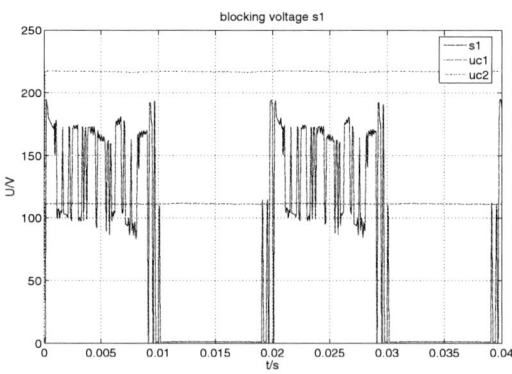

Fig. 9. blocking voltage S_1 without parallel resistors

an example for the change of the blocking voltages by using the parallel resistors. S_2 reaches in both cases a blocking voltage of 222 V ($\frac{2}{3}$ of intermediate circuit voltage). The blocking voltage of S_1 increases without the resistor through the undefined voltage division (like shown in figure 3). The blocking voltage can be reduced to 111 V ($\frac{1}{3}$ of intermediate circuit voltage)) by defining the voltage division. The blocking voltage can therefor be dimensioned through the high ohmic parallel resistors (57 kΩ / 2 * 57 kΩ used) to reduce (neglect) the internal losses.

IV. EXPANSION

An additional branch has to be installed in the circuit to reach the next expansion stage. The number of expansion steps is therby unlimited. A configuration example for an seven - point - converter is shwon in figure 12. The additional branches are marked. Examples for the configuration of the capacitors in the intermediate circuite are shown in table IV. The capacitors should be distributed for example in the fifth expansion stage like $C_1 = 2C_x$, $C_2 = 6C_x$, $C_3 = C_x$, $C_4 = 3C_x$ und $C_5 = C_x$, where C_x is a reference capacity.

The simulation result of the seven - point - converter (figure 12) is shown in figure 13, using the switching configuration like in table V.

The resistors for the voltage distribution (like shown with the four - point - converter) have to be calculated to

TABLE IV
NUMBER OF POTENTIALS (SERO - POTENTIAL IS MENTIONED AS +1)

expansion stage	number of potentials	distribution of capacitors
7	22+1	2,2,2,7,3,5,1
6	17+1	1,4,4,4,2,1
5	13+1	2,6,1,3,1
4	9+1	2,3,3,1
3	6+1	2,3,1
2	3+1	2,1

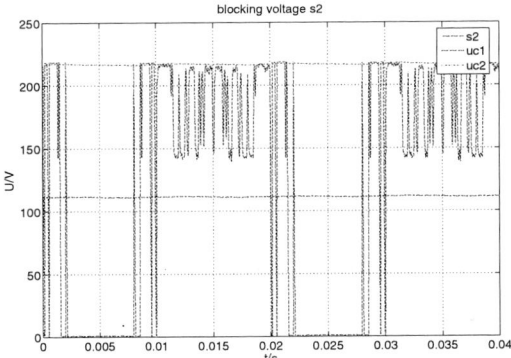

Fig. 10.   blocking voltage S_2 with parallel resistors

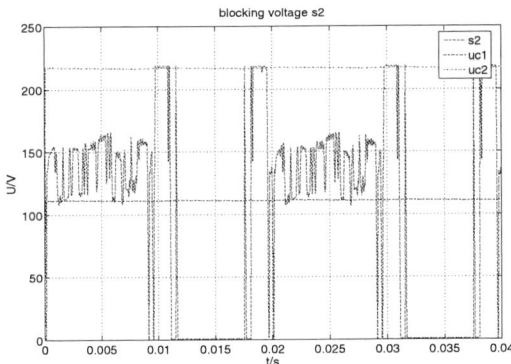

Fig. 11.   blocking voltage S_2 without parallel resistors

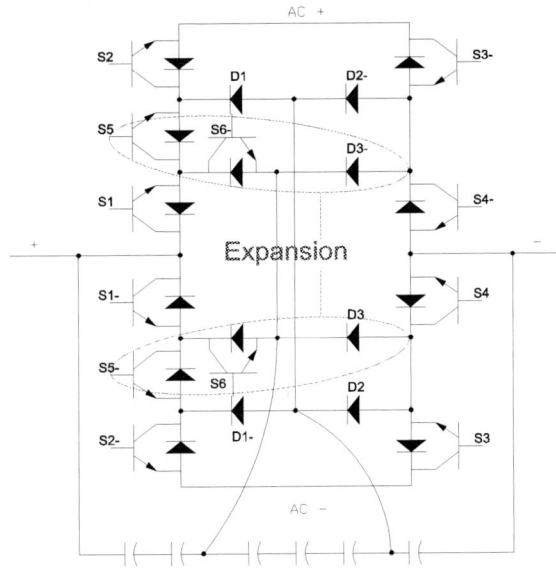

Fig. 12.   seven - point - converter

TABLE V
SWITCHING CONFIGURATION, SEVEN - POINT - CONVERTER

U_{AC}	$[S1, S2, S3, S4, S5, S6]$
0	$[1, 1, 0, 0, 1, 0]$
$\frac{1}{6}U_d$	$[0, 1, 1, 1, 0, 0]$
$\frac{2}{6}U_d$	$[1, 1, 0, 0, 1, 1]$
$\frac{3}{6}U_d$	$[0, 1, 1, 0, 1, 0]$
$\frac{4}{6}U_d$	$[0, 1, 1, 1, 1, 0]$
$\frac{5}{6}U_d$	$[1, 1, 1, 0, 1, 0]$
U_d	$[1, 1, 1, 1, 1, 0]$

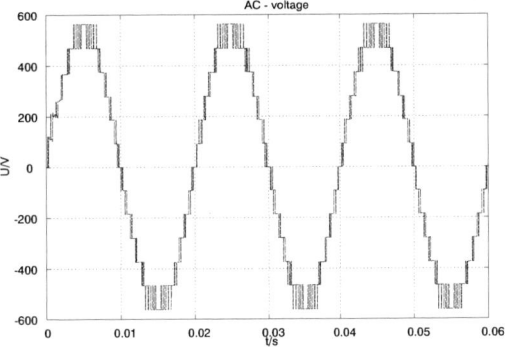

Fig. 13.   AC - voltage, simulation, seven - point - converter

$2R_{S1} = 3R_{S5} = R_{S2}$ ($U_{S1} = \frac{2}{6}U_d$, $U_{S5} = \frac{3}{6}U_d$ and $U_{S2} = \frac{1}{6}U_d$). The voltage of the elements S_3 and S_4 can be calculated to $\frac{1}{6}U_d$ and $\frac{5}{6}U_d$, which results in a resistor ratio of 1:5.

The three - phase to three - phase converter has passed the stage of simulation. The commonly known prinziples of voltage- and current - control of inverter and inverse inverter are thereby implemented.

V.   SWITCHING FREQUENCY

One of the problems of multi level - converter is the increased switching frequency of some valves. The switching tables are shown in table III and V. To compare the different configurations a defined pulse sequency or scaling has to be used. The comparism will be done by reaching all possible voltage steps one by one. The necessary switching pulses will be counted and scaled to the number of voltage steps.

TABLE VI
SWITCHING FREQUENCY

voltage division	$[S_1, S_2, S_3, S_4, S_5, S_6]$
$\frac{1}{2}$	$[1, 1, 1, 1]$
$\frac{1}{3}$	$[2, 1, 1, 1]$
$\frac{1}{6}$	$[3, 1, 2, 3, 2, 1]$

According to table III S_1 has to be switched on two times. The other elements are only switched one time. The notation is $[2, 1, 1, 1]$. Only the scaling is missed which is corrected in table VII. Only S_1 at the four - level - converter reaches a scaled switching frequency which is higher than the one of the three - level - converter. With higher expansion stage only some valves will work at the limit of the switching frequency. Therefor the switching frequency of the whole system can be increased by dimensioning these valves separately (thermal and electrical).

TABLE VII

SWITCHING FREQUENCY

voltage division	$[S_1, S_2, S_3, S_4, S_5, S_6]$					
$\frac{1}{2}$		$\frac{1}{2}$,	$\frac{1}{2}$,	$\frac{1}{2}$,	$\frac{1}{2}$	
$\frac{1}{3}$		$\frac{2}{3}$,	$\frac{1}{3}$,	$\frac{1}{3}$,	$\frac{1}{3}$	
$\frac{1}{6}$	$\frac{1}{2}$,	$\frac{1}{6}$,	$\frac{1}{3}$,	$\frac{1}{2}$,	$\frac{1}{3}$,	$\frac{1}{6}$

The practical pulse sequency is only in some special cases identical with such a standard sequency. In normal working conditions the higher voltage level will be prefered. By switching between level $\frac{5}{6}$ and $\frac{6}{6}$ only S_4 will be switched. A voltage step from $\frac{4}{6}$ and $\frac{5}{6}$ will influence S_1 and S_4. These „standard" switching elements has to be dimensioned as critical elements.

VI. CONCLUSION

A multi - level - converter can be established with the new topology without using an input transformer. The amount of necessary material (switching elements and transformer) can thereby be reduced compared with actual topologies. The increase of levels can be reached by using an asymmetric intermediate circiut. The reverse blocking voltage of the valves can be dimensioned through resistors. The output characteristic can be calculated like the one of any other multi - level - converter.

Frank Grundmann received the M.S. degree in electrical engineering from University of Magdeburg in 2002. He is currently preparing for the Ph.D. degree at the University of Ulm.

Jian Xie received his B.Sc. degree at Jiao - Tong - university Shanghai. He received his M.S. and Ph.D. from University of Darmstadt. He was system engineer for the frequency converter station Jübeck and project engineer at adtranz railway systems. He became a Professor at university of ulm in 1998.

REFERENCES

[1] J. R. et al., "Multilevel interters: A survey of topologies, controls and applications," *IEEE Transactions on Insutrial Electronics*, vol. 49, no. 4, August 2002.

[2] A. L. R. Marquardt, "A new modular voltage source inverter topology," *EPE 2003*.

[3] F. Peng, "A generalized multilevel inverter topology with self voltage balancing," *IEEE Transactions on industry applications*, vol. 37, no. 2, März 2001.

[4] M. V. A. Rufer, "Control of a hybrid asymmetric multilevel inverter for competitive medium - voltage industrial drives," *IEEE Transactions on Industry Applications*, vol. 41, no. 2, März / April 2005.

[5] M. F. E. et al., "Flying capacitor multileven inverters and dtc motor drive applications," *IEEE Transactions on Industrial Electronics*, vol. 49, no. 4, August 2002.

[6] R. T. et al., "Multilevel inverter by cascading industial vsi," *IEEE Transactions on Insutrial Electronics*, vol. 49, no. 4, August 2002.

[7] T. et al., "Charge balance control scheme for cascade multilevel converter in hybrid electric vehicles," *IEEE Transactions on industrial electronics*, vol. 49, no. 5, Oktober 2002.

[8] K. C. Y. Familiant, "A new cascaded multilevel h-bride drive," *IEEE Transactions on Power Electronics*, vol. 17, no. 1, Januar 2002.

[9] B. P. M. et al., "Multicarrier pwm strategies for multilevel inverters," *IEEE Transactions on Insutrial Electronics*, vol. 49, no. 4, August 2002.

[10] S. S. et al., "Optimum harmonic reduction with a wide range of modulation indexes for multilevel converter," *IEEE Transactions on Industrial Electronics*, vol. 49, no. 4, August 2002.

[11] D. Hasenkopf, "Regelverfahren für einen umrichter zur symmetrierung einphasiger lasten in drehstromnetzen," Ph.D. dissertation, Universität Ulm, 2005.

2006 5th International Power Electronics and Motion Control Conference

PWM Based Sensing and Control of Magnetic Bearings

Shuliang Lei[*] and Ralph Jansen[**]

[*] NASA Glenn Research Center, Cleveland, OH 44135, U.S.A.

slei@neo.tamu.edu

[**] University of Toledo, NASA Glenn Research Center, Cleveland, OH 44135, U.S.A.

Abstract—This paper develops a PWM Based Position Sensing method for motion control of active magnetic bearings. A practical method to determine the rotor position in terms of rotor-stator magnetic properties was investigated. An analog signal processing circuit which conditions the PWM voltage and current signals is used to implement the method. Analog biquad band pass filters, as well as low pass filters, demodulators, isolators and dividers were built. Preliminary tests were conducted using a four pole homopolar magnetic bearing by manually varying the rotor position and measuring the effectiveness of the signal processing module. The closed loop magnetic bearing motion control based on PWM sensing module was simulated using Matlab/Simulink. The simulation model includes 4-axis radial magnetic bearings with the dynamics of a high speed flywheel. A PD controller along with notch filters and lead compensators were employed in the closed loop control system. The simulation results demonstrated converged orbit trajectories of the rotors, which proved the effectiveness of the sensing scheme and the control strategy.

Keywords-coil inductance; magnetic bearings; pulse-width modulation (PWM); position sensing; signal processing

I. INTRODUCTION

Pulse-width modulation (PWM) has found wide applications in the area of motion control. PWM controlled H-bridges have demonstrated very low power loss due to their switching property which leads to negligible power dissipation. Most of the power stages in magnetic bearing motion control utilize PWM controlled H-bridges as power amplifiers in closed loop systems. A PWM Based Position Sensing algorithm is simulated and implemented with analog signal processing circuits. Position sensing technology utilizing the PWM signal itself in magnetic bearing applications eliminates the use of physical sensors, which greatly reduces the structural complexity of the mechanical system and lowers the system cost.

In the magnetic bearing control, the PWM controlled H-bridge power amplifier provides current to the coils controlling the motion of magnetic bearings. A PWM signal generator produces the pulse width-modulated square wave as the gate signal to the H-Bridge. The creation of the gate signal is based on the comparison of

the command and a reference triangle wave at a carrier frequency. The carrier frequency is usually very high as compared to that of the control signal. We will process the high frequency current and voltage signals measured at the coils to determine the rotor positions relative to the magnetic bearing stators.

For the motion control of magnetic bearings, a large variety of methods were discussed in literature concerning the sensing of the rotor position without physical sensors. In [1], a self-sensing scheme was formulated and implemented on homopolar magnetic bearings with a field programmable gate array (FPGA). However, the specific configuration of the 'homopolar' magnetic bearing method was not applicable to other categories of magnetic bearings; for example, to heteropolar magnetic bearings with opposing C-cores. The formulas in [1] can not be directly applied to heteropolar magnetic bearings. Some other authors used modern control theory to estimate the rotor position with a Luenberger observer [2] or parameter estimation [3]. Whereas these approaches are more theoretical, the implementations for practical industrial applications are costly and complicated. Motivated by the modulation approach proposed in [4], the PWM Based Position Sensing method developed in this paper utilizes the PWM coil current and voltage signals to implement the self-sensing strategy.

In this paper the closed loop magnetic bearing motion control in terms of PWM Based Position Sensing was simulated using Matlab/Simulink. A model of the system, including the levitated shaft, the power amplifiers, the PWM drive circuit and the feedback control system was created. PID controllers in corporation with notch filters and lead compensators were employed in the feedback control loop. A model of the PWM Based Position Sensing system, driven by the PWM modulation algorithm, was first built and simulated. Later this position sensing model was put into the magnetic bearing control loop to replace the physical sensor models. Stable levitation was achieved with the self-sensing models. The orbits of the rotor positions at upper and lower magnetic bearings converged to circles, similar to the results obtained from the physical sensing models, which verified the effectiveness of the PWM Based Position Sensing method.

In engineering applications, it is very important that the proposed method is feasible. Generally there are three hardware approaches for implementation of the position

---

This work is sponsored by the NASA Glenn Research Center, Cleveland, OH 44135, U. S. A.

1-4244-0448-7/06/$25.00 ©2006 IEEE   875

sensing algorithms: FPGAs, microprocessors and analog circuits. The first two fall into the digital signal processing category. With the rapid advances in recent IC technology, the cost-effectiveness of systems made up of analog microchips is improving, making analog signal processing more convenient and affordable. In addition, analog signal processing needs less supporting tools and therefore much less investment to start up the project, whereas DSPs and FPGAs need sophisticated and expensive software development packages.

The proposed PWM Based Position Sensing approach has the advantage of eliminating the use of an external modulation signal to obtain the position information of the rotor in the magnetic bearings. The hardware design objective is to create circuit realization suitable for industrial applications. Individual analog isolator, band pass filter, demodulator and divider signal processing circuits were built and integrated. The circuit was tested with a four pole homopolar magnetic bearing in which the shaft was manually positioned in various locations.

II. PWM BASED POSITION SENSING METHOD

A heteropolar magnetic bearing actuator can be schematically described as two opposing C-cores shown in Fig. 1.

The basic principle of magnetic bearing is similar to that in motor control. Referring to Fig. 1, and neglecting the iron reluctance, the magnetic flux ϕ^+ in the positive C-core is

$$\phi^+ = \frac{(N_B i_B + N_c i)}{2w^+} \mu_0 A_a = \frac{(N_B i_B + N_c i)}{2(g_0 - w)} \mu_0 A_a$$

(1)

The magnetic flux ϕ^- in the negative C-core is

$$\phi^- = \frac{(N_B i_B - N_c i)}{2w^-} \mu_0 A_a = \frac{(N_B i_B - N_c i)}{2(g_0 + w)} \mu_0 A_a \quad (2)$$

where g_0 is the air gap of the magnetic bearing and w is the displacement of the rotor in positive direction. Also

$\mu_0 = 4\pi \times 10^{-7}$, *permeability*

A_a , *area per pole*

i_B , *bias current*

N_B, *turns per C – core, bias coil*

N_c, *turns per C – core, control coil*

According to Faraday's law, the coil voltage in each C-core can be written as follows:

Positive C-core:

$$V_L^+ = N_c \frac{d\phi^+}{dt} = \frac{N_c^2 \mu_0 A_a}{2} \frac{d}{dt}(\frac{i}{w^+}), \quad (3)$$

negative C-core:

$$V_L^- = N_c \frac{d\phi^-}{dt} = \frac{N_c^2 \mu_0 A_a}{2} \frac{d}{dt}(\frac{-i}{w^-}) \quad (4)$$

where $w^+ = g_0 - w$, $w^- = g_0 + w$.

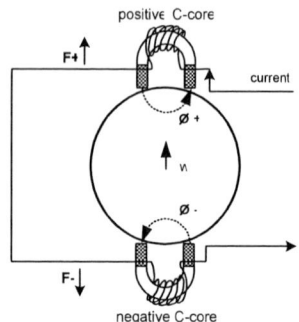

Figure 1. Schematic of two opposing C-cores

Let $\qquad K_\alpha = \frac{N_c^2 \mu_0 A_a}{2}$,

we have

$$V_L^+ = K_\alpha \frac{d}{dt}(\frac{i}{w^+}), \quad V_L^- = K_\alpha \frac{d}{dt}(\frac{-i}{w^-}) \quad (5)$$

From these results we can see that with the air gap change due to the radial displacement of the rotor, the inductance changes consequently. We will use this as the basis of the algorithm.

Fig. 2 depicts the block diagram of the PWM amplifier driving magnetic bearing coils. In this diagram, the nonlinear inductances vary with the air gap between the rotor and the stator. The inductance change results in changes of the coil voltage and current. The PWM Based Position Sensing algorithm determines rotor position based on this information. Ideally, the magnitude of the coil current is maximized when the magnetic bearing is centered because the inductance is the smallest at this position. When the rotor moves in the positive direction, the air gap in the positive C-core is reduced, the inductance and the voltage drop in the coil is reduced whereas the inductance and voltage drop in the opposing coil increases.

III. SIGNAL PROCESSING

Consider the PWM modulation signal with angular frequency ω_c and apply a band pass filter to extract the high frequency signal. The voltage signal V_L^+ after the band pass filter can be written as $V_{L(bpf)}^+ = V_{dem} \cos \omega_c t$ [4]. Referring to (5):

$$i = \frac{w^+}{K_\alpha} \int V_L^+ dt$$

The filtered coil current can thus be obtained

876

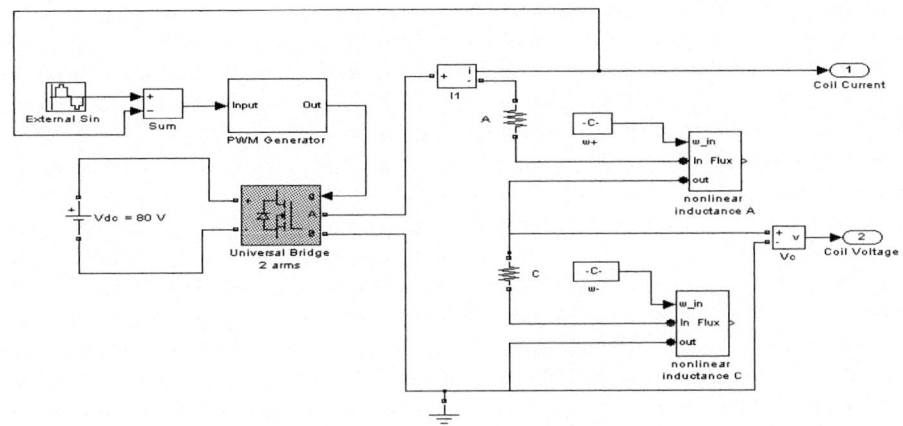

Figure 2. PWM power amplifier driving the magnetic bearing coils

$$i_{bpf} = \frac{w^+}{K_\alpha} \int V_{L(bpf)}^+ dt = \frac{w^+}{K_\alpha \omega_c} V_{dem} \sin \omega_c t.$$

Use demodulation to obtain $i_{dem} = \frac{w^+}{K_\alpha \omega_c} V_{dem}$.

Finally we have $w^+ = K_\alpha \omega_c \dfrac{i_{dem}}{V_{dem}}$. From $w^+ = g_0 - w$,

we can obtain the rotor displacement w. This algorithm was first simulated according to the block diagram in Fig. 3. Fig. 4 shows the simulation result with ramp input signal. After the transient initial condition period the self-sensing output follows the ramp input. The output can be trimmed to exactly match the real displacement. The implementation of the analog signal processing also followed the block diagram in Fig. 3. The filters employ biquad topology with operational amplifiers [6], as shown in Fig. 5 for the band pass filter. The isolator is needed because the signal processing circuits are single-ended reference whereas the coil voltage is floating.

IV. EXPERIMENTAL RESULTS

The coil PWM voltage, rectifier output, and coil current measured on an oscilloscope are depicted in Fig 6. The coil PWM voltage after the isolator is not rectangular and has overshoots because of an RLC power filter utilized to suppress electro-magnetic noise. The rectifier signal is fed to a low pass filter for demodulation. The demodulated output is a DC voltage whose level depends on the air gap magnitude.

Preliminary tests using a four pole homopolar magnetic bearing with the air gap position changed by manually locating the shaft yields the following results. When the rotor is located in the maximum positive Y position, the measured inductance is 1.3 mH and the output DC voltage from the processing module is -0.669 V. When the shaft is located in the maximum negative Y direction, the inductance is 1.4 mH and the output DC voltage from the processing module is -0.530 V. We have 25%

voltage change from the air gap change. These test result prove that it is possible to effectively measure the rotor position using the PWM Based Self Sensing Method.

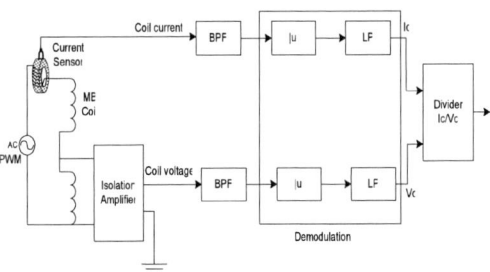

Figure 3 Block diagram for simulation and implementation

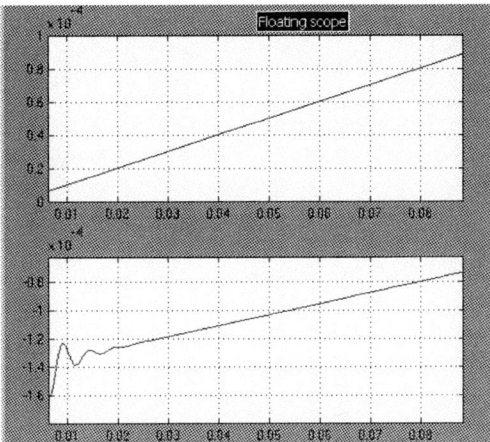

Figure 4. Simulation result: The upper plot is the real displacement of the rotor; the lower plot is the self-sensing output

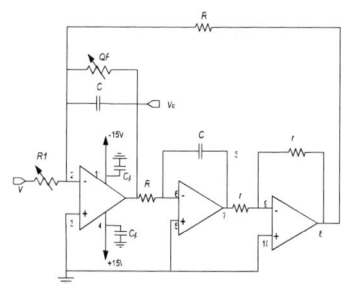

Figure 5. Biquad topology in signal processing

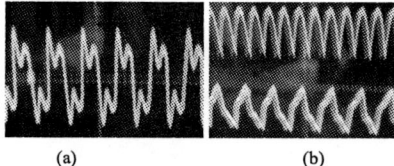

(a)   (b)

Figure 6. Waveforms of the processing circuit outputs: (a) The PWM voltage output. (b) The upper curve is the rectifier output; the lower curve is the coil current signal

V.   CLOSED LOOP CONTROL SYSTEM SIMULATION

The control of magnetic suspension in high speed rotating machinery requires closed loop feedback to support the rotating shaft and to suppress vibrations. In most industrial applications, the feedback system includes a linear PID controller with compensators, power amplifiers, magnetic actuators and sensors [5]. The PWM Based Position Sensing technique eliminates the need for physical sensors, such as eddy current or optical sensors. In the design of magnetic bearing suspension systems, there are two kinds of configurations: collocated and non-collocated sensors-actuators. In most cases the magnetic bearing system utilizes the non-collocated sensors due to the physical constraints in the mechanical design. However, since the sensors and the actuators are not at the same spot of the shaft, instability occurs because of the phase shift between the sensed position and the actual magnetic bearing position, which makes the controller design more difficult. When the shaft spins at very high speed and flexible modes are excited, the shaft at the sensor locations may have reversed phase angles with

those of the actuators. The self-sensing techniques remove the risk of instability caused by non-collocated sensors and actuators.

The mechanical system model of the test rig used for closed loop simulation consists of a spinning shaft with flywheel. Radial magnetic bearings installed on the upper and lower part support the shaft during high speed operations. Instead of using physical sensors, the signal processing module shown in Fig. 3 was employed in each channel. The control loop utilized PID control with notch filters and lead compensators to appropriately stabilize the magnetically suspended flywheel.

Fig. 7 is the schematic diagram of the closed loop system for magnetic bearing control. The error signals are difference of the position target and the feedback position signals which are generated by the self-sensing processing circuit. A coordinate transformation converts the motion in each axis to the mass center displacements and rotations about the mass center. The purpose of doing this is to consider the cross couplings between axes due to gyroscopic effect at high rotational speed. The controller includes a proportional-derivative (PD) feedback, lead compensators and notch filters. In the PD stage, cross couplings are implemented between the orthogonal radial axes. The output transformation converts the mass center based coordinates back to each magnetic bearing axis coordinates. The power amplifier on each magnetic bearing axis receives the controller signal and generates the current in the magnetic bearing coil. Based on the block diagram, two Simulink models were built for comparison: one using physical sensors and the other using the PWM Based Position Sensing model.

The parameters used in the simulations are: Fc=20000 Hz (PWM carrier frequency); K_a=6.22644e-7 (magnetic bearing parameter); g0=0.02/39.37 m (air gap); Vdc=80 V (DC voltage source of the H-bridge).

The orbits of the upper and lower rotor of the magnetic bearings are shown in Fig. 8. The displacements of the upper and the lower part of the rotor versus time are shown in Fig. 9. The resulting stable orbits exhibit the effectiveness of the PWM Based Position Sensing method. Fig. 9 compares the real displacements of the

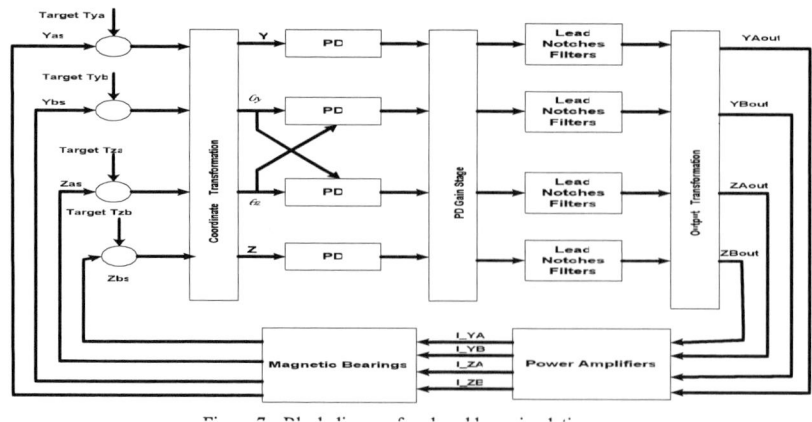

rotors with the self-sensing results of the radial magnetic bearings. It is important to note that, although the PWM Based Position Sensing signal does not quite coincide with the actual displacement signal, the phase angles between these two are the same. If these signals were out of phase the system would be unstable.

VI. CONCLUSION

In this paper the PWM Based Position Sensing approach of self-sensing and magnetic bearing motion control is investigated. First, a signal processing method calculating the rotor position in terms of the rotor-stator magnetic relations was formulated. Simulation and circuit realizations were performed. In the simulations, the PWM generator and H-bridge with inductance loads were modeled as switches, coils and resistors. The processing circuit including band pass filters, low pass filters, demodulators, isolators and dividers were built and tested. The change of air gap between the rotor and the stator was successfully measured with a four pole homopolar magnetic bearing by manually varying shaft positions. Closed loop simulation of magnetic bearing control was carried out. The self-sensing module was applied to closed loop motion control of magnetic bearings. The simulation model includes the radial 4-axis magnetic bearings and the dynamics of the rotational

flywheel. A PD controller with notch filters and lead compensators were employed for feedback control. The simulation results exhibited stable rotor orbit trajectory , which proved the effectiveness of the PWM Based Position Sensing scheme.

REFERENCES

[1] P. Tsao, *et al.* "A self-sensing homopolar magnetic bearing: analysis and experimental results," *IEEE –IAS 34ᵗ Annual Meeting,* vol. 4, No. 1, pp. 2560-2565, Oct. 1999

[2] S. Overbo, *et al.* "New self sensing scheme based on INFORM, heterodyning and Luenberger observer," *IEEE Int. Electric Machines and Drives Conf* ,vol. 3, pp. 1819-1825, June 2003

[3] M. Noh, *et al.* "Self-sensing magnetic bearings using parameter estimation," *IEEE Trans Instrumentation and Measurement Systems Technology,* Vol. 46, No. 1, pp. 45-50, Feb. 1997

[4] A. Schammass, *et al.* "New results for self-sensing active magnetic bearings using modulation approach," *IEEE Trans. Control Syst. Technology,* Vol. 13, No. 4, pp. 509-516, July 2005

[5] C. Kim, A. B. Palazzolo, "Eddy current effects on the design of rotor-magnetic bearing systems" 1995 *Trans. of the ASME* 117, pp. 162-170.

[6] A. Sedra and K. Smith, *Microelectronic Circuits*, 4ᵗʰ Edition, Oxford University Press, Inc. 1998

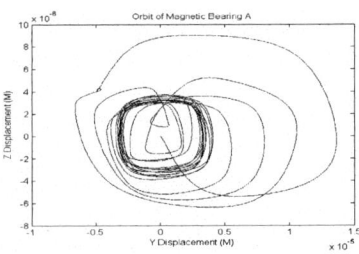 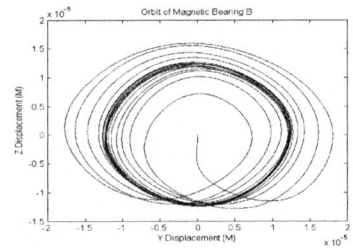

Figure 8 Orbits of the magnetic bearing: upper bearing (left) and lower bearing (right)

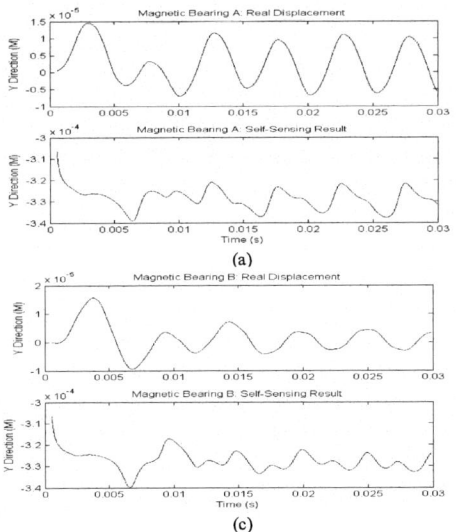 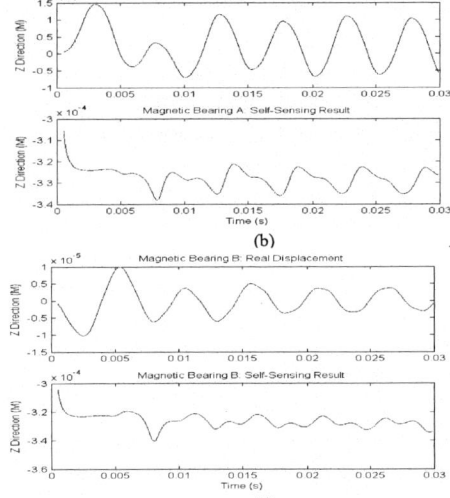

Figure 9. Comparison of the real displacements of the rotors with the self-sensing results (a) Upper MB in Y direction (b) Upper MB in Z direction (c) Lower MB in Y direction (d) Lower MB in Y direction. Top figures are real displacements; bottom figures are self sensing results

Position Sensorless Direct Torque Control of Synchronous Reluctance with Permanent Magnet Motor

Jiang Dong   Zhao Zhengming   Duan Yao   Guo Wei

State Key Laboratory of Power Systems, Department of Electrical Engineering, Tsinghua University, Beijing, P.R.China

jd@mails.tsinghua.edu.cn

Abstract—This paper introduces the structure of synchronous reluctance with permanent magnet (SR-PM) motor and its principle of direct torque control (DTC), especially focuses on characters of DTC in SR-PM motor. Then the (position/speed) sensorless strategy of SR-PM motor is studied. The key point of sensorless control is estimating the torque angle δ precisely.   This paper presents two different compensating methods and simulates the results, and proves that they can achieve sensorless DTC when feedback is sent to the position and speed loop, both in speed control and position control.

Keywords: SR-PM motor; Direct torque control; Torque angle; Sensorless

I.   Introduction

With the development of permanent magnetic materials, Synchronous Reluctance with Permanent Magnet (SR-PM) motor is becoming widely used in industrial world. Its structure contains the features of both conventional permanent magnetic synchronous motor (PMSM) and synchronous reluctance motor (SRM). It includes PMSM's high power-quality and SRM's high power-density.

In recent years, several scholars study the high-performance controlling strategies of permanent magnetic motors especially PMSM. The strategies include direct torque control (DTC) and position/speed sensorless algorithms [2],[3],[4],[5],[6]. These strategies perform well in conventional PMSM's control.

DTC strategy of conventional PMSM including internal PMSM (IPMSM) is developed from DTC strategy of induction motor, and its main principle is to use bang-bang controller to control stator flux of the motor and determine the voltage vector by flux position and torque tendency.

There are several algorithms for position/speed sensorless control on AC motors. One of the most widely used sensorless algorithms is directly calculating flux by back-EMF integral. Also, in PMSM's position estimation, torque angle's influence is very important, which is showed in figure-2. Stator flux could be easily calculated in $\alpha - \beta$ coordinate. But the rotor position should compensate the torque angle δ on the basis of stator position, that is $\theta = \phi - \delta$. Traditional δ estimating method is based on function (7), which is in section II. In section III two δ estimating algorithms are introduced respectively. Traditional speed estimating algorithm is based on flux rotating speed of function (1)[2]. In section IV the fact that flux-rotating speed cannot replace rotor speed in transient process will be proved and a new algorithm of speed estimator based on precise position estimator will be introduced and compared.

$$\omega = \frac{d(arctg\frac{\psi_\beta}{\psi_\alpha})}{dt} = \frac{\psi_\alpha \frac{d\psi_\beta}{dt} - \psi_\beta \frac{d\psi_\alpha}{dt}}{\psi_\alpha{}^2 + \psi_\beta{}^2} = \frac{\psi_\alpha e_\beta - \psi_\beta e_\alpha}{|\psi|^2} \quad (1)$$

II.   Motor structure and its mathematical model

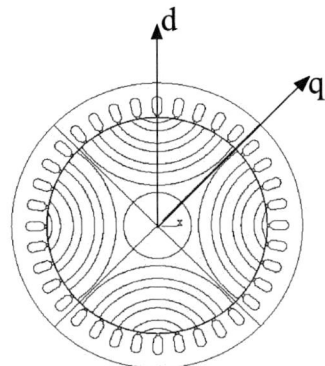

Figure-1 SR-PM motor's structure

The structure of SR-PM motor is showed by figure-1. The rotor of SR-PM motor is constituted by multi-layers of permanent magnetic materials. This structure makes the motor contain three characteristics [1].

(1) To allow q-axis flux to flow across the whole pole surface in order to obtain maximum q-axis inductance;
(2) To enhance d-axis air gap length to minimize d-axis inductance;
(3) To insert permanent magnets into rotor laminations to produce assistance torque.

The mathematical model of SR-PM motor almost approximates that of normal PMSM. But just because of the special structure, it contains a higher salient effect—a higher ratio of Lq and Ld, which will enhance its reluctance torque. The mathematical model is described in the following functions.

$$\psi_d = \psi_f + L_d i_d \qquad (2)$$

$$\psi_q = L_q i_q \qquad (3)$$

$$u_d = r_s i_d + \frac{d\psi_d}{dt} - p\omega\psi_q \qquad (4)$$

$$u_q = r_s i_q + \frac{d\psi_q}{dt} + p\omega\psi_d \qquad (5)$$

$$T_e = \frac{3}{2}p(\psi_f i_q + (L_d - L_q)i_d i_q)$$
$$= \frac{3}{2}p(\psi_f i_s \sin\delta + \frac{(L_d - L_q)i_s^2 \sin 2\delta}{2}) \qquad (6)$$

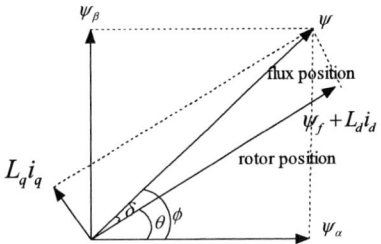

Figure-2 SR-PM motor's space flux

From function (2) to (6), the parameters are defined as follows:

ψ_f —permanent flux linkage, p —pole number,

L_d, L_q —inductance of d-axis and q-axis,

ψ_d, ψ_q , u_d, u_q , i_d, i_q —flux linkage, voltage and current of d-axis and q-axis,

r_s, i_s —stator resistance and current and δ —torque angle, the angle of flux in d-q coordinate. (Showed in figure-2)

III.   A Novel DTC Algorithm of SR-PM

In DTC strategy, flux linkage is restricted in a bang-bang controlled loop. By limiting the width of the loop can make the flux linkage approaching to the circle of settled radius. The controller samplings the torque error between real and reference value in order

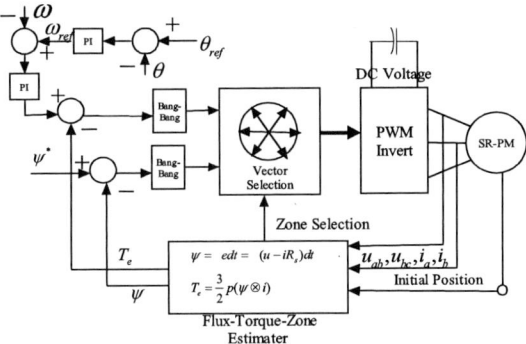

Figure-3 Position Controlled DTC Scheme of SR-PM Motor

to determine whether to increase or decrease motor's torque, then actualize the torque's control by accelerating or decelerating flux rotating speed.   And with a position controller and a speed controller, a scheme of position feedback controlled DTC is showed in figure-3.

For a long time people neglected the reluctance torque's existence, accelerating flux speed cannot always increase motor's torque. The essence of accelerating flux speed is increasing the torque angle δ, as showed in figure-2. Function (7) is calculated to determine whether increasing torque angle could increase motor's torque.

$$\frac{dT_e}{d\delta} = \frac{3p|\psi|}{4L_dL_q}[2\psi_fL_q\cos\delta - 2|\psi|(L_q - L_d)\cos 2\delta] \quad (7)$$

Just because of the reluctance torque in T_e, $\dfrac{dT_e}{d\delta} > 0$ is restricted by (7), and the result of the equation is (8):

$$\frac{a/|\psi| - \sqrt{(a/|\psi|)^2 + 8}}{4} < \cos\delta < \frac{a/|\psi| + \sqrt{(a/|\psi|)^2 + 8}}{4} \quad (8)$$

$$\left(a = \frac{\psi_f L_q}{L_q - L_d} \right)$$

In torque bang-bang controller, a three level table is used instead of conventional two-level bang-bang controller, to eliminate the torque ripple by adding zero voltage vectors.

IV.   A Novel Position and Speed Estimating Algorithm of SR-PM

As discussed in part I, estimated rotor position of SR-PM motor should include both flux position and torque angle. As showed in figure-2:

$$\theta = \phi - \delta \quad (9)$$

For flux position angle ϕ, a simple integrator can easily get its estimated value. In order to eliminate the effect of integral error people introduced several filters to get an ideal flux estimated value. For torque angle, as discussed in part I, the former researchers acquire δ from Eq. (6), which needs torque's value. But precisely estimated torque value is difficult to acquire. Also because function (6) is a non-linear function, its solving is complex.

New algorithm to estimate δ is directly based on figure-2. As showed in (10):

$$\delta(k) = arctg\frac{L_q i_q(k-1)}{L_d i_d(k-1) + \psi_f} \quad (10)$$

When the position of the nearest period k-1 is

estimated, the torque angle of period k can be gained by Eq.(10).

Another algorithm to calculate δ is also based on figure-2, but it does not need to be iterative. This algorithm is directly calculated from following algebraic transformation based on figure-2:

$$\sin\delta = \frac{L_q i_q}{\psi}$$

$$\Rightarrow \sin^2\delta = \frac{L_q^2(i_s^2 - i_d^2)}{\psi^2}$$

$$\Rightarrow \psi^2\sin^2\delta = L_q^2(i_s^2 - (\frac{\psi\cos\delta - \psi_f}{L_d})^2)$$

$$\Rightarrow \psi^2(1 - \cos^2\delta) = L_q^2 i_s^2 - \frac{L_q^2}{L_d^2}(\psi^2\cos^2\delta - 2\psi\psi_f\cos\delta + \psi_f^2)$$

$$\Rightarrow (\frac{L_q^2}{L_d^2} - 1)\psi^2\cos^2\delta - \frac{2L_q^2\psi\psi_f}{L_d^2}\cos\delta + (\frac{L_q^2}{L_d^2}\psi_f^2 + \psi^2 - L_q^2 i_s^2) = 0$$

$$(11)$$

In the end of the transformation (11), a quadratic equation is derived. Its coefficients are known motor parameters such as L_d, L_q, ψ_f and the combined flux and current values ψ, i_s that can be easily calculated. Directly solve function (11) can get $\cos\delta$ and δ.

For a long time people are using simple flux rotating speed (by function (1)) to replace rotor-rotating speed in synchronous motor. But it is untenable in transient process like starting period and torque varying period. Because in this period, i_d and i_q are changing and δ is also varying. So that flux speed is not equal to rotor speed. In fact we use a speed-demodulation module to calculate rotor speed from rotor position, as showed in figure-4. It has a minus feedback process as Eq.(12) which can track real speed adaptively.

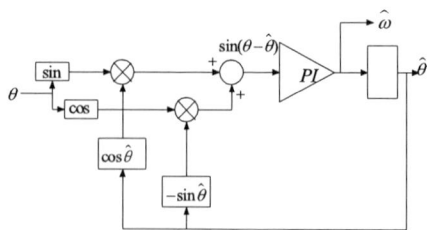

Figure-4 speed-demodulation module

882

$$\widehat{\theta}\uparrow\rightarrow(\theta-\widehat{\theta})\downarrow\rightarrow\sin((\theta-\widehat{\theta}))\downarrow\rightarrow\widehat{\omega}\downarrow\rightarrow\widehat{\theta}\downarrow \quad (12)$$

V.　　Simulating Results

A SR-PM motor for experiment has parameters in the end of part V. Following simulations have been done to check the effects of the control strategy discussed above.

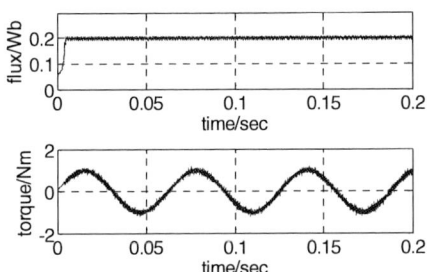

Figure-5 Flux and torque response of DTC under sinusoidal

reference torque

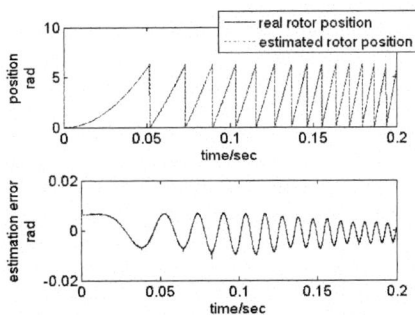

Figure-6 Real and estimated position's comparisons and

estimated errors (direct calculated method)

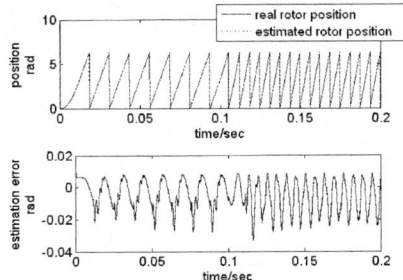

Figure-7 Real and estimated position's comparisons and

estimated errors (iterative method)

In order to testify the torque and flux controlling effect of DTC with the new algorithm, a toque-settled simulation is done, with sinusoidal torque reference.

From figure-5, SR-PM motor controlled by DTC can gain an ideal flux circle; even with sinusoidal reference torque, the real torque of motor can track the

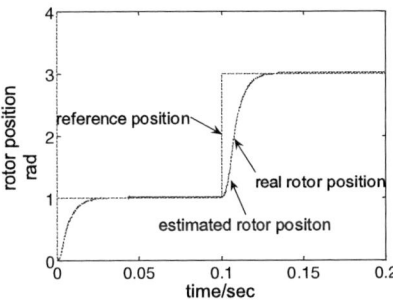

Figure-8 reference position, real position and estimated

position in sensorless DTC in position servo system—

rectangle reference

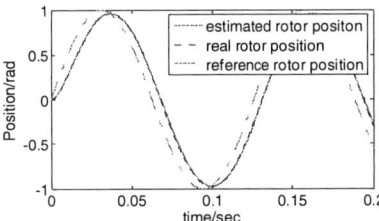

Figure-9 reference position, real position and estimated

position in sensorless DTC in position servo system—

sinusoidal reference

reference torque quickly. Adding zero voltage vectors can weaken the torque ripple effectively.

From figure-6 and figure-7, the simulation results show that both the two algorithms can track the real position effectively. Because the iterative method is based on approximate angle of the nearest sampling period substituting real angel, its position error (less than 0.04rad) is more obvious than direct calculated method (less than 0.02rad). Both of the errors are mainly caused by flux-estimated errors.

Then, after validating the new sensorless DTC's effect in speed feedback loop, a simulation of position servo system without position sensor is made and the results are showed by figure-8 and figure-9. In figure-8, although without position sensor, rotor can quickly arrive at the reference position. In figure-9, in order to check the ability of position servo system, a sinusoidal position signal is set, and real position

883

output signal's attenuation is smaller than 0.5dB.

Simulating parameters include rated speed 8000rpm, rated current 4.2A, rated voltage 270V, and rated torque 1Nm. And the other parameters are showed in Table.1.

Table.1. Motor parameters

Rs（ohm）	1.4
Lq（H）	0.0222758
Ld（H）	0.0027113
ψ_f（Wb）	0.05389
Rotor inertia（kg.m^2）	0.74e-4
Friction coefficient	0
Poles	4

VI.   Conclusion

In this paper, a new kind of permanent motor—SR-PM motor and its mathematical model is introduced. Although SR-PM motor's mathematical model is almost the same as conventional PMSM, its controlling effect has its own characteristics because of its special structure. For its DTC scheme, an accessory limited Eq.8 should be satisfied. And adding zero voltage vectors could effectively weaken the torque ripples. Understanding these we get a fast and precisely tracking torque response of SR-PM motor under DTC showed in figure-5. Then the key point of SR-PM motor's sensorless DTC is to estimate the torque angle δ precisely has been validated. This paper presents two new algorithms to estimate δ, and to estimate a more precise speed by δ than traditional speed estimating Eq.8. It has ideal results of both speed and position servo system by simulation.

References

[1] Zhao Zhenming, El-Antably Alumed, Optimization of Added Permanent Magnet Amount in Synchronous Reluctance Machines for High Performance, TSINGHUAHUA SCIENCE AND TECHNOLOGY ISSN 1007-0214 21/23 pp1137-1142 Volume 3, Number 3, September 1998

[2] Peter.Vas. Sensorless vector and direct torque control. Oxford Press. 1998

[3] Muhammed.Fazlur.Rahman, L.Zhong, Khiang Wee Lim. A direct torque-controlled interior permanent magnet synchronous motor drive incorporating field weakening. IEEE Transaction on Industry Applications, 1998, Vol 34, No.6: 1246~1253.

[4] Muhammed.Fazlur.Rahman, L Zhong, Md.Enamul.Haque, M.A.Rahman.   A   direct   torque-controlled   interior permanent-magnet synchronous motor drive without a speed sensor.   IEEE Transaction on Energy Conversion 2003, No.1: 17~22

[5] Lixin Tang, Limin Zhong, Muhammed Fazlur Rahman, Yuwen Hu   A   Novel   Direct   Torque   Control   for   Interior Permanent-Magnet Synchronous Machine Drive With Low Ripple in Torque and Flux---A Speed-Sensorless Approach IEEE Transaction on Industry Application Vol.39 No.6 2003:1748~1755

[6] Hu Yuwen, Tian cun, Gu yikang, You Zhiqing, L.X.Tang, M.F.Rahman   In-depth Research on Direct Torque Control of Permanent Magnet Synchronous Motor IEEE IECON 02: 1060 - 1065 vol.2

2006 5th International Power Electronics and Motion Control Conference

Counter-Rotating Permanent Magnet Brushless DC Motor for Underwater Propulsion

Jianqi Qiu, Cenwei Shi, Mengjia Jin, Ruiguang Lin

College of Electrical Engineering, Zhejiang University，Hangzhou 310027, China

Email: motor@zju.edu.cn

Abstract—This paper presents a novel counter-rotating permanent magnet BLDCM (CRBLDCM) for underwater propulsion. This kind of motor is different from the traditional BLDCM in that both the rotor and the armature are rotating in the opposite directions, which is particularly suitable for underwater propulsion system. In this paper, the mathematical model of CRBLDCM is established based on its operation principle, and the operation performances are analyzed, from which the counter-rotating condition for steady state operation is deduced. Furthermore, the control strategy for counter-rotating using two sets of position sensors is proposed. The simulation and the experimental results with prototype machines are also shown in this paper, which confirm that the control method proposed in this paper not only simplify the construction of the control system, but also ensure the reliable start and precise commutation of the motor.

Keywords-Counter-rotating, Permanent magnet, Brushless dc motor, Position Sensor

I. INTRODUCTION

Nowadays, electrical propulsion systems are drawing more and more interests in underwater marine application because of their considerable advantages such as high efficiency, low noise and the elimination of peg-top effect due to the use of conventional propulsion systems [1-2]. In this application, permanent magnet brushless motors are the most competitive choice because of their attractive characteristics in such key categories as power density, torque to inertia, and electrical efficiency. A kind of PM synchronous motor with axial-flux was presented being used in the ship propulsion system [3]. With one stator and two independent rotors to drive the main and the counter-rotating propellers, this kind of motor usually has a complicated construction and manufacture. A counter-rotating ring thruster for underwater vehicle was reported in [4], which shows that the ring thrusters are essentially two permanent magnet brushless motors with a large diameter hollow rotor.

This paper presents a novel counter-rotating permanent magnet BLDCM (CRBLDCM) for marine propulsion applications. Compared with the formerly presented counter-rotating motors, this kind of motor has relatively simpler structure like traditional BLDCM, except that

both the rotor and the armature with windings embedded in it are rotating in the counter directions. In this paper, the mathematical model of CRBLDCM is established based on its operation principle, and the operation performances are analyzed, from which the counter-rotating condition for steady state operation is deduced. Furthermore, the control technology for counter-rotating using two sets of position sensors is proposed. The simulation and the experimental results with a prototype machine are also shown in this paper, which confirm that the control method in this paper not only simplify the construction of the system, but also ensure the reliable start and precise commutation of the motor.

II. CONSTRUCTION AND MATHEMATICAL MODEL OF CRBLDCM

The schematic diagram of the CRBLDCM construction is shown in Fig.1. The outer PM rotor, marked with number 1, is rotating with the bearing set marked with 3 and 4, while the armature rotor marked with number 2 is rotating in the opposite direction with the bearing set marked by 5 and 6, which is accomplished by the hollow shaft with polyphase winding leads, marked by 8, going through it, and the terminals connected to three slip rings marked by 9.

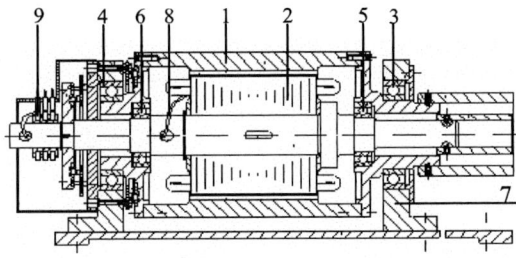

Fig.1 Schematic Diagram of the CRBLDCM Construction

The voltage equations of CRBLDCM can be expressed as

$$\begin{cases} u_a = Ri_a + pi_a(L-M) + e_a \\ u_b = Ri_b + pi_b(L-M) + e_b \\ u_c = Ri_c + pi_c(L-M) + e_c \end{cases} \quad (1)$$

1-4244-0448-7/06/$25.00 ©2006 IEEE   885

where u_a, u_b, u_c are the phase voltages, i_a, i_b, i_c are the phase currents, e_a, e_b, e_c are the phase back-EMF, L and M represents the phase self inductance and mutual inductance respectively.

The phase back-EMF of the motor can be described as follows

$$\begin{cases} e_a = k_b f_a(\theta_{r1} + \theta_{r2})(\omega_{r1} + \omega_{r2}) \\ e_b = k_b f_b(\theta_{r1} + \theta_{r2})(\omega_{r1} + \omega_{r2}) \\ e_c = k_b f_c(\theta_{r1} + \theta_{r2})(\omega_{r1} + \omega_{r2}) \end{cases} \quad (2)$$

where ω_{r1}, ω_{r2} are electrical angular velocities of the PM rotor and the armature rotor respectively, θ_{r1}, θ_{r2} are the angles of the two rotors with respect to the static parts, $f_a(\theta_{r1} + \theta_{r2})$, $f_b(\theta_{r1} + \theta_{r2})$, $f_c(\theta_{r1} + \theta_{r2})$ are the three phase back-EMF shape functions.

The motion equation and the electromagnetic torque can be expressed as (3) and (4)

$$\begin{cases} J_1 p\omega_{r1} = T_{em} - T_{L1} - B_1\omega_{r1} \\ J_2 p\omega_{r2} = T_{em} - T_{L2} - B_2\omega_{r2} \end{cases} \quad (3)$$

$$T_{em} = (e_a i_a + e_b i_b + e_c i_c)/(\omega_{r1} + \omega_{r2}) \quad (4)$$

where J_1, J_2 are the moment of inertia of the PM rotor and the armature rotor, T_{L1}, T_{L2} represent the load torques of the two rotors respectively, and B_1, B_2 are the damping coefficients of the two rotors respectively.

III. COUNTER-ROTATING CONDITION FOR STEADY STATE OPERATION

For counter-rotating BLDCM, the electromagnetic torques acting on the two rotors are of the same magnitude, but opposite in directions. However, the same magnitude of electromagnetic torques does not usually mean the same speed of the two rotors in steady state, because their inertia, coefficients, and mechanical load torques are different. If the parameters of the two rotors are unmatched extremely, one of the rotors will stop slowly and the counter-rotating BLDCM will become a normal single-rotating BLDCM. In order to analyze the influences of these parameters to the speed characteristics, the simulation studies are carried out on two prototype motors with different parameters. The parameters of the

two motors are shown in table I, where T_{01}, T_{02} represent the no-load torques of the inner and outer rotors respectively, and T_{m1}, T_{m2} represent the mechanical loads of the two rotors. The total load torques can be expressed as (5)

$$\begin{cases} T_{L1} = T_{01} + T_{m1} \\ T_{L2} = T_{02} + T_{m2} \end{cases} \quad (5)$$

The simulation results of the speed characteristics with different mechanical loads of two prototype motors are shown in Fig.2 and Fig.3. The main conclusions may be summarized as follows.

- Counter-rotating condition for steady state operation can not be obtained under no-load condition if the no-load torques of the two rotors are unequal. The rotor with higher no-load torque will stop slowly, as shown in Fig.2-(a). By forcing an additional mechanical load on the rotor with lower no-load torque, the total load torques of the two rotors become identical, both rotors will run with unequal speed in steady state, as shown in Fig.2-(b).
- For the mechanical loads with constant torque, the no-load torques of the two rotors should be of no difference, otherwise the counter-rotating will be failed, even if the same constant mechanical load torques act on them, which are shown in Fig.3-(a).
- If the magnitude of the mechanical loads vary with the rotor speed, such as fan or pump type, the counter-rotating can be achieved easily. In this case, the same no-load torques of the two rotors is not the indispensable condition for counter-rotating, the steady state speed of the two rotors can be adjusted by the variation of the mechanical loads, which are shown in Fig.3-(b) and Fig.3-(c).
- The speed of the two rotors in steady state are also influenced by the damping coefficients of the two rotors under the same load torques($T_{L1} = T_{L2}$). The moment of inertia of the PM rotor and the armature rotor determine the dynamic performance, and do not influence the steady state speed of the two rotors.

TABLE I

PARAMETERS OF THE TWO PROTOTYPE MOTORS FOR SIMULATION

	J_1	J_2	B_1	B_2	T_{01}	T_{02}	T_{m1}	T_{m2}	Fig. No.
	Kgm^2		Nm/(rad/s)		Nm		Nm		
I	13.2e-3	15.0e-3	0.01	0.03	0.5	2.5	0.0	0.0	2-(a)
							2.0	0.0	2-(b)
							$0.0003 \cdot \omega_r^2$	$0.0003 \cdot \omega_r^2$	2-(c)
II	13.2 e-3	15.0e-3	0.001	0.003	0.2	0.8	2.0	2.0	3-(a)
							$0.0003 \cdot \omega_r^2$	$0.0003 \cdot \omega_r^2$	3-(b)
							$0.000325 \cdot \omega_r^2$	$0.000275 \cdot \omega_r^2$	3-(c)

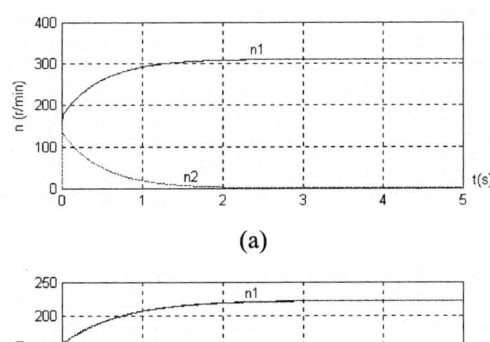

(a)

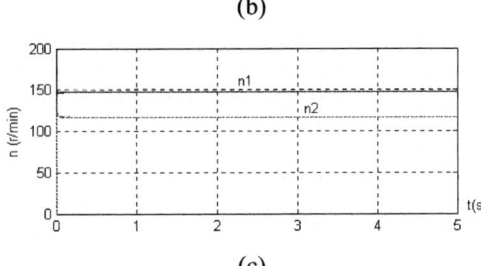

(b)

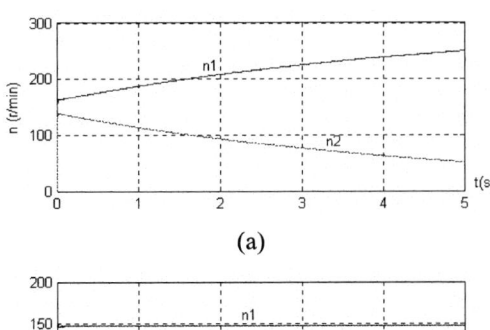

(c)

Fig.2 Speed Characteristics of prototype motor I

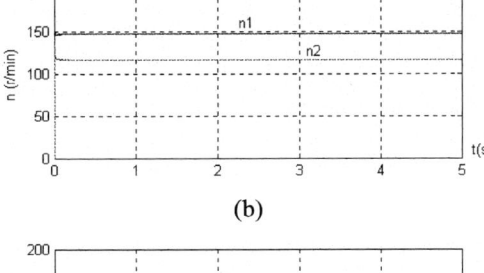

(a)

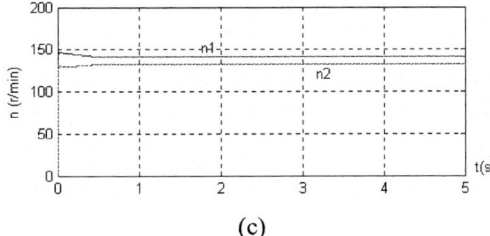

(b)

(c)

Fig.3 Speed Characteristics of prototype motor II

IV. CONTROL TECHNIQUES OF CRBLDCM

A. Position sensors of CRBLDCM

In this paper, two sets of position sensors are employed to detect the position signals of both the PM rotor and the armature rotor with respect to the static parts, and the relative position signal of the two rotors will be obtained by further calculation, thus precise commutation signals can be obtained.

The arrangement of Hall sensors in the experimental motor is shown in Fig.4, the six Hall sensors which are displaced by 30° electrical degree provides a 12-step rotor position information relative to the static parts, and the waveforms of the signals produced by the sensors are shown in Fig.5. The combination of the signals from two set of Hall position sensors forms a two dimension logical table(12×12), and it can be used to produce commutation sequence of the windings directly.

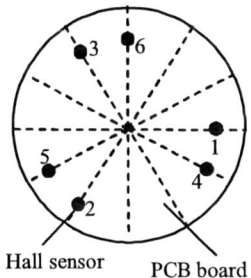

Fig.4 Arrangement of Hall sensors in PCB board

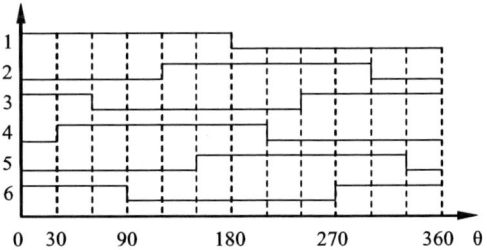

Fig.5 Signals from Hall sensors

The proposed position sensor method simplifies the construction of the system by eliminating the use of slip rings for sensor terminals. The CRBLDCM can start from an arbitrary initial position reliably, and by further commutation correction, precise commutation will always be obtained.

B. Relative position estimation of CRBLDCM

The counter-rotating BLDCM can be run properly using the commutation signals from two sets of position sensors directly, but the commutation instants will be advanced or delayed uncertainly because of the unequal speed of the two rotors, which distort the current waveforms and produce torque ripple. The precise commutation instants depend on the relative position information which can be estimated as follows:

887

$$\theta_{r1} = \theta_{r1(0)} + \int_0^t \omega_{r1} dt$$

$$\theta_{r2} = \theta_{r2(0)} + \int_0^t \omega_{r2} dt \qquad (6)$$

$$\theta = \theta_{r1} + \theta_{r2}$$

Where θ is the relative angle between the two rotors, and is calculated every PWM period. The rotor speed ω_{r1} or ω_{r2} is estimated twelve times each cycle, and it is an average speed in one cycle. The relative angle θ varies between $0°$ elec to $360°$ elec, every $60°$ elec, a commutation signal is triggered to energy the corresponding winding, total six commutation signals is produced in an electrical period. The experimental results are shown in Fig.6, though the speed of the two rotors are unequal, the commutation interval of the six conduction mode are almost the same with the proposed estimated method.

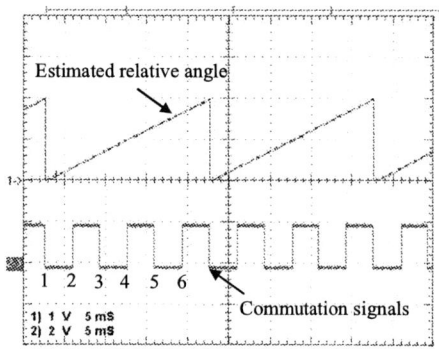

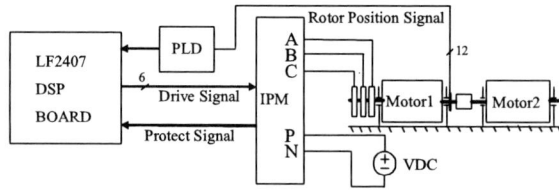

Fig.6 Estimated relative angle and commutation signals

V. EXPERIMENTAL RESULTS

The experiment is carried out on a counter-rotating BLDCM set with one of it acts as a motor and the other acts as a generator. the electric energy produced by the generator is consumed through a resistance load. The entire drive system is controlled by a low-cost, fixed-point DSP (TMS320LF2407). The schematic diagram of the control system is shown in Fig.7.

Fig.7 Schematic diagram of the control system

The experimental results are shown in Fig.8 to Fig.11. Fig.8 shows the Hall sensor signals from the PM rotor and the armature rotor, indicating the speed of the two rotors are unequal. Fig.9 shows the current waveforms using the commutation signal from two set of Hall position sensors directly, the current waveforms distort greatly without relative position estimation. It also indicates the commutation intervals of the six conduction mode are different from each other, which is caused by the unequal speed of the two rotors. Fig.10 shows the continuous change of the current waveforms during switching procedure. The switching procedure is accomplished by control system automatically at the starting stage of the motor. By comparing Fig.9 with Fig.11, it is obviously that the current distortion will be significantly reduced with relative position estimation, which proves the accuracy of the control strategy.

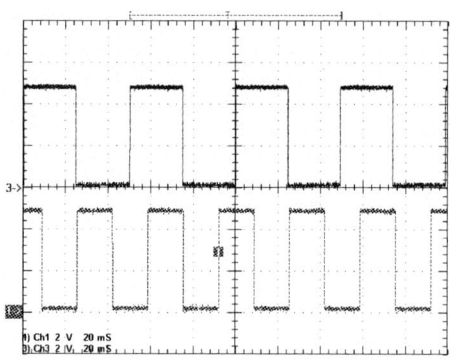

Fig.8 Position sensor signals from the PM rotor and the armature rotor (top: the PM rotor, bottom: the armature rotor)

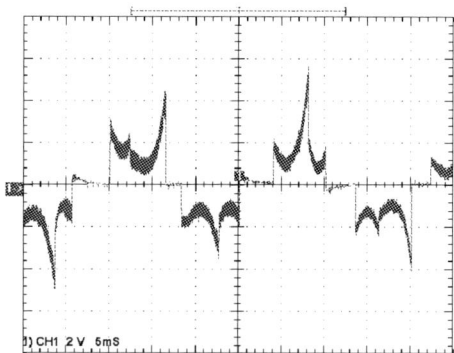

Fig.9 Current waveform without relative position estimation

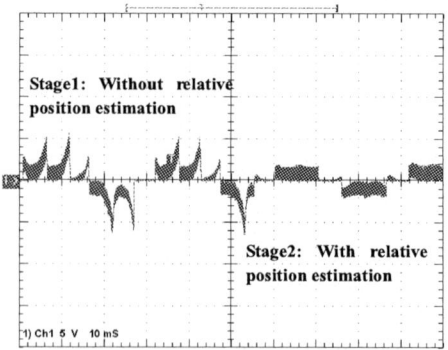

Fig.10 Current waveform during switching procedure

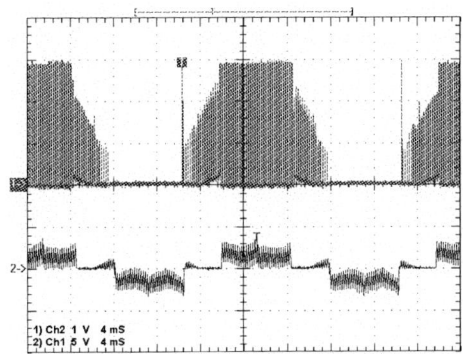

Fig.11 Terminal voltage waveform and Current waveform

with relative position estimation

VI. CONCLUSIONS

This paper has introduced a novel counter-rotating permanent magnet brushless dc motor for underwater propulsion. By the analysis of the mechanical characteristic of the motor, it can be concluded that the steady state counter-rotating operation condition depends on the parameters of the PM and armature rotors such as no-load torque characteristic, moment of inertia, damping coefficient, and the mechanical load torque, therefore the parameters of the two rotors should be designed as close as possible. The simulation and the experimental results confirm that control method using two sets of position sensors not only simplify the construction of the system, but also ensure the reliable start and precise commutation of the motor.

REFERENCES

[1] Mei Duanjing, "Electrical thrust in opposite direction of aggressive submarine" *Marine Electric & Electronic Technology*, No.2, 1995, pp.30-35.

[2] P. Pillay and R. Krishnan, "Modeling of permanent magnet motor drives". *IEEE Transactions on Industrial Electronics*, Vol.35, No.4, 1998, pp.537-541.

[3] F.Caricchi, F.Crescimbini, and E.Santini, "Basic Principle and design Criteria of Axial-Flux PM machines having Counterrotating Rotors", *IEEE Transactions on Industry Applications*, Vol.31, No.5, 1995, pp. 1062-1068.

[4] J.K.Holt, and D.G.White, "High efficiency, counter-rotating ring thruster for underwater vehicles", *Autonomous Underwater Vehicle Technology*, 1994, pp. 337-339.

2006 5th International Power Electronics and Motion Control Conference

A Special Flux-weakening Control Scheme of PMSM – Incorporating and Adaptive to Wide-Range Speed Regulation

Song Chi, Student Member, and Longya Xu, Fellow IEEE
Dept. of Electrical and Computer Engineering
The Ohio State University
2015 Neil Avenue
Columbus, OH 43210 USA
chi.36@osu.edu, xu.12@osu.edu

Abstract— This paper presents a special flux-weakening control scheme for permanent magnet synchronous machines (PMSM) over wide speed range. In contrast to conventional two-loop (d-, q-axis) control methods, the proposed control scheme achieves both flux-weakening and speed control simultaneously using only one speed/flux-weakening controller. The controller automatically generates both the required demagnetizing and torque-producing currents based on the crossing-coupling effects between d- and q-axes. Additional futures of the proposed controller include 1) not requiring knowledge of motor parameters and dc bus voltage of power inverter, and 2) preventing saturation of the current regulators under any load conditions. Therefore, this scheme is adaptive to the variation of motor parameters and load levels. The effectiveness of the proposed control scheme is verified by both computer simulation and experimental results.

Keywords-flux weakening; PMSM; adaptive

I. INTRODUCTION

Permanent magnet synchronous machine (PMSM) drives have been increasingly used in a wide variety of industrial applications due to their high power density and efficiency, high torque to inertia ratio and high reliability. In high-performance applications, PMSM can readily meet sophisticated requirements such as fast dynamic response, high power. Recently, the continuous cost reduction of magnetic materials with high energy density and coercivity makes the ac drives based on PMSM more attractive and competitive. This has opened up new possibilities for large-scale application of PMSM. A continuous increase in the use of PMSM drives will be witnessed in the near future [1-3].

The advantages of PMSM recently make them highly attractive candidates for traction and residential drive applications, such as hybrid electrical vehicles (HEV) or electrical vehicles (EV) and washing machines. The PMSM drive systems for such applications normally require high starting torque and wide speed range as well as high efficiency and power density. In order to satisfy these requirements, PMSMs are operated not only in the constant torque region when the speed is below the base speed but also in the constant power region over a wide speed range. In this way, the cost and size of the motor drive can be significantly reduced. The constant torque operation can be easily achieved in PMSM drives by conventional vector control. However, when the speed is above the base speed, the back-EMF of PMSM is much larger than the line voltage so that the PMSM suffers from the difficulty to produce torque due to the voltage constraints. Thanks to the flux-weakening technology, the operating speed range can be extended by applying negative field stator current component to weaken the air-gap flux [4, 5].

In general, permanent magnet synchronous machines with approximate sinusoidal back-EMF (electromotive force) can be broadly categorized into two types [5]: 1) interior (or buried) permanent motors (IPM) with saliency and 2) surface-mounted permanent motors (SPM) without saliency. Both IPM and SPM can be operated in the constant power region by flux-weakening technologies to an extended speed range. Various control algorithms for flux weakening have been published.

Macminn and Jahns presented two control techniques to enhance the performance of the IPM drive over an extended speed range. Although the proposed feed-forward compensation and flux-weakening algorithms were combined to improve the torque production capability of the IPM in high speeds, full effectiveness of the techniques heavily depended on accurate machine parameters, and the performance degraded noticeably as errors between the programmed and actual parameters increased [6].

Dhaoudi and Mohan researched a current-regulated flux-weakening method by introducing a negative current component to create a d-axis flux in opposition to that of the rotor permanent magnets, resulting in a decreased air-gap flux. This armature reaction was used to extend the operating speed range of a PMSM and relieve the current regulator from saturation in high speeds [7]. Similarly, a

1-4244-0448-7/06/$25.00 ©2006 IEEE   890

current vector control to expand the operating limits under the constant inverter capacity and the improvement by decoupling feed forward compensation were proposed in [8, 9] respectively by Morimoto et al. With these flux-weakening schemes, the required demagnetizing current component was calculated based on the mathematical model of PMSMs and, consequently, the performance of the PMSM drive system was seriously affected by the system parameters and sensitive to operating conditions.

Sozer and Torrey [10] presented an approach for adaptive control of the surface mounted PM motor over its entire speed range. The adaptive flux-weakening scheme was able to determine the right amount of direct-axis current at any operating conditions without knowing the load torque and inverter parameters. The level of demagnetizing current was obtained by using the current error between the actual and reference currents. Integration of the error with a proper forgetting factor was used to drive the direct-axis current.

Y. S. Kim et al, J. M. Kim et al and J. H. Song et al respectively proposed a flux-weakening control algorithm based on a voltage regulator using the voltage error signals between the maximum voltage and the voltage command [11-13]. The output of the voltage regulator determined the required mount of the demagnetizing current. In addition, the onset of flux weakening could be adjusted to prevent the saturation of the current regulators required by the vector control of PMSM.

Both current-error- and voltage-error-based flux-weakening methods require an additional PI regulator or integrator to generate the demagnetizing current command. However, the added regulator could only operate properly in the tuned conditions, which is not easily reached, resulting in the increased complexity of the overall control system.

Conventionally, two current regulators are always required to achieve torque (q-axis current) and flux (d-axis current) control as in [14]. Unfortunately the d- and q-axis currents cannot be truly controlled independently due to the cross-coupling effects inside the PMSM. The cross-coupling effects increase with the speed and become dominant in the high-speed flux-weakening region. As a result, the performance of current and torque response is degraded without good decoupling control.

In the paper, a novel flux-weakening control of PMSM incorporating wide-range speed regulation is presented. The cross-coupling effect between d- and q-axis current is re-examined and a speed/flux-weakening controller proposed. The controller is able to achieve the closed-loop speed control and flux weakening control simultaneously, which simplifies the control algorithm by using only one current regulator. The proposed control scheme does not require the knowledge of motor parameters and dc bus voltage of the power inverter. In addition, saturation of current regulation is prevented under any load conditions, showing control robustness in high speeds. Computer simulation results are presented to show the control

performance and robustness with respect to disturbances from load and dc bus voltage. An experimental setup is built based on a PMSM without saliency. Experimental results are used to demonstrate the effectiveness of the proposed approach.

II. MATHEMATICAL MODEL OF PMSM IN THE SYNCHRONOUS REFERENCE FRAME

The dynamic equations of a PMSM in the synchronous reference frame with d-axis aligned to the actual rotor magnets can be expressed in the matrix form as

$$\begin{bmatrix} v_{ds} \\ v_{qs} \end{bmatrix} = \begin{bmatrix} R_s + pL_s & -\omega_r \cdot L_q \\ \omega_r \cdot L_d & R_s + pL_s \end{bmatrix} \cdot \begin{bmatrix} i_{ds} \\ i_{qs} \end{bmatrix} + K_e \cdot \omega_r \cdot \begin{bmatrix} 0 \\ 1 \end{bmatrix}. \quad (1)$$

In steady state, the equations are reduced to

$$\begin{bmatrix} v_{ds} \\ v_{qs} \end{bmatrix} = \begin{bmatrix} R_s & -\omega_r \cdot L_q \\ \omega_r \cdot L_d & R_s \end{bmatrix} \cdot \begin{bmatrix} i_{ds} \\ i_{qs} \end{bmatrix} + K_e \cdot \omega_r \cdot \begin{bmatrix} 0 \\ 1 \end{bmatrix}. \quad (2)$$

where

v_{ds}, v_{qs}	d-axis, q-axis stator voltage;
i_{ds}, i_{qs}	d-axis, q-axis stator current;
L_d, L_q	d-axis, q-axis stator inductance;
R_s	stator resistance;
K_e	back-EMF constant, or magnetic flux linkage per phase;
ω_r	rotor speed.

The developed electromagnetic torque T_e in terms of stator currents is expressed as

$$T_e = \tfrac{3}{2} \cdot p \cdot [K_e \cdot i_{qs} + (L_d - L_q)i_{qs}i_{ds}], \quad (3)$$

where p is the number of pole pairs.

From (2), we can find the cross coupling effect between i_{ds} and i_{qs}, which is expressed in

$$i_{qs} = -\frac{\omega_r \cdot L_d}{R_s} i_{ds} + \frac{v_{qs} - K_e \cdot \omega_r}{R_s}. \quad (4)$$

Equation (4) clearly shows that the i_{qs}-i_{ds} cross coupling becomes stronger when the rotor speed, ω_r, is higher.

When $L_q = L_d = L_s$, referring to surface-mounted PMSM without saliency, (3) and (4) can be rewritten as

$$T_e = \tfrac{3}{2} \cdot p \cdot K_e \cdot i_{qs} \quad (5)$$

$$i_{qs} = -\frac{\omega_r \cdot L_s}{R_s} i_{ds} + \frac{v_{qs} - K_e \cdot \omega_r}{R_s}. \quad (6)$$

As shown in (3), T_e consists of the torque components contributed by the permanent magnets and reluctance. When only the magnet-contributed torque is produced as in SPM, (5) applies. In order to increase the efficiency of overall PMSM drive system, it is obvious that the maximum torque per ampere can always be achieved under the current and voltage constraints.

III. CURRENT AND VOLTAGE CONSTRAINTS IN FLUX-WEAKENING REGION

In general, the required terminal voltage of a PMSM increases as its speed goes high. However, the maximum available voltage V_{smax} applied to the PMSM is always limited by the fixed dc bus voltage. Also the maximum current I_{smax} is limited by the thermal ratings of both the inverter and the stator windings of the PMSM. Such constraints can be expressed in

$$\begin{cases} v_{qs}^{2} + v_{ds}^{2} \le V_{s\max}^{2} \\ i_{qs}^{2} + i_{ds}^{2} \le I_{s\max}^{2} \end{cases}. \qquad (7)$$

Neglecting the stator resistance, we can rewrite the voltage limit equation as

$$i_{qs}^{2} + (\frac{L_d}{L_q})^2 (i_{ds} + \frac{K_e}{L_d})^2 \le (\frac{V_{s\max}}{\omega_r L_q})^2. \qquad (8)$$

From (7) and (8), we can see that the current limit equation determines a circle with a radius of I_{smax} while the voltage limit equation determines a series of nested circles ($L_q = L_d$) or ellipses ($L_q > L_d$).

Fig.1 shows the current-limit circle CLMT and the voltage-limit ellipses/circles VLMT1, 2 in the i_d-i_q plane. The voltage-limit circles centering at Point "A" ($-K_e/L_s$, 0) with the radius become smaller as the speed ω_r increases. The maximum torque-per-ampere trajectory is also illustrated by a bold and piecewise curve from Origin "O" to Point "A". When a PMSM is operated from the start up to the base speed ω_1 in the constant torque region, the voltage limit ellipse/circle meets the intersection of the maximum torque-per-ampere curve and the current limit circle CLMT. The PMSM cannot be operated beyond the base speed without flux-weakening control. To extend the speed range, a proper demagnetizing current has to be applied according to its operating speed so that the current vector trajectory can move along the maximum torque-per-ampere curve from B to A, corresponding to the speed changing from ω_1 to infinite, assuming $K_e/L_s < I_{smax}$.

IV. ADAPTIVE FLUX-WEAKENING CONTROLLER

Define $V_{FWC} = v_{qs} = const.$, and consider (4), we get

$$i_{qs} = -\frac{\omega_r \cdot L_d}{R_s} i_{ds} + \frac{V_{FWC} - K_e \cdot \omega_r}{R_s}. \qquad (9)$$

Examining (9), we can see that at the specific speed ω_r, there exists a linear relationship between i_{qs} and i_{ds}, showing the i_{qs}-i_{ds} cross coupling. This equation suggests

a control strategy that the q-axis current i_{qs}, or torque, can be controlled by means of controlling i_{ds}, adaptive to the load and operating speed. Fig.2 shows a speed control block diagram including an adaptive speed/flux-weakening controller based on the above control concept. The speed/flux-weakening controller is but not limited to a PI regulator. The input signals are actual speed and speed error. And the outputs are i_{ds} command and an enabling signal to the FWC Mux. The actual speed determines the onset of flux weakening, while the speed error determines the i_d command.

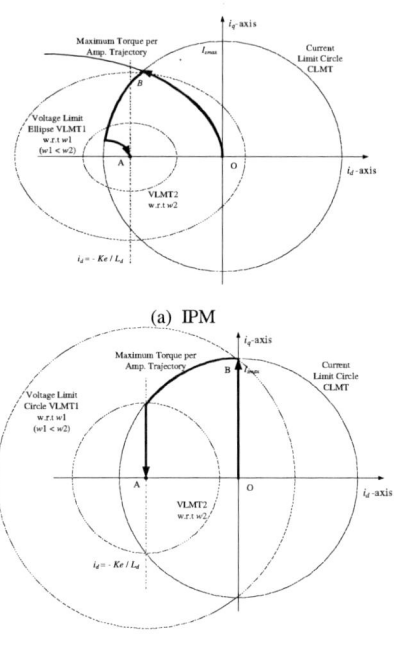

(a) IPM

(b) SPM

Figure 1. Current-limit CLMT, voltage-limit ellipse/circle VLMT1, VLMT2 and maximum torque per ampere trajectory in the i_d-i_q plane.

In the constant torque region, one speed regulator and two current regulators (referring to i_{qs}-and i_{ds}-regulator) are used in the synchronous reference frame. The three regulators work together for the speed control below the base speed as in conventional vector control systems [6]-[8]. When the speed increases near but still less than the base speed, the system starts to smoothly switch into flux-weakening operation using the proposed adaptive controller. Thereafter, speed control is achieved by the speed/flux-weakening controller and the i_{ds} current regulator. The i_{ds} current command (i.e. i_d^* in Fig. 2) generated by the speed/flux-weakening controller comprises not only the required demagnetizing current component but also the torque-related component defined by (9). In this way, the cross-coupling effect is utilized to control the torque as well as flux weakening, instead of being purposely eliminated in conventional systems.

It should be noticed that there is a tradeoff between the maximum torque capability and efficiency of the PMSM as using the proposed flux-weakening controller.

According to the load torque profile, V_{FWC} can be properly selected to meet the requirements of torque and efficiency.

In the paper, $V_{FWC} = 0.5$ (p.u.) is selected for the computer simulation and experimental testing.

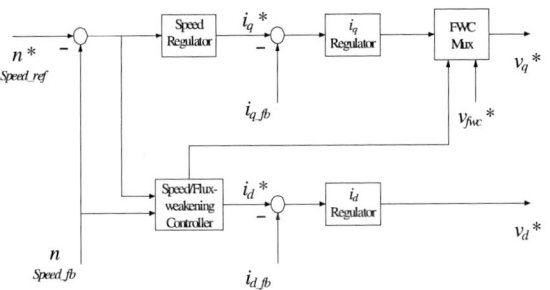

Figure 2.   Block diagram of the adaptive flux-weakening control

V.   SIMULATION AND EXPERIMENTAL RESULTS

A.   Simulation results

A PMSM drive system has been simulated by Simulink/MatLab. The parameters of PMSM were: $R_s =$ 16 ohm, $L_s = 60$ mH and $K_e = 0.22$ Vs/rad. The base speed was 250 rpm. The onset of the flux-weakening operation was 0.2 in per unit. The speed base was 1250 rpm and current base was 7A.

Figs.3 and Fig.4 show the simulation results when the motor was running up from 0 to 1000 rpm and stayed for 0.5 s and then slowed down to 0 with a constant load of 1 Nm.

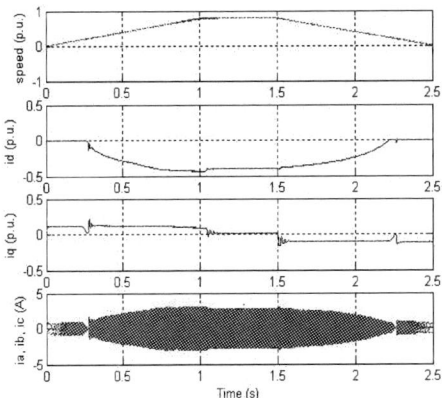

Figure 3.   Speed command and feedback (up), i_{ds} (2nd), i_{qs} (3rd) and stator phase current i_{as} (5.0 A/div, bottom)

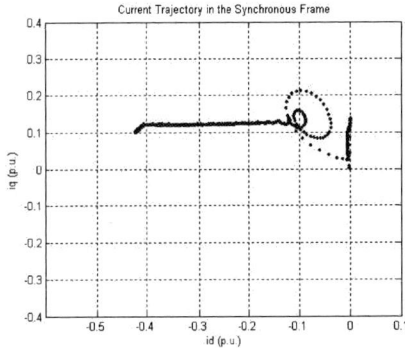

Figure 4.   Current vector trajectory in the synchronous i_d-i_q plane during speeding up

The dc bus voltage was 310V. We can observe the automatically generated demagnetizing current i_{ds} and good performance of speed control within the entire speed range.

Fig.5 shows the robustness of the system with respect to the disturbances from load and dc bus voltage. In the simulation the motor was accelerated from 0 to 500 rpm and stayed after. The dc bus voltage was initially 310V and ramped to 320V in 2 s. and then down from 320V to 280V. The load was initially 0 Nm and stepped to 1 Nm. in 2.5 s. We can observe that the speed regulation in the flux-weakening region functioned very well. It shows that the proposed scheme is adaptive to the variation of dc bus voltage and load condition.

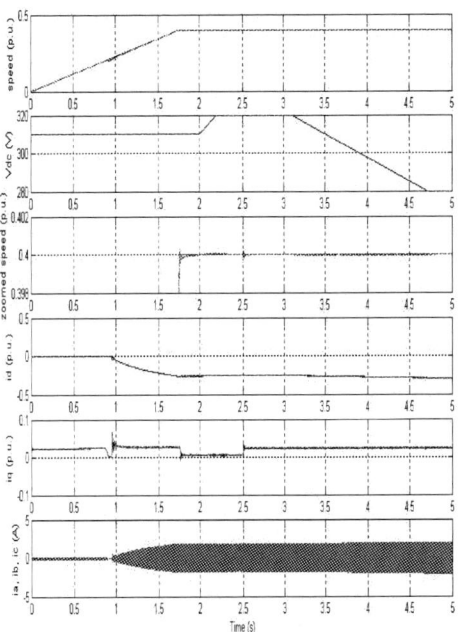

Figure 5.   Speed command and feedback (up), V_{dc} (2nd), zoomed speed command and feedback (3rd), i_{ds} (4th), i_{qs} (5th) and stator phase current i_{as} (5.0 A/div, bottom)

893

B. Experimental results

The proposed flux-weakening control scheme was verified by an experimental PMSM drive system including: 1) a 48-pole non-saliency outer-rotor PMSM with its base speed of 250 rpm, 2) a DSP controller based on an eZdspF2812 DSP board, 3) a three-phase power inverter and 4) a dynamometer coupled with the shaft of the PMSM as load. The switching frequency of power inverter was 20kHz. Space vector PWM was used for the PWM generation. The dc bus voltage of the power inverter was 310V and the maximum current was 7A. The sampling frequency of the current and voltage measurement was 20 kHz. The parameters of PMSM were same as in the computer simulation.

Figs.6 and 7 show the experimental results when the motor was accelerated from 50 to 1025 rpm and stayed for 25 s and then down to 50 rpm. The dc bus voltage varied between 320V and 280V due to operating conditions. The load was 0.5 Nm. We can observe the automatically generated demagnetizing current i_{ds} and the good performance of speed control within the wide speed range including flux-weakening region. Comparing the results of computer simulation and those of experimental testing, it is clearly seen that the experimental results agree with the simulation results very well, indicating that the proposed adaptive flux-weakening control scheme is valid and the real-time implementation of the speed/flux-weakening controller is successful.

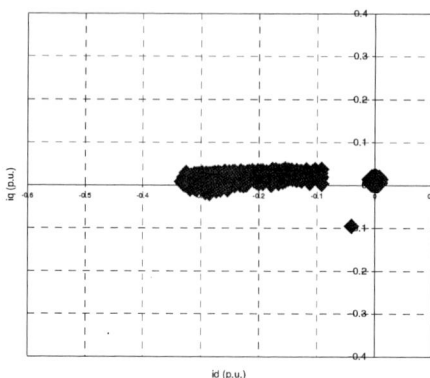

Figure 7. Current vector trajectory in the synchronous i_d-i_q plane during speeding up

VI. CONCLUSIONS

In this paper, a special adaptive flux-weakening control scheme incorporating wide-range speed regulation is presented. No knowledge of motor parameters and dc bus voltage of power inverter is required, indicating that this scheme is adaptive to the variation of system parameters and load levels. The automatically generated demagnetizing current by the proposed speed/flux-weakening controller satisfies both flux-weakening operation and torque control based on the cross-coupling inherent to PMSMs. The robustness and stability of the system has been demonstrated. In addition, the two-controller structure for both flux and speed control reduces computation time in real-time implementation. The effectiveness of the proposed flux-weakening control scheme and the speed control performance of the speed/flux-weakening controller have been verified by both computer simulation and experimental results.

ACKNOWLEDGMENT

This work was supported by the Research and Engineering Center of Whirlpool.

REFERENCES

[1] B. K. Bose, B. Bose, "Power electronics and motion control—Technology status and recent trends," *IEEE Trans. Ind. Applicat.,* vol. 29, pp. 902–909, Sept./Oct. 1993.

[2] W. Leonhard, "Adjustable speed ac drives," *Prod. IEEE,* vol. 76, pp. 455-471, Apr. 1988.

[3] B. K. Bose, "Variable frequency drives-technology and applications," in *Proc. ISIE 93(Budapest),* June, 1993, pp 1-18.

[4] T. M. Jahns and V. Blasko, "Recent advances in power electronics technology for industrial and traction machine drives," *Proc. IEEE,* vol. 89, pp. 963–975, June 2001.

[5] Thomas M. Jahns, "Motion control with permanent-magnet ac machines," in *Proc. IEEE,* vol. 82, Aug. 1994, pp. 1241-1252.

[6] S. R. Macminn and T. M. Jahns, "Control techniques for improved high-speed performance of interior PM synchronous motor

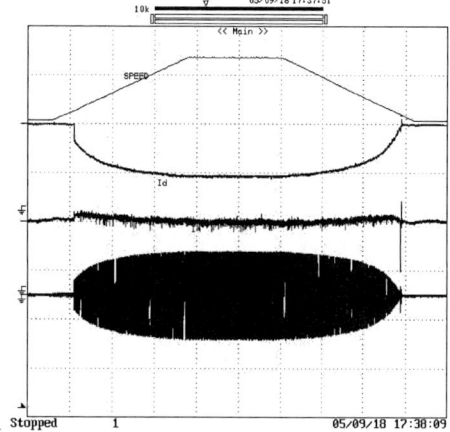

Figure 6. Speed command and feedback (757 rpm/div, up), i_{ds} (1.7 A/div, 2nd), i_{qs} (1.7 A/div, 3rd) and stator phase current i_{as} (1.0 A/div, bottom), 10 s/div

drives," *IEEE Trans. Ind. Applicat.,* vol. 2, pp. 997-1004, Sept./Oct. 1991.

[7] R. Dhaouadi and N. Mohan, "Analysis of current-regulated voltage-source inverters for permanent magnet synchronous motor drives in normal and extended speed ranges," *IEEE Trans. Energy Conv.,* vol. 5, pp. 137-144, Mar. 1990.

[8] S. Morimoto, M. Sanada and K. Takeda, "Wide-speed operation of interior permanent magnet synchronous motors with high-performance current regulator," *IEEE Trans. Ind. Applicat.,* vol. 30, pp. 920-926, July/Aug. 1994.

[9] S. Morimoto, Y. Takeda, T. Hirasa, and K. Taniguchi, "Expansion of operating limits for permanent magnet by current vector control considering inverter capacity," *IEEE Trans. Ind. Applicat.,* vol. 26, pp. 866-871, Sept./Oct. 1990.

[10] Y. Sozer and D. A. Torrey, "Adaptive Flux weakening control of permanent magnet synchronous motors," in *Conf. Rec. IEEE-IAS Annu. Meeting,* vol. 1, St. Louis, MO, 1998, pp. 475–482.

[11] Y. S. Kim, Y. K. Choi and J. H. Lee, "Speed-sensorless vector control for permanent-magnet synchronous motors based on instantaneous reactive power in the wide-speed region," *IEE Proc-Electr. Power Appl.,* vol. 152, No. 5, pp. 1343-1349, Sept. 2005.

[12] J. M. Kim and S. K. Sul, "Speed control of interior permanent magnet synchronous motor drive for the flux weakening operation," *IEEE Trans. Ind. Applicat.,* vol. 33, pp. 43-48, Jan./Feb. 1997.

[13] J. H. Song, J. M. Kim, and S. K. Sul, "A new robust SPMSM control to parameter variations in flux weakening region," *IEEE IECON,* vol. 2, pp. 1193-1198, 1996.

[14] J. J. Chen and K. P. Chin, "Automatic flux-weakening control of permanent magnet synchronous motors using a reduced-order controller," *IEEE Trans. Power Electron.,* vol. 15, pp. 881-890, Sept. 2000.

2006 5th International Power Electronics and Motion Control Conference

Model-based Disturbance Attenuation for Linear Motor Servo System

Guiqiu Liu, Qingding Guo

School of Electrical Engineering, Shenyang University of Technology, Shenyang, P.R. China
Liugq_shy@sina.com, Guoqd@sut.edu.cn

Abstract—This paper presents a method using Model-based Disturbance Attenuation (MBDA) to reject external disturbances, such as frictional forces, cutting forces and load variations for a permanent magnet linear synchronous motor (PMLSM) with high precision and micro-feed for CNC machining centers. In this method, a nominal plant is designed in parallel with the plant. Through the output feedback, the error signal, which is the difference between the plant output and the nominal plant output, is achieved by means of a compensator controller in the velocity loop. Accordingly, disturbance attenuation can be attained. In addition, an integral and proportional (IP) controller is designed directed toward the velocity loop to cater the fast tracking and resist disturbance greatly. Simulation results show that this proposed method has quick-response speed and strong disturbance attenuation. Moreover, machining precision can be improved remarkably.

Keywords- linear motor; MBDA; IP speed controller

I. INTRODUCTION

In recent years, there has been significant interest in the permanent magnet linear synchronous motor (PMLSM). For the reliability advantage of eliminating mechanically-drive chain of rotating electrical motor from rotation to linear motion, the PMLSM became increasingly popular in the high precision and micro-feed linear servo system. In many applications, the PMLSM has become a preferred choice, due to its characteristics: high efficiency, low noise, long life expectancy and being controlled easily [1].

But the plant is connected directly with the active cell of the PMLSM, causing that the linear servo system performance is impacted by load variation and external disturbance. In the linear servo system, disturbance is a main factor reducing the performance. Disturbance attenuation has been an important topic, especially in the high precision and micro-feed linear servo system [1][3][4]. In this paper, an adaptive control method is presented, using integral and proportional (IP) controller[2] in the velocity loop. It is rapidly for the plant to track the command with the proposed algorithm. The disturbance can be rejected stronger as well.

In order to suppress disturbance effectively, Model-based Disturbance Attenuation (MBDA)[3][4] method is adopted. The MBDA utilizes the nominal plant in parallel

with the plant, through an IP controller to compensate the disturbance in the linear servo system. Accordingly, disturbance can be attenuated stronger. The simulation results illustrate the validity of our approach.

II. MATHEMATICAL MODEL OF PMLSM

The PMLSM is a thrust device converting AC to linear motion directly. When the fundamental component is only taken into account, d- and q-axis model can be applied in the PMLSM. In the d- and q-axis model, the mover voltages are characterized by the following equation.

$$\begin{bmatrix} u_d \\ u_q \end{bmatrix} = \begin{bmatrix} R_s & 0 \\ 0 & R_s \end{bmatrix} \begin{bmatrix} i_d \\ i_q \end{bmatrix} + \begin{bmatrix} -v & p \\ v & p \end{bmatrix} \begin{bmatrix} \lambda_d \\ \lambda_q \end{bmatrix} \quad (1)$$

The flux linkages equations are given by

$$\lambda_d = L_d i_d + \lambda_{PM1} \quad (2)$$

$$\lambda_q = L_q i_q \quad (3)$$

The mover current equations are expressed as following.

$$\begin{bmatrix} i_d \\ i_q \end{bmatrix} = \frac{2}{3} \begin{bmatrix} A_1 & B_1 & C_1 \\ A_2 & B_2 & C_2 \end{bmatrix} \begin{bmatrix} i_a \\ i_b \\ i_c \end{bmatrix} \quad (4)$$

Where:

$A_1 = \cos(-\theta_r)$,
$B_1 = \cos(-\theta_r + 2\pi/3)$,
$C_1 = \cos(-\theta_r - 2\pi/3)$
$A_2 = \sin(-\theta_r)$,
$B_2 = \sin(-\theta_r + 2\pi/3)$,
$C_2 = \sin(-\theta_r - 2\pi/3)$

The electromagnetic thrust equation can be defined as following

$$F_e = \frac{3\pi}{2\tau} [\lambda_{PM} i_q + (L_d - L_q) i_d i_q] \quad (5)$$

In this paper, the mover current vector is perpendicular to permanent magnet of the stator. In the inner loop of current, the excitation component $i_d = 0$ is adopted. Accordingly, the electromagnetic thrust equation can be expressed as

1-4244-0448-7/06/$25.00 ©2006 IEEE   896

$$F_e = \frac{3\pi}{2\tau}\lambda_{PM}i_q \qquad (6)$$

In (6), it is obvious that the electromagnetic thrust is determined by i_q. The motive equation of the PMLSM can be attained.

$$F_e = K_T i_q = F_L + Dv + M\frac{dv}{dt} \qquad (7)$$

Where:

M -- the mass of the mover and load.

D -- the viscous frictional coefficient.

F_e -- the electromagnetic thrust.

F -- the load resistance.

F_L – the load resistance (including the equivalent resistance caused by end effects.)

K_T -- the thrust coefficient.

v -- the liner speed of mover.

λ_{PM} – the excitation chain caused by permanent magnet of stator.

τ -- the polar distance.

III. THE DESIGN OF IP SPEED CONTROLLER

In this paper, the PMLSM can be thought as the plant after it is modulated in the inner loop. The structural pattern is shown in Fig.1.

In order to acquire the zero steady-state error and expectable transient state response, a proportional and integral (PI) controller is adopted in the velocity loop in the traditional control system. From ω_r and F_L to ω_m, the transfer function can be expressed as

$$T(s) = G_v(s)\omega_r + G_p(s)F_L \qquad (8)$$

Where :

$$G_v(s) = \frac{K_T K(s)}{Ms + D + K_T K(s)}$$

$$G_p(s) = \frac{1}{Ms + D + K_T K(s)}$$

In the traditional control system, the $K(s)$ can reject disturbance F_L.

In this paper, IP controller is adopted in the velocity loop. IP controller has those advantages, such as speed quick response and strong retrain disturbance. From ω_r and F_L to ω_m, the transfer function can be expressed as (9)

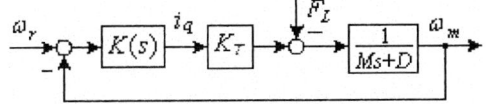

Fig.1.Block diagram of the control system

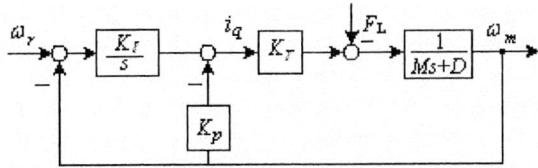

Fig.2. Block diagram of IP control system

$$T(s) = G_v(s)\omega_r + G_p(s)F_L \qquad (9)$$

$$G_v(s) = \frac{K_I K_T}{Ms^2 + (D + K_p K_T)s + K_I K_T}$$

$$G_p(s) = \frac{-s}{Ms^2 + (D + K_p K_T)s + K_I K_T}$$

IV. MODEL-BASED DISTURBANCE ATTENUATION METHOD

A. The General Design of MBDA

In this paper, Model-based Disturbance Attenuation method is proposed, which is shown in Fig.3. It is obvious that the velocity loop is thought of as the plant. In this structure, a nominal plant is designed in parallel with the plant. Through the output feedback, the error signal, which is the difference between the nominal plant output and the plant output, is fed back to input of the plant $G_v(s)$ by means of a compensation controller. Accordingly, the disturbance can be restrained. Then, from ω_r and F_L to ω_m, the transfer function can be expressed as.

$$T(s) = G_m(s)\omega_r + D_m(s)G_p(s)F_L \qquad (10)$$

Where:

$$G_m(s) = \frac{1 + K_m(s)G_n(s)}{1 + K_m(s)G_v(s)}G_v(s)$$

$$D_m(s) = \frac{1}{1 + K_m(s)G_v(s)}$$

$$G_n(s) = \frac{K_I K_T}{Ms^2 + (D + K_p K_T)s + K_I K_T}$$

B. The Design of PID Compensator

When the nominal plant is close to the plant, the speed vibration response of the plant is a little faster than the nominal plant. In equation (8), it is shown that $G_p(s)$ reduces the disturbance from F_L to ω_r. In equation (10), $D_m(s)$ can reject disturbance further for the MBDA. In Fig.3, the compensation may have any type, and in this paper, PID controller was adopted.

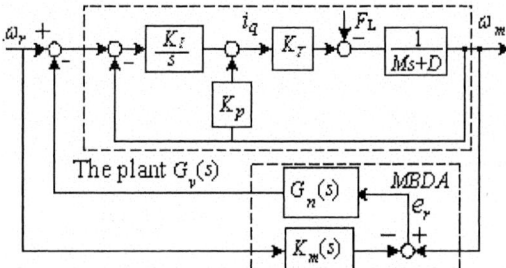

Fig.3. Diagram block of MBDA control system

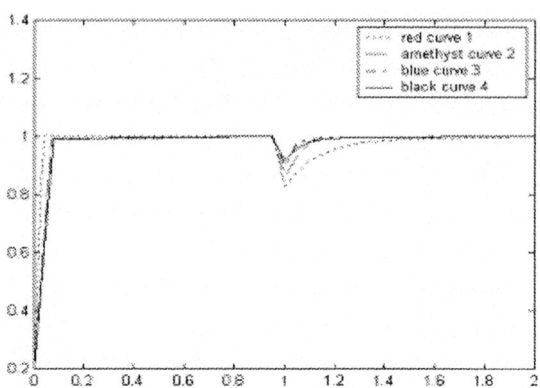

Fig.4. The curves of simulation results

It is adaptable for IP controller to be implemented in the compensation. Where, the proportional gain can reject all kinds of frequency. The integral gain is applied to attenuate the disturbance of low frequency, so as to increase the steady-state performance of the plant. The differential component is implemented in median frequency, in order to improve the dynamic performance of the plant. It is obvious that the gain is higher, while the disturbance attenuation is more effective.

Whereas, when the gain is too high, the plant is unstable for the model being determinate.

V. SIMULATION RESULTS

The MBDA method has been investigated using computer simulation. Matlab/Simulink is used for the simulation. In order to give a meaningful idea of the system performance, simulation results are presented considering the plant, whose parameters are adopted in $M = 10kg$, $D = 1.2N \cdot S / m$, $K_T = 25N / A$. The parameters of the PI controller are selected as $K_{spp} = 60$, $K_{si} = 250$. The parameters of the IP controller are selected as $K_i = 1800$, $K_p = 50$. In the MBDA, the parameters of IP controller are $K_i = 1800$, $K_p = 50$. In the feedback compensator, the parameters of PI controller are $K_{mi} = 0.01$, $K_{mp} = 0.8$, and the parameters of IP controller are $K_{mi} = 0.01$, $K_{mp} = 0.8$, $K_{md} = 0.02$. When unit step command and time is $t = 1S$, the load disturbance $F_L = 200N$ is put in abruptly. The simulation results are shown in Fig.3. It is seen that the PID compensate curve has the fast tracking and strong disturbance attenuation.

VI. CONCLUSION

This paper has introduced a technique using MBDA method for linear servo system control. It is suitable for the requirement to high precision and micro-feed for CNC machining centre. The simulation results have demonstrated that this proposed method has quick-response speed and strong disturbance attenuation. Moreover, machining precision can be improved remarkably.

Where: 1- PI control, 2- IP control, 3- MBDA PID compensation control, 4- MBDA PI compensation control.

REFERENCES

[1] Qingding Guo and Chengyuan Wang., *Precision control for linear AC servo system*. Beijing: China Machine Press, 2000.

[2] Zhuo Yue, *The On-line Design of IP Position Controller of the Permanent Magnet Linear Synchronous Motor*, The master degree. thesis, School of Electrical Engineering, *Shenyang University of Technology*, 1999.

[3] Beyong-Kap Choi, Chong-Ho choi, and Hyuk Lim, "Model-based disturbance attenuation for CNC machining centers in Cutting Process, " *IEEE/ASME Trans. Mechatronics*, vol .4,no.2, pp.157-178, 1999.

[4] Bong Keun Kim,and Wan Kyun Chung, "Unified analysis and design of robust disturbance attenuation algorithms using inherent structural equivalence," in *Proceedings of the American Control Conference*, pp. 4046-4051, 2001.

A Fuzzy-Wavelet-Network-Based Position Control for PMSM

Wang Jun [*], Peng Hong [**], Xia Ling [*]

[*] School of Electric and Information, Xihua University, Chengdu, China
[**] School of Mathematics & Computer Science, Xihua University, Chengdu, China
jiayu@mail.sc.cninfo.net

Abstract—A robust position controller with variable structure control (VSC) and dynamic recurrent fuzzy wavelet network (DRFWN) technique for permanent magnet synchronous motor (PMSM) is presented in this paper. Based on the VSC method, the closed-loop system can system dynamics with an invariance property to uncertainties. However, the sliding controller based the assumption of known uncertainty bounds often cause the chattering phenomena in the control system. DRFWN identification with adaptive learning rates is implemented to evaluate the uncertainty bounds of PMSM which are caused by the variable of system parameters and the external load disturbance. The performance of the proposed drive is investigated both in experiment and simulation at different condition.

Keywords-Fuzzy wavelet network; permanent magnet synchronous motor

I. INTRODUCTION

In the past decade, variable structure control (VSC) with sliding mode is one of the effective nonlinear robust control approaches since it provides system dynamics with an invariance property to uncertainties once the system dynamics are controlled in the sliding mode[1][2]. So, sliding-mode control system were developed for the rotor position control of the field-oriented permanent magnet synchronous motor (PMSM) drive [3][4]. However, the sliding controllers which are the assumption of known uncertainty bounds often cause the chattering phenomena in the control system so that system can become instability. Therefore, in the design of a sliding mode controller the bound of uncertainties should be obtained in advance.

The combination of wavelet analysis, fuzzy system and neural network is a new idea and technology, which have the advantages of better accuracy, generalization capability, fast convergence and multi-resolution. Fuzzy wavelet network (FWN) was proposed in [5] and [6]. However, the proposed FWN is static network due to the inherent feedforward network structure. Since, If the dynamic system like PMSM is identified by the static FWN neural network, dynamic time modeling question actually will become static space modeling question, which exist many problems. Especially, with the

increasing the order of the system, a large network size will cause slow learning rate. In the meantime, the large number of inputs should leads sensitive to output noise of the system.

This paper will focus on the motivation and implementation of identification schemes for the uncertainty bounds of PMSM using dynamic recurrent fuzzy wavelet network (DRFWN). Cheng-Jian Lin and Cheng-Chung Chin[7] presents a wavelet network to the consequents parts of the fuzzy rules. Their methods have complicated structure and a large subset of the network parameters. First, VSC method for PMSM is described in Section 2. Second, based on our previous research [8], a DRFWN model is presented to evaluate the uncertainty bounds of PMSM which are caused by the variable of system parameters and the external load disturbance in Section 3. The simulation examples are given to illustrate the performance in Section 4. Finally, a brief conclusion is drawn in Section 5.

II. THE SLIDING CONTROL FOR PMSM

The basic principle in controlling a PMSM drive is based on field orientation through decoupling the quadrate axis (q axis) and direct axis (d axis). For a permanent magnet machine, if d axis current i_d is at zero, PMSM becomes a reduced-order linear time-invariant system as follows.

$$T_e = k_t i_q^*(t) \tag{1}$$

$$\ddot{\theta} = \frac{k_t}{J} i_q^*(t) - \frac{B}{J}\dot{\theta} - \frac{1}{J}T_L = A\dot{\theta} + Bi_q^*(t) + DT_L \tag{2}$$

where $K_t=3P.\lambda_M/4$, i_q is the q-axis stator current, T_e is the electromagnetic torque, θ is the rotor position, respectively. In (1), since K_t is fixed, the electromagnetic torque T_e is then proportional to $i_q^*(t)$, which is determined by closed-loop control. Since the generated motor torque is linearly proportional to the q-axis current, the maximum torque per ampere can be achieved.

Considering the existing of disturbance, the mathematic model can be rewritten as

$$\ddot{\theta} = (A + \Delta A)\dot{\theta} + (B + \Delta B)i_q^*(t) + (D + \Delta D)T_L \tag{3}$$

$$= A\dot{\theta} + Bi_q^*(t) + W(t)$$

Where $W(t)$ is the total uncertainty disturbances which are defined as

$$W(t) = \Delta A\,\dot{\theta} + \Delta Bi_q^{\,*}(t) + (D + \Delta D)T_L \quad (4)$$

Assuming the bound of the total uncertainty disturbances is expressed by

$$W(t) \le \beta \quad (5)$$

In (5), β is positive. The objective is to find control law $i_q^{\,*}(t)$ so that the rotor position θ can track any the command position θ_r. Therefore, define the error $e(t) = \theta_r - \theta$.

$$S(t) = \dot{e}(t) + 2\lambda e(t) + \lambda^2 \int_0^t e(\tau)dt \quad (6)$$

In (6), λ is positive. Differential (9)

$$\dot{S}(t) = \ddot{e}(t) + 2\lambda\,\dot{e}(t) + \lambda^2 e(t) \quad (7)$$

$$= \ddot{\theta}_r - [A\,\dot{\theta} + Bi_q^{\,*}(t) + DT_L] + 2\lambda\,\dot{e}(t) + \lambda^2 e(t)$$

Conclusion: When the sliding control law is designed as

$$i_q^{\,*}(t) = B^{-1}[\ddot{\theta}_r - A\,\dot{\theta} + 2\lambda\,\dot{e} + \lambda^2 e + \rho\,\mathrm{sgn}(S(t))] \quad (8)$$

If $\rho \ge \beta$, the system is global stability.

Proof: Defined Lyapunov function as

$$V[S(t)] = \frac{1}{2}S^2(t) \quad (9)$$

Differential (9)

$$\dot{V}[S(t)] = S(t)\dot{S}(t)$$

$$= S(t)[\ddot{\theta}_r - A\,\dot{\theta} - Bi_q^{\,*}(t) - DT_L + 2\lambda\,\dot{e}(t) + \lambda^2\,e(t)] \quad (10)$$

$$= S(t)[-DT_L - \rho\,\mathrm{sgn}(S(t))]$$

$$= -S(t)DT_L - \rho|S(t)| \le |S(t)||DT_L| - \rho|S(t)|$$

$$= -|S(t)|[\rho - |DT_L|] < 0$$

If select control gain $\rho \ge \beta$, the stability of the sliding control system can be guaranteed during control domain.

The block diagram of control system based VSC is shown in Fig.1. However, based on the assumption of known uncertainty bounds, it is very difficult to decide β in advance which decide the selection of the ρ. In the meantime, if selected β is inexact, system can cause the chattering phenomena. In this paper, we propose to identify the uncertainty bounds of PMSM using dynamic recurrent fuzzy wavelet network (DRFWN) so that the stability and accuracy of system can be increased.

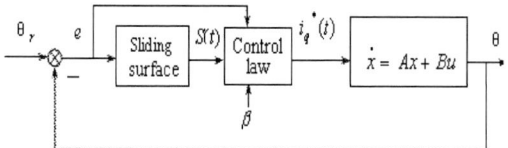

Figure 1. The block diagram of control system based VSC

III. THE SLIDING CONTROL WITH DRFWN FOR PMSM

The block diagram of control system based VSC with the DRFWN identification is shown in Fig.2. The value of the ρ can be decided online through the DRFWN identification. The input of the FWN are S(t) and S(t)(1-Z^{-1}). The output is $\hat{\rho}$ which can be expressed by

$$\hat{\rho} = \theta^T\Gamma \quad (11)$$

DRNN provides an effective way for the identification of dynamic process. Through storing past state into delay units, networks have the memory capability. So, they can process the object related to time. The recurrent property is achieved in the proposed RCFWN. As a result, it not only utilizes its previous knowledge, but also has better response for the dynamic system.

The structure of the RCFWN is illustrated in Fig.3.

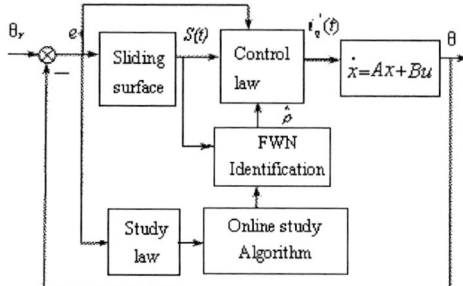

Figure 2. The block diagram of control system based VSC with the DRFWN identification

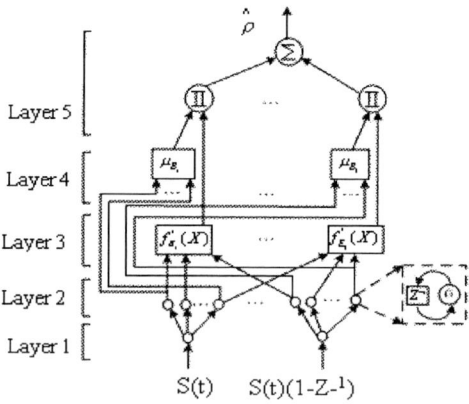

Figure 3. The architecture of the RCFWN

Layer 1. The node number is n. The ith node of the net output can be denoted by

$$O_j^{(1)} = I_j^{(1)} = x_j \quad (12)$$

where $j=1,...,n$.

Layer 2. In this layer, every node represents a linguistic variable, which compute fuzzy membership function. The B-spline wavelet membership function, a particular function, is adopted here as the membership function. Then

$$O_{ij}^{(2)} = \phi_{P_{i,j}, l_{P_{i,j}+k}}(I_{ij}^{(2)}(k)) \tag{13}$$

$$= 2^{-p_i/2} \phi(2^{-p_i} I_{ij}^{(2)}(k) - l_{P_{i,j}+k})$$

where p_i and $l_{P_{i,j}+k}$ are dilation and translation of the scaling function, respectively. $i = 1,..., M$, and M is the fuzzy rule number. For the approximation function network, the input of this layer is determined by

$$I_{ij}^{(2)}(k) = O_j^{(1)}(k) + d_{ij}.O_{ij}^{(2)}(k-1) \tag{14}$$

where d_{ij} is the weight of the dynamic feedback. Clearly, the memory terms $O_{ij}^{(2)}(k-1)$ stores the past information of the network, so, it can realize the dynamic reflection.

Layer 3. Every node in this layer implements the fuzzy subsystem $f_{E_i}'(X)$ under the resolution factor p_i. Assuming the rule number of the fuzzy subsystem with resolution p_i is I. The output can be expressed by

$$O_i^{(3)} = f_{E_i}'(X) = \frac{\sum_{k=1}^{I_{P_i}} w_{ik} \prod_{j=1}^{n} \phi_{P_{i,j}, l_{P_{i,j}+k}}(I_{ij}^{(2)}(k))}{\sum_{k=1}^{I_{P_i}} \prod_{j=1}^{n} \phi_{P_{i,j}, l_{P_{i,j}+k}}(I_{ij}^{(2)}(k))}$$

$$= \frac{\sum_{k=1}^{I_{P_i}} w_{ik} \Phi_{P_i, L_k}(X)}{\sum_{k=1}^{I_{P_i}} \Phi_{P_i, L_k}(X)} \tag{15}$$

where $\Phi_{P_i, L_k}(X) = \prod_{j=1}^{n} \phi_{P_{i,j}, l_{P_{i,j}+k}}(I_{ij}^{(2)}(k))$.

Layer 4. In layer 4, the membership function in frequency domain can be denoted by

$$O_i^{(4)} = \mu_{B_i}(P_i, X) \tag{16}$$

where

$$\mu_{B_i}(P_i, X) = \prod_{j=1}^{n} \mu_{B_i}(p_{i,j}, x_j) \tag{17}$$

where $P_i = (p_{i,1}, p_{i,2}, \cdots, p_{i,n})$.

Layer 5. This layer is output layer which denote fuzzy system with multiresolution capability. The output can be calculated using the output $O_i^{(3)}$ of the third layer and the output $O_i^{(4)}$ of the fourth layer, which can be written by

$$O^{(5)} = \hat{\rho} = \frac{\sum_{i=1}^{I} \mu_{B_i}(P_i, X) f_{E_i}'(X)}{\sum_{i=1}^{I} \mu_{B_i}(P, X)}$$

$$= \sum_{i=1}^{I} \hat{\mu}_{B_i}(P_i, X) \frac{\sum_{k=1}^{I_{P_i}} w_{ik} \Phi_{P_i, L_k}(X)}{\sum_{k=1}^{I_{P_i}} \Phi_{P_i, L_k}(X)} \tag{18}$$

where

$$\hat{\rho} = \frac{\mu_{B_i}(P_i, X)}{\sum_{i=1}^{I} \mu_{B_i}(P_i, X)} \tag{19}$$

$\hat{\mu}_{B_i}(P_i, X)$ represents the when the membership function in frequency domain is the ratio of the corresponding scale function for x_j in time domain.

RCFWN is composed of five parts: input layer, fuzzy layer, fuzzy subsystem layer, frequency membership function layer and output layer. In layer 2, recurrent neural unit has inherent feedback connection which function store past information of the network and can achieve dynamic response of the system, so, the network can more strongly dynamic system with better response and small number of tuning parameters.

Owing to the simplicity of the algorithm and easy implementation using hardware, the backpropagation learning algorithm (BP) has been used for training recurrent neural network. So, in this paper, we adopt BP algorithm to train the free parameters of RCFWN network. The cost function E is defined as

$$E(k) = \frac{1}{2p} \sum_{l=1}^{p} [\theta_r(k) - \hat{\theta}(k)]^2 \tag{20}$$

where $\theta_r(k)$ is lth desired output and p denotes the number of output nodes. When the BP learning algorithm is used, the weighting vector of the RCFWN model is adjusted such that the error defined in (20) is less than the desired threshold value after a given number of training cycles.

IV. SIMULATION RESULTS

To demonstrate the control performance of the proposed controller, the simulation results are compared under the VSC controller and the VSC controller with DRFWN. The simulating conditions are the two cases with unload and load. In experiment, Node numbers of every layer from input layer to output are 2, 6, 9 and 1. The values of weights initialization are between 0 and 1.

The simulation conditions are the following two cases.
Case 1. T_L =0N.m, B' =B, J' =J.
Case 2. T_L =5N.m, B'=5B, J' =5J.

Fig.4 shows the position responses and control effort under VSC controller. Fig.5 shows the position responses and control effort under VSC controller with DRFWN. Fig. 4(a) and (b) and Fig. 5(a) and (b) are done when unload, the square input signal and the initial value $\hat{\rho}$ is same which express the two control method both have better response.

Fig. 4(c) and (d) and Fig. 5(c) and (d) is position responses and control effort when the sinusoid input signal and the load torque step changes from 0 to 1.5N.m at t=3.16s. From the simulation results, compared the two controllers, the proposed algorithm has smaller overshoot and shorter transient time. It can be already seen under the control of the proposed controller that the degenerated and smooth responses under inertia variations and load disturbances are much improved with the augmentation of the proposed controller.

V. CONCLUSIONS

901

A controller based VSC and DRFWN techniques for PMSM are theoretically analyzed and simulated in this paper. Major characters of the proposed controller are described on the following: (1) It makes system stronger robustness. (2) It has on-liner learning capability that can deal with a larger range of parameter variations, and external disturbances. (3) It reduces the chattering under VSC controller. More smooth responses can be obtained.

Appendix: The nominal parameters of a PMSM are: Rated power is 700w, rated speed is 1500rpm, number of poles is 4, magnetic flux is 0.175wb; Stator resistance is 2.875Ω, stator induction is 0.085H.

ACKNOWLEDGMENT

This paper is supported by the importance project fund of the education department of Sichuan province, China (No. 2005A117) .

REFERENCES

[1] Shyu K.K., Kim J.H., "A New Switching surface Sliding-mode Speed Controller for Induction Motor Drive Systems," *IEEE Trans. Power Electron*, vol. 11, no. 4, pp. 660-667,1996.

[2] Guo Xingding, Wang Qingyuan, "Study of Sliding Mode Variable Structure Acceleration Control of AC Servo System", *Transactions of China Electrotechnical Society*, vol. 11, No. 2, pp. 34-37, 1996.

[3] Rong-Jong Wai, Jia-Ming Chang, et al, "Implementation of Robust Wavelet-Neural-Network Sliding-Mode Control for Induction Servo Motor Drive," *IEEE Trans. Industrial Electronics*, vol. 50, no. 6, pp. 1317-1334, 2003.

[4] Kuo-Kai Shyu, et al, "A Newly Robust Controller Design for the Position Control of Permanent-Magnet Synchronous Motor," *IEEE Trans. Industry Electronic*, vol. 49, no. 3, pp. 558-564, 2002.

[5] Daniel J., HO W. C., Ping-An Zhang, and Jinhua Xu. "Fuzzy wavelet networks for function learning," *IEEE Trans. on Fuzzy Systems*, vol.9, no. 1, pp. 200-211, 2001.

[6] Jun Wang, Jian Xiao, Dan Hu, " Fuzzy Wavelet Network Modeling with B-Spline Wavelet," *The fourth International Conference on Machine Learning and Cybernetics*, pp. 889-898, 2005.

[7] Cheng-Jian Lin and Cheng-Chung Chin, "Prediction and Identifaction Using Wavelet-Based Recurrent Fuzzy Neural Networks," *IEEE Trans. on Systems, Man and Cybenetics-Part B: Cybenetics*, vol. 34, no. 5, pp. 2144-2154, 2004.

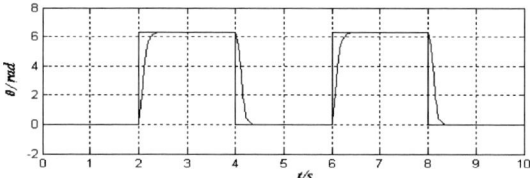

(*a*) The position responses when the square input signal

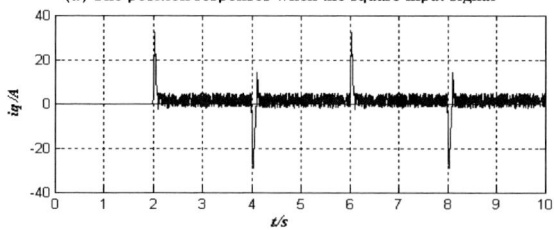

(*b*) The control effort when the square input signal

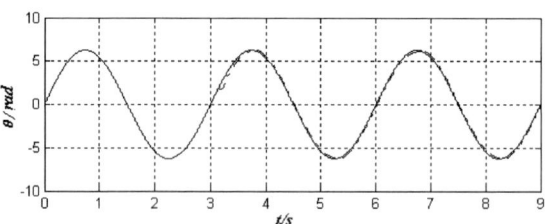

(*c*) The position responses when the sinusoid input signal

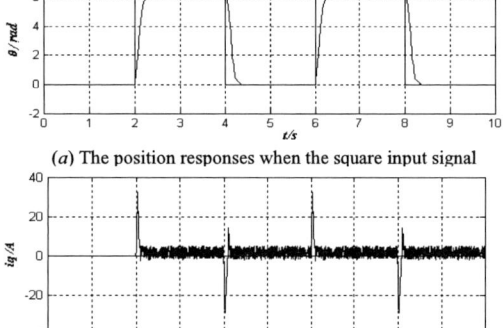

(*a*) The position responses when the square input signal

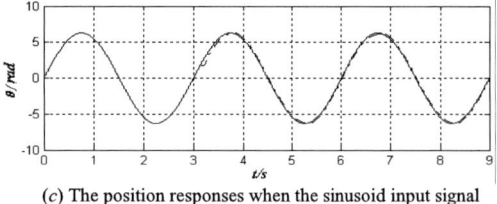

(*b*) The control effort when the square input signal

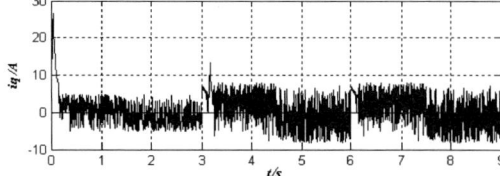

(*c*) The position responses when the sinusoid input signal

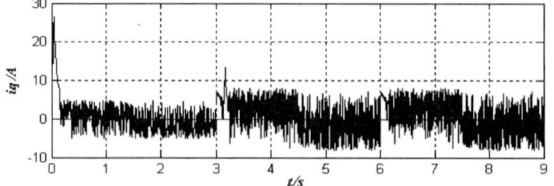

(*d*) The control effort when the sinusoid input signal

Figure 5. The simulation position responses and control under VSC controller and DRFWN identification

(*d*) The control effort when the sinusoid input signal

Figure 4. The simulation position responses and control under VSC controller

2006 5th International Power Electronics and Motion Control Conference

Stability Analysis of Magnetic Bearing with Resonance Circuit

Zong Ming, Wang Fengxiang, Sun Yidan, Wang Jiqiang
School of Electrical Engineering Shenyang University of Technology, Shenyang 110023, China

Abstract —In this paper, a structure of a radial magnetic bearing with resonance circuit is proposed, which dose not need separate displacement sensors. The operating principle of the radial magnetic bearing is also introduced. The stability of the radial magnetic bearing is analyzed by the control system theory. The analysis shows that the proposed radial magnetic bearing with resonance circuit is instable in dynamic state process. The simulation study on an example shows that the radial magnetic bearing with resonance circuit can levitate stably by means of a PD controller. But it is of steady-state error.

Keywords- magnetic bearing; PD controller; simulation, magnetic levitation; resonance circuit

I. INTRODUCTION

Magnetic bearing are designed to support rotating and linear moving machinery elements without coming into contact with motion parts. This is accomplished by applying the principle which an electromagnet will attract a ferromagnetic material. Using this principle the motion part can be suspended in a magnetic field which is generated by the bearing.

A typical radial active magnetic bearing consists of stator, displacement sensors mounted above the rotor and control system. The rotor position can be controlled by the control system according to measuring results of the displacement sensor. In the radial magnetic bearing with resonance circuit proposed in this paper, the separate displacement sensors are not needed. But the capacitors in series with stator windings of the magnetic bearing and high frequency alternating supply voltage are required. So the control system design of magnetic bearing with resonance circuit can be simplified. Because of the dynamic state instability, PD controller is required. The information of the rotor position can be gotten by measuring the current in stators winding.

II. STRUCTURE OF RADIAL MAGNETIC BEARING WITH RESONANCE CIRCUIT

The structure of the radial magnetic bearing with resonance circuit proposed in this paper is shown in Figure.1.

The characteristics of the radial magnetic bearing in

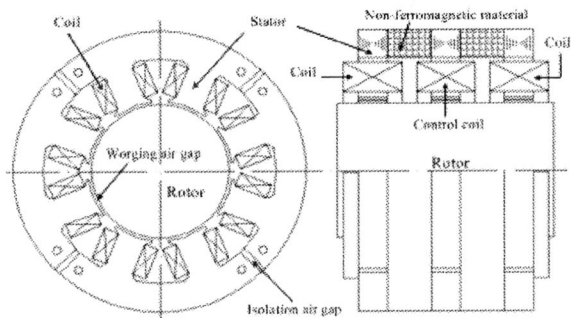

Fig.1 Structure of the radial active magnetic bearing proposed

structure are

1). 12 pairs of the magnetic poles are insulated by the isolation air gaps to reduce magnetic coupling between different poles.

2). In order to eliminate variation of the electromagnetic force caused by AC excitation, the twins0 magnetic circuits insulated by non-ferromagnetic material are adopted.

3). A set of the control winding is used to improve the dynamic response of the magnetic bearing with resonance circuit.

III. OPERATION PRINCIPLE OF MAGNETIC BEARING WITH RESONANCE CIRCUIT

Fig.2 is the schematic diagram of the radial magnetic bearing with resonance circuit in one degree of freedom (DOF).

S_{UL}, S_{UR}, S_{DL} and S_{DR} represent the top left electromagnet, top right electromagnet, bottom left electromagnet and bottom right electromagnet respectively. The $u1$ is an AC power supply for left electromagnets and the $u2$ is for right electromagnets. C is the capacitor in series with the electromagnet winding of the magnetic bearing. It can be selected from Eq.1.

$$C = \frac{1}{\omega^2 \times L_{\min} \times K_c} \quad (0.7 < K_c < 1) \quad (1)$$

Where ω is the angular frequency of the alternating supply voltage; L_{min} is minimum inductance within the bound of rotor moving.

$$L = \frac{\mu_0 \times S}{2\delta} \times N^2 \quad (2)$$

Where S is the cross-sectional area of the stator core,

1-4244-0448-7/06/$25.00 ©2006 IEEE   903

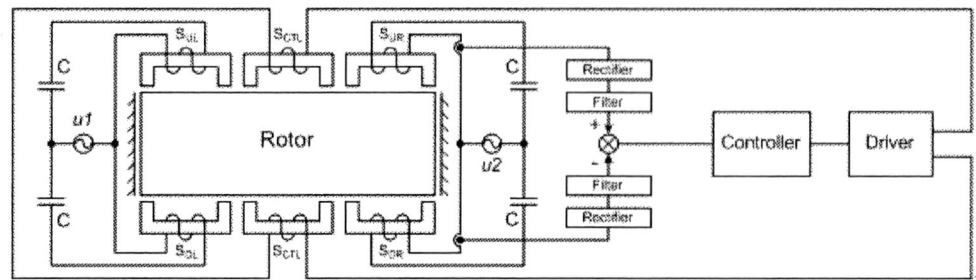

Fig.2. Schematic diagram of the radial magnetic bearing proposed

N is the winding turns, δ is working air gap length.

The peak amplitude of the steady-state current in any electromagnet winding is given by

$$I_m = \frac{U_m}{\sqrt{R^2 + \left(\omega L - \dfrac{1}{\omega C}\right)}} \tag{3}$$

Where U_m is the voltage peak amplitude of the alternating supply voltage, R is the resistance of the winding.

From Eq.3, the relationship between inductance L (i.e. air gap length, see Eq.2) and current in any electromagnet winding is shown in Fig.3.

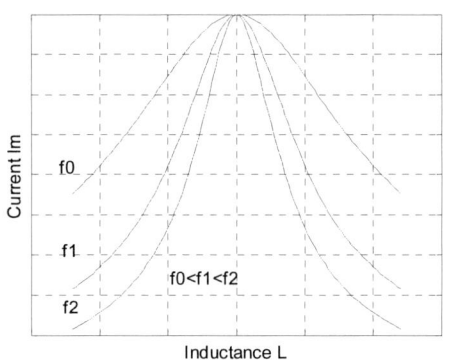

Fig.3 Relationship curve between inductance and current in an electromagnet

Where $f_0 \sim f_2$ are frequency of the AC power supply in Hz.

The peak amplitude current through an electromagnet winding occurs when $\omega L = 1/\omega C$. From Fig.3, as long as the inductance variation is limited in the bound of right hand side of Fig.3, the steady-state current amplitude through an electromagnet winding will be reduced with inductance increasing rapidly.

If the working air gap length is smaller, the magnetic attracting force to rotor produced by the top left electromagnet is

$$F_{UL} = \frac{\Phi_m^2}{2\mu_0 S} - \frac{\Phi_m^2}{2\mu_0 S}\cos 2\omega t = F_{ULm} - F_{ULm}\cos 2\omega t \tag{4}$$

Where φ is the alternating magnetic flux; ϕ_m is the peak amplitude of the alternating magnetic flux.

Owing to the magnetic attracting force, F_{UL}, is fluctuating, the rotor pulsate is occurred. In order to eliminate the variation, another alternating magnetic flux with 90 degree phase difference is applied to the top right

electromagnet. So the magnetic attracting force to rotor produced by this electromagnet is

$$F_{UR} = \frac{\Phi_m^2}{2\mu_0 S} + \frac{\Phi_m^2}{2\mu_0 S}\cos 2\omega t = F_{URm} + F_{URm}\cos 2\omega t \tag{5}$$

The resultant force acting on the rotor produced by top electromagnet can be written as

$$F_U = F_{UR} + F_{UL} = \frac{\Phi_m^2}{\mu_0 S} = \frac{\mu_0 S}{4\delta_u^2} N^2 I_{um}^2 \tag{6}$$

Where δ_u is the upper working air gap length, I_{um} is the peak amplitude of the steady-state current in upper electromagnet windings.

Eq.6 shows the resultant force produced by top electromagnets is invariable.

According to the structure symmetry, the resultant force acting on the rotor produced by bottom electromagnets can be written as

$$F_D = F_{DR} + F_{DL} = \frac{\Phi_m^2}{\mu_0 S} = \frac{\mu_0 S}{4\delta_d^2} N^2 I_{dm}^2 \tag{7}$$

Where δ_d is the subjacent working air gap length, I_{dm} is the peak amplitude of the steady-state current in subjacent electromagnet windings.

Under no excitation in the control winding, S_{CTR}, the resultant force acting on the rotor can be written

$$F = F_D - F_U = \frac{1}{4}\mu_0 SN^2\left(\frac{I_{um}^2}{\delta_u^2} - \frac{I_{dm}^2}{\delta_d^2}\right)$$
$$= K_0\left[\frac{1}{K_1(a+x)^2 + K_2(b-x)^2} - \frac{1}{K_1(a-x)^2 + K_2(b+x)^2}\right] \tag{8}$$

Where $K_0 = \mu_0 S(NK_c U_m \delta_{max})^2$; $K_1 = 4K_c^2 R^2 \delta_{max}^2$; $K_2 = \mu_0^2 \omega^2 S^2 N^4$; $b = K_c \delta_{max} - a$; a is the nominal working air gap; x is rotor displacement; δ_{max} is the maximum of working air gap.

From Eq.8, the steady-state resultant force acting on the rotor is shown in Fig.4. It shows that the working air gap increasing results in the steady-state resultant force augmenting. If some reason causes the rotor leave its original position, the increasing magnetic force will draw it back. So the radial magnetic bearing proposed is stable in the steady-state condition,

In order to improve the dynamic state stability, the control winding, SCTR, and a controller are adopted.

IV. STABILITY ANALYSIS OF RADIAL MAGNETIC BEARING WITH RESONANCE CIRCUIT

If the rotor displacement influence to stator current is taken into account, the balance of voltage in R, L, C loop

can be written as

$$u(t) - L(t)\frac{d}{dt}i(t) - i(t)\frac{d}{dt}L(t) - \frac{1}{C}\int i(t)dt = i(t)R \quad (9)$$

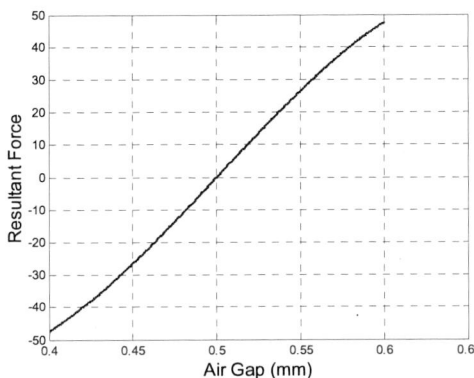

Fig.4 Relationship between working air gap and
resultant force acting on rotor

The motion equation of the rotor is expressed as

$$F = m\frac{d^2}{dt^2}x(t) \quad (10)$$

According to the Eq.8, 9 and 10, the dynamic model of the radial magnetic bearing proposed can be built with SIMULINK under no excitation in the control winding. See Fig.5.

For an example, a=0.5mm, R=0.05Ω, f=1750Hz, N=42T, K_C=0.9, U_m=30V, m=2kg and S=137mm^2, the dynamic model of Fig.5 can be integrated into Fig.6 based on step response [1]. The transfer function of Fig.6 system is as follows

$$G(s) = \frac{6.17\times10^{-7}S^2 + 1\times10^{-3}S + 1}{1.234\times10^{-6}S^4 + 2\times10^{-3}S^3 + 2S^2 + 2.1\times10^6} \quad (11)$$

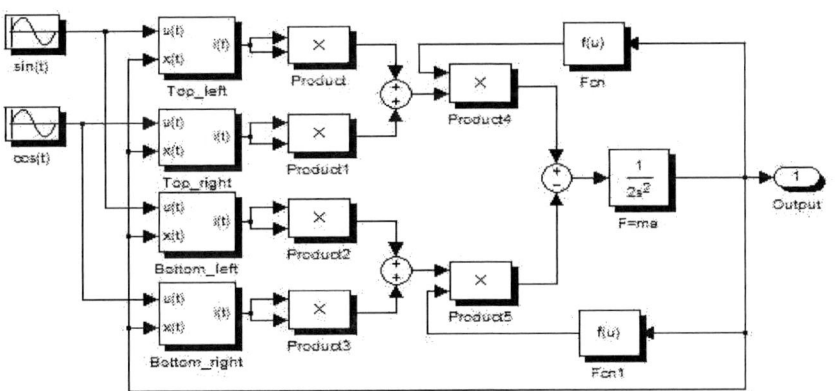

Fig.5 Dynamic model of the radial magnetic bearing proposed with SIMULINK

For the transfer function G(s), Routh array[2] can be calculated as follows

$$a_0 = 1.234\times10^{-6}$$

$$a_1 = 2\times10^{-3}$$

$$b_1 = 2$$

$$c_1 = -2.1\times10^3$$

Because c_1 is negative, the system dynamic state response is instability according to Routh criterion[2].

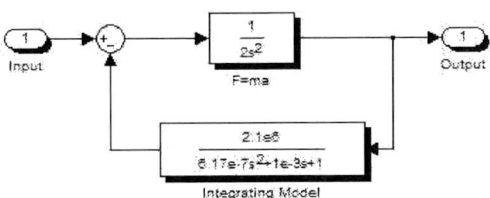

Fig.6 Simplified model of Fig.5 system

rectifiers and 2nd-order Butterworth lowpass filters are adopted. See Fig.2.

The transfer function between working air gap and difference of currents for the example mentioned above can be integrated as follows

$$G_i(s) = \frac{1.7\times10^5}{1.95\times10^{-6}S^2 + 3.88\times10^{-3}S + 1} \quad (12)$$

The system structure inserted into a controller is shown in Fig.7.

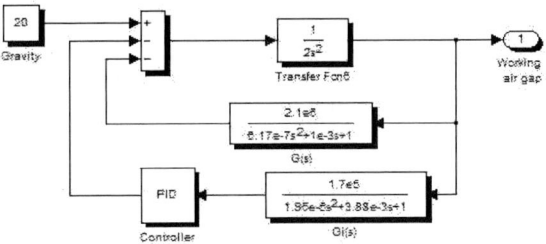

Fig.7 Control system of the radial magnetic bearing proposed

V. SIMULATION RESULTS OF CONTROL SYSTEM

In order to enable the system stabilize, a controller is used in the system. The variation of working air gap is detected by measuring the difference between top right and bottom right stator winding current. Full wave

The simulation results are seen from Fig.8 to Fig.10.

Fig.8 is the simulation results of the system shown in Fig.6. Fig.8a indicates the displacement variation of the rotor with time in the case of gravity acting. Fig.8b gives the gravity and the changing of the resultant

905

electromagnetic force. Because the resultant electromagnetic force or displacement of rotor is divergent, the system shown in Fig.6 is instable.

Fig.9 is the simulation results of the system with PD controller shown in Fig.7.

Fig.9a indicates the variation of the rotor displacement with time in the case of gravity acting. Fig.9b gives the gravity, the resultant electromagnetic force and the compensation force of PD controller. Because the PD control strategy is used in the system shown in Fig.7, the rotor displacement is leveled off. The system shown in Fig.7 becomes stable. The PD controller only used for compensating variation of disturbance force in the transient process. So it is a system with static error.

Fig.10 illustrates the simulation results in the case of the gravity and a sinusoidal disturbance force with a frequency of 1Hz and amplitude of 10N acting simultaneously.

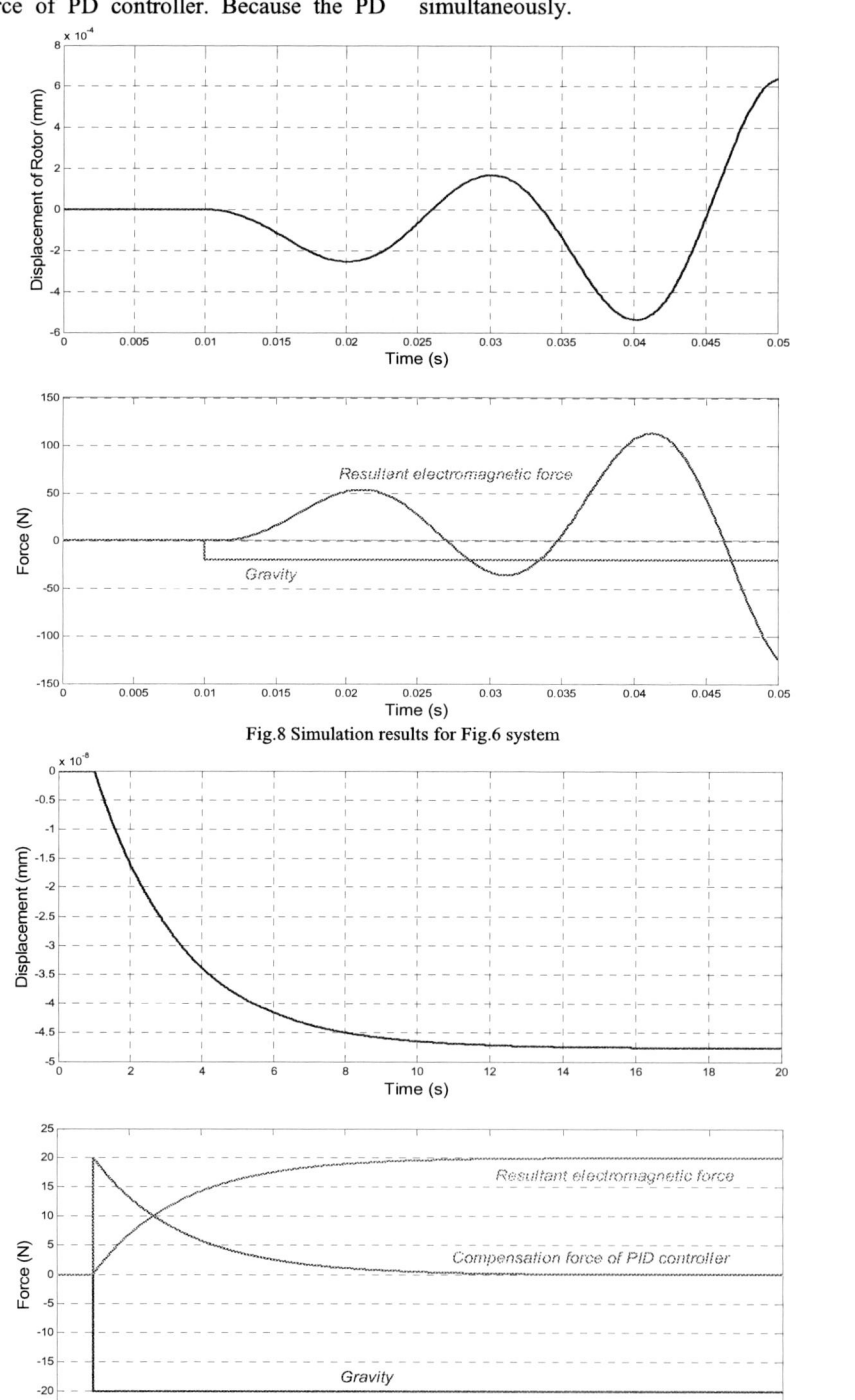

Fig.8 Simulation results for Fig.6 system

Fig.9 Simulation results for Fig.7 system

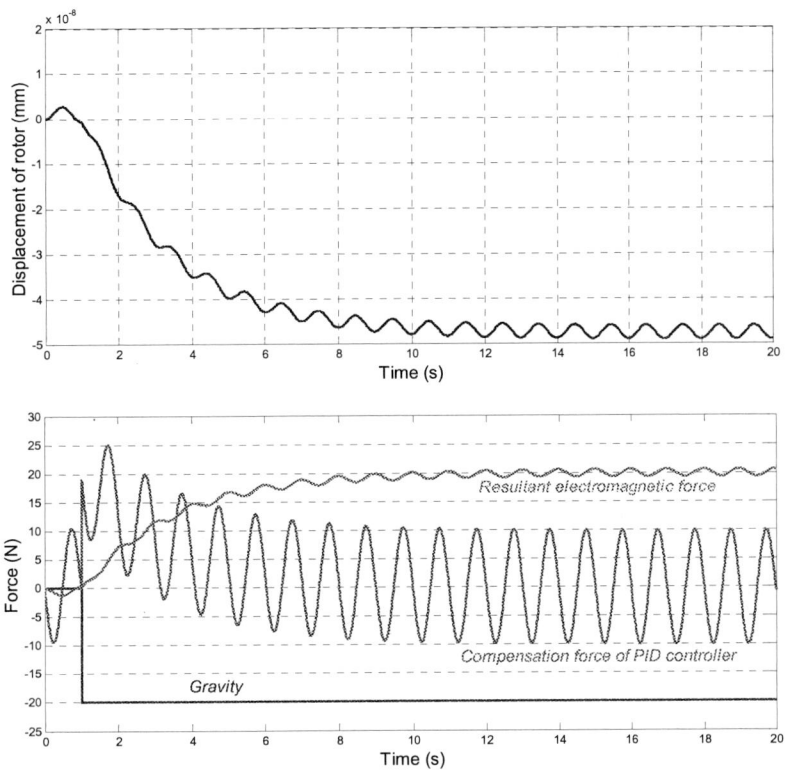

Fig.10 Simulation results for Fig.7 system with PD controller

VI. CONCLUSION

Through studying on a novel radial magnetic bearing proposed in this paper, the following conclusions can be drawn.

The proposed system without PD controller is instable though it is stable in the steady-state.

Using PD control strategy, the system becomes stable in the dynamic state process, but it is a static error system.

Because no displacement sensor needed, the structure of the radial magnetic bearing proposed is simplified and easy to control.

The proposed radial magnetic bearing system is feasible.

ACKNOWLEDGMENT

The work reported in this paper was supported by the National Natural Science Foundation of China(50437010).

REFERENCES

[1] Manfred Morari. N. Lawrence Ricker. Model Predictive Control Tollbox User's Guide. MathWorks. Inc. 1998.

[2] P. B. Deshpande. R. H. Ash. Computer Process Control with Advanced Control Applications, 2nd ed. ISA, 1988

[3] Betschon, F. Knospe, C.R. Reducing magnetic bearing currents via gain scheduled adaptive control Mechatronics, IEEE/ASME Transactions on, Volume: 6 Issue: 4, Dec. 2001 Page(s): 437 -443.

[4] Mukhopadhyay, S.C. Ohji, T. Iwahara, M. Yamada, S. Modeling and control of a new horizontal-shaft hybrid-type magnetic bearing. Industrial Electronics, IEEE Transactions on, Volume: 47 Issue: 1, Feb. 2000 Page(s): 100 -108.

[5] Kasarda, M.E. Clements, J. Wicks, A.L.; Hall, C.D. Kirk, R.G. Effect of sinusoidal base motion on a magnetic bearing. Control Applications, 2000. Proceedings of the 2000 IEEE International Conference on, 25-27 Sept. 2000, Page(s): 144 -149

[6] McMullen P T, S.Huynh C, Hayes R J. Combination radial-axial magnetic bearing [C]. In: Proc. 6th Int. Symp. Magnetic bearings, ETH Zurich, Switzerland, 2000：473-478.

2006 5th International Power Electronics and Motion Control Conference

Flux-Weakening Characteristics of Trapezoidal Back-EMF Machines in Brushless DC and AC Modes

Z.Q. Zhu[1], J.X. Shen[1,2], D. Howe[1]

[1]Department of Electronic & Electrical Engineering, University of Sheffield, Mappin Street, Sheffield S1 3JD, UK
[2]College of Electrical Engineering, Zhejiang University, Hangzhou, P. R. China, 310037

Abstract—**The performance of a brushless motor which has a surface-mounted magnet rotor and a trapezoidal back-emf waveform when it is operated in BLDC and BLAC modes is evaluated, in both constant torque and flux-weakening regions, assuming (a) the same torque, (b) the same peak current, and (c) the same rms current. It is shown that although the motor has an essentially trapezoidal back-emf waveform, the output power and torque when operated in the BLAC mode in the flux-weakening region are significantly higher than that can be achieved when operated in the BLDC mode due to the influence of the winding inductance and back-emf harmonics.**

Keywords-brushless ac, brushless dc, commutation advance, flux weakening, permanent magnet machine

I. INTRODUCTION

Permanent magnet brushless motors are generally classified according to their back-emf waveform, as being either sinusoidal or trapezoidal back-emf machines, as well as by their control strategy, which is usually classified as being either brushless DC (BLDC), in which case the phase current waveforms are essentially rectangular, or brushless AC (BLAC), in which case the phase current waveforms are essentially sinusoidal. Thus, in order to minimize torque pulsations, a machine with a trapezoidal back-emf waveform should be operated in BLDC mode, while a machine with a sinusoidal back-emf waveform should be operated in BLAC mode. For surface-mounted magnet, trapezoidal back-emf machines, maximum torque per ampere and extended speed operation can realized by advancing the commutation angle for both 2-phase, 120° and 3-phase, 180° BLDC conduction modes, as reported in [1][2].

For sinusoidal back-emf machines, it is relatively easier to realize maximum torque per ampere control and extended speed operation since the optimal relationship between d- and q-axis currents can be analytically determined by employing vector control and flux-weakening control strategies [3][4]. However, in practice, it is inevitable that harmonics exist in the back-emf waveform. Various design features may be employed to obtain a sinusoidal back-emf waveform. For example, the stator slots and/or rotor magnets may be skewed, a

distributed stator winding may be employed, the magnets might be appropriately shaped or magnetised, etc. However, while such methods reduce the harmonic content in the back-emf waveform, they also reduce the average torque and increase manufacturing complexity and cost. Therefore, a machine with a non-sinusoidal back-emf waveform may be operated in BLAC mode [5], although its performance, in terms of efficiency and torque ripple, for example, may then be compromised.

In this paper, the performance of a motor having surface-mounted magnets and a trapezoidal back-emf waveform when operated in both BLDC and BLAC modes, Fig. 1, is evaluated theoretically and experimentally, with particular reference to the flux-weakening performance. To complement this investigation, the performance of a motor having interior magnets and a sinusoidal back-emf waveform when operated in BLDC and BLAC modes is compared in a companion paper [6].

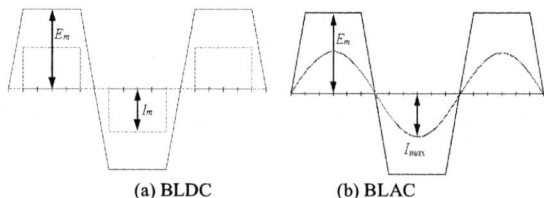

(a) BLDC   (b) BLAC

Figure 1. Schematic illustrating BLDC and BLAC operation of PM motor having trapezoidal back-emf waveform (below base-speed).

II. THEORETICAL COMPARISON OF TORQUE CAPABILITY

It is well known that if both the phase current and back-emf waveforms are ideal, as shown in Fig. 1(a), i.e. the back-emf waveform is trapezoidal with a flat top of at least 120° elec., and the current waveform is rectangular with a conduction angle of 120° elec., the electromagnetic torque of a BLDC motor will be ripple free, and can be expressed as:

$$T_{m\_BLDC} = \frac{2E_m I_m}{\Omega} = \frac{2pE_m I_m}{\omega} \tag{1}$$

1-4244-0448-7/06/$25.00 ©2006 IEEE   908

where p is the number of pole-pairs, and Ω and ω are the mechanical and electrical angular velocity, respectively. Both the phase current and back-emf waveforms are rich in harmonics, as listed in Table 1, interaction between current and back-emf harmonics of the same order resulting in electromagnetic torque. The relative magnitudes of the component torques are also given in Table I. From Table I, it can be seen that:

- There are no third or other triplen harmonics in the phase current waveforms of a machine with a star-connected winding. However, other high order harmonics are significant.

- The most significant harmonic in the phase back-emf waveform is the third harmonic, higher harmonics being relatively low. However, the third-harmonic back-emf does not contribute to the production of electromagnetic torque.

- The electromagnetic torque results predominantly from the interaction between the fundamental components of the phase back-emfs and currents.

TABLE I.   HARMONICS IN IDEALISED BACK-EMF AND PHASE CURRENT WAVEFORMS OF BLDC MACHINE, AND ELECTROMAGNETIC TORQUE DUE TO INTERACTION HARMONICS OF SAME-ORDER

Order of harmonic (v)	Amplitude of back-emf harmonic (E_{mv}/E_m)	Amplitude of phase current harmonic (I_{mv}/I_m)	Torque due to harmonics of same-order (T_{mv}/T_m)
1	121.6%	110.3%	100.6%
3	27.0%	0	0
5	4.9%	-22.1%	-0.8%
7	-2.5%	-15.8%	0.3%
...	...	...	...

TABLE II.   CURRENT AND TORQUE OF TRAPEZOIDAL BACK-EMF MOTOR IN BLDC AND BLAC MODES

Amplitude of 120°-trapezoidal back-emf waveform	E_m		
BLDC mode	Amplitude of rectangular phase current waveform	I_m		
	Electromagnetic torque	$2E_m I_m / \Omega$		
BLAC mode	Amplitude of sinusoidal phase current waveform, I_{max}	$1.096 I_m$	I_m	$1.155 I_m$
	Electromagnetic torque	$2E_m I_m / \Omega$	$1.825 E_m I_m /\Omega$	$2.107 E_m I_m / \Omega$
	Condition	Same torque in BLDC mode	Same peak current as in BLDC mode	Same RMS current as in BLDC mode

The electromagnetic torque in BLAC mode when the motor has a non-sinusoidal back-emf waveform can be expressed as:

$$T_{m\_BLAC} = \frac{3E_{m1}I_{max}}{2\Omega} = \frac{3pE_{m1}I_{max}}{2\omega} \qquad (2)$$

where E_{m1} is the amplitude of fundamental back-emf.

The torque and current of a trapezoidal back-emf motor when operated in ideal BLDC and BLAC modes is given in Table II. In BLAC mode, three cases are considered, viz. (a) the same torque, (b) the same peak current, and (c) the same RMS current as in BLDC mode.

It can be seen that to produce the same torque, the amplitude of the phase current in BLAC mode is 9.6% higher than that for the BLDC mode, while for the same peak current the torque which results in BLAC mode is 17.5% lower. For the same RMS current, i.e. the same copper loss, the phase current in BLAC mode can be increased by 15.5%, which results in the torque being increased by 10.7%.

III.   FLUX-WEAKENING CONTROL AS BLDC DRIVE

The 3-phase, 6-pole, 18-slot, surface-mounted PM brushless motor whose specification is given in Table III, and which has a full-pitched overlapping stator winding, is considered. The phase back-emf waveforms are approximately trapezoidal, as shown in Fig. 2, which also shows the measured third-harmonic voltage between the star-point and neutral. The DC link voltage is 200V, and the rated peak phase current for BLDC operation is $I_m = 3.30A$.

TABLE III.   SPECIFICATION OF BLDC MOTOR

Number of poles:	6	Number of slots:	18
Winding:	Overlapping	Stator skew:	1 slot-pitch
Magnets:	Surface-mounted sintered ferrite	Self- & Mutual-inductances:	20.68mH, -2.81mH
Phase resistance:	3.37 ohm	Rated speed:	830rpm
DC link voltage:	200 V	Rated torque:	4.5Nm
Fundamental back-emf constant (k_{Em1}):		287.3 mV/(elec-rad/s)	
3rd harmonic back-emf constant (k_{Em3}):		64.5 mV/(elec-rad/s)	
5th harmonic back-emf constant (k_{Em5}):		15.6 mV/(elec-rad/s)	
7th harmonic back-emf constant (k_{Em7}):		2.5 mV/(elec-rad/s)	

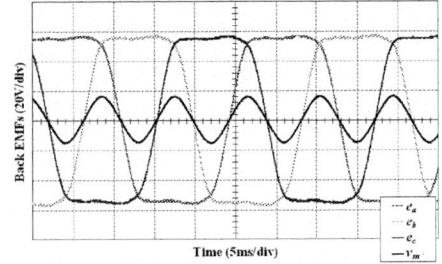

Figure 2.   Back-emf waveforms of 3-phase, 6-pole, 18-slot, BLDC motor.

Flux-weakening control of a BLDC motor is achieved by advancing the commutation, measured torque-speed curves which result with different commutation advance angles being shown in Fig. 3(a), from which the optimal commutation advance angle for maximum torque at any speed is obtained, Fig. 3(b).

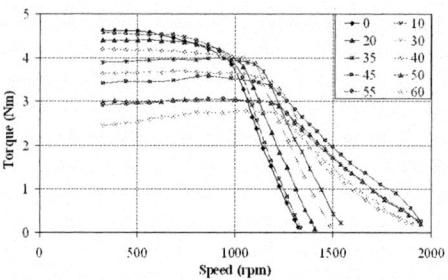

(a) Torque-speed curves for different commutation advance angles

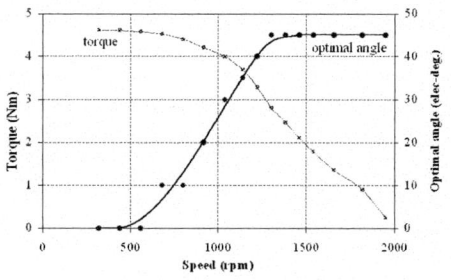

(b) Optimal commutation advance angle for maximum torque

Figure 3. Torque-speed performance of BLDC motor.

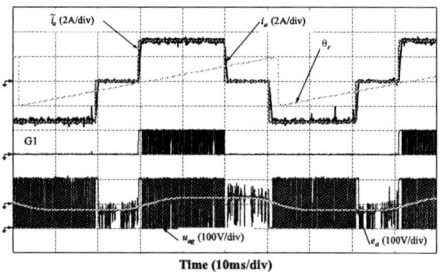

(a) Low speed, 320rpm, zero commutation advance

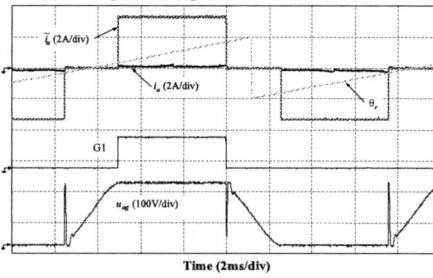

(b) 1320rpm, zero commutation advance

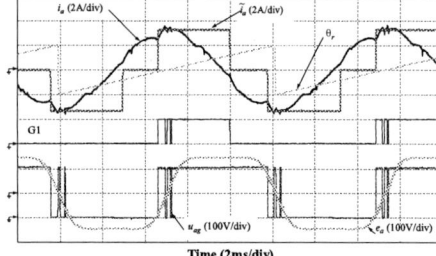

(c) 1950rpm, optimal commutation advance

Figure 4. Measured waveforms in BLDC mode: rotor position (θ_r), reference phase current ($\tilde{i}_a$), actual phase current (i_a), inverter switching signal (G1), terminal voltage (u_{ag}), simulated back-emf (e_a).

As can be seen, at low speed, the optimal commutation advance angle is zero, i.e. the phase current is in phase with the back-emf for maximum torque per ampere, Fig. 4(a). As can be seen, the phase current waveform is essentially rectangular, and the same as the demanded current $\tilde{i}_a$. However, above the base-speed the phase current which results with zero commutation advance is very much lower than the demanded current due to the increase in the back-emf, as shown in Fig. 4(b). With optimal commutation advance, however, the phase current can again be regulated to have an essentially rectangular waveform. However, at yet higher speeds in the flux-weakening mode, the phase current waveform deteriorates and becomes almost sinusoidal due to the influence of winding inductance since the higher order current harmonics are suppressed by their higher reactance as the speed increases, as shown in Fig. 4(c), in which it will also be noted that the peak-to-peak amplitude of the phase back-emf is significantly higher than the DC link voltage. When optimal commutation advance is employed, however, the maximum achievable output power increases to 450W which compares to 400W without commutation advance, as will be shown later, while the maximum speed increases from 1320rpm to 1950rpm.

IV. FLUX-WEAKENING CONTROL AS BLAC DRIVE

Although the flux-weakening control of BLAC machines has been studied extensively [3][4], generally it has been applied to machines which have an essentially sinusoidal back-emf waveform, not a trapezoidal waveform as is the case for the motor under consideration. It is easy to implement, since simple analytical equations exist for the optimal d- and q-axis currents, according to the speed and the motor parameters [4]. The maximum DC link voltage, $U_{max}=2U_{dc}/\pi$, is utilized in the flux-weakening mode. However, in order to compare the relative torque and speed capabilities when the motor under consideration is operated in BLDC and BLAC modes, both with and without optimal flux-weakening, the three criteria cited in section II are used to determine the phase current amplitude I_{max}, viz.: (a) the same electromagnetic torque is produced in the constant torque region in both BLDC and BLAC modes, (b) in both modes, the amplitude of the phase currents are the same, and (c) in both modes the copper loss, and, hence, the rms phase current is the same. The amplitude of the phase currents which corresponds to the foregoing conditions are given in Table IV.

TABLE IV.   AMPLITUDE OF PHASE CURRENTS

Amplitude of phase current of BLDC motor (I_m)	Condition	Amplitude of phase current of BLAC motor (I_{max})
3.30 A	Same electromagnetic torque	3.56 A
	Same phase current amplitude	3.30 A
	Same RMS phase current	3.81 A

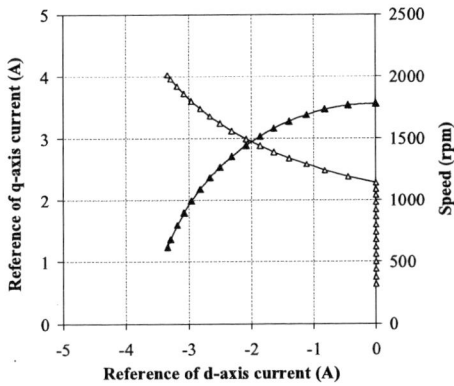

Figure 5. Variation of optimal d-q axis current components with speed.

Fig. 5 shows the variation of the optimal d-and q-axis currents with speed, when I_{max}=3.56A. It can be seen from Figs. 6(a) and (b) that when flux-weakening control is not employed the phase current waveform is essentially sinusoidal when the motor operates below base-speed, but becomes very distorted above base-speed. However, with optimal flux-weakening control the phase current waveform is essentially sinusoidal throughout the speed range, Fig 6(c), and, as can be seen from Fig. 7(a), the maximum attainable speed increases from ~1480rpm to ~2010rpm. Fig. 7 also shows the measured maximum torque-speed curves which result when the motor is operated in BLDC and BLAC modes under the criteria which were cited earlier, viz. I_m = 3.30A, and I_{max}=3.56A, 3.30A and 3.81A, the performance being summarized in Table V.

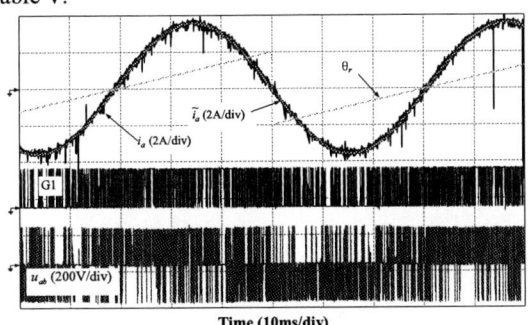

(a) 320rpm, without flux-weakening

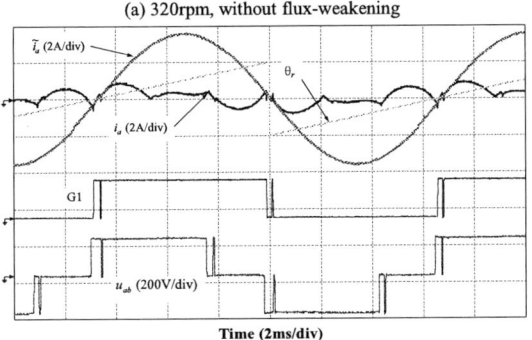

(b) 1480rpm, without flux-weakening

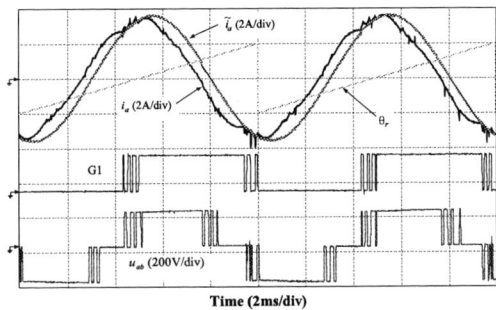

(c) 2010rpm, with optimal flux-weakening

Figure 6. Measured waveforms in BLAC mode when I_{max}=3.56A.: rotor position (θ_r), reference phase current ($\tilde{i}_a$), actual phase current (i_a), inverter switching signal (G1), and terminal voltage (u_{ag})

TABLE V. SUMMARY OF PERFORMANCE WHEN OPERATED IN BLDC AND BLAC MODES

Mode and phase current amplitude	Measured maximum torque	Measured max. speed
		With phase advance /flux-weakening
BLDC, I_m=3.30A	4.6Nm	1950rpm
BLAC, I_{max}=3.56A (+9.6%)	4.6Nm	2010rpm
BLAC, I_{max}=3.30A	4.3Nm (-6.5%)	1940rpm
BLAC, I_{max}=3.81A (+15.5%)	5.0Nm (+8.7%)	2080rpm

Note: Values in () are with reference to BLDC mode

V. COMPARISON OF BLDC AND BLAC MODES OF OPERATION

Fig. 7(a) shows the performance which is achieved when the amplitude of the idealized phase currents waveforms in BLDC and BLAC modes correspond to the same torque capability below base-speed, viz. 3.30A and 3.56A, respectively. As will be seen, above base-speed the torque and power which result in BLAC mode are significantly higher, due to the influence of the winding inductance and back-emf harmonics on the current waveform. Thus, for example, at the rated speed of 830rpm, while the phase current in BLAC mode is close to the reference value, in BLDC mode it is somewhat lower than the reference value. Fig. 7(b) shows the maximum torque/power-speed curves which result when the amplitude of the phase currents is the same in both BLDC and BLAC modes, viz. 3.30A. When flux-weakening is employed above base-speed, both BLDC and BLAC modes exhibit almost the same performance. Below base-speed, however, BLDC operation results in a higher output power for the same RMS phase current. Fig. 7(c) shows the maximum achievable torque/power-speed curves which result with the same RMS phase current in both BLDC and BLAC modes. As will be seen, the power and torque capability in the BLAC mode is significantly higher than that in BLDC mode. As can also be seen in Table IV, since the phase back-emf waveform of the motor under consideration does not have the ideal 120° elec. flat top which was assumed in the derivation of Table II, the variation in the maximum torque capability with the mode of operation differs slightly from that which was predicted earlier.

911

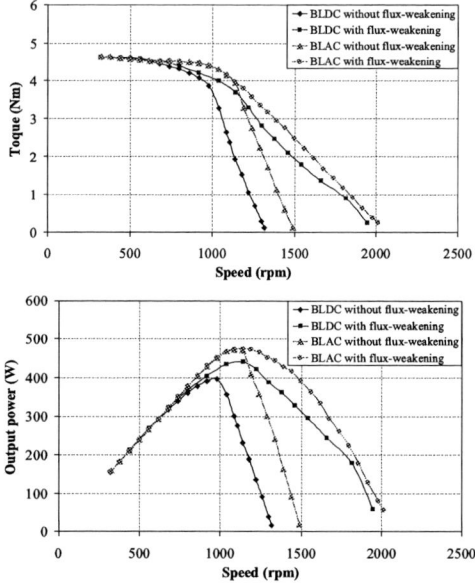

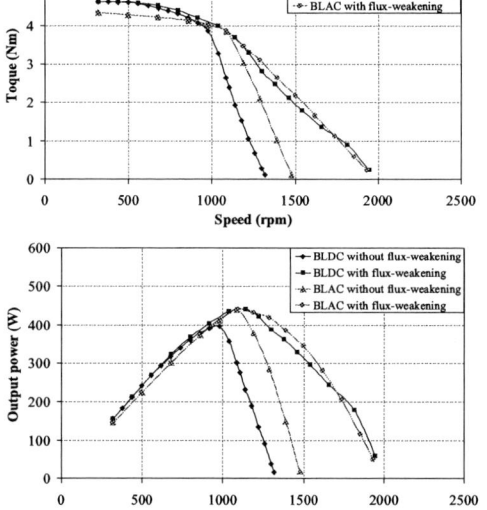

(a) Phase current amplitude in BLDC mode I_m=3.30A and in BLAC mode I_{max}=3.56A

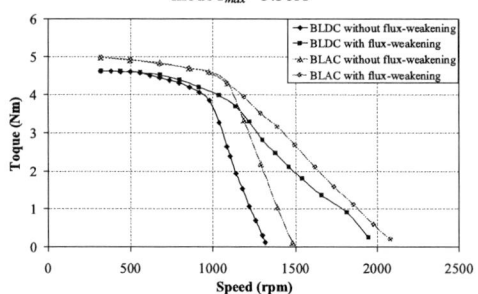

(b) Phase current amplitude in BLDC mode I_m =3.30A, and BLAC mode I_{max} =3.30A

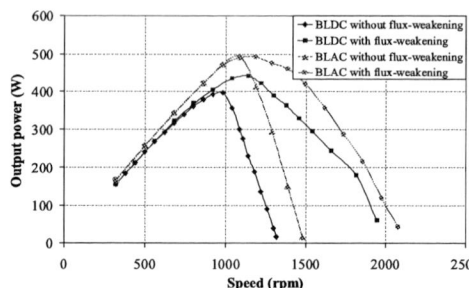

(c) Phase current amplitude in BLDC mode I_m =3.30A, and BLAC mode I_{max} =3.81A

Figure 7.   Performance comparisons when motor is operated in BLDC and BLAC modes.

VI.   CONCLUSIONS

The performance of a permanent magnet brushless motor having a surface-mounted magnet rotor and an essentially trapezoidal back-emf waveform has been determined, when it is operated in both BLDC and BLAC modes in the constant torque and flux-weakening regions, assuming (a) the same torque, (b) the same peak current, (c) the same rms current. The results show that in the flux-weakening region the output power and torque in the BLAC mode are higher than those in the BLDC mode for all cases although the motor has a trapezoidal back-emf waveform.

However, it should be recognized that while a permanent magnet brushless motor with a trapezoidal back-emf waveform can be operated in either BLDC or BLAC mode, other aspects of performance, in terms of efficiency and torque ripple, should also be considered. Nevertheless, it is advantageous to operate such a motor in BLAC mode in the flux-weakening region, in terms of maximizing the torque and speed range.

REFERENCES

[1]   T. M. Jahns, "Torque production in permanent magnet synchronous motor drives with rectangular current excitation," *IEEE Trans. Industry Applications*, Vol. 20, No. 4, 1984, pp. 803-813.

[2]   S.K. Safi, P.P. Acarnley, A.G. Jack, "Analysis and simulation of the high-speed torque performance of brushless DC motor drives," Proc. IEE –EPA, Vol.142, No.3, 1995, pp.191-200.

[3]   T. M. Jahns, "Flux-weakening regime operation of an interior permanent magnet synchronous motor drive," *IEEE Trans. Industry Applications*, Vol. 23, No. 4, 1987, pp. 681-689.

[4]   S. Morimoto, M. Sanada, and Y. Takeda, "Wide-speed operation of interior permanent magnet synchronous motors with high-performance current regulator," *IEEE Trans. Industry Applications*, Vol. 30, No. 4, 1994, pp. 920-926.

[5]   Y. Liu, Z.Q. Zhu, and D. Howe, "Direct torque control of PM brushless AC motors having non-sinusoidal back-emf waveforms," *Proc. 3rd Int. Conf. on Power Electronics, Machines, and Drives, PEMD 2006*, 4-6 April 2006, Dublin, Ireland, pp.425-429.

[6]   Y. F. Shi, Z.Q. Zhu, and D. Howe, "Torque-speed characteristics of interior-magnet machine in brushless AC and DC modes, with particular reference to their flux-weakening performance," *Proc. Int. Power Electronics and Motion Control Conf., IPEMC 2006*, 13-16 August, 2006, Shanghai, China.

A Cost Effective Sensorless Control Method for Permanent Magnet Synchronous Motors Based on Average Terminal Voltage

Cheng-Hu Chen, Wei-Chih Tai, and Ming-Yang Cheng
Department of Electrical Engineering, National Cheng Kung University
E-Mail: mycheng@mail.ncku.edu.tw

Abstract—This paper presents the design, analysis, and implementation of a high performance and cost effective sensorless control scheme for the extensively used Brushless DC Motors (BLDCM), Permanent Magnet Synchronous Motors (PMSM), and Interior Permanent Magnet Synchronous Motors (IPMSM). In the proposed approach, instead of sensing the non-excited back EMF or injecting the additional high frequency switching signals, the commutation signals are extracted directly from the specific average line to line voltages with simple RC circuits and comparators. As a result, the proposed approach is particularly suitable for rectangular current commutation. For the case of sinusoidal commutation, which is desired for PMSMs and IPMSMs, the required continuous position can be obtained from the speed information which is calculated and updated every 60 electric degrees. Moreover, the speed-dependent commutation error caused by the low pass filter and inherent armature inductance is analyzed. An optimal commutation strategy is proposed to keep the armature current as small as possible. Compared with the conventional methods, complex calculation and sensitive machine parameters are not required in the proposed approach. Because of the inherent low cost property, the proposed control algorithm is particularly suitable for air purifiers, air blowers, cooling fans, air conditioners, and related home appliances, etc. Theoretical analysis and various experiments have been conducted to evaluate the effectiveness of the proposed method.

Keywords- sensorless control; permanent magnet synchronous motors; average terminal voltage

I. INTRODUCTION

Taking into consideration torque/speed controllability, future maintenance, and dynamic response, BLDCMs, PMSMs, and IPMSMs are popular choices for various applications [1]-[10]. In fact, due to their excellent performance, Permanent Magnet AC Motors (PMACMs) are considered the best candidates for most small to midsized variable speed applications. To obtain the commutation signals for various PMACMs, the Hall effect sensor is the most popular sensor since it is much more economic than the costly optical encoders or resolvers. However, there are limitations on the installation of Hall effect sensors in some applications [5]-[6]. Firstly, the Hall effect sensors are very sensitive to heat. They become ineffective if the working temperature is higher than 75-100 $^\circ C$. The other limitation of the Hall effect sensors is its size, the installation of the Hall effect sensors will substantially increase the volume of the system.

However, space/volume constraints are important issues for computer and consumer electronics related products. Therefore, there is a strong demand for low cost and easily implemented sensorless control drives.

During the last two decades, much research on sensorless control techniques has been conducted which can be divided into two categories: rectangular current commutation [2]-[6] and sinusoidal current commutation [7]-[10]. Among the various techniques, the back EMF based method is one of the most popular solutions, in which it has been extensively applied in rectangular and sinusoidal commutation [2], [4], [7]. Unfortunately, there are several practical implementation problems concerning the back EMF detection. 1.) In most rectangular commutation methods, the neutral voltage and phase delay circuit are required so that the complexity of the algorithm will be increased. 2.) For the case of sinusoidal commutation, sensitive machine parameters, complex model based calculation, and current feedback are indispensable, in which they will increase the implementation complexity and cost.

In order to cope with the aforementioned problems, a novel sensorless control algorithm based on average terminal voltage was proposed in [6], in which it is particularly suitable for rectangular commutation. In this study, a new sensorless sinusoidal commutation based on the average terminal voltage is proposed. The continuous position required over one electric cycle is estimated with the speed information which is updated every 60 electric degrees. Moreover, the speed-dependent commutation error caused by the low pass filter and inherent armature inductance is analyzed, and the compensation angle is provided to keep the armature current as small as possible. Compared with the back EMF methods and the high frequency injection methods, complex calculation and sensitive machine parameters are not required in the proposed approach. In addition, since the estimation of inductance is not needed, it can be operated in various types of permanent AC motors, e.g., BLDCMs, PMSMs, and IPMSMs. Because of the inherent low cost property, the proposed control algorithm is particularly suitable for cost sensitive products such as home appliances, computer peripherals, automotive components, etc. Theoretical analysis and various experiments have been conducted to demonstrate the effectiveness of the proposed method.

II. PRINCIPLE OF THE NEW SENSORLESS COMMUTATION

A. Rectangular Commutation

Fig. 1 illustrates the back EMF waveform with 120 degree flat top of an ideal BLDCM and the corresponding conduction current. Fig. 2 shows the equivalent circuit of a BLDCM and the inverter topology. In order to regulate the conducting current so that the motor will follow the given velocity or torque command faithfully, the power switches are generally controlled via a high frequency PWM signal. In order to reduce the cost of the gate drive circuit, the boost strap circuit accompanying with the high frequency switching signal (10-20 KHz) imposed on the high side of the inverter is widely used in many home appliances and industry applications. According to the polarity of the armature current as illustrated in Fig. 1, the terminal voltage of each phase can be divided into three sub-sections, i.e., positive conduction current, negative conduction current, and zero conduction current. It should be noted that each motor terminal is inserted between the upper and lower diodes, and each is connected to the positive and negative sides of the DC source. Therefore it can be expected that the maximum and minimum terminal voltage will be fixed between V_{dc} and 0. If the terminal voltages are expressed in the average form, namely in terms of duty ratio, then the switching states can be eliminated. The average terminal voltage of phase "a" is derived in the follows. The same results can be obtained for the other phases with 120 and 240 degrees lagging [6].

$$V_a = D \cdot V_{dc} \qquad \theta_e = 30^o \sim 150^o \qquad (1)$$

$$V_a = \frac{(\theta_e + 30)}{60} \cdot (V_{dc} \cdot D) \qquad \theta_e = -30^o \sim 30^o \qquad (2)$$

$$V_a = (V_{dc} \cdot D) - \frac{(\theta_e - 150)}{60} \cdot (V_{dc} \cdot D) \qquad \theta_e = 150^o \sim 210^o$$

$$V_a = 0 \qquad \theta_e = 210^o \sim 330^o \qquad (3)$$

According to (1)-(3), the average line to line voltage can be expressed as (4) and (5).

$$V_a = D \cdot V_{dc} \qquad \theta_e = 90^o \sim 150^o \qquad (4)$$

$$V_a = -D \cdot V_{dc} \qquad \theta_e = 270^o \sim 330^o$$

$$V_a = (-V_{dc} \cdot D) + \frac{(\theta_e + 30)}{120} \cdot (2V_{dc} \cdot D) \qquad \theta_e = -30^o \sim 90^o \qquad (5)$$

$$V_a = (V_{dc} \cdot D) - \frac{(\theta_e - 150)}{120} \cdot (2V_{dc} \cdot D) \qquad \theta_e = 150^o \sim 270^o$$

According to (4) and (5), the zero crossing points of the average line to line voltage will occur at 30 and 210 degrees, respectively. Fig. 3 shows the phase relationship between the ideal back EMF waveform, the average terminal voltage, and the line to line voltage according to (1)-(5). It is easy to find that the output of the average line to line voltage has an inherent lagging angle of 30 electric degrees compared with the back EMF, namely the zero crossing points of the line to line voltage are in phase with the ideal commutation points. According to the properties of the average terminal voltage, the commutation points can be obtained directly from the three terminal voltages without including the neutral voltage. Moreover, the zero crossing points are obtained in phase with the ideal commutation points, that is, the phase shift circuit can be eliminated.

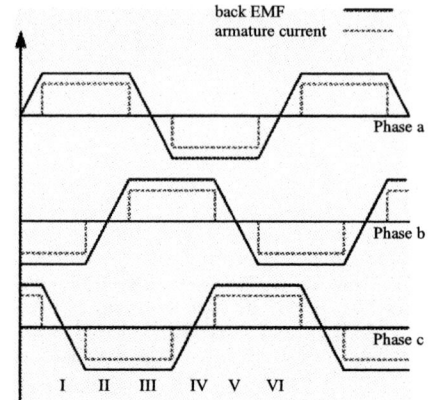

Figure 1. Back EMF and excited current of an ideal BLDCM

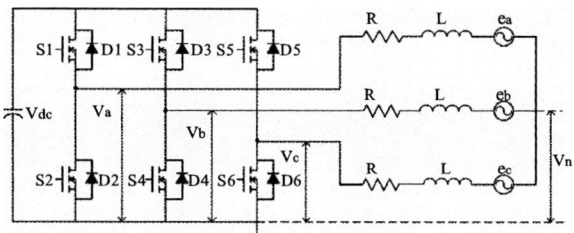

Figure 2. Inverter topology and equivalent circuit of a BLDCM

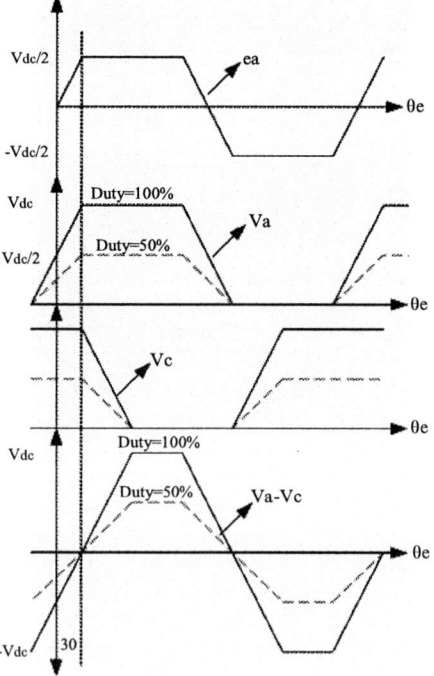

Figure 3. Phase relationship between the back EMF, the average terminal voltage, and the average line to line voltage

B. Sinusoidal Commutation

The rectangular current commutation is known for its simplicity and ease of implementation. However, because of the time delay of the power switches and the effect of armature reaction, the armature current cannot rise/fall instantaneously. Consequently, the periodic noise and torque ripple are generated in each 60 degree commutation process [1]. In high precision motion control applications or in the cases where the torque ripple and noise are the issues of concern, sinusoidal current commutation will be adopted. Compared with the rectangular current commutation, precise and continuous rotor position is required in continuous commutation. Sinusoidal Pulse Width Modulation (SPWM) is one of the most popular modulation schemes to generate the gate signals for the inverter as shown in Fig. 2. When the current command is higher than the carrier signal, the upper legs of the inverter will be turned on. In contrast, the lower legs will be turned on if the command is less than the carrier signal. The instantaneous output voltage of the motor terminal is switched between V_{dc} and ground level. If the high frequency terms are filtered with a proper low pass filter, the average terminal voltage and the average line to line voltage can be expressed in terms of duty ratio as expressed in (6) and (7),

$$V_a = \frac{1}{2} D \cdot V_{dc} + \frac{1}{2} D \cdot V_{dc} \sin(\omega t)$$

$$V_b = \frac{1}{2} D \cdot V_{dc} + \frac{1}{2} D \cdot V_{dc} \sin(\omega t - 120)$$

$$V_c = \frac{1}{2} D \cdot V_{dc} + \frac{1}{2} D \cdot V_{dc} \sin(\omega t - 240)$$

$$V_{ac} = \sqrt{3}/2 \cdot D \cdot V_{dc} \sin(\omega t - 30)$$

$$V_{ba} = \sqrt{3}/2 \cdot D \cdot V_{dc} \sin(\omega t - 150) \qquad (7)$$

$$V_{cb} = \sqrt{3}/2 \cdot D \cdot V_{dc} \sin(\omega t - 270)$$

According to (7), the zero crossing points of the average line to line voltage V_{ac}, will occur at 30 and 210 degrees respectively, which is same as the rectangular current commutation as shown in (5). According to the characteristics of the average terminal voltage as discussed in (1)-(7), the useful commutation signals can be generated with simple RC filters and comparators without the neutral voltage and the phase delay circuit as shown in Fig. 4. The measured commutation signals generated from the average terminal voltage illustrated in Fig. 4 is shown in Fig. 5. It can be seen that each commutation signal is 120 electric degrees apart. Moreover, there is a fixed lagging angle, namely 30 electric degrees compared with the corresponding back EMF as mentioned previously. These commutation signals are particularly useful for rectangular commutation without any additional computation. In what follows, the continuous position of the rotor required for sinusoidal commutation will be calculated between each rising and falling edge of the estimated commutation signals.

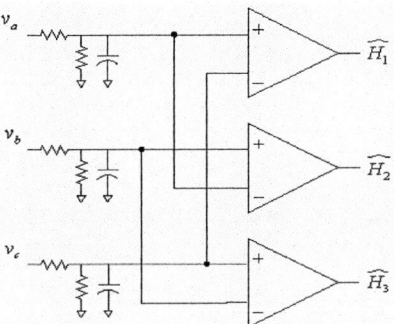

Figure 4. Low cost sensorless commutation signal generator

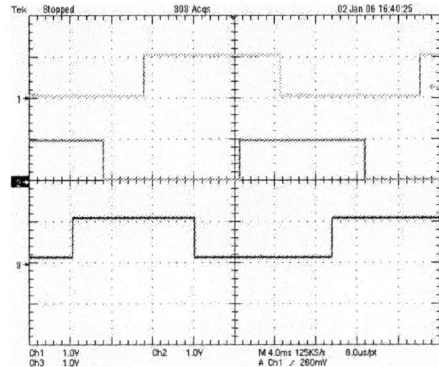

Figure 5. Output of the proposed commutation signal generator

III. POSITION INTERPOLATION AND COMPENSATION

A. Position Interpolation

According to the estimated commutation signals shown in Fig. 5, it can be seen that there are three rising edges and three falling edges between phases over one electric cycle. Namely, there is a fixed 60 electric degrees apart between each rising and falling edge. Therefore, the average angular speed ω_r over each 60 electric degrees can be estimated using (8) if the number of pole pairs P of the rotor is given.

$$\omega_r = \frac{60}{2\pi \cdot P \cdot T_{rf}} (rad/s) \qquad (8)$$

where T_{rf} is the time between every rising and falling edge as shown in Fig. 5.

If the speed does not vary significantly during each T_{rf}, the position between each rising and falling edge can be estimated by (9).

$$\theta_r = \theta_n + \omega_r \cdot \Delta T_s \qquad (9)$$

where $n=1\sim6$, $\theta_1 = 30^0$, $\theta_2 = 90^0$..., $\theta_6 = 330^0$, ΔT_s is the interpolation time period which is determined by the performance of the micro-controller or the digital signal processor used. Equation (9) reveals that the continuous position of the rotor can be obtained if the angular speed between each 60 electric degrees does not change significantly. This assumption is reasonable if the inertia of the system or the number of pole pairs of the rotor is large enough.

B. Commutation error and compensation

According to the position information estimated using (9), the initial position will be reset in each 60 electric degrees. This will improve the estimation accuracy significantly, namely the commutation error will not be accumulated. However the position relationship for each rising and falling edge between the back EMF will change if the rotor speed increases significantly, which is mainly caused by adding the low pass filter as shown in Fig. 4 and Fig. 6 (a). Equation (10) shows the induced phase delay angle by the RC filter in terms of the machine's frequency.

$$\phi_1 = -\tan^{-1}\left(\frac{CR_1R_2}{R_1+R_2}\omega_e\right) \tag{10}$$

In addition, in order to make the copper losses as low as possible, the current lagging angle caused by the armature inductance is estimated using (11).

$$\phi_2 = -\tan^{-1}\left(\frac{L_a}{R_a}\omega_e\right) \tag{11}$$

Both the position errors induced by the low pass filter and the armature impedance in (10) and (11) depend on the fundamental frequency ω_e of the terminal voltage. Therefore, appropriate compensation can be performed if the angular speed of the rotor is known, which can be obtained from (8). According to (8)-(11), the rotor position can be modified using (12) to obtain the optimal commutation process, namely the copper losses will be kept at the minimum level.

$$\theta_r = \theta_n + \omega_r \cdot \Delta T_s + \phi_1 - \phi_2 \tag{12}$$

(a) Low pass filter (b) Armature impedance

Figure 6. Commutation error by the RC filter and armature impedance

IV. EXPERIMENTAL EVALUATION

Fig. 7 shows the experimental setup which consists of a 400W, 4 poles, 3000RPM PMSM with sinusoidal back EMF. The proposed rectangular and sinusoidal sensorless algorithm is implemented by a cost effective 16 bit micro-controller, Microchip dsPIC30F4012. The power stage consists of a 600V/10A IGBT module, Cyntec IM11400, which can be directly interfaced with the micro-controller. Fig. 8 shows the measured instantaneous and average terminal voltages in rectangular and sinusoidal commutation. The waveforms are closely related to the switching signals as analyzed in section III and IV. The oscilloscope channels from top to bottom in Fig. 9 shows the measured average terminal voltage for phase "a" (trace 1) and "c" (trace 2), the average line to line voltage V_{ac} (trace3), and the commutation signal obtained from the proposed circuit shown in Fig. 4. The measured current waveforms are

shown in Fig. 10. It can be seen that both the rectangular and sinusoidal commutation can be well achieved with the proposed algorithm.

Figure 7. Prototype of the proposed cost effective sensorless drive

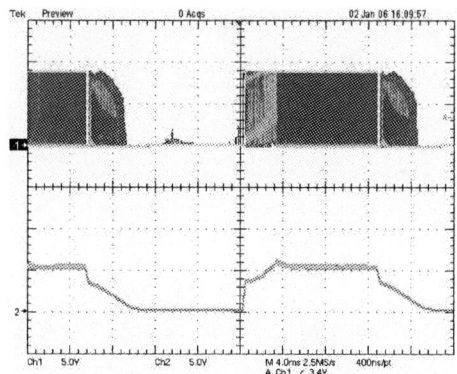

(a) Instantaneous (Ch1) and average (Ch2) terminal voltage, rec. commutation

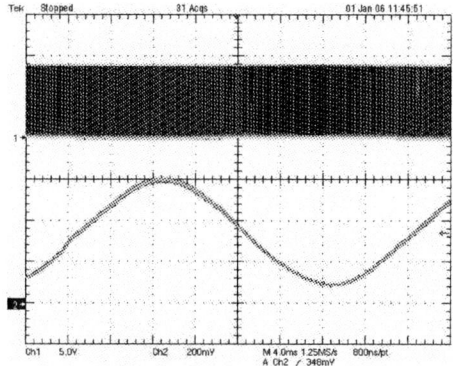

(b) Instantaneous (Ch1) and average (Ch2) terminal voltage, sin. commutation

Figure 8. Measured terminal voltage

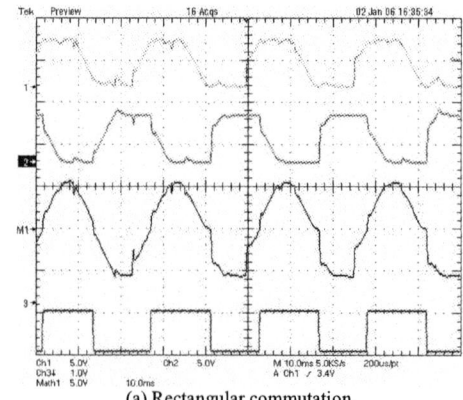

(a) Rectangular commutation

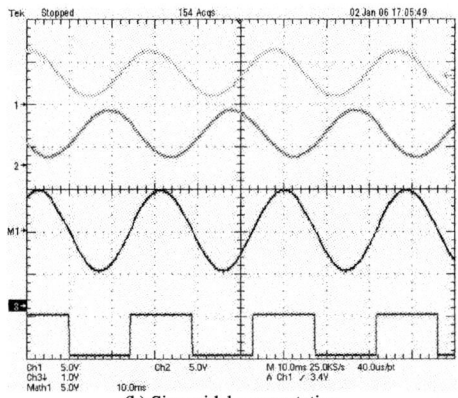

(b) Sinusoidal commutation

Figure 9. Measured terminal voltage and commutation signal
(From top to bottom: average voltage Va, Vc, Vac, commutation signal)

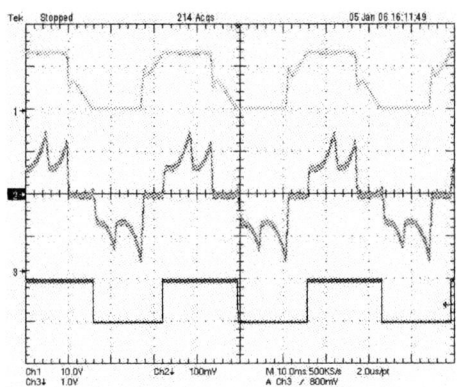

(a) Rectangular commutation

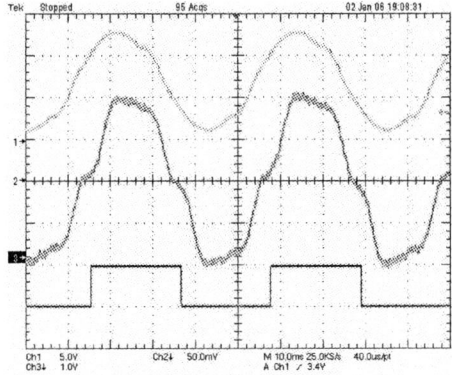

(b) Sinusoidal commutation

Figure 10. Measured current waveforms with rec. and sin. commutation
(From top to bottom: average terminal voltage, armature current,
commutation signal)

V. CONCLUSION

In this study, a novel mechanical and current sensorless control algorithm based on the average terminal voltage is proposed. The presented cost effective circuit can provide useful commutation signals which are well in phase with the output of real Hall effect sensors. Unlike the conventional solutions, the complex phase delay circuits and neutral voltage are not required in the proposed approach. It is particularly useful for BLDC motors with rectangular current commutation. Furthermore, the precise position interpolation based on the estimated commutation signals is exploited in the sinusoidal current commutation to reduce the commutation noises and torque ripples. The commutation error caused by the low pass filter and armature impedance is analyzed, and an optimal commutation process is derived to keep the copper losses at the lowest level. Theoretical analysis and experimental results has verified that satisfactory performance is achieved with the proposed approach.

ACKNOWLEDGMENT

The authors would like to thank the Electric Motor Technology Center of National Cheng Kung University, for their kind support, and providing a number of testing instruments. Special thanks are dedicated to Bryan Liao and Ray Hung for their assistances with this work.

REFERENCES

[1] R. Krishnan, *Electric Motor Drives, Modeling, Analysis, and Control*, Prentice Hall, USA, 2001.

[2] K. Hzuka, H. Uzuhashi, M. Kano, T. Endo, and K. Mohri, "Microcomputer Control for Sensorless Brushless Motor," *IEEE Trans. on Ind. Applicat.*, Vol. 21, No. 4, pp. 595-601, May/June 1985.

[3] J. Moreira, "Indirect Sensing for Rotor Flux Position of Permanent Magnet AC Motors Operating Over a Wide Speed Range," *IEEE Trans. on Ind. Applicat.*, Vol. 32, No. 6, pp. 1394-1401, Nov./Dec. 1996.

[4] J. P. Jahnson, M. Ehsani, and Y. Guzelaunler, "Review of Sensorless Methods for Brushless DC," *in Proc. of IEEE IAS Annu. Meeting Conf.*, Phoenix, USA, pp. 143-150, 1999.

[5] J. Shao, D. Nolan, M. Teissier, and D. Swanson, "A Novel Microcontroller-Based Sensorless Brushless DC (BLDC) Motor Drive for Automotive Fuel Pumps," *IEEE Trans. on Ind. Applicat.*, Vol. 39, No. 6, pp. 1734-1740, Nov./Dec. 2003.

[6] C.-H. Chen, and M.-Y. Cheng, "A new sensorless control scheme for brushless DC motors without phase shift circuit," in *Proc. of the 6th IEEE International Conference on Power Electronics and Drive Systems*, K.L, Malaysia, pp. 1084-1089, 2005.

[7] R. Wu, and G. R. Slemon, "A Permanent Magnet Motor Drive Without a Shaft Sensor," *IEEE Trans. on Ind. Applicat.*, Vol. 27, No. 5, pp. 1005-1011, Sep./Oct. 1991.

[8] M. J. Corley and R. D. Lorenz, "Rotor Position and Velocity Estimation for a Salient-Pole Permanent Magnet Synchronous Machine at Standstill and High Speed," *IEEE Trans. on Ind. Applicat.*, Vol. 34, No. 4, pp. 784-789, July/Aug. 1998.

[9] B. Bae, S. Sul, J. Kwon, and J. Byeon, "Implementation of Sensorless Vector Control for Super-High-Speed PMSM of Turbo-Compressor," *IEEE Trans. on Ind. Applicat.*, Vol. 39, No. 3, pp. 811-818, May/June 2003.

[10] C. Wang, and L. Xu, "A Novel Approach for Sensorless Control of PM Machine Down to Zero Speed Without Signal Injection or Special PWM Technique," *IEEE Trans. on Power Electron.*, Vol. 19, No. 6, pp. 1601-1607, Nov. 2004.

2006 5th International Power Electronics and Motion Control Conference

DSP-based Discrete-Time Reaching Law Control of Switched Reluctance Motor

Ge Baoming and Zhao Nan

School of Electrical Engineering, Beijing Jiaotong University, Beijing 100044, China

Abstract—In this paper, a robust switched reluctance motor (SRM) drive using the speed closed-loop with the discrete time reaching law control and the current closed-loop with nonlinear conversion function from torque to average current is proposed. Firstly, the current closed-loop is implemented by employing the average value of three phase-currents and the digital proportion and integrator, which greatly simplifies the computational procedure. The nonlinear conversion function from the torque to the average current is obtained using experimental measures, which compensates the nonlinear characteristics between the average torque and current. And the discrete time reaching law control as an effective approach to sliding mode control is applied to the speed control of SRM drive, since the discrete time analysis and design are more appropriate for the digital control applications. The design equations are derived in the discrete time and the robustness of the control approach is analyzed. The experiments are carried out on a SRM drive consisting of 5kW SRM and the DSP-based controller, and experimental results verify the proposed drive.

Keywords- reaching law control; sliding mode control; SRM; digital motion control

I. Introduction

Switched reluctance motor (SRM) gradually becomes an important drive motor in the motion control [1] [2]. In comparison with DC motor, asynchronous motor, and synchronous motor, SRM has many specific advantages such as robustness, simple structure, and a wide speed range etc. However, the accuracy and nonlinearity of parameters in the motor model directly influences the drive characteristics when using the traditional control methods. An approximate model will result in a degradation of the drive performances or even an unstable response. In fact, it is very difficult to get the accurate model due to the nonlinearities and uncertainties.

Sliding mode control (SMC) is a powerful technique to control the nonlinear and uncertain systems [3-5], which do not excessively depend on the precision of motor model. As a strongly robust control method, it can be ap-

plied in the presence of model uncertainties, parameter fluctuations and external disturbances and can effectively compensate for the nonlinearity and uncertainties of the plant. Reaching law control (RLC), which is a new SMC design technique, was introduced by Gao and Hung in [6]. This approach not only establishes a reaching condition to the sliding line (or surface) directly but also specifies the dynamic characteristics of the system during the reaching phase. Additional merits of the RLC approach include simplification of the solution for SMC and providing a measure for the reduction of chattering.

However, the nonlinear relationship between the torque and the phase current makes it difficult to effectively control the motor torque. The control system will be very complicated if the torque share function and the instantaneous phase current control are employed, where three current closed-loops are necessary. In fact, we can simplify the current closed-loop using feedback of average value of phase-currents, and as a result a current closed-loop is enough. The function table of average torque versus average current is derived using practical measures. Thus, the nonlinear function links the speed controller to the average current closed-loop system, which improves the performances of whole SRM drive. The control system based on DSP TMS320LF2407 is setup to finish the experimental study. The measured results verify the perfect performance of the designed drive.

II. Digital Speed Control Based on RLC

A. Plant Model

The mechanical dynamic of SRM can be written as

$$\frac{d\omega_r}{dt} = (T_e - B\omega_r - T_l)/J \qquad (1)$$

where T_e is the electromagnetic torque produced by SRM, B denotes the damping coefficient, T_l is the load torque, J is the moment of inertia, and ω_r is the rotor speed.

Let $T_l = 0$, then we obtain

$$G_P(s) = \frac{\omega_r(s)}{T_e(s)} = \frac{1}{Js + B} \qquad (2)$$

When sampling time is T_s, the pulse transfer function is

$$G_p(z) = Z\{G_h(s)G_p(s)\} = \frac{C_p}{z - P_p} \qquad (3)$$

This work is partially supported by the Key Project of Chinese Ministry of Education (2004104051), Delta Science & Technology Educational Development Program (DREG2005006), and in part by the Special Science Foundation of Beijing Jiaotong University (2003SM013).

1-4244-0448-7/06/$25.00 ©2006 IEEE

where $P_p = \exp(-BT_s/J)$ and $C_p = (1-P_p)/B$, $G_h(s)$ represents the zero-order hold.

Obviously the mechanical dynamic is of the first order and SMC is not directly applicable [7]. The system is made the second order by integral compensation, namely, adding an integrator before the mechanical dynamic. Then the plant model is

$$G'_p(z) = \frac{\omega_r(z)}{u(z)} = \frac{C_p T_s z}{(z-1)(z-P_p)} \qquad (4)$$

B. Discrete Time RLC

Because the direct implementation of sgn(.) function in the general SMC results in a chattering problem in the discrete time systems, let us consider a discrete time reaching law without the sgn(.) function:

$$\Delta S(k+1) = -\alpha S(k) \qquad (5)$$

where k is the sampling instant, α is a positive constant, Δ operator is defined as $\Delta x(k+1) = (x(k+1) - x(k))/T_s$, which is supplemented with the condition $\Delta x(0) = 0$.

The switching function is given by

$$S(k) = \lambda e(k) + \Delta e(k) \qquad (6)$$

where

$$e(k) = \omega_{ref}(k) - \omega_r(k) \qquad (7)$$

ω_{ref} and ω_r are the reference and actual speed respectively.

Note that the reaching law (5) basically implies that the switching function S exponentially reduces to zero with a desired dynamics defined by α.

From (6), the dynamic difference equation for ΔS is derived as

$$\Delta S(k+1) = \frac{1+\lambda T_s}{T_s} \Delta e(k+1) - \frac{1}{T_s}\Delta e(k)$$

Similarly, using (4) and (7) the change in the error term is written as

$$\Delta e(k+1) = P_p \Delta e(k) - C_p u(k) + \Delta \omega_{ref}(k+1) - P_p \Delta \omega_{ref}(k)$$

and thus,

$$\Delta S(k+1) = \frac{(1+\lambda T_s)}{T_s}((P_p - \frac{1}{(1+\lambda T_s)})\Delta e(k) - C_p u(k) \\ + \Delta \omega_{ref}(k+1) - P_p \Delta \omega_{ref}(k)) \qquad (8)$$

Setting (8) equal to $-\alpha S(k)$ from (5), and solving for $u(k)$ gives the control law

$$u(k) = \left(\frac{T_s \alpha}{(1+\lambda T_s)C_p}\right)S(k) + \left(\frac{(1+\lambda T_s)P_p - 1}{(1+\lambda T_s)C_p}\right)\Delta e(k) \\ + \frac{1}{C_p}\left(\Delta \omega_{ref}(k+1) - P_p \Delta \omega_{ref}(k)\right) \qquad (9)$$

An advance term $\Delta \omega_{ref}(k+1)$ is seen on the right hand side of (9), but this is not a problem since $\omega_{ref}(k)$ is a known reference input. For simplicity, let us assume that ω_{ref} is a step demand (i.e. $\Delta \omega_{ref}(k+1) = \Delta \omega_{ref}(k) = 0$). The control law then becomes

$$u(k) = KS(k) + K_{eq}\Delta e(k) \qquad (10)$$

where

$$K = \frac{T_s \alpha}{(1+\lambda T_s)C_p} \qquad (11)$$

$$K_{eq} = \frac{(1+\lambda T_s)P_p - 1}{(1+\lambda T_s)C_p} \qquad (12)$$

The second term of (10), $K_{eq}\Delta e(k)$, actually corresponds to the equivalent control, which can be interpreted as the control law that would maintain $\Delta S(k) = 0$ if the dynamics were exactly known [3][4].

If (9) is substituted in (8), a discrete equation for the switching function is obtained

$$S(k+1) = (1-\alpha T_s)S(k) \qquad (13)$$

Since a sufficient condition for the stability [8] is

$$|S(k+1)| < |S(k)| \qquad (14)$$

this requires

$$|1-\alpha T_s| < 1 \qquad (15)$$

It should be noted that if

$$\frac{1}{T_s} < \alpha < \frac{2}{T_s} \qquad (16)$$

then $1-\alpha T_s$ becomes a negative number and S will have damped chattering.

S will exponentially reduce to zero if

$$0 < \alpha \le \frac{1}{T_s} \qquad (17)$$

which requires

$$0 < K \le K_m \qquad (18)$$

$$K_m = \frac{1}{(1+\lambda T_s)C_p} \qquad (19)$$

Both (18) and (19) clearly show the restriction on K. The robustness is inversely proportional to T_s, as expected.

III. DIGITAL CURRENT CONTROL

The control $u(k)$ represents an incremental torque, and the desired total torque can be obtained by the digital integrator. The current loop control ensures the actual average torque following the desired torque if the feedback current of SRM with three phase-windings is

$$I_{sum} = \frac{1}{3}\sum_{j=1}^{3} i_j = F(T_e) \qquad (20)$$

where the function $F(T_e)$ is derived from the experimental measures, and i_j denotes the current of phase-j.

PI current controller with fixed switching frequency PWM cooperates with a suitable switching scheme to achieve the current tracking around the optimum current reference. The output of the PI current controller, after being normalized, is used as the duty cycle for PWM signal of the converter. PI controller in Fig.1 is most suitable for regulation problems, where zero steady-state error is the main performance criterion. Its digital implementation is written as

$$U_k = K_p e_{I,k} + K_i e_{I,k} + K_i \sum_{n=0}^{k-1} e_{I,n} \qquad (21)$$

with

$$e_{I,k} = I_{ref}(k) - I_{sum}(k)$$

$$K_i = K_p T / T_I$$

where T is the sample time of the current loop, T_I is the time constant of integrator, and K_p is the proportion coefficient.

When the maximum and minimum limited values are used to PI control, the current controller in Fig.1 is obtained.

The control algorithm of PI controller is summarized as
a) Input reference and feedback signals I_{ref} and I_{sum};
b) Calculate error $e_{I,k} = I_{ref} - I_{sum}$;
c) Calculate control $U_0 = x_i + K_p \, e_{I,k}$;
d) Set $U_k = U_0$;

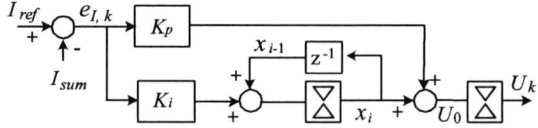

Fig.1. PI controller with the limited saturation.

e) If $U_0 > U_{max}$, then $U_k = U_{max}$;
f) If $U_0 < U_{min}$, then $U_k = U_{min}$;
g) Output U_k;
h) Calculate integral action $x_i = x_{i-1} + K_i e_{I,k}$;
i) If $x_i > x_{max}$, then $x_i = x_{max}$;
j) If $x_i < x_{min}$, then $x_i = x_{min}$;
k) Return to a).

where U_{max} and U_{min} are the limited values of controller output, and x_{max} and x_{min} are the limited values of integrator.

IV. CONTROL SYSTEM AND EXPERIMENTS

The proposed control system of SRM is shown in Fig. 2, which contains the speed and current closed-loops. RLC control is applied to the speed controller, and PI control is useful in the current controller. The data table of conversion function $F(T_e)$ is stored in the memory. In our experiments, the prototype is a 5 kW SRM with 12/8 poles, three phases, which is fed by a classic asymmetrical half-bridge IGBT inverter, and the current and speed controllers are implemented using DSP TMS320LF2407. The hardware of controller provides all kinds of feedback signals such as phase currents, dc bus voltage, temperature, position for commutation, and given reference, as shown in Fig. 3. Current protection will limit the current value within an allowable scope. The over current signal will interrupt the work of controller if it is larger than an upmost value. Fig. 4 shows picture of the controller hardware.

In the experiments, a controlled dc machine is used as the load to provide the required load torque. The parameters of prototype SRM include the rated power of 5kW, the rated speed of 3000 r/min, the rated dc voltage of 200V, and the moment of inertia J=0.003833 kg.m^2.

The experimental results are shown in Figs. 5-8, respectively. Figs. 5 and 6 show the SRM currents when the control parameters of current closed-loop are different. We can find that the increasing of K_p will improve the dynamic response of current, and the phase current of Fig. 5 (b) rises more quickly than that of Fig. 5 (a). In addition, the phase current cannot reach the reference value in Fig. 5 due to the integral action K_i =0, moreover the offset is larger for the smaller K_p in the case of K_i =0, as shown in Fig. 5 (a) and (b). The actual current tracks the reference

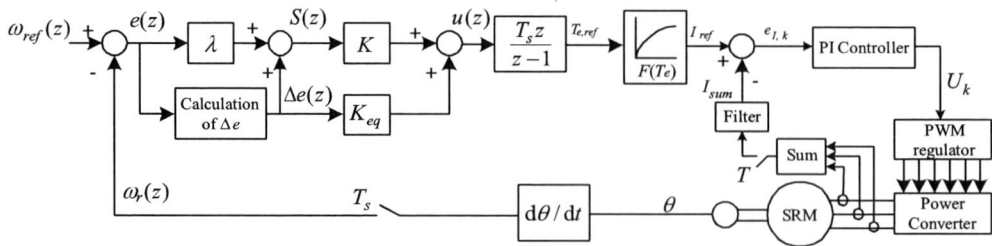

Fig. 2 . SRM control system based on RLC.

920

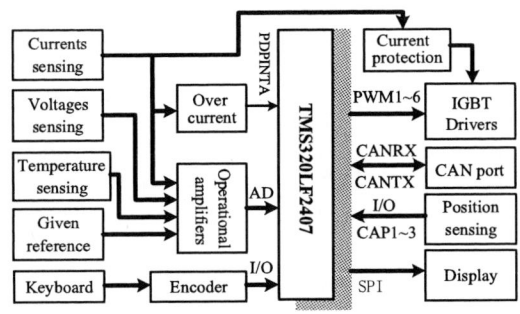

Fig. 3. Structure of controller.

Fig. 4. Controller hardware based on DSP.

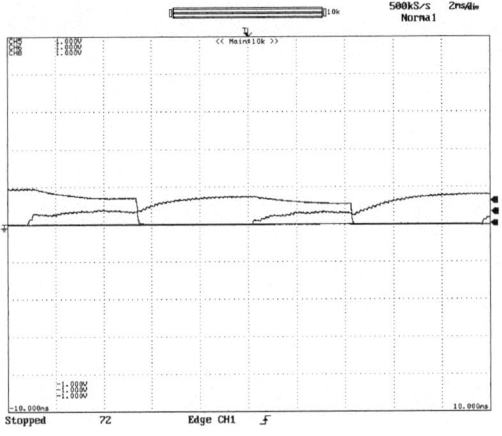

(a) K_p=0.0625

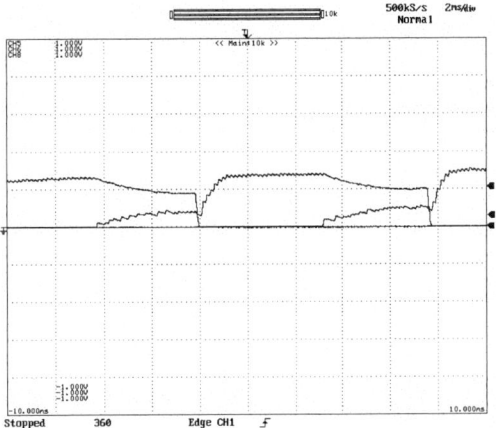

Fig. 6. Measured current waves when K_i =0.004 and K_p =0.25.

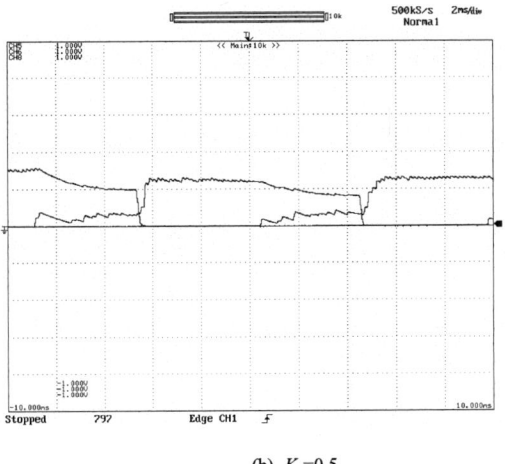

(b) K_p=0.5

Fig. 5. Measured current waves when K_i=0.

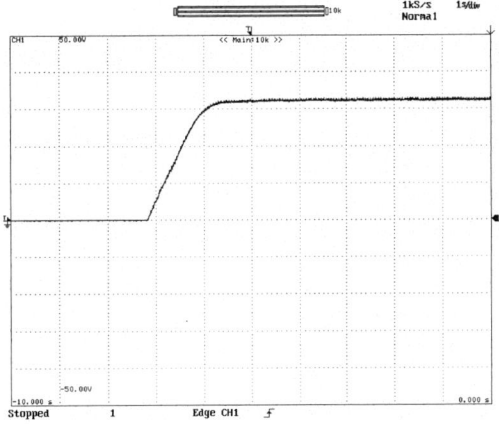

Fig. 7. Speed response of the proposed SRM drive.

value in the zero offset when the integral action is effective in Fig. 6.

The performance of speed closed-loop is tested when the given speed is 1188r/min and the load torque is 30N.m,

and Fig. 7 shows us the speed response of the proposed drive. For the comparison with the conventional PI control, Fig. 8 presents the result when PI control is used as the speed controller. The RLC controller provides a better speed response than PI control. As a result, there is not

921

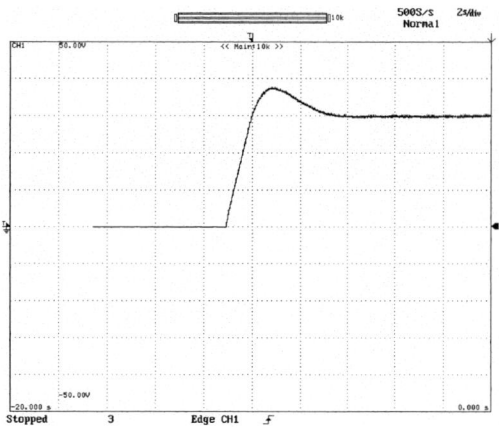

Fig. 8. Speed of SRM drive using conventional PI control.

overshoot in Fig. 7, and it takes about one second from the starting to the steady state. However, PI control results in a big overshoot and the motor needs about four seconds from the starting to the steady operation. A lot of results show that the increase of K makes the drive dynamic quick, and the increasing λ can reduce the speed response time. However, for $K>2K_m$, the system becomes unstable.

V. CONCLOSION

In this paper, a discrete time RLC method, which is an approach to SMC design, is applied to the speed control of a SRM drive system since the discrete time analysis and design are more appropriate for the practical implementations. The average current control is applied to the current control of the drive and a nonlinear conversion function between the torque and the average current is derived using experimental measures, which not only simplifies the design procedure of the controllers, but also improves the performance of the drive. DSP-based control system was constructed in order to implement the proposed algorithms. The experimental SRM prototype of 5kW is employed for the purpose of verification. The perfect dynamic and static performances of the designed drive were verified by using experimental tests. It is shown that the RLC provides better performances than the conventional PI control. The increase of K makes the drive dynamic quick, and the increasing λ can reduce the speed response time. However, for $K>2K_m$, the system becomes unstable.

REFERENCES

[1] Ge Baoming, Wang Xiangheng, Su Pengsheng, and Jiang Jingping, "Nonlinear internal model control for switched reluctance drives," *IEEE Trans. on Power Electronics*, vol. 17, pp.379-388, May 2002.

[2] P.C. Kjaer, J.J. Gribble, and T.J.E. Miller, "High-grade control of switched reluctance machines," *IEEE Trans. Ind. Applicat.*, vol.33, pp.1585-1593, Nov. 1997.

[3] J.J.E. Slotine and W. Li, *Applied Nonlinear Control*. Prentice-Hall, 1991.

[4] J.Y. Hung, W.B. Gao, and J.C. Hung, "Variable structure control: A survey," *IEEE Trans. Ind. Electron.*, vol. 40, no. 1, pp. 2-22, 1993.

[5] M. Morari and E. Zafiriou, *Robust Process Control*. Englewood Cliffs, NJ: Prentice-Hall, 1989.

[6] W.B. Gao and J.C. Hung, "Variable structure control of nonlinear systems: A new approach," *IEEE Trans. Ind. Electron.*, vol. 40, no. 1, pp. 45-55, 1993.

[7] E. Y. Y. Ho and P. C. Sen, "Control dynamics of speed drive systems using sliding mode controllers with integral compensation," *IEEE Trans. Ind. Applicat.*, vol.27, pp.883-892, 1991.

[8] S.Z. Sarpturk, Y. Istefanopulos, and O. Kaynak, "On the stability of discrete-time sliding mode control systems," *IEEE Trans. Automat. Contr.*, vol. 32, no. 10, pp. 930-932, 1987.

2006 5th International Power Electronics and Motion Control Conference

Digital Control System on Bearingless Permanent Magnet-type Synchronous Motors

Jianming Deng, Huangqiu Zhu and Yang Zhou

Jiangsu University/School of Electrical and Information Engineering, Zhenjiang 212013, China

jxdjm@sohu.com

Abstract—The bearingless permanent magnet synchronous motor (BPMSM) is an innovational type of motor, which has all excellence of magnetic bearings. The BPMSM is a high-order, nonlinear and strong coupling multivariable system. Bearingless motors have a set of torque windings and a set of additional suspension force windings in the stator slots. In order to realize the rotor suspending and motor operation steadily, decoupling control between the radial forces and torque force is the chief precondition. In the paper, based on the basic working principle of BPMSM, the mathematics models of the radial forces and rotation part on BPMSM are deduced by adopting rotor magnetic field oriented control (FOC) strategy. A digital vector control system according to the demand of decoupling is designed by using DSP (TMS320LF2407) technique. The hardware and the software of the digital control system are developed in this paper. The experiment results have shown that the steady suspension of the rotor is realized and the speed of the rotor can be continuously adjusted within the range of 0 - 5 000 r/min.

Keywords-bearingless motor; permanent magnet-type synchronous motor; mathematics model; decoupling control; digital control

I. INTRODUCTION

As bearingless motors have all the advantages of magnetic bearings, such as no mechanical contact, no friction, no need of lubrication, high speed and high precision, they have good application prospects in life science, chemical processing industry, semiconductor manufacturing industry, and so on [1]-[7]. The BPMSM is a new kind of motor on which the principle of magnetic bearing is applied. In order to realize the rotor suspension and speed regulation for BPMSM, independent control of the radial forces and torque force based on the decoupling algorithm is required. So the key technology is how to realize the independent control of the radial forces and torque force for BPMSM. In the paper, decoupling algorithm dealing with torque force and radial forces is deduced by adopting rotor magnetic field oriented control

The project was supported by National Natural Science Foundation of China (50275067), High technology research of Jiangsu Province (BG2005027), and SRF for ROCS, SEM.

strategy and digital control system using TMS320LF2407 is designed on the basis of the algorithm. The hardware and software of the digital control system are designed to meet the need and experiments are completed.

II. WORKING PRINCIPLE OF BPMSM AND MATHEMATICS MODEL

A. Working Principle of BPMSM

Aside from torque windings, the BPMSM has a set of additional suspension force windings in the stator slots. In order to produce controllable suspension forces, the pole pairs relationship between torque windings P_1 and suspension force windings P_2 should be $P_2=P_1\pm1$[2], [3]. As shown in Fig. 1, the torque windings consist of N_a and N_b with pole pairs 2, the suspension force windings consist of N_x and N_y with the pole pairs 1. When the rotor is located in the center with no current in the suspension force windings, symmetrical 4-pole flux ϕ_p can be produced by permanent magnet and the flux density in the each airgap is equal. So no suspension forces are produced. However, according to electromagnetic field theory, when the rotor has a displacement towards negative direction of *x*-axis, the symmetrical distributes of the flux will be broken in the motor airgap. So a Maxwell-Force which is positive direction in *x*-axis is generated. This magnetictractive force will increase as rotor displacement increases. In order to insure the rotor come back to balanced position, the magnetic flux in airgap 1 and 3 must be regulated. If the current of suspension force windings N_x is positive, 2-pole flux ϕ_x is

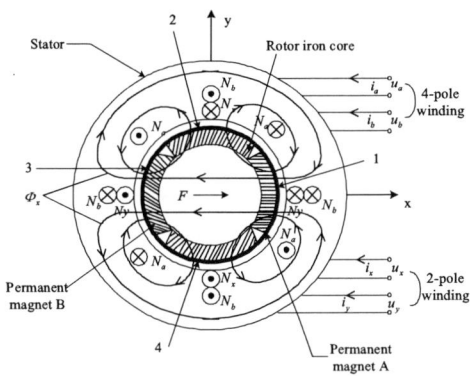

Figure 1.   Principles of radial force production

generated. As a result, the flux density in airgap 1 is increased while the flux density in airgap 3 is decreased. Therefore, a positive directional force is produced to draw the rotor back to the center. When the rotor has a displacement towards negative direction of x-axis and the current of the suspension force windings N_x is negative, the flux density in airgap 1 is decreased and the flux density in airgap 3 is increased. The radial forces in the negative direction of x-axis are produced. Similarly, the radial forces can be produced by controlling the current of the suspension force windings N_y. So the direction and magnitude of the radial forces can be adjusted by the currents of N_x and N_y.

B. Mathematics Model of BPMSM

In d-q coordinate system, the current and flux linkage of BPMSM can be expressed by the components of 2-phase d-axis and q-axis coordinate, respectively. When the suspension force windings input the currents, the mathematical expressions between the radial forces and the currents of suspension force windings can be written as

$$\begin{cases} F_{ix} = (K_M \pm K_L) \cdot (i_{2d} \cdot \psi_{1d} + i_{2q} \cdot \psi_{1q}) \\ F_{iy} = (K_L \pm K_M) \cdot (i_{2q} \cdot \psi_{1d} - i_{2d} \cdot \psi_{1q}) \end{cases} \quad (1)$$

where F_{ix}, F_{iy} are the radial forces which are composed of Maxwell-Force and Lorentz-Force produced by the currents of the suspension force windings. K_M is Maxwell-Force constant. K_L is Lorentz-Force constant. i_{2d}, i_{2q} are the currents of suspension force windings in the 2-phase d-axis and q-axis coordinate, respectively. ψ_{1d}, ψ_{1q} are the airgap flux linkage of torque windings in the 2-phase d-axis and q-axis coordinate, respectively.

In addition, according to electromagnetic field theory, when the rotor has an off-center displacement, the Maxwell-Force F_{sx}, F_{sy} which are in proportion to off-center displacement will be put on the rotor. They are inherent forces and can be written as

$$\begin{cases} F_{sx} = k_s x \\ F_{sy} = k_s y \end{cases} \quad (2)$$

where $k_s = k \cdot \dfrac{\pi r l B^2}{\mu_0 \delta}$. k_s is the force-displacement coefficient. μ_0 is the vacuum permeability. δ is the airgap length. k is the attenuation factor, $k \approx 0.3$ [2].

So the radial forces F_x and F_y of x- and y-direction can be expressed as

$$\begin{cases} F_x = F_{ix} + F_{sx} \\ F_y = F_{iy} + F_{sy} \end{cases} \quad (3)$$

Substituting (1), (2) into (3), while (3) can be written as

$$\begin{cases} F_x = (K_M \pm K_L) \cdot (i_{2d} \cdot \psi_{1d} + i_{2q} \cdot \psi_{1q}) + k_s \cdot x \\ F_y = (K_L \pm K_M) \cdot (i_{2q} \cdot \psi_{1d} - i_{2d} \cdot \psi_{1q}) + k_s \cdot y \end{cases} \quad (4)$$

When $P_2 = P_1 + 1$, (4) can be written as

$$\begin{cases} F_x = (K_M + K_L) \cdot (i_{2d} \cdot \psi_{1d} + i_{2q} \cdot \psi_{1q}) + k_s \cdot x \\ F_y = (K_L + K_M) \cdot (i_{2q} \cdot \psi_{1d} - i_{2d} \cdot \psi_{1q}) + k_s \cdot y \end{cases} \quad (5)$$

When $P_2 = P_1 - 1$, (4) can be written as

$$\begin{cases} F_x = (K_M - K_L) \cdot (i_{2d} \cdot \psi_{1d} + i_{2q} \cdot \psi_{1q}) + k_s \cdot x \\ F_y = (K_L - K_M) \cdot (i_{2q} \cdot \psi_{1d} - i_{2d} \cdot \psi_{1q}) + k_s \cdot y \end{cases} \quad (6)$$

As for the rotary part, in order to concentrate on the main problem, assumed the motor is linear, parameter cannot be changed with temperature and some minor factors need to be neglected such as hysteresis, vortex wastage, and so on. In d-q coordinate system, stator voltage equation, stator flux linkage equation and torque equation can be obtained [8].

The stator flux linkage equation is as follows

$$\begin{cases} \psi_{1d} = L_d i_{1d} + \psi_r \\ \psi_{1q} = L_q i_{1q} \end{cases} \quad (7)$$

where ψ_{1d} and ψ_{1q} are airgap flux linkage. ψ_r is rotor flux linkage. L_d and L_q are the self-inductance of motor windings in the 2-phase d-axis and q-axis coordinate, respectively.

The stator voltage equation can be written as follows

$$\begin{cases} u_{1d} = p\psi_{1d} - \omega\psi_{1q} + r_1 i_{1d} \\ u_{1q} = p\psi_{1q} + \omega\psi_{1d} + r_1 i_{1q} \end{cases} \quad (8)$$

where u_{1d} and u_{1q} are the voltages of motor windings in the 2-phase d-axis and q-axis coordinate, respectively. r_1 is resistance of stator torque windings. ω is mechanical rotational angular speed.

The torque equation can be written as follows

$$T_{em} = p_1(\psi_{1d} i_{1q} - \psi_{1q} i_{1d}) = \frac{J}{p_1} \cdot \frac{d\omega}{dt} + T_L \quad (9)$$

where T_{em} is the electromagnetic torque. J is moment of inertia. P_1 is pole pairs of torque windings.

As for the BPMSM, the position of rotor magnetic field can be obtained by detecting the position of rotor. In the case of rotor magnetic field oriented control ($i^*_{1d}=0$), as shown in (7) and (9), due to rotor flux linkage invariablenes, controlled current i_{1q} in q-axis can be used to control electromagnetic torque T_{em}. By coordinate transformation, the current of stator windings can be

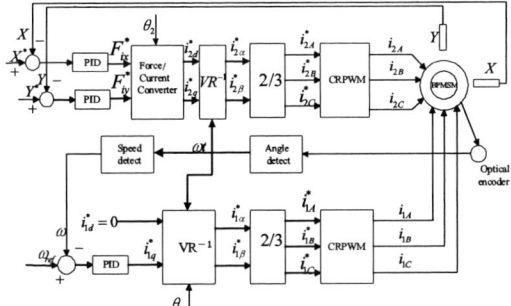

Figure 2. Vector control system of bearingless permanent Magnet-type synchronous motor

decomposed into the rotation coordinate system where the d-axis is fixed up the position of rotor flux linkage ψ_r. When rotor magnetic field oriented control strategy ($i_{1d}^*=0$) is adopted, the mathematics model of the rotary part in motor can be written as

$$\begin{cases} \psi_{1d} = \psi_r \\ \psi_{1q} = L_{qs}i_{1q} \\ u_{1d} = -\omega\psi_{1q} = -\omega L_{qs}i_{1q} \\ u_{1q} = \omega\psi_r + r_1 i_{1q} + L_{qs}pi_{1q} \\ T_{em} = p_1\psi_r i_{1q} \end{cases} \tag{10}$$

According to the mathematics model, a vector control system block diagram is given as shown in Fig. 2.

In Fig. 2, θ_1 is the initial angle between d-axis and A-phase windings of stator，and θ_2 is the load angle.

III.   CONFIGURATION OF DIGITAL CONTROL SYSTEM

The digital control system of the BPMSM consists of hardware system and software system.

A.   Hardware of Control System

The hardware of control system consists of a BPMSM, DSP controller TMS320LF2407 [9]-[10], CRPWM (current regulate pulse width modulation) converter, sensors and interface circuits, and so on. As the control system requires high precision and high performance, TMS320LF2407 is one of the fix-point DSPs TMS320C2000™ platform, which has advantages of low-cost, low-power, and high performance in processing capability. It has been widely used in the motor control fields. Speed sensor is the photoelectric encoder who can generate two pulse signals of variable frequency and 90° fixed phase difference. When rotor rotates a circle, speed sensor will output a Z pulse signal. The pulse numbers and frequency can be used for calculating the rotor mechanical position and speed. In order to get the precise radial displacement information, two displacement sensors with high precision are built in each direction to carry out differential measurement.

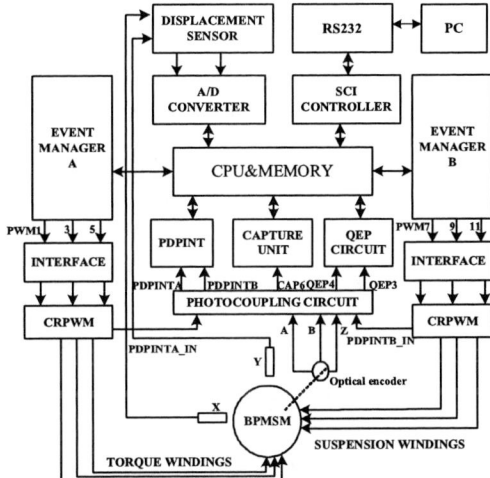

Figure 3.   Digital control configuration diagram

Fig. 3 shows the digital control configuration diagram. The converting time of 10-bit A/D converter with sample and hold circuit is 375 ns. ADCIN01 and ADCIN02 receive the output signal of the displacement sensor. The quadrature encoder pulse (QEP) circuit detects two pulse signals generated by the photoelectric encoder. After CRPWM fault signals are isolated, they will input power device protection interrupt pins (PDPINTA and PDPINTB). When overtemperature, overcurrent or power fault occurs, system interrupt is triggered. Because the typical working voltage of DSP is 3.3 V, interface circuit and photocoupler are designed to realize the isolated connection between different voltage levels.

B.   Software of Control System

As a real time control system needs precision and appropriate answer time, software design plays a very important role in the performance of the whole control system. Applied modular programming has greatly facilitated development and verification of complex programs.

The program is based on two modules: the initialization module and the interrupt module. As shown in Fig. 4, the initialization module executes only once when system starts to perform the following tasks: hardware initialization, variables initialization, interrupt vectors definition, rotor initial position and waiting loop. The interrupt module performs the protection and control of the whole control system.

According to Shannon Theory, considering the operation time of the software as well as the count and control precision, the sampling period including algorithm time and background loop time is set to 136 μs. The sampling period is established by setting the timer period T2PER to 2048 when its working frequency is 30 MHz. This timer is set up up-down count mode and generates a periodical interrupt when underflow event happens.

TMS320LF2407 CPU supports one non-maskable interrupt (NMI) and six prioritized maskable interrupt request (INT1-INT6) at the core level. Since C240X CPU does not have sufficient capacity of generating one or more interrupt requests at the core level, a peripheral interrupt controller (PIE) is required to arbitrate the interrupt request from various sources. When INT1 which is power device protection interrupt is triggered, the corresponding interrupt service program can lock the output of PWM pins. When T_2 underflow event happens, INT3 performs the control program of torque windings and radial suspension force windings. INT4 is capture interrupt triggered by the Z signal of photoelectric encoder.

Fig. 5 shows INT3 flowchart of interrupt service routine. When T_2 underflow event happens and necessary context is protected, pdpinta_flg and pdpintb_flg decide whether the program will perform next step or not. Their initialization value is zero. When INT1 is triggered, the

value of pdpinta_flg and pdpintb_flg is set to 1. In calculating rotation speed subroutine, the rotation speed is calculated one time every other 60 sampling periods. The variable *speedstep* counts the sampling period from 60 to 0.

Incremental speed proportional integral regulator with saturation and amplitude limit produces the stator torque current i^*_{1q}. The inverse Park coordinate transformation and the inverse Clark coordinate transformation is applied to obtain pulse width modulate wave of 3-phase stator currents (i^*_{1A}, i^*_{1B}, i^*_{1C}). For the suspension force windings, ADCIN01 and ADCIN02 receive the x and y displacement output signals of the displacement sensor and convert x and y displacement signals. A digital filter eliminated error and interference signal has been designed. Relying on the fast dynamic characteristic of radial displacement, incomplete differential PID control is introduced. Force-current conversion model is applied to obtain two-phase currents (i^*_{α}, i^*_{β}). After inverse Clark transformation, two-phase currents (i^*_{α}, i^*_{β}) can be converted into 3-phase currents (i^*_{2A}, i^*_{2B}, i^*_{2C}). After 3-phase currents input the CRPWM, they will be used to drive 3-phase suspension force windings.

IV. EXPERIMENT RESULTS

The parameters of the prototype motor are given as follows: the nominal power P_N=1 kW, the rated speed n=5 000 r/min, pole pairs of torque windings P_1=2 and pole pairs of suspension windings P_2=3, mass of rotor m=2.85 kg, stator resistance r_1=2.01 Ω, moment of inertia J=0.00769 kg·m^2, stator inductance $L_d = L_q$=0.008 H, radial airgap of outboard bearing δ=0.25 mm.

Fig. 6 and Fig. 7 show the experimental waveforms of x- and y- direction displacements, current of suspension force windings i_{1A} and current of torque windings i_{2A} when the rotor speed is n=1 200 r/min and n=3 000 r/min under no-load condition, respectively. The results show that the steady suspension of rotor is realized and the vibration amplitude in x and y direction is less than 20 μm.

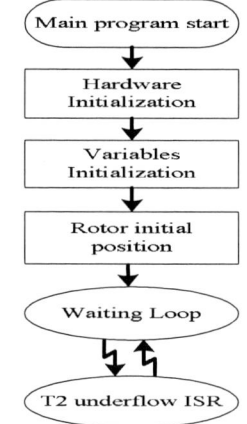

Figure 4. Flowchart of system initialization

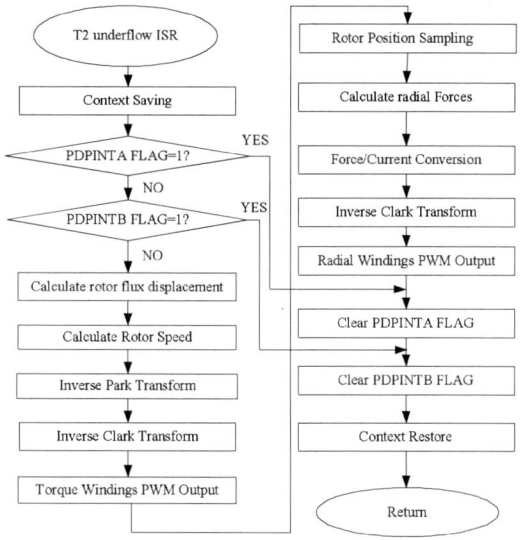

Figure 5. INT3 flowchart of BPMSM

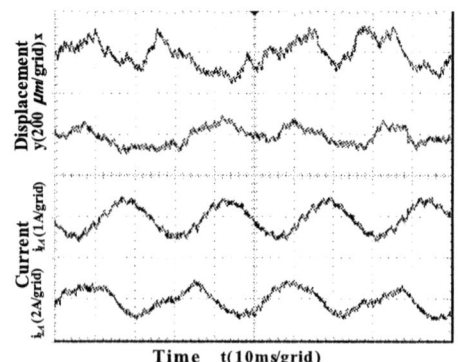

Figure 6. Experiment waveforms of BPMSM when the speed is 1 200 r/min

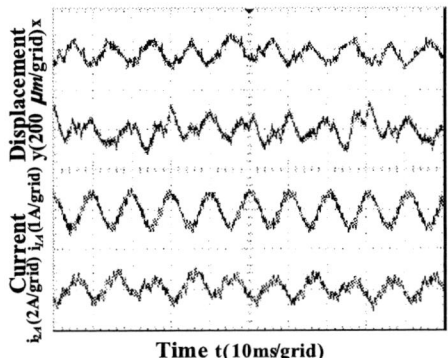

Figure 7. Experiment waveforms of BPMSM when the speed is
3 000 r/min

V. CONCLUSIONS

In this paper, mathematics model of BPMSM is deduced by adopting rotor magnetic field oriented control strategy. A digital vector control system is designed by using DSP technique, and the hardware and the software of the digital control system are developed. The experiment results have shown that the designed digital vector control system based on decoupling algorithm can ensure rotor suspension steadily, and content demand of dynamic performance of the motor, and the speed of the motor can be continuously adjusted within the range of 0 - 5 000 r/min.

REFERENCES

[1] Huangqiu Zhu, Zhiquan Deng, Yangguang Yan, and Shouqi Yuan, "Principles of bearingless motors and research status," *Micromotors*, vol. 33, no. 6, pp. 29-31, June 2000.

[2] J. Bichsel, "The bearingless electrical machine," in *Proc. Int. Symp. Magn. Suspension Technol.* NASA Langley Res. Center, Hampton, 1991, pp. 561-573.

[3] R. Schob and J. Bichsel, "Vector control of the bearingless motor," in *Proc. 4th Int. Symp. Magnetic Bearings.* ETH Zürich, 1994, pp.327-332.

[4] M. Ooshima, A. Chiba, T. Fukao, and et al, "Design and analysis of permanent magnet-type bearingless motors," *IEEE Trans. Industrial Electronics*, vol. 43, no. 2, pp. 292-299, March/April 1996.

[5] M. Ooshima, A. Chiba, and T. Fukao, "Characteristics of a permanent magnet type bearingless motor," *IEEE Trans, Industry Applications*, vol. 32, no. 2, pp. 363-370, March/April 1996.

[6] M. Ooshima, S. Miyazawa, A. Chiba, and et al, "Performance evaluation and test results of a 11000 rpm, 4kW surface-mounted permanent magnet-type bearingless motor," in *7th International Symposium on Magnetic Bearings,* ETH Zürich, 2000, pp.23-25.

[7] Huangqiu Zhu and Tao Zhang, "Finite element analysis for bearingless permanent magnet-type synchronous motor," *Proceeding of the CSEE*, vol. 26, no. 3, pp. 136-140, February 2006.

[8] Yongdong Li, *Digital Control System of AC Motors.* Beijing: China Machine Press, 2002.

[9] Antai Han, Zhifei Liu, Hai Huang, and et al, *DSP Controller Principle and Application in the motion control system.* Beijing: Tsinghua University Press, 2003.

[10] Simin Jiang, *TMS320LF240X DSP Hardware Developer Textbook.* Beijing: China Machine Press, 2003.

2006 5th International Power Electronics and Motion Control Conference

Practical Issues in Sensorless Control of PM Brushless Machines Using Third-Harmonic Back-EMF

J.X. Shen[*], Z.Q. Zhu[**] and D. Howe[**]

[*] College of Electrical Engineering, Zhejiang University, Hangzhou, 310027, China
[**] Department of Electronic and Electrical Engineering, Sheffield University, Sheffield, S1 3JD, UK
J.X.Shen@ieee.org, Z.Q.Zhu@ieee.org, s.j.gawthorpe@sheffield.ac.uk

Abstract—This paper presents some practical issues in the utility of third-harmonic back-EMF in sensorless control of PM brushless machines, including detection methods of the third-harmonic EMF, open-loop starting and normal operation of the machines in both BLDC and BLAC modes, as well as restrictions of the sensorless control.

Keywords-PM brushless machines, sensorless control, third harmonic, back-EMF

I. INTRODUCTION

Numerous sensorless control techniques have been developed for permanent magnet (PM) brushless dc (BLDC) and brushless ac (BLAC) machines. Methods which utilise the 3rd harmonic component of the back-EMF are attractive since they are relatively simple and potentially low-cost. In the authors' previous paper [1], the utility of the 3rd harmonic back-EMF method is demonstrated for the sensorless flux weakening control of both BLDC and BLAC drives. In this paper, some practical issues in implementing the sensorless control are described, and the influence of the machine design parameters will also be investigated.

II. DETECTION OF THIRD-HARMONIC BACK-EMF

Fig.1 shows the schematic of a PM brushless machine drive and the 3rd harmonic EMF detecting circuit. It was proposed in [2,3] that the 3rd harmonic could be extracted between points "s" and "n", points "s" and "h", or points "h" and "n". However, as proved in [1], only between the points "s" and "n" can the 3rd harmonic EMF be extracted for both BLDC and BLAC operations, both with and without PWM. The voltage between these two points, denoted as u_{sn}, contains the 3rd harmonic information. However, it usually contains some noise, hence, it should be preprocessed with band-pass filters. Then, its zero-crossings which represent some particular rotor positions are detected with a voltage comparator, resulting in a digital position signal (denoted as SGN). The rising and falling edges of the signal SGN correspond to the zero-crossings of the filtered 3rd harmonic EMF. Moreover, SGN may also contains noise, and the noise can be eliminated with a software-based digital filter. Fig.2 shows the position signal SGN detected from the 3rd harmonic EMF when a surface-mounted permanent magnet machine (referred to as Motor-I) is operated in BLDC and BLAC modes, respectively. The upper figure illustrates the machine terminal voltage (u_{ag}), the third harmonic EMF before band-pass filtering (u_{sn}), and the motor phase current (i_a). The lower figure illustrates rotor position signal (SGN) both before and after digital filtering. It is seen that the 3rd harmonic EMF detection works well in both modes. It should be pointed out that the experimental waveforms in Figs. (a) and (b) were obtained with the same surface-mounted permanent magnet machine, but different operation modes (BLDC and BLAC) were applied.

III. MOTOR OPERATIONS

Usually, BLDC operation requires trapezoidal phase EMF in the machine winding, which contains the 3rd harmonic component. On the other hand, BLAC operation needs sinusoidal phase EMF. However, if the 3rd harmonic exists in the phase EMF, the line-to-line EMF can still be sinusoidal, hence, the machine can be operated in BLAC mode. Therefore, if the machine is appropriately designed, the 3rd harmonic EMF-based sensorless control can be applied to both BLDC and BLAC operations.

A. Normal Operation

6 particular rotor positions are essential for BLDC operation. However, continuous rotor position information with high-resolution will be required if phase-advancing control is applied in BLDC drive, such that flux-weakening control can be realized [1].

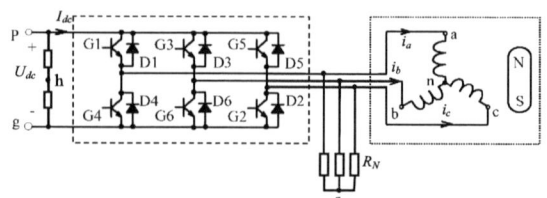

Figure1. Schematic of PM brushless drive and third-harmonic back-EMF detecting circuit.

1-4244-0448-7/06/$25.00 ©2006 IEEE

$$\theta_{re(k)} = \theta_{re(k-1)} + \omega_{rm} \times \Delta t_c \qquad (1)$$

where ω_{rm} is the machine speed which can be calculated from the time cycle of the signal SGN, whilst Δt_c is the time interval of position estimation. The estimated position is corrected with the particular rotor position when an edge of the signal SGN appears. Fig. 3 shows the estimated and actual rotor positions when the machine operates in both BLDC and BLAC modes, the actual position being measured with an encoder. Clearly, the sensorless position estimation is workable for both modes, and has sufficient accuracy and resolution.

Moreover, as verified in Fig. 4, the 3rd harmonic EMF-based sensorless control is reliable even if the machine operation is abnormal, such as one or two MOSFETs in the inverter are open-circuited, as long as the winding terminals are connected to the detection circuit.

B. Open-Loop Starting

Like all other EMF-based sensorless control, the motor has to start in the position open-loop mode until reaching sufficient speed. It is clear that the open-loop start in BLAC mode (i.e., with sine-wave phase currents) has lower risk of losing-step than in BLDC mode (i.e., with square-wave currents), since the stator field rotates smoothly in the former mode but steps forward in the latter, thus the electromagnetic torque is much smoother in the former mode. In other EMF-based sensorless control, such as that with the zero-crossing detection of the non-energized terminal voltage [4], the BLAC starting mode is not suitable, since the zero-crossing cannot be detected in the BLAC mode. However, the 3rd harmonic EMF-based sensorless control is always workable in BLAC operation, therefore, the machine can be started in BLAC mode. This is one of the unique advantages of the 3rd harmonic-based sensorless control.

IV. LIMITATIONS OF THIRD-ARMONIC BACK-EMF-BASED SENSORLESS CONTROL

It has been seen that the 3rd harmonic EMF-based sensorless control is rather attractive. However, it has some limitations, which are analysed below.

A. Requirement of Neutral Line

As the 3rd harmonic EMF is obtained from the voltage between the star point of resistor network (point "s" in Fig. 1) and the winding neutral (point "n" in Fig. 1), a neutral line is therefore required. However, the neutral line can be thin, since there is no power current in it.

B. Absence of Third-Harmonic Back-EMF

The 3rd harmonic back-EMF component must be high enough. The amplitude of the 3rd harmonic back-EMF (E_{m3}) is:

$$E_{m3} \propto \omega \cdot B_3 \cdot k_{w3} \qquad (2)$$

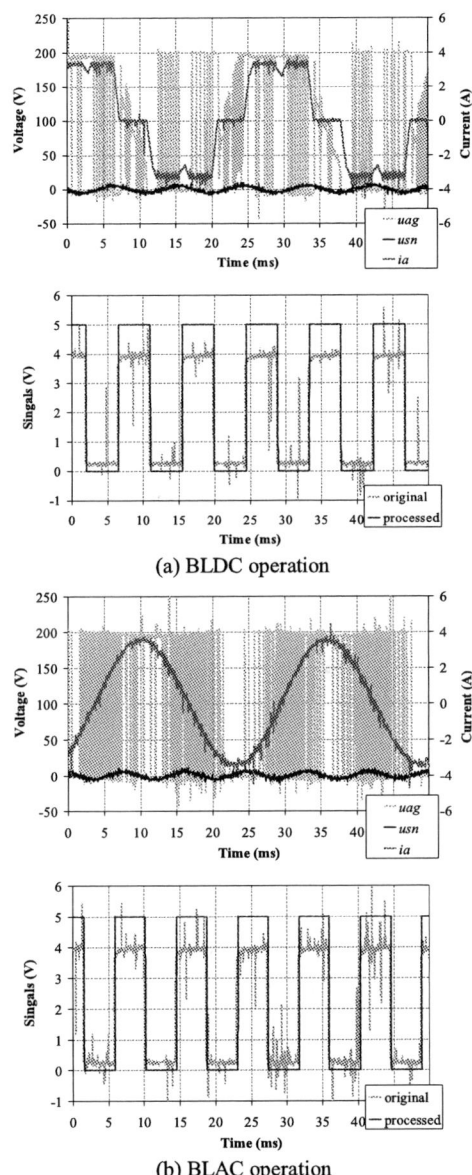

(a) BLDC operation

(b) BLAC operation

Figure 2. Detection of 3rd harmonic EMF and rotor position signal SGN for Motor-I with different operation modes. (u_{ag}: terminal voltage; u_{sn}: 3rd harmonic EMF; i_a: phase current; original: position signal SGN before digital filtering; processed: SGN after digital filtering)

On the other hand, high-resolution rotor position is inherently required for BLAC operation. Therefore, in this paper, the high-resolution rotor position is derived from the above-mentioned rotor position signal SGN. The rising and falling edges of SGN correspond to the EMF zero-crossings and particular rotor positions, with a phase shift due to the band-pass filers. However, the phase shift can be calculated with the software according to the machine speed and the filter specification. Thus, the high-resolution position is estimated as:

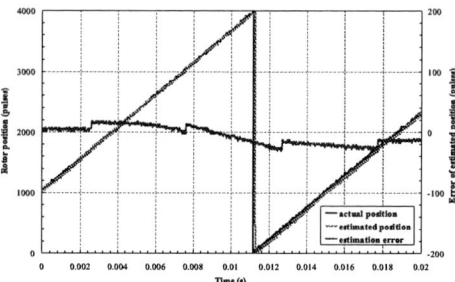

(a) BLDC operation,
only falling edges of SGN being used

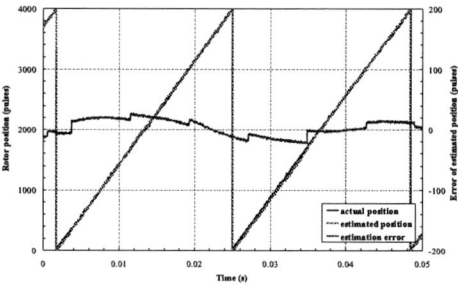

(b) BLAC operation,
both falling and rising edges of SGN being used

Figure 3. Comparison of estimated and actual rotor position of
Motor-I during normal operation.

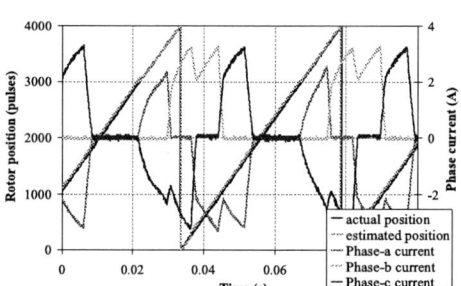

(a) Lower MOSFET of Phase-b open-circuited,
BLDC operation

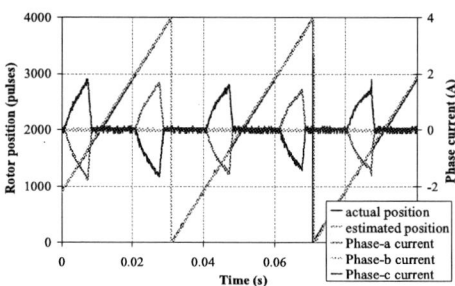

(b) Both MOSFET of Phase-b open-circuited,
BLDC operation

Figure 4. Robustness of sensorless rotor position estimation
with Motor-I under abnormal operating conditions.

where ω is the machine speed and B_3 is the amplitude of the 3rd harmonic component of the excitation flux density, whilst k_{w3} is the winding factor for the third-harmonic. k_{w3} is calculated as:

$$k_{w3} = k_{p3} \cdot k_{d3} \cdot k_{s3} \qquad (3)$$

where k_{p3}, k_{d3} and k_{s3} are the coil pitch factor, distribution factor and skew factor, respectively. If either B_3, k_{p3}, k_{d3} or k_{s3} are zero or very small, the third-harmonic back-EMF will be zero or too small to sense. Furthermore, (2) also shows that the 3rd harmonic back-EMF-based sensorless control is also unsuitable for zero or low speed operation.

In some PM brushless machines, the 3rd harmonic excitation field is very weak. For example, if a diametrically magnetised surface-mounted magnet ring is used in a 2-pole machine, the airgap field distribution is essentially sinusoidal [5]. Similarly, in some machines the pole-arc is optimised to ~2π/3 elec-rad in order, for example, to reduce the cogging torque, and hence, the 3rd harmonic airgap field is also very small.

Machines with non-overlapping windings and a 3:2 ratio of slot number to pole number is very popular for BLDC drives. However, the 3rd harmonic winding factor of such machines is zero. Therefore, in such machines, even if the 3rd harmonic component exists in the airgap field and the waveform of the phase back-EMF is non-sinusoidal, the 3rd harmonic back-EMF is zero.

By the way, it was stated in [3] that the condition for the existence of a third-harmonic back-EMF was that the coil pitch must be greater than 2π/3 elec-rad. However, this is not the case.

C. Influence of Rotor Saliency

It has been stated in Section II that the third harmonic back-EMF can be obtained from the voltage u_{sn} (see Fig. 1). However, this is correct only when the winding inductance is constant. In salient machines, such as those with interior/inserted magnets, the winding inductance varies with the rotor position. Thus, the expression of u_{sn} can be derived as below.

The winding inductances are expressed as:

$$\begin{cases} L_{aa} = \sum_k L_k \cos k\theta_r \\ L_{bb} = \sum_k L_k \cos k(\theta_r - 2\pi/3) \\ L_{cc} = \sum_k L_k \cos k(\theta_r + 2\pi/3) \\ M_{bc} = M_{cb} = \sum_k M_k \cos k\theta_r \\ M_{ca} = M_{ac} = \sum_k M_k \cos k(\theta_r - 2\pi/3) \\ M_{ab} = M_{ba} = \sum_k M_k \cos k(\theta_r + 2\pi/3) \\ \quad k = 0, 2, 4, 6, 8, 10, \cdots \end{cases} \qquad (4)$$

whilst the motor voltage equations are:

$$\begin{bmatrix} u_{an} \\ u_{bn} \\ u_{cn} \end{bmatrix} = R \begin{bmatrix} i_a \\ i_b \\ i_c \end{bmatrix} + \begin{bmatrix} e_a \\ e_b \\ e_c \end{bmatrix} + p \left\{ \begin{bmatrix} L_{aa} & M_{ab} & M_{ac} \\ M_{ba} & L_{bb} & M_{bc} \\ M_{ca} & M_{cb} & L_{cc} \end{bmatrix} \cdot \begin{bmatrix} i_a \\ i_b \\ i_c \end{bmatrix} \right\} \qquad (5)$$

Moreover, the back-EMFs are expressed as:

$$
\begin{cases}
\begin{aligned}
e_a &= e_{a1} + e_3 + e_{a5} + e_{a7} + e_9 + \cdots \\
&= -E_{m1}\sin\theta_r - E_{m3}\sin 3\theta_r - E_{m5}\sin 5\theta_r \\
&\quad - E_{m7}\sin 7\theta_r - E_{m9}\sin 9\theta_r - \cdots \\
e_b &= e_{b1} + e_3 + e_{b5} + e_{b7} + e_9 + \cdots \\
&= -E_{m1}\sin(\theta_r - 2\pi/3) - E_{m3}\sin 3\theta_r \\
&\quad - E_{m5}\sin 5(\theta_r - 2\pi/3) - E_{m7}\sin 7(\theta_r - 2\pi/3) \\
&\quad - E_{m9}\sin 9\theta_r - \cdots \\
e_c &= e_{c1} + e_3 + e_{c5} + e_{c7} + e_9 + \cdots \\
&= -E_{m1}\sin(\theta_r + 2\pi/3) - E_{m3}\sin 3\theta_r \\
&\quad - E_{m5}\sin 5(\theta_r + 2\pi/3) - E_{m7}\sin 7(\theta_r + 2\pi/3) \\
&\quad - E_{m9}\sin 9\theta_r - \cdots
\end{aligned}
\end{cases} \tag{6}
$$

Therefore, from (4)~(6) and the circuit shown in Fig. 1, the voltage u_{sn} can be derived as:

$$
\begin{aligned}
u_{sn} &= (e_3 + e_9 + e_{15} + \cdots) + \frac{p}{3} \times \sum_j \Big\{ (L_j - M_j) \\
&\quad \times \big[i_a \cos j\theta_r + i_b \cos j(\theta_r - 2\pi/3) \\
&\quad + i_c \cos j(\theta_r + 2\pi/3) \big] \Big\} \\
&\quad j = 2, 4, 8, 10, \cdots
\end{aligned} \tag{7}
$$

Clearly, if there is no rotor saliency, i.e., both L_j and M_j in (7) are 0, u_{sn} will mainly represents the 3rd harmonic back-EMF e_3. However, when the rotor saliency exists, neither L_j nor M_j is 0, and usually $(L_j - M_j)$ is not 0, either. Therefore, as long as currents flow in the machine windings, the second part on the right side of (7) can cause significant distortion in the measured u_{sn}.

The winding inductances of Motor-I, which has been used in experiments shown in Figs. 2~4, can be regarded as constant. However, the inductances vary in an interior permanent magnet machine (Motor-II), which also has the 3rd harmonic back-EMF. Experimentally captured waveforms of u_{sn} for these two machines are given in Figs. 5 and 6, respectively. Clearly, u_{sn} with the interior permanent magnet machine is significantly distorted no matter how large the phase current, but it becomes very clean with only the zero-sequence EMF component remaining when the machine windings are disconnected from the inverter. In contrast, u_{sn} with the surface-mounted permanent magnet machine contains a detectable third-harmonic back-EMF, although it also includes noise when PWM is applied. Therefore, if the rotor saliency exists, the 3rd harmonic back-EMF-based sensorless control technique is inapplicable.

D. Influence of Unbalance between the Three Phases

In some surface-mounted permanent magnet machines, the three phases are unbalanced on the aspects of EMF amplitude, winding resistance and inductance. Although the unbalance is not too serious, it can cause a distortion in u_{sn}. Moreover, in the practical implementation, u_{sn} usually contains some noise. Hence, u_{sn} is processed with band-pass filters, resulting in a modified signal (u_{xn}), which is actually used for the rotor position estimation. Clearly, distortion due to phase unbalance also occurs in u_{xn}. Furthermore, distortion in both u_{sn} and u_{xn} is influenced by the motor current. By way of example,

experimental results for Motor-III with surface-mounted magnets and unbalance are given in Fig. 7. Since u_{sn} is a relative small signal, its distortion is not very observable. However, the distortion in u_{xn} is very significant if the motor current is large. In conclusion, the 3rd harmonic back-EMF-based sensorless control is inapplicable to machines with unbalance between the three phases. In contrast, the three phases of Motor-I are well balanced.

It will be noted that the machine parameters, such as resistance, inductance and EMF constant, may vary due, for example, to temperature rise or other factors. However, if the three phases are originally balanced, such variations occur simultaneously in all phases and will not cause unbalancing. Furthermore, the 3rd harmonic back-EMF-based sensorless control does not use the exact values of the machine parameters, instead, it detects the EMF zero-crossings. Therefore, the sensorless control is insensitive to the machine parameter variations.

Therefore, the 3rd harmonic back-EMF-based sensorless control is inapplicable in the following cases:
(i) The winding neutral point is not accessible;
(ii) The third-harmonic component of the excitation field is zero or very small;
(iii) The third-harmonic winding factor is zero or very small;
(iv) The winding inductances are not constant due to rotor saliency or other factors;
(v) The three phases are not well balanced.

However, in practical applications, many brushless machines are not subject to the foregoing restrictions. Therefore, the development of such a sensorless technique is rather useful.

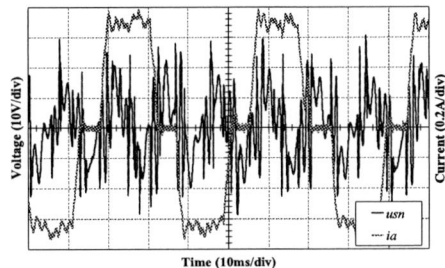

(a) BLDC operation with PWM

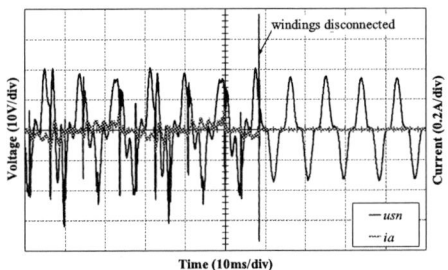

(b) BLDC operation without PWM;
then windings disconnected from inverter

Figure 5. Measured voltage (u_{sn}) with Motor-II.

931

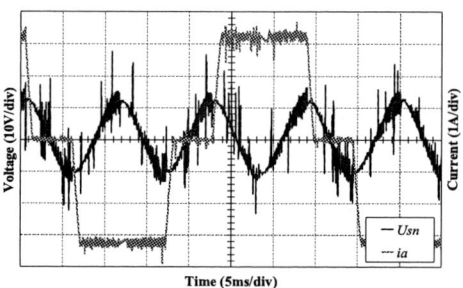

(a) BLDC operation with PWM

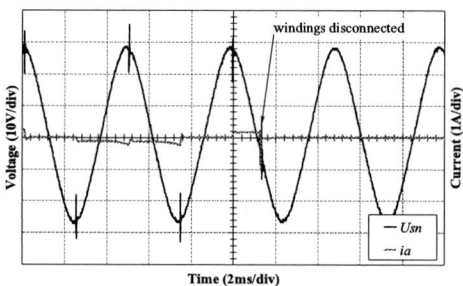

(b) BLDC operation without PWM;
then windings disconnected from inverter

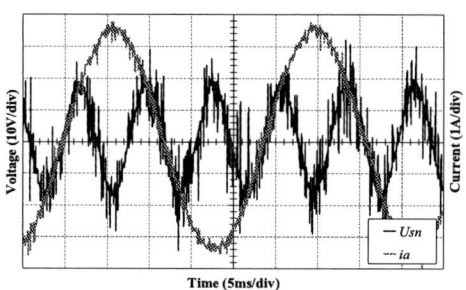

(c) BLAC operation with PWM

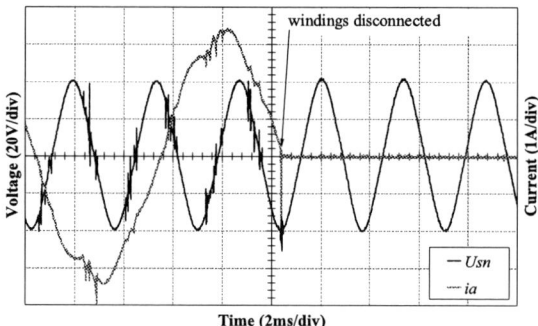

(d) BLAC operation with PWM;
then windings disconnected from inverter

Figure 6. Measured voltage (u_{sn}) with Motor-I.

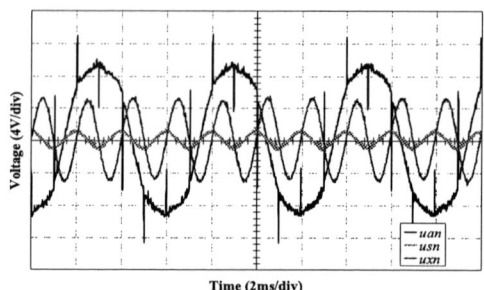

(a) BLDC operation without PWM, 18.3Vdc, 0.76Adc

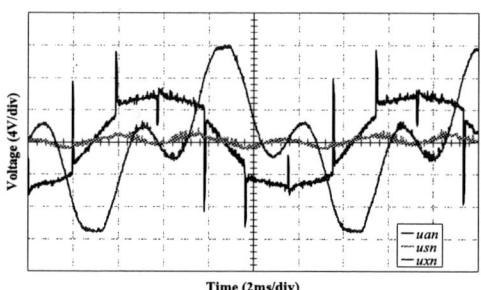

(d) BLDC operation without PWM, 17.9Vdc, 3.9Adc

Figure 7. Influence of phase unbalance on voltages u_{sn} and u_{xn} in Motor-III.

V. CONCLUSIONS

The 3rd harmonic back-EMF-based sensorless control is workable for both BLDC (phase-advancing can be applied) and BLAC operations. The motor can be started in the open-loop BLAC mode, which has a smooth electromagnetic torque and consequently low risk of losing-step. Implementation of the sensorless control is simple, whilst the accuracy is sufficiently high. Furthermore, restrictions of the sensorless control have been highlighted, such that the control strategy can be utilized for appropriate machines.

REFERENCES

[1] J. X. Shen, *et al*, Sensorless Flux-Weakening Control of Permanent Magnet Brushless Machines Using Third-Harmonic Back-EMF, *IEEE Trans. Industry Applications*, vol.40, no.6, 2004, pp.1629-1636.

[2] F. Profumo, *et al*, Universal Field Oriented Controller Based on Air Gap Flux Sensing via Third Harmonic Stator Voltage, *IEEE Trans. Industry Applications*, vol.30, no.2, 2994, pp.448-455.

[3] J. C. Moreira, Indirect Sensing for Rotor Flux Position of Permanent Magnet AC Motors Operating in a Wide Speed Range, *IEEE Trans. Industry Applications*, vol.32, no.6, 1996, pp.1394-1401.

[4] K. Iizuka, *et al*, Microcomputer Control for Sensorless Brushless Motor, *IEEE Trans. Industry Applications*, vol.21, no.4, 1985, pp.595-601.

[5] Z. Q. Zhu, et al, Desig and Analysis of High-Speed Brushless Permanent Magnet Motors, *Proc. 1997 IEE Int'l Conf. Electrical Machines and Drives (EMD'97)*, pp. 381-385.

2006 5th International Power Electronics and Motion Control Conference

Switched Reluctance Motors Drive for the Electrical Traction in Shearer

H. Chen

College of Information and Electrical Engineering
China University of Mining & Technology, Xuzhou 221008, China
chenhaocumt@tom.com

Abstract—The paper presented the double Switched Reluctance motors parallel drive system for the electrical traction in shearer. The system components, such as the Switched Reluctance motor, the main circuit of the power converter and the controller, were described. The control strategies of the closed-loop rotor speed control with PI algorithm and balancing the distribution of the loads with fuzzy logic algorithm were given. The tests results were also presented. It is shown that the relative deviation of the average *DC* supplied current of the power converter in the Switched Reluctance motor 1 and in the Switched Reluctance motor 2 is within $\pm 10\%$.

Keywords- switched reluctance; motor control; shearer; coal mine; electrical drive

I. INTRODUCTION

The underground surroundings of the coal mines are very execrable. One side, it is the moist, high dust and inflammable surroundings. On the other side, the space of roadway is limited since it is necessary to save the investment of exploiting coal mines so that it is difficult to maintain the equipments. In the modern coal mines, the automatization equipments could be used widely. The faults of the automatization equipments could affect the production and the benefit of the coal mines. The shearer is the mining equipment that coal could be cut from the coal wall. The traditional shearer was driven by the hydrostatic transmission system. The fault ratio of the hydrostatic transmission system is high since the fluid in hydrostatic transmission system could be polluted easily. The faults of the hydrostatic transmission system could affect the production and the benefit of the coal mines directly. The fault ratio of the motor drive system is lower than that of the hydrostatic transmission system, but it is difficult to cool the motor drive system in coal mines since the motor drive system should be installed within the flameproof enclosure for safety protection. The motor drive system is also one of the pivotal parts in the automatization equipments. The development of the novel types of the motor drive system had been attached importance to by the coal mines. The Switched Reluctance motor drive could become the main equipments for adjustable speed electrical drive system in coal mines [1],

because it has the high operational reliability and the fault tolerant ability [2]. The Switched Reluctance motor drive made up of the double-salient pole Switched Reluctance motor, the unipolar power converter and the controller is firm in the motor and in the power converter. There is no brush structure in the motor and no fault of ambipolar power converter in the power converter [3][4]. The Switched Reluctance motor drive could be operated at the condition of lacked phases fault depended on the independence of each phase in the motor and the power converter [5]. There is no winding in the rotor so that there is no copper loss in the loss and there is only little iron loss in the rotor. It is easy to cool the motor since it is not necessary to cool the rotor. The shearer driven by the Switched Reluctance motor drive had been developed. The paper presented the developed prototype.

II. SYSTEM COMPONENTS

The developed Switched Reluctance motors drive for the electrical traction in shearer is a type of the double Switched Reluctance motors parallel drive system. The system is made up of two Switched Reluctance motors, a control box installed the power converter and the controller. The adopted two Switched Reluctance motors are all three-phase 12/8 structure Switched Reluctance motor, which were shown in Figure 1. The two Switched Reluctance motors were packing by the explosion-proof enclosure, respectively. The rated output power of one motor is *40 KW* at the rotor speed *1155 r/min*, and the adjustable speed range is from *100 r/min* to *1500r/min*.

Figure 1. Photograph of the two three-phase 12/8 structure Switched Reluctance motor

1-4244-0448-7/06/$25.00 ©2006 IEEE

The power converter consists of two three-phase asymmetric bridge power converter in parallel. The IGBTs were used as the main switches. Three-phase 380V AC power source was rectificated and supplied to the power converter. The main circuit of the power converter was shown in Figure 2.

In the controller, there were the rotor position detection circuit, the commutation circuit, the current and voltage protection circuit, the main switches' gate driver circuit and the digital controller for rotor speed closed-loop and balancing the distribution of the loads.

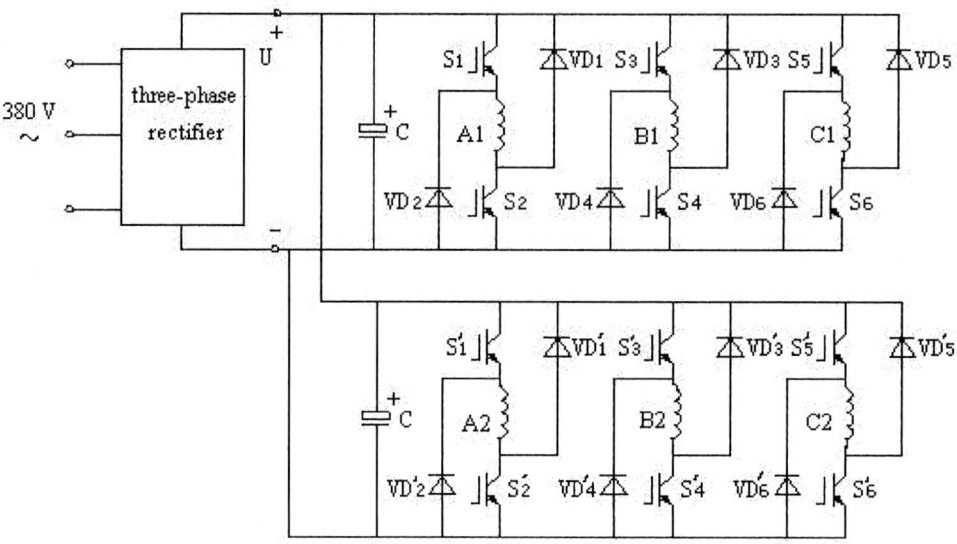

Figure 2. Main circuit of the power converter

III. CONTROL STRATEGY

The two Switched Reluctance motor could all drive the shearer by the transmission outfit in the same traction guide way so that the rotor speed of the two Switched Reluctance motors could be synchronized.

The closed-loop rotor speed control of the double Switched Reluctance motors parallel drive system could be implemented by PI algorithm. In the Switched Reluctance motor 1, the triggered signals of the main switches in the power converter are modulated by PWM signal, the comparison of the given rotor speed and the practical rotor speed are made and the duty ratio of PWM signal are regulated as follows,

$$e = n_g - n_f \tag{1}$$

$$\Delta D_{1(k)} = K_i \cdot e_k + K_p \cdot (e_k - e_{k-1}) \tag{2}$$

$$D_{1(k)} = D_{1(k-1)} + \Delta D_{1(k)} \tag{3}$$

where, n_g is the given rotor speed, n_f is the practical rotor speed, e is the difference of the rotor speed, $\Delta D_{1(k)}$ is the increment of the duty ratio of PWM signal of the Switched Reluctance motor 1 at k time, K_i is the integral coefficient, K_p is the proportion coefficient, e_k is the difference of the rotor speed at k time, e_{k-1} is the difference of the rotor speed at $k-1$ time, $D_{1(k)}$ is the duty ratio of PWM signal of the Switched Reluctance motor 1 at k

time, and $D_{1(k-1)}$ is the duty ratio of PWM signal of the Switched Reluctance motor 1 at $k-1$ time.

The output power of the Switched Reluctance motor drive system is approximately in proportion to the average DC supplied current of the power converter as follows,

$$P_2 \propto I_{in} \tag{4}$$

where, P_2 is the output power of the Switched Reluctance motor drive system, I_{in} is the average DC supplied current of the power converter.

In the Switched Reluctance motor 2, the triggered signals of the main switches in the power converter are also modulated by PWM signal. The balancing the distribution of the loads between the two Switched Reluctance motors could be implemented by fuzzy logic algorithm. In the fuzzy logic regulator, there are two input control parameters, one is the deviation of the average DC supplied current of the power converter between the two Switched Reluctance motors, and the other is the variation of the deviation of the average DC supplied current of the power converter between the two Switched Reluctance motors. The output control parameter is the increment of the duty ratio of the PWM signal of the Switched Reluctance motor 2. The block diagram of the double Switched Reluctance motors parallel drive system for the electrical traction in shearer was shown in Figure 3.

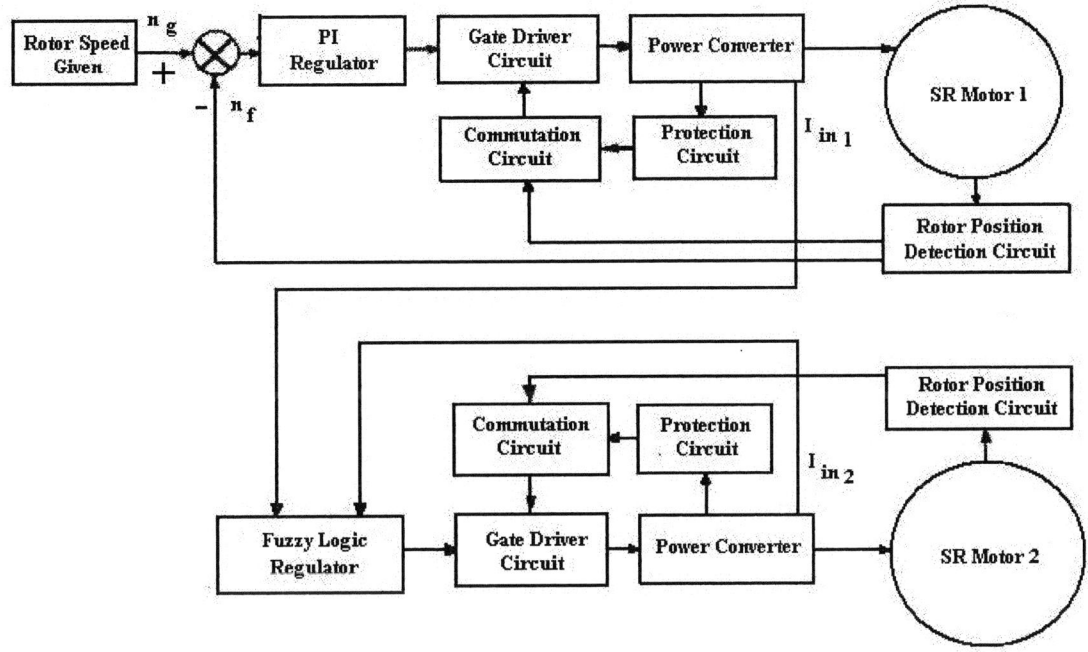

Figure 3. Block diagram of the double Switched Reluctance motors parallel drive system for the electrical traction in shearer

The deviation of the average DC supplied current of the power converter between the two Switched Reluctance motors at the moment of t_i is

$$e_i = I_{in1} - I_{in2} \qquad (5)$$

where, I_{in1} is the practical average DC supplied current of the power converter in the Switched Reluctance motor 1 at the moment of t_i, I_{in2} is the practical average DC supplied current of the power converter in the Switched Reluctance motor 2 at the moment of t_i.

The variation of the deviation of the average DC supplied current of the power converter between the two Switched Reluctance motors at the moment of t_i is

$$\dot{e}_i = e_i - e_{i-1} \qquad (6)$$

where, e_{i-1} is the deviation of the average DC supplied current of the power converter between the two Switched Reluctance motors at the moment of t_{i-1}.

The duty ratio of the PWM signal of the Switched Reluctance motor 2 at the moment of t_i is

$$D_{2(i)} = D_{2(i-1)} + \Delta D_{2(i)} \qquad (7)$$

where, $\Delta D_{2(i)}$ is the increment of the duty ratio of the PWM signal of the Switched Reluctance motor 2 at the moment of t_i and $D_{2(i-1)}$ is the duty ratio of the PWM signal of the Switched Reluctance motor 2 at the moment of t_{i-1}.

The fuzzy logic algorithm could be expressed as follows,

$$\text{if } \widetilde{E} = \widetilde{E}_i \text{ and } \widetilde{E}C = \widetilde{E}C_j \text{ then } \widetilde{U} = \widetilde{U}_{ij} \qquad (8)$$

$$i = 1,2, \cdots, m, j = 1,2, \cdots, n$$

where, $\widetilde{E}$ is the fuzzy set of the deviation of the average DC supplied current of the power converter between the two Switched Reluctance motors, $\widetilde{E}C$ is the fuzzy set of the variation of the deviation of the average DC supplied current of the power converter between the two Switched Reluctance motors, and $\widetilde{U}$ is the fuzzy set of the increment of the duty ratio of the PWM signal of the Switched Reluctance motor 2.

The continuous deviation of the average DC supplied current of the power converter between the two Switched Reluctance motors could be changed into the discrete amount at the interval [-5, +5], based on the equations as follows,

$$e = INT[K_e \cdot e_i] \qquad (9)$$

$$K_e = \frac{10}{220} \qquad (10)$$

The continuous variation of the deviation of the average DC supplied current of the power converter between the two Switched Reluctance motors could also be changed into the discrete amount at the interval [-5, +5], based on the equations as follows,

$$\dot{e} = INT[K_{\dot{e}} \cdot \dot{e}_i] \qquad (11)$$

$$K_{\dot{e}} = \frac{10}{40} \qquad (12)$$

The discrete increment of the duty ratio of PWM signal of the Switched Reluctance motor 2 at the interval [-5, +5] could be changed into the continuous amount at the interval [-1.0%, +1.0%], based on the equations as follows,

$$\Delta D_{2(i)} = INT^{-1}[K_{\Delta D} \cdot \Delta D] \tag{13}$$

$$K_{\Delta D} = \frac{10}{0.02} \tag{14}$$

There is a decision forms of the fuzzy logic algorithm based on the above principles, which was stored in the programme storage cell of the controller.

While the difference of the distribution of the loads between the two Switched Reluctance motors could be got, the duty ratio of PWM signal of the Switched Reluctance motor 2 will be regulated based on the decision forms of the fuzzy logic algorithm and the distribution of the loads between the two Switched Reluctance motors could be balanced.

IV. TESTED RESULTS

The developed double Switched Reluctance motors parallel drive system prototype had been tested experimentally. Table I gives the tests results, where σ_1 is the relative deviation of the average DC supplied current of the power converter in the Switched Reluctance motor 1, σ_2 is the relative deviation of the average DC supplied current of the power converter in the Switched Reluctance motor 2, and,

$$\sigma_1 = \frac{I_{in1} - \dfrac{I_{in1} + I_{in2}}{2}}{\dfrac{I_{in1} + I_{in2}}{2}} \times 100\% \tag{15}$$

$$\sigma_2 = \frac{I_{in2} - \dfrac{I_{in1} + I_{in2}}{2}}{\dfrac{I_{in1} + I_{in2}}{2}} \times 100\% \tag{16}$$

TABLE I.
TESTS RESULTS OF PROTOTYPE

Rotor speed (r/min)	I_{in1} (A)	I_{in2} (A)	σ_1 (%)	σ_2 (%)
153	43.8	51.5	-8.1	+8.1
500	95.1	109.8	-7.2	+7.2
700	90.6	98.6	-4.2	+4.2
1155	80.5	89.0	-5.0	+5.0
1500	68.8	75.1	-4.4	+4.4

It is shown that the relative deviation of the average DC supplied current of the power converter in the Switched Reluctance motor 1 and in the Switched Reluctance motor 2 is within $\pm10\%$.

V. CONCLUSION

The paper presented the double Switched Reluctance motors parallel drive system for the electrical traction in shearer. The novel type of the shearer in coal mines driven by the Switched Reluctance motors drive system contributes to reduce the fault ratio of the shearer, enhance the operational reliability of the shearer and increase the benefit of the coal mines directly. The drive type of the double Switched Reluctance motors parallel drive system could also contribute to enhance the operational reliability compared with the drive type of the single Switched Reluctance motor drive system.

ACKNOWLEDGMENT

The authors would like to thank for the project supported in part by the 333 Engineering Training Programme Foundation for New Century Science & Technology Leaders of Jiangsu Province Grant No.2003-16, the Indigo Blue Engineering Training Programme Foundation for Middle-Young Academic Leaders by the Education Department of Jiangsu Province Grant No.[2002]60, the Higher College or University of Jiangsu Province High & New Technology Industry Development Project Grant No.JH03-002 and the Science & Technology Project of Xuzhou City Grant No.X20052392.

REFERENCES

[1] H. Chen, G. Xie, "A Switched Reluctance Motor Drive System for Storage Battery Electric Vehicle in Coal Mine," *Proceedings of the 5th IFAC Symposium on Low Cost Automation*, pp.95-99, Sept. 1998.

[2] H. Chen, X. Meng, F. Xiao, T. Su, G. Xie, "Fault tolerant control for switched reluctance motor drive," *Proceedings of the 28 Annual Conference of the IEEE Industrial Electronics Society*, pp.1050-1054, Nov. 2002.

[3] R. M. Davis, W. F. Ray, R. J. Blake, "Inverter drive for switched reluctance motor: circuit and component ratings," *IEE Proc. B*, vol.128, no.3, pp. 126-136, Sept. 1981.

[4] D. Liu, et al., *Switched Reluctance Motor Drive*. Beijing: Mechanical Industry Press, 1994.

[5] H. Chen, J. Jiang, C. Zhang, G. Xie, "Analysis of the four-phase switched reluctance motor drive under the lacking one phase fault condition," *Proceedings of IEEE 5th Asia-Pacific Conference on Circuit and Systems*, pp.304-308, Dec. 2000.

2006 5th International Power Electronics and Motion Control Conference

Research on Three-level Inverter of Six-phase Synchronous Motor

Yao Wenxi, Hu Haibing, Lu Zhengyu, Xu Haijie

State Key Laboratory of Power Electronics of ZheJiang University, Hangzhou 310027, China

ywxi@zju.edu.cn

Abstract: Using adjustable speed scheme in high power motor drive can not only save energy greatly but also improve the dynamic performance of the motor. Recently, medium-voltage inverter becomes the most popular scheme for high-power motor drive, but in some especial fields multi-phase motor drives have more advantages. In this paper a vector control method of six-phase synchronous motor (SM) based on three-level inverter is studied. And some key techniques are analyzed in detail such as air gap flux-oriented vector control of six-phase SM, Space vector pulse-width modulation (SVPWM) of six-phase three-level inverter and its neutral-point control. Then a 690V/200kW three-level inverter of six-phase SM is developed, and experimental results verify the techniques presented in the paper.

Keyword: Six-phase SM, Air gap flux-oriented, SVPWM, Neutral-point balance

I. INTRODUCTION

The power rating of the inverter should meet the required level for the machine and driven load. However, the inverter ratings can not be increased over a certain range due to the limitation on the power rating of semiconductor devices. One solution to this problem is using multi-level inverter where switches of reduced rating are employed to develop high power level inverters. The advent of inverter fed motor drives also removed the limits of the number of motor phases. This fact made it possible to design machine with more than three phases and brought about the increasing investigation and applications of multi-phase (more than three phases) motor drives.[1][2]

The main advantages of multi-phase motor drives can be summarized as:

1. Multi-phase motor drives achieve the high power by enhancing the number of phases, which makes it possible to use low voltage or low current inverters.

2. Some low orders of air gap harmonics can be eliminated in multi-phase motor drives, so the torque propulsion is reduced greatly.

3. The current harmonics to DC-link capacitor of multi-phase inverter is reduced.

4. Multi-phase improves the reliability of system, when one or several phases are at fault, multi-phase system can keep running at lower load.

In this paper, the techniques of motor drive based on multi-level and multi-phase is studied, and a 690V/200kW three-level inverter for six-phase SM is developed.

I. AIR GAP FLUX-ORIENTED VECTOR CONTROL OF SIX-PHASE SM

Air gap flux-oriented vector control is one of the most popular control schemes of SM. The air gap flux-oriented vector control scheme of six-phase SM can be developed from that of three-phase SM, which has been widely used[3], as shown in Figure1, where 'n' is rotor speed, $\cos \phi_e$ is power factor, λ is rotor position angle, ϕ_s is the angle from rotor d axis to stator α axis, i_a, i_b, i_c are the stator currents, i_{sM}, i_{sT} are the excitation component and torque component of stator currents, u_a, u_b, u_c are the stator voltages. The subscript of the variables in Figure1, 1 and 2, denote the first three-phase stator coils and the second three-phase stator coils respectively, and the superscript * suppose the variables are given ones.

The vector control system of six-phase SM with three-level inverter consists of two parts: one is the manage of SM, which includes rotor position and speed detection, speed control, power factor setting, air-gap flux observation and so on; the other is the manage of three-level inverter, which includes voltage and current sampling, current control, SVPWM of three-level inverter and so on. In the control scheme, the leading difference between three-level six-phase SM's drive and normal two-level three-phase SM's drive is air gap flux observing of six-phase SM and SVPWM of three-level inverter, which will be discussed in the following text.

II. AIR-GAP FLUX OBSERVATION OF SIX-PHASE SM

The key technique of air gap flux-oriented vector

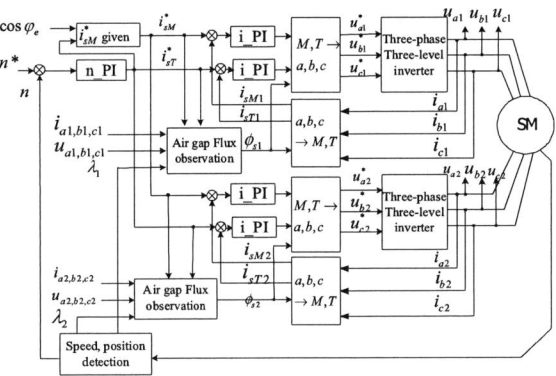

Figure 1. Vector control diagram of Three-level six-phase SM

1-4244-0448-7/06/$25.00 ©2006 IEEE

control is the air gap flux observation. Since it is difficult and low reliable to detect the air-gap flux by sensors directly, current model and voltage model are generally used to observe the air-gap flux. The air gap flux observation of three-phase SM has been used widely, based on which the flux observation of six-phase SM will be discussed in this section.

Current model gets the air gap flux by a closed loop[3]. Because it neglects the leakage inductance of stator, the current model of six-phase SM is almost the same as that of three-phase SM. It can be achieved by dividing the six-phase stator coils into two three-phase stator coils and adding up the currents of the two three-phase stator coils in M-T coordinates.

However, voltage model of six-phase SM can not be approached from three-phase SM simply. Equ.(1) is the expression of air gap flux of six-phase, where L_s is the leakage inductance of stator, L_{al} is the mutual leakage inductance of two stator coils, which expresses the flux coupling through two stator coils but not coupling through rotor coils.

$$
\begin{cases}
\psi_{\alpha 1} = \int(u_{s\alpha 1} - r_s i_{s\alpha 1})dt - (L_{sl} + L_{al})i_{s\alpha 1} - \dfrac{\sqrt{3}}{2}L_{al}i_{s\alpha 2} + \dfrac{1}{2}L_{al}i_{s\beta 2} \\[2mm]
\psi_{\beta 1} = \int(u_{s\beta 1} - r_s i_{s\beta 1})dt - (L_{sl} + L_{al})i_{s\beta 1} - \dfrac{\sqrt{3}}{2}L_{al}i_{s\beta 2} - \dfrac{1}{2}L_{al}i_{s\alpha 2} \\[2mm]
\psi_{\alpha 2} = \int(u_{s\alpha 2} - r_s i_{s\alpha 2})dt - (L_{sl} + L_{al})i_{s\alpha 2} - \dfrac{\sqrt{3}}{2}L_{al}i_{s\alpha 1} - \dfrac{1}{2}L_{al}i_{s\beta 1} \\[2mm]
\psi_{\beta 2} = \int(u_{s\beta 2} - r_s i_{s\beta 2})dt - (L_{sl} + L_{al})i_{s\beta 2} + \dfrac{\sqrt{3}}{2}L_{al}i_{s\beta 1} - \dfrac{1}{2}L_{al}i_{s\alpha 1}
\end{cases}
$$
（1）

It is difficult to acquire the air gap flux via Equ.(1) directly, because the initial values of integral are uncertain, and voltages sampling are inaccurate in low rotor speed. Actually a feedback scheme is developed to approach the voltage model, which uses flux value acquired from current model as the initial value of integral. Figure2 shows the block diagram of the first three-phase stator coils, of which the transfer function in $\alpha 1$ axis is Equ.(2), and it is similar in $\beta 1$ axis.

$$
\psi_{\alpha 1}(j\omega) = \frac{a}{1 + aj\omega}e_{s\alpha 1} + \frac{1}{1 + aj\omega}\psi_{\alpha 1}^{*}
$$
（2）

Where $\psi_{\alpha 1}^{*}(\psi_{\beta 1}^{*})$ is the flux acquired from current model, 'a' is the feedback factor that satisfied 0<a<1, $e_{s\alpha 1}$ is the electromotive force:

$$
e_{s\alpha 1} = u_{s\alpha 1} - r_s i_{s\alpha 1} - (L_{sl} + L_{al})\frac{di_{s\alpha 1}}{dt} - \frac{\sqrt{3}}{2}L_{al}\frac{di_{s\alpha 2}}{dt} + \frac{1}{2}L_{al}\frac{di_{s\alpha 2}}{dt}
$$
（3）

From the transfer function Equ.(2), model show in Figure2 has the following features: At low speed, ajω<<1, $\psi_{\alpha 1}$ is mainly decided by $\psi_{\alpha 1}^{*}$, and current model works; At high speed ajω>>1, $\psi_{\alpha 1}$ mainly comes from the integral of $e_{s\alpha 1}$, and voltage model works.

III. SVPWM OF SIX-PHASE THREE-LEVEL INVERTER

The three-level inverter of six-phase SM is composed of two three-phase three-level inverters which have the same DC-link capacitors, as show in Figure3. The control methods of this inverter are various due to its multi-phase and multi-level, such as multilevel Carrier PWM, three-phase SVPWM, six-phase SVPWM and so on. In this paper, the six-phase inverter is divided into two three-phase inverters, and the SVPWM scheme is dealt in each three-phase inverter separately. As an example, one of three-phase inverters composed of phase A, phase B and phase C will be discussed in the following.

Each three-level leg can output three statues [-1, 0, 1], so three phases can form 27 vectors, in which there are 19 independent vectors, as show in Figure4, where V_0 is zero vector, V_1-V_6 are short vectors, V_7, V_9, V_{11}, V_{13}, V_{15}, V_{17} are long vectors, V_8, V_{10}, V_{12}, V_{14}, V_{16}, V_{18} are middle vectors.

The vectors can be divided into six major triangular sectors (I to VI), and the details of sector I is given in Figure5. Assume that reference vector $\vec{V}_{ref}$ lies in sector I, and suppose t_a, t_b, t_c are the time duration of three nearest vectors of $\vec{V}_{ref}$ in one PWM cycle, as show in Figure5, which can be calculated by building equations of $\alpha - \beta$ coordinates[4]. The results are shown in Table I.

When reference vector lies in sector II-VI, the time duration of vectors can be calculated by building equations of $\alpha - \beta$ coordinates likewise. However, a more simple method is to establish a mapping from sector II-VI to sector I which can be achieved by the following procedure.

1) The reference vector lied in sector II-VI should be converted to sector I by coordinates rotating. For instance, when reference vector lied in sector II should be converted to sector I by rotating the reference vector 60° clockwise. Then it is possible to calculate the PWM of the reference vector in any sector all in sector I.

2) The relationship of switch-sets should be decided between sector I and others. It is easy to find that the difference of sector I, sector III and sector V is the sequence of phases, so the only thing is to regroup the

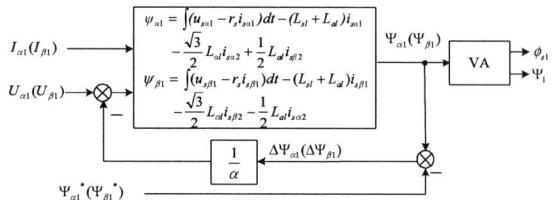

Figure 2. Air-gap flux observation using voltage, current model

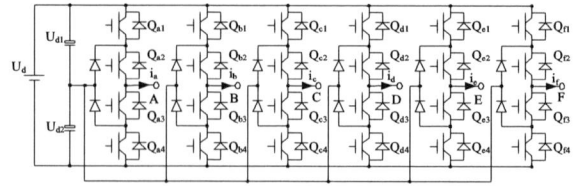

Figure 3. Main circuit of six-phase three-level inverter

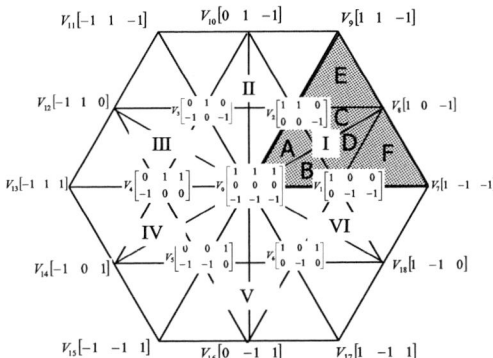

Figure 4. Vector diagram of three-level inverter

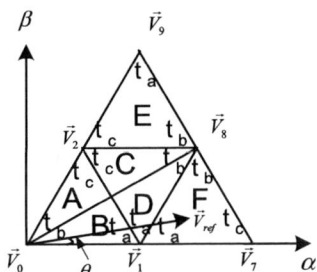

Figure 5. Vectors detail in sector I

sequence of phase A, B, C in these sectors. Furth more, the switch sets in sector II are the negative of those in sector V. For example, of middle vector $V_{10}[0\ 1\ -1]$ in sector II, the corresponding vector in sector V is $V_{16}[0\ -1\ 1]$. In the same way, switch sets in sector IV and sector VI are the negative ones in sector I and sector III respectively. Suppose the scheme decomposed the three-level SVPWM into two-level carrier PWMs, proposed in reference [4], is used to achieve the three-level SVPWM, and define P_{a1}, P_{a2}, P_{b1}, P_{b2}, P_{c1}, P_{c2} are the duty-rations of Q_{a1}, Q_{a4}, Q_{b1}, Q_{b4}, Q_{c1}, Q_{c4} in Figure3, and P_{a1}', P_{a2}', P_{b1}', P_{b2}', P_{c1}', P_{c2}' are the duty-rations of them acquired in sector I, then the rotated angle of reference vector and the relationship of PWMs in each sector are shown in Table II.

This section proposed a method to achieve the six-phase SVPWM by two three-phase SVPWMs, and generalized the SVPWM of three-level inverter. Then a simplified scheme of three-phase three-level SVPWM was proposed by establishing a mapping from sector II-VI to sector I.

TABLE I.
Time duration of Vectors in sector I

Area	t_a	t_b	t_c
AB	$2k\sin(\frac{\pi}{3}-\theta)$	$T_s-2k\sin(\frac{\pi}{3}+\theta)$	$2k\sin\theta$
F	$2T_s-2k\sin(\frac{\pi}{3}+\theta)$	$2k\sin\theta$	$2k\sin(\frac{\pi}{3}-\theta)-T_s$
CD	$T_s-2k\sin\theta$	$2k\sin(\frac{\pi}{3}+\theta)-T_s$	$T_s-2k\sin(\frac{\pi}{3}-\theta)$
E	$2k\sin\theta-T_s$	$2k\sin(\frac{\pi}{3}-\theta)$	$2T_s-2k\sin(\frac{\pi}{3}+\theta)$

TABLE II.
Rotated angle of reference vector and relationship of PWM in each sector

Sector	Rotated angle of reference vector	P_{a1}	P_{b1}	P_{c1}	P_{a2}	P_{b2}	P_{c2}
I	$0°$	P_{a1}'	P_{b1}'	P_{c1}'	P_{a2}'	P_{b2}'	P_{c2}'
II	$60°$ clockwise	P_{c2}'	P_{a2}'	P_{b2}'	P_{c1}'	P_{a1}'	P_{b1}'
III	$120°$ clockwise	P_{b1}'	P_{c1}'	P_{a1}'	P_{b2}'	P_{c2}'	P_{a2}'
IV	$180°$ clockwise	P_{a2}'	P_{b2}'	P_{c2}'	P_{a1}'	P_{b1}'	P_{c1}'
V	$240°$ clockwise	P_{c1}'	P_{a1}'	P_{b1}'	P_{c2}'	P_{a2}'	P_{b2}'
VI	$300°$ clockwise	P_{a2}'	P_{b2}'	P_{c2}'	P_{b1}'	P_{c1}'	P_{a1}'

IV. NEUTRAL-POINT CONTROL OF NPC THREE-LEVEL INVERTER

Neutral-point unbalance is an inherent problem in NPC three-level inverter. There are two factors lead to unbalance: 1) The inverter inject or extract current from neutral-point when one or several phases output the neutral-point potential, which made the neutral-point have ripple of 3 times output frequency; 2) A fixed bias may be occurred when there are unbalance factors in the inverter, such as the unbalance load, unbalance components and so on.

The schemes of neutral-point balance mainly include: using hardware balance circuits in addition[5], zero-sequence voltage injection[6] and distribution of short vectors' time duration[7]. In this paper a BANG-BANG control scheme will be proposed based on distribution of short vectors' duration time.

Only short vectors have the ability of balance in the 19 vectors of three-level inverter[4]. For sector I in Figure4, if the reference vector lies in area B, D, F, the nearest short vector V_1 will be used to balance the neutral-point. Suppose the distribution of time duration of V_1 in one PWM cycle show as Equ.4, where t_a is the duration of short vector V_1 in one PWM cycle, t_{a1} is the duration of positive switch-set $[1\ 0\ 0]$ of V_1, t_{a2} is the duration of negative switch-set $[0\ -1\ -1]$ of V_1, Δt is the time bias acquired by the principle of BANG-BANG control, as shown in Equ.(5), where ΔU_o is the bias of neutral-point potential which defined as $\Delta U_o=(U_{d2}-U_{d1})$, i_a is the current of phase A, sgn is direction function, k is balance factor that satisfied $0<k<1$. In the same way, short vector V_2 is used in area A, C, E of sector I, as shown in Equ.(6) and Equ.(7).

$$\begin{cases} t_{a1}=\dfrac{t_a}{2}-\Delta t \\ t_{a2}=\dfrac{t_a}{2}+\Delta t \end{cases} \quad (4)$$

$$\Delta t=k\,\mathrm{sgn}(\Delta U_o)\,\mathrm{sgn}(i_a)t_a \quad (5)$$

$$\begin{cases} t_{c1}=\dfrac{t_c}{2}-\Delta t \\ t_{c2}=\dfrac{t_c}{2}+\Delta t \end{cases} \quad (6)$$

$$\Delta t=k\,\mathrm{sgn}(\Delta U_o)\,\mathrm{sgn}(-i_c)t_c \quad (7)$$

From Table II, the PWM calculation can be simplified greatly by establishing a mapping form sector II-VI into sector I. Accordingly, neutral-point balance can also be simplified similarly. From Equ.(5) and Equ.(7), the differentia in different sectors is which phase current can be used to balance the neutral-point in each sector which can be decided by the short vectors of each sector, as shown in Table III. When the reference vector lies in sector II, IV and VI, it should be noted that t_{a1} and t_{a2} express negative and positive short vectors respectively.

The scheme of BANB-BANG control achieves neutral-point balance only using the direction of phase current and the direction of neutral-point potential bias, so it is easy to be realized.

TABLE III.
Relationship between sectors and output currents in neutral-point balance

Sectors	I	II	III	IV	V	VI
Area A,C,E	$-i_c$	$-i_b$	$-i_a$	$-i_c$	$-i_b$	$-i_a$
Area B,D,F	i_a	i_c	i_b	i_a	i_c	i_b

V. EXPERIMENT RESULTS

The proposed control scheme is tested in 690V/200kW six-phase three-level SM drive system. The diagram of the system is shown in Figure6, of which the main features are:

1) 12-pulse uncontrolled rectifier
2) Six-phase NPC three-level inverter
3) Full digital control system based on multi-DSP
4) 690V/200kW six-phase SM
5) With load of direct current dynamo

First, the drive system is tested in stable state with full load, 2/3 load, half load, 1/3 load and no load respectively, and the results are shown in Figure7 and Figure8. Figure7 is the relationship between stator currents and rotor speeds, which shows that the stator current is in direct ratio with rotor speed and output power. Figure8 is the relationship between power factor and rotor speed, which shows that the power factor is closed to 1 at high load that verified the scheme of air gap flux observation proposed in this paper, the power factor is low with light load because the stator current contains more harmonics relatively such as 5th and 7th harmonics. Figure9 and Figure10 are the stator currents in 200kW, 450r/min and 58kW, 250r/min respectively. Figure11 is the stator voltage in 200kW, 450r/min. Figure12 is the wave of neutral-point potential, which shows that the amplitude of neutral-point potential is less than 10V, and also verified the scheme of neutral-point BANG-BANG control.

Fig 13 is four-quadrant run of SM at light load, where wave 1 is the torque part of stator voltage, and wave 2 is the torque part of stator current, which are calculated by

DSP and acquired by DA converter. In this Figure, from A to B, SM accelerates to highest positive speed, where stator current and stator voltage both are positive, SM runs in quadrant 1. From B to C, SM decelerates, where stator current is negative and the stator voltage is positive, SM runs in quadrant 2. From C to D, SM accelerates and runs at highest negative speed, where stator current and stator voltage both are negative, SM runs in quadrant 3. From D to E, SM decelerates, where stator current is positive and stator voltage is negative, SM runs in quadrant 4.

VI. CONCLUSION

In this paper, a new high-power drive scheme based on six-phase SM and three-level inverter is proposed, of which the key techniques such as air-gap flux oriented

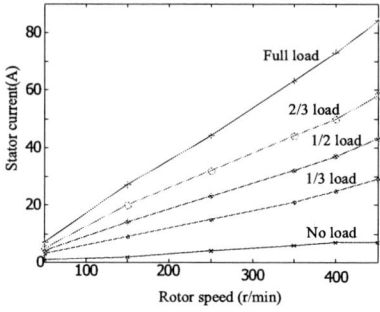

Figure 7. Relationship between stator currents and rotor speeds

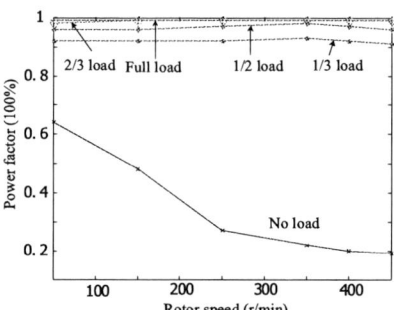

Figure 8. Relationship between power-factors and rotor speeds

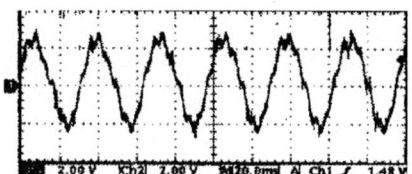

Figure 9. Stator current of SM in 200kW, 450r/min
(100A/div, 20ms/div)

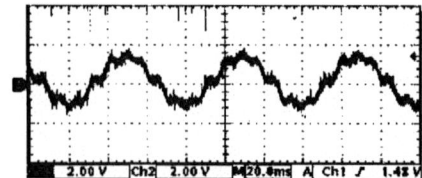

Figure 10. Stator current of SM in 58kW, 250r/min
(100A/div, 20ms/div)

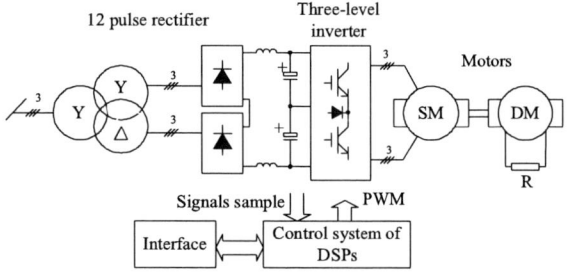

Figure 6. Diagram of SM drive system

940

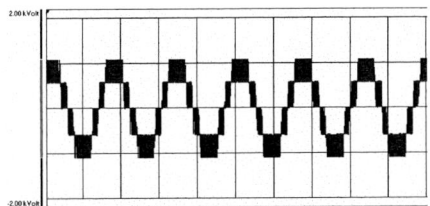

Figure 11. Stator voltage or SM in 200kW, 450r/min
(1kV/div, 20ms/div)

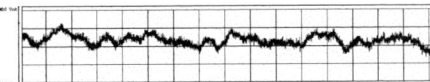

Figure 12. Neutral-point potential (2.5V/div, 10ms/div)

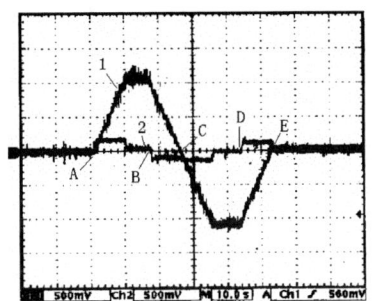

Figure 13. Four-quadrant run of SM with light load
(1:Voltage 10s/div, 300V/div, 2: Current 10s/div, 60A/div)

vector control of SM, the SVPWM of three-level inverter and neutral-point balance of NPC three-level inverter are studied. Finally, the proposed control scheme is tested in 690V/200kW six-phase three-level SM drive system.

REFERENCES

[1] Leila Parsa, "On Advantages of Multi-Phase Machines", IECON'05, pp.1574-1579., Nov. 2005

[2] Willems, J. "The analysis of high-phase-order power transmission systems", Circuits and Systems, IEEE Transactions on Volume 29, Issue 11, Nov 1982 Page(s):786 – 789

[3] Ma Xiaoliang, "High Power Cycloconverter and Vector Control System", China Machine Press

[4] W. Yao, Z. Lu, W. Fei, Z. Qiao and Y. Gu, "Three-Level SVPWM Method Based on Two-Level PWM Cell in DSP", IEEE APEC 2004, 2004.2, pp.1720-1724

[5] Von Jouanne, A.; Dai, S.; Zhang, H., "A multilevel inverter approach providing DC-link balancing, ride-through enhancement, and common-mode voltage elimination", Industrial Electronics, IEEE Transactions on Volume 49, Issue 4, Aug. 2002 Page(s):739 – 745

[6] Song Qiang, Liu Wen-hua, Yan Gan-gu etc, "a neutral-point potential for three-level algorithm for NPC inverter by using analytically injected zero-sequence voltage", Proceedings of the CSEE, Vol.24 No.5 May 2004

[7] Wei Lixiang, Liu congwei, SunXudong, Li Fahai, "DC Voltage Balance control method for three-level converter", J Tsinghua Univ (Sci & Tech), 2002,Vol.43, No.9

2006 5th International Power Electronics and Motion Control Conference

Doubly-Salient Permanent-Magnet Machine with Skewed Rotor and Six-State Commutating Mode

Yongbin Li[*, **], Chris Mi[**]

* School of Electrical Engineering, Shandong University, Ji'nan, Shandong, China

** Department of Electrical and Computer Engineering, University of Michigan-Dearborn, USA

Email: ybli@eee.hku.hk; mi@ieee.org

Abstract— New concepts are proposed for the design, optimization and control of doubly-salient permanent-magnet machine (DSPM). A uniform pole-width is employed in which the width of rotor pole is designed to be the same as that of the stator salient pole. This design is targeted to minimum the cogging torque. Secondly, a skewed rotor at one half of pole width is used to improve the commutation. Lastly, based on the flux-linkage of the skewed-rotor, a six-state commutating mode is used to control the motor. The feasibility of these new design concepts is validated by simulations, field analysis, and experimental results of the prototype.

Keywords - doubly-salient permanent-magnet machines, DSPM; skewed-rotor; pole-width; commutating-mode; cogging torque; simulation

I. INTRODUCTIONS

Doubly-salient permanent magnet machine (DSPM), a new kind of inverter-fed electrical traction motor first proposed in the early 1990's [1], is becoming more and more attractive because of its distinct features, such as high efficiency, high power density and simple structure. Progresses have been made on the design of DSPM. For example, Liao discussed the basic principal in 1992, 1993 and 1995 respectively [1-3]. Cheng and Chau studied the steady state characteristics and performance using the method of nonlinear varying-network magnetic circuit analysis [4-8].

Fig.1 shows the typical geometry of a DSPM with 6/4 pole-pair. It bears a resemblance to the structure of switched reluctance machines (SRM), except that permanent magnets are inserted in the stator. Therefore some common techniques used in SRM can be used in the design and control of DSPM. For example, wider rotor pole arc, advanced shut-off angle control, and lagged firing angle control can all be used in the design and control of DSPM. Due to existence of permanent magnets in the stator, the behavior of DSPM is different from that of DSPM. Therefore, new design and control concepts need to be explored to optimize the performance of DSPM. In this paper, a novel DSPM with skewed-rotor and six-state commutating mode is investigated. Simulations and experimental data of the prototype verified the validity of the design concepts.

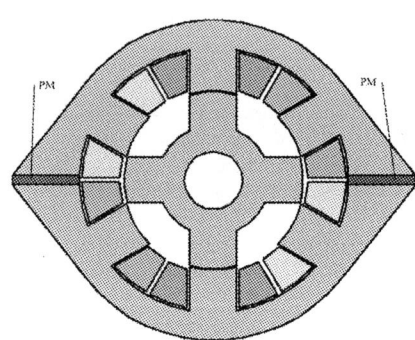

Fig. 1. Typical DSPM geometry with 6/4 pole pairs.

II. SKEWED-ROTOR DESIGN CONCEPT

In the design practices of SRM, in order to ensure the winding commutation and self-start capability at any rotor position and either rotating direction, there should be a small overlap between the adjacent stator and rotor salient poles when the axes of the stator pole is aligned with that of the rotor pole. Therefore the width of the rotor pole width is usually larger than that of the stator. This technology is also used in DSPM [4], because of the structure similarity. Due to the existence of permanent magnets in the stator, cogging torque exists in DSPM. According to the Flux-MMF diagram of PM machines, cogging torque will reach its minimum if the resultant gap reluctance is uniform at any rotor position [9]. Therefore, if the rotor pole width is larger than the rotor pole width, the cogging torque will be significant for 6/4 pole-paired DSPM because the gap reluctance will not be uniform as the rotor position varies. Cogging torque is one of the most important issues of DSPM.

In order to minimize the cogging torque of DSPM, the width of rotor pole-width is designed to be the same as that of the stator, and both equal to half of pole-pitch,

$$\theta_s = \theta_r = \frac{\pi}{N_s}. \qquad (1)$$

When neglecting saturation and fringe effects, cogging torque is eliminated in a 6/4 pole-paired DSPM design if the width of rotor pole-width equals to that of stator and their width is a half of the stator's pitch, or $30°$

1-4244-0448-7/06/$25.00 ©2006 IEEE

942

mechanic degree [3].

Secondly, skewed-rotor is used to ensure the capability of self-starting at any rotor position and either rotating direction. From the curve of flux linkage, skewed-rotor can also lead to overlap between the adjacent of stator and rotor salient poles, which is the same effect as that of the larger rotor pole-width used in SRM; and in order to obtain the largest output, the skew-angle of the rotor is chosen to be half of stator salient pole-width,

$$\theta_{skew} = \frac{\theta_r}{2} = \frac{\pi}{2N_s} \qquad (2)$$

For a 6/4 pole-paired DSPM, the skew-angle is 15°.

The flux linkage of the skewed rotor DSPM is shown in Fig. 2(a). For comparison, the flux linkage of the DSPM with un-skewed rotor is shown in Fig. 2(b).

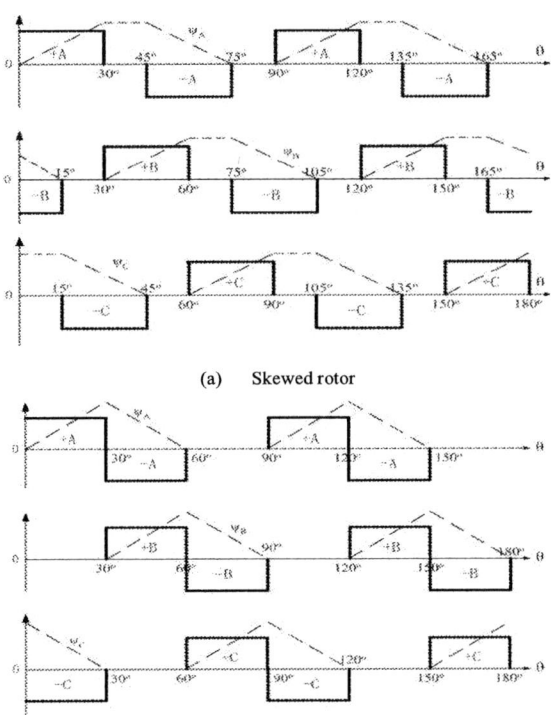

(a)   Skewed rotor

(b)   Un-skewed rotor

Fig. 2. Flux linkage (dashed lines) and commutating mode (solid lines). (a) skewed rotor DSPM; (b) un-skewed-rotor DSPM

It can be seen from Fig.2 that the flux linkages of the skewed rotor DSPM are different from that of the un-skewed rotor DSPM. The magnitude of flux linkage of the skewed rotor DSPM decreased, and the coverage of flux linkage increased from 120° to 150°. As a result, the six-state commutating mode was employed.

III. SIX-STATE COMMUTATING MODE

In the conventional control of DSPM, three-state commutating mode was used according to its flux linkage. The conducting sequence are: +A-C, -A+B and –B+C respectively as shown in Fig.2(b). This simple commutating mode generally results poor performance of the DSPM. It can be seen from Fig. 3 that the current commutating from positive to negative is in sequence. Because of the current continuity, the switch-off angle of the positive current should be advanced, or the flywheel current will create a reverse torque, as shown in Fig. 3. Fig.3 also shows the asymmetrical current and fluctuated torque without advanced switch-off angle control.

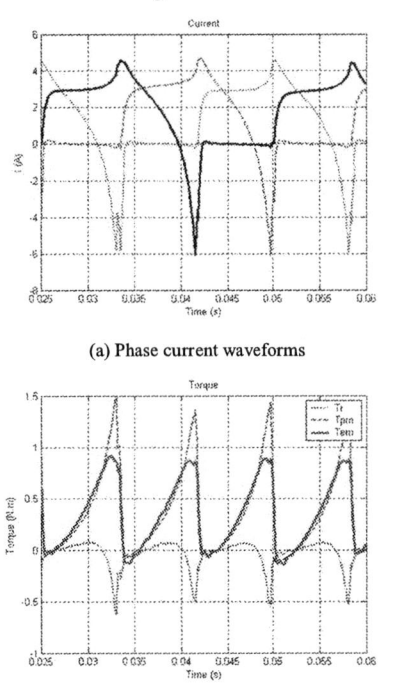

(a) Phase current waveforms

(b) Output Torque waveform

Fig. 3.   Steady state waveforms under conventional conducting mode

According to the same commutating principle as conventional DSPM motor, i.e., positive current conducts at the rising slope of flux linkage, while negative conducts at the falling slope, a new commutating mode for the novel skewed rotor DSPM can be developed as shown in Fig.2(a). Six-state commutating mode is used. The conducting sequences are +A-B, +A-C, +B-C, +B-A, +C-A and +C-B respectively. Each state will be conducted for 60° electrical degree continuously, and there is a 60° interval between the positive and negative current commutation. This commutating mode makes it be possible to neglect the control of switch-off angle during the commutating moment. Therefore commutating performance can be improved and the reliability enhanced.

943

IV. SIMULATION

In order to confirm the proposed design, the performance of the proposed DSPM is simulated. The mathematic model of DSPM can be expressed as following:

$$\begin{cases} \dfrac{d}{dt}[\Psi] = [u] - [R]\cdot[i] \\[2mm] \dfrac{d\omega}{dt} = \dfrac{1}{J}\cdot(T_e - T_l - k_\omega\cdot\omega) \end{cases} \tag{3}$$

and

$$\begin{cases} [i] = \dfrac{1}{[L]}\big([\Psi] - [\Psi_{PM}]\big) \\[2mm] T_e = \dfrac{1}{2}[i]^T\left(\dfrac{\partial}{\partial\theta}[L]\right)[i] + \left(\dfrac{\partial}{\partial\theta}[\Psi]_{PM}^T\right)[i] \end{cases} \tag{4}$$

Where $[u]=[u_A\ u_B\ u_C]^T$ is the stator voltage, $[i]=[i_A\ i_B\ i_C]^T$ is the stator current, $[\psi]=[\psi_A\ \psi_B\ \psi_C]^T$ is the flux linkage of the stator winding, $[R]=\text{diag}(R_A, R_B, R_C)$ is the stator resistance, $[\Psi_{PM}]$ is flux linkage of the permanent magnets, and $[L]$ is the stator winding inductance matrix. The permanent magnet flux and the stator winding inductance can be obtained from finite element analysis (FEA) [12].

The simulated steady-state performance of un-skewed rotor under three-state commutating mode is shown in Fig. 4. Compared with Fig. 3, the performance of DSPM has been improved by the advanced shut-off angle. The positive conducting time has been shortened from 120° to 108°. But the output toque ripple is still notable. The torque also crosses over t-axis as shown in Fig. 4(b). Therefore the motor can't self-start.

The simulated steady state performance of the skewed rotor under the six-state commutating mode is shown in Fig. 5. Obviously, the magnitudes of positive and negative current are more identical than those shown in Fig.4. The output toque was increased and the torque ripple was significantly decreased. Self-start capability can be easily achieved.

Six-state commutating mode has been widely used in Brushless DC motors (BLDCM) control [10], but to date it was seldom discussed in the control of DSPM.

V. PROTOTYPE AND EXPERIMENTATION

In order to validate the proposed methods, two prototypes were designed and built, one with un-skewed rotor, the other with skewed rotor as shown in Fig. 6.

The flux distributions at different rotor positions are shown in Fig. 7. The winding flux-linkages for the skewed rotor and un-skewed rotor are shown in Fig. 8.

A hardware control system based on ADMCF34X was developed and the current waveform at steady-state is shown in Fig. 9. The starting performance is shown in Fig.

10, which illustrated its self-starting ability. They are consistent with the simulation and FEA results.

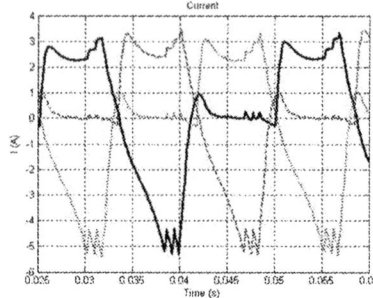

(a) Phase current waveforms

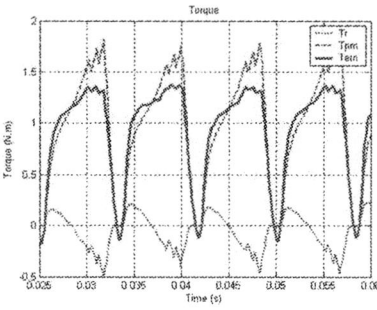

(b) Output torque waveforms

Fig. 4 Steady state performance of DSPM with un-skewed rotor

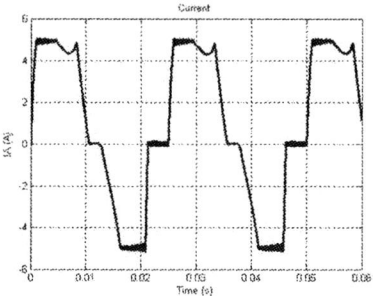

(a) Current waveform

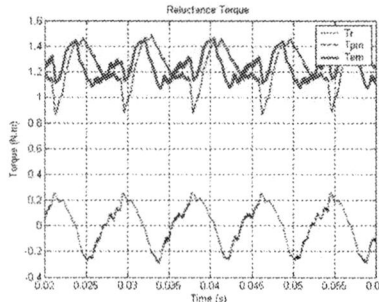

(b) Output torque waveforms

Fig. 5. Steady state performance of DSPM with skewed-rotor

VI. CONCLUSION

A new skewed-rotor design concept for 6/4 pole-paired DSPM, which refers to the rotor with the same pole-arc as that of the stator and skews at a half pole-arc, is proposed in this paper. And based on the flux-linkage of the skewed-rotor, a six-state commutating mode for DSPM is presented. There are some advantages of these design concepts, such as minimum cogging torque, small torque ripple, simplified control parameters and self-starting ability, et al. The feasibility of these ideas is verified by digital simulation and experimental results.

Fig. 6. Photo of the prototype rotors. Top: un-skewed rotor. Bottom: skewed rotor

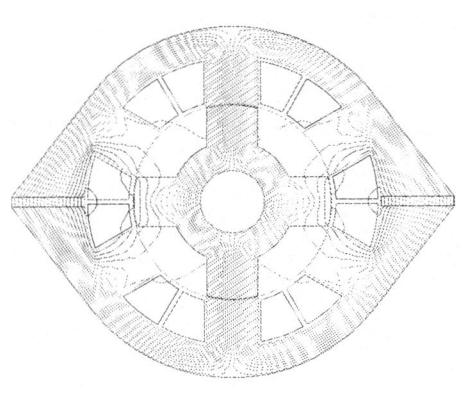

(a) 0°

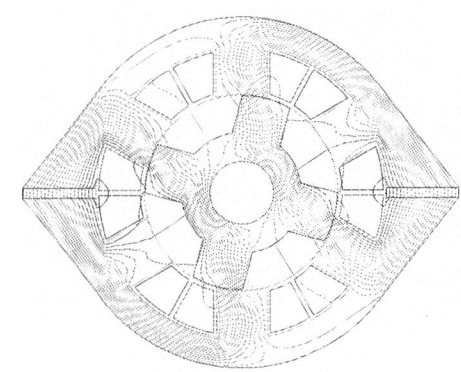

(b) 15°

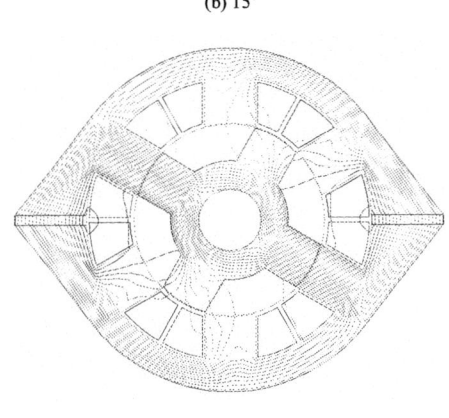

(c) 30°

Fig. 7. Flux distributions at different rotor positions

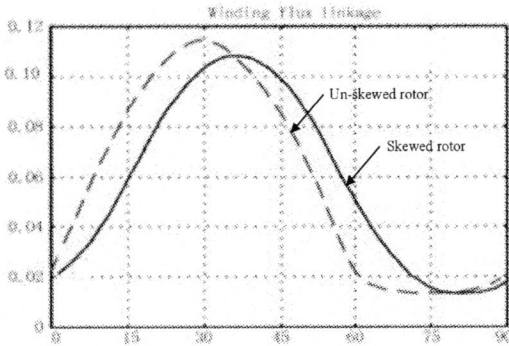

Fig. 8. Flux waveforms of the two rotor designs

945

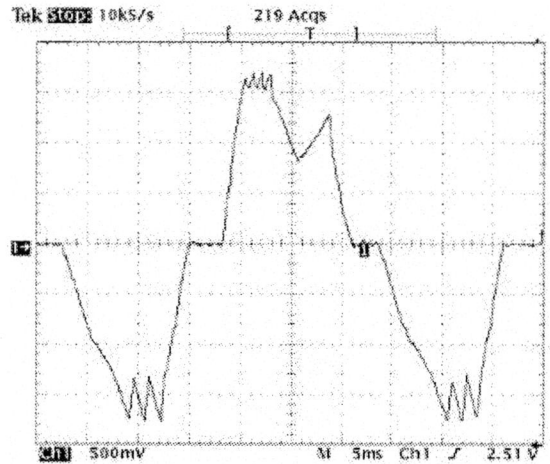

Fig,.9. Measured steady current waveform

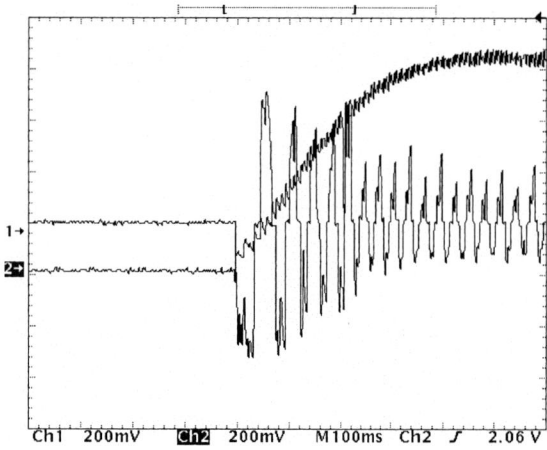

Fig. 10. Measured starting speed (2) and current (1) waveforms

REFERENCES

[1] Y. Liao, F. Liang, T. A. Lipo, "A novel permanent magnet motor with doubly salient structure", *Industry Applications Society Annual Meeting, Oct. 1992*, pp.308 -314.

[2] Y. Liao, T. A. Lipo, "Sizing and optimal design of doubly salient permanent magnet motors", *Sixth International Conference on Electrical Machines and Drives*, Sep 1993, pp.452 –456.

[3] Y. Liao, F. Liang and T. A. Lipo, "A Novel Permanent Magnet Motor with Doubly Salient Structure", *IEEE Trans. on IA*, 31 (5), 1995, pp.1069 –1078.

[4] M. Cheng, K.T. Chau, C.C. Chan, "Nonlinear varying-network magnetic circuit analysis for DSPM motors", *IEEE Trans. on Magnetics*, 36(1), 2000, pp.339 –348.

[5] M. Cheng, K. T. Chau, C. C. Chan, "Static characteristics of a new doubly salient permanent magnet motor", *IEEE Trans. on Energy Conversion*, 16 (1), 2001, pp. 20 –25.

[6] M. Cheng, K.T. Chau, C.C. Chan, Qiang Sun; "Control and operation of a new 8/6-pole doubly salient permanent-magnet motor drive", *IEEE Transactions IA*, 39(5), 2003, pp.1363 – 1371

[7] M. Cheng, K.T. Chau, C.C. Chan, " Design and analysis of a new doubly salient permanent magnet motor", *IEEE Trans. on Magnetics*, 37(4), 2001, pp.3012 – 3020

[8] K.T Chau,.; Qiang Sun; Ying Fan; Ming Cheng, "Torque ripple minimization of doubly salient permanent-magnet motors", *IEEE Trans. On EC*, 20(2), 2005 pp.352 - 358

[9] Deodhar R. P., Staton D. A., Miller T. J. E., "Prediction of cogging torque using the flux-MMF diagram technique", *IEEE Trans. on IA*, 1996, 32(6), pp.569 – 576.

[10] J. R. Hendershot Jr., T.J.E Miller, *Design of Brushless Permanent-Magnet Motors.* Magna Physics Publishing and Clarendon Press, Oxford, 1994.

[11] Yongbin Li, Jianzhong Jiang and K. T. Chau. "Research on a Novel Doubly-fed, Doubly-Salient Permanent Magnet Machine", CSEE，2005, 25(1), pp198-203.

[12] Martis, C.; Radulescu, M.M.; Biro, K., "On the dynamic model of a doubly-salient permanent-magnet motor", MELECON 98, 1998 pp. 410 - 414

[13] Hu Qinfeng; Yan Yangguang; "Steady Characteristics Analysis Method of Doubly Salient Permanent Magnet Motor System" , *ICEMS 2005*, Volume 3, 2005, pp.1917 – 1921

[14] Sekhar babu, A.R.C.; Rajagopal, K.R., "FE analysis of multi-phase doubly salient permanent magnet motors", *INTERMAG Asia 2005*. pp.733 – 734

[15] Y. Fan, K.T. Chau, M. Cheng, "A New Three-Phase Doubly Salient Permanent Magnet Machine for Wind Power Generation", *IEEE Trans. On IA*, 42(1), 2006, pp.53 – 60

2006 5th International Power Electronics and Motion Control Conference

Sensorless Control and PMSM Drive System for Compressor Applications

Dongsheng Li[*], Takahiro Suzuki[*], Kiyoshi Sakamoto[*], Yasuo Notohara[*], Tsunehiro Endo[**],
Chikara Tanaka[**] and Tatsuo Ando[***]
[*] Hitachi Research Laboratory, Hitachi, Ltd., JAPAN
[**] Hitachi, Ltd. Power Systems, JAPAN
[***] Hitachi Appliances, Inc., JAPAN
dongli@gm.hrl.hitachi.co.jp

Abstract—**Sensorless control strategies and a PMSM drive system for compressor applications are presented. The drive system consists of a boost PFC rectifier and a three-phase inverter controlled by one Micro-Controller-Unit (MCU: SH7046, Renesas Tech.). To realize low-cost, high performance, and high reliability, sensorless (current and position) control and a novel vector control strategy are proposed to satisfy the requirements of different compressors. The effectiveness and validity of the proposed control strategies are confirmed by experimental results.**

Keywords-PMSM; sensorless control; vector control; inverter

I. INTRODUCTION

Since the early 1990's, air-conditioners with inverter driven Permanent Magnet Synchronous Motors (PMSMs) have become increasingly popular due to comfort and high efficiency [1]. In fact, from 1998, the ratio of inverter driven room air-conditioners sold in Japan is above 90 percent. With a worldwide shortage of energy, it is clear that the other countries will follow the trend in the future [2].

In the PMSM drive system of a compressor, trapezoidal current (120-degree commutation) drive mode is widely used because of its simplicity and low cost. However, the vibration of the compressor and the losses of the motor are relatively large because of the pulsating torque and distorted motor currents.

Recently, with the remarkable progresses of sensorless control technology and the improvements of low-cost Micro-Controller-Units (MCUs), sinusoidal current (180-degree commutation) drive mode has become available for these applications.

In this paper, a PMSM drive system for compressor applications is presented. The drive system consists of a boost Power-Factor Correction (PFC) rectifier and a three-phase inverter controlled by only one MCU.

To realize low cost, high performance, and high reliability, current and position sensorless control technologies are adopted. Moreover, a novel vector control strategy is proposed to satisfy the requirements of compressors.

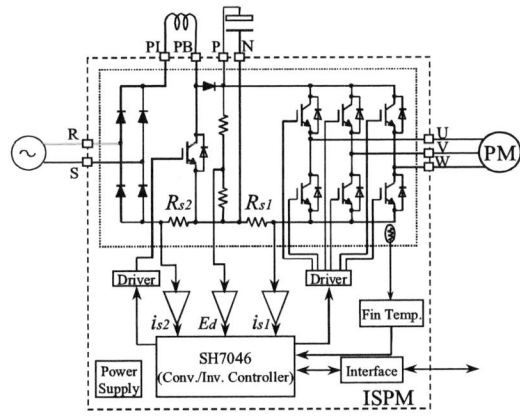

Figure 1. Configuration of proposed PMSM drive system

To satisfy the guidelines of harmonics and enlarge the rated input ac voltage range, a boost type PFC rectifier is added. The control of the rectifier is carried out by the same MCU, also with sensorless control.

The proposed drive system has the characteristics of:

Low cost: There are no current/voltage/position sensors in the system. The inverter and rectifier are controlled by one MCU.

High efficiency: The input source current and motor currents are almost sinusoidal; and currents of the d-q axis can be individually controlled to minimize the motor currents.

High performance: With a boost PFC rectifier, the input voltage range can be significantly enlarged and the input current can be kept sinusoid; with the proposed vector control strategy, the speed range can be enlarged even with a low-cost microcomputer.

High reliability and small size: All power devices (IGBTs and diodes), driver circuits, control circuits, interface circuit and the power supply are enclosed in one power module, named the Inverter System Power Module (ISPM).

In this paper, the configuration of the proposed drive system, the algorithms of the sensorless control and PFC rectifier control are introduced firstly. Then, a simplified

1-4244-0448-7/06/$25.00 ©2006 IEEE   947

vector control strategy suitable for the PMSM drive for compressor applications is investigated in detail. Finally, experimental results are given to verify the effectiveness of the proposed drive system.

II. PMSM DRIVE SYSTEM

A. Configuration of Drive System

The proposed PMSM drive system consists of a three-phase inverter, a boost PFC rectifier and a digital control unit (MCU) as shown in Fig. 1. The inverter and rectifier are controlled by the same MCU (SH7046, Renesas Tech.). The currents of the inverter and rectifier are determined through the shunt resistance R_{s1} and R_{s2} instead of current sensors. It should be noted that an ac source voltage signal is not necessary in the control of the PFC rectifier due to the proposed control. Therefore, there are no current/voltage sensors besides the two shunt resistances in this drive system. Furthermore, with the position sensorless control, the position or speed sensors of a regular PMSM can also be omitted.

B. Current Sensorless control

The technology to determine (reconstruct) the motor currents from the dc-link current was reported by Green and Williams [3] and other researchers. Fig. 2 shows the waveforms of the carrier, three-phase reference signals, corresponding dc-link current and equal circuits.

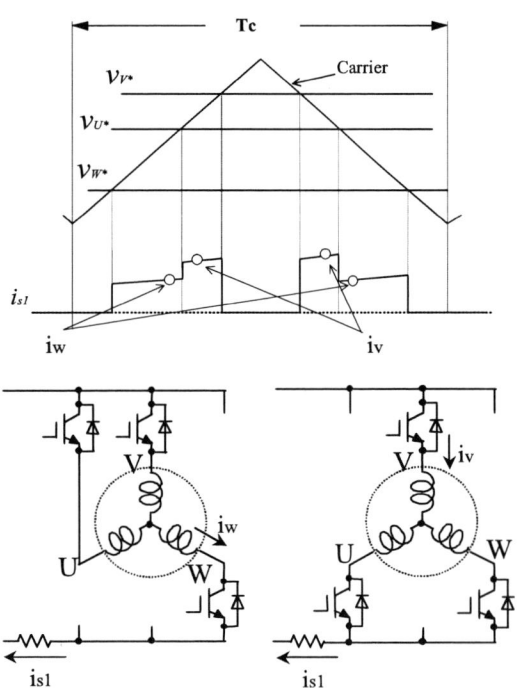

Figure 2. Three-phase reference, carrier and dc-link current (upper) and equal circuits (lower)

As shown in the figure, during one PWM period, there are two chances for sampling the motor current for each phase.

C. Position sensorless control

The position sensorless control is realized by using the direct position error estimation approach presented by author [4].

With the direct position error estimation approach, the position error of the rotor can be obtained from Equation (1):

$$\Delta \theta_c = \tan^{-1}\left(\frac{v_{dc}^* - r \cdot i_{dc} + \omega_1 L_q \cdot i_{qc}}{v_{qc}^* - r \cdot i_{qc} - \omega_1 L_q \cdot i_{dc}} \right) \quad (1)$$

where $\Delta \theta_c$ is the estimated position error between the assumed d-q axis and the real d-q axis of the motor as shown in Fig. 3; r is the stator winding resistance; L_q is the q-axis inductance; ω_1 is the reference of the inverter angular velocity; and v_{dc}, v_{qc} are the d-q axis motor voltages (which can be approximated from the output reference); and i_{dc} and i_{qc} are the determined motor currents.

As shown in Fig. 6, a phase-locked loop (PLL) is employed to adjust the assumed position θ_{dc} in the proposed control. In other words, the PLL controller adjusts the assumed rotor speed ω_1 using $\Delta \theta_c$, and the assumed rotor position θ_{dc} is obtained by integrating with ω_1. One of the benefits of this scheme is that the controller is very stable.

D. Control of PFC rectifier

To improve the power factor of a single-phase rectifier, a boost type PFC rectifier composed with a full-bridge diode rectifier and a boost dc-dc converter is widely used. Usually, the control of the PFC rectifier can be carried out by an analog IC such as UC3854.

In the proposed drive system, control of the boost PFC

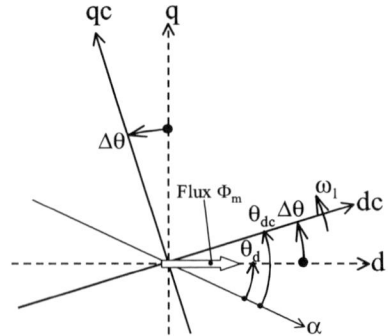

Figure 3. Definition of the rotationary reference frames

948

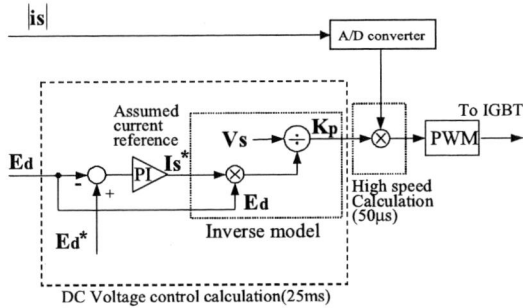

Figure 4. Block diagram of PFC rectifier with stabilization of dc voltage control

rectifier is carried out by the same MCU to simplify the composition and reduce costs. In addition, a novel control method that can provide a sinusoidal input current without using an input voltage sinusoidal standard waveform is adopted.

In a steady state, the switch duty ratio d of the boost PFC rectifier can be expressed as

$$d = 1 - \frac{|v_s|}{E_d} = 1 - K_p \cdot |i_s| \qquad (2)$$

where $v_s = \sqrt{2} \cdot V_s \cdot \sin \omega t$ is the source voltage, i_s is the source current, and K_p is the current control gain.

From Equation (2), i_s can be expressed as

$$i_s = \frac{\sqrt{2} \cdot V_s \cdot \sin \omega t}{K_p \cdot E_d} \qquad (3)$$

Equation (3) indicates that if the dc voltage E_d and control gain K_p are kept steady, the source current will be a sine wave. Furthermore, the amplitude of the source current can be controlled by adjusting K_p. Accordingly, the dc voltage can be regulated though feedback control of K_p.

The block diagram with stabilization of dc voltage control is shown in Fig. 4. To deal with the nonlinear relationship between the dc voltage error $\Delta E_d = E_d^* - E_d$ and the control gain K_p, an inverse model is used.

It should be noted that only the calculation of the multiplication of i_s and K_p is needed in every PWM carrier period as shown in Fig. 4. The other processes can be done in a much longer period, for example 25[ms]. Therefore, the switching frequency can be up to 20[kHz] or higher without a tremendous increase in the process load of the MCU.

III. VECTOR CONTROL STRATEGIES

The vector control was widely used in servo and ac induction motors. To obtain high performance, Auto-Speed-Regulators (ASRs) and Auto-Current-Regulators (ACRs) are commonly used. An example of a conventional vector control is shown in Fig. 5. However, with the limitation in the processing ability of low-cost MCUs, control in the high-speed region (>5000[rpm]) becomes very difficult with conventional vector control. To solve this problem, a novel vector control strategy is proposed.

Since the needs of the responses of speed and torque are not as strict as for servo motors in compressor applications, it is possible to omit some feedback control loops to

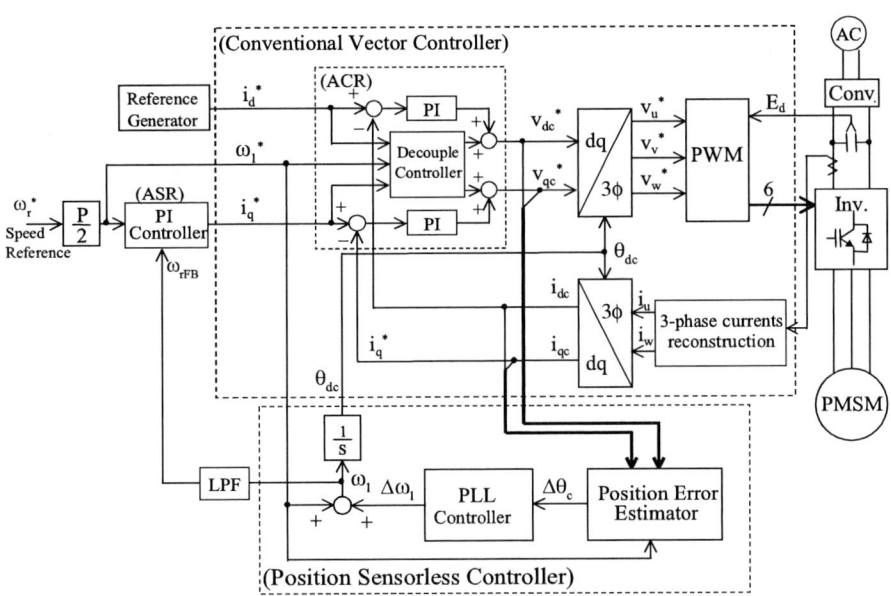

Figure 5. An example of conventional vector control for sensorless PMSM drives

949

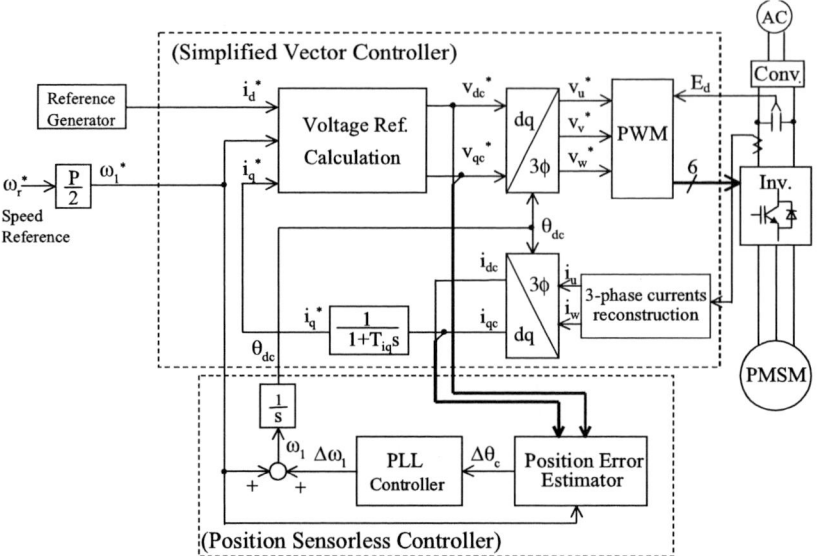

Figure 6. SMART vector control for sensorless PMSM drive

simplify the whole control system.

Fig. 6 shows the proposed simplified vector control strategy, named SMART vector control. Unlike conventional vector controllers, the SMART vector control eliminates the speed regulator and the current regulator in order to realize a simple control system.

The current reference of the q-axis, i_q*, is obtained by the determined motor current, i_{qc}, using Equation (4).

$$i_q^* = \frac{1}{1 + T_{iq} s} i_{qc} .$$ (4)

where T_{iq} is the low-pass filter time constant.

The current reference of the d-axis, i_d*, is set to minimize the amplitude of the motor current in order to decrease losses.

IV. EXPERIMENTAL RESULTS

Experiments were carried out with proposed drive system. The specifications of the proposed drive module and the parameters used in experiment are listed in Table I.

Fig. 7 shows the input source current and motor current at 3600[rpm]. One can see that both currents are almost sine waves.

Fig. 8 shows the speed response of PMSM with the proposed SMART vector control. This result confirms that the speed response can satisfy the requirements of compressor applications.

V. CONCLUSIONS

An integrated drive system and a novel vector control strategy of PMSM for compressor applications were introduced in this paper. Fig. 9 shows the outside of the proposed Inverter System Power Module (ISPM).

TABLE I. Specifications of ISPM and parameters used in experiment

Rated Input Voltage	1Φ-200/220V
Rated Input Current	30A(rms, max.)
Rated Output Current	28A(rms, max.)
Carrier Freqs.	20kHz (Conv.)/7kHz (Inv.)
Dimension	W138*D110*H148 mm³
Weight	610g
Rated Power of Motor	3.7 kW

To reduce costs and improve reliability, both motor

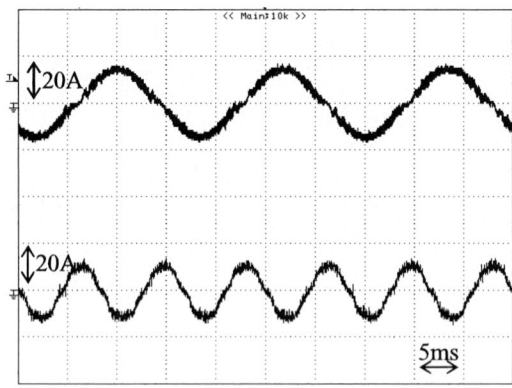

Figure 7. Source current (upper, 20A/div) and motor current (lower, 20A/div) with SMART vector control

current and position sensors are estimated using sensorless control. In addition, a PFC rectifier controlled by the same MCU is added to reduce input current harmonics and enlarge input voltage range.

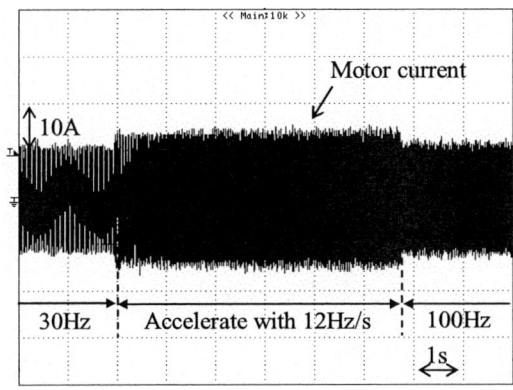

Figure 8. Speed response of PMSM with SMART vector control from 30[Hz] to 100[Hz]

Figure 9. Outside of proposed Inverter System Power Module (ISPM)

The PFC rectifier is controlled with a unique control strategy, which can be realized simply using a digital controller. In the proposed control strategy, a source voltage signal is not necessary and the switching frequency can be above 20[kHz] without the problem of overloading the processing ability.

The SMART vector control, which can be accomplished by a very low-cost MCU due to its simplified configuration, shows a good speed response and steady-state characteristics even without ACR and ASR control loops.

The proposed drive system has been successfully used in room air-conditioners and package air-conditioners.

REFERENCES

[1] K. Ohyama, and S. Matsuno, "Advances of Power Electronics Technology in Air Conditioners," *Journal of IEE Japan*, vol. 125, no. 12, pp. 772–775, 2005. (in Japanese)

[2] V. R. Stefanovic, "Trends in ac Drive Applications," *IPEC'2005*, pp. 72-80, 2005.

[3] T. C. Green, and B. W. Williams, "Derivation of Motor Line-current Waveforms from the dc-link Current of an Inverter," *Proc. Inst. Elect. Eng.*, vol. 136, no. 4, pp. 196–203, July 1989.

[4] K. Sakamoto, Y. Iwaji, T. Endo, and Y. Takakura, "Position and Speed Sensorless Control for PMSM Driver Using Direct Position Error Estimation," *IECON'01*, pp. 1680–1685, 2001.

[5] K. Iizuha, H. Uzuhashi, M. Kano, T. Endo, and K. Mohri, "Microcomputer Control for Sensorless Brushless Motor," *IEEE Trans. Industry Applications*, vol. IA-21, no. 4, pp. 595–601, 1985.

[6] F. Blaabjerg, J. K. Pedersen, U. Jaeger, and P. Thoegersen, "Single Current Sensor Technique in the DC Link of Three-Phase PWM–VS Inverters: A Review and a Novel Solution," *IEEE Trans. Ind. Applicat.*, vol. 33, no. 5, pp. 1241-1253, 1997.

[7] K. Sakamoto, Y. Iwaji, and T. Endo, "A Simplified Vector Control of Position Sensorless Permanent Magnet Synchronous Motor for Electrical Household Appliances," *T. IEE Japan, vol.124, no.11*, pp. 1133–1140, 2004. (in Japanese)

[8] Y. Notohara, T. Endo, T. Suzuki, K. Sakamoto, K. Murayama, and S. Furusawa, "Examination of All Digital Single Phase Converter," *IEE Japan, Annual Meeting*, 2003. (in Japanese)

[9] T. Suzuki, Y. Notohara, T. Endo, C. Tanaka, D. Kawase, and H. Umeda, "Development of the Inverter System Power Module Equipped with a Converter-inverter Controller Realized with One Microprocessor," *IEE Japan, Annual Meeting*, 2004. (in Japanese)

2006 5th International Power Electronics and Motion Control Conference

Analysis and Experimental Study of Slot Effect in Synchronous Reluctance Permanent Magnet Motors

Wei Guo，Zhengming Zhao，Yingchao Zhang
Department of Electrical Engineering , Tsinghua University
State Key Laboratory of Control and Simulation of Power System and Generation Equipment
Beijing, P.R.China
Guo-w@mail.tsinghua.edu.cn

Abstract—**Slot effect in Synchronous Reluctance Motors with Permanent Magnet assisted (SR-PM) are deeply analyzed, reduced and verified by experiments in this paper. The three different type sample SR-PM motors are introduced firstly. Then how the rotor layer numbers affect the torque characteristics is deeply analyzed, including the torque-angle, steady operating torque and position-setting torque. Then the reduction method is presented with simulation results. The detail experiment results are given in the final. The results shows that both more layers and stator slot skewing can reduce the teeth harmonics. However for the rotor with sintered NdFeB magnet, stator slot skewed is the best way because it only has few layers due to manufacture reason.**

Keywords-Reluctance motors; Permanent magnet motors; slot effect

I. INTRODUCTION

A synchronous reluctance motor with permanent magnet assisted (SR-PM) is a combination of a synchronous reluctance motor (SynRM) and a permanent magnet synchronous motor (PMSM). Because a SR-PM motor has many good features, such as high power density and power factor, excellent efficiency and a wide constant-power speed range, it has been a focus of research recently. The rotor structure of this motor has many different types. The rotor can radial-laminated or axially-laminated. Fig.1 shows a SR-PM motor [1] with a axially-laminated rotor with the permanent material added in between and mounted on the steel bracket. The stator is the same as that of a common induction motor (IM).

Since SR-PM machines share both the characteristics of SynRM and PMSM, the research on SynRM and PMSM is a good start [2][3][4][5][6][7][13][12]. For example the analysis of the axially-laminated SynRM by Staton D.A [3], the analysis and design of axially-laminated interior permanent magnet motor with the shape of sandwich by Soong W.L. [2] and the optimization of the added permanent magnetic amount.

Usually slot effect exists in the SR-PM motors due to the interaction between the stator slots and rotor layers. This effect acts as torque ripple and position-setting torque which are very harmful to servo application. Unlike IM, SR-PM motors have no slots on rotor but many layers, the number of layers N_L plays an

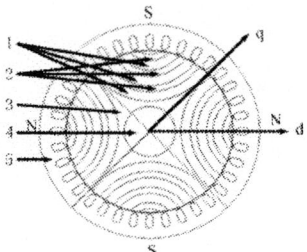

1-PM, 2-Silicon-steel sheet, 3-Bracket, 4-Shaft, 5-Stator

Figure 1.　SR-PM motor structure

important role in slot effect, which is similar to the number of rotor slots in IM. So how N_L affect the torque-angle characteristics, steady operating torque and position setting torque are deeply analyzed. Then the reduction methods of slot effect are also introduced. The proof tests have been carried out on three sample motors which are listed in TABLE I. Motor 1 (M1) and 2 (M2) have been analyzed in ref [14]. Though M1 has reached the required power yet, the temperature rise and the position setting torque make the dynamic response worst. To solve these problems, the third motor (M3) is developed. The lower field density, more high speed , stator slot skewing and radial-laminated rotor makes it lower loss, high efficiency, little harmonics and good dynamic characteristics.

TABLE I.　PARAMETERS

	Motor 1	Motor 2	Motor 3
Structure	axially-laminated	axially-laminated	radial laminated
Magnet	sintered NdFeB magnet	rubber NdFeN magnet	sintered NdFeB magnet
Rated power （kW）	1.0	1.0	1.0
Rated speed （rpm）	8000	8000	11000
Number of poles	4	4	4
Number of stator slots	36	36	36
Stator Slot skewing	no	No	yes

It is must be noted that the sintered NdFeB magnet (SNM) has been used in M1 which has axially-laminated rotor. The build method is introduced in [14]. The rotor of M3 has been modified to radial-laminated for

1-4244-0448-7/06/$25.00 ©2006 IEEE
952

structure-strengthen in high speed and simplifying working process.

The experiment also verify that the skewed stator slots can eliminate the harmonics without degrade the performance and a good design of radial-laminated motor can also lead to good performance with feasible manufacture technology.

II. SLOT EFFECT ANALYSIS AND REDUCTION

A. Torque-current Angle Curves and Harmonic Analysis

In control model (vector control and direct torque control) of the SR-PM motors, the following linear torque equation is adapted generally,

$$T_e = \frac{3}{2} p(\psi_{PM} i_q + (L_d - L_q) i_d i_q) \tag{1}$$
$$= \frac{3}{2} p(\psi_{PM} i_s \sin \delta + \frac{(L_d - L_q) i_s^2 \sin 2\delta}{2})$$

This Equation shows that torque lies on amplitude and phase angle of stator current and is irrelevant to rotor position. In fact, δ is defined in synchronize coordinate, the relation to the fixed coordinate is

$$\delta = \theta_s - \theta_r \tag{2}$$

Where, θ_s is the angle between i_s and a-axis, θ_r is the angle between d-axis and a-axis. So to calculate the torque-current angle curves, we can vary the phase of i_s with the constant θ_r and amplitude of i_s or vary θ_r with constant i_s. The slot effect is ignored in the former method because relative position is not changed while the latter includes slot effect. Fig.2(a) is the $T - \delta$ curves acquired by the latter method with different rotor layers N_L. There are apparently ripple in torque. The corresponding FFT analyses are shown in Fig.2 (b). Except for large fundamental wave component, the times of master harmonic is multiple of 9. The period of 9th harmonic is 10 mechanical degrees, which is same as teeth pitch. It verifies that stator slots have relation to harmonic. For further study of relationship of harmonic and N_L, the harmonic amplitude are shown in Fig.2(c). It can be seen that 9th and 18th harmonic reach its maximum at 3 layers and 6 layers respectively while 27th harmonic does not reach its maximum at 9 layers. The harmonic effect will decrease with the increase of N_L on the whole.

The above analysis shows that rational choice of N_L will improve motor performance with decreased manufacturing complex.

B. Torque Ripple at Steady-state Operation

Except that slot effect affects the $T - \delta$ characteristics, it has more influences on torque ripple

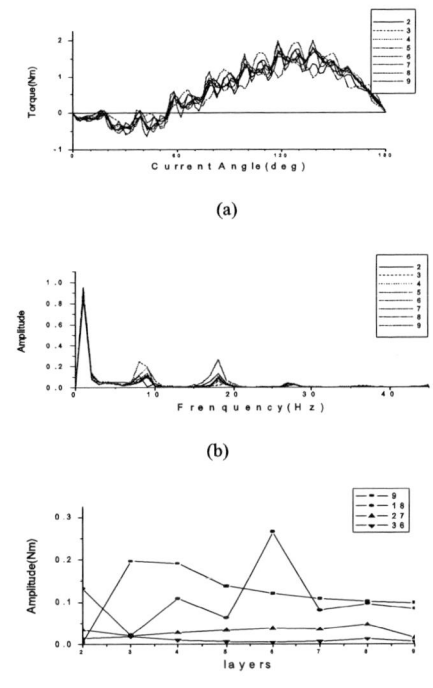

(a)

(b)

(c)

Figure 2. $T - \delta$ torque curves. (a) Torque-current angle characteristics with different N_L. (b) FFT analysis. (c) Harmonic amplitude vs. N_L.

at steady-state operation. The torque ripple causes the acoustic vibration of motor. The vibration and acoustic noise will aggravate when the harmonic torque frequency is close to the resonant frequency of the motor. Furthermore the control model of motor in practice is linear or nonlinear one ignored of the slot effect. The torque will be constant with constant current amplitude and constant δ from control aspect. However the exist of slot effect make this false. This effect is also caused by interaction of stator slots and rotor layers. The torque profile versus the rotor position are showed in Fig.3(a), with constant i_s and δ. The corresponding FFT analyses are showed in Fig.3 (b). In calculation, the rated current is applied and δ is optimized against with layers.

From Fig.3 (a) and (b) it can be seen that the torque ripple are deeply impressed by N_L. The third harmonic has the maximum amplitude and others are very lower oppositely. The harmonic amplitude is becoming more and more small with the harmonic times increase. FFT analysis does not only open out the harmonic component and trends, but also reveal the relation between N_L and harmonic torque. Fig.3(c) is the curves of harmonic torque amplitude vs. N_L. It is clear that the third and sixth harmonic reached maximum at 3 layers and 6 layers respectively while the ninth harmonic is very small

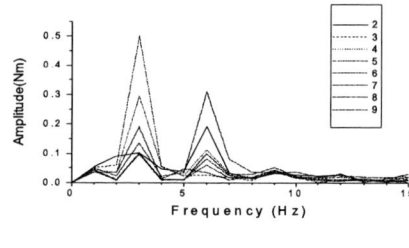

(b)

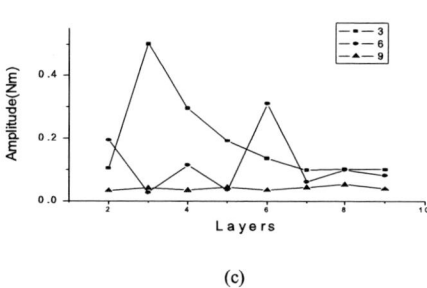

(c)

Figure 3. Steady-state torque. (a)Torque waveform with different N_L. (b) FFT analysis. (c) Harmonic amplitude vs. N_L.

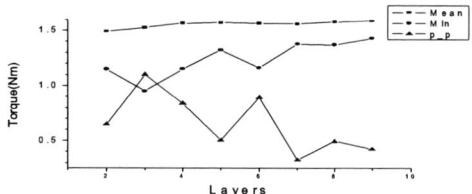

Figure 4. Mean, minimum and peak to peak torque characteristics

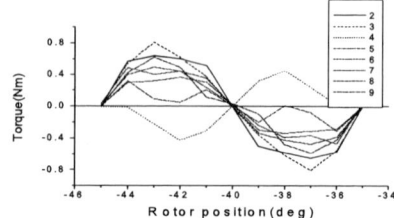

Figure 5. Position-setting torque (Br=1.1T)

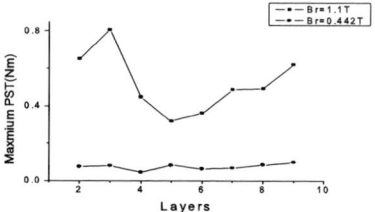

Figure 6. Position-setting torque vs. N_L and PM material

comparatively. It implies that the influence of slot effect will become more and more feebleness with harmonic times increasing, which is similar to torque-current angle analysis.

Except for harmonic analysis, the mean torque T_{av}, minimum torque T_{min} and peak to peak value T_{p_p} are get more attention to in practice. The results are shown in Fig.4. T_{av} is independent of N_L if the computational error is ignored. The fact that T_{min} is minimum at $N_L = 3$ indicates that start-up ability is the worst in this condition. T_{min} will approach to T_{av} as N_L increasing. T_{p_p} curves indicates that the torque ripple are severe at 3 and 6 layers.

In general, more layers means lower torque ripple. So if it is possible, more layers are best choice which not only can reduce torque ripple, but also decrease iron loss. It must be avoid selecting 3 layers and 6 layers when manufacturing limits exist for this sample motor.

C. Slot Effect on Position-setting Torque

Similar to PMSM, Position-Setting Torque (PST) of SR-PM motor is produced by rotor PM and stator slots.

Because PST causes the motor can not be positioned accurately, the reduction of PST is very important for SR-PM motors used for position control servo system. The calculation of PST will be restricted in one periodic (10 mechanical degrees) because of its periodicity. Fig.5 shows the PST curves respect to rotor layers which indicate that PST is the function of rotor position and rotor layers. The rotor layers do not only affect the PST value, but also affect the profile of PST. Except for rotor layers, permanent magnet material characteristics also have important impact on PST value and profile. Fig.6 shows maximum PST ($T_{ot\max}$) vs. rotor layers for two different PM material added to rotor layers. All $T_{ot\max}$ with residual flux density $B_r = 1.1T$ are larger than $B_r = 0.442T$. The variety regulation is different for different PM material. So for the reduction of PST, rational N_L select is also a good method.

D. Reduction of Slot Effect

1) N_L optimization

From above analysis, it can be seen that $N_L = 3$ $N_L = 6$ must be avoided for this these motor with 36 stator slots. If possible, $N_L >= 9$ is a good choice.

2) Enlargement of air gap

The air gap length is the most important parameter in motor design. For reluctance motor design, it must be very small for maximizing the magnetic inductivity and inductance to produce larger torque and diminish the excitation current. The reluctance torque is very sensitive to air gap length. In fact, the harmonic torque produced by slot effect is reluctance torque in chief. So to diminish the torque ripple, enlarging the air gap is an alternative. This reduction technology has been verified in common motor.

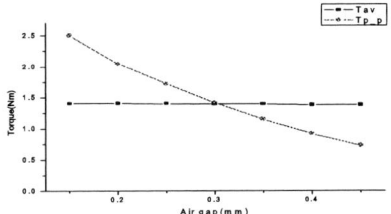

Figure 7. Enlargement of air gap

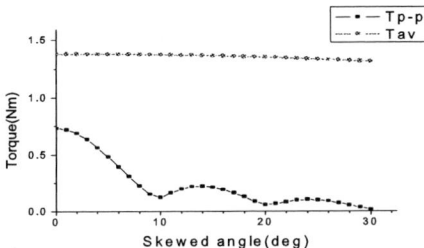

Figure 8. Skewed slots effect

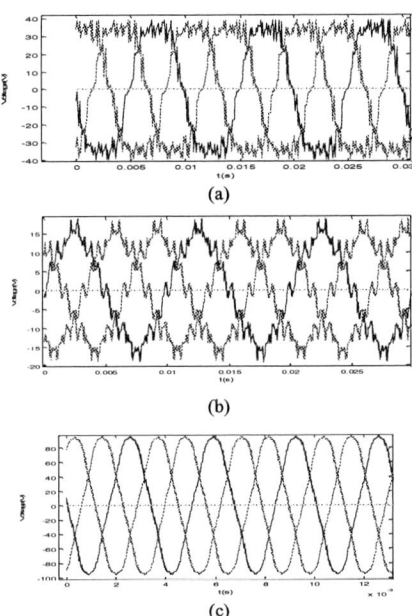

Figure 9. EMF waveforms. (a) M1. (b)M2. (c) M3.

For traditional SynRM, the main torque is also reluctance torque, so it is strongly restrict of this method. But for the SR-PM motor, both PM torque and reluctance torque exist. The air gap can be enlarged with only slight degrade in mean torque. This can be indicated from Fig.7. With the increase of air gap, T_{p_p} decreases remarkably while T_{av} is only a minor decline. These results must be own to the added PM material. It is well known that the magnetic inductivity of PM material is closed to air and the PM torque is not sensitive to air gap length. So for SR-PM motor with high magnetic energy product (MEP) PM material, enlargement of air gap is a rational choice. For SR-PM motor with added low maximum energy product magnet, it must be carefully to increase the air gap to avoid decreasing the reluctance torque excessively.

3) Skewed stator slots

To reduce the harmonic torque caused by slot effect, the skewed slots technology can be adopted, similarly to general motor. Because the rotor of SR-PM is too complex to be skewed, to skew the stator slots is a feasible means. The effect is calculated by skew integration method. Fig.8 shows T_{av} and T_{p_p} varied with skewed angle. It is very clear that skewed slots can reduce harmonic torque markedly with only a little decreasing in average torque. The best skewed angle is close to the integral multiple of slot pitch, That is to say $\theta_{skew\_opt} = k\tau_{pitch}$, k is positive integer. For the sample motor, $\tau_{pitch} = 10\deg(mech)$.

III. EXPERIMENTS

A. Electromotive force(EMF)

The back EMF is the most important parameters of the motor. Figure 9. are the waveforms of phase voltage. M1 has a trapezoid waveform with some harmonics due to slot effect. M2 also have some harmonics but approach to sinusoidal because of more layers. M3 indicate that slot skewing can not only eliminate harmonica, but also improve the distribution of EMF apparently.

B. Parameters Measurement

The inductance characteristics were measured using the instantaneous flux linkage method described in [10]. The measured inductances are shown in Table II. It gives a comparison of different designed parameters. The measured inductances are all larger than designed. This is because the FEA method does not include the winding end effect and flux leakage. L_d of M1 and M3 have much leakage because the SNM makes the magnetic circuit more saturated.

The measured ψ_{PM} are also listed in TABLE II along with the design values for comparison. All actual ψ_{PM}'s are lower than design values. The main reason for these phenomena is because the end effect and flux leakage.

TABLE II.   COMPARISON OF PARAMETERS

Items		ψ_{PM}(Wb)	Ld(mH)	Lq(mH)
Motor 1	Designed	0.0684	0.41	1.70
	Measured	0.0628	0.51	1.97
Motor 2	Designed	0.0348	1.88	6.16
	Measured	0.0222	2.06	6.41
Motor 3	Designed	0.0566	1.02	7.80
	Measured	0.0503	1.22	8.28

C. Statict torque test

The static torque-current angle curves are shown in Figure 10. In this test, a constant current is applied to stator winding with $i_A = -2i_B = -2i_C$. The torque is

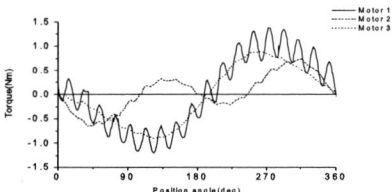

Figure 10. Torque-current angle curve

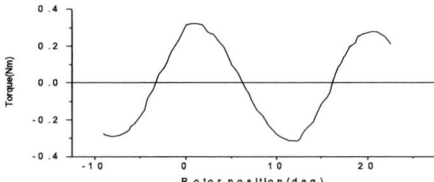

Figure 11. Position-setting torque of M1

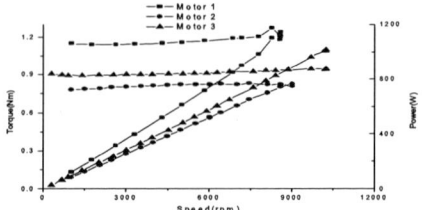

Figure 12. Power and torque tests

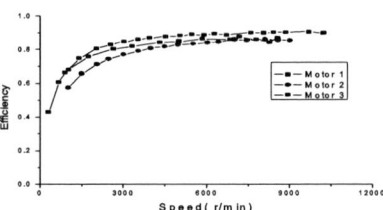

Figure 13. Efficiency results

recorded along with the rotor position. The torque ripples of M2 and M3 are very fewer than M1 because of their different structure. Because the rubber NdFeN magnet has low MEP (Br=0.442) compare to SNM(Br=1.1), the reluctance characteristics of M2 has been improved to increase power density and lead to more salient effect than M1 and M3.

The position-setting torques is also tested with the currents set to zero. Only the result of M1 is shown in Fig.11. The test is not done on the other two motors because the torque is too small to be measured for the torque transducer. This test also verifies that N_L and permanent material greatly affect the torque characteristic. Because N_L of the SNM motor is very small, the stator slot must be skewed and δ must be enlarged to reduce the cogging torque for future design.

D. Load test

To verify the design, the full load operation tests are done on test bed. The load is a PMSM motor operating in generating mode. An adjustable resistor is connected to PMSM motor to adjust the load. The speed, torque output and power output are measured with torque transducer. The input power, efficiency and current waveform are recorded by the digital power meter YOKOGAWA WT1600.

Fig.12 and Fig.13 show efficiency, power and torque results of these three motors. The results show that motor 3 has the highest efficiency and reaches the power and torque requirement of the primary design though its rotor is radial-laminated.

IV. CONCLUSIONS

The slot effect in SR-PM motors are analyzed and verified with experiments of three different structure motors. The emphases of the analyses are to open out the effect of rotor layers on torque ripple. The simulation results indicates that N_L affect the slot effect

intensively. The reduction methods of slot effect are also introduced with simulation results.

The final experiments verify that slot effect can be reduced by more rotor layers or skewed slots.

REFERENCES

[1] Zhao Zhengming, "Review for the development of a novel Synchronous reluctance and Permanent Magnet Machine". *Advanced Technology of Electrical Engineering and Energy,* vol.17, issue.3, pp.22-25 1998.

[2] Soong, W.L.; Staton, D.A.; Miller, T.J.E.; "Design of a new axially-laminated interior permanent magnet motor". *Industry Applications, IEEE Transactions on ,* vol.31, issue.2, pp.358- 367, 1998.

[3] Staton, D.A.; Miller, T.J.E.; Wood, S.E.; "Maximising the saliency ratio of the synchronous reluctance motor". *Electric Power Applications, IEE Proceedings B,* vol.140, issue.4, pp.249 – 259, 1998.

[4] Schiferl, R.F.; Lipo, T.A.; "Power capability of salient pole permanent magnet synchronous motors in variable speed drive applications," *Industry Applications, IEEE Transactions on,* vol. 26 , issue: 1, pp. 115–123, 1990.

[5] Zhao Zhengming, El-Antably Ahmed, "Optimization of added permanent magnet amount in synchronous reluctance machines for high performance," *Tsinghua Science and Technology,* vol.3, pp. 1137-1142, 1998.

[6] Zhao Zhengming, El-Antably Ahmed, "Advanced Computer-aided Design and Analysis for Synchronous Reluctance-Permanent Magnet Machines," *Tsinghua Science and Technology,* vol.3, pp. 1143-1148, 1998.

[7] Soong W.L., Miller T.J.E., "Theoretical limitations to the field-weakening performance of the five classes of brushless synchronous AC motor drive," *Electrical Machines and Drives, Sixth International Conference on,* vol.376, pp. 127–132, 1993.

[8] Cristian Lascu, Ion Boldea, and Frede Blaabjerg, "A modified direct torque control for induction motor sensorless drive", *Industry Applications, IEEE Transactions on,* vol. 36, issue. 1, pp. 122–130, 2000.

[9] Rahman M.F., Zhong L., Hu W.Y., Lim K.W., Rahman M.A., "A direct torque controller for permanent magnet synchronous motor drives," *Electric Machines and Drives Conference Record, IEEE International,* TD1/2.1 - TD1/2.3, 1997.

[10] W. L. Soong, D. A. Staton, and T. J. E. Miller, "Validation of lumped-circuit and finite-element modelling of axiallylaminated motors", *IEEE Electrical Machiner and Driver Conference,* 1993.

[11] Soong, W.L.; Ertugrul, N.; "Field-weakening performance of interior permanent-magnet motors", *Industry Applications, IEEE Transactions on,* Vol.38, Issue. 5, pp. 1251-1258,2002

[12] Isaac, F.N.; Arkadan, A.A.; El-Antably, A.; "Characterization of axially laminated anisotropic-rotor synchronous reluctance motors", *Energy Conversion, IEEE Transactions on,* Vol.14, Issue 3, pp. 506 – 511,1999

[13] Gu, C.L.; Li, L.R.; Shao, K.R.; Xiang, Y.Q.; "Anisotropic finite element computation of high density axially-laminated rotor reluctance machine" *Magnetics, IEEE Transactions on,* Vol. 30, Issue 5, pp. 3679 – 3682,1994

[14] Guo Wei, Zhao Zhengming, "Design and Experiments of two Glued Axially-Laminated Synchronous Reluctance Permanent Magnetic Motors", *the Sixth IEEE International Conference on Power Electronics and Drive Systems,* pp. 1374-1379, 2005

2006 5th International Power Electronics and Motion Control Conference

A New BLDC Motor Drives Method Based on BUCK Converter for Torque Ripple Reduction

Zhang Xiaofeng ,Lu Zhengyu
Department of Electrical Engineering
Zhejiang University, Hangzhou,310027, P.R.China
Zxfeng001@163.com

Abstract —This paper presents a comprehensive analysis on torque ripples of brushless dc motor drives in conduction region and commutation region. A novel method for reducing the torque ripple in brushless dc motors with a single current sensor has been proposed by adding BUCK converter in the front of 3-phase inverter.In such drives, torque ripple suppression technique is theoretically effective in commutation region as well as conduction region. Effectiveness and feasibility of the proposed control method is verified through experiments.

Keywords —*Brushless dc motor； torque ripple； conduction region； commutation region*

I .INTRODUCTION

The brushless direct current (BLDC) motor has high torque,compact size, and high efficiency. Therefore, the BLDC motor is widely used in computers, household, industrial products and automobiles.However, the BLDC motor has a disadvantage of high cost compared with the direct current (DC) motor because it is necessary to use an inverter and controller to remove a brush of DC motor.

As known to all,brushless dc motor with trapezoidal back-EMF have been widely used due to their high power density and easy control method. Further, basic trapezoidal brushless dc motors make it possible to use a single dc-link current sensor to regulate the phase current flowing through two motor phases.But torque ripple generated in conduction region and commutation region is the main drawback of BLDCM, however, which deteriorates the precision of BLDCM[1].In this paper, the proposed control scheme eliminates torque ripples in conduction region and attenuates torque ripples in commutation region .Meanwhile, this scheme improves the veracity of a single dc-link current sensor.

II .PRINCIPLE AND ANALYSIS

Generally a BLDCM has two operation region: conduction region and commutation region.In the conduction region,with position of rotor selected 2-phases are conducted.On the other hand commutation region is to be transient region which converts from the current conduction into next one,is relatively shorter than

conduction region,and 3-phases(rising phase,decaying phase, non-commutation phase)are all conducted. Conduction and commutation appears six times per one electrical rotating of the rotor[2].

The novel proposed circuit consists of the step-down BUCK converter that regulates the amplitude of the current and the inverter which is controlled in such a way as to supply 3-phase rectangular current with a pulse width of $120°$ electrical degree to the motor. Fig.1shows the conventional and new proposed circuit configuration respectively[3][4][5]. The motor is assumed symmetrical and salient effect is neglected. The phase inductance denoted by L_a , L_b , L_c are constant. E_a , E_b , E_c represent three phase back EMF respectively. i_a,i_b,i_c are stator current in phase A,B,C respectively. R_a , R_b , R_c are phase resistance respectively. N denotes the neutral node of the motor windings with reference to groud. R_{sample} is the current sampling resistance[6].

Fig.2 shows the normal and new type of PWM pulse patterns in 2-phase feeding scheme waveforms.The following analysis is based on the region of 6-1 shown in Fig.2 .

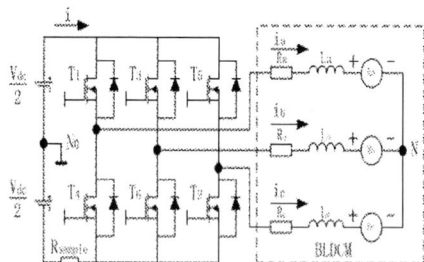

Fig1(a). The conventional circuit configuration

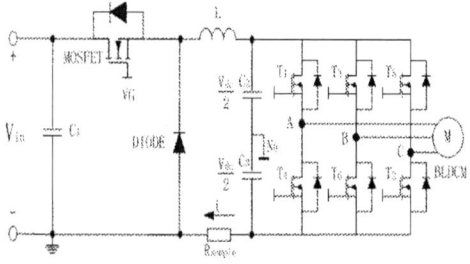

Fig1(b). The new proposed circuit configuration

1-4244-0448-7/06/$25.00 ©2006 IEEE   958

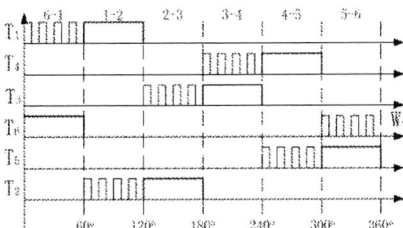

Fig2(a). The normal modulation of PWM-ON pattern

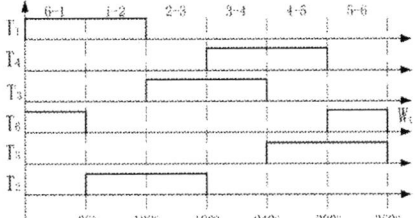

Fig2(b).   The new modulation pattern

The torque is given by Eq.(1).

$$T_e = \frac{1}{\omega_m}\left(e_a i_a + e_b i_b + e_c i_c\right) \qquad (1)$$

where T_e is the belectromagnetic torque of motor, ω_m is mechanical angular velocity of rotor. e_a, e_b, e_c are the BEMF of 3-phase, i_a, i_b, i_c are the current of 3-phase. In order to only show off the torque ripple due to supply, ideal trapezoidal EMFs with a $120°$ constant plateau are considered.

According to the switching conditions of the inverter switches, the voltage equations related to the switch-on and switch-off intervals in the normal mode can be described as Eq.(2).

$$\begin{bmatrix} V_{ka} \\ \dfrac{V_{dc}}{2} \\ V_{kc} \end{bmatrix} = \begin{bmatrix} R & 0 & 0 \\ 0 & R & 0 \\ 0 & 0 & R \end{bmatrix} \bullet \begin{bmatrix} i_a \\ i_b \\ i_c \end{bmatrix} + \frac{d}{d} \begin{bmatrix} L & 0 & 0 \\ 0 & L & 0 \\ 0 & 0 & L \end{bmatrix} \bullet \begin{bmatrix} i_a \\ i_b \\ i_c \end{bmatrix} + \begin{bmatrix} e_a \\ e_b \\ e_c \end{bmatrix} + \begin{bmatrix} V_{NN_0} \\ V_{NN_0} \\ V_{NN_0} \end{bmatrix}$$

$$(2)$$

Where V_{dc} is the dc-link voltage, $L = L_s - M$ (L_s is self inductance and M is mutual inductance), V_{ka}, V_{kc} are the voltages of phase A and C, S is the switching function(1 denotes switch-on,0 denotes switch-off), N_0 and N are shown in Fig.1. V_{NN_0} is derived from (2) as follows.

$$V_{NN_0} = \frac{1}{3}\left(V_{ka} - V_{kc} + \frac{1}{2}V_{dc}\right) - \frac{1}{3}\left(e_a + e_b + e_c\right) \qquad (3)$$

Combining (2) and (3),the currents of 3-phase can be described as Eq.(4).

$$\begin{cases} i_a = \left(\dfrac{2}{3}V_{ka} + \dfrac{1}{3}V_{kc} - \dfrac{1}{6}V_{dc} - \dfrac{4}{3}E_m\right)\dfrac{t}{L} + i_{a0} \\[2mm] i_b = \left(-\dfrac{1}{3}V_{ka} + \dfrac{1}{3}V_{kc} + \dfrac{1}{3}V_{dc} + \dfrac{2}{3}E_m\right)\dfrac{t}{L} + i_{b0} \\[2mm] i_c = \left(-\dfrac{1}{3}V_{ka} - \dfrac{2}{3}V_{kc} - \dfrac{1}{6}V_{dc} + \dfrac{2}{3}E_m\right)\dfrac{t}{L} + i_{c0} \end{cases}$$

$$(4)$$

Where E_m is BEMF constant and i_{a0}, i_{b0}, i_{c0} are steady-state value of phase-current in the conduction region. It is assumed that the motor winding resistance is neglected,and e_a maintains the value of E_m and e_b, e_c hold $-E_m$.

In the conduction region(taking"6-1" for example),the torque equation is derived from(1).In principle, the torque in BLDCM is proportional to the current amplitude in the non-commutation phase winding.

$$T_e = \frac{1}{\omega_m}\left[E_m \bullet I_a + (-E_m)\bullet(-I_a) + e_c \bullet 0\right] = \frac{2E_m}{\omega_m}I_a$$

$$(5)$$

In the commutation region(taking from "6-1" to "1-2" for example), the torque equation is expressed as Eq.(6) derived from (1) and (4). Correspondingly the torque ripples which are caused by the inductance L is Eq.(7).

$$T_e = \frac{2E_m I_a}{\omega_m} + \frac{2E_m t}{3\omega_m L}\left(2V_{ka} + V_{kc} - \frac{1}{2}V_{dc} - 4E_m\right)$$

$$(6)$$

$$\Delta T_e = \frac{2E_m t}{3\omega_m L}\left(2V_{ka} + V_{kc} - \frac{1}{2}V_{dc} - 4E_m\right) \qquad (7)$$

Where t is time of commutation region.

In conduction region, the dc-link current sensor can not reflect the real phase-current when the switching device turns off in the conventional PWM-ON modulation method, then the torque ripples are generated because of these.However in the new proposed method, BUCK converter which adopts PWM modulation transforms V_{in} into V_{dc},accordingly changing constant voltage-source into quasi-current-source which can provide the satisfactory waveform matched to the induced EMF waveform in the stator windings.So, the dc-link current sensor can reflect the phase-current exactly,conclusively eliminating the torque ripples in conduction region.

In commutation region, the conventional control method adopts PWM-ON modulation.The torque ripples are expressed as Eq.(8) during commutation interval ($S_a = 1, S_c = 0$ or 1).

$$\begin{cases} \Delta T_e = \dfrac{2E_m t}{3\omega_m L}\left(V_{dc} - 4E_m\right)\ldots\ldots S_C = 1 \\[4mm] \Delta T_e = \dfrac{2E_m t}{3\omega_m L}\left(-4E_m\right)\ldots\ldots S_C = 0 \end{cases} \qquad (8)$$

The expression shows the commutation torque ripples in trapezoidal BLDCM, including torque spikes in the low speed range($V_{dc} > 4E_m$) and torque dips in the high speed range($V_{dc} < 4E_m$).

But using the new proposed method that adds Buck converter between the battery and the inverter,the converter can regulate the amplitude of the out-voltage and the current.The duty ratio is denoted by D(t),the following equation is gained:

$$V_{dc} = D(t) \bullet V_{in} \qquad (9)$$

So torque spikes decreases with decreasing the duty ratio D in the low speed range($V_{dc} > 4E_m$) and torque dips decreases with increasing the duty ratio D in the high speed range($V_{dc} < 4E_m$).Because of $S_a = S_c = 1$ in the proposed scheme,the torque ripples are expressed as Eq.(10) during commutation interval.In this way,the case of $V_{dc} < 4E_m$ exists over the entire speed range , consequently attenuating the commutation torque ripples.

$$\Delta T_e = \frac{2E_m t}{3\omega_m L}\left[D(t) \bullet V_{in} - 4E_m\right] \qquad (10)$$

In the low speed range, the torque ripple ΔT_e can be eliminated by making the $D(t) \bullet V_{in} = 4E_m$.

In the high speed, the torque ripple ΔT_e can be attenuated by making the $D(t) = 1$.

In summary,the new proposed method makes drive to produce smooth torque and linear torque with current.

III.EXPERIMENTAL RESULTS

Experiments are carried out to verify the feasibility and effectiveness of the proposed method. The parameters of BLDCM prototype are shown in table.1.

TABLE1:PARAMETERS OF BLDCM PROTOTYPE

U_{rated} (V)	24
P_{rated} (W)	146
Poles	4
$R_{phase}\,(\Omega)$	0.402
$L_{phase}\,(mH)$	0.185
$f_{buck}\,(KHz)$	30

To implement the new control method, the TMS320LF2407A DSP is employed in the prototype. Digtal PID speed control, constant current control and constant voltage control operate every 33.3 μs sampling time.Fig.3 and Fig.4 show the experiment

waveforms of two-phase currents and commutation currents respectively in the case of conventional control method and in the case of the new proposed scheme.Fig.3 shows that torque ripples during conduction region are eliminated effectively by the new proposed control technique.Fig.4 notes that the commutation current slopes of the incoming and outgoing phases balanced, so the resulting commutation torque ripples are effectively suppressed with the help of the new proposed scheme.

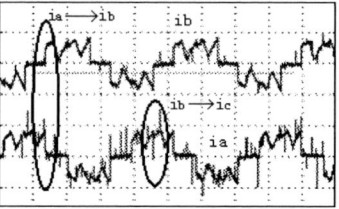

i: 10A/div   time: 3ms/div

Fig.3.(a) The 2-phase current-waveforms
of conventional modulation mode

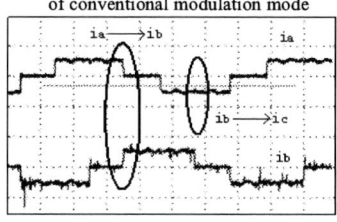

i: 10A/div   time: 2ms/div

Fig.3.(b) The 2-phase current-waveforms
of new proposed modulation mode

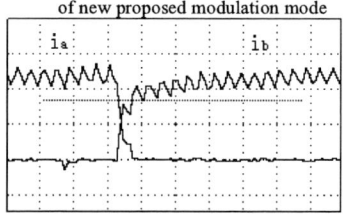

i: 2A/div   time: 100us/div

Fig.4.(a)The commutation current-waveforms
of conventional modulation mode

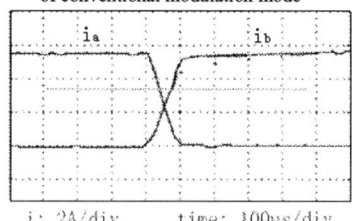

i: 2A/div   time: 100us/div

Fig.4.(b)The commutation current-waveforms
of new proposed modulation mode

IV.CONCLUSIONS

In this paper,a new torque ripple reduction method based on buck converter has been proposed for brushless dc motor drives using a single dc current sensor.In such control method, the dc-link current sensor can give correct information corresponding to the motor phase currents to eliminate torque ripples in conduction region. Meanwhile, torque ripples have been attenuated effectively during commutation region.Subsequently effectiveness and feasibility of the proposed control method are verified through experiments.

REFERENCE

[1] Joong-Ho Song and Ick Choy, "Commutation torque ripple reduction in brushless DC motor drives using a single DC current sensor,"*IEEE Trans. on Power Electronics*,vol. 19, No.2 ,pp.312-319,March 2004.

[2] Byoung-Hee Kang,Choel-Ju Kim,Hyung-Su Mok and Gyu-Ha Choe, "Analysis of torque ripple in BLDC motor with commutation time,"*Proceedings of IEEE*,vol.2,pp.1044-1048, June 2001.

[3] Carlson R,Lajoie-Mazenc M and Fagundes J.C.d.S, "Analysis of torque ripple due to phase communtation in brushless DC machines,"*IEEE Trans. on Industry Applications*,vol.28,no.3, pp.632-638,May-June 1992.

[4] Luk P.C.K and Lee C.K, "Efficient modeling for a brushless DC motor drive,"*International Conference on Industrial Electronics,Control and Instrumentation*,vol.1,pp.188-191, September 1994.

[5] Lei Hao,Toliyat,H.A, "BLDC motor full speed range operation including the flux-weakening region,"*IEEE-IAS Annual Meeting*,vol.1,pp.618-624, Octorber 2003.

[6] Wei Kun,Lou Zhenli,Zhang Zhongchao, "Estimate of rotor position of BDCM based on the third harmonic component"*Power Electronics and Motion Control Conference,2004, IPEMC* ,vol.3,pp.1306-1310, August 2004.

2006 5th International Power Electronics and Motion Control Conference

Performance Investigation of a Fault-Tolerant Brushless Permanent Magnet AC Motor Drive

Jingwei Zhu, Nesimi Ertugrul and Wen Liang Soong
School of Electrical & Electronic Engineering
The University of Adelaide, Adelaide, Australia, 5005
jingwei@eleceng.adelaide.edu.au

Abstract—**Fault-tolerant motor drives are required in various safety critical applications. Using special motor design and inverter topology, brushless permanent magnet AC motor drives can have a fault-tolerant capability. In this paper, a dual motor drive configuration has been proposed which offer high robustness and reliability. This paper studies various critical performance characteristics of the dual motor drive including drive loss and system efficiency. The paper aims to address some of important performance issues in the motor drive, which has not been addressed previously. The paper presents the details of the experimental setup and explains the measurement methods for the motor parameters and for the open-circuit torque loss. The efficiency prediction method and corresponding results are also presented and verified by the computer simulation study both under healthy and various faulty conditions.**

Keywords-fault-tolerant drive; brushless permanent magnet motor; efficiency calculation.

I. INTRODUCTION

Safety critical motor drive systems are becoming more important in many areas such as aerospace, transportation, medical and military applications, and nuclear power plants. In such applications, any failure of motor drives may result in catastrophic loss of property and human life. Therefore, the motor drives utilized in such applications must be fault tolerant, and they should continue to operate until the operation of system is ceased safely.

It is well known that the Switched Reluctance Motor (SRM) is inherently fault tolerant [1]. However, such motors are not desirable in many applications due to higher acoustic noise and lower torque density.

A number of studies have been reported in the literature that investigates fault tolerant motor drives. For example, brushless Permanent Magnet (PM) AC motors can be specially designed to be fault-tolerant [2-4]. A comparative study carried out in [5] suggested that brushless PM machines can achieve similar degrees of fault-tolerance as SRM while offering higher torque density and lower acoustic noise.

However, since brushless PM motors require an inverter to operate, a number of faults may also occur in such motor drives. Inverter faults and remedial strategies

in a brushless PM motor drive were investigated in [6]. Winding turn-to-turn faults and control methods were presented in [7]. In addition, in order to improve the robustness and reliability of the fault-tolerant motor drive, a dual fault tolerant motor drive configuration has been proposed in [8] and its electromagnetic torque performance described in [9]. However, in these earlier studies, performance characteristics such as power loss and efficiency in a dual fault-tolerant motor drive have not been investigated, which can significantly affect on the operation and control strategies of the motor drive.

This paper considers a dual fault tolerant motor drive configuration that can offer redundancy and presents results to demonstrate its performance both under healthy and faulty operating conditions. In addition, the parameter measurements and the efficiency prediction of the drive are presented in the paper. The efficiency tests results are also verified by the computer simulations under various fault scenarios.

II. PROPOSED FAULT-TOLERANT MOTOR DRIVE

In order to obtain a fault-tolerant brushless PMAC motor, it is important to minimize or eliminate electrical, magnetic and thermal coupling between the motor windings [2]. Thus, the failure in one winding will not affect the operation of other windings. This may be achieved physically by separating the motor windings and driving each winding using a separate single-phase H-bridge inverter circuit.

Fig. 1 shows a three-phase fault-tolerant brushless PMAC motor drive configuration, where each phase winding occupies two different slots around a single tooth.

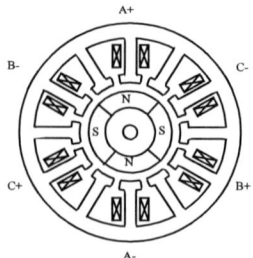

Figure 1. Winding arrangement of a fault-tolerant brushless PMAC motor with three phases and 4 poles

1-4244-0448-7/06/$25.00 ©2006 IEEE

It should be reported here that although a fault tolerant drive system with a single motor may overcome most of the problems in safety critical applications, it is still possible to result in the loss of the entire drive function due to some potential faults.

Therefore, this paper considers a dual motor drive system on the same shaft to introduce redundancy. In this system each motor module has also three isolated phases driven by separate H-bridge inverters, and each motor module is powered via a separate power supply and associated controller. The controllers can provide the primary functions for each motor module, and may be linked to communicate with each other to determine the status of the operation.

Fig. 2 illustrates the test setup that is used to investigate the fault tolerant operation concept, which includes a DC machine and two fault-tolerant PMAC motor modules on a common shaft.

It should be reported here that the motors used in the setup are based on Soft Magnetic Composite cores with concentrated windings as shown in Fig.1. In addition, each motor module has identical outer dimensions and similar torque ratings, but has different winding parameters. This is due to different number of winding turns and different gauge wire used in the design of each motor module.

Table 1 lists the measured motor parameters of the PMAC motor modules shown in Fig.2. As can be seen in the table, the Motor 2 has a winding inductance about four times and a back-EMF constant about two times greater than the values of Motor 1.

Fig. 3 is given to illustrate the back-EMF waveforms of one of the phases of Motor 1 and Motor 2, which were measured at a constant rotor speed of 870 rpm. As can be seen in the figure, two motor modules have similar back-EMF profiles that can be approximated to ideal sinusoidal waveforms. Therefore, for simplicity in the simulation study, it is assumed that the back-EMFs and reference current waveforms are sinusoidal waveforms for both motor modules.

Figure 2. The test setup including a dual fault-tolerant PMAC motor drive and a dynamometer.

Table 1: Parameters of the PMAC Motor Modules

Parameters	Motor 1	Motor 2
Winding resistance (Ω)	0.2535	0.7173
Winding inductance (mH)	0.3867	1.7933
Back-EMF constant (V/rad/s)	0.0404	0.0931

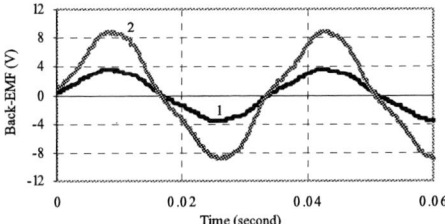

Figure 3. Phase A back-EMF waveforms of Motor 1 (curve 1) and Motor 2 (curve 2) at 870 rpm

III. PERFORMANCE ESTIMATION BASED ON MEASUREMENTS

A. Open-Circuit Power Loss Tests

The open-circuit losses (iron and mechanical loss) of the dual brushless PMAC machine were performed using the test setup mentioned above. Firstly, the DC motor was tested under no-load to obtain its torque and mechanical loss characteristics, which are a function of the rotor speed. Secondly, the DC motor was used to drive the dual brushless PMAC motors to obtain the total torque loss of the test setup. Finally, the open-circuit torque loss of the dual PMAC motors was obtained, which is equal to the difference between the total torque loss of the setup system and the DC motor torque loss.

It can be noted here that the open-circuit torque loss of Motor 1 and Motor 1+ Motor 2 were measured using the identical techniques. However, during the test of Motor 1, Motor 2 was decoupled from the common shaft to eliminate the associated windage and friction losses.

Fig. 4 shows the measured open-circuit torque loss for each motor module and the total torque loss. In the figure, the open-circuit torque loss of Motor 2 was calculated by subtracting Motor 1 torque loss from the total torque loss. This figure reveals that both motor modules have similar torque loss profiles but different values. The open-circuit torque loss of Motor 1 was found greater than Motor 2.

The torque loss curves given above can be used to obtain the open-circuit power loss in each motor module and in the entire motor system by multiplying the open circuit torque loss with the rotor speed.

B. Copper Loss Calculation

As well known, the copper loss in brushless PMAC motors is a function of torque, hence can be calculated using stator resistance and torque constant.

In this study, due to the difference between the parameters of the motor modules, it is assumed that both motors are excited by different values of current, but they generate equal electromagnetic torque. Therefore, the total copper loss in two PMAC motors can be given as.

$$P_{copperloss} = (\frac{R_1}{6K_1^2} + \frac{R_2}{6K_2^2}) \cdot T_s^2 \qquad (1)$$

Here R_1, R_2 are the winding resistances and K_1, K_2 are the back-EMF constants of Motor 1 and Motor 2 respectively, and T_s represents the shaft torque of the motor drive that

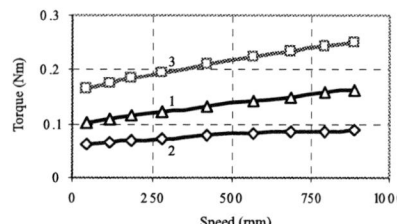

Figure 4. PMAC motor drive torque loss versus speed curves
1 - Motor 1 iron and mechanical torque loss
2 - Motor 2 iron and mechanical torque loss
3 - Total iron and mechanical torque loss

is equal to the sum of the open-circuit torque loss and the load torque.

C. Motor Efficiency Prediction

The open-circuit power loss and copper loss obtained above can be used to determine the efficiency characteristics of the dual fault-tolerant PMAC motor drive. The efficiency of the motor drive is calculated by

$$\eta = \frac{P_{out}}{P_{out} + P_{copperloss} + P_{openloss}} \times 100\% \quad (2)$$

Where

$$P_{out} = T_L \cdot \omega \quad (3)$$

Here P_{out}, $P_{copperloss}$ and $P_{openloss}$ are the output power, copper loss and open-circuit power loss respectively, and ω is the rotor speed and T_L is the load torque.

Fig. 5 displays the contour plot of the efficiency calculations. It was observed from this figure that the efficiency of the motor drive increases with output load torque. In addition, the efficiency at low rotor speeds is smaller than the efficiency at high speed.

Fig. 6 illustrates the change of output power, input power, copper loss and open-circuit power loss as a function of the rotor speed. As can be seen in the figure, if the output torque is kept constant, the copper loss almost remains constant while the open-circuit power loss increases with speed. It can also be observed that due to the higher open-circuit losses in the drive, the total efficiency of the motor drive is relatively low. Therefore, to increase the efficiency it is important to decrease the open-circuit power loss, which can be achieved by improving bearings and by utilizing lower iron loss magnetic material in the motor structure.

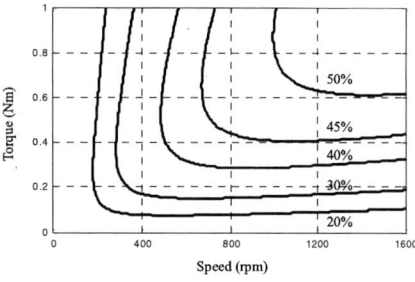

Figure 5. Efficiency contour of the dual PMAC motor drive

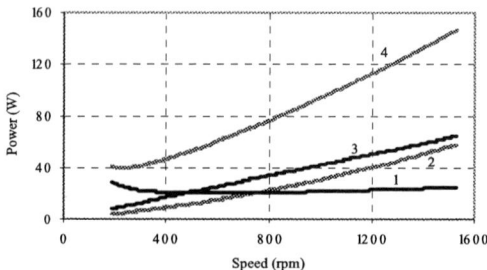

Figure 6. Power curves of the motor drive versus rotor speed at 0.4Nm load torque
1 – copper loss   2 – open-circuit loss power
3 – output power   4 – input power

IV. COMPUTER SIMULATION STUDIES

The simulation study in this paper was performed by modeling the entire drive system using Matlab/Simulink, which primarily include two motor modules, inverter and controller. The principal block diagram of the simulation tool developed is given in Fig. 7.

A. Mathematical Model

To model the entire motor drive system, it is necessary to model the motors first. The windings of brushless PMAC motors are modeled by a series circuit consisting of a resistance R, an inductance L (equivalent winding inductance in magnetically coupled windings) and a speed dependent back-EMF voltage e(t).

The simulation study included four equations that describe the dynamic behavior of a motor drive. These are the voltage equation, the electromagnetic torque equation, the equation of motion and the motor speed equation as given below.

$$v_i(t) = R\,i_i(t) + L\frac{di_i(t)}{dt} + e_i(t) \quad (4)$$

$$T_e(t) = \frac{1}{\omega(t)}\sum_{i=1}^{n} e_i(t) i_i(t) \quad (5)$$

$$T_e(t) = J\frac{d\omega(t)}{dt} + B\omega(t) + T_L \quad (6)$$

$$\omega(t) = \frac{d\theta(t)}{dt} \quad (7)$$

Where i represents the number of phases in the motor drive, i =1, 2, 3, 4, 5, 6 for the dual three phase motors, $v_i(t)$ is the phase voltage, $i_i(t)$ is the phase current, $T_e(t)$ is the total electromagnetic torque developed by the dual motor system, $\omega(t)$ is the angular speed of the rotor that is common for both motor modules, $\theta(t)$ is the rotor position, J is the polar moment of inertia of the entire system, and B is the damping coefficient.

The calculated parameters in the drive simulation was utilized to estimate the phase currents and voltages in each motor module, the total electromagnetic torque, the rotor speed and the efficiency under healthy as well as faulty operating conditions.

964

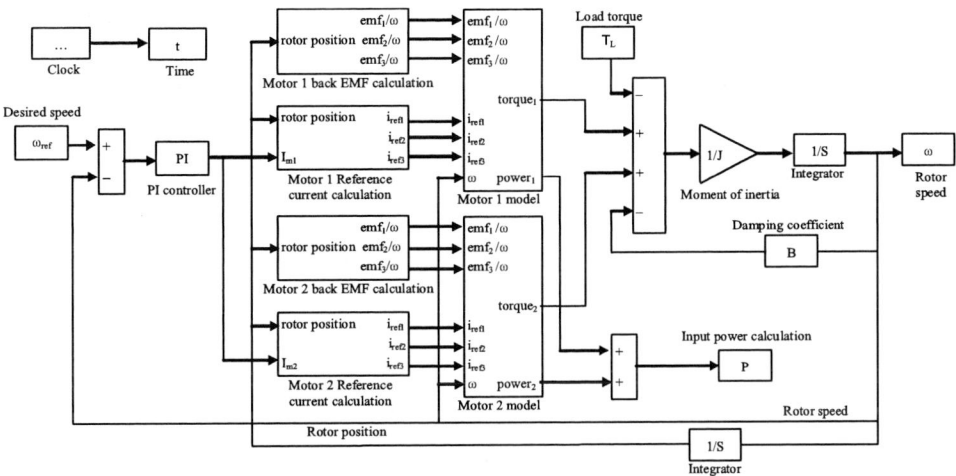

Figure 7 : Matlab/Simulink model of the dual fault tolerant PMAC motor drive

Since the efficiency is equal to the ratio of the output power and the input power, the simulation utilizes Eq. (3) to calculate the output power, and the input power is obtained by the following equation.

$$P_{in} = \sum_{j=1}^{6} \frac{1}{T} \int_0^T V_{DC} \, i_j(t) dt \qquad (8)$$

Here V_{DC} is the DC link voltage of the inverter, $i_j(t)$ is the phase current and T is the electrical period of the phase current.

B. No-Fault Operation

As stated above the mathematical equations given above were utilized to simulate the entire motor drive. The control parameters of the simulation system are the rotor speed and the load torque. The simulation study is used to examine the efficiency as a function of load torque.

Fig. 8 display a set of selected steady state phase current waveforms of Motor 1 while the load torque is 0.4Nm at a rotor speed of 382rpm. Motor 2 has the same phase current form but different current value. It can be noted here that despite the difference in the amplitude of phase currents in each motor module, the output torque is same, which is due to the corresponding back EMF value at a given speed.

Fig. 9 is produced to illustrate the efficiency curves of the motor drive as a function of rotor speed and load torque. In the figure the predicted values that are based on the motor characteristics obtained in the previous section are also shown, which indicate a good correlation.

C. Effects of Faults on Efficiency Characteristics

Although the dual fault-tolerant motors offer the capability of operation under a fault, it is necessary to investigate the efficiency of the entire motor drive under various faulty scenarios.

The potential faults that can occur in a dual fault tolerant PMAC motor drive include winding short circuit, winding open circuit, switch open circuit, switch short circuit, power supply open and short circuit and one motor complete fault. Among these, the most common faults in a motor drive are winding open circuit and short circuit faults.

In a dual PMAC motor drive setup with six (since each motor have three phases) identical and electrically and magnetically independent phases, if the rotor speed and the output load torque are constant, the open circuit power loss remains the same value both in healthy and faulty operation with winding open circuit fault.

In a faulty operating condition, in order to keep the output torque constant, it is assumed that the currents in healthy phases are increased proportionally. However,

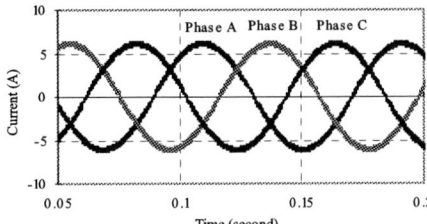

Figure 8.   Steady-state phase current waveforms of Motor 1 under T_L=0.4Nm, ω =382rpm

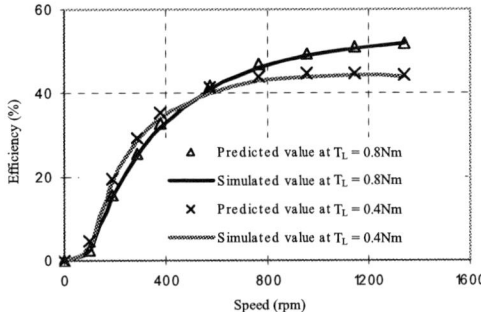

Figure 9.   Predicted and simulated efficiency versus speed characteristics under different loading conditions

the copper loss in the healthy phases, which is proportional to the current square, also increase under this condition. Therefore, the total copper loss of the motor drive should be increased accordingly, which can be given by

$$P_{copperloss} = \frac{6}{6-m} \cdot P_{copperloss0} \qquad (9)$$

$$P_{copperloss0} = \frac{R}{3K^2} \cdot T_s^2 \qquad (10)$$

Here m is the number of phases with winding open circuit fault ($0 \leq m \leq 5$), R is the phase resistance, K is the back-EMF constant, and $P_{copperloss0}$ and $P_{copperloss}$ are the copper loss in healthy and faulty operations respectively.

Fig. 10 displays the changing coefficient of copper loss versus the number of phases, m. The figure illustrates that the copper loss increases with the number of faulty phases. Therefore, the efficiency of the motor drive under open circuit faults will be low.

Under winding short circuit faults, however, the efficiency becomes worst. This is due to the fact the winding short circuit fault generates a negative torque by the induced short circuit current in a failed phase. Therefore, to keep the same output torque and rotor speed as in the healthy operation, it is necessary to increase the current in healthy phases significantly (much more than the value under open circuit fault). Thus, the efficiency of the motor drive under short circuit faults is much lower than the value under open circuit fault operation.

Fig. 11 shows the simulation results of efficiency under conditions of health, one phase open circuit fault (phase C in motor 1), one phase short circuit fault (phase C in motor 1), and motor 2 complete open circuit fault when the load torque is kept in 0.8Nm. The simulation results are consistent with the analysis results.

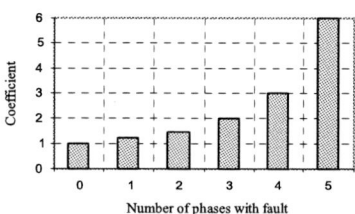

Figure 10. Increasing coefficient of copper loss versus number m under winding open circuit fault.

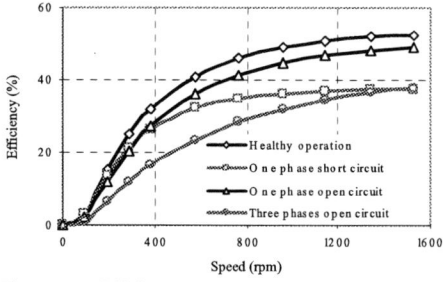

Figure 11. Efficiency versus speed curves obtained from the simulation, under different operating conditions at T_L=0.8Nm.

V. CONCLUSION

This paper examined the performance of a fault-tolerant dual brushless permanent magnet AC motor drive configuration, which offers redundancy against complete motor failure. A prototype motor drive was built and key motor parameters were determined experimentally. A method of efficiency estimation is described and verified by the simulation results.

In addition, the analysis and simulation results were obtained to study the effect of faults on to the efficiency of the motor drive. It was demonstrated that total copper loss of the motor drive increases under different faults, which assumed the output torque and the rotor speed of the motor drive was kept constant.

The future work in this research will include the real time implementation of the entire fault tolerant system and potential fault detection and elimination methods in an actual motor drive.

REFERENCES

[1] C. M. Stephens, "Fault detection and management system for fault-tolerant switched reluctance motor drives", IEEE Transactions on Industry Applications, Vol. 27(6), Nov.-Dec. 1991, pp. 1098-1102.

[2] B. C. Mecrow, A. G. Jack, J. A. Haylock, J. Coles, "Fault-tolerant permanent magnet machine drives", IEEE Proceedings-Electric Power Applications, Vol. 143(6), Nov. 1996, pp. 437-442.

[3] J. A. Haylock, B. C. Mecrow, A. G. Jack, and D. J. Atkinson, "Operation of a fault tolerant PM drive for an aerospace fuel pump application", IEEE Proceedings on Electric Power Applications, Vol. 145(5), September 1998, pp. 441-448.

[4] B. C. Mecrow, A. G. Jack, D. J. Atkinson and S. R. Green, "Design and testing of a four-phase fault-tolerant permanent-magnet machine for an engine fuel pump", IEEE Transactions on Energe Conversion, Vol. 19(4), December. 2004, pp. 671-678.

[5] A. G. Jack, B. C. Mecrow, and J. A. Haylock, "A comparative study of permanent magnet and switched reluctance motors for high performance fault-tolerant applications", IEEE Transactions on Industry Applications, Vol. 32(4), July-Aug. 1996, pp. 889-895.

[6] S. Bolognani, M. Zordan, and M. Zigliotto, "Experimental fault-tolerant control of a PMSM drive", IEEE Transactions on Industrial Electronics, Vol. 47(5), Oct. 2000, pp. 1134-1141.

[7] C. Gerada, K. Bradley, and M. Sumner, "Winding turn-to-turn faults in permanent magnet synchronous machine drives." Proceeding of the 40th IAS Annual Meeting, Hong Kong, Oct. 2005.

[8] N. Ertugrul, W. Soong, G. Dostal, and D. Saxon, "Fault tolerant motor drive system with redundancy for critical application", IEEE 33rd Power Electronics Specialists Conference, Cairns, Australia, 23-27, June 2002.

[9] J. Zhu, N. Ertugrul, and W. Soong, "Modeling and Simulation of Electromagnetic Torque in a Fault Tolerant Brushless PMAC Motor Drive", Proceeding of AUPEC Conference, Hobart, Australia, Vol.2, pp. 635-640, September 2005.

Current sensorless integral variable structure controller of synchronous reluctance motor

Huann-Keng Chiang[1], Chien-An Chen[2], Bor-Ren Lin[1] and Kai-Sheng Hsu[1]

[1] Department of Electrical Engineering, National Yunlin University of Science & Technology, Douliu, Yunlin 640, Taiwan, China
[2] Graduate School of Engineering Science & Technology, National Yunlin University of Science & Technology, Douliu, Yunlin 640, Taiwan, China
Email: chianghk@yuntech.edu.tw Fax: +886-5-5312065

Abstract—A synchronous reluctance motor driven by an integral variable structure controller for current sensorless is presented. The mathematical model for the synchronous reluctance motor includes core loss. We develop a maximum torque control strategy. The integral variable structure controller is used to reject the uncertain bounded disturbances and parameter variations. The voltage reference equation generates the require voltage from the torque command and the motor speed. The proposed controller is implemented using dSPACE DS1102 processor board. This system has a fast response and a good disturbance rejection capability. Experimental results show that the proposed controller is valid for the synchronous reluctance motor.

Keywords-synchronous reluctance motor; core loss; integral variable structure control; maximum torque control.

I. Introduction

The Synchronous Reluctance Motor (SynRM) has a mechanically simple and robust structure. It can rotate at high speeds in high temperature environments. Since last two decades, many researchers have focus their attention on the speed control of SynRM using sliding mode control strategies[1-3]. There are also a few researches focusing their attention on the current sensorless using different control strategies[4-7]. In [4], an ideal model of SynRM speed control with current sensorless and low resolution position sensor was discussed. In [5], an ideal model of SynRM current sensorless speed control was discussed for the maximum power factor control (MPFC) strategy. In [6], an ideal model of SynRM current sensorless speed control using sliding mode controller was proposed for maximum torque control (MTC), MPFC, maximum rate of change of torque (MRCTC) and constant current in inductive control (CCIAC) strategies. In [7], ideal model of SynRM for PI controller in optimal efficiency control without current sensor was discussed .

In general, we only consider the copper loss in motor modeling analysis [8,9]. The core losses of hysteresis and eddy currents are neglected. Core loss makes the torque decrease in acceleration and increase during braking [10-13]. Core loss also decreases the efficiency, at high switching frequencies, of pulse width modulation.

Therefore, the SynRM model including core loss is considered in this paper.

One of the popular methods about robust control is the so-called variable structure control [14,15]. It has been proven as an effective and robust control technology in SynRM [1-3,13]. The integral variable structure control can offer fast dynamic response, insensitivity to parameter variations and external disturbances rejection without reaching condition. Hence, a current sensorless of the SynRM model including core loss using integral variable structure control is considered for MTC strategy in this paper.

SynRM modeling, including the core loss, is discussed in Section II. In Section III, the integral variable structure control for MTC strategy is used to reject uncertain bounded disturbances and system parameter variations. We propose the current sensorless speed control in Section IV. In Section V, the simulation and experimental results are presented to validate the proposed current sensorless integral variable structure controller. Our conclusions are presented in Section VI.

II. Modeling of the SynRM

Fig. 1 shows the d-q equivalent circuits of a SynRM in a synchronously rotating rotor reference frame, including core losses. The voltage equations of the SynRM are represented as [10,16]

$$V_d = R_s i_{ds} - \omega_r L_q i_{qt} + L_d \frac{di_{dt}}{dt} \qquad (1)$$

$$V_q = R_s i_{qs} + \omega_r L_d i_{dt} + L_q \frac{di_{qt}}{dt} \qquad (2)$$

where
V_d and V_q : the direct axis and quadrature axis
      terminal voltages
i_{ds} and i_{qs} : the direct axis and quadrature axis
      terminal currents
i_{dt} and i_{qt} : the direct axis and quadrature axis
      torque producing currents

L_d and L_q : the direct axis and quadrature axis
magnetizing inductances

R_s and R_c : the stator resistance and core loss per
phase

ω_r : the speed of the rotor

The electromagnetic torque T_e and motor dynamic equation are stated as

$$T_e = \frac{3}{4}P(L_d - L_q)i_{dt}i_{qt} \qquad (3)$$

$$T_e - T_L = J_m \frac{d\omega_r}{dt} + B_m\omega_r \qquad (4)$$

where

T_L : torque of load

J_m : inertia moment of the rotor

B_m : viscous friction coefficient

P : number of poles

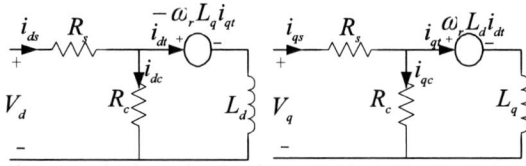

Fig. 1 Equivalent circuit of SynRM including core loss

The torque producing currents i_{dt} and i_{qt} differ from the stator currents i_{ds} and i_{qs} , respectively. The terminal currents i_{ds} and i_{qs} can be measured but the torque producing currents i_{dt} and i_{qt} must be calculated. The core loss resistance R_c is difficult to measure in the transition state. Hence, we ignore the inductance transition voltage in the steady state, and the relationship between stator currents and torque currents are represented as

$$i_{ds} = i_{dt} - \frac{1}{R_c}(\omega_r L_q i_{qt}) \qquad (5)$$

$$i_{qs} = i_{qt} + \frac{1}{R_c}(\omega_r L_d i_{dt}) \qquad (6)$$

Substituting (5) and (6) into (1) and (2), we obtain the voltage equations as

$$V_d = R_s i_{dt} - \omega_r(1 + \frac{R_s}{R_c})L_q i_{qt} + L_d \frac{di_{dt}}{dt} \qquad (7)$$

$$V_q = R_s i_{qt} + \omega_r(1 + \frac{R_s}{R_c})L_d i_{dt} + L_q \frac{di_{qt}}{dt} \qquad (8)$$

III. THE MTC STRATEGY

The losses concentrated in the stator and the rotor due to the flux ripple are assumed to be negligible. From (5)

and (6), we know $i_{ds} \neq i_{dt}$ and $i_{ds} \neq i_{qt}$. First, setting the current angle as

$$\phi = \tan^{-1}\frac{i_{qs}}{i_{ds}} \qquad (9)$$

The electromagnetic torque can be represented as

$$T_e = \frac{3}{4}P(L_d - L_q)R_c^2\left[\frac{(R_c + \omega_r L_q \tan\phi)(R_c \tan\phi - \omega_r L_d)}{(1 + \tan^2\phi)(R_c^2 + \omega_r^2 L_d L_q)^2}\right]I_s^2$$

$$(10)$$

where $I_s = \sqrt{i_{ds}^2 + i_{qs}^2}$. The optimal current angle of maximum torque control is obtained by setting $\frac{dT_e}{d\phi} = 0$. We get the optimal maximum torque per ampere current angle as

$$\phi_{opt} = \tan^{-1}\left[\frac{\omega_r R_c(L_d + L_q)}{R_c^2 - \omega_r^2 L_d L_q} + \sqrt{(\frac{\omega_r R_c(L_d + L_q)}{R_c^2 - \omega_r^2 L_d L_q})^2 + 1}\right] , \quad \text{for}$$

$$T_e \geq 0 \qquad (11a)$$

and

$$\phi_{opt} = \tan^{-1}\left[\frac{\omega_r R_c(L_d + L_q)}{R_c^2 - \omega_r^2 L_d L_q} - \sqrt{(\frac{\omega_r R_c(L_d + L_q)}{R_c^2 - \omega_r^2 L_d L_q})^2 + 1}\right] , \quad \text{for}$$

$$T_e < 0 \qquad (11b)$$

If the core loss of (11) is ignored, i.e. $R_c \to \infty$, the maximum torque current angle is $\phi_{opt} = \pm\pi/4$. From (3), (5), (6) and (11), we can get the torque current i_{dt} and i_{qt} as

$$i_{dt} = \frac{\sqrt{\dfrac{|T_e|}{\frac{3}{4}P(L_d - L_q)R_i^2 K_a}\cos\phi_{opt}(R_i^2 + R_i\omega_r Lq\tan\phi_{opt})}}{R_i^2 + \omega_r^2 L_d L_q} = i_{dt\_MTC} \qquad (12)$$

$$i_{qt} = \pm\frac{\sqrt{\dfrac{|T_e|}{\frac{3}{4}P(L_d - L_q)R_i^2 K_a}\cos\phi_{opt}(R_i^2 + R_i\omega_r Lq\tan\phi_{opt})}}{R_i^2 + \omega_r^2 L_d L_q} = i_{qt\_MTC} \qquad (13)$$

where $K_a = \dfrac{(R_i + \omega_r L_q \tan\phi_{opt})(R_i \tan\phi_{opt} - \omega_r L_d)}{(1 + \tan^2\phi_{opt})(R_i^2 + \omega_r^2 L_d L_q)^2}$. In (13), i_{qt} is positive when $T_e > 0$ and i_{qt} is negative when $T_e < 0$.

IV. THE VOLTAGE REFERENCE CALCULATION OF CURRENT SENSORLESS FOR INTEGRAL VARIABLE STRUCTURE CONTROLLER(IVSC)

A current sensorlesss control scheme includes the voltage reference calculator which generates the required

By inserting (11) into (1) and (2), the required voltages V_d and V_q are

$$V_d = R_s i_{dt\_MTC} - \omega_r \left(1 + \frac{R_s}{R_c}\right) L_q i_{qt\_MTC} + L_d \frac{di_{dt\_MTC}}{dt}$$

(14)

$$V_q = R_s i_{qt\_MTC} + \omega_r \left(1 + \frac{R_s}{R_c}\right) L_d i_{dt\_MTC} + L_q \frac{di_{qt\_MTC}}{dt}$$

(15)

Defining the velocity error $e(t) = \omega_r - \omega_{ref}$, ω_{ref} is the velocity command. Assume the time required to change the velocity command is much longer than the velocity response time (i.e. $d\omega_{ref}/dt = 0$). The velocity error differential equation of SynRM can be expressed as follows

$$\frac{de(t)}{dt} = (-\frac{B_m}{J_m})\omega_r + (\frac{1}{J_m})T_e - (\frac{1}{J_m})T_L$$

$$= \tilde{a}e(t) + \tilde{a}\omega_{ref} + \tilde{b}u(t) - \tilde{b}T_L(t)$$

(16)

where

$$\tilde{a} \equiv -\frac{B_m}{J_m} = a_0 + \Delta a$$

$$\tilde{b} \equiv \frac{1}{J_m} = b_0 + \Delta b$$

$$u \equiv T_e$$

The subscript index "o" indicates nominal system value; "Δ" symbol indicates uncertainty. To have complete robustness, the sliding function S is combined with the integration of the state [1,15] as

$$S = e(t) + c\int_{-\infty}^{t} e(\tau)d\tau, \quad c > 0$$

(17)

If we want the system to be stable, the velocity error integration and the velocity error must be reversed. Therefore, the sliding line $S = 0$ must be designed in the second and fourth quadrants. The larger c value of (17) shows that the system has a faster dynamic response.

The sliding mode controller has different dynamic responses, depending on the initial condition of the integrator in the sliding line. In other words, the system dynamic response can be controlled by selecting the initial condition of the integrator. At $t = 0$, (17) can be expressed as

$$S = e(0) + cI(0)$$

(18)

where $e(0)$ is the initial condition of the error e and $I(0)$ is the initial condition of the integrator which is defined as

$$I(0) = \int_{-\infty}^{0} e(\tau)d\tau$$

(19)

$$I(0) = \frac{\omega_{ref} - \omega_r}{c}$$

(20)

The sliding mode control satisfies the hitting condition in the sliding surface and no reaching time when $t = 0$, and $e \to 0$ for $t \to \infty$.

V. EXPERIMENTAL RESULTS

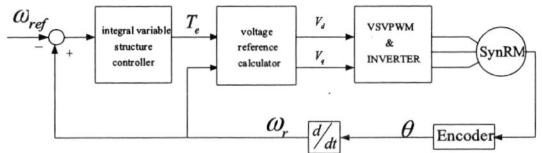

Fig. 2 Experiment structure of current sensorless SynRM

The block diagram of the experimental SynRM system shown in Fig. 2 has the hardware system, SynRM, mechanical loads and auxiliary circuitry for control and measurement. The controller is adopted a dSPACE DS1102 control board. The 0.25hp three-phase SynRM, whose nominal parameters are shown in Table 1, is driven using IGBTs by a three-phase voltage space vector pulsewidth modulation (VSVPWM) inverter. The sampling period of control rules is set as 200 μ s.

Fig. 3 and 4 show the speed response and electromagnetic torque for 300rev/min and 1200rev/min reference command, respectively. In Fig. 3(a) and 4(a), this proposed scheme shows the good speed responses. In Fig. 5, the reference commands are changed from 500rev/min to -500rev/min. Fig. 5(a) shows a good response under no disturbance. Fig. 6 shows the speed response and output torque for 700rev/min reference command with the addition of a 0.3Nt-m machine load at the beginning. The disturbance is changed from 0.3Nt-m to 0.6Nt-m in t =1.9sec. Fig. 6(a) shows this scheme has a good response when external load disturbance is added.

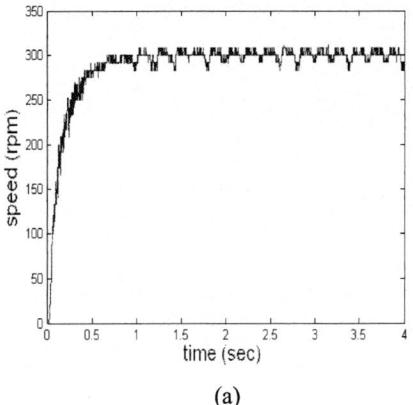

(a)

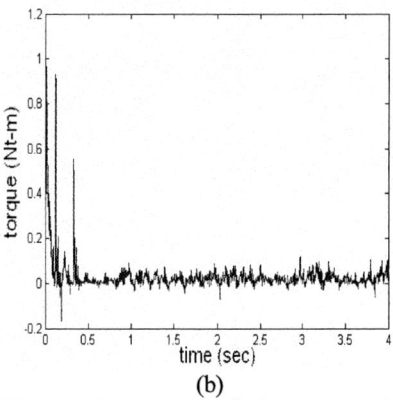

(b)

Fig. 3 The speed response and electromagnetic torque for speed command 300rpm.

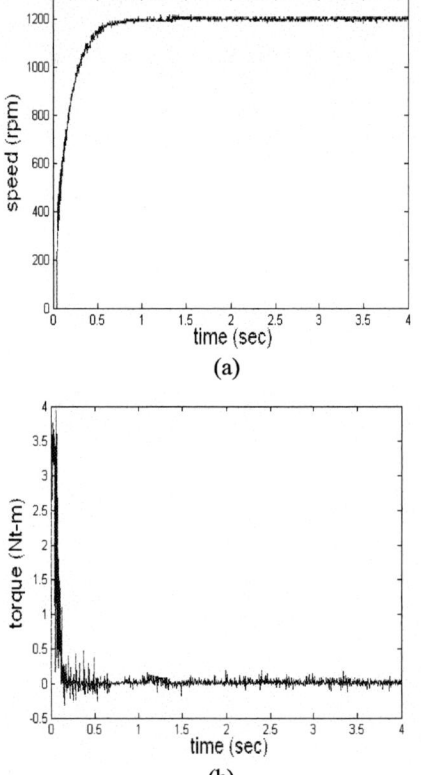

(a)

(b)

Fig. 4 The speed response and electromagnetic torque for speed command 1200rpm.

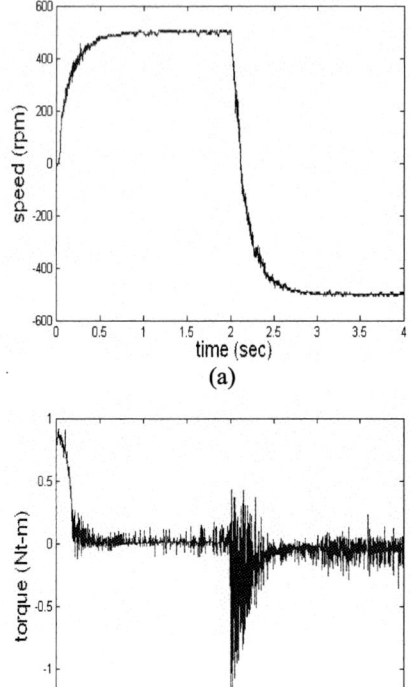

(a)

(b)

Fig. 5 The speed response and electromagnetic torque for speed command from 500rpm to -500rpm.

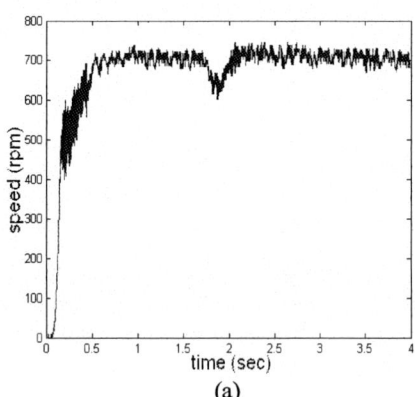

(a)

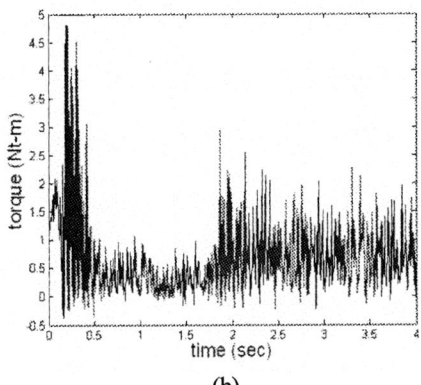

(b)

Fig. 6 The speed response and electromagnetic torque for speed command 700rpm under 0.3Nt-m to 0.6Nt-m load torque step disturbance.

VI. CONCLUSIONS

A complete model development and analysis for the current sensorless integral variable structure control of synchronous reluctance motor is presented in this paper. The mathematical model of synchronous reluctance motor includes the core loss. We develop a MTC strategy of the current sensorless integral variable structure controller. This control scheme does not use current sensor which reduces the system cost. Finally, we employ the experiments to validate the proposed method.

Table 1 The parameters of SynRM (0.37KW)

R_s	R_c	L_d	L_q
$4.2\,\Omega$	$50\,\Omega$ (f=60Hz)	$0.328\,H$	$0.181\,H$

J_m	B_m	P
$0.00076\,kg-m^2$	$0.00012\,Nt-m.s$	2

ACKNOWLEDGMENT

This research was supported by the National Science Council, Taiwan, China, under grant NSC-94-2213-E-224-031.

REFERENCES

[1] T. H. Liu and M. T. Lin, "A fuzzy sliding-mode controller design for a synchronous reluctance motor drive", *IEEE Transactions on Aerospace And Electronic Systems*, vol. 32, no. 3, pp. 1065-1076, 1996.

[2] K. K. Shyu, C. K. Lai, and Y. W. Tsai, "Optimal position control of synchronous reluctance motor via totally invariant variable structure control", *IEE Proceeding-Control Theory and Applications*, vol. 147, no. 1, pp. 28-36, 2000 .

[3] K. K. Shyu and C. K. Lai, "Incremental motion control of synchronous reluctance motor via

multisegment sliding mode control method", *IEEE Transactions on Control Systems Technology*, vol. 10, no. 2, pp. 169-176, 2002.

[4] S. Morimoto, M. Sanada, and Y. Takeda, "High-performance current-sensorless drive for PMSM and SynRM with only low-resolution position position sensor", *IEEE Transactions on Industry Applications*, vol. 39, no. 3, pp. 792-801, 2003.

[5] T. Matsuo and T. A. Lipo "Current sensorless field oriented control of synchronous reluctance motor", *IEEE Conference on Industry Applications*, pp. 672-678, 1993.

[6] S. Morimoto, M. Sanada, and Y. Takeda, "Optimum efficiency operation of synchronous reluctance motor without current-sensor", *IEE Conference on Power Electronics and Variable Speed Drives*, pp. 506-511, 2000.

[7] J. Soltani and H. A. Zarchi, "Robust optimal speed tracking control of a current-sensorless synchronous reluctance motor drive using a new sliding mode controller", *The Fifth International Conference on Power Electronics and Drive Systems, 2003 PEDS*, pp. 474-479, 2003.

[8] R. E. Betz, "Theoretical aspects of control of synchronous reluctance machines", *IEE Proceedings B*, . vol. 139, no. 4, pp. 355-364, 1992.

[9] M. G., Jovanovic and R. E. Betz, "Effects of uncompensated vector control on synchronous reluctance motor performance", *IEEE Transactions on Energy Conversion*, vol. 14, no. 3, pp. 532-537, 1999 .

[10] L. Xu, X. Xu, T. A. Lipo, and D. W. Novotny, "Vector control of a synchronous reluctance motor including saturation and iron loss", *IEEE Transactions on Industry Applications*, vol. 27, no. 5, pp. 977-985, 1991.

[11] T. M. A. El-Antably and T. A. Lipo, "A new control strategy for optimum-efficiency operation of a synchronous reluctance motor", *IEEE Transaction on Industry Applications*, vol. 33, no. 5, pp. 1146-1153, 1997.

[12] H. D. Lee, S. J. Kang, and S. K. Sul, "Efficiency-optimized direct torque control of synchronous reluctance motor using feedback linearization", *IEEE Transactions on Industrial Electronics*, vol. 46, no. 1, pp. 192-198, 1999 .

[13] T. Sharf-Eldin, M. Dunnigan, J. E. Fletche, and B.W. Williams, "Nonlinear robust control of a vector-controlled synchronous reluctance machine", *IEEE Transactions on Power Electronics*, vol. 14, no. 6, pp. 1111-1121, 1999.

[14] U. Itkis, *Control System of Variable Structure,* New York, Wiley, 1976.

[15] V. I. Utkin, J. Guldner, and J. Shi, *Sliding Mode Control in Electromechanical Systems*, Taylor & Francis, 1999 .

[16] S. J. Kang and S. K. Sul, "Highly dynamic torque control of synchronous reluctance motor", *IEEE Transactions on Power Electronics*, vol. 13, no. 4, pp.793-798,1998.

2006 5th International Power Electronics and Motion Control Conference

An Improved Sliding Mode Observer for Speed Sensorless Vector Control Drive of PMSM

K. Paponpen and M. Konghirun

Department of Electrical Engineering , King Mongkut's University of Technology Thonburi, Bangkok, Thailand

Abstract—This paper proposes a speed sensorless control scheme for permanent magnet synchronous motor (PMSM) drive using an improved sliding mode observer (SMO). A variable frequency or cutoff frequency in low pass filter is not essential to use in the improved SMO. Since the product of the improved SMO gain and the control action of sigmoid function, which replaces the Bang-Bang control or discontinuous control, commonly found in the conventional SMO, can determine the equivalent back emfs. The estimated rotor position and speed are obtained directly from them. Therefore, the low pass filter can be eliminated and the improved SMO could simplify the conventional SMO. Also, the cutoff frequency tuning is not necessary. A DSP based digital controller using the TMS320F2812 from the Texas Instruments has been employed to realize the proposed sensorless control scheme. Experimental results show that the proposed control scheme can achieve robust sensorless requirement.

Keywords-permanent magnet synchronous motors; sliding mode observer; sigmoid function

I. INTRODUCTION

The PMSMs have found in several applications such as machine tools and robotics, due to their high ratio of power to weight and ease of control. Speed control of PMSM usually requires the mechanical sensor for sensing rotor position to achieve drive system. However, the position sensors such as resolver, absolute position encoder, and QEP sensor increase the overall cost and reduce reliability of the system. From disadvantages of the mechanical sensors, the sensorless operation of PMSM has been receiving wide attention.

The sensorless speed control of PMSM is presented in [1]. In this paper, the rotor position is estimated from calculating the flux linkage that is obtained by integrator processing. However, the suffering from the effects of integrator drift is evident. The sensorless speed control using extended Kalman filter is present to avoid this problem [2]-[3]. However, the system model used in Kalman filter greatly depends on the changing of parameters of motor and complex calculation.

The SMO is proposed to overcome these disadvantages [4]-[5]. It is robust to variation of motor parameters. Since the switching signals of the SMO contain the induced voltages of the motor, it is possible to obtain the rotor

speed and position of the motor directly from the switching signals. However, the chattering problem due to the discontinuous control in SMO is the major factor to make the high system oscillation. Generally, the low pass filter is used to reduce the chattering problem but it produces the delay time. So, the calculated rotor position is normally added with offset position. Moreover, to achieve for speed sensorless control, the cut off frequency of low pass filter have to be varied according to the rotor speed. It is a task of tuning to obtain the accurate estimated position and speed over a wide range of operating speeds.

In this paper, the improved SMO is proposed. It replaces the discontinuous control, Bang-Bang control, by using the sigmoid function [6] in order to reduced chattering problem. The ripples of the state variables are reduced. So the low pass filter and the back emf calculation are not essentially used to reject them. Therefore, the cut off frequency tuning is not required and the improved SMO simplifies its structure.

II. PMSM MODEL

The PMSM model in the stationary reference frame ($\alpha\beta$-axis) is shown below:

$$\dot{i}_\alpha = -\frac{R}{L}i_\alpha - \frac{1}{L}e_\alpha + \frac{1}{L}u_\alpha$$
$$\dot{i}_\beta = -\frac{R}{L}i_\beta - \frac{1}{L}e_\beta + \frac{1}{L}u_\beta \qquad (1)$$
$$e_\alpha = -\lambda_0\omega_e\sin\theta_e \,; e_\beta = \lambda_0\omega_e\cos\theta_e$$

where R is the stator resistance (ohm), L is stator self inductance (H), $i_\alpha, i_\beta, u_\alpha, u_\beta$ and e_α, e_β are the phase currents (amp), phase voltages (volt) and back emf (volt) in the stationary reference frame, respectively. The ω_e is electrical angular velocity (rad/sec), λ_0 is the flux linkage of permanent magnet (volt.sec/rad) and θ_e is the electrical rotor position (rad).

Assuming the rotor speed changes slowly, i.e., $\dot{\omega}_e \approx 0$, then the back emf can be simply rewritten as

$$\dot{e}_\alpha = -\omega_e e_\beta$$
$$\dot{e}_\beta = \omega_e e_\alpha \qquad (2)$$

1-4244-0448-7/06/$25.00 ©2006 IEEE

III. THE CONVENTIONAL SMO

Fig.1 shows the system configuration of the speed sensorless control of PMSM using the SMO. Two phase currents are measured and transformed to the αβ-axis. Then, these currents and two computed phase voltages in αβ-axis are fed to the SMO. It calculates the estimated rotor position and speed that are used to transform the variables in the stationary reference frame into the synchronous rotating reference frame and to be the state feedback for closed-loop system drive, respectively.

The conventional SMO shown in Fig.2 consists of the sliding mode current observer, Bang-Bang control, low pass filter, back emf observer, rotor position and speed calculation, rotor position correction, and cutoff frequency tuning. The estimated currents are computed by the sliding mode current observer. The current error between the estimated currents and the actual one fed into the Bang-Bang control. Since it is the discontinuous control, the chattering problem from this action control always appears in the state variables. Therefore, the low pass filter is used to reduce this problem. However, the delay time caused by the low pass filter maintains the obstacle in the estimated rotor position. The calculated rotor position is normally added with offset position. To achieve speed sensorless control at any operating speeds, the cutoff frequency has been therefore adjusted according to the operating speed. The cut off frequency to be tuned is formulated as expressed in (3).

$$\omega_c = \omega_{c\_history} + \left(\frac{\omega_{c2} - \omega_{c1}}{\omega_2 - \omega_1}\right)\omega^* - \left(\frac{\omega_{c2} - \omega_{c1}}{\omega_2 - \omega_1}\right)\omega_1 \quad (3)$$

where $\omega_c = 2\pi f_0$ (rad/s), and f_0 = cutoff frequency of the filter (Hz), $\omega_{c\_history}$ = previous value of ω_c, ω_{c2}, ω_{c1} are the angular frequency at the rotor speed ω_2 and ω_1, and ω^* is the rotor speed command (rad/s).

According to the equation (3), the cutoff frequency tuning is used to achieve the accurate estimated position in the conventional SMO. Thus, the improved SMO is proposed to eliminate this tuning process.

IV. THE IMPROVED SMO

The block diagram of the improved SMO can be depicted in Fig.3. It primarily consists of only three major parts, i.e., current observer, continuous sigmoid function and rotor position and speed calculator, which explain as follows:

Current observer model:

$$\dot{\hat{i}}_\alpha = -\frac{R}{L}\hat{i}_\alpha + \frac{1}{L}u_\alpha - \frac{1}{L}k_1 H\left(\hat{i}_\alpha - i_\alpha\right)$$
$$\dot{\hat{i}}_\beta = -\frac{R}{L}\hat{i}_\beta + \frac{1}{L}u_\beta - \frac{1}{L}k_1 H\left(\hat{i}_\beta - i_\beta\right) \quad (4)$$

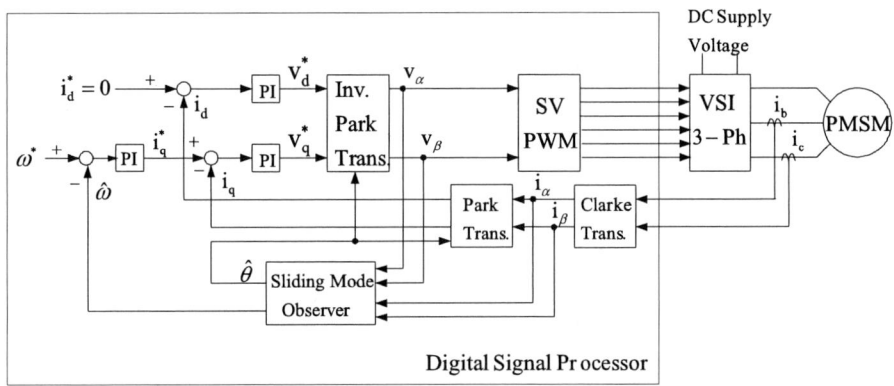

Figure 1. The overall configuration of the proposed system

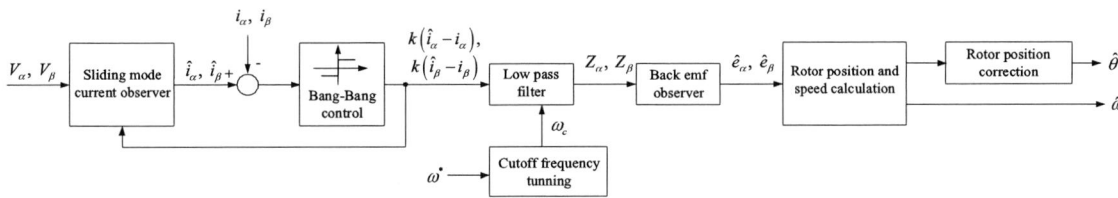

Figure 2. The conventional SMO

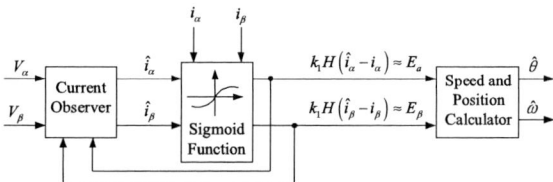

Figure 3. The improved SMO

where superscript " $\wedge$ " represents the estimated quantities, k_1 is constant current observer gain, and H is the sigmoid function [6] which will replace the conventional discontinuous control. This function can be formulated as the following:

$$
\begin{bmatrix} H\left(\bar{i}_\alpha\right) \\ H\left(\bar{i}_\beta\right) \end{bmatrix} = \begin{bmatrix} \left(\dfrac{2}{1+\exp(-a\bar{i}_\alpha)}\right)-1 \\ \left(\dfrac{2}{1+\exp(-a\bar{i}_\beta)}\right)-1 \end{bmatrix} \tag{5}
$$

where a is a positive constant that can be adjusted the slope of the sigmoid function. The $\bar{i}_\alpha = \hat{i}_\alpha - i_\alpha, \bar{i}_\beta = \hat{i}_\beta - i_\beta$ are the current errors between estimated currents and actual ones in the α- and β-axis, respectively.

The sliding mode surface is defined as

$$
s_n = \begin{bmatrix} s_\alpha & s_\beta \end{bmatrix}^T \tag{6}
$$

where $s_\alpha = \bar{i}_\alpha$ and $s_\beta = \bar{i}_\beta$. When the estimation error trajectories reach the sliding surface, i.e., $s_n = 0$. Obviously, the observed currents will eventually converge to the actual ones, i.e., $\hat{i}_\alpha = i_\alpha$ and $\hat{i}_\beta = i_\beta$. It is important that on the sliding surface, the observer will not be affected by any system parameters or any disturbance.

The sliding mode exists when $s_n \dot{s}_n < 0$, that is $s_n \to 0$ as $t \to \infty$. Let's design the Lyapunov function to find such condition of sliding mode existence.

$$
V = \frac{1}{2} s_n^T s_n = \frac{1}{2}\left(s_\alpha^2 + s_\beta^2\right) \tag{7}
$$

The error equation is obtained by subtracting (4) from (1) as

$$
\dot{\bar{s}}_\alpha = \dot{\bar{i}}_\alpha = \dot{\hat{i}}_\alpha - \dot{i}_\alpha = -\frac{R}{L}\bar{i}_\alpha + \frac{1}{L}e_\alpha - \frac{1}{L}k_1 H\left(\bar{i}_\alpha\right)
$$

$$
\dot{\bar{s}}_\beta = \dot{\bar{i}}_\beta = \dot{\hat{i}}_\beta - \dot{i}_\beta = -\frac{R}{L}\bar{i}_\beta + \frac{1}{L}e_\beta - \frac{1}{L}k_1 H\left(\bar{i}_\beta\right) \tag{8}
$$

The sliding mode exists if $\dot{V} = s_n^T \dot{s}_n < 0$, i.e.,

$$
\begin{aligned}
s_n^T \dot{s}_n = &-\frac{R}{L}\left(\bar{i}_\alpha^2 + \bar{i}_\beta^2\right) + \frac{1}{L}\left(e_\alpha \bar{i}_\alpha - k_1 \bar{i}_\alpha H\left(\bar{i}_\alpha\right)\right) \\
&+ \frac{1}{L}\left(e_\beta \bar{i}_\beta - k_1 \bar{i}_\beta H\left(\bar{i}_\beta\right)\right) < 0
\end{aligned} \tag{9}
$$

As a result,

$$
k_1 \geq \max\left(\left|e_\alpha\right|, \left|e_\beta\right|\right) \tag{10}
$$

Once the sliding mode occur with selecting large enough k_1, then the sliding surface become as

$$
\begin{bmatrix} \dot{s}_\alpha & \dot{s}_\beta \end{bmatrix}^T = \begin{bmatrix} s_\alpha & s_\beta \end{bmatrix}^T = \begin{bmatrix} 0 & 0 \end{bmatrix} \tag{11}
$$

and (11) can be rewritten as

$$
\begin{aligned}
\left(k_1 H\left(\bar{i}_\alpha\right)\right)_{eq} &= \hat{e}_\alpha \\
\left(k_1 H\left(\bar{i}_\beta\right)\right)_{eq} &= \hat{e}_\beta
\end{aligned} \tag{12}
$$

The chattering problem could be rejected by the sigmoid function which is the continuous control. Back emf from equation (12) can be used to calculate the rotor speed and position directly as follows.

$$
\hat{\theta}_e = -\tan^{-1}\left(\frac{\hat{e}_\alpha}{\hat{e}_\beta}\right) \tag{13}
$$

$$
\hat{\omega}_e = \frac{d\hat{\theta}_e}{dt} \tag{14}
$$

V. EXPERIMENTAL RESULTS

The configuration of a DSP-based experimental system is shown in Fig.4. The fixed-point TMS320F2812 DSP with a clock frequency of 150 MHz is employed as the digital controller. The PMSM is supplied by a three phase voltage source PWM inverter with a switching frequency of 20 kHz. A Quadratic Encoder Pulse (QEP) sensor is used to merely detect the rotor position and compare with the estimated rotor position. Two motor phase currents are sensed, rescaled, and converted to digital values by on-- chip ADC with 12-bit resolution. The PWM gate firing signals from the desired phase voltage commands are generated by means of the space vector modulation technique. The sampling period of the control system is set at 50 μs. The motor parameters are given in the appendix.

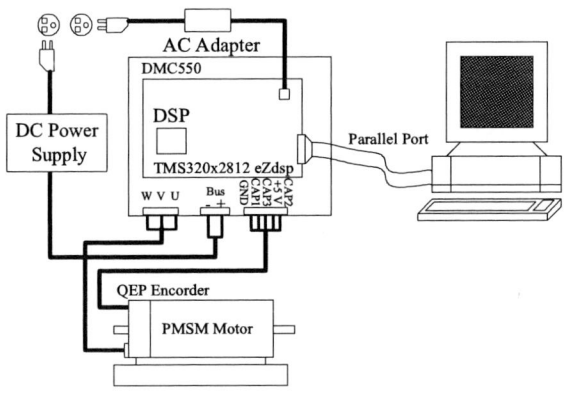

Figure 4. Drive System

Fig.5 shows the speed responses of sensorless drive system of PMSM using the conventional SMO with fixed cutoff frequency and the offset of rotor position is added. A ramp speed command is applied at time = 0.4 sec from 0.2 PU (180 rpm) to 0.8 PU (720 rpm) at the no load condition. In Fig. 5(a) shows the speed response that can be converged to the speed command. Although the estimated rotor position is compensated by the rotor position offset, the position error between the estimated rotor position and the actual one is still evident as shown in Fig. 5(d). The fixed cutoff frequency should be adjusted according to the rotor speed to reduce this position error.

Fig.6 shows the responses of sensorless drive system using the conventional SMO with adjustable cutoff frequency according to the equation (3) and the offset of rotor position is also added. For the same experimental conditions, the speed response can also be converged to the speed command as seen in Fig. 6(a). Clearly, the position error is smaller than the position error when the fixed cutoff frequency is used. In this test, the cutoff frequency has to be tuned to achieve drive system.

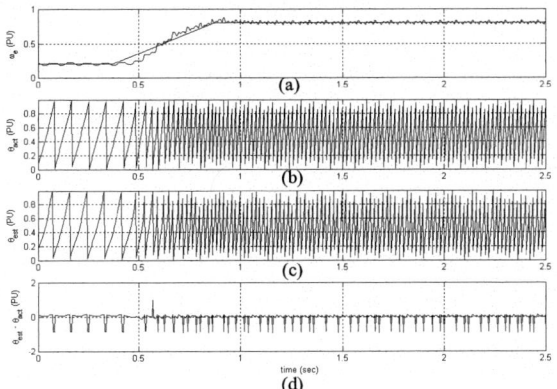

Figure 5. The responses speed sensorless control using the conventional SMO with fixed cutoff frequency at 0.0004 PU (0.15 rad/s) (a) speed response to the ramp speed command from 0.2 – 0.8 PU (b) rotor position from QEP (c) estimated rotor position (d) the position error between the estimated rotor position and the actual one

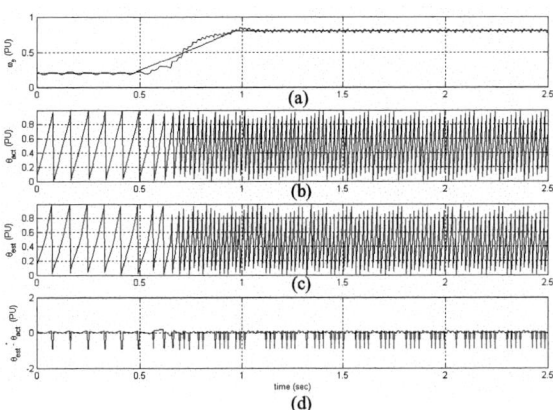

Figure 6. The responses of speed sensorless control using the conventional SMO with cutoff frequency tuning according to the equation (3) (a) speed response to the ramp speed command from 0.2 – 0.8 PU (b) rotor position from QEP (c) estimated rotor position (d) the position error between the estimated rotor position and the actual one

Now the proposed SMO with sigmoid function is tested. The successful speed control can also be expected in Fig. 7(a). By using this improved SMO, the estimated rotor position is nearly coincides with the actual one as seen in Fig.7(d). Consequently, the proposed SMO improve the accuracy of rotor position estimation over a wide range of speeds and the cutoff frequency tuning process is eliminated.

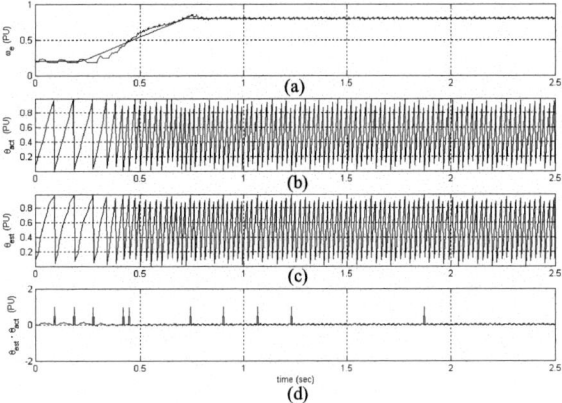

Figure 7. The responses of speed sensorless control using the improved SMO   (a) speed response to the ramp speed command from 0.2 – 0.8 PU (b) rotor position from QEP (c) estimated rotor position (d) the position error between the estimated rotor position and the actual one

VI. CONCLUSION

The proposed SMO has been presented to estimate the rotor speed and rotor position of the PMSM. This observer is very easy to implement and is not required for tuning the cutoff frequency in the SMO-based sensorless drive system of PMSM. The low pass filter is not necessary once the sigmoid function replaces the discontinuous function. According to the experimental results, the estimated rotor speed is nearly identical to the actual ones. The proposed SMO greatly simplifies the algorithm, comparing with the conventional SMO. The obstacle of cutoff frequency tuning is significantly lessened when using this proposed observer.

VII. APPENDIX

PARAMETERS OF THE 24 V APPLIED PMSM MOTOR

Stator resistance	0.79 Ω
Stator inductance	1.17 mH
Permanent magnet flux	17.666×10^{-3} (volt.sec/rad)
Number of poles	8
Slope of sigmoid function	a = 4
Sling mode observer gain	K = 0.1

REFERENCES

[1] S. Tomonobu, S. Tsuyoshi and U. Katsumi, "Vector control of permanent magnet synchronous motors without position and speed sensors," *in PESC'95,* pp.759 – 765.

[2] Q. Albert., B. Wu, K. Hassan, "Sensorless control of permanent magnet synchronous motor using extended Kalman filter," *Electrical and computer Engineering Conference 2004,* pp.1557 – 1562.

[3] G. Garcia Soto, E. Mendes and A. Razek, " Reduced order observers for rotor flux, rotor resistance and speed estimation for vector controlled induction motor drive using the extended Kalman filter technique," *Electric Power Applications, IEE Proceeding volume 146,* 1999, pp.282-288.

[4] T. Furuhashi, , S. Sangwongwanich, and S. Okuma, "A position – and – velocity sensorless control for brushless DC motors using an adaptive sliding mode observer", *IEEE Trans. Industrial Electronics, Vol.39,* pp. 89 – 95, Apr. 1992.

[5] Z.M.A. Peixoto, Sa F.M Freitas, P.F. Seixas, and B.R. Menezes, "Application of sliding mode observer for induced e.m.f., position and speed estimation of permanent magnet motors", *in 1995 Proc. Power Electronics and Drive Systems Int. Conf.,* pp.599 – 604.

[6] M. Ertugrul, , O. Kaynak, , A. Sabanovic, and K. Ohnishi, "A generalized approach for Lyapunov design of sliding mode controllers for motion control applications", *in 1996 Proc. AMC'96-MIE Conf.,* pp.407 – 412.

2006 5th International Power Electronics and Motion Control Conference

Analysis of an AC fed direct converter for a switched reluctance machine in aerospace applications

S. J. Forrest, J. Wang, G. W. Jewell, C. M. Johnson and S.D. Calverley
University of Sheffield / Dept. Electronic and Electrical Engineering, Sheffield, United Kingdom

Abstract - This paper proposes an N-phase, AC fed, switched reluctance drive for aerospace applications. The proposed converter is capable of direct AC conversion and eliminates the unreliable DC link capacitor present in conventional DC supply converters. It is shown that, by Space Vector Modulation of output current, control of active and reactive power can be achieved whilst three sinusoidal AC input phase-currents with the required power factor can be synthesised. A comparison between numerically simulated and analytically predicted DC output current ripple is also presented.

Keywords – Switched Reluctance Machine, Current Source Inverter, Active Rectifier, Power Quality, Current Control.

I. Introduction

Increasing commercial and environmental pressures have resulted in considerable interest and associated research activities into the so called 'more-electric aircraft'[1]. It is envisaged that in the future generations of aircraft, controlled electric drives will be adopted in preference to established hydraulic and mechanical systems to perform functions such as control and surface actuation, braking actuation embedded electrical starter-generators [2][3] etc.

Of the various machine topologies chosen for these emerging applications, the switched-reluctance (SR) machine remains a leading candidate due to its potential fault-tolerance capability and a simple rotor design that allows higher operating speeds and operation within harsh environments [4].

However, the viability of adopting more–electric technologies to perform key functions on an aircraft is also reliant upon realising power converters that have both the desired volumetric power density and the reliability required for safety critical systems. The power density of electrical machines continues to be improved, however it is widely recognised that the power electronic converter often poses the greatest difficulty in terms of gaining end user acceptance. Conventional drives for controlling bi-directional power flow between an SR machine and a 3-phase AC supply require a large DC link capacitance, usually provided by an array of electrolytic capacitors. However, whilst electrolytic capacitors are widely used in industrial drives, they are generally regarded as unsuitable for aerospace applications, where

Rolls-Royce PLC *(sponsor)*

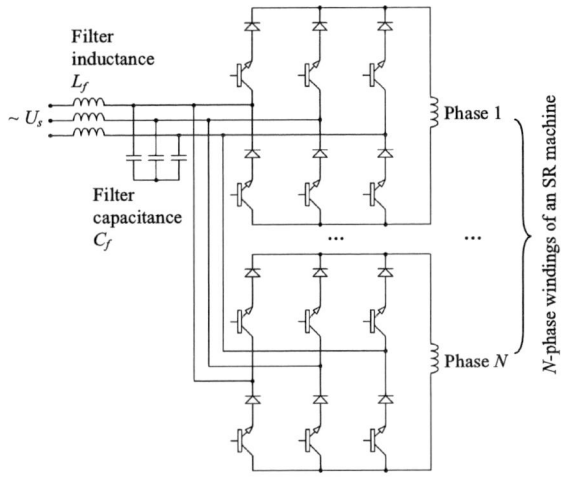

Fig. 1. **Low Energy Storage Converter.**

weight and reliability are key considerations. In this paper an AC fed direct converter for a Switched Reluctance machine is proposed. The current control strategy is developed and power quality and performance is analysed.

II. Modulation and Control

The circuit configuration of the proposed low energy storage converter topology for controlling bi-directional power flow between an N-phase SR machine and a 3-phase AC supply, is shown schematically in Fig. 1. L_f and C_f are the inductance and capacitance of the supply filter and U_s the supply voltage. The rectifier bridge consists of six controllable switches (IGBT's), each with a series diode, which provides the reverse blocking capability. These diodes also provide an inherent fault tolerance that, in the event of IGBT failure, the converter reverts to the form of an uncontrolled rectifier. By using reverse blocking IGBT's, the series diode can be removed. However, in this case, the fault tolerant feature of this converter design may be compromised in the event of device failure.

The converter allows full four-quadrant input operation, providing only uni-polar output currents, whilst having the capability of blocking voltage of either polarity. These operational features are exactly what are required for controlling an SR machine. A switching strategy using Space Vector Modulation has been adopted in order

1-4244-0448-7/06/$25.00 ©2006 IEEE   977

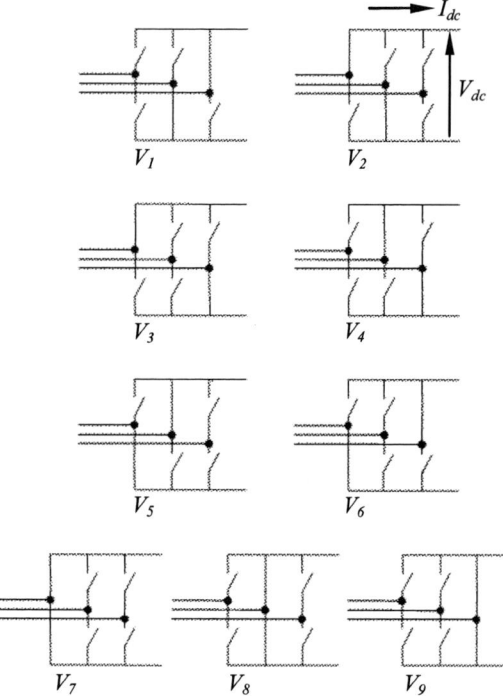

Fig. 2: Switching States for the Three-Phase Rectifier.

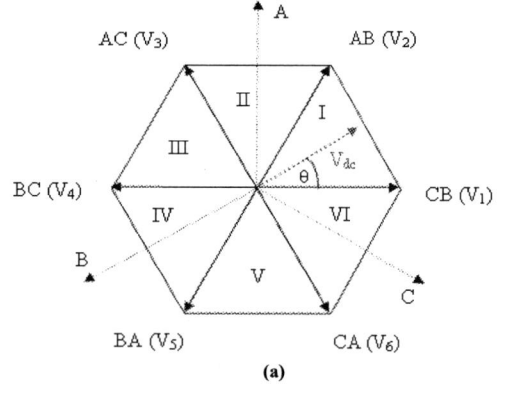

(a)

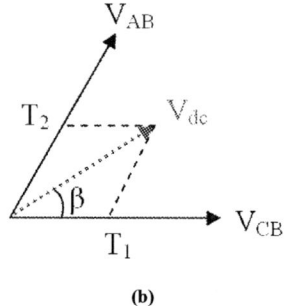

(b)

Fig. 3: Voltage Space Vector Diagram showing Active Vectors.

to control current in the phase winding. When the sum of the net winding current for N-output phases is held at a constant value for a given output torque, the converter behaves as a current source rectifier/inverter, allowing for three sinusoidal input-phase currents, with the required power factor to be synthesised. In order to maintain system power quality within acceptable limits, the control system of the N-phase converter endeavours to synthesise a net sinusoidal current at the input terminals under all load conditions.

For the space vector modulation scheme, there are nine switching states: six active states and three zero states, as shown in Fig. 3(a) [5]. Each switching state is represented as the space vector for the input of the rectifier. V_{dc} is the reference vector of which the rectifier output voltage is composed by these active switching states. During one sampling interval, T_s, the output voltage vector can be written as:

$$V_{dc}(t) = \delta_n V_n + \delta_{n+1} V_{n+1} + \delta_7 V_7 \,(\text{or } \delta_8 V_8 \text{ or } \delta_9 V_9), \qquad (1)$$

where $V_7 \sim V_9$ are the three zero vectors, V_n, V_{n+1} (n =1,2,..,5) are two active vectors, and $\delta_0, \delta_n ..., \delta_9$ are the duty cycles for each state, $V_0, V_n...V_9$. The composition of V_{dc}, by the available switching states, can be achieved in many ways. However in order to minimise switching events and maximise the active voltage vector on-time, the V_{dc} vector is commonly split between two adjacent active voltage vectors and zero voltage vectors, for an arbitrary sector. Fig. 2 and Fig. 3 show the distribution of voltage vectors and the sequence of switching states to minimise switching frequency and output current ripple. By way of an example, if V_{CB} and V_{AB} are the two active line voltages in sector I (Fig. 3(b)), the average voltage V_{dc} applied to the phase winding over T_s period is given by (2). The rotating reference vector, V_{dc}, is synchronised with the supply voltage, i.e., $\theta = \omega t$ corresponds to the electrical angular frequency of the supply voltage times the time.

$$V_{dc}(t) = \frac{T_1 V_{cb} + T_2 V_{ab} + T_8 V_8}{T_s} \qquad (2)$$

T_1, T_2, and T_8 are given by:

$$T_1 = mT_s \sin\left(\frac{\pi}{3} - \beta\right), \qquad (3)$$

$$T_2 = mT_s \sin \beta, \qquad (4)$$

$$T_8 = \left[T_s - (T_1 + T_2)\right] \qquad (5)$$

And the modulation index by:

$$m = \frac{2V_{dc}}{3V_m}, \qquad (6)$$

where, V_m is the peak phase voltage of the AC supply.

From (6), it follows that the per-cycle average voltage across the phase winding is independent of time, and can be controlled by the modulation index m. Therefore by directly varying the modulation index of the converter branch, the phase voltage and thus the active power of an SR machine, can be directly controlled. Conversely it is possible to verify that, by introducing a phase shift angle between the synchronous modulation reference and the supply-voltage based reference, the reactive power can be controlled[6]. Therefore, it is possible to compensate for the leading power factor which results from the capacitive currents drawn by the supply filter. By expressing the supply filter equations in the synchronous rotating reference frame, the appropriate phase shift correction angle can be calculated. The equations of the supply filter are formed using the equivalent circuit shown in Fig. 4. The equations governing the equivalent circuit in the stationary reference frame are:

$$u_s = \frac{L_1 R \frac{d}{dt} i_s}{R + L_1 + L_2} + \frac{L_1 L_2 \frac{d}{dt} i_s}{R + L_1 + L_2} + u_c, \tag{7}$$

$$u_c = \frac{1}{C_f} \int i_c dt, \tag{8}$$

$$i_s = i_c + i_r. \tag{9}$$

Therefore, substituting (8) and (9) into (7), gives the following expression:

$$u_s = \frac{L_1 R \frac{d}{dt} i_s}{R + L_1 + L_2} + \frac{L_1 L_2 \frac{d}{dt} i_s}{R + L_1 + L_2} + \frac{1}{C_f} \int (i_s - i_r) dt. \tag{10}$$

Solving for i_r and considering the direct and quadrature components in the supply-voltage based x-y reference frame, a simplified expression for reactive power compensation can be derived as follows [7]:

$$i_{ys}^* \approx -C_f \omega_s V_m. \tag{11}$$

By considering the time average of the AC phase current over a switching period, with reference to the switching scheme for sector I shown in Fig. 3(b), the following relationship between the demand for the AC current magnitude i_{sx}^* and the required sum of the SR winding currents, I_{dc}, can be derived as:

$$i_{sx}^* = I_{dc} * m_p. \tag{12}$$

Where, m_p is the modulation index when the power factor correction is not considered. Equation (12) is obtained by considering power balance, assuming no losses in the converter. If a power factor correction angle, α, between the modulation reference and the supply-voltage based

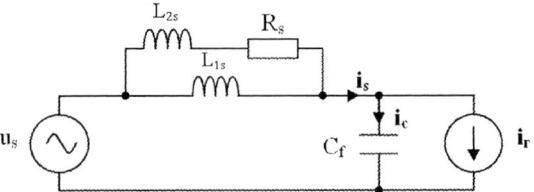

Fig. 4: Equivalent Circuit of the Supply Filter.

reference is introduced, the average voltage across the phase winding of the SR machine is given by (13).

$$V_{dc} = \frac{3}{2} m V_m \cos \alpha \tag{13}$$

As α changes, the modulation index is required to vary in order to maintain the desired V_{dc} and hence the output current I_{dc}. Thus, the resulting modulation index when correcting the input power factor is given by (14).

$$m = \frac{m_p}{\cos \alpha} \tag{14}$$

Active power control is obtained from the x-axis component (12) and reactive power from the y-axis component (11) of the capacitor current vector [7]. From these components, the correction angle, α, can be obtained.

To control I_{dc}, the modulation index m_p, is obtained via the current feedback PI controller. The relationship between modulation index and the average voltage across the SR machine phase winding is linear which allows for linear current control.

III. Analysis of Output Current Ripple

Power quality is a major issue in the design of high power electrical systems for any application, particularly aerospace. By modulation of the converter in order to realise a desired output current, harmonics are introduced at integer multiples of the switching frequency and at side bands of all these frequencies [8] [9]. These harmonics determine the line current total harmonic distortion (THD) and ultimately affect copper losses and torque ripple of an SR machine. Hence, the current harmonics drawn from, and returned to, the supply will have a significant bearing on a given converter / control combination. Therefore, it is important that the harmonic content of the converter output is analysed. Further to this, if the harmonic content of the SR phase current is known, more optimised input filter design can be easily achieved.

In order to evaluate the output current spectral content, the output harmonic flux (time integral of harmonic voltage) is considered [8-10]. In accordance with the space vector modulation scheme, within each carrier cycle, the harmonic voltage vectors, V_{1h}, V_{2h} and V_{0h} exist

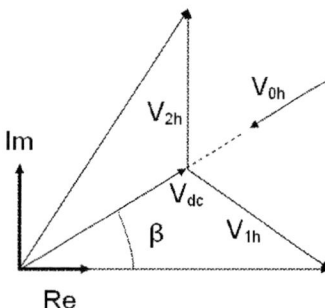

Fig. 5: Harmonic Voltage Vectors in the first segment of the space vector hexagon.

and are both space and modulation index dependent, as shown in Fig. 5. The harmonic flux in a carrier cycle is calculated by (15):

$$\psi_T(\theta,m,t) = \int_0^t (V_k - V_{dc})dt. \qquad (15)$$

where, V_k is the instantaneous voltage across SR machine phase windings for a given switching state, V_{dc} is the desired output voltage and T_s is the carrier period. The per-carrier harmonic flux is then normalised to the nominal output voltage-second magnitude.

$$\psi_n(\theta,m,d) = \frac{\psi_T(\theta,m,t)}{\frac{3}{2}V_m T_s}, \qquad (16)$$

where, $d = t/T_s$. Harmonic current and harmonic flux are related to the inductance of the SR machine phase winding by $\psi_T = L_s i$.

The current ripple over one carrier cycle can then be expressed as:

$$\Delta I_{dc} = \frac{3V_m}{2f_s L_s}\psi_n(\theta,m,d). \qquad (17)$$

The characteristic SVPWM method has six-fold symmetry; hence the per-fundamental cycle rms flux harmonic can be calculated by evaluating the per-carrier harmonic flux over a 60° inverter segment.

IV. Simulation Results

Rather than focussing on a specific SR machine and a set of particular operating conditions from the outset, the research program has been structured to progressively and logically address the various features of the SR machine and converter. This will ultimately be achieved by considering the converter of Fig. 1 operating in a

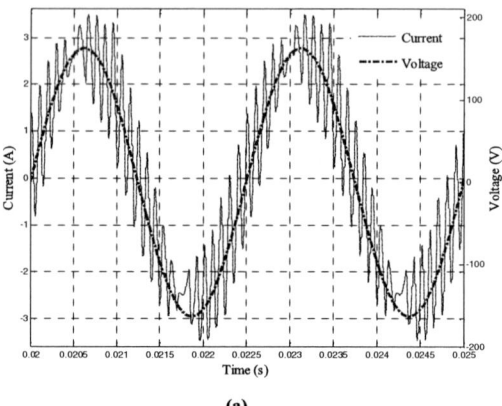

(a)

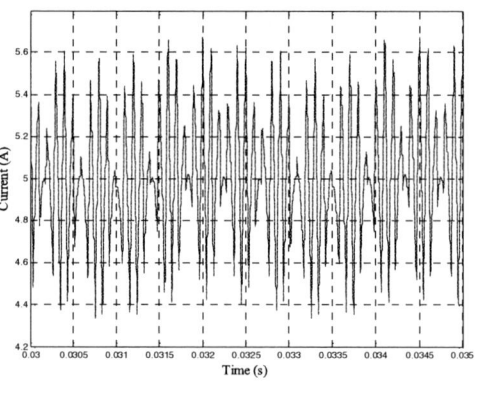

(b)

Fig. 6: (a) Phase A Input Current and Voltage waveforms and (b) Output Current Waveform for a single phase converter.

series of different modes and with different load conditions.

Reconfigurable system models consisting of one phase and five phases of the converter, has been developed utilising MATLAB / Simulink and SimPowerSystems Toolbox. Each model implements current control and power factor correction. The modulation algorithm employs the principles outlined in the previous sections. An appropriate filter design was chosen to minimise the input current THD [10].

The initial investigation considered simulation of the single phase converter model with inductive load with a constant value inductance. This demonstrates bi-directional power flow control capability of the converter, through a fixed value inductor. Fig. 6 (a) and (b) show the simulated input AC voltage and current waveforms and DC output current waveform, respectively, in steady-state for fixed inductive load (5mH), 400Hz, 115V AC supply and a controlled DC output current of 15A. Although this is a relatively simple mode of operation, it provides a mechanism to establish a relationship between switching frequency, load current filter parameter and power quality. Fig. 7 compares the analytically predicted and numerically simulated DC output current ripples over a switching cycle. As will be seen, the two predictions are in good agreement.

The methodologies developed for the single output phase converter have been extended to five independent branches. This allowed the supply side power quality and filter performance to be explored when tracking multiple current demands in a five phase SR machine, each having identical active (emf) and passive components. The current demand waveform profiles of each phase are sequentially offset by 72 electrical degrees. Given that a time averaged sinusoidal input phase current is only possible when drawing a constant load current, it follows that the sum of the phase currents in the SR machine must be held constant in order to maintain the necessary power quality on the supply side. Thus, coordinated control of phase currents in the SR machine is of great importance, particularly during the commutation of current between one phase and another. The current demand waveform profiles of the SR machine were determined by finite element analysis based optimisation with the constraints that the sum of phase currents drawn by all five converters remains constant for a given torque demand. Fig. 8 and Fig. 9 show simulated three-phase input current and voltage waveforms as well as the

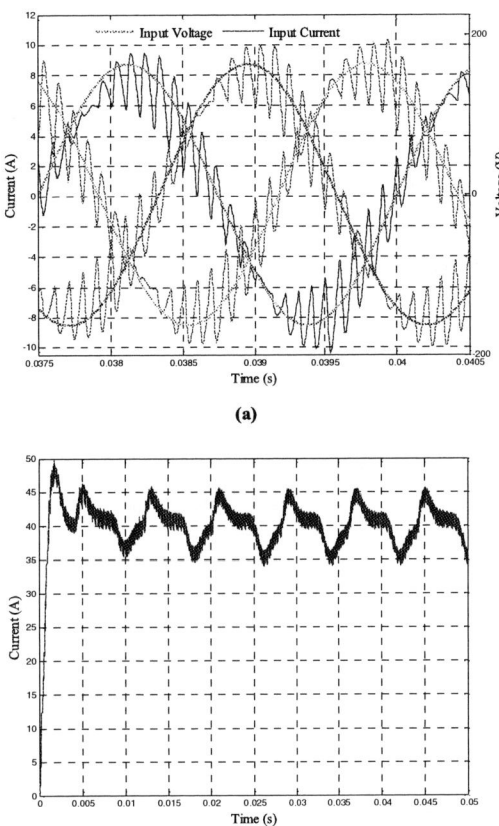

(a)

(b)

Fig. 8: (a) Simulated Three-Phase Input Voltage and Current for a 5-Phase Converter and (b) Sum of Phase Currents Drawn by 5-Phase Converters in Simulation with power factor correction.

waveforms of the sum of phase currents drawn by the 5-phase converters with and without the power factor correction, respectively. Fig. 10 shows the current demand and simulated output current for phase 1 of the 5-phase converter. As will be seen, due to the dynamic delay of the current tracking, the total current drawn by the 5-phase converters is not exactly constant. This will in turn affect the input power quality.

It should be noted that, in order to track the current demand under all operating conditions, the polarity of the voltage across each phase branch needs to be bidirectional. The negative output voltage can be achieved by adding an 180^0 phase shift in the SVPWM modulator.

V. Conclusions

In this paper, the SVPWM technique is proposed for the three-phase AC fed direct converter for SR machines. By switching between the six space vectors and three zero vectors, the reference output voltage, V_{dc}, is realised. Using PI control, the desired output (load) current can be

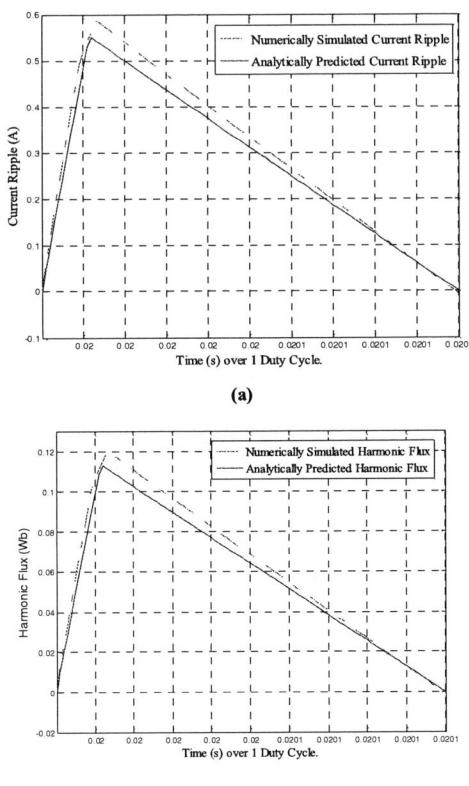

(a)

(b)

Fig. 7: Numerical Simulation and Analytical Prediction of (a) DC Output Current Ripple and (b) Harmonic Flux.

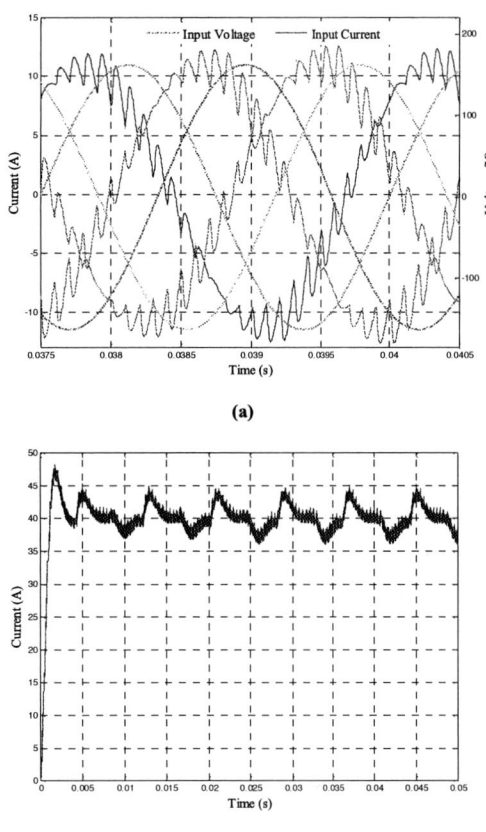

(a)

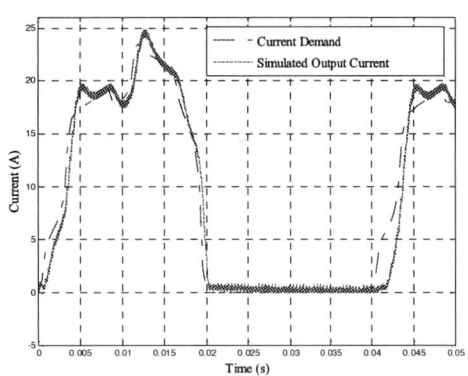

(b)

Fig. 9: (a) Simulated Three-Phase Input Voltage and Current for a 5-Phase Converter and (b) Sum of Phase Currents Drawn by 5-Phase Converters in Simulation without power factor correction.

Fig. 10: Phase 1 Current Demand and Simulated Output Current of a 5 Phase Converter.

realised by appropriate adjustment of the voltage reference. The modulation scheme has been extended by adopting a technique to correct the power factor of the converter by compensation of the reactive power drawn by the supply filter capacitor. Analysis of the output current ripple has been carried out to assess the output power quality and performance of the converter. The initial simulation results of a 5-phase converter have been presented. It has been shown that by co-ordinated control of phase currents in the SR machine, harmonic distortion on 3-phase AC input currents can be minimised for a given output torque demand.

Acknowledgment

This work is supported by Rolls-Royce PLC.

References

[1] Emadi, A., Ehsani, M., "Aircraft power systems: Technology, state of the art future trends", *IEEE AS Systems magazine*, January 2000, p28-32.

[2] Guanghai Gong, Marcelo Lobo Heldwein, Uwe Drofenik, Johann Minibock, Kazuaki Mino, Johann W. Kolar. "Comparative evaluation of three-phase high-power-factor AC-DC converter concepts for application in future more electric aircraft". *IEEE Trans. On Ind. Electronics.* Vol. 52, No. 3, June 2005, pp. 727 – 737.

[3] Provost, M., "A general overview from an engine manufacturer", *IMechE seminar 'the more electric aircraft and beyond',* January 2000.

[4] Powell, D.J., Jewell, G.W., Howe, D., Atallah, K., "Rotor topologies for a switched reluctance machine for the 'more-electric' aircraft engine*", IEE Proc. Electric Power Applications,* Vol. 150(3), 2003, pp.311-318.

[5] Ma, J.D., Wu, B., Zargari, N.R., Rizzo, S.C., " A space vector modulated CSI-based AC drive for multimotor applications", *IEEE Trans. Power Electronics,* Vol. 16, No. 4, July 2001.

[6] Salo, M., Tuusa, H., "A vector-controlled PWM current-source-inverter-fed induction motor drive with a new stator current control method", *IEEE Trans. Ind. Electronics,* Vol. 52, No. 2, April 2005.

[7] Miko Salo and Heikki Tuusa, " A vector controlled current-source PWM rectifier with a novel current damping method", *IEEE Trans. Power Electronics,* Vol. 15, pp. 464-470, No. 3, May 2000.

[8] Ahmet M. Hava, Russel J. Kerkman and Thomas A. Lipo, "Simple analytical and graphical methods for carrier-based PWM – VSI drives", *IEEE Trans. Power Electronics,* Vol. 14, No.1, pp. 49-61, January 1999.

[9] Casadei, D., Serra, G., Tani, A., "A general approach for the analysis of the input power quality in matrix converters", *IEEE Trans Power Electronics",* Vol. 13, No. 5, September 1998.

[10] Evelyn Matheson and Kamiar Karimi, "Power quality specification development for more electric airplane architectures" The Boeing Company.

2006 5th International Power Electronics and Motion Control Conference

Direct Torque Control of an Interior Permanent Magnet Synchronous Machine fed by a Direct AC-AC Converter

D. Xiao and M. F. Rahman

School of Electrical Engineering and Telecommunications
The University of New South Wales, Sydney, Australia

Abstract—**This paper presents a novel direct torque control (DTC) scheme of an Interior Permanent Magnet Synchronous Motor (IPMSM) fed by a matrix converter. In the proposed scheme, a sliding mode torque and stator flux controller is integrated with a simplified Venturini's modulation algorithm to achieve low torque and flux ripples, sinusoidal input/output currents and unity fundamental displacement factor on the input side regardless of the load power factor while maintaining constant and controllable switching frequency and preserving the fast response and robustness. Numerical simulations are carried out for the classical and modified DTC schemes in both steady-state and transient conditions, verifying the effectiveness of this new proposed scheme which is superior to the classical one.**

Keywords—direct torque control (DTC), Interior Permanent Magnet Synchronous Machine (IPMSM), matrix converter

I. INTRODUCTION

Recently, the three-phase to three-phase matrix converter (MC) has emerged to become a viable alternative to the conventional voltage-source inverter (VSI). The matrix converter, as a member of AC-to-AC direct converter family, provides sinusoidal input/output waveforms and allows inherent four-quadrant operation and the adjustment of input power factor on the main side. Furthermore, the absence of bulky dc-link electrolytic capacitors for energy storage allows long lifetime, high integration capability, extreme temperature and critical volume/weight applications [1]-[4].

The basic DTC scheme for matrix converter was initially proposed to apply to induction motor drives [5]. However, some drawbacks, such as large flux and torque ripples and switching frequency variation according to the change of the motor speed and the hysteresis band amplitudes, still exist.

In this paper, a new DTC scheme for IPMSM drives fed by matrix converter is presented as shown in Fig. 1. The variable-structure controller produces the most appropriate stator voltage vectors to track the reference torque and flux. The control stator voltage signals have been limited and converted to the three-phase target voltages before

proceeding to the Venturini's optimum modulation module. The three-phase voltages and instantaneous main voltages are used as the inputs for the modulation algorithm to generate the required duty cycle for each switch in the matrix converter. This method has more advantages over the classical DTC, such as significantly reduced torque and flux ripples and constant switching frequency while preserving the fast response and robustness of DTC. In addition, the quality of input and stator currents can be improved by accurate synthesis of the desired output voltages and input currents.

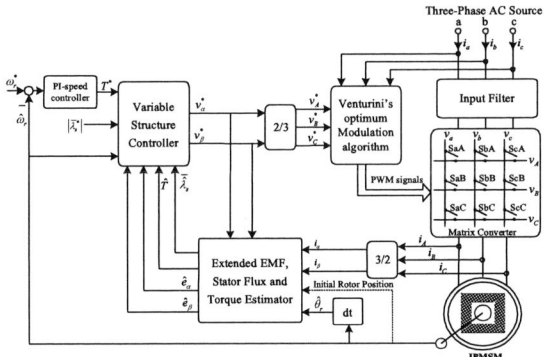

Figure 1. Proposed DTC of the matrix converter fed IPMSM drive

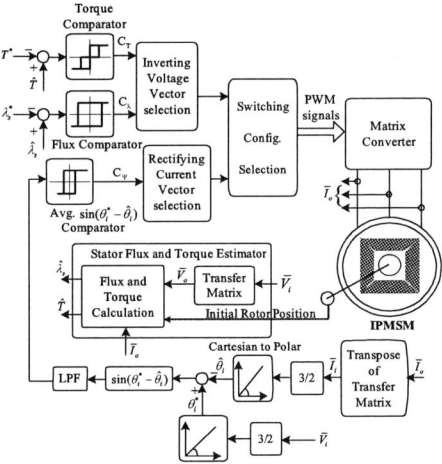

Figure 2. Basic DTC scheme with matrix converter

II. Modulation Algorithm for Matrix Converter

The main task of the modulation algorithm for the matrix converter is to find a modulation matrix $M(t)$ from a set of input voltages and an assumed set of output currents to synthesize the target output voltages and input currents [2], [3]. An injection of a third harmonic of the input and output voltage was proposed in order to obtain a higher voltage transfer ratio [2], [4]. A simplified form of this algorithm with unity input displacement factor is adopted in this work [6]. The nine modulation functions for three-phase outputs are calculated in each sampling interval by measuring any two of three input line-to-line voltages. Then, the magnitudes and positions of the input and target voltage vectors are calculated as

$$V_{im} = \frac{2}{3}(v_{ab}^2 + v_{bc}^2 + v_{ab}v_{bc})^{1/2} \tag{1}$$

$$V_{om} = \sqrt{\frac{2}{3}}(v_A^2 + v_B^2 + v_C^2)^{1/2} \tag{2}$$

$$\omega_i t = \arctan(\frac{\sqrt{3}v_{bc}}{2v_{ab} + v_{bc}}) \tag{3}$$

$$\omega_o t = \arctan(\frac{v_B - v_C}{\sqrt{3}v_A}) \tag{4}$$

Assuming the input power factor is set to one, three triple harmonic terms are defined as

$$K_{31} = \frac{2}{9}\frac{V_{om}/V_{im}}{\sqrt{3}/2}\sin(\omega_i t)\sin(3\omega_i t) \tag{5}$$

$$K_{32} = \frac{2}{9}\frac{V_{om}/V_{im}}{\sqrt{3}/2}\sin(\omega_i t + \frac{2\pi}{3})\sin(3\omega_i t) \tag{6}$$

$$K_{33} = -V_{im}\left[\frac{1}{6}\cos(3\omega_o t) - \frac{1}{2\sqrt{3}}\cos(3\omega_i t)\right] \tag{7}$$

Then, the modulation functions for output phase j ($j = A, B, C$) are given as

$$M_{aj} = \frac{1}{3} + K_{31} + \frac{2}{9V_{im}^2}(v_j + K_{33})(2v_{ab} + v_{bc}) \tag{8}$$

$$M_{bj} = \frac{1}{3} + K_{32} + \frac{2}{9V_{im}^2}(v_j + K_{33})(v_{bc} - v_{ab}) \tag{9}$$

$$M_{cj} = 1 - (M_{aj} + M_{bj}) \tag{10}$$

III. Direct Torque Control for Matrix Converter

A. Basic DTC Using Matrix Converter

The conventional DTC for matrix converter drive can be realized by two stages as represented in Fig. 2. In the inverter stage, the hysteresis flux and torque control can be achieved by selecting one of the six active and two zero voltage vectors with respect to the virtual dc-link voltage, exactly as in VSI. In addition, the direct control of the average value of $\sin(\theta_i^* - \hat{\theta}_i)$ can be completed by choosing a single current vector in the rectifier stage. The stator flux, torque and displacement angle estimators are realized by only measuring the input voltages and stator currents, since the other qualities are calculated on the basis of the switching states in each sampling period.

B. VS-DTC Using Matrix converter

In order to drive the torque and flux to track their desired trajectories, the VS controller is designed in more details in [7]. The procedure can be exhibited by the following equations.

The switching surface is defined as $S = [S_1 \quad S_2]^T$,

$$\begin{aligned} S_1 &= e_T(t) + K_T\int_0^t e_T(\tau)d\tau - e_T(0) \\ S_2 &= e_\lambda(t) + K_\lambda\int_0^t e_\lambda(\tau)d\tau - e_\lambda(0) \end{aligned} \tag{11}$$

Differentiating the switching surface vector gives

$$\begin{aligned} \dot{S}_2 &= \dot{e}_\lambda + K_\lambda e_\lambda = (\dot{\lambda}^* - \dot{\hat{\lambda}}) + K_\lambda(\lambda^* - \hat{\lambda}) \\ \dot{S}_1 &= \dot{e}_T + K_T e_T = (\dot{T}^* - \dot{\hat{T}}) + K_T(T^* - \hat{T}) \end{aligned} \tag{12}$$

where $e_T = T^* - \hat{T}$, $e_\lambda = \lambda^* - \hat{\lambda} = \lambda_s^{*2}/2 - (\hat{\lambda}_\alpha^2 + \hat{\lambda}_\beta^2)/2$ K_T and K_λ are positive control gains of flux and torque.

Substituting for $\hat{T}$, $\hat{\lambda}$, $\dot{\hat{T}}$ and $\dot{\hat{\lambda}}$ into (12) leads to $\dot{\underline{S}} = \underline{F} + \underline{D} \cdot [u_\alpha, u_\beta]^T$.

$$\underline{D} = \begin{bmatrix} -1.5P(i_\beta - \hat{\lambda}_\beta/L_d) & -1.5P(\hat{\lambda}_\alpha/L_d - i_\alpha) \\ -\hat{\lambda}_\alpha & -\hat{\lambda}_\beta \end{bmatrix}$$

$$\begin{aligned} F_1 = K_T e_T &- 1.5P\{\hat{\lambda}_\alpha[\omega_{re}(L_d - L_q)/L_d \cdot i_\alpha - R_s i_\beta/L_d - e_\beta/L_d] \\ &- \hat{\lambda}_\beta[-R_s i_\alpha/L_d - \omega_{re}(L_d - L_q)/L_d \cdot i_\beta - e_\alpha/L_d]\} \end{aligned}$$

$$F_2 = K_\lambda e_\lambda + \hat{\lambda}_\alpha R_s i_\alpha + \hat{\lambda}_\beta R_s i_\beta$$

where R_s is the stator resistance, p is the differential operator, L_d, L_q are direct and quadrature inductances, ω_{re} is the rotor speed in electrical rad/s and e_α and e_β are extended EMFs.

The VS controller generates the command output voltage vector components which are converted to three-phase voltages and provided for the modulation module. The switching control law can be selected as (13) according to Lyapunov approach [7].

$$u_{\alpha,\beta}^* = -D^{-1}\begin{bmatrix} \mu_1 & 0 \\ 0 & \mu_2 \end{bmatrix}\begin{bmatrix} sign(S_1) \\ sign(S_2) \end{bmatrix} \tag{13}$$

The chattering problem can be remedied by (14). The switching function is replaced by a continuous function around the sliding surface neighborhood by introducing smoothing factors.

$$sign(S_i) = \begin{cases} 1, & if \ S_i > \lambda_i \\ -1, & if \ S_i < -\lambda_i \quad (Smoothing \ factors \ \lambda_i > 0) \\ S_i/\lambda_i, & if \ |S_i| < \lambda_i \end{cases} \tag{14}$$

IV. Numerical Simulation

Two Simulink models are built to test the proposed scheme. The steady state and dynamic responses are compared with the basic DTC scheme. The parameters of the tested IPMSM and models are given in Tables I, II and III. The simulations have been carried out assuming three-phase balanced input voltages and ideal switching devices.

Rated torque reversal at -6 Nm and +6 Nm was tested with stator flux linkage maintained at 0.55 Wb. Figs. 3-6 show the dynamic responses of estimated torque, three-phase stator currents and one of three-phase filtered input currents against the corresponding input phase voltage, respectively. It is obvious that the variable-structure DTC drive system operates better than the classical DTC under transient conditions. The torque and flux ripples are significantly reduced by proposed scheme. The stator and filtered input currents are sinusoidal with nearly unity input power factor and smoother than basic DTC as shown in Figs. 3 and 4. During motoring operation (from 0 to 0.1 sec.), the filtered input current is in phase with the corresponding phase voltage. Then the input current suddenly reverses the phase 180°, opposite that of the input voltage, during the regenerative breaking. Figs. 5-7 give the comparison of flux response of the two schemes. It can be noticed that the magnitude of flux stays at rated value immediately after startup. The flux components and extended EMFs exhibit smoother sinusoidal waveforms under VS-DTC. The manifolds of torque and flux are shown in Fig. 6. The states of torque and flux are driven towards their sliding surfaces by discontinuous control signals. The flux is greatly reduced without influencing dynamic response speed under VS-DTC than classical DTC scheme as shown in Fig. 7. Fig. 9 shows that torque steady-state waveform of VS-DTC contains less ripples than that of basic DTC. The proposed scheme still preserves the fast dynamic response as shown in Fig. 8.

Fig. 10 shows the four-quadrant operation characteristics of the proposed drive system. This is examined by applying ±1000 rpm speed step reference and 1 Nm load. Fig. 11 shows stator flux magnitude, flux components in the stationary reference frame, extended back EMFs and manifolds of torque and flux for torque and speed step commands. The torque-speed curve and flux locus for all four-quadrant regions are demonstrated by Fig. 12. The steady-state behavior has been investigated at 1000 rev/min and 3 Nm load under two schemes. From Figs. 13 and 15, high ripples and distortion in currents can be observed and harmonics are distributed over a large frequency span undesirably. The harmonics in the currents are greatly reduced under proposed DTC. The dominant harmonics in currents, around 10 kHz, are less than 0.2% and 3.3% of those of their fundamentals, respectively as shown in Figs. 14 and 16. Figs. 17 and 18 show the output line-to-line voltages and frequency spectrums for these two methods in the range from 0 to 11 kHz. The output line-to-line voltage of VS-DTC contains dominant harmonics around switching frequency while more harmonics are scattered from fundamental to 10 kHz for basic DTC. In this case, a fixed switching frequency 10 kHz is achieved for the VS-DTC.

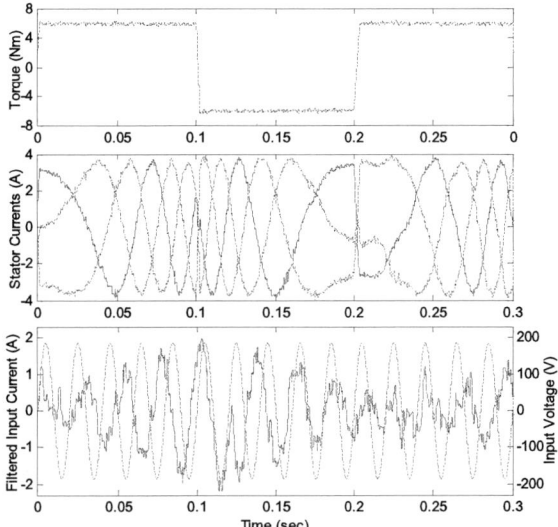

Figure 3.   Dynamic response with ±6 Nm command (basic DTC)

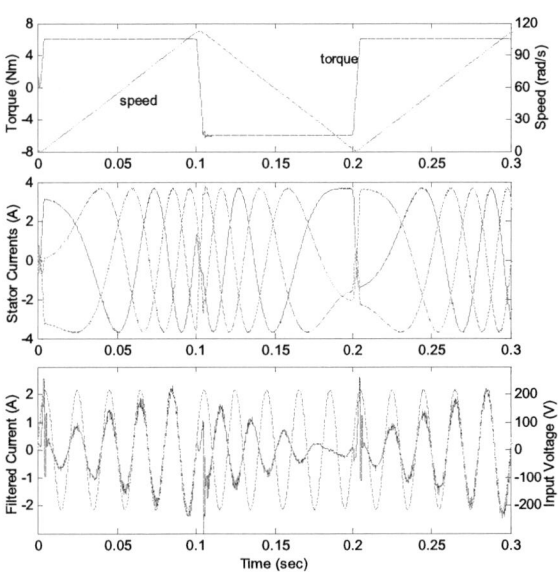

Figure 4.   Dynamic response of the VS-DTC with ±6 Nm command

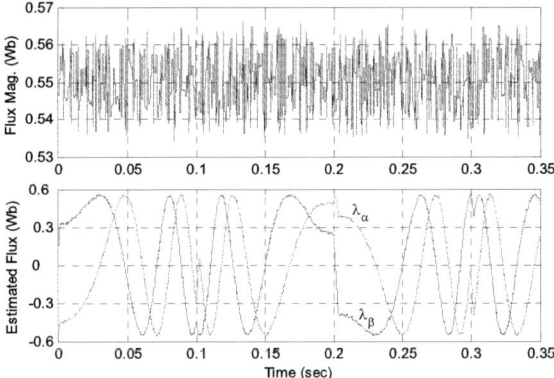

Figure 5.   Flux of the basic DTC with ±6 Nm command

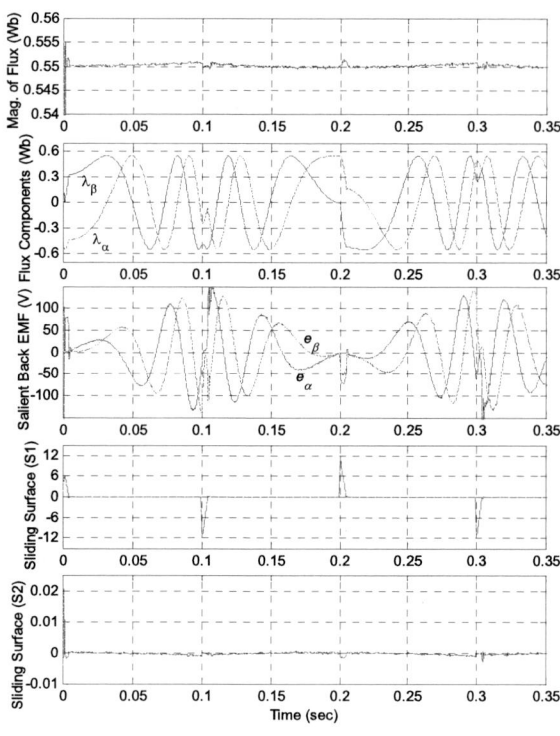

Figure 6. Stator flux, EEMF and sliding surfaces with ±6 Nm step

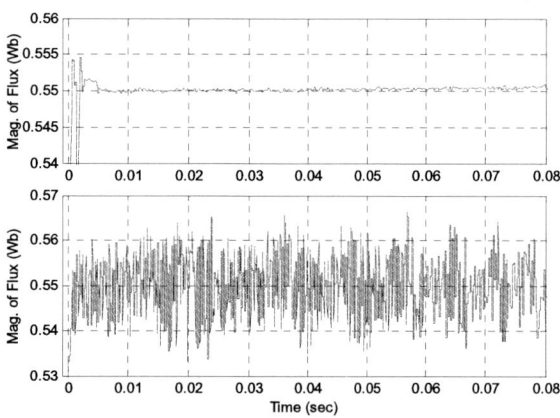

Figure 7. Flux response under VS (a) and basic DTC (b) schemes

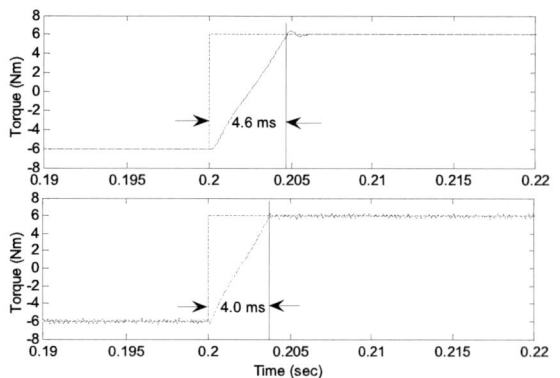

Figure 8. Torque response under VS (a) and basic DTC (b) schemes

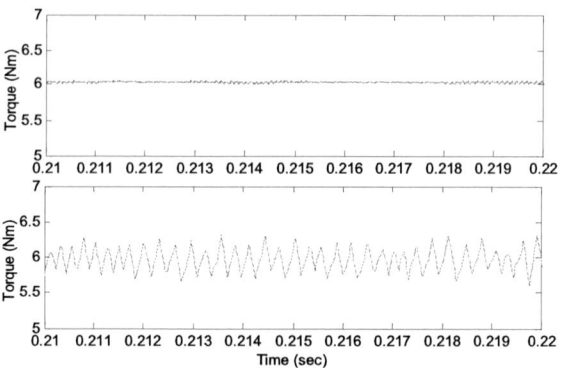

Figure 9. Steady-state torque under VS (a) and basic DTC (b) schemes

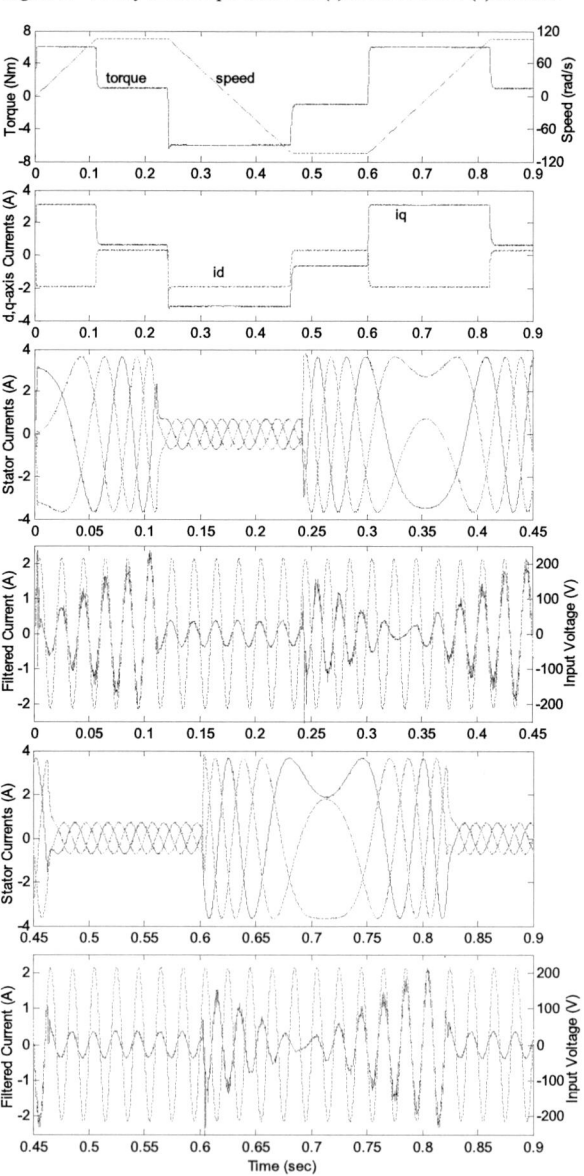

Figure 10. Four-quadrant operation waveforms with light load

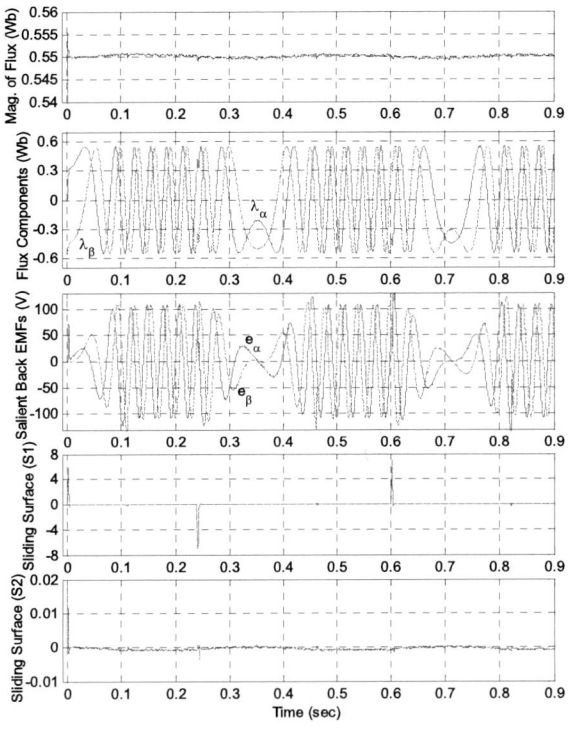

Figure 11. Stator flux, EEMF and sliding surfaces with speed step

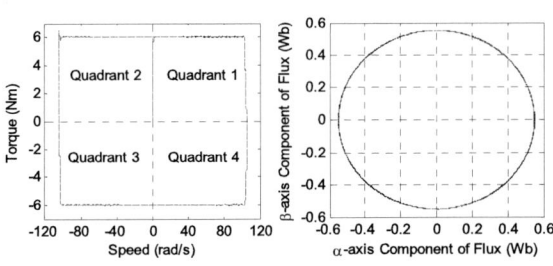

Figure 12. Torque-speed characteristic and flux locus

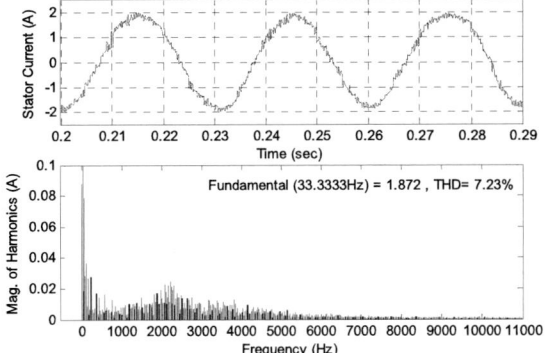

Figure 13. Stator current and spectrum at 1000 r/min and 3 Nm (basic)

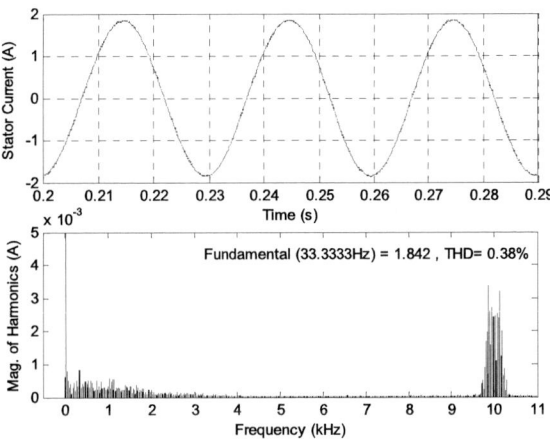

Figure 14. Stator current and its spectrum at 1000 r/min and 3 Nm (VS)

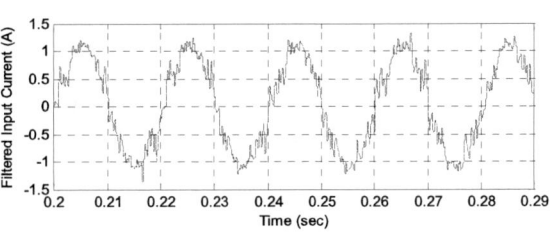

Figure 15. Filtered current and spectrum at 1000 r/min, 3 Nm (basic)

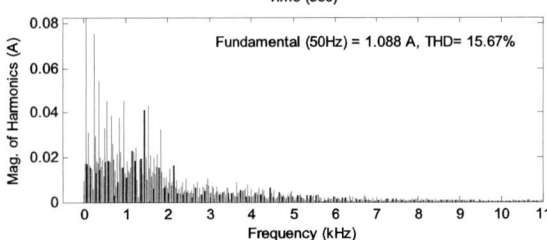

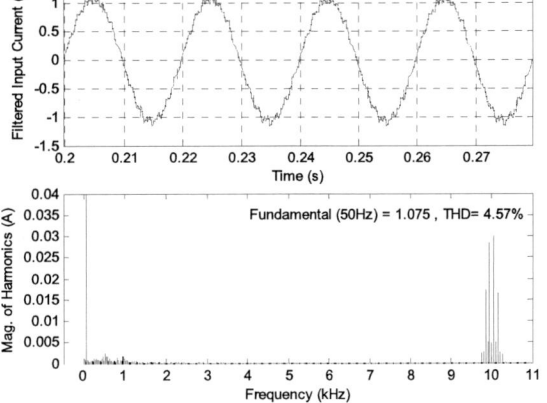

Figure 16. Filtered current and spectrum at 1000 r/min, 3 Nm (VS)

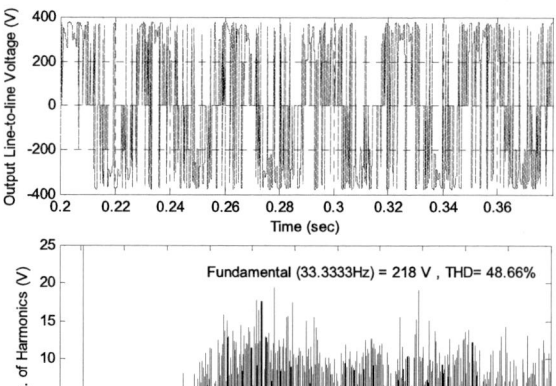

Figure 17. Output voltage and spectrum at 1000 r/min, 3 Nm (basic)

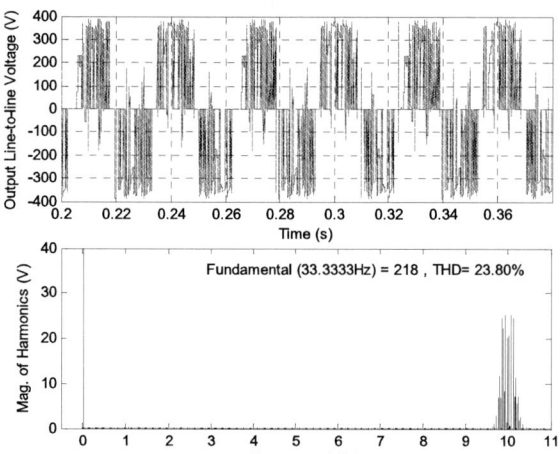

Figure 18. Output voltage and spectrum at 1000 r/min, 3 Nm (VS)

V. CONCLUSION

In this paper, a new matrix converter fed IPMSM drive scheme using DTC technique has been proposed. The Venturini's simplified modulation algorithm is employed to achieve the target output voltages of the matrix converter on the basis of the control requirements of the motor side with unity input power factor on the grid side.

The proposed scheme has been tested in steady-state and transient conditions under rated torque reversal and four-quadrant operation, carrying out some numerical simulations. The numerical simulation results of the proposed drive system are compared with those of the classical DTC. The torque and flux ripples are significantly reduced, and also the switching frequency remains constant independent of operating conditions. In comparison with the basic DTC, the filtered input currents, the stator currents and output line-to-line voltages are improved, of which the dominant harmonics are around a fixed frequency 10 kHz determined by the modulation sampling period.

TABLE I. PARAMETERS OF THE IPMSM USED IN THIS PAPER

Rated output power (Watt)	P_r	1000
Rated phase voltage/current (V/A)	V / I	132/3
Magnetic flux linkage (Wb.)	λ_f	0.533
Number of pole pairs	P	2
Rated torque (Nm)	T_b	6
Stator resistance (Ω)	R_s	5.8
dq-axis inductances (mH)	L_d , L_q	44.8, 102.7
Rotor inertia (Kg.m^2)	J	0.00529
Friction coefficient (Nm/rad/s)	D	0.0006

TABLE II. PARAMETERS OF BASIC DTC SCHEME

Stator flux linkage reference (Wb.)	λ_s^*	0.55
Maximum output of speed PI controller (Nm)	T_{max}^*	±6.0
Proportional gain of speed PI controller	K_{SP}	2.0
Integral gain of speed PI controller	K_{SI}	0.05
Electromagnetic torque band (Nm)	B_T	±0.02
Stator flux band (Wb)	B_λ	±0.001
Average $\sin(\theta_i^* - \hat{\theta}_i)$ band	B_ψ	0
Sampling period (μs)	T_s	40

TABLE III. PARAMETERS OF VS-DTC SCHEME

Positive control gains of VS controller	μ_1 , μ_2	3200, 200
Smoothing factors of sliding surfaces	λ_1 , λ_2	0.4, 0.04
Control gains of VS controller	K_T , K_λ	3.0, 5.0
Initial stator flux components (Wb)	$\lambda_{\alpha0}$, $\lambda_{\beta0}$	0, -0.533
Sampling period (μs)	T_s	100

REFERENCES

[1] O.Simon, J. Mahlein, M. N. Muenzer, and M. Bruckmann, "Modern solutions for industrial matrix-converter applications," *IEEE Trans. Ind. Electron.*, vol. 49, no. 2, pp. 401–406, Apr. 2002.

[2] P. W. Wheeler, J. Rodríguez, J. C. Clare, L. Empringham, and A. Weinstein, "Matrix converter: a technology review," *IEEE Trans. Ind. Electron.*, vol. 49, no. 2, pp. 276–288, Apr. 2002.

[3] M. Venturini, "A new sine wave in sine wave out, conversion technique which eliminates reactive elements," in *Proc. POWERCON 7*, 1980. pp. E3_1–E3_15.

[4] M. Venturini, and A. Alesina, "Analysis and design of optimum-amplitude nine-switch direct AC-AC converters," *IEEE Trans. Power Electron.*, vol. 4, no. 1, pp. 101–112, Jan. 1989.

[5] D. Casadei, G. Serra, and A. Tani, "The use of matrix converters in direct torque control of induction machines," *IEEE Trans. Ind. Electron.*, vol. 48, no. 6, pp. 1057–1064, Dec. 2001.

[6] H. Altun and S. Sunter, "A vector controlled matrix converter induction motor drive," PhD Thesis, Department of Electrical and Electronic Engineering, The University of Nottingham, UK, 1993.

[7] Z. Xu and M. F. Rahman, "A variable structure torque and flux controller for a DTC IPM synchronous motor drive", in *Proc. IEEE PESC'04*, 2004, vol. 1, pp. 445–450.

A Novel Modular Permanent Magnet Drive System Design

Wen Ouyang, Nicholas Lemberg, Ruoping Yao*, T.A.Lipo

Department of Electrical and Computer Engineering
University of Wisconsin-Madison
1415 Engineering Drive
Madison, WI 53706
Eamil: ouyang@cae.wisc.edu , lipo@engr.wisc.edu

Department of Electrical Engineering*
Shanghai Jiao Tong University
1954 HuaShan Road
Shanghai, 200030
Eamil: raoyrp@sjtu.edu.cn

Abstract — With improvement in magnetic property of soft magnetic composites (SMC), electrical machines can be fabricated with more flexible structure at low cost, which results in more options of the drive system design different from the conventional technology, for example, 3D machine structure compared with the 2D structure enforced by the lamination sheets. In this paper, a modular permanent drive system based on the independent machine module and drive circuit are proposed. With independent control, the potential of fault tolerant capability of the drive system is discussed as well. The experiment results from the prototype system will be provided for the verification.

Keywords — Permanent Magnet, Modular Design, Fault Tolerant

I. INTRODUCTION

With the successive technology improvement of powder metallurgy, the soft magnetic composites (SMC) offers more options in the machine fabrication. The machine iron section which provides path for the main alternating flux can be fabricated from the composites easily by a single punching action and heat curing. Compared with the conventional laminations used in the machine iron, the fabrication process is greatly simplified. Moreover, machine structure is no longer limited in 2D as enforced by lamination sheets. The SMC offers flexible design options in 3D [1][2][3][4]. With higher operation frequency, the SMC material exhibits better loss property compared with typical silicon steel laminations. All these merits make SMC a competitive candidate for the motor drive applications. In this paper, a modular permanent magnet drive system design is proposed with independent control unit for each SMC based module. The independent control offers the potential of fault tolerant capability of the drive system. Preliminary simulations based on the general equations for the description of machine behavior are provided. The experiment results from the prototype system will be presented in a future

paper.

II. SOFT MAGNETIC COMPOSITES

Laminations have been the dominant choice for the iron components in electrical machines subject to alternating magnetic fields, since it is possible to choose the thickness and treatment of steel sheets to keep hysteresis and eddy losses low, which satisfies the typical motor application under 50~60 Hz excitation. With the introduction of soft magnetic composite (SMC) material, the iron component design of an electrical machine is no longer limited to the laminations. The frequency barrier set by the steel laminations is overcome by the more acceptable high frequency property of SMC, which especially benefits the machines operated in high speed range.

The SMC material is based on the powder metallurgy technology. Each iron particle is insulated by a chemical coating to avoid eddy currents as illustrated by Figure 1. Finer base powder particle size distribution yields better high frequency performance while coarser particles optimize low frequency behavior. The content of the binder governs the mechanical property of the components. SMC is quite different from traditional laminations in the unsaturated permeability, isotropic property, iron loss, and production process.

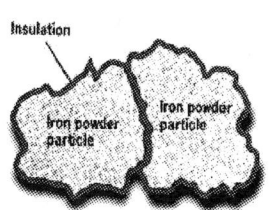
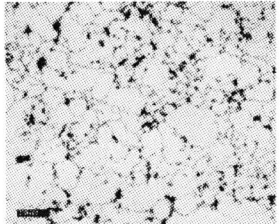

Figure 1: Microstructure of SMC500 (courtesy of Hoganas AB)

The SMC magnetic property is improved with higher compaction pressure. For the electrical machine design, the magnetic property in the linear and starting saturation range is of more interest for the final performance of the machine. Although with relatively low unsaturated permeability of Somaloy500 ($\mu_r \approx 500$), compared to the excellent permeability of steel laminations (μ_r as high as 3000), the applications which

---

This work was supported in part by the ERC program of the National Science Foundation under award number EEC-9731677 for the Center for Power Electronic Systems.

require armature magnetization may suffer from the high magnetizing current, but for the applications in the PM machines, especially for the surface PM (SPM) design, it is not a critical problem due to the large effective air gap, which makes the machine performance non-sensitive to the permeance of the iron components.

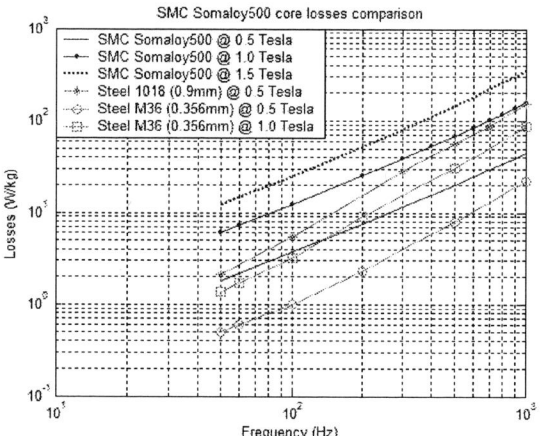

Figure 2: Loss Comparison between steel and SMC

As observed from Figure 2, all the loss curves exhibit frequency dependant property as the excitation frequency sweeps from 50 Hz to 1,000 Hz. At low frequency range, the Somaloy500 has higher core losses compared with steel M36 (0.356mm) in different induction levels mainly due to the higher hysteresis loss. However, if compared with steel 1018 (0.9mm), the SMC behaves well above the steel under 0.5 T induction. Further comparison of the loss data comes to the conclusion that SMC has excellent loss property associated with eddy and excess loss, while the hysteresis loss is still significant compared with standard steel laminations.

For the machine design with SMC modules, the machine winding structures can be simplified. The concentric windings are preferred for this single module mounted with single winding. Furthermore, the winding can even be pre-pressed [3], which greatly reduced the manual efforts in the insertion of windings. The winding fill factor can be significantly improved and the winding thermal resistance can be reduced as well by this compact structure, which benefits the machine cooling. When the machine can be effectively cooled, the designed current density can be enhanced, making a higher power/torque density design possible.

However, the tradeoff between the overall machine performance and the cost always exists in a practical design. Although the SMC material has better loss performance compared with steel laminations in wide frequency range, it is low in the unsaturated permeability, which will impact the machine design with large armature inductance. Although the thin laminations (<0.35mm) may also exhibits excellent core losses in hundreds to kilo-Hz frequency range, the fabrication cost of such thin steel sheets and the corresponding stacking

may require careful consideration. Although the SMC material simplifies the machine iron core fabrication process by punched modules and pre-wound windings, the magnetic stability and mechanical strength of the material and the required curing process still needs further practical experiences. Before the large scale usage of SMC material, the availability and expenses compared with traditional steel laminations is always an important tradeoff before a successful design can be turned into reality.

III. MACHINE DRIVE SYSTEM DESCRIPTION

For modular structure design, the proper shape of each module is of interest for its significant impact on the while drive system performance. In [5], various module shape and machine configuration are fully discussed. The air gap MMF and the corresponding winding factor with the consideration of harmonics are evaluated. In this paper, a trapezoidal module shape design is evaluated by the fundamental and harmonic component of its winding factor as illustrated in Figure 3.

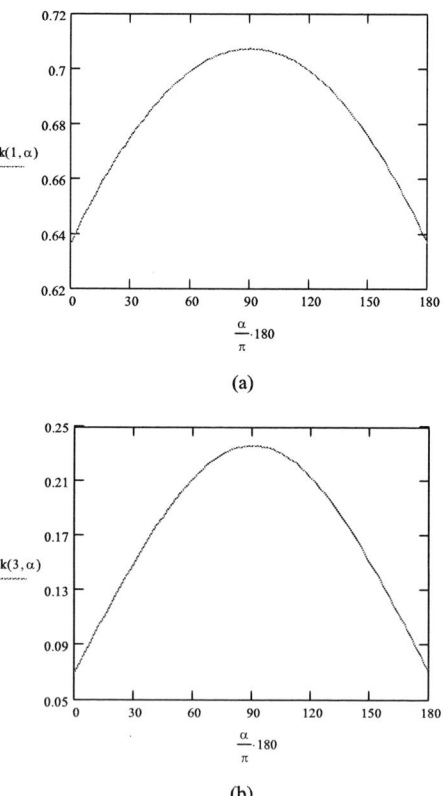

Figure 3: Normalized Winding Factor: (a)1st ; (b) 3rd

In Figure 3, angle α denotes the bottom angle of the symmetric trapezoid. When α reaches 90 degree, a square shape will be formed. And the trapezoid will be shaped upside down when α goes beyond 90 degree. It is obvious that the winding factor reaches maximum when the square pole shape is reached. But square shape suffers from higher order harmonics exhibited from the similar dependence on the shape

In general, if stator leakage inductance and resistance can be neglected, the output power for any electrical machine can be expressed as:

$$P_R = \eta \frac{m}{T} \int_0^T e(t)\, i(t)dt = \eta\, m\, K_{pw} E_{pk} I_{pk} \qquad (1)$$

With K_{pw}, E_{pk}, and I_{pk} defined as power wave factor, peak back EMF and peak current value.

The back EMF for the machines is given by (2):

$$e(t) = \frac{d\lambda_g}{dt} = 2\pi K_w N_t B_{gmax} \frac{f}{p} L_e\, \lambda\, D_o f_e(t) \qquad (2)$$

Where λ_g is the air-gap flux linkage per phase, N_t is the number of turns per phase, $f_e(t)$ is the normalized function for the alternate field. The ratio λ is defined as (3):

$$\lambda = \frac{D_i}{D_o} \qquad (3)$$

Where D_i is the stator inner diameter of the machine and D_o the stator outer diameter. From Eq. (2) it is apparent that:

$$E_{pk} = 2\pi K_w N_t B_{gmax} \frac{f}{p} L_e \lambda\, Do \qquad (4)$$

With the factor K_i defined as (5):

$$K_i = \frac{I_{pk}}{I_{rms}} = \left[\frac{1}{T} \int_0^T \left(\frac{i(t)}{I_{phmax}} \right)^2 dt \right]^{-1/2} \qquad (5)$$

Where I_{rms} is the rms phase current which is related to the stator electric loading A_s:

$$A_s = 2m_1 N_t \frac{I_{rms}}{\pi D_i} \qquad (6)$$

In the general case, the total electric loading A should include both the stator electric loading A_s and rotor electric loading A_r:

$$A_s = A - A_r = \frac{A}{1+K_\phi} \qquad (7)$$

By equation (4), (5) and (6), an expression for the peak current is:

$$I_{pk} = \frac{1}{1+K_\phi} K_i A\, \pi \frac{D_g}{2m_1 N_t} \qquad (8)$$

Substituting (4) and (8) into (1), the general purpose $D_o^2 L_e$ sizing equation for the dual-rotor radial flux machines takes the form of (9):

$$P_R = \frac{1}{1+K_\phi} \frac{m}{m_1} \pi^2 K_w K_i K_{pw} \eta\, B_{gmax} A \frac{f}{p} \lambda\, D_o^{\,2} L_e \qquad (9)$$

angle at the same time.

The overall power density can be defined as (10):

$$\xi = \frac{P_R}{\frac{\pi}{4} D_t^2 L_t} \qquad (10)$$

Where D_t is the total outer diameter of the stator, and L_t is the total machine length including the stack length and the protrusion of the end winding from the iron stack.

In different application background, the selection of K_L favors quite different values. For example, when the machine is design in low speed range with high torque requirement, a small K_L is preferred, which converges to a disk shape structure; while for the high speed application, larger K_L is selected for a slim design with smaller machine diameter and longer stacking length. It is a major design parameter that has significant effect on the characteristic of the machine. When combined with the optimization goals, for instance, efficiency, the optimal value of K_L is usually not same for different rated power, pole pairs, power supply frequency, etc. Furthermore, if different materials or different structures are involved, the optimal K_L will have the different value.

In this study, a surface-mounted permanent magnet (SPM) motor with five phase concentric winding configuration is selected for the investigation of high power density design with fault tolerant capability. For the surface mounted PM structure, the machine dimension of inner and outer diameters can be described by:

$$D_i = D_o - 2d_{ss} - 2d_{cs} \qquad (11)$$

Where d_{cs} and d_{ss} are back iron core (yoke) depth and stator slot depth respectively.

For the slot depth, a proper estimation based on the stator electrical load A_s and copper current density J_c by:

$$d_{ss}\, \pi\, d_s\, K_{slot} K_{ds} K_{cu} J_c = 2m N_t I_{rms} \qquad (12)$$

Where K_{slot} is the ratio between the slot width to the span of a pair of slot and tooth, K_{ds} is the slot shape factor that converts between the irregular slot shape and rectangular shape, and K_{cu} is the bare copper slot fill factor.

Thus, with (6), the slot depth can be derived as:

$$d_{ss} = \frac{A_s}{K_{slot} K_{ds} K_{cu} J_c} \qquad (13)$$

If B_{cs} is defined as the flux density in the stator core, then the back core depth can be found as:

$$d_{cs} = \frac{\pi}{4p} \frac{\alpha_i B_{gmax} D_i}{K_{fe} B_{cs}} \qquad (14)$$

Where α_i is the ratio between the averaged and maximal air gap flux density, and K_{fe} is the lamination stacking factor, $K_{fe} = 1$ for SMC module.

Substituting (13) and (14) into (11), it yields:

$$\lambda = \frac{D_i}{D_o} = (1 - \frac{2A_s}{D_o K_{slot} K_{ds} K_{cu} J_c}) / (1 + \frac{\pi \, \alpha_i B_{gmax}}{2p \, K_{fe} B_{cs}}) \qquad (15)$$

Table 1 lists the λ variations with different pole pair p and different air gap flux densities with typical value for ferrite (0.3 Tesla) and NdFeB (0.65 Tesla) permanent magnets used in the surface mounted rotor structure.

Table 1: λ variations with respect to the pole pair numbers p and magnet material @ $B_{cs} = 1.4$ Tesla, $\alpha_i = 0.75$, $A_s = 19{,}500$ A/m, $J_c = 6$A/mm²

p	1	2	3	4
B_{gmax}=0.3 T	0.556	0.618	0.642	0.655
B_{gmax}=0.65 T	0.450	0.547	0.589	0.613

As observed from Table 1, with multi-pole design, the inner stator diameter is increased due to the less back iron core (yoke) used to provide the flux path at the proper level. While the ferrite PMs can enhance λ further with the penalty of lower torque/power density, since the machine air gap flux level is reduced significantly.

From (9), for the modular radial flux PM (MRFPM) machines, $K_\Phi = 0$ for the PM excitation on rotor, if trapezoidal waveform is considered, $K_i K_{pw} = 0.88$, then the machine sizing equations can be further simplified as:

$$P_{R(MRFPM)} = 8.69 \, K_w \, \eta \, B_{gmax} \, A \frac{f}{p} \lambda \, D_o^2 \, L_e \qquad (16)$$

With the SMC module, the end winding section is embedded between the iron back core and module tooth tip section, if constant flux distribution is assumed for the whole air gap area, which means $L_e = L_t$, then the power density can be simplified as (17):

$$\xi_{(MRFPM)} = 11.07 \, K_w \, \eta \, B_{gmax} \, A \frac{f}{p} \lambda \qquad (17)$$

For the IPM rotor structure design, the general analysis above is still valid. However, the existence of the flux ribs and bridges provides saturated path for the inserted permanent magnets, which leads to significant flux leakage and lower air gap flux density, if compared with the SPM structure. While the less air gap length and the L_q/L_d ratio by the configuration of PM slots provides the possibility of field weakening capability for this type of machine.

With the general sizing and power equations discussed in the previous sections, the power density comparisons between the induction machine (IM) and MRFPM machine family are provided as illustrated in Figure 4.

A standard induction machine with P_R=75 kW, A=60kA/m, J_c=6.2×10⁶ A/m² was selected. For the MRFPM machine family, the dimension and power rating are reduced for the fabrication consideration, with A = 19.5 kA/m, J_c=6×10⁶A/m², and P_R=2.2kW. With higher power rating, the power density of the induction machine is enhanced as well compared with low horse power induction machines. However, the power density of induction machines is much lower compared with the MRFPM machine family, which is an expected result for the PM machines due to the existence of strong flux source.

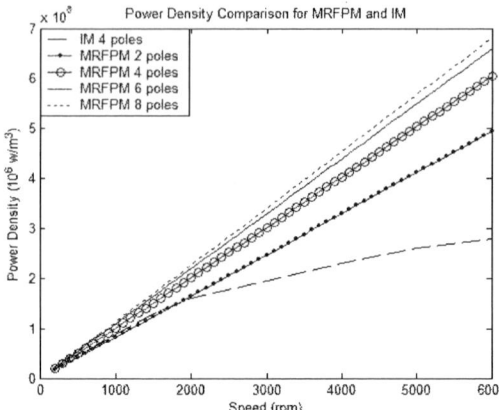

Figure 4: Power density comparison between IM and MRFPM machines

With more poles, the power density of MRFPM is increased at a slowing down pace. The pole pitch is reduced with higher pole number design, which results in less contribution of the torque production of a single pole. While manufacture cost of the high pole number design is increased as well for the much more complicated machine structure. By playing with these tradeoffs, the machine design converges to an optimal point with weighted considerations of all the factors.

In order to achieve higher power density design of the drive system, a five phase modular PM motor based on independent square pole module design is developed as shown in Figure 5.

Figure 5: Module and Prototype PM motor

The main parameters of the machine are compared with standard induction machine as listed in Table 2.

Table 2: Five-phase Modular SPM Motor Parameters

Parameters	IM (GE/3HP)	5 Phase SPM Motor
OD	190mm	120mm
ID	120mm	72mm
RPM	1750	1800
Machine Length	70mm(iron)/150mm(full)	140mm
Torque (T)	11.87Nm	13.26Nm
Effective Volume	4.2529×10⁻³ m³	1.5834×10⁻³ m³
Torque Density	2.791×10³ Nm/m³	8.374×10³ Nm/m³
Cogging Torque	0	2.5% of rated
Torque Density Ratio	5 phase PM / IM = 3.0	

The machine performance can be described by the equations [18] ~ [21] based on the general electrical characteristics of SPM motor.

$$[v] = [e] + [r][i] + \frac{d}{dt}[\lambda] \qquad (18)$$

where v, e, i, λ denotes the vector of phase terminal voltage, back EMF (emf), current, and flux linkage.

$$[\lambda] = [L] \cdot [i] \tag{19}$$

$$[L] = \begin{bmatrix} L_{11} & L_{12} & L_{13} & L_{14} & L_{15} \\ L_{21} & L_{22} & L_{23} & L_{24} & L_{25} \\ L_{31} & L_{32} & L_{33} & L_{34} & L_{35} \\ L_{41} & L_{42} & L_{43} & L_{44} & L_{45} \\ L_{51} & L_{52} & L_{53} & L_{54} & L_{55} \end{bmatrix} \tag{20}$$

With balanced energy conversion, the machine torque and speed production equals the power input from the inverter side as (21), if the resistance and other losses are neglected.

$$T_e \cdot \omega_m = [e]^T \cdot [i] \tag{21}$$

Due to the large effective air gap, the armature reaction is negligible compared with the machines with significant phase inductance, for instance, interior PM (IPM) design. Thus, the back EMF wave form impacts the machine torque characteristics significantly, for example, the torque pulsation. In this system, a control strategy based on the current regulation is proposed for the mitigation of the torque pulsation and the phase fault compensation, which will be covered in the future paper.

A standard inverter topology based on independent control unit for each module is illustrated in Figure 6.

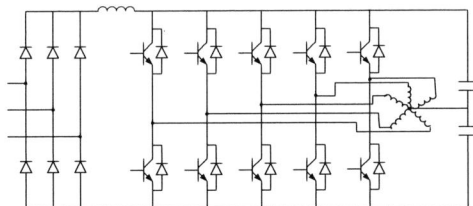

Figure 6: Inverter Topology

In Figure 6, half bridge leg is designed for the independent module control. When the machine is working with healthy modules, the neutral point of this drive circuit provides relative stiff voltage reference. While under phase fault condition, the neutral point suffers from the unbalanced phase current and the voltage fluctuation makes it challenge to control the rest independent healthy modules properly. More discussion of the control strategy and the drive circuit topology is still under investigation, and the results will be reported in a future report.

IV. SIMULATION RESULTS

With the system described above, some preliminary system simulations are provided with fan load characteristics as illustrated in Figure 7 ~11.

The starting of the drive system for the position open loop and closed loop are shown in Figure 7 and 8.

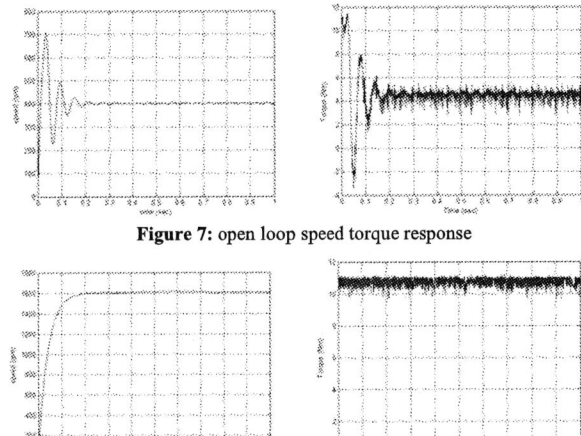

Figure 7: open loop speed torque response

Figure 8: Closed loop speed torque response

With the rotor feedback (close loop), the machine can be started very quickly compared with the transient process for the open loop system simulation under certain frequency.

For the phase open circuit, the machine torque suffers from the phase loss and torque pulsation is simulated by the FEM analysis as shown in Figure 9 (a). Due to the design with small number of modules (five modules in this design), the machine torque exhibits significant cogging torque pulsation under healthy operation, which may be mitigated by the proper design from the stator or rotor side, for instance, skewed PMs as illustrated in the actual machine construction.

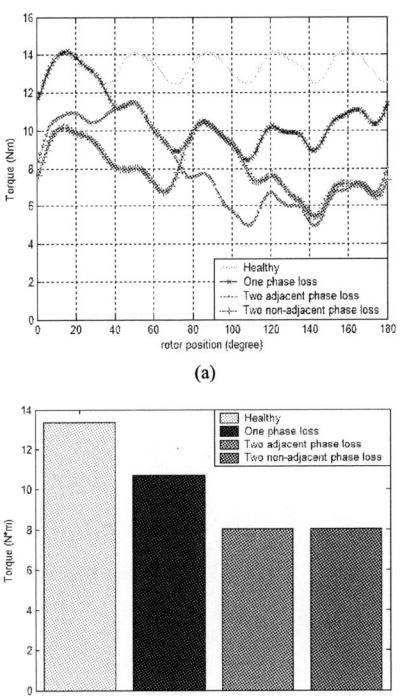

Figure 9: Machine torque characteristics under phase open circuit fault by FEM analysis. (a) transient. (b) average

The averaged torque under open circuit fault is directly dependent on the healthy phase numbers as shown in Figure 9 (b). The system simulations under phase open circuit fault are provided in Figure 10 ~ 11.

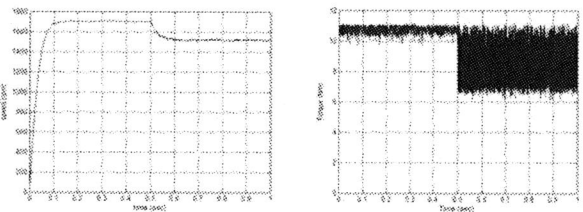

Figure 10: One phase open circuit fault

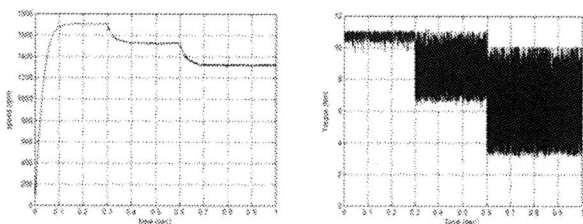

Figure 11: Two phase open circuit fault

The single phase short circuit fault is also simulated as shown in Figure 12. The machine torque suffers most due to the induced current in the fault phase.

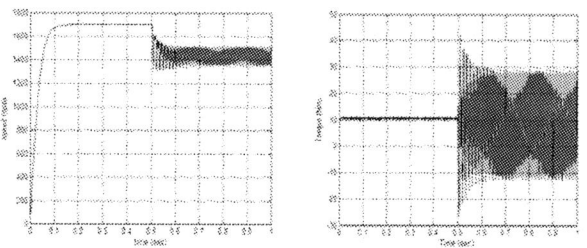

Figure 12: Single phase short circuit fault

The modular drive system proposed in this paper benefits from it natural fault tolerant capability due to the independently controlled modules. While the torque pulsation can be further mitigated by the control strategies embedded by the control program, which will be specifically discussed in the future paper.

V. EXPERIMENT SETUP AND RESULTS

The prototype motor and inverter drive system is illustrated in Figure 13.

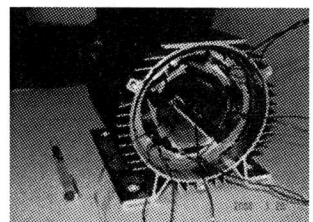 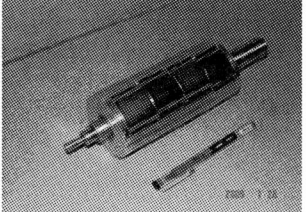

(a) modular SPM motor

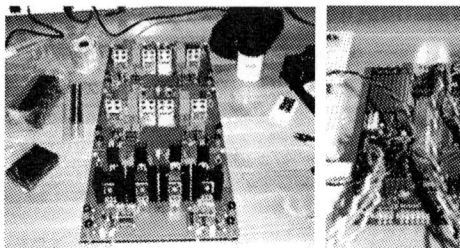 

(b) Inverter
Figure 13: System Prototype

The experiment setup work is still under way. The test results will be provided in the future report.

VI. CONCLUSIONS

In this paper, a modular PM drive system is proposed and developed. With new machine fabrication methodology, the PM machine can be designed with independent modules, which offers the option with independent control for each module as well. The proposed SPM machine topology can achieve high power/torque density with potentials for fault tolerant operations. The test validation for this proposed drive system will be provided in future report.

REFERENCES

[1] A. Jack, *Experience with using soft magnetic composites for electrical machines*, New Magnetic Materials - Bonded Iron, Lamination Steels. Sintered Iron and Permanent Magnets, *IEE* Colloquium on 28, May,1998, pp:3/1 - ¾

[2] A. G. Jack, B. C. Mecrow, C. P. Maddison, and N. A. Wahab, *Claw pole armature permanent magnet machines exploiting soft iron powder metallurgy*, Electric Machines and Drives Conference Record, 1997, *IEEE* International, pp.18-21, May 1997

[3] A. G. Jack, B. C. Mecrow, P. G. Dickinson, D. Stephenson, J. S. Burdess, N. Fawcett, and J. T. Evans, *Permanent-magnet machines with powdered iron cores and prepressed windings*, Industry Applications, IEEE Transactions on, vol. 36, pp. 1077, 2000

[4] Y. G. Guo, J. G. Zhu, P. A. Watterson, and W. Wu, *Comparative study of 3D flux electrical machines with soft magnetic composite cores*, Industry Applications Conference, 2002. 37th IAS Annual Meeting, pp.1147 vol.2

[5] T.A. Lipo, S. Madani and R.J.White, "Soft Magnetic Composites for AC Machines – A Fresh Perspective", in Proc. of EPE Conf. on Power Electronics and Motion Control, Sept. 2004, Riga, Latvia.

2006 5th International Power Electronics and Motion Control Conference

Research on Digital Control Systems for Large Power AC-DC-AC Converters with Synchronous Motor Load

Xiaotan Zhao [*,**], Chongjian Li [***], Weihui Sheng [***], Yaohua Li [*]

[*]Institute of Electrical Engineering Chinese Academy of Science, Beijing, China
[**]Graduate University of the Chinese Academy of Science, Beijing, China
[***]Automation Research and Design Institute of Metallurgical Industry, Beijing, China

Abstract— **Large power AC converters with synchronous motor load are widely used in field projects. In this article, hardware modules with VME-bus and graphical developing environment are used to construct a digital control system. This system is used to control a 5MVA 3-level PWM AC-DC-AC IGCT converter with synchronous motor load. The open-loop control experiment result proves that using such a set of platform to construct the digital control system for large power AC-DC-AC converters with synchronous motor load is feasible. And the close-loop control experiment is in process.**

Keywords- Large Power; AC-DC-AC converter; VME-bus; Synchronous motor

I. INTRODUCTION

Along with the rapid development of power electronics, micro-electronics technique and modern control theory, the trend of AC drives replacing DC ones has appeared in large power speed-regulating field. Large power AC converters with synchronous motor load have been widely adopted in rolling mill drive of metallurgy industry. And the economic benefit is evident. This kind of converters has also been used for mine hoist drive in coal and non-ferrous metal industry. The large power compressor drive used in "transmitting gas in the west area to the east area" project, the marine propulsion system, the traction power supply system used for linear synchronous motor in high-speed maglev project, et al. use large power AC converters with synchronous motor load too.

The large power AC converter with synchronous motor load has high capacity. And the system is very complex. Usual universities and colleges don't have the experiment devices. Thus the research and experiment on the digital control system for such converters are rarely done. It is some famous electrical companies that promote the development of the digital control system for large power AC drives.

In this article, a digital control system is constructed using hardware modules with VME-bus and softwares developed in graphical programming environment. In the experiment, this system is used to control a 5MVA 3-level PWM AC-DC-AC IGCT converter with synchronous motor load. The power of the synchronous motor is 3kW. In the experiment, the synchronous motor runs well under open-loop control algorithm.

II. DIGITAL CONTROL SYSTEMS EXISTING FOR LARGE POWER AC CONVERTERS WITH SYNCHRONOUS MOTOR LOAD

As everyone knows, the digital control system for large power converters is very complex. For the convenience of system expanding, using and maintaining, the digital control systems for large power AC converters with synchronous motor load used in rolling mill, mine hoist and high-speed maglev systems all adopt backplane bus structure [3][4]. The functions these control systems need to fulfill are complex, besides for the control algorithms, other functions such as process control, communicating with other systems, also need to be implemented. One rack or one CPU module can't meet all these functions alone. Usually, there are at least two racks with backplane bus in such a control system. And there are multi-CPU modules and other types of modules, such as analog input and output modules, digital input and output modules, communication interface modules, et al. in one rack. All the functions the control system to fulfill are divided into many parts. Different CPU modules implement different function parts. Multi-CPU modules in one rack exchange data through backplane bus. And CPU modules in different racks exchange data through hardware modules that connect racks together. The medium for communication among racks is usually optical fiber.

The programming languages for such kind of control systems are usually graphical programming languages that conform to IEC61131-3 standard, such as Ladder and Function Block. Graphical programming languages are convenient for developing and reading application software. Generally, the programming environment has some basic algorithm modules existing and allows users to develop their own algorithm modules. This function makes it easy to implement modularized application

---

1-4244-0448-7/06/$25.00 ©2006 IEEE

softwares, which is convenient for integration of softwares developed by different developers.

To sum up, the multi-CPU system with backplane bus structure is easy to expand. If certain function is needed by the system, adding appropriate hardware module or rack is enough. And such a system is convenient for fault handling. If certain hardware module is in trouble, all we need to do is to replace this module. Thus it has little bad effect on the performance of the whole system.

To use multi-CPU modules in one rack, a kind of backplane bus supporting multi-CPUs is needed. The backplane bus controls reasonable data flow. Now there are many kinds of backplane bus supporting multi-CPUs in market. And the VME-bus is the most widely used in embedded field, such as industry control, spaceflight, telecommunication, medical devices, et al. The VME-bus has many advantages, such as 64-bit addressing and data transmitting ability, matching IEC 297 EURO card standard, reliable mechanical performance, et al. Thousands of products supporting VME-bus standard have been produced since the VME-bus appears. And now there are hundreds of Printed Circuit Board, hardware, software and bus interface manufactures that are producing VME-bus products. Thus, we have more freedom to choose hardware modules and developing environment when choosing the VME-bus as backplane bus.

III. CONTROL SYSTEM BASED ON VME-BUS CONSTRUCTED IN THIS ARTICLE

A. Framework of AC-DC-AC converters with synchronous motor load

The AC-DC-AC voltage-fed converter is widely used. The devices used in such a converter usually are IGBTs, IGCTs or IEGTs, et al.

Fig.1 shows the block diagram of the AC-DC-AC converter with synchronous motor load. The flux observation unit computes θ, the inclination between the flux axis and the still axis, using practical voltages and currents. The I/U unit computes the stator voltage components u_{sm}, u_{st} through current components i_{sm}, i_{st}. The current model Mi uses magnetizing current i_{μ} and torque current i_{st} to obtain rotor field current i_f and current component i_{sm}.

From Fig.1, we can see that converters with DC-link are similar to those without DC-link. The difference is the trigger signal. For converters with DC-link, the trigger signals are voltage module $|u|$ and rotation angle θ_u. For converters without DC-link, those are 3-phase AC voltages.

The AC-DC-AC voltage-fed converter used in the experiment of this article contains rectification cabinet,

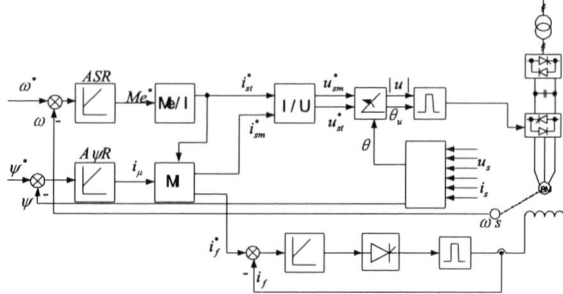

Figure 1. Block diagram of AC-DC-AC converter with synchronous motor load

DC filter cabinet, inversion cabinet, field excitation cabinet and purified-water cooling system. The power of this converter is 5MVA and the devices are IGCTs.

B. Construction of the digital control system used in this article

Based on the example existing and in consideration of system expanding requirements, hardware modules with VME-bus are used to construct the control system in this article. All the modules in this system support VME-bus standard. Among them modules interfacing with drive system are developed independently. Embedded operating system core VxWorks is installed in CPU modules to guarantee the real-time performance.

Fig.2 shows the block diagram of the digital control system constructed in this article. Both Rack1 and Rack2 have VME backplane buses. Rack1 acts as upper rack to monitor the whole system. Rack2 acts as controlling rack to implement system control and protection functions. The CPU module, A/D module, D/A module, digital I/O module (shown as DI/DO in Fig.2), Profibus interface module (shown as Profi in Fig.2) and Reflective Memory module (shown as RM in Fig.2) are standard VME-bus modules.

The Reflective Memory module is used to communicate between Rack1 and Rack2. It uses optical fiber as communication medium. There are at most 256 nodes in a whole system and the transmission distance is 2km at most. The communication speed can reach 174Mbyte/s, which can satisfy the requirements of fast communication. Furthermore, the reflective memory needs little software configuration and easy to use.

The Profibus interface module is used to communicate with field equipments with Profibus-DP interface, such as ET200, PLC. This module collects field fault signals, feedback signals and outputs command signals. The fastest communication speed of Profibus-DP can reach 12Mbit/s and it is suited to field application.

The Pulses Output module (shown as PO in Fig.2) is developed independently to interface with drive system. This module generates 3-level PWM pulses and outputs these pulses to IGCTs in the inversion circuit.

996

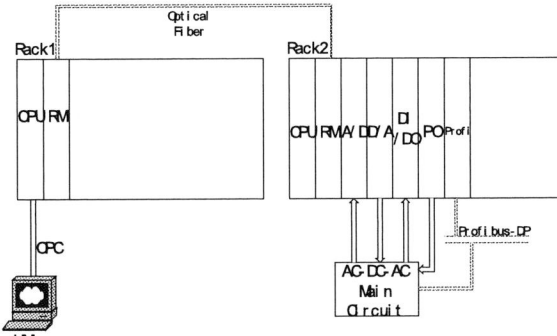

Figure 2.   Block diagram of constructed digital control system

The programming software supports OPC (OLE for Processing Control) agreement. Thus, it is easy to communicate with HMI software supporting OPC agreement to monitor the system.

Fig.3 shows the software-developing environment of this system.

The software-developing environment supports Ladder, Function Block and C languages. It has some basic algorithm modules existing and allows users to develop algorithms using C language and encapsulate them into software modules. By using graphical programming language, the developing software is easy to use and the application softwares are easy to read. This makes it possible to implement modularized application softwares, which is convenient for expanding application softwares.

C.Experiment

The control system constructed in this article is used to control a 5MVA IGCT AC-DC-AC converter with a 3kw synchronous motor load. The open-loop control algorithm is adopted.

Fig.4 shows the 3-level PWM waveforms output by the Pulse Output module. Curve 1 is the waveform of the output voltage and Curve 2 is that of the output current.

The open-loop experiment result shows that the speed-regulating performance of the whole system is good, which approves that using such a platform to construct a digital control system for large power AC-DC-AC converters with synchronous motor load is feasible.

According to practical situation, the hardware and software modules can both be expanded due to the excellent expanding performance of this system.

IV.   CONCLUSION AND FURTHER TASKS

The existing digital control systems for large power AC drives are introduced in this article. A set of control system platform is constructed using hardware modules with VME backplane bus and graphical software-developing environment. In the experiment, this system controls a 5MVA IGCT AC-DC-AC converter with a 3kw synchronous motor. The open-loop experiment result proves that it is feasible to use such a kind of platform to

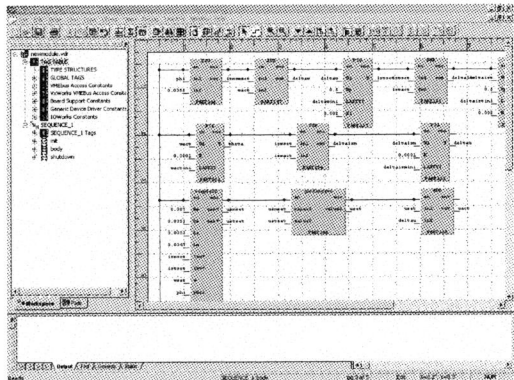

Figure 3.   Software-developing environment of the constructed digital control system

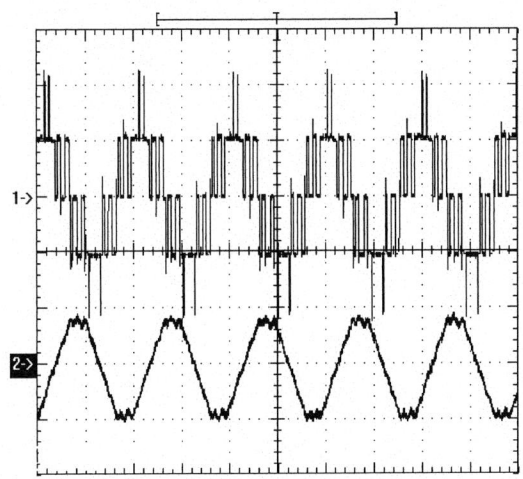

Figure 4.   Waveforms of output voltage and current

construct a digital control system for large power AC-DC-AC converters with synchronous motor load.

Further tasks to do are as follows:

(a) Expand the hard wares of this system to satisfy requirements of other functions needed by large power AC-DC-AC converters with synchronous motor load.

(b) Develop the close-loop algorithm using graphical programming language.

(c) Commission in the lab and field.

The goal is to implement a set of digital control system for large power AC-DC-AC converters with synchronous load and to use this system in practice.

REFERENCES

[1] Li chongjian, *Adjusting Speed System of Synchronous Machine*. Science Press, in press.

[2] Bimal K. Bose, *Modern Power Electronics and AC Drives*. China Machine Press, 2003.1.

[3] SIEMENS, *Digital Control System SIMADYN D Configuring Guide*, 1995.

[4] SIEMENS, *Control System SIMATIC TDC*, 1999.

[5] Ma xiaoliang, *Large Power Cycloconverter-fed and Vector Control Technique*. China Machine Press, 2003.10.

[6] Li Chongjian, Zhu Chunyi and Li Yaohua, "Modeling and Simulation of AC/AC Field-oriented Control System of Synchronous Motor," IPEMC94, P507.

[7] Li Chongjian, Wang Xiangheng, Li Fahai, Gao Jingde, Ding Yunshi and KuligTS, "Dynamic Performance of the Field Oriented Control Cycloconverter-Fed Synchronous Machine," IPEMC94, P510.

[8] Yongge Gan, Wen Wang, Chongjian Li, Fahai Li, "New Method of Reactive Power Compensation of Field-oriented Control of Cycloconverter-fed Synchronous Motor Driving Mine Hoist," *The 24th Annual Conference of the IEEE Industrial Electronics Conference*, Aachen-Germany, pp 473-478.

2006 5th International Power Electronics and Motion Control Conference

About the Prediction of Undesired Higher Current and Torque Harmonics of Inverter Driven Motors with Numerical Methods

C. Grabner

Research and Development, Siemens AG, Bad Neustadt an der Saale, Germany
grabner.christian@siemens.com

Abstract — A certain method for the direct coupling of electrical circuits and two-dimensional finite element calculations is used for the transient analysis of a complete drive system in the time-domain. Thereby, a novel and deeper insight into the invoked harmonic spectrum of the electrical stator current and the mechanical torque at the rotating shaft could be achieved.

Keywords — Finite element method; coupled circuits; inverter topology; induction motor.

I. INTRODUCTION

The application of variable speed induction drive systems requires a variable frequency and voltage supply including a fast and high-efficient electronic control [1]. Fig.1 shows an example of an optimized industrial drive system.

Figure 1. Speed-variable induction drive system.

One main problem arises with the prediction of mechanical torque fluctuations at the rotating shaft even from unsuitable machine designs or inappropriate power converter strategies [2,3]. Moreover, a deeper knowledge about the prediction of additional losses in converter driven motors is still lacking [4].

The proposed numerical calculation treats the complete system, concerning the converter which is built up of discrete electronic devices, as well as the real 3D motor design represented by iron laminations, insulation material, stator winding and squirrel cage with finite elements in the time-domain [5]. The method overcomes previous insufficiencies and delivers the knowledge about novel interrelations between electrical phase voltages, electrical phase currents, and mechanical quantities in dependency on various electrical control methods with good correctness, because the real motor geometry and nonlinear material effects as well as minor changes in the shaft speed due to torque ripple are regarded [6].

II. DRIVE SYSTEM TOPOLOGY

The power conversion from the public three-phase ac grid of constant frequency and voltage amplitude into an arbitrary three-phase ac system with variable settings is performed by means of the power converter topology in Fig.2. In a first step, the input ac to dc rectification is done by a classical B6 bridge. Thereby, the diodes D1 to D6 depicted in Fig.2 cause a distinct voltage ripple within the dc voltage link U_{ZK}. For this reason, the capacitor C smoothes these fluctuations. The implemented inductance L in Fig.2 restricts current peaks during unexpected operational states [7]. The conversion of the dc link to three-phase output ac power is exclusively performed in the switched mode. Power semiconductor switches T1 to T6 of Fig.2 effectuate temporary connections at high repetition rates between the two dc terminals and the three phases of the ac drive motor [8]. The actual power flow, which is determined by $u(t), i(t), \varphi(t)$, is controlled by the on/off ratio, or duty cycle, of the respective switches T1 to T6 by various control strategies [9,10].

1-4244-0448-7/06/$25.00 ©2006 IEEE

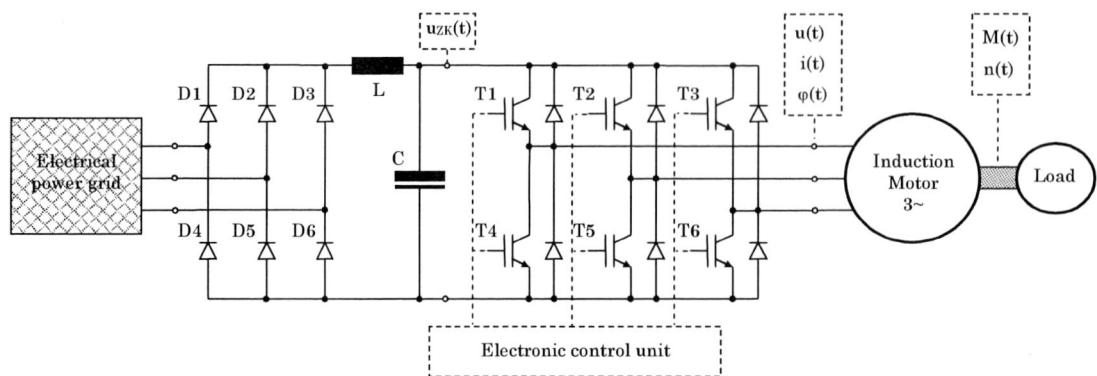

Figure 2. The investigated power converter consists of the input B6 rectifier with diodes D1 to D6, the DC voltage link, three half-bridges formed by semiconductors T1-T4, T2-T5 and T3-T6 and an electronic control unit. The electrical motor supply is $u(t), i(t), \varphi(t)$, whereas the resulting mechanical characteristics at the load are $M(t), n(t)$.

III. ELECTROMECHANICAL FINITE ELEMENT APPROACH WITH DIRECT COUPLED CIRCUITS

The local field quantities of the applied 2D finite element algorithm must be coupled to external circuits in order to include the arbitrary voltage waveforms of the power converter. Moreover, concentrated circuit elements are used for modeling of 3D motor end-region effects. Due to the non-linear iron material properties, commonly used linearization as well as time-discretization methods are necessary in order to solve such kind of coupled electromechanical problem in the time domain.

A. Finite element implementation

The three dimensional transient magnetic field within the domain Ω can be fully described in case of a gauged magnetic vector potential formulation $\vec{\nabla} \cdot \vec{A}(t) = 0$ with

$$\vec{\nabla} \times \frac{1}{\mu} \vec{\nabla} \times \vec{A}(t) = \vec{J}_b(t), \qquad (1)$$

whereby the used nonlinear permeability $\mu = B / H$ is directly accessible from Tab.1. Due to the assumed independency of the local field quantities along the axial $\vec{e}_z$ direction in Fig.3, the electrical current density in (1) can be written with an isotropic electrical conductivity γ as

$$\vec{J}_b(t) = \frac{\gamma}{l_b} U_b(t)\vec{e}_z - \gamma \frac{\partial}{\partial t} \vec{A}(t), \qquad (2)$$

whereas only one component $\vec{A}(t) = A(t)\vec{e}_z$ of the magnetic vector potential is necessary. The uniqueness of the field problem (1), (2) is established by means of Dirichlet boundary conditions $A(t) = 0$ along certain sections of Ω in Fig.3 [11].

TABLE I
NON–LINEAR IRON MAGNETIZATION CHARACTERISTIC.

B [T]	0	0.4	0.8	1.2	1.6	2.0
H [A/m]	0	140	190	260	1370	20500

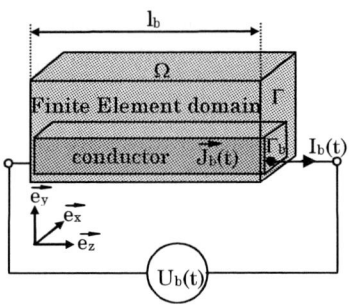

Figure 3. Finite element region with coupled circuit.

The finite element method is based on the principle of Galerkin. Thereby, it is assumed that a selected approximation $\hat{A}(t) \neq A(t)$ of the exact solution $A(t)$ causes an undesired residuum $\Re\langle \hat{A}(t) \rangle \neq 0$. Following the ideas of Galerkin, the minimization of this residuum is performed with an arbitrarily chosen weighting function W

$$\int_\Gamma W \Re \langle \hat{A}(t) \rangle d\Gamma = 0 \qquad (3)$$

across the 2D-domain Γ [12]. A very practicable choice for the weighting function in (3) is given due to the usage of the simple shape function $W = N$. Thus, a reduced system (1), (2) and (3) can be rewritten due to partial integration as

$$\int_\Gamma \frac{1}{\mu} \left[\frac{\partial}{\partial x} N \frac{\partial}{\partial x} \hat{A}(t) + \frac{\partial}{\partial y} N \frac{\partial}{\partial y} \hat{A}(t) \right] d\Gamma +$$
$$+ \int_\Gamma \gamma N \frac{\partial}{\partial t} \hat{A}(t) d\Gamma - \int_\Gamma \frac{\gamma}{l_b} N U_b(t) d\Gamma = 0 \quad . \qquad (4)$$

The finite element approach takes account of the fact that an integral over the connected domain (4) can alternatively be evaluated from contributions of integrals over sub-domains. Thus, the domain Γ has to be discretizised into a distinct number of sub-domains Γ_e. The finite element representation of the iron-made squirrel cage induction motor parts by small triangular finite elements is exemplarity shown in Fig.4.

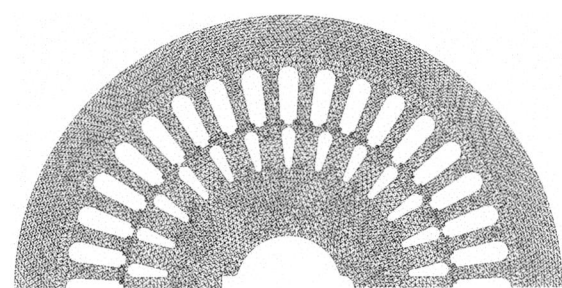

Figure 4. Finite element mesh of iron-made stator and rotor.

The proposed numerical method allows the summation across sub-areas formed by a number of e single elements according to

$$\sum_e \left[\int_{\Gamma_e} \frac{1}{\mu} \left[\frac{\partial}{\partial x} N \frac{\partial}{\partial x} \hat{A}^e(t) + \frac{\partial}{\partial y} N \frac{\partial}{\partial y} \hat{A}^e(t) \right] d\Gamma \right. $$
$$\left. + \int_{\Gamma_e} \gamma N \frac{\partial}{\partial t} \hat{A}^e(t) d\Gamma - \int_{\Gamma_e} \frac{\gamma}{l_b} N U_b(t) d\Gamma \right] = \{0\} . \quad (5)$$

The most favorable finite element approach uses triangular finite elements with a number of six nodes. The element approximation

$$\hat{A}^e(t) = \sum_{i=1}^{6} N_i \tilde{A}_i^e(t) \quad (6)$$

of the primarily unknown six time-dependent nodal vector potential values $\tilde{A}_i^e(t)$ per element is therein performed by quadratic shape functions N_i. Fortunately, the chosen approximation (6) avoids some well-known difficulties along magnetic saturable material boundaries. Moreover, (6) allows the diction of the equation (4) by means of primarily unknown local quantities $\left\{ \tilde{A}^e(t) \right\} = \left(\tilde{A}_1^e(t) \quad \cdots \quad \tilde{A}_6^e(t) \right)^T$ in a more favorable way as

$$\sum_e \frac{1}{\mu} [S^e] \left\{ \tilde{A}^e(t) \right\} + \gamma [T^e] \frac{\partial}{\partial t} \left\{ \tilde{A}^e(t) \right\} - \left\{ Q^e \right\} \frac{\gamma}{l_b} U_b(t) = \{0\} \quad (7)$$

whereby the element matrices and vectors

$$S_{ij}^e = \iint \left[\frac{\partial}{\partial x} N_i \frac{\partial}{\partial x} N_j + \frac{\partial}{\partial y} N_i \frac{\partial}{\partial y} N_j \right] dxdy , \quad (8)$$

$$T_{ij}^e = \iint N_i N_j dxdy , \quad (9)$$

$$Q_i^e = \iint N_i dxdy , \quad (10)$$

are introduced to characterize each element.

B. Calculation of the total electrical current

In order to couple the circuit and field equation as shown in Fig.3, it is necessary to calculate the total current flow in $\vec{e}_z$ direction through the surface Γ_b. The electrical bar current is found by integrating (2) over the associated domain as

$$I_b(t) = \int_{\Gamma_b} \vec{e}_z \cdot \vec{J}_b(t) d\Gamma_b . \quad (11)$$

With the electrical current density (2), the rewritten form of (11) is given as

$$I_b(t) = \iint \left[\frac{\gamma}{l_b} U_b(t) - \gamma \frac{\partial}{\partial t} A(t) \right] dxdy . \quad (12)$$

Regarding the finite element notation (6), we derive with (10) and the surface Δ^e of one single element the governing relation for the electrical current in one single bar as

$$I_b(t) = \sum_{e \in \Gamma_b} \left[\Delta^e \frac{\gamma}{l_b} U_b(t) - \gamma \{Q^e\}^T \frac{\partial}{\partial t} \left\{ \tilde{A}^e(t) \right\} \right] . \quad (13)$$

C. Coupling of finite element stator domains with external circuits

The modeling of the stator winding system demands in general a distinct number of n series connected bars in order to form one single coil c, as it is schematically shown in Fig.5 [13]. The small wire diameter and the high number of windings avoid any eddy currents in the stator windings themselves. Each bar component in the current vector $\left\{ I_b(t) \right\} = \left(I_{b1}(t) \quad \cdots \quad I_{bn}(t) \right)^T$ of the coupled system is only consisting of contributions due to the first term in (13). Moreover, each separate bar in Fig.5 carries the same magnitude of the total current $I_c(t)$, but successive bars 1 to n carry local bar currents $I_{b1}(t)$ to $I_{bn}(t)$ in opposite directions. The leads of the coil are brought out of the finite element region and are connected to a voltage source. With the usage of a bar voltage vector $\left\{ U_b(t) \right\} = \left(U_{b1}(t) \quad \cdots \quad U_{bn}(t) \right)^T$ we obtain the electrical circuit relation

$$U_c(t) = \{d_b\}^T \{U_b(t)\} + R_\sigma I_c(t) + L_\sigma \frac{\partial}{\partial t} I_c(t). \quad (14)$$

The values of the resistor R_σ and the inductance L_σ in (14) represent both three-dimensional end-winding effects of one stator phase within the two-dimensional finite element calculation. The important equation (14) serves to couple certain finite element regions, represented by the components $U_{b1}(t)$ to $U_{bn}(t)$, to external circuits and sources. Thereby, the changing polarity is regarded in (14) by an introduced weighting coefficient vector $\{d_b\}$ with the entries $\pm w/2$, which takes account of the total acting winding number w per phase. As shown in Fig.5, bar 1 and bar (n-1) give a 'positive' contribution, whereas bar 2 and bar (n) are counted with a 'negative' sign. The used values of the stator resistor $R_\sigma = 7.74\Omega$ and the inductance $L_\sigma = 6.4mH$ are therefore evaluated at operating temperature.

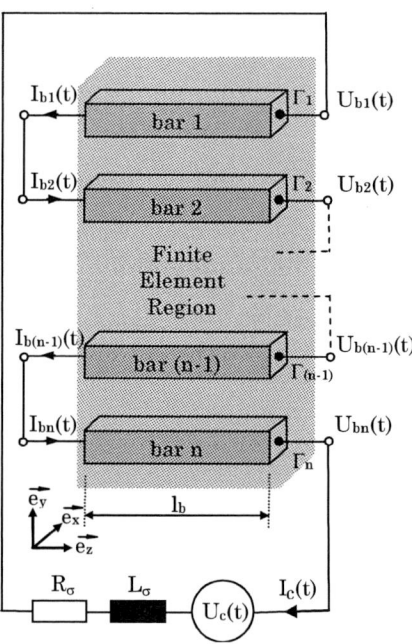

Figure 5. A number of n-serial connected bars are forming one stator coil driven by an external voltage source.

D. Coupling of finite element rotor domains with external circuits

The substituted winding schema for the squirrel cage rotor is shown in Fig.6. It is assumed that one coil (14) is only consisting of one single bar. The resistors R_σ in (14) stands now for the fictive resistance R_b, whereas the inductance L_σ in (14) is now approximated by L_b. The used 2D finite element model is not suitable to take account of 3D end-ring effects of the rotor. Thus, R_b and L_b in Fig.6 stands for a combined parameter, which also represents the 3D end-ring influences. A number of m different coils are parallel connected in Fig.6 in order to form the squirrel cage. The bar components $I_{b1}(t)$ to $I_{bm}(t)$ within the vector $\{I_b(t)\} = \left(I_{b1}(t) \quad \cdots \quad I_{bm}(t)\right)$ have to fulfill the condition $\{d_b\}^T \{I_b(t)\} = 0$ at each connection point in Fig.6, whereby all entries in the vector $\{d_b\}$ are 1. Furthermore, the contributions $U_{c1}(t)$ to $U_{cm}(t)$ at these connection points are forced to be equal. The resistance and inductance of both end-rings are included in the rotor bar parameters $R_b = 1.74\mu\Omega$ and $L_b = 22.5\text{nH}$ as an additional contribution.

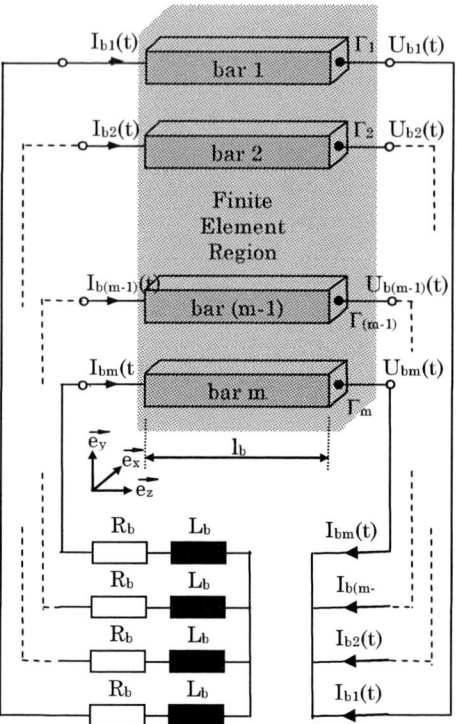

Figure 6. A number of m-parallel connected coils build up the complete squirrel cage rotor.

E. Forces and movement in the finite element analysis

The basic relation of motion can be written with the mass m of the moving part, the actual position $x(t)$, the damping coefficient λ and the acting electromagnetic forces $F_m(t)$ and the mechanical load $F_l(t)$ as

$$m\frac{\partial^2}{\partial t^2}x(t) + \lambda\frac{\partial}{\partial t}x(t) = F_m(t) - F_l(t). \qquad (15)$$

Thereby, the damping coefficient is assumed to have a speed independent value of $\lambda = 0.00045$. The mechanical boundary condition for the line start behavior is given by $x(t_0) = x_0$. The finite element approach (6), (8) allows the computation of the magnetic force, which is acting on a solid body in (15), by 'virtual motion' as

$$F_m(t) = -\frac{1}{2}l_b\frac{1}{\mu}\{\tilde{A}(t)\}^T\left[\frac{\partial S}{\partial x}\right]\{\tilde{A}(t)\}. \qquad (16)$$

The differentiation of the contributions in the assembled matrix (8) is straightforward, since the entries are simple functions of the local directions.

F. Time discretization and linearization

The field equation, the total current, the circuit equations and the equation of motion have to be discretized in the time domain [14]. This is based on the Crank-Nicholson Method according to

$$\frac{1}{2}\left[\frac{\partial f(t)}{\partial t}\right]^{t+\Delta t} + \frac{1}{2}\left[\frac{\partial f(t)}{\partial t}\right]^{t} = \frac{\{f(t)\}^{t+\Delta t} - \{f(t)\}^{t}}{\Delta t} \quad (17)$$

Moreover, the field equation (7) and the acceleration equation (15), (16) are non-linear functions of the unknown vector potential $\tilde{A}(t)$ and/or the component displacement $x(t)$. Thus, these interrelations have to be linearized in spite of the non-linear behavior by applying the Newton-Raphson method

$$[J]\left\{\xi^{(k+1)} - \xi^{(k)}\right\} + \left\{f(\xi^{(k)})\right\} = \{0\} \quad (18)$$

with the Jacobean-Matrix

$$[J] = \begin{bmatrix} \dfrac{\partial f_1(\xi)}{\partial \xi_1}, & \dfrac{\partial f_2(\xi)}{\partial \xi_2}, & \cdots, & \dfrac{\partial f_n(\xi)}{\partial \xi_m} \\ \vdots & & & \vdots \\ \dfrac{\partial f_m(\xi)}{\partial \xi_1}, & \cdots, & \cdots, & \dfrac{\partial f_m(\xi)}{\partial \xi_m} \end{bmatrix} \quad (19)$$

before they can be combined with the other equations in order to form the global system matrix.

G. Assembling of the global system

With the changes of the potential vector of each node $\{\Delta\tilde{A}\}$, the change of the voltage vector across each bar $\{\Delta U_b\}$, the change in the current vector in each bar $\{\Delta I_b\}$, the change in the terminal voltage vector $\{\Delta U_c\}$ and finally the change in the position vector of the nodes of the movable rotor part $\{\Delta x\}$, the global system matrix can be assembled by using (7) to (19) as

$$\begin{bmatrix} M_{11} & M_{12} & & & M_{15} \\ M_{12}^T & M_{22} & M_{23} & & \\ & M_{23}^T & M_{33} & M_{34} & \\ & & M_{34}^T & M_{44} & \\ M_{15}^T & & & & M_{55} \end{bmatrix} \begin{Bmatrix} \{\Delta\tilde{A}\} \\ \{\Delta U_b\} \\ \{\Delta I_b\} \\ \{\Delta U_c\} \\ \{\Delta x\} \end{Bmatrix}^{t+\Delta t}_{k+1} = \begin{Bmatrix} \{C_1\} \\ \{C_2\} \\ \{C_3\} \\ \{C_4\} \\ \{C_5\} \end{Bmatrix}. \quad (20)$$

Taking account of the necessary boundary conditions, the global system (20) is solved with (18), (19) for each time step (17).

IV. VOLTAGE WAVEFORMS FOR VARIOUS ELECTRONIC CONTROL STRATEGIES

The control unit in Fig.2 has the task to generate different duty cycles for the semiconductor devices T1 to T6. So on, the three half-bridges are able to create nine distinct voltage levels within the phase-to-neutral motor voltage as it is obvious in Fig.7. The invoked terminal voltage spectrum is thereby mainly governed by the chosen internal carrier frequency of the control unit.

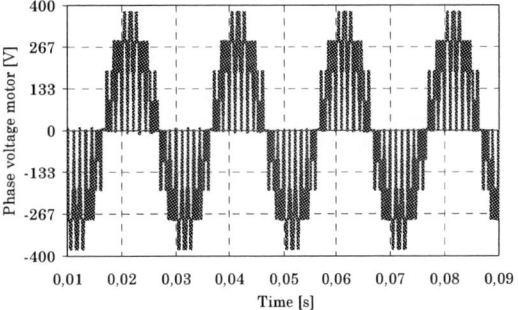

Figure 7. Switched phase-to-neutral voltage of the star-connected motor for a carrier frequency of 450Hz.

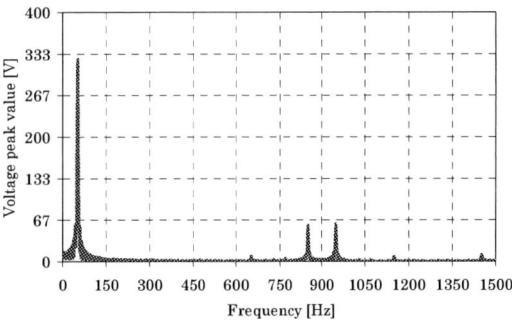

Figure 8. Spectrum of the motors phase-to-neutral voltage for a carrier frequency of 450Hz.

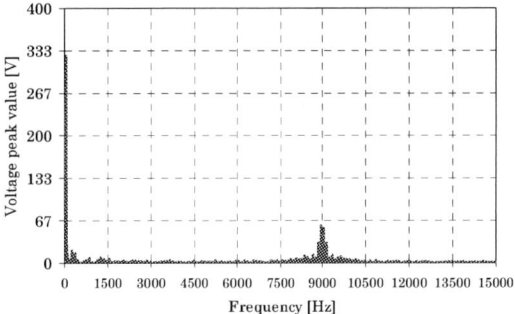

Figure 9. Spectrum of the motors phase-to-neutral voltage for a carrier frequency of 4500Hz.

The periodical discrete voltage u_n in Fig.7 is analyzed with the aid of the discrete Fourier-transformation at a number of N samples as

$$\hat{U}_\upsilon = \sum_{n=0}^{N-1} u_n e^{-j(2\pi n/N)\upsilon} \ , \ \upsilon = 0,1,2\dots N-1. \qquad (21)$$

Due to the star connection of the stator winding, the derived spectrum in Fig.8 does never contain the distinct components 3υ with $\upsilon \in \mathbb{N}_{Odd}$. Thus, it is almost favorable to take carrier frequencies for the control schema in accordance to

$$f_P = 3gf_N \ , \ g \in \mathbb{N}, \qquad (22)$$

in order to eliminate thus characteristic contributions at the motor. Other undesired harmonic magnitudes could be implicitly shifted to higher ordinal numbers by increasing the carrier frequency up to $f_p = 4500\text{Hz}$, as it is exemplarity shown in Fig.9. Following this strategy, some attention has to be given to eventually unacceptable increasing switching losses inside the semiconductors [15].

V. STATOR CURRENT AND TORQUE SPECTRUM IN DEPENDECY ON THE CARRIER FREQUENCY

A. Invoked higher harmonics in the electrical stator current

In the case of the lower carrier frequency of $f_p = 450\text{Hz}$ the depicted voltage waveform in Fig.7 is processed as input voltage of the motor within the coupled finite element analysis. The derived phase current waveform $i(t)$ is shown in Fig.10. When the Fourier analysis procedure (21) is applied to $i(t)$ in Fig.10, the total harmonic current is governed by the important components $\hat{I}_1 = 2.82\text{A}$ at 50Hz, $\hat{I}_{17} = 0.25\text{A}$ at 850Hz and $\hat{I}_{19} = 0.23\text{A}$ at 950Hz. Other harmonic components are only of secondary interest. Unfortunately, a main discrepancy to the desired sinusoidal distribution is established by using such low carrier frequency. This undesired effect could be easily overcome by applying the much higher carrier frequency of $f_p = 4500\text{Hz}$. Thereby, the applied source voltage spectrum of Fig.9 leads to the current wave-shape $i(t)$ shown in Fig.11. Thereby, the harmonic analysis delivers a very distinct fundamental component $\hat{I}_1 = 2.89\text{A}$ at 50Hz, which allows a very good sinusoidal approximation.

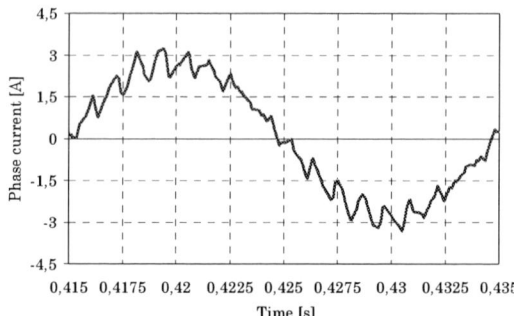

Figure 10. Calculated phase current $i(t)$ of the motor at the carrier frequency 450Hz.

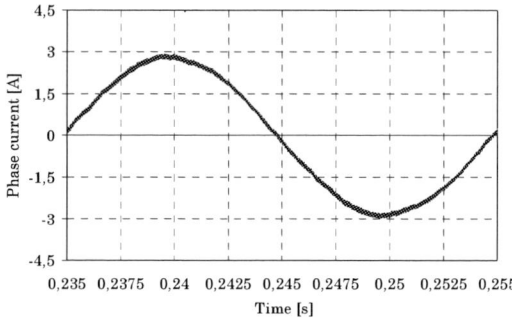

Figure 11. Calculated phase current $i(t)$ of the motor at the carrier frequency of 4500Hz.

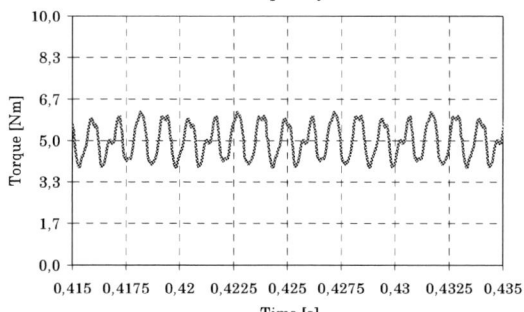

Figure 12. Calculated mechanical torque $M(t)$ at the carrier frequency of 450Hz.

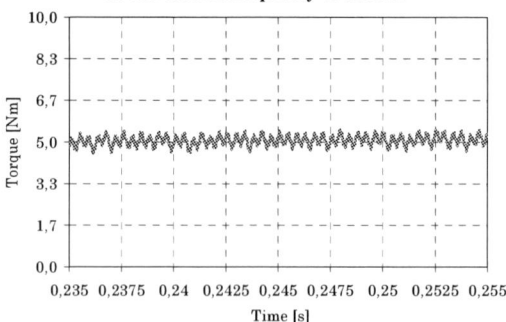

Figure 13. Calculated mechanical torque $M(t)$ at the carrier frequency of 4500Hz.

B. Invoked higher harmonics in the mechanical torque

The mechanical torque wave shape $M(t)$ is depicted in Fig.12 in case of a control strategy based on the carrier frequency of $f_p = 450 \text{Hz}$. It results from the Fourier analysis (21) that there exists the desired constant contribution $\hat{M}_0 = 5 \text{Nm}$. The well-known torque fluctuation in case of the converter topology of Fig.2 is thereby caused by the very dominant component $\hat{M}_{18} = 0.95 \text{Nm}$ at 900Hz. Other contributions to the torque ripple due to $\upsilon = 6$ at 300Hz or $\upsilon = 12$ at 600Hz are obviously suppressed in Fig.12. However, undesired torque pulsation effects could be drastically reduced by applying the much higher carrier frequency of $f_p = 4500 \text{Hz}$. The calculated time-dependent torque $M(t)$ is shown in Fig.13. Therein only one major disturbing contribution $\hat{M}_{48} = 0.25 \text{Nm}$ exists at 2400Hz. Thus, the previously caused torque ripple component $\upsilon = 18$ at 900Hz in Fig.12 is successfully shifted to a much higher ordinal number $\upsilon = 48$ at 2400Hz in Fig.13. Moreover, the torque ripple magnitude could be significant decreased. Thus, undesired effects concerning the quality of the true shaft motion of the drive system due to improper converter control could be avoided by using a higher carrier frequency.

VI. CONCLUSION

The prediction of undesired current and torque harmonics within converter fed induction motors is of crucial interest in order to guarantee a high quality level of the drive system. Thereby, the complex interaction of the converter and the squirrel cage induction motor has to be considered. This is done by using the 2D transient electromagnetic-mechanical finite element method with additionally coupled external circuits. Fortunately, the generated arbitrary time-dependent output voltage waveforms of the converter are directly processed within the non-linear finite element analysis in the time-domain. Effects of minor changes in the mechanical rotor true running due to the torque ripple as well as fluctuations in the electrical current consumption are thereby also regarded. Summarized, the proposed coupled numerical method is very suitable for a fast and accurate analysis of the complete speed-variable squirrel cage induction drive system.

REFERENCES

[1] K. Heintze, H. Tappeiner and M. Weibelzahl, *Pulswechselrichter zur Drehzahlsteuerung von Asynchronmaschinen*, Siemens Zeitschrift 45, Heft 3, 1971.

[2] T.A. Lipo, P.C. Krause and H.E. Jordan, *Harmonic Torque and Speed Pulsation in a Rectifier-Inverter Induction Motor Drive*, IEEE Transactions on Power Apparatus and Systems, Vol. PAS-88, No.5, 1969.

[3] D.T. Stuart and K.M. Hebbar, *Torque Pulsation in Induction Motors with Inverter Drives*, IEEE Transactions on Industry and General Applications, Vol. IGA-7, No. 2, 1971.

[4] A. Heimbrock and H.O. Seinsch, *Neue Erkenntnisse über Oberschwingungsverluste in Umrichtergespeisten Käfigläufern*, Elektrotechnik und Informationstechnik, 122. Jahrgang, Heft 7/8, 2005.

[5] P. P. Silvester, *Finite Elements for Electrical Engineers*, Kluwer Academic Publisher: Boston/London, 1995.

[6] J. S. Salon, *Finite Element Analysis of Electrical Machines*, Cambridge University Press: Cambridge, 1996.

[7] A.W. Krieger and J.C. Salmon, *Hysteresis-Based Current Control at Fixed Frequency with a Resonating Integrator to Eliminate the Steady State Error*, IEEE Canadian Conference on Electrical and Computer Engineering, CD Rom, May 2005.

[8] G.S. Buja and G.B. Indri, *Optimal Pulse Width Modulation for Feeding AC Motors*, IEEE Transactions on Industry Applications, Vol. IA-13, No.1, 1977.

[9] G. B. Kliman and A.B. Plunkett, *Development of a Modulation Strategy for a PWM Inverter Drive*, IEEE Transactions on Industry Applications, Vol. IA-15, No. 1, 1979.

[10] J.M.D. Murphy and M.G. Egan, *A Comparison of PWM Strategies for Inverter-Fed Induction Motors*, IEEE Transactions on Industry Applications, Vol. IA-19, No. 3, 1983.

[11] O. Biro, K. Preis and K. R. Richter, *On the Use of the Magnetic Vector Potential in the Nodal and Edge Finite Element Analysis of 3D Magnetostatic Problems*, IEEE Transactions on Magnetics, vol. 32, no. 5, 1996.

[12] K. J. Bins, P. J. Lawrenson and C. W. Trowbridge, *The Analytical and Numerical Solution of Electric and Magnetic Fields*, John Wiley & Sons: Chichester, 1992.

[13] Basim Istfan, *Extensions to the Finite Element Method for Nonlinear Magnetic Field Problems*, PhD Thesis Rensselaer Polytechnic Institute, New York, 1987.

[14] J. W. Kolar, H. Ertl and F.C. Zach, *Efficiency Optimal Control for AC Drives with PWM Inverters*, IEEE Transactions on Industry Applications, Vol. IA-21, No.4, 1985.

[15] F.C. Zach and H. Ertl, *Influence of the Modulation Method on the Conduction and Switching Losses of a PWM Converter System*, IEEE Transactions on Industry Applications, Vol. 27, No.6, 1991.

2006 5th International Power Electronics and Motion Control Conference

A Method of Stator Voltage Error Compensation in MRAS Sensorless Vector Control of Induction Motor

Wen Xuhui*, Chen Guilan*, Han Li**

* Institute of Electrical Engineering Chinese Academy of Science, Beijing, P. R. China 100080
** Graduate School of Chinese Academy of Science, Beijing, P. R. China 100080

Abstract—**In sensorless vector control of induction motor using voltage-fed inverter, the flux estimation error caused by stator voltage error due to the dead time and switch voltage drop and so on can deteriorate the performance of vector control .This paper proposes a method to compensate the stator voltage error by utilizing the error value between the measured current and estimated current. The dc component and fundamental component of reconstructed stator voltage errors have been compensated.**

Keywords-sensorless; vector control; stator voltage error; MRAS; flux estimation

I. INTRODUCTION

The vector control technique has been widely used in the induction motor control at present. The rotor-flux-oriented control is usually employed since it can realize the decoupling between the flux component and torque component. To obtain accurate field orientation in sensorless vector control, the key technology is flux estimation[1].

With the fast developing of DSP technology, Model Reference Adaptive system (MRAS) approach based on state observer is receiving more attention due to its highly robustness[2]. However, in practice the dead time and the switch voltage drop and so on can cause both fundamental magnitude-phase error and dc component error of reconstructed stator voltage in pulse-wide-modulation inverter system. All these can cause flux estimation error.

For current-fed inverter, the stator voltage can be compensated directly because the voltage waveform is sinusoidal. Kubota had proposed some methods of voltage error compensation [3][4][5], but those methods have the problem of gain coefficient and calculation complexity.

For voltage-fed inverter, [6] proposed method to estimate the stator voltage error while the flux is estimated at the same time. The voltage error item is added to full-order state observer to estimate and compensate the dc error component and ac magnitude error of stator voltage. The algorithm, which depends on the selection of matrix L, is difficult to be realized because the matrix L is hard to calculate.

This paper presents the analysis of influence on full-order state observer flux estimation from stator voltage error when the motor speed is known. A method is proposed to compensate the dc component and fundamental component of stator voltage reconstructed error by utilizing the error value between measured current and estimated current.

Both simulation and experimental results have verified the proposed method of stator voltage error compensation.

II. MRAS FLUX ESTIMATOR MODEL

For the induction motor it is convenient to consider the stator current and the rotor flux as the state vector components. The induction motor can be described by the following state equations in the stationary reference frame:

$$p \begin{bmatrix} i_1 \\ \psi_2 \end{bmatrix} = \begin{bmatrix} \mathbf{A}_{11} & \mathbf{A}_{12} \\ \mathbf{A}_{21} & \mathbf{A}_{22} \end{bmatrix} \begin{bmatrix} i_1 \\ \psi_2 \end{bmatrix} + \begin{bmatrix} \mathbf{B}_1 \\ 0 \end{bmatrix} u_1 \quad (1)$$

The full-order state observer which estimates the stator current and the rotor flux together in stationary reference frame is written by the following equation:

$$p \begin{bmatrix} \hat{i}_1 \\ \hat{\psi}_2 \end{bmatrix} = \begin{bmatrix} \hat{\mathbf{A}}_{11} & \hat{\mathbf{A}}_{12} \\ \hat{\mathbf{A}}_{21} & \hat{\mathbf{A}}_{22} \end{bmatrix} \begin{bmatrix} \hat{i}_1 \\ \hat{\psi}_2 \end{bmatrix} + \begin{bmatrix} \mathbf{B}_1 \\ 0 \end{bmatrix} u_1^* + \mathbf{G}(\hat{i}_1 - i_1)$$

$$(2)$$

where

$$\mathbf{B}_1 = b_1 \mathbf{I} = \mathbf{I} / (\sigma L_1), \quad u_1^* = u_1 - \Delta u_1$$

$$\mathbf{A}_{11} = a_{r11} \mathbf{I}, \qquad \mathbf{A}_{12} = a_{r12} \mathbf{I} + a_{i12} \mathbf{J},$$

$$\mathbf{A}_{21} = a_{r21} \mathbf{I}, \quad \mathbf{A}_{22} = a_{r22} \mathbf{I} + a_{i22} \mathbf{J}$$

$$a_{r11} = -\left[1/(\sigma \tau_1) + (1-\sigma)/(\sigma \tau_2) \right] < 0$$

$$a_{r12} = -1/\rho (1/\tau_2) > 0$$

$$a_{i12} = 1/\rho \omega_r < 0$$

$$a_{r21} = L_m / \tau_2 > 0$$

$$a_{r22} = -1/\tau_2 = \rho a_{r12} < 0$$

$$a_{i22} = \omega_r = \rho a_{i12} > 0$$

$$\tau_1 = L_1 / R_1, \tau_2 = L_2 / R_2,$$

1-4244-0448-7/06/$25.00 ©2006 IEEE

$$\sigma = 1 - L_m{}^2 /(L_1 L_2) , \quad \rho = -(\sigma L_1 L_2)/L_m$$

$$\mathbf{I} = \begin{bmatrix} 1 & 0 \\ 0 & 1 \end{bmatrix}, \quad \mathbf{J} = \begin{bmatrix} 0 & -1 \\ 1 & 0 \end{bmatrix},$$

$$\mathbf{G} = \begin{bmatrix} g_1 \mathbf{I} + g_2 \mathbf{J} & 0 \\ g_3 \mathbf{I} + g_4 \mathbf{J} & 0 \end{bmatrix} = \begin{bmatrix} \mathbf{G}_1 & \mathbf{0} \\ \mathbf{G}_2 & \mathbf{0} \end{bmatrix}$$

$$g_1 = (k-1)(a_{r11} + a_{r22})$$

$$g_2 = (k-1)a_{i22}$$

$$g_3 = (k^2 - 1)(a_{r21} - \rho a_{r11}) + \rho g_1$$

$$g_4 = \rho g_2$$

R_1、R_2 : Stator and rotor resistance.

L_1、L_2 : Stator and rotor self-inductance.

L_m : Mutual inductance.

ω_r : Motor angular velocity.

i_1、u_1 : Stator current and voltage vector

ψ_2 : Rotor flux vector

p : Ddifferential operator

Δu_1 : Stator voltage error

α、β : Stationary reference frame

In the full-order adaptive state observer based on the Model Reference Adaptive System (MRAS) approach , the reference model is the motor itself. The adaptive model is the observer. The motor speed is estimated by the following adaptive scheme. This scheme is deduced by the Popov's hyperstability theorem[7][8].

$$\hat{\omega}_r = (k_p + k_I/p)|e_i \times \hat{\psi}_2| = (k_p + k_I/p)(e_{i\alpha}\hat{\psi}_{2\beta} - e_{i\beta}\hat{\psi}_{2\alpha}) \quad (3)$$

III. INFLUENCE OF VOLTAGE ERROR ON FLUX ESTIMATION

When the system is linear and the parameters and motor speed are clearly known, the influence of stator voltage error on full-order state observer can be analyzed in stationary reference frame. When stator voltage error exists, the error equation is derived by subtraction of (2) from (1).

$$\begin{bmatrix} \dot{e}_i \\ \dot{e}_\psi \end{bmatrix} = \begin{bmatrix} \mathbf{A}_{11} + \mathbf{G}_1 & \mathbf{A}_{12} \\ \mathbf{A}_{21} + \mathbf{G}_2 & \mathbf{A}_{22} \end{bmatrix} \begin{bmatrix} e_i \\ e_\psi \end{bmatrix} + \begin{bmatrix} \mathbf{B} \\ 0 \end{bmatrix} \Delta u_1 \quad (4)$$

The stator voltage error, which is the input of observer, can be separated into dc component and harmonic components. The rotor flux estimation error and stator current estimation error due to the stator voltage error can also be separated into dc component and harmonic components.

The dc component of stator voltage error and flux estimation error and current estimation error can be expressed as the following :

$$\Delta u_{1-D} = \mathbf{c}_0 \quad (5)$$

$$e_{\psi-D} = a_0 \quad (6)$$

$$e_{i-D} = b_0 \quad (7)$$

The relationship among the dc component of voltage error, flux estimation error and current estimation error can be derived as the (8) from (4).

$$\begin{bmatrix} 0 \\ 0 \end{bmatrix} = \begin{bmatrix} \mathbf{A}_{11} + \mathbf{G}_1 & \mathbf{A}_{12} \\ \mathbf{A}_{21} + \mathbf{G}_2 & \mathbf{A}_{22} \end{bmatrix} \begin{bmatrix} e_{i-D} \\ e_{\psi-D} \end{bmatrix} + \begin{bmatrix} \mathbf{B} \\ 0 \end{bmatrix} \Delta u_{1-D} \quad (8)$$

The fundamental and harmonic components of voltage error, flux estimation error and current estimation error can be expressed as (9)-(11). Where $n = 1, 2, 3 \cdots\cdots$.

$$\Delta u_{1-A(n)} = \mathbf{c}_n e^{Jn\omega_1 t + \theta_n} \quad (9)$$

$$e_{\psi-A(n)} = a_n e^{Jn\omega_1 t + \alpha_n} \quad (10)$$

$$e_{i-A(n)} = b_n e^{Jn\omega_1 t + \beta_n} \quad (11)$$

According to (4), the relationship among the ac component of voltage error, flux estimation error and current estimation error is derived as (12). Where $n = 1, 2, 3 \cdots\cdots$

$$\begin{bmatrix} \mathbf{J} n\omega_1 e_{i-A(n)} \\ \mathbf{J} n\omega_1 e_{\psi-A(n)} \end{bmatrix} = \begin{bmatrix} \mathbf{A}_{11} + \mathbf{G}_1 & \mathbf{A}_{12} \\ \mathbf{A}_{21} + \mathbf{G}_2 & \mathbf{A}_{22} \end{bmatrix} \begin{bmatrix} e_{i-A(n)} \\ e_{\psi-A(n)} \end{bmatrix} + \begin{bmatrix} \mathbf{B} \\ 0 \end{bmatrix} \Delta u_{1-A(n)}$$

$$(12)$$

The relationship among the components of voltage error, flux estimation error and current estimation error is obtained by (12) and (8). Where $n = 0, 1, 2, 3 \cdots\cdots$. When n=0, all the errors are dc components.

$$\begin{bmatrix} 0 \\ 0 \end{bmatrix} = \begin{bmatrix} \mathbf{A}_{11} + \mathbf{G}_1 - \mathbf{J} n\omega_1 & \mathbf{A}_{12} \\ \mathbf{A}_{21} + \mathbf{G}_2 & \mathbf{A}_{22} - \mathbf{J} n\omega_1 \end{bmatrix} \begin{bmatrix} e_{i-A(n)} \\ e_{\psi-A(n)} \end{bmatrix} + \begin{bmatrix} \mathbf{B} \\ 0 \end{bmatrix} \Delta u_{1-A(n)}$$

$$(13)$$

The relationship between current error and flux estimation error can be derived as (14) from the second row of (13). The relationship between voltage error and flux estimation error can be obtained as (15) by substituting (14) into the first row of (13).

$$e_{i-A(n)} = -(\mathbf{A}_{21} + \mathbf{G}_2)^{-1}(\mathbf{A}_{22} - \mathbf{J} n\omega_1)e_{\psi-A(n)}$$

$$(14)$$

$$e_{\psi-A(n)} = [-\mathbf{A}_{12} + (\mathbf{A}_{11} + \mathbf{G}_1 - \mathbf{J} n\omega_1)(\mathbf{A}_{21} + \mathbf{G}_2)^{-1} (\mathbf{A}_{22} - \mathbf{J} n\omega_1)]^{-1} \mathbf{B} \Delta u_{1-A(n)}$$

$$= \mathbf{C}_v \Delta u_{1-A(n)}$$

$$(15)$$

Where $\mathbf{C}_v$ is a 2 by 2 matrix and defined as flux-voltage error coefficient due to voltage error. The matrix can be expressed as $\mathbf{C}_v = \begin{bmatrix} C_v(1) & C_v(3) \\ C_v(2) & C_v(4) \end{bmatrix}$ .After calculation, the components of the matrix have the following constrains.

$$C_v(1) = C_v(4), \quad C_v(2) = -C_v(3).$$

Flux-voltage dc error coefficient varies with the motor speed is shown in fig.1 by matrix calculation without pole placement （K=1）.

The flux-voltage ac error coefficient varies with motor speed is shown in fig.2. When motor speed $\omega_r = 0.02$, the trend of flux-voltage ac error coefficient vs. harmonic number n is shown in fig.3. It can be known that the influence of voltage error on flux estimation error is mainly concentrated in low speed region and decreases quickly with the increasing of harmonic number n.

The parameters of the induction motor used in the analysis are listed in table 1. All the variables and parameters of motor are expressed in per unit system.

IV. THE COMPENSATION OF STATOR VOLTAGE FUNDAMENTAL COMPONENT MAGNITUDE ERROR AND DC COMPONENT ERROR

In practice, the current estimation error can be obtained and used to obtain the information of voltage error. When n=0，the stator voltage error dc component estimation is derived as (16) by （13）.

$$\Delta \boldsymbol{u}_{1-D}^* = \frac{1}{b_1} \left\{ -\mathbf{A}_{11} + \mathbf{A}_{12}\mathbf{A}_{22}^{-1}\mathbf{A}_{21} \right\} \boldsymbol{e}_{i-D} \quad (16)$$

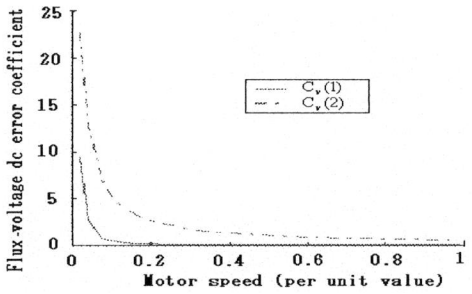

Figure 1. Dc error coefficient vs. motor speed

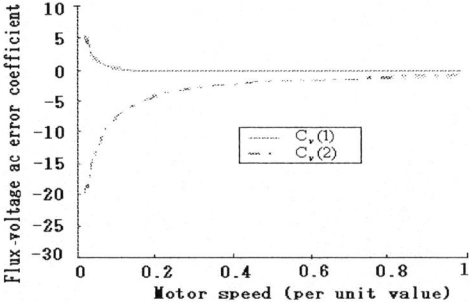

Figure 2. Ac error coefficient vs. motor speed

At the same time, under the assumption of equivalence between stator leakage inductance and rotor leakage inductance, the （16）can be simplified as (17).

$$\Delta \boldsymbol{u}_{1-D}^* = (R_1 + R_2 - \frac{\sigma L_1}{L_2}R_2)\boldsymbol{e}_{i-D} - \frac{L_m^2}{L_2^2}R_2\,\boldsymbol{e}_{i-D} \quad （17）$$

$$= R_1\,\boldsymbol{e}_{i-D}$$

The ac component of stator voltage error can be estimated as (18) by (13).

$$\Delta \boldsymbol{u}_{1-A}^* = \frac{1}{b_1}\{-(\mathbf{A}_{11} - \mathbf{J}\,n\omega_1)$$

$$+ \mathbf{A}_{12}(\mathbf{A}_{22} - \mathbf{J}\,n\omega_1)^{-1}\mathbf{A}_{21}\}\boldsymbol{e}_{i-A}$$

$$= \boldsymbol{C}_{v/i}\boldsymbol{e}_{i-A} \quad （18）$$

where $\boldsymbol{C}_{v/i} = \begin{bmatrix} C_{v/i}(1) & C_{v/i}(3) \\ C_{v/i}(2) & C_{v/i}(4) \end{bmatrix}$,

$$C_{v/i}(1) = C_{v/i}(4), \quad C_{v/i}(2) = -C_{v/i}(3).$$

According to (18), the relationship between voltage error and estimated current error is defined as $\boldsymbol{C}_{v/i}$. The magnitude of $\boldsymbol{C}_{v/i}$ is

$$F_{v/i} = \sqrt{2(C_{v/i}(1)^2 + C_{v/i}(2)^2)}.$$

When n=1, the magnitude of fundamental component error is derived as (19)

$$\left\| \Delta \boldsymbol{u}_{1-A(1)}^* \right\| = \left\| \boldsymbol{C}_{v/i}\boldsymbol{e}_{i-A(1)} \right\| \quad （19）$$

The fundamental current error is expressed in (20).

$$\boldsymbol{e}_{i-A(1)} = \left\| \boldsymbol{e}_{i-A(1)} \right\| \begin{bmatrix} \cos\theta_{ei} \\ \sin\theta_{ei} \end{bmatrix} \quad （20）$$

Substituting (20) into (19), （21）is obtained.

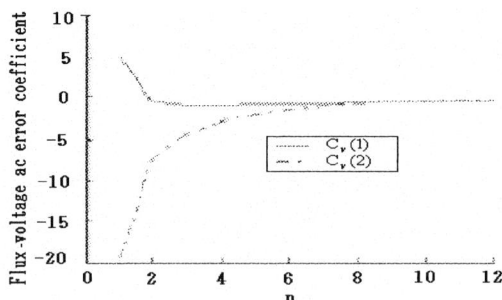

Figure 3. Ac error coefficient vs. harmonic number n

Table 1

Motor parameters	value	unit
Rated Output Power	20	kw
Rated Voltage	180	v
Rated Frequency	120	Hz
Stator Resistance	0.0365	Ohm
Rotor Resistance	0.0202	Ohm
Stator Leakage Inductance	0.155	mH
Rotor Leakage Inductance:	0.155	mH
Magnetizing Inductance	3.1	mH
Poles	2	

$$\left\|\Delta\boldsymbol{u}^{*}_{1-A(1)}\right\|=\left\|\boldsymbol{e}_{i-A(1)}\right\|\cdot$$

$$\left\|\begin{bmatrix}C_{v/i-A(1)}(1)cos\theta_{ei}-C_{v/i-A(1)}(2)sin\theta_{ei}\\C_{v/i-A(1)}(2)cos\theta_{ei}+C_{v/i-A(1)}(1)sin\theta_{ei}\end{bmatrix}\right\|$$

$$=\left\|\boldsymbol{e}_{i-A(1)}\right\|\sqrt{C_{v/i-A(1)}(1)^{2}+C_{v/i-A(1)}(2)^{2}}$$

$$=\frac{1}{\sqrt{2}}F_{v/i(1)}\left\|\boldsymbol{e}_{i-A(1)}\right\|$$

(21)

In the condition of rated load, the variation of error coefficient magnitude $F_{v/i(1)}$ with motor speed is shown in fig.4 by matrix calculation. With the increasing of motor speed, the $F_{v/i(1)}$ increases at a certain proportional rate, i.e. the fundamental component magnitude error of stator voltage is proportional to fundamental component magnitude error of estimated current. In addition, the proportional coefficient is proportional to motor speed.

Thus, the relationship between the fundamental component magnitude error of stator voltage and the fundamental component magnitude error of estimated current can be expressed as (22)

$$\left\|\Delta\boldsymbol{u}^{*}_{1-A(1)}\right\|=k_{s}\omega_{r}\left\|\boldsymbol{e}_{i-A(1)}\right\|\qquad(22)$$

In practical control, $\boldsymbol{e}_i$ contains both dc component and ac components. The dc component $\boldsymbol{e}_{i-D}$ can be obtained by filtering of $\boldsymbol{e}_i$, i.e.

$$\boldsymbol{e}_{i-D}=\frac{1}{\tau s+1}\boldsymbol{e}_{i}\qquad(23)$$

The dc component of stator voltage error $\Delta\boldsymbol{u}_{1-D}{}^{*}$ can be obtained by (17).

$$\Delta\boldsymbol{u}_{1-D}{}^{*}=\frac{-R_{1}}{\tau s+1}\boldsymbol{e}_{i}\qquad(24)$$

The harmonic components can be ignored, the fundamental component of current estimation error is shown in (25).

$$\boldsymbol{e}_{i-A(1)}=\boldsymbol{e}_{i}-\frac{1}{\tau s+1}\boldsymbol{e}_{i}\qquad(25)$$

The stator voltage phase can be obtained by the current value of stator voltage in（26）.

$$\theta_{u}=arcth(\frac{u_{1\beta}}{u_{1\alpha}})\qquad(26)$$

According to (22), the stator voltage magnitude error compensation is obtained in the following.

$$\begin{cases}\Delta u^{*}_{1\alpha}=k_{s}\omega_{r}\left\|\Delta\boldsymbol{u}^{*}_{1-A(1)}\right\|cos\theta_{u}\\\Delta u^{*}_{1\beta}=k_{s}\omega_{r}\left\|\Delta\boldsymbol{u}^{*}_{1-A(1)}\right\|sin\theta_{u}\end{cases}\qquad(27)$$

where k_{s} is a constant .

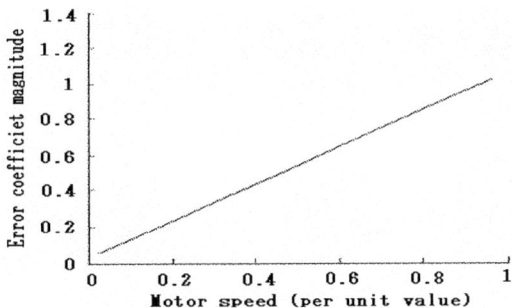

Figure 4.　Error coefficient magnitude vs. motor speed

The relationship among stator voltage real value, stator voltage reference value and magnitude error compensation is shown in (28).

$$\begin{cases}u_{1\alpha}=u^{*}_{1\alpha}+\Delta u^{*}_{1\alpha}\\u_{1\beta}=u^{*}_{1\beta}+\Delta u^{*}_{1\beta}\end{cases}\qquad(28)$$

The system with stator voltage dc component error compensation and stator voltage fundamental component magnitude error compensation is shown in fig.5.

V.　SIMULATION RESULTS

A.　Simulation analysis without stator voltage error compensation

When the dead time is 3e-6s, the voltage drop of power switch Vce is 0.8V and the load is 40N.m，the sensorless vector control without stator voltage fundamental component magnitude error compensation is simulated and the results are shown in the following.

1）The stator voltage reference is aberrant from the real value with the current loop control because of the influence of dead time and voltage drop of switch. Error exists between the fundamental component magnitude of reference valued and real value of stator voltage as shown in fig.6.

2）There are serious magnitude and phase error between the estimated flux value and the real flux value.

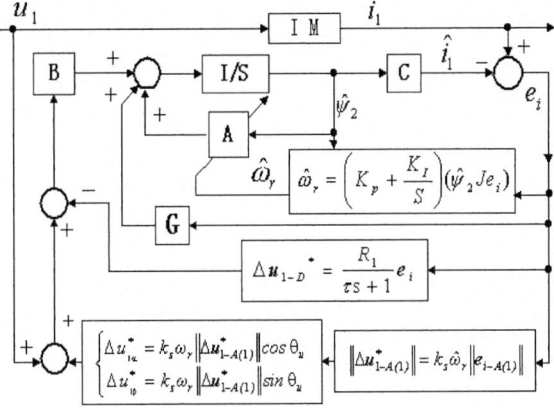

Figure 5.　Block diagram of system with stator voltage error compensation

1009

3）The motor cannot be started up because of the inaccuracy of the field orientation. The output torque of motor decreases so quickly that the error between estimated value and real value of motor speed increases quickly.

B. Simulation analysis with stator voltage error compensation

According to（28）, the sensorless vector control with stator voltage fundamental component magnitude error compensation is simulated and the results are shown in the following.

1）The fundamental component of observer stator voltage input coincides with fundamental component of motor stator voltage real input value after compensation of fundamental component magnitude error compensation as shown in fig.7.

2）After the compensation of fundamental component magnitude error, the magnitude error of flux is removed . The accurate electromagnetic torque is obtained and the motor can be started up. The compensation value of stator voltage magnitude error is shown in fig.8. This makes the estimated motor speed to tend toward real value as shown in fig.9.

In the same simulation system, 0.1v dc component is added to stator voltage in stationary reference frame. Both the dc component error and fundamental component magnitude error of stator voltage are compensated. The result of stator voltage dc component error estimation is shown in fig.10.

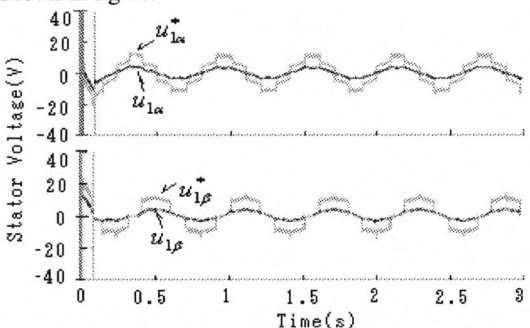

Figure 6.   Voltage reference value and real value

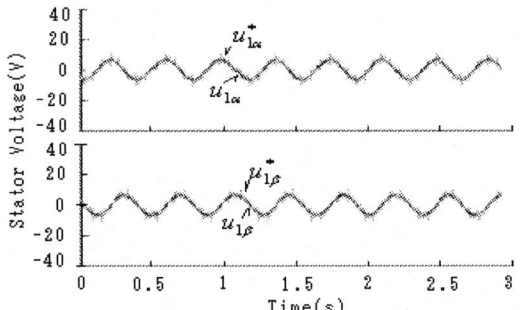

Figure 7.   The voltage reference value and real value after compensation

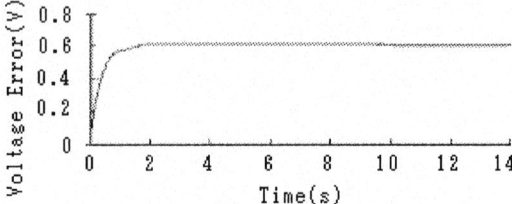

Figure 8.   Compensation value of stator voltage fundamental component magnitude error

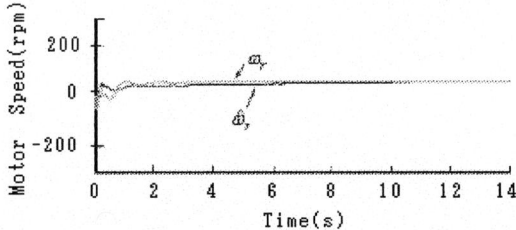

Figure 9.   Estimated value and real value of motor speed

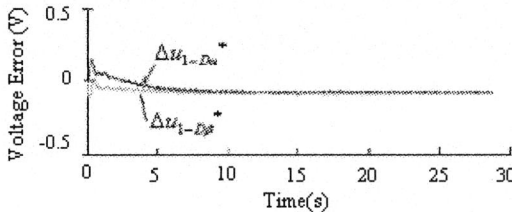

Figure 10.   Dc component error of stator voltage

It is verified from simulation that the estimated values of flux and motor speed approach the real values after stator voltage magnitude error compensation. This improves the accuracy of field orientation and favors the start-up performance in low speed region with heavy load. Thus the system can run steadily in low speed region.

VI.   EXPERIMENTAL RESULTS AND ANALYSIS

In the experimental system, a DC generator and an adjustable resistance act as system load. The induction motor is used in the experimental tests, whose parameters are reported in the Table2. The induction motor is magnetized by constant current with the rated value 8.6A. The dead time in program is 3e-6s. According to (28). k_s can be adjusted properly to realize the stabilized speed loop control.

The speed loop control result is obtained in fig.11.when $\omega_r = 18$ rpm. The compensation value of magnitude error is shown in fig. 12.

When k_s is too small, the start-up procedure will oscillate or even fail. The estimated and real value of motor speed in this case is shown in fig.13.

When k_s is too large, the estimated value of motor speed is much less than the real value. This causes the deteriorated performance as shown in fig.14.

The experimental results show that steady speed loop control in wide speed region can be realized by the compensation of stator voltage fundamental component magnitude error.

By adjusting the magnitude error coefficient k_s properly, the start-up performance of motor control system can be improved. No dc component of voltage error exists in the experiment.

VII. CONCLUSIONS

For voltage-fed inverter, when the reference value of stator voltage is the input of full-order state observer, the dead time, switch voltage drop and so on can cause stator voltage error which causes the error of flux estimation. Thus the system cannot start up and the control performance is deteriorated. This paper presents the sensorless vector control of induction motor in experimental platform based on TMS320F2407A DSP. The experimental results have verified that the proposed method can effectively improve the system start-up performance and wide speed range steady performance.

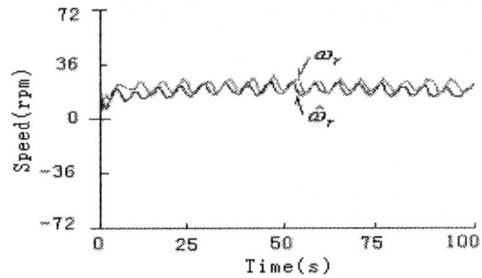

Figure 11. Estimated speed and real speed with proper k_s

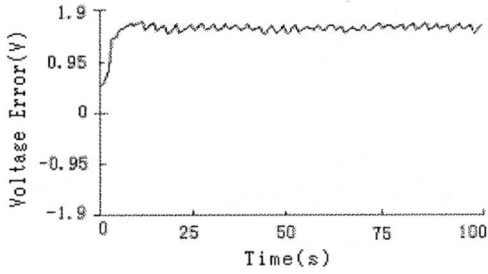

Figure 12. The compensation value of stator voltage fundamental component magnitude error

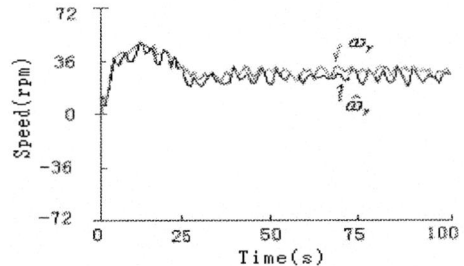

Figure 13. Estimated speed and real speed with very small k_s

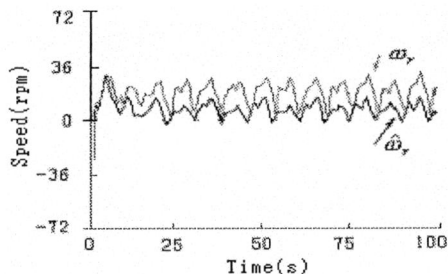

Figure 14. Estimated speed and real speed with very large k_s

Table 2

Motor parameters	value	unit
Rated Output Power	1.8	kw
Rated Voltage	200	v
Rated Frequency	80	Hz
Stator Resistance	0.2221	Ohm
Rotor Resistance	0.2092	Ohm
Stator Leakage Inductance	1.4962	mH
Rotor Leakage Inductance	1.7829	mH
Magnetizing Inductance	28.96	mH
Poles	4	

REFERENCES

[1] Li Yongdong, Li Mingcai. Speed Sensorless Control of Induction Motor with High Performance.Electric Transmission .2004,(1): 4-10.

[2] C.Ilas, A.Bettini,L.Ferraris,G.Griva,F.Profumo. Comparison of different Schemes without Shaft Sensors for Field Oriented Control Drives .Proc.of IEEE IECON,1994 1579-1588.

[3] Hisao Kubota, Yukio Kataoka, Hisayoshi Ohta, et al. Sensorless vector controlled induction machine drives with fast stator voltage error compensataion.Pro.of IEEE IAS Ann.Mtg.,1999,(4)2321-2324

[4] Kubota, H., Sato, I., Tamura, Y., etal. Regenerating-Mode Low-Speed Operation of Sensorless Induction Motor Drive With Adaptive Observer .IEEE Tran.Ind. Applicat.2002 38(4) 1081-1086.

[5] Kubota,H.,Matsuse,K..The improvement of performance at low speed by error compensation of stator voltage in sensorless vector controlled induction machines. Thirty-First IAS Annual Meeting,IAS '96,Conference Record of the 1996 IEEE (1):257-261

[6] Kozo Ide, Koji Hazama, Teruo Tsuji,Ryuichi Oguro.Speed sensorless Field-Oriented Controlled IM with Stator Voltage Error Compensator. T.IEE Japan. 1996 116-D ,No.8：835-843.

[7] Geng Yang, Tung-Hai Chin.Hyperstability of the Full Order Adaptive Observer for Vector Controlled-Induction Motor Drive without Speed-Sensor，T.IEE Japan. 1992，112-D,No.11：1047-1055.

[8] Hisao Kubota, Kouki Matasuse. Adaptive Flux Observer of Induction Motor and its stability. T.IEE Japan. 1991，111-D,No.3：188-194..

2006 5th International Power Electronics and Motion Control Conference

Systematic Design of Fuzzy Logic Based Hybrid On-Line Minimum Input Power Search Control Strategy for Efficiency Optimization of IM

Zhang Liwei*, Liu Jun**, Wen Xuhui**, *Member, IEEE*, and Trillion Q. Zheng*

* School of Electrical Engineering, Beijing Jiaotong University, Beijing, China
** Institute of Electrical Engineering, Chinese Academy of Sciences, Beijing, China

Abstract—Operated under rated conditions, the induction motor (IM) is highly efficient, but the efficiency is greatly reduced with the normal conditions changed. So efficiency optimization has significance for economic saving and environment protecting. This paper summarizes the main efficiency optimization control strategies appearing recently years, and presents a new fuzzy logic based hybrid on-lion minimum input power search control strategy. Unlike other fuzzy methods that need simulation calculation to get the coefficients of scaling factors, the gains derivative method of this strategy utilizes the research results of loss model control to do on-line scaling factor calculation, which can mark out the optimization course in advance. Meanwhile, the systematic design for new fuzzy sets and Membership Functions (MFs) can make control system converge fast, with avoiding the oscillation around the optimal flux. The simulation and experiment results confirm the validity and usefulness of the proposed techniques.

Keywords- induction motor; efficiency optimization control; fuzzy logic

I. NOMENCALTURE

R_s	Equivalent stator copper loss resistance.
R_r	Equivalent rotor copper loss resistance.
R_m	Equivalent motor iron loss resistance.
L_m	Magnetizing inductance.
i_{ds}^*, i_{qs}^*	d- and q-axes stator reference currents.
i_{dse}^*, i_{qse}^*	d- and q-axes stator rated reference currents.
i_{dso}^*, i_{qso}^*	d- and q-axes stator optimized reference currents of LMC.
i_{ds}, i_{qs}	d- and q-axes stator currents.
ω_r	Electrical angular speed of rotor.
T_e	The developed torque of IM.
p	Pole pairs.

II. INTRODUCTION

It is estimated that electric machines consume more than 50% of the world electric energy generated. So it is important to improve the efficiency of electric drives, mainly for two reasons: economic saving and environment protecting. The main losses of IM are usually split into 5 components: stator copper losses, rotor copper losses, iron losses, mechanical and stray losses, but cop-

per loss and iron losses constitute the major percentage of IM losses and are controllable. As a rule, IM is highly efficient operated under the rated conditions [1]. But at light loads, the iron losses increase dramatically, with the efficiency reducing considerably [2]. Loss minimization is possible by an optimal balance of the core and copper losses in machine [3].

In vector control system, the efficiency optimization can easily be realized by controlling the reference flux current. Basically, there exist two different approaches to improve the efficiency [4]:

A. Loss Model Control (LMC)

Based on the IM loss model [5-9] the optimal flux is computed analytically, and the convergence speed is fast. Without extra hardware, LMC can be conveniently realized. However, it must need an accurate knowledge of motor parameters, which change considerably with temperature, saturation, skin effect, etc.

B. On-Lion Power Measure Search Control (SC)

Based on minimum input power control, SC uses particular search algorithms to find the optimal flux [10-16]. This approach does not require the knowledge of motor parameters, and the optimal efficiency can be found ultimately. But the optimal flux search time is longer than that of LMC.

In these search methods, the fuzzy logic based search method (FLSC) is the more successful one [13]. But the need of simulation calculation to get the coefficients of the fuzzy logic scaling factors, no doubt, will limit the application of this method. Ref.[16] presents a new FLSC strategy, which have both good characteristics of LMC and SC. Unlike the simulation method [13], the new on-line gain calculation method, with simple and effective characteristics, makes full use of the research results of LMC [5], and can mark out the optimization course in advance. On the bases of its initial simulation studies [16], this paper makes further researches on this FLSC strategy.

This paper reveals the rather specific relationship between the increment of Δi_{sd}^* and the increment of ΔP_d during FLSC optimization course. Based on this relationship, the systematic design method for new fuzzy sets and MFs is summarized, which can make the control system

1-4244-0448-7/06/$25.00 ©2006 IEEE

converge fast with avoiding the oscillations around the optimal flux. Moreover, the robust studies show that this FLSC strategy is completely insensitive to rotor resistance (R_r). Even if R_r changes greatly for the same steady output state, the optimal efficiency still can be derived, keeping almost unchanged. Simulations and experiments verify that this strategy, congregating good characteristics of LMC and SC, is very fast and highly precise, and can be applied for any steady state of IMs.

III. A NEW FUZZY LOGIC SEARCH CONTROLLER

For simple analyses, the following simplifications have been made:

i)   The iron saturation can be neglected since FLSC is valid only at light loads.

ii)   At a specified motor speed, the iron loss resistance can be assumed constant and independent of the small variations of the slip frequency.

iii)   It only considers controllable losses (iron and copper losses).

iv)   The following researches only adapt to steady states of IMs.

Fig.1 is the proposed configuration of this proposed FLSC for IM drives, and Fig.2 is the core of this method. One input of FLSC is the dc power increment $\Delta P_{d(k)}$, which is the power increment between this sampled value and the previous one of dc power. The other input is the last flux current increment $\Delta i_{ds}^*{}_{(k-1)}$. On these bases [17], the new flux current increment $\Delta i_{ds}^*{}_{(k)}$ is generated.

For IM steady states, the currents i_{ds} and i_{qs} can be assumed to be equal to the reference currents i_{ds}^* and i_{qs}^*. So in the following sections, i_{ds}^* and i_{qs}^* are chosen as the control variables. In the following simulation studies, the FLSC is activated every 0.3s only after dc power settles down corresponding to the change of i_{ds}^*.

A New Derivative Method of Scaling Factor

Fig.3 is the equivalent circuit of the induction motor. With the help of the research results of LMC [5], the following (1)-(4) expressions can be derived easily.

In the vector control state, the total loss of motor is [5]

$$P_l = R_q i_{qs}^*{}^2 + R_d i_{ds}^*{}^2 \qquad (1)$$

where

$$R_q = R_s + \frac{R_m R_r}{R_m + R_r} \text{ and } R_d = R_s + \frac{L_m^2 \omega_r^2}{R_m + R_r} \qquad (2)$$

So, for the normal rated flux control, the total loss can be expressed as

$$P_{l\_nor} = R_q i_{qse}^*{}^2 + R_d i_{dse}^*{}^2 \qquad (3)$$

Furthermore, in efficiency optimization control state, the relationship between i_{dso}^* and i_{qso}^* is [5]

$$i_{dso}^* = K_{min} \left| i_{qso}^* \right| \ (K_{min} = \sqrt{R_q / R_d}) \qquad (4)$$

For the same steady output state of IM, there is

$$T_e = p L_m i_{dse}^* i_{qse}^* = p L_m i_{dso}^* i_{qso}^* \qquad (5)$$

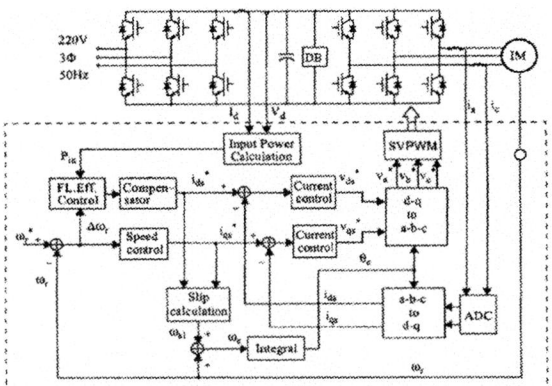

Fig.1. Proposed configuration of FLSC for IM drives

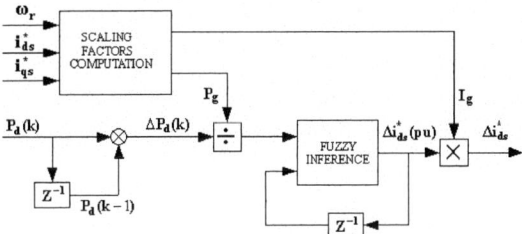

Fig.2. Efficiency optimization control block diagram.

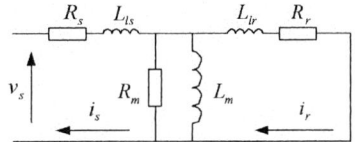

Fig.3. The equivalent circuit of IM drives

So i_{dso}^* can be expressed as the function of i_{dse}^* and i_{qse}^*

$$i_{dso}^* = \sqrt{K_{min} i_{dse}^* i_{qse}^*} \qquad (6)$$

From (1) to (6), the total loss of efficiency optimization control state can be expressed as

$$P_{l\_opt} = 2 R_d i_{dso}^*{}^2 \qquad (7)$$

Based on above analyses, the input gain P_g and output gain I_g of FLSC for each steady state can be made as

$$P_g = P_{l\_nor} - P_{l\_opt} = R_q i_{qse}^*{}^2 + R_d i_{dse}^*{}^2 - 2 R_d i_{dso}^*{}^2 \qquad (8)$$

$$I_g = i_{dse}^* - i_{dso}^* \qquad (9)$$

B. Systematic design of fuzzy sets and MFs

The fuzzy rule base for fuzzy control is described as a series of IF-THEN rules. There are 14 rules in this rule base and they are designed as in Table I, where $\Delta P_{d(k)}(pu)$ is defined as $\Delta P_{d(k)}/P_g$, and $\Delta i_{ds}^*{}_{(k-1)}(pu)$ is defined as $\Delta i_{ds}^*{}_{(k-1)}/I_g$, as well as $\Delta i_{ds}^*{}_{(k1)}(pu)$ defined as $\Delta i_{ds}^*{}_{(k)}/I_g$.

One example of the fuzzy IF-THEN rules is given as:

R1: "IF power increment $\Delta P_{d(k)}(pu)$ is Positive Big (PB) and the last flux current increment $\Delta i_{ds}^*{}_{(k-1)}(pu)$ is Negative (N), THEN the new flux current increment $\Delta i_{ds}^*{}_{(k)}(pu)$ is Positive Medium (PM)."

According to the abundant previous experiment data and other references statistic analyses[7], a rather specific relationship, as Table II shows, is existed between the increment of $\Delta i_{sd}^{*}{}_{(k-1)}(pu)$ and the increment of $\Delta P_{d(k)}(pu)$.

In Table II, the numbers of the first line represent the optimization control order, and the numbers of the second line represent the assumed $\Delta i_{ds}^{*}{}_{(k-1)}(pu)$ value of each optimization control, and the numbers of the third line represent the corresponding $\Delta P_{d(k)}(pu)$ value with the change of $\Delta i_{ds}^{*}{}_{(k-1)}(pu)$.

In order to realize the fast converge of FLSC in the practical control, the FLSC is designed to finish the efficiency optimization with 4-step change, and the value of $\Delta i_{ds}^{*}{}_{(k)}(pu)$ in each step is assumed to be -0.4, -0.3, -0.2 and -0.1. According as Table II shows, the corresponding $\Delta P_{d(k)}(pu)$ value should be <-0.4, ≈-0.3, >-0.2, >-0.1, and such value should belong to NB, NM, NS and ZE sets of $\Delta P_{d(k)}(pu)$ individually. Let $\mu_A(x)$ denote the degree of membership of a given element x in the universe of discourse X (denoted by $x \in X$). The following are the systematic design method of the fuzzy sets and MFs of $\Delta P_{d(k)}(pu)$ and $\Delta i_{ds}^{*}{}_{(k)}(pu)$:

i) The first change of $\Delta i_{ds}^{*}{}_{(k)}(pu)$ is set as -0.4, so the corresponding $\Delta P_{d(k)}(pu)$ change is between -0.4 and -0.6, belonging to NB and NM sets of $\Delta P_{d(k)}(pu)$. For convenience, NB set of $\Delta P_{d(k)}(pu)$ is defined by trapezoidal shape function (-1,-1,-0.6,-0.4), and NM set of $\Delta P_{d(k)}(pu)$ is defined by triangular shape function (-0.6,-0.4, X1), Here, X1 is an undetermined value needing to be calculated, as well as the following Y1, Y2, etc.. Supposing first change of $\Delta P_{d(k)}(pu)$ is -0.45, it is easy to calculate $\mu_{NB}(\Delta P_{d(k)}(pu)) = 0.25$ and $\mu_{NM}(\Delta P_{d(k)}(pu)) = 0.75$.

ii) For fuzzy controller inputs ($\Delta P_{d(k)}(pu) = -0.45$) and ($\Delta i_{ds}^{*}{}_{(k-1)}(pu) = -0.4$), according to the fuzzy rules of Table I, the output $\Delta i_{ds}^{*}{}_{(k)}(pu)$ should be -0.3, with $\mu_{NB}(\Delta i_{ds}^{*}{}_{(k)}(pu)) = 0.25$ and $\mu_{NM}(\Delta i_{ds}^{*}{}_{(k)}(pu)) = 0.75$. So it can be deduced that the center of $\Delta i_{ds}^{*}{}_{(k)}(pu)$ NB set is on the left of -0.3, and the center of $\Delta i_{ds}^{*}{}_{(k)}(pu)$ NM set is on the right of -0.3. Considering the first change of $\Delta i_{ds}^{*}{}_{(k)}(pu)$ is -0.4, so -0.4 is chosen as the NB center, and the NB set can be defined by trapezoidal shape function (-1,-1,-0.4, Y1). With the fuzzy implication methods [17], it is easy to get Y1= 0.267, here Y1= 0.25 is chosen for programming convenience ultimately. Now the NM set of $\Delta i_{ds}^{*}{}_{(k)}(pu)$ can be defined by triangular shape function (-0.4,-0.25, Y2). According to such calculation methods, X1= -0.2 and Y2= -0.1 can be gotten, etc.

iii) In order to avoid the oscillations around the optimal flux, the ZE set of $\Delta P_{d(k)}(pu)$ is defined by trapezoidal shape function. Considering practical application, the ZE set is defined by (-0.2, -0.08, 0.08, 0.2).

iv) The previous experiment data and other references statistic data show that the total motor loss decrease with the decrease of the rotor flux. But under the optimal rotor flux, the total motor loss will get rapider increment than its decrease for the rotor flux is above the optimal one. So in this FLSC strategy, the fuzzy set of $\Delta i_{ds}^{*}{}_{(k)}(pu)$ is desi-

gned to be an asymmetry shape, which can improve the system robust.

Based on the above analyses, the MFs of these fuzzy sets for the inputs and output are defined by triangular /trapezoidal shape functions, as fig. 4 shows. The control surface of this FLSC controller is plotted in fig.5.

TABLE I.
FUZZY RULES FOR POWER EFFICIENCY OPTIMIZATION

$\Delta P_{d(k)}(pu)\backslash\Delta i_{ds}^{*}{}_{(k-1)}(pu)$	N	P
PB	PM	NM
PM	PS	NS
PS	PS	NS
ZE	ZE	ZE
NS	NS	PS
NM	NM	PM
NB	NB	PB

TABLE II.
THE RELATIONSHIP BETWEEN THE INCREMENT OF $\Delta i_{SD}^{*}{}_{(k-1)}(PU)$ AND THE INCREMENT OF $\Delta P_{D(K)}(PU)$

Opt. Order	1	2	3	4
$\Delta i_{ds}^{*}{}_{(k-1)}(pu)$	-0.4	-0.3	-0.2	-0.1
$\Delta P_{d(k)}(pu)$	>-0.6 & <-0.4	≈-0.3	>-0.2	>-0.1

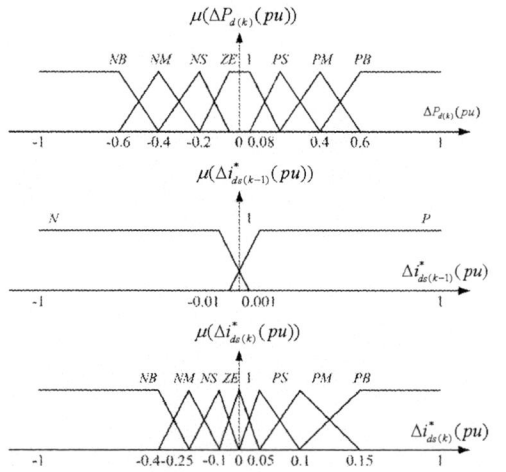

Fig.4. Membership functions for efficiency controller. (a) Change of dc link power ($\Delta P_{d(k)}(pu)$). (b) Last change in flux current ($\Delta i_{ds}^{*}{}_{(k-1)}(pu)$). (c) Flux current control increment ($\Delta i_{ds}^{*}{}_{(k)}(pu)$).

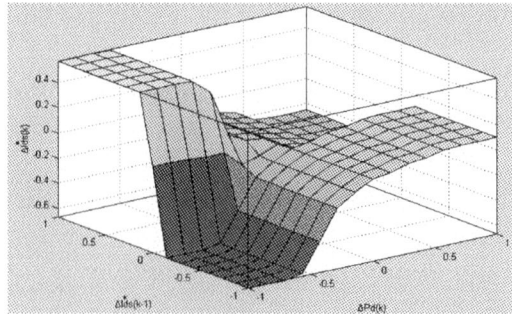

Fig.5. Crisp I/O map of FLSC strategy.

IV. SIMULATION RESULTS

The performance of this strategy has been studied through computer simulations. Fig.6 is the time responses when applying a constant load of 6N.m to the IM driven at 750r/min, and the output of IM is 471W. The parameters of IM are as Appendix shows.

Fig.6 (a) and (b) show that the flux current conference i_{ds}^* and input power P_d are remarkably decreased with the FLSC control, if compared to the conventional rated flux control. The i_{ds}^* changes from the rated 9A to 4.1A, as well as P_d from 910W to 694W. The efficiency of IM changes from 51.8% to 67.9%. Fig.6 (c) shows the i_{qs}^* change during the FLSC efficiency optimization control.

Fig.6 (d) and (e) illustrate that the same regulated torque and motor speed are assured with FLSC of IM while minimizing the total losses.

Furthermore, this paper makes robust simulation research. In the simulation system and in the above (1)-(9) expressions, keeping the same output (6N.m load and 750r/min motor speed), the rotor resistance R_r is made to reduce 50% or increase 50% from the normal value, and the time responses of i_{ds}^* and P_d can be derived. Fig.7 (a) and (b) are the results answering to 50% reduction of R_r, as well as Fig.8 answering to 50% increase of R_r.

Fig.7 shows that i_{ds}^* changes from the rated 9A to 2.7A, and P_d is from 1000W to 696W, with 67.8% optimal efficiency. Fig.8 shows that i_{ds}^* changes from the rated 9A to 5.8A, and P_d is from 820W to 690W, with 68.3%

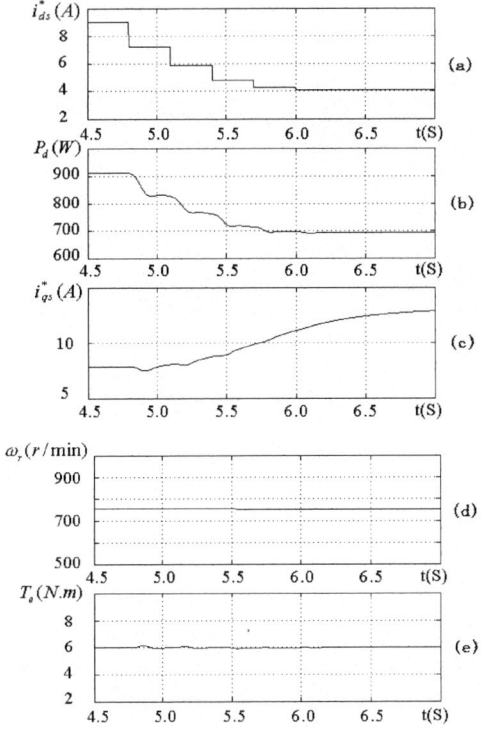

Fig.6. Time responses of FLSC simulation. (a) The changes of i_{ds}^*. (b) The response of P_d. (c) The changes of i_{qs}^*. (d) The developed torque T_e of IM in FLSC. (e) The motor speed ω_r.

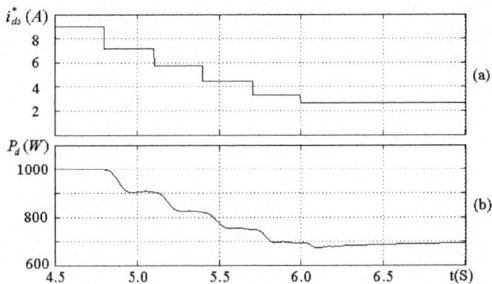

Fig.7. Time responses of FLSC simulation for 50% reduction of R_r. (a) The changes of i_{ds}^*. (b) The response of P_d.

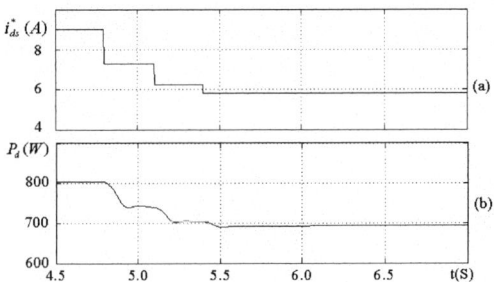

Fig.8. Time responses of FLSC simulation for 50% increase of R_r. (a) The changes of t i_{ds}^*. (b) The response of P_d.

optimal efficiency. Therefore, the changes of R_r only affect the optimal i_{ds}^* and the rated input P_d, but the optimal efficiency, which is about 68%, keeps unchanged.

V. EXPERIMENT RESULTS

Fig.9 is the experiment system with a digital signal processor (DSP) board, featuring a fixed-point TMS320-LF2407A DSP. An insulated gate bipolar transistor(IGBT) PWM inverter is chosen, and the switching frequency is 10 kHz. The load T_L is 1.4Nm, and the IM is driven at 750r/min. The parameters of IM are as appendix shows.

With the help of Controller Area Network (CAN), the results, as Fig.10 shows, are displayed on a PC. Fig.10 (a) and (b) show that the flux current conference i_{ds}^* and input power P_d are remarkably decreased with the FLSC if compared to the conventional rated flux control. i_{ds}^* changes from the rated 9A to 4A, as well as P_d from 400W to 250W. The efficiency of such IM changes from the 27.5% to 44.0%. During the efficiency optimization, the motor speed keeps 750r/min, as Fig.10 (c) shows.

Moreover, this FLSC strategy only uses three or fours steps to finish the optimization course. The convergence speed is very faster than that of other search control methods.

VI. CONCLUSIONS

This paper summarizes the main efficiency optimization control strategies appearing recently years, and presents a new FLSC strategy. Such strategy congregates the good characteristics of LMC and SC. The systematic design for new fuzzy sets and MFs can get over the oscillations around the optimal flux. Unlike other fuzzy

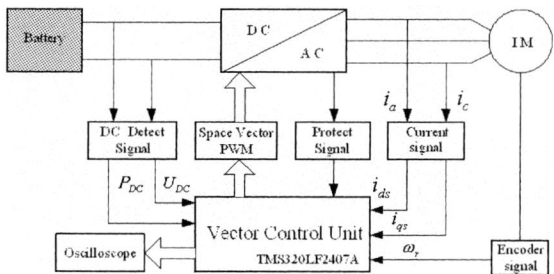

Fig.9. Experiment system

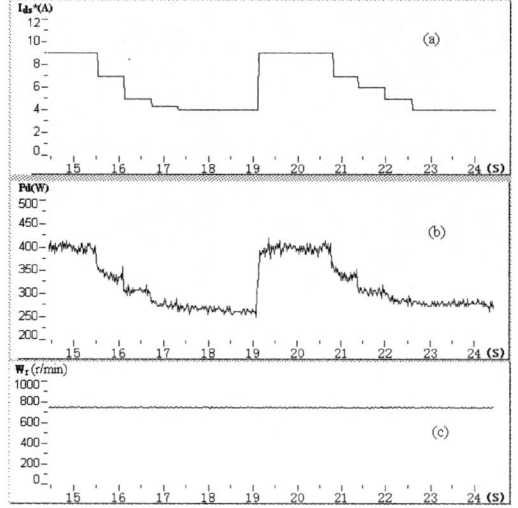

Fig.10. Time responses of FLSC experiment. (a) The changes of the reference flux current i_{ds}^*. (b) The response of the input power P_d. (c) The motor speed ω_r of IM in FLSC.

methods that need simulation calculation to get the coefficients of scaling factors, the gains derivative method of this strategy is simple and effective. The gains of FLSC for every steady state can be on-line derived automatically, and it can mark out the optimization course in advance. Simulation and experiment results of the proposed strategy on a 1.8 kW induction motor drive demonstrate the validity of the described methods. Such strategy should be widely used in efficiency optimization control for IMs.

VII. APPENDIX

Induction motor rating: 200V three-phase, 80Hz, 14Nm, 1.8kW, eight-pole, 1200r/min. Motor parameters:

$R_s = 0.161\ \Omega, R_r = 0.316\ \Omega, R_m = 13.625\ \Omega$

$L_m = 27.926\ mH, L_s = L_r = 29.308\ mH$

The rotor inertia J of the IM is 0.36 $kg.m^2$.

REFERENCES

[1] A.Bonet, "Understanding efficiency in squirrel cage induction motors", *IEEE Trans. IA.*, vol. 16, No. 4, pp. 476-483, Jul./Aug. 1980.

[2] A. Kusko, D. Galler, "Control means for minimization of losses in ac and dc motor drives", *IEEE Trans. IA.*, vol. 19, no. 4, pp. 561-570, Jul./Aug. 1983.

[3] B. K.Bose, *Power Electronics and AC Drives.* Englewood Cliffs, NJ: Prentice Hall, 1986.

[4] J. Abrahamsen, J.K. Pedersen, F. Blaabjerg, "State-of-the-art of optimal efficiency control of low cost induction motor drives", *PEMC'96*, vol. 2, pp. 163-170, Budapest, Hungry, 2-4 Sep. 1996.

[5] G.O.Garcia, J.C. Mendes Luis, R.M.Stephan, and E.H.Watanabe, "An efficient controller for an adjustable speed induction motor drive", *IEEE Transactions on Industrial Electronics*, vol. 41, No. 5, pp. 533-539, Oct. 1994.

[6] I. Kioskeridis, N. Margaris, "Loss minimization in induction motor adjustable-speed drives", *IEEE Transactions on Industrial Electronics*, vol. 43, no. 1, pp. 226-231, February 1996.

[7] F.Abrahamsen, F.Blaabjerg, "On the energy optimized control of standard and high-efficiency induction motors in CT and HVAC applications", *IEEE Trans. IA.*, vol. 34, No. 4, pp. 822-831, Jul. /Aug. 1998.

[8] F.F.Bernal, A.G.Cerrada, and R.Faure, "Model-based loss minimization for DC and AC vector-controlled motors including core saturation", *IEEE Trans. IA.*, vol. 36, No. 3, pp. 755-763, May. /Jun. 2000.

[9] A.A.Hassan, "Improving the power efficiency of a rotor flux-oriented induction motor drive", *Electric Power Components and Systems*, 30: 431-442, 2002.

[10] D.Kirschen, D.Novotny, and T.Lipo, "Optimal efficiency control of an induction motor drive", *IEEE Transactions on Energy Conversion*, vol. EC-2, No. 1, Mar. 1987.

[11] J.C. Moreira, T.A. Lipo, V. Blasko, "Simple efficiency maximizer for an adjustable frequency induction motor drive", *IEEE Transactions on IA.*, vol. 27, no. 5, pp. 940-945, Sep./Oct. 1991.

[12] G. Kim, I. Ha, M. Ko, "Control of induction motors for both high dynamic performance and high power efficiency", *IEEE Transactions on Industrial Electronics*, vol. 39, no. 4, pp. 323-333, Aug. 1992.

[13] G.C.D. Sousa, B.K. Bose, J.G. Cleland, "Fuzzy logic based on-line efficiency optimization control of an indirect vector-controlled induction motor drive", *IEEE Transactions on Industrial Electronics*, vol. 42, no. 2, pp. 192-198, Apr. 1995.

[14] I. Kioskeridis, N. Margaris, "Loss minimization in scalar-controlled induction motor drives with search controllers", *IEEE Transactions on Power Electronics*, vol. 11, no. 2, pp. 213-220, Mar. 1996.

[15] Cao-Minh Ta, Yoichi Hori, "Convergence improvement of efficiency optimization control of induction motor drives", *IEEE Trans. IA.*, vol. 37, No. 6, pp. 1746-1753, Nov./Dec. 2001.

[16] Zhang Liwei, Liu Jun, Wen Xuhui, "A new fuzzy logic based search control for efficiency optimization of induction motor drives", *Proc. of IPEC'05*, Singapore, Nov., 2005.

[17] B.K.Bose, "Expert system, fuzzy logic, and neural network applications in power electronics and motion control", *Proc. of the IEEE*, vol. 82, pp. 1303-1323, Aug. 1994.

2006 5th International Power Electronics and Motion Control Conference

Research on an AC Variable-frequency Power Dynamometer Based on PWM Rectifier and Fuzzy Direct Torque Control

Jia-qiang Yang, Jin Huang

College of Electrical Engineering Zhejiang University, Hangzhou, 310027, P.R. China
yjq1998@163.com, ee_huangj@emb.zju.edu.cn

Abstract—The traditional dynamometers consume more fuel or electric energy while working, so they not only waste energy, but also have lower dynamic performance. An AC variable-frequency power dynamometer scheme is presented based on PWM Rectifier and Fuzzy Direct Torque Control (DTC) of induction motor in this paper. The dynamometer can operate in both motoring and generating states. The bidirectional flow of the energy with unit power factor between the power grids and the dynamometer is realized by voltage-oriented control (VOC) three-phase PWM rectifier. The dynamometer absorbs the energy from the power grids in motoring state, and feeds back the energy generated by the dynamometer to the power grids in generating state. In order to improve dynamic response capability and decrease the torque ripple of the dynamometer, the fuzzy logic idea is adopted based on conventional DTC. The stator flux angle, the stator flux error, and the torque error are reasonably fuzzified into several fuzzy subsets to optimize selection of the voltage vector. The proposed scheme is implemented with a single board microcomputer that uses TMS320LF2407A DSP and intelligent power model (IPM) to validate its feasibility. Experimental results show that the dynamometer has outstanding dynamic and static performance. The application of this novel dynamometer both saves the energy source and decreases the disturbances to power grids caused by rectification.

Keywords-power dynamometer; PWM rectifier; direct torque control; fuzzy control; induction motor

I. INTRODUCTION

The dynamometer, which is core equipment of motor performance test station, mechanical drive test station and engine performance test station, must have the characteristics of better dynamic performance, little inertia, good stability so as to meet the demands of examination. Traditional dynamometers, such as eddy current dynamometers, hysteretic dynamometers, mechanic dynamometers and waterpower dynamometers, do not have ideal dynamic performances. They can only operate static test, in which generated energy is converted into heat energy. Thus they not only waste energy, but also

need additional equipment to dissipate heat energy [1][2][3][4]. However, the power dynamometer can achieve the circulation of energy, and doesn't need additional equipment to dissipate heat energy. Therefore, power dynamometer attracts many researchers' focus in this field. Nowadays, the DC motor is mainly used in the system of power dynamometer, because of its excellent speed regulation capability and facile control ability. But due to the influence of its commutators, the DC motor is not able to run in high speed. In the case of high rotate speed, a mechanic speed-reducer must be installed, which increases the complexity and noise of the system [5][6]. Since the AC motor does not have the problem of the commutators as the DC one, brief structure and high reliability are its characteristics, so it can be used in the system of AC power dynamometer. References [7] and [8] adopt the synchronous scheme that dynamometer feeds back the energy while testing the power. But the synchronous rotate speed limits the highest rotate speed of the induction motor, when the dynamometer runs in motoring state. In [9],[10],and [11], an AC variable-frequency scheme is proposed, which not only realizes the bi-directional energy flow but also improves its static and dynamic performance. But the double-controllers device complicates the whole control system, and field oriented vector-control technology is prone to change with the alteration of the motor parameters. References [12] adopts genetic algorithms (GA) to optimize the performance of the dynamometer, now justly simulation results, it is difficult to practice. This paper proposes a novel scheme about the power dynamometer. Fuzzy direct torque control (DTC) is used to improve dynamic performance and decrease the torque ripple, and DTC is little dependent on the model parameters. Three-phase PWM technology is adopted to commute the DC link voltage, and feed back the electrical energy generated by the dynamometer to the power grids with unit power factor. Therefore, it can reduce the harmonic disturbances to the power grids and save the energy sources.

II. STRUCURE AND PRINCIPLE OF DYNAMOMETER

Fig.1 shows the structure of the dynamic power dynamometer proposed in the paper. The whole system consists of industrial control computer, control board

1-4244-0448-7/06/$25.00 ©2006 IEEE   1017

based on TMS320LF2407 DSP, inverters, AC induction motor and detecting circuits. Industrial control computer communicates with the DSP control board through RS232 serial port. In this way, control software sends DSP board start or stop signal to set rotate speed and torque, while DSP control board sends the detected rotate speed, torque, voltage and flux signals to the computer simultaneously through RS323 serial port, thus both real-time analysis and real-time control are achieved.

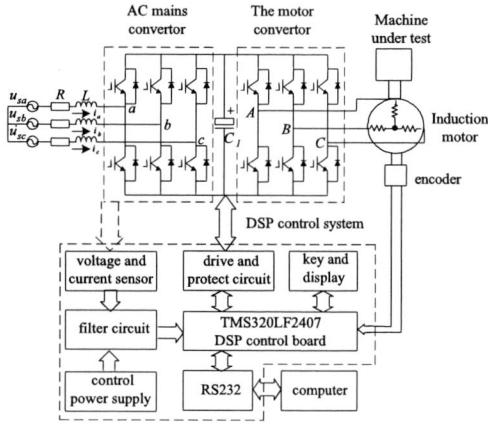

Figure 1. The basic structure diagram of the dynamometer

The proposed dynamometer can operate in both motoring and generating states. The dynamometer runs in the motoring state, when machine under test (MUT) is passive machine or using for start-up, grinding, and measure friction power. In this case, three-phase AC voltage of the power grids is commuted to DC link voltage u_{dc} with unit power factor under control of DSP board, and u_{dc} is inverted for dynamometer to drive machine under test (MUT) according to needed speed or torque by DTC system. The dynamometer runs in the generating state, when machine under test (MUT) is prime mover and its rotated speed exceed the synchronous speed of dynamometer. The dynamometer commutes random frequency generated-voltage to DC link voltage, and then inverts DC link voltage into 50Hz sine wave voltage to feed back the power grid with same phase. The dynamometer implements the bidirectional flow of the energy by automatic conversion between motoring and generating states. The proposed dynamometer not only has a full function dynamometer, but also decreases the harmonic disturbances to the power grids, saves the energy sources, and cuts the cost of purchasing radiator.

III. CONTROL STRATEGY OF THE DYNAMOMETER

A. VOC Three-Phase PWM Rectifier Mathematics Mode And Control

Traditional non-controlled rectifier and phase-controlled rectifier may bring the power grids serious pollution. To resolve the problem, an ordinary method is to install a filter that can compensate reactive power and eliminate harmonic waves, but this method can only be implemented afterwards. However, three-phase PWM technology can not only decrease current distortion and promote power factor, but also achieve energy circulation. There are two familiar PWM technologies, phase and amplitude control [13], and hysteresis current control [14]. In phase and amplitude control, the power factor relies on the accuracy of the main circuit's resistances, thus its dynamic performance is not very perfect. Although the hystersis current control can gain excellent results, its switch frequency is rather high and unfixed. Therefore, this paper provides an easy digital arithmetic method to realize voltage-oriented control (VOC) PWM rectifier based on volt-age and current double closed loop control.

From Fig.1, we can acquire the mathematical model of the three-phase rectifier in static ABC coordinate system, shown as "(1)".

$$\begin{bmatrix} L\frac{di_a}{dt} \\ L\frac{di_b}{dt} \\ L\frac{di_c}{dt} \\ C\frac{du_{dc}}{dt} \end{bmatrix} = \begin{bmatrix} -R & 0 & 0 & -S_a+\frac{1}{3}(S_a+S_b+S_c) \\ 0 & -R & 0 & -S_b+\frac{1}{3}(S_a+S_b+S_c) \\ 0 & 0 & -R & -S_c+\frac{1}{3}(S_a+S_b+S_c) \\ S_a & S_b & S_c & \frac{-1}{R} \end{bmatrix} \begin{bmatrix} i_a \\ i_b \\ i_c \\ u_{dc} \end{bmatrix} + \begin{bmatrix} 1 & 0 & 0 \\ 0 & 1 & 0 \\ 0 & 0 & 1 \\ 0 & 0 & 0 \end{bmatrix} \begin{bmatrix} u_{sa} \\ u_{sb} \\ u_{sc} \end{bmatrix} \quad (1)$$

Where, i_a, i_b, i_c are the current of a,b,c phases, respectively. u_{sa}, u_{sb}, u_{sc} are three phase power grids voltages. u_{dc} is DC link voltage. R is equivalent resistance. L is equivalent inductance. S_a, S_b, S_c are the switch function of different bridges. '1' denotes to turn-on in upper bridge and turn-off in lower bridge, while '0' denotes to turn-off in upper bridge and turn-on in lower bridge.

Equation (1) means that all three phase currents are interrelated with each other. Because the change of one phase current not only is related with the switch states itself, but also closely connects with the other two. In static ABC coordinate system, since the three phase currents couple with each other closely, it is difficult to decouple them. Therefore, a swift and accurate response of the system could not be acquired easily.

In order to decouple the phase current, "(1)" is converted to "(2)" in d-q coordinate system.

$$\begin{bmatrix} L\frac{di_d}{dt} \\ L\frac{di_q}{dt} \\ L\frac{du_{dc}}{dt} \end{bmatrix} = \begin{bmatrix} -R & \omega L & -S_d \\ -\omega L & -R & -S_q \\ \frac{3S_dL}{2C} & \frac{3S_qL}{2C} & 0 \end{bmatrix} \begin{bmatrix} i_d \\ i_q \\ u_{dc} \end{bmatrix} + \begin{bmatrix} 1 & 0 & 0 \\ 0 & 1 & 0 \\ 0 & 0 & -\frac{L}{C} \end{bmatrix} \begin{bmatrix} u_{sd} \\ u_{sq} \\ i_L \end{bmatrix} \quad (2)$$

Where, ω is radian frequency of the AC power grids. "(3)" can be deduced from "(2)".

$$\begin{cases} L\dfrac{di_d}{dt} = -R\,i_d + \omega L i_q + u_{sd} - S_d u_{dc} \\ L\dfrac{di_q}{dt} = -R\,i_q - \omega L i_d + u_{sq} - S_q u_{dc} \end{cases} \quad (3)$$

Equation (3) displays that, besides the influence of $S_d u_{dc}$ and $S_q u_{dc}$, d-q current is related with both coupling voltage $\omega L i_q$ and $-\omega L i_d$, and main voltage u_{sd} and

u_{sq}. S_d and S_q in the "(3)" can be regulated to ensure the correctness of "(4)".

$$\begin{cases} S_d u_{dc} = u_{rd} + u_{sd} + \omega L i_q \\ S_q u_{dc} = u_{rq} + u_{sq} - \omega L i_d \end{cases} \quad (4)$$

Where, u_{rd} and u_{rq} are only related with i_d and i_q, respectively. Putting "(4)" into "(3)", "(5)" can be acquired.

$$\begin{cases} L \dfrac{di_d}{dt} + R i_d = -u_{rd} \\ L \dfrac{di_q}{dt} + R i_q = -u_{rq} \end{cases} \quad (5)$$

Equation (5) shows that the two axis current are totally decoupled. Fig.2 displays the double closed-loop control system, which fulfills the current decoupling of the PWM rectifier.

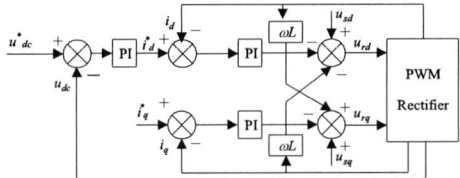

Figure 2. Theory diagram of PWM rectifier with current decoupling

Due to the introduction of current feedback, the process of decoupling, in fact, is a course that each of the PI regulated axis current result is injected with the other axis currents components. These injected current components are equal to the coupling value produced by the controlled object, but in an opposite direction. At the same time, AC main voltages u_{sd} and u_{sq} compensate it for voltage.

When the voltages u_{rd} and u_{rq} are acquired, $u_{r\alpha}$ and $u_{r\beta}$, two components of the referential voltage u_{ref} in $\alpha - \beta$ coordinate system, can also be acquired through coordinate transformation. Taking advantage of the voltage space vector, the converter on the side of AC power grids can trace the current command exactly, then the u_{dc}^{*} can be fixed.

B. Fuzzy Direc Torque Control

Direct torque control (DTC) is a new-style high performance AC speed regulation technology that was proposed by M.Depenbrock and I.Takahashi in 1985 [15][16]. The high dynamic torque performance and robust to the machine parameter variation are its advantages [17]. The conventional DTC is based on two discrete hysteretic comparators in which the torque and the stator flux are controlled directly. Because its selective range is too small,it is easy to select the same voltage vector, when the error of torque and the stator flux are big or small. So it will lead to high torque ripple and reduce torque response. In order to further improve static and dynamic torque response performance of DTC, and decrease torque ripple, a fuzzy logic idea was bring into

DTC. The stator flux phase angle, the stator flux error, and the torque error are reasonably fuzzified into several fuzzy subsets to optimize selection of the voltage vector.The framework of fuzzy DTC system, as shown Fig.3.

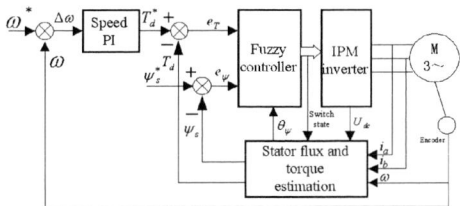

Figure 3. The framework of fuzzy DTC system

In 3-phase voltage-source inverter, "1" denotes to turn-on in above bridge, and "0" denotes to turn-on in below bridge, so we can get 8 voltage vector, $u_1 \sim u_6$ is 6 nonzero voltage vector ,and u_0、u_7 is two zero voltage vector in the origin, as Fig.4.

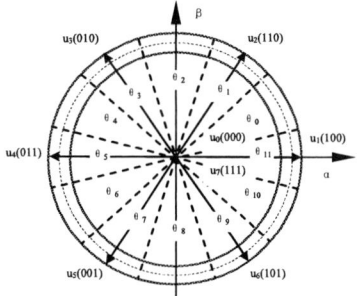

Figure 4. Voltage vector of 3-phase voltage-source inverter

The electromagnet torque formula of induction motor in static $\alpha - \beta$ coordinate is shown as "(6)"

$$T_d = \frac{3}{2} P_n (\psi_s \otimes i_s) \quad (6)$$

Where, ψ_s is the stator flux, i_s is the stator current, P_n is the pole-pair number, the stator current i_s can be detected by current sensor. The expression of stator flux is shown as "(7)".

$$\psi_s = \int (u_s - R_s i_s) dt \quad (7)$$

Where, u_s is the stator voltage space vector, which can easily be measured by the voltage sensor. R_s is the stator resistance. But noise disturbances and integral bias may result in inaccurate flux estimation in a relative lower speed. To solve the problem, we must establish a lowspeed model [18], in which the rotor resistance R_r, leakage inductance L_σ and main inductance L should be confirmed in advance. Consequently, this model is relatively complicated and its independence on the model parameters is weakened. Therefore, the paper replaces the integrator with a lowpass filter, as "(8)".

$$\hat{\psi}_s = \frac{u_s - R_s i_s}{s + a} \quad (8)$$

Where, a is the cut-off frequency of the lowpass filter. Compared to the integrator, lowpass filter may leads to the alteration of the stator flux amplitude and the problem

1019

that the flux no longer lags the counter voltage 90 degree. Therefore, an amendment for the amplitude and phase is necessary. The modified stator flux is shown as "(9)".

$$\begin{cases} \phi = \tan^{-1}(\omega_e / a) \\ \dfrac{(u_s - R_s i_s)}{s + a} \dfrac{\sqrt{\omega^2_e + a^2}}{a} \exp(-j(\tfrac{\pi}{2} - \phi)) \end{cases} \quad (9)$$

Where, ω_e is the synchronous radiant speed.

The fuzzy controller is a system with 3-input, 1-output. e_ψ, e_T, and θ_n are input variable, voltage vector u_s is output variable. e_ψ is fuzzified into E_ψ, including 3 fuzzy subsets {P,N,Z}, e_T is fuzzified into E_T, including 5 fuzzy subsets {PL,PS,ZE,NS,NL}, their membership function as Fig.5 (a) (b), respectively. The θ_n has 12 components in 2π range, as "(10)", it is fuzzified into 12 fuzzy subsets $\{\theta_0, \theta_1, \cdots \theta_{11}\}$, the membership function as Fig.5(c).

$$\frac{2n-1}{12}\pi \le \theta_n \le \frac{2n+1}{12}\pi \quad n = 0,1,\cdots 11 \quad (10)$$

Because output variable u_s is only 8 selective voltage vectors, it isn't essential to subdivide fuzzy subsets, so it is fuzzified into point fuzzy subsets{ $u_0, u_1, u_2, u_3, u_4, u_5, u_6, u_7$}, its membership function as Fig.5(d).

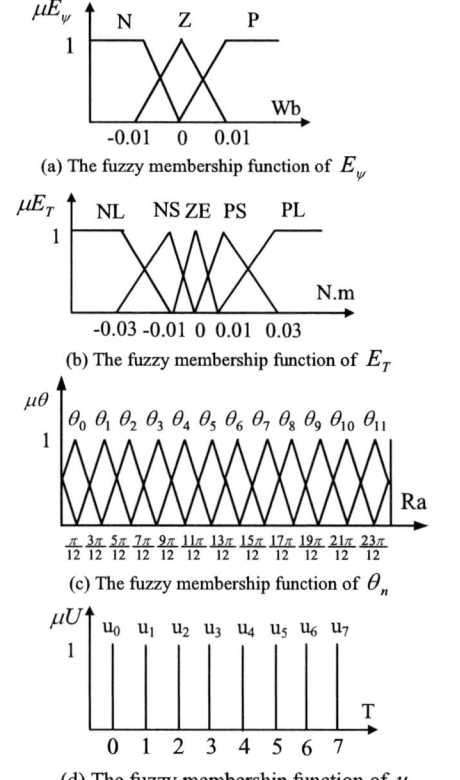

(a) The fuzzy membership function of E_ψ

(b) The fuzzy membership function of E_T

(c) The fuzzy membership function of θ_n

(d) The fuzzy membership function of u

Figure 5. The fuzzy membership function of input-output variable

The fuzzy control rule R_i as "(11)"

$$R_i : if\ E_\psi = A\ and\ E_T = B\ and\ \theta = \theta_j\ then\ U = u_k \quad (11)$$

Where, variables A, B, θ_j, u_i are fuzzy subsets of flux error, torque error, flux phase angle, and voltage vector. i=1~180, j=0~11, k=0~7. There are 180 rules in the strategy. Adopting Mamdani's minimum operation rule[19], which leads to control decision, and the membership function $\mu U(n)$ of the output is pointwise given, as "(12)".

$$\begin{cases} \mu U'(n) = \max_{i=1}^{180} (\min\ (\min\ (\mu E_{\psi i},\ \mu E_{Ti},\ \mu\theta_i,\ \mu N_i(n)) \\ \mu U(n) = \max_{N=0}^{7} (\mu N'(n)) \end{cases} \quad (12)$$

Because the fuzzy subsets of output variable are point subsets, so it doesn't need to defuzzified. the output variable u_i as "(13)".

$$u_i = u_N \quad (13)$$

Where, N is corresponding variable of "(12)".
According to the all rules and Mamdani's organon, and all the variable membership function, a fuzzy control table can gain, as shown Table1. The fuzzy control table can be queried in real time, deposited the table into memory of microcomputer.

TABLE I
FUZZY CONTROL TABLE

$E\psi$	ET	θ_0	θ_1	θ_2	θ_3	θ_4	θ_5	θ_6	θ_7	θ_8	θ_9	θ_{10}	θ_{11}
	PL	u_1	u_2	u_2	u_3	u_3	u_4	u_4	u_5	u_5	u_6	u_6	u_1
	PS	u_1	u_2	u_2	u_3	u_3	u_4	u_4	u_5	u_5	u_6	u_6	u_1
P	ZE	u_0	u_7	u_7	u_0	u_0	u_7	u_7	u_0	u_0	u_7	u_7	u_0
	NS	u_1	u_1	u_2	u_2	u_3	u_3	u_4	u_4	u_5	u_5	u_6	u_6
	NL	u_6	u_1	u_1	u_2	u_2	u_3	u_3	u_4	u_4	u_5	u_5	u_6
	PL	u_2	u_3	u_3	u_4	u_4	u_5	u_5	u_6	u_6	u_1	u_1	u_2
	PS	u_2	u_3	u_3	u_4	u_4	u_5	u_5	u_6	u_6	u_1	u_1	u_2
Z	ZE	u_7	u_0	u_0	u_7	u_7	u_0	u_0	u_7	u_7	u_0	u_0	u_7
	NS	u_7	u_0	u_0	u_7	u_7	u_0	u_0	u_7	u_7	u_0	u_0	u_7
	NL	u_6	u_1	u_1	u_2	u_2	u_3	u_3	u_4	u_4	u_5	u_5	u_6
	PL	u_3	u_3	u_4	u_4	u_5	u_5	u_6	u_6	u_1	u_1	u_2	u_2
	PS	u_4	u_4	u_5	u_5	u_6	u_6	u_1	u_1	u_2	u_2	u_3	u_3
N	ZE	u_7	u_7	u_0	u_0	u_7	u_7	u_0	u_0	u_7	u_7	u_0	u_0
	NS	u_5	u_5	u_6	u_6	u_1	u_1	u_2	u_2	u_3	u_3	u_4	u_4
	NL	u_5	u_6	u_6	u_1	u_1	u_2	u_2	u_3	u_3	u_4	u_4	u_5

IV. EXPERIMENTAL RESULT AND ANALYSIS

In order to validate the proposed scheme about the power dynamometer, we design a set of experimental device. its structure is shown as Fig.6.

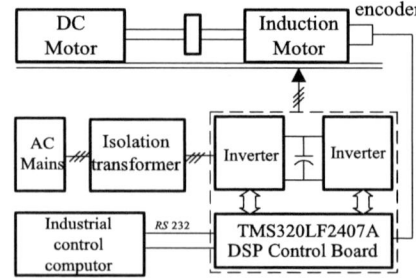

Figure 6. The framework diagram of experimental setup

The experimental device consists of an industrial control computer, a DSP control board, two IPM intelligent power modules, a photoelectrical encoder, an induction motor, voltage and current sensors and a DC motor.

The parameters of induction motor are listed as follows:P_N=370W,T_N=2.5N•m,U_N=220V,I_N=0.89A,n_N=1 440rpm,Pn=2,Rs=46Ω,f=50Hz.The parameters of the DC motor are listed as follows: PN=185W, U_N=220V, I_N=1.1A , n_N=1600rpm, P_n=2, the rated exiting current less than 0.16A.

In the experiment, the working frequency of the VOC three-phase PWM rectifier is 5 KHz, while the working frequency of the fuzzy DTC inverter is 10 KHz.

Fig.7, Fig.8 and Fig.9 display the experimental results, when the proposed dynamometer operates in motoring state. Fig.10, Fig.11 and Fig.12 display the experimental results, when the proposed dynamometer operates in generating state.

Fig.7 shows the A-phase voltage and current waveform on the side of the power grids in motoring states. Although the power grids' voltage waveform is not a perfect sinusoid, A-phase current waveform is close to sinusoid, and its voltage and current are in the same phase. Thus, the device achieves PWM rectification with unit power factor.

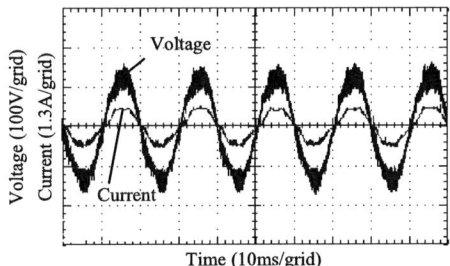

Figure 7. u_a and i_a waveform of the power grid in motoring state

Fig.8 shows the A-phase current waveform of dynamometer at 30r/min. the current aberrance is faintness, and order harmonic waves are smaller, so the fuzzy logical improves DTC performance at low speed range.

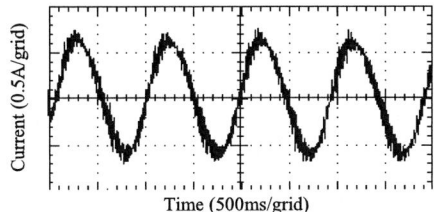

Figure 8. i_a waveform of the dynamometer at 30r/min

Fig.9 shows dynamometer's electromagnetic torque waveform of conventional DTC and fuzzy DTC at 1.0N.m. The curve 1 is based on conventional DTC; the curve 2 is based on fuzzy DTC. Torque ripple is smaller curve 2 than curve 1, so proposed fuzzy DTC decease the torque ripple of dynamometer.

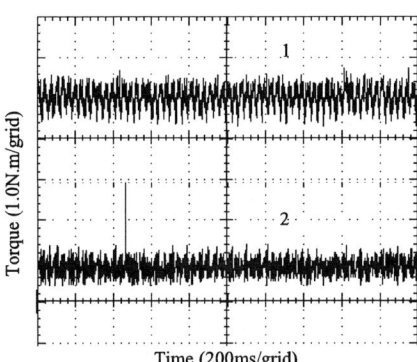

Figure 9. Torque response waveform of conventional DTC and fuzzy DTC at 1.0N.m

Fig.10 shows the A-phase voltage and current waveform of dynamometer, when the dynamometer operates in generating state. The figure displays that the voltage and current waveforms are close to sinusoids, there are few harmonic contents in the waveforms, voltage and current waveforms are in the opposite direction. Thus, the unit power factor energy could be feed back to the power grids and the disturbances to the power grids are reduced.

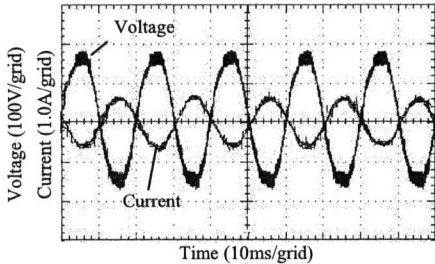

Figure 10. u_a and i_a waveform of the power grid in generating state

Fig.11 shows the torque response waveform of dynamometer in generating state, when the set torque increases from 0 to 1.0N.m. It takes 5ms for the dynamometer to reach the set torque, there is no obvious overshooting, and the dynamic capability is excellent. Therefore, it is indicated that the dynamometer has an outstanding dynamic performance with fuzzy control.

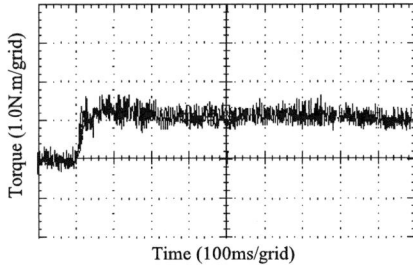

Figure 11. Torque response from 0.0N.m to 1.0N.m in generating state

Fig.12 shows the phase current waveform of dynamometer from 1.0N.m to 2.0N.m in generating state. Amplitude of current increase and variety placidly. The period is 100ms at 1.0N.m, however the period is 120ms

1021

at 2.0N.m.That to say, the dynamometer reduces synchronous rotate speed to increase the torque.

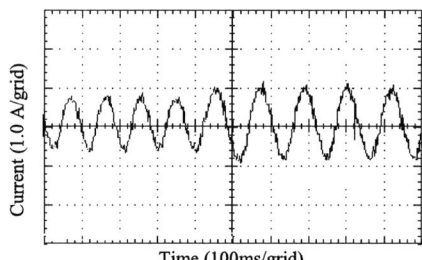

Figure 12. i_a waveform of the dynamometer from 1.0N.m to 2.0N.m in generating state

V. CONCLUSION

The paper proposes a variable-frequency power dynamometer scheme based on the PWM rectification technology and fuzzy DTC technology of induction motor. PWM rectifier is unlike the traditional AC/DC converter, which can run in both rectification and inversion states, The bidirectional flow of the energy with unit power factor between the power grids and the dynamometer is realized by voltage-oriented control three-phase PWM rectifier. The DTC technology not only enables the dynamometer to run in motoring state, but also in generating state. The fuzzy logic optimizes selection of the voltage vector, and improves dynamic response capability and decrease the torque ripple of the dynamometer. The proposed dynamometer operating in generating state provides the dynamic load torque for machine under test, which the one operating in motoring state drives machine under test with the set torque or the set rotate speed. The dynamometer is little dependent on the parameters of the induction motor model, realizes bidirectional flow of the energy with unit power factor, and reduces the disturbances to the power grids, saves the energy sources. Its practical value is attractive.

REFERENCES

[1] Z.D Yan,*Testing Technology of Internal Combustion Engine.*HZ: Zhejiang University Press,1993,pp. 10-15.

[2] S.S Yu and J Huang,"T-n Curve Auto-test for Asynchronous Motor Based on Hysteres is Power Tester," *S&M Electric Machines,* vol. 28, no. 1, pp. 48–52, Jan. 2001.

[3] J.B Byron, A. F Matthew, and E. T Bruce, "Robust multivariable control of an engine-dynamometer system," *IEEE Trans. Control Systems Technology*, vol.5, no. 2,pp. 189-199,1997.

[4] F Aghili, M Buehler, and J.M. Hollerbach, "Motion Control Systems with H$^\infty$ Positive Join torque feedback," *IEEE Trans.Control Systems Technology*, vol.9, no. 5, pp. 685-695, Sept. 2001.

[5] S.J Zhang, J.L Ren, and X.D Ji, "Development of Vertical DC Dynamometer," *Small Internal Combustion Engine And Motorcycle*,vol.30,no.4,pp.44-45,2001.

[6] C.R Hewson, G..M Asher, and M Sumner, "Dynamometer control for emulation of mechanical loads," *in Proc. 33rd IAS Annu. Meeting. Industry Applications*,vol.2,pp.1511-1518,Oct. 1998.

[7] A.C Williamson, and K.M.S Al-Khalidi, "An improved engine-testing dynamometer," *in Fourth International Conference. Electrical Machines and Drives*, pp.374-378, Sept. 1989.

[8] H Zhuo, "Exploration of Enlarging the Capacity and Range of Application for Synchronous-Dynamometer," *Journal of Fuzhou University (Natural Science))*, vol.26,no.5,pp.43-47,1998.

[9] E.R Collins, and Y Huang, "A programmable dynamometer for testing rotating machinery using a three-phase induction machine," *IEEE Trans. Energy Conversion*, vol.9,no.3,pp.521-527, Sept.1994.

[10] G.C.D Sousa, and D.R Errera, "A high performance dynamometer for drive systems testing," *the 23rd International Conference. Industrial Electronics, Control and Instrumentation*, vol.2, pp.500-504, Nov.1997.

[11] R.S Wieser , and A Lechner, "Four-quadrant drive as fast operating and precise dynamometric brake,", *in Record Thirty-First IAS Annual Meeting, Industry Applications*, IAS '96, vol.1,pp.180-185,Oct. 1996.

[12] L Weng, Z.Y. Dong. "Optimal design of a regenerative dynamic dynamometer using genetic algorithms," *The Congress on Evolutionary Computation*, CEC'03 , vol.4,pp. 2665-2672,2003.

[13] W Rusong, S.B Dewan, and G.. R Slemon, "Analysis of an AC to DC voltage source converter using PWM with phase and amplitude control," *IEEE Trans. Industry Applications*, vol.27,no.2,pp.355-364, Mar. 1991.

[14] B. D Min, J.H Youm, and B.H Kwon, "SVM-based hysteresis current controller for three-phase PWM rectifier,"*in Proc. Electric Power Applications*, vol.146,no.2,pp.225-230, Mar. 1999.

[15] M. Depenbrock, "Direct self-control (DSC) of inverter-fed induction machine," *IEEE Trans. Power Electronics*,vol.3,no.4,pp.420-429,1988.

[16] I Takahashi, and Y Ohmori, "High-performance direct torque control of an induction motor," *IEEE Trans. Industry Applications*, vol. 25, no. 2, pp. 257–264, 1989.

[17] J.Q Yang and J Huang, "Research on direct torque control of induction machine based on torque prediction," *Journal of Zhejiang University:Engineering Science*,vol.39,no.9,pp.1277-1281, Sep.2005.

[18] J.Q Yang, J Huang, and X.P Xu, "Research on Full-speed Model Direct Torque Control System of Asynchronous motor Based on the DSP," *Proceeding of EPSA*,vol.16,no.2,pp.28-32, Apr.2004.

[19] J Zhu. *Fuzzy control theory and application.* Beijing: Mechanical Industry Press, 2001.

Author Index

A

Abbasian, M.A. 1043
Abdelhamid, T. H. 411
Abedini, A. 224
Abjadi, N. R. 1917
Abo-Khalil, Ahmed G. 1477
Abramovitz, A. 1412
Agarwal, Anant K. 157
Agarwal, Vivek 281
Ahmed, Nabil A. 242
Ahn, C. H. 1198
Aide, Xu 1162
Ai-Juan, Jin 1421, 1426
Ait-Amirat, Y. 1882
Ajjarapu, Venkataramana 505
Akagi, Hirofumi 23
Akimasa, Koji 1613
Andersen, Henrik Rosendal 1032
Ando, Tatsuo 947
Arpilliere, M. 249
Ashida, M. 2000
Askari, J. 1917

B

Badica, M. 1751
Bai, Haijun 219, 826
Bai, Zhifeng 1581
Baihua, ...161
Balda, J.C. 1353
Banaei, M. R. 759, 764
Bao, G.Q. 813
Baocheng, Wang 569, 1991
Baoming, Ge 918
Barsoum, ...1148
Bendjedia, M. 1882
Berthon, A. 515, 1882, 2005
Bhattacharya, Subhashish 1450
Bin, Su 406
bin, Wu 1368
Binder, A. 842
Bing, Chen 1401
Bisogno, F.E. 1117
Blaaberg, F. 46, 1107, 2029
Bo, Chen 122
Böcker, Joachim 1112
Bodson, M. 1912
Bojoi, R. 1651
Boroyevich, D. 249
Boroyevich, Dushan 1836
Bréhaut, Stéphane 92
Brouji, H. El 1663

C

C, Sreekumar 281
Cailin, Wang 1167

Calderon-Lopez, G. 1328
Calverley, S.D. 977
Camara, M.B. 515
Câmpeanu, A. 1751
Cao, Binggang. 1581
Cao, R. X. 510
Cao, Yanjie 2015
Carazo, A. V. 1117
Cartes, David 774
Cen, Yuwan 1986
Chan, C.C. 57
Chang, Chung-Hsing 291
Chang, Duan Qi 1551
Chang, H.-H. 1343
Chang, Jie (Jay) 102
Chang, Lon-Kou 417
Chang, Yuan 1722
Changhong, Wang 1793
Changzheng, Zhang 489, 739
Chau, K. T. 1788
Chen, Bin 1450
Chen, C.-C. 117
Chen, Cheng-Hu 913
Chen, Chern-Lin 332
Chen, Chien-An 967
Chen, Guiyou 1202
Chen, Guocheng 1560
Chen, Guozhu 794
Chen, H. 933
Chen, H. G. 1129
Chen, J. 194
Chen, Jian 1218
Chen, Jiann-Fuh 361, 1178
Chen, Jiaxin 346, 831
Chen, Jie 113
Chen, Jun-Ning 199, 286, 1283, 1392
Chen, Min 442
Chen, Qiaoliang 433, 642
Chen, Rui 1171
Chen, Ruijuan 1253, 1877
Chen, Tso-Min 332
Chen, Wei 171, 1081
Chen, Xi 1507
Chen, Xiangjun 607
Chen, XuWu 438
Chen, Y.-M. 108, 117
Chen, Yao 1454
Chen, Yen-Ming 1763
Chen, Yuan-rui 1397
Chen, Yunpeng 236
Chen, Z. 49, 499, 1773, 2029
Chen, Zongxiang 142
Chenchen..386
Cheng, Chun-An 1178
Cheng, Ming 1746, 1815
Cheng, Ming-Yang 913

Author Index

Cheng-ning, Zhang 1027
Chengsheng, Wang 589
Cherifi, A. 574
Chi, Song 890, 1825
Chiang, Huann-Keng 967
Chiasson, J. N. 1703, 1912
Chiu, Huang-Jen 291
Cho, Yun-hyun 1238, 1784
Choi, E.S. 1382
Chongjian, Li 589
Chun, Dong 1623
Chun, YonDo 1784
Chung, Jung Kee 1736
Chunjiang, Zhang 554, 559, 1473, 1618
Corzine, Keith A. 637
Costa, François 92
Crausaz, A. 2005
Cui, Bo 798
Cui, Jiefan 657
Cui, Junwei 1436
Cvetkovski, G. 254

D

Dai, Ke 789
Dai, Renchang 1122
Dai, Yue-Hua 199
Da-ming, Liu 1674
Danhe, Li 1991
De Doncker, R. W. 31
Deng, Jianming 923
Deng, Yan 1931
Dianguo, Xu 301, 1713
Ding, Xiaoyu 1560
Ding, Ye 1223
Divan, D. 16
Divan, Deepak M. 2010
Doi, Toshimitsu 356, 1302, 1307
Dong, Jiang 880
Donghua, Luo 1634
Dongsheng, Zuo 468
Dong-Shoutian, ...484
DongYu, ...1623
Dou, Sen 537
Du, Guiping 316
Du, Zhong 1450
Duan, Baoxing 70
Duan, Huijuan 798
Duan, Shan Xu 1522
Duan, Shanxu 1218
Duarte, Jorge L. 784

E

Ebrahimi, Yousef 779
Eiuo, Bin 1358
El Din, Ashraf Salah El Din Zein 847
Elbanhawy, Alan 342. 1967

Endo, Tsunehiro 947
Ertugrul, Nesimi 147, 962

F

Fa, Naiguang 1253, 1877
Fang, Liang 817
Fang, Xin 789
Fang, Xu-Peng 166
Fang, Yu 1406
Fang, Zhuo 122, 1542
Fathy, Khairy 356, 1302, 1307, 1358, 1363
Fei, Wanmin 1138
Fei-peng, Xu 647
Feng, D. 499
Feng, Huang 1401
Feng, L. 842
Feng, Zhao 622, 679
Feng, Zheng 585
Fengxiang, Wang 449, 903
Feyzi, M. Reza 204, 1228
Forrest, S. J. 977
Forsyth, A. J. 1323, 1328
Francis, Jerry 1836
Friedrichs, Peter 132
Fröhleke, Norbert 1112
Fuchs, F.-W. 325
Fujita, Kouetsu 1971
Fukuda, S. 1468
Fukushima, Kentaro 1333
Funian, Hua 449
Furuya, Atsushi 1598
Fu-sheng, Wang 463
Futami, M. 1468

G

Gallay, R. 2005
Gang, Ma 378
Gao, F. 1107
Gao, Yan 157
Gao, Yang 113, 1159
Gao, Yong 1198
Gao, Z. Y. 1071
Garinto, Dodi 306
Ge, L.S. 1458
Ge, Lu-sheng 1368, 1576
Ge, Qiongxuan 1171
Geng, Pan 789
Goharrizi, A. Yazdanpanah 1697
Gong, Yu 1223, 1788
Grabner, C. 999
Graczkowski, J. J. 1096
Grantham, Colin 1207, 1858
Gruenberger, Hans Pert 219, 826
Grundmann, Frank 870, 1442
Gu, G. 175
Gu, Herong 473, 1585

Author Index

Gu, Yilei 171, 276
Gualous, H. 515, 2005
Guan, Xiaohan 688
Guang, Zeng 734, 1669
Guangzheng, NI 1091
Guenther, D. 842
Guilan, Chen 554, 1006, 1049
Gui-xin, Shao 1027
Guiyou, Chen 1630, 1634
Guo, Hongche 612, 853
Guo, Qingding 612, 853, 896
Guo, Wei 952
Guo, Xin 337
Guo, Youguang 346, 831
Guobiao, Gu 808
Guocheng, San 1473
Guojun, Lu 802
Guoxin, Zhu 1802
Gustin, F. 515

H

Habetler, Thomas G. 836
Haibing, Hu 937
Haijie, Xu 937
Haiping, Xu 684, 1298
Haitao, Zhang 161
Halász, S. 693
Han, B.D. 1143
Han, Chong 1450
Han, Chong Zhao 652
Han, F. T. 1071
Han, S.K. 1382
Hang-Tian, Li 1421, 1426
Harley, Ronald G. 836, 2010
Hartavi, A.E. 2018
Hashimoto, Takayoshi 1333
He, Guofeng 657
He, Junping 1081
He, Shijie 1463
He, Xiangning 83, 1931
He, Zhongyi 1537
Hemin, Wang 1849
Heming, Li 458
Hendrix, Marcel A. M. 784
Herong, Gu 559, 1618
Hirao, Mitsuhiro 1768
Ho, Chien-Yeh 1527, 1995
Ho, S. L. 1901
Hong, Peng 899
Hong, Shen 559, 1273, 1618
Hong-mei, LI 463
Hongren, Yin 401
Hori, Yoichi 1797
Hosseini, S. H. 759, 764, 1697
Hosseini, Seyyed Hossein 753, 779, 1679
Howe, D. 908. 928, 1841

Hsu, Kai-Sheng 967
Hsu, Ken-Chuan 1957
Hsu, W.P. 718
Hu, D.Q. 1143
Hu, Haibing 127, 1183
Hu, Jiangang 703
Hu, Qing 1806
Hu, Qingbo 526
Hu, Songqin 351
Hu, Weihao 321, 585
Hu, Wenhua 397
Hu, Xuezhi 438
Hu, Y. 2029
Hu, Z. L. 1708
Hu, Zongbo 316
Hua, Li 729, 1571
Hua, Wei 1746, 1758
Huade, Li 401
Huang, Zhenyue 1213
Huang, Alex Q. 113, 157, 1159, 1450
Huang, Chien-Lan 748, 1278
Huang, Congsheng 1288
Huang, Jin 1017
Huang, Xuwen 1288
Huang, Yafeng 542
Huang, Yi 1076
Huang, Yuehui 580, 1512
Hui, Li 729
Hui, Wu 862
Hui, Zhang 1032, 1492
Hui-jie, Xiang 1942

I

Ichinose, M. 1468
Inoue, Kaoru 1233, 1613
Inoue, Shigenori 23
Iov, F. 46
Iwanski, G. 494

J

Jang, Jeong-Ik 1482
Jangwanitlert, A. 1353
Járdán, R.K. 1338
Jeon, K. S. 1198
Jewell, G. W. 977
Ji, Yanchao 627, 1507
Jia, C. 1323
Jia, Y. 2000
Jia, Y.P. 1143
Jiag, Maoh-Chin 1527
Jian, Chen 74, 769
Jian, Cui 1431
Jian, Liu 272
Jian, Wu 1713
Jiang , Chang 1213
Jiang-Hui, Chen 1401

Author Index

Jiang, J. Z. 1788
Jiang, J.G. 1458, 1952
Jiang, J.J. 152
Jiang, J.Z. 813
Jiang, Jianguo 1081
Jiang, Jianzhong 1223
Jiang, Xianglong 1608
Jiang, Xiaochun 1896
Jianguo, Jiang 468
Jianlin, Zhu 1557
Jianru, Wan 1273
Jian-Ru, Wan 1431
Jian-wen, Zhang 1657
Jianze, Wang 1693
Jiarong, Kan 1532
JiaYi, Yuan 808
Jie, Shuo 1806
Jie, Wang 569
Jiefan, Cui 862, 1849
Jin, Jianxun 831
Jin, Mengjia 885, 1872
Jin, Shun 617
Jin, Tianjun 1183
Jin, Wenxi 127
Jin, Xin Min 1454
Jing, Liu 88
Jin-gang, Li 549
Jing-Gang, Zhang 1669
Jinjun, Liu 1061, 1492, 1722
Jinlong, Zhang 401
Jinupun, P. 1887
Jiqiang, Wang 903
Jiuhe, Wang 401
Jo, WonYoung 1784
Johal, H. 16
Johnson, C. M. 977
Joseph, Alan 1076
Jou, H.L 718
Jun, Liu 1012
Jun, Wang 899
Jung, Kun-seok 1238
Junjuan, Sun Xiaofeng Wu 569
Junmin, Zhang 1684
Junzhu, Wan 1849
Jwo, Ko-Wen 1590

K

Kaijie, Feng 1741
Kaipei, Liu 209, 1684
Kang, B.W. 1143
Kang, Ju-Sung 1358
Kang, Y. 194
Kang, Yong 97, 564, 789, 1218, 1522, 1981
Karimi, E. 1697
Kato, Tomohiko 1971
Kato, Toshiji 1233, 1613

Ke, Dao-Ming 199
Ke, Fu-Jing 1154
Ke, Yi-Jing 1392
Kerkman, Russel J. 1054
Kesong, Ye 699, 1832
Khaehintung, Noppadol 137, 368
Khajee, M. Darkalee 759
Khan, Mahamnad Mansoor 386
Kim, E. D. 1198
Kim, Jang-Hwan 662
Kim, Joo Han 1736
Kim, Young-Sin 1482
Kimura, Noriyuki 1768
Kiranon, Wiwat 137
Kita, H. 1468
Koczara, W. 494
Konghirun, M. 972
Koo, DaeHyun 1784
Kou, X. 1096
Krishnaswami, Sumi 157
Ku, Chung-Ping 1590
Kumar, Pavan 537
Kun, Li 1447
Kunakorn, Anantawat 368
Kuo, J. -S. 1343

L

L., M. 1148
Lai, Ching-Ming 1590
Lai, Stephen L. 1502
Lai, Y. M. 296
Lang, Yongqiang 708
Lee, Chi-Yang 1192
Lee, Dong-Choon 1477, 1482
Lee, Fred C. 1
Lee, Hyun Woo 392, 1302, 1307, 1358, 1363, 1372, 1377
Lee, Se-Hyun 1477
Lemberg, Nicholas 989
Li, Chongjian 995
Li, Dong 1202
Li, Dongsheng 947
Li, F. 152
Li, H. 1773
Li, Han 1006, 1049
Li, Hongtao 688, 1248
Li, M. 1703, 1912
Li, Ma 88, 209
Li, Mingzhu 1537
Li, Min-zu 97
Li, Qi 79
Li, Qunzhan 423
Li, Rongyuan 1112
Li, Shijie 1171
Li, Tianbo 1947
Li, Wen 1797

Author Index

Li, Wenguang 1815
Li, Xia 1674
Li, Y.W. 1101
Li, Yaohua 995
Li, Yong 428
Li, Yongbin 942
Li, Yongdong 1892
Li, Zhanlong 1416
Li, Zhaoji 70, 79
Li, Zheng-Ping 286
Li, Zhou 1630, 1634
Liang, L. 1129
Liang, Tsorng-Juu 361, 1178
Liang, Zhonghua 607
Liao, Changming 102
Li-Jiahui, ... 484
Lijie, Chen 1595
Li-jun, Hang 406
Lijun, Zhao 862, 1849
Lili, Jiang 1849
Liming, Liu 74
Lin, Bor-Ren 748, 967, 1278
Lin, Chang-Hua 1957
Lin, Fei 184, 1976
Lin, Liangrui 1243
Lin, Li-Wei 291
Lin, Ray-Lee 361
Lin, Ruan 808
Lin, Ruiguang 885, 1872
Lin, W.-C. 108
Lin, Yang-Sheng 1178
Lin, Ying-De 1154
Lin, Yu-Tzung 1192
Ling, Xia 899
Lingjie, Meng 674
Lipo, T.A. 989
Liqiang, Yuan 161
Liu, Cheng-Tsung 1763
Liu, Ching-Hsiung 361
Liu, Guiqiu 896
Liu, Hongwei 798
Liu, Hsing-Fu 417
Liu, Jian 1267
Liu, Jianqiang 184
Liu, Jingbo 703
Liu, Jinjun 713, 1726
Liu, K. 1071
Liu, Kaipei 453
Liu, Shu-Lin 1267
Liu, Tien-Shuo 1957
Liu, Wei-Shih 361, 1178
Liu, Wenhua 542
Liu, Wenji 1248
Liu, Xiang 229
Liu, Xiaodong 351
Liu, Xinhua 1223

Liu, Yuanchao 236, 1248
Liu, Zhengang 688
Liu-Xueli, ... 484
Liwei, Zhang 1012
Loh, P. C. 1107
Lorenz, L. 39
Lou, Z. L. 373
Lu, Bin 836
Lu, Bing 1
Lu, Cheng 489, 739
Lu, Haihui 1054
Lu, P.-C. 117
Lu, Shuai 637
Lu, Xiaodong 83
Lu, Zhengyu 127, 171, 276, 526, 1183
Luk, P.C.K 478, 1872, 1887
Luo, Fang 789, 1522
Lyons, James 1122

M

Ma, Hao 1312, 1637
Ma, Hongfei 708
Ma, Wenchuan 627
Ma, Xiangfei 1836
Ma, Xuejun 438, 1288
Maeda, Toshihiro 1971
Manmek, Thip 1207
Mansouri, O. 574
Mao, Hong 1267
Mathew, Anu 442
Matsumoto, Shuji 1598
Matsuse, Kouki 1598
Mayor, J. Rhett 2010
Meghriche, K. 574
Member, Student 1825
Meng, Zheng 236
Mi, Chris 942
Miao, Guan 744
Miao, Zhao 549
Miller, Nicholas 1122
Ming, Cheng 1758
Ming, Zhou 1431
Ming, Zong 449, 903
Ming-fu, Zhao 1623
Mingli, Ding 1793
Min-qian, Ke 734
Miyatake, Masafumi 242
Moghbelli, H. 597
Mohr, M. 325
Molinas, Marta 63
Moon, G.W. 1382
Morimoto, Keiki 356, 1302, 1307
Morizane, Toshimitsu 1768
Mou, Shann-Chyi 291
Mu, Gang 542
Mudannayake, Chathura P. 1207

Author Index

N

N., N. 1148
Na, He 1713
Nagy, I. 1338
Naidu, S. R. 1731
Nakaoka, Mutsuo 356, 392, ...1302, 1307, 1358, 1363, 1372, 1377
Nakayama, Y. 1468
Nan, C. H. 1708
Nan, Liu 214, 1942
Nan, Zhao 918
Nasiri, A. 224
Neff, K. L. 1096
Niasar, A. Halvaei 597
Ning, Gaidi 1463
Ninomiya, Tamotsu 1333
Nishimae, Kazuya 1233
Nittayarumphong, S. 1117
Niu, Shuangxia 1788
Nolle, Eugen 219, 826
Nondahl, Thomas A. 1054
Notohara, Yasuo 947
Nozawa, Yusuke 1598
Nuttall, D. R. 1328

O

Ogiwara, Hiroyuki 1307, 1358
Ohara, S. 1468
Oka, Kazuo 1598
Okude, Takaaki 1363
Oleschuk, V. 1651
Omata, Ryuji 1598
Omori, Hideki 392, 1358, 1363, 1372, 1377
Ou, Chung-Lun 1957
Ouyang, Wen 989

P

Pan, Junmin 142, 267, 1348
Pan, Ming-Ho 1590
Pan, Sanbo 1348
Pang, Da-Chen 1763
Paponpen, K. 972
Park, J. D. 1198
Paweletz, A. 842
Payam, A. Farrokh 1906
Pedersen, John K. 1773
Pei, Yunqing 321
Peng, Fang Z. 1076
Pengcheng, Zhu 74
Petchjatuporn, Panom 137
Petkovska, L. 254
Piwko, Richard 1122
Poon, N. K. 1502
Poure, P. 1663
Prado, R. N. do 1117

Pratt, Annabelle 537
Profumo, F. 1651

Q

Qi, Feng 1637
Qi, Wang 1793
Qian, Lewei 774
Qian, Zhaoming 127, 171, 276, 1076, 1183
Qiang, Li 1853
Qiang, Mei 1926
Qiao, Ermin 337
Qiao, Wei 836
Qiaofu, Chen 489, 739
Qi-gang, Fu 734
Qing, Sun 862
Qingding, Guo 1802, 1846
Qingdong, Zhou 1793
Qingfan, Zhang 1634
Qinglin, Zhao 532, 1387, 1517
Qingyu, Yang 699, 1832
Qinmu, Wu 1820
Qiu, Dongyuan 316, 1293
Qiu, Jianqi 885, 1872
Qiu, Zhiling 794
Qizhi, Zhan 1542

R

Radecker, M. 1117
Rafik, F. 2005
Ragon, S. 249
Rahman, M. F. 983, 1646, 1867, 2023
Rahman, M. Faz 1858
Rahnavard, Reza 779
Rajagopalan, Satish 2010
Ren, Hai Peng 652
Ren, Shi 699, 1832
Rentschler, A. 842
Rhyu, Se Hyun 1736
Rosado, Sebastian 1836
Rosario, L.C. 478
Ruan, L. 175
Ruan, Xinbo 1936
Ruixia, Wang 1853
Ruliang, Zhang 1167
Ruxi, Wang 1061

S

Saadate, S. 1663
Sabahi, Mehran 753, 1228, 1679
Sabzali, A. 411
Saha, Bishwajit 392, 1372
Sahinkaya, M.N. 2018
Sakamoto, Kiyoshi 947
Sanchez-Gasca, Juan 1122
Scozzie, Charles 157
Segawa, Takeshi 1333

Author Index

Shancheng, Xing 1023
Shanxu, Duan 769
Shao, Changhong 607
Shao-De, Zhang 1447
Shaojun, Xie 1532
Shao-Long, Li 1421, 1426
Sharifian, M. B. B. 204
Shen, Guoqiao 1566
Shen, Hong 1603
Shen, J.X. 908, 928
Shen, Miaosen 1076
Shen, W. 249
Sheng, K. 1188
Sheng, Weihui 995
Shergin, V. V. 1133
Shi, Cenwei 885, 1872
Shi, Y. F. 1841
Shiang, J. -Z. 1086
Shibata, R. 2000
Shi-feng, Zhang 1447
Shiri, A. 821
Shoulaie, A. 821
Shu, Mantang 1560
Shu, Zhibing 1811
Shun, Jin ....
Shutong, Qiao 468
Shyu, Kuo-Kai 1590
Sibo, Ge 699, 1832
Sirisuk, Phaophak 137, 368
Skorokhod, Y. Y. 1133
Sneineh, Anees Abu 1318
Soltani, J. 1038, 1043, 1906, 1917
Song, Wenchao 1450
Song, Wenxiang 1560
Songboonkaew, J. 1353
Songhua, Shen 744
Soong, Wen Liang 962
Souza, E. V. N. 1731
Stankovic, A.M. 1651
Stefanovic, V. 249
Su, Hongsheng 423
Sugimoto, Hidehiko 1258
Sugimura, Hisayuki 392, 1372, 1377
Sul, Seung-Ki 662
Sun, Chin 417
Sun, Jia-E 199
Sun, Jian 442
Sun, Sizhou 351
Sun, Wei 866
Sun, Xiaofeng 674
Sun, Yuxin 179
Sunat, Khamron 137
Sung, Ha Kyeong 1736
Suul, Jon Are 63
Suzuki, Takahiro 947

T

Tai, Wei-Chih 913
Takahashi, Toshio 428
Tan, Guang-Hui 627, 1507
Tan, Ruimin 1032
Tan, Siew-Chong 296
Tanaka, Chikara 947
Tang, Yan 866
Tang, Yupeng 1416
Tang, Yu-peng 669
Taniguchi, Katsunori 1768
Tao, Haimin 784
Tao, Liu 122
Tenconi, A. 1651
Teodorescu, R. 46
Tezcan, Ibrahim 1546
Thammasiriroj, W. 1353
Tian, Kai 1318
Toba, Akio 1971
Tolbert, L. M. 1703, 1912
Tongjing, Sun 1630
Tsai, C.-T. 108
Tsai, Ming-Fa 1154
Tsay, Shuh-Chuan 1278
Tse, Chi K. 296, 580, 1512
Tseng, S. -Y. 1086, 1343
Tseng, S.-H. 1086
Tuncay, R.N. 2018
Tzou, Ying-Yu 1192

U

Undeland, Tore 63

V

Vahedi, A. 597, 821
van der Broeck, Heinz 1546
Vansencc, Flalph 875
Varjasi, I. 693
Venkataramanan, Giri 259
Volskiy, S. I. 1133

W

Walther, B. 1882
Wan, Deyu 1585
Wan, Shuyun 1608
Wang, Bin 674
Wang, Bingsen 259
Wang, Changkun 1258
Wang, Chengxue 2015
Wang, Chien-Ming 1527, 1995
Wang, Deyu 473
Wang, F. 249
Wang, Fred 1836
Wang, Gang 1986
Wang, Hua 1507

Author Index

Wang, J. 977
Wang, J.K. 813
Wang, Jianhui 1436
Wang, Jian-quan 229
Wang, Juan 798
Wang, Jui-Kum 1154
Wang, Li-Li 1283
Wang, Linbing 311
Wang, Liqiao 632, 724
Wang, Ming-Yan 1318
Wang, Pei-zhen 1576
Wang, Qingyi 1608
Wang, Qun-jing 857
Wang, Shuo 1
Wang, Xiaofeng 1931
Wang, Xiaoyu 713, 1726
Wang, Y.F. 1458
Wang, Yaonan 866
Wang, Yue 1463
Wang, Yunfei 1896
Wang, Z. A. 1708
Wang, Z. S. 373, 1901
Wang, Zhaoan 321, 433, 642, 713, 1463, 1726
Wang, Zhixin 520, 1487
Watkins, S. J. 1551
Wei, Dong 189
Wei, Guo 880
Wei, Liu 1473
Wei, Shi 1061
Wei, Wen 684, 1298
Wei, Xueliang 789, 1522
Weibin, Cheng 602
Weiguo, Liu 679
Wei-ping, Zhou 1674
Weiyang, Wu 189, 532, 554, 559, 569, 1273, 1387, 1473, 1517, 1618, 1991
Wei-Yang, Wu 1926
Wen , Z. 175
Wen, H.-T. 1343
Wen, Xuhui 337, 622
Wenjuan, Dong 1542
Wenlang, Deng 1557
Wenlong, Qu 378
Wenqing, Shi 684, 1298
Wenxi, Yao 937
Wetzel, Hermann 1112
Wiseman, J. 1101
Wu, B. 1101
Wu, Bin 397
Wu, C.-Y. 117
Wu, Chih-Yu 417
Wu, Hongxia 438, 1288
Wu, J.C 718
Wu, Jiaju 1258
Wu, Jiande 83
Wu, Li 1487

Wu, Li 520
Wu, Q. P. 1071
Wu, Shanshan 1892
Wu, T.-F. 108, 117
Wu, Tao 1936
Wu, Wei-yang 1603
Wu, Weiyang 473, 632, 674, 724, 1585
Wu, Wilson 537
Wu, Yong 1608

X

Xi, Zhai 122
Xia, Kun 857
Xiangjun, Zhang 301
Xiangrong, Li 301
Xiangyun, Fu 1693
Xianmin, Ma 1922
Xianmin, Mu 1693
Xiao, D. 983, 1867, 2023
Xiao, G. C. 1708
Xiao, Lei 209
Xiao, Wenxun 1293
Xiao, Zheng 1447
Xiaobo, Yang 1273
Xiaofeng, Sun 189, 1991
Xiaofeng, Zhang 958, 1626
Xiaohuan, Wang 1273
Xiaojie, Wu 468
Xiao-ping, Yang 1718
Xiaoqiang, Guo 532
Xiaotan, Zhao 589
Xiaoxia, Wei 1693
Xiaoyi, Jin 189
Xiaoyu, Wang 1722
Xie, Jian 870, 1442
Xie, S.S. 1143
Xie, Yong 1406
Ximei, Zhao 1802, 1846
Xindong, Tian 808
Xing, Yan 1406, 1537
Xinming, Huang 1492
Xinxin, Wang 1162
Xu, Cai 1657
Xu, D. 1101
Xu, Dehong 1566
Xu, Dianguo 708
Xu, Jianping 1066
Xu, Jia-peng 669
Xu, Jinbang 1608
Xu, Longya 703, 890, 1779
Xu, Ming 1
Xu, Wancai 627
Xu, Y. N. 194
Xu, Yanping 1863
Xu, Longya 1825
Xuan-fang, Yang 1674

Author Index

Xue, H. 1952
Xue, Shan 622
Xuhui, Wen 684, 1006, 1012, 1049, 1298
Xun, Li 769

Y

Yabin, LI 458
Yamamura, N. 2000
Yan, Caizhong 1811
Yan, Chen 1623
Yan, Gangui 542
Yan, Wang 1718
Yanchao, Ji 1693
Yanfeng, Wu 729
Yang, Bo 83, 311
Yang, Chun-Sheng 1278
Yang, Geng 1896
Yang, Hui 1863
Yang, J.J. 718
Yang, Jia-qiang 1017
Yang, Junyou 657, 1253, 1877
Yang, R. 152
Yang, S. Y. 510
Yang, Sheng 505
Yang, X.J. 1458
Yang, Xiao-bo 1603
Yang, Xi-jun 229
Yang, Xing-hua 229
Yang, Xu 433, 642
Yang, Zhaoning 1450
Yang, Zhongping 184
Yang, Zilong 473
Yanhui, He 1061
Yanliang, Xu 1741
Yan-min, Su 734, 1669
Yan-ru, Zhong 382, 549, 602, 1718
Yansong, Hou 1571
Yao, Duan 880
Yao, Lei 1463
Yao, Ruoping 989
Yao, Tianjun 127
Yao, Yue-feng 1397
Yaogang, ...386
Yaohua, Li 589
Yao-Xin, ...484
Yazdanpanah, R. 1038
Ye, Min 1581
Ye, Pengsheng 142
Ye, Peng-sheng 229
Yeic, Zhuliang 875
Yesong, Li 1820
Yeung, Heidi H.T. 1502
Yi, Qin 1820
Yi, Wen 1387
Yi, Zhang 862
Yidan, Sun 449, 903

YII, ...1148
Yin, Qiang 1054
Ying, Jiang 1942
Ying, Li 147
Yinhai, Fan 1162
Yong, Gao 88, 1167, 1595
Yong, Kang 74, 769
Yong, Wang 744
Yongchang, Zhang 161
Yonglong, Peng 458
Yoon, H.K. 1382
You, Keping 1646
You, Xiaojie 1976
Youbin, Zhao 739
Yougui, Guo 1557
Youn, M.J. 1382
Yu, Dongmei 1806
Yu, Hongxiang 627
Yu, L.C. 1188
Yu, Y. H. 1129
Yu, Zhang 489
Yuan, Chang 713, 1726
Yuan, Xiaoming 593, 1122, 1566
Yuan, Yang 1595
Yuan, Zhou 647
Yuanbin, Wang 272
Yuanfang, Wen 802
Yuanyuan, Liu 378, 1162
Yuda, Chen 739
Yue, Wang 1061
Yue-feng, Yang 406
Yu-fan, Xi 1669
Yu-gang, Yang 1942
Yugang, Yang 214
Yun-qing, Pei 585
Yun-Xiang, Xie 1401

Z

Zargari, N. 1101
Zeng, Fanpeng 1507
Zeng, Guohong 1689
Zhai, Xiaohua 866
Zhang , B. 152
Zhang, Bo 70, 316, 505, 1293
Zhang, C. L. 1198
Zhang, C. W. 510
Zhang, D. 813
Zhang, Dong 1788
Zhang, Dongyan 236, 1243
Zhang, Fengge 219, 826
Zhang, Hairong 1811
Zhang, Handong 1986
Zhang, Hongyan 794
Zhang, Hui 453
Zhang, Jia 1689
Zhang, Jianzhong 1746

Author Index

Zhang, Jun 1858
Zhang, Kai 564, 1981
Zhang, L. 1551
Zhang, Luan-guo 229
Zhang, Qian 1976
Zhang, Qiang 774
Zhang, Shifu 219, 826
Zhang, Tengchao 179
Zhang, Weiping 236, 688, 1243, 1248
Zhang, X. 510
Zhang, Xi 267
Zhang, Xianmiao 1183
Zhang, Xiaoqiang 1248
Zhang, Xueguang 708
Zhang, Yanli 1138
Zhang, Yingchao 952
Zhang, Yingqi 593
Zhang, Yonggao 564, 1981
Zhang, Yongping 316
Zhang, Yu 1218
Zhang, Yuan 1779
Zhao, Wenxiang 1746, 1815
Zhao, Xiaotan 995
Zhao, Xusen 688, 1243
Zhao, Y. 194
Zhao, Zhengming 952
Zhaoan, Wang 122, 1061, 1542, 1722
Zhaomin, Fanyinhai 1962
Zhao-ming, Qian 406
Zhao-yong, Zhou 647
Zhao-Yulin, ... 484
Zhe, Chen 802, 1387
Zhe, Zhang 559, 1618
Zheng- guo, Wu 1674
Zheng, Shi-cheng 1576
Zheng, Trillion Q. 184, 1012, 1976
Zheng, Zedong 1892
Zhengfeng, Ming 382, 1091
Zhengguo, Wu 729, 1023
Zhengming, Zhao 161, 880
Zheng-Na, ... 484
Zhengyu, Lu 406, 937, 958, 1626
Zhen-lin, Xu 1926
Zhi, Na 1198
Zhili, Tan 769
Zhi-Qiang, Wei 1431
Zhi-yuan, Zhang 1623
Zhong, Yanru 1863
Zhong, Yan-ru 617
Zhongmin, Wang 1630
Zhongnan, Guo 554
Zhongying, Chen 1517
Zhou, Qian-zhi 1368
Zhou, Qianzhi 397
Zhou, Tao 1066
Zhou, Wenqi 1312, 1637

Zhou, Y. M. 1129
Zhou, Yang 923, 1947
Zhou, Yu-Fei 286, 1283, 1392
Zhou, Yunbin 564, 1981
Zhu, Guo-rong 97
Zhu, Huangqiu 179, 817, 923, 1213, 1947
Zhu, Jianguo 346, 831
Zhu, Jingwei 962
Zhu, Xiaoyong 1746, 1815
Zhu, Yuran 798
Zhu, Yu-wu 1238
Zhu, Z.Q. 908, 928, 1841
Zou, X. D. 194